S0-BMA-530

Calcium: the molecular basis of calcium action in biology and medicine

Calcium:
The molecular basis of calcium action in biology and medicine

Edited by

ROLAND POCHET
Université Libre de Bruxelles, Belgium

Co-edited by

ROSARIO DONATO, JACQUES HAIECH,
CLAUS HEIZMANN and VOLKER GERKE

Kluwer Academic Publishers
DORDRECHT / BOSTON / LONDON

Library of Congress Cataloging-in-Publication Data

ISBN: 0-7923-6421-X

Published by Kluwer Academic Publishers,
P.O. Box 17, 3300 AA Dordrecht, The Netherlands

Sold and distributed in North, Central and South America
by Kluwer Academic Publishers,
101 Philip Drive, Norwell, MA 02061, U.S.A.

In all other countries, sold and distributed
by Kluwer Academic Publishers,
P.O. Box 322, 3300 AH Dordrecht, The Netherlands.

Printed on acid-free paper

Printed in the Netherlands

Contents

Calcium Signalling in Extracellular Matrix

Calcium Storage and Release

Calcium and Apoptosis

Calcium and Cellular Ageing

Calcium and Growth Factor

Part Two: Calcium Related Diseases

Cancer

Liver Diseases

Neurodegeneration

Neurodegenerative Diseases

Part Three: Methodological Chapters

Calcium Imaging, Fluorescence Microscopy, Confocal Microscopy

List of Contributors

Mei Bai • Endocrine-Hypertension Division and Membrane Biology Program, Department of Medicine, Brigham and Women's Hospital and Harvard Medical School, 221 Longwood Ave., Boston, MA 02115, U.S.A. [Contribution, pp. 415–442]

Frank Barletta • Department of Biochemistry and Molecular Biology, UMDNJ, New Jersey Medical School, Newark, NJ 07103, U.S.A. [Contribution, pp. 259–275]

Virginia Barone • DIBIT, Istituto Scientifico San Raffaele, Via Olgettina 58, I-20132 Milan, Italy [Contribution, pp. 205–219]

Greg J. Barritt • Department of Medical Biochemistry, School of Medicine, Faculty of Health Sciences, Flinders University, G.P.O. Box 2100, Adelaide, South Australia 5001, Australia [Contribution, pp. 73–94]

Karl-Heinz Braunewell • Signal Transduction Research Group, Department of Molecular Biology and Neurochemsitry, Leibniz-Institute for Neurobiology, P.O. Box 1860, D-39008 Magdeburg, Germany [Contribution, pp. 129–149]

David Brown • Department of Pharmacology, University College London, Gower Street, London WC1E 6BT, U.K. [Contribution, pp. 27–44]

Edward M. Brown • Endocrine-Hypertension Division and Membrane Biology Program, Department of Medicine, Brigham and Women's Hospital and Harvard Medical School, 221 Longwood Ave., Boston, MA 02115, U.S.A. [Contribution, pp. 415–442]

Geert Callewaert • Laboratorium voor Fysiologie, Campus Gasthuisberg O/N, Katholieke Universiteit Leuven, Herestraat 49, B-3000 Leuven, Belgium [Contribution, pp. 179–190]

France Carrier • University of Maryland, School of Medicine, Department of Biochemistry and Molecular Biology, 108 N. Greene Street, Baltimore, MD 21201, U.S.A. [Contribution, pp. 521–539]

Mounia Chami • Unité INSERM 370, Institut Pasteur/Necker, Faculté de Médécine Necker, F-75730 Paris Cédex 15, France [Contribution, pp. 505–519]

Naibedya Chattopadhyay • Endocrine-Hypertension Division and Membrane Biology Program, Department of Medicine, Brigham and Women's Hospital and Harvard Medical School, 221 Longwood Ave., Boston, MA 02115, U.S.A. [Contribution, pp. 415–442]

Sylvia Christakos • Department of Biochemistry and Molecular Biology, UMDNJ, New Jersey Medical School, Newark, NJ 07103, U.S.A. [Contribution, pp. 259–275]

Kelly Y. Chun • Department of Pathology, Brigham and Women's Hospital and Harvard Medical School, Thorn 530, 75 Francis Street, Boston, MA 02115, U.S.A. [Contribution, pp. 541–563]

Caroline Clair • INSERM U 442, UPS, bat 443, F-91405 Orsay, France [Contribution, pp. 95–108]

Michel Claret • INSERM U 442, UPS, bat 443, F-91405 Orsay, France [Contribution, pp. 95–108]

Laurent Combettes • INSERM U 442, UPS, bat 443, F-91405 Orsay, France [Contribution, pp. 95–108]

Vadim N. Dedov • The School of Pharmacy, The University of Sydney, New South Wales, Australia 2006 [Contribution, pp. 697–713]

Humbert De Smedt • Laboratorium voor Fysiologie, Campus Gasthuisberg O/N, Katholieke Universiteit Leuven, Herestraat 49, B-3000 Leuven, Belgium [Contribution, pp. 179–190]

Patrick De Smet • Laboratorium voor Fysiologie, Campus Gasthuisberg O/N, Katholieke Universiteit Leuven, Herestraat 49, B-3000 Leuven, Belgium [Contribution, pp. 179–190]

Rosario Donato • Department of Experimental Medicine & Biochemical Sciences, University of Perugia, Via del Giochetto, C.P. 81, Succ. 3, I-06126 Perugia, Italy

Ada Dormann • Department of Neurobiology, Life Science Institute, The Hebrew University of Jerusalem, Israel/The Interuniversity Institute for Marine Sciences, Eilat, Israel [Contribution, pp. 589–603]

Vincent J. Dupriez • Euroscreen S.A., 802 Route de Lennik, B-1070 Brussels, Belgium [Contribution, pp. 647–659]

Paul Eggleton • MRC Immunochemistry Unit, Department of Biochemistry, University of Oxford, Oxford OX1 3QH, U.K. [Contribution, pp. 317–331]

Jürgen Engel • Biozentrum der Universität, CH-4056 Basel, Switzerland [Contribution, pp. 151–164]

Gareth J.O. Evans • The Physiological Laboratory, Department of Physiology, University of Liverpool, P.O. Box 147, Crown Street, Liverpool L69 3BX, U.K. [Contribution, pp. 683–696]

Jean-Luc Galzi • IFR Gilbert LAUSTRIAT, Faculté de Pharmacie/Ecole Supérieure de Biotechnologie, 74 Route du Rhin, F-67401 Illkirch, France [Contribution, pp. 641–646]

Ricardo L. Gee • Department of Ophtalmology and Visual Sciences and the Department of Biomolecular Chemistry, University of Wisconsin, Madison, WI 53792, U.S.A. [Contribution, pp. 493–504]

Volker Gerke • Clinical Research Group for Endothelial Cell Biology, University of Münster, Von Esmarckstrasse 56, D-48149 Münster, Germany

Dani Gitler • Department of Neurobiology, Life Science Institute, The Hebrew University of Jerusalem, Israel/The Interuniversity Institute for Marine Sciences, Eilat, Israel [Contribution, pp. 589–605]

Miguel A. Gorriti • Instituto Universitario de Drogodependencias (Departamento de Psicobiología, Facultad de Psicología, Universitad Complutense, E-28223 Madrid, Spain [Contribution, pp. 465–476]

Devrim Gozuachik • Unité INSERM 370, Institut Pasteur/Necker, Faculté de Médécine Necker, F-75730 Paris Cédex 15, France [Contribution, pp. 505–519]

Eckart D. Gundelfinger • Signal Transduction Research Group, Department of Molecular Biology and Neurochemistry, Leibniz-Institute for Neurobiology, P.O. Box 1860, D-39008 Magdeburg, Germany [Contribution, pp. 129–149]

Andreas H. Guse • University of Hamburg, University Hospital Eppendorf, Institute for Medical Biochemistry and Molecular Biology, Division of Cellular Signal Transduction, Grindelallee 117, D-20146 Hamburg, Germany [Contribution, pp. 109–128]

Gerry Hagens • Clinique de Dermatologie and DHURDV, University Hospital Geneva, CH-1211 Geneva 14, Switzerland [Contribution, pp. 477–492]

Jacques Haiech • IFR Gilbert LAUSTRIAT, Faculté de Pharmacie/Ecole Supérieure de Biotechnologie, 74 Route du Rhin, F-67401 Illkirch, France [Contribution, pp. 1–6]

Shoji Hata • Laboratory of Molecular Structure and Function, Department of Molecular Biology, Institute of Molecular and Cellular Biosciences,

University of Tokyo, Yayoi 1-1-1, Bunkyo-ku, Tokyo 113-0032, Japan [Contribution, pp. 443–464]

Brigitte Hayek • Department of Pathophysiology, Vienna General Hospital, University of Vienna, Medical School, Währinger Gürtel 18-20, A-1090 Vienna, Austria [Contribution, pp. 365–377]

Johan W.M. Heemskerk • Departments of Biochemistry (CARIM) and Human Biology (NUTRIM), University of Maastricht, P.O. Box 616, NL-6200 MD Maastricht, The Netherlands [Contribution, pp. 45–71]

Claus W. Heizmann • Department of Clinical Chemistry, Kinderspital Zürich, Steinwiesstrasse 75, CH-8032 Zürich, Switzerland

Marcel Hibert • IFR Gilbert LAUSTRIAT, Faculté de Pharmacie/Ecole Supérieure de Biotechnologie, 74 Route du Rhin, F-67401 Illkirch, France [Contribution, pp. 1–6]

Michael Huening • Department of Biochemistry and Molecular Biology, UMDNJ, New Jersey Medical School, Newark, NJ 07103, U.S.A. [Contribution, pp. 259–275]

Mikio Iijima • Department of Biochemistry, Faculty of Medicine, Kagoshima University 8-35-1, Sakuragaoka, Kagoshima 890-8520, Japan [Contribution, pp. 565–587]

Shahidul Islam • Department of Molecular Medicine, Karolinska Insitutet, Karolinska Hospital L1:02, S-171 76 Stockholm, Sweden[Contribution, pp. 401–413]

H. Kawamichi • Department of Pharmacology, Gunma University School of Medicine, Maebashi, Gunma 371-8511, Japan [Contribution, pp. 221–244]

Bernhard U. Keller • Zentrum Psychologie und Pathophysologie, Universität Göttingen, Humboldallee 23, D-37073 Göttingen, Germany [Contribution, pp. 625–637]

Keiko Kobayashi • Department of Biochemistry, Faculty of Medicine, Kagoshima University 8-35-1, Sakuragaoka, Kagoshima 890-8520, Japan [Contribution, pp. 565–587]

Alexander Koch • Department of Biochemistry and Molecular Biology, Mt. Sinai School of Medicine, New York, NY 10029, U.S.A. [Contribution, pp. 151–164]

Kazuhiro Kohama • Department of Pharmacology, Gunma University School of Medicine, Maebashi, Gunma 371-8511, Japan [Contribution, pp. 221–244]

Jody Kohut • Department of Biochemistry and Molecular Biology, UMDNJ, New Jersey Medical School, Newark, NJ 07103, U.S.A. [Contribution, pp. 259–275]

Dietrich Kraft • Department of Pathophysiology, Vienna General Hospital, University of Vienna, Medical School, Währinger Gürtel 18-20, A-1090 Vienna, Austria [Contribution, pp. 365–375]

Hongbing Li • Instituto de Investigaciones Biomédicas, Consejo Superior de Investigaciones Científicas/Universidad Autónoma de Madrid, Arturo Duperier 4, E-28029 Madrid, Spain [Contribution, pp. 287–303]

David Llewellyn • Department of Medical Biochemistry, University of Wales, College of Medicine, Cardiff, Wales, U.K. [Contribution, pp. 317–331]

Nathalie Macrez • Laboratoire de Physiologie Cellulaire et Pharmacologie Moleculaire, CNRS UMR 5017, Université de Bordeaux II, 146 Rue Leo Saignat, F-33076 Bordeaux Cédex, France [Contribution, pp. 9–25]

Masatoshi Maki • Department of Applied Molecular Biosciences, Graduate School of Bioagricultural Sciences, Nagoya University, Furo-Cho, Chikusa-ku, Nagoya 464-8601, Japan [Contribution, pp. 245–257]

Steve J. Marsh • Department of Pharmacology, University College London, Gower Street, London, WC1E 6BT, U.K. [Contribution, pp. 27–44]

José Martín-Nieto • División de Genética, Departamento de Fisiología, Genética y Microbiología, Universidad de Alicante, Apartado de Correos 99, E-03080 Alicante, Spain [Contribution, pp. 287–303]

Patrick Maurer • Institut für Biochemie, D-50931 Köln, Germany [Contribution, pp. 151–164]

Marek Michalak • Medical Research Council Group in Molecular Biology of Membranes and the Department of Biochemistry, University of Alberta, Edmonton, Alberta, Canada T6G 2H7 [Contribution, pp. 191–204]

Jean Mironneau • Laboratoire de Physiologie Cellulaire et Pharmacologie Moleculaire, CNRS UMR 5017, Université de Bordeaux II, 146 Rue Leo Saignat, F-33076 Bordeaux Cédex, France [Contribution, pp. 9–25]

Ludwig Missiaen • Laboratorium voor Fysiologie, Campus Gasthuisberg O/N, Katholieke Universiteit Leuven, Herestraat 49, B-3000 Leuven, Belgium [Contribution, pp. 179–190]

Mervyn Monteiro • Medical Biotechnology Center, Department of Neurology and Division of Human Genetics, University of Maryland, 725 Lombard Street, Baltimore, MD 21201, U.S.A. [Contribution, pp. 607–623]

Gregory R. Monteith • The School of Pharmacy, The University of Queensland, St. Lucia, Queensland, Australia 4072 [Contribution, pp. 697–713]

Angel Nadal • Institute of Bioengineering and Department of Physiology, Miguel Hernandez University, San Juan Campus, E-03550 Alicante, Spain [Contribution, pp. 661–671]

Akio Nakamura • Department of Pharmacology, Gunma University School of Medicine, Maebashi, Gunma 371-8511, Japan [Contribution, pp. 221–244]

Kimitoshi Nakamura • Medical Research Council Group in Molecular Biology of Membranes and the Department of Biochemistry, University of Alberta, Edmonton, Alberta, Canada T6G 2H7 [Contribution, pp. 191–204]

Miguel Navarro • Instituto Universitario de Drogodependencias (Departamento de Psicobiología, Facultad de Psicología), Universitad Complutense, E-28223 Madrid, Spain [Contribution, pp. 465–476]

Yasuko Ono • Laboratory of Molecular Structure and Function, Department of Molecular Biology, Institute of Molecular and Cellular Biosciences, University of Tokyo, Yayoi 1-1-1, Bunkyo-ku, Tokyo 113-0032, Japan [Contribution, pp. 443–464]

Michal Opas • Department of Anatomy and Cell Biology, University of Toronto, Toronto, Canada M5S 1A8 [Contribution, pp. 191–204]

Ruthi Oren • Department of Neurobiology, Life Science Institute, The Hebrew University of Jerusalem, Israel/The Interuniversity Institute for Marine Sciences, Eilat, Israel [Contribution, pp. 589–605]

Paloma I. Palomo-Jiménez • Instituto de Investigaciones Biomédicas, Consejo Superior de Investigaciones Científicas/Universidad Autónoma de Madrid, Arturo Duperier 4, E-28029 Madrid, Spain [Contribution, pp. 287–303]

Sylvia Papp • Department of Anatomy and Cell Biology, University of Toronto, Toronto, Canada M5S 1A8 [Contribution, pp. 191–204]

Jan B. Parys • Laboratorium voor Fysiologie, Campus Gasthuisberg O/N, Katholieke Universiteit Leuven, Herestraat 49, B-3000 Leuven, Belgium [Contribution, pp. 179–190]

Patricia Paterlini-Bréchot • Unité INSERM 370, Institut Pasteur/Necker, Faculté de Médécine Necker, 75730 Paris Cédex 15, France [Contribution, pp. 505–519]

Arlette Pesty • INSERM Unité 355, "Maturation Gamétique et Fécondation", 32 Rue des Carnets, F-92140 Clamart, France [Contribution, pp. 673–682]

Daniela Pietrobon • Department of Biomedical Sciences, University of Padova, V. le G. Colombo 3, I-35121 Padova, Italy [Contribution, pp. 379–400]

Roland Pochet • Laboratory of Histopathology CP 620, School of Medicine, ULB, 808 Route de Lennik, B-1070 Brussels, Belgium

Jennifer Pocock • Cell Signalling Laboratory, Department of Neurochemistry, Institute of Neurology, University College London, 1 Wakefield Street, London WC1N 1PJ, U.K. [Contribution, pp. 683–696]

Arthur S. Polans • Department of Ophtalmology and Visual Sciences and the Department of Biomolecular Chemistry, University of Wisconsin, Madison, WI 53792, U.S.A. [Contribution, pp. 493–504]

Harvey B. Pollard • Departments of Anatomy & Cell Biology and Physiology/Institute for Molecular Medicine, USU School of Medicine (USUHS), Bethesda, MD 20814, U.S.A. [Contribution, pp. 307–316]

Mihali Raval-Pandya • Department of Biochemistry and Molecular Biology, UMDNJ, New Jersey Medical School, Newark, NJ 07103, U.S.A. [Contribution, pp. 259–275]

Carsten Reissner • Signal Transduction Research Group, Department of Molecular Biology and Neurochemistry, Leibniz-Institute for Neurobiology, P.O. Box 1860, D-39008 Magdeburg, Germany [Contribution, pp. 129–149]

Daniela Riccardi • School of Biological Sciences, University of Manchester, G38 Stopford Building, Oxford Road Manchester M13 9PT, U.K. [Contribution, pp. 165–177]

Fernando Rodríguez de Fonseca • Instituto Universitario de Drogodependencias (Departamento de Psicobiología, Facultad de Psicología), Universitad Complutense, E-28223 Madrid, Spain [Contribution, pp. 465–476]

Daniela Rossi • Section of Molecular Medicine, Department of Neurosciences, University of Siena, Italy [Contribution, pp. 205–219]

Basil D. Roufogalis • The School of Pharmacy, The University of Sydney, New South Wales, Australia 2006 [Contribution, pp. 697–711]

Maria José Ruano • Instituto de Investigaciones Biomédicas, Consejo Superior de Investigaciones Científicas/Universidad Autónoma de Madrid, Arturo Duperier 4, E-28029 Madrid, Spain [Contribution, pp. 287–303]

Richard R. Rustandi • University of Maryland, School of Medicine, Department of Biochemistry and Molecular Biology, 108 N. Greene Street, Baltimore, MD 21201, U.S.A. [Contribution, pp. 521–539]

David B. Sacks • Department of Pathology, Brigham and Women's Hospital and Harvard Medical School, Thorn 530, 75 Francis Street, Boston, MA 02115, U.S.A. [Contribution, pp. 541–563]

Takeyori Saheki • Department of Biochemistry, Faculty of Medicine, Kagoshima University 8-35-1, Sakuragaoka, Kagoshima 890-8520, Japan [Contribution, pp. 565–587]

Stephen W. Scherer • Department of Molecular and Medical Genetics, University of Toronto, Toronto, Ontario M55 1A8, Canada/Department of Genetics, The Hospital for Sick Children, Toronto, Ontario M5G 1X8 [Contribution, pp. 565–587]

Alexander A. Selyanko • Department of Pharmacology, University College London, Gower Street, London WC1E 6BT, U.K. [Contribution, pp. 27–44]

Georges Siegenthaler • Clinique de Dermatologie and DHURDV, University Hospital Geneva, CH-1211 Geneva 14, Switzerland [Contribution, pp. 477–492]

Ilse Sienaert • Department of Molecular Medicine, University of Texas Health Science Center, Institute of Biotechnology, San Antonio, TX 78245, U.S.A. [Contribution, pp. 179–190]

Ilenia Simeoni • Section of Molecular Medicine, Department of Neurosciences, University of Siena, Italy [Contribution, pp. 205–219]

David S. Sinasac • Department of Molecular and Medical Genetics, University of Toronto, Toronto, Ontario M55 1A8, Canada/Department of Genetics, The Hospital for Sick Children, Toronto, Ontario M5G 1X8, Canada [Contribution, pp. 565–587]

Bernat Soria • Institute of Bioengineering and Department of Physiology, Miguel Hernandez University, San Juan Campus, E-03550 Alicante, Spain [Contribution, pp. 661–671]

Hiroyuki Sorimachi • Laboratory of Molecular Structure and Function, Department of Molecular Biology, Institute of Molecular and Cellular Biosciences, University of Tokyo, Yayoi 1-1-1, Bunkyo-ku, Tokyo 113-0032, Japan [Contribution, pp. 443–464]

Vincenzo Sorrentino • DIBIT, Istituto Scientifico San Raffaele, Via Olgettina 58, I-20132 Milan, Italy [Contribution, pp. 205–219]

Micha E. Spira • Department of Neurobiology, Life Science Institute, The Hebrew University of Jerusalem, Israel/The Interuniversity Institute for Marine Sciences, Eilat, Israel [Contribution, pp. 589–605]

Susanne Spitzauer • Department of Medical and Clinical Chemistry, Vienna General Hospital, A-1090 Vienna, Austria [Contribution, pp. 365–377]

Meera Srivastava • Departments of Anatomy & Cell Biology and Physiology/ Institute for Molecular Medicine, USU School of Medicine (USUHS), Bethesda, MD 20814, U.S.A. [Contribution, pp. 309–316]

Stacy M. Stabler • Medical Biotechnology Center, Department of Neurology and Division of Human Genetics, University of Maryland, 725 Lombard Street, Baltimore, MD 21201, U.S.A. [Contribution, pp. 607–623]

Lalita Subramanian • Department of Ophtalmology and Visual Sciences and the Department of Biomolecular Chemistry, University of Wisconsin, Madison, WI 53792, U.S.A. [Contribution, pp. 493–504]

Koichi Suzuki • Institute of Molecular and Cellular Biosciences, University of Tokyo, 1-1-1 Yayoi, Bunkyo-ku, Tokyo 113-0032, Japan [Contribution, pp. 443–464]

Ines Swoboda • Department of Medical and Clinical Chemistry, Vienna General Hospital, A-1090 Vienna, Austria [Contribution, pp. 365–377]

Emil Toescu • Department of Physiology, Birmingham University, Birmingham B15 2TT, U.K. [Contribution, pp. 277–286]

Thierry Tordjmann • INSERM U 442, UPS, bat 443, F-91405 Orsay, France [Contribution, pp. 95–108]

Lap-Chee Tsui • Department of Molecular and Medical Genetics, University of Toronto, Toronto, Ontario M55 1A8, Canada/Department of Genetics, The Hospital for Sick Children, Toronto, Ontario M5G 1X8, Canada [Contribution, pp. 565–587]

Anna Twardosz • Department of Pathophysiology, Vienna General Hospital, University of Vienna, Medical School, Währinger Gürtel 18–20, A-1090 Vienna, Austria [Contribution, pp. 365–377]

Rudolf Valenta • Department of Pathophysiology, Vienna General Hospital, University of Vienna, Medical School, Währinger Gürtel 18–20, A-1090 Vienna, Austria [Contribution, pp. 365–377]

Paul R. van Ginkel • Department of Ophtalmology and Visual Sciences and the Department of Biomolecular Chemistry, University of Wisconsin, Madison, WI 53792, U.S.A. [Contribution, pp. 493–504]

Sara Vanlingen • Laboratorium voor Fysiologie, Campus Gasthuisberg O/N, Katholieke Universiteit Leuven, Herestraat 49, B-3000 Leuven, Belgium [Contribution, pp. 179–190]

Pieter A. van Zwieten • Departments of Pharmacotherapy, Cardiology and Cardiopulmonary Surgery, Academic Medical Centre, University of Amsterdam, Meibergdreef 15, NL-1105 AZ Amsterdam, The Netherlands [Contribution, pp. 333–363]

Alex Verkhratsky • Manchester University, School of Biological Sciences, 1.124 Stopford Building, Oxford Road, Manchester M13 9PT, U.K. [Contribution, pp. 277–286]

Antonio Villalobo • Instituto de Investigaciones Biomédicas, Consejo Superior de Investigaciones Científicas/Universidad Autónoma de Madrid, Arturo Duperier 4, E-28029 Madrid, Spain [Contribution, pp. 287–303]

Teresa M. Walker • Department of Ophtalmology and Visual Sciences and the Department of Biomolecular Chemistry, University of Wisconsin, Madison, WI 53792, U.S.A. [Contribution, pp. 493–504]

Nicolas J. Wanaverbecq • Department of Pharmacology, University College London, Gower Street, London WC1E 6BT, U.K. [Contribution, pp. 27–44]

Donald T. Ward • School of Biological Sciences, University of Manchester, Oxford Road, Manchester M13 9PT, U.K. [Contribution, pp. 165–177]

David J. Weber • University of Maryland, School of Medicine, Department of Biochemistry and Molecular Biology, 108 N. Greene Street, Baltimore, MD 21201, U.S.A. [Contribution, pp. 521–539]

Naoki Yamaguchi • Department of Biochemistry, Faculty of Medicine, Kagoshima University 8-35-1, Sakuragaoka, Kagoshima 890-8520, Japan [Contribution, pp. 565–587]

Tomotsugu Yasuda • Department of Biochemistry, Faculty of Medicine, Kagoshima University 8-35-1, Sakuragaoka, Kagoshima 890-8520, Japan [Contribution, pp. 565–587]

Danna B. Zimmer • University of South Alabama, College of Medicine, Department of Pharmacology, Mobile, AL 36688, U.S.A. [Contribution, pp. 521–539]

Noam Ziv • Department of Anatomy and Cell Biology, Bruce Rappaport Faculty of Medicine, The Technion, Israel [Contribution, pp. 589–605]

Editorial

The enormous and varied role of calcium in living systems is now widely appreciated by both cell biologists and clinicians. The identification and characterisation of new calcium binding proteins and regulatory pathways is matched by the recognition of the involvement of calcium binding proteins in an growing number of disease states. This book is intended to fulfil a long-standing ambition of mine, namely to succeed in introducing clinicians to fundamental biological research, whilst at the same time attracting researchers to the clinical world.

The publication of the book coincides with the elucidation of the complete Human Genomic Sequence. As a result of this, scientists now have access to an unprecedented array of data, from which new calcium binding proteins and hence new regulatory pathways will undoubtedly be discovered. It is a further aim of this book to provide a "key" to open the door to the new postgenomic era.

The book is divided into in three parts. The first section introduces the reader to the role of calcium in cell biology, allowing him to appreciate how this small, simple, non-metabolisable agent can move rapidly and silently through the different cellular compartments, thereby influencing and controlling the fate of the cell. This section also illustrates and dissects the often complex interplay between calcium and numerous agents in muscle and endocrine cells, neurons, hepatocytes and platelets.

In the second section the reader will discover the role of calcium and its partners in common diseases such as migraine and drug dependence. New classes of diseases such as annexinopathies, channelopathies, calcium-sensing disorders and cittrulinemia are discussed, and the authors give many new insights into the molecular mechanisms of the diseases, thereby explaining how and why they occur. Such information is clearly of primary importance for the pharmaceutical industry. New ideas and concepts on neurodegenerative diseases are introduced, which should stimulate new approaches. Clinicians will also have access, in a comprehensive and authoritative yet highly readable chapter to data from recent large-scale clinical studies on the numerous and widely prescribed calcium antagonists.

The final section gives information on new methods and devices for calcium imaging, and illustrates how calcium movement and change can be monitored and ingeniously utilised as a fast, cheap and accurate drug screening instrument.

The transformation of an idea into a book is always a difficult process, and I was aided immeasurably with this by the active network of the European Calcium Society. I would also like to thank the Université Libre de Bruxelles who encouraged and supported me with this endeavour.

Roland Pochet
Université Libre de Bruxelles, Belgium

Colourplate Section

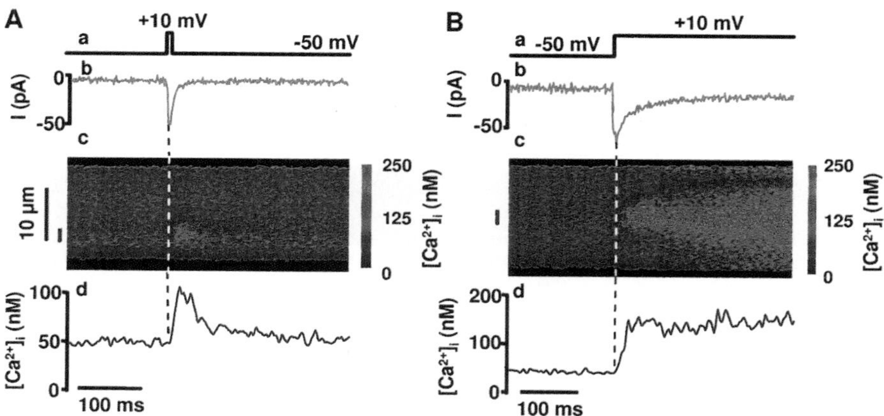

Figure 1. Ca^{2+} sparks and Ca^{2+} waves activated by membrane depolarizations in isolated vascular myocytes. (A) A depolarization from −50 to +10 mV for 10 ms (a) activates a brief Ca^{2+} current (b) and a Ca^{2+} spark illustrated in the line-scan image (c) and shown as spatial averaged fluorescence (d) from a 2 μm-region of the line-scan image. (B) A longer depolarization to +10 mV (a) activates a durable Ca^{2+} current (b) and a propagated Ca^{2+} wave (c, d). Myocytes are loaded with Fluo-3/Fura red mixture and held at −50 mV. Modified from Arnaudeau et al. (1997) with permission from the publisher, Churchill Livingstone.
This figure appears in black/white on page 13.

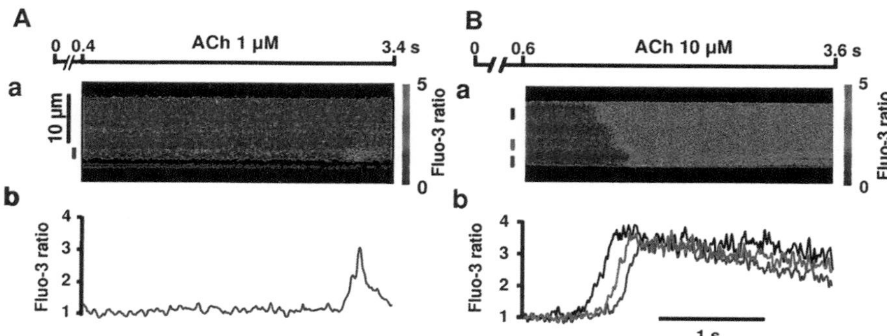

Figure 3. Localized Ca^{2+} signals ("Ca^{2+} puffs") and Ca^{2+} waves in rat ureteric myocytes. (A) Typical Ca^{2+} responses evoked by 1 μM acetylcholine, shown as line-scan image (a) and averaged fluorescence from a 2-μm region (b), indicated by the vertical bar on the line-scan image. (B) Propagated Ca^{2+} wave evoked by 10 μM acetylcholine, shown as line-scan image (a) and averaged fluorescence from three 2-μm regions (b), indicated by vertical bars on the line-scan image. Myocytes are loaded with Fluo-3/AM and not patch-clamped. No change in fluorescence corresponds to Fluo-3 ratio = 1.
This figure appears in black/white on page 16.

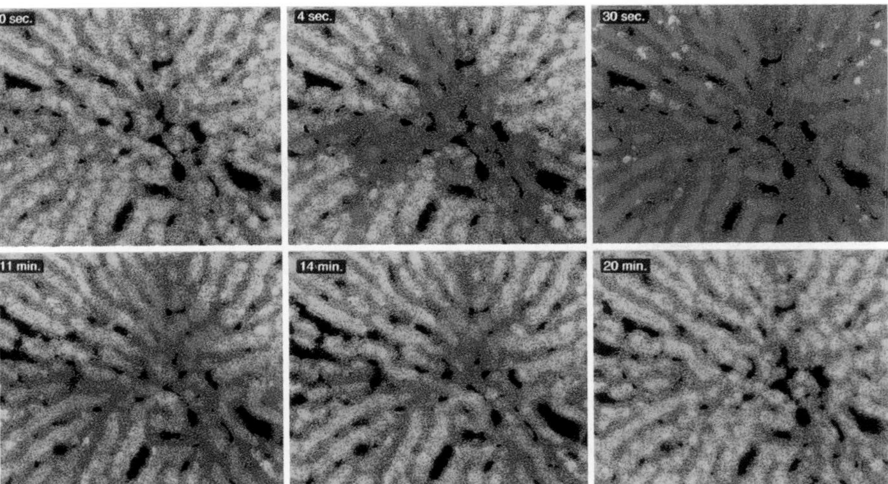

Figure 1. Continued. (B) A wave of increased $[Ca^{2+}]_{cyt}$ moving outwards along liver cell plates from the central vein to the bile duct in a perfused rat liver in which the hepatocytes are loaded with fluo-3. The wave was initiated by the infusion of vasopressin for 1–2 min beginning at time 0 sec. The panels at 11, 14 and 20 min represent the decline in $[Ca^{2+}]_{cyt}$. The central vein is located at the centre of each frame, and hepatic plates radiate outwards in all directions The resting $[Ca^{2+}]_{cyt}$ (100 nM) is represented by yellow and the maximum value of $[Ca^{2+}]_{cyt}$ (400 nM) by red. Reproduced from Motoyama, K., Karl, I.E., Flye, M.W., Osborne, D.F. and Hotchkiss, R.S., 1999, Effect of Ca^{2+} agonist in the perfused liver: determination via laser scanning confocal microscopy, *Am. J. Physiol.* 276, R575–R585, by copyright permission of The American Physiological Society.
This figure appears in black/white on page 76.

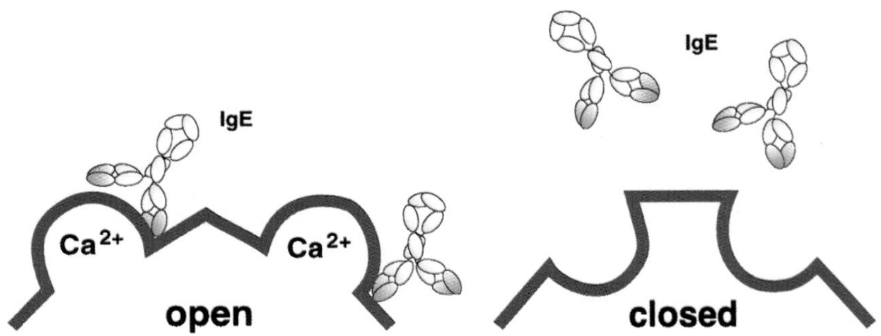

Figure 2. IgE antibodies recognize preferentially epitopes on the open (i.e., calcium-bound) form but not on the closed (i.e., apo-) form of calcium-binding allergens.
This figure appears in black/white on page 372.

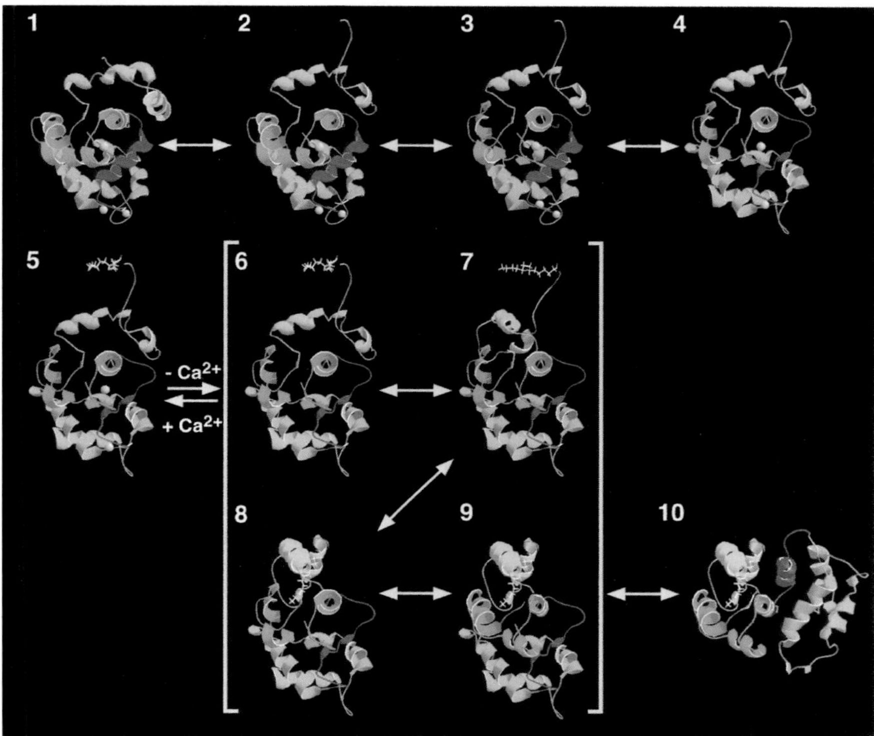

Figure 3. Model of calcium-dependent motions and a myristoyl switch of VILIP-1. The structure is modeled using the coordinates of neurocalcin δ (model 1), calcium-bound recoverin (5) and calcium-free recoverin (10). Models 1 to 4 illustrate the possible motion of the N-terminal domain (yellow) upon decrease of calcium concentration. This motion induces a response in the central hydrophobic domain (green) by deforming secondary structural elements (3), and is finally supported by a deformation of helix F (red) (4). The C-terminal domain shows minor changes. Recoverin is not capable of binding calcium at EF-hand 4, but VILIP-1 most likely is. This calcium ion may be released independently (4) and may be available for calcium-dependent target interactions. Models 5-10 illustrate the calcium-myristoyl switch from the "open" to the "closed" form of VILIP-1. Structures 6–9 are intermediates. After calcium removal, the N-terminal domain can flip over the buried helix E (green) (6 to 7). This opens the way for the "induced fit" of the myristoyl group into the hydrophobic binding pocket provided by the N-terminal and central domains (7 to 9). Finally the helical-linker (red) covers the remaining cavity to the myristoyl group and this movement is stabilized by a rotation of the rigid helical-linker at Ser-90. The C-terminal domain (blue) follows that movement and adapts exactly to the surface formed by helices D, E, and F (10). Calcium binding to the "closed" form of VILIP-1 must induce the reverse order of movements since the helical-linker and C-terminal domain complex in the "closed" calcium-free form occupy the same position as the N-terminal domain in the "open" calcium-bound structure (5). The motions are also available as movies at http://www.synprot.de. Details of the methods used for modeling and RMS values are also presented at this site.

This figure appears in black/white on page 134.

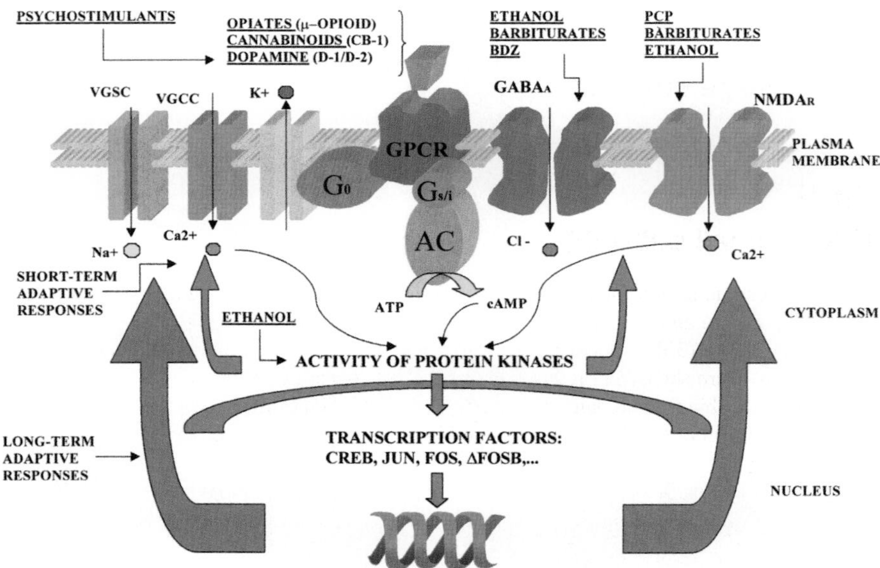

Figure 1. Place of calcium in molecular events associated with acute and chronic drug exposure. Four different types of cellular targets have been described for abused drugs: uptake systems, G-protein-coupled receptors (GPCR), ligand-gated ion channels and cytoplasmic protein kinases. Psychostimulants (cocaine, amphetamine) affect the functionality of monoamine uptake systems, increasing the synaptic concentrations of these transmitters, specially of dopamine, in reward-relevant synapses. Dopamine, as well as opiates or cannabinoids activate specific GPCR such as dopamine D-1 and D-2 families of receptors, the μ-opioid or the cannabinoid CB-1 receptor respectively. GPCR activation results in dynamic changes in second messengers such as cAMP and Ca^{2+}, through the G-protein mediated regulation of effectors such as adenylate cyclase (AC) or Ca^{2+} and K^+ channels. Ethanol, barbiturates, phencyclidine and benzodiazepines (BDZ) affect the functionality of $GABA_A$ or glutamate NMDA receptors ($NMDA_R$), leading to cellular hyperpolarization and to an elevation in cytoplasmic Ca^{2+} levels, respectively. Changes in cAMP and Ca^{2+} levels activate protein kinases, the later effect also directly elicited by ethanol. Protein kinases modulate the function of plasma membrane (i.e. voltage-gated calcium channels (VGCC) and membrane receptors such as GCPR, $GABA_A$ or $NMDA_R$) and transcription factors. Protein-kinase-mediated phosphorylation of plasma membrane proteins constitute the first stage in cellular homeostatic responses counteracting acute effects of drugs abuse. For instance, short-term adaptive cellular responses underlying acute ethanol tolerance are mediated by protein kinases (PK) such as PKC, PKA or Fyn-kinase, which directly modulate $GABA_A$ receptor activity and the desensitization of $NMDA_R$. Activation of transcription factors lead to long-term cellular changes responsible for the neuroadaptions associated with chronic tolerance, sensitization or dependence. These cellular neuroadaptions associated to long-term exposure to abused drugs include, among others, the up-regulation of the cAMP-PKA signaling pathway in opiate and cannabinoid-dependent animals or the up-regulation of L-type VGCC and modifications in subunit composition of glutamate $NMDA_R$ induced by chronic ethanol and opiates. Additional cellular proteins such as voltage-gated sodium channels (VGSC), G-proteins, neurotransmitter receptors, enzymes involved in transmitter synthesis, etc., can also be affected by chronic drug exposure. Other abused drugs such as nicotine or methylxantines (caffeine, etc.), not depicted here to improve clarity of the scheme, similarly modulate the activaty of specific endogenous signaling systems (cholinergic and purinergic respectively).

This figure appears in black/white on page 468.

Figure 1. Schematic representation of the molecular structure of calpain. Conventional calpains, μ-calpain and m-calpain, are composed of a large catalytic subunit (μCL or mCL) and a small subunit (30K). Domain II is a cysteine protease domain. IIa and IIb correspond to the crystal structure shown in Figure 2. Cys, His, and Asn are catalytic residues. Domains IV and VI are Ca^{2+}-binding domains containing five EF-hand motifs.
This figure appears in black/white on page 444.

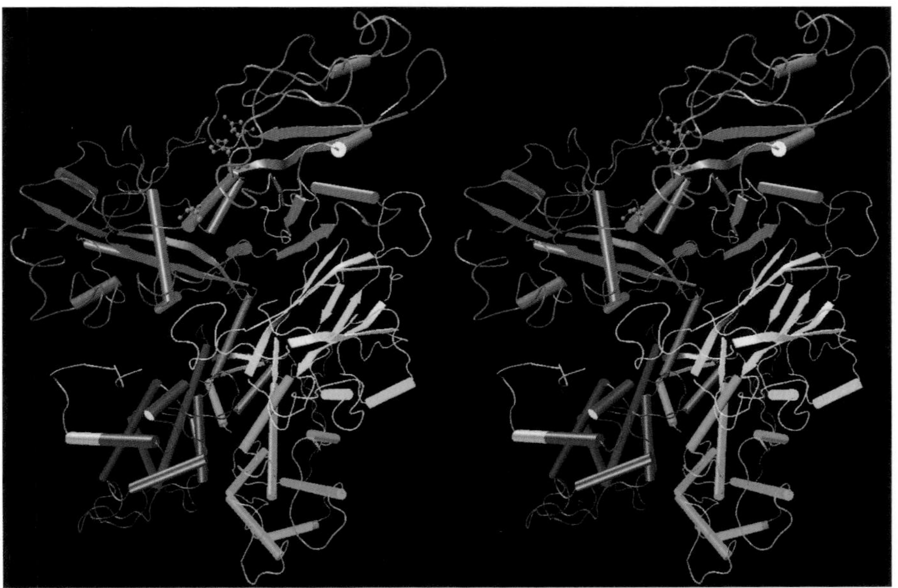

Figure 2. Ribbon structure of human m-calpain in the absence of Ca^{2+}. The colour of each domain corresponds to that in Figure 1. The cylinders represent helical structures. The N-terminus of mCL resides in the center of the molecule.
This figure appears in black/white on page 446.

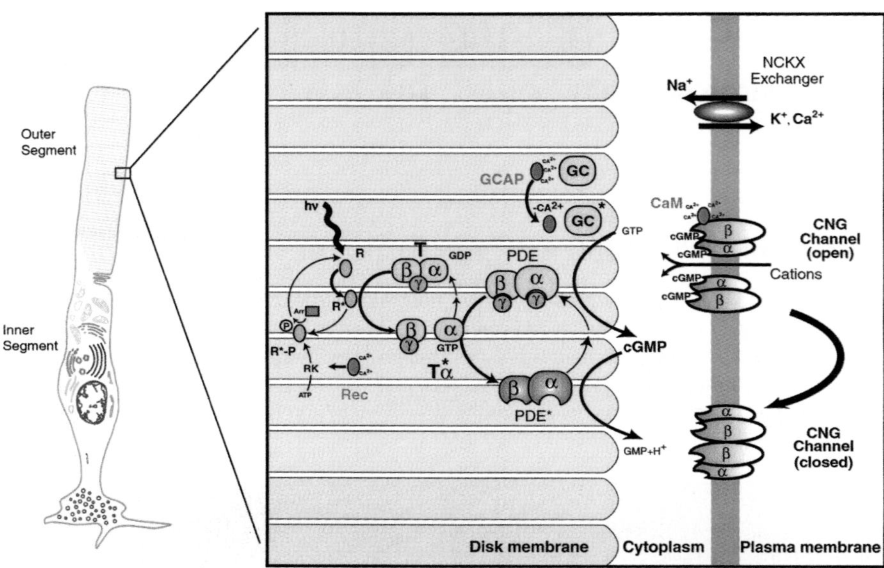

Figure 1. Phototransduction and the role of calcium-binding proteins. The outer segment of the photoreceptor is the primary site of phototransduction. The outer segment consists of a series of flattened, double-membranous disks surrounded by a plasma membrane. Light causes the isomerization of the visual chromophore, 11-cis retinal, to its all-trans configuration. 11-cis retinal is normally coupled to the protein opsin, the complex referred to as rhodopsin (R), and isomerization leads to the activation of rhodopsin (R*) in the disk membranes and the subsequent biochemical reactions depicted in the figure. A trimeric G protein called transducin (T) is activated (T*), causing the stimulation of a phosphodiesterase (PDE*) by removing two inhibitory gamma subunits; this leads to the hydrolysis of cGMP. Cyclic nucleotide gated (CNG) channels in the outer segment plasma membrane are open in the dark when cGMP levels are elevated and close upon illumination as the concentration of cGMP decreases, thus altering the flux of cations into the photoreceptor. The change in membrane potential diminishes the release of chemical transmitter at the synapse so that the reception of light in the outer segment is conveyed to second order neurons in the retina. While sodium ions carry most of the current entering the outer segment, calcium ions also flow through the CNG channels. The flux of calcium ions through the channels is reduced by illumination, while their extrusion via the sodium-calcium-potassium exchanger (NCKX) continues, thus reducing the intracellular concentration of calcium in the light. This reduction stimulates a guanylate cyclase (GC) through the action of GCAP (guanylate cyclase activating protein) in order to resynthesize cGMP, reopen the CNG channels, and reestablish the dark current. Calcium also modifies the phototransduction cycle through the action of recoverin (Rec), which is thought to interact with rhodopsin kinase, inhibiting the phosphorylation of rhodopsin, which otherwise is a major step required for the inactivation of R*. Calmodulin (CaM) also modulates the phototransduction cycle by modifying the affinity of the CNG channels towards cGMP. As light lowers the concentration of intracellular calcium, the binding of calmodulin to the channels changes, increasing their affinity to cGMP, thereby enhancing the return to the dark-adapted state. The light-dependent changes in the concentration of calcium, mediated by calcium-binding proteins, play a critical role in determining the adaptation mechanisms of the photoreceptor. In the absence of such changes in the concentration of calcium, photoreceptor cells could not respond to the wide range of light intensities that we normally experience (Palczewski et al., 2000).

This figure appears in black/white on page 494.

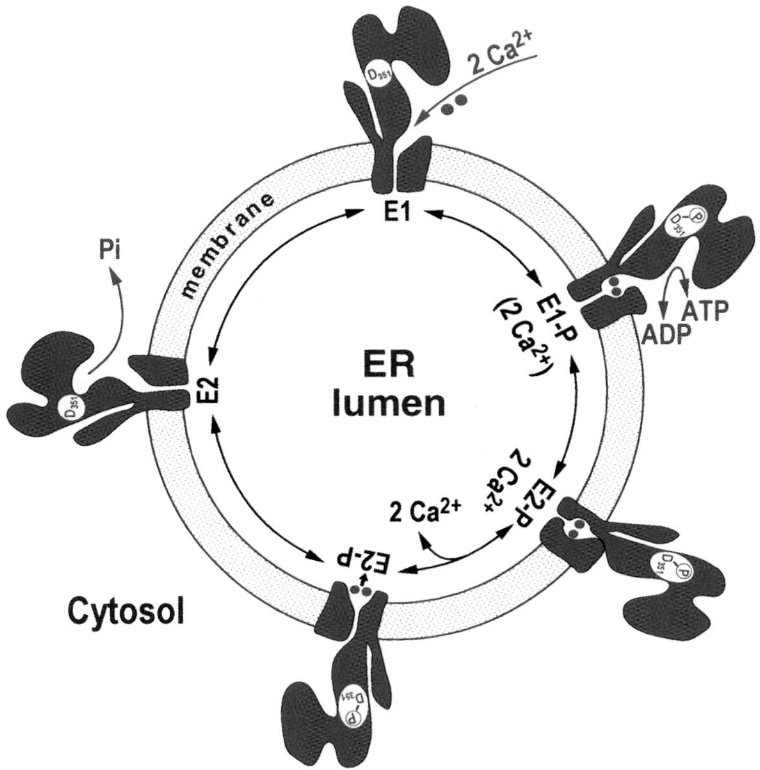

Figure 2. Reaction cycle for Ca^{2+} transport by SERCA pumps. In the E1 conformation, high affinity Ca^{2+} binding sites located near the center of helices M4, M5, and M6 are accessible to cytoplasmic Ca^{2+} but not to lumenal Ca^{2+}. Phosphorylation from ATP, following the occupation of both sites by Ca^{2+} (E1-P ($2Ca^{2+}$)), leads to linked movements of both cytoplasmic and transmembrane domains, resulting in occlusion through closure of the entry gate (E2-P $2Ca^{2+}$). Ca^{2+} is now contained in a polar cavity formed near the center of the transmembrane domains. Further conformational changes open the gate allowing the exit of two Ca^{2+} ions to the lumen (E2-P). This conformation, which Ca^{2+} affinity is very low, also activates dephosphorylation (E2) and returns the pump to the high Ca^{2+} affinity form (E1), completing the cycle (MacLennan et al., 1997; Yonekura et al., 1997).
This figure appears in black/white on page 508.

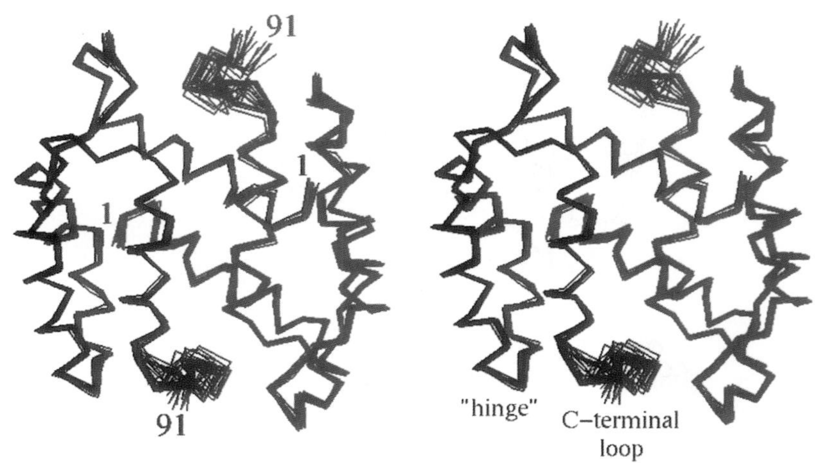

Figure 2. Stereoview of 20 apo-S100B($\beta\beta$) structures refined to high resolution using dipolar coupling restraints as described previously (Drohat et al., 1999). Residues in the C-terminus of apo-S100B($\beta\beta$) and Ca^{2+}-loaded S100B($\beta\beta$) (Drohat et al., 1998) (Figure 3) are in conformational exchange, and are not well defined. The conformation in the C-terminal region of S100B($\beta\beta$) is stabilized when bound to a peptide derived from p53 (Rustandi et al., 1999). The proximity of non-conserved residues in loop 2 (the "hinge") and the C-terminal loop ultimately confer specificity to S100-target protein interactions.
This figure appears in black/white on page 527.

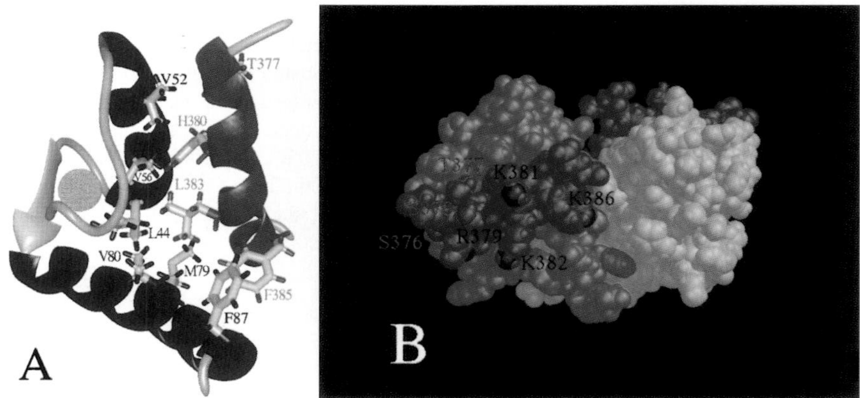

Figure 4. Ribbon diagram and space-filling model showing the positioning of sidechain residues of S100B($\beta\beta$) and p53. (A) Ribbon diagram displaying the sidechains involved in a hydrophobic binding pocket. These residues include L44, V52, V56, M79, V80, F87 (colored in blue) from S100($\beta\beta$) and H380, L383, and F385 (colored in green) from the p53 peptide. (B) Space-filling model of Ca^{2+}-loaded S100B($\beta\beta$) bound to the p53 peptide illustrating the positions of the PKC-phosphorylation sites, S376, T377, and S378 (red) and several p300-acetylation sites (in blue) on p53.
This figure appears in black/white on page 532.

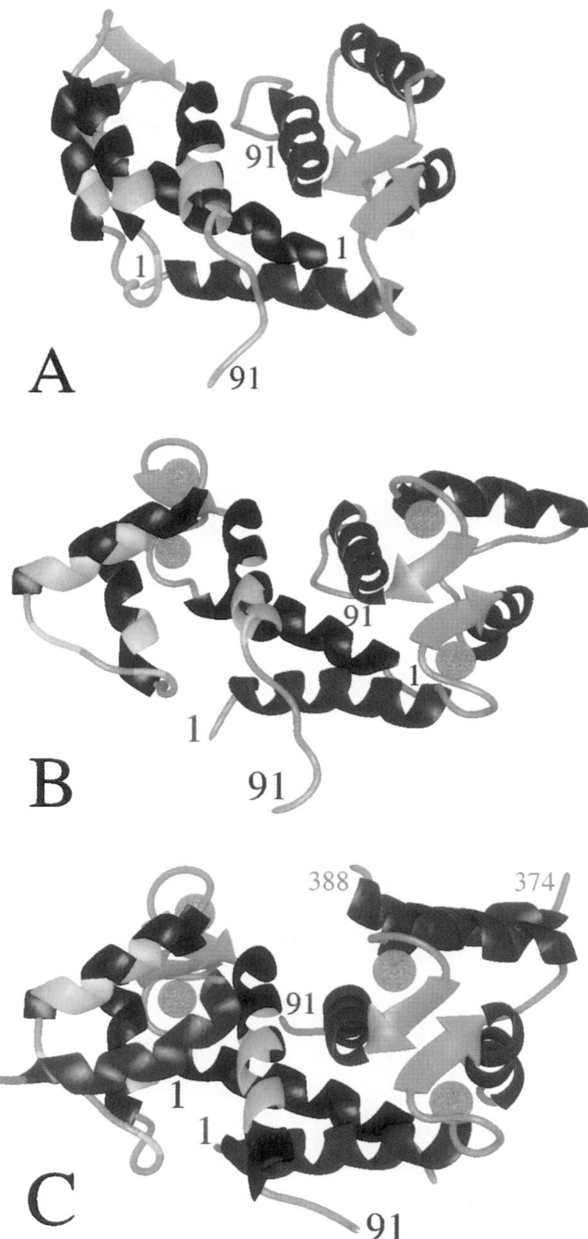

Figure 3. Ribbon diagrams illustrating the 3D structures of (A) apo-S100B($\beta\beta$), (B) Ca^{2+}-bound S100B($\beta\beta$), and (C) p53 peptide-bound S100B($\beta\beta$). Residues that interact with the p53 peptide (in C) are shown in purple on one subunit of the apo- and Ca^{2+}-bound S100B($\beta\beta$) structures to illustrate the Ca^{2+}-dependence of the p53-S100B interaction. Residues that interact with p53 are buried in (A) apo-S100B, but are exposed to solvent after a Ca^{2+}-dependent conformational change (in B). This change in protein structure is required for p53 binding (Rustandi et al., 1998, 1999).

This figure appears in black/white on page 529.

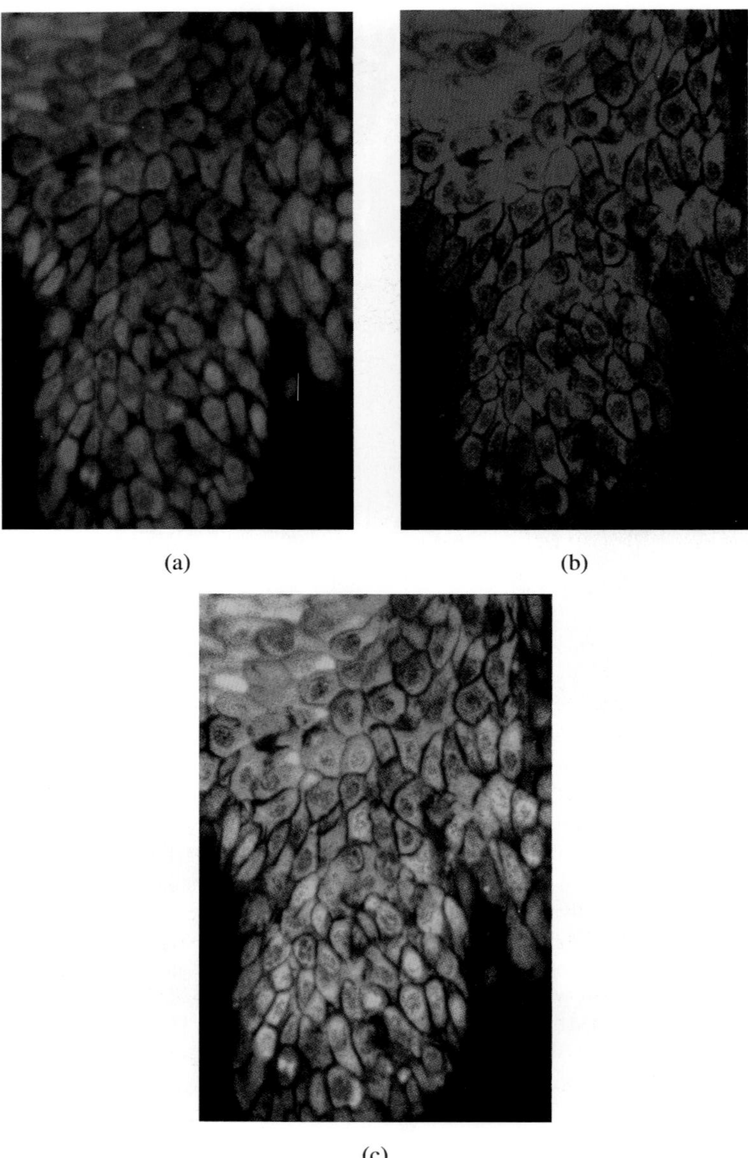

(a) (b)

(c)

Figure 1. Confocal microscopical analysis of S100A7 (a), E-FABP (b) and double staining (c) of lesional psoriatic skin. Picture (c) shows that S100A7 and E-FABP co-localize in all suprabasal cell layers (yellow fluorescence). The predominant green coloured basal layer (c) confirms the very low expression of E-FABP in basal keratinocytes (b).
This figure appears in black/white on page 485.

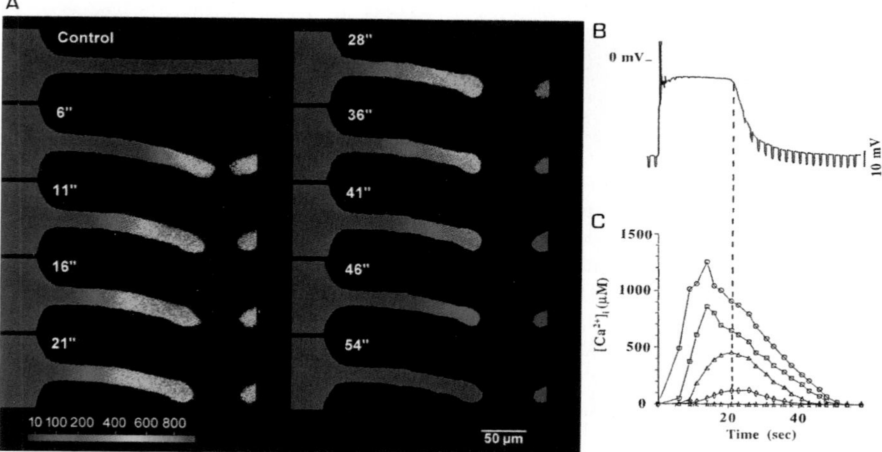

Figure 1. (A) Mag-fura-2 ratio images of the $[Ca^{2+}]_i$ after axotomy of a cultured metacerebral *Aplysia* neuron. Ca^{2+} is observed to diffuse from the cut end towards the cell body (6, 11, 16 seconds), elevating $[Ca^{2+}]_i$ to more than 1000 μM along segments near the cut end of the axon. Following the resealing of the ruptured membrane (between 21 and 28 seconds), the $[Ca^{2+}]_i$ recovers to near control levels. Time is given in seconds from axotomy. Calibration of $[Ca^{2+}]_i$ is given in μM. Correlation between the electrophysiological manifestation of axotomy and the alterations in $[Ca^{2+}]_i$ as obtained from the images shown in (A) are detailed in (B) and (C). (B) Resting potential and the transmembrane voltage drops in response to constant hyperpolaryzing pulses (input resistance measurements) during axonal transection and throughout the recovery period. (C) The $[Ca^{2+}]_i$ as a function of time post axotomy at five points along the axon (from top to bottom: 36 μm from the point of transection, 73, 110, 147 and 211). Note that the recovery of $[Ca^{2+}]_i$ began before the recovery of the membrane potential (dashed line connecting (B) and (C)). From Ziv and Spira (1995).
This figure appears in black/white on page 591.

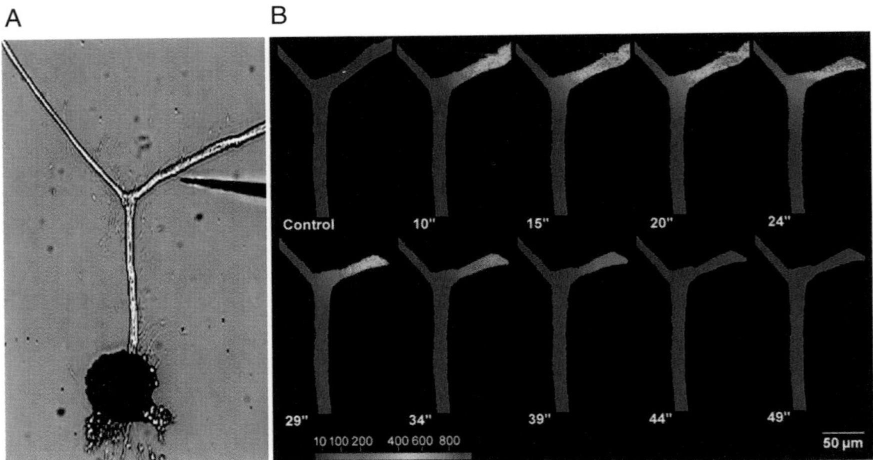

Figure 2. The spatiotemporal distribution of $[Ca^{2+}]_i$ following axotomy at a point of axonal bifurcation. (A) A brightfield image of the bifurcated axon of a cultured *Aplysia* metacerebral neuron. (B) Mag-fura-2 pseudocolor images of the spatiotemporal distribution of $[Ca^{2+}]_i$ following the transection of the right branch. The front of elevated $[Ca^{2+}]_i$ is observed to reach the branch point, and a significant $[Ca^{2+}]_i$ gradient is formed on both sides of the bifurcation point. The resealing of the membrane occurred 25 seconds after axotomy (not shown). Time is given in seconds from axotomy. Calibration of $[Ca^{2+}]_i$ is given in μM. From Ziv and Spira (1995).

This figure appears in black/white on page 593.

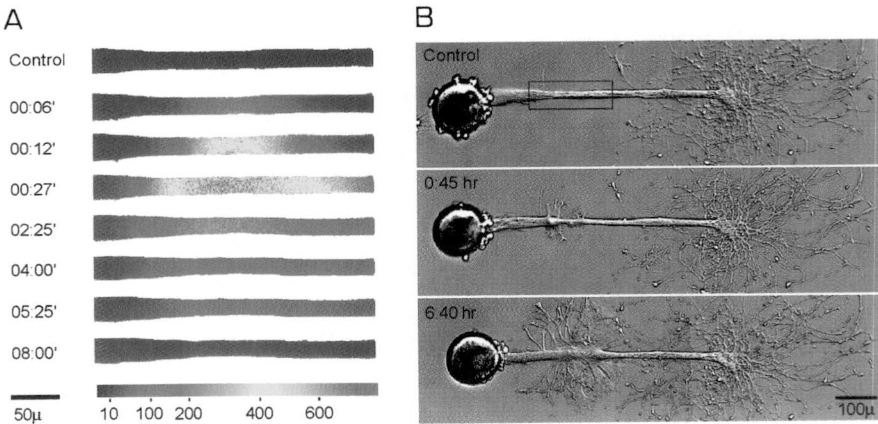

Figure 4. Transient and localized elevation of the $[Ca^{2+}]_i$ to 300–500 μM induces growth cone formation and neuritogenesis in an axon of intact neuron. Mag-fura-2 ratiometric fluorescence microscopy was used to determine the intra-axonal $[Ca^{2+}]_i$ required to induce the transformation of an intact axonal segment into a growth cone. (A) The spatiotemporal alterations in the axonal $[Ca^{2+}]_i$ induced by a focal application of ionomycin. The region shown corresponds to the rectangle in (B), upper panel. $[Ca^{2+}]_i$ is given in μM. (B) The resulting changes in axonal morphology. The transient increase of $[Ca^{2+}]_i$ to $\sim$ 500 μM induced the formation of a growth cone at the application site that subsequently developed into a new neuritic tree. From Ziv and Spira (1997).

This figure appears in black/white on page 598.

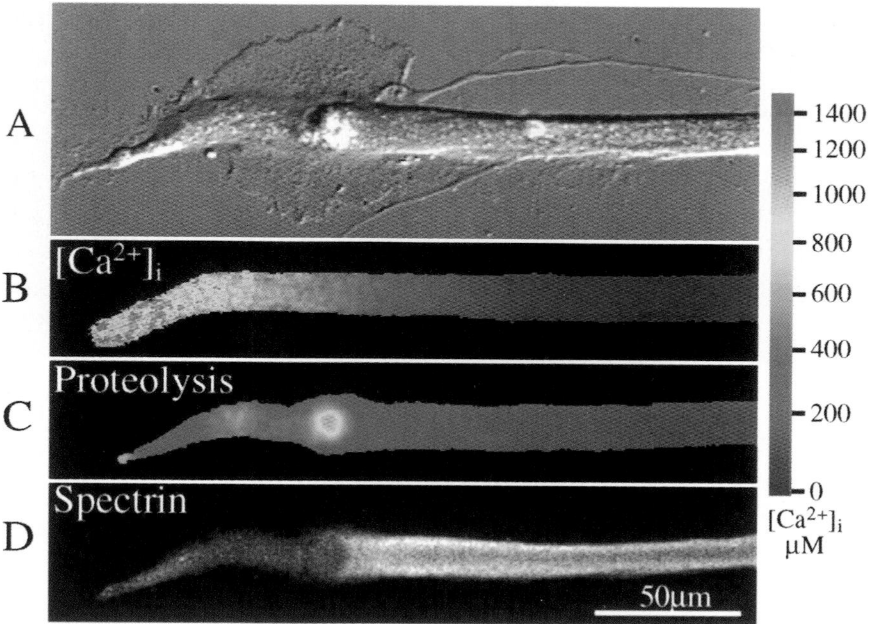

Figure 5. $[Ca^{2+}]_i$, proteolytic activity and spectrin distribution after axotomy. A cultured *Aplysia* buccal neuron was transected while $[Ca^{2+}]_i$ and proteolytic activity were measured. The neuron was fixed and immunolabeled for spectrin 25 minutes after axotomy. (A) Image of the axon prior to fixation. The neuron extended a sizable growth cone lamellipodium following axotomy. (B) The maximal levels of $[Ca^{2+}]_i$ measured during axotomy. The image was acquired 25 seconds after the axotomy was performed. (C) The levels of proteolytic activity, as measured just prior to fixation, 25 minutes after axotomy. (D) Spectrin density is seen to be drastically decreased from the center of the growth cone and up to the transected tip. Notice that the sites from which spectrin was removed correlate very well with the locations in which highest proteolytic activity was measured (C). There exists a very sharp border between the regions in which spectrin density was reduced and those in which spectrin density appears to be unaffected. In contrast, notice that the distribution of spectrin does not conform to the shape of the $[Ca^{2+}]_i$ gradient (B). The location where the growth cone's center was formed correlates to those areas in which $[Ca^{2+}]_i$ was elevated to approximately 350 μM. A $[Ca^{2+}]_i$ scale bar is given to the right. From Gitler and Spira (1998).
This figure appears in black/white on page 600.

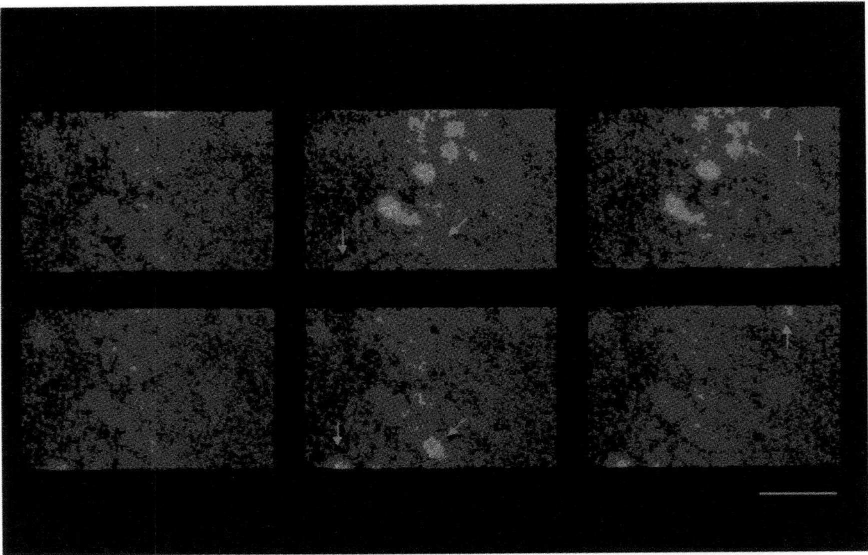

Figure 2. Confocal image of $[Ca^{2+}]_i$ from a fluo-3 loaded cortical brain slice obtained from a neonatal rat. Upper panel: Application of 100 μM NMDA in 0 Mg^{2+} and 10 μM glycine induced Ca^{2+} elevation in a population of cells within the cortical brain slice. From left to right: basal level of fluorescence, 10 s after and 30 s after NMDA application. Lower panel: Effect of 20 mg/ml BPA on the same field of cells. Note that cells indicated by arrows respond to BPA but not to NMDA. Scale bar is 50 μm. After Nadal et al. (1998), reproduced with permission.
This figure appears in black/white on page 665.

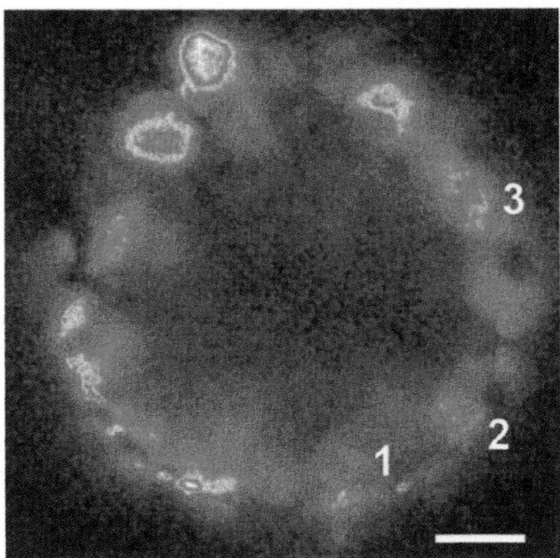

Figure 3. Fluorescence changes measured from individual cells within an intact islet of Langerhans. Colour image of a fluo-3 loaded islet exposed to 3 mM glucose, blue corresponds to low and red to high fluorescence intensity. Scale bar at bottom right represents 15 μm. After Nadal et al. (1999), reproduced with permission.
This figure appears in black/white on page 667.

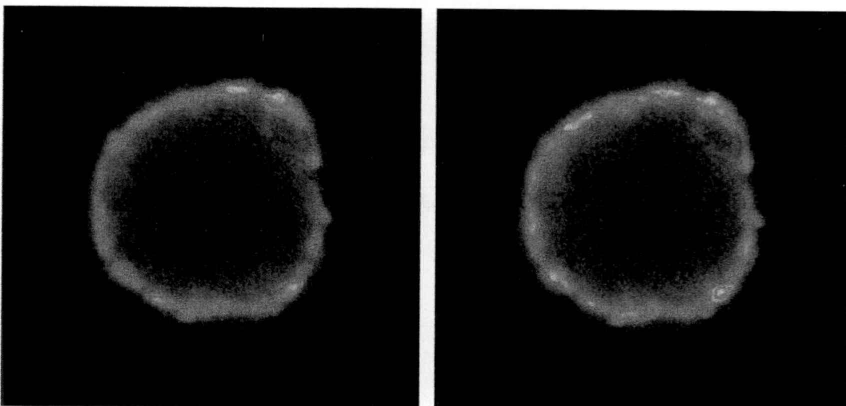

Figure 4. Synchronicity between pancreatic β-cells within an intact islet of Langerhans. Time sequence of a $[Ca^{2+}]_i$ oscillation in an islet of Langerhans exposed to 11 mM glucose imaged with a confocal microscope. From left to right: at the beginning of a $[Ca^{2+}]_i$ oscillation and at the peak of the oscillation.
This figure appears in black/white on page 668.

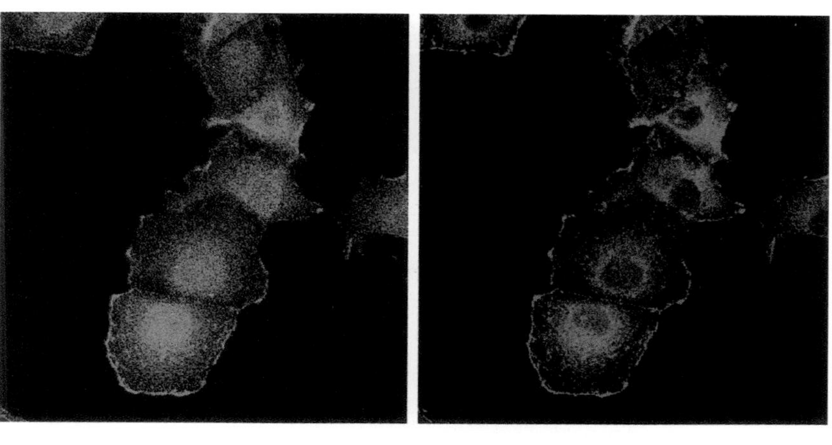

Figure 5. Subcellular location of calmodulin and IQGAP1 by confocal microscopy. MCF-7 cells were grown on plastic slides, fixed and probed with anti-calmodulin and anti-IQGAP1 antibodies. Primary antibodies were visualized with the appropriate fluorescent labeled secondary antibodies. Calmodulin is depicted in green and IQGAP1 in red.
This figure appears in black/white on page 558.

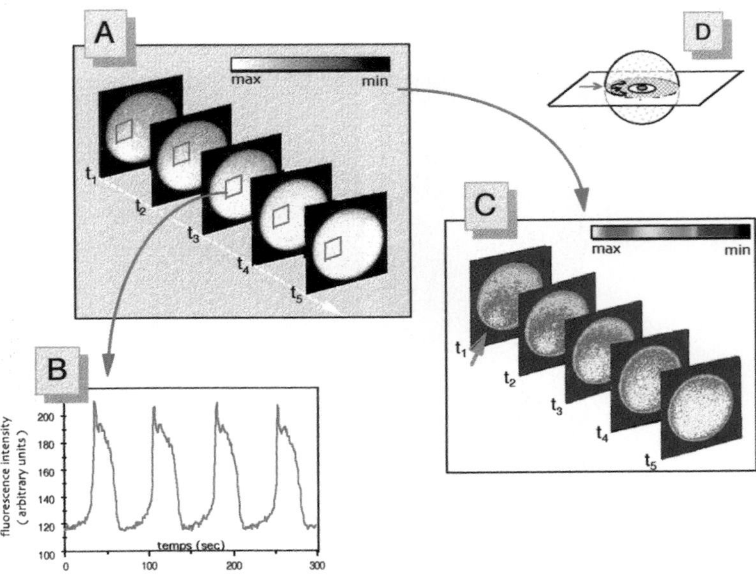

Figure 2. Acquisition and analysis of data recorded in confocal microscopy: (A) sequential grey scale digitalized images are recorded for later analysis; then (B) either a time-course curve of calcium indicator fluorescence emission is constructed on a selected zone of the image to study the kinetics of the Ca^{2+} signal; or (C) a pseudo-colors scale is calibrated on the grey scale of the recorded images, to improve the readability of imaging the spatial progression of the signal. The inset (D) shows the progression of the Ca^{2+} signal in the selected optical plane, through the GV and nucleolus. *This figure appears in black/white on page 677.*

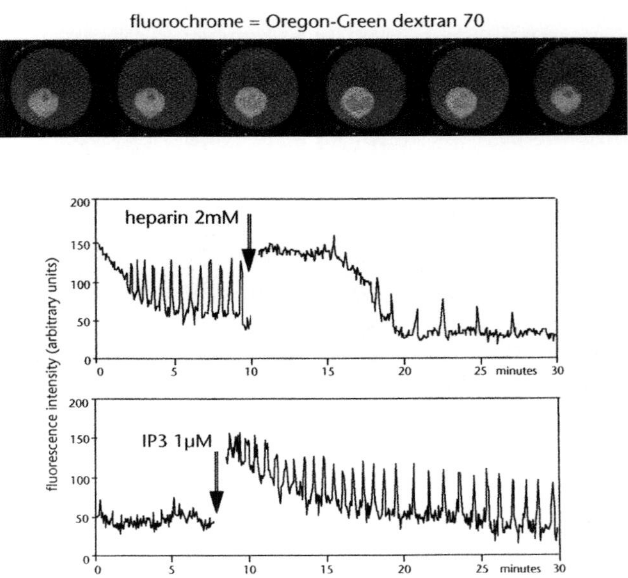

Figure 3. Pseudo-color imaging of nuclear calcium oscillations after microinjection of Oregon Green 488 BAPTA-1 Dextran into the nucleus of a whole oocyte. Below, the time-course curves show the effect of a second microinjection of phosphoinositide cycle antagonists on the spontaneous calcium train of oscillations. *This figure appears in black/white on page 678.*

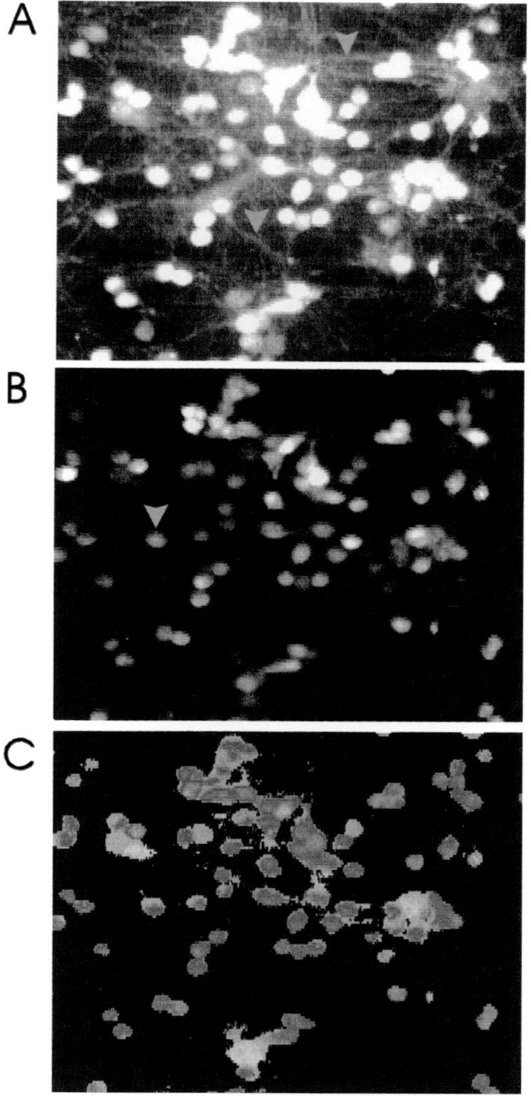

Figure 2. Nine DIV Cerebellar granule neurons plated on small coverslips at low density and loaded for 30 min at 37°C in the presence of 16 μM BSA. The cells were washed and mounted into a heated holder with 150 μM of incubation medium, and placed on the stage of an Olympus IX70 inverted fluorescence microscope. Images were captured using a 12-bit cooled digital CCD camera, Life Science Resources SpectraMASTER High Speed Monochromator, controller and xenon UV lamp, and the output displayed using Life Science Resources Merlin software. Data were collected and analysed off-line to produce (A), (B) and (C). (A) Emission from 380 nm excitation showing field of neurons in which the grey level fluorescence scale has been set so that whilst the somatic response appears saturated (i.e. white), the neurites (arrowheads) are visible. B. The same field of cells in which the fluorescence from the somata is now within scale. (C) 340/380 ratio image of the same cells in (A) and (B) but with a grey level scale set to observe somatic response only. Yellow fluorescence is likely to derive from astrocytes.

This figure appears in black/white on page 688.

From Genes to Drugs

High throughput screening on genome targets

Jacques Haiech and Marcel Hibert

1. INTRODUCTION: PRE- AND POST-GENOME ERA

The working draft of the human genome (95% of the whole genome) has now been published. The whole set of human genes will soon be known and therefore, all the "putative" proteins involved in calcium signal transduction will be accessible.

A new paradigm is appearing in biology: the possibility of exhaustively studying the transcription (transcriptome) and the expression (proteome) of a family or a sub-family of genes. These tools are or will be available through centers of competency in genomics. This avalanche of data is going to force the biologist to develop new strategies and methods as well as to be more intuitive and comprehensive in his approach. In particular, it will be necessary to develop biological hypotheses and to test them not only *in vitro* or *in vivo* but also *in silico*.

One of the main tasks the biologist already has to face is to analyze the putative functional role of orphan proteins. Three strategies might be considered:

– a genetic approach (knock-in and knock-out mice, transcriptome, proteomics);
– a structural approach (production and structure determination of the protein);
– a pharmacological approach (screening to find new molecules acting at the target protein and allowing the exploration of its physio-pathological function both *in vitro* and *in vivo*).

These three approaches are complementary and the last one may lead to new pharmaceutical drugs. It is this screening approach, which might revolutionize drug discovery, that we will describe here.

Jacques Haiech and Marcel Hibert • Institut Gilbert Laustriat, Faculté de Pharmacie, Illkirch, France.

R. Pochet, R. Donato, J. Haiech, C. Heizmann and V. Gerke (eds.): Calcium: The Molecular Basis of Calcium Action in Biology and Medicine, 1–6.

In pharmaceutical chemistry, one of the main aims of academic and industrial research is to accelerate the discovery of molecules active in biological systems in order to decipher vital processes and to lead to new and innovative therapeutic applications.

Four strategies have been used to reach these objectives.

Since our origin, human beings have tested the natural compounds on themselves. Discoveries were the consequences of trial and error (often deadly errors). By analyzing in a systematic manner today the traditional medicines, we are trying to capitalize on the potential of this evolutionary screening.

At the beginning of the industrial era, systematic screening has been established, mainly on animals, without trying to characterize the biological targets and the underlying molecular mechanisms. However, this approach has led to the main known therapeutic classes of synthetic compounds. Generally, until the end of the 1970s, the drugs were developed without any knowledge of the identity and chemical structure of their targets.

The progresses in enzymology, molecular pharmacology and molecular biology have progressively allowed us to link specific pathologies with specific biological macromolecules. Furthermore, structural biology has opened the road to the rational design of new drugs and to the optimization of some active compounds. This strategy has been impaired by the difficulty of linking a chemical structure to a target undoubtedly associated with a given pathology. Such a problem is going to be overcome with the global deciphering of the human genome. Indeed, with this avalanche of data, the sequence of thousands of proteins is going to be characterized. The main challenge is to associate orphan proteins with a function and to evaluate their putative implications in specific pathologies. It becomes crucial to develop pharmacological tools that are active on a given target protein in order to characterize its physio-pathological role. Such compounds may lead to new commercial drugs. Awaiting progress in gene therapy, the pharmaceutical firms have invested in high throughput screening, combinatorial chemistry and drug design in order to exploit as much as possible the knowledge issuing from the human sequencing data.

2. SCREENING: VALIDATED TARGETS, AUTOMATED ASSAYS, COMPOUND LIBRARIES, SCREENING METHODS AND PHARMACO-INFORMATICS

The basic idea behind the screening strategy is to obtain all the interesting biological targets defined through a bioinformatic analysis, to set up a screening test and to test a huge number of molecules in order to find an original compound that interacts with a given biological target. It is thus possible to obtain pharmacological tools and hits that could be optimized into commercial drugs. This strategy is theoretically feasible due to the genome knowledge and the progress in robotics. In practical terms, numerous scientific, technical, methodological and managerial problems have to be solved, namely the choice of the targets, the management of the huge amount of data, the choice of smart drug libraries, the rationale behind screening and the use of good bioinformatics and finally the follow up and the exploitation of the hits. These different aspects are briefly detailed below.

2.1. Validated targets

The human genome possesses the capacity to code more than 100 000 proteins. It is not reasonable to test all proteins. It is necessary to select a subset of interesting validated targets. What are the targets we should choose? Two kinds of targets may be considered:

- the known targets, partially characterized through their functions, but without potent or specific ligands;
- the "orphan" targets for which we know only the sequence.

Obviously most known targets are important. For instance, among the different calcium binding proteins, few active synthetic agonists or antagonists have been synthesized. It will be of paramount importance, in order to identify the different cellular calcium pathways, to discover specific inhibitors.

Among the orphan targets, which ones should we choose? How should we identify the most interesting ones? What are the best therapeutic targets? To address these issues, it will be necessary to develop and use genomic bioinformatics, the transcriptome, the proteome and the clinical study of patients and transgenic animals.

2.2. Automated assays

Once targets have been chosen, relevant and cost effective automated assays have to be developed. In most simple cases, one might transfer classical binding or enzymatic assays from the bench to a robot. However, a number of problems become acute in high throughput screening: quantity and cost of consumables and biological material, cost and safety issues associated with the use of radioactivity, the duration of the assay, etc. This necessitates additional fundamental and technical research and development to set up more sensitive and cost effective assays. Miniaturized assays in microtiterplates (96, 384, 1536 wells) are now operational as well as assays on solid support (eg chips). Sophisticated screening techniques are under investigation showing advantages but also new limitations compared to radio- or fluorescence detection (for instance, mass spectrometry, Nuclear Magnetic Resonance, microcalorimetry). Using molecular biology, smart biological constructs allow the development of alternative detection methods based for instance on Fluorescence Resonance Energy Transfer, luminescence, membrane potential variations, etc.

Finally, let us mention the development of automated assays aiming at screening compounds in order to evaluate their absorption, distribution, metabolism and toxicity at a very early stage, before entering into the most expensive drug development phases.

The design of miniaturized automated assays is not a trivial task. It still necessitates a lot of effort in fundamental research and technology transfer.

2.3. Compound libraries

In order to obtain hits on a validated biological target, one must have access to large libraries of compounds to be screened. What are the different sources of molecules? The first one consists in natural or synthetic compounds that are already available in academic or industrial laboratories. Pharmaceutical companies typically had from 40 000 to 200 000 products in their archives. Over the past five years, an active purchasing and exchange policy has led to corporate libraries containing up to one million compounds that could be screened on different targets. However, the "limited" number of compounds available and the very high throughput of the screening robots led chemists to develop new strategies and methods to speed up synthesis. This resulted in the emergence of "combinatorial chemistry" that now makes possible the synthesis of medium size to huge libraries (hundred to a one million compounds per campaign) in a few days. Millions of molecules from these two sources have thus been screened based on the idea that, statistically, testing one million molecules should lead to hits for any target. However, one has rapidly observed that this number was much too high for some targets (when hundreds of hits were found) and too low for others (when no hit could be found). Hence, the current trend is to develop smarter (cost effective) screening to improve the ratio: number of hits *versus* number of compounds being tested. This might be achieved in two ways.

- First, in identifying the sub-library with the minimal size which best represents the molecular diversity of the parent library. This subset will be screened first to evaluate the hit rate on any new target and to tune the screening strategy.
- The second approach consists in rationalizing the selection of compounds to be screened in taking into account existing structure-activity relationships. For example, bioinformatics and modeling methods might make it possible to link the target protein to a well known structural class of homologous proteins. Three dimensional modelling, pharmacophore searching, and virtual screening techniques might then provide useful information about the chemicals that are the most likely to interact with the target protein. These molecules will be screened with priority.

These two strategies (molecular diversity based-sampling and rational design and selection of biased libraries for a given target protein) are currently being extensively explored but their efficiency has still to be further established.

2.4. Pharmaco-informatics

Bioinformatics can be defined as the set of computer techniques used to analyze biological data. It covers many areas of biology and might intervene at different steps in the process of discovery of bioactive molecules. We will discuss here the specificity of "drug bioinformatics", or "pharmaco-informatics".

Pharmaco-informatics can be defined as the *in silico* strategies and methods to be set up in order to facilitate the understanding of protein physiological function and the discovery of bioactive ligands. Downstream of genomic bioinformatics, it aims at going from the annotated sequences to drug design and optimization. Its different components include the following:

- Refinement of the classification of proteins into structural and functional classes using sequence comparison programs (in one, two or three dimensions);
- Definition of a three-dimensional model of the target protein by homology with structural congeners, when possible;
- Docking study of known ligands and simulation of the molecular mechanisms underlying the protein function;
- Design and diversity evaluation of combinatorial chemistry libraries;
- Virtual screening of compound libraries;
- Database mining;
- Structure-activity relationship study and pharmacophore characterization in order to optimize the potency and selectivity of existing ligands;
- Prediction of physicochemical parameters (solubility, partition coefficient, etc.) and pharmacokinetic properties (absorption, metabolism, distribution, toxicity, etc.).

Several programs and packages are currently available to perform some of these tasks. However, extensive development and experimental validation are still necessary to reinforce the contribution of pharmaco-informatics to the study of protein function and to drug design.

3. PERSPECTIVES OF THE POST-GENOMIC ERA IN THE CALCIUM FIELD

One of the main challenges in the calcium field is to trace the signal transduction pathway in normal and tumor cells. The exhaustive knowledge of the complete genome has led A. Gilman to launch an interdisciplinary initiative to describe G-protein associated pathways in two model systems, namely the cardiac cell and the B lymphocyte (Alliance for cell signaling: http://afcs.swmed.edu). This type of study needs the use of new technologies to analyze the transcriptome (the expressed genes), the proteome (the expressed proteins) in a given cell or the metabolome (macromolecules involved in a given metabolic pathway). The establishment of national genomic centers allows the propagation of such techniques and a dramatic increase in the production of experimental data. We are urged to organize and to structure such an avalanche of data into usable models.

The knowledge of the human genome must make it possible to describe the complete set of putative calcium binding proteins and then to look for their targets. Using technologies (see the chapter by Nadal and Soria in this book) that allow us to see the calcium signals and the macromolecules involved in such pathways in a given cell at a given time, we now have the opportunity to detect the differences between a normal cell and a tumoral cell.

This set of macromolecules is now amenable to screening in a rational manner. In the near future, we will have thousands of new targets that are going to be described. We need to be intelligent and rational in order not to be overwhelmed by the huge number of possibilities that are now opened up.

PART ONE

Basic Mechanisms of Action of Calcium

Ca^{2+} Release in Muscle Cells

Nathalie Macrez and Jean Mironneau

1. INTRODUCTION

It is well established that Ca^{2+} release from intracellular stores is required for muscle contraction. This process involves Ca^{2+} release channels such as the ryanodine receptor (RYR) and the inositol 1,4,5-trisphosphate receptor ($InsP_3R$). Confocal imaging has recently made possible the visualization of localized Ca^{2+} release through RYRs and $InsP_3Rs$ leading to the existence of a hierarchy of Ca^{2+} signals from the smallest Ca^{2+} response corresponding to the opening of a single channel to the propagating Ca^{2+} wave. Both RYRs and $InsP_3Rs$ show some structural and functional similarities (both are tetramers regulated by Ca^{2+}) but they also display discrepancies and distinct tissue distribution. Although both channel families are activated by Ca^{2+} in a mechanism known as "Ca^{2+}-induced Ca^{2+} release" (CICR), the different classes and isoforms display distinct Ca^{2+}-dependencies. Some structure/function relationships underlying these differential regulation by Ca^{2+} ions have been reported elsewhere (Ehrlich, 1995) and in this book (see the chapter by Parys et al., this book). Here, it is our intention to review the different isoforms of RYRs and $InsP_3Rs$ and their role in inducing localized Ca^{2+} responses and propagated Ca^{2+} waves in muscle cells. In addition, we will report how the local control of excitation-contraction coupling may be used to propose new subcellular mechanisms to explain modifications of the contractility in physiopathological conditions.

Nathalie Macrez and Jean Mironneau • Laboratoire de Physiologie Cellulaire et Pharmacologie Moléculaire, CNRS UMR 5017, Université de Bordeaux II, Bordeaux, France.

R. Pochet, R. Donato, J. Haiech, C. Heizmann and V. Gerke (eds.): Calcium: The Molecular Basis of Calcium Action in Biology and Medicine, 9–25.

2. DIVERSITY OF Ca^{2+} RELEASE CHANNELS IN MUSCLE CELLS

For each one of the RYR and $InsP_3R$ family of Ca^{2+} release channels, three subtypes have been described and cloned. Further structural diversity of Ca^{2+} release channels arises from alternative splicing of the mRNA encoding them and from assembly of heterotetrameric complexes of IP3Rs (Joseph et al., 1995). Because they show different patterns of expression in different tissues and during development, each subtype of Ca^{2+} release channel might have a specific role, and their diversity may be responsible for various Ca^{2+} signals controlling cell functions.

2.1. Expression and role of RYR isoforms in muscle cells

The function of RYR as Ca^{2+}-activated Ca^{2+} release channel has been widely studied in both skeletal and cardiac muscles, but much less in smooth muscles, although members of the RYR family have been shown to be expressed in all muscle cell types. Moreover, we have recently demonstrated the functional importance of RYRs in vascular smooth muscle by showing that all Ca^{2+} release processes (initiated either by Ca^{2+} or by $InsP_3$) are amplified by a Ca^{2+}-induced Ca^{2+} release mechanism through RYRs (Boittin et al. 1998, 1999). Subtypes 1 and 3 of RYR have been shown to be expressed in early stages of skeletal muscle development. In adult animals, RYR1 has been shown to be expressed in all skeletal muscles while RYR3 display a differential distribution, having been observed in slow-twitch muscles such as the diaphragm but not in fast-twitch muscles such as abdominal skeletal muscle (Conti et al., 1996). In cardiac myocytes, both RYR subtypes 2 and 3 have been shown to be expressed (Ledbetter et al., 1994). RYR subtypes are differentially expressed in vascular and visceral smooth muscles. In intestinal, arterial and venous myocytes, all three subtypes of RYR have been reported (Neylon et al., 1995; Coussin et al., 2000); while in ureteric or non-pregnant myometrial smooth muscle mainly RYR3 have been detected by RT-PCR (Awad et al., 1997).

RYR subtypes have attracted much interest as they offer one of the structural bases supporting the difference between the skeletal and cardiac excitation-contraction (EC) coupling. The skeletal EC coupling specifically occurs through the RYR1 subtype as shown by the lack of voltage sensor-gated Ca^{2+} release in mutant mice lacking RYR1 (Takeshima et al., 1994). In spite of a recent report showing that in embryonic cardiac myocytes isolated from mutant mice lacking RYR2, the CICR mechanism was not significantly affected (Takeshima et al., 1998), the EC coupling in mature cardiac cells is thought to occur through RYR2 which is the main RYR subtype expressed in these myocytes. Up to now, the role of RYR3 in muscle cells remains unclear and the knockout mice lacking RYR3 ($RYR3^{-/-}$ show apparently normal growth and EC coupling, although a reduced skeletal muscle contraction was observed. RYR3 has been shown to be activated by high concentrations of caffeine, by high Ca^{2+}-containing external solutions and by cADPR (Sonnleitner et al., 1998; Guse et al., 1999 and this book). The normal growth of the RYR3$^{-/-}$ mice is somewhat amazing since RYR3 is widely expressed, suggesting its important role, in murine skeletal muscles during the post-natal phase of muscle development (Conti et al., 1996). Because specific inhibitors of RYR subtypes do not exist, all studies of RYR subtype functions have been performed on genetically modified mice expressing mutants of one or two RYR subtypes. However, since the mutant mice lacking functional RYR1 or RYR2 die perinatally, these models are not suitable for studying EC couplings in mature cells.

We have used an antisense oligonucleotide strategy allowing the transient knockout of RYR subtypes in mature venous myocytes. We have identified both RYR1 and RYR2 as Ca^{2+} release channels required for Ca^{2+}-induced Ca^{2+} release mechanism in vascular myocytes; while RYR3 does not seem to be involved in this process (Coussin et al., 2000). These results assign a physiological role for RYR1 as a non-voltage sensor gated Ca^{2+} release channel and provide a new piece in the puzzling debate concerning the molecular composition of RYR Ca^{2+} release units by suggesting that, in venous myocytes, RYR1 and RYR2 are grouped within the same units. We propose several hypotheses for the respective roles of RYR1 and RYR2 within these Ca^{2+} release units. A first possibility would be that the clusters are composed of RYR1 and RYR2 homotetramers which can similarly be activated by Ca^{2+}. In agreement with the concept that functional RYRs exist as homotetramers (Lai et al., 1989), we have reported existing, although decreased, global Ca^{2+} responses to caffeine when either RYR1 or RYR2 expression has been suppressed. A second possibility would be that RYR1 may be responsible for addressing of RYR2 clusters to the proper plasma membrane sites, since RYR1 has been shown to be tightly bridged to the plasma membrane through an associated protein network in skeletal myocytes (Franzini-Armstrong and Protasi, 1997). A third possibility would be that the Ca^{2+} release unit is reduced to a single Ca^{2+} release channel able to generate a Ca^{2+} spark. Under these conditions, in venous myocytes, this channel might be an heterotetramer made from RYR1 and RYR2 proteins. This third possibility seems unlikely since RYRs have been suggested to be organized as homotetramers only (Lai et al., 1989), whereas both homo- and heterotetramers have been reported for $InsP_3Rs$ (Joseph et al., 1995). Biochemical reconstitutions are needed to elucidate how each RYR subtype contributes to the Ca^{2+} release units in vascular myocytes.

2.2. Expression and role of $InsP_3R$ isoforms in muscle cells

The increase in cytoplasmic Ca^{2+} ($[Ca^{2+}]_i$) mediated by $InsP_3Rs$ is important for various cellular processes such as neurotransmission, plasticity of synaptic response, secretion and smooth muscle contraction (Berridge, 1993; Parys et al., this book). Skeletal and cardiac muscles are special tissues where the ryanodine receptor is the main channel for Ca^{2+} release from sarcoplasmic reticulum (SR). However, although not clearly demonstrated, many studies suggest that $InsP_3Rs$ may play a role in these striated muscles as well. For example, some hormones that activate the $InsP_3$ cascade enhance the cardiac contraction (Vigne et al., 1989); and $InsP_3Rs$ have been reported to be expressed in cardiac and skeletal myocytes (Moschella and Marks, 1993).

To date, the coding regions of three mammalian $InsP_3R$ genes have been sequenced and type 1 of $InsP_3R$ ($InsP_3R1$) mRNA has been shown to undergo alternative splicing in three distinct regions designated as S1–S3. Based on partial sequence analysis, it has been suggested that additional $InsP_3Rs$ isoforms ($InsP_3R4$ and $InsP_3R5$) are expressed in mouse cells; but further RT-PCR study only detects three $InsP_3Rs$ in various mouse cells and the putative types 4 and 5 are supposed to be mouse counterparts of rat $InsP_3Rs$ (De Smedt et al., 1997). In a wide range of cell types tested, the expression patterns reveal simultaneous expression of $InsP_3R1$, $InsP_3R2$ and $InsP_3R3$. However, the subtypes are expressed in different amounts in different cell types. $InsP_3R1$ is particularly predominant in the central nervous system, in oocytes and eggs and in vascular smooth muscle (De Smedt et al., 1997; Parys et al., this book). The splice variant of $InsP_3R1$ expressed in smooth muscle (as well as in most of the peripheral tissues) is thought to belong to the shortest form (S1–S3 regions deleted) of $InsP_3R1$; the S1+ and the S3+ variants have been found in brain and nervous system while the S2+ variant seems to be related to T-lymphocytes and fibroblasts. Heart and skeletal muscles predominantly express $InsP_3R2$ (De Smedt et al., 1997). Skeletal muscle also express significant $InsP_3R3$ levels. Higher levels of $InsP_3R3$ *versus* $InsP_3R1$ have been immunodetected in vessels from neonatal rats, while this ratio is inverted in fully developed animals. Messenger RNAs encoding types 1, 2 and 3 of $InsP_3R$ have been detected in visceral smooth muscles (Morgan et al., 1996), but their presence has not been checked in vascular smooth muscles. Interestingly, high levels of $InsP_3R3$ have been observed in all cultured cell types, suggesting a relation between cell proliferation and $InsP_3R3$ expression (De Smedt et al., 1997).

The various $InsP_3R$ subtypes have been shown to display distinct $InsP_3$-, ATP- and Ca^{2+}-sensitivities leading to generation of different Ca^{2+} signals within the same cellular expression system (Miyakawa et al., 1999). Moreover, co-expression of several $InsP_3R$ subtypes may generate heterotetramers (Joseph et al., 1995), thus producing additional degrees of variability for $InsP_3R$ function and $InsP_3$-mediated Ca^{2+} signalling.

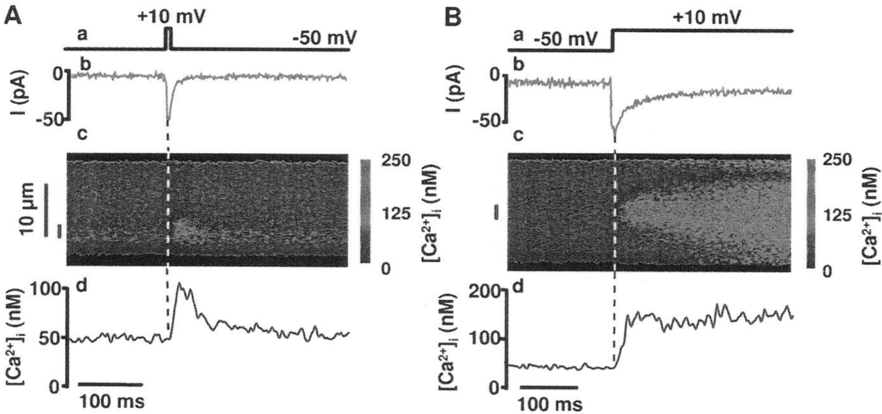

Figure 1. Ca^{2+} sparks and Ca^{2+} waves activated by membrane depolarizations in isolated vascular myocytes. (A) A depolarization from −50 to +10 mV for 10 ms (a) activates a brief Ca^{2+} current (b) and a Ca^{2+} spark illustrated in the line-scan image (c) and shown as spatial averaged fluorescence (d) from a 2 μm-region of the line-scan image. (B) A longer depolarization to +10 mV (a) activates a durable Ca^{2+} current (b) and a propagated Ca^{2+} wave (c, d). Myocytes are loaded with Fluo-3/Fura red mixture and held at −50 mV. Modified from Arnaudeau et al. (1997) with permission from the publisher, Churchill Livingstone. For a colour version of this figure, see page xxiii.

Taking into account the diversity of $InsP_3Rs$ and RYRs, their co-expression in many cell types may provide structural supports for the hypothesis that the combination of various Ca^{2+} release channels may be involved in the regulation of specific cellular functions.

3. Ca^{2+} SIGNALS ACTIVATED BY Ca^{2+} RELEASE CHANNELS

Recent data obtained by confocal imaging have suggested that the intracellular Ca^{2+} signals are organized as a hierarchy (Lipp and Niggli, 1996; Berridge, 1997). The opening of individual Ca^{2+} release channels gives rise to fundamental events referred to as blips in the case of $InsP_3R$ or quarks for the RYR. The next level of organization is represented by small groups of $InsP_3Rs$ or RYRs releasing Ca^{2+} ions as localized units to give puffs and sparks, respectively. Typical examples of homologous hierarchical Ca^{2+}-signalling systems have been described in non-excitable and excitable cells (Lipp and Niggli, 1996; Bootman et al., 1997).

3.1. Ryanodine-sensitive Ca^{2+} release channels

Elementary Ca^{2+} signals have been detected in skeletal, cardiac and smooth muscles (Cheng et al., 1993; Nelson et al., 1995; Tsugorka et al., 1995; Arnaudeau et al., 1996). Although amplitude of Ca^{2+} sparks may differ in different muscle cells, the kinetic parameters remain relatively similar. For example, the mean time to reach peak Ca^{2+} spark, the half-time of decay and the mean full width at half-maximal amplitude (FWHM) are around 20 ms, 25 ms and 2–5 μm, respectively. Since Ca^{2+} sparks are spatially restricted Ca^{2+}-release events, their amplitude may depend on the concerted activity of a cluster or group of RYRs in a specific area of the sarcoplasmic reticulum. In vascular myocytes, a quasi-maximal L-type Ca^{2+} current evokes a single Ca^{2+} spark (Figure 1A), in contrast to cardiac cells where partial inhibition of L-type Ca^{2+} channels is needed to trigger isolated Ca^{2+} sparks. This observation suggests that the detection of Ca^{2+} sparks implies a system with a low positive feedback control and that the size of Ca^{2+} sparks is below the critical threshold for propagation. In addition, Ca^{2+} sparks appear to be localized in special areas of the cells (T-tubule or caveolae) and cannot be triggered in other areas. As a consequence, Ca^{2+} sparks are induced by activation of L-type Ca^{2+} channels, whereas homogeneous Ca^{2+} trigger signals (e.g. UV-flash photolysis of DM-nitrophen) evoke apparently homogeneous Ca^{2+} transients in cardiac myocytes because they recruit all the RYRs (isolated and clustered Ca^{2+} release channels). However, in vascular myocytes, UV-flashes of reduced intensity appear to activate isolated Ca^{2+} sparks in the same areas where L-type Ca^{2+} currents are effective (Arnaudeau et al., 1997).

The existence of a close association between sarcolemmal L-type Ca^{2+} channels and ryanodine-sensitive Ca^{2+} release channels is supported by both immunological and ultrastructural studies in muscle cells. Opening of a single L-type Ca^{2+} channel could be sufficient to trigger Ca^{2+} release from the SR (Lopez-Lopez et al., 1995) as suggested by the voltage-dependency of Ca^{2+} spark triggering. However, similar whole-cell Ca^{2+} currents obtained at -10 mV and $+30$ mV produce different Ca^{2+} signals in vascular myocytes (Arnaudeau et al., 1997). At -10 mV, the L-type Ca^{2+} channel opens infrequently; but the single-channel current is large because of the important electrochemical gradient for Ca^{2+} ions and it then triggers Ca^{2+} sparks. In contrast, at $+30$ mV, the Ca^{2+} channel opens frequently but the single-channel current is smaller as the reversal potential is approached and is unable to trigger a Ca^{2+} spark. Thus, the amplitude of the single-channel current could be crucial for triggering either Ca^{2+} sparks or homogeneous Ca^{2+} response. In recent years, cloning and sequencing of 3 genes encoding different RYR subtypes and 10 genes encoding dihydropyridine receptor (DHPR) subtypes have provided structural bases for the understanding of the different types of RYR activation by membrane depolarization. The Ca^{2+} release in skeletal muscle depends on RYR1 activation; although RYR3 may contribute to Ca^{2+} sparks in embryonic skeletal muscle (Conklin et al., 1999). In cardiac muscle, the Ca^{2+} release is thought to occur through RYR2. A recent paper shows that the cardiac RYR2 expressed in CHO cells forms functional Ca^{2+} release channels (Bhat et al., 1999). However, the lack of Ca^{2+} sparks in these cells supports the idea that Ca^{2+} sparks observed in native cells involve cooperative gating of a group of RYRs and/or their interaction with

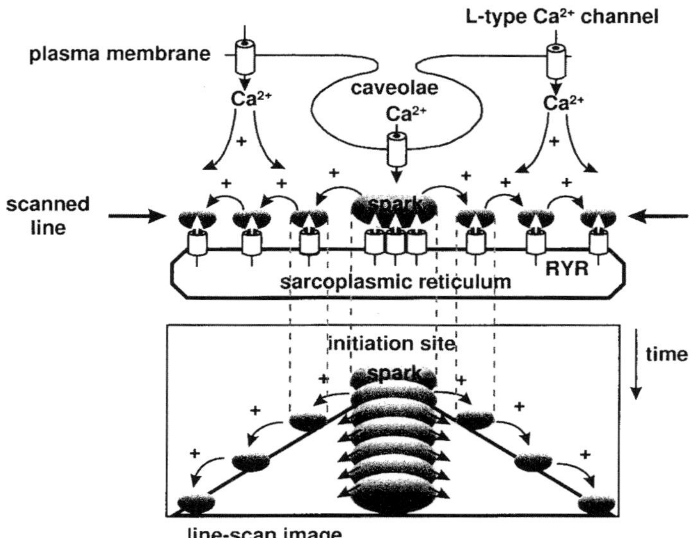

Figure 2. Propagation of Ca^{2+} wave initiated by the prolonged activation of L-type Ca^{2+} channels in vascular myocytes. A single L-type Ca^{2+} channel controls the opening of a Ca^{2+} release unit producing a Ca^{2+} spark. Repetitive activation of the Ca^{2+} spark recruits neighbouring isolated ryanodine channels by Ca^{2+} diffusion. L-type Ca^{2+} channels not coupled with ryanodine channels may increase the $[Ca^{2+}]_i$ in the vicinity of isolated ryanodine channels.

specific proteins. In vascular myocytes where the three RYR subtypes are expressed, selective deletion of one RYR subtype by an antisense oligonucleotide strategy reveals that both RYR1 and RYR2 subtypes are required for Ca^{2+} sparks activated by voltage-gated Ca^{2+} channels (Coussin et al., 2000). Different hypotheses concerning RYR assembling in Ca^{2+} release units are detailed in Section 2.1.

Recruitment of a varying number of Ca^{2+} release units is one mechanism which enables the myocyte to generate smoothly graded Ca^{2+} responses or propagated Ca^{2+} waves. In cardiac cells, Ca^{2+} waves have been reported to be due to the recruitment and summation of Ca^{2+} sparks (saltatory transmission) which can be detected in the wave front (Cheng et al., 1996). In contrast, in vascular myocytes, propagated Ca^{2+} waves start from a Ca^{2+} spark as initiation site which is repetitively activated during the depolarization and then progressively recruits neighbouring single Ca^{2+} release channels, probably not coupled with L-type Ca^{2+} channels (Figures 1B and 2). Interestingly, transmission of the wave in vascular myocytes appears as fast as that measured in cardiac cells (50–70 μm.s^{-1}).

Spontaneous Ca^{2+} sparks have been recorded in all types of muscle cells with amplitude and kinetic parameters similar to Ca^{2+} sparks triggered by L-type Ca^{2+} channel activation. It is likely that this similarity is linked to the micro-architecture of the SR, i.e. clustering of RYRs. The most plaus-

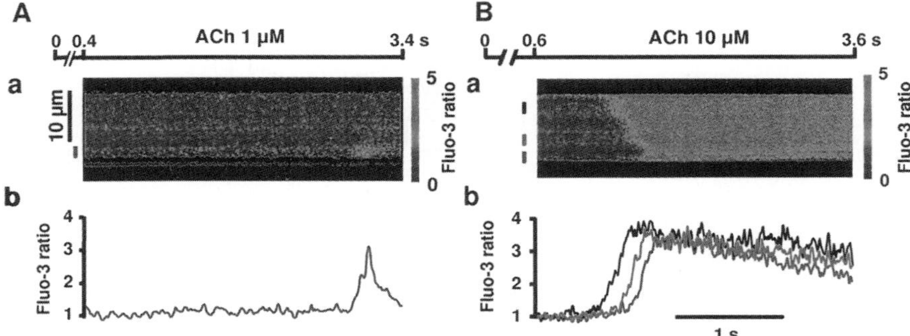

Figure 3. Localized Ca^{2+} signals ("Ca^{2+} puffs") and Ca^{2+} waves in rat ureteric myocytes. (A) Typical Ca^{2+} responses evoked by 1 μM acetylcholine, shown as line-scan image (a) and averaged fluorescence from a 2-μm region (b), indicated by the vertical bar on the line-scan image. (B) Propagated Ca^{2+} wave evoked by 10 μM acetylcholine, shown as line-scan image (a) and averaged fluorescence from three 2-μm regions (b), indicated by vertical bars on the line-scan image. Myocytes are loaded with Fluo-3/AM and not patch-clamped. No change in fluorescence corresponds to Fluo-3 ratio = 1. For a colour version of this figure, see page xxiii.

ible mechanism responsible for spontaneous Ca^{2+} sparks is the loading state of the SR (i.e. intraluminal [Ca^{2+}]). It has been reported that the luminal [Ca^{2+}] controls the gating behaviour of the Ca^{2+} release channels by shifting the sensitivity of the channel towards lower $[Ca^{2+}]_i$ with increasing luminal [Ca^{2+}]. This notion implies that every mechanism shifting the Ca^{2+}-dependency of RYR (associated proteins, phosphorylation, second messenger) will influence the probability of Ca^{2+} spark triggering. Spontaneous Ca^{2+} sparks may exert a key role in controlling the basal electrical activity of vascular myocytes since they can activate both Ca^{2+}-dependent K^+ channels and Ca^{2+}-dependent non-selective cation channel (Section 4.2).

3.2. $InsP_3$-gated Ca^{2+} release channels

Ca^{2+} signals evoked by activation of $InsP_3$Rs are poorly understood in muscle cells. In smooth muscles, there are indications that $InsP_3$Rs and RYRs are present on the SR membrane and even that they can be localized on the same intracellular Ca^{2+} store (e.g. in portal vein myocytes, see Boittin et al., 1999). However, immunostaining of $InsP_3$Rs indicates that these channels are not organized in Ca^{2+} release units, and elementary Ca^{2+} signals, i.e. Ca^{2+} puffs, have not been detected in these myocytes (Boittin et al., 1998).

It has recently been suggested that rat ureteric myocytes may possess only an $InsP_3$-sensitive Ca^{2+} store (Burdyga et al., 1998). Application of acetylcholine or photorelease of caged $InsP_3$ by UV flash pulses on rat ureteric myocytes evoke localized Ca^{2+} signals which appear to vary in amplitude, time course and spatial spread (Figure 3A). The largest Ca^{2+} signals are characterized by a time-to-peak of about 50–90 ms, a half-time decay of

100–150 ms and a FWHM of 1–2.5 μm. These kinetic parameters are similar to those reported for Ca^{2+} puffs in non-excitable cells. These observations are compatible with a scheme where $InsP_3$-gated Ca^{2+} channels would exist in clusters containing variable numbers of channels and that within these clusters any number of channels could be recruited. This proposal is supported by immunodetection of $InsP_3Rs$ performed with an anti-$InsP_3$ antibody showing that $InsP_3Rs$ are distributed in the whole confocal section, with several marked spots of fluorescence (clusters) in areas corresponding to the cell periphery or to infoldings of the plasma membrane in close association with the SR (unpublished data).

Global Ca^{2+} responses propagate throughout the cell by alternating between regeneration at several Ca^{2+} release sites along the scanned line and Ca^{2+} diffusion between the release sites. Generally, Ca^{2+} waves start from the edge of the cell, in agreement with the location of clusters of $InsP_3Rs$ near the plasma membrane, and they propagate at a velocity of 50–60 $\mu m.s^{-1}$ (Figure 3B). It is likely that Ca^{2+} puffs contribute to a slow rise in basal free $[Ca^{2+}]_i$ which further increases puff frequency and finally sensitized $InsP_3Rs$ by binding $InsP_3$ and Ca^{2+}. Triggering of a Ca^{2+} wave may depend in principle on recruitment of Ca^{2+} puffs in frequency, amplitude and spatial domains (Bootman et al., 1997).

Existence of elementary and propagated Ca^{2+} signals in ureteric myocytes is related to the high expression of $InsP_3Rs$, as revealed by binding experiments with $[^3H]InsP_3$, and this suggests a micro-architecture of Ca^{2+} release channels. The complex expression pattern of $InsP_3R$ subtypes is believed to be responsible for the generation of cell-type specific Ca^{2+} signalling. Differential expression of $InsP_3R$ subtypes has revealed that Ca^{2+}-signalling patterns differ in their responses to $InsP_3$, Ca^{2+} and ATP. $InsP_3R2$ is the most sensitive to $InsP_3$ and is required for Ca^{2+} oscillations; $InsP_3R1$ is highly sensitive to ATP and mediates less regular Ca^{2+} oscillations; $InsP_3R3$ is the least sensitive to $InsP_3$ and Ca^{2+} and generates monophasic Ca^{2+} responses (Miyakawa et al., 1999).

3.3. Cooperativity between ryanodine- and $InsP_3$-sensitive Ca^{2+} release channels

In myocytes when both $InsP_3Rs$ and RYRs coexist, a cooperativity between these receptors has been reported which represents a mechanism to amplify Ca^{2+} release from the same intracellular store and to give rise to large propagated Ca^{2+} waves (Boittin et al., 1998). In vascular myocytes, the density of RYRs is 3–4 times higher than that of $InsP_3Rs$, whereas the opposite is observed in intestinal smooth muscle (Wibo and Godfraind, 1994). In muscles displaying a higher density of RYRs (such as vascular myocytes), release of $InsP_3$, either from a caged compound by UV flashes or activation of muscarinic receptors leading to $InsP_3$ production, evokes homogeneous

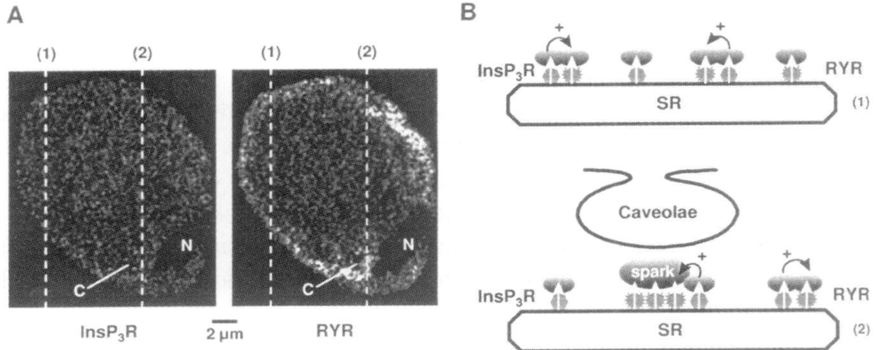

Figure 4. Subcellular localization of Ca^{2+} release channels in vascular myocytes. (A) Immunolocalization of $InsP_3$ receptors ($InsP_3R$) and ryanodine receptors (RYR) in confocal cell section prepared by double staining protocol. N: nucleus; C: caveolae. (B) Schematic representation of the micro-architecture of the sarcoplasmic reticulum (SR) in areas where both RYRs and $InsP_3Rs$ are homogeneously distributed (dotted line 1) or where RYRs are clustered in the vicinity of caveolae of the plasma membrane (dotted line 2). Opening of isolated $InsP_3R$ may initiate Ca^{2+} sparks via a recruitment process, and repetitive activation of these Ca^{2+} sparks may lead to a propagated Ca^{2+} wave. Reprinted from Boittin et al. (1998) with permission from the publisher, Churchill Livingstone.

Ca^{2+} responses of low amplitude and slow upstroke velocity when RYRs are blocked by high concentrations of ryanodine or intracellular application of an anti-ryanodine receptor antibody (Boittin et al., 1999). These functional experiments are supported by immunostaining showing the lack of high fluorescence spots in the confocal cell sections. When both RYRs and $InsP_3Rs$ are functional, Ca^{2+} sparks can be detected in isolation in response to low concentrations of noradrenaline or low intensity UV flashes. Maximal Ca^{2+} responses are obtained with high production of $InsP_3$ and correspond to the addition of $InsP_3$- and Ca^{2+}-activated Ca^{2+} components. The $[Ca^{2+}]_i$ threshold for activation of RYR is estimated to be around 75–95 nM; whereas the Ca^{2+}-dependent positive feedback of $InsP_3$-induced Ca^{2+} release is around 150–180 nM in various cell types (Iino et al., 1993). Thus, the small increase in $[Ca^{2+}]_i$ due to the opening of $InsP_3$-gated channels is able to activate the RYRs located in the vicinity of these $InsP_3Rs$, then producing Ca^{2+} sparks and subsequently the propagated Ca^{2+} waves. Such a cooperation between RYRs and $InsP_3Rs$ implies a micro-architecture where $InsP_3Rs$ are distributed homogeneously in the SR, even within the specialized areas showing clusters of RYRs (Figure 4). The physiological role of this Ca^{2+}-amplifying mechanism is supported by contraction experiments showing that the noradrenaline-induced contraction is reduced by 40–50% after blockade of RYRs.

Cooperativity between $InsP_3Rs$ and RYRs may be considered as a mechanism functionally equivalent to the Ca^{2+}-dependent positive feedback control of $InsP_3$-induced Ca^{2+} release, observed in some cell types (Iino and Endo, 1992).

4. CELLULAR MECHANISMS ACTIVATED BY Ca^{2+} SIGNALS

In muscle cells, an increase in cytosolic Ca^{2+} is generally associated with a contraction process. However, description of hierarchies of Ca^{2+} release signals, showing various spatio-temporal patterns, raises the question which Ca^{2+} pattern is efficient for inducing contraction and for activating ion channels sensitive to $[Ca^{2+}]_i$.

4.1. Ca^{2+} release events and contraction

The description of localized Ca^{2+} release events raised the question of their functional role in muscle contraction. Are puffs and sparks only trigger signals, giving rise to more global Ca^{2+} responses responsible for the cellular responses, or, are sparks and puffs sufficient to trigger cellular responses by themselves? The relation between a single Ca^{2+} release event and contraction has been studied only for RYRs. To date, no link has been described between the pattern of $InsP_3R$-gated Ca^{2+} events and cellular responses.

In all muscle types, spontaneous Ca^{2+} sparks have been observed in resting conditions suggesting that these sparks are not sufficient to mediate contraction but that they may participate to the resting Ca^{2+} level. In a pressurized mesenteric artery, the myogenic tone is under the control of L-type Ca^{2+} channels and the subsequent Ca^{2+} sparks. In this smooth muscle, inhibition of L-type Ca^{2+} channels causes nearly full vasodilatation, although Ca^{2+} oscillations may persist (Miriel et al., 1999). A relaxing effect of spontaneous Ca^{2+} sparks has been suggested on the basis of the simultaneous recordings of both Ca^{2+} sparks and STOCS (Spontaneous Transient Outward Currents) in resting cells (Nelson et al., 1995). However, in all muscle types, stimulation by caffeine or membrane depolarization gives rise to Ca^{2+} sparks initiating global Ca^{2+} responses and contractions, so that evoked Ca^{2+} sparks can be considered as contracting rather than relaxing events. Some recent studies support the hypothesis that different types of sparks may exist within a cell and that some of these sparks are activated spontaneously in "frequent discharge sites" (Bolton and Gordienko, 1998). However, since we have shown that depolarization or caffeine evoke Ca^{2+} sparks and waves arising from the same sites as spontaneous Ca^{2+} sparks (Mironneau et al., 1996), the same spark-generating sites could be involved in both relaxing and contracting processes. In the case of evoked sparks, repetitive stimulation of RYRs by L-type Ca^{2+} current may produce a durable activation of a RYR cluster, activating the neighbouring isolated RYRs and leading to a Ca^{2+} wave underlying the contraction.

In all muscle cell types, the global Ca^{2+} responses arising from the synchronized release of Ca^{2+} from a large number of Ca^{2+} release channels produce contractions. Depending on the cell type, different mechanisms of synchronization have been described involving simultaneous stimulation of several clusters of Ca^{2+} release channels, synchronization of several clusters through a bridging protein (like FKBP12), and local regeneration of the Ca^{2+} release process by the CICR mechanism (Berridge, 1997; Marx et al., 1998).

It is obvious that the respective location of Ca^{2+} release units and ion channels in the plasma membrane as well as the timing and extent of the Ca^{2+} release events determine whether they contribute to a global Ca^{2+} signal leading to contraction, or to a local signal which may lead to relaxation.

4.2. Ion channels of the plasma membrane regulated by Ca^{2+} release events

As mentioned above, the relation between Ca^{2+} release events and ion channels is probably a key component regulating cellular activity. Several interactions between RYRs or RYR-mediated Ca^{2+} release and ion channels of the plasma membrane have been reported, thus establishing an indirect control of the membrane potential by RYRs.

Since the description of STOCs as K^+ channels transiently activated in resting smooth muscle cells (Benham and Bolton, 1986), the Ca^{2+}-dependence of STOC activation has been well documented. More recently, Nelson et al., (1995) suggested that STOC could be activated by Ca^{2+} sparks. Several groups have now reported simultaneous activation of Ca^{2+} sparks and STOCs on the same cell (Mironneau et al., 1996; Perez et al., 1999); however, in the whole-cell recording mode, the location of STOC towards Ca^{2+} sparks has not been resolved. Among the other conductances regulated by Ca^{2+}, the cationic STICs (Spontaneous Transient Inward Currents) have also been shown to be activated simultaneously with Ca^{2+} sparks in vascular myocytes (Mironneau et al., 1996). Activation of such depolarizing current could counteract the hyperpolarizing K^+ current also activated by sparks, thus creating a balance of membrane conductances regulating the resting potential. However, since the average density of K^+ channels has been estimated to be 1000–3000 channels per cell whereas the density of cationic channels is around 1–5 channels per cell, it is likely that the dominant function of Ca^{2+} sparks remains to hyperpolarize the cell membrane.

The Ca^{2+}-dependent chloride channels are not observed during Ca^{2+} spark activation but they are activated during the global Ca^{2+} responses mediated either by RYRs or $InsP_3Rs$ (Mironneau et al., 1996; Machaca and Hartzell, 1999). This observation suggests that the chloride channels have a higher threshold of activation by Ca^{2+} or that they are not co-localized with Ca^{2+} sparks. Thus, the Ca^{2+}-activated chloride channels are thought to participate in the large global Ca^{2+} responses rather than regulating myogenic vascular tone.

Recently, $InsP_3Rs$ have been described to specifically interact with SOCs (Store-Operated Ca^{2+} channels) which are thought to be involved in the maintained phase of long-lasting Ca^{2+} signals or in the refilling of the emptied Ca^{2+} stores (Kiselyov et al., 1998). These data indicate that $InsP_3Rs$ may probably regulate muscle contraction via ways other than only the Ca^{2+} release process.

5. PHYSIOPATHOLOGICAL IMPLICATIONS

Fundamental knowledge of the local control mechanism of excitation-contraction coupling can help with better understanding of some dysfunctions or modifications by experimental conditions of cardiac and vascular myocytes. Two examples are illustrated.

5.1. Cardiac dysfunction

The local control mechanism of excitation-contraction coupling in heart myocytes is impaired in two forms of cardiac dysfunction. In hypertension, cardiac myocytes compensate for the increased pressure by a cell hypertrophy but demonstrate an impaired contraction. By studying the different steps involved in the local-control mechanism, it has been proposed that an impaired coupling between adjacent L-type Ca^{2+} channels and RYRs is the primary defect in the overall contractile failure, without any alterations in L-type Ca^{2+} current and Ca^{2+} sparks (Gomez et al., 1997). If hypertension persists, a congestive heart failure develops, which shows a similar defect in coupling. Thus, variable forms of cardiac dysfunction may depend on the same defect at the fundamental cellular level. One possible explanation for the impaired coupling in cardiac dysfunction is that it may be related to changes in the micro-architecture of local response elements in specialized areas where L-type Ca^{2+} channels and RYRs are co-localized (Figure 5).

5.2. Venous dysfunction

Simulated weightlessness elicits many of the alterations that occur during inactivity and old age in the cardiovascular system, e.g. decreased contractile activity. In venous myocytes of rats suspended for 14 days, the decrease in contractility is related to smaller Ca^{2+} responses evoked by neuromediators and hormones. L-type Ca^{2+} currents, Ca^{2+} sparks as well as Ca^{2+} sensitivity of RYRs remain unchanged during hindlimb suspension (Morel et al., 1997). In contrast, the propagated Ca^{2+} waves are significantly reduced in parallel

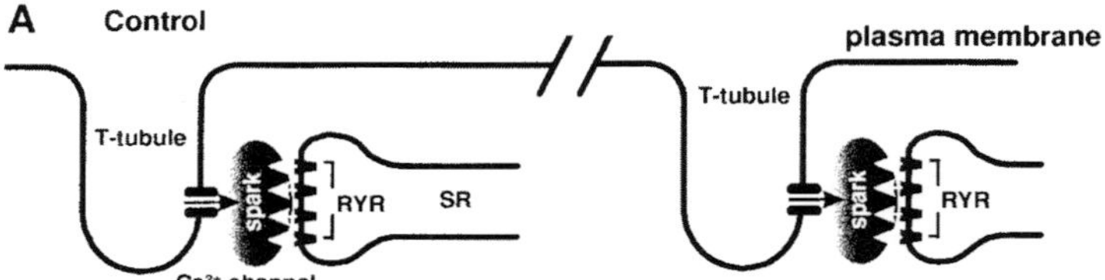

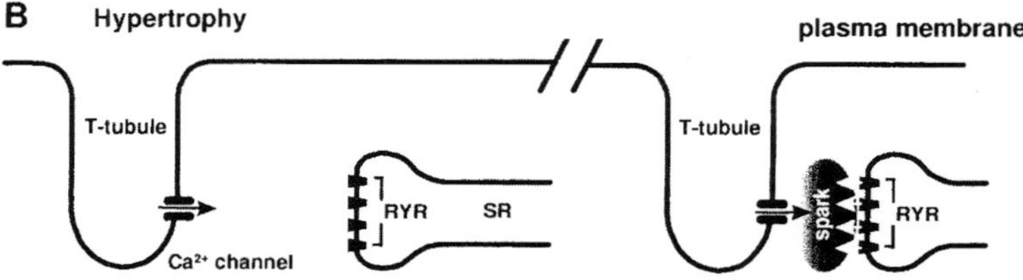

Figure 5. Local control of Ca^{2+} release in control (A) and hypertrophied cardiac myocytes (B). In the latter case, the defect would result in increased spacing between L-type Ca^{2+} channels of the T-tubule and RYRs. SR: sarcoplasmic reticulum.

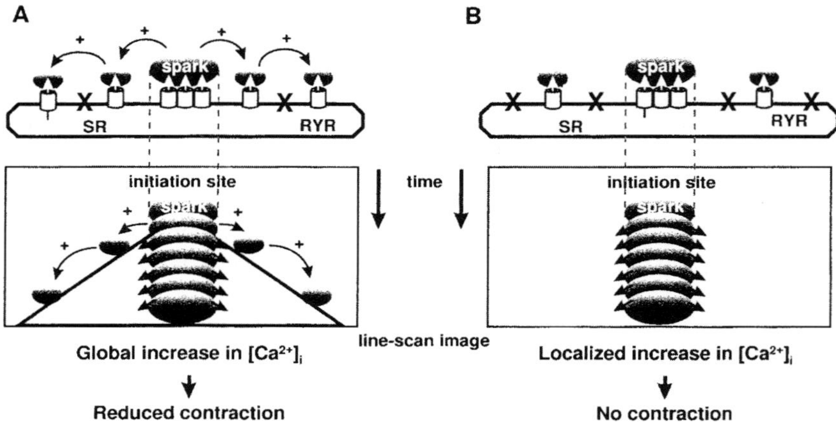

Figure 6. Local control of Ca^{2+} release in venous myocytes of hindlimb suspended rats (B). (A) In myocytes where the RYR density is slightly reduced (X) a global increase in $[Ca^{2+}]_i$ triggers a reduced contractile response. (B) In myocytes where the RYR density is strongly reduced, Ca^{2+} sparks are recorded but Ca^{2+} wave is not triggered because isolated RYRs are too distant to be activated by Ca^{2+} diffusion. SR: sarcoplasmic reticulum.

with a noticeable inhibition of [^{3}H]ryanodine binding. Therefore, the reduction of the number of functional RYRs (and $InsP_3Rs$) in the SR of venous myocytes may explain the impaired contractility without any alterations in the micro-architecture of the SR (Figure 6).

REFERENCES

Arnaudeau, S., Macrez-Leprêtre, N. and Mironneau, J., 1996, Activation of calcium sparks by angiotensin II in vascular myocytes, *Biochem. Biophys. Res. Commun.* 222, 809–815.

Arnaudeau, S., Boittin, F.X., Macrez, N., Lavie, J.L., Mironneau, C. and Mironneau, J., 1997, L-type and Ca^{2+} release channel-dependent hierarchical Ca^{2+} signalling in rat portal vein myocytes, *Cell Calcium* 22, 399–411.

Awad, S.S., Lamb, H.K., Morgan, J.M., Dunlop, W. and Gillespie, J.I., 1997, Differential expression of ryanodine receptor RYR2 mRNA in the non-pregnant and pregnant human myometrium, *Biochem. J.* 322, 777–783.

Benham, C.D. and Bolton, T.B., 1986, Spontaneous transient outward currents in single visceral and vascular smooth muscle cells of the rabbit, *J. Physiol. (Lond.)* 381, 385–406. Berridge, M.J., 1993, Inositol trisphosphate and calcium signalling, *Nature* 361, 315–325.

Berridge, M.J., 1997, Elementary and global aspects of calcium signalling, *J. Physiol. (Lond.)* 499, 291–306.

Bhat, M.B., Hayek, S.M., Zhao, J., Zang, W., Takeshima, H., Wier, W.G. and Ma, J., 1999, Expression and functional characterization of the cardiac muscle ryanodine receptor Ca^{2+} release channel in Chinese hamster ovary cells, *Biophys. J.* 77, 808–816.

Boittin, F.X., Coussin, C., Macrez, N., Mironneau, C. and Mironneau J., 1998, Inositol 1,4,5-trisphosphate- and ryanodine-sensitive Ca^{2+} release channel-dependent Ca^{2+} signalling in rat portal vein myocytes, *Cell Calcium* 23, 303–311.

Boittin, F.X., Macrez, N., Halet, G. and Mironneau J., 1999, Norepinephrine-induced Ca^{2+} waves depend on $InsP_3$ and ryanodine receptor activation in vascular myocytes, *Am. J. Physiol.* 277, C139–C151.

Bolton, T.B. and Gordienko, D.V., 1998, Confocal imaging of calcium release events in single smooth muscle cells, *Acta Physiol. Scand.* 164, 567–575.

Bootman, M., Berridge, M. and Lipp, P., 1997, Cooking with calcium: The recipes for composing global signals from elementary events, *Cell* 91, 367–373.

Burdyga, T.V., Taggart, M.J., Crichton, C., Smith, G.L. and Wray, S., 1998, The mechanism of Ca^{2+} release from the SR of permeabilised guinea-pig and rat ureteric smooth muscle, *Biochim. Biophys. Acta* 1402, 109–114.

Cheng, H., Lederer, W.J. and Cannell, M.B., 1993, Calcium sparks: Elementary events underlying excitation-contraction coupling in heart muscle, *Science* 262, 740–744.

Cheng, H., Lederer, M.R., Lederer, W.J. and Cannell, M., 1996, Calcium sparks and $[Ca^{2+}]_i$ waves in cardiac myocytes, *Am. J. Physiol.* 270, C148–C159.

Conklin, M.W., Barone, V., Sorrentino, V. and Coronado, R., 1999, Contribution of ryanodine receptor type 3 to Ca^{2+} sparks in embryonic mouse skeletal muscle, *Biophys. J.* 77, 1394–1403.

Conti, A., Gorza, L. and Sorrentino, V., 1996, Differential distribution of ryanodine receptor type 3 (RYR3) gene product in mammalian skeletal muscles, *Biochem. J.* 316, 19–23.

Coussin, F., Macrez, N., Morel, J.-L. and Mironneau, J., 2000, Requirement of both ryanodine receptor subtypes 1 and 2 for Ca^{2+}-induced Ca^{2+} release in vascular smooth muscle, *J. Biol. Chem.* 275, 9596–9603.

De Smedt, H., Missiaen, L., Parys, J.B., Henning, R.H., Sienaert, I., Vanlingen, S., Gijsens, A., Himpens, B. and Casteels, R., 1997, Isoform diversity of the inositol trisphosphate receptor in cell types of mouse origin, *Biochem. J.* 322, 575–583.

Ehrlich, B.A., 1995, Functionnal properties of intracellular calcium release channels, *Curr. Opinion Neurobiol.* 5, 304–309.

Franzini-Armstrong, C. and Protasi, F., 1997, Ryanodine receptors of striated muscles: A complex channel capable of multiple interactions, *Physiol. Rev.* 77, 699–729.

Gomez, A.M., Valdivia, H., Cheng, H., Lederer, M.R., Santana, L.F., Cannell, M.B., McCune, S.A., Altschuld, R.A. and Lederer, W.J., 1997, Defective excitation-contraction coupling in experimental cardiac hypertrophy and heart failure, *Science* 276, 800–806.

Guse, A.H., da Silva, C.P., Berg, I., Skapenko, A.L., Weber, K., Heyer, P., Hohenegger, M., Ashamu, G.A., Schulze-Koops, H., Potter, B.V. and Mayr, G.W., 1999, Regulation of Ca^{2+} signalling in T lymphocytes by the second messenger cyclic ADP-ribose, *Nature* 398, 70–73.

Iino, M. and Endo, M., 1992, Calcium-dependent immediate feedback control of inositol 1,4,5-triphosphate-induced Ca^{2+} release, *Nature* 360, 76–78.

Iino, M., Yamazawa, T., Miyashita, Y., Endo, M. and Kasai, H., 1993, Critical intracellular Ca^{2+} concentration for all-or-none Ca^{2+} spiking in single smooth muscle cells, *EMBO J.* 12, 5287–5291.

Joseph, S.K., Lin, C., Pierson, S., Thomas, A.P. and Maranto, A.R., 1995, Heteroligomers of type-I and type-III inositol trisphosphate receptors in WB rat liver epithelial cells, *J. Biol. Chem.* 270, 23310–23316.

Kiselyov, K., Xu, X., Mozhayeva, G., Kuo, T., Pessah, I., Mignery, G., Zhu, X., Birnbaumer, L. and Muallem, S., 1998, Functional interaction between $InsP_3$ receptors and store-operated Htrp3 channels, *Nature* 396, 478–482.

Lai, F.A., Misra, M., Xu, L., Smith, H.A. and Meissner, G., 1989, The ryanodine receptor-Ca^{2+} release channel complex of skeletal muscle sarcoplasmic reticulum. Evidence for a cooperatively coupled, negatively charged homotetramer, *J. Biol. Chem.* 264, 16776–16785.

Ledbetter, M.W., Preiner, J.K., Louis, C.F. and Mickelson, J.R., 1994, Tissue distribution of ryanodine receptor isoforms and alleles determined by reverse transcription polymerase chain reaction, *J. Biol. Chem.* 269, 31544–31551.

Lipp, P. and Niggli, E., 1996, A hierarchical concept of cellular and subcellular Ca^{2+}-signalling, *Prog. Biophys. Mol. Biol.* 65, 265–296.

Lopez-Lopez, J.R., Shacklock, P.S., Balke, C.W. and Wier, W.G., 1995, Local Ca^{2+} transients triggered by single L-type Ca^{2+} channel currents in cardiac cells, *Science* 268, 1042–1045.

Machaca, K. and Hartzell, H.C., 1999, Adenophostin A and inositol 1,4,5-trisphosphate differentially activate Cl-currents in Xenopus oocytes because of disparate Ca^{2+} release kinetics, *J. Biol. Chem.* 274, 4824–4831.

Marx, S.O., Ondrias, K. and Marks, A.R., 1998, Coupled gating between individual skeletal muscle Ca^{2+} release channels, *Science* 281, 818–821.

Miriel, V.A., Mauban, J.R., Blaustein, M.P. and Wier, W.G., 1999, Local and cellular Ca^{2+} transients in smooth muscle of pressurized rat resistance arteries during myogenic and agonist stimulation, *J. Physiol. (Lond.)* 518, 815–824.

Mironneau, J., Arnaudeau, S., Macrez-Lepretre, N. and Boittin, F.X., 1996, Ca^{2+} sparks and Ca^{2+} waves activate different Ca^{2+}-dependent ion channels in single myocytes from rat portal vein, *Cell Calcium* 1996, 20, 153–160.

Miyakawa, T., Maeda, A., Yamazawa, T., Hirose, K., Kurosaki, T. and Iino, M., 1999, Encoding of Ca^{2+} signals by differential expression of IP_3 receptor subtypes, *EMBO J.* 18, 1303–1308.

Morel, J.L., Boittin, F.X., Halet, G., Arnaudeau, S., Mironneau, C. and Mironneau, J., 1997, Effect of a 14-day hindlimb suspension on cytosolic Ca^{2+} concentration in rat portal vein myocytes, *Am. J. Physiol.* 273, H2867–H2875.

Morgan, J.M., De Smedt, H. and Gillespie, J.I., 1996, Identification of three isoforms of the $InsP_3$ receptor in human myometrial smooth muscle, *Pflügers Arch.* 431, 697–705.

Moschella, M.C. and Marks, A.R., 1993, Inositol 1,4,5-trisphosphate receptor expression in cardiac myocytes, *J. Cell Biol.* 120, 1137–1146.

Nelson, M.T., Cheng, H., Rubart, M., Santana, L.F., Bonev, A.D., Knot, H.J. and Lederer, W.J., 1995, Relaxation of arterial smooth muscle by calcium sparks, *Science* 270, 633–637.

Neylon, C.B., Richards, S.M., Larsen, M.A., Agrotis, A. and Bobik, A., 1995, Multiple types of ryanodine receptor/Ca^{2+} release channels are expressed in vascular smooth muscle, *Biochem. Biophys. Res. Commun.* 215, 814–821.

Parys, J.B., Sienaert, I., Vanlingen, S., Callewaert, G., De Smet, P., Missiaen, L. and De Smet, H., 2000, Regulation of inositol 1,4,5-trisphosphate-induced Ca^{2+} release by Ca^{2+}, this book.

Perez, G.J., Bonev, A.D., Patlak, J.B. and Nelson, M.T., 1999, Functional coupling of ryanodine receptors to KCa channels in smooth muscle cells from rat cerebral arteries, *J. Gen. Physiol.* 113, 229–238.

Sonnleitner, A., Conti, A., Bertocchini, F., Schindler, H. and Sorrentino, V., 1998, Functional properties of the ryanodine receptor type 3 (RyR3) Ca^{2+} release channel, *EMBO J.* 17, 2790–2798.

Takeshima, H., Iino, M., Takekura, H., Nishi, M., Kuno, J., Minowa, O., Takano, H. and Noda, T., 1994, Excitation-contraction uncoupling and muscular degeneration in mice lacking functional skeletal muscle ryanodine-receptor gene, *Nature* 369, 556–559.

Takeshima, H., Komazaki, S., Hirose, K., Nishi, M., Noda, T. and Iino, M., 1998, Embryonic lethality and abnormal cardiac myocytes in mice lacking ryanodine receptor type 2, *EMBO J.* 17, 3309–3316.

Tsugorka, A., Rios, E. and Blatter, L.A., 1995, Imaging elementary events of calcium release in skeletal muscle cells, *Science* 269, 1723–1726.

Vigne, P., Lazdunski, M. and Frelin, C., 1989, The inotropic effect of endothelin-1 on rat atria involves hydrolysis of phosphatidylinositol, *FEBS Lett.* 249, 143–146.

Wibo, M. and Godfraind, T., 1994, Comparative localization of inositol 1,4,5-trisphosphate and ryanodine receptors in intestinal smooth muscle: An analytical subfractionation study, *Biochem. J.* 297, 415–423.

Calcium Signalling in Neurons Exemplified by Rat Sympathetic Ganglion Cells

S.J. Marsh, N. Wanaverbecq, A.A. Selyanko and D.A. Brown

1. INTRODUCTION

Calcium (Ca^{2+}) is essential for neural function. The most obvious requirement is to trigger the release of transmitter when the action potential arrives at the axon terminals. However, Ca^{2+} can also affect neuronal function in other ways. The mechanisms for Ca^{2+} entry, release, sequestration and extrusion vary somewhat from one type of nerve cell to another, but follow certain principles in common. These can be helpfully illustrated with reference to neurons in the rat superior cervical sympathetic (SCG) ganglion. These are well-studied peripheral neurons, which are particularly convenient for simultaneous recording of membrane ion channel currents and intracellular Ca^{2+} changes using fluorescent indicators.

In this chapter we consider first the elements of Ca^{2+} homeostasis – that is, the mechanisms of Ca^{2+} entry, sequestration and extrusion – and second some of the effects of changing intracellular Ca^{2+} concentration ($[Ca^{2+}]_i$) on ion channels in the outer cell membrane and their consequences for cell excitability.

2. CALCIUM HOMEOSTASIS

The resting concentration of Ca^{2+} recorded in these neurons normally lies between 50 and 125 nM (e.g., Thayer et al., 1988; Beech et al., 1991;

S.J. Marsh, N. Wanaverbecq, A.A. Selyanko and D.A. Brown • Department of Pharmacology, University College London, London, U.K.

R. Pochet, R. Donato, J. Haiech, C. Heizmann and V. Gerke (eds.): Calcium: The Molecular Basis of Calcium Action in Biology and Medicine, 27–44.

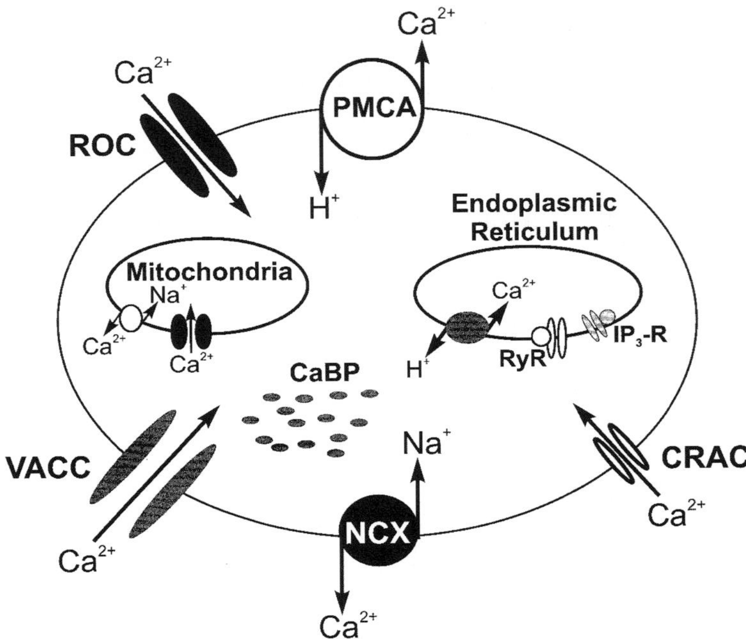

Figure 1. Calcium homeostatic mechanisms in rat sympathetic neurons.

Trouslard et al., 1993; Foucart et al., 1995). This is maintained by the balance between Ca^{2+} entry, intracellular buffering and sequestration, and Ca^{2+} extrusion (Figure 1).

2.1. Ca^{2+} entry

2.1.1. Voltage-activated Ca^{2+} channels (VACC)

These cells possess (principally) two types of voltage-activated Ca^{2+} channels (VACCs) – a functionally-distinct isoform of the α_{1B} (ω-conotoxin GVIA-sensitive N-type) channel (Lin et al., 1997) and dihydropyridine-sensitive L-type channels. Both may be activated by depolarizing the somatic cell membrane (Hirning et al., 1988; Plummer et al., 1989; Regan et al., 1991) and are opened during the normal action potential (Davies et al., 1996) with 95% of the Ca^{2+} charge being carried through N-type channels (Toth and Miller, 1995). In keeping with this, N-type channels are overwhelmingly responsible for triggering the release of transmitter (noradrenaline) from SCG neuron processes (Hirning et al., 1988; Toth et al., 1993; Koh and Hille, 1997), though releasing effects mediated by L-type channels have been detected in cultured cells (Przywara et al., 1993). A significant feature of both N- and L-type channels is that they can be inhibited (and hence Ca^{2+} influx reduced) by neurotransmitters acting via G protein-coupled receptors (Hille, 1994). Although a remarkable variety of transmitters and hormones can do this, of most significance physiologically are the effects of the normal preganglionic transmitter (acetylcholine – ACh) on the somato-dendritic channels and of the normal postganglionic transmitter (noradrenaline – NE) on channels in axonal processes and their terminals. The effect of ACh on the somato-dendritic N- and L-type channels (mediated by M_4 and M_1 muscarinic receptors) is to increase excitability and facilitate high-frequency repetitive firing, through the consequential reduction of the Ca^{2+}-activated K^+ current (see Section 3 below). The effect of NE on the processes and their terminals (mediated by α_2-adrenoceptors) is to inhibit Ca^{2+} entry through N-type channels and thereby reduce transmitter release (Koh and Hille, 1997). This is probably the principal mechanism underlying auto-inhibition of NE release from these axons (Boehm and Huck, 1996; see also Toth et al., 1993).

2.1.2. Receptor operated channels (ROC)

ACh is the natural transmitter released from preganglionic fibres onto these neurons, and gates cation-conducting nicotinic channels to produce an excitatory post-synaptic current (epsc) generating the excitatory post-synaptic potential (epsp). The precise molecular composition of the ganglionic nicotinic receptor is not certain. The most abundant mRNAs are those for $\alpha 3/\beta 4$, followed by $\alpha 7$, $\beta 2$, and possibly $\alpha 4$ and $\alpha 5$ (McGehee and Role, 1995). Pharmacologically, $\alpha 3/\alpha 4$ mixes provide the best match (Luetje and Patrick, 1991; Covernton et al., 1994). However, co-precipitation (Vernallis et al.,1995) and recent single channel conductance measurements (Sivilotti et al., 1997) suggest a possible triplet combination of $\alpha 3/\alpha 5/\beta 4$.

The channels are noticeably permeant to Ca^{2+} (P_{Ca}:P_{Na} 2.6–3.8; Trouslard et al., 1993). As a result, there is a substantial inward flux of Ca^{2+} during nicotinic receptor activation. Thus, by comparing the ratio of the Ca^{2+}/Indo-1 fluorescence signals to the total charge transfer induced by activating nicotinic receptors with the equivalent ratio following activation of the Ca^{2+} channels (Trouslard et al., 1993), it may be calculated that about 4.4% of the nicotinic inward current is carried by Ca^{2+} ions at 2.5 mM external $[Ca^{2+}]$. Rogers and Dani (1995) using saturating Fura-2 buffering have deduced a concordant value of 4.7 %. This places the ganglionic nicotinic receptor somewhere between the low-permeability AMPA receptor and the high permeability NMDA receptor. Notwithstanding, it is unlikely that sufficient Ca^{2+} enters during a single epsp to raise cytosolic Ca^{2+} by more than a few nM, which is probably negligible compared with that entering through the voltage-gated Ca^{2+} channels during the ensuing action potential.

Nevertheless, the role of Ca^{2+} entry during the epsp probably requires further study, since experiments on frog sympathetic neurons (Tokimasa and North, 1984) suggest that the sub-membrane rise might be sufficient to activate Ca^{2+}-dependent K^+ channels and accelerate recovery of the depolarization.

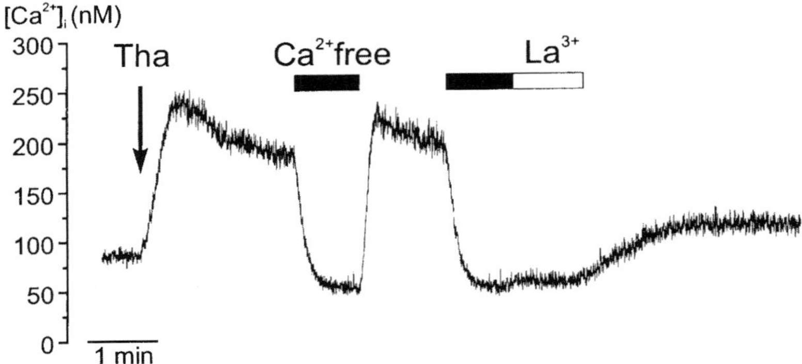

Figure 2. Role of intracellular Ca^{2+} store in the regulation of Ca^{2+} permeability of the plasma membrane. Thapsigargin (Tha, 1 μM, 5 sec) applied by pressure ejection to an Indo-1 loaded SCG neuron caused an increase in $[Ca^{2+}]_i$. The initial rise in $[Ca^{2+}]_i$ was unaffected in Ca^{2+}-free extracellular solution (data not shown), however the later phase of Ca^{2+} entry was inhibited by removing extracellular Ca^{2+} or by the addition of Lanthanum (La^{3+} 1 μM). This suggests that store depletion by thapsigargin opens a CRAC-type conductance in the plasma membrane resulting in Ca^{2+} entry.

2.1.3. Calcium release-activated calcium (CRAC) channels

There is some evidence that these neurons might posses some form of "CRAC" channel, in that the delayed component of the thapsigargin-induced elevation of $[Ca^{2+}]_i$ (see below) depends on an influx of Ca^{2+} and is inhibited on removing Ca^{2+} from the external solution or by adding 10 μM La^{3+} (Figure 2). However the significance of this for neuronal function is not yet clear.

2.2. Calcium release

2.2.1. Calcium-induced calcium release (CICR)

Caffeine when applied to rat sympathetic neurons induces a rapid and pronounced change in $[Ca^{2+}]_i$ by releasing calcium from internal stores, which is blocked by pre-treating the cells with ryanodine (see, e.g., Thayer et al., 1988; Hernandez-Cruz et al., 1995). In fact, the density of ryanodine receptors (RyR) in these cells approaches that in cardiac muscle (Hernandez-Cruz et al., 1995), so these neurons have a substantial potential capacity for amplification of Ca^{2+} signals by CICR. However, although there is evidence that a component of the rise in $[Ca^{2+}]_i$ produced by caffeine itself results from CICR-mediated amplification (Hernandez-Cruz et al., 1997), CICR seems to play a surprisingly modest role in amplifying signals mediated by Ca^{2+} entry. Thus, the increase in intracellular $[Ca^{2+}]_i$ during voltage-activated Ca^{2+} currents of varying amplitude and duration showed a linear (or sublinear) dependence on charge entry (Figure 3), suggesting little or no amplification (Trouslard et al., 1993). This contrasts with the CICR-mediated amplification

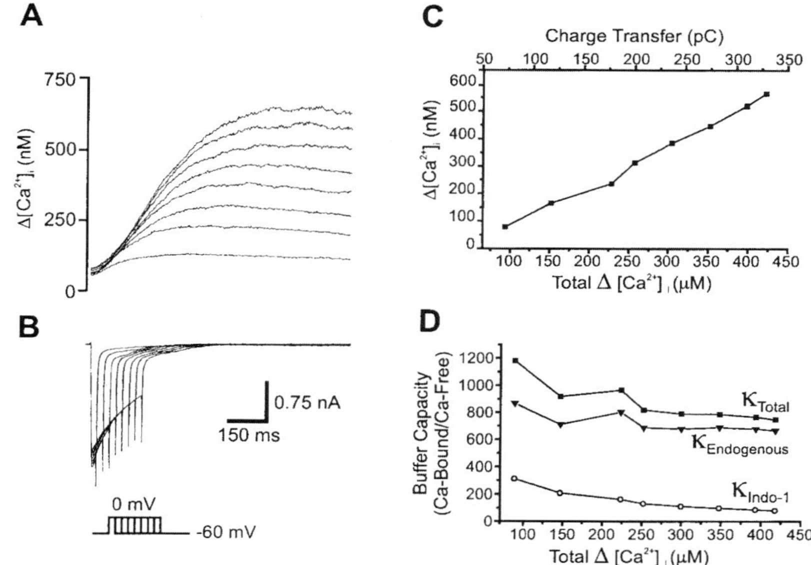

Figure 3. The determination of the endogenous Ca^{2+} binding capacity of rat sympathetic neurons. A single cell (~ 4 pl volume) was subjected to a whole-cell voltage-clamp experiments whilst simultaneously monitoring $[Ca^{2+}]_i$ using Indo-1 (see Trouslard et al., 1993). Transient rises in $[Ca^{2+}]_i$ (A) were induced by activation of voltage-activated calcium channels with voltage steps of various durations from a holding potential of −60 to 0 mV (B). In (C), the variation in $[Ca^{2+}]_i$ ($\Delta[Ca^{2+}]_i$ in nM) was plotted against the charge transfer (top), measured from the integral of the inward Ca2+ current, and the total change in $[Ca^{2+}]_i$ (bottom), determined as described in the text. Note that very little of the Ca^{2+} that enters this neuron is available as the free ion. In (D), calculation of the calcium binding capacity: the buffer capacity of Indo-1 was determined as described by Palecek et al. (1999), using the equation: $\kappa_{\text{Indo}-1} = [\text{Indo}] * K_d/(([Ca^{2+}]_{\text{Rest}} + K_d) * ([Ca^{2+}]_{\text{Peak}} + K_d))$ with a concentration of Indo-1 of 100 μM and the K_d determined within our whole-cell calibration procedure. The endogenous buffer ($\kappa_{\text{Endogenous}}$) was then estimated by subtracting the Indo-1 binding capacity from the total calcium binding capacity (κ_{Total}).

in cerebellar Purkinje cells (see Llano et al., 1994). On the other hand, treatment with ryanodine has been reported to produce a modest, though clear, abbreviation of the Ca^{2+}-dependent spike after-hyperpolarization, suggesting some contribution to the maintenance of the Ca^{2+} signal by CICR (Kawai and Watanabe, 1989; Davies et al., 1996). Even so, the effects of ryanodine were far less dramatic than those seen in guinea-pig vagal neurons that exhibit much more long-lasting Ca^{2+}-dependent K^+ currents (see, e.g., Sah and McLachlan, 1991). The reason may be that the ryanodine-sensitive stores are too remote from the outer membrane (Hernandez-Cruz et al., 1997).

2.2.2. IP_3-Receptors (IP_3-R)

Intact rat sympathetic ganglia show a robust phosphoinositide response to musarinic receptor agonists (e.g., Bone et al., 1984; Horwitz et al., 1984), and this has recently been confirmed in single dissociated SCG neurons (del

Rio et al., 1999). Indeed, experiments on these ganglia played a key role in the early formulation of inositol phospholipid theories of signal transduction (Larrabee et al., 1963). It is rather surprising, therefore, that activation of muscarinic receptors in these cells produces only small and inconsistent IP_3-mediated $[Ca^{2+}]_i$ rises (e.g., Beech et al., 1991; Marsh et al., 1995), and that these are only apparent when the intracellular stores are filled by prolonged depolarization (del Rio et al., 1999). Bradykinin produces a more consistent IP_3-induced rise in $[Ca^{2+}]_i$, and this appears to be the transducer responsible for M current inhibition by this agonist (Cruzblanca et al., 1998). Even so, the changes in $[Ca^{2+}]_i$ produced by either agent are much less pronounced than those in some related neural cell lines such as NG108-15 cells (see Robbins et al., 1993), for example, unlike the latter, the rises in SCG neurons are insufficient to activate K_{Ca} channels. Thus, the receptor-coupled IP_3-gated release mechanism seems to be relatively weakly-developed in these neurons compared (for example) to certain central neurons such as cerebellar granule cells that show quite strong Ca^{2+} signals in response to "metabotropic" receptor stimulation (Irving et al., 1992). The reasons for this weak Ca^{2+} signalling are not yet clear.

2.3. Intracellular sequestration

2.3.1. Endoplasmic reticulum

As both caffeine and IP_3 increase $[Ca^{2+}]_i$ levels in SCG neurons it is apparent that these cells also have the ability to sequester Ca^{2+} into these same intracellular compartments. Uptake into these stores is mediated by the sarco(endo)plasmic reticulum Ca^{2+}-ATPase (SERCA) that is sensitive to thapsigargin (Tha) (Cruzblanca et al., 1998). Interestingly thapsigargin itself can induce release from these stores (Figure 2). It is believed that these compartments are subject to a constant "leak" e.g. the channels have a high probability of opening under non-stimulated conditions. Whether this "leak" represents basal levels of IP_3 or even cyclic-ADP-ribose is still not clear but it has been suggested that thapsigargin induced release is mediated by the failure of the now inhibited SERCA to recycle the tonic release of calcium back into the store. The "leak" component probably also means that, at rest, the intracellular store is not full and thus might act as a temporary sequestration site during large Ca^{2+} transients.

2.3.2. Mitochondrial uptake

The mitochondria of peripheral neurons have been shown to accumulate large amounts of Ca^{2+} under certain experimental conditions but whether this represents a pathological or physiological response is still a contentious issue (Miller, 1991). However, with the recent discovery of intramitochondrial enzymes that are regulated by Ca^{2+}, the former "calcium sink" hypothesis that mitochondrial uptake might occur only under cytotoxic conditions, may have to be re-evaluated. To obtain significant calcium uptake into mitochondria it usually requires high levels of free Ca^{2+} (> 500 nM) as influx via the low affinity calcium uniporter is constantly being opposed by Na^+/Ca^{2+} or H^+/Ca^{2+} exchangers (Duchen, 1999). Nevertheless, there is now considerable evidence that mitochondrial uptake plays a major role in the temporary sequestration of large Ca^{2+} loads following activation of voltage-activated Ca^{2+} channels in bovine adrenal chromaffin cells (Herrington et al., 1996). We also have evidence to suggest a significant contribution of mitochondrial uptake to the clearance of large Ca^{2+} loads in rat SCG neurons (Wanaverbecq and Marsh, unpublished).

2.3.3. Calcium binding proteins (CaBP)

Sympathetic neurons have a large capacity for buffering $[Ca^{2+}]_i$. Thus, using whole-cell voltage-clamp recording while simultaneously measuring $[Ca^{2+}]_i$ using Indo-1, we found that less than 0.2% of all calcium that entered the cell via voltage-activated Ca^{2+} channels was available in the intracellular compartment as free Ca^{2+}. A simple calculation using the integral of the Ca^{2+} current to determine transmembrane charge transfer in Coulombs (Q) allowed us to estimate the predicted change in $[Ca^{2+}]_{\mathrm{Total}}$ in the absence of an endogenous buffer: $\Delta[Ca^{2+}]_{\mathrm{Total}} = (Q/z\mathrm{F})/V$, were z is the valency of calcium (2), F is Faradays constant (96485) and V is the volume of the cell. Comparing these predicted changes in $[Ca^{2+}]_i$ to measurements of $[Ca^{2+}]_i$ by Indo-1, showed that these cells had a Ca^{2+} binding capacity ($\kappa_{\mathrm{Endogenous}} = [Ca^{2+}]_{\mathrm{Bound}}/[Ca^{2+}]_{\mathrm{Free}}$) between 600–800 (Figure 3). This is far greater than that previously described for spinal cord motoneurons ($\kappa_{\mathrm{Endogenous}} \cong 50$, see Palecek et al., 1999), adrenal chromaffin cells ($\kappa_{\mathrm{Endogenous}} \cong 75$, see Neher and Augustine, 1992) or nerve endings ($\kappa_{\mathrm{Endogenous}} \cong 175$, see Stuenkel, 1994) but less than that determined in cerebellar Purkinje cells ($\kappa_{\mathrm{BUFFER}} \cong$ 900–2000, see Fierro and Llano 1996). The nature of the intracellular calcium buffer in SCG neurons is not known but several of the EF-hand type CaBP such as calbindin D_{28k} (Sanchez-Vivas et al., 1994), parvalbumin (Endo and Onaya, 1988) and calmodulin (Setoohshima et al., 1987) have all been identified within the soma of these cells. The influence of a combination of these or other candidates on Ca^{2+}-dependent regulation of excitability should not be underestimated.

2.4. Calcium efflux

2.4.1. Plasma membrane calcium ATPase (PMCA)

The Plasma Membrane Calcium ATPases (PMCA, 130–140 kDa) belong to the P-type family (Carafoli, 1987; Carafoli and Sauffer, 1994) and are products of a multigene family with four isogenes undergoing several types of alternative splicing (Carafoli, 1997). The principal isoforms of the PMCA, named PMCA1 to 4, differ in the structure of their N- and C-terminal regions, i.e. the regulatory sites, and in their tissue distribution. Thus it has been demonstrated that PMCA1 and 4 are expressed in large amount in most tissues and therefore are thought to represent the house-keeping isoforms. The PMCA2 and 3 isoforms have a more restricted distribution and are believed to have specialized functions (Carafoli, 1994; Carafoli and Sauffer, 1994). The PMCA affinity (K_m) for Ca^{2+} ranges from about 10–20 μM (low-affinity state) to 0.5 μM or less (high-affinity state). It has been demonstrated that the binding of calmodulin, to a specific site on the C-terminal loop, induces the transition to the high-affinity state. There are very few pharmacological tools with which we can selectively inhibit the PMCA. However, broad spectrum inhibition of PMCA using ortho-vanadate and lanthanum which inhibit P-type ATPase's, 5-(and 6-)-carboxyeosin (the actual most potent Ca^{2+}-ATPase inhibitor) and extracellular alkalization (reduction of extracellular $[H^+]$) all have significant effects on Ca^{2+} homeostasis and suggest a dominant role for PMCA in $[Ca^{2+}]_i$ regulation.

In SCG neurons, very little is known about the nature of the PMCA nor about its importance in Ca^{2+} extrusion. However, a depolarization under whole-cell voltage clamp conditions induces a Ca^{2+} transient with a decay time constant about 10–15 s at 30°C. When ortho-vanadate (100 μM) or carboxyeosin (50 μM) are added to the intracellular medium the decay time constant is dramatically increased (to several minutes) (Figure 4A). The PMCA is an obligatory Ca^{2+}:H^+ exchanger (Carafoli, 1994; Carafoli and Sauffer, 1994), therefore reducing extracellular proton concentration by alkalization to pH 9 ($[H^+] = 10^{-9}$ M) inhibits its activity. In SCG neurons this results in both an increased basal $[Ca^{2+}]_i$ and a prolongation of the calcium transient induced by voltage-activated Ca^{2+} entry (Figure 4B) (see also Benham et al., 1992). These experiments suggest that the PMCA plays a substantial role not only in regulating calcium transients but also in maintaining the resting level $[Ca^{2+}]_i$. Preliminary results using reverse transcription PCR and immunocytochemistry have demonstrated the presence of the PMCA in SCG neurons but the exact isoform has not yet been identified.

2.4.2. Sodium/calcium exchanger (NCX)

Along with the PMCA, the sodium-calcium exchanger (NCX) is thought to be involved in neuronal Ca^{2+} transport (Blaustein and Hodgkin, 1969) and three genes (NCX1 to 3) coding for NCX have been cloned in mammalian tissues. These proteins have an ubiquitous distribution and a high homology. Most of the isoforms (over 30 splice variants) are related to the NCX1 form and their differences mainly lay in the structure of the intracellular loop, which is involved in the modulation of NCX activity (Blaustein and Lederer, 1999). NCX is electrogenic, the rate of Ca^{2+} transport and its direction are dependent upon the electrochemical gradient of the transported ions and the membrane potential (V_m). Until recently it was believed that the stoichiometry of the NCX was $3Na^+$:$1Ca^{2+}$ and the "motivational force" (Δv) determined by the following equation: $\Delta v = V_m - (3E_{Na} - 2E_{Ca})$. Therefore, in SCG neurons

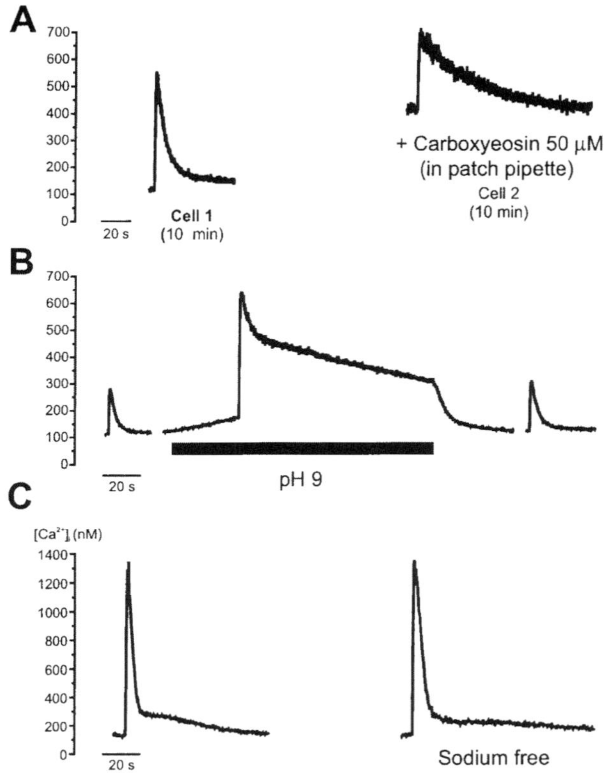

Figure 4. Role of PMCA and NCX in Ca^{2+} homeostasis in SCG neurons. Cultured rat SCG neurons were loaded with Indo-1-AM and voltage-clamped at a holding potential of −60 mV in the perforated patch (amphotericin B) configuration or in the whole-cell configuration using 100 μM Indo-1 free acid. Transient intracellular calcium rises were then elicited by activating voltage-activated calcium channels at 0 mV for 60 or 500 ms. (A) Transient calcium rises were elicited ten minutes after obtaining the whole-cell recording configuration. In the presence of the PMCA inhibitor carboxyeosin (50 μM added to the intracellular pipette solution – see cell2) there is a substantial rise in the resting calcium levels and the calcium transient is prolonged. (B) Alkalinization of the extracellular medium (pH 9) also reversibly increases the resting level of calcium and slows the rate of decay of the calcium transient recorded under perforated patch recording conditions. (C) Replacement of extracellular sodium with N-methyl-D-glucamine does not effect the resting $[Ca^{2+}]_i$ or significantly change the time-course of an intracellular calcium transient induced by a 500 ms depolarizing voltage step.

at 30°C and a resting potential of −60 mV, with E_{Na} set at +67 mV and E_{Ca} at +132 mV, Δv was only −3 mV. Thus at the resting membrane potential, NCX would have minimal effect on Ca^{2+}-transport. However, it was envisaged that during prolonged hyperpolarization like that induced during activation of Ca^{2+}-activated K^+ channels (see below Section 3) one might expect an enhanced contribution of NCX to $[Ca^{2+}]_i$ regulation. Recent experiments have suggested that the stoichiometry may be as high as $4Na^+{:}1Ca^{2+}$

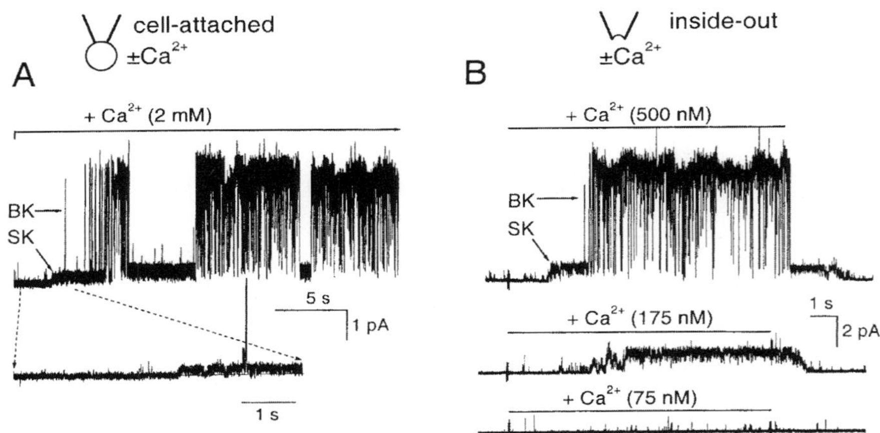

Figure 5. Large (BK) and small (SK) Ca^{2+}-activated potassium channels in SCG neurones. Activities of single BK and multiple SK channels were recorded in cell-attached (A) and inside-out (B) patch configurations in two different experiments. In (A) a cell was pre-depolarized to ~ -30 mV by adding 25 mM K^+ to the bath and the patch was further depolarized by 10 mV. Channels were then activated by adding 2 mM Ca^{2+} to the bath; presumably Ca^{2+} entered the cell through voltage-dependent Ca^{2+} channels. In (B) different concentrations of Ca^{2+} were applied directly to the inside face of the patch (at 0 mV). Note that SK channels required elevation in $[Ca^{2+}]_i$ above its resting level ($\sim$ 75 mM) and (at this potential) a BK channel was less sensitive to Ca^{2+} than the SK channels. Pipettes contained low (2.5 mM) $[K^+]$.

(Fujioka et al., 2000) which set the motivational force for NCX more negative ($\Delta v = -64$ mV) and so the direction of the NCX remains predominantly in Ca^{2+} efflux mode under physiological conditions ($4E_{\text{Na}} - 2E_{\text{Ca}} = +4$ mV).

There are no selective inhibitors available against NCX, but the exchanger, highly selective to Na^+, can be inhibited by extracellular Na^+ removal. In SCG neurons, the substitution of extracellular Na^+ with N-methyl D-glucamine, induces a small increase in the decay time of Ca^{2+} transient, especially for large rises in free $[Ca^{2+}]_i$ (Figure 4C), without affecting the resting $[Ca^{2+}]_i$, suggesting that the NCX plays only a minor role in Ca^{2+} extrusion.

3. CALCIUM AND CELL EXCITABILITY

The major immediate role for Ca^{2+} in the nervous system is to trigger the release of the neurotransmitter at all synapses. However, changes in intracellular Ca^{2+} can have other effects on neuronal excitability, by modifying the activity of membrane ion channels that regulate excitability and activity. Thus, in the rat SCG neuron, elevation of intracellular Ca^{2+} can activate two classes of Ca^{2+}-dependent K^+ channels (K_{Ca} channels), inhibit M-type voltage-gated K^+ channels (K_{M} channels) and activate Cl^- channels.

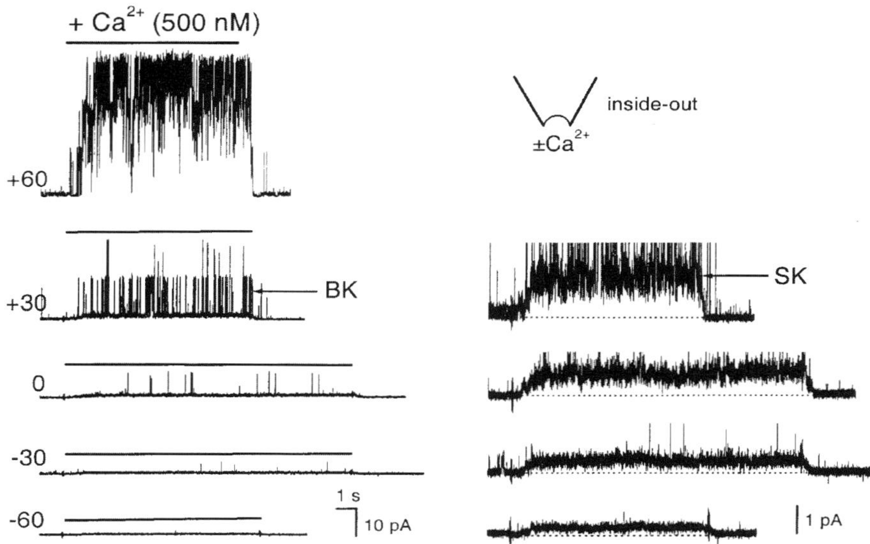

Figure 6. Differential voltage sensitivities of BK and SK channels. Activities of multiple BK and SK channels produced by 500 nM Ca^{2+} were recorded from the same inside-out patch at different membrane potentials (indicated near each record). Note that activation of BK but not SK channels was voltage-dependent: the activity of BK channels was reduced by membrane hyperpolarization and completely eliminated at the resting membrane potential level (−60 mV). The same records, shown at 10 times higher gain (right panel), revealed the activity of SK channels which persisted at all negative membrane potentials.

3.1. Ca^{2+}-dependent K^+ channels

As in many other types of neuron, two classes of Ca^{2+}-dependent K^+ channels can be detected in membrane patches from rat sympathetic neurons: large-conductance ("BK") channels, with a conductance of $\sim$ 200 pS in symmetric (150/150) K solution (Smart, 1987) and $\sim$ 120 pS at 0 mV in asymmetric (2/150) K (Figure 5) and small-conductance ("SK") channels, of conductance $\sim$ 2 pS in asymmetric K (Selyanko, 1996; Figure 5). The former are also voltage-dependent (Smart, 1987; see Figure 6) and presumably belong to the mSlo family of K_{Ca} channels. These channels are blocked by external TEA, with an IC_{50} $\sim$ 0.2 mM at 0 mV (Smart, 1987), and by charybdotoxin (Davies et al. 1996). The SK channels are insensitive to voltage (Figure 6) or TEA, but are blocked by $\leq$ 100 nM apamin (Kawai and Watanabe, 1986; Selyanko, 1996). Recent evidence suggests that the channel is composed (in part, at least) of products of the SK3 gene (Hosseini et al., 1999).

Both types of channel are opened by Ca^{2+} entry during an action potential (Davies et al., 1996). Opening of BK channels contributes to spike repolarization and the early spike after-hyperpolarization (see also Marsh and Brown, 1991), whereas SK channels are responsible for the slow (long-lasting) after-hyperpolarization (sAHP: see also Kawai and Watanabe, 1986). This resembles the situation in many central neurons such as hippocampal pyramidal cells (see Storm, 1990), except that the SK channel-driven I_{sAHP} in sympathetic neurons appears to be generated without the pronounced delay (or build-up) seen in central neurons. The function of the sAHP is also similar to that in central neurons, in that it acts as a negative-feedback inhibitor of repetitive discharges. Thus, suppression of the sAHP with apamin promotes tonic firing during sustained depolarization (Kawai and Watanabe,1986; Sacchi et al., 1995; Davies et al., 1996).

Interestingly, BK and SK channels in rat SCG neurons also differ in respect of the source of Ca^{2+} required to open them. Whereas the BK channels are opened by Ca^{2+} entering through somatic L-type Ca^{2+} channels, the SK channels are selectively activated following opening of N-type Ca^{2+} channels (Davies et al., 1996). This implies some quite discrete topographic co-localization of Ca^{2+} and K_{Ca} channels, since both channel types can coexist in excised somatic membrane patches and can readily be opened by applying Ca^{2+} to the inside face of such patches (Figure 5). The dependence of SK channel opening on entry through N-type Ca^{2+} channels also has the consequence that the sAHP can be modulated by neurotransmitters. Thus, opening of N-type Ca^{2+} channels can be inhibited, or its gating slowed, by several transmitters, including the natural transmitter, ACh, acting through M_1 and M_4 muscarinic receptors (see Hille, 1994, and references therein; also Delmas et al., 1998). It is this action, rather than an action on the SK channels themselves, that appears to be responsible for the reduced I_{sAHP} in these cells (Brown and Selyanko, 1995). Inhibition of the sAHP co-operates with inhibition of the voltage-gated K_M current (see below) in generating the increased excitability seen in these neurons following cholinergic stimulation of muscarinic receptors (Brown, 1988); and inhibition of both currents can lead to the most dramatic spike activity (Haley et al., 1998). Superficially, this resembles the situation in some central neurons, such as hippocampal neurons, the excitability of which are also enhanced following cholinergic stimulation (Cole and Nicoll, 1983). However, in the latter case, it is thought that the activated muscarinic receptors affect the response of the K_{Ca} channels to Ca^{2+}, rather than the entry of Ca^{2+} (e.g., Knoepfel et al., 1990), though an effect on hippocampal Ca^{2+} channels (L-type, in this case) cannot be discounted (Marrion and Tavalin, 1998).

3.1.1. K_M channels

These are subthreshold, slowly-gated channels that open during depolarization and serve to limit the ability of the neuron to sustain repetitive spike discharges (see Brown, 1988). They are inhibited by a variety of neurotransmitters, most notably by ACh acting through M_1 muscarinic receptors: this has the effect of producing a small depolarization (the "slow epsp") and enhancing repetitive spiking activity. The channels belong to the KCNQ family

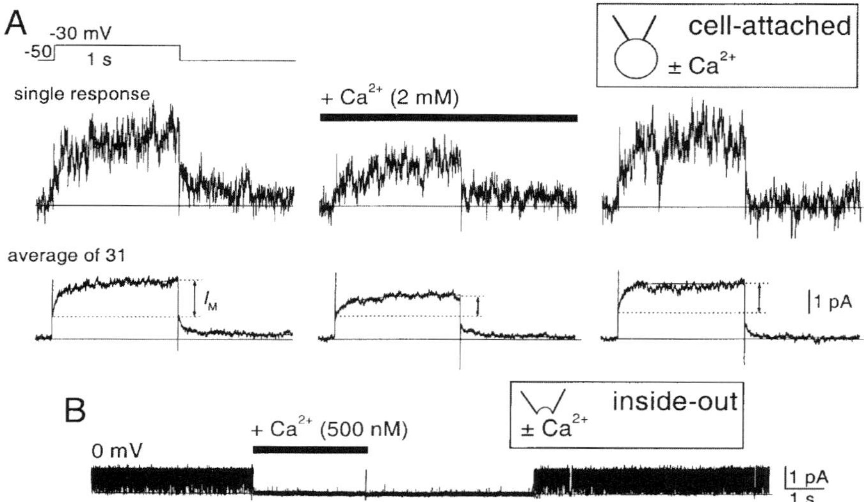

Figure 7. Intracellular Ca^{2+} inhibits voltage-dependent M-type potassium channels in SCG neurones. Activities of multiple (A) and single (B) M-channels were recorded in cell-attached and inside-out patch configurations, respectively. In A a cell was bathed in a low (2.5 mM) K^+ solution and M-channels were activated by 1 s depolarizing voltage steps (from −50 to −30 mV); the averaged response (lower row) matched the macroscopic M-current (not shown) in its slow activation (denoted by I_M) and deactivation. Channels were then inhibited by adding 2 mM Ca^{2+} to the bath; presumably Ca^{2+} entered the cell through some non-voltage-dependent Ca^{2+}-permeable channels. In (B) the M-channel activity (recorded at 0 mV) was completely inhibited by direct application of 500 nM Ca^{2+} to the inside face of the patch. Data in (A) and (B) were obtained in two different experiments.

of K channels, and are probably composed of a KCNQ2+3 heteromer (Wang et al., 1998).

Unlike the SK channels, they are not directly gated by Ca^{2+}. Instead, in isolated patches, their activity is *reduced* by intracellular Ca^{2+}, with an IC_{50} around 100 nM (Selyanko and Brown, 1996). This implies that they are likely to be under some degree of tonic inhibition at the normal resting Ca^{2+} concentrations of 50–125 nM (see above). Supporting evidence for this is provided by the effect of removing external Ca^{2+}, which leads to a pronounced increase in the activity of K_M channels recorded from intact neurons with cell-attached patch electrodes (Figure 7). Also, a reduction in activity can result when intracellular Ca^{2+} is raised above normal – for example, following application of a nicotinic agonist (Marsh and Trouslard, unpublished). This apparent inhibitory effect of intracellular Ca^{2+} runs counter to the activating effect of Ca^{2+} on SK channels. Since both SK and K_M channels have similar effects on excitability (and indeed co-operate: see Brown, 1988), the changes in excitability produced by elevating intracellular Ca^{2+} may be more complex than originally thought, and need further study.

$I_{K(M)}$ is inhibited by transmitters and hormones acting on receptors coupled to G proteins of the G_q family, such as M_1 muscarinic receptors

(Haley et al., 1998) and B_2 bradykinin receptors (Jones et al., 1995). Since activation of G_q also activated phospholipase C (PLC) and hence (potentially) raises intracellular Ca^{2+} through the formation of IP_3, it might plausibly be argued from the observed sensitivity of K_M channels to Ca^{2+} that Ca^{2+} might be the "final messenger" for receptor-mediated inhibition of $I_{K(M)}$. There is indeed evidence to support this in the case of the bradykinin B_2 receptor (Cruzblanca et al., 1998). However (and notwithstanding the fact that activating muscarinic receptors can clearly stimulate PLC: see del Rio et al., 1999), the release of Ca^{2+} on stimulating M_1 muscarinic receptors in these cells is small and inconsistent, and appears not to explain cholinergic inhibition of $I_{K(M)}$ (Cruzblanca et al., 1998; del Rio et al., 1999). The reason for this difference between bradykinin and acetylcholine is not yet known.

3.2. Activation of Cl^- currents

The Cl^- equilibrium potential (E_{Cl}) is normally well positive to the membrane potential in rat sympathetic neurons so opening of Cl^- channels gives an inward (depolarizing) current. Two such inward Cl^- currents have been described following elevation of intracellular Ca^{2+} by (e.g.) opening of voltage-activated Ca^{2+} channels. The first is a rapid transient current, giving an after-depolarization following an action potential (Sanchez-Vivas and Gallego, 1994), and probably results from a fairly "direct" gating of the underlying channels by Ca^{2+}. This current is normally very small or negligible, but becomes prominent after axotomy. The second is a delayed current that starts several seconds after the priming Ca^{2+} trigger and lasts some 20 seconds or so (Marsh et al., 1995). This appears not to be due to direct gating of the underlying Cl^- channels but instead to the Ca^{2+}-induced membrane-translocation and activation of protein kinase C (PKC). Thus, it is enhanced (that is, its Ca^{2+} threshold is reduced) on co-activation of PKC and suppressed by inhibition or down-regulation of PKC. This means that it is essentially a PKC-activated current, though requiring a priming Ca^{2+} charge to initiate the translocation of sufficient PKC to a site in the membrane where it can be activated. It is also enhanced following activation of receptors linked to PLC, such as B_2 bradykinin and M_1 muscarinic receptors, which activate PKC through formation of diacylglycerols. Further, the priming charge of Ca^{2+} can be delivered just as readily through the Ca^{2+}-permeable nicotinic receptors (see above) as through the voltage-gated Ca^{2+} channels. This means that the natural transmitter ACh is capable of inducing the current since it can produce both the priming (nicotinic) Ca^{2+} charge and the enhancing (muscarinic) PKC activation. The normal function of this current is not yet clear, but it might explain some of the delayed inward currents previously reported following strong presynaptic stimulation in these neurons (see, e.g., Brown and Selyanko, 1985).

REFERENCES

Beech, D.J., Bernheim, L., Mathie, A. and Hille, B., 1991, Intracellular Ca^{2+} buffers disrupt muscarinic suppression of Ca^{2+} current and M current in rat sympathetic neurons, *Proc. Natl. Acad. Sci., USA* 88, 652–656.

Benham, C.D., Evans, M.L. and McBain, C.J., 1992, Ca^{2+} efflux mechanisms following depolarization evoked calcium transients in cultured rat sensory neurones, *J. Physiol. Lond.* 455, 567–583.

Blaustein, M.P. and Hodgkin, A.L., 1969, The effect of cyanide on the efflux of calcium from squid axons, *J. Physiol. Lond.* 200, 497–527.

Blaustein, M.P. and Lederer, W.J., 1999, Sodium/calcium exchange: Its physiological implications, *Physiol. Rev.* 79, 763–854.

Boehm, S. and Huck, S., 1996, Inhibition of N-type calcium channels: The only mechanism by which presynaptic α_2-autoreceptors control sympathetic transmitter release, *Eur. J. Neurosci.* 8, 1924–1931.

Bone, E.A., Fretten, P., Palmer, S., Kirk, C.J. and Michell, R.H., 1984, Rapid accumulation of inositol phosphates in isolated rat superior cervical ganglia exposed to V_1-vasopressin and muscarinic cholinergic stimuli, *Biochem. J.* 221, 803–811.

Brown, D.A., 1988, M currents, in *Ion Channels*, Vol. 1, T. Narahashi (ed.), Plenum, New York, pp. 55–99.

Brown, D.A. and Selyanko, A.A., 1985, Membrane currents underlying the cholinergic slow excitatory post-synaptic potential in the rat sympathetic ganglion, *J. Physiol.* 365, 365–387.

Carafoli, E., 1987, Intracellular calcium homeostasis, *Annu. Rev. Biochem.* 56, 395–433.

Carafoli, E., 1994, Biogenesis: Plasma membrane calcium ATPase: 15 years of work on the purified enzyme, *FASEB J.* 8, 993–1002.

Carafoli, E., 1997, Plasma membrane calcium pump: Structure, function and relationships, *Basic. Res. Cardiol.* 92 (Suppl. 1), 59–61.

Carafoli, E. and Sauffer, T., 1994, The plasma membrane calcium pump: Functional domains, regulation of the activity, and tissue specificity of isoform expression, *J. Neurobiol.* 25, 312–324.

Cole, A.E. and Nicoll, R.A., 1983, Acetylcholine mediates a slow synaptic potential in hippocampal pyramidal cells, *Science* 221, 1299–1301.

Covernton, P.J.O., Kojima, H., Sivilotti, L., Gibb, A.J. and Colquhoun, D., 1994, Comparison of neuronal nicotinic receptors in rat sympathetic neurones with subunit pairs expressed in *Xenopus* oocytes, *J. Physiol.* 481, 27–34.

Cruzblanca, H., Koh, D.-S. and Hille, B., 1998, Bradykinin inhibits M-current via phospholipase C and Ca^{2+} release from $InsP_3$-sensitive stores in rat sympathetic neurons, *Proc. Natl. Acad. Sci., USA* 95, 7151–7156.

Davies, P.J., Ireland, D.R. and McLachlan, E.M., 1996, Sources of Ca^{2+} for different Ca^{2+}-activated K^+ conductances in neurones of the rat superior cervical ganglion, *J. Physiol.* 495, 353–366.

Delmas, P., Abogadie, F.C., Dayrell, M., Haley, J.E., Milligan, G., Caulfield, M.P., Brown, D.A. and Buckley, N.J., 1998, G-proteins and G-protein subunits mediating cholinergic inhibition of N-type calcium currents, *Eur. J. Neurosci.* 10, 1654–1666.

del Rio, E., Bevilacqua, J.A., Marsh, S.J., Hallet, P.M. and Caulfield, M.P., 1999, Muscarinic M_1 receptors activate phosphoinositide turnover and Ca^{2+} mobilization in rat sympathetic neurones, but this signalling pathway does not mediate M-current inhibition, *J. Physiol. Lond.* 520, 101–111.

Duchen, M.R., 1999, Contributions of mitochondria to animal physiology: From homeostatic sensor to calcium signalling and cell death, *J. Physiol. Lond.* 516, 1–17.

Endo, T. and Onaya, T., 1988, Immunohistochemical localization of parvalbumin in rat and monkey autonomic ganglia, *J. Neurocytol.* 17, 73–77.

Fierro, L. and Llano, I., 1996, High endogenous calcium buffering in Purkinje cells from rat cerebellar slices, *J. Physiol. Lond.* 496(3), 617–625.

Foucart, S., Gibbons, S.J., Brorison, J.R. and Miller, R.J., 1995, Increases in [Ca^{2+}] in adult rat sympathetic neurons are not dependent on intracellular Ca^{2+} pools, *Am. J. Physiol.* 268, C829–C837.

Fujioka, Y., Komeda, M. and Matsuoka, S., 2000, Stoichiometry of Na^{+}-Ca^{2+} exchange in inside-out patches excised from guinea-pig ventricular myocytes, *J. Physiol. Lond.* 523, 339–351.

Haley, J.E., Abogadie, F.C., Delmas, P., Dayrell, M., Vallis, Y., Milligan, G., Caulfield, M.P., Brown, D.A. and Buckley, N.J., 1998, The α subunit of G_q contributes to muscarinic inhibition of the M-type potassum current in sympathetic neurons, *J. Neurosci.* 18, 4521–4531.

Hernandez-Cruz, A., Diaz-Munoz, M., Gomez-Chavaron, M., Canedo-Merino, R., Protti, D.A., Escobar, A.L., Sierralta, J. and Suarez-Isla, B.A., 1995, Properties of the rynaodine-sensitive release channels that underlie caffeine-induced Ca^{2+} mobilization from intracellular stores in mammalian sympathetic neurons, *Eur. J. Neurosci.* 7, 1684–1699.

Hernandez-Cruz, A., Escobar, A.L. and Jimenez, N., 1997, Ca^{2+}-induced Ca^{2+} release phenomena in mammalian sympathetic neurons are critically dependent on the rate of rise of trigger Ca^{2+}, *J. Gen. Physiol.* 109, 147–167.

Herrington, J., Park, Y.B., Babcock, D.F. and Hille B., 1996, Dominant role of mitochondria in clearance of large Ca^{2+} loads from rat adrenal chromaffin cells, *Neuron* 16, 219–228.

Hille, B., 1994, Modulation of ion channel function by G-protein-coupled receptors, *Trends Neurosci.* 17, 531–536.

Hirning, L.D., Fox, A.P., McLeskey, E.W., Olivera, B.M., Thayer, S.A., Miller, R.J. and Tsien, R.W., 1988, Dominant role of N-type Ca^{2+} channels in evoked release of norepinephrine from sympathetic neurons, *Science* 239, 57–61.

Horwitz, J., Tsymbalow, S. and Perlman, R.L., 1984, Muscarine increases tyrosine-3-mono-oxyegnase activity and phospholipid metabolism in the superior cervical ganglion of the rat, *J. Pharmacol. Exper. Ther.* 239, 577–582.

Hosseini, R., Benton, D.C., Haylett, D.G. and Moss, G.W.J., 1999, Cloning of an SK channels from rat sympathetic neurones, *J. Physiol. Lond.*, 518P, 133P.

Irving, A.J., Collingridge, G.L. and Schofield, G., 1992, Interactions between Ca^{2+} mobilizing mechanisms in cultured rat cerebellar granule cells, *J. Physiol.* 456, 667–680.

Jones, S., Brown, D.A., Milligan, G., Willer, E., Buckley, N.J. and Caulfield, M.P., 1995, Bradykinin excites rat sympathetic neurons by inhibition of M current through a mechanism involving B_2 receptors and $G_{\alpha q/11}$, *Neuron* 14, 399–405.

Kawai, T. and Watanabe, M., 1986, Blockage of Ca-activated K-conductance by apamin in rat sympathetic neurones, *Br. J. Pharmacol.* 87, 225–232.

Kawai, T. and Watanabe, M., 1989, Effects of ryanodine on the spike after-hyperpolarization in sympathetic neurones of the rat superior cervical ganglion, *Pflueg. Arch.* 413, 470–475.

Knoepfel, T., Vranesic, I., Gaehwiler, B.H. and Brown, D.A., 1990, Muscarinic and beta-adrenergic depressioin of the slow Ca^{2+}-activated potassium conductance in hippocampal CA3 pyramidal cells is not mediated by a reduction of depolarization-induced cytosolic Ca^{2+} transients, *Proc. Natl. Acad. Sci., USA* 87, 4083–4087.

Koh, D.-S. and Hille, B., 1997, Modulation by neurotransmitters of catecholamine secretion from sympathetic ganglion neurons detected by amperometry, *Proc. Natl. Acad. Sci., USA* 94, 1506–1511.

Larrabee, M.G., Klingman, J.D. and Leicht, W.S., 1963, Effects of temperature, calcium and activity on phospholipid metabolism in a sympathetic ganglion, *J. Neurochem.* 12, 1–13.

Lin, Z., Haus, S., Edgerton, J. and Lipscombe, D., 1997, Identification of functionally distinct isoforms of the N-type Ca^{2+} channel in rat sympathetic ganglia and brain, *Neuron* 18, 153–166.

Llano, I., DiPolo, R. and Marty, A., 1994, Calcium-induced calcium release in cerebellar Purkinje cells, *Neuron* 12, 663–673.

Luetje, C.W. and Patrick, J., 1991, Both α- and β-subunits contribute to the agonist sensitivity of neuronal nicotinic acetylcholine receptors, *J. Neurosci.* 11, 837–845.

Marrion, N.V. and Tavalin, S.J., 1998, Selective activation of Ca^{2+}-activated K^+ channels by co-localized Ca^{2+} channels in hippocampal neurons, *Nature* 395, 900–905.

Marsh, S.J. and Brown, D.A., 1991, Potassium currents contributing to action potential repolarization in dissociated cultured rat superior cervical sympathetic neurones, *Neurosci. Lett.* 133, 298–302.

Marsh, S.J., Trouslard, J., Leaney, J.L. and Brown, D.A., 1995, Synergistic regulation of a neuronal chloride current by intracellular calcium and muscarinic receptor activation: A role for protein kinase C, *Neuron* 15, 729–737.

McGehee, D.S. and Role, L.W., 1995, Physiological diversity of nicotinic acetylcholine receptors expressed by vertebrate neurons, *Annu. Rev. Physiol.* 57, 521–546.

Miller, R.J., 1991, The control of neuronal Ca^{2+} homeostasis. *Prog. Neurobiol.* 37, 255–285.

Neher, E. and Augustine, G.J., 1992, Calcium gradients and buffers in bovine chromaffin cells, *J. Physiol. Lond.* 450, 273–301.

Palecek, J., Lips, M.B. and Keller, B.U., 1999, Calcium dynamics and buffering in motoneurones of the mouse spinal cord, *J. Physiol. Lond.* 520, 485–502.

Plummer, M.R., Logothetis, D.E. and Hess, P., 1989, Elementary properties and pharmacological sensitivities of calcium channels in mammalian peripheral neurons, *Neuron* 2, 1453–1463.

Przywara, D.A., Bhave, S.V., Chowdhury, P.S., Wkade, T.D. and Wakade, A.R., 1993, Sites of transmitter release and relation to intracellular Ca^{2+} in cultured sympathetic neurons, *Neuroscience* 52, 973–988.

Regan, L.J., Sah, D.W. and Bean, B.P., 1991, Ca^{2+} channels in rat central and peripheral neurons: High-threshold current resistant to dihydropyridine blockers and ω-conotoxin, *Neuron* 6, 269–280.

Robbins, J., Marsh, S.J. and Brown, D.A., 1993, On the mechanism of M-current inhibition by muscarinic m1 receptors in DNA-transfected rodent neuroblastoma x glioma cells, *J. Physiol. Lond.* 469, 153–178.

Rogers, M. and Dani, J.A., 1995, Comparison of quantitative calcium flux through NMDA, ATP and ACh receptor channels, *Biophys. J.* 68, 501–506.

Sacchi, O., Rossi, M.L. and Canella, R., 1995, The slow Ca^{2+}-activated K^+ current, I_{AHP}, in the rat sympathetic neurone, *J. Physiol. Lond.* 483, 15–27.

Sah, P. and McLachlan, E.M., 1991, Ca^{2+}-activated K^+ currents underlying the afterhyperpolarization in guinea pig vagal neurons: A role for Ca^{2+}-activated Ca^{2+} release, *Neuron* 7, 257–264.

Sanchez-Vivas, M.V. and Gallego, R., 1994, Calcium-dependent chloride current induced by axotomy in rat sympathetic neurons, *J. Physiol. Lond.* 475, 391–400.

Sanchez-Vivas, M.V., Valdeolmillos, M., Martinez, S. and Gallego, R., 1994, Axotomy-induced changes in Ca^{2+} homeostasis in rat sympathetic-ganglion cells, *Eur. J. Neurosci.* 6, 9–17.

Selyanko, A.A., 1996, Single apamin-sensitive, small conductance calcium-activated potassium channels (SK_{Ca}) in membrane patches from rat sympathetic neurones, *J. Physiol. Lond.* 494P, 52P.

Selyanko, A.A. and Brown, D.A., 1996, Intracellular calcium directly inhibits potassium M channels in excised membrane patches from rat sympathetic neurons, *Neuron* 16, 151–162.

Seto-Ohshima, A., Sano, M., Kitajima, S., Kawamura, N., Yamazaki, Y. and Nagata, Y., 1987, The effect of axotomy and denervation on calmodulin content in the superior cervical sympathetic-ganglion of the rat, *Brain Res.* 410, 292–298.

Sivilotti, L.G., McNeil, D.K., Lewis, T.M., Nassar, M.A., Schoepfer, R. and Colquhoun, D., 1997, Recombinant nicotinic receptors expressed in Xenopus oocytes do not resemble native rat sympathetic ganglion receptors in single-channel behaviour, *J. Physiol. Lond.* 500, 123–138.

Smart, T.G., 1987, Single calcium-activated potassium channels recorded from cultured rat sympathetic neurones, *J. Physiol. Lond.* 389, 337–360.

Storm, J.F., 1990, Potassium currents in hippocampal pyramidal cells, *Progr. Brain Res.* 83, 161–187.

Stuenkel, E.L., 1994, Regulation of intracellular calcium and calcium buffering properties of rat isolated neurohypophyseal nerve-endings, *J. Physiol. Lond.* 481, 251–271.

Thayer, S.A., Hirning, L.D. and Miller, R.J., 1988, The role of caffeine-sensitive calcium stores in the regulation of the intracellular free calcium concentration in rat sympathetic neurons in vitro, *Mol. Pharmacol.* 34, 664–673.

Tokimasa, T. and North, R.A., 1984, Calcium entry through acetylcholine-channels can activate potassium conductance in bullfrog sympathetic neurons, *Brain Res.* 295, 364–367.

Toth, P.T. and Miller, R.J., 1995, Calcium and sodium currents evoked by action potential waveforms in rat sympathetic neurones, *J. Physiol.* 485, 43–57.

Toth, P.T., Bindokas, V.P., Bleakman, D., Colmers, W.F. and Miller, R.J., 1993, Mechanism of presynaptic inhibition by neuropeptide Y at sympathetic nerve terminals, *Nature* 364, 635–639.

Trouslard, J., Marsh, S.J. and Brown, D.A., 1993, Calcium entry through nicotinic receptor channels and calcium channels in cultured rat superior cervical ganglion cells, *J. Physiol. Lond.* 468, 53–71.

Vernallis, A.B., Conroy, W.G. and Berg, D.K., 1995, Neurons assemble acetylcholine receptors with as many as three kinds of subunits while maintaining subunit segregation among receptor subtypes, *Neuron* 10, 451–563.

Wang, H.-S., Pan, Z., Shi, W., Brown, B.S., Wymore, R.S., Cohen, I.S., Dixon, J.E. and McKinnon, D., 1998, KCNQ2 and KCNQ3 potassium channel subunits: Molecular correlates of the M-channel, *Science* 282, 1890–1893.

Calcium and Platelets

J.W.M. Heemskerk

1. INTRODUCTION

Platelets have an essential role in primary haemostasis to arrest bleeding at a site of vessel wall disruption. Insufficient platelet activation or a reduced platelet number give rise to bleeding syndromes. On the other hand, high platelet activation contributes to thrombosis, which often starts with platelet deposition on a damaged atherosclerotic plaque. The formation of vaso-occlusive platelet thrombi or aggregates, usually with repeated embolization, is a major cause of arterial thrombosis, as occurring in the heart and brains, and resulting in myocardial infarction and so-called cerebrovascular accidents (stroke). Although there is little doubt that coronary and cerebral artery diseases are multifactorial disorders, where also abnormal coagulation, fibrinolysis, vessel wall function and blood flow dynamics can play a role, intervention studies show that in particular antiplatelet drugs provide a risk reduction of one out of every four cases.

Platelets can be stimulated to adhere or aggregate by a large number of agents. Under conditions of high flow, as in arteries, initial arrest of platelets is achieved by their interaction with von Willebrand factor (vWF), which is present in a damaged vessel wall. Irreversible adhesion, e.g. to collagen fibers in the subendothelial matrix, is then accomplished by the platelet integrin $\alpha_2\beta_1$ and $\alpha_{IIb}\beta_3$ receptors. Platelet aggregation will start on top of the bound cells, and the platelets in an aggregate are held together by tight bridges of the integrin $\alpha_{IIb}\beta_3$ receptors with fibrinogen in the plasma. In addition to aggregating, platelets show various other responses, such as shape change (rounding-off), development of pseudopods, secretion of activation-promoting substances, and phospholipid modifications at their surface to support blood coagulation. Most of these responses appear to be essential

J.W.M. Heemskerk • Departments of Biochemistry (CARIM) and Human Biology (NUTRIM), University of Maastricht, Maastricht, The Netherlands.

R. Pochet, R. Donato, J. Haiech, C. Heizmann and V. Gerke (eds.): Calcium: The Molecular Basis of Calcium Action in Biology and Medicine, 45–71.

in accelerating and extending the overall platelet activation process. For instance, the platelet products ADP and thromboxane A_2 are potent platelet agonists by themselves. Similarly, thrombin that is formed as a result of the coagulation is a strong stimulator of platelet activation and aggregation.

This chapter evaluates the importance of the intracellular calcium signal for these various platelet responses. It reviews the current understanding of signaling pathways involved in the generation and abolition of this signal. This information is used to evaluate the signaling defects and functional changes in platelets from a number of patients that have been described with platelet abnormalities.

2. PLATELET AGGREGATION AND THE CALCIUM RESPONSE

Many functional platelet responses are directly dependent on an increase in cytosolic $[Ca^{2+}]_i$. These include shape change, secretion, formation of prostaglandins and thromboxanes, exposure of a coagulation-promoting surface and shedding of microvesicles from the platelet body. However, adhesion and aggregation of platelets are not necessarily dependent on generation of a calcium signal, although these processes are promoted by elevated $[Ca^{2+}]_i$, and are tightly connected to the other, Ca^{2+}-dependent activation events. As indicated below, many of the so-called anti-aggregatory agents that are currently in clinical or pre-clinical use, such as aspirin, abciximab, clopidogrel and ticlopidine, will act in part by influencing the platelet calcium signal.

Platelet aggregation is a direct consequence of activation of the fibrinogen receptor, integrin $\alpha_{IIb}\beta_3$ (glycoprotein IIb/IIIa) on the platelet surface (Shattil et al., 1998). This integrin is the most abundant receptor on platelets with about 50,000 copies present per cell. Activation of integrin $\alpha_{IIb}\beta_3$ can be accomplished by many platelet agonists, and is often referred to as *inside-out signaling*. The activation is accompanied by clustering of the integrin molecules within the plasma membrane, and generates high-affinity binding sites for fibrinogen, vWF and collagen. Many signaling pathways have been implicated in the integrin activation process, involving serine and protein kinases (e.g. protein kinase C and phosphoinositide 3-kinase), protein tyrosine kinases (Src, Syk and Pyk2) and Ras-related G proteins (Hers et al., 1998; Shattil et al., 1998). Increases in $[Ca^{2+}]_i$ may stimulate one or more of these activation routes. On the other hand, platelet-derived products that are released in a Ca^{2+}-dependent way (e.g. ADP and thromboxane A_2) stimulate the activation of integrin $\alpha_{IIb}\beta_3$.

The most common screening test for platelet function *in vivo* is measurement of the bleeding time after incision. However, this test suffers from a high variance and lack of predictive value for post-operative bleeding. Measurement of the platelet aggregation tendency has for long been used to test platelet function in the laboratory. Because of standarization problems, the aggregation method is gradually being replaced by flow-cytometric methods and platelet adhesion tests, which appear to give reproducible results (Marshall et al., 1997; Fressinaud et al., 1998).

In recent years, much achievement has been made in elucidating the manyfold signal transduction routes in platelets. Principal routes of generating and controling the calcium response are described below. Given vast amount of literature and the high complexity of signaling patterns in platelets, however, many subjects can only be mentioned here in brief in sometimes an oversimplified way.

3. GENERATION OF CALCIUM SIGNAL IN PLATELETS

Various types of agonists are capable to raise $[Ca^{2+}]_i$ in platelets. In addition to phospholipase C (PLC)-stimulating receptor agonists, inducing production of inositol 1,4,5-trisphosphate ($InsP_3$), these are mainly inhibitors of the sarco/endoplasmic reticulum Ca^{2+}-ATPases (SERCAs) and sulfhydryl reagents sensitizing the platelet $InsP_3$ receptors (Sargeant et al., 1992, Authi et al., 1993; Heemskerk et al., 1993). The sulfhydryl reagents act in the absence of InsP3 production by promoting the process of Ca^{2+}-induced Ca^{2+} release (CICR, see below).

Like other cell types, platelets contain various phosphoinositide-specific PLC isoforms, which all cleave phosphatidylinositol 4,5-bisphosphate into $InsP_3$ and diacylglycerol. At least seven different PLC forms are detected in human platelets, of which PLCγ2 and PLCβ2/3 appear to be the most abundant ones (Lee et al., 1996; Banno et al., 1998). The canonical pathway of $InsP_3$ action is highly operative in platelets. Thus, the formed $InsP_3$ binds to its receptor on the endoplasmic reticulum (in platelets also indicated as dense tubular system), which results in the release of Ca^{2+} through the $InsP_3$ receptor channels from store compartments to the cytosol (Authi et al., 1986). Ryanodine receptors have not been found in platelets. Calcium store depletion triggers the so-called store-regulated Ca^{2+} influx pathway, by which means also extracellular Ca^{2+} enters the cell. The diacylglycerol that is formed upon PLC action activates protein kinase C isoforms, which phosphorylate many platelet proteins and, thereby, stimulate aggregation and evoke downregulation of the calcium signal generation. Other predominant ways of suppressing Ca^{2+} mobilization in platelets are activation of protein kinase A and protein kinase G, caused by rises in the cytosolic levels of cAMP and cGMP, respectively. Details of these classical signaling routes are extensively described in earlier reviews (Siess, 1989; Heemskerk and Sage, 1994; Authi, 1997; Sage, 1997). Some newly described mechanisms are indicated below.

Members of the seven transmembrane domain class of receptor provide a major way of achieving mobilization of Ca^{2+} ions. In the last few years, various of these receptors have been identified in platelets. These include receptors of the Ca^{2+}-mobilizing agonists, thrombin, platelet-activating factor receptor, ADP and thromboxane A_2 (Brass et al., 1997; Coughlin, 1999).

Platelets from mice with deficient G_q α subunits lack $InsP_3$ formation and Ca^{2+} mobilization in response to thrombin, ADP and thromboxane A_2, indicating that G_q is involved in the stimulation of PLC via all three receptor types (Offermanns et al., 1998).

Thrombin, which is formed during the coagulation process by cleavage of its parent protein prothrombin, is a potent platelet agonist. In human platelets two protease-activated receptors, PAR1 and PAR4, for thrombin have been identified (Vu et al., 1991; Xu et al., 1999). Together, these two receptors may account for most or all effects of thrombin on platelets, including increases in $[Ca^{2+}]_i$, shape change, secretion and aggregation (Kahn et al., 1999). On the other hand, there is evidence that PAR1 is the major activator of human platelets, at least at low concentrations of thrombin (Brass et al., 1992; Kahn et al., 1999). Note that in mouse platelets the PAR3 receptor is responsible for part of the thrombin-induced activation events, while the PAR1 receptor is inactive here. In human platelets, the thrombin-induced cleavage of PAR1 results in auto-activation of the receptor, followed by its coupling to G_q and stimulation of PLCβ isoforms. In addition, thrombin is proposed to elicit the membrane translocation and the stimulation of PLCγ2 (Banno et al., 1996). Although the latter effect may point to a cross-talk between thrombin receptor- and collagen receptor-mediated signaling pathways (Watson, 1999), it is still unclear how PLCγ2 stimulation is achieved. Thrombin also binds to the platelet glycoprotein Ib/V/IX complex, and it can cleave glycoprotein V. However, whether this cleavage indeed results in calcium signaling, as has been proposed (Greco et al., 1996), needs confirmation.

The nature of the purinergic (ADP and ATP) receptors on platelets has been a subject of much debate (Mills, 1996). Nowadays, three different ADP receptors on the platelet surface are recognized. Two of these, the $P2Y_1$ and the P2T receptors, appear to be responsible for most of the ADP-evoked platelet activation events (Jin and Kunapuli, 1998; Hechler et al., 1998; Kunapuli, 1998). The $P2Y_1$ receptor is coupled to G_q and, thus, stimulates PLCβ with consequent Ca^{2+} mobilization (Heemskerk et al., 1993; Daniel et al., 1998; Hechler et al., 1998). The importance of this receptor form can be deduced from the observation that mice deficient in $P2Y_1$ receptors show an increased bleeding time and a reduced thromboembolic reaction. The mutated animals have platelets that fail to respond to ADP by a rise in $[Ca^{2+}]_i$, and are reduced in ADP-evoked aggregation (Fabre et al., 1999). The P2T ($P2T_{AC}$) receptor, on the other hand, is coupled to G_i and thus mediates inhibition of adenylate cyclase. This signaling route is considered to trigger platelet aggregation, although it is unclear whether the aggregation is a direct consequence of the resulting decrease in cAMP level (Jin and Kunapuli, 1998). The P2T receptor is also the principal target of new antiplatelet drugs, such as ticlopidine and clopidogrel (Cattaneo and Gachet, 1999; Geiger et al., 1999).

In addition, a third purinergic receptor, with ionotropic action, has been identified in human platelets. This is the $P2X_1$ receptor, which in fact is a receptor-operated cation ion channel, being responsible for the initial phase of ADP-evoked Ca^{2+} influx into platelets (MacKenzie et al., 1996; Vial et al., 1997; Sun et al., 1998). Earlier observations indicates that, under normal conditions, the calcium signal elicited by stimulation of this receptor is rather small and becomes rapidly desensitized (Sage, 1997). However, the channel activity is relatively insensitive to activation of protein kinases A or G (i.e. elevation in cAMP or cGMP), and it has been shown that under such conditions the receptor-mediated Ca^{2+} influx significantly contributes to the total ADP-evoked calcium response (Geiger et al., 1992; Sage et al., 2000).

Other platelet serpentine receptors, whose stimulation results in PLCβ activation and Ca^{2+} mobilization, are the TPα receptors for thromboxane A_2, which bind to G_q (Offermanns et al., 1998; Paul et al., 1999b), and the receptors for platelet-activating factor and lysophosphatidic acid, which interact with still undefined G proteins.

Various receptor agonists, such as thrombin, ADP and epinephrine, are potent stimulators of G_i activity (Siess, 1989). For long, it was thought that this stimulation is somehow linked to PLC activation (Brass et al., 1986). This idea was supported by the finding that, in a variety of cells other than platelets, G_i stimulation leads to PLCβ stimulation via G$\beta\gamma$ subunits that are released upon receptor-G_i coupling. However, current data indicate that the G$\beta\gamma$ signaling pathway does not play a prominent role in platelets. Evidence comes from the observation that stimulation of the α_{2A}-adrenergic receptor with epinephrine or other specific agents does not result in $InsP_3$ formation or a significant calcium signal, but can potentiate the Ca^{2+} mobilization by other agonists (Keularts et al., 2000). Epinephrine-stimulated activation of G_i appears to reduce the "basal" level of cytosolic cAMP, which then results in downregulation of the suppressive effect of protein kinase A on the platelet calcium response. This would imply that platelets in the circulation have a (slightly) elevated level of cAMP, which acts as activation-suppressive.

Platelets contain various collagen-binding proteins (see below). Interaction with glycoprotein VI appears to be both essential and sufficient to mediate platelet activation by collagen. Glycoprotein VI is thus considered as the main signaling collagen receptor (Watson, 1999). Recent findings indicate that it is a member of the immunoglobulin family of receptors (Clemetson et al., 1999). Cross-linking of glycoprotein VI by collagen results in the assembly of a large complex of several protein tyrosine kinases, protein serine/threonine kinases, adapter proteins and cytoskeletal elements (Watson, 1999). Crucial steps in the myriad of signaling events occurring in this complex are phosphorylation at tyrosine of the Fc receptor γ-chain in the plasma membrane, and the subsequent activation of the tyrosine kinase Syk. This somehow results in phosphorylation and activation of PLCγ2, in a consequent increase in $[Ca^{2+}]_i$ (Daniel et al., 1994; Ichinohe et al., 1995; Asselin et al., 1997), but also in platelet aggregation (Kehrel et al., 1998).

Evidence for this pathway comes from observations on platelets from patients lacking glycoprotein VI and mice platelets lacking either the Fc receptor γ-chain or Syk, which are all severely compromised in collagen-induced platelet activation (Moroi and Jung, 1997; Ichinohe et al., 1997; Poole et al., 1997). Furthermore, quite similar signaling events are triggered by other glycoprotein VI-binding agents, such as a collagen-related peptide and the snake venom convulxin (Jandrot-Perrus et al., 1997; Polgár et al., 1997; Gibbins et al., 1997). As stated above, Syk appears to play an important role in glycoprotein VI-induced platelet activation. Its binding to the Fc receptor γ-chain leads to auto-phosphorylation and activation of other protein tyrosine kinases (Lyn and Fyn), which phosphorylate other proteins in the complex. Other factors required for (full) activation of PLCγ2 are the adapter proteins, SLP-76 and LAT, and also phosphoinositide 3-kinase which forms 3-phosphorylated phosphoinositides (Gross et al., 1999; Lagrue et al., 1999; Watson, 1999). The latter effect agrees well with the general idea that the 3-kinase and its product, phosphatidylinositol 3,4,5-trisphosphate, are important for sustained PLCγ functioning (Scharenberg and Kinet, 1998). In platelets, activation of phosphoinositide 3-kinase appears also to play an important role in the stimulation of PLCγ2 elicited by immunoglobulin G, which binds to the platelet FcγRIIA receptors (Gratcap et al., 1998).

An important though variable part of the calcium response in platelets consists of effects mediated by stimulating agents that are released by the platelets themselves (Siess, 1989). The most potent autocrine agents are thromboxane A_2 and adenosine nucleotides. Activation of the 85 kDa cytosolic phospholipase A_2 (cPLA$_2$) leads to liberation of arachidonic acid from the various phospholipids. cPLA$_2$ stimulation is supposed to be dependent on an increase in $[Ca^{2+}]_i$ and to be facilitated by phosphorylation by stress-activated protein kinases (Kramer et al., 1993; Börsch-Haubold et al., 1999). Arachidonate is converted in the labile, platelet-activating agents, prostaglandin H_2 and thromboxane A_2 by cyclooxygenase. The latter enzyme is inhibitable by acetyl salicylic acid (aspirin) and other anti-inflammatory drugs. Once produced, thromboxane A_2 interacts with the TPα receptor and causes Ca^{2+} mobilization, shape change and secretion. Secretion of platelet agonists (mainly ADP) and subsequently activation of the G_i-coupled receptors is thought to be required for full thromboxane A_2-induced platelet aggregation (Paul et al., 1999b). In resting platelets, the adenosine nucleotides ADP and ATP are stored together with Ca^{2+} within secretory granules, the so-called dense granules. Many platelet agonists are capable to induce the secretion reaction, which results in the release of ADP that can then exert its effects by interacting with the platelet purinergic receptors. Thereby, ADP acts as a potentiating agent of many responses (Siess, 1989).

Suppression of the calcium and aggregation responses is mainly achieved by endothelial-derived compounds. One of the most potent of these is prostaglandin I_2 (prostacyclin), which binds to the G_s-coupled IP receptor on platelets and thereby stimulates adenylate cyclase. The cAMP thus produced has one principal way of action in platelets, so far known, i.e. activation of protein kinase A (Hallbrügge and Walter, 1993; Brass et al., 1997). Another endothelial product is nitric oxide (NO), which stimulates platelet guanylate cyclase and raises the cGMP level. Elevated cGMP may act, for a substantial part, by inhibiting a cAMP phosphodiesterase, which reduced the degradation of cAMP and, thereby, causes a net increase in cAMP level (Hallbrügge and Walter, 1993). Especially protein kinase A, but probably also protein kinase G, phosphorylate various proteins that are involved in calcium signal generation (Authi, 1997).

4. COMPOSITION OF THE CALCIUM SIGNAL IN PLATELETS

Activation of phosphoinositide-specific PLC in platelets occurs in two common ways, through GTP-bound G_q (e.g. involving PLCβ3) and by stimulating the tyrosine kinase Syk (e.g. PLCγ2) (Brass et al., 1997). The alternative mechanism of activation of PLCβ via $\beta\gamma$ subunits derived from (pertussis toxin-sensitive) G_i can be mimicked *in vitro* using platelet-derived PLC and purified $\beta\gamma$ subunits of brain G proteins (Banno et al., 1998). However, this pathway has not yet been proven to work in intact platelets (Keularts et al., 2000).

Platelets contain at least two different $InsP_3$ receptor isotypes among which the type 1 receptor (Quinton et al., 1996; Authi, 1997). It is most probably this type that is greatly downregulated by protein kinase A-mediated phosphorylation (Cavallini et al., 1996; Keularts et al., 2000). The observation that $InsP_3$-generating agonists and SERCA inhibitors potentiate each other in mobilizing Ca^{2+} from the endoplasmic reticulum, has led to the idea that platelets contain different calcium store compartments, where the various types of $InsP_3$ receptors and SERCAs are segregated (Authi et al., 1993; Authi, 1997). This agrees with the evidence from immunochemical studies that the various SERCA isoforms present in platelets (the house-keeping SERCA2b and SERCA3 forms) have a somewhat different subcellular localization (Kovàcs et al., 1997). The functional consequences of such subcellular heterogeneity are not clear. SERCAs or associated proteins may be a target of phosphorylation by protein kinase A. However, the literature is still ambiguous on the involvement of (phosphorylated) Rap1B in increasing SERCA activity (Authi, 1997). Extrusion of Ca^{2+} out of the platelets is achieved by a plasma membrane Ca^{2+}-ATPAse (PMCA4b), which also seems to be increased in activity upon phosphorylation by protein kinase A (Dean et al., 1997).

Platelets do not exhibit voltage-gated Ca^{2+} channels. The most prominent way of stimulating Ca^{2+} influx across the plasma membrane is by Ca^{2+} store depletion, according to the mechanism of store-regulated Ca^{2+} entry (Authi, 1997; Sage, 1997). Thus, mobilization of Ca^{2+} from internal stores, by either receptor agonists or SERCA inhibitors, is coupled to Ca^{2+} influx and, accordingly, inhibition of Ca^{2+} mobilization by protein kinase A/G stimulation suppresses the Ca^{2+} influx (Alonso et al., 1991; Heemskerk et al., 1994). As for other cell types, the mechanism regulating this influx has not been clarified, although there is evidence that it involves protein tyrosine phosphorylation, specific inositolphosphates and/or cytoskeletal reorganization (Sargeant et al., 1993; Authi, 1997; Vuist et al., 1997; Rosado et al., 2000). Cation channels of the Trp family are nowadays considered as the most likely candidates for store-regulated calcium channels (Parekh and Penner, 1997). These channel proteins are probably also present in platelets and platelet-like cells (Berg et al., 1997).

Measurements of the calcium signal in single, immobilized platelets show that, in general, stimulation of the G_q-linked receptors results in a series of repetitive $[Ca^{2+}]_i$ spikes, occurring at irregular frequency and amplitude (Heemskerk et al., 1993, Heemskerk and Sage, 1994; Ariyoshi and Salzman, 1996; Hussain and Mahaut-Smith, 1999). This holds for the agonists thrombin, PAR1-activating peptides, thromboxane A_2 mimics and ADP (e.g., see Figure 1B). Typically for thrombin stimulation is that the $[Ca^{2+}]_i$ spiking is persistent for up to 20 min. This seems to be due to continuous receptor cleavage by thrombin, in which each newly cleaved (PAR1) receptor adds to the calcium signal (Heemskerk et al., 1997a). The prolonged thrombin-evoked calcium response appears to rely heavily on influx of external Ca^{2+}. When the assumption is made that, with these receptor agonists, the average half-life time of every mobilized Ca^{2+} ion is equal (i.e. the rate of Ca^{2+}-pumping remains constant), the Ca^{2+}-mobilizing potency of the agonists can be estimated from the time-integral of the total calcium response. By this approach, it is shown that the Ca^{2+}-mobilizing potency of thrombin on platelets is much higher than that of PAR1 peptide, ADP or thromboxane (Heemskerk et al., 1997a). This is an indication that several primary (PAR1- and PAR4-mediated) and secondary (ADP and thromboxane A_2-mediated) signaling events synergize in the thrombin-evoked the Ca^{2+} response.

The mechanism underlying the generation of $[Ca^{2+}]_i$ spikes is relatively well understood. As in other cell types, the rapid Ca^{2+} mobilization in each spike is controled by the mechanism of Ca^{2+}-induced Ca^{2+} release (CICR), based on a Ca^{2+}-promoting effect of $InsP_3$ receptor functioning (Heemskerk et al., 1993; Authi, 1997; Keularts et al., 2000). Accordingly, Ca^{2+} mobilization can occur even at basal $InsP_3$ levels, provided that $[Ca^{2+}]_i$ is already slightly elevated.

5. PLATELET ADHESION AND CALCIUM

Recent investigations shed more light on the calcium-signaling effects of platelet adhesion to physiological receptor ligands (vWF, fibrinogen and collagen). Platelet adhesion to immobilized vWF is considered to be a shear-dependent process. vWF is secreted by endothelial cells and then deposited in the subendothelial matrix of the vessel wall. Other sources of vWF are blood plasma and platelets themselves, which also secrete this substance. Next to serving as a coagulation factor, vWF is considered to play an important role

A Activation of platelet on collagen surface

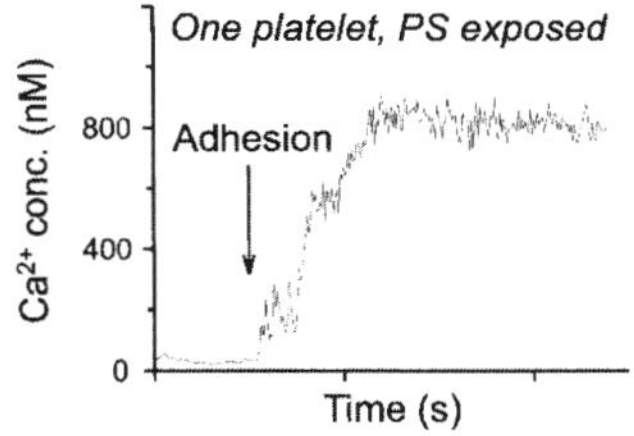

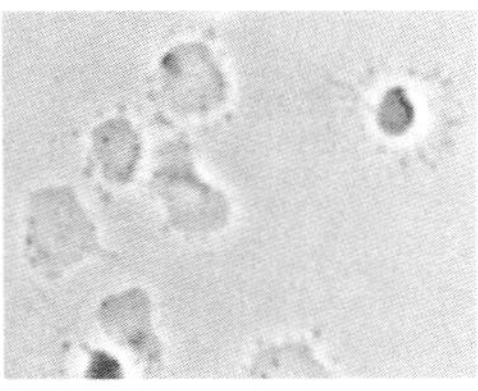

B Thrombin activation of platelet on fibrinogen surface

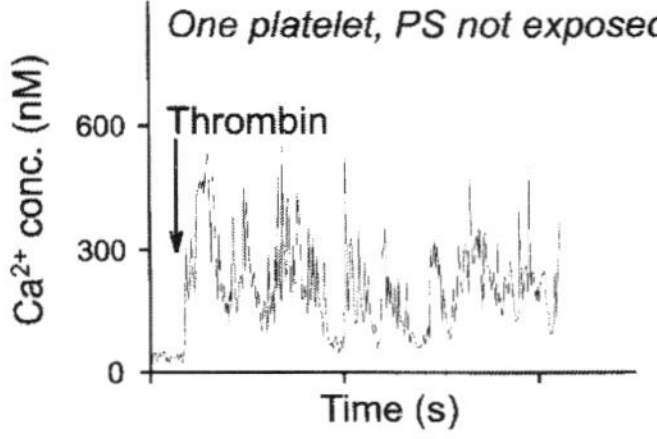

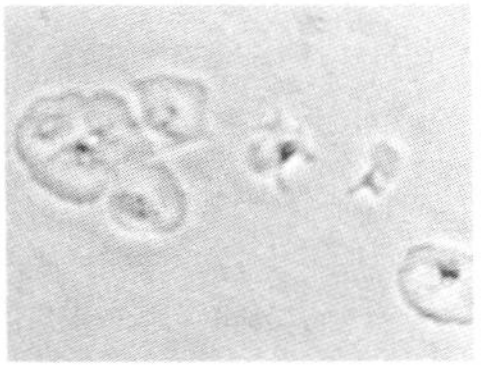

C Phosphatidylserine exposure of platelets on various adhesive surfaces

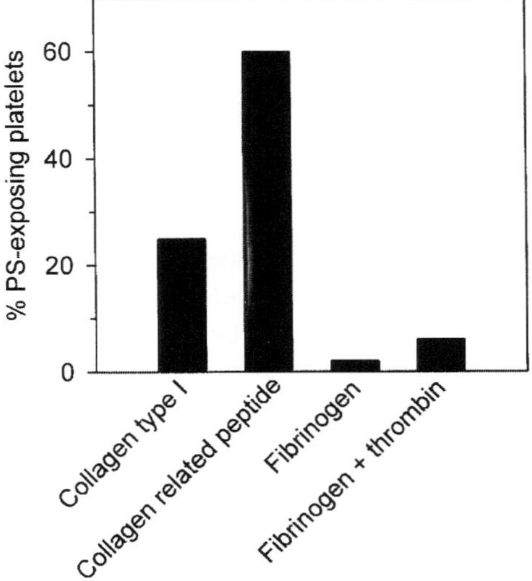

Figure 1. Procoagulant response of platelets adhering to a collagen, but not to a fibrinogen surface. (A) Single, fura-2-loaded platelets show a prolonged calcium response soon after the adhesion to collagen. This effect is accompanied by phosphatidylserine (PS) exposure and blebbing of the plasma membrane. (B) Fura-2-loaded platelets adhering to a fibrinogen surface need to be stimulated with thrombin to respond by repetitive spiking in $[Ca^{2+}]_i$. This calcium response promotes spreading of the platelets, but does not lead to PS exposure or bleb formation. (C) Platelet adhesion to a surface covered with a glycoprotein VI-binding collagen-related peptide is a most efficient trigger of PS exposure.

in both the physiologic and pathologic formation of platelet thrombi under conditions of high shear rate (Sixma and Wester, 1977; Sadler, 1991; Kroll et al., 1996). *Ex vivo* experiments where blood was flowing over surfaces coated with vWF have confirmed this idea (Goto et al., 1998). Platelets appear firstly to interact with surface-immobilized vWF through the glycoprotein Ib/V/IX complex (25,000 copies/platelet) (Tsuji et al., 1996; Kroll et al., 1996). This interaction is reversible, leads to a transient capture of flowing platelets, and is antagonized upon lowering of the shear rate. Subsequent, firm attachment to the vWF surface and contact with other platelets to form an aggregate requires the involvement of integrin $\alpha_{IIb}\beta_3$ receptors, which become gradually activated and bind irreversibly to vWF (Lankhof et al., 1995; Goto et al., 1998). Various authors have proposed that stable platelet interaction with vWF, preferably via integrin $\alpha_{IIb}\beta_3$, results in a calcium signal, particularly consisting of Ca^{2+} influx (Ikeda et al., 1993; Bertolino et al., 1995; Kroll et al., 1996). Recent studies, using reconstituted blood with fluorescent-labeled platelets, indicate that the stable phase of platelet adhesion (integrin $\alpha_{IIb}\beta_3$-dependent) is accompanied by a rise in $[Ca^{2+}]_i$, whereas the initial (putatively glycoprotein Ib/V/IX-dependent) adhesion is not (Kuwahara et al., 1999). However, it is unclear from this study whether integrin $\alpha_{IIb}\beta_3$ activation itself mobilizes Ca^{2+}.

The signaling events underlying vWF-dependent calcium signal generation are unclear. Work with specific glycoprotein Ib-binding agents, among which mutated forms of vWF, indicates that the vWF-glycoprotein Ib/V/IX interaction results in Fc receptor γ-chain phosphorylation as well as in activation of the tyrosine kinases Fyn, Lyn and Syk. In a way resembling the glycoprotein VI-induced signaling system, this may result in activation of PLCγ2 (Falati et al., 1999). Others report that glycoprotein Ib/V/IX is connected to the membrane skeleton via actin-binding protein to form a complex with adapter proteins like 14-3-3ζ (Calverley et al., 1998). This complex can assist in recruitment to the cytoskeleton of signaling elements, such as the tyrosine kinases Src, Lyn and Syk, the tyrosine phosphatase PTP1B and focal adhesion kinase. Ultimately, this may lead to activation of the Ca^{2+}-dependent enzymes, cPLA$_2$ and μ-calpain (Calverley et al., 1998; Yuan et al., 1997).

Many studies have been performed with platelets adhering to a fibrinogen-coated surface through their integrin $\alpha_{IIb}\beta_3$ receptors. Fibrinogen-bound platelets tend to spread over the surface in a way that is promoted by but is not necessarily dependent on elevation in $[Ca^{2+}]_i$ (Jen et al., 1996; Heemskerk et al., 1997b). Platelets spreading over fibrinogen (i.e. developing lamellipodia) often show incidental spiking rises in $[Ca^{2+}]_i$, which is partly due to secretion of ADP by the cells (Heemskerk et al., 1993). However, some of the spike formation may be secondary to the other, extensive signaling events reported for spreading platelets. These events, triggered by integrin $\alpha_{IIb}\beta_3$ occupation, are known as *outside-in* signaling. First reports of the signaling pathways during the spreading process point to extensive cytoskeletal reorganization and tyrosine phosphorylation (Haimovich et al., 1993). Evidence that integrin binding leads to phosphorylation of Src and Syk (Clark et al., 1994) may suggest that this leads to PLC stimulation, because Syk is considered to be an upstream activator of PLCγ2. Conversely, for Syk-null murine platelets it has been observed that Syk plays a significant role in activation of integrin $\alpha_{IIb}\beta_3$ (Law et al., 1999). Nowadays, it is clear that integrin $\alpha_{IIb}\beta_3$-dependent spreading also results in the cytoskeletal recruitment of protein kinases, such as focal adhesion kinase and Tec, and, eventually, in protein degradation by μ-calpain and activation of protein tyrosine phosphatases (PTP1B and SHP1), which results in a late decrease in tyrosine phosphorylation (Shattil et al., 1998). The focal adhesion kinase provides another assembly site of signaling and cytoskeletal elements. In addition, Rho-family G-proteins are probably involved both in spreading (lamellipod formation) and in the extrusion of pseudopods (filopods). A particular role of Ca^{2+} has been suggested in such morphological changes. The necessary association of myosin with actin filaments is promoted by myosin light-chain phosphorylation. This phosphorylation appears to be achieved by two independent pathways, one Ca^{2+}/calmodulin-sensitive involving myosin light-chain kinase, and another Ca^{2+}-insensitive mediated by the small G protein Rho and by Rho-kinase (p160^{ROCK}) (Bauer et al., 1999; Paul et al., 1999a).

Platelet adhesion to and spreading on collagens of various types is currently seen as a two-step process. The initial interaction is mediated by the integrin $\alpha_2\beta_1$ (glycoprotein Ia/IIa) receptor (Coller et al., 1989; Polanowska-Grabowska and Gear, 1992; Saelman et al., 1994). By itself, this integrin binding results in little signaling, although this receptor might be brought into a higher affinity binding state by previous activation of the platelets (Jung and Moroi, 1998). On the other hand, integrin $\alpha_2\beta_1$-binding appears to play an important role in the subsequent, glycoprotein VI-dependent platelet activation responses, leading to Syk and PLCγ2 activation and a high calcium signal (Keely and Parise, 1996; Heemskerk et al., 1999). In flowing blood, initial platelet-collagen interaction also involves the glycoprotein Ib/V/IX-mediated binding to vWF that becomes attached to the collagen (Moroi et al., 1997; Verkleij et al., 1998). There is still ambiguity whether glycoprotein IV (CD36) also plays a role in platelet adhesion under such conditions (Saelman et al., 1994).

6. CALCIUM AND THE COAGULATION-STIMULATING EFFECT OF PLATELETS

It has been known for almost 20 years that activated platelets can facilitate the coagulation of blood plasma by promoting thrombin formation. Initially, it was recognized that Ca^{2+}-ionophores such as A23187 and ionomycin are efficient triggers of the so-called platelet procoagulant response, suggesting that this reaction is dependent on an increase in $[Ca^{2+}]_i$ (Bevers et al.,

1982). The coagulation-promoting effect appears to consist of two series of responses that usually occur simultaneously. First, the Ca^{2+}-ionophores cause abolition of the normal phospholipid transbilayer asymmetry of the plasma membrane. In resting platelets, the aminophospholipids phosphatidylserine and phosphatidylethanolamine are inwardly transported by a putative aminophospholipid translocase. However, their inside location becomes lost in the presence of Ca^{2+}-ionophore, putatively due to the activation of a Ca^{2+}-dependent phospholipid scramblase (Williamson et al., 1995; Zwaal and Schroit, 1997). This loss of phospholipid asymmetry is thus accompanied by the exposure of aminophospholipids at the platelet outer membrane surface. Especially the exposed phosphatidylserine has a potent coagulation-stimulating effect, by serving as a site for the assembly and activation of the tenase and prothrombinase complexes, which are responsible for the formation of coagulation factor Xa and thrombin, respectively (Zwaal et al., 1993). The second ionophore effect is formation of membrane blebs at the platelet surface, which may shed into the circulation as platelet-derived microvesicles (Sims et al., 1989; Comfurius et al., 1990; Zwaal et al., 1992). Since these microvesicles also have phosphatidylserine exposed (Tans et al., 1991) and do not to stick to platelet aggregates, they may disperse the coagulation process and thus be associated with thrombotic conditions (Zwaal and Schroit, 1997).

When examined with platelets in suspension, not only Ca^{2+}-ionophores but also combinations of other activating agents appear to induce phosphatidylserine exposure and microvesiculation. These include collagen with thrombin (Zwaal et al., 1992), SERCA inhibitors whether or not in combination with thrombin (Smeets et al., 1993; Dachary-Prigent et al., 1995), and complement proteins C5-9 (Sims et al., 1988). Typically, all these agents cause prominent increases in $[Ca^{2+}]_i$ in platelets. By using fluorescence microscopic imaging techniques, it has recently become possible to study these calcium responses in combination with phosphatidylserine exposure and bleb formation in platelets adhering to immobilized collagen or fibrinogen. From such studies it is concluded that only conditions that elicit a prolonged, non-spiking increase in $[Ca^{2+}]_i$ lead to phosphatidylserine exposure and membrane blebbing (Heemskerk et al., 1997; Briedé et al., 1999). For instance, most platelets adhering to fibrillar collagen or to a collagen-related peptide show this combination of responses, while platelets adhering to fibrinogen are mostly unable to do so (Figure 1). Even when the fibrinogen-bound platelets are stimulated with thrombin or other agonists of G-protein-coupled receptors, this results in no more than little phosphatidylserine exposure. On the other hand, these agonists can potentiate the effect of adhesion to collagen.

Principal receptors and signaling pathways involved in the procoagulant response are gradually being elucidated. Platelet-collagen adhesion via glycoprotein VI is a most effective way of triggering this response, as apparent from studies with a specific collagen-related peptide and convulxin. Platelet-collagen binding via integrin $\alpha_2\beta_1$ is inactive by itself, but can prime for the procoagulant response elicited by glycoprotein VI ligandation (Heemskerk et al., 1999). Although thrombin is only a very weak activator of the procoagulant response, this low activity has been attributed to signaling through the PAR1 receptor (Andersen et al., 1999). A prominent increase in $[Ca^{2+}]_i$ is essential in causing procoagulant activity. For instance, the glycoprotein VI-mediated procoagulant response is inhibited by all interventions aimed to reduce the calcium response, e.g. by inhibitors of protein tyrosine kinases (Syk), by elevating cAMP or by abolition of thromboxane A_2 formation (Heemskerk et al., 1997b). Furthermore, the procoagulant response evoked by collagen adhesion, and also that by SERCA inhibition, appears to require influx of external Ca^{2+}, probably to reach sufficiently high levels of $[Ca^{2+}]_i$ (Pasquet et al., 1996; Heemskerk et al., 1999). The elevated $[Ca^{2+}]_i$ seems to have at least two effects: activation of the putative scramblase and activation of the protease μ-calpain, which is autolysed at elevated $[Ca^{2+}]_i$ and thereby constitutively activated (Saido et al., 1993). Where investigated, with the exception of complement C5b-9-activated platelets, microvesiculation appeared to be preceded by μ-calpain activation (Wiedmer et al., 1990; Dachary-Prigent et al., 1995). Calpain is probably responsible for the protein degradation necessary to weaken interactions of the plasma membrane with the membrane skeleton and cytoskeleton. It has also been proposed that (unspecified) protein tyrosine phosphatases are involved in bleb/microvesicle formation (Pasquet et al., 1998).

Additional clues as to the processes involved in the procoagulant response come from studies with platelets from patients with a rare inherited bleeding disorder, the Scott syndrome. Platelets from these patients are deficient in procoagulant activity and in vesiculation, when stimulated with either Ca^{2+}-ionophore or collagen/thrombin (Rosing et al., 1985; Sims et al., 1989). This finding has led to the idea that the putative scramblase activity is reduced in this patient, and also that its activation is linked to the shedding of membrane vesicles (Zwaal and Schroit, 1997; Bettache et al., 1998). However, there is also evidence that the platelets from a Scott patient are reduced in protein tyrosine phosphorylation, which suggests that the alteration underlying this syndrome is rather a signaling defect involving tyrosine kinase or phosphatase activity (Dachary-Prigent et al., 1997).

Since membrane blebbing and phosphatidylserine expression also occur in nucleated cells during apoptosis, there has been some interest in finding "apoptotic" signaling pathways during the platelet procoagulant response. Platelets indeed contain relevant elements of the apoptotic signaling system, such as caspases, caspase-activators and Bcl-2 family proteins (Vanags et al., 1997; Wolf et al., 1999). However, it is also clear that, in platelets, the Ca^{2+}-activated μ-calpain prevents caspase-mediated proteolysis and, thus, the onset of apoptotic-like protein degradation (Wolf et al., 1999).

Evidence for the importance of platelet adhesion in the procoagulant response comes from a series of experiments, in which platelet-containing plasma is triggered to coagulate by recalcification and then the coagulation process is continuously examined. The triggering results in a resonance loop in which (i) small amounts of thrombin formed activate platelets and convert fibrinogen to fibrin, (ii) fibrin in combination with thrombin stimulates platelets to expose procoagulant phosphatidylserine, and (iii) the exposed phosphatidylserine facilitates the generation of more thrombin (Béguin and Kumar, 1997). Experiments with antibodies and blocking peptides indicate that glycoprotein Ib/V/IX as well as integrin $\alpha_{IIb}\beta_3$ in conjunction with their ligands, vWF and fibrin(ogen), are involved in the platelet-dependent stimulation of the coagulation (Béguin and Kumar, 1997; Béguin et al., 1999). Because of the requirement of an increase in $[Ca^{2+}]_i$ for this response and because thrombin alone is a bad inducer of phosphatidylserine exposure, it is likely that these adhesive receptors increase or extend the platelet calcium response.

Given the contribution of integrin $\alpha_{IIb}\beta_3$ in this process, it may not be a surprise that the aggregation-inhibiting agent, abciximab (ReoPro, being a Fab fragment of a mouse/human chimeric antibody 7E3 against integrin $\alpha_{IIb}\beta_3$), is well-effective in inhibiting the platelet-dependent stimulation of coagulation (Reverter et al., 1996). Consequently, integrin $\alpha_{IIb}\beta_3$ antagonists appear to act as anti-thrombotic agents not only by reducing thrombus formation, but also by inhibiting thrombin generation (Coller, 1997).

7. PLATELET CALCIUM SIGNAL IN HEALTH AND DISEASE

In the past, numerous patients have been identified with mild to severe bleeding problems that associate with a reduced platelet function. The most important aberrations, as far as the platelet calcium signal is altered, are described here. A rare bleeding disorder that is accompanied by altered expression of platelet adhesive proteins is the Bernard–Soulier syndrome. These patients mostly have abnormally large (giant) platelets, which are characterized by absent or decreased expression of the vWF receptor, the glycoprotein Ib/V/IX complex (Konkle, 1997). The stabilizing and anchoring function of glycoprotein Ib/V/IX for the actin filaments of the membrane skeleton probably explains, why its reduced expression influences platelet size and structure. Platelets from these patients clearly lack the vWF-evoked adhesion and intracellular responses, including generation of a calcium signal. Various genetic defects are detected that may cause the decreased expression of the glycoprotein Ib/V/IX complex or influence its interaction with the membrane skeleton. In most cases, these are located in the genes encoding for glycoprotein Ibα and IX chains (Konkle, 1997; López et al., 1998), although mutations in the glycoprotein Ibβ chain are also known (Kunishima et al., 1997).

An extensively described inherited, sometimes severe, bleeding disorder is Glanzmann's thrombasthenia, which is accompanied by quantitative or qualitative defects in integrin $\alpha_{IIb}\beta_3$ expression on the platelet surface. The number of functionally active integrin molecular is highly variable from patient to patient, but is related to the type of molecular defect (Perutelli and Mori, 1992). More than 15 different mutations have been detected in the glycoprotein IIb as well as the glycoprotein IIIa chain of the integrin molecule (Nurden, 1999). The major consequence of thrombasthenia is a reduced or defective aggregation of the platelets in response to all agonists. Compatible with the proposed function of integrin $\alpha_{\text{IIb}}\beta_3$ in the platelet procoagulant response, it is found that platelets from thrombasthenic patients have a reduced "secondary" calcium response and a decreased production of microvesicles upon stimulation (Gemmell et al., 1993; Weiss and Lages, 1997).

Few patients are known, again with recurrent bleeding, whose platelets lack the collagen receptor, glycoprotein VI, or who carry auto-antibodies against glycoprotein VI (Moroi and Jung, 1997). Platelets from these patients cannot aggregate with collagen, and these have been used to elucidate the glycoprotein VI signaling effects (Kehrel et al., 1998; Watson, 1999). Another patient is known with selective absence of collagen-induced adhesion and aggregation due to a deficiency in integrin $\alpha_2\beta_1$ (Nieuwenhuis et al., 1986). Further, the many identified subjects with deficient glycoprotein IV expression on platelets (i.e. 3–10% of the Japanese population) do not seem to exhibit bleeding abnormalities, and their platelets respond normally to collagen (Moroi and Jung, 1997). This questions the proposed function of glycoprotein IV as an essential platelet collagen receptor.

A limited number of patients appear in the literature, again with mild to moderate bleeding disorders and normal platelet count, whose platelets are defective in aggregation in response to agonists of G-coupled receptors. In platelets from some of the patients, PLC activation and Ca^{2+} mobilization are reduced in response to both thrombin and ADP (Rao et al., 1993). One of the subjects appeared to have platelets with a three-fold reduced amount of PLCβ2 (Lee et al., 1996). Other subjects carry platelets with defective PLC activation and calcium mobilization in response to only thromboxane A_2, where the thromboxane binding is normal (Fuse et al., 1996; Mitsui et al., 1997). In addition, subjects from two families are known who have platelets, where the ADP-evoked aggregation and/or calcium responses are reduced, sometimes in combination with reduced ADP-binding characteristics. Defects have thus been established in P2T or $P2Y_1$ purinergic receptor signaling (Cattaneo et al., 1992; Nurden et al., 1995). Importantly, platelets with a defective P2T signaling resemble in responsiveness to platelets from humans that are treated with the drug, clopidogrel (Cattaneo and Gachet, 1999). This category closes with a family of patients, whose platelets are putatively reduced in α_{2A}-adrenergic receptor expression, are defective in epinephrine-induced aggregation, but are unchanged in G_i signaling (Rao et al., 1988). Taken together, it appears that the rare receptor-related (calcium) signaling defect in platelets can give rise to increased bleeding, but not to severe pathological situations.

Among the most common congenital disorders of platelet function are the so-called secretion defects, also known as storage pool diseases (Bennett,

1995; Nurden, 1999). The patients usually suffer from prolonged bleeding. The platelets are defective in the release of one or more granular constituents. Although this type of disorder is heterogeneous in appearance, in many cases the potentiating effects of secreted ADP on the aggregation and calcium responses are lacking. For example, patients with a severe form of storage pool deficiency, as prominently present in the Hermansky–Pudlak syndrome (HPS), have somewhat decreased calcium responses with thrombin and also a decreased (Ca^{2+}-dependent) prothrombinase-converting activity. Addition of ADP restores these responses (Weiss and Lages, 1997). The defective HPS gene is known, and the protein may have a function in the biogenesis or trafficking of the granules (Nurden, 1999). Intriguingly, somewhat similar defects have been noted for integrin $\alpha_{IIb}\beta_3$-deficient platelets from thrombasthenic patients (Weiss and Lages, 1997), which suggests that the integrin participates in maintaining a high calcium signal and in triggering the procoagulant response. This work recalls earlier provocative studies that integrin $\alpha_{IIb}\beta_3$ is actively involved in Ca^{2+} influx (see Heemskerk and Sage, 1994).

In recent years, research is concentrating on the identification of factors responsible for increased platelet activity, reasoning that hyperactive platelets may be an independent risk factor of arterial thrombosis. To estimate platelet activation from the amount of calcium signal generation, in the author's laboratory, platelets from 50 healthy subjects were triggered with various agonists (thrombin, collagen or SERCA inhibitors), and the calcium responses were measured (Feijge et al., 1998). In addition to the normal inter-individual variation in platelet responsiveness, which could partly be explained by a variation in SERCA expression levels, it appeared that a few subjects had platelets showing relatively high calcium responses upon triggering with one or more agonists. The increased reactivity was a persistent phenomenon and was not influenced by aspirin intake. Figure 2 gives an example of the ADP- and thrombin-evoked calcium responses of a subject with hyperactive platelets.

Increased levels of basal $[Ca^{2+}]_i$ have been noted in platelets from patients with a variety of (prethrombotic) diseases, e.g. with diabetes type II (Standley et al., 1993) or severe arteriosclerosis (Vicari et al., 1994). This is suggestive for a chronic alteration of the intracellular calcium homeostasis, but the underlying mechanism is still unclear. Increased $[Ca^{2+}]_i$ and a higher activation tendency are also observed for platelets from hypertensive individuals, possibly due to reduced activity of the plasma membrane Ca^{2+}-ATPase (Dean et al., 1994). In addition, patients have been described with a variety of thrombotic disorders, such as transient ischemic attack or myocardial infarct, who have increased amounts of circulating microvesicles that had probably shed from procoagulant platelets (Lee et al., 1993; Jy et al., 1995). Accordingly, several indications can be found in the literature that arterial thrombosis may associate with increased platelet activity.

Genetic screening now provides additional tools to identify platelet-dependent risk factors for arterial disease. In the last few years, various genetic polymorphisms of platelet adhesion receptors are identified, which

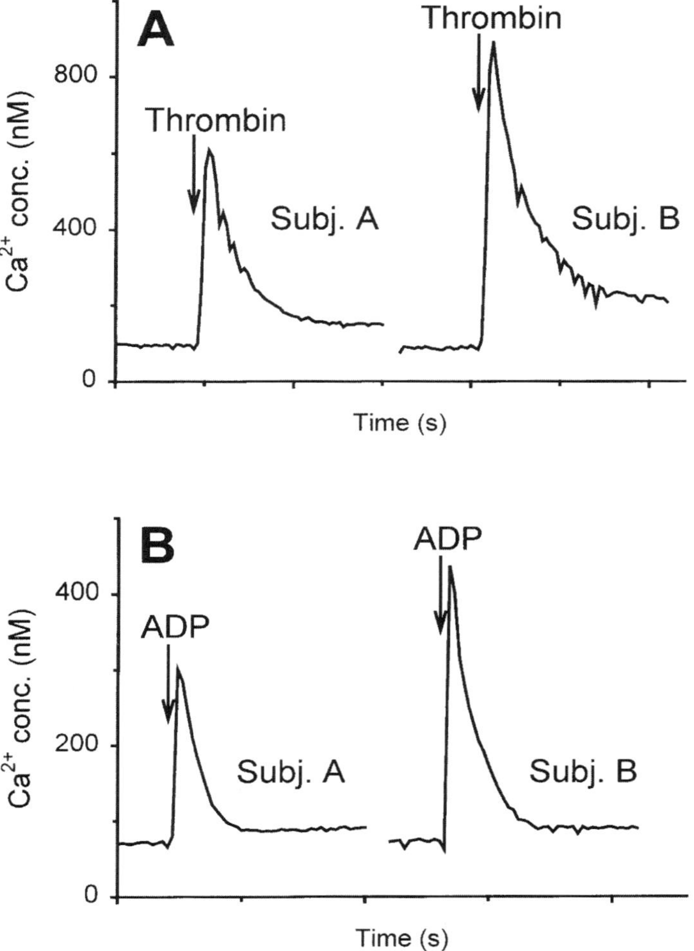

Figure 2. Platelet calcium responses of a normo-active (subject A) and a hyperactive (subject B) individual. Suspensions of fura-2-loaded platelets were triggered in the presence of 1 mM $CaCl_2$ with 4 nM thrombin (A) or 20 μM ADP (B) under standard conditions.

associate with an increased risk of cerebrovascular accidents or myocardial infarction. In a number of cases, this concerns an alloantigen form of glycoproteins that is responsible for alloimmune thrombocytopenia for example after transfusion. Best-studied is the so-called Pl^{A2} (HPA-1) polymorphism of the glycoprotein IIIa chain in integrin $\alpha_{IIb}\beta_3$ (Weiss et al., 1996; Carter et al., 1998; Feng et al., 1999). In general, Pl^{A2}-positive platelets seem to be more prone towards aggregation, but they also more sensitive to inhibition by aspirin or abciximab (Bray, 1999). Three polymorphisms concern the glycoprotein Ibα chain of the vWF receptor: the so-called Kozak sequence, variable number tandem repeat (VNTR) polymorphism, and a M/T145 (HPA-2) polymorphism (Gonzalez-Conejero et al., 1998; Carter et al., 1998; Afshar-Kharghan et al., 1999). Another risk factor is found in the 807C/T polymorphism of the glycoprotein Ia chain from integrin $\alpha_2\beta_1$ (Carlsson et al., 1999; Santoso et al., 1999). As general trend, these genetic variations are accompanied by modest changes in glycoprotein-ligand interaction and, thus, in platelet adhesion or aggregation. In summary, the little evidence as yet suggests that increased platelet activation can indeed be detected among patients with thrombo-arterial diseases. Such hyperactivity may appear as increased adhesion, aggregation or procoagulant activity. Given the direct or indirect contribution of the calcium signal in these responses, it is not unlikely that increased Ca^{2+} mobilization is indeed a significant risk factor of arterial thrombosis. Experiments are under way to demonstrate this.

ACKNOWLEDGEMENT

I acknowledge many colleagues for sharing experimental data and for discussions, particularly Drs. J.-W. Akkerman, E. Bevers, R. Farndale, C. Nieuwenhuys, S. Sage and R. van Gorp.

REFERENCES

Afshar-Kharghan, V., Li, C.Q., Khoshnevis-Asl, M. and Lopez, J.A., 1999, Kozak sequence polymorphism of the glycoprotein (GP) Ibα gene is a major determinant of the plasma membrane levels of the platelet GP Ib-IX-V complex, *Blood* 94, 186–191.

Alonso, M.T., Alvarez, J., Montero, M., Sanchez, A. and Garcia-Sancho, J., 1991, Agonist-induced Ca^{2+} influx into human platelets is secondary to the emptying of intracellular Ca^{2+} stores, *Biochem. J.* 280, 783–789.

Andersen, H., Greenberg, D.L., Fujikawa, K., Xu, W., Chung, D.W. and Davie, E.W., 1999, Protease-activated receptor 1 is the primary mediator of thrombin-stimulated platelet procoagulant activity, *Proc. Natl. Acad. Sci. USA* 96, 11189–11193.

Ariyoshi, H. and Salzman, E.W., 1996, Association of localized Ca^{2+} gradients with redistribution of glycoprotein IIb-IIIa and F-actin in activated human blood platelets, *Arterioscler. Thromb. Vasc. Biol.* 16, 230–235.

Asselin, J., Gibbins, J.M., Achison, M., Lee, Y.H., Morton, L.F., Farndale, R.W., Barnes, M.J. and Watson, S.P., 1997, A collagen-like peptide stimulates tyrosine phosphorylation of Syk and phospholipase Cγ2 in platelets independent of the integrin $\alpha_2\beta_1$, *Blood* 89, 1235–1242.

Authi, K.S., 1997, Ca^{2+} homeostasis in human platelets, in *Handbook of Experimental Pharmacology*, Vol. 126, *Platelets and Their Factors*, F. von Bruchhausen and U. Walter (eds.), Springer, Berlin, pp. 325–370.

Authi, K.S., Evenden, B.J. and Crawford, N., 1986, Metabolic and functional consequences of introducing inositol 1,4,5-trisphosphate into saponin-permeabilized human platelets, *Biochem. J.* 233, 707–718.

Authi, K.S., Bokkala, S., Patel, Y., Kakkar, V.V. and Munkonge, F., 1993, Ca^{2+} release from platelet intracellular stores by thapsigargin and 2,5-di-(t-butyl)-1,4-benzohydroquinone: Relationship to Ca^{2+} pools and relevance in platelet activation, *Biochem. J.* 294, 119–126.

Banno, Y., Nakashima, S., Ohzawa, M. and Nozawa, Y., 1996, Differential translocation of phospholipase C isozymes to integrin-mediated cytoskeletal complexes in thrombin-stimulated human platelets, *J. Biol. Chem.* 271, 14989–14994.

Banno, Y., Asano, T. and Nozawa, Y., 1998, Stimulation by G protein $\beta\gamma$ subunits of phospholipase C-β isoforms in human platelets, *Thromb. Haemostas.* 79, 1008–1013.

Bauer, M., Retzer, M., Wilde, J.I., Maschberger, P., Essler, M., Aepfelbacher, M., Watson, S.P. and Siess, W., 1999, Dichotomous regulation of myosin phosphorylation and shape change by Rho-kinase and calcium in intact human platelets, *Blood* 94, 1665–1672.

Béguin, S. and Kumar, R., 1997, Thrombin, fibrin and platelets: A resonance loop in which von Willebrand factor is a necessary link, *Thromb. Haemostas.* 78, 590–594.

Béguin, S., Kumar, R., Keularts, I., Seligsohn, U., Coller, B.S. and Hemker, H.C., 1999, Fibrin-dependent platelet procoagulant activity requires GPIb receptors and von Willebrand factor, *Blood* 93, 564–570.

Bennett, J.S., 1995, Hereditary disorders of platelet function, in *Hematology, Basic Principles and Practice*, R. Hoffman, E.J. Benz, S.S. Shattil, B. Furie, H.J. Cohen and L.E. Silberstein (eds.), Churchill Livingstone, New York, pp. 1909–1925.

Berg, L.P., Shamsher, M.K., El-Daher, S.S., Kakkar, V.V. and Authi, K.S., 1997, Expression of human TRPC genes in the megakaryocytic cell lines MEG01, DAMI and HEL, *FEBS Lett.* 403, 83–86.

Bertolino, G., Noris, P., Spedini, P. and Balduino, C.L., 1995, Ristocetin-induced platelet agglutination stimulates GPIIb/IIIa-dependent calcium influx, *Thromb. Haemostas.* 73, 689–692.

Bettache, N., Gaffet, P., Allegre, N., Maurin, L., Toti, F., Freyssinet, J.M. and Bienvenüe, A., 1998, Impaired redistribution of phospholipids with distinctive cell shape change during Ca^{2+}-induced activation of platelets from a patient with Scott syndrome, *Br. J. Haematol.* 101, 50–58.

Bevers, E.M., Comfurius, P., van Rijn, J.L.M.L., Hemker, H.C. and Zwaal, R.F.A., 1982, Generation of prothrombin-converting activity and the exposure of phosphatidylserine at the outer surface of platelets, *Eur. J. Biochem.* 122, 429–436.

Börsch-Haubold, A.G., Ghomashchi, F., Pasquet, S., Goedert, M., Cohen, P., Gelb, M.H. and Watson, S.P., 1999, Phosphorylation of cytosolic phospholipase A_2 in platelets is mediated by multiple stress-activated protein kinase pathways, *Eur. J. Biochem.* 265, 195–203.

Brass, L.F., Laposata, M., Singh Banga and Rittenhouse, S.E., 1986, Regulation of the phosphoinositide hydrolysis pathway in thrombin-stimulated platelets by pertussis toxin-sensitive guanine nucleotide-binding protein, *J. Biol. Chem.* 261, 16838–16847.

Brass, L.F., Vasallo, R.R., Belmonte, E., Ahuka, M., Cichowski, K. and Hoxie, J.A., 1992, Structure and function of the human platelet thrombin receptor. Studies using monoclonal antibodies directed against a defined domain within the receptor N terminus, *J. Biol. Chem.* 267, 13793–13798.

Brass, L.F., Manning, D.R., Cichowski, K. and Abrams, C.S., 1997, Signaling through G proteins in platelets: To the integrins and beyond, *Thromb. Haemostas.* 78, 581–589.

Bray, P.F., 1999, Integrin polymorphisms as risk factors for thrombosis, *Thromb. Haemostas.* 82, 337–344.

Briedé, J.J., Heemskerk, J.W.M., Hemker, H.C. and Lindhout, T., 1999, Heterogeneity in microparticle formation and exposure of anionic phospholipids at the plasma membrane of single adherent platelets, *Biochim. Biophys. Acta* 1451, 163–172.

Calverley, D.C., Kavanagh, T.J. and Roth, G.J., 1998, Human signaling protein 14-3-3ζ interacts with platelet glycoprotein Ib subunits Ibα and Ibβ, *Blood* 91, 1295–1303.

Carlsson, L.E., Santoso, S., Spitzr, C., Kessler, C. and Greinacher, A., 1999, The α_2 gene coding sequence T807/A873 of the platelet collagen receptor integrin $\alpha_2\beta_1$ might be a genetic risk factor for the development of stroke in younger patients, *Blood* 93, 3583–3586.

Carter, A.M., Catto, A.J., Bamford, J.M. and Grant, P.J., 1998, Platelet GP IIIa Pl^A and GP Ib variable tandem repeat polymorphisms and markers of platelet activation in acute stroke, *Arterioscler. Thromb. Vasc. Biol.* 18, 1124–1131.

Cattaneo, M. and Gachet, C., 1999, ADP receptors and clinical bleeding disorders, *Arterioscler. Thromb. Vasc. Biol.* 19, 2281–2285.

Cattaneo, M., Lecchi, A., Randi, A.M., McGregor, J.L. and Mannucci, P.M., 1992, Identification of a new congenital defect of platelet function characterized by severe impairment of platelet responses to adenosine diphosphate, *Blood* 80, 2787–2796.

Cavallini, L., Coassin, M., Borean, A. and Alexandre, A., 1996, Prostacyclin and sodium nitroprusside inhibit the activity of the platelet inositol 1,4,5-trisphosphate receptor and promote its phosphorylation, *J. Biol. Chem.* 271, 5545–5551.

Clark, E.A., Shattil, S., Ginsberg, M.H., Bolen, J. and Brugge, S.J., 1994, Regulation of the protein tyrosine kinase, $pp72^{syk}$, by platelet agonists and the integrin $\alpha_{IIb}\beta_3$, *J. Biol. Chem.* 269, 28859–28864.

Clemetson, J.M., Polgár, J., Magnenat, E.M., Wells, T.N.C. and Clemetson, K.J., 1999, The platelet collagen receptor glycoprotein VI is a member of the immunoglobulin superfamily closely related to FcαR and the natural killer receptors, *J. Biol. Chem.* 274, 29019–29024.

Coller, B.S., 1997, Platelet GPIIb/IIIa antagonists: The first anti-integrin receptor therapeutics, *J. Clin. Invest.* 9, 1467–1471.

Coller, B.S., Beer, J.H., Scudder, L.E. and Steinberg, M.H., 1989, Collagen-platelet interactions: Evidence for a direct interaction of collagen with platelet GPIa/IIa and an indirect interaction with platelet GPIIb/IIIa mediated by adhesive proteins, *Blood* 74, 182–192.

Comfurius, P., Senden, J.M., Tilly, R.H., Schroit, A.J., Bevers, E.M. and Zwaal, R.F.A., 1990, Loss of membrane phospholipid asymmetry in platelets and red cells may be associated with calcium-induced shedding of plasma membrane and inhibition of aminophospholipid translocase, *Biochim. Biophys. Acta* 1026, 153–160.

Coughlin, S.R., 1999, Protease-activated receptors and platelet function, *Thromb. Haemostas.* 82, 353–356.

Dachary-Prigent, J., Pasquet, J.M., Freyssinet, J.M. and Nurden, A.T., 1995, Calcium involvement in aminophospholipid exposure and microparticle formation during platelet activation: A study using Ca^{2+}-ATPase inhibitors, *Biochemistry* 34, 11625–11634.

Dachary-Prigent, J., Pasquet, J.M., Fressinaud, E., Toti, F. and Freyssinet, J.M., 1997, Aminophospholipid exposure, microvesiculation and abnormal protein tyrosine phosphorylation in the platelets of a patient with Scott syndrome: A study using physiologic agonists and local anaesthetics, *Br. J. Haematol.* 99, 959–967.

Daniel, J.L., Dangelmaier, C. and Smith, J.B., 1994, Evidence for a role for tyrosine phosphorylation of phospholipase Cγ2 in collagen-induced platelet cytosolic calcium mobilization, *Biochem. J.* 302, 617–622.

Daniel, J.L., Dangermaier, C., Jin, J., Ashby, B., Smith, J.B. and Kunapuli, S.P., 1998, Molecular basis for ADP-induced platelet activation. Evidence for three distinct ADP receptors on human platelets, *J. Biol. Chem.* 273, 2024–2029.

Dean, W.L., Pope, J.E., Brier, M.E. and Aronoff, G.R., 1994, Platelet calcium transport in hypertension, *Hypertension* 23, 31–37.

Dean, W.L., Chen, D., Brandt, P.C. and Vanaman, T.C., 1997, Regulation of platelet plasma membrane Ca^{2+}-ATPase by cAMP-dependent and tyrosine phosphorylation, *J. Biol. Chem.* 272, 15113–15119.

Fabre, J.E., Nguyen, M.T., Latour, A., Keifer, J.A., Audoly, L.P., Coffman, T.M. and Koller, B.H., 1999, Decreased platelet aggregation, increase bleeding time and resistance to thromboembolism in $P2Y_1$-deficient mice, *Nature Medic.* 10, 1199–1202.

Falati, S., Edmead, C.E. and Poole, A.W., 1999, Glycoprotein Ib-V-IX, a receptor for von Willebrand factor, couples physically and functionally to the Fc receptor γ-chain, Fyn, and Lyn to activate human platelets, *Blood* 94, 1648–1656.

Feijge, M.A.H., van Pampus, E.C.M., Lacabaratz-Porret, C., Hamulyàk, K., Lévy-Toledano, S., Enouf, J. and Heemskerk, J.W.M., 1998, Inter-individual varability in Ca^{2+} signalling in platelets from healthy volunteers: Relation with expression of endomembrane Ca^{2+}-ATPases, *Br. J. Haematol.* 102, 850–859.

Feng, D.L., Lindpaintner, K., Larson, M.G., Rao, V.S., O'Donnell, C.J., Lipnska, I., Schmitz, C., Sutherland, P.A., Silbershatz, H., D'Agostino, R.B., Muller, J.E., Myers, R.H., Levy,

D. and Tofler, G.H., 1999, Increased platelet aggregability associated with platelet GPIIIa Pl^{A2} polymorphism. The Framingham offspring study, *Arterioscler. Thromb. Vasc. Biol.* 19, 1142–1147.

Fox, J.E., Austin, C.D., Reynolds, C.C. and Steffen, P.K., 1991, Evidence that agonist-induced activation of calpain causes the shedding of procoagulant-containing microvesicles from the membrane of aggregating platelets, *J. Biol. Chem.* 266, 13289–13295.

Fressinaud, E., Veyradier, A., Truchaud, F., Martin, I., Boyer-Neumann, C., Trossaert, M., and Meyer, D., 1998, Screening for von Willebrand disease with a new analyzer using high shear stress: A study of 60 cases, *Blood* 91, 479–483.

Fuse, I., Hattori, A., Mito, M., Higuchi, W., Yahata, K., Shibata, A. and Aizawa, Y., 1996, Pathogenetic analysis of five cases with a platelet disorder characterized by the absence of thromboxane A_2 (TxA_2)-induced platelet aggregation in spite of normal TxA_2 binding activity, *Thromb. Haemostas.* 76, 1080–1085.

Geiger, J., Nolte, C., Butte, E., Sage, S.O. and Walter, U., 1992, Role of cGMP and cGMP-dependent protein kinase in nitrovasidilator inhibition of agonist-evoked calcium elevation in platelets, *Proc. Natl. Acad. Sci. USA* 89, 1031–1035.

Geiger, J., Brich, J., Hönig-Liedl, P., Eigenthaler, M., Schanzenbächer, P., Herbert, J.M. and Walter, U., 1999, Specific impairment of human platelet $P2Y_{AC}$ ADP receptor-mediated signaling by the antiplatelet drug clopidogrel, *Arterioscler. Thromb. Vasc. Biol.* 19, 2007–2011.

Gemmell, C.T., Sefton, M.V. and Yeo, E.L., 1993, Platelet-derived microparticle formation involves glycoprotein IIb-IIIa. Inhibition by RGDS and a Glanzmann's thrombastenia defect, *J. Biol. Chem.* 268, 14586–14589.

Gibbins, J.M., Okuma, M., Farndale, R., Barnes, M. and Watson, S.P., 1997, Glycoprotein VI is the collagen receptor in platelets which underlies tyrosine phosphorylation of the Fc receptor γ-chain, *FEBS Lett.* 413, 255–259.

Goldstein, R.E., Andrews, M., Hall, W.J. and Moss, A.J., 1996, Marked reduction in long-term cardiac deaths with aspirin after a coronary event, *J. Am. Coll. Cardiol.* 28, 326–330.

Gonzalez-Conejero, R., Lozano, M.L., Rivera, J., Corral, J., Iniesta, J.A., Moraleda, J.M. and Vicente, V., 1998, Polymorphisms of platelet membrane glycoprotein Ibα associated with arterial thrombotic disease, *Blood* 92, 2771–2776.

Goto, S., Ikeda, Y., Saldivar, E. and Ruggeri, Z.M., 1998, Distinct mechanisms of platelet aggregation as a consequence of different shearing conditions, *J. Clin. Invest.* 101, 479–486.

Gratacap, M.P., Payrastre, B., Viala, C., Mauco, G., Plantavid, M. and Chap, H., 1998, Phosphatidylinositol 3,4,5-trisphosphate-dependent stimulation of phospholipase C-γ2 is an early key event in FcγRIIA-mediated activation of human platelets, *J. Biol. Chem.* 273, 24314–24321.

Greco, N., Jones, G.D., Tandon, N.N., Kornhauser, R., Jackson, B. and Jamieson, G.A., 1996, Differentiation of the two forms of GPIb functioning as receptors for α-thrombin and von Willebrand factor: Ca^{2+} responses of protease-treated human platelets activated with α-thrombin and the tethered ligand peptide, *Biochemistry* 35, 915–921.

Gross, B.S., Melford, S.K. and Watson, S.P., 1999, Evidence that phospholipase Cγ2 interacts with Slp-76, Syk, Lyn, LAT and the Fc receptor γ-chain after stimulated of the collagen receptor glycoprotein VI in human platelets, *Eur. J. Biochem.* 263, 612–623.

Haimovich, B., Lipfert, L., Brugge, J.S. and Shattil, S.J., 1993, Tyrosine phosphorylation and cytoskeletal reorganization in platelets are triggered by interaction of integrin receptors with their immobilized ligands, *J. Biol. Chem.* 268, 15868–15877.

Hallbrügge, M. and Walter, U., 1993, The regulation of platelet function by protein kinases, in *Protein Kinases in Blood Cell Function*, C.K. Huang and R.I. Sha'afi (eds.), CRC Press, Boca Raton, FL, pp. 245–298.

Hechler, B., Léon, C., Vial, C., Vigne, P., Frelin, C., Cazenave, J.P. and Gachet, C., 1998, The $P2Y_1$ receptor is necessary for adenosine 5′-diphosphate-induced platelet aggregation, *Blood* 92, 152–159.

Heemskerk, J.W.M. and Sage, S.O., 1994, Calcium signalling in platelets and other cells, *Platelets* 5, 295–316.

Heemskerk, J.W.M., Vis, P., Feijge, M.A.H., Hoyland, J., Mason, W.T. and Sage, S.O., 1993, Roles of phospholipase C and Ca^{2+}-ATPase in calcium responses of single, fibrinogen-bound platelets, *J. Biol. Chem.* 268, 356–363.

Heemskerk, J.W.M., Feijge, M.A.H., Sage, S.O. and Walter, U., 1994, Indirect regulation of Ca^{2+} entry by cAMP-dependent and cGMP-dependent protein kinases and phospholipase C in rat platelets, *Eur. J. Biochem.* 223, 543–551.

Heemskerk, J.W.M., Feijge, M.A.H., Henneman, L., Rosing, R. and Hemker, H.C., 1997a, The Ca^{2+}-mobilizing potency of α-thrombin and thrombin-receptor-activating peptide on human platelets. Concentration and time effects of thrombin-induced Ca^{2+} signaling, *Eur. J. Biochem.* 249, 547–555.

Heemskerk, J.W.M., Vuist, W.M.J., Feijge, M.A.H., Reutelingsperger, C.P.M. and Lindhout, T., 1997b, Collagen but not fibrinogen surfaces induce bleb formation, exposure of phosphatidylserine and procoagulant activity of adherent platelets: Evidence for regulation by protein tyrosine kinase-dependent Ca^{2+} responses, *Blood* 90, 2615–2625

Heemskerk, J.W.M., Siljander, P., Vuist, W.M.J., Breikers, G., Reutelingsperger, C.P.M., Barnes, M.J., Knight, C.G., Lassila, R. and Farndale, R.W., 1999, Function of glycoprotein VI and integrin $\alpha_2\beta_1$ in the procoagulant response of single, collagen-adherent platelets, *Thromb. Haemostas.* 81, 78–92.

Hers, I., Donath, J., van Willigen, G. and Akkerman, J.W.N., 1998, Differential involvement of tyrosine and serine/threonine kinases in platelet integrin $\alpha_{IIb}\beta_3$ exposure, *Arterioscler. Thromb. Vasc. Biol.* 18, 404–414.

Hussain, J.F. and Mahaut-Smith, M.P., 1999, Reversible and irreversible intracellular Ca^{2+} spiking in single isolated human platelets, *J. Physiol.* 514, 713–718.

Ichinohe, T., Takayama, H., Ezumi, Y., Yanagi, S., Yamamura, H. and Okuma, M., 1995, Cyclic AMP-insensitive activation of c-Src and Syk protein-tyrosine kinases through platelet membrane glycoprotein VI, *J. Biol. Chem.* 270, 28029–29036.

Ichinohe, T., Takayama, H., Ezumi, Y., Arai, M., Yamamoto, N., Takahashi, H. and Okuma, M., 1997, Collagen-stimulated activation of Syk but not c-Src is severely compromised in human platelets lacking membrane glycoprotein VI, *J. Biol. Chem.* 272, 63–68.

Ikeda, Y., Handa, M., Kamata, T., Kawano, K., Kawai, Y., Watanabe, K., Kawakami, K., Sakai, K., Fukuyama, M., Itagaki, I., Yoshioka, A. and Ruggeri, Z.M., 1993, Transmembrane calcium influx associated with von Willebrand factor binding to GP Ib in the initiation of shear-induced platelet aggregation, *Thromb. Haemostas.* 69, 496–502.

Jandrot-Perrus, M., Lagrue, A.H., Okuma, M. and Bon, C., 1997, Adhesion and activation of human platelets induced by convulxin involve glycoprotein VI and integrin $\alpha_2\beta_1$, *J. Biol. Chem.* 272, 27035–27041.

Jen, C.J., Chen, H.I., Lai, K.C. and Usami, S., 1996, Changes in cytosolic calcium concentrations and cell morphology in single platelets adhered to fibrinogen-coated surface under flow, *Blood* 87, 3775–3782.

Jin, J. and Kunapoli, S.P., 1998, Coactivation of two different G protein-coupled receptors is essential for ADP-induced platelet aggregation, *Proc. Natl. Acad. Sci. USA* 95, 8070–8074.

Jung, S.M. and Moroi, M., 1998, Platelets interact with soluble and insoluble collagens through characteristically different reactions, *J. Biol. Chem.* 273, 14827–14837.

Jy, W., Horstman, L.L., Wang, F., Duncan, R. and Ahn, Y.S., 1995, Platelet factor 3 in plasma fractions: Its relation to microparticle size and thromboses, *Thromb. Res.* 80, 471–482.

Kahn, M.L., Nakanishi-Matsui, M., Shapiro, M.J., Ishihara, H. and Coughlin, S.R., 1999, Protease-activated receptors 1 and 4 mediate activation of human platelets by thrombin, *J. Clin. Invest.* 103, 879–887.

Keely, P.J. and Parise, L.V., 1996, The $\alpha_2\beta_1$ integrin is a necessary co-receptor for collagen-induced activation of syk and the subsequent phosphorylation of phospholipase Cγ2 in platelets, *J. Biol. Chem.* 271, 26688–26676.

Kehrel, B., Wierwille, S., Clemetson, K.J., Anders, O., Steiner, M., Knight, C.G., Farndale, R.W., Okuma, M. and Barnes, M.J., 1998, Glycoprotein VI is a major collagen receptor for platelet activation; it recognizes the platelet-activating quaternary structure of collagen, whereas CD36, glycoprotein IIb/IIIa, and von Willebrand factor do not, *Blood* 91, 491–499.

Keularts, I.M.L.W., van Gorp, R.M.A., Feijge, M.A.H., Vuist, W.M.J. and Heemskerk, J.W.M., 2000, α_{2A}-Adrenergic receptor stimulation potentiates calcium release in platelets by modulating cAMP levels, *J. Biol. Chem.* 275, 1763–1772.

Konkle, B., 1997, The Bernard–Soulier syndrome, *Trends Cardiovasc. Medic.* 7, 239–244.

Kovàcs, T., Berger, G., Corvazier, E., Pàszty, K., Brown, A., Bobe, R., Papp, B., Wuytack, F., Cramer, E.M. and Enouf, J., 1997, Immunolocalization of the multi-sarcoendoplasmic reticulum Ca^{2+}-ATPase system in human platelets, *Br. J. Haematol.* 97, 192–203.

Kramer, R.M., Roberts, E.F., Manetta, J.V., Hyslop, P.A. and Jakubowski, J.A., 1993, Thrombin-induced phosphorylation and activation of Ca^{2+}-sensitive cytosolic phospholipase A_2 in human platelets, *J. Biol. Chem.* 268, 26796–26804.

Kroll, M.H., Hellums, J.D., McIntire, L.V., Schafer, A.I. and Moake, J.L., 1996, Platelets and shear stress, *Blood* 88, 1525–1541.

Kunapuli, S.P., 1998, Multiple P_2 receptor subtypes on platelets: A new interpretation of their function, *Trends Pharmacol. Sci.* 19, 391–394.

Kunishima, S., Lopez, J.A., Kobayashi, S., Imai, N., Kamiya, T., Saito, H. and Naoe, T., 1997, Missense mutations of the glycoprotein (GP) Ibα gene impairing the GPIb α/β disulfide linkage in a family with giant platelet disorder. Blood, 89, 2402–2412.

Kuwahara, M., Sugimoto, M., Tsuji, S., Miyata, S. and Yoshioka, A., 1999, Cytosolic calcium changes in a process of platelet adhesion and cohesion on a von Willebrand factor-coated surface under flow conditions, *Blood* 94, 1149–1155.

Lacabaratz-Porret, C., Corvazier, E., Kovàcs, T., Bobe, R., Bredoux, R., Launay, S., Papp, B. and Enouf, J., 1998, Platelet sarco/endoplasmic reticulum Ca^{2+}-ATPase isoform 3b and Rap1b: Interrelation and regulation in physiopathology, *Biochem. J.* 332, 173–181.

Lagrue, A.H., Francischetti, I.M.B., Guimaraes, J.A. and Jandrot-Perrus, M., 1999, Phosphatidylinositol 3′-kinase and tyrosine-phosphatase activation positively modulate convulxin-induced platelet activation. Comparison with collagen, *FEBS Lett.* 44, 95–100.

Lankhof, H., Wy, Y.P., Vink, T., Schiphorst, M.E., Zerwes, H.G., de Groot, P.G. and Sixma, J.J., 1995, Role of the glycoprotein Ib-binding A1 repeat and the RGDS sequence in platelet adhesion to human recombinant von Willebrand factor, *Blood* 86, 1035–1042.

Law, D.A., Nannizzi-Alaimo, L., Ministri, K., Hughes, P.E., Forsyth, J., Turner, M., Shattil, S.J., Ginsberg, M.H., Tybulewicz, V.L.J. and Phillips, D.R., 1999, Genetic and pharmacological analyses of Syk function in $\alpha_{IIb}\beta_3$ signaling in platelets, *Blood* 93, 2645–2652.

Lee, Y., Jy, W., Horstman, L.L., Janania, J., Kelley, R. and Ahn, Y.S., 1993, Elevated platelet microparticles in multi-infarct dementias and transient ischemic attacks, *Thromb. Res.* 72, 295–304.

Lee, S.B., Rao, A.K., Lee, K.H., Yang, X., Bae, Y.S. and Rhee, S.G., 1996, Decreased expression of phospholipase C-β2 isozyme in human platelets with impaired function, *Blood* 88, 1684–1691.

López, J.A., Andrews, R.K., Afshar-Kharghan, V. and Berndt, M.C., 1998, Bernard–Soulier syndrome, *Blood* 91, 4397–4418.

MacKenzie, A.B., Mahaut-Smith, M.P. and Sage, S.O., 1996, Activation of receptor-operated cation channels via $P2X_1$ not P2T purinoceptors in human platelets, *J. Biol. Chem.* 271, 2879–2881.

Marshall, P.W., Williams, A.J., Dixon, R.M., Growcott, J.W., Warburton, S., Armstrong, J. and Moores, J., 1997, A comparison of the effects of aspirin on bleeding time measured using the Simplate method and closure time measured using the PFA-100 in healthy volunteers, *Br. J. Clin. Pharmacol.* 44, 151–155.

Mills, D.C., 1996, ADP receptors on platelets, *Thromb. Haemostas.* 76, 835–856.

Mitsui, T., Yokoyama, S., Shimizu, Y., Katsuura, M., Akiba, K. and Hayasaka, K., 1997, Defective signal transduction through the thromboxane A_2 receptor in a patient with a mild bleeding disorder: Deficiency of the inositol 1,4,5-trisphosphate formation despite normal G-protein activation, *Thromb. Haemostas.* 77, 991–995.

Moroi, M. and Jung, S.M., 1997, Platelet receptors for collagen, *Thromb. Haemostas.* 78, 439–444.

Moroi, M., Jung, S.M., Nomura, S., Sekiguchi, S., Ordinas, A. and Diaz-Ricart, M., 1997, Analysis of the involvement of the von Willebrand factor-glycoprotein Ib interaction in platelet adhesion to a collagen-coated surface under flow conditions, *Blood* 90, 4413–4424.

Nieuwenhuis, H.K., Sakariassen, K.S., Houdijk, W.P.M., Nievelstein, P.F.E.M. and Sixma, J.J., 1986, Deficiency of platelet membrane glycoprotein Ia associated with a decreased platelet adhesion to subendothelium: A defect in platelet spreading, *Blood* 68, 692–695.

Nurden, A.T., 1999, Inherited abnormalities of platelets, *Thromb. Haemostas.* 82, 468–480.

Nurden, P., Savi, P., Heilmann, E., Bihour, C., Herbert, J.M., Maffrand, J.P. and Nurden, A., 1995, An inherited bleeding disorder linked to a defective interaction between ADP and its receptor on platelets. Its influence on glycoprotein IIb-IIIa complex function, *J. Clin. Invest.* 95, 1612–1622.

Offermanns, S., Toombs, C.F., Hu, Y.H. and Simon, M.I., 1998, Defective platelet activation in $G\alpha_q$-deficient mice, *Nature* 389, 183–186.

Parekh, A.B. and Penner, R., 1997, Store depletion and calcium influx, *Physiol. Rev.* 77, 902–930.

Pasquet, J.M., Dachary-Prigent, J. and Nurden, A.T., 1996, Calcium influx is a determining factor of calpain activation and microparticle formation in platelets, *Eur. J. Biochem.* 239, 647–654.

Pasquet, J.M., Dachary-Prigent, J. and Nurden, A.T., 1998, Microvesicle release is associated with extensive protein tyrosine dephosphorylation in platelets stimulated by A23187 or a mixture of thrombin and collagen, *Biochem. J.* 333, 591–599.

Paul, B.Z.S., Daniel, J.L. and Kunapuli, S.P., 1999a, Platelet shape change is mediated by both calcium-dependent and -independent signaling pathways, *J. Biol. Chem.* 274, 28293–28300.

Paul, B.Z.S., Jin, J. and Kunapuli, S.P., 1999b, Molecular mechanism of thromboxane A_2-induced platelet aggregation, *J. Biol. Chem.* 274, 29108–29114.

Perutelli, P. and Mori, P.G., 1992, Biochemical and molecular basis of Glanzmann's thrombasthenia, *Haematologica* 77, 421–426.

Polanowska-Grabowska, R. and Gear, A.R.L., 1992, High-speed platelet adhesion under conditions of rapid flow, *Proc. Natl. Acad. Sci. USA* 89, 5754–5758.

Polgár, J., Clemetson, J.M., Kehrel, B.E., Wiedemann, M., Magnenat, E.M., Wells, T.N.C. and Clemetson, K.J., 1997, Platelet activation and signal transduction by convulxin, a C-type lectin from Crotalus durissus terrificus (tropical rattlesnake) venom, via the p62/GPVI collagen receptor, *J. Biol. Chem.* 272, 13576–13583.

Poole, A., Gibbins, J.M., Turner, M., van Vugt, M.J., van de Winkel, J.G.J., Tybulewicz, V.L.J. and Watson, S.P., 1997, The Fc receptor γ-chain and the tyrosine kinase Syk are essential for activation of mouse platelets by collagen, *EMBO J.* 16, 2333–2341.

Quinton, T.M., Brown, K.D. and Dean, W.T., 1996, Inositol 1,4,5-trisphosphate-mediated Ca^{2+} release from platelet internal membranes is regulated by differential phosphorylation, *Biochemistry* 35, 6865–6871.

Rao, A.K., Kowalska, M.A., Wachtfogel, Y.T. and Colman, R.W., 1988, Differential requirements for platelet aggregation and inhibition of adenylate cyclase by epinephrine. Studies of a familial platelet α_{2A}-adrenergic receptor defect, *Blood* 71, 494–501.

Rao, A.K., Disa, J. and Yang, X., 1993, Concomitant defect in internal release and influx of calcium in patients with congenital platelet dysfunction and impaired agonist-induced calcium mobilization, *J. Lab. Clin. Med.* 121, 52–63.

Reverter, J.C., Béguin, S., Kessels, H., Kumar, R., Hemker, H.C. and Coller, B.S., 1996, Inhibition of platelet-mediated, tissue-factor-induced thrombin generation by the mouse/human chimeric 7E3 antibody. Potential implications for the effect of c7E3 Fab treatment on acute thrombosis and 'clinical restenosis', *J. Clin. Invest.* 98, 863–874.

Rosado, J.A., Jenner, S. and Sage, S.O., 2000, A role for the active cytoskeleton in the initiation and maintenance of store-mediated calcium entry in human platelets: Evidence for conformational coupling, *J. Biol. Chem.* 275, 7527–7533.

Rosing, J., Bevers, E.M., Comfurius, P., Hemker, H.C., van Dieijen, G., Weiss, H.J. and Zwaal, R.F.A., 1985, Impaired factor X and prothrombin activation associated with decreased phospholipid exposure in platelets from a patient with a bleeding disorder, *Blood* 65, 1557–1561.

Sadler, J.E., 1991, Von Willebrand factor, *J. Biol. Chem.* 266, 22777–22780.

Saelman, E.M., Kehrel, B., Hese, K.M., de Groot, P.G., Sixma, J.J. and Nieuwenhuis, H.K., 1994, Platelet adhesion to collagen and endothelial cell matrix under flow conditions is not dependent on platelet glycoprotein IV, *Blood* 83, 3240–3244.

Sage, S.O., 1997, Calcium entry mechanisms in human platelets, *Exp. Physiol.* 82, 807–823.

Sage, S.O., Yamoah, E.H. and Heemskerk, J.W.M., 2000. The roles of $P2X_1$ and $P2T_{AC}$ receptors in ADP-evoked calcium signalling in human platelets, *Cell Calcium*, in press.

Saido, T.C., Suzuki, H., Yamazaki, H., Tanoue, K. and Suzuki, K., 1991, In situ capture of μ-calpain activation in platelets, *J. Biol. Chem.* 268, 7422–7426.

Santoso, S., Kunicki, T.J., Kroll, H., Haberbosch, W. and Gardemann, A., 1999, Association of the platelet glycoprotein Ia C807T gene polymorphism with nonfatal myocardial infarction in younger patients, *Blood* 93, 2449–2453.

Sargeant, P., Clarkson, W.D., Sage, S.O. and Heemskerk, J.W.M., 1992, Calcium influx in fura-2-loaded human platelets is evoked by thapsigargin and 2,5-di-(*t*-butyl)-1,4-benzohydroquinone and reduced by inhibitors of cytochrome *P*-450, *Cell Calcium* 13, 553–564.

Sargeant, P., Farndale, R.W. and Sage, S.O., 1993, ADP- and thapsigargin-evoked Ca^{2+} entry and protein tyrosine phosphorylation are inhibited by the tyrosine kinase inhibitors genistein and methyl-2,5-dihydroxycinnamate in fura-2-loaded human platelets, *J. Biol. Chem.* 268, 18151–18156.

Scharenberg, A.M. and Kinet, J.P., 1998, Ptd-3,4,5-P_3, a regulatory nexus between tyrosine kinases and sustained calcium signals, *Cell* 94, 5–8.

Shattil, S.J., Kashiwagi, H. and Pampori, N., 1998, Integrin signaling: The platelet paradigm, *Blood* 91, 2645–2657.

Siess, W., 1989, Molecular mechanisms of platelet activation, *Physiol. Rev.* 69, 58–178.

Sims, P.J., Faioni, E.M., Wiedmer, T. and Shattil, S.J., 1988, Complement proteins C5-9 cause release of membrane vesicles from the platelet surface that are enriched in the membrane receptor for coagulation factor Va and express prothrombinase activity, *J. Biol. Chem.* 263, 18205–18212.

Sims, P.J., Wiedmer, T., Esmon, C.T., Weiss, H.J. and Shattil, S.J., 1989, Assembly of the platelet prothrombinase complex is linked to vesiculation of the platelet plasma membrane. Studies in Scott syndrome: An isolated defect in platelet procoagulant activity, *J. Biol. Chem.* 264, 17049–17057.

Sixma, J.J. and Wester J., 1977, The hemostatic plug, *Semin. Hematol.* 14, 265–299.

Smeets, E.F., Heemskerk, J.W.M., Comfurius, P., Bevers, E.M. and Zwaal, R.F.A., 1993, Thapsigargin amplifies the platelet procoagulant response caused by thrombin, *Thromb. Haemostas.* 70, 1024–1029.

Standley, P.R., Ali, S., Bapna, C. and Sowers, J.R., 1993, Increased platelet cytosolic calcium responses to low density lipoprotein in type II diabetes with and without hypertension, *Am. J. Hypertens.* 6, 938–943.

Sun, B., Li, J., Okahara, K. and Kambayashi, J.I., 1998, $P2X_1$ purinoceptor in human platelets. Molecular cloning and functional characterization after heterologous expression, *J. Biol. Chem.* 273, 11544–11547.

Tans, G., Rosing, J., Thomassen, M.C.L.G.D., Heeb, M.J., Zwaal, R.F.A. and Griffin, J.H., 1991, Comparison of anticoagulant and procoagulant activities of stimulated platelets and platelet-derived microparticles, *Blood* 77, 2641–2648.

Tsuji, S., Sugimoto, M., Kuwahara, M., Nishio, K., Takahashi, Y., Fujimura, Y., Ikeda, Y. and Yoshioka, A., 1996, Role and initiation mechanism of the interaction of glycoprotein Ib with surface-immobilized von Willebrand factor in a solid-phase platelet cohesion process, *Blood* 88, 3854–3861.

Vanags, D.M., Orrenius, S. and Aguilar-Santelises, M., 1997, Alterations in Bcl-2/Bax protein levels in platelets form part of an ionomycin-induced process that resembles apoptosis, *Br. J. Haematol.* 99, 824–831.

Verkleij, M.W., Morton, L.F., Knight, C.G., de Groot, P.G., Barnes, M.J. and Sixma, J.J., 1998, Simple collagen-like peptides support platelet adhesion under static but not under flow conditions: Interaction via $\alpha_2\beta_1$ and von Willebrand factor with specific sequences in native collagen is a requirement to resist shear forces, *Blood* 91, 3808–3816.

Vial, C., Hechler, B., Léon, C., Cazenave, J.P. and Gachet, C., 1997, Presence of $P2X_1$ purinoceptors in human platelets and megakaryoblastic cell lines, *Thromb. Haemostas.* 78, 1500–1504.

Vicari, A.M., Monzani, M.L., Pellegatta, F., Ronchi, P., Galli, L. and Folli, F., 1994, Platelet calcium homeostasis is abnormal in patients with severe arteriosclerosis, *Arterioscler. Thromb.* 14, 1420–1424.

Vu, T.K.H., Hung, D.T., Wheaton, V.I. and Coughlin, S.R., 1991, Molecular cloning of a functional thrombin receptor reveals a novel proteolytic mechanism of receptor activation, *Cell* 64, 1057–1068.

Vuist, W.M.J., Feijge, M.A.H. and Heemskerk, J.W.M., 1997, Kinetics of store-operated Ca^{2+} influx evoked by endomembrane Ca^{2+}-ATPase inhibitors in human platelets, *Prostagland. Leukotr. Essential Fatty Acids* 57, 447–550.

Watson, S.P., 1999, Collagen receptor signaling in platelets and megakaryocytes, *Thromb. Haemostas.* 82, 365–376.

Weiss, H.J. and Lages, B., 1997, Platelet prothrombinase activity and intracellular calcium responses in patients with storage pool deficiency, glycoprotein IIb-IIIa deficiency, or impaired platelet coagulant activity. A comparison with Scott syndrome, *Blood* 89, 1599–1611.

Weiss, E.J., Bray, P.F., Tayback, M., Schulman, S.P., Kickler, T.S., Becker, L.C., Weiss, J.L., Gerstenblith, G. and Goldschmidt-Clermont, P.J., 1996, A polymorphism of a platelet glycoprotein receptor as an inherited risk factor for coronary thrombosis, *New Engl. J. Med.* 334, 1090–1094.

Wiedmer, T., Shattil, S.J., Cunningham, M. and Sims, P.J., 1990, Role of calcium and calpain in complement-induced vesiculation of the platelet plasma membrane and in the exposure of the platelet factor Va receptor, *Biochemistry* 29, 623–632.

Williamson, P., Bevers, E.M., Smeets, E.F., Comfurius, P., Schlegel, R.A. and Zwaal, R.F.A., 1995, Continuous analysis of the mechanism of activated transbilayer lipid movement in platelets, *Biochemistry* 34, 10448–10455.

Wolf, B.B., Goldstein, J.C., Stennicke, H.R., Beere, H., Amarante-Mendes, G.P., Salvesen, G.S. and Green, D.G., 1999, Calpain functions in a caspase-independent manner to promote apoptosis-like events during platelet activation, *Blood* 94, 1683–1692.

Xu, W.F., Andersen, H., Whitmore, T.E., Presnell, S.R., Yee, D.P., Ching, A., Gilbert, T., Davie, E.W. and Foster, D.C., 1999, Cloning and characterization of human protease-activated receptor 4, *Proc. Natl. Acad. Sci. USA* 95, 6642–6646.

Yuan, Y., Dopheide, S.M., Ivanidis, C., Salem, H.H. and Jackson, S.P., 1997, Calpain regulation of cytoskeletal signaling complexes in von Willebrand factor-stimulated platelets. Distinct roles for glycoprotein Ib-V-IX and glycoprotein IIb-IIIa (integrin $\alpha_{IIb}\beta_3$) in von Willebrand factor-induced signal transduction, *J. Biol. Chem.* 272, 21847–21854.

Zwaal, R.F.A. and Schroit, A.J., 1997, Pathophysiological implications of membrane phospholipid asymmetry in blood cells, *Blood* 89, 1121–1132.

Zwaal, R.F.A., Comfurius, P. and Bevers, E.M., 1992, Platelet procoagulant activity and microvesicle formation in hemostasis and thrombosis, *Biochim. Biophys. Acta* 1180, 1–8.

Zwaal, R.F.A., Comfurius, P. and Bevers, E.M., 1993, Mechanism and function of changes in membrane-phospholipid asymmetry in platelets and erythrocytes, *Biochem. Soc. Trans.* 21, 248–253.

Calcium Signalling in Liver Cells

Greg J. Barritt

1. SUMMARY

In hepatocytes, increases in the concentration of Ca^{2+} in the cytoplasmic space and mitochondrial matrix play central roles in the regulation by hormones, neurotransmitters and growth factors of metabolic pathways, the secretion of bile acids, cell growth, and the metabolism of drugs and toxic agents. One of several important forms of intracellular Ca^{2+} signal is the hormone-induced wave of increased cytoplasmic Ca^{2+} concentration which begins at the canalicular membrane and moves to the sinusoidal membrane in an individual hepatocyte, and travels between neighbouring hepatocytes via gap junctions. The nature of this Ca^{2+} signal is, in part, a product of the polarised structure of the hepatocyte. This structure also has an important bearing on the process by which the Ca^{2+} wave is generated. The maintenance of successive waves of increased cytoplasmic Ca^{2+} concentration involves the inositol 1,4,5-trisphosphate and ryanodine Ca^{2+} channels in the endoplasmic reticulum, Ca^{2+} channels and transporters in the mitochondria, a variety of plasma membrane Ca^{2+} channels and the plasma membrane (Ca^{2+} + Mg^{2+})ATP-ase, GTP-binding regulatory proteins and the cytoskeleton. Glucagon, in the presence of a hormone which generates inositol 1,4,5-trisphosphate, causes a profound stimulation of Ca^{2+} inflow and increases the mitochondrial Ca^{2+} concentration. The latter contributes to the regulation of mitochondrial ATP synthesis. Changes or abnormalities in the intracellular Ca^{2+} signalling pathways in hepatocytes may underlie a number of diseased states, including some forms of cholestasis, and the response of the liver to toxic insults.

Greg J. Barritt • Department of Medical Biochemistry, School of Medicine, Faculty of Health Sciences, Flinders University, Adelaide, Australia.

R. Pochet, R. Donato, J. Haiech, C. Heizmann and V. Gerke (eds.): Calcium: The Molecular Basis of Calcium Action in Biology and Medicine, 73–94.

2. INTRODUCTION

Increases in the concentration of free (ionised, unbound) Ca^{2+} in the cytoplasmic space ($[Ca^{2+}]_{cyt}$), the mitochondrial matrix ($[Ca^{2+}]_m$), the lumen of the endoplasmic reticulum (ER) ($[Ca^{2+}]_{er}$) and the nucleus, play a central role in regulating the functions of liver parenchymal cells (hepatocytes) in response to hormones, neurotransmitters and growth factors (Graf and Häussinger, 1996; Robb-Gaspers et al., 1998; Tordjmann et al., 1998; Patel et al., 1999). The cytoplasmic Ca^{2+} signal is defined by the concentration of Ca^{2+} at a given location in the cytoplasmic space, the region of the cytoplasmic space in which the concentration of Ca^{2+} is increased, and the time over which this increase occurs. The hepatocyte cytoplasmic Ca^{2+} signal is often composed of repetitive waves or oscillations of increased $[Ca^{2+}]_{cyt}$. It is thought that signalling information is encoded in both the frequency and amplitude of the repetitive changes in $[Ca^{2+}]_{cyt}$ (Robb-Gaspers et al., 1998; Tordjmann et al., 1998; Patel et al., 1999). Regulation of the concentrations of free Ca^{2+} in the cytoplasmic space and other intracellular compartments requires control of the inflow and outflow of Ca^{2+} across the plasma membrane and across the membranes of intracellular organelles. Several diseases affecting the liver and digestive system involve changes in Ca^{2+} signalling in the liver (Frost et al., 1996; Graf and Häussinger, 1996; Crenesse et al., 1999).

The aims of this chapter are to summarise current knowledge of the nature and physiological functions of intracellular Ca^{2+} signals in hepatocytes and to describe the mechanisms by which these signals are generated. Since the nature of the signals and the mechanisms by which they are generated depend very closely on the structure and polarity of hepatocytes, the structure and physiological functions of these polarised epithelial cells will first be briefly reviewed. The chapter will then principally focus on recent developments in understanding how intra- and inter-cellular Ca^{2+} waves are generated in hepatocytes and the nature of the plasma membrane Ca^{2+} channels.

3. THE STRUCTURE AND PHYSIOLOGICAL FUNCTIONS OF HEPATOCYTES

Hepatocytes are multifunctional cells which play a major role in regulating the normal physiology of the whole body. They perform the interconversion of carbohydrates, lipids, proteins, and amino acids, the synthesis and secretion of lipids (chiefly as low density lipoproteins), the synthesis and secretion of albumin and other proteins, the metabolism of xenobiotic compounds, the synthesis and transcellular movement of bile acids and associated anions and cations, and provide energy to support all these functions (Graf and Häussinger, 1996; Robb-Gaspers et al., 1998; Tordjmann et al., 1998; Patel et al., 1999). Hepatocytes are arranged in plates (the hepatic cords) which are approximately 20 cells in length (Figure 1A). These form the "acinus". The

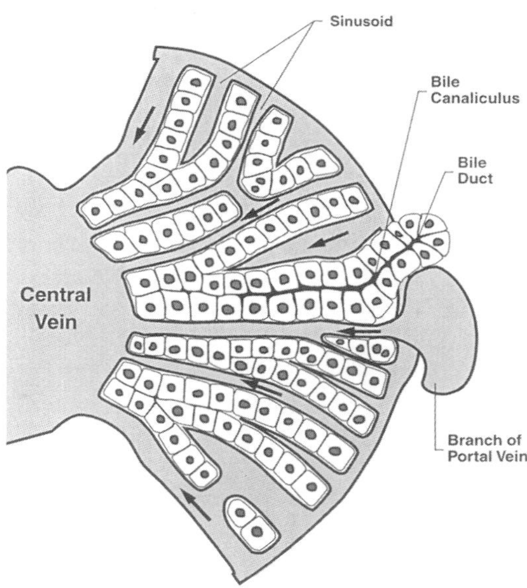

Figure 1. The arrangement of hepatocytes in an acinus (part of a liver lobule), and the generation and decline of a wave of increased cytoplasmic Ca^{2+} concentration which moves along liver plates. (A) A schematic representation of the structure of an "acinus" in a liver lobule, showing plates of hepatocytes lying between the central vein and bile duct. Re-drawn from Ham, A.W., 1965, *Textbook of Histology*, 5th edition, J.B. Lippincott, Philadelphia, by permission of J.B. Lippincot Co.

portal vein and the bile duct are located at one end of the hepatic plate and the central vein at the other end (Figure 1A). Each hepatocyte is in contact with interstitial fluid and blood at the sinusoidal surface and at the basolateral surface, and with bile acids at the bile canaliculus at the canalicular surface (Ham, 1965).

Most studies of Ca^{2+} signalling in hepatocytes have been conducted with the perfused liver, freshly-isolated hepatocytes, or with hepatocytes in primary culture. Since the hepatocyte couplet, in which two hepatocytes enclose part of the bile canalicular space (Figure 2), is an important functional unit of the liver, and hepatocyte couplets can readily be isolated by collagenase digestion of the perfused liver, studies with hepatocyte couplets have played a central role in understanding intracellular Ca^{2+} signalling (Spray et al., 1986; Nathanson et al., 1994; Tran et al., 1999). Recently, the perfused liver and confluent monolayers of hepatocytes in primary culture have been used to characterise intercellular Ca^{2+} waves and to relate these to the physiological functions of the liver (Tordjmann et al., 1998; Patel et al., 1999).

Hepatocytes are highly polarised epithelial cells (Figure 2). Tight junctions, composed of connexin proteins, between the two cells are, in part, responsible for maintaining cell polarity, including the separate canalicular and basolateral membrane domains (Figure 2B). Gap junctions located on the basolateral surface permit the movement of low molecular weight molecules

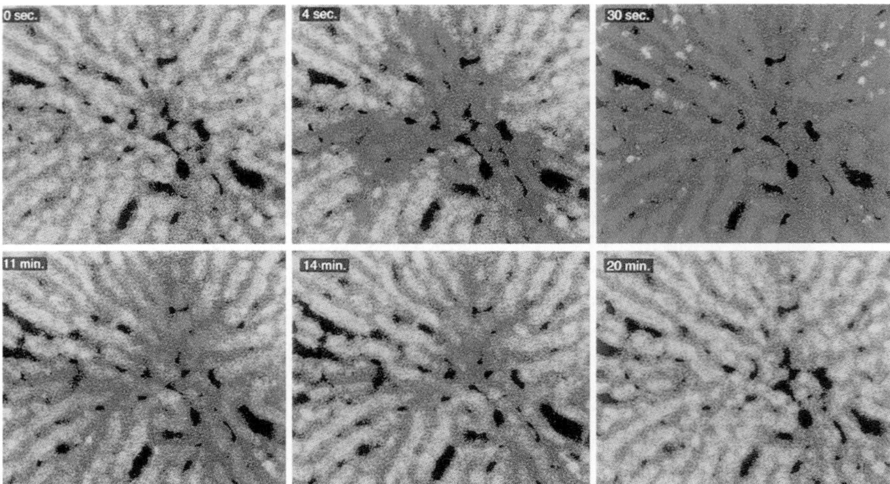

Figure 1. Continued. (B) A wave of increased $[Ca^{2+}]_{cyt}$ moving outwards along liver cell plates from the central vein to the bile duct in a perfused rat liver in which the hepatocytes are loaded with fluo-3. The wave was initiated by the infusion of vasopressin for 1–2 min beginning at time 0 sec. The panels at 11, 14 and 20 min represent the decline in $[Ca^{2+}]_{cyt}$. The central vein is located at the centre of each frame, and hepatic plates radiate outwards in all directions The resting $[Ca^{2+}]_{cyt}$ (100 nM) is represented by yellow and the maximum value of $[Ca^{2+}]_{cyt}$ (400 nM) by red. Reproduced from Motoyama, K., Karl, I.E., Flye, M.W., Osborne, D.F. and Hotchkiss, R.S., 1999, Effect of Ca^{2+} agonists in the perfused liver: Determination via laser scanning confocal microscopy, *Am. J. Physiol.* 276, R575–585, by copyright permission of The American Physiological Society. For a colour version of this figure, see page xxiv.

across the basolateral membranes between neighbouring cells. Like most animal cells, hepatocytes are dynamic structures. A constant flow of vesicles is responsible for the movement of proteins and lipids from the ER and Golgi to the apical canalicular membrane and to the sinusoidal membrane (for release into the blood) (reviewed in Fernando and Barritt, 1996). Proteins and sphingolipids synthesised in the Golgi are conveyed in secretory vesicles to a sub-apical compartment and from there to the canalicular (apical) membrane. Other proteins and lipids originating from the Golgi take an indirect route, via the basolateral membrane and early endosomes, to the canalicular membrane.

The ER, mitochondria and cytoskeleton (Figure 2) are key elements of the hepatocyte structure which both determine the nature and direction of intracellular Ca^{2+} signals and play a central role in the processes by which Ca^{2+} is moved from one region to another. Since one of their major functions is the synthesis and secretion of proteins, hepatocytes are enriched in rough and smooth ER. At low magnification, the ER can be detected by light microscopy as basophilic bodies (Ham, 1986). At higher magnification, fluorescence microscopy of cells stained with DiOC6 and electron microscopy reveals that the ER is distributed throughout the cell (Figure 2A). Mitochondria are also distributed throughout the cytoplasmic space (Figure 2A). A characteristic of

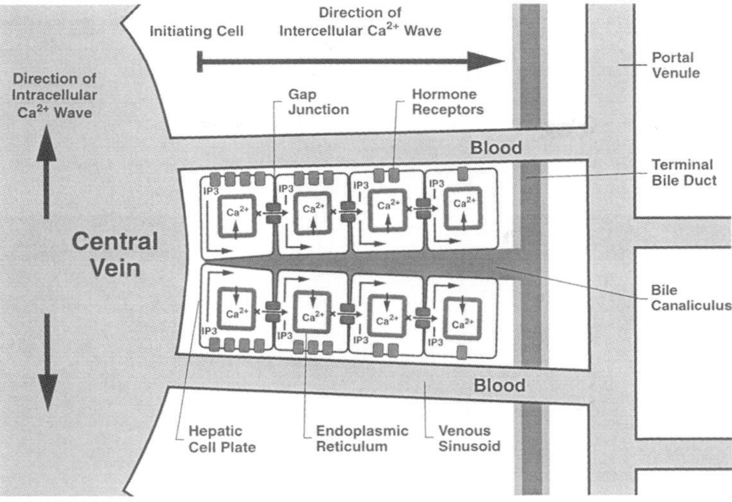

Figure 1. Continued. (C) A schematic representation of one hepatic plate showing the proposed roles of hormone receptors, $InsP_3$, $InsP_3$ receptors (Ca^{2+} channels), and the release of Ca^{2+} from the ER in the generation of a Ca^{2+} wave which moves from the bile canalicular to the sinusoidal side of each individual cell. The scheme also shows the roles of an intracellular messenger (indicated X, possibly $InsP_3$), gap junctions, and hormone receptors in the generation of an intercellular wave which moves from the central vein to the terminal bile duct. Re-drawn from Tordjmann, T., Berthon, B., Jacquemin, E., Clair, C., Stelly, N., Guillon, G., Claret, M. and Combettes, L., 1998, Receptor-oriented intercellular calcium waves evoked by vasopressin in rat hepatocytes, *EMBO J.* 17, 4695–4703, by copyright permission of Oxford University Press.

hepatocyte couplets and of hepatocytes *in situ* is the arrangement of the actin cytoskeleton around the cortex with an enrichment of actin filaments in the bile canalicular region (Figures 2B and 3D). A close association of actin with the plasma membrane has been confirmed by the analysis of purified liver cell plasma membranes. In addition to actin, these preparations have been shown to contain ER membranes, and there is evidence that the ER is attached to the plasma membrane by actin filaments (reviewed in Gregory et al., 1999).

4. THE ROLE OF Ca^{2+} AS AN INTRACELLULAR MESSENGER IN HEPATOCYTES

Hormones, neurotransmitters and growth factors which use changes in $[Ca^{2+}]_{cyt}$ and $[Ca^{2+}]_m$ as intracellular signals in regulating the functions of hepatocytes include adrenaline, noradrenaline, insulin, glucagon, vasopressin, angiotensin II, epidermal growth factor and ATP (Graf and Häussinger, 1996; Robb-Gaspers et al., 1998; Tordjmann et al., 1998; Patel et al., 1999). Osmotically-induced changes in cell volume may also cause changes in $[Ca^{2+}]_{cyt}$ (Graf and Häussinger, 1996). The main targets for the action

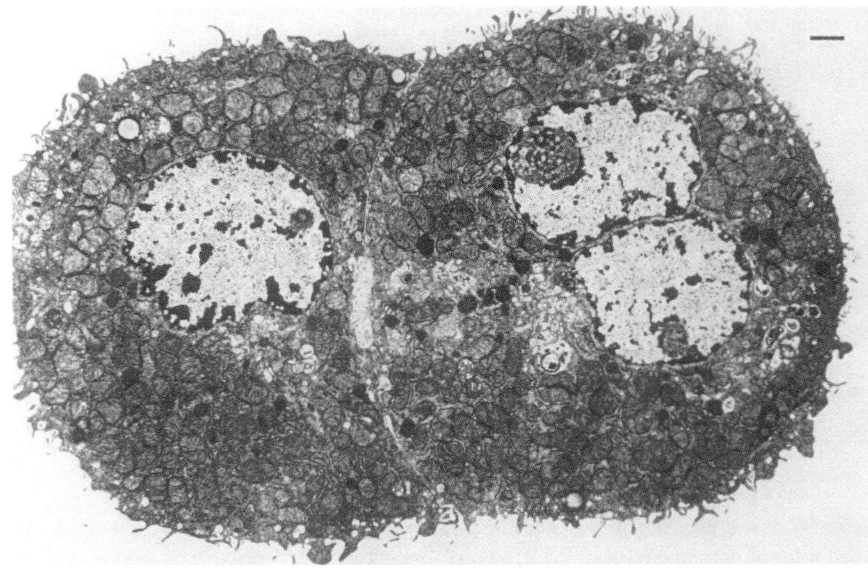

Figure 2. Some morphological features of polarised hepatocytes in an hepatocyte couplet. (A) A transmission electron micrograph of an isolated rat hepatocyte couplet showing the bile canaliculus in the centre, located between the two cells. The scale bar represents 1 μm. Reproduced from Spray, D.C., Ginzberg, R.D., Morales, E.A., Gatmaitan, Z. and Arias, I.M., 1986, Electrophysiological properties of gap junctions between dissociated pairs of rat hepatocytes, *J. Cell Biol.* 103, 135–144, by copyright permission of the Rockefeller University Press.

of increased $[Ca^{2+}]_{cyt}$ are Ca^{2+}-dependent regulatory enzymes in metabolic pathways (e.g. glycogen phosphorylase kinase), proteins which regulate the shape of the actin cytoskeleton (e.g. in regulation of the contraction of the bile canaliculus), plasma membrane ion channels and transporters (e.g. Ca^{2+}-activated K^+ channels which contribute to the regulation of membrane potential) and channels and transporters involved in the regulation of cell volume and cell growth (Graf and Häussinger, 1996). Changes in $[Ca^{2+}]_m$, which are transmitted to the mitochondrial matrix by changes in $[Ca^{2+}]_{cyt}$ regulate the activity of the mitochondrial dehydrogenases involved in ATP synthesis (Robb-Gaspers et al., 1998).

Numerous channels and transporters for ions other than Ca^{2+} are present in the hepatocyte plasma membrane (Graf and Häussinger, 1996). In the sinusoidal membrane, these include proteins which facilitate the exchange of Na^+ for an amino acid, Na^+ for H^+, and Na^+ for taurocholate; co-transporters for Na^+ and HCO_3^-; Na^+, Cl^- and K^+; and channels for Cl^-, Na^+ and H^+. The canalicular membrane possesses a Cl^- channel; an exchanger for Cl^- and HCO_3^-; an ATP-dependent taurocholate transporter; and several isoforms of ATP-dependent multidrug resistant proteins which transport phospholipids, conjugated bilirubin and several other molecules. Many of these ion channels and transporters are involved in the transport of bile salts and in the regulation of liver cell volume in response to changes in the intracellular concentrations

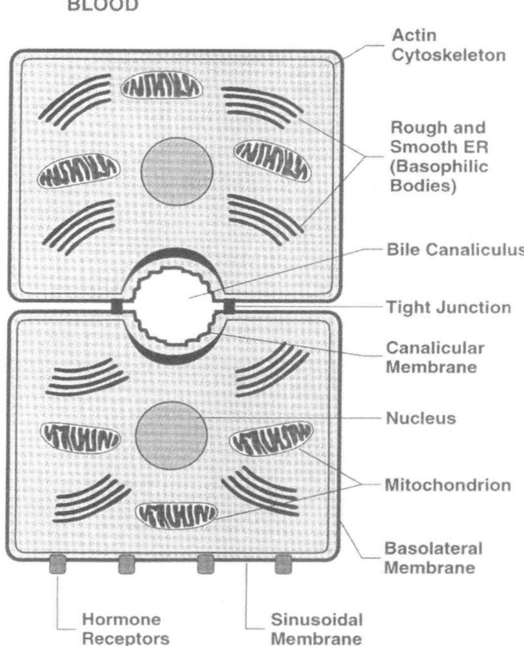

Figure 2. Continued. (B) A schematic representation of some of the main morphological features of an hepatocyte couplet.

of bile salts and ions caused by the rapid transport of these molecules from the blood to the bile canaliculus (Graf and Häussinger, 1996). While these channels and transporters do not facilitate the movement of Ca^{2+} and do not directly affect $[Ca^{2+}]_{cyt}$, knowledge of their properties and functions is valuable in understanding the role of Ca^{2+} as an intracellular messenger in hepatocytes, and in interpreting the results of experiments directed towards elucidation of mechanisms which regulate $[Ca^{2+}]_{cyt}$.

5. HORMONE-INDUCED WAVES OF INCREASED CYTOPLASMIC FREE Ca^{2+} CONCENTRATION IN HEPATOCYTES AND THE LIVER

Hepatocytes loaded with the luminescent protein, aequorin, provided the first demonstration that, at sub-maximal concentrations, hormones, growth factors and neurotransmitters can induce oscillations in $[Ca^{2+}]_{cyt}$ in non-excitable animal cells (reviewed in Woods et al., 1999). The frequency and amplitude were shown to depend on the nature of the hormone and the intracellular signalling pathway(s) employed. Using fura-2 to measure $[Ca^{2+}]_{cyt}$, primary cultures of rat hepatocytes attached to a collagen-coated coverslip, and image analysis, the hormone-induced oscillations in $[Ca^{2+}]_{cyt}$ were shown to be

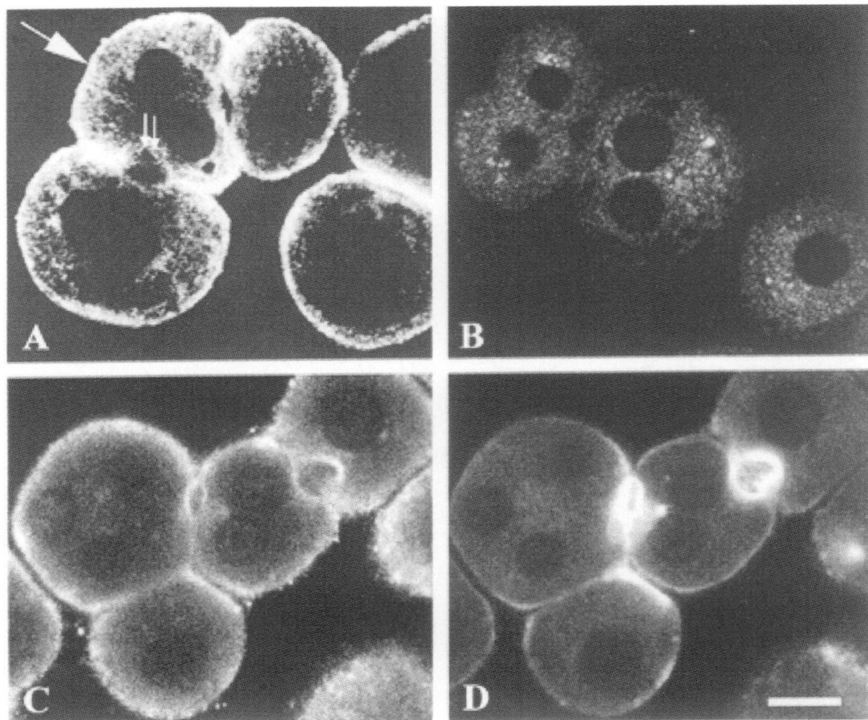

Figure 3. The subcellular distribution of type 1 inositol 1,4,5-trisphosphate receptors and F-actin in an hepatocyte couplet. (A) The subcellular distribution of type 1 $InsP_3$ receptors determined using an anti-type 1 $InsP_3$ receptor antibody and immunofluorescence. (B) The control for A in which the primary antibody was omitted. (C) and (D) Double labelling with an anti-type 1 $InsP_3$ receptor antibody (C) and an anti-F-actin antibody (D) showing the partial co-localisation of type 1 $InsP_3$ receptors and F-actin. Reproduced from Tran, D., Stelly, N., Tordjmann, T., Durroux, T., Dufour, M.N., Forchioni, A., Seyer, R., Claret, M. and Guillon, G., 1999, Distribution of signaling molecules involved in vasopressin-induced Ca^{2+} mobilisation in rat hepatocyte multiplets, *J. Histochem. Cytochem.* 47, 601–616, by copyright permission of The Histochemical Society Inc.

waves of increased $[Ca^{2+}]_{cyt}$. These are initiated by an increase in $[Ca^{2+}]_{cyt}$ in a specific region of the hepatocyte and move across most regions of the cytoplasmic space (reviewed in Patel et al., 1999). Other studies by Nathanson and colleagues using hepatocyte couplets showed that hormone-induced waves of increased $[Ca^{2+}]_{cyt}$ often begin in the canalicular region (Nathanson et al., 1994). They also found the majority of type 1 $InsP_3$ receptors in this region. In this respect, the behaviour of hepatocyte couplets may be similar to that of groups of pancreatic acinar cells in which waves of increased $[Ca^{2+}]_{cyt}$ are observed to originate in the acinar region and move to the basal region (reviewed in Petersen et al., 1999).

The application of fluorescent Ca^{2+} indicators such as fluo-3 and image analysis techniques to the perfused liver has permitted researchers to address the question of whether waves of increased $[Ca^{2+}]_{cyt}$ move from one cell to another in liver plates. These experiments have yielded some quite fascinating results on the cooperative behaviour of cells in a tissue (Tordjmann et al., 1998; Motoyama et al., 1999; Patel et al., 1999). Hormones such as vasopressin, adrenaline and ATP induce intercellular waves of increased $[Ca^{2+}]_{cyt}$ in liver "acini" which move along the hepatocytes in liver plates. Such waves can be observed moving through the same cells, and in the same direction, upon removal and subsequent re-addition of the hormone. The waves are often observed to move along the liver cell plate from the central vein to the portal vein as shown in Figure 1B. However, the direction and location of the increase in $[Ca^{2+}]_{cyt}$ depend on the concentration of the applied hormone (Motoyama et al., 1999). On the basis of experiments conducted with the perfused liver and with groups of confluent hepatocytes in primary culture which have re-established gap junctions, it has been concluded that both the binding of the hormone to plasma membrane receptors on each cell in the liver cell plate and the diffusion of an intracellular messenger (possibly $InsP_3$) through gap junctions from one cell to another are necessary for the propagation of these intercellular waves (Tordjmann et al., 1998). Moreover, other experiments indicate that the concentration of vasopressin receptors (the number of receptors per cell) is highest for cells near the central vein (reviewed in Tordjmann et al., 1998). This may explain why the wave of increased $[Ca^{2+}]_{cyt}$ is initiated at this end of the liver cell plate. These ideas are shown schematically in Figure 1C.

Since an increase in $[Ca^{2+}]_{cyt}$ has been shown to cause a contraction of the bile canaliculus, it has been proposed that one of the functions of intercellular waves of increased $[Ca^{2+}]_{cyt}$ which move along liver cell plates from the central vein to the bile duct is to move bile fluid excreted from individual hepatocytes along the bile canaliculus to the bile duct (reviewed in Tordjmann et al., 1998; Motoyama et al., 1999). Within a given hepatocyte, initiation of the hormone-induced increase in $[Ca^{2+}]_{cyt}$ at the canalicular membrane may reflect an important regulatory role for a high $[Ca^{2+}]_{cyt}$ in the canalicular region (Nathanson et al., 1994).

6. INTRACELLULAR DISTRIBUTION OF KEY PROTEINS INVOLVED IN GENERATING Ca^{2+} SIGNALS IN HEPATOCYTES

Some of the key proteins involved in initiating and propagating Ca^{2+} signals in hepatocytes, and their likely distribution in hepatocyte couplets are shown schematically in Figure 4. Receptors for hormones and growth factors, some trimeric GTP-binding proteins and phosphoinositide phospholipase C (phospholipase C), which catalyses the hydrolysis of phosphatidylinositol 4,5-bisphosphate to form diacylglycerol and $InsP_3$, are principally located on the sinusoidal and basolateral membranes (reviewed in Tran et al., 1999). There is evidence that the plasma membrane $(Ca^{2+} + Mg^{2+})$ATP-ase is prin-

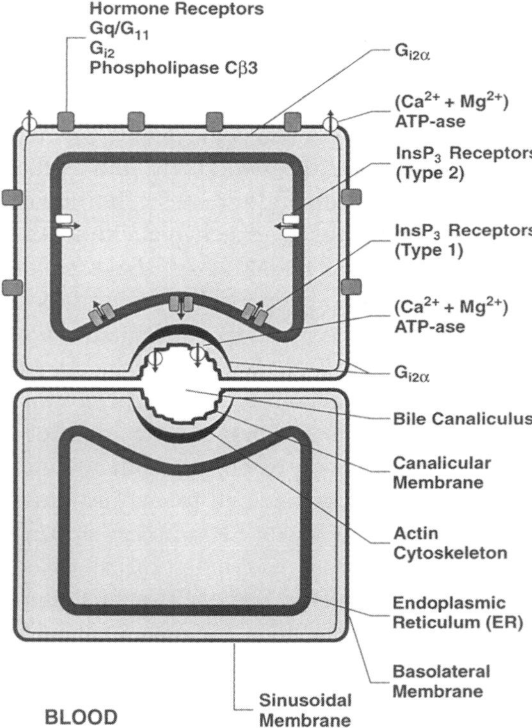

Figure 4. A schematic representation of the intracellular distribution of key proteins involved in generating and regulating intracellular Ca^{2+} signals in hepatocytes.

cipally located at the sinusoidal membrane and at the canalicular membrane (reviewed in Tran et al., 1999).

Hepatocytes express type 1 (about 20% of the total $InsP_3$ receptors) and type 2 (about 80% of the $InsP_3$ receptors), with negligible expression of type 3 $InsP_3$ receptors (reviewed in Gregory et al., 1999 and chapter on "Calcium storage and release" by Parys et al., in this book). Immunolocalisation has shown that, in hepatocyte couplets and in hepatocytes *in situ*, the type 1 $InsP_3$ receptors are located near the plasma membrane at most regions of the cell surface and near the nucleus, but are especially concentrated in the canalicular region (Nathanson et al., 1994) (Figures 3A and 3C). Presently, little is known about the location of type 2 $InsP_3$ receptors in hepatocytes. It is likely that the observed intracellular location of $InsP_3$ receptors is influenced by the hepatocyte preparation examined (e.g. whether whole liver, single hepatocytes or hepatocyte couplets) and the conditions under which hepatocytes are cultured, since these factors may determine the cell architecture and intracellular targetting of $InsP_3$ receptors.

Ryanodine receptors may also be present in hepatocytes and most likely play an important role in the propagation of waves of increased $[Ca^{2+}]_{cyt}$

by mediating Ca^{2+}-induced Ca^{2+} release (reviewed in Patel et al., 1999). Ryanodine receptors, which are located in the ER, are Ca^{2+} channels activated by the binding of cyclic ADP-ribose and by Ca^{2+} itself. While there are several reports indicating the presence of ryanodine receptors in hepatocytes, others suggest that these receptors cannot be detected. Komazaki et al. (1998) have recently utilised transgenic mice in which the type 1 or the type 3 ryanodine receptor was ablated. Ablation of either of these ryanodine receptors was associated with liver hypertrophy and excessive accumulation of liver glycogen. These results may suggest that the ryanodine receptor is required for the normal function of hepatocytes such as hormone-initiated, $[Ca^{2+}]_{cyt}$-mediated glycogen hydrolysis.

7. PLASMA MEMBRANE Ca^{2+} CHANNELS IN LIVER CELLS

The maintenance of repeated waves of hormone-induced increases in $[Ca^{2+}]_{cyt}$ requires increased Ca^{2+} inflow across the plasma membrane since some Ca^{2+} is transported out of the cell through the plasma membrane (Ca^{2+} + Mg^{2+})ATP-ases during each wave of increased $[Ca^{2+}]_{cyt}$ (reviewed in Tordjmann et al., 1998; Patel et al., 1999; Woods et al., 1999). A large part of this Ca^{2+} outflow occurs at the canalicular region. An increase in plasma membrane Ca^{2+} inflow is required to replenish Ca^{2+} in the ER. Furthermore, it is likely that there are other forms of intracellular Ca^{2+} signal involving an increase in $[Ca^{2+}]_{cyt}$ in specific regions of the cytoplasmic space generated principally by enhanced plasma membrane Ca^{2+} inflow.

Hepatocytes process several different types of receptor-activated Ca^{2+} channel (RACC) (a plasma membrane channel, the opening of which is initiated by the binding of a hormone to its receptor protein which is separate from the channel protein) and at least one type of ligand-gated Ca^{2+} channel (a plasma membrane Ca^{2+} channel which is part of the multiprotein complex which constitutes the receptor protein). Extensive experiments, conducted using a variety of techniques, have provided no evidence for the expression of voltage-operated Ca^{2+} channels (VOCCs) in hepatocytes (reviewed in Brereton et al., 1997; Woods et al., 1999).

The main types of plasma membrane Ca^{2+} channels found in hepatocytes are listed in Table 1. There is at least one type of store-operated Ca^{2+} channel (SOC) and several types of intracellular messenger-activated non-selective cation channels which admit Ca^{2+}. At least one ligand-gated Ca^{2+} channel, the P2X purinergic receptor which can bind ATP, has also been identified (Capiod, 1998) (Table 1). SOCs are defined as a plasma membrane Ca^{2+} channels the opening of which is initiated by a decrease in the concentration of Ca^{2+} in the lumen of the ER (reviewed in Petersen et al., 1999). Under physiological conditions this involves the binding of $InsP_3$ to $InsP_3$ receptors or, possibly, the binding of cyclic ADP ribose to ryanodine receptors. SOCs can also be activated artificially using thapsigargin or 2,5-di-(*t*-butyl)-1,4-

Table 1. The types of plasma membrane Ca^{2+} channels detected in the perfused liver, freshly-isolated hepatocytes, and hepatocytes in primary culture

Channel	Liver cell preparation	Agents used to initiate Ca^{2+} inflow	Technique employed to measure Ca^{2+} inflow	References
Store-operated Ca^{2+} channel (some permeability to Mn^{2+} reported)	Rat, isolated hepatocytes	Hormones, ATP (P2Y), $InsP_3$, thapsigargin, DBHQ	Intracellular fluorescent Ca^{2+} sensor	Reviewed in Kass et al., 1994; Fernando and Barritt, 1995; Ikari et al., 1997
		Hormones, ATP, $InsP_3$	Patch-clamp recording	Reviewed in Fernando et al., 1997
	Rat, perfused liver	Hormones, DBHQ	Decrease in perfusate $[Ca^{2+}]$	Applegate et al., 1997
Receptor (intracellular messenger-activated) non-selective cation channel(s). (admission of Mn^{2+} often observed)	Rat, isolated hepatocytes	Hormones, ATP, $InsP_3$, thapsigargin, DBHQ, maitotoxin	Intracellular fluorescent Ca^{2+} sensor or aequorin	Reviewed in Kass et al., 1994; Fernando and Barritt, 1995; Woods et al., 1999
		Hormones	Patch-clamp recording	Reviewed in Lidofsky et al., 1997
	Rat, perfused liver	Hormones	Decrease in perfusate $[Ca^{2+}]$	Reviewed in Applegate et al., 1997
Stretch (ATP)-activated non-selective cation channel	Rat, isolated hepatocytes	Stretch	Patch-clamp recording	Reviewed in Fernando and Barritt, 1995

Continued on next page

Table 1. Continued

Channel	Liver cell preparation	Agents used to initiate Ca^{2+} inflow	Technique employed to measure Ca^{2+} inflow	References
Cyclic AMP-activated Ca^{2+} channel	Axolotl, isolated hepatocytes	Cyclic AMP analogues	Intracellular fluorescent Ca^{2+} sensor	Lenz and Kleineke, 1997
	Rat, isolated hepatocytes	Glucagon, cyclic AMP analogues, parathyroid hormone	Intracellular fluorescent Ca^{2+} sensor	Reviewed in Kass et al., 1994; Ikari et al., 1997; Applegate et al., 1997
	Rat liver and isolated hepatocytes	Cyclic AMP	Northern blot to detect expression of mRNA encoding the CNGCα pore-forming subunit of cyclic nucleotide-gated channel Intracellular fluorescent Ca^{2+} sensor	Feng et al., 1996
Purinergic (P2X) ligand-gated non-selective cation channel	Guinea pig, isolated hepatocytes	ATP	Patch-clamp recording	Capiod, 1998

hydroquinone (DBHQ), inhibitors of the (Ca^{2+} + Mg^{2+})ATP-ase of the ER (reviewed in Petersen et al., 1999; Gregory et al., 1999). While there is good evidence for the presence of a cyclic AMP-activated plasma membrane Ca^{2+} channel in axolotl hepatocytes (Lenz and Kleineke, 1997), further experiments are required to clearly define cyclic nucleotide-activated Ca^{2+} channels which may be present in rat and human hepatocytes. In addition to the plasma membrane channels listed in Table 1, there is also evidence that some Ca^{2+} enters hepatocytes by the process of pinocytosis (Fernando and Barritt, 1996). This may account, in part, for the relatively large basal rate of Ca^{2+} inflow observed in isolated hepatocytes and in the perfused liver.

It is likely that the action of a hormone on hepatocytes often leads to the activation of more than one type of plasma membrane Ca^{2+} channel (Fernando and Barritt, 1995). Thapsigargin may also activate other RACCs in addition to SOCs (Fernando and Barritt, 1995; Ikari et al., 1997).

A number of RACCs admit Mn^{2+} as well as Ca^{2+}. Since Mn^{2+} can quench the fluorescence of intracellular fura-2, this has been used as a convenient assay for RACC activity. However, the introduction of fura-2 to the cytoplasmic space of hepatocytes by incubation with the acetoxymethyl ester of fura-2 often leads to the accumulation of fura-2 in the ER (reviewed in Kass et al., 1994). Thus results obtained for Mn^{2+} inflow to hepatocytes loaded with fura-2 in this way have been difficult to interpret. An additional complication is the evidence that the maintenance of rat hepatocytes in primary culture for 24 h leads to the appearance of a RACC which admits Mn^{2+}. This channel is not observed, or is not pronounced, in freshly-isolated rat hepatocytes (reviewed in Kass et al., 1994).

Liver cell lines have provided a number of advantages in studies of hepatocyte plasma membrane Ca^{2+} channels. They are more robust, can be attached more easily to a polished glass pipette for patch-clamp recording experiments, and can be grown in secondary culture under controlled conditions (Lidofsky et al., 1997). However, liver cell lines do have the disadvantage that they are genetically partially transformed cells and, consequently, their state of differentiation and phenotype can differ substantially from those of normal hepatocytes (Brereton et al., 1997). Notwithstanding these potential difficulties, studies with liver cell lines have led to the identification of a number of plasma membrane Ca^{2+} channels (Table 2). These include SOCs and a hormone-activated non-selective cation channel which admits Ca^{2+} as well as Na^{+}. This channel is also observed in freshly-isolated hepatocytes. Use of the patch-clamp recording technique has identified a Ca^{2+}-activated non-selective cation channel in HTC and HepG2 cells (Lidofsky et al., 1997).

Liver cell SOCs are inhibited by La^{3+}, Gd^{3+} and other lanthanides in the concentration range 0.1 to 10 μM and by SK&F 96365 at about 10–50 μM (Fernando and Barritt, 1995). Intracellular messenger-activated non-selective cation channels in liver cells are generally inhibited by higher concentrations of lanthanides and by SK&F 96365 (Fernando and Barritt, 1995). A number of laboratories have reported that hormone-stimulated Ca^{2+} inflow

Table 2. Types of plasma membrane Ca^{2+} channels detected in liver cell lines (transformed cells grown in secondary culture)

Channel	Liver cell line	Agents used to initiate Ca^{2+} inflow	Technique employed to measure Ca^{2+} inflow	References
Store-operated Ca^{2+} channel	H4-IIE (Reuber hepatoma)	ATP (P2Y), thapsigargin, $InsP_3$	Intracellular fluorescent Ca^{2+} sensor	Brereton et al., 1997
			Patch-clamp recording	Rychkov, Brereton, Harland, and Barritt, unpublished results
Receptor (intracellular messenger)-activated Na^+ channel (non-selective cation channel)	HTC rat hepatoma	ATP	Patch-clamp recording Intracellular fluorescent Na^+ sensor	Reviewed in Lidofsky et al., 1997; Fernando et al., 1997
Ca^{2+}-activated non-selective cation channel	HTC rat hepatoma, HepG2	Intracellular Ca^{2+}	Patch-clamp recording	Lidofsky et al., 1997
Stretch (ATP)-activated non-selective cation channel	Rat hepatoma	Stretch, ATP	Patch-clamp recording	Reviewed in Fernando et al., 1997

in liver cells is inhibited by verapamil, nifedipine and other inhibitors of L-type VOCCs (reviewed in Brereton et al., 1997). In most cases, the type of RACC affected was not unequivocally identified. Cyclic AMP-activated Ca^{2+} inflow has been shown to be inhibited by Ca^{2+} antagonists (Lenz and Kleineke, 1997; reviewed in Applegate et al., 1997). Verapamil and other Ca^{2+} channel antagonists inhibit a number of liver cell functions which are dependent on extracellular Ca^{2+} (reviewed in Crenesse et al., 1999), although it is not known whether the inhibition is due to an effect on plasma membrane Ca^{2+} channels, on other channels, transporters and proteins responsible for the intracellular movement of Ca^{2+}, or an indirect consequence of inhibition of other proteins.

8. STRUCTURES OF LIVER CELL PLASMA MEMBRANE Ca^{2+} CHANNELS

While the nature and number of subunits, the amino acid sequence and the topology of the polypeptide chain for each subunit of any hepatocyte RACC has not yet been determined, a number of hypotheses are being tested. Using an RT-PCR strategy, mRNA encoding isoforms of the pore-forming α_{1A} subunit of L-type VOCCs was detected in freshly-isolated rat hepatocytes and in the H4-IIE rat liver cell line. While some of these isoforms are potentially interesting, including those in which the sequence encoding the positively-charged voltage- sensing amino acids in the S4 domain was deleted, it was concluded that the isoforms are likely to have arisen from mutations in the gene encoding the α_1 subunit of L-type VOCCs during the transformation of rat hepatocytes to H4-IIE cells (Brereton et al., 1997). In another approach, an RT-PCR strategy has been used to identify in liver an mRNA species encoding the cyclic nucleotide-activated non-selective cation channel (Feng et al., 1996). Further studies are required to determine whether this mRNA species encodes a functional cyclic nucleotide-activated cation channel in liver.

It has been proposed that a family of non-selective Ca^{2+} channels called the TRP Ca^{2+} channels are animal cell RACCs (reviewed in Harteneck et al., 2000). The TRP Ca^{2+} channels are homologous of the *Drosophila melanogaster* TRP Ca^{2+} channel which is present in the *Drosophila photoreceptor* cell. TRP was named after the phenotype of a *Drosophila* TRP mutant (Transient Receptor Potential) in which the function of the TRP Ca^{2+} channel is abolished. Seven TRP channels have so far been identified in animal cells, chiefly in nerve cells of the brain (Harteneck et al., 2000). RT-PCR and Northern blot analysis have detected mRNA encoding TRP-1 in liver, indicating that this TRP Ca^{2+} channel is expressed in hepatocytes (Wang et al., 1999). While further experiments are required to elucidate the roles of TRP channels as RACCs in hepatocytes (and in other animal cells) it is likely that TRP-1 and possibly other TRP channels are the molecular forms of one or more of the hepatocyte non-selective cation channels listed in Tables 1 and 2.

9. THE ROLES OF GTP-BINDING PROTEINS AND THE CYTOSKELETON IN THE ACTIVATION OF PLASMA MEMBRANE Ca^{2+} CHANNELS IN HEPATOCYTES

Studies of the mechanisms of activation of SOCs and other RACCs in freshly-isolated hepatocytes by hormones and thapsigargin have provided evidence that the function of a trimeric GTP-binding regulatory protein (G-protein) is required in the mechanism by which SOCs are activated (Fernando et al., 1997; Gregory et al., 1999). These experiments have identified the G-protein as G_{i2} and have shown that, in hepatocytes, G_{i2_α} is ADP-ribosylated quite slowly upon the treatment of livers in vivo with pertussis toxin. A key experimental result was the observation that the microinjection of an inhibitory anti-G_{i2_α} antibody, or an inhibitory peptide corresponding to a region of the carboxy terminus of G_{i2_α} inhibits the activation of SOCs (reviewed in Gregory et al., 1999). The results of other experiments suggest that a brefeldin A-sensitive protein, tentatively identified as an Arf monomeric G-protein, is also required for the activation of SOCs in hepatocytes (Fernando et al., 1997).

While there is considerable evidence that a functional G-protein is required in the activation of SOCs in some other cell types, there has been no clear demonstration that G_{i2} is required for the activation of SOCs in any cell type other than hepatocytes (reviewed in Gregory et al., 1999). Thus hepatocytes appear somewhat unique in exhibiting a requirement for G_{i2} in the activation of SOCs. G_{i2_α} (and possibly a brefeldin A-sensitive protein) may be required to maintain communication within the lumen of the ER and/or the correct intracellular location of a region of the ER which might be directly involved in the activation of SOCs in hepatocytes (Gregory et al., 1999). Such a role for G_{i2} may involve the actin cytoskeleton since, in some cells, there is evidence for the association of Gi2 with actin (reviewed in Gregory et al., 1999). Another possibility is that Gi2 and a brefeldin A-sensitive protein are required for the insertion of SOCs into the plasma membrane by exocytosis (cf. mechanisms reviewed in Harteneck et al., 2000), although the time scale of this process may be too slow to account for the observed rate of activation of SOCs.

10. PHYSIOLOGICAL FUNCTIONS OF PLASMA MEMBRANE Ca^{2+} CHANNELS IN POLARISED HEPATOCYTES

As described earlier, hormone-induced waves of increased $[Ca^{2+}]_{cyt}$, which often originate at the canalicular region of each hepatocyte and move to the sinusoidal region of the same hepatocyte and also spread to the neighbouring hepatocytes on the portal side through gap junctions, constitute an important intracellular Ca^{2+} signal in hepatocytes. The maintenance of successive waves of increased $[Ca^{2+}]_{cyt}$ depends on the presence of extracellular Ca^{2+}.

This ensures an adequate supply of intracellular Ca^{2+} by replacing the Ca^{2+} which is lost from the cytoplasmic space through the action of the plasma membrane (Ca^{2+} + Mg^{2+})ATP-ases each time $[Ca^{2+}]_{cyt}$ is increased (reviewed in Petersen et al., 1999). Since the distance over which Ca^{2+} can effectively diffuse in the cytoplasmic space is a few microns, regenerative Ca^{2+} release from the ER mediated by $InsP_3$ and ryanodine receptors provides the mechanism by which the Ca^{2+} signal (the wave of increased $[Ca^{2+}]_{cyt}$) is transmitted through the cytoplasmic space (reviewed in Patel et al., 1999). Thus the ER, acting as a reservoir of Ca^{2+} which is widely distributed through the cytoplasmic space, plays an essential role in generating the $[Ca^{2+}]_{cyt}$ signal. Therefore, the maintenance of repetitive waves of increased $[Ca^{2+}]_{cyt}$ requires the constant replenishment of Ca^{2+} in the ER with extracellular Ca^{2+}. A major function of SOCs in hepatocytes is thought to replenish the Ca^{2+} which is lost from the ER in this manner.

As indicated in Figure 4, hormone receptors are most likely located principally on the sinusoidal membrane of hepatocytes while the majority of the type 1 $InsP_3$ receptors are located in the canalicular region where hormone-initiated waves of increased $[Ca^{2+}]_{cyt}$ often originate (Nathanson et al., 1994). SOCs are likely to be principally located on the sinusoidal and basolateral membranes since studies with the perfused liver indicate that, after the administration of a Ca^{2+}-mobilising hormone, considerable amounts of Ca^{2+} move from the blood to hepatocytes (reviewed by Applegate et al., 1997). Thus, by analogy with ideas developed for Ca^{2+} signalling in the pancreatic acinar cell (Petersen et al., 1999), it is likely that the binding of a hormone to receptors on the sinusoidal domain of the hepatocyte plasma membrane leads to an increase in the concentration of $InsP_3$ in this region of the cell, the diffusion of $InsP_3$ through all regions of the cytoplasmic space, the initiation of a wave of increased $[Ca^{2+}]_{cyt}$ in the canalicular region, and the advance of the wave of increased $[Ca^{2+}]_{cyt}$ from the canalicular to the sinusoidal region. The movement of $InsP_3$ and direction of the Ca^{2+} wave in individual hepatocytes are shown schematically in Figure 1C.

Ca^{2+} which flows through intracellular messenger-activated non-selective cation channels most likely has different roles compared with the role of Ca^{2+} which enters through SOCs. These may involve the regulation of cell volume in response to the transcellular movement of bile acids and other ions (Graf and Häussinger, 1996). non-selective cation channels may deliver a local high concentration of Ca^{2+} to specific regions of the cytoplasmic space in the vicinity of the channels. In addition, Na^+ which enters the cytoplasmic space through intracellular messenger-activated non-selective cation channels may also play a role in the regulation of cell volume and shape.

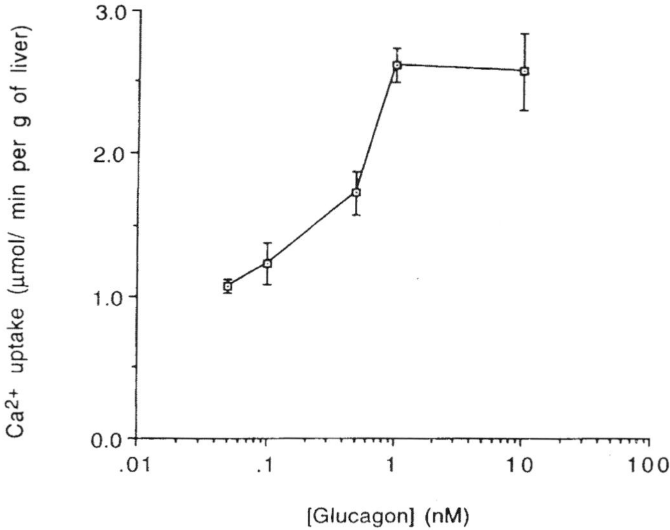

Figure 5. The dramatic enhancement by glucagon of 2,5-di-(t-butyl)-1,4-hydroquinone (DBHQ)-stimulated Ca^{2+} inflow to hepatocytes. A concentration-response curve for the enhancement by glucagon of Ca^{2+} inflow in the perfused rat liver. The rate of Ca^{2+} inflow was estimated by measuring the decrease in the concentration of Ca^{2+} in the perfusion medium using a Ca^{2+}-selective electrode. Reproduced from Applegate, T.L., Karjalainen, A. and Bygrave, F.L., 1997, Rapid Ca^{2+} influx induced by the action of dibutyl-hydroquinone and glucagon in the perfused rat liver, *Biochem. J.* 323, 463–467, by copyright permission of the Biochemical Society.

11. THE ROLE OF MITOCHONDRIA IN GLUCAGON-ENHANCED PLASMA MEMBRANE Ca^{2+} INFLOW

An unexpected and most interesting aspect of Ca^{2+} signalling in hepatocytes concerns the action of glucagon on Ca^{2+} inflow and mitochondrial Ca^{2+} concentrations. The exposure of hepatocytes to glucagon in the presence of a hormone such as vasopressin, which increases the concentration of $InsP_3$ and activates Ca^{2+} inflow, causes a pronounced enhancement of Ca^{2+} inflow and a large accumulation of Ca^{2+} in the mitochondria (reviewed in Applegate et al., 1997; Fernando et al., 1998). This effect of glucagon is also observed when an inhibitor of the (Ca^{2+} + Mg^{2+})ATP-ase such as DBHQ is used in place of the $InsP_3$-generating hormone (Figure 5). The mechanism by which glucagon enhances Ca^{2+} inflow is not fully understood. It most likely involves a cyclic AMP-mediated increase in the rate at which mitochondria remove Ca^{2+} from the cytoplasmic space (Fernando et al., 1998). The physiological functions of this action of glucagon are to permit the activation of mitochondrial enzymes involved in oxidative phosphorylation and ATP synthesis. These include pyruvate, α-ketoglutarate and isocitrate dehydrogenases (Robb-Gaspers et al., 1998).

In the absence of glucagon, the increase in $[Ca^{2+}]_{cyt}$ induced by hormones such as vasopressin also causes an increase in the concentration of Ca^{2+} in the mitochondrial matrix and activates mitochondrial dehydrogenases (Robb-Gaspers et al., 1998). Moreover, in hepatocytes (and in many other cell types), mitochondria are often spatially very closely associated with the ER so that Ca^{2+} is released from the ER by the action of $InsP_3$ which plays a major role in inducing an increase in the mitochondrial Ca^{2+} concentration (Robb-Gaspers et al., 1998).

12. Ca^{2+} SIGNALLING AND LIVER DISEASE

Several diseased states, such as some forms of cholestasis or toxic insults to the liver, lead to alterations in intracellular Ca^{2+} signalling (Frost et al., 1996; Crenesse et al., 1999; and chapter by Kobayashi et al., in this book). Potential abnormalities in Ca^{2+} signalling may involve plasma membrane and intracellular Ca^{2+} channels and transporters. Some forms of liver damage are prevented by treatment with verapamil and other Ca^{2+} antagonists which inhibit L-type voltage-operated Ca^{2+} channels (Crenesse et al., 1999). The effects of these Ca^{2+} antagonists may reflect their actions on either the multidrug resistant proteins, which bind verapamil, or on plasma membrane Ca^{2+} channels.

13. FURTHER DEVELOPMENTS

Future developments in understanding Ca^{2+} signalling in hepatocytes are likely to involve the following areas. (i) A more clear understanding of the nature of waves of increased $[Ca^{2+}]_{cyt}$ within an individual hepatocyte and within the whole liver, especially their mechanism of generation, transmission between hepatocytes, and their role in regulating the physiological functions of the liver as a whole. (ii) The cellular functions, structures and mechanism of activation of the different plasma membrane Ca^{2+} channels. Of particular interest are the SOCs and the TRP non-selective cation channels. (iii) How changes in Ca^{2+} signalling are involved in, or cause, various diseased states or liver damage, especially following the reperfusion of newly-transplanted livers and in liver cells in liver replacement therapy.

ACKNOWLEDGEMENTS

Research conducted in the applicant's laboratory on Ca^{2+} signalling in the liver was supported by grants from the National Health and Medical Research Council and the Australian Research Council.

REFERENCES

Applegate, T.L., Karjalainen, A. and Bygrave, F.L., 1997, Rapid Ca^{2+} influx induced by the action of dibutyl-hydroquinone and glucagon in the perfused rat liver, *Biochem. J.* 323, 463–467.

Brereton, H.M., Harland, M.L., Froscio, M., Petronijevic, T. and Barritt, G.J., 1997, Novel variants of voltage-operated calcium channel α_1-subunit transcripts in a rat liver-derived cell line: Deletion in the IVS4 voltage-sensing region, *Cell Calcium* 22, 39–52.

Capiod, T., 1998, ATP-activated cation currents in single guinea-pig hepatocytes, *J. Physiol.* 507.3, 795–805.

Crenesse, D., Hugues, M., Ferre, C., Poiree, J.C., Benoliel, J., Dolisi, C. and Gugenheim, J., 1999, Inhibition of calcium influx during hypoxia/reoxygenation in primary cultured rat hepatocytes, *Pharmacol.* 58, 160–170.

Feng, L., Subbaraya, I., Yamamoto, N., Baehr, W. and Kraus-Friedmann, N., 1996, Expression of photoreceptor cyclic nucleotide-gated cation channel α subunit (CNGCα) in the liver and skeletal muscle, *FEBS Lett.* 395, 77–81.

Fernando, K.C. and Barritt, G.J., 1995, Characterisation of the divalent cation channels of the hepatocyte plasma membrane receptor-activated Ca^{2+} inflow system using lanthanide ions, *Biochim. Biophys. Acta* 1268, 97–106.

Fernando, K.C. and Barritt, G.J., 1996, Pinocytosis in 2,5-di-tert-butylhydroquinone-stimulated hepatocytes and evaluation of its role in Ca^{2+} inflow, *Mol. Cell. Biochem.* 162, 23–29.

Fernando, K.C., Gregory, R.B., Katsis, F., Kemp, B.E. and Barritt, G.J., 1997, Evidence that a low-molecular-mass GTP-binding protein is required for store-activated Ca^{2+} inflow in hepatocytes, *Biochem. J.* 328, 463–471.

Fernando, K.C., Gregory, R.B. and Barritt, G.J., 1998, Protein kinase A regulates the disposition of Ca^{2+} which enters the cytoplasmic space through store-operated Ca^{2+} channels in rat hepatocytes by diverting inflowing Ca^{2+} to mitochondria, *Biochem. J.* 330, 1179–1187.

Frost, L., Mahoney, J., Field, J. and Farrell, G.C., 1996, Impaired bile flow and disordered hepatic calcium homeostasis are early features of halothane-induced liver injury in guinea pigs, *Hepatol.* 23, 80–86.

Graf, J. and Häussinger, D., 1996, Ion transport in hepatocytes: Mechanisms and correlations to cell volume, hormone actions and metabolism, *J. Hepatol.* 24, 53–77.

Gregory, R.B., Wilcox, R.A., Berven, L.A., van Straten, N.C.R., van der Marel, G.A., van Boom, J.H. and Barritt, G.J., 1999, Evidence for the involvement of a small subregion of the endoplasmic reticulum in the inositol trisphosphate receptor-induced activation of Ca^{2+} inflow in rat hepatocytes, *Biochem. J.* 341, 401–408.

Ham, A.W., 1965, *Textbook of Histology*, 5th edition, J.B. Lippincott, Philadelphia.

Harteneck, C., Plant, T.D. and Schultz, G., 2000, From worm to man: Three subfamilies of TRP channels, *Trends Neurosci.* 23, 159–166.

Ikari, A., Sakai, H. and Takeguchi, N., 1997, ATP, thapsigargin and cAMP increase Ca^{2+} in rat hepatocytes by activating three different Ca^{2+} influx pathways, *Japanese J. Physiol.* 47, 235–239.

Kass, G.E.N., Webb, D.-L., Chow, S.C., Llopis, J. and Berggren, P.-O., 1994, Receptor-mediated Mn^{2+} influx in rat hepatocytes: Comparison of cells loaded with fura-2 ester and cells microinjected with fura-2 salt, *Biochem. J.* 302, 5–9.

Komazaki, S., Ikemoto, T., Takeshima, H., Iino, M., Endo, M. and Nakamura, H., 1998, Morphological abnormalities of adrenal gland and hypertrophy of liver in mutant mice lacking ryanodine receptors, *Cell Tissue Res.* 294, 467–473.

Lenz, T. and Kleineke, J.W., 1997, Hormone-induced rise in cytosolic Ca^{2+} in axolotl hepatocytes: Properties of the Ca^{2+} influx channel, *Am. J. Physiol.* 273, C1536–C1532.

Lidofsky, S.D., Sostman, A. and Fitz, J.G., 1997, Regulation of cation-selective channels in liver cells, *J. Membr. Biol.* 157, 231–236.

Motoyama, K., Karl, I.E., Flye, M.W., Osborne, D.F. and Hotchkiss, R.S., 1999, Effect of Ca^{2+} agonists in the perfused liver: Determination via laser scanning confocal microscopy, *Am. J. Physiol.* 276, R575–R585.

Nathanson, M.H., Burgstahler, A.D. and Fallon, M.B., 1994, Multistep mechanism of polarized Ca^{2+} wave patterns in hepatocytes, *Am. J. Physiol.* 267, G338–G349.

Patel, S., Robb-Gaspers, L.D., Stellato, K.A., Shon, M. and Thomas, A.P., 1999, Coordination of calcium signalling by endothelial-derived nitric oxide in the intact liver, *Nature Cell Biol.* 1, 467–471.

Petersen, O.H., Burdakov, D. and Tepikin, A.V., 1999, Regulation of store-operated calcium entry: Lessons from a polarized cell, *Eur. J. Cell Biol.* 78, 221–223.

Robb-Gaspers, L.D., Rutter, G.A., Burnett, P., Hajnóczky, G., Denton, R.M. and Thomas, A.P., 1998, Coupling between cytosolic and mitochondrial calcium oscillations: Role in the regulation of hepatic metabolism, *Biochim. Biophys. Acta* 1366, 17–32.

Spray, D.C., Ginzberg, R.D., Morales, E.A., Gatmaitan, Z. and Arias, I.M., 1986, Electrophysiological properties of gap junctions between dissociated pairs of rat hepatocytes, *J. Cell Biol.* 103, 135–144.

Tordjmann, T., Berthon, B., Jacquemin, E., Clair, C., Stelly, N., Guillon, G., Claret, M. and Combettes, L., 1998, Receptor-oriented intercellular calcium waves evoked by vasopressin in rat hepatocytes, *EMBO J.* 17, 4695–4703.

Tran, D., Stelly, N., Tordjmann, T., Durroux, T., Dufour, M.N., Forchioni, A., Seyer, R., Claret, M. and Guillon, G., 1999, Distribution of signaling molecules involved in vasopressin-induced Ca^{2+} mobilisation in rat hepatocyte multiplets, *J. Histochem. Cytochem.* 47, 601–616.

Wang, W., O'Connell, B., Dykeman, R., Sakai, T., Delporte, C., Swaim, W., Zhu, X., Birnbaumer, L. and Ambudkar, I.S., 1999, Cloning of Trp1β isoform from rat brain: Immunodetection and localisation of the endogenous Trp1 protein, *Am. J. Physiol.* 276, C969–C979.

Woods, N.M., Dixon, C.J., Yasumoto, T., Cuthbertson, K.S.R. and Cobbold, P.H., 1999, Maitotoxin-induced free Ca changes in single rat hepatocytes, *Cell. Signal.* 11, 805–811.

Intercellular Calcium Signaling in "Non-Excitable" Cells

Thierry Tordjmann, Caroline Clair, Michel Claret and Laurent Combettes

1. BACKGROUND

Communication is the governing word of our epoch, growth of networks such as the internet or mobile phone are tangible proof of our need to communicate. These human developments follow the normal evolution of life. Indeed, the first life forms were single cells that did not communicate with each other. During evolution, prokaryotic cells, and later eukaryotes, began to associate into multicellular colonies. The key advantages of this "multicellularity" are intercellular cooperation and cellular specialization. Communication between cells is a fundamental requirement for the social behavior of cells, facilitating cooperation between cells of the same specialized group (tissue or organ) and between different cell groups, thereby ensuring the coordination of complex and intricate functions. Cells communicate with each other by several non-exclusive pathways, which may be direct or indirect. Direct communication involves intercellular junctions (known as gap junctions), connecting the cytosols of the adjacent cells and facilitating the transfer of low-molecular weight molecules. Thus, the molecules that act as intracellular messengers (cAMP, $InsP_3$, Ca^{2+}) may also act on other cells, by passing from one cell to another via the gap junctions. Indirect communication involves the emission by certain cells of a biological messenger into the extracellular environment. Embryogenesis, tissue ontogeny, cell growth and regeneration and the coordination of many tissue and cell functions could not occur in the absence of direct or indirect communication between cells.

The concentration of free Ca^{2+} in the cytosol ($[Ca^{2+}]_i$) is actively kept much lower (100 to 200 nM) than extracellular (1 to 2 mM) and intrareticular

Thierry Tordjmann, Caroline Clair, Michel Claret and Laurent Combettes • INSERM U442, UPS, Orsay, France.

R. Pochet, R. Donato, J. Haiech, C. Heizmann and V. Gerke (eds.): Calcium: The Molecular Basis of Calcium Action in Biology and Medicine, 95–108.

(0.5 mM) Ca^{2+} concentrations. The cytosol, at the interface of these two very calcium-rich environments, is a site of major, rapid variations in $[Ca^{2+}]_i$ in response to the transfer of small quantities of Ca^{2+} from the extracellular medium or intracellular storage compartments (Berridge, 1997). These variations are induced by hormones and neurotransmitters and are described as "calcium signals". It is becoming increasingly evident that the calcium signals produced by these agonists are extraordinarily well organized in both space and time, from the subcellular to whole tissue level (Miyazaki, 1995; Thomas et al., 1996; Berridge, 1997).

Fifteen years ago, Woods et al. (1986) showed that the Ca^{2+} signals in response to hormonal stimulation generally consists of a series of peaks in $[Ca^{2+}]_i$ (oscillations), with a period of a few seconds to a few minutes, depending on agonist concentration, since $[Ca^{2+}]_i$ oscillations have been observed in numerous cell types (for a review, see Thomas et al., 1996; Berridge, 1997). It has also been shown in various cell types that each Ca^{2+} peak is organized spatially: Ca^{2+} concentration first increases locally, then the increase is propagated in all the cell as a sort of wave, traveling at a speed of 10 to 20 $\mu m.s^{-1}$ (Miyazaki, 1995; Thomas et al., 1996; Berridge, 1997).

These intracellular movements of Ca^{2+}, induced by hormones and neurotransmitters, may be propagated from cell to cell creating an apparent intercellular wave (see Sanderson et al., 1994; Sanderson, 1995). We will focused this review on intercellular Ca^{2+} signaling in non-excitable cells with a particular attention to hepatocytes and the liver. Indeed, there is a wide range of experiments performed on Ca^{2+} signaling in isolated hepatocytes and the liver is the sole tissue where intercellular Ca^{2+} waves have been observed in the intact organ with no invasive techniques (see below).

2. INTERCELLULAR CALCIUM SIGNALING

Intercellular calcium signaling, also called intercellular calcium waves, were first observed in cells in culture, but have more recently been demonstrated in more highly integrated systems such as sections of cerebral tissue (Dani et al., 1992), preparations of intact retina (Newman and Zahs, 1997), the ocular ciliary epithelium (Hirata et al., 1998), the intestinal crypts (Lindqvist et al., 1998) and the isolated intact perfused liver (Nathanson et al., 1995; Robb-Gaspers and Thomas, 1995; Motoyama et al., 1999). The precise functions initiated or regulated by these waves in the various tissues in which they have been observed are still mostly a matter for speculation. In addition, the mechanisms by which they are propagated and coordinated, which are beginning to be elucidated for certain cell types only, involve several factors that cannot be generalized and applied to all tissues. As described above, the cells of a multicellular organism may communicate directly via gap junctions (the "junctional coupling" pathway) or indirectly via a chemical messenger that is released by the cell into the extracellular medium, where it stimulates a target cell (paracrine, endocrine and synaptic pathways). Intercellular waves may be propagated via one or several of these pathways (Figure 1).

3. JUNCTIONAL COUPLING PATHWAY

This has been the most studied pathway in early investigations of the intercellular propagation of calcium signals. The gap junctions are made up of an assembly of several "hemichannels", each consisting of six transmembrane proteins (connexins) associated in a unit called a "connexon". The connexons of adjacent cells align to form true channels, providing a means of direct communication between the cytosols of the two cells (for more details on structure and functions of connexins and gap junction, see Bruzzone et al., 1996). Molecules – less than 1.5 nm diameter or 1200 Da – that pass through the junction are ions (facilitating the electrical coupling of cells), second messengers (facilitating synchronization of cell function), metabolites and substrates (which may be involved in intercellular "metabolic collaboration"). In excitable cells, a major function of gap junctions is the rapid relay of currents generated at the plasma membrane by the passage of K^+, Na^+ or Cl^- ions between cells. This electrical coupling is involved in the synchronization of calcium signals in certain types of cell (Steinberg et al., 1998). In unexcitable cells, the principal role of the gap junctions is the exchange of signals and metabolites.

Although it has not been formally demonstrated, it is widely believed that intercellular calcium waves mediated by the diffusion of a messenger through gap junctions occur in hepatocytes (Nathanson and Burgstahler, 1992; Thomas et al., 1996) and in many other types of cell (Sanderson, 1995; Thomas et al., 1996). Several lines of evidence support the existence of an intercellular calcium wave mediated by the diffusion of a messenger across junctions. As pointed out by Sanderson et al. (1994) and Sanderson (1995), the propagation of intercellular calcium waves is often discontinuous. The intracellular wave propagates continuously, stops at the boundary separating the two cells and then starts again in the adjacent cell, close to the intercellular boundary, suggesting the involvement of a direct intercellular communication pathway in the propagation of calcium waves. Moreover, gap junction inhibitors such as octanol or 18α-glycyrrhetinic acid, reduce or abolish the propagation of these intercellular calcium waves. In the same way, the distance traveled by intercellular calcium waves from a mechanically stimulated cell is correlated with the quantity of functional connexins expressed in the cells (Enkvist and McCarthy, 1992; Charles et al., 1992; Toyofuku et al., 1998).

Most of these experimental results come from studies investigating the propagation of calcium waves in monolayers of cultured cells following the mechanical stimulation of one cell. This situation mimics only very specific physiological situations, such as stimulation of epithelial ciliate cells of the

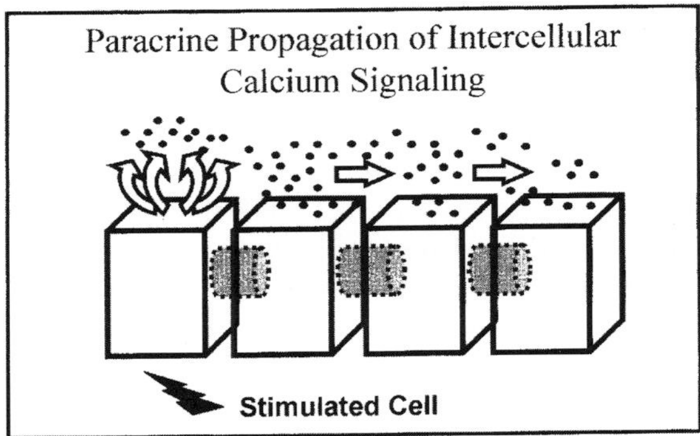

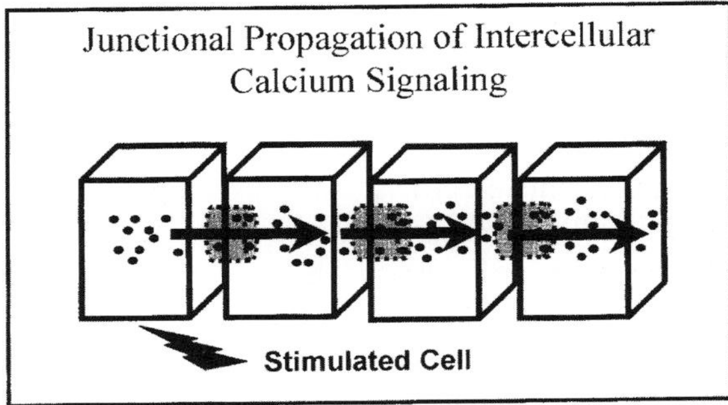

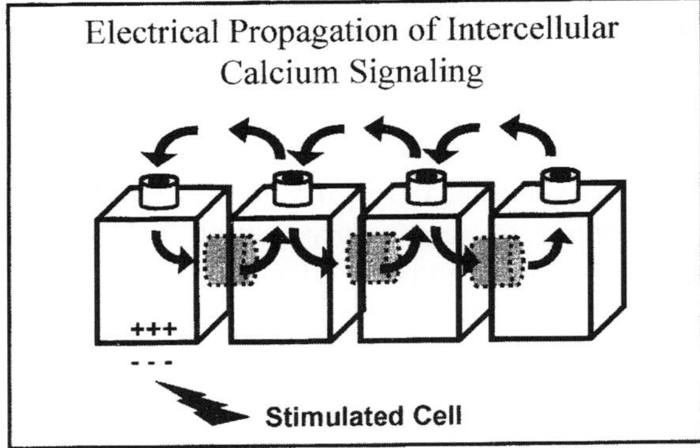

Figure 1. Schematic representation of different possible mechanisms for intercellular propagation of calcium signaling. Gray boxes represent gap junction, white arrows represent the release of active molecules in the external medium (paracrine pathway). Black arrows represent molecules (junctional propagation) or ions (electrical propagation) moving through gap junctions or membrane channels.

upper respiratory tract, osteoblasts and chondrocytes in bone and endothelial cells in the bloodstream. However, most of the time in physiological conditions, tissues are entirely bathed with hormones. In studies which have been carried out with cells globally stimulated by Ca^{2+}-mobilizing agonists, the inhibition of junctional coupling results in the desynchronization of the calcium signals (Nathanson and Burgstahler, 1992; Stauffer et al., 1993; Tordjmann et al., 1997).

4. PARACRINE PATHWAY

A paracrine pathway may be involved in the propagation of intercellular calcium signals in many cell types and lines, independently of, or in association with, junctional coupling. The principal characteristic of this type of propagation is that the distance over which the wave can travel is limited by the extracellular diffusion of the messenger and that the intercellular wave follows the direction of rinsing. Several candidates for extracellular paracrine messengers have been suggested. Osipchuk and Cahalan (1992) first proposed that ATP is involved in the propagation of the Ca^{2+} signal because the addition of suramine, an antagonist of ATP receptors, prevents propagation of the Ca^{2+} signal in rat basophilic leukemia cells. Studies with cell cultures have shown that this mechanism may also account for the propagation of Ca^{2+} signals in many other different cell types such as mammary epithelial cells (Enomoto et al., 1994), astrocytes (Charles, 1998; Giaume and Venance, 1998) or hepatocytes (Schlosser et al., 1996). Other extracellular messengers like substance P and/or endothelin in endothelial cells (Himmel et al., 1993) or glutamate in astrocytes (Parpura et al. 1994), may also be involved in the propagation of intercellular calcium signals.

5. THE ROLES OF Ca^{2+} AND $InsP_3$ IN THE PROPAGATION OF INTERCELLULAR CALCIUM SIGNALS

The intercellular exchange of second messengers is considered to be a fundamental property of gap junctions. It has been shown that cAMP diffuses across gap junctions (Hempel et al., 1996). For intercellular calcium waves, Ca^{2+} and $InsP_3$ are the two most likely candidates for the intercellular messenger.

5.1. Role of Ca^{2+}

In 1989, Saez and colleagues showed that the injection of Ca^{2+} or $InsP_3$ into the cells of a hepatocyte doublet induced an increase in $[Ca^{2+}]_i$ in the adjoining cell and that this effect was inhibited by octanol (Saez et al., 1989). Many studies have since shown that intracellular messengers such as $InsP_3$ and Ca^{2+} can cross gap junctions in physiological conditions. However, there is considerable evidence against the primary involvement of Ca^{2+} in the propagation of intercellular waves. Ca^{2+} diffuses very poorly in the cytosol due to the abundance of immobile Ca^{2+} buffers (Allbritton et al., 1992), thus the intercellular Ca^{2+} wave, like the intracellular Ca^{2+} wave, cannot by accounted for by the simple diffusion of Ca^{2+} alone. Moreover, in numerous cell types, the calcium oscillations may be restricted to one cell, with no propagation to adjacent coupled cells (Stauffer et al., 1993; Tordjmann et al. 1997) and asynchronous calcium oscillations may occur within a system of coupled cells, independently of any intercellular calcium wave (Sanderson, 1995; Giaume and Venance 1998). These observations demonstrate that the simple diffusion of Ca^{2+} across gap junctions is not essential to the development of intercellular calcium waves.

5.2. Role of $InsP_3$

$InsP_3$ diffuses much more readily than Ca^{2+} in the cytosol (Allbritton et al., 1992) and is able to cross gap junctions and drive propagation of the intercellular calcium wave. The injection of $InsP_3$ into a cell may induce a calcium response not only in the cell injected, but also in the adjacent coupled cell or cells (Saez et al., 1989; Sanderson et al., 1990) and although it is not possible to follow directly the intercellular $InsP_3$ diffusion, it has been shown that caged $InsP_3$ can cross gap junctions in certain types of cell (Carter et al., 1996; Tordjmann et al., 1998; Hirata et al., 1998). Finally, studies using inhibitors of the $InsP_3$ pathway, such as heparin, which inhibits the binding of $InsP_3$ to its receptor, and U73122, which inhibits phospholipase C, have provided indirect experimental evidence for the involvement of a mechanism relayed by $InsP_3$ in the propagation of intercellular calcium waves (Giaume and Venance, 1998; Domenighetti et al., 1998).

Sneyd et al. (1998) developed a mathematical model formalizing Ca^{2+} wave propagation in which the "passive diffusion" of $InsP_3$ drives propagation of the intercellular wave. In this model, the $InsP_3$ generated by the mechanical stimulation of a cell diffuses across gap junctions, triggering the release of intracellular Ca^{2+} successively in adjacent cells (Figure 2). As these authors pointed out, the passive diffusion of $InsP_3$, which has been shown to occur in certain epithelial cells following mechanical stimulation, cannot account for the propagation of calcium waves in all types of cell. Indeed, the large distances covered by intercellular calcium waves in certain tissues and cell cultures and the tendency for the speed and amplitude of the Ca^{2+} signal to remain constant suggest that there is an autocatalytic process in the various cells "traversed" by the waves. Two mechanisms which are thought to be involved in the propagation of intracellular waves, may be implicated in this regenerative process. Firstly, $InsP_3$ may be regenerated via the stimulation of

Hypothesis for Initiation and Propagation of Intercellular Calcium Signaling

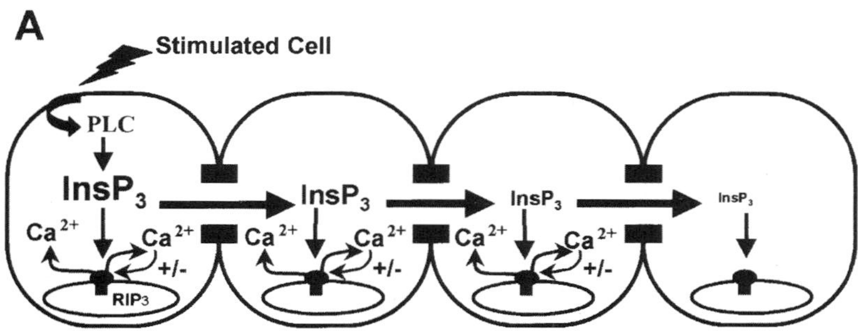

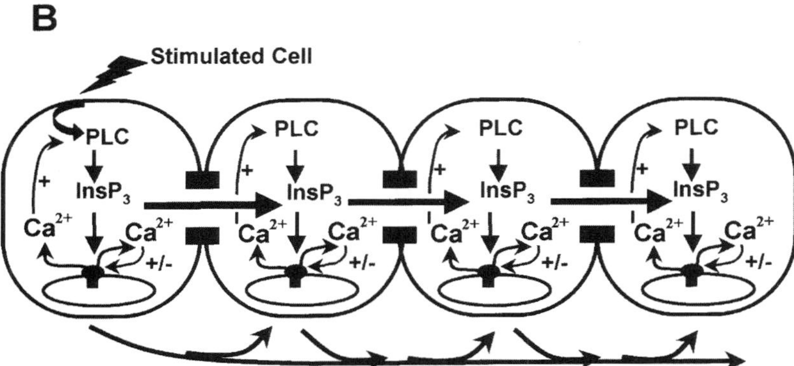

Figure 2. Hypothesis for the propagation of intercellular calcium signaling. (A) Passive propagation of intercellular calcium signaling between adjacent cells through gap junctions. After $InsP_3$ is produced in the stimulated cell, it induces the release of calcium from internal stores, in the form of an *intracellular* calcium wave. This wave is propagated by the combined action of $InsP_3$ and Ca^{2+} which transforms the cytoplasm into an "excitable" environment favoring the development of calcium waves. $InsP_3$ is believed to move through gap junctions, initiating similar calcium waves in adjacent cells. This phenomenon reproduces itself in successive cells, as long as enough $InsP_3$ enters the cell to induce a calcium wave. This model accounts for the propagation of a restricted intercellular calcium wave after focal stimulation of a single cell as described by Sneyd et al. (1994). (B) Regenerative propagation of intercellular calcium signaling. In this case, in adjunction to the $InsP_3$ diffusion descibed above (A), an "autocatalytic process" occurs and may involve regeneration of $InsP_3$,via the stimulation of phospholipase C (PLC) by Ca^{2+}, or the release of a Ca^{2+}-mobilizing substance into the extracellular medium, or both mechanisms.

phospholipase C by Ca^{2+}. Evidence for such a mechanism has been obtained in experiments with astrocytes, showing that a large increase in $[Ca^{2+}]_i$ leads to the activation of phospholipase C, resulting in $InsP_3$ production and that the inhibition of phospholipase C results in the inhibition of regenerative intercellular calcium waves (Giaume and Venance, 1998). Secondly, Ca^{2+} is known to be the principal regulator of its own release (for a review, see Taylor, 1998). Ca^{2+} may therefore play a major role in the propagation of regenerative intercellular calcium waves via the stimulation of $InsP_3$ production, and/or through an increase in the effect of $InsP_3$ on its receptor (Figure 2). Recent experiments have shown that both these mechanisms are required to account for this type of propagation (Domenighetti et al., 1998). Moreover, in many cell types, a combination of the junctional pathway and the paracrine pathway may be involved in regenerative intercellular calcium signals (Charles, 1998; Giaume and Venance, 1998).

Finally, intracellular messengers other than Ca^{2+} and $InsP_3$ may be involved in the propagation of intercellular calcium signals. For example, cAMP modifies both the permeability of gap junctions (Lowenstein, 1985) and the sensitivity of $InsP_3R$ to $InsP_3$ (Taylor, 1998). These two effects may control the distance traveled by intercellular Ca^{2+} signals. Other molecules meeting particular criteria (size and capacity to induce the release of intracellular Ca^{2+}), such as cyclic ADP ribose, may interfere more directly in the propagation of these signals (Churchill and Louis, 1998).

6. INTERHEPATOCYTIC CALCIUM SIGNALS: A RECEPTOR-ORIENTED WAVE

As we have seen, most studies of the propagation of intercellular calcium signals have been carried out with cultured cells which clearly do not accurately reflect the situation in tissues. The most important aspects of tissue organization and of the connections between cells *in situ* are not taken into account and so the results obtained cannot be directly extrapolated to the intact tissue or organ. Studying intercellular calcium signals in the liver has made it possible to demonstrate a direct relationship between the organization of the liver tissue and of interhepatocytic calcium signals.

The various studies carried out with freshly isolated hepatocyte doublets and triplets, which can be considered as fragments of hepatocyte plates, and with intact perfused liver, have provided important evidence that Sneyd's model (1998) cannot be transposed to interhepatocytic calcium signals and that a paracrine pathway cannot be responsible for the propagation of these waves (Tordjmann et al., 1997; Tordjmann et al., 1998). The calcium responses induced by $InsP_3$-dependent agonists (noradrenaline and vasopressin) in these cells are typically coordinated and sequential (Nathanson and Burgstahler, 1992; Combettes et al. 1994; Nathanson et al., 1995; Robb-Gaspers and Thomas, 1995; Motoyama et al., 1999). The order in which the

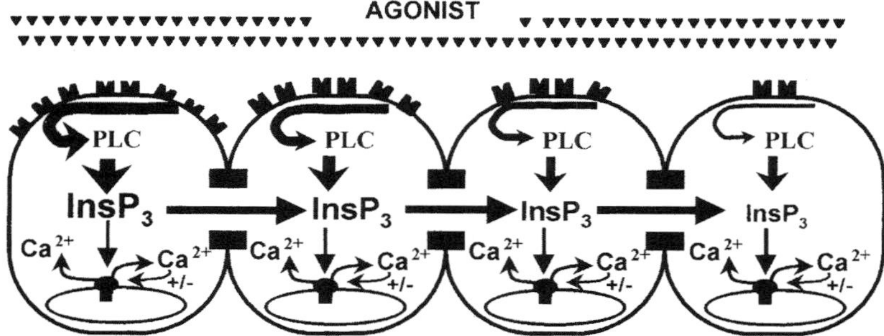

Figure 3. Schematic model of the receptor-oriented and coordinated intercellular Ca^{2+} waves in rat hepatocytes. Gradient of agonist receptors along the liver cell plate leads to a gradient in the $InsP_3$ production (represented by letter sizes) and thus, determines the direction of the wave. $InsP_3$, via the $InsP_3$ receptor, releases Ca^{2+} from the endoplasmic reticulum (ER). Diffusion of small amounts of $InsP_3$ through gap junctions reduces the cell-to-cell differences in the levels of $InsP_3$ and induces coordinated Ca^{2+} oscillations.

cells respond to the agonist is the same at each oscillation, over prolonged periods of recording, and appears to be conserved regardless of agonist concentration (Combettes et al., 1994).

The authors have shown that this pattern of Ca^{2+} response results from an heterogeneity in $InsP_3$ production between hepatocytes which creates the conditions necessary for the intercellular diffusion of $InsP_3$, which Sneyd et al. (1998) described as the driving force behind the calcium wave. Even if the diffusion across gap junctions of one or several messengers is essential for the coordination of oscillating calcium signals between various coupled cells, it cannot in itself generate the propagation of calcium signals from one hepatocyte to another (Tordjmann et al., 1997). In fact, gradient in $InsP_3$ production is due to a gradient in the number of hormone receptors across the hepatocyte plate (Tordjmann et al., 1998). The sequential nature of calcium responses and thus the direction of the apparent intercellular calcium wave results from this heterogeneity, creating, as shown in Figure 3, a "receptor-oriented" Ca^{2+} wave (Tordjmann et al., 1998; Dupont et al. 2000).

Such an organization should be of major functional advantage. In certain tissues, such as the liver, intercellular gradients may themselves support one or several functions (see below). The capacity of the various cells to respond to a stimulus (globally applied to all the cells) in a particular order, despite the existence of junctional coupling, may make it possible to regulate the direction of the hormonal response, from cell to cell. These functional characteristics may be of importance, not only in the liver, but also in other epithelial tissues, for the fine regulation of intercellular communication.

7. RECEPTOR ORIENTED INTERCELLULAR CALCIUM SIGNALS IN OTHER TISSUES

It is possible that gradients in the sensitivity of cells to agonists exist in non-hepatic tissues, particularly the pancreas and intestinal crypts.

The pancreatic parenchyma, which is both exocrine and endocrine in nature, is composed of functionally heterogeneous cells (Moitoso de Vargas et al., 1997). The heterogeneity of these cells, which appears to be less organized than that of hepatocytes, may account for the directional nature of the propagation of the intercellular calcium waves sometimes observed in the acini and islets (Bertuzzi et al., 1996).

In the intestinal villi, enterocytes mature along the cryptovillous axis, which may account for the functional differences observed between adjacent cells. Intercellular calcium waves have recently been reported in this cell model (Lindqvist et al., 1998). These waves begin in the crypts and migrate towards the tips of the villi. It is possible that a gradient in the expression of certain hormone receptors along the cryptovillous axis is responsible for these intercellular calcium waves.

8. FUNCTIONS OF INTERCELLULAR CALCIUM SIGNALS

Although our understanding of the mechanisms by which intercellular Ca^{2+} signals are propagated is increasing, the functions of these waves are still largely unknown. The most simple view is that the "goal" of the propagation of a calcium signal is the generalization to all neighboring cells of a response generated locally in an individual cell. The calcium wave would thus coordinate the execution of a particular function by this group of cells. For example, the clearance of the mucus covering the ciliated epithelium of the upper respiratory airways may be regulated by intercellular calcium waves. It has been demonstrated that the beating frequency of the cilia at the apical pole of these epithelial cells and the coordination of their movement in a cell monolayer depend on mechanically generated intercellular calcium waves (Sanderson et al., 1990). Intercellular calcium signals may be also involved in the regulation of blood flow by endothelial cells, the regulation of lactation, the osteogenesis and during the initial steps of tissue repair (Sanderson et al., 1994, 1995). The intercellular calcium waves observed in the glial cells may regulate certain neuronal functions and play a key role in the development of migraines (Giaume and Venance, 1998; Charles, 1998). Ca^{2+} signaling has been shown to modulate cell proliferation in a number of tissues (Berridge 1995) and so the transfer between cells of calcium signals may be involved in the normal cell growth and in the proliferation of cells in culture (Charles et al., 1992). Finally, oscillations in $[Ca^{2+}]_i$ have been shown to have a direct effect on the expression of numerous developmentally regulated genes in cultured cells (Hardingham and Bading, 1999). In this context, intercellular calcium signals are one possible mechanism for the coordination of cellular processes during embryogenesis (Gilland et al., 1999).

In the liver, intercellular calcium waves may be thought of as a means of transmitting a signal received in one zone of the plate to other hepatocytes. In the rat, following stimulation of hepatic nerves, the release of neurotransmitters (particularly noradrenaline and ATP) induces calcium signals in the periportal zone that are transmitted to the perivenous zone exclusively via the gap junctions (Nelles et al., 1996). Propagation of stimulation to the entire lobe leads in particular to the production of large amounts of glucose induced by noradrenaline and ATP. Interhepatocytic calcium waves may also themselves be direct signals for Ca^{2+}-dependent functions that require the directional stimulation of the hepatocyte plate. Canalicular contraction, a Ca^{2+}-dependent phenomenon, may be propagated in the hepatic lobule, from the perivenous zones to the portal spaces, via interhepatocytic calcium signals induced by certain agonists, such as vasopressin, noradrenaline and angiotensin II (Tordjmann et al., 1998). The same may also be true for biliary secretion, certain steps of which are dependent on Ca^{2+}.

9. CONCLUSION

Calcium signals generated by $InsP_3$-dependent agonists are a means of transduction widely used in a large number of cell types. These signals reflect a very complex organization in both space and time, at the cellular and multicellular levels. In response to agonists, oscillating intracellular and intercellular calcium waves are propagated from a precise initiation point (or in a precise cell) and may progress in a non-random direction. Mechanisms of propagation and coordination of intercellular calcium signals in non-excitable cells are beginning to be understood but the functions they regulate, remain largely unclear. There is no doubt however that these signals, which appear to be a general phenomenon observed virtually in all cell types, are fundamental elements of cell communication.

ACKNOWLEDGEMENTS

We thank J. Knight and C. Cruttwell for their help in editing the manuscript. Our work is supported by the Association pour la Recherche sur le Cancer (ARC 5457).

REFERENCES

Allbritton, N.L., Meyer, T. and Stryer, L., 1992, Range of messenger action of calcium ion and inositol 1,4,5-trisphosphate, *Science* 258, 1812–1815.

Berridge, M.J., 1995, Calcium signalling and cell proliferation, *Bioessays* 17, 491–500.

Berridge, M.J., 1997, Elementary and global aspects of calcium signalling, *J. Physiol.* 499, 291–306.

Bertuzzi, F., Zacchetti, D., Berra, C., Socci, C., Pozza, G., Pontiroli, A.E. and Grohovaz, F., 1996, Intercellular Ca^{2+} waves sustain coordinate insulin secretion in pig islets of Langerhans, *FEBS Lett.* 379, 21–25.

Bruzzone, R., White, T.W. and Paul, D.L., 1996, Connections with connexins: The molecular basis of direct intercellular signaling, *Eur. J. Biochem.* 238, 1–27.

Carter, T.D., Chen, X.Y., Carlile, G., Kalapothakis, E., Ogden, D. and Evans, W H., 1996, Porcine aortic endothelial gap junctions: Identification and permeation by caged $InsP_3$, *J. Cell Sci.* 109, 1765–1773.

Charles, A.C., Naus, C.C., Zhu, D., Kidder, G.M., Dirksen, E.R. and Sanderson, M.J., 1992, Intercellular calcium signaling via gap junctions in glioma cells, *J. Cell Biol.* 118, 195–201.

Charles, A., 1998, Intercellular calcium waves in glia, *Glia* 24, 39–49.

Churchill, G. and Louis, C., 1998, Roles of Ca^{2+}, inositol trisphosphate and cyclic ADP-ribose in mediating intercellular Ca^{2+} signaling in sheep lens cells, *J. Cell Sci.* 111, 1217–1225.

Combettes, L., Tran, D., Tordjmann, T., Laurent, M., Berthon, B. and Claret, M., 1994, Ca^{2+}-mobilizing hormones induce sequentially ordered Ca^{2+} signals in multicellular systems of rat hepatocytes, *Biochem. J.* 304, 585–594.

Dani, J.W., Chernjavsky, A. and Smith, S.J., 1992, Neuronal activity triggers calcium waves in hippocampal astrocyte networks, *Neuron* 8, 429–440.

Domenighetti, A.A., Beny, J.L., Chabaud, F. and Frieden, M., 1998, An intercellular regenerative calcium wave in porcine coronary artery endothelial cells in primary culture, *J. Physiol. (Lond.)* 513, 103–116.

Dupont, G., Tordjmann, T., Clair, C., Swillens, S., Claret, M. and Combettes, L., 2000, Mechanism of receptor-oriented intercellular calcium wave propagation in hepatocytes, *FASEB J.* 14, 279–289.

Enkvist, M.O.K. and McCarthy, K.D., 1992, Activation of protein kinase C blocks astroglial gap junction communication and inhibits the spread of calcium waves, *J. Neurochem.* 59, 519–526.

Enomoto, K., Furuya, K., Yamagishi, S., Oka, T. and Maeno, T., 1994, The increase in the intracellular Ca^{2+} concentration induced by mechanical stimulation is propagated via release of pyrophosphorylated nucleotides in mammary epithelial cells, *Pflügers Arch.* 427, 533–542.

Giaume, C. and Venance, L., 1998, Intercellular calcium signaling and gap junctional communication in astrocytes, *Glia* 24, 50–64.

Gilland, E., Miller, A.L., Karplus, E., Baker, R. and Webb, S.E., 1999, Imaging of multicellular large-scale rhythmic calcium waves during zebrafish gastrulation, *Proc. Natl. Acad. Sci. USA* 96, 157–161.

Hardingham, G.E. and Bading, H., 1999, Calcium as a versatile second messenger in the control of gene expression, *Microsc. Res. Tech.* 46, 348–355.

Hempel, C.M., Vincent, P., Adams, S.R., Tsien, R.Y. and Selverston, A.I., 1996, Spatio-temporal dynamics of cyclic AMP signals in an intact neural circuit, *Nature* 384, 166–169.

Himmel, H.M., Whorton, A.R. and Strauss, H.C., 1993, Intracellular calcium, currents, and stimulus-response coupling in endothelial cells, *Hypertension* 21, 112–127.

Hirata, K., Nathanson, M.H. and Sears, M.L., 1998, Novel paracrine signaling mechanism in the ocular ciliary epithelium, *Proc. Natl. Acad. Sci. USA* 95, 8381–8386.

Lindqvist, S.M., Sharp, P., Johnson, I., Satoh, Y. and Williams, M., 1998, Acetylcholine-induced calcium signaling along the rat colonic crypt axis, *Gastroenterology* 115, 1131–1143.

Lowenstein, W.R., 1985, Regulation of cell-to-cell communication by phosphorylation, *Biochem. Soc. Symp.* 50, 43–58.

Miyazaki, S., 1995, Inositol trisphosphate receptor-mediated spatiotemporal calcium signalling, *Curr. Opin. Cell Biol.* 7, 190–196.

Moitoso De Vargas, L., Sobolewski, J., Siegel, R. and Moss, G.L., 1997, Individual β cells within the intact islet differentially respond to glucose, *J. Biol. Chem.* 272, 26573–26577.

Motoyama, K., Karl, I.E., Flye, M.W., Osborne, D.F. and Hotchkiss, R.S., 1999, Effect of Ca^{2+} agonists in the perfused liver: Determination via laser scanning confocal microscopy, *Am. J. Physiol.* 276, R575–R585.

Nathanson, M.H. and Burgstahler, A.D., 1992, Coordination of hormone-induced calcium signals in isolated rat hepatocyte couplets – Demonstration with confocal microscopy, *Mol. Biol. Cell.* 3, 113–121.

Nathanson, M.H., Burgstahler, A.D., Mennone, A., Fallon, M.B., Gonzalez, C.B. and Saez, J.C., 1995, Ca^{2+} waves are organized among hepatocytes in the intact organ, *Am. J. Physiol.* 32, G167–G171.

Nelles, E., Butzler, C., Jung, D., Temme, A., Gabriel, H.D., Dahl, U., Traub, O., Stumpel, F., Jungermann, K., Zielasek, J., Toyka, K.V., Dermietzel, R. and Willecke, K., 1996, Defective propagation of signals generated by sympathetic nerve stimulation in the liver of connexin32-deficient mice, *Proc. Natl. Acad. Sci. USA* 93, 9565–9570.

Newman, E.A. and Zahs, K.R., 1997, Calcium waves in retinal glial cells, *Science* 275, 844–847.

Osipchuk, Y. and Cahalan, M., 1992, Cell-to-cell spread of calcium signals mediated by ATP receptors in mast cells, *Nature* 359, 241–244.

Parpura, V., Basarsky, T.A., Liu, F., Jeftinija, K., Jeftinija, S. and Haydon, P.G., 1994, Glutamate-mediated astrocyte-neuron signalling, *Nature* 369, 744–747.

Robb-Gaspers, L.D. and Thomas, A.P., 1995, Coordination of Ca^{2+} signaling by intercellular propagation of Ca^{2+} waves in the intact liver, *J. Biol. Chem.* 270, 8102–8107.

Saez, J.C., Connor, J.A., Spray, D.C. and Bennett, M.V.L., 1989, Hepatocyte gap junctions are permeable to the 2nd messenger, inositol 1,4,5-trisphosphate, and to calcium ions, *Proc. Natl. Acad. Sci. USA* 86, 2708–2712.

Sanderson, M.J., 1995, Intercellular calcium waves mediated by inositol trisphosphate, *Ciba Found Symp.* 188, 175–189.

Sanderson, M.J., Charles, A.C. and Dirksen, E.R., 1990, Mechanical stimulation and intercellular communication increases intracellular Ca^{2+} in epithelial cells, *Cell. Regul.* 1, 585–596.

Sanderson, M.J., Charles, A.C., Boitano, S. and Dirksen, E.R., 1994, Mechanisms and functions of intercellular calcium signaling, *Mol. Cell. Endocrinol.* 98, 173–187.

Schlosser, S.F., Burgstahler, A.D. and Nathanson, M.H., 1996, Isolated rat hepatocytes can signal to other hepatocytes and bile duct cells by release of nucleotides, *Proc. Natl. Acad. Sci. USA* 93, 9948–9953.

Sneyd, J., Wilkins, M., Strahonja, A. and Sanderson, M.J., 1998, Calcium waves and oscillations driven by an intercellular gradient of inositol (1,4,5)-trisphosphate, *Biophys. Chem.* 72, 101–109.

Stauffer, P.L., Zhao, H., Luby-Phelps, K., Moss, R.L., Star, R.A. and Muallem, S., 1993, Gap junction communication modulates $[Ca^{2+}]_i$ oscillations and enzyme secretion in pancreatic acini, *J. Biol. Chem.* 15, 268, 19769–19775.

Steinberg, T.H., Civitelli, R., Beyer, E.C., Jorgensen, N.R., Cao, D., Geist, S.T. and Lin, G., 1998, Multiple mechanisms for intercellular calcium waves, in *Gap Junctions*, R. Werner (ed.), IOS Press, pp. 271–275.

Taylor, C.W., 1998, Inositol trisphosphate receptors: Ca^{2+} modulated intracellular Ca^{2+} channels, *Biochim. Biophys. Acta* 1436, 19–33.

Thomas, A.P., Bird, G.S.J., Hajnoczky, G., Robb-Gaspers, L.D. and Putney, J.W., 1996, Spatial and temporal aspects of cellular calcium signalling, *FASEB J.* 10, 1505–1517.

Tordjmann, T., Berthon, B., Claret, M. and Combettes, L., 1997, Coordinated intercellular calcium waves induced by noradrenaline in rat hepatocytes: Dual control by gap junction permeability and agonist, *EMBO J.* 16, 5398–5407.

Tordjmann, T., Berthon, B., Jacquemin, E., Clair, C., Stelly, N., Guillon, G., Claret, M. and Combettes, L., 1998, Receptor-oriented intercellular calcium waves evoked by vasopressin in rat hepatocytes, *EMBO J.* 17, 4695–4703.

Toyofuku, T., Yabuki, M., Otsu, K., Kuzuya, T., Hori, M. and Tada, M., 1998, Intercellular calcium signaling via gap junction in connexin-43-transfected cells, *J. Biol. Chem.* 273, 1519–1528.

Woods, N.M., Cuthbertson, K.S. and Cobbold, P.H., 1986. Repetitive transient rises in cytoplasmic free calcium in hormone-stimulated hepatocytes, *Nature* 329, 719–721.

The Ca^{2+}-Mobilizing Second Messenger Cyclic ADP-Ribose

Andreas H. Guse

ABBREVIATIONS

ADPR	adenosine diphosphoribose;
$[Ca^{2+}]_i$	free intracellular Ca^{2+}-concentration;
cADPR	cyclic adenosine diphosphoribose;
cArisDPR	cyclic aristeromycin diphosphoribose;
cATPR	cyclic adenosine triphosphoribose;
cGDPR	cyclic guanosine diphosphoribose;
cGMP	$3',5'$-cyclic guanosine monophosphate;
IC_{50}	halfmaximal inhibitory concentration;
$InsP_3$	D-*myo*-inositol 1,4,5-trisphosphate;
ER	endoplasmic reticulum;
$NAADP^+$	nicotinic acid adenine dinucleotide phosphate;
NADase	NAD^+-glycohydrolase;
$NAD(P)^+$	nicotinamide adenine dinucleotide (phosphate);
NGD^+	nicotinamide guanine dinucleotide;
PLC	phospholipase C;
RyR	ryanodine receptor(s);
TCR/CD3 complex	T cell receptor/CD3 complex.

1. INTRODUCTION

Signal transduction via elevation of the intracellular free Ca^{2+}-concentration ($[Ca^{2+}]_i$) is one of the fundamental events observed in cell regulation. Cellular systems as diverse as oocytes, muscle cells, neurons and cells of the immune system, but also plant cells and protozoa utilize Ca^{2+}-signaling to

Andreas H. Guse • Division of Cellular Signal Transduction, Institute for Medical Biochemistry and Molecular Biology, University Hospital Eppendorf, University of Hamburg, Hamburg, Germany.

R. Pochet, R. Donato, J. Haiech, C. Heizmann and V. Gerke (eds.): Calcium: The Molecular Basis of Calcium Action in Biology and Medicine, 109–128.

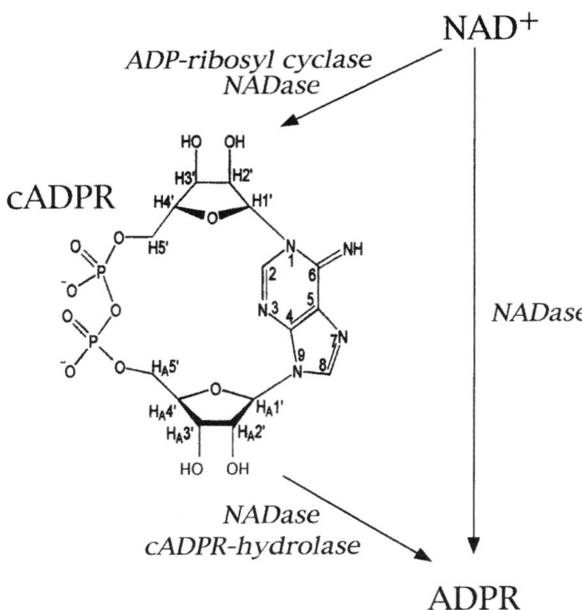

Figure 1. Structure and metabolic pathways of cADPR. NAD^+ can be converted to cADPR by both ADP-ribosyl cyclases and NADases, although the known NADases produce only a small quantity of cADPR (usually < 10%). Catabolism of cADPR proceeds by cADPR-hydrolase and/or NADase.

fulfill their diverse physiological functions. This fundamental importance implies that the principle of Ca^{2+}-signaling – an elevation of the concentration of free Ca^{2+}-ions in the cytosol, the nucleus and perhaps also inside specific organelles – has been conserved during evolution. On the other hand, research in the field of Ca^{2+}-signaling during the last years has shown that multiple cellular mechanisms with a multitude of molecular players are involved in the regulation of $[Ca^{2+}]_i$. The main cellular mechanisms that directly increase $[Ca^{2+}]_i$ are Ca^{2+}-release from intracellular Ca^{2+}-stores and Ca^{2+}-entry across the plasma membrane. Re-uptake into Ca^{2+}-stores, active transport of Ca^{2+}-ions from the cytosol into the extracellular space and buffering by Ca^{2+}-binding proteins are the mechanisms to decrease $[Ca^{2+}]_i$.

Cyclic adenosine diphosphoribose (cADPR) is a potent calcium-mobilizing, endogenous compound which has been discovered by Lee and co-workers (Clapper et al., 1987; Lee et al., 1989). The structure of cADPR has been elucidated by NMR and mass spectroscopy in 1989 (Lee et al., 1989), and has been corrected by X-ray crystallography in 1994 (Lee, 1994; Figure 1). The involvement of cADPR in the regulation of intracellular Ca^{2+}-signaling has been reported for different cell types from many species covering protozoa, invertebrates and vertebrates, humans, and plants (reviewed in Lee, 1997, 1999; Dousa et al., 1996; Guse, 1999; Zhang et al., 1999). However, cADPR is not the only calcium-mobilizing second messenger. D-*myo*-inositol 1,4,5-trisphosphate ($InsP_3$; Berridge, 1993) and nicotinic acid adenine dinucleotide phosphate ($NAADP^+$, Aarhus et al., 1995) are also involved in the regulation of intracellular Ca^{2+} signaling. The biochemistry and physiology of $InsP_3$ have been investigated in great detail; the data obtained indicated that all or almost all aspects of receptor-mediated Ca^{2+}-signaling may be explained by the action of $InsP_3$ (Berridge, 1993). However, the discovery of further Ca^{2+}-mobilizing compounds, such as cADPR and $NAADP^+$, made clear that several molecular mechanisms appear to be involved in the regulation of Ca^{2+}-signaling. However, the exact roles of each of these messengers in the generation of the complex elementary and global pattern of Ca^{2+} signals observed in cells, e.g. spatial events like Ca^{2+} waves or temporal events like Ca^{2+} oscillations, have not yet been clearly defined.

The understanding of these complex Ca^{2+}-signaling pattern is further complicated, because some aspects of the biochemistry and physiology of cADPR remain to be clarified. This is even more the case for $NAADP^+$; the precise function and metabolism of this molecule are only beginning to be understood.

Looking at cADPR, the main issues currently discussed in a controverse manner are the identity and regulation of its synthesizing enzyme(s) and its intracellular receptor. Thus, in Section 2 of this chapter I shall discuss the current knowledge of metabolism of cADPR by ADP-ribosyl cyclases and cADPR-hydrolases, with special emphasis on the activation of ADP-ribosyl cyclases via stimulation of receptors in the plasma membrane. It is important to note here, that this part is not intended to merge different results into one unifying model. I feel it is much too early to judge whether individual models should be taken into account, or not.

Further parts of this chapter are the biochemistry and pharmacology of Ca^{2+}-release by cADPR via ryanodine receptor/Ca^{2+}-channels (RyR), the involvement of cADPR in Ca^{2+}-entry, and the physiological role of cADPR in cellular signal transduction, with special emphasis on fertilization of oocytes and activation of T-lymphocytes.

Although I will concentrate on results obtained in non-excitable cells, I will mention results from excitable cells as well.

2. METABOLISM OF cADPR BY ADP-RIBOSYL CYCLASES AND cADPR-HYDROLASES

The enzymes involved in metabolism of cADPR are ADP-ribosyl cyclases, cADPR-hydrolases and NAD^+-glycohydrolases (NADase; Table 1, Figure 1).

ADP-ribosyl cyclases from ovotestis of *Aplysia californica* (Hellmich and Strumwasser, 1991) and *A. kurodai* so far are the first and only enzymes exhibiting almost pure cyclase activity. The enzyme from *A. californica* has been used as a prototypic ADP-ribosyl cyclase, and many of its characteristics

Table 1. Enzymes involved in metabolism of cADPR

Enzyme name	Substrate	Product	By-product	Example
NAD^+-glycohydrolase	NAD^+ cADPR	ADPR ADPR	cADPR ?	from canine spleen (Kim et al., 1993)
ADP-ribosyl cyclase	NAD^+	cADPR		from *A. californica* (Hellmich and Strumwasser, 1991)
cADPR-hydrolase	cADPR	ADPR	?	*

*An enzyme displaying pure cADPR-hydrolase activity has not been purified so far. However, cADPR-hydrolase activity is present in many tissues, but may turn out to be performed by NAD^+-glycohydrolases.

have been investigated, e.g. it was cloned, recombinantly expressed and its 3D-structure and active site were determined (Prasad et al., 1996; Munshi et al., 1997, 1999). ADP-ribosyl cyclase from *A. californica* displays a relatively low substrate specificity. Besides NAD^+ other endogenous nucleotides, such as NADP+ are converted to the corresponding cyclic compounds (Aarhus et al., 1995; Zhang et al., 1995; Vu et al., 1996, Guse et al., 1997a).

Since ADP-ribosyl cyclase from *A. californica* can additionally catalyze a base-exchange reaction using $NADP^+$ as substrate, Lee recently introduced a "unified mechanism of enzymatic synthesis of two calcium messengers: cADPR and NAADP" based on structural and biochemical data obtained with the ADP-ribosyl cyclase from *A. californica* (Lee, 1999). The model suggests that ADP-ribosyl cyclase cyclizes NAD+ to cADPR, and in addition, in the presence of nicotinic acid, and preferentially at acidic pH, via a base exchange reaction forms $NAADP^+$ (from $NADP^+$; Lee, 1999). Although the cyclase from *A. californica* certainly can catalyze these reactions *in vitro*, the *in vivo* role of this enzyme is much less clear, e.g. it is so far unknown whether the enzyme can be activated upon stimulation of the cells in the ovotestis, and whether this would result in intracellular Ca^{2+}-signaling. However, at present it is unknown whether other soluble ADP-ribosyl cyclases for which stimulatory mechanisms, such as heterotrimeric G-protein-, Tyr-phosphorylation-, cGMP- or ATP-dependent activation (see below), have been proposed may share the bi-functional properties with the cyclase from *A. californica*. Purification of such stimulatable ADP-ribosyl cyclases is urgently required to answer these important questions in the future.

Except for the ADP-ribosyl cyclases from the two *Aplysia* species all other known cyclases display as their main enzyme activity a NADase activity which converts β-NAD^+ to adenosine diphosphoribose (ADPR).

A soluble NADase with remarkable properties was purified to homogeneity from bovine brain cytosol (Matsumura and Tanuma, 1998). Like the classical membrane-bound NADases, this NADase synthesized mainly

ADPR, but also relatively small amounts of cADPR were obtained (Matsumura and Tanuma, 1998). Once a threshold concentration of cADPR was reached, there was a massive increase in hydrolysis of cADPR indicating that the by-product cADPR itself may be an allosteric modulator that activates the intrinsic cADPR hydrolyzing activity of this NADase (Matsumura and Tanuma, 1998).

Other soluble ADP-ribosyl cyclase activities have been reported from sea urchin egg and human T-lymphocyte cytosolic preparations (Graeff et al., 1998; Guse et al., 1999).

However, soluble ADP-ribosyl cyclases or NADases are, at least up to now, a minority: most of such enzymes purified or characterized so far are membrane-bound, either found in intracellular membranes or the plasma membrane. Membrane-bound NADases which also exhibit ADP-ribosyl cyclase activity have been found in different spleen preparations, from membranes prepared from brain, smooth or cardiac muscle, from total membranes of NG 108-15 neuroblastoma x glioma hybrid cells, or from mitochondrial membranes (for references, see Guse, 1999). Another class of membrane-bound NADases which also produce small amounts of cADPR are the ectoenzymes CD38 (Howard et al., 1993) and CD157 (BST-1; Hirata et al., 1994), which are expressed on activated T- and B-lymphocytes, erythrocytes and natural killer cells or bone marrow stromal cells, respectively.

Different models have been proposed to explain receptor-mediated synthesis of cADPR in the context of Ca^{2+}-signaling in cells.

The soluble ADP-ribosyl cyclases from neurosecretory PC 12 cells and sea urchin eggs can be activated by extracellular addition of NO (Galione et al., 1993; Willmott et al., 1996; Clementi et al., 1996; Graeff et al., 1998). NO diffuses through the plasma membrane and results in formation of cGMP by activated guanylyl cyclase, activation of cGMP-dependent protein kinase and subsequent activation of the ADP-ribosyl cyclase (Figure 2).

Stimulation of the T cell receptor/CD3 (TCR/CD3) complex in human Jurkat T-lymphocytes activated a soluble ADP-ribosyl cyclase (Guse et al., 1999). Since tyrosine kinases play a major role in transducing signals via the TCR/CD3 complex, and since activation of the ADP-ribosyl cyclase was more pronounced in the presence of the Tyr-phosphatase inhibitor dephostatin, it is likely that Tyr-phosphorylation is involved in this process (Figure 2).

Another model involves the ectoenzyme CD38 (Figure 2). Overexpression of CD38 in pancreatic β-cells stimulated an increased production of a calcium mobilizing compound, likely cADPR, and increased insulin secretion (Kato et al., 1995). It is well accepted that high extracellular concentrations of glucose are taken up into pancreatic β-cells by the non-insulin dependent glucose transporter GLUT2. Subsequent aerobic metabolism of glucose leads to increased ATP production which in the classical model inhibits an ATP-dependent K^+-channel (Ashcroft and Ashcroft, 1992). Okamoto and co-workers have shown that high concentrations of ATP modulated the

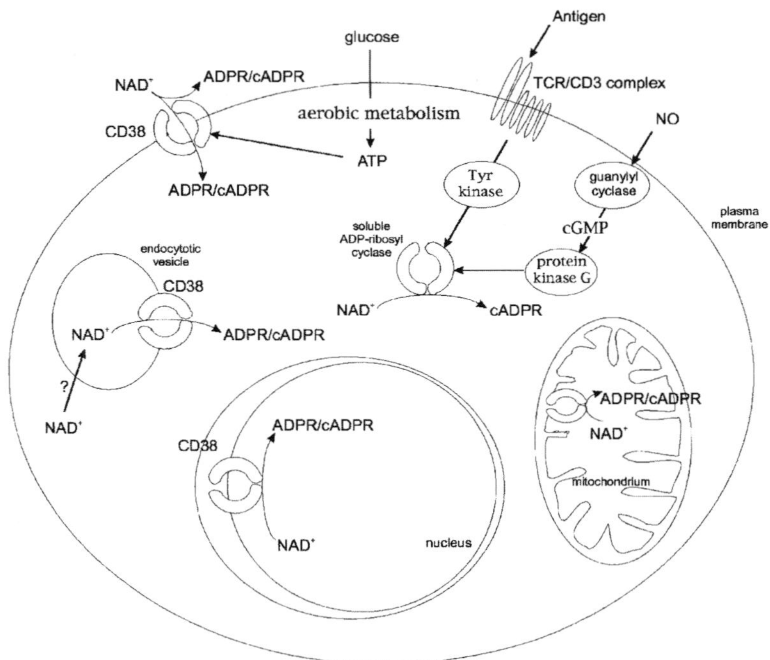

Figure 2. Activation pathways of ADP-ribosyl cyclases. In this schematical presentation different cell types have been merged into one cell to display postulated activation pathways of membrane-bound and soluble ADP-ribosyl cyclases. Individual activation pathways can be found in T-lymphocytes (via the TCR/CD3 complex), in pancreatic β-cells (via uptake of glucose), in sea urchin eggs and PC12 cells (via stimulation by nitric oxide [NO]), or in CD38 expressing cells. In individual cell types, CD38 has also been detected in endocytotic vesicles and in the inner nuclear membrane. An additional ADP-ribosyl cyclase has been found in mitochondria.

cyclase-to-hydrolase-ratio of CD38 in pancreatic β-cells towards the cyclase activity (Kato et al., 1995; Takasawa et al., 1993a) indicating an alternative route for Ca^{2+}-signaling stimulated by intracellularly elevated ATP. Most interestingly, autoantibodies against CD38 were detected in patients with non-insulin dependent diabetes; the anti-CD38 autoantibodies inhibited the ADP-ribosyl cyclase activity of CD38 as well as glucose-induced secretion of insulin from pancreatic islets (Ikehata et al., 1998). These results indicate an involvement of CD38 in cADPR formation in pancreatic β-cells.

However, three major problems with CD38 (and CD157) as cADPR-producing enzymes exist: (i) besides their ADP-ribosyl cyclase activity they display a major NAD^+-glycohydrolase and cADPR-hydrolase activity resulting mainly in formation of ADPR, (ii) both CD38 and CD157 are ectoenzymes with their active sites located in the extracellular space, and since cADPR is not membrane-permeant, it is not clear how the extracellularly produced cADPR would reach its intracellular target site(s), and (iii) the substrate NAD^+ is found in sufficient quantity within, but not outside of the cell.

The first problem may, at least in pancreatic β-cells, be solved by inhibition of the hydrolase activity of CD38 by increased ATP; in fact, an increase in ATP from 2 to 4 mM resulted in about 60% inhibition of the cADPR hydrolase activity (Kato et al., 1995). However, it remains to be established whether other cell types display comparable differences in intracellular ATP upon stimulation of membrane receptors.

A model to solve the second problem has been suggested by de Flora and co-workers (Zocchi et al., 1998, 1999; Franco et al., 1998). Transfection or retroviral infection of a cDNA for CD38 into CD38-negative 3T3- or HeLa cells resulted in intracellular accumulation of cADPR accompanied by a small increase in the basal $[Ca^{2+}]_i$ indicating that CD38 may constitute a transport system across the plasma membrane (Figure 2; Zocchi et al., 1998). Evidence for the latter was obtained in resealed membranes of CD38-positive human erythrocytes and in CD38 reconstituted proteoliposomes (Franco et al., 1998). Furthermore, it was shown in CD38-transfected HeLa cells that internalization of CD38 resulted in activation of influx of cytosolic NAD^+ into the endocytotic vesicles by a so far unknown dinucleotide transport system, and importantly, in intravesicular CD38-catalyzed conversion of NAD^+ into cADPR and its subsequent transport into the cytosol (Zocchi et al., 1999). The same stimuli which induced internalization (NAD^+, glutathione or N-acetylcysteine) also induced a somewhat increased intracellular Ca^{2+}-concentration (increase from about 20 to 80 nM; Zocchi et al., 1999).

However, these results could not be reproduced in other cell systems. In CD38-positive Jurkat T cells addition of NAD^+ to the extracellular medium did not result (i) in internalization of CD38, (ii) in significant increases in $[Ca^{2+}]_i$, or (iii) in significant increases in the intracellular cADPR concentration (da Silva et al., 1998a). Furthermore, no significant transport of extracellularly produced cADPR (or cGDPR as the product of NGD^+) into the Jurkat cells could be observed (da Silva et al., 1998a). It is important to note that there was substantial production of cADPR (and cGDPR) on the surface of the Jurkat cells as measured by HPLC (da Silva et al., 1998b), but extensive washing revealed that almost all of the cyclic products were on the outside and were not transported into the cell (da Silva et al., 1998a). In a further attempt to clarify any potential role of CD38 in T cell Ca^{2+}-signaling, recently we used different methods to inactivate or to inhibit CD38 on the surface of intact Jurkat cells without any significant inhibitory effect on TCR/CD3-mediated Ca^{2+}-signaling (Schweitzer, Deterre and Guse, unpublished results). Since an essential role for cADPR in TCR/CD3-mediated Ca^{2+}-signaling has been confirmed (Guse et al., 1999), these data suggest that CD38 does not play any major role in receptor-mediated cADPR formation in T cells.

In conclusion, there is experimental evidence for a variety of different models for receptor-mediated formation of cADPR. It may well turn out in the next years that different isoenzymes of ADP-ribosyl cyclase and/or NADase are involved in the different biological systems. More knowledge of the enzymology of ADP-ribosyl cyclases and NADases will hopefuly help to integrate the aspects discovered so far into a unifying model.

Attempts for pharmacological intervention in the cADPR signaling pathway have concentrated so far on the development of antagonists for cADPR (see below Section 3). However, also the second messenger producing ADP-ribosyl cyclases appear to be valuable targets to intervene in this important signaling pathway. ADP-ribosyl cyclase inhibitors have been recently introduced by Potter and co-workers (Migaud et al., 1999). Both C1′m-benzamide ribose adenine dinucleotide and C1′-m-benzamide ribose nicotinamide dinucleotide inhibited *A. californica* ADP-ribosyl cyclase at nanomolar concentrations (Migaud et al., 1999). The usefulness of ADP-ribosyl cyclase inhibitors in complex cellular systems should be investigated in the future.

3. BIOCHEMISTRY AND PHARMACOLOGY OF Ca^{2+}-RELEASE BY cADPR VIA RYANODINE RECEPTOR/Ca^{2+}-CHANNELS

Ca^{2+}-release mediated by cADPR has been discovered in sea urchin eggs in 1987 (Clapper et al., 1987). The responsiveness towards cADPR has been demonstrated in an increasing number of cell systems in the last 12 years (for comprehensive listing, the interested reader may refer to Lee, 1997). Most importantly, cADPR is not only active in invertebrates. More and more mammalian and human cells have been shown to respond to cADPR with Ca^{2+}-release from intracellular stores. In addition, cADPR-mediated Ca^{2+}-release was also found in one protozoon (*Euglena gracilis*; Masuda et al., 1997) and in plants (Wu et al., 1997) demonstrating the fundamental importance of the cADPR/Ca^{2+} pathway as an intracellular signaling system among living organisms.

To date it is not completely understood how cADPR exerts its Ca^{2+}-mobilizing effect inside the cell. In the majority of systems studied, there is pharmacological evidence that the Ca^{2+}-channel involved is the ryanodine receptor (RyR; Figure 3). It has been demonstrated that Ca^{2+}-induced and caffeine-induced Ca^{2+}-release are potentiated by cADPR (Lee, 1993), while ruthenium red and high Mg^{2+} concentrations inhibited cADPR-mediated Ca^{2+}-release (Galione et al., 1991; Guse et al., 1996). However, experimental data obtained so far have not shown direct binding of cADPR to one (or more) of the 3 RyR subtypes, and therefore, the possibility remains that cADPR-binding proteins mediate its effect at the RyR (Figure 3). Photoaffinity labeling experiments with [^{32}P]-8-N_3-cADPR showed the existence of such binding proteins of 100 and 140 kD in sea urchin egg homogenates (Walseth et al., 1993). However, there is also the possibility that these two

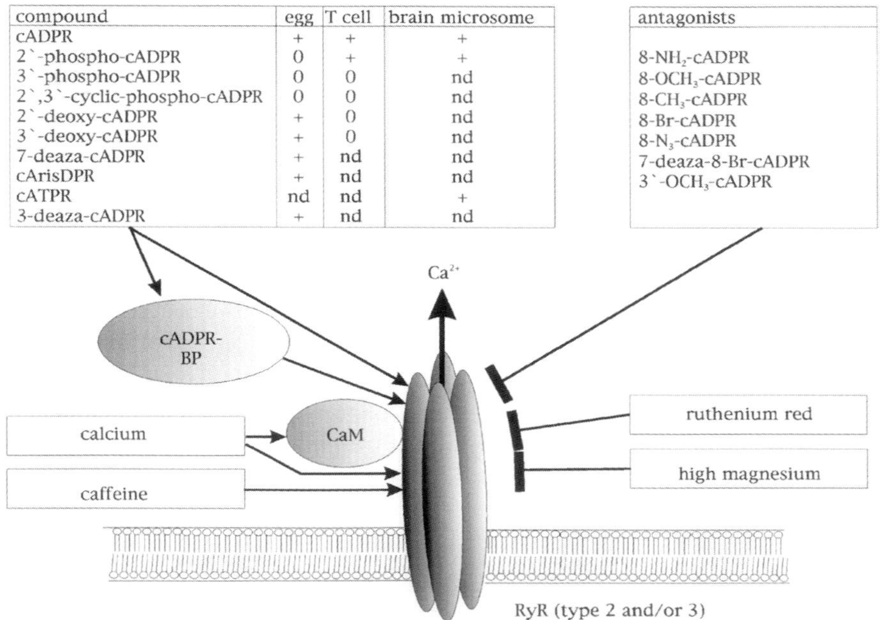

Figure 3. Ca^{2+}-release by cADPR. The cADPR derivatives listed in the box on the left upper side display different properties (+ agonistic; o no effect) in different cell systems. For the antagonists shown in the right upper box such contrary effects (as for 2′-deoxy-cADPR) have not yet been described. Calcium and caffeine are generally known as agonists at the RyR, whereas ruthenium red and high magnesium ion concentrations are inhibitory. Abbreviations: cADPR-BP, cADPR-binding protein; CaM, calmodulin; nd, not determined.

proteins are proteolytical fragments of the RyR from sea urchin eggs. In addition, it was shown that calmodulin markedly enhanced the sensitivity of sea urchin egg microsomes towards cADPR (Lee et al., 1994).

In some mammalian cell types, effects of cADPR on RyR have been described. The type 3 RyR is expressed in the brain, but also in peripheral tissues. Examples for the latter are human Jurkat T-lymphocytes (Guse et al., 1999) and rat diaphragm muscle (Sonnleitner et al., 1998). In brain, T cells and diaphragm muscle, specific effects of cADPR on the RyR have been detected, either (i) as an increase in single channel openings in reconstituted lipid planar bilayers (Meszaros et al., 1993), (ii) as release of [$^{45}Ca^{2+}$] from brain microsomes (Meszaros et al., 1993), (iii) as a stimulatory effect on the velocity of [^{3}H]ryanodine binding to T cell membranes (Guse et al., 1999), or (iv) as shift of the open probability of the RyR/Ca^{2+}-channel towards lower Ca^{2+}-concentrations in diaphragm muscle (Sonnleitner et al., 1998).

The type 2 RyR, mainly found in cardiac muscle, was also stimulated by cADPR in lipid planar bilayer experiments and in release experiments in [$^{45}Ca^{2+}$] microsomes (Meszaros et al., 1993). Evidence for a role of cADPR in cardiac myocytes was also obtained by Terrar and co-workers showing (i) that the cADPR antagonist 8-NH_2-cADPR could reduce electrically stimulated contractions, Ca^{2+}-oscillations and Ca^{2+}-activated currents (Rakovic et al., 1996, 1999), and (ii) photoreleased cADPR had specific effects on the magnitude of whole-cell Ca^{2+} transients and on the frequency of Ca^{2+} sparks in cardiac myocytes (Cui et al., 1999). Also in PC12 cells, which mainly express the type 2 RyR, Ca^{2+}-release by cADPR was observed in permeabilized cell preparations (Clementi et al., 1996).

For the type 1 RyR the data are conflicting: either no effect of cADPR was observed (Meszaros et al., 1993), or the effect was not specific for cADPR because similar results were produced with NAD^+ or ADPR (Sitsapesan and Williams, 1995). Furthermore, channel opening was observed under special conditions only, e.g. 10 μM cytosolic Ca^{2+}-concentration and 1 mM luminal Ca^{2+}-concentration (Sitsapesan and Williams, 1995).

In conclusion, the types 2 and 3 RyR are the best candidates for the cADPR receptor; however, the possibility remains that additional binding proteins are required to mediate the Ca^{2+}-mobilizing effect of cADPR at the Ca^{2+} channel protein.

Wherever the binding site for cADPR is located, at a so far unidentified cADPR-binding protein or directly at the RyR, the use of a number of cADPR derivatives obtained by chemo-enzymatic synthesis using *Aplysia californica* ADP-ribosyl cyclase already showed that there are common features, but that there must also be differences in the binding pocket for cADPR in the different cell systems studied.

The majority of work has been performed in three cell systems: the invertebrate sea urchin egg, human T-lymphocytes and brain microsomes. In addition to cADPR, six other compounds were active as full or partial agonists (Figure 3), namely 2′-phospho-cADPR (Zhang et al., 1995; Vu et al., 1996; Guse et al., 1997a), cyclic adenosine triphosphate ribose (cATPR; Zhang et al., 1996), 2′-deoxy-cADPR (Ashamu et al., 1997), 7-deaza-cADPR (Bailey et al., 1997), cyclic asteromycin diphosphate ribose (cArisDPR), a carbocyclic analogue of cADPR (Bailey et al., 1996), and 3-deaza-cADPR (Wong et al., 1999). Interestingly, two out of these five derivatives were so far tested as agonists in the vertebrate cell systems, namely 2′-phospho-cADPR in T cells and both 2′-phospho-cADPR and cATPR in brain microsomes (Zhang et al., 1995, 1996; Vu et al., 1996; Guse et al., 1997a). In the invertebrate sea urchin egg system, the four other compounds have been identified as full or partial agonists (Figure 3). 7-Deaza-cADPR, cArisDPR and 3-deaza-cADPR have not been investigated in vertebrate cell systems, 2′-deoxy-cADPR and 2′-phospho-cADPR were tested in sea urchin eggs as well as in human T cells (Ashamu et al., 1997; Guse et al., 1997a). 2′-Phospho-cADPR was also investigated in brain microsomes (Zhang et al., 1995; Vu et al., 1996). Most interestingly, 2′-deoxy-cADPR was fully active in the invertebrate system (Ashamu et al., 1997), while 2′-phospho-cADPR was inactive (Aarhus et al., 1995). In contrast, in the vertebrate cell systems, results were the opposite: 2′-deoxy-cADPR was almost inactive in T cells (Potter and Guse, unpublished results) and 2′-phospho-cADPR was almost as potent as cADPR (Zhang et al., 1995; Vu et al., 1996; Guse et al., 1997a). These data indicate that different receptor proteins for cADPR are used by vertebrate and invertebrate cells. Binding proteins (100 and 140 kD) for cADPR have been labeled in sea urchin egg extracts in photoaffinity experiments (Walseth et al., 1993); in line with the pharmacological characterization, so far we were unable to identify any protein of a similar size in extracts from human Jurkat T cells (Walseth and Guse, unpublished results).

The majority of cADPR derivatives synthesized so far are antagonists of cADPR-mediated Ca^{2+}-signaling. In human T cells, five compounds tested were inhibitory (Figure 3), namely the 8-substituted 8-NH_2-, 8-OCH_3-, 8-CH_3-, 8-Br-cADPR and 7-deaza-8-Br-cADPR (Guse et al., 1995, 1997a, 1999). Similarly, in the invertebrate sea urchin egg the 8-substituted compounds 8-NH_2-, 8-Br-, 8-N_3-cADPR and 7-deaza-8-Br-cADPR were antagonists (Walseth and Lee, 1993; Sethi et al., 1997). These data indicate that alterations at carbon 8 of the adenine ring, even those leading to structurally very different molecules, convert the cADPR derivative into an antagonist. In contrast, the published agonists all have the same natural small ligand at carbon 8, the hydrogen atom. However, recent investigations indicate that the hydrogen atom bonded to carbon 8 is not essential for agonistic activity since the new derivative 8-aza-9-deaza-cADPR (cyclic formycin diphosphoribose) is a relatively potent agonist in permeabilized T cells (Guse and Potter, unpublished results).

Derivatives of cADPR at the 3′-position of the ribose either were without any activity (3′-phospho-cADPR and 2′,3′-cyclic-phospho-cADPR (Ashamu et al., 1997; Guse et al., 1997)), or were antagonists (3′-OCH_3-cADPR; Ashamu et al., 1997) in sea urchin eggs. This highlights the importance of the hydroxyl group at the 3′-position (Figure 3).

In addition to these compounds which all contain adenine or a closely related derivative of adenine as the base, further cyclic compounds have been produced, such as cyclic etheno-cytosine diphosphoribose, cyclic guanosine diphosphoribose, cyclic inosine diphosphoribose and 1,N^6-etheno-cADPR (reviewed in Zhang et al., 1999). None of these compounds appears to release Ca^{2+} or block cADPR-mediated Ca^{2+}-release; however, because of their fluorescent properties some of them are useful to monitor the enzymatic activity of ADP-ribosyl cyclases or cADPR hydrolases.

4. Ca^{2+}-ENTRY MEDIATED BY cADPR

In the majority of systems studied, cADPR has been shown to release Ca^{2+} from an intracellular store. Ca^{2+} stores sensitive to cADPR are very often also sensitive to caffeine and/or ryanodine, because the RyR/Ca^{2+} channel is activated by these pharmacological compounds.

In general, the molecular mechanism by which Ca^{2+} ions enter non-excitable cells in response to activation of plasma membrane bound receptors

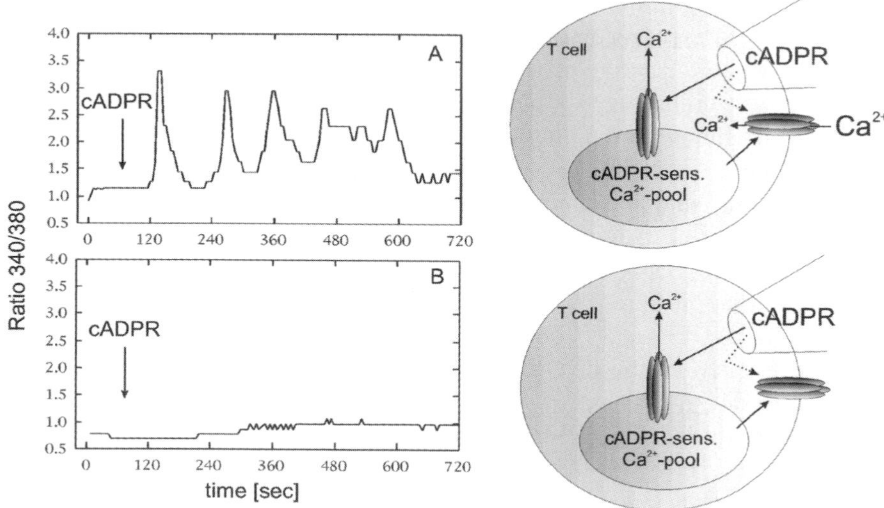

Figure 4. Involvement of cADPR in Ca^{2+}-entry. Ratiometric Ca^{2+}-imaging was performed in single Fura-2-loaded Jurkat T cells. During the experiment, the cells were microinjected with cADPR (10 μM) in the presence (A) or absence (B) of 1 mM extracellular Ca^{2+} (methods described in detail in Guse et al., 1997c). The right panel indicates schematically potential pathways of activation of Ca^{2+}-entry: (i) secondary to Ca^{2+}-release from cADPR-sensitive stores, or (ii) directly by activation of Ca^{2+}-channels in the plasma membrane.

is still not clarified (reviewed in Guse, 1998). Depletion of Ca^{2+} stores is one of the mechanisms proposed to be responsible for receptor-mediated Ca^{2+} entry. This model was termed "capacitative Ca^{2+} entry model" by Putney (1986; reviewed in Favre et al., 1996). When the "capacitative Ca^{2+} entry model" was introduced in 1986, cADPR was not even discovered as a Ca^{2+} mobilizing second messenger. Today, $InsP_3$ still is regarded as the most important Ca^{2+} mobilizer, perhaps, because receptor-mediated formation of $InsP_3$ and Ca^{2+} release mediated via the $InsP_3$ receptor have been more intensively studied. However, both cADPR and $InsP_3$ do release Ca^{2+} from internal stores, and thereby may both contribute to subsequent Ca^{2+} entry via the capacitative mechanism.

Evidence for a role for cADPR in Ca^{2+}-entry was obtained in human Jurkat T cells where it was shown that microinjection of cADPR into intact cells specifically stimulated sustained Ca^{2+} signaling, while in the absence of extracellular Ca^{2+} the signals were largely abolished (Figure 4; Guse et al., 1997b). Since cADPR mediated Ca^{2+} release from intracellular stores has been demonstrated in permeabilized T cells (Guse et al., 1995), it is likely that the observed effect was due to capacitative Ca^{2+} entry mediated by depletion of the cADPR sensitive Ca^{2+} pool (Figure 4). However, a direct effect of cADPR on Ca^{2+} entry cannot completely be excluded at present (Figure 4). The fact that a membrane-permeant antagonist of cADPR inhibited TCR/CD3 complex-mediated long-lasting Ca^{2+}-signaling (Guse et al., 1999) provides additional evidence for the involvement of cADPR in Ca^{2+} entry in T cells.

Less direct evidence for a role of cADPR in Ca^{2+} influx was obtained in submandibular gland acinar cells, anococcygeus smooth muscle cells, and PC12 cells (for references, see Guse, 1999). In all these cells, it was shown that the depletion of caffeine-sensitive Ca^{2+}-stores resulted in subsequent Ca^{2+}-entry. Since the caffeine- and the cADPR-sensitive Ca^{2+} stores are thought to be identical, or at least to overlap largely, these results may indicate that cADPR could be similarly active.

In conclusion, cADPR appears to be involved also in the regulation of Ca^{2+}-entry in certain cell types. The data obtained so far indicate that a mechanism based on Ca^{2+}-store depletion as suggested for the effect of $InsP_3$ on Ca^{2+}-entry may also exist for cADPR.

5. THE PHYSIOLOGICAL ROLE OF THE cADPR/Ca^{2+}-SIGNALING SYSTEM IN CELLULAR SIGNAL TRANSDUCTION

In a growing number of non-excitable and excitable cells important elements of the cADPR/Ca^{2+}-signaling pathway, such as (i) cADPR-mediated Ca^{2+}-signaling, (ii) inhibition of receptor-mediated Ca^{2+}-signaling by antagonists of cADPR, (iii) receptor-mediated formation of cADPR, and (iv) the presence of ADP-ribosyl cyclase or cADPR-hydrolase activities, have been observed. Despite these advances, in the majority of cell systems studied only limited information on the role of cADPR is available, mainly because only one of the points listed above has been investigated for each individual cell system in detail. However, in some cell systems more definitive evidence for an involvement of cADPR in signal transduction has been obtained.

In longitudinal muscle cells cholecystokinin mediated activation of an ADP-ribosyl cyclase, an increase in $[Ca^{2+}]_i$, and contraction (Kuemmerle and Makhlouf, 1995). Cholecystokinin also appears to activate the cADPR signaling pathway in pancreatic acinar cells: (i) cADPR stimulated Ca^{2+}-release in permeabilized and Ca^{2+}-dependent current oscillations in intact acinar cells (Schulz et al., 1999; Thorn et al., 1994), and (ii) 8-NH_2-cADPR antagonized Ca^{2+}-signaling mediated by cholecystokinin (Cancela and Petersen, 1998).

Evidence for coupling of the muscarinic cholinergic receptor to the cADPR system was obtained in several cell types: (i) in tracheal smooth muscle cells 8-NH_2-cADPR inhibited Ca^{2+}-oscillations mediated via the muscarinic cholinergic receptor (Prakash et al., 1998), (ii) ligation of the muscarinic acetylcholine receptor in adrenal chromaffin cells stimulated a membrane-bound ADP-ribosyl cyclase (Morita et al., 1997), and (iii) cholera-toxin sensitive activation of a membrane-bound ADP-ribosyl cyclase was reported for NG 108-15 neuroblastoma x glioma hybrid cells in response to stimulation of the muscarinic cholinergic receptor (Higashida et al., 1997). A cholera-toxin sensitive ADP-ribosyl cyclase was also detected in cardiac myocytes stimulated by isoproterenol (Higashida et al., 1999).

In alveolar macrophages, ATP-mediated activation of an outward potassium current was inhibited by intracellularly applied 8-NH_2-cADPR (Ebihara et al., 1997).

In a hippocampal model evidence for the involvement of cADPR in presynaptic, NO-triggered induction of long-term synaptic depression was obtained (Reyes-Harde et al., 1999).

Pancreatic β-cells are important regulators of the blood glucose concentration. Okamoto and co-workers have shown that glucose induced an increase in cADPR in pancreatic islets (Takasawa et al., 1998). The ADP-ribosyl cyclase involved appears to be CD38 which is expressed in β-cells and which can be modulated towards a higher ADP-ribosyl cyclase to cADPR-hydrolase ratio by ATP (Takasawa et al., 1993a; Kato et al., 1995). It was also shown that cADPR stimulated Ca^{2+}-release and insulin secretion in permeabilized β-cells (Takasawa et al., 1993b). Despite these results, the problems accompanied with the ectoenzyme CD38 as discussed above (see Section 2) have not yet been solved satisfactorily for β-cells.

Non-excitable cell systems studied in detail are sea urchin eggs and human T cells. Fertilization of sea urchin eggs requires a cytosolic Ca^{2+} transient, which is induced by binding of the sperm to the egg. Specific inhibition of cADPR-mediated or $InsP_3$-mediated Ca^{2+}-release alone was not sufficient to inhibit the fertilization of the egg, whereas combined inhibition of the two Ca^{2+}-release mechanisms efficiently antagonized the cortical reaction, which is associated with fertilization (Lee et al., 1993). Synthesis of cADPR can be stimulated by cGMP (Galione et al., 1993), and the gaseous messenger NO was shown to trigger guanylyl cyclase and cADPR production in sea urchin eggs (Willmott et al., 1996). These data suggest that both the cADPR- and the $InsP_3$-dependent Ca^{2+}-release mechanism are involved in fertilization of sea urchin eggs, likely in a redundant fashion (Figure 5).

In human T-lymphocytes, the involvement of both the $InsP_3$- and cADPR-mediated Ca^{2+}-release in the signal transduction process via the TCR/CD3 complex have been demonstrated. An antisense approach against the type 1 $InsP_3$ receptor resulted in complete inhibition of TCR/CD3 complex-mediated Ca^{2+}-signaling and IL-2 production in human Jurkat T cells (reviewed in Guse, 1998). Similarly, inhibition of $InsP_3$-mediated Ca^{2+}-release by the membrane-permeant $InsP_3$-antagonist xestospongin C resulted in marked inhibition of TCR/CD3 complex-mediated Ca^{2+}-signaling (Guse et al., 1999). In contrast, a membrane-permeant antagonist of cADPR, 7-deaza-8-Br-cADPR (Sethi et al., 1997), mainly blocked the second, sustained phase of TCR/CD3 complex-mediated Ca^{2+}-signaling (Guse et al., 1999). Additionally, the expression of activation antigens and proliferation was effectively inhibited by 7-deaza-8-Br-cADPR (Guse et al., 1999). This indicates that in T cells the 2 second messengers $InsP_3$ and cADPR appear to act in a

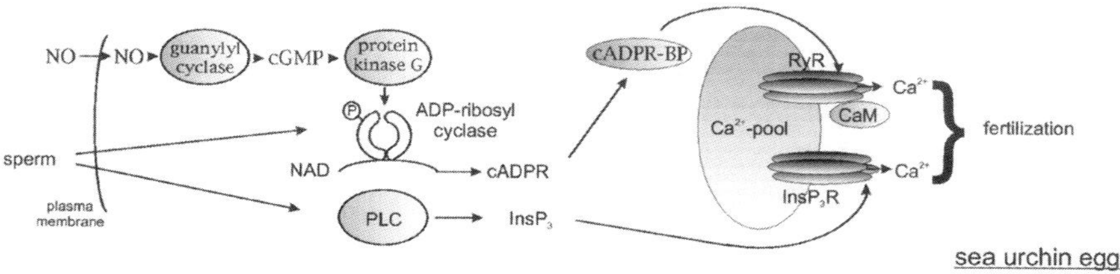

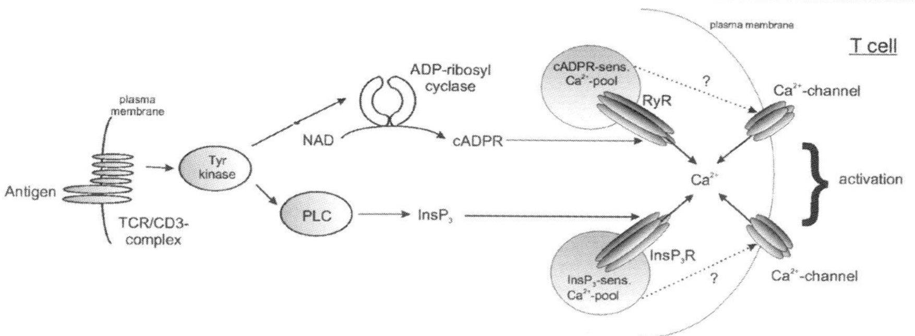

Figure 5. Involvement of cADPR in Ca^{2+}-signaling in sea urchin eggs and human T cells. Schematical presentation of Ca^{2+}-signaling events leading to fertilization of sea urchin eggs (upper panel) and activation and proliferation of T cells (lower panel). Abbreviations: cADPR-BP, cADPR-binding protein; CaM, calmodulin; PLC, phospholipase C.

temporally co-ordinated fashion. While $InsP_3$ releases Ca^{2+} and promotes subsequent Ca^{2+}-influx in the initial phase, cADPR acts, perhaps in concert with pre-elevated $[Ca^{2+}]_i$, during the second, long-lasting phase of T cell Ca^{2+}-signaling (Figure 5). Such a model is also in accordance with the kinetics of $InsP_3$- and cADPR-formation: $InsP_3$ is highly elevated mainly in the first seconds or minutes after cell stimulation (Ng et al., 1990), whereas the intracellular concentration of cADPR increased slowly within the first minutes, but then remained elvated for at least 60 minutes (Guse et al., 1999). Since quantitatively most of the Ca^{2+}-signaling observed in T cells results from Ca^{2+}-entry, the decrease in the intraluminal Ca^{2+}-concentrations of the $InsP_3$-sensitive store during the initial phase, and of the cADPR-sensitive store during the sustained phase may be the trigger for rapid and sustained Ca^{2+}-entry in T cells (Figure 5).

6. CONCLUSION

Ten years ago the world of receptor-mediated Ca^{2+}-signaling was governed by $InsP_3$. However, knowledge accumulated in the last decade indicates that the network of intracellular Ca^{2+}-signaling includes many more players, such as the novel endogenous Ca^{2+}-mobilizing compounds cADPR and $NAADP^+$. The role of $NAADP^+$ is still unclear in many aspects, but recent data obtained in pancreatic acinar cells may indicate that $NAADP^+$ evokes an initial Ca^{2+} elevation, which is subsequently needed for full activity of $InsP_3$ and cADPR (Cancela et al., 1999). In contrast, the results for cADPR obtained in several phylogenetically very different cell systems, such as invertebrate and mammalian cells, protozoa and plants, suggest a role as second messenger for this molecule. Future research aimed at the molecular characterization of the enzymes and receptor protein(s) involved in the action of cADPR and $NAADP^+$ will help to more accurately understand the role of these emerging new signaling molecules within the framework of intracellular signal transduction.

ACKNOWLEDGEMENTS

I am grateful to my co-workers who significantly contributed to the understanding of the cADPR system in T-lymphocytes. Thanks are also expressed to my collaboration partners for their engagement and support, and to Dr. Cristina P. da Silva for critically reading the manuscript. I apologize for not having cited many important articles; this is solely due to limited space for references in this volume.

Research in my lab was supported by grants from the Deutsche Forschungsgemeinschaft (grant No. Gu 360, jointly with G.W. Mayr, Hamburg), the Alexander-von-Humboldt-Stiftung, the Wellcome Trust (research collaboration grant No. 051326 jointly with B.V.L. Potter, Bath, U.K.) and the Deutsche Akademische Austauschdienst (grant No. D/9822786, jointly with P. Deterre, Paris, France).

REFERENCES

Aarhus, R., Graeff, R.M., Dickey, D.M., Walseth, T.F. and Lee, H.C., 1995, ADP-ribosyl cyclase and CD38 catalyze the synthesis of a calcium-mobilizing metabolite from $NADP^+$, *J. Biol. Chem.* 270, 30327–30333.

Ashamu, G.A., Sethi, J.K., Galione, A. and Potter, B.V.L., 1997, Roles for adenosine ribose hydroxyl groups in cyclic adenosine 5′-diphsophate ribose-mediated Ca^{2+}-release, *Biochemistry* 36, 9509–9517.

Ashcroft, F.M. and Ashcroft, S.J.H., 1992, Mechanism of insulin secretion, in *Insulin – Molecular Biology to Pathology*, F.M. Ashcroft and S.J.H. Ashcroft (eds.), Oxford University Press, Oxford, pp. 97–150.

Bailey, V.C., Fortt, S.M., Summerhill, R.J., Galione, A. and Potter, B.V.L., 1996, Cyclic aristeromycin diphosphate ribose: A potent and poorly hydrolysable Ca^{2+}-mobilising mimic of cyclic adenosine diphosphate ribose, *FEBS Lett.* 379, 227–230.

Bailey, V.C., Sethi, J.K., Fortt, S.M., Galione, A. and Potter, B.V.L., 1997, 7-Deaza cyclic adenosine 5′-diphosphate ribose: First example of a Ca^{2+}-mobilizing partial agonist related to cyclic adenosine 5′-diphosphate ribose, *Chem. Biol.* 4, 51–61.

Berridge, M.J., 1993, Inositol trisphosphate and calcium signaling, *Nature* 361, 315–325.

Cancela, J.M. and Petersen, O.H., 1998, The cyclic ADP-ribose antagonist 8-NH_2-cADP-ribose blocks cholecystokinin-evoked cytosolic Ca^{2+} spiking in pancreatic acinar cells, *Pflügers Arch. Eur. J. Physiol.* 435, 746–748.

Cancela, J.M., Churchill, G.C. and Galione, A., 1999, Coordination of agonist-induced Ca^{2+}-signaling patterns by NAADP in pancreatic acinar cells, *Nature* 398, 74–76.

Clapper, D.L., Walseth, T.F., Dargie, P.J. and Lee, H.C., 1987, Pyridine nucleotide metabolites stimulate calcium release from sea urchin egg microsomes desensitized to inositol trisphosphate, *J. Biol. Chem.* 262, 9561–9568.

Clementi, E., Riccio, M., Sciorati, C., Nistico, G. and Meldolesi, J., 1996, The type 2 ryanodine receptor of neurosecretory PC 12 cells is activated by cyclic ADP-ribose, *J. Biol. Chem.* 271, 17739–17745.

Cui, Y., Galione, A. and Terrar, D.A., 1999, Effects of photoreleased cADP-ribose on calcium transients and calcium sparks in myocytes isolated from guinea-pig and rat ventricles, *Biochem. J.* 342, 269–273.

da Silva, C., Heyer, P., Schweitzer, K., Malavasi, F., Mayr, G.W. and Guse, A.H., 1998a, Ectocellular CD38-catalyzed synthesis and intracellular Ca^{2+}-signaling activity of cyclic ADP-ribose in T-lymphocytes are not functionally related, *FEBS Lett.* 439, 291–296.

da Silva, C.P., Potter, B.V.L., Mayr, G.W. and Guse, A.H., 1998b, Quantification of intracellular levels of cyclic ADP-ribose by high-performance liquid chromatography, *J. Chromatogr. B* 707, 43–50.

Dousa, T.P., Chini, E.N. and Beers, K.W., 1996, Adenine nucleotide diphosphates: Emerging second messengers acting via intracellular Ca^{2+} release, *Am. J. Physiol.* 271 (*Cell Physiol.* 40), C1007–C1024.

Ebihara, S., Sasaki, T., Hida, W., Kikuchi, Y., Oshiro, T., Shimura, S., Takasawa, T., Okamoto, H., Nishiyama, A., Akaike, N. and Shirato, K., 1997, Role of cyclic ADP-ribose in ATP-activated potassium currents in alveolar macrophages, *J. Biol. Chem.* 272, 16023–16029.

Favre, C.J., Nüsse, O., Lew, D.P. and Krause, K.-H., 1996, Store-operated Ca^{2+} influx: What is the message from the stores to the membrane?, *J. Lab. Clin. Med.* 128, 1–9.

Franco, L., Guida, L., Bruzzone, S., Zocchi, E., Usai, C. and de Flora, A., 1998, The transmembrane glycoprotein CD38 is a catalytically active transporter responsible for generation and influx of the second messenger cADPR across membranes, *FASEB J.* 12, 1507–1520.

Galione, A., Lee, H.C. and Busa W.B., 1991, Ca^{2+}-induced Ca^{2+}-release in sea urchin egg homogenates: Modulation by cyclic ADP-ribose, *Science* 253, 1143–1146.

Galione, A., White, A., Willmott, N., Turner, M., Potter, B.V.L. and Watson, S.P., 1993, cGMP mobilizes intracellular Ca^{2+} in sea urchin eggs by stimulating cyclic ADP-ribose synthesis, *Nature* 365, 456–459.

Graeff, R.M., Franco, L., De Flora, A. and Lee, H.C., 1998, Cyclic GMP-dependent and -independent effects on the synthesis of the calcium messengers cyclic ADP-ribose and nicotinic acid adenine dinucleotide phosphate, *J. Biol. Chem.* 273, 118–125.

Guse, A.H., da Silva, C.P., Emmrich, F., Ashamu, G.A., Potter, B.V.L. and Mayr, G.W., 1995, Characterization of cyclic adenosine diphosphate-ribose-induced Ca^{2+}-release in T-lymphocyte cell lines, *J. Immunol.* 155, 3353–3359.

Guse, A.H., da Silva, C.P., Weber, K., Ashamu, G.A., Potter, B.V.L. and Mayr, G.W., 1996, Regulation of cyclic ADP-ribose-induced Ca^{2+}-release by Mg2+ and inorganic phosphate, *J. Biol. Chem.* 271, 23946–23954.

Guse, A.H., da Silva, C.P., Weber, K., Armah, C., Schulze, C., Potter, B.V.L., Mayr, G.W. and Hilz, H., 1997a, 1-(5-Phospho-β-D-ribosyl)2′-phosphoadenosine 5′-phosphate cyclic anhydride-induced Ca^{2+}-release in human T cell lines, *Eur. J. Biochem.* 245, 411–417.

Guse, A.H., Berg, I., da Silva, C.P., Potter, B.V.L. and Mayr, G.W., 1997b, Ca^{2+}-entry induced by cyclic ADP-ribose in intact T-lymphocytes, *J. Biol. Chem.* 272, 8546–8550.

Guse, A.H., 1998, Ca^{2+}-signaling in T-lymphocytes, *Crit. Rev. Immunol.* 18, 419–448.

Guse, A.H., 1999, Cyclic ADP-ribose: A novel Ca^{2+}-mobilising second messenger, *Cell. Signal.* 5, 309–316.

Guse, A.H., da Silva, C.P., Berg, I., Weber, K., Heyer, P., Hohenegger, M., Ashamu, G.A., Skapenko, A.L., Schulze-Koops, H., Potter, B.V.L. and Mayr, G.W., 1999, Regulation of Ca^{2+}-signaling in T-lymphocytes by the second messenger cyclic ADP-ribose, *Nature* 398, 70–73.

Hellmich, M.R. and Strumwasser, F., 1991, Purification and characterization of a molluscan egg-specific NADase, a second messenger enzyme, *Cell Reg.* 2, 193–202.

Higashida, H., Yokoyama, S., Hashii, M., Taketo, M., Higashida, M., Takayasu, T., Ohshima, T., Takasawa, S., Okamoto, H. and Noda, M., 1997, Muscarinic receptor-mediated dual regulation of ADP-ribosyl cyclase in NG108-15 neuronal cell membranes, *J. Biol. Chem.* 272, 31272–31277.

Higashida, H., Egorova, A., Higashida, C., Zhong, Z.G., Yokohama, S., Noda, M. and Zhang, J.S., 1999, Sympathetic potentiation of cyclic ADP-ribose formation in rat cardiac myocytes, *J. Biol. Chem.* 274, 33348–33354.

Hirata, Y., Kimuram, N., Sato, K., Ohsugi, Y., Takasawa, S., Okamoto, H., Ishikawa, J., Kaisho, T., Ishihara, K. and Hirano, T., 1994, ADP-ribosyl cyclase activity of a novel bone marrow stromal cell surface molecule, BST-1, *FEBS Lett.* 356, 244–248.

Howard, M., Grimaldi, J.C., Bazan, J.F., Lund, F., Santos-Argumedo, L., Parkhouse, R.M.E., Walseth, T.F. and Lee, H.C., 1993, Formation and hydrolysis of cyclic ADP-ribose catalyzed by lymphocyte antigen CD38, *Science* 262, 1056–1059.

Ikehata, F., Satoh, J., Nata, K., Tohgo, A., Nakazawa, T., Kato, I., Kobayashi, I., Akiyama, T., Takasawa, S., Toyota, T. and Okamoto, H., 1998, Autoantibodies against CD38 (ADP-ribosyl cyclase/cyclic ADP-ribose hydrolase) that impair glucose-induced insulin secretion in noninsulin-dependent diabetes patients, *J. Clin. Invest.* 102, 395–401.

Kato, I., Takasawa, S., Akabane, A., Tanaka, O., Abe, H., Takamura, T., Suzuki, Y., Nata, K., Yonekura, H., Yoshimoto, T. and Okamoto, H., 1995, Regulatory role of CD38 (ADP-ribosyl cyclase/cyclic ADP-ribose hydrolase) in insulin secretion by glucose in pancreatic β cells, *J. Biol. Chem.* 270, 30045–30050.

Kim, H., Jacobson, E.L. and Jacobson, M.K., 1993, Synthesis and degradation of cyclic ADP-ribose by NAD glycohydrolases, *Science* 261, 1330–1333.

Kuemmerle, J.F. and Makhlouf, G.M., 1995, Agonist-stimulated cyclic ADP-ribose, *J. Biol. Chem.* 270, 25488–25494.

Lee, H.C., Walseth, T.F., Bratt, G.T., Hayes, R.N. and Clapper, D.L., 1989. Structural determination of a cyclic metabolite of NAD^{+} with intracellular Ca^{2+}-mobilizing activity, *J. Biol. Chem.* 264, 1608–1615.

Lee, H.C., 1993, Potentiation of calcium- and caffeine-induced calcium release by cADPR, *J. Biol. Chem.* 268, 293–299.

Lee, H.C., Aarhus, R. and Walseth, T.F., 1993, Calcium mobilization by dual receptors during fertilization of sea urchin eggs, *Science* 1993, 352–355.

Lee, H.C., 1994, The crystal structure of cyclic ADP-ribose, *Nature Struct. Biol.* 1, 143–144.

Lee, H.C., Aarhus, R., Graeff, R., Gurnack, M.E. and Walseth, T.F., 1994, Cyclic ADP-ribose activation of the ryanodine receptor is mediated by calmodulin, *Nature* 370, 307–309.

Lee, H.C., 1997, Mechanisms of calcium signaling by cyclic ADP-ribose and NAADP, *Physiol. Rev.* 77, 1133–1164.

Lee, H.C., 1999, A unified mechanism of enzymatic synthesis of two calcium messengers: cADPR and NAADP, *Biol. Chem.* 380, 785–793.

Masuda, W., Takenaka, S., Tsuyama, S., Tokunaga, M., Yamaji, R., Inui, H., Miyatake, K. and Nakano, Y., 1997, Inositol 1,4,5-trisphosphate and cyclic ADP-ribose mobilize Ca^{2+} in a protist, *Euglena gracilis*, *Comp. Biochem. Physiol.* 118, 279–283.

Matsumura, N. and Tanuma, S., 1998, Involvement of cytosolic NAD^{+} glycohydrolase in cADPR metabolism, *Biochem. Biophys. Res. Commun.* 253, 246–252.

Meszaros, L.G., Bak, J. and Chu, A., 1993, Cyclic ADP-ribose as an endogenous activator of the non-skeletal type ryanodine receptor Ca^{2+} channel, *Nature* 364, 76–79.

Migaud, M.E., Pederick, R.L., Bailey, V.C. and Potter, B.V.L., 1999, Probing Aplysia californica adenosine 5′-diphosphate-ribosyl cyclase for substrate binding requirements: Design of potent inhibitors, *Biochemistry* 38, 9195–9214.

Morita, K., Kitayama, S. and Dohi, T., 1997, Stimulation of cyclic ADP-ribose synthesis by acetylcholine and ist role in catecholamine release in bovine adrenal chromaffin cells, *J. Biol. Chem.* 272, 21002–21009.

Munshi, C., Fryxell, K.B., Lee, H.C. and Branton, W.B., 1997, Large-scale production of human CD38 in yeast fermentation, *Meth. Enzymol.* 280, 318–330.

Munshi, C., Thiel, D.J., Mathews, I.I., Aarhus, R., Walseth, T.F. and Lee, H.C., 1999, Characterization of the active site of ADP-ribosyl cyclase, *J. Biol. Chem.* 274, 30770–30777.

Ng, J., Gustavsson, J., Jondal, M. and Andersson, T., 1990, Regulation of calcium influx across the plasma membrane of the human T-leukemic cell line, JURKAT: Dependence on a rise in cytosolic free calcium can be dissociated from formation of inositol phosphates, *Biochim. Biophys. Acta* 1053, 97–105.

Prakash, Y.S., Kannan, M.S., Walseth, T.F. and Sieck, G.C., 1998, Role of cyclic ADP-ribose in the regulation of $[Ca^{2+}]_i$ in porcine tracheal smooth muscle, *Am. J. Physiol.* 274 (*Cell Physiol.* 43), C1653–C1660.

Prasad, G.S., McRee, D.E., Stura, E.A., Levitt, D.G., Lee, H.C. and Stout, C.D., 1996, Crystal structure of Aplysia ADP ribosyl cyclase, a homologue of the bifunctional ectoenzyme CD38, *Nature Struct. Biol.* 3, 957–964.

Putney, J.W., Jr., 1986, A model for receptor-regulated calcium entry, *Cell Calcium* 7, 1–12.

Rakovic, S., Galione, A., Ashamu, G.A., Potter, B.V.L. and Terrar, D.A., 1996, A specific cyclic ADP-ribose antagonist inhibits cardiac excitation-contraction coupling, *Curr. Biol.* 6, 989–996.

Rakovic, S., Ciu, Y., Iino, S., Galione, A., Ashamu, G.A., Potter, B.V.L. and Terrar, D.A., 1999, An antagonist of cADP-ribose inhibits arrythmogenic oscillations of intracellular Ca^{2+} in heart cells, *J. Biol. Chem.* 274, 17820–17827.

Reyes-Harde, M., Empson, R., Potter, B.V.L., Galione, A. and Stanton, P.K., 1999, Evidence of a role for cyclic ADP-ribose in long-term synaptoc depression in hippocampus, *Proc. Natl. Acad. Sci. USA* 96, 4061–4066.

Schulz, I., Krause, E., Gonzalez, A., Göbel, A., Sternfeld, L. and Schmid, A., 1999, Agonist-stimulated pathways of calcium signaling in pancreatic acinar cells, *Biol. Chem.* 380, 903–908.

Sethi, J.K., Empson, R.M., Bailey, V.C., Potter, B.V.L. and Galione, A., 1997, 7-Deaza-8-bromo-cyclic ADP-ribose, the first membrane-permeant, hydrolysis-resistant cyclic ADP-ribose antagonist, *J. Biol. Chem.* 272, 16358–16363.

Sitsapesan, R. and Williams, A.J., 1995, Cyclic ADP-ribose and related compounds activate sheep skeletal sarcoplasmic reticulum Ca^{2+} release channel, *Am. J. Physiol.* 268 (*Cell Physiol.* 37), C1235–C1240.

Sonnleitner, A., Conti, A., Bertocchini, F., Schindler, H.G. and Sorrentino, V., 1998, Functional properties of the ryanodine receptor type 3 (RyR3) Ca^{2+} release channel, *EMBO J.* 17, 2790–2798.

Takasawa, S., Tohgo, A., Noguchi, N., Koguma, T., Nata, K., Sugimoto, T., Yonekura, H. and Okamoto, H., 1993a, Synthesis and hydrolysis of cyclic ADP-ribose by human leukocyte antigen CD38 and inhibition of the hydrolysis by ATP, *J. Biol. Chem.* 268, 26052–26054.

Takasawa, S., Nata, K., Yonekura, H. and Okamoto, H., 1993b, Cyclic ADP-ribose in insulin secretion from pancreatic β cells, *Science* 259, 370–373.

Takasawa, S., Akiyama, T., Nata, K., Kuroki, M., Tohgo, A., Noguchi, N., Kobayashi, S., Kato, I., Katada, T. and Okamoto, H., 1998, Cyclic ADP-ribose and inositol 1,4,5-trsiphosphate as alternate second messengers for intracellular Ca^{2+} mobilization in normal and diabetic β-cells, *J. Biol. Chem.* 273, 2497–2500.

Thorn, P., Gerasimenko, O. and Petersen O.H., 1994, Cyclic ADP-ribose regulation of ryanodine receptors involved in agonist evoked cytosolic Ca^{2+}-oscillations in pancreatic acinar cells, *EMBO J.* 13, 2038–2043.

Vu, C.Q., Lu, P.-J., Chen, C.S. and Jacobson, M.K., 1996, 2′-Phospho-cyclic ADP-ribose, a calcium-mobilizing agent derived from NADP, *J. Biol. Chem.* 271, 4747–4754.

Walseth, T.F., Aarhus, R., Kerr, J.A. and Lee, H.C., 1993, Identification of cyclic ADP-ribose-binding proteins by photoaffinity labeling, *J. Biol. Chem.* 268, 26686–26691.

Walseth, T.F. and Lee, H.C., 1993, Synthesis and characterization of antagonists of cyclic-ADP-ribose-induced Ca^{2+}-release, *Biochim. Biophys. Acta* 1178, 235–242.

Willmott, N., Sethi, J., Walseth, T.F., Lee, H.C., White, A.M. and Galione, A., 1996, Nitric oxide-induced mobilization of intracellular calcium via the cyclic ADP-ribose signaling pathway, *J. Biol. Chem.* 271, 3699–3705.

Wong, L., Aarhus, R., Lee, H.C. and Walseth, T.F., 1999, Cyclic 3-deaza-adenosine diphosphoribose: A potent and stable analog of cyclic ADP-ribose, *Biochim. Biophys. Acta* 1472, 555–564.

Wu, Y., Kuzma, J., Marechal, E., Graeff, R., Lee, H.C., Foster, R. and Chuah, N.H., 1997, Abscisic acid signaling through cyclic ADP-ribose in plants, *Science*, 278, 2126–2130.

Zhang, F.-J., Gu, Q.-M., Jing, P. and Sih, C.J., 1995, Enzymatic cyclization of nicotinamide adenine dinucleotide phosphate (NADP), *Bioorg. Med. Chem. Lett.* 19, 2267–2272.

Zhang, F.-J., Yamada, S., Gu, Q.-M. and Sih, C.J., 1996, Synthesis and characterization of cyclic ATP-ribose: A potent mediator of calcium release, *J. Bioorg. Med. Chem. Lett.* 6, 1203–1208.

Zhang, F.-J., Gu, Q.-M. and Sih, C.J., 1999, Bioorganic chemistry of cyclic ADP-ribose (cADPR), *Bioorg. Med. Chem.* 7, 653–664.

Zocchi, E., Daga, A., Usai, C., Franco, L., Guida, L., Bruzzone, S., Costa, A., Marchetti, C. and De Flora, A., 1998, Expression of CD38 increases intracellular calcium concentration and reduces doubling time in HeLa and 3T3 cells, *J. Biol. Chem.* 273, 8017–8024.

Zocchi, E., Usai, C., Guida, L., Franco, L., Bruzzone, S., Passalacqua, M. and de Flora, A., 1999, Ligand-induced internalization of CD38 results in intracellular Ca^{2+} mobilization: Role of NAD transport across cell membranes, *FASEB J.* 13, 273–283.

Visinin-Like Proteins (VILIPs) – Emerging Role in Cross-Talk between Cellular Signaling Pathways

Karl-Heinz Braunewell, Carsten Reissner and Eckart D. Gundelfinger

1. INTRODUCTION: THE VISININ-LIKE PROTEINS (VILIPS), A SUBFAMILY OF THE INTRACELLULAR NEURONAL CALCIUM SENSOR (NCS) PROTEINS

For the interaction of calcium with calcium-binding proteins, which is a prerequisite for the translation of calcium signals into physiological answers, different calcium-binding motifs and thus different families of intracellular calcium-binding proteins have evolved. These families include the C2 domain-containing proteins, the annexins and the EF-hand calcium-binding proteins (Heizmann and Hunziker, 1991). The latter group constitutes a superfamily of proteins with more than 300 members. Within this superfamily, a major discrimination has been made between calcium-buffering and calcium-sensing proteins (Ikura, 1996).

In addition to the ubiquitous calcium sensor protein calmodulin, a variety of calcium-sensing proteins have been identified in the nervous system reflecting the importance of calcium and its regulative function in nerve cells. These molecules have been termed intracellular neuronal calcium sensor (NCS) proteins (Nef, 1996; Braunewell and Gundelfinger, 1999). To date, more than 40 NCS protein family members are known from various species. Based on their sequence similarity, they can be grouped into different subfamilies: the Recoverins, the Frequenins, the guanylyl cyclase-activating proteins (GCAPs) and the VILIPs – also named NVPs for "neural visinin-like

Karl-Heinz Braunewell, Carsten Reissner and Eckart D. Gundelfinger • Signal Transduction Research Group, Department of Molecular Biology and Neurochemistry, Leibniz-Institute for Neurobiology, Magdeburg, Germany.

R. Pochet, R. Donato, J. Haiech, C. Heizmann and V. Gerke (eds.): Calcium: The Molecular Basis of Calcium Action in Biology and Medicine, 129–149.

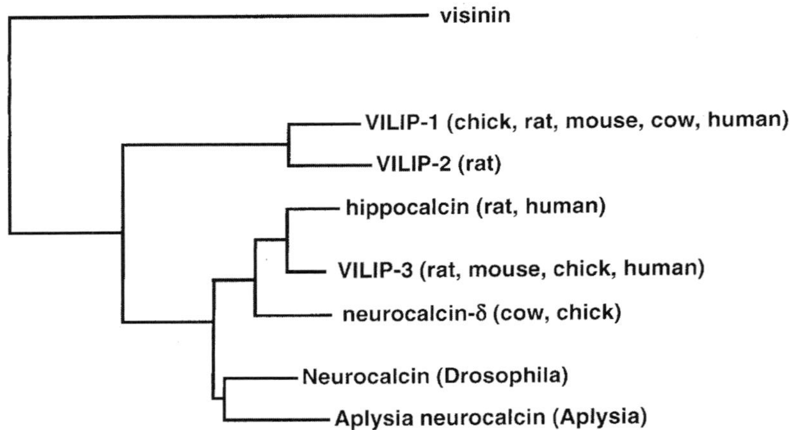

Figure 1. Dendrogram showing the relationships between the VILIP subfamily of NCS proteins based on amino acid similarity. Visinin, the founder member of the family, is used as a reference member for the VILIP subfamily of EF-hand calcium-binding proteins. The dendrogram was generated using sequences from the SwissProt database and the clustalX and NJplot program.

proteins" (Polans et al., 1996; Braunewell and Gundelfinger, 1999). Interestingly, calcium sensors may also occur as functional domains of larger proteins as exemplified by calsenilin/DREAM, which form a new subfamily of NCS proteins and may function in the regulation of presenilin function and/or as a calcium-dependent transcriptional repressor (Buxbaum et al., 1999; Carrion et al., 1999). This chapter will be focussed on the VILIP subfamily of intracellular NCS proteins, which includes VILIP-1, VILIP-2, VILIP-3, hippocalcin and several neurocalcins.

2. STRUCTURE of VILIPS

2.1. Primary structure

Sequence analyses of the NCS protein family illustrates, that VILIPs form a subfamily of closely related proteins (Figure 1). Whereas subfamily members share between 70 to 95% identical amino acids, the degree of identity between the VILIPs and other NCS subfamilies, i.e. Recoverins, Frequenins or GCAPs, are in the range of 30–60% (Figure 2). VILIP subfamily members have been cloned and sequenced from various species (Table 1). They are typically 190–200 amino acid residues long, have a consensus sequence for N-terminal myristoylation and posses two pairs of EF-hand calcium-binding motifs. The first EF-hand appears not to be capable of calcium binding due to a Cys-Pro amino acid pair in the loop region, which is conserved among most NCS proteins (Cox et al., 1994). EF-hand 1 also forms the most variable

Table 1. Members of the VILIP subfamily of intracellular NCS proteins

Protein	Ortholog	Species
VILIP-1		chick
	NVP-1	rat, mouse
	Neurocalcin α	cow
	VSNL 1	human
VILIP-2	NVP-2	rat
VILIP-3	Rem-1	chick
	NVP-3	rat, mouse
	hHLP2	human
Hippocalcin		rat, human
Neurocalcin δ		cow, chick
D-neurocalcin		Drosophila
Aplysia neurocalcin		Aplysia

part in the sequence of NCS proteins (Figure 2) and is therefore discussed as a possible interaction site with target proteins.

2.2. Tertiary structure and mechanics of the calcium myristoyl switch

First data on the higher order structure of NCS proteins were available for recoverin, the most intensively studied family member, which was analyzed in its calcium-bound and calcium-free myristoylated forms (Ames et al., 1997). Myristoylated recoverin, the prototype NCS protein, is able to perform a so-called calcium-myristoyl switch (for a review, see Ames et al., 1996). For this molecule the mechanism of the switch has been analyzed at the molecular level in detail (Ames et al., 1996, 1997). Binding of calcium to recoverin induces a conformational change leading to the surface exposure of hydrophobic parts of the polypeptide and of the myristoyl side chain, thereby making these structures available for interaction with cellular membranes and/or target proteins. More recently, also the three dimensional structure of a VILIP subfamily member, neurocalcin δ, was analyzed in its calcium-bound and non-myristoylated form and compared to recoverin (Vijay-Kumar and Kumar, 1999). Based on these two sets of structural data the molecular structure and dynamics of other NCS proteins, e.g. a similar molecular switch mechanism for VILIP-1, can now be modeled (Figures 3 and 4). The studies on recoverin and neurocalcin focused on the two double EF-hand motifs as separate structural elements, which divide the NCS structure in two domains. With respect to protein motions, EF-hands 1, 2, and 3 behave not as rigid bodies. Calculations of internal protein backbone distances for all structures

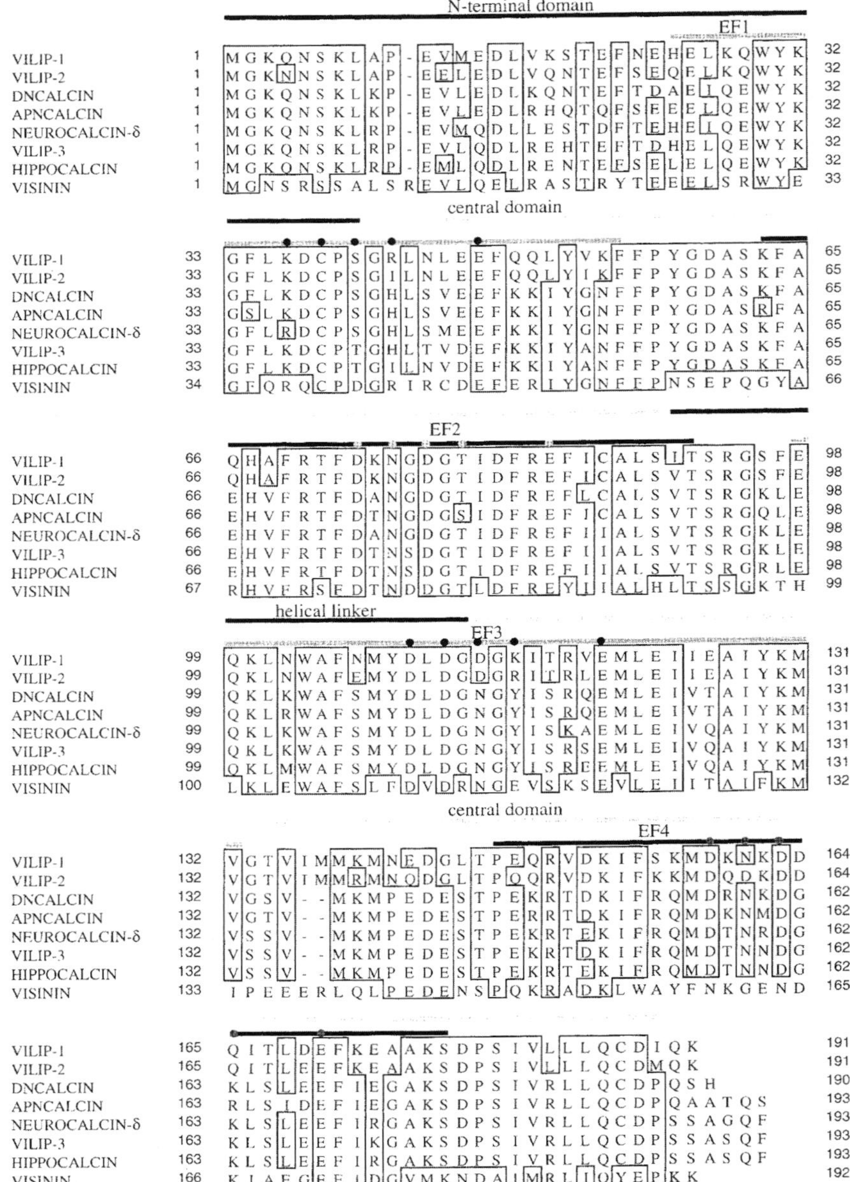

Figure 2. Alignment of visinin-like proteins. Sequences of the VILIP subfamily members VILIP-1, VILIP-2, *Drosophila* neurocalcin (DNcalcin), *Aplysia* neurocalcin (ApNcalcin), neurocalcin δ, VILIP-3 and hippocalcin were taken from the SwissProt database and aligned. In addition visinin, as an NCS protein of the Recoverin subfamily, was included in the alignment. Black dots in EF-hands 2, 3 and 4 mark residues that in the case of recoverin and neurocalcin reportedly are involved in calcium binding. In the case of the disabled EF-hand 1 they mark the potential calcium binding residues. Individual subdomains used for modeling in Figure 3 are indicated.

against each other (data not shown) gave result to a precise description of protein fragments and regions in space that move upon binding of calcium or the myristoyl group (Figure 3). VILIP-1 is modeled using the coordinates from recoverin (1JSA, 1IKU) and neurocalcin δ (1BJF). The main differences of VILIP-1 to the template structures remain on side chains. Hence, Figure 3 represents a mechanism applying to all of the three proteins. The structures are aligned on the central fragment comprising residues 41 to 90 of VILIP-1 (green). The N-terminal domain (yellow, residues 1–40), the helical linker (red, residues 91–114), and the C-terminal domain (cyan, residues 115–188) each behave as rigid bodies (root-mean-square-deviation, rmsd < 0.13 nm) and move in relation to the central domain (green). Calcium-bound neurocalcin δ (Figure 3.1) and calcium-free recoverin (Figure 3.10) are geometrically well-defined structures, while calcium-bound recoverin (Figure 3.5) may represent an intermediate state with deformed helices A, D, E and F. This possibly results from a lower occupancy of bound calcium at EF-hands 2 and 3. Calcium-bound recoverin cannot bind calcium at EF-hand 4 (Flaherty et al., 1993), but this is not likely to be responsible for the deformation of the helices (Vijay-Kumar and Kumar, 1999). The EF-hands 1, 2, and 3 bridge the moveable domains and function as hinges (Figure 3). Because EF-hand 1 is not capable to bind calcium in any one of the NCS proteins (Ames et al., 1997), changes at EF-hand 1 are calcium-independent, while changes in EF-hand 2 and 3 are calcium-concentration dependent. The structures suggest, that hinges at EF-hands 2 and 3 are rigid, when calcium is bound, getting weaker, when the calcium concentration is lowered (Figures 3.1–3.4), and become highly flexible in a calcium-free environment.

A comparison of the calcium-free with the calcium-bound structures show that the N-terminal domain is making a shear motion above helix E. Due to steric hinderance, this movement is not allowed with the myristoyl group bound to the N-terminal domain. Hence, the N-terminal domain has to move with the myristoyl group extruded (Figures 3.6–3.7). The motion will also not take place in the calcium saturated state, as helix E has to be partly unfolded to allow the crossing. A decrease in calcium concentrations increases the flexibility of EF-hands 2 and 3, which in turn allows the motion of the N-terminal domain (Figures 3.6–3.7) and finally leads to the "induced fit" of the myristoyl group into the hydrophobic pocket (Figures 3.7–3.8). The binding pocket is partly constructed by the N-terminal (yellow), the central (green), and the helical linker domain (red). The central domain is likely to move immediately (Figures 3.8–3.9) followed by the helical linker domain. The C-terminal domain moves in concert with the helical linker, but finally makes an additional adjustment to complement the surface built by helices D, E, and F (Figures 3.9–3.10).

From the structural point of view, the mechanism of the calcium-myristoyl-switch is reversible. The unfolding of the compact myristoylated structure, therefore the described "classical" calcium-induced myristoyl-switch (Figure 3.10) has to start with the movement of the C-terminal domain

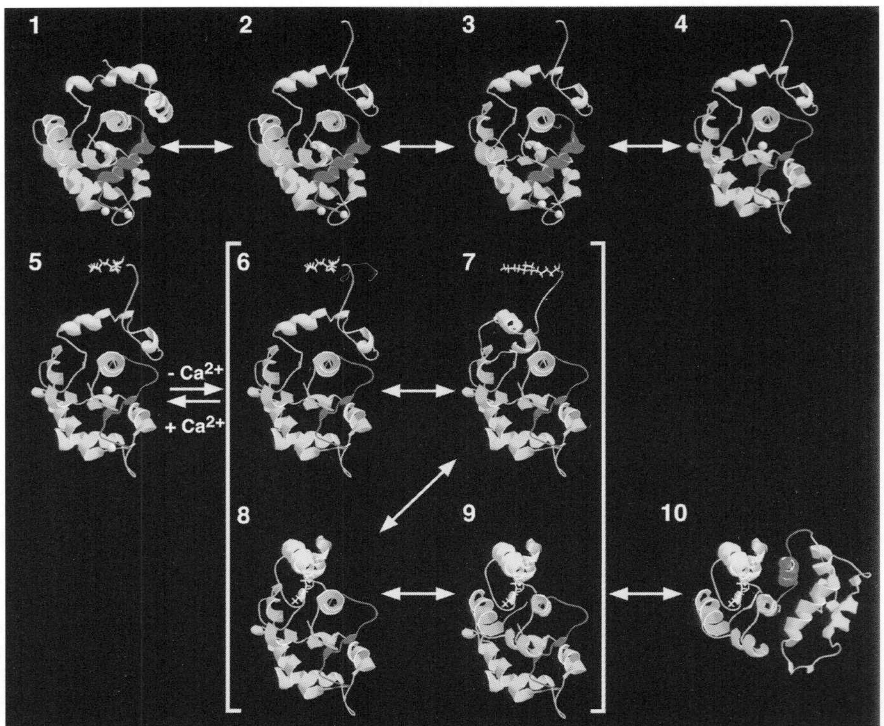

Figure 3. Model of calcium-dependent motions and a myristoyl switch of VILIP-1. The structure is modeled using the coordinates of neurocalcin δ (model 1), calcium-bound recoverin (5) and calcium-free recoverin (10). Models 1 to 4 illustrate the possible motion of the N-terminal domain (yellow) upon decrease of calcium concentration. This motion induces a response in the central hydrophobic domain (green) by deforming secondary structural elements (3), and is finally supported by a deformation of helix F (red) (4). The C-terminal domain shows minor changes. Recoverin is not capable of binding calcium at EF-hand 4, but VILIP-1 most likely is. This calcium ion may be released independently (4) and may be available for calcium-dependent target interactions. Models 5-10 illustrate the calcium-myristoyl switch from the "open" to the "closed" form of VILIP-1. Structures 6–9 are intermediates. After calcium removal, the N-terminal domain can flip over the buried helix E (green) (6 to 7). This opens the way for the "induced fit" of the myristoyl group into the hydrophobic binding pocket provided by the N-terminal and central domains (7 to 9). Finally the helical-linker (red) covers the remaining cavity to the myristoyl group and this movement is stabilized by a rotation of the rigid helical-linker at Ser-90. The C-terminal domain (blue) follows that movement and adapts exactly to the surface formed by helices D, E, and F (10). Calcium binding to the "closed" form of VILIP-1 must induce the reverse order of movements since the helical-linker and C-terminal domain complex in the "closed" calcium-free form occupy the same position as the N-terminal domain in the "open" calcium-bound structure (5). The motions are also available as movies at http://www.synprot.de. Details of the methods used for modeling and RMS values are also presented at this site. For a colour version of this figure, see page xxv.

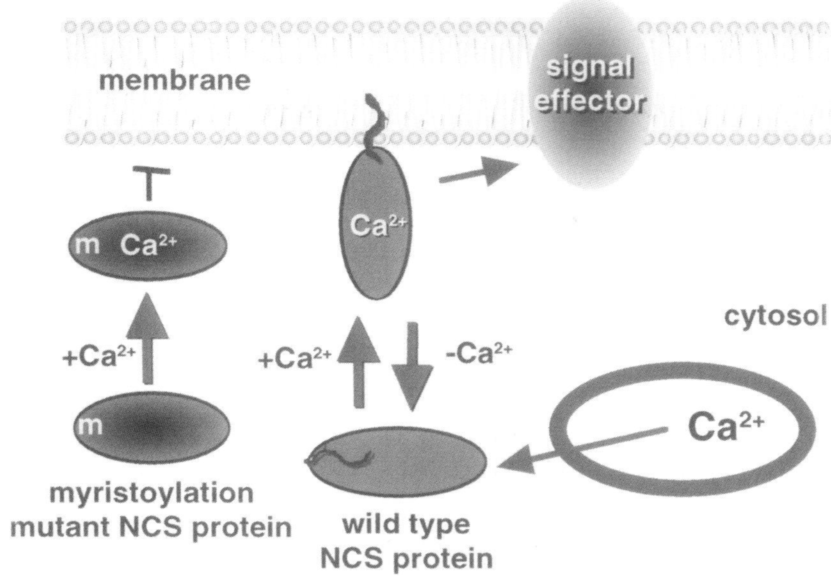

Figure 4. Model of the calcium-myristoyl switch of NCS proteins. Following binding of cytosolic calcium a calcium-induced conformation change of NCS proteins occurs leading to the exposure of the hydrophobic myristoyl side chain. In turn the NCS protein can interact with the cellular membrane. The NCS protein shuttles from the cytosol to the membrane and vice versa. The calcium- and myristoylation-dependent interaction of NCS proteins with cell membranes may be a prerequisite for the interaction with membrane-bound target molecules from various cellular signaling cascades.

(blue), as this domain together with helix F (red) sterically block the motion of the N-terminal domain, and results in the extrusion of the myristoyl group. The binding of calcium induces the formation of EF-hand 3 as shown in Figure 3.5. The binding of calcium further reorientates EF-hand 2 and destabilizes the hydrophobic interaction with the myristoyl group. Interestingly, it may be possible, that a calcium-bound intermediate structure folded as shown in Figure 3.9 becomes stable, if no acceptor protein for the myristoyl group is present. In summary, calcium functions as mediator of NCS protein flexibility, while the myristoyl group triggers the motion of the N-terminal domain via an "induced fit" mechanism. The N-terminal domain and helical-linker to C-terminal domain complex are competitive as they try to occupy the same space in the calcium-bound and calcium-free structure, respectively (compare Figures 3.4 and 3.10).

3. DISTRIBUTION OF VILIPS IN THE CENTRAL NERVOUS SYSTEM

In the nervous system VILIPs show a widespread but distinct expression pattern primarily in nerve cells. In some instances, expression in glial cells and in cells outside of the nervous system has been reported. For example in chicken, VILIP-3/Rem-1 transcripts have been detected not only in eye and brain, but also in bone marrow and gut (Kraut et al., 1995). Our knowledge about expression patterns of NCS proteins and their transcripts have been reviewed in detail previously (Braunewell and Gundelfinger, 1999). Here we will focus on the comparative examination of NCS protein expression in distinct brain areas.

Several comparative expression studies for NCS proteins, which include VILIP subfamily members, have been performed. These include an analysis of the differential expression of neurocalcin isoforms in rat spinal cord, dorsal root ganglia and muscle spindle (Okazaki et al. 1994); the comparison of the distribution of recoverin, NCS-1 and VILIP-1 in various vertebrate retinae (De Raad et al., 1995); of recoverin and neurocalcins in the rat olfactory epithelium and hippocampus (Bastianelli et al., 1995a, b); of neurocalcins α and δ in the rat cerebellum (Kato et al., 1998); of VILIP-1 and VILIP-3 in the rat cerebellum and hippocampus (Spilker et al., 2000), and of VILIP-1 and VILIP-3 in human brain (Bernstein et al., 1999). It should be noted that the immunohistochemical data have to be taken with some caution because, due to the high similarity of subfamily members, antibodies may recognize several NCS proteins.

Many detailed immunohistochemical studies on the distribution of NCS proteins in the brain have focussed on the hippocampus and the cerebellum. For VILIP-1 and VILIP-2, studies were carried out in the hippocampus of rat and gerbil, respectively (Saitoh et al., 1995; Lenz et al., 1996b). The two proteins appear to be differentially distributed. Whereas VILIP-2 expression is most prominent in CA1/CA2 regions and the dentate gyrus of the rat hippocampus, VILIP-1 immunoreactivity is strongest in the CA3 region of the gerbil. Hippocalcin is strongly expressed, though with varying intensities, in the different parts of the rat hippocampus, but is also present in some other brain regions including the cerebellum (Saitoh et al., 1994; Grant et al., 1996). At the transcript level regional differences in the expression levels of hippocalcin and VILIP-3 can be observed in the rat hippocampus. As compared to other hippocampal subregions, a relatively high expression level of VILIP-3 and a relatively low level of hippocalcin, is observed in the dentate gyrus (Spilker et al., 2000).

Co-immunolocalization studies in the rat hippocampus have suggested that all neurocalcin immunoreactive neurons are GABAergic. They comprise about 19% of GABAergic neurons, suggesting that neurocalcin is a specific marker for a subpopulation of GABAergic interneurons (Martinez-Guijarro et al., 1998). However, it is not clear which of the at least 4 rodent neurocalcins were recognized, since the neurocalcin antibodies reacted with several NCS proteins in the bovine brain (Martinez-Guijarro et al., 1998).

The second brain structure which has been studied in more detail is the cerebellum. Northern blot data indicate that VILIP-2 is not expressed (Kajimoto et al., 1993), whereas VILIP-3 is prominently expressed in the rat cerebellum (Kajimoto et al., 1993). Immunohistochemical and *in situ* hybridization data revealed that at the cellular level, VILIP-3 is mainly expressed in Purkinje cells, but also a few scattered granule cells show VILIP-3 immunoreactivity (Spilker et al., 2000). In comparison, VILIP-1 transcripts are primarily found in the granule cell layer but are absent from Purkinje cells in chicken and rat cerebellum (Lenz et al., 1992; for neurocalcin α: Kato et al., 1998). Consistently VILIP-1 (and its ortholog neurocalcin α) is absent from Purkinje cells. Immunoreactivity is restricted to granule cells and to the molecular layer of the chicken and rat cerebellum (Lenz et al., 1996a; Kato et al., 1998). As detected by immunohistochemistry, hippocalcin, neurocalcin δ and the Frequenin subfamily member NCS-1 also appear to be expressed at different levels in Purkinje cells of the rat cerebellum (Saitoh et al., 1994; Kato et al., 1998; Schaad et al., 1996).

Studies in the human brain have been performed with antibodies against VILIP-1 and VILIP-3. Generally both proteins are expressed in subsets of neurons in virtually all brain regions. However, VILIP-1 displays a much more intense immunoreactivity than VILIP-3 (Bernstein et al., 1999).

Members of the VILIP-group of NCS proteins are also expressed in neurons of sensory pathways including the retina and the olfactory system. In the retina of various species, VILIP-1 or neurocalcin α is found in subsets of bipolar, amacrine and retinal ganglion cells, but not in photoreceptors (Lenz et al., 1992; Nakano et al., 1992; De Raad et al., 1995). Moreover a subset of olfactory receptor neurons in the rat olfactory epithelium were shown to express VILIP-1 (Boekhoff et al., 1997; Bastianelli et al., 1995a; Braunewell and Gundelfinger, 1999). Studies on the distribution of other VILIP subfamily members in these sensory systems are still incomplete and therefore it is unclear whether subsets of neurons expressing individual VILIPs are complementary or overlap to some extent.

4. FUNCTIONAL ASPECTS

4.1. Calcium-myristoyl switch of NCS proteins – A mechanism for signal transduction from and to the cell membrane?

N-terminal acylation seems to be a common feature of nearly all intracellular NCS proteins. For several members of the VILIP subfamily including hippocalcin (Kobayashi et al., 1993), neurocalcin δ (Ladant, 1995), VILIP-1 (Braunewell et al., 1997) and VILIP-3 (Spilker et al., 2000) as well as for the Frequenin subfamily member NCS-1 (McFerran et al., 1999) it has been shown that they can be myristoylated at their N-termini. Myristoylated recoverin, the prototype NCS protein, is able to perform a so-called calcium-myristoyl switch (for a review, see Ames et al., 1996 and Figure 3). Based on molecular modeling studies, a similar switch mechanism can be predicted for VILIP-1 (Figures 3 and 4). The functional significance of fatty acylation has been demonstrated for several polypeptides. For example, N-terminal myristoylation of the non-receptor protein tyrosine kinase src and of the myristoylated alanine-rich protein kinase C substrate MARCKS is essential for their functional activities (McLaughlin and Aderem, 1995).

Though most NCS proteins share the N-terminal myristoylation site, the significance and mechanism of the calcium-myristoyl switch may vary between individual family members. In GCAP-2, for example, the calcium-myristoyl switch acts in an opposite direction, i.e. after fractionation of retinal outer segments *in vitro* a dissociation from the membrane at high calcium concentrations was observed (Olshevskaya et al., 1997). However, this phenomenon was not seen for GCAP-1 and has been discussed as only occurring *in vitro* (Rudnicka-Nawrot et al., 1998). In contrast, the ability of NCS-1 to influence neurotransmitter release in permeabilized cells, was independent of myristoylation. NCS-1 was not found to perform a calcium-myristoyl switch following calcium binding, but did interact with cellular membranes in a calcium-independent manner (McFerran et al., 1999).

For VILIP-1 and VILIP-3 a calcium-dependent membrane association has been observed in cerebellar homogenates and in several transfected cell lines (Lenz et al., 1996c; Spilker et al., 1997, 2000). Furthermore, in subcellular fractionation experiments *in vitro*, VILIP-1 and VILIP-3 have been shown to interact with membranes in a calcium- and myristoylation-dependent manner (Braunewell et al., 1997; Spilker et al., 2000). However, the membrane association of VILIP-1 and VILIP-3 in the cerebellum in vivo appears to follow different mechanisms. Whereas VILIP-1 shows a tight membrane association in granule cells, VILIP-3 is evenly distributed in Purkinje cells, although it is well able to perform a calcium-myristoyl switch *in vitro*. Whether these differences between the two NCS proteins reflect differences in their calcium-binding abilities, in their conformational change in response to calcium binding, or properties of the cellular context remains to be investigated.

Myristoylated, but not myristoylation-deficient, VILIP-1 has a stimulating effect on Forskolin-stimulated cAMP accumulation in a stably transfected glioma cell model (Braunewell et al., 1997). Suppression of fatty acylation by mutation of the myristoylation site has even a dominant negative effect on Forskolin-induced cAMP production, indicating that myristoylation is of functional significance. The exact physiological role and the actual occurrence of the calcium-myristoyl switch of NCS proteins in living neurons *in vivo*, however, still remain to be clarified. The switch has been suggested to serve as a mechanism for compartmentalization of signaling cascades and/or for the specific transfer of calcium signals to or from the membrane compartment. The latter may be achieved by either calcium-mediated activation of membrane proteins or calcium-mediated recruitment of target proteins to the membrane. For each NCS protein these mechanisms may function differently and synergistically. A central issue will be to find out whether myristoylation of NCS proteins does serve actual signaling functions *in vivo*.

4.2. Modulation of different intracellular neuronal signaling cascades by VILIPs

A variety of extracellular signals including light, odorants, hormones, growth factors, electrical activity and neurotransmitters evoke highly specific increases of intracellular calcium levels. Apparently NCS proteins serve as effector or modulator proteins to transduce these calcium signals into appropriate physiological answers. In the past it has turned out to be very difficult to pinpoint the specific function of individual NCS proteins. One possible explanation for this failure may be that NCS proteins, like calmodulin, may modulate multiple intracellular targets. This "pleiotropy" of action appears to also hold true for VILIP subfamily members. It may be reflected by the widespread expression of VILIPs, in particular of VILIP-1, in subsets of neurons throughout the nervous system including sensory pathways such as the visual and olfactory systems (see above). Here we will summarize the still very patchy picture of the effects of VILIP subfamily proteins on various signaling components and pathways, and compare them with known functions of other NCS proteins. In particular functional studies on NCS proteins of retinal photoreceptor cells have stimulated the functional investigation of VILIPs.

4.2.1. Influence on G-protein-coupled receptor signaling

The well accessible signaling cascades of photoreceptor cells have been advantageous for the discovery of functions of recoverin, which initially served as the model NCS protein. Activation of the rhodopsin receptor by light leads to a G-protein-mediated increase in cGMP phosphodiesterase (PDE) activity which then decreases the intracellular cGMP concentration. The decrease in cGMP causes the closure of the cGMP-gated ion channels. The resulting hyperpolarization and decrease of intracellular calcium trigger the recovery of the dark state and adaptation of the system to light intensity. These latter signaling processes include up-regulation of retinal guanylyl cyclase (retGC) activity and turning off the light-induced signaling by rhodopsin phosphorylation and subsequent binding of arrestin to the light receptor. Both processes are calcium-dependent and modulated by recoverin, which has been reported to prolong PDE activation and to inhibit phosphorylation of rhodopsin at high calcium levels *in vitro*, possibly by physical interaction with rhodopsin kinase. The exact role of recoverin in the visual system is, however, still under investigation (for a review, see Polans et al., 1996; Braunewell and Gundelfinger, 1999).

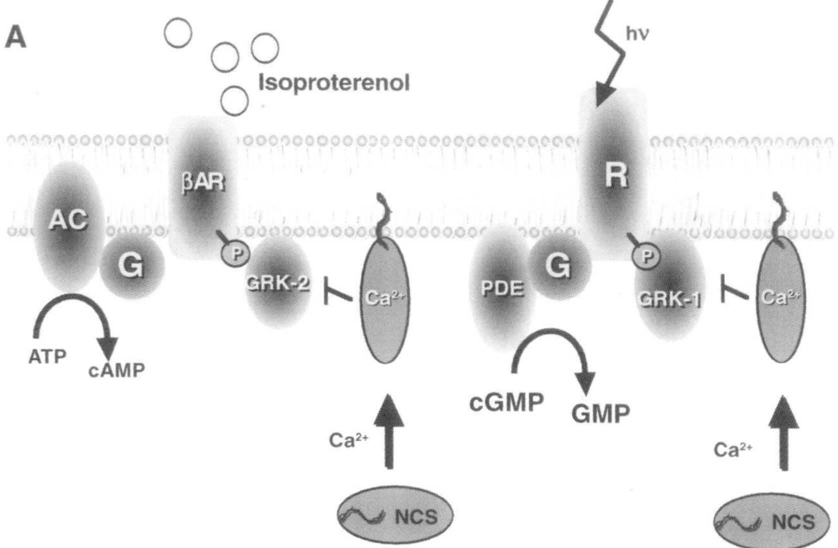

Figure 5A. NCS proteins from the VILIP-subfamily as modulators of G-protein coupled receptor kinases (GRK) signaling. Influence on β-adrenergic (left, GRK2, 3) and rhodopsin kinase (right, GRK1) subfamily of GRKs. Ligand binding (isoproterenol or light, hv) to the G-protein-coupled receptors (β-adrenerg βAR or light receptor R) activates an adenylyl cyclase (AC) or a cGMP-phosphodiesterase (PDE) via a G-protein (G). NCS-proteins inhibit receptor phosphorylation and therefore receptor desensitization by inhibiting the receptor kinases (GRK-1, GRK-2) in a calcium-dependent manner.

Various intracellular NCS proteins are expressed in retinal neurons other than photoreceptor cells. These include recoverin itself, NCS-1, neurocalcins, hippocalcin and VILIP-1 (see Section 3). NCS-1 and VILIP-1 show recoverin-like activity on the G-protein coupled receptor kinase 1 (GRK1, rhodopsin kinase) *in vitro* (DeCastro et al., 1995; Figure 5A), suggesting a potential interaction of NCS proteins with different GRKs in neurons. Neurocalcin α, the bovine ortholog of VILIP-1, was shown to inhibit the phosphorylation activity of GRK2 (β-adrenergic receptor kinase), another member of the GRK family (Kato et al., 1998). In contrast, a comparable effect of VILIP-1 on GRK3 activity in olfactory membranes (Boekhoff et al., 1997) or on β-adrenergic receptor kinase (GRK2) activity in stimulated C6 cells (Braunewell et al., 1997) has not been observed. Generally it appears that the family of GRKs is differentially affected by NCS proteins (reviewed by Iacovelli et al., 1999). The GRK family consists of six cloned members, which form three sub-families: the rhodopsin kinase subfamily (GRK1), the β-adrenergic subfamily (GRK2, 3) and the GRK4 subfamily (GRK 4α-δ, 5, 6). Whereas the GRK4 subfamily is preferentially influenced by calmodulin and not by NCS proteins, GRK1 can be inhibited by hippocalcin, VILIP-1 and NCS-1. As discussed above, VILIP-1 may not influence GRK2 and 3, the receptor kinases expressed in C6 cells and olfactory neurons, respectively.

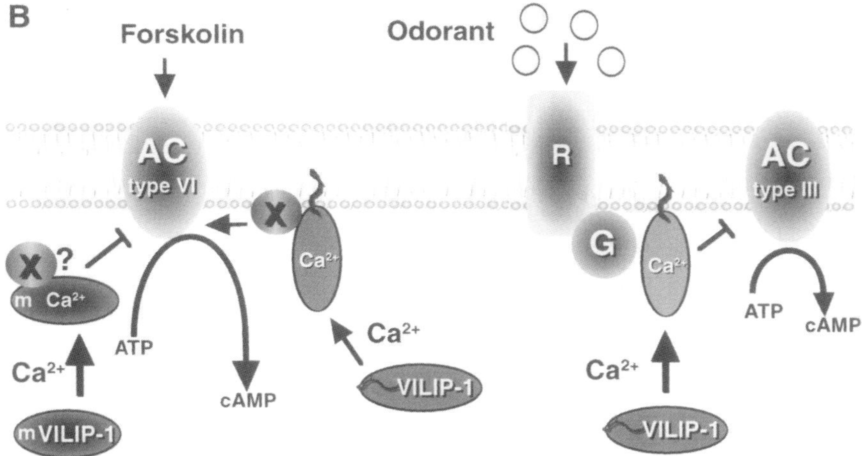

Figure 5B. NCS protein VILIP-1 as modulator of cyclase enzymes cascades in olfactory receptor neurons and C6 cells. Ligand binding (odorant) of the G-protein-coupled odor receptor (R) activates an adenylyl cyclase (olfactory AC type III) via a G-protein (G). Alternatively the adenylyl cyclase (AC type VI) can be directly activated via Forskolin in C6 glioma cells. Depending on the adenylyl cyclase isoform, VILIP-1 inhibits (AC type III) or enhances (AC type VI) cyclase activity. In the case of AC type VI wild type VILIP-1 leads to the enhancement of cAMP accumulation in VILIP-1 transfected C6 cells, whereas the myristoylation mutant (m) of VILIP-1, which does not show a calcium-myristoyl-switch, leads to a reduction of AC activity, possibly by trapping an as yet unknown factor x.

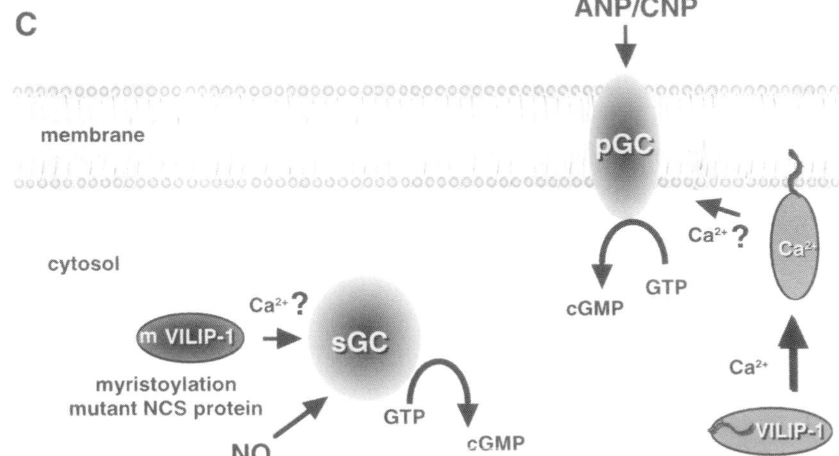

Figure 5C. NCS protein VILIP-1 as modulator of guanylyl cyclase enzymes cascades in transfected cells. Depending on the subcellular localization of VILIP-1 in PC12 cells the cytosolic myristoylation mutant of VILIP-1 influences soluble guanylyl cyclase and the membrane-localized wild-type VILIP-1 affects the particular guanylyl cyclases GC-A and GC-B. The natural ligand ANP and CNP for the particulate guanylyl cyclase receptors pGC-A and pGC-B leads to the increased production of cGMP. Membrane localized wild type VILIP-1 increases the particulate guanylyl cyclase (pGC) activity. In contrast the cytosolically localized myristoylation mutant (m) of VILIP-1 influences the NO-stimulated soluble guanylyl cyclase (sGC).

4.2.2. Influence on adenylyl cyclase signaling systems

A second primary sensory system in which calcium-dependent regulation mechanisms are essential for signal transduction, recovery of the receptive state and adaptation to stimulus intensity is the olfactory system (for a review, see Anholt, 1993). The binding of odorants to appropriate receptors in the olfactory knobs of sensory neurons, activates a trimeric G-protein, which in turn up-regulates adenylyl cyclase activity (Figure 5B). The olfactory adenylyl cyclase type III produces cAMP which opens a cyclic nucleotide-gated ion channel and leads to an influx of sodium and calcium ions and a subsequent depolarization of the sensory neuron. An odor-induced elevation of cGMP levels has also been observed although its precise role in olfactory signaling is not yet understood. It has been postulated that cGMP acts on cGMP-gated olfactory ion channels to produce odor adaptation (Zufall and Leinders-Zufall, 1997).

Analogous to the visual system, a calcium-dependent adaptation process is described for olfactory neurons. The type III adenylyl cyclase expressed in olfactory neurons is controlled by calcium ions, and the elevated concentrations of calcium following odor-stimulation abrogate the odorant-induced cAMP response in olfactory cilia (Anholt, 1993). In addition, elevated calcium levels mediated by calmodulin lead to an increased PDE activity and subsequently the closure of cyclic nucleotide-gated cation channels. This cascade contributes to a decrease in responsiveness of olfactory neurons. Besides calmodulin other calcium-binding proteins have been identified in olfactory neurons including VILIP-1 (Boekhoff et al., 1997). A strong accumulation of VILIP-1 immunoreactivity is observed in olfactory knobs containing the olfactory signaling machinery (Braunewell and Gundelfinger, 1999). VILIP-1 inhibits adenylyl cyclase type III activity following odor stimulation in a calcium-dependent manner in olfactory membranes, which is proposed as an additional mechanism of olfactory adaptation (Boekhoff et al., 1997).

Regulation of cAMP levels by VILIP-1 has also been observed in VILIP-1-transfected C6 glioma cells, which express the adenylyl cyclase type VI (Braunewell et al., 1997). After stimulation of β-adrenergic receptors by isoproterenol or of adenylyl cyclase type VI by Forskolin, the cAMP level in these cells were increased several-fold in wild-type VILIP-1 transfected cells as compared to untransfected C6 cells. A myristoylation-deficient mutant of VILIP-1, which does not show membrane association in these cells, has the opposite effect. Interestingly, also the basic cAMP levels appear to be elevated in VILIP-1-transfected C6 cells, which induces differentiation processes in these glioma cells (Braunewell and Gundelfinger, 1997).

Together, these results obtained in different cellular systems suggest a possible modulatory role for VILIP-1 on the activity of distinct adenylyl cyclase isoforms either by a direct physical interaction or by involving an additional, yet unknown, factor (Figure 5B).

4.2.3. Influence on guanylyl cyclase signaling systems

As mentioned above, calcium triggers the recovery of the dark state and adaptation to light intensity in photoreceptor cells. These processes include the up-regulation of retinal guanylyl cyclase (retGC) activity. NCS proteins expressed in the retina, i.e. recoverin, GCAP-1 and GCAP-2, have been found to act as calcium-dependent modulators of retGC. GCAPs inhibit the retGC at resting calcium concentrations and activate the enzyme after lowering calcium by photoexcitation (for a review, see Polans et al., 1996). Similarly, in the olfactory system a member of the NCS family which is closely related to GCAP-1, rather than to GCAP-2, was immunohistochemically identified in olfactory neurons and was shown to influence one of two biochemically defined olfactory GCs (Moon et al., 1998). In contrast to what was observed in photoreceptor cells the GCAP-like molecule stimulates the olfactory GC at high calcium concentrations. cGMP- and cAMP-dependent signaling appear to be coupled in olfactory neurons (Moon et al., 1998). As expression of various GC enzymes outside the olfactory and visual system has been described, it is conceivable that other NCS proteins may have the various GCs as physiological targets in the nervous system. First clues came from studies on neurocalcin δ, showing that neurocalcin can substitute for GCAPs in retGC activation (Olshevskaya et al., 1999; Kumar et al., 1999; Vijay-Kumar and Kumar, 1999), as it has been suggested earlier for recoverin (Pongs et al., 1993). In contrast to GCAPs, neurocalcin δ and recoverin do activate retGC at high calcium concentrations. An influence of VILIP-1 on non-retinal GCs, namely the receptor GCs A and B and soluble cyclases, has been observed *in vitro*. In several transfected neural cell lines, including C6, PC12 and Neuro2A cells, VILIP-1 affects cGMP-levels. Following stimulation of receptor GCs with natriuretic peptides or the soluble GC with nitric oxide an enhanced cGMP-accumulation was observed in all three cell lines when transfected with VILIP-1. Interestingly, the effect on soluble and particulate GCs depends on the subcellular localization of VILIP-1. The membrane-localized wildtype VILIP-1 mainly affects the particulate receptor cyclases GC-A and GC-B, whereas the cytosolically localized myristoylation mutant of VILIP-1 influences the soluble cyclase enzyme (Figure 5C, Braunewell et al., 2000).

These data suggest that one cellular function of VILIPs is to interact with and modulate different types of cyclase enzymes, including receptor GC, soluble cyclases and certain isoforms of adenylyl cyclases in the nervous system. These NCS proteins may confer calcium-dependency to these cyclase enzymes.

4.2.4. Other functional activities of VILIPs

Influence of NCS proteins on two other cellular signaling processes, i.e. on growth factor and cytokine signaling pathways has been suggested. In an attempt to identify proteins that affect the activity and localization of MLK2/3 (mixed-lineage Ser/Thr kinases) a yeast two-hybrid cDNA library was screened and hippocalcin was identified as a possible interaction partner (Nagata et al., 1998). MLKs are closely related to the mitogen-activated protein kinase kinase kinase (MAPKKK) family (Tibbles and Woodgett, 1999). MLK2 and 3 interact with the small GTPases Rac and Cdc42, with motor proteins of the kinesin superfamily and colocalize with the tubulin cytoskeleton. This suggests that they are involved in regulation of cytoskeleton dynamics in cells. After activation, MLK2/3 activate the JNK (c-Jun N-terminal kinase)/MAP kinase cascade, but also activation of ERK (extracellular signal-regulated kinase) and p38 by MLKs has been reported (Tibbles et al., 1996). However, to date the interaction of hippocalcin with MLK has not been shown to be calcium-dependent and more importantly no influence on kinase activity or microtubule localization has been observed. Therefore the biological significance of the MLK-hippocalcin interaction remains unclear.

Another signaling pathway, which may be affected by NCS proteins, is the neurotrophin pathway involving neurotrophic factors such as BDNF or NGF and their receptors, the trks (tyrosine receptor kinases). In an attempt to identify novel proteins binding double-strand (ds) RNAs, a dsRNA binding domain has been identified in VILIP-1, which shows homology to the conserved amino acids of the binding domains of classical dsRNA binding proteins. Furthermore, a specific calcium-dependent interaction of VILIP-1 with the 3′-untranslated region of trkB has been shown (Mathisen et al., 1999). Therefore, an indirect effect of VILIP-1 on trkB signaling by influencing the expression and/or localization of the corresponding transcripts has been postulated. Notably in this context, it has been shown that VILIP-1 expression is regulated by neurotrophin receptor activation (Friedel et al., 1997). Since VILIP-1 and trkB are both expressed in hippocampal neurons a cooperation on the cellular level seems possible. At the subcellular level, trkB is localized in dendrites in hippocampal neurons and VILIP-1 is expressed throughout the neuron including dendrites and synaptic structures (Lenz et al., 1996b). Interestingly, neuronal activity localizes trkB to the dendritic compartment in hippocampal neurons (Tongiorgi et al., 1997) and for VILIP-1 an increase in expression in the hippocampus following neuronal stimulation has been shown (Manahan-Vaughan and Braunewell, 1999). Therefore, a functional interaction of the two molecules *in vivo* is conceivable; in a feedback regulation VILIP-1 may affect trkB expression and VILIP-1 expression may be regulated by trkB activation. However, further studies are needed to substantiate this conjecture and to elucidate the signal cascade from calcium to neurotrophins in the hippocampus.

Members of the VILIP subfamily of NCS proteins have also been implicated in the pathophysiology of neurodegenerative diseases. A serum against neurocalcins from bovine brain, showed a decreased staining in the neuropil surrounding neocortical neurons in Alzheimer patients (Shimohama et al., 1996). The pathologic mechanism underlying the alteration of the immunoreactivity of neurocalcins in Alzheimers disease is presently unclear. It was speculated that the reduction in immunoreactivity of neurocalcins simply reflects synaptic pathology known to occur in Alzheimers disease. Changes in calcium sensor proteins may therefore be secondary to a perturbed calcium homeostasis, as it is common for Alzheimers disease.

It was recently shown that VILIP-1 and VILIP-3 immunoreactivity is reduced in the cerebral cortex in Alzheimers disease in comparison to control brains. The number of VILIP-1 containing neurons was reduced in the temporal cortex and both proteins were found to be closely associated with typical neuropathologic hallmarks (Braunewell et al., 2000). The facts that there were fewer VILIP-1 containing nerve cells in the cerebral cortex of Alzheimers disease patients and that VILIP-1 immunoreactivity is associated with plaques and tangles may point to a certain susceptibility of VILIP-expressing neurons in Alzheimers disease. In contrast, although being discussed controversially, there are hints that cortical neurons that contain calcium-buffering proteins, such as calbindin-D28k, appear to be relatively resistant to cellular changes in Alzheimers disease (Iacopino and Christakos, 1990). One may argue that changes in calcium-sensor and calcium-buffer proteins may be the Yin and Yang of disturbed calcium homeostasis in Alzheimers disease. Therefore, detailed co-localization studies for "neuroprotective" and "neurotoxic" calcium-binding proteins will be necessary to substantiate this hypothesis.

5. SUMMARY: AN EMERGING ROLE FOR VILIPS IN MULTIPLE SIGNALING PROCESSES

Members of the VILIP subfamily of NCS proteins are involved in a variety of calcium-dependent signaling cascades in nerve cells. The calcium-myristoyl switch mechanism may enable NCS proteins to shuttle information from the cytoplasm to the cell membrane and vice versa in a calcium-dependent manner. NCS proteins may therefore act as mediators to confer calcium-dependence to key players of intracellular signaling pathways including G-protein-coupled receptor kinases and members of the superfamily of cyclase enzymes. Furthermore, there are recent reports on an implication of NCS proteins on different levels of receptor tyrosine kinase signaling pathways including neurotrophin and MAPK pathways. Generally, NCS proteins may play an essential role in the cross-talk between second messenger systems, such as calcium and cAMP and cGMP or calcium and MAPK pathways in neurons and accordingly may affect the fine-tuning of neuronal signaling and/or participate in novel neuronal coincidence detection systems.

ACKNOWLEDGEMENTS

Work in the authors' laboratories is supported by grants from Deutsche Forschungsgemeinschaft, Kultusministerium des Landes Sachsen-Anhalt and by the Fonds der Chemischen Industrie.

REFERENCES

Ames J.B., Tanaka, T., Stryer, L. and Ikura, M., 1996, Portrait of a myristoyl switch protein, *Curr. Opin. Struct. Biol.* 6, 432–438.

Ames, J.B., Ishima, R., Tanaka, T., Gordon, J.I., Stryer, L. and Ikura, M., 1997, Molecular mechanics of calcium-myristoyl switches, *Nature* 389, 198–202.

Anholt, R.R., 1993, Molecular neurobiology of olfaction, *Crit. Rev. Neurobiol.* 7, 1–22.

Bastianelli, E., Polans, A.S., Hidaka, H. and Pochet, R., 1995a, Differential distribution of six calcium-binding proteins in the rat olfactory epithelium during postnatal development and adulthood, *J. Comp. Neurol.* 354, 395–409.

Bastianelli, E., Takamatsu, K., Okazaki, K., Hidaka, H. and Pochet, R., 1995b, Hippocalcin in rat retina. Comparison with calbindin-D28k, calretinin and neurocalcin, *Exp. Eye Res.* 60, 257–266.

Bernstein, H.-G., Baumann, B., Danos, P., Diekmann, S., Bogerts, B., Gundelfinger, E.D. and Braunewell, K.-H., 1999, Regional and cellular distribution of neural visinin-like protein immunoreactivities (VILIP-1 and VILIP-3) in human brain, *J. Neurocytol.*, in press.

Boekhoff, I., Braunewell, K.-H., Andreini, I., Breer, H. and Gundelfinger, E.D., 1997, The calcium-binding protein VILIP in olfactory neurons: Regulation of second messenger signaling, *Eur. J. Cell Biol.* 72, 151–158.

Braunewell, K.-H. and Gundelfinger, E.D., 1997, Low level expression of calcium-sensor protein VILIP induces cAMP-dependent differentiation in rat C6 glioma cells, *Neurosci. Lett.* 234, 139–142.

Braunewell, K.-H. and Gundelfinger, E.D., 1999, Intracellular neuronal calcium sensor (NCS) proteins – A family of EF-hand calcium-binding proteins in search for function, *Cell Tissue Res.* 299, 1–12.

Braunewell, K.-H., Spilker, C., Behnisch, T. and Gundelfinger, E.D., 1997, The neuronal calcium-sensor protein VILIP modulates cyclic AMP accumulation in stably transfected C6 glioma cells: Amino-terminal myristoylation determines functional activity, *J Neurochem.* 68, 2129–2139.

Braunewell, K.-H., Riederer, P., Spilker, C., Gundelfinger, E.D., Bogerts, B. and Bernstein, H.-G., 2000, Abnormal localization of two neuronal calcium sensor proteins, Visinin-Like Proteins (VILIPs)-1 and -3, in neocortical brain areas of Alzheimer Disease patients, *Dementia*, in press.

Buxbaum, J.D., Choi, E.K., Luo, Y., Lilliehook, C., Crowley, A.C., Merriam, D.E. and Wasco, W., 1998, Calsenilin: A calcium-binding protein that interacts with the presenilins and regulates the levels of a presenilin fragment, *Nat. Med.* 4, 1177–1181.

Carrion, A.M., Link, W.A., Ledo, F., Mellstrom, B. and Naranjo, J.R., 1999, DREAM is a Ca^{2+}-regulated transcriptional repressor, *Nature* 398, 80–84.

Cox, J.A., Durussel, I., Comte, M., Nef, S., Nef, P., Lenz, S.E. and Gundelfinger, E.D., 1994, Neuron-specific calcium-binding proteins, *J. Biol. Chem.* 269, 32807–32813.

De Castro, E., Nef, S., Fiumelli, H., Lenz, S.E., Kawamura, S. and Nef, P., 1995, Regulation of rhodopsin phosphorylation by a family of neuronal calcium sensors, *Biochem. Biophys. Res. Commun.* 216, 133–140.

De Raad, S., Comte, M., Nef, P., Lenz, S.E., Gundelfinger, E.D. and Cox, J.A., 1995, Distribution pattern of three neural calcium-binding proteins (NCS-1, VILIP and recoverin) in chicken, bovine and rat retina, *Histochem. J.* 27, 524–535.

Flaherty, K.M., Zozulya, S., Stryer, L. and McKay, D.B., 1993, Three-dimensional structure of recoverin, a calcium-sensor in vision, *Cell* 75, 709–716.

Friedel, R.H., Schnurch, H., Stubbusch, J. and Barde, Y.A., 1997, Identification of genes differentially expressed by nerve growth factor- and neurotrophin-3-dependent sensory neurons, *Proc. Natl. Acad. Sci. USA* 94, 12670–12675.

Grant, A.L., Jones, A., Thomas, K.L. and Wisden, W., 1996, Characterization of the rat hippocalcin gene: The 5′ flanking region directs expression to the hippocampus, *Neuroscience* 75, 1099–1115.

Heizmann, C.W. and Hunziker, W., 1991, Intracellular calcium-binding proteins, more sites than insights, *Trends Biochem.* 16, 98–103.

Iacopino, A.M. and Christakos, S., 1990, Specific reduction of calcium-binding protein (28-kilodalton calbindin-D) gene expression in aging and neurodegenerative diseases, *Proc. Natl. Acad. Sci. USA* 87, 4078–4082.

Iacovelli, L., Sallese, M., Mariggio, S. and de Blasi, A., 1999, Regulation of G-protein-coupled receptor kinase subtypes by calcium sensor proteins, *FASEB J.* 13, 1–8.

Ikura, M., 1996, Calcium binding and conformational response in EF-hand proteins, *Trends Biochem.* 21, 14–17.

Kajimoto, Y., Shirai, Y., Mukai, H., Kuno, T. and Tanaka, C., 1993, Molecular cloning of two additional members of the neural visinin-like Ca(2+)-binding protein gene family, *J. Neurochem.* 61, 1091–1096.

Kato, M., Watanabe, Y., Iino, S., Takaoka, Y., Kobayashi, S., Haga, T. and Hidaka, H., 1998, Cloning and expression of a cDNA encoding a new neurocalcin isoform (neurocalcin alpha) from bovine brain, *Biochem. J.* 331, 871–876.

Kobayashi, M., Takamatsu, K., Saitoh, S. and Nogushi, T., 1993, Myristoylation of hippocalcin is linked to its membrane association properties, *J. Biol. Chem.* 268, 18898–18904.

Kraut, N., Frampton, J. and Graf, T., 1995, Rem-1, a putative direct target gene of the Myb-Ets fusion oncoprotein in haematopoietic progenitors, is a member of the recoverin family, *Oncogene* 10, 1027–1036.

Kumar, V.D., Vijay-Kumar, S., Krishnan, A., Duda, T. and Sharma, R.K., 1999, A second calcium regulator of rod outer segment membrane guanylate cyclase, ROS-GC1: Neurocalcin, *Biochemistry* 38, 12614–12620.

Ladant, D., 1995, Calcium and membrane binding properties of bovine neurocalcin delta expressed in Escherichia coli, *J. Biol. Chem.* 270, 3179–3185.

Lenz, S.E., Henschel, Y., Zopf, D., Voss, B. and Gundelfinger, E.D., 1992, VILIP, a cognate protein of the retinal calcium binding proteins visinin and recoverin, is expressed in the developing chicken brain, *Brain Res. Mol. Brain Res.* 15, 133–140.

Lenz, S.E., Jiang, S., Braun, K. and Gundelfinger, E.D., 1996a, Localization of the neural calcium-binding protein VILIP (visinin-like protein) in neurons of the chick visual system and cerebellum, *Cell Tissue Res.* 283, 413–424.

Lenz, S.E., Zuschratter, W. and Gundelfinger, E.D., 1996b, Distribution of visinin-like protein (VILIP) immunoreactivity in the hippocampus of the Mongolian gerbil (Meriones unguiculatus), *Neurosci. Lett.* 206, 133–136.

Lenz, S.E., Braunewell, K.-H., Weise, C., Nedlina-Chittka, A. and Gundelfinger, E.D., 1996c, The neuronal EF-hand Ca(2+)-binding protein VILIP: Interaction with cell membrane and actin-based cytoskeleton, *Biochem. Biophys. Res. Commun.* 225, 1078–1083.

Manahan-Vaughan, D. and Braunewell, K.-H., 1999, Metabotropic glutamate receptor activation regulates expression of the neuronal calcium-sensor protein VILIP-1 in the hippocampus of freely moving rats: Implication in long-term potentation, *Neuropharmacology* 396, A27.

Martinez-Guijarro, F.J., Brinon, J.G., Blasco-Ibanez, J.M., Okazaki, K., Hidaka, H. and Alonso, J.R., 1998, Neurocalcin-immunoreactive cells in the rat hippocampus are GABAergic interneurons, *Hippocampus* 8, 2–23.

Mathisen, P.M., Johnson, J.M., Kawczak, J.A. and Tuohy, V.K., 1999, Visinin-like protein (VILIP) is a neuron-specific calcium-dependent double-stranded RNA-binding protein, *J. Biol. Chem.* 274, 31571–31576.

McFerran, B.W., Weiss, J.L. and Burgoyne, R.D., 1999, Neuronal Ca(2+) sensor 1. Characterization of the myristoylated protein, its cellular effects in permeabilized adrenal chromaffin cells, Ca(2+)-independent membrane association, and interaction with binding proteins, suggesting a role in rapid Ca(2+) signal transduction, *J. Biol. Chem.* 274, 30258–30265.

McLaughlin, S. and Aderem, A., 1995, The myristoyl-electrostatic switch: A modulator of reversible protein-membrane interactions, *Trends Biochem.* 20, 272–276.

Moon, C., Jaberi, P., Otto-Bruc, A., Baehr, W., Palczewski, K. and Ronnett, G.V., 1998. Calcium-sensitive particulate guanylyl cyclase as a modulator of cAMP in olfactory receptor neurons, *J. Neurosci.* 18, 3195–3205.

Nagata, K., Puls, A., Futter, C., Aspenstrom, P., Schaefer, E., Nakata, T., Hirokawa, N. and Hall, A., 1998, The MAP kinase kinase kinase MLK2 co-localizes with activated JNK along microtubules and associates with kinesin superfamily motor KIF3, *EMBO J.* 17, 149–158.

Nakano, A., Terasawa, M., Watanabe, M., Usuda, N., Morita, T. and Hidaka, H., 1992, Neurocalcin, a novel calcium binding protein with three EF-hand domains, expressed in retinal amacrine cells and ganglion cells, *Biochem. Biophys. Res. Commun.* 186, 1207–1211.

Nef, P., 1996, Neuron-specific calcium sensors (the NCS subfamily), in *Guidebook to the Calcium-Binding Proteins*, M.R. Celio (ed.), Oxford University Press, New York, pp. 94–98.

Okazaki, K., Iino, S., Inoue, S., Kobayashi, S. and Hidaka, H., 1994, Differential distribution of neurocalcin isoforms in rat spinal cord, dorsal root ganglia and muscle spindle, *Biochim. Biophys. Acta* 1223, 311–317.

Olshevskaya, E.V., Hughes, R.E., Hurley, J.B. and Dizhoor, A.M., 1997, Calcium binding, but not a calcium-myristoyl switch, controls the ability of guanylyl cyclase-activating protein GCAP-2 to regulate photoreceptor guanylyl cyclase, *J. Biol. Chem.* 272, 14327–14333.

Olshevskaya, E.V., Ermilov, A.N. and Dizhoor, A.M., 1999, Dimerization of guanylyl cyclase-activating protein and a mechanism of photoreceptor guanylyl cyclase activation, *J. Biol. Chem.* 274, 25583–25587.

Polans, A., Baehr, W. and Palczewski, K., 1996, Turned on by Ca^{2+}! The physiology and pathology of Ca(2+)-binding proteins in the retina, *Trends Neurosci.* 19, 547–554.

Pongs, O., Lindemeier, J., Zhu, X.R., Theil, T., Engelkamp, D., Krah-Jentgens, I., Lambrecht, H.-G., Koch, K.-W., Schwerner, J., Rivosecchi, R., Mallart, A., Galceran, J., Canal, I., Barbas, J.A. and Ferrus, A., 1993, Frequenin – A novel calcium-binding protein that modulates synaptic efficacy in the Drosophila nervous system, *Neuron* 11, 15–28.

Rudnicka-Nawrot, M., Surgucheva, I., Hulmes, J.D., Haeseleer, F., Sokal, I., Crabb, J.W., Baehr, W. and Palczewski, K., 1998, Changes in biological activity and folding of guanylate cyclase-activating protein 1 as a function of calcium, *Biochemistry* 37, 248–257.

Saitoh, S., Takamatsu, K., Kobayashi, M. and Noguchi, T., 1994, Expression of hippocalcin in the developing rat brain, *Brain Res. Dev. Brain Res.* 80, 199–208.

Saitoh, S., Kobayashi, M., Kuroki, T., Noguchi, T. and Takamatsu, K., 1995, The development of neural visinin-like Ca(2+)-binding protein 2 immunoreactivity in the rat neocortex and hippocampus, *Neurosci Res.* 23, 383–388.

Schaad, N.C., De Castro, E., Nef, S., Hegi, S., Hinrichsen, R., Martone, M.E., Ellisman, M.H., Sikkink, R., Rusnak, F., Sygush, J. and Nef, P., 1996, Direct modulation of calmodulin targets by the neuronal calcium sensor NCS-1. *Proc. Natl. Acad. Sci., USA* 93, 9253–9258.

Shimohama, S., Chachin, M., Taniguchi, T., Hidaka, H. and Kimura, J., 1996, Changes of neurocalcin, a calcium-binding protein, in the brain of patients with Alzheimer's disease, *Brain Res.* 716, 233–236.

Spilker, C., Gundelfinger, E.D. and Braunewell, K.-H., 1997, Calcium- and myristoyl-dependent subcellular localization of the neuronal calcium-binding protein VILIP in transfected PC12 cells, *Neurosci. Lett.* 225, 126–128.

Spilker, C., Richter, K., Smalla, K.-H., Manahan-Vaughan, D., Gundelfinger, E.D. and Braunewell, K.-H., 2000, The neuronal EF-hand calcium-binding protein VILIP-3 is expressed in cerebellar Purkinje cells and shows a calcium-dependent membrane association, *Neurosci.* 96, 121–129.

Tibbles, L.A. and Woodgett, J.R., 1999, The stress-activated protein kinase pathways, *Cell. Mol. Life Sci.* 55, 1230–1254.

Tibbles, L.A., Ing, Y.L., Kiefer, F., Chan, J., Iscove, N., Woodgett, J.R. and Lassam, N.J., 1996, MLK-3 activates the SAPK/JNK and p38/RK pathways via SEK1 and MKK3/6, *EMBO J.* 15, 7026–7035.

Tongiorgi, E., Righi, M. and Cattaneo, A., 1997, Activity-dependent dendritic targeting of BDNF and TrkB mRNAs in hippocampal neurons, *J. Neurosci.* 17, 9492–9505.

Vijay-Kumar, S. and Kumar, V.D., 1999, Crystal structure of recombinant bovine neurocalcin, *Nat. Struct. Biol.* 6, 80–88.

Zufall, F. and Leinders-Zufall, T., 1997, Identification of a long-lasting form of odor adaptation that depends on the carbon Monoxide/cGMP second-messenger system, *J. Neurosci.* 17, 2703–2712.

Calcium Binding to Extracellular Matrix Proteins, Functional and Pathological Effects

Alexander W. Koch, Jürgen Engel and Patrik Maurer

1. INTRODUCTION

Calcium concentrations in the extracellular space are 4 to 5 orders of magnitude larger than inside cells. These concentrations are highly controlled by a unique receptor system (Brown et al., 1995; Ward and Ricardi, this book; Brown et al., this book). Free calcium concentration in serum has been determined to be 1.2 mM, however, spatial and time-dependent fluctuations have been detected (for a review, see Maurer et al., 1996). Many different proteins in the extracellular space bind calcium ions and a variety of calcium-binding motifs have been identified in these proteins, e.g. the EF-hand motif, which is common in cytosolic proteins (Maurer et al., 1996). Calcium ions bound to extracellular proteins may serve various functions, but primarily stabilize protein structure which explains tight calcium binding (micromolar or smaller K_D values). In addition, sites for weakly bound calcium ions may sense variations in extracellular calcium levels and may be involved in regulation (Maurer et al., 1996; Koch et al., 1997). This chapter will focus on relevant findings concerning functional and pathological effects of calcium binding to two extracellular matrix proteins (fibrillin-1 and COMP) and the cell adhesion protein E-cadherin. Possible molecular mechanisms for these effects will be discussed as well.

For a number of matrix proteins, of which fibrillin-1 and cartilage oligomeric matrix protein (COMP) are the best studied examples, mutations in calcium binding domains lead to severe diseases, Marfan syndrome (MFS) for fibrillin-1 and pseudoachondroplasia for COMP. The genetic links between the diseases and the proteins are well established, but correlations

Alexander W. Koch • Department of Biochemistry and Molecular Biology, Mt. Sinai School of Medicine, New York, U.S.A. **Jürgen Engel** • Biozentrum der Universität, Basel, Switzerland. **Patrik Maurer** • Institut für Biochemie, Köln, Germany.

R. Pochet, R. Donato, J. Haiech, C. Heizmann and V. Gerke (eds.): Calcium: The Molecular Basis of Calcium Action in Biology and Medicine, 151–164.

between the mutations and the observed phenotypes are understood only in a few cases.

The structure and function of the cell adhesion protein E-cadherin strongly depends on calcium binding. The ectodomain of this essential molecule exhibits an unusual calcium-binding mode with up to three calcium ions binding with different affinity between pairs of modular units (domains). It was recently demonstrated that a weakly bound calcium ion between domains 1 and 2 is substantial for adhesion and most likely involved in regulation of this process (Koch et al., 1997; Pertz et al., 1999). At present, familial gastric cancer is the only inherited disease known to be caused by mutations in the E-cadherin gene (Guilford et al., 1998), but it appears likely that other such diseases exist.

2. MARFAN SYNDROME AND OTHER GENETIC DISORDERS CAUSED BY MUTATIONS IN THE CALCIUM-BINDING EGF DOMAINS OF FIBRILLIN-1

Marfan syndrome (MFS) is a heritable connective tissue disorder characterized by cardiovascular, skeletal and ocular abnormalities. It is caused by mutations in the gene coding for fibrillin-1 (Ramirez and Pereira, 1999). This extracellular matrix protein is an integral component of fibrillar structures called microfibrils, which are abundant in elastic tissues. Fibrillin-1 (molar mass about 350 kDa) consists of numerous modules, many of which are homologous and repeated along the polypeptide chain (Sakai and Keene, 1994). Calcium-binding EGF modules (cbEGF) are the largest group of homologous modules with 43 copies per fibrillin-1 molecule.

The domain organization of fibrillin-1 is schematically shown below:

$$xe_3hE_2cpeE_4cE_3hE_1cE_{12}cE_2cE_7cE_5cE_7z$$

with E: *cbEGF* motif, e: *EGF* motif, c: 8-cysteine motif, h: hybrid motif, p: proline rich region, and x, y, z: unique regions. Subscripts indicate the number of modules with internal homology repeated in uninterrupted sequence.

A cbEGF module (E) itself is characterized by the following amino acid sequence pattern:

$$D/N - X - D/N - E/Q - X_m - D^*/N^* - X_n - Y/F$$

in which X stands for any residue, n and m are variable integers and the stars denote potential β-hydroxylation sites. The three-dimensional structure of cbEGF modules has been elucidated by NMR (Figure 1) and the mechanism of calcium binding is well understood (Downing et al., 1996). Dissociation equilibrium constants of Ca^{2+} ions were determined for a number of cbEGF domains of fibrillin-1 (Knott et al., 1996), factor X (Persson et al., 1989)

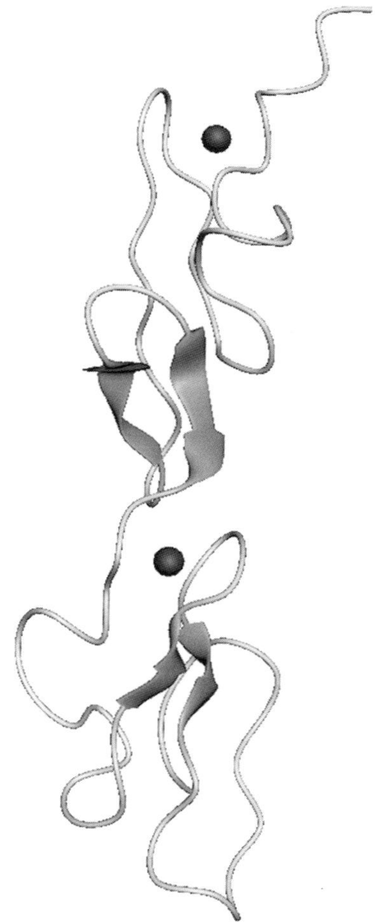

Figure 1. The structure of a pair of cbEGF domains 32 and 33 of fibrillin-1 as solved by NMR (Dowling et al., 1996). Beta-strands are depicted by arrows and calcium ions are shown as spheres. The figure was rendered from Molscript (Kraulis, 1991).

and of Notch protein (Rand et al., 1997). Values are in the range of 10^{-6} to 10^{-4} M demonstrating that the cbEGF domains are calcium saturated at physiological calcium levels. Removal of calcium by addition of chelators leads to a large increase in proteolytic susceptibility of fibrillin (Handford et al., 1995; Reinhard et al., 1997).

Approximately 120 fibrillin-1 mutations have been identified so far which lead to MFS or related diseases (Ramirez and Pereira, 1999). Most of these mutations occur within cbEGF modules and affect residues involved in calcium binding. These mutations are expected to lead to local structural changes in the fibrillin protein and possibly to increased proteolytic degradation because of the stabilizing role of calcium. For example, in the fibrillin mutation E2447K an increased disruption of the protein by matrix metalloproteinases was indeed observed (Ashworth et al., 1999). In a model system using two MFS-causing mutations in cbEGF modules (N548I, E1073K), the mutated polypeptides were degraded by various proteases at sites close to the mutations, but not always within the mutated module (Reinhardt et al., 2000). In this context it may be recalled that calcium binding is also influenced by domain linkages (Smallridge et al., 1999). A loss or modification of fibrillin by degradation is a possible reason for MFS and other fibrillin-related genetic disorders, but more direct effects like the disturbance of microfibril formation from mutated molecules may also play a role.

3. CALCIUM BINDING TO CARTILAGE OLIGOMERIC MATRIX PROTEIN IS AFFECTED IN PSEUDOACHONDROPLASIA AND MULTIPLE EPIPHYSEAL DYSPLASIA

Cartilage oligomeric matrix protein (COMP) is a secreted pentameric glycoprotein and belongs to the thrombospondin protein family (Hedbom et al., 1992; Mörgelin et al., 1992; Oldberg et al., 1992). It is built from modular units with a coiled-coil domain at the amino terminus which is responsible for the pentamerization of COMP (Mörgelin et al., 1992; Malashkevich et al., 1996). This assembly domain is followed by four epidermal growth factor (EGF)-like domains, a region encompassing eight so-called type 3 repeats, and a carboxy-terminal domain. In electron microscopy a bouquet-shaped form of COMP was seen with five arms held together by the small coiled-coil domain. The arms are formed by thin filaments with globular ends (Figure 2A, Mörgelin et al., 1992). COMP is found in cartilage, ligaments, tendons and synovial fluid (Hedbom et al., 1992; DiCesare et al., 1994, 1997). COMP has been proven to be a prognostic marker for cartilage destruction in early rheumatoid arthritis, osteoarthritis and after traumatic injury (Saxne and Heinegard, 1992; Forslind et al., 1992; Neidhard et al., 1997), however, its precise function is not established.

Pseudoachondroplasia (PSACH) and Multiple Epiphyseal Dysplasia (MED) constitute two autosomal dominantly inherited forms of osteochondrodysplasias, which are characterized by severe-to-moderate short-limb dwarfism with normal skull. Their clinical features also include pronounced joint laxity and limitation of the movement of joints. The major clinical complication is caused by premature osteoarthritis of the weight-bearing joints (International Working Group on Constitutional Diseases of Bone, 1998). Both syndromes exhibit considerable clinical heterogeneity and show a broad phenotypical overlap with PSACH at the severe and MED at the mild end of the disease spectra (International Working Group on Constitutional Diseases of Bone, 1998 and references therein). Genetic linkage to chromosome 19 was recently demonstrated for mild and severe forms of osteochondrodysplasias and was soon followed by the identification of mutations in COMP

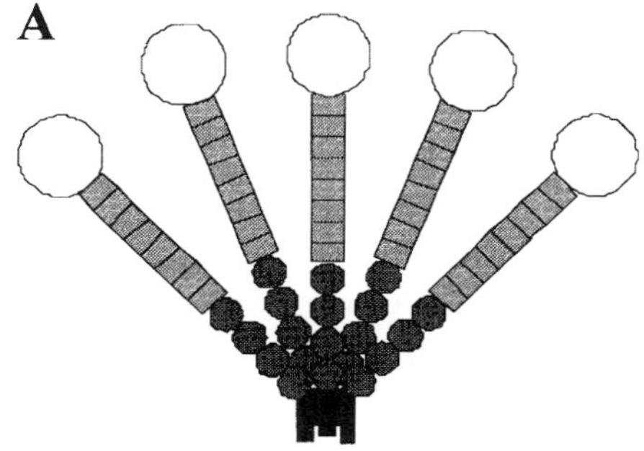

B

```
R1          .....RDTDLDGFPDEKLRC.....PEPQCRKDNCVTV
R2          PNSGQEDVDRDGIGDAC..DPDADGDGVPNEKDNCPLV
R3          RNPDQRNTDEDKWGDAC...............DNCRSQ
R4          KNDDQKDTDQDGRGDAC..DDDIDGDRIRNQADNCPRV
R5          PNSDQKDSDGDGIGDAC...............DNCPQK
R6          SNPDQADVDHDFVGDACDSDQDQDGDGHQDSRDNCPTV
R7          PNSAQEDSDHDGQGDAC..DDDDDNDGVPDSRDNCRLV
R8          PNPGQEDADRDGVGDVC..QDDFDADKVVDKIDVCPEN

Consensus   .N..Q.D.D.DGhGDhC..D.D.D.D.h.D..DNCP.h
```

Figure 2. Domain structure of COMP and sequence of the calcium-binding type 3 repeats. (A) The coiled-coil domain are depicted as rectangles, the epidermal growth factor-like domains as black circles, the type 3 repeats as gray rectangles and the carboxy-terminal globular domain as an open circle. (B) The sequences of the eight calcium-binding type 3 repeats of human COMP (residues 268-523) are aligned. Letters shown on gray backgrounds indicate conserved residues; residues shown in a white font on dark gray background are mutated in PSACH and MED patients, respectively.

responsible for both PSACH and MED (Hecht et al., 1995; Briggs et al., 1995).

All mutations in the COMP gene that are known to cause PSACH or MED result either in a single amino acid substitution, a small deletion or an insertion of one or two amino acids (Delot et al., 1999; Deere et al., 1999, and references therein). No premature stop codon has been found so far. Most mutations in COMP affect residues in the type 3 repeats and only a few mutations are located in the C-terminal domain. The type 3 repeats encompass 256 amino acids and can be subdivided into eight homologous parts. These are characterized by the presence of two cysteines and one or two conserved patterns of aspartic acid residues, DxDxDxxxDxxD (Figure 2). This pattern is reminiscent of calcium-coordinating residues in EF-hands, however, no indications of helices flanking the acidic residues are found in COMP and the related thrombospondins. For thrombospondin-1, it was postulated that the region encompassing the type 3 repeats is responsible for calcium binding although definite proof is still lacking (Misenheimer and Mosher, 1995). Interestingly, most of the mutations causing PSACH and MED affect the acidic residues of COMP (Figure 2). Thus, it was speculated that the mutations of acidic residues in the type 3 repeats may disturb calcium binding, folding and stability of COMP.

In the chondrocytes of PSACH and MED patients, large inclusions within the rough endoplasmic reticulum were detected and they were shown to contain COMP, as well as collagen IX and proteoglycans (Maynard et al., 1972; Maddox et al., 1997). However, tenocytes of patients do not show inclusions. In cell culture, chondrocytes, tenocytes and ligament cells of patients have been shown to secrete COMP (Hecht et al., 1998; Delot et al., 1998). This suggests that the PSACH and MED phenotypes are presumably caused by the combined effect of reduced amounts of secreted wild-type COMP (and other proteins), and a dominant malfunction of secreted mutated COMP.

In order to understand the consequences of mutations in COMP which cause PSACH and MED, several fragments of COMP as well as the full-length protein were expressed recombinantly. It was demonstrated that the type 3 repeats of COMP are responsible for highly cooperative calcium binding and that they undergo a conformational change upon calcium binding. Two mutations were introduced in recombinant COMP. One mutation resulted in the deletion of Asp470 which causes PSACH with severe dwarfism, and in the second mutation Asp361 was substituted with Tyr which causes MED with early onset osteoarthritis. Both of the mutated COMP proteins were secreted by the mammalian cells following recombinant expression, and both had folds similar to wild-type COMP. This demonstrates that the mutations in the type 3 repeats do not dramatically affect folding and stability of COMP. However, calcium binding was affected by these mutations. Normal type 3 repeats bind calcium with a $K_d = 0.3$ mM and a Hill coefficient of 3.7. In the mutated type 3 repeats 80% of the conformational change following calcium binding shows the same affinity and cooperativity, but 20% reveals a five to nine-fold reduction in affinity (Thur, Paulsson and Maurer, manuscript in preparation). Similar results were obtained for a Asp446 to Asn mutation which causes PSACH (Maddox et al., 1997). The number of calcium ions bound was reduced from 17 calcium ions per wild-type peptide to 8 calcium ions per mutated type 3 repeat, and structural differences were observed by electron microscopy and ultracentrifugation (Maddox and Bächinger, personal communication). The distinct structural changes in the mutated type 3 repeats may alter binding sites of COMP for its ligands such as collagens and may thus interfere with the structure and stability of cartilage, tendons and ligaments.

4. CALCIUM DEPENDENT CELL TO CELL ADHESION BY CADHERINS

Cadherins ("**Ca**lcium dependent **adhe**rent prote**ins**") were initially identified and characterized as glycoproteins responsible for calcium-dependent cell to cell adhesion (Takeichi, 1988). To date, a large family of cadherin proteins has been identified, with at least 60 different family members. The recent discovery of 52 brain specific so-called protocadherins (Wu and Maniatis, 1999) indicates that the number of cadherins and cadherin-related proteins is even much larger. The cadherin superfamily diverges into several subfamilies, namely classical cadherins, e.g. E- and N-cadherin, desmosomal cadherins, protocadherins and other cadherin-related proteins. In cadherin nomenclature, the initial letter refers to the tissue in which the corresponding cadherin was first detected, e.g. E for epithelial and N for neuronal, but cadherins are generally not restricted to any one tissue type. All the cadherins are membrane associated – most are type I transmembrane proteins – and their extracellular regions are built up by repeated modules, so-called cadherin repeats or extracellular cadherin domains (CA domains). Besides these two features, the superfamily is quite diverse as different cadherins can have variable numbers of cadherin domains and other modules can be inserted between CA domains. Conversely, classical cadherins, which are by far the best characterized members of the superfamily, are defined by their conserved cytosolic domain and have five such CA domains. Besides strict calcium dependence, adhesion mediated by classical cadherins is generally homophilic, that is, a cadherin molecule of a certain type interacts preferentially with another cadherin of the same type, thus enabling differential sorting of cells which express different cadherins.

4.1. Biological function of cadherins

The ability to perform homophilic adhesion is of great importance for the role of cadherins in early development. For instance, it has been shown that the very first adhesion event in early mouse development, the so-called blastomere compaction, is mediated by E-cadherin (Hyafil et al., 1980; Vestweber and Kemler, 1984). Also later in development morphogenetic events can often be directly associated with the modulation of cadherin expression. For example, in gastrulation and neurulation, two major morphogenetic events during embryogenesis, cadherin expression within a particular cell layer changes from E- to N-cadherin (Takeichi, 1988). However, the function of cadherins is not restricted to developmental processes and cadherins are in general crucial for the initiation, maintenance and proper functioning of tissue architecture in both vertebrates and invertebrates. Cadherin mediated cell to cell adhesion is not only regulated by the expression of different cadherin subtypes (with different adhesive specificities), but also by interactions of cadherins with cytosolic proteins. These cytosolic binding partners, which are called catenins, not only physically anchor cadherins to the cytoskeleton but also involve them in signaling events.

In many cases, loss of E-cadherin function can be correlated with the invasive behavior of tumor cells. A transgenic mouse model demonstrated that E-cadherin loss-of-function plays a causal role in the multifactorial process of tumorigenesis (Perl et al., 1998). E-cadherin has therefore a clear tumor suppressor function (Semb and Christofori, 1998). Interestingly, in at least one specific type of tumors – the so-called diffuse type gastric tumors – many mutations were found in the E-cadherin gene, namely skipping of exon 8 or 9 and substitutions of aspartic acid residues, which all destroy calcium-binding sites (Becker et al., 1994). Structural studies (see below) provide an explanation of why an amino acid substitution of only a single calcium-binding residue can lead to a drastic effect on the structure and function of cadherins.

Cadherins are also exploited as portals of entry for some pathogens. For instance, the entry of the gram-positive bacterium *Listeria monocytogenes* into epithelial cells involves the specific interaction of a bacterial surface protein (internalin) with human E-cadherin (Marra and Isberg, 1996). Another gram-positive bacterium, *Shigella flexneri*, uses cadherins to spread between epithelial cells (Sansonetti et al., 1994). Cadherins have also been implicated in inflammatory bowel disease, e.g. E-cadherin is downregulated in ulcerative colitis and Crohn's disease (Karayiannakis et al., 1998), whereas, blocking N-cadherin function in transgenic mice results in a chronic inflammatory bowel disease resembling Crohn's disease (Hermiston and Gordon, 1995).

4.2. Role of calcium in cadherin mediated cell to cell adhesion

Early observations have revealed that Ca^{2+} provides protection against proteolysis of cadherins (Hyafil et al., 1981). In the absence of Ca^{2+}, cadherins are cleaved off from the cell surface and eventually further fragmentated. As a result, dissociation occurs in cells, which are held together by cadherin mediated adhesion. Conserved aspartate-rich regions in cadherin sequences have been previously predicted to be involved in calcium binding (Ringwald et al., 1987). More recently, high resolution structures and electron microscopy unveiled molecular details of calcium binding as well as the three dimensional structure of CA domains. CA domains have a compact barrel-like fold consisting of seven antiparallel β-strands ("β-sandwich") and a rather flexible loop region at the C-terminus, which connects two consecutive cadherin domains (Shapiro et al., 1995; Nagar et al., 1996). In the presence of Ca^{2+} the entire extracellular region (ectodomain) of classical cadherins adopts an elongated rod-like structure which enables a cadherin molecule to interact in a parallel fashion with a neighboring cadherin molecule. These so-called cis-dimers interact in an antiparallel fashion with identical cis-dimers protruding

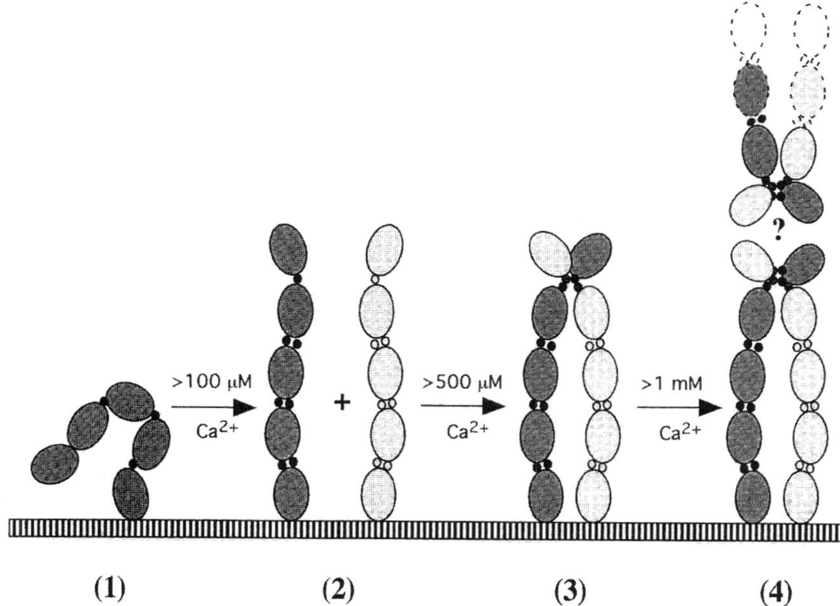

Figure 3. Model for cadherin homophilic interactions. Cadherin domains are depicted as elongated globes and calcium ions as small spheres. According to calcium-binding data and structural studies the following model for the calcium dependence of homophilic cadherin interactions has been proposed (Koch et al., 1999). (1) In the absence of Ca^{2+} or at very low calcium concentrations ($\ll$ 100 μM) the interdomain linkages are flexible. (2) At 100 μM most of the calcium-binding sites between domains 3, 4 and 5 are saturated leading to a rod-like shape of the entire ectodomain. (3) Cis dimerization of cadherin ectodomains is observed at medium calcium concentrations ($\sim$ 500 μM) whereas calcium concentrations of 1 mM or higher are necessary for antiparallel, adhesive interactions between activated cis-dimers (4).

from an opposing cell surface, thus establishing adhesive cell to cell contacts (Figure 3, Koch et al., 1999). Without Ca^{2+}, the rod-like structure collapses to a more condensed globular structure (Pokutta et al., 1994) rendering parallel and antiparallel interactions impossible. Binding of Ca^{2+} to the ectodomain leads to a major conformational change, because calcium binding rigidifies the otherwise flexible loop regions between single CA domains. Crystal structures of pairs of domains from both N- and E-cadherin finally have nicely shown that a large array of aspartate and glutamate residues from the first and the second CA domain form a calcium-binding pocket (Nagar et al., 1996; Tamura et al., 1998; Pertz et al., 1999). This calcium-binding pocket, which has no similarities to other known calcium-binding motifs, is formed by 11 residues which provide the coordination sphere for a total of three calcium ions in the interdomain loop region. This calcium-binding mode does not only render the loop region less flexible but also orients the two consecutive CA domains towards each other. The presence of four similar calcium-binding pockets between the five cadherin domains, therefore, readily explains the rod-like structure of the whole extracellular region. Furthermore, potential cleavage sites for proteases in the loop regions are not accessible in the calcium-bound state, thus explaining the increased protease resistance in the presence of Ca^{2+}.

In conclusion, the structural details of calcium binding illustrate why cadherins do not function in the absence of Ca^{2+}. Calcium-binding measurements, however, indicate that the role of Ca^{2+} might go beyond a solely structural one. Binding studies have revealed that the four calcium-binding pockets in E-cadherin are not identical in terms of numbers of bound Ca^{2+} and their dissociation constants (Koch et al., 1997). The first N-terminal binding pocket of E-cadherin – between CA domains 1 and 2 – binds three calcium, whereas, the four calcium-binding pockets of the entire ectodomain bind 9 Ca^{2+}. Moreover, binding of Ca^{2+} to the first calcium-binding pocket is considerably weaker than to the other three pockets. Especially one of the three calcium ions exhibits rather low binding affinity in the millimolar range. These different calcium-binding affinities may be of importance since the first calcium-binding pocket is located in the vicinity of proposed parallel and antiparallel interaction sites. Electron microscopy studies of E-cadherin ectodomains provided a model of how different calcium concentrations could influence cadherin-mediated adhesion (Pertz et al., 1999): Rather low calcium concentrations ($\sim$ 100 μM) are necessary and sufficient to maintain the rod-like structure of the entire ectodomain (Figure 3) which is in perfect agreement with earlier measured dissociation constants (Koch et al., 1997). No parallel or antiparallel interactions are observed at these calcium concentrations, instead the formation of parallel dimers requires calcium concentrations of about 500 μM and the antiparallel interaction between cis-dimers requires even higher concentrations of 1 mM Ca^{2+} or more (Figure 3). This is the calcium concentration at which the weakest of the three binding sites between domain 1 and 2 is becoming saturated. At present, the structural basis for this calcium-dependent activation of cadherins has not been totally scrutinized, but it is striking, that the most N-terminal domain of cadherins has been implicated in parallel and antiparallel (head-to-head) interactions (Nose et al., 1990; Shapiro et al., 1995), and that the neighboring calcium-binding sites have to be saturated in order to obtain these interactions.

The physiological significance of calcium-mediated activation of cadherins is not immediately apparent. In vertebrates, the free calcium concentration in blood is tightly regulated at $\sim$ 1.2 mM and is, therefore, assumed to be invariant in all other extracellular fluid. At this calcium concentration, most of the calcium-binding sites in extracellular proteins including cadherins are saturated and, consequently, should not be involved in activation or regulation processes. Conversely, there is increasing evidence that the dogma of an invariant extracellular calcium concentration has to be modified, at least with respect to certain tissues and certain developmental stages (Maurer et al., 1996). For example, drastic changes in extracellular calcium concentrations has been detected in bones, skin, placenta, lung alveoli of fetuses, and in the adult brain (Brown et al., 1995). In addition, the discovery of the so-called calcium-sensing receptor serves as the first example of an extracellular protein regulated by Ca^{2+} and demonstrates that Ca^{2+} can act as an extracellular messenger (Brown et al., 1998). Based on calcium-binding data an activation of cadherins by Ca^{2+} fluctuations has been proposed (Koch et al., 1997; Pertz et al.,1999) however, this has yet to be confirmed *in vivo*.

ACKNOWLEDGEMENTS

This work was financially supported by grants of the Swiss National Science Foundation to J.E. and to A.W.K. The authors are grateful to K. Manzur for critically reading the manuscript.

REFERENCES

Ashworth, J.L., Murphy, G., Rock, M.J., Sherratt, M.J., Shapiro, S.D., Shuttleworth, C.A. and Kielty, C.M., 1999, Fibrillin degradation by matrix metalloproteinases: Implications for connective tissue remodelling, *Biochem. J.* 340, 171–181.

Becker, K.F., Atkinson, M.J., Reich, U., Becker, I., Nekarda, H., Siewert, J.R. and Hofler, H., 1994, E-cadherin gene mutations provide clues to diffuse type gastric carcinomas, *Cancer Res.* 54, 3845–3852.

Briggs, M.D., Hoffman, S.M., King, L.M., Olsen, A.S., Mohrenweiser, H., Leroy, J.G., Mortier, G.R., Rimoin, D.L., Lachman, R.S., Gaines, E.S. et al., 1995, Pseudoachondroplasia and multiple epiphyseal dysplasia due to mutations in the cartilage oligomeric matrix protein gene, *Nat. Genet.* 10, 330–336.

Brown, E.M., Vassilev, P.M. and Hebert, S.C., 1995, Calcium ions as extracellular messengers, *Cell* 83, 679–682.

Brown, E.M., Pollak, M. and Hebert, S.C., 1998, The extracellular calcium-sensing receptor: Its role in health and disease, *Annu. Rev. Med.* 49, 15–29.

Deere, M., Sanford, T., Francomano, C.A., Daniels, K. and Hecht, J.T., 1999, Identification of nine novel mutations in cartilage oligomeric matrix protein in patients with pseudoachondroplasia and multiple epiphyseal dysplasia, *Am. J. Med. Genet.* 85, 486–490.

Delot, E., Brodie, S.G., King, L.M., Wilcox, W.R. and Cohn, D.H., 1998, Physiological and pathological secretion of cartilage oligomeric matrix protein by cells in culture, *J. Biol. Chem.* 273, 26692–26697.

Delot, E., King, L.M., Briggs, M.D., Wilcox, W.R. and Cohn, D.H., 1999, Trinucleotide expansion mutations in the cartilage oligomeric matrix protein (COMP) gene, *Hum. Mol. Genet.* 8, 123–128.

DiCesare, P., Hauser, N., Lehman, D., Pasumarti, S. and Paulsson, M., 1994, Cartilage oligomeric matrix protein (COMP) is an abundant component of tendon, *FEBS Lett.* 354, 237–240.

DiCesare, P., Carlson, C., Stollerman, E., Chen, F., Leslie, M. and Perris, R., 1997, Expression of cartilage oligomeric matrix protein by human synovium, *FEBS Lett.* 412, 249–252.

Downing, A.K., Knott, V., Werner, J.M., Cardy, C.M., Campbell, I.D. and Handford, P.A., 1996, Solution structure of a pair of calcium-binding epidermal growth factor-like domains: Implications for the Marfan syndrome and other genetic disorders, *Cell* 85, 597–605.

Forslind, K., Eberhardt, K., Jonsson, A. and Saxne, T., 1992, Increased serum concentrations of cartilage oligomeric matrix protein. A prognostic marker in early rheumatoid arthritis, *Br. J. Rheumatol.* 3, 593–598.

Guilford, P., Hopkins, J., Harraway, J., McLeod, M., McLeod, N., Harawira, P., Taite, H., Scoular, R., Miller, A. and Reeve, A.E., 1998, E-cadherin germline mutations in familial gastric cancer, *Nature* 392, 402–405.

Handford, P., Downing, A.K., Rao, Z., Hewett, D.R., Sykes, B.C. and Kielty, C.M., 1995, The calcium binding properties and molecular organization of epidermal growth factor-like domains in human fibrillin-1, *J. Biol. Chem.* 270, 6751–6756.

Hecht, J.T., Deere, M., Putnam, E., Cole, W., Vertel, B., Chen, H. and Lawler, J., 1998, Characterization of cartilage oligomeric matrix protein (COMP) in human normal and pseudoachondroplasia musculoskeletal tissues, *Matrix Biol.* 17, 269–278.

Hecht, J.T., Nelson, L.D., Crowder, E., Wang, Y., Elder, F.F., Harrison, W.R., Francomano, C.A., Prange, C.K., Lennon, G.G., Deere, M., et al., 1995, Mutations in exon 17B of cartilage oligomeric matrix protein (COMP) cause pseudoachondroplasia, *Nat. Genet.* 10, 325–329.

Hedbom, E., Antonsson, P., Hjerpe, A., Aeschlimann, D., Paulsson, M., Rosa-Pimentel, E., Sommarin, Y., Wendel, M., Oldberg, A. and Heinegard, D., 1992, Cartilage matrix proteins. An acidic oligomeric protein (COMP) detected only in cartilage, *J. Biol. Chem.* 267, 6132–6136.

Hermiston, M.L. and Gordon, J.I., 1995, Inflammatory bowel disease and adenomas in mice expressing a dominant negative N-cadherin, *Science* 270, 1203–1207.

Hyafil, F., Morello, D., Babinet, C. and Jacob, F., 1980, A cell surface glycoprotein involved in the compaction of embryonal carcinoma cells and cleavage stage embryos, *Cell* 21, 927–934.

Hyafil, F., Babinet, C. and Jacob, F., 1981, Cell-cell interactions in early embryogenesis: A molecular approach to the role of calcium, *Cell* 26, 447–454.

International Working Group on Constitutional Diseases of Bone, 1998, International nomenclature and classification of the osteochondrodysplasias, 1997, International Working Group on Constitutional Diseases of Bone, *Am. J. Med. Genet.* 79, 376–382.

Karayiannakis, A.J., Syrigos, K.N., Efstathiou, J., Valizadeh, A., Noda, M., Playford, R.J., Kmiot, W. and Pignatelli, M., 1998, Expression of catenins and E-cadherin during epithelial restitution in inflammatory bowel disease, *J. Pathol.* 185, 413–418.

Knott, V., Downing, A.K., Cardy, C.M. and Handford, P., 1996, Calcium binding properties of an epidermal growth factor-like domain pair from human fibrillin-1, *J. Mol. Biol.* 255, 22–27.

Koch, A.W., Pokutta, S., Lustig, A. and Engel, J., 1997, Calcium binding and homoassociation of E-cadherin domains, *Biochemistry* 36, 7697–7705.

Koch, A.W., Bozic, D., Pertz, O. and Engel, J., 1999, Homophilic adhesion by cadherins, *Curr. Opin. Struct. Biol.* 9, 275–281.

Kraulis, P.J., 1991, MOLSCRIPT, a program to produce both detailed and schematic plots of protein structure, *J. Appl. Cryst.* 24, 946–950.

Maddox, B.K., Keene, D.R., Sakai, L.Y., Charbonneau, N.L., Morris, N.P., Ridgway, C.C., Boswell, B.A., Sussman, M.D., Horton, W.A., Bachinger, H.P. and Hecht, J.T., 1997, The fate of cartilage oligomeric matrix protein is determined by the cell type in the case of a novel mutation in pseudoachondroplasia, *J. Biol. Chem.* 272, 30993–30997.

Malashkevich, V., Kammerer, R., Efimov, V., Schulthess, T. and Engel, J., 1996, The crystal structure of a five-stranded coiled coil in COMP: A prototype ion channel?, *Science* 274, 761–765.

Marra, A. and Isberg, R.R., 1996, Bacterial pathogenesis: Common entry mechanisms, *Curr. Biol.* 6, 1084–1086.

Maurer, P., Hohenester, E. and Engel, J., 1996, Extracellular calcium binding proteins, *Curr. Opin. Cell Biol.* 8, 609–617.

Maynard, J.A., Cooper, R.R. and Ponseti, I.V., 1972, A unique rough surfaced endoplasmic reticulum inclusion in pseudoachondroplasia, *Lab. Invest.* 26, 40–44.

Misenheimer, T.M. and Mosher, D.F., 1995, Calcium ion binding to thrombospondin 1, *J. Biol. Chem.* 270, 1729–1733.

Mörgelin, M., Heinegard, D., Engel, J. and Paulsson, M., 1992, Electron microscopy of native cartilage oligomeric matrix protein purified from the Swarm rat chondrosarcoma reveals a five-armed structure, *J. Biol. Chem.* 267, 6137–6141.

Nagar, B., Overduin, M., Ikura, M. and Rini, J.M., 1996, Structural basis of calcium-induced E-cadherin rigidification and dimerization, *Nature* 380, 360–364.

Neidhart, M., Hauser, N., Paulsson, M., Di Cesare, P.E., Michel, B.A. and Hauselmann, H.J., 1997, Small fragments of cartilage oligomeric matrix protein in synovial fluid and serum as markers for cartilage degradation, *Br. J. Rheumatol.* 36, 1151–1160.

Nose, A., Tsuji, K. and Takeichi, M., 1990, Localization of specificity determining sites in cadherin cell adhesion molecules, *Cell* 6, 147–155.

Oldberg, A., Antonsson, P., Lindblom, K. and Heinegard, D., 1992, COMP (cartilage oligomeric matrix protein) is structurally related to the thrombospondins, *J. Biol. Chem.* 267, 22346–22350.

Perl, A.K., Wilgenbus, P., Dahl, U., Semb, H. and Christofori, G., 1998, A causal role for E-cadherin in the transition from adenoma to carcinoma, *Nature* 392, 190–193.

Persson, E., Selander, M., Linse, S., Drakenberg, T., Oehlin, A.-K. and Stenflo, J., 1989. Calcium binding to the isolated β-hydroxyaspartic acid-containing epidermal growth factor-like domain of bovine factor X, *J. Biol. Chem.* 264, 16897–16904.

Pertz, O., Bozic, D., Koch, A.W., Fauser, C., Brancaccio, A. and Engel, J., 1999, A new crystal structure Ca^{2+} dependence and mutational analysis reveal molecular details of E-cadherin homoassociation, *EMBO J.* 18, 1738–1747.

Pokutta, S., Herrenknecht, K., Kemler, R. and Engel, J., 1994, Conformational changes of the recombinant extracellular domain of E-cadherin upon calcium binding, *Eur. J. Biochem.* 223, 1019–1026.

Ramirez, F. and Pereira, L., 1999, The fibrillins, *Int. J. Biochem. Cell. Biol.* 31, 255–259.

Rand, M.D., Lindblom, A., Carlson, J., Villoutreix, B.O. and Stenflo, J., 1997, Calcium binding to tandem repeats of EGF-like modules. Expression and characterization of the EGF-like modules of human Notch-1 implicated in receptor-ligand interactions, *Protein Sci.* 6, 2059–2071.

Reinhardt, D.P., Ono, R.N. and Sakai, L.Y., 1997, Calcium stabilizes fibrillin-1 against proteolytic degradation, *J. Biol. Chem.* 272, 1231–1236.

Reinhardt, D.P., Ono, R.N., Notbohm, H., Mueller, P.K., Bächinger, H.P. and Sakai, L.Y., 2000, Mutations in calcium-binding EGF modules render fibrillin-1 susceptible to proteolysis: A potential disease-causing mechanism in Marfan syndrom, *J. Biol. Chem.* 275, 12339–12345.

Ringwald, M., Schuh, R., Vestweber, D., Eistetter, H., Lottspeich, F., Engel, J., Dolz, R., Jahnig, F., Epplen, J., Mayer, S., Müller, C. and Kemler, R., 1987, The structure of cell adhesion molecule uvomorulin. Insights into the molecular mechanism of Ca^{2+}-dependent cell adhesion, *EMBO J.* 6, 3647–3653.

Sakai, L.Y. and Keene, D.R., 1994, Fibrillin: Monomers and microfibrils, *Methods Enzymol.* 245, 29–52.

Sansonetti, P.J., Mounier, J., Prevost, M.C. and Mege, R.M., 1994, Cadherin expression is required for the spread of Shigella flexneri between epithelial cells, *Cell* 76, 829–839.

Saxne, T. and Heinegard, D., 1992, Cartilage oligomeric matrix protein: A novel marker of cartilage turnover detectable in synovial fluid and blood, *Br. J. Rheumatol.* 31, 583–591.

Semb, H. and Christofori, G., 1998, The tumor-suppressor function of E-cadherin, *Am. J. Hum. Genet.* 63, 1588–1593.

Shapiro, L., Fannon, A.M., Kwong, P.D., Thompson, A., Lehmann, M.S., Grubel, G., Legrand, J.F., Als-Nielsen, J., Colman, D.R. and Hendrickson, W.A., 1995, Structural basis of cell-cell adhesion by cadherins, *Nature* 374, 327–337.

Smallridge, R.S., Whiteman, P., Doering, K., Handford, P.A. and Downing, A.K., 1999, EGF-like domain calcium affinity modulated by N-terminal domain linkage in human fibrillin-1, *J. Mol. Biol.* 286, 661–668.

Takeichi, M., 1988, The cadherins: Cell-cell adhesion molecules controlling animal morphogenesis, *Development* 102, 639–655.

Tamura, K., Shan, W.-S., Hendrickson, W.A., Colman, D.R. and Shapiro, L., 1998, Structure-function analysis of cell adhesion by neural (N-) cadherin, *Neuron* 20, 1153–1163.

Vestweber, D. and Kemler, R., 1984, Rabbit antiserum against a purified surface glycoprotein decompacts mouse preimplantation embryos and reacts with specific adult tissues, *Exp. Cell. Res.* 52, 169–178.

Wu, Q. and Maniatis, T., 1999, A striking organization of a large family of human neural cadherin-like cell adhesion genes, *Cell* 97, 779–790.

The Extracellular Calcium-Sensing Receptor: Molecular Features, Distribution and Its Role in Physiology and Disease

Donald T. Ward and Daniela Riccardi

1. OVERVIEW

Extracellular Ca^{2+} levels in blood and other biological fluids can be sensed by cells of the Ca^{2+} homeostatic system that possess a G protein-coupled receptor (GPR) known as the extracellular calcium-sensing receptor (CaR). The discovery that extracellular Ca^{2+} (Ca_0^{2+}) can invoke intracellular signaling responses similarly to GPR agonists such as epinephrine and vasopressin, has revealed that, in addition to being a fundamental intracellular second messenger, Ca^{2+} also represents an important extracellular "first" messenger in its own right.

CaRs are activated by physiological concentrations of extracellular Ca^{2+} and Mg^{2+} and also by pharmacological agents such as Gd^{3+} and neomycin. CaRs were first discovered in parathyroid glands and later kidney but are now known to be expressed in multiple tissues throughout the body both within and without the calcium homeostatic system. Accordingly, the proposed physiological roles of the CaR include both modulation of calcium homeostasis and a range of other functions not directly related to regulation of blood calcium levels.

For example, CaR activation inhibits parathyroid hormone (PTH) formation and secretion in parathyroid gland. In the kidney, CaRs modulate vasopressin-dependent and independent divalent cation excretion and water reabsorption, protecting the tubules from the risk of stone formation. Further functions of extracellular Ca^{2+} concentration have been proposed for other cell types including regulation of differentiation and apoptosis.

Donald T. Ward and Daniela Riccardi • School of Biological Sciences, University of Manchester, Manchester, U.K.

R. Pochet, R. Donato, J. Haiech, C. Heizmann and V. Gerke (eds.): Calcium: The Molecular Basis of Calcium Action in Biology and Medicine, 165–177.

Pharmacological and biochemical studies of CaRs have revealed that they share certain structural and functional features that are both common to and distinct from other members of the GPR superfamily. In common with other GPRs, CaRs contain a seven-transmembrane region and an intracellular domain that mediates coupling to heterotrimeric G-proteins. However, the CaR differs from classic GPRs in three major respects: (i) CaRs sense ions rather than modified amino acids or other organic molecules, (ii) ligand recognition occurs in the extracellular domain rather than the transmembrane region, and (iii) CaRs exist in the membrane as dimers, i.e. two CaR molecules bound together, rather than as monomers.

In this chapter we describe how the ability of a cell to sense the concentration of surrounding Ca^{2+} levels can have a profound effect on its physiological activity and how these effects of $[Ca^{2+}]_0$ vary widely from cell-type to cell-type.

2. FUNCTIONS OF CaR IN THE Ca^{2+} HOMEOSTATIC SYSTEM

Free ionized Ca^{2+} levels represent one of the body's most tightly regulated parameters with little fluctuation permitted beyond a narrow range of normal values. The organs that control this are the parathyroids, kidney, bone and intestine and each in turn possess Ca^{2+}-sensing cells.

Levels of circulating PTH are regulated by changes in $[Ca^{2+}]_0$. If plasma Ca^{2+} levels fall, PTH secretion occurs which stimulates renal and intestinal Ca^{2+} reabsorption and bone resorption thereby restoring normocalcemia (Figure 1; Brown, 1991). In contrast, hypercalcemia induces activation of the parathyroid CaR causing inhibition of the formation and secretion of PTH. There is an inverse sigmoidal relationship between $[Ca^{2+}]_0$ and PTH secretion in which $[Ca^{2+}]_0$ exhibits a half-maximal inhibition on PTH release at about 1.1–1.3 mM in humans (Brown, 1991). This represents the normal range of free ionized calcium in the blood and correlates very closely with the EC_{50} (concentration of half-maximal response) of $[Ca^{2+}]_0$ acting on CaR protein exogenously expressed in artificial expression systems supporting the notion that the CaR is a central player in calcium homeostasis (Brown et al., 1993).

Following the onset of hypercalcemia, urinary calcium excretion rises sharply. In thick ascending limb (TAL) cells, CaR activation inhibits the NaCl reabsorption and K^+ recycling mechanism that produces the lumen positive-potential difference that in turn drives the positively charged Ca^{2+} and Mg^{2+} ions back into the bloodstream (Hebert et al., 1997). Thus $[Ca^{2+}]_0$ has the ability to regulate not only its own reabsorption in TAL but also that of NaCl.

Several bone-forming, osteoblast-derived cell lines express CaR and exhibit chemotaxis and proliferation in response to CaR agonists (Yamaguchi et al., 1998a; Godwin and Soltoff, 1997). The CaR has also been found in bone-resorbing osteoclasts, both in the mature cell (Kameda et al., 1998) but

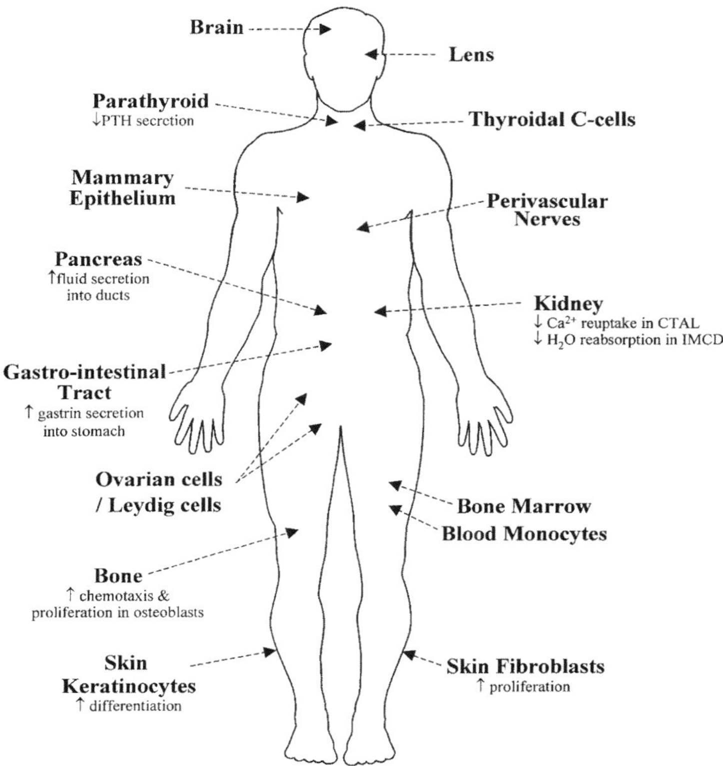

Figure 1. Localization of extracellular Ca^{2+}-sensing receptors throughout the body. CaR is located in multiple tissues and cell types and exerts a diverse range of functions. This schematic diagram indicates the locations of CaR expression that are currently known and where there is evidence of a specific function of CaR in that tissue, the figure is annotated accordingly.

also in the osteoclast-precursor stromal cells (Yamaguchi et al., 1998b). In stromal cells the CaR may participate in their proliferation and migration to sites of bone resorption as a result of local, osteoclast-mediated release of Ca^{2+}.

Another effect of hypercalcemia is the stimulation of calcitonin release from thyroidal C cells. Calcitonin acts on osteoclasts to inhibit bone resorption, so reducing the release of calcium into the bloodstream (Austin and Heath, 1981). Thyroidal C cells express CaR (Garrett et al., 1995), however it is not clear whether the C cell CaR directly regulates calcitonin secretion since there is also evidence that a voltage-sensitive Ca^{2+} channel may be responsible (Scherubl et al., 1991).

Finally, the other tissue important for calcium homeostasis is the intestine since it regulates dietary calcium absorption under the control of $1,25(OH)_2D_3$. In the gastrointestinal tract the receptor has been identified in both small and large intestine, where it has been suggested to modulate absorptive functions (Chattopadhyay et al., 1998a). Thus the CaRs present in parathyroid gland, kidney, intestine and bone work in concert to ensure normocalcemia. When mutations that reduce or enhance receptor responsiveness to Ca_0^{2+} occur, these cause inherited disorders of Ca^{2+} metabolism such as familial hypocalciuric hypercalcemia, neonatal severe hyperparathyroidism and autosomal dominant hyperparathyroidism (see the chapter "Inherited and acquired disorders of extracellular calcium-sensing" in this book).

3. FUNCTIONS OF CaR OUTSIDE OF THE Ca^{2+} HOMEOSTATIC SYSTEM

In addition to its localization within the Ca^{2+} homeostatic system, CaR is also expressed in a wide variety of tissues throughout the body apparently serving a number of diverse purposes. However, of the non-Ca^{2+} homeostatic functions of CaR so far elucidated, three particular types of effect seem to be emerging: (i) reduction of lithogenic potential, (ii) modulation of secretion, and (iii) regulation of cell fate.

During antidiuresis, vasopressin-elicited water permeability in inner medullary collecting ducts (IMCDs) raises luminal calcium concentrations to levels above those associated with the formation of calcium-containing urinary precipitates. In IMCD, CaRs are localized to the apical membrane where they can sense the rise in $[Ca^{2+}]_0$ in the urine as it is concentrated and attenuate the vasopressin-elicited water reabsorption so protecting against possible stone formation (Sands et al., 1997). CaRs exert a similar anti-lithogenic action in pancreas. In the exocrine pancreas the balance between the concentrations of free ionized Ca^{2+} and bicarbonate is of critical importance in preventing calcium carbonate precipitation in the pancreatic juice. In the pancreatic duct, CaRs monitor $[Ca^{2+}]_0$ in the pancreatic juice and regulate fluid and HCO_3 secretion thereby preventing the risk of calcium stone formation (Bruce et al., 1999). CaR is also found in insulin-secreting β-cells and interestingly previous studies have demonstrated that elevating $[Ca^{2+}]_0$ (2–18 mM) stimulates insulin secretion from β-cells even in the absence of other secretagogues.

The region of the brain where CaR mRNA abundance is at its highest is in the subfornical organ or "thirst-center" and this indicates a possible locale for the integration of calcium and H_2O homeostasis (reviewed in Chattopadhyay et al., 1998b). In addition, many other areas of the brain such as hippocampus, cerebellum, olfactory bulbs and ependymal areas of the cerebral ventricles also express the receptor. In all these regions the function of the receptor is unclear, but has been associated with regulation of local ionic homeostasis, long-term potentiation, cognitive function, regulation of neurotransmitter release and neuronal excitability. CaR has also been found in neurons located outside of the brain itself. Bukoski et al. (1997) showed that the perivascular nerve (PVN) network that innervates artery smooth muscle contains CaR and that increasing $[Ca^{2+}]_0$ above 1.5 mM causes a dose-dependent, endothelium-independent, relaxation of rat mesenteric arteries pre-constricted with norepinephrine. The putative vasorelaxant effect of PVN-CaR could be caused by an alteration in the secretion of vasoconstrictor or vasodilatory agents.

Cultured human lens epithelial cells express a CaR that is functionally coupled to K^+ channels (Chattopadhyay et al., 1997). Since hypocalcemia has been associated with initiation of cataract formation, it has been suggested that the receptor could play a role in ionic homeostasis of the lens. In the stomach the CaR is expressed in antral gastrin cells, where it modulates gastrin secretion and thus could mediate the acid rebound phenomenon associated with calcium-containing antacid preparations (Ray et al., 1997). More recently CaR has also been found in epithelial cells of the gastric mucosa and in cells of the gastric submucosal and myenteric nervous systems (Cheng et al., 1999). Indeed the small and large intestines also express the receptor (Chattopadhyay et al., 1998a) where it could be involved in absorptive and/or secretomotor functions.

Human bone marrow is enriched with bone cell precursors and several haematopoietic cell lineages and is exposed to substantial alteration in $[Ca^{2+}]_0$ due to bone turnover. House et al. (1997) reported finding abundant CaR mRNA and protein in murine erythroid precursors, megakaryocytes and platelets although not in peripheral blood erythrocytes. It has also since been found in human peripheral blood monocytes (Yamaguchi et al., 1998c).

A series of studies have also investigated whether CaR has any influence over the cell cycle including proliferation, differentiation and apoptosis in a variety of cell types. In the human colonic cell line CACO2, a reduction in luminal $[Ca^{2+}\ ]$ results in an up-regulation of the proto-oncogene c-*myc* and this signal is depressed by stimulation of a luminal CaR (Kallay et al., 1997). Taken together with an epidemiological observation that dietary Ca^{2+} load may inhibit cell proliferation (Garland et al., 1991), it is possible that extracellular Ca^{2+} acting via the CaR could regulate cell proliferation. Extracellular Ca^{2+} actually induces proliferation in cell types such as fibroblasts (McNeil et al., 1998) and osteoblasts (Huang et al., 1999). Lin et al. (1998) have proposed that CaR may regulate the balance between cell survival and cell death. These investigators have shown that in prostate carcinoma cells, CaR stimulation prevents apoptosis so providing a novel mechanism by which Ca^{2+} ions modulate cell survival. Finally, extracellular Ca^{2+} has also been shown to induce differentiation in cultured mammary cells (Cheng et al., 1998), keratinocytes (Filvaroff et al., 1994) and cells of the placental cytotrophoblast (Bradbury et al., 1998). Human keratinocytes contain both a full length CaR and an alternatively spliced CaR and it has been shown that upon differentiation, there is an increase in the relative amounts of the splice variant compared to the full length receptor (Oda et al., 2000). The splice variant CaR does not appear to respond to Ca_0^{2+} suggesting that the full-length CaR is required to mediate calcium signaling in the keratinocytes.

4. PHYSIOLOGICAL, PHARMACOLOGICAL AND PATHOLOGICAL AGONISTS OF THE CaR

The CaR is a glycoprotein that shares limited but significant overall homology with the metabotropic glutamate receptors (mGluRs) (Brown et al., 1993) and pheromone receptors (Ryba and Tirindelli, 1997). The extracellular domain of the CaR contains putative low affinity Ca^{2+} binding sites comprising clusters of negatively charged amino acid residues. Consistent with its physiological role, the CaR does not possess any high affinity Ca^{2+}-binding sites (such as EF hands that are sensitive to micromolar quantities of Ca^{2+}) as otherwise it would be permanently activated by the millimolar quantities of free ionized Ca^{2+} found in the blood and other extracellular fluids.

In addition to Ca^{2+}, the CaR also senses other divalent cations such as Mg^{2+}, albeit with a lower affinity. More potent CaR agonists include trivalent lanthanides (e.g., Gd^{3+}) and certain polyvalent cations (e.g. aminoglycoside antibiotics such as neomycin, and some highly basic peptides) (Brown et al., 1993). As membrane impermeant CaR agonists, Gd^{3+} and neomycin are now commonly used for the study of receptor pharmacology and function, particularly since Ca^{2+} may freely permeate the plasma membrane and effect non-CaR specific intracellular responses.

Given the problems of selectivity surrounding the current CaR agonists and the need to develop "calcimimetic" drugs for the treatment of some diseases involving disturbed calcium homeostasis, a class of compounds known as the phenylalkylamines (e.g. NPS R-568) have been studied for their usefulness as pharmacological probes of the CaR (Nemeth et al., 1998). These compounds act in a stereoselective fashion as positive allosteric modulators to enhance CaR sensitivity to Ca_0^{2+}. Currently these compounds are in clinical use where they have been shown to lower plasma levels of Ca^{2+} and PTH levels in normal humans with secondary hyperparathyroidism (Nemeth, 1996) (see the chapter "Inherited and aqcuired disorders of extracellular calcium-sensing" in this book).

The polycationic CaR agonists include non-endogenous compounds such as polyarginine and neomycin but also some naturally occurring compounds such as spermine and protamine, although whether these are actual physiological agonists is unclear (Brown et al., 1991; Quinn et al., 1997). Interestingly, amyloid-β peptides, proteins that are produced in massive quantities in Alzheimer's disease, are also CaR agonists and this could explain the sustained elevation of intracellular Ca^{2+} that produces the neuronal dysfunction and degeneration seen in the disease (Ye et al., 1997).

5. STRUCTURAL FEATURES OF THE CaR DETERMINING AGONIST SENSITIVITY AND RECEPTOR FUNCTION

It is known that the CaR binds Ca^{2+} within its large amino-terminal extracellular domain (ATED) since chimeric constructs that either lack the ATED, or that have had it replaced by the ATED of mGluR1, are incapable of sensing changes in Ca_0^{2+} (Hammerland et al., 1999). Interestingly Gd^{3+} activates both of these mutant CaR constructs, indicating that the binding sites of Ca^{2+} and Gd^{3+} on the CaR are not necessarily the same and that some or all of gadolinium's binding sites may be located in the extracellular loops of the transmembrane domain.

Two amino acid residues located in the ATED that are crucial for CaR agonist binding and receptor function are Ser-147 and Ser-170. Mutant CaRs containing alanine residues substituted for these two residues exhibit reduced sensitivity to Ca^{2+} (Bräuner-Osborne et al., 1999).

Another important factor for the Ca_0^{2+}-sensitivity of CaR is the ionic strength of the surrounding media. Quinn et al. (1998) reported that the Ca_0^{2+}-sensitivity of CaRs expressed in HEK cells is increased when $[Na^+]_0$ is reduced. If this is also true of endogenously expressed CaRs in the body then they could potentially detect fluctuations in $[Na^+]_0$ levels under conditions where $[Ca^{2+}]_0$ levels remain static.

6. DISULFIDE-LINKED HOMODIMERIZATION OF THE CALCIUM RECEPTOR

The disulfide-linked dimerization of CaR protein has been demonstrated by a number of biochemical techniques including immunoblotting under non-reducing/reducing conditions, estimation of native mass by sucrose density gradient ultracentrifugation and use of covalent crosslinking reagents (Ward et al., 1998; Bai et al., 1998a). In order to prove that the putative CaR dimer truly represents two CaRs bound together, i.e. a homodimer, rather than one CaR bound to a different protein(s) of similar molecular mass, Bai et al. (1998a) successfully coprecipitated flagged and non-flagged CaRs from membranes of HEK cells cotransfected with FLAG epitope-tagged and wild type CaRs.

It is proposed that extracellular cysteine residues that are highly conserved between CaRs and mGluRs mediate CaR dimerization (Figure 2; Ward et al., 1998). When the ATED of CaR is expressed in HEK cells, the translated protein still forms homodimers (Goldsmith et al., 1999) despite the absence of the intracellular or transmembrane domains. Ray et al. (1999) and Pace et al. (1999) have between them identified four extracellular cysteine residues (Cys-101, 129, 131 and 236) that are likely candidates for mediating this disulfide-linked dimerization.

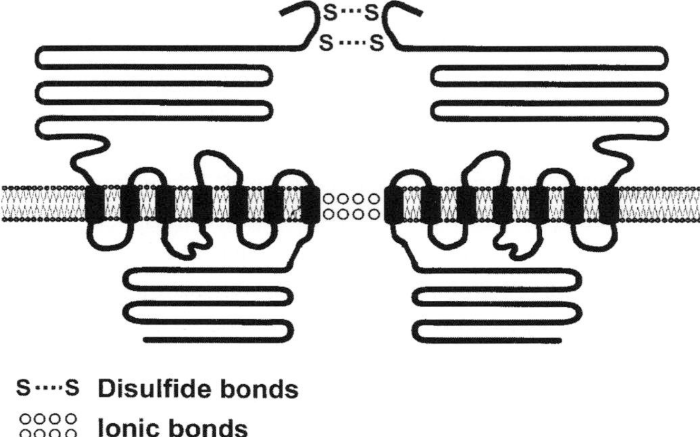

Figure 2. Model of extracellular Ca^{2+}-sensing receptor dimerization. CaR exists in the cell membrane as a dimeric molecule that is linked to a second CaR molecule both by intermolecular disulfide bonds in the extracellular domain and ionic bonds in the transmembrane region.

In addition to the disulfide-linked associations described above, CaR molecules also associate via non-covalent interactions. Triton X-100-solubilized CaR sediments in sucrose gradients as for a dimer both in the presence and absence of reducing agents (Ward et al., 1998). Thus, reduction alone is insufficient to separate the putative CaR dimer. It is only upon subsequent SDS denaturation (in the presence of reducing agent) that monomeric CaR molecules are obtained. Triton X-100 is a mild non-ionic detergent that tends to preserve ionic interactions whereas SDS is a stronger ionic detergent that disrupts them. This suggests that the putative CaR dimer is held together both by disulfide bonds and by ionic bonds (Ward et al., 1998; Bai et al. 1998a). This was investigated by Pace et al. (1999) who successfully coprecipitated two CaR mutants, that had been differentially flagged using green fluorescent protein and the FLAG epitope, and that additionally had had cysteines 101 and 236 substituted for alanine. Thus, assuming that the CaR-C101S and C236S mutations did successfully prevent formation of intermolecular disulfide bonds between the CaR monomers, then coprecipitation could only be achieved if the CaR molecules were also bound non-covalently.

It is important to note regarding CaR dimerization that unlike receptor tyrosine kinases which dimerize upon agonist binding to produce a functional response, the CaR dimer appears to pre-exist agonist treatment (Ward et al., 1998; Bai et al., 1998a). Therefore, if dimerization precedes agonist binding then what is its functional significance or does it represent anything other than a convenient structural conformation? Addressing this question Bai et al. (1999) reported that when inactivating CaR mutants are co-expressed with wild type CaRs in HEK cells, there is a partial recovery of signaling function. However, the presence of an abnormal domain in each mutant monomer substantially impairs the function of the CaR heterodimer (Bai et al., 1999).

7. INTRACELLULAR SIGNALING MECHANISMS DOWNSTREAM OF CaR ACTIVATION

There exist in the body a wide variety of endocrine effector molecules that are capable of evoking a range of cell-specific responses. Since each cell type contains a variety of membrane receptors, the cell must somehow differentially respond to each agonist without confusion. To manage this a plethora of distinct intracellular signaling mechanisms are employed.

The first level of control over GPR signaling, concerns which heterotrimeric G protein(s) actually couple to the receptor since distinct G proteins effect different responses. CaR associates with both $Gq/_{11}\alpha$ and $Gi_{2-3}\alpha$ proteins in thyroid parafollicular cells (Tamir et al., 1996) and in Madin–Darby canine-kidney (MDCK) cells (Arthur et al., 1997).

Another important class of intracellular signaling mediators are the phospholipid-derived second messengers that are generated by the action of phospholipases. Kifor et al. (1997) reported that in response to CaR agonists, both bovine parathyroid cells and CaR-transfected HEK cells exhibited increased the activities of phospholipases C, A_2 and D causing increased production of inositol trisphosphate, arachidonic acid and phosphatidic acid respectively.

CaR signaling has also been shown to involve signaling via protein kinases that phosphorylate serine/threonine (Kifor et al., 1997; Bai et al., 1998b) and tyrosine (Filvaroff et al., 1994) residues. Indeed, the activity of the CaR may itself be modulated by protein kinase C (PKC). The cloned CaR contains at least four putative PKC phosphorylation sites (Brown et al., 1993; Riccardi et al., 1995) and their phosphorylation greatly reduces the affinity of CaR for Ca^{2+}, that is, inhibits the receptor (Bai et al., 1998b). In contrast, protein kinase A (PKA) does not seem to play an important role in regulation of receptor function since bovine parathyroid CaR (Brown et al., 1993) does not contain a PKA consensus sequence despite the rat kidney CaR containing one (Riccardi et al., 1995).

CaR agonists induce protein tyrosine phosphorylation in mouse keratinocytes (Filvaroff et al., 1994) and rat fibroblasts (McNeil et al., 1998) and thus tyrosine kinases may be employed to mediate certain responses such as CaR-mediated differentiation. However, the CaR itself cannot be directly regulated by tyrosine phosphorylation, as it contains no intracellular tyrosine residues. In fibroblasts it was specifically shown that receptor stimulation activates c-SRC kinase and induces proliferation (McNeil et al., 1998). Mitogen activated protein kinase (MAPK) activation is another important signaling mechanism involved in proliferation that is downstream of CaR activation.

CaR agonists have been reported to induce the p42/p44 MAPKs in fibroblasts (McNeil et al., 1998) and osteoblasts (Huang et al., 1999).

8. CONCLUDING REMARKS

The existence of a G-protein coupled extracellular Ca^{2+}-sensing receptor in multiple cell-types means that extracellular Ca^{2+}extracellular Ca^{2+} may be thought of an agonist or "first" messenger in its own right. Indeed, the promiscuity of CaR in sensing other metal ions and polycations renders such agents first messengers as well. Given the array of putative CaR agonists it is possible that in certain cell-types the physiological agonist of CaR may not even be extracellular Ca^{2+}. The knowledge that CaR agonists include the aminoglycoside antibiotics, which can cause nephrotoxicity, and the β-amyloid peptides that are overexpressed in Alzheimer's disease, may also shed some light on these pathologies. Ultimately, it is our understanding of the regulation of calcium homeostasis itself that has been most improved by the discovery of CaR.

NOTE ADDED IN PROOF

An association between $[Ca^{2+}]_0$ and insulin secretion has been confirmed by Squires et al. (2000), however this recently published study suggests that the relationship is actually inversely proportional.

Reference: Squires, P.E., Harris, T.E., Persaud, S.J., Curtis, S.B., Buchan, A.M. and Jones, P.M., 2000, The extracellular calcium-sensing receptor on human beta-cells negatively modulates insulin secretion, *Diabetes* 49, 409–417.

REFERENCES

Arthur, J.M., Collinsworth, G.P., Gettys, T.W., Quarles, L.D. and Raymond, J.R., 1997, Specific coupling of a cation-sensing receptor to G protein u-subunits in MDCK cells, *Am. J. Physiol.* 273, F129–F135.

Austin, L.A. and Heath, H., 3rd, 1981, Calcitonin: Physiology and pathophysiology, *New Engl. J. Med.* 304, 269–278.

Bai, M., Trivedi, S. and Brown, E.M., 1998a, Dimerization of the extracellular calcium-sensing receptor (CaR) on the cell surface of CaR-transfected HEK293 cells, *J. Biol. Chem.* 273, 23605–23610.

Bai, M., Trivedi, S., Lane, C.R., Yang, Y., Quinn, S.J. and Brown, E.M., 1998b, Protein kinase C phosphorylation of threonine at position 888 in Ca_0^{2+}-sensing receptor (CaR) inhibits coupling to Ca^{2+} store release, *J. Biol. Chem.* 273, 21267–21275.

Bai, M., Trivedi, S., Kifor, O., Quinn, S.J. and Brown, E.M., 1999, Intermolecular interactions between dimeric calcium-sensing receptor monomers are important for its normal function, *Proc. Natl. Acad. Sci. USA* 96, 2834–2839.

Bradbury, R.A., Sunn, K.L., Crossley, M., Bai, M., Brown, E.M., Delbridge, L. and Conigrave, A.D., 1998, Expression of the parathyroid Ca^{2+}-sensing receptor in cytotrophoblasts from human term placenta, *J Endocrinol.* 156, 425–430.

Bräuner-Osborne, H., Jensen, A.A., Sheppard, P.O., O'Hara, P. and Krogsgaard-Larsen, P., 1999, The agonist-binding domain of the calcium-sensing receptor is located at the amino-terminal domain, *J. Biol. Chem.* 274, 18382–18386.

Brown, E.M., 1991, Extracellular Ca^{2+} sensing, regulation of parathyroid cell function, and role of Ca^{2+} and other ions as extracellular (first) messengers, *Physiol. Rev.* 71, 371–411.

Brown, E.M., Katz, C., Butters, R. and Kifor, O., 1991, Polyarginine, polylysine, and protamine mimic the effects of high extracellular calcium concentrations on dispersed bovine parathyroid cells, *J. Bone Miner. Res.* 6, 1217–1225.

Brown, E.M., Gamba, G., Riccardi, D., Lombardi, M., Butters, R., Kifor, O., Sun, A., Hediger, M.A., Lytton, J. and Hebert, S.C., 1993, Cloning and characterization of an extracellular Ca^{2+}-sensing receptor from bovine parathyroid, *Nature* 366, 575–580.

Bruce, J.I., Yang, X., Ferguson, C.J., Elliott, A.C., Steward, M.C., Case, R.M. and Riccardi, D., 1999, Molecular and functional identification of a Ca^{2+} (polyvalent cation)-sensing receptor in rat pancreas, *J. Biol. Chem.* 274, 20561–20568.

Bukoski, R.D., Bian, K., Wang, Y. and Mupanomunda, M., 1997, Perivascular sensory nerve Ca^{2+} receptor and Ca^{2+}-induced relaxation of isolated arteries, *Hypertension* 30, 1431–1439.

Chattopadhyay, N., Ye, C., Singh, D.P., Kifor, O., Vassilev, P.M., Shinohara, T. and Chylack, L.T., Jr. and Brown, E.M., 1997, Expression of extracellular calcium-sensing receptor by human lens epithelial cells, *Biochem. Biophys. Res. Commun.* 233, 801–805.

Chattopadhyay, N., Cheng, I., Rogers, K., Riccardi, D., Hall, A., Diaz, R., Hebert, S.C., Soybel, D.I. and Brown, E.M., 1998a, Identification and localization of extracellular Ca^{2+}-sensing receptor in rat intestine, *Am. J. Physiol.* 274, G122–G130.

Chattopadhyay, N., Yamaguchi, T. and Brown, E.M., 1998b, Ca^{2+} receptors from brain to gut: Common stimulus, diverse actions, *TEM* 9, 354–359.

Cheng, I., Klingensmith, M.E., Chattopadhyay, N., Kifor, O., Butters, R.R., Soybel, D.I. and Brown, E.M., 1998, Identification and localization of the extracellular calcium-sensing receptor in human breast, *J. Clin. Endocrinol. Metab.* 83, 703–707.

Cheng, I., Qureshi, I., Chattopadhyay, N., Qureshi, A., Butters, R.R., Hall, A.E., Cima, R.R., Rogers, K.V., Hebert, S.C., Geibel, J.P., Brown, E.M. and Soybel, D.I., 1999, Expression of an extracellular calcium-sensing receptor in rat stomach, *Gastroenterology* 116, 118–126.

Filvaroff, E., Calautti, E., Reiss, M. and Dotto, G.P., 1994, Functional evidence for an extracellular calcium receptor mechanism triggering tyrosine kinase activation associated with mouse keratinocyte differentiation, *J. Biol. Chem.* 269, 21735–21740.

Garland, C.F., Garland, F.C. and Gorham, E.D., 1991, Can colon cancer incidence and death rates be reduced with calcium and vitamin D?, *Am. J. Clin. Nutr.* 54, 193S–201S.

Garrett, J.E., Tamir, H., Kifor, O., Simin, R.T., Rogers, K.V., Mithal, A., Gagel, R.F. and Brown, E.M., 1995, Calcitonin-secreting cells of the thyroid express an extracellular calcium receptor gene, *Endocrinology* 136, 5202–5211.

Godwin, S.L. and Soltoff, S.P., 1997, Extracellular calcium and platelet-derived growth factor promote receptor-mediated chemotaxis in osteoblasts through different signaling pathways, *J. Biol. Chem.* 272, 11307–11312.

Goldsmith, P.K., Fan, G.F., Ray, K., Shiloach, J., McPhie, P., Rogers, K.V. and Spiegel, A.M., 1999, Expression, purification, and biochemical characterization of the amino-terminal extracellular domain of the human calcium receptor, *J. Biol. Chem.* 274, 11303–11309.

Hammerland, L.G., Krapcho, K.J., Garrett, J.E., Alasti, N., Hung, B.C., Simin, R.T., Levinthal, C., Nemeth, E.F. and Fuller, F.H., 1999, Domains determining ligand specificity for Ca^{2+} receptors, *Mol. Pharmacol.* 55, 642–648.

Hebert, S.C., Brown, E.M. and Harris, H.W., 1997, Role of calcium-sensing receptor in divalent mineral homeostasis, *J. Exper. Med.* 200, 295–302.

House, M.G., Kohlmeier, L., Chattopadhyay, N., Kifor, O., Yamaguchi, T., Leboff, M.S., Glowacki, J. and Brown, E.M., 1997, Expression of an extracellular calcium-sensing receptor in human and mouse bone marrow cells, *J. Bone Miner. Res.* 12, 1959–1970.

Huang, Z., Gong, Q., Lu, Y., Brown, A., Dusso, A., Cheng, S.-L. and Slatopolsky, E., 1999, Selective activation of MAP kinases (p44/p42) and stimulation of cell proliferation in human osteoblast cell culture by extracellular calcium, *J. Am. Soc. Nephrol.* 10, A3090.

Kallay, E., Kifor, O., Chattopadhyay, N., Brown, E.M., Bischof, M.G., Peterlik, M. and Cross, H.S., 1997, Calcium-dependent c-myc proto-oncogene expression and proliferation of Caco-2 cells: A role for a luminal extracellular calcium-sensing receptor, *Biochem. Biophys. Res. Commun.* 232, 80–83.

Kameda, T., Mano, H., Yamada, Y., Takai, H., Amizuka, N., Kobori, M., Izumi, N., Kawashima, H., Ozawa, H., Ikeda, K., Kameda, A., Hakeda, Y. and Kumegawa, M., 1998, Calcium-sensing receptor in mature osteoclasts, which are bone resorbing cells, *Biochem. Biophys. Res. Commun.* 245, 419–422.

Kifor, O., Diaz, R., Butters, R. and Brown, E.M., 1997, The Ca^{2+}-sensing receptor (CaR) activates phospholipases C, A2, and D in bovine parathyroid and CaR-transfected, human embryonic kidney (HEK293) cells, *J. Bone Miner. Res.* 12, 715–725.

Lin, K.I., Chattopadhyay, N., Bai, M., Alvarez, R., Dang, C.V., Baraban, J.M., Brown, E.M. and Ratan, R.R., 1998, Elevated extracellular calcium can prevent apoptosis via the calcium-sensing receptor, *Biochem. Biophys. Res. Commun.* 249, 325–331.

McNeil, S.E., Hobson, S.A., Nipper, V. and Rodland, K.D., 1998, Functional calcium-sensing receptors in rat fibroblasts are required for activation of SRC kinase and mitogen-activated protein kinase in response to extracellular calcium, *J. Biol. Chem.* 273, 1114–1120,

Nemeth, E.F., 1996, *Principles of Bone Biology*, Academic Press, London, pp. 1019–1035.

Nemeth, E.F., Steffey, M.E., Hammerland, L.G., Hung, B.C., Van Wagenen, B.C., DelMar, E.G. and Balandrin, M.F., 1998, Calcimimetics with potent and selective activity on the parathyroid calcium receptor, *Proc. Natl. Acad. Sci. USA* 95, 4040–4045.

Oda, Y., Tu, C.L., Chang, W., Crumrine, D., Komuves, L., Mauro, T., Elias, P.M. and Bikle, D.D., 2000, The calcium sensing receptor and its alternatively spliced form in murine epidermal differentiation, *J. Biol. Chem.* 275, 1183–1190.

Pace, A.J., Gama, L. and Breitwieser, G.E., 1999, Dimerization of the calcium-sensing receptor occurs within the extracellular domain and is eliminated by Cys to Ser mutations at Cys101 and Cys236, *J. Biol. Chem.* 274, 11629–11634.

Quinn, S.J., Ye, C.P., Diaz, R., Kifor, O., Bai, M., Vassilev, P. and Brown, E.M., 1997, The Ca^{2+}-sensing receptor: A target for polyamines, *Am. J. Physiol.* 273, C1315–C1323.

Quinn, S.J., Kifor, O., Trivedi, S., Diaz, R., Vassilev, P. and Brown, E., 1998, Sodium and ionic strength sensing by the calcium receptor, *J. Biol. Chem.* 273, 19579–19586.

Ray, K., Hauschild, B.C., Steinbach, P.J., Goldsmith, P.K., Hauache, O. and Spiegel, A.M., 1999, Identification of the cysteine residues in the amino-terminal extracellular domain of the human Ca^{2+} receptor critical for dimerization. Implications for function of monomeric Ca^{2+} receptor, *J. Biol. Chem.* 274, 27642–27650.

Riccardi, D., Park, J., Lee, W., Gamba, G., Brown, E.M. and Hebert, S.C., 1995, Cloning and functional expression of a rat kidney extracellular calcium/polyvalent cation-sensing receptor, *Proc. Natl. Acad. Sci. USA* 92, 131–135.

Ryba, N.J. and Tirindelli, R., 1997, A new multigene family of putative pheromone receptors, *Neuron* 19, 371–379.

Sands, J.M., Naruse, M., Baum, M., Jo, I., Hebert, S.C., Brown, E.M., Harris, H.W., 1997, Apical extracellular calcium/polyvalent cation-sensing receptor regulates vasopressin-elicited water permeability in rat kidney inner medullary collecting duct, *J. Clin. Invest.* 99, 1399–1405.

Scherubl, H., Schultz, G. and Hescheler, J., 1991, Electrophysiological properties of rat calcitonin-secreting cells, *Mol. Cell Endocrinol.* 82, 293–301.

Tamir, H., Liu, K.P., Adlersberg, M., Hsiung, S.C. and Gershon, M.D., 1996, Acidification of serotonin-containing secretory vesicles induced by a plasma membrane calcium receptor, *J. Biol. Chem.* 271, 6441–6450.

Ward, D.T., Brown, E.M. and Harris, H.W., 1998, Disulfide bonds in the extracellular calcium-polyvalent cation-sensing receptor correlate with dimer formation and its response to divalent cations in vitro, *J. Biol. Chem.* 273, 14476–14483.

Yamaguchi, T., Kifor, O., Chattopadhyay, N. and Brown, E.M., 1998a, Expression of extracellular calcium (Ca^{2+})-sensing receptor in the clonal osteoblast-like cell lines, UMR-106 and SAOS-2, *Biochem. Biophys. Res. Commun.* 243, 753–757.

Yamaguchi, T., Chattopadhyay, N., Kifor, O. and Brown, E.M., 1998b, Extracellular calcium (Ca_0^{2+})-sensing receptor in a murine bone marrow-derived stromal cell line (ST2): Potential mediator of the actions of Ca^{2+} , on the function of ST2 cells, *Endocrinology* 139, 3561–3568.

Yamaguchi, T., Olozak, I., Chattopadhyay, N., Butters, R.R., Kifor, O., Scadden, D.T., Brown, E.M., 1998c, Expression of extracellular calcium (Ca_0^{2+})-sensing receptor in human peripheral blood monocytes, *Biochem. Biophys. Res. Commun.* 246, 501–506.

Ye, C., Ho-Pao, C.L., Kanazirska, M., Quinn, S., Rogers, K., Seidman, C.E., Seidman, J.G., Brown, E.M. and Vassilev, P.M., 1997, Amyloid-beta proteins activate Ca^{2+}-permeable channels through calcium-sensing receptors, *J. Neurosci. Res.* 47, 547–554.

Regulation of Inositol 1,4,5-Trisphosphate-Induced Ca^{2+} Release by Ca^{2+}

Jan B. Parys, Ilse Sienaert, Sara Vanlingen, Geert Callewaert, Patrick De Smet, Ludwig Missiaen and Humbert De Smedt

1. INTRODUCTION

Activation of cells by extracellular stimuli like hormones, growth factors or neurotransmitters leads to a controlled release of Ca^{2+} ions from intracellular Ca^{2+} stores into the cytosol. This release can take place locally or produce a global cellular response whereby the release takes the form of complex spatio-temporal signals, such as Ca^{2+} oscillations and Ca^{2+} waves (Berridge, 1997). The intracellular Ca^{2+} stores are part of the endoplasmic reticulum and their function is governed by three major types of proteins: (1) the ATP-driven Ca^{2+} pumps responsible for store filling, (2) intraluminal Ca^{2+}-binding proteins (see Michalak et al., this book), and (3) one or both of the two major types of intracellular Ca^{2+}-release channels, the inositol 1,4,5-trisphosphate receptors (IP_3Rs) and the ryanodine receptors (RyRs) (see Rossi et al., this book; Macrez and Mironneau, this book).

The relation between cell activation by extracellular agonists and opening of the IP_3R is fairly well elucidated (Berridge, 1993). The agonists interact with specific plasma-membrane receptors that are coupled either to heterotrimeric G proteins or to tyrosine kinase activity. Both pathways lead to activation of a different phospholipase C isoform (β or γ, respectively) and subsequent inositol 1,4,5-trisphosphate (IP_3) production. IP_3 diffuses in the cytosol and activates the IP_3R, thereby leading to Ca^{2+} release. The IP_3R is a large protein (about 1200 kDa) composed of four subunits. Each monomer

---

Jan B. Parys, Ilse Sienaert, Sara Vanlingen, Geert Callewaert, Patrick De Smet, Ludwig Missiaen, Humbert De Smedt • Laboratorium voor Fysiologie, Katholieke Universiteit Leuven, Leuven, Belgium.

R. Pochet, R. Donato, J. Haiech, C. Heizmann and V. Gerke (eds.): Calcium: The Molecular Basis of Calcium Action in Biology and Medicine, 179–190.

Table 1. Important features of the three mammalian IP_3R isoforms. Question marks indicate that not all observations concur on that point. The formation of IP_3R heterotetramers or the expression of cell-type dependent regulatory proteins may be responsible for part of the contradictory results. For further information the interested reader is referred to the text or to a recent review (Patel et al., 1999)

	IP_3R1	IP_3R2	IP_3R3
Cloned human IP_3R (length in amino acids)	2695–2743*	2701	2671
Chromosomal localization in humans	3p25–26	12p11	6p21
Homology with human IP_3R1	100%	77%	72%
Predominant expression	Cerebellum, oocytes, smooth muscle	Hepatocytes, heart, RBL-2H3 cells	Epithelial cells proliferative cells, inclus. most cell lines
Affinity for IP_3	Intermediate	Highest	Lowest
Effects of Ca^{2+}:			
– Stimulation of Ca^{2+} release	Yes	Yes	Yes
– Inhibition of Ca^{2+} release	Yes	Yes ?	Yes ?
– Effect of IP_3 binding	Inhibition	Stimulation	?
Effects of ATP:			
– Stimulation of Ca^{2+} release	High sensitivity	None	Low sensitivity
– Effect on IP_3 binding	Inhibition	Inhibition	Inhibition
Effects of SH-reagent thimerosal:			
– Effect on Ca^{2+} release	Biphasic	Stimulation	Inhibition
– Effect on IP_3 binding	Biphasic	Stimulation ?	None

*Dependent on splicing events.

binds a single IP_3 molecule and consists of an N-terminal IP_3-binding domain (about 600 amino acids), which is coupled by means of a large regulatory domain (about 1500 amino acids) to the C-terminal channel domain (about 600 amino acids). The latter contains the pore region, six transmembrane stretches, a small luminal loop region, and a small cytosolic tail.

Different cytosolic factors modulate IP_3-induced Ca^{2+} release (e.g. Ca^{2+}, nucleotides, pH, protein kinases, associated proteins) of which Ca^{2+} is the most important (Taylor, 1998; Patel et al., 1999). Three IP_3R isoforms have been cloned from human cells (IP_3R1, IP_3R2, IP_3R3) and the subunits encoded by the various genes can assemble to either homo- or heterotetramers. The three IP_3R isoforms, although having a similar structure, differ in characteristics, and those characteristics are combined in the heterotetrameric receptors. The most important properties of the three IP_3R isoforms are summarized in Table 1.

The regulation by Ca^{2+} is particularly important for setting up complex spatio-temporal Ca^{2+} signals in the cell. It is the aim of this chapter to discuss the current views on the regulation of the various IP_3R isoforms by Ca^{2+}, thereby focusing in particular on the molecular pathways underlying this regulation.

2. BIPHASIC REGULATION OF IP_3-INDUCED Ca^{2+} RELEASE BY CYTOSOLIC Ca^{2+}

A fundamental observation, reported exactly a decade ago, was the biphasic modulation of IP_3-induced Ca^{2+} release by Ca^{2+}. Such behavior was observed in intestinal smooth muscle (Iino, 1990), in neurons (Bezprozvanny et al., 1991; Finch et al., 1991), in *Xenopus* oocytes (Parys et al., 1992) and subsequently in almost all other tissues investigated. This biphasic effect was observed in permeabilized cells, as well as in microsomal preparations and in IP_3Rs reconstituted in lipid bilayers. The basic observation (Figure 1) is the rapid increase in IP_3-induced Ca^{2+} release at increasing cytosolic Ca^{2+} concentrations thereby providing a strong positive feedback on the channel activity, up to a certain concentration (0.3–1 μM Ca^{2+}) where the Ca^{2+} concentration becomes inhibitory. Reaching this concentration, Ca^{2+} induces a negative feedback on the IP_3R, which leads to a decrease and an eventual inhibition of the release (Figure 1). Stimulation and inhibition are rapid processes, but the latter develops more slowly than the former (Finch et al., 1991). This kinetic difference, indicating a difference in underlying mechanism, is important for the initiation and the propagation of intracellular Ca^{2+} signals (*vide infra*). The regulation of the IP_3R by cytosolic Ca^{2+} is further modulated by various cellular factors, including IP_3, luminal Ca^{2+}, ATP, Mg^{2+}, pH, and regulatory proteins. Hereby is the cytosolic Ca^{2+}-binding protein calmodulin an important, recently identified, regulator of this process. It strongly potentiates the inhibitory phase of IP_3-induced Ca^{2+} release (Missiaen et al., 1999) (Figure 1C) and decreases, at high Ca^{2+} concentrations, the open probability of the purified IP_3R1 incorporated in lipid bilayers (Michikawa et al., 1999).

The biphasic regulation of the type 1 IP_3R by Ca^{2+} has been extensively studied, but it is not yet clear whether a similar phenomenon occurs for the type 2 and the type 3 IP_3Rs. Activation of IP_3-induced Ca^{2+} release by cytosolic Ca^{2+} is a universal feature, occurring in cell types with vastly different IP_3R-expression patterns. However, it is not established whether the same applies for the Ca^{2+}-dependent inhibition of IP_3-induced Ca^{2+} release (Taylor, 1998). The existence of heterotetrameric IP_3Rs and the role of cell-type dependent regulatory factors, as e.g. calmodulin, may be responsible for some of the conflicting results obtained thus far.

The complex interplay between IP_3, Ca^{2+} and regulatory factors, the fast kinetics of channel activation and the IP_3R isoform diversity make it difficult to establish the sequence of events occurring at a molecular level. Rapid kinetic analysis of IP_3-induced Ca^{2+} release in permeabilized hepatocytes (expressing IP_3R1 and IP_3R2 in a 1:4 ratio) provided compelling evidence that on each subunit, subsequent to IP_3 binding, a Ca^{2+}-binding site becomes accessible which must be occupied within a narrow time frame in order to allow channel activation. Binding of either IP_3 or Ca^{2+} alone would lead to an inactive state (Marchant and Taylor, 1997; Adkins and Taylor, 1999).

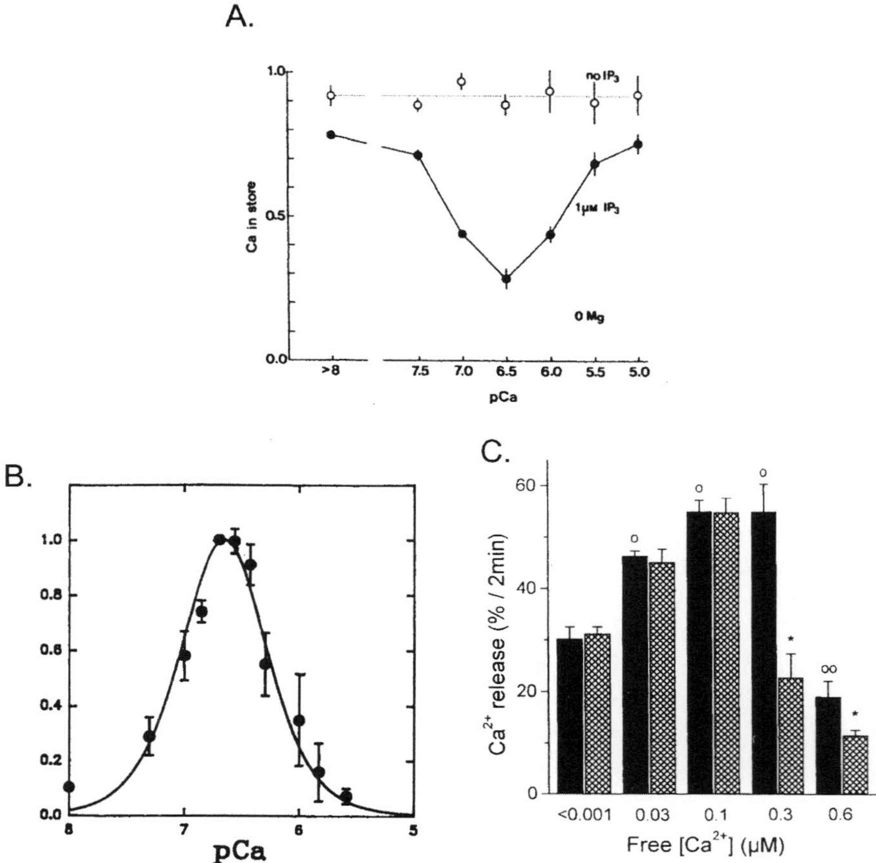

Figure 1. Biphasic dependence of IP_3-induced Ca^{2+} release on the free cytosolic Ca^{2+} concentration. (A) Ca^{2+} release was measured in permeabilized intestinal smooth muscle cells. The Ca^{2+} remaining in the stores is expressed as a function of the free cytosolic Ca^{2+} concentration and represents the reciprocal of the amount of Ca^{2+} released. In the absence of IP_3 no Ca^{2+} release is observed (open symbols) while the release evoked by 1 μM IP_3 (closed symbols) is clearly dependent on the Ca^{2+} concentration (after Iino 1990, reproduced from *The Journal of General Physiology*, 1990, Vol. 95, pp. 1103–1122 by copyright permission of The Rockefeller University Press). (B) An identical biphasic relationship is observed for the cerebellar IP_3R incorporated in artificial lipid bilayer. The open probability of the channel is plotted against the free Ca^{2+} concentration (arbitrary units, maximal opening corresponds to an open probability of about 15%) (modified after Bezprozvanny et al., 1991; reproduced with permission from *Nature*, Vol. 351, pp. 751–754, Copyright 1991, Macmillan Magazines Limited). (C) A recent and important finding is the fact that the modulation of the IP_3R is strongly affected by calmodulin. The fractional Ca^{2+} loss from the Ca^{2+} stores in permeabilized smooth muscle cells is expressed as a function of the free cytosolic Ca^{2+} concentration in the absence (filled bars) or in the presence of 10 μM calmodulin (hatched bars). Calmodulin does not affect the activation of the release by Ca^{2+} but strongly potentiates the inhibitory phase (after Missiaen et al., 1999; reproduced from *The Journal of Biological Chemistry*, Vol. 274, pp. 13748–13751, with permission of The American Society for Biochemistry & Molecular Biology). In all cases maximal Ca^{2+} release occurs at about 0.3 μM free Ca^{2+}.

This observation may be related to the steady-state analysis performed on the single channel properties of the *Xenopus* oocyte IP_3R1 in its natural membrane environment. This analysis suggested that the inhibition of the channel by Ca^{2+} is a fundamental property of the receptor. IP_3 would activate the receptor by decreasing the affinity of the inhibitory Ca^{2+} site, thereby suppressing the inhibitory phase (Mak et al., 1998).

3. STIMULATION OF IP_3-INDUCED Ca^{2+} RELEASE BY LUMINAL Ca^{2+}

The IP_3R is not only modulated by cytosolic Ca^{2+}, but also by the Ca^{2+} present in the lumen of the stores. In contrast with the regulation by cytosolic Ca^{2+}, the effect of luminal Ca^{2+} appears to be monophasic. The sensitivity of Ca^{2+} stores to submaximal IP_3 concentrations is strongly reduced at low levels of store loading (e.g., Missiaen et al., 1992, 1994; Tanimura and Turner, 1996). Higher concentrations of IP_3 (or Ca^{2+}-ionophores) lead to a complete emptying of the stores independently of the degree of store loading (Missiaen et al., 1992, 1994; Tanimura and Turner, 1996). The effect of luminal Ca^{2+} is itself dependent on the cytosolic Ca^{2+} concentration (Missiaen et al., 1994) suggesting an at least partially common underlying mechanism (Swillens et al., 1994). One possibility is that the activation of IP_3-induced Ca^{2+} release by luminal Ca^{2+} is related to binding of released Ca^{2+} ions to the stimulatory cytosolic site. Although there is still some controversy (Taylor, 1998), the differences in characteristics between both phenomena make it likely that two separate regulatory pathways are involved (Parys et al., 1996; Sienaert et al., 1997). It can therefore be envisaged that the interaction of Ca^{2+} to either a cytosolic or a luminal binding site lead to an equivalent, active conformational state.

4. MODULATION OF IP_3 BINDING BY CYTOSOLIC Ca^{2+}

Several effects of cytosolic Ca^{2+} on IP_3 binding were described. The nature of the effects were quite variable and may actually depend on experimental conditions, the isoform(s) expressed, and/or the presence of accessory proteins (Taylor, 1998). Most information on the modulation of IP_3 binding by Ca^{2+} concerns the type 1 IP_3R. Binding of IP_3 to endogenous (e.g. cerebellum, smooth muscle) or recombinant IP_3R1 is inhibited by Ca^{2+} (Joseph et al., 1989; Yoneshima et al., 1997; Cardy et al., 1997; Picard et al., 1998; Sipma et al., 1999). Most, but not all, studies agree that the inhibition by Ca^{2+} of IP_3 binding to IP_3R1 is characterized by an about 3-fold decrease in affinity for IP_3, a submicromolar sensitivity for Ca^{2+} (IC_{50} 100–300 nM Ca^{2+}) and a saturation of the effect at about 40% inhibition, even at Ca^{2+} concentrations well above the physiological range. Depending on the cooperativity of IP_3-induced Ca^{2+} release, a partial inhibition of IP_3 binding can still lead to a complete inhibition of the Ca^{2+} release (Coquil et al., 1999). Although the link between the inhibitory effects on IP_3 binding and channel activity is not clear, the selectivity for divalent cations ($Ca^{2+} > Mn^{2+} >> Sr^{2+} >>> Ba^{2+}$) is the same, suggesting that the same binding site is involved (Coquil et al., 1999).

The effects of Ca^{2+} on IP_3 binding to the type 2 or type 3 IP_3R are much less investigated. In contrast to cell types expressing predominantly IP_3R1, the affinity of the IP_3R for IP_3 increased in the presence of cytosolic Ca^{2+} in hepatocytes, which express mainly IP_3R2. This increase would represent the formation of a high-affinity desensitized state and is probably mediated by a regulatory protein (Picard et al., 1998). In agreement with these results, physiologically relevant Ca^{2+} concentrations (100 nM–25 μM) did not modulate IP_3 binding to the bacterially expressed IP_3-binding domain of IP_3R2 (Vanlingen et al., 2000). In fact, IP_3 binding to the latter domain was already significantly lower at 100 nM Ca^{2+} than in the complete absence of Ca^{2+}. This suggests that a Ca^{2+}-binding site is conserved in the IP_3-binding domain of the type 2 IP_3R, but that under physiological conditions the site is always occupied by Ca^{2+} (Vanlingen et al., 2000).

The effect of Ca^{2+} on IP_3 binding to IP_3R3 was studied in Sf9 insect cells after overexpression of mammalian IP_3R3 (Yoneshima et al., 1997; Cardy et al., 1997). Both studies show that the major effect of increased Ca^{2+} on IP_3 binding is inhibitory for IP_3R1 and stimulatory for IP_3R3, but the results were partially contradictory. Concerning IP_3R3, Ca^{2+} increased in one study the affinity without affecting the total number of binding sites (Yoneshima et al., 1997). In the other study, Ca^{2+} initially increased the number of binding sites but at higher concentrations subsequently decreased the affinity (Cardy et al., 1997). The reason for the discrepancy is not clear and may lie in the different levels of expression or the experimental conditions (Cardy et al., 1997). Ca^{2+} also inhibits IP_3 binding to the bacterially expressed IP_3-binding domain of IP_3R3 (Vanlingen et al., 2000) and to IP_3R3 heterologously expressed in COS cells (Lin et al., 2000). The basic characteristics of the inhibition of IP_3 binding are similar for the IP_3R3 expressed in Sf9 cells and for its ligand-binding domain expressed in *E. coli*, i.e. a partial inhibition (about 40%) occurring with a lower sensitivity for Ca^{2+} than for IP_3R1 (IC_{50} 500–700 nM Ca^{2+}) (Cardy et al., 1997; Vanlingen et al., 2000).

5. STRUCTURAL DETERMINANTS FOR THE REGULATION OF THE IP_3R BY Ca^{2+}

At the functional level three separate processes (activation by cytosolic Ca^{2+}, activation by luminal Ca^{2+} and inhibition by cytosolic Ca^{2+}) can be discerned, implying the existence of various regulatory sites for Ca^{2+}. Moreover,

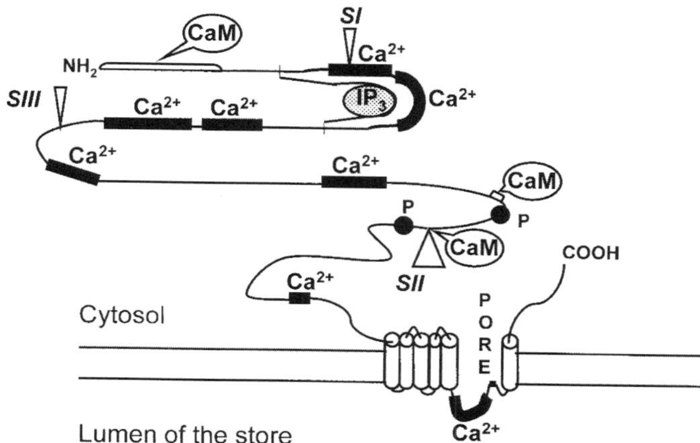

Figure 2. Ca^{2+}- and calmodulin-binding sites on IP_3R1. Scale model of a single IP_3R1 subunit with indication of the binding sites for Ca^{2+} and calmodulin (CaM). The seven cytosolic and the one luminal Ca^{2+}-binding sites are indicated, as well as the three calmodulin-binding sites already identified on IP_3R1. The coordinates of the various sites for Ca^{2+} are amino acids 304–381, 378–450, 660–745, 741–849, 994–1059, 1347–1426 (Sienaert et al., 1997), 2124–2146 and 2463–2528 (Sienaert et al., 1996). For calmodulin, the binding sites are 1–159 (Adkins et al., 2000), 1564–1585 (Yamada et al., 1995) and near the S2-splice region (Lin et al., 2000). In addition, the IP_3-binding domain, the two identified protein kinase A-sensitive phosphorylation sites (P), the six transmembrane stretches, the pore region and the localization of the splice domains (SI, SII, SIII) are indicated.

each of these processes can have one or multiple causes. For example, the Ca^{2+}-dependent inhibition of IP_3-induced Ca^{2+} release could, at least in permeabilized smooth muscle cells, already be divided in two separate effects. Indeed, a different sensitivity to divalent cations allowed to discriminate between an effect on the cooperativity of the release (whereby Sr^{2+} mimics Ca^{2+}, although with lower affinity) and an effect on the IP_3 sensitivity (strictly specific for Ca^{2+}) (Sienaert et al., 1997). Finally, the time-dependence of the activation and inhibition processes indicate that protein-protein interactions or kinase-mediated reactions could also be involved.

A first mechanism whereby Ca^{2+} can modulate the IP_3R is by direct interaction. The type 1 IP_3R contains 7 cytosolic Ca^{2+}-binding sites and 1 intraluminal Ca^{2+}-binding site, none of them containing an EF-hand motif (Sienaert et al., 1996, 1997). The role of these different Ca^{2+}-binding sites has not yet been ascertained, but several of those sites are located in potentially important domains of the receptor (Figure 2). In particular, Ca^{2+}-binding sites are localized in the IP_3-binding region, near the putative binding sites for regulatory proteins as e.g. calmodulin and in the vicinity of the channel pore (Sienaert et al., 1996, 1997).

The inhibition by Ca^{2+} of IP_3 binding to IP_3R1 was originally proposed to be mediated by an accessory protein, named calmedin (Danoff et al., 1988). In several studies the effect of Ca^{2+} on IP_3 binding was lost during receptor purification, suggesting a role for an accessory protein. On the other hand, Ca^{2+} still inhibited IP_3 binding to immuno-precipitated IP_3R1, to IP_3R1 from which associated proteins were removed (Picard et al., 1998) and to its bacterially expressed IP_3-binding domain (Sipma et al., 1999). Therefore, it is conceivable that one of the two Ca^{2+}-binding sites located in the IP_3-binding domain (Figure 2), may mediate this effect. Particularly, the domain between amino acids 378 and 450 is an attractive candidate because it is conserved between the three isoforms, and an effect of Ca^{2+} on the bacterially expressed IP_3-binding domains occurs for the three IP_3R isoforms (Sipma et al., 1999; Vanlingen et al., 2000).

A functional role can also be expected for the luminal binding site (amino acids 2463-2528), localized in the immediate vicinity of the pore region, between the fifth and the sixth transmembrane stretches (Sienaert et al., 1996) and which is conserved in the different IP_3R isoforms. This Ca^{2+}-binding site may be involved in the luminal regulation of the IP_3R by Ca^{2+} and/or the accumulation of Ca^{2+} ions near the channel pore.

Regulation of the IP_3R by Ca^{2+}-binding proteins is also documented. In particular evidence exist for an important role of calmodulin in the process of Ca^{2+}-dependent inhibition (Missiaen, et al., 1999; Michikawa, et al., 1999). A Ca^{2+}-dependent calmodulin-binding site was identified in the regulatory region of IP_3R1 and IP_3R2 (Yamada et al., 1995), but there is now evidence for the existence of at least two other binding sites (Figure 2). Indeed, a low affinity calmodulin-binding site is located in the N-terminal first 159 amino acids of IP_3R1 (Adkins et al., 2000), which may be responsible for the effect of calmodulin on IP_3 binding (Sipma et al., 1999; Vanlingen et al., 2000). Finally, evidence for a third calmodulin-binding site at or near the S2 splice region has been proposed for peripheral tissues (Lin et al., 2000). Interestingly, protein-kinase A-mediated phosphorylation inhibits calmodulin binding to IP_3R1 (Lin et al., 2000), thereby providing a link between cAMP- and Ca^{2+}-dependent signaling pathways and offering an additional mechanism for fine-tuning of the IP_3R.

The possibility that the IP_3R is regulated from its luminal side by Ca^{2+}-binding proteins, in analogy with the RyR, has been proposed, but no such protein has yet been positively identified (Parys et al., 1996).

As mentioned above, Ca^{2+}-dependent phosphorylation and dephosphorylation processes may be involved in the regulation of the IP_3R by Ca^{2+}, but up to now neither the functional role nor the actual phosphorylation sites for Ca^{2+}-dependent protein kinases (protein kinase C, Ca^{2+}/calmodulin kinase II) have been identified (Patel et al., 1999). Moreover, crucial events in the Ca^{2+}-dependent regulation of the IP_3R, such as effects on IP_3 binding and on Ca^{2+} release, occur even in the absence of ATP (Taylor, 1998).

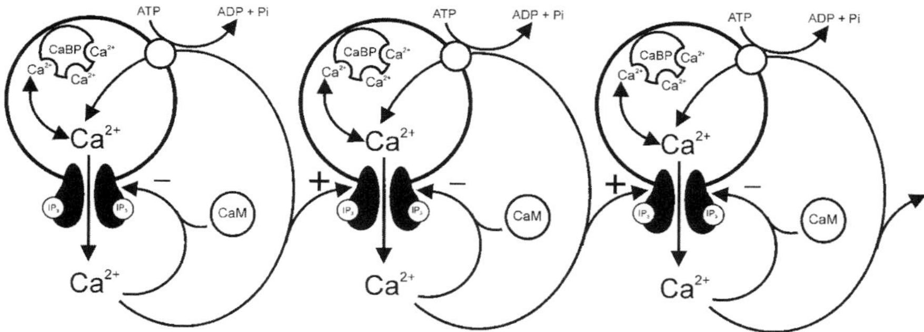

Figure 3. Propagation of intracellular Ca^{2+} waves. The model incorporates the effect of cytosolic Ca^{2+}, luminal Ca^{2+} and of the cytosolic Ca^{2+}-binding protein calmodulin on the regulation of IP_3R1. Different Ca^{2+}-store units are shown. Arrows indicate the flux of Ca^{2+} ions and the Ca^{2+} wave propagates from left to right in the figure. −, inhibition; +, stimulation. See text for details.

6. FUNCTIONAL IMPLICATIONS OF THE COMPLEX REGULATION OF THE IP_3R BY Ca^{2+}

Intracellular Ca^{2+} signals often take the form of oscillations or waves. The physiological importance of these complex spatio-temporal Ca^{2+} signals is recognized e.g. in oocytes at fertilization (Stricker, 1999), in exocrine pancreas (Petersen et al., 1994) or in liver (see Barritt, this book). Recently, a distinction has been introduced between elementary events, which play a role in setting up the resting Ca^{2+} concentration in the cytosol and which can be involved in local Ca^{2+} signaling, and global events which can evolve from the former and which affect the whole cell (Berridge, 1997).

It was recognized quite early that the biphasic regulation of IP_3-induced Ca^{2+} release by Ca^{2+} was instrumental for the propagation of intracellular Ca^{2+} waves (Lechleiter and Clapham, 1992; DeLisle and Welsh, 1992). Furthermore, both cytosolic and luminal Ca^{2+} concentrations are important factors determining the level of sensitization of the IP_3R to the level of ambient IP_3. Subsequent to the activation of IP_3R channels at a given intracellular localization, the released Ca^{2+} ions will diffuse to neighboring IP_3Rs and initiate subsequent release (Figure 3). Meanwhile, the increase in Ca^{2+} concentration in the cytosol, leading to activation of calmodulin, combined with the decline in the luminal Ca^{2+} concentration, will lead to a closure of the channels, and eventually to a refractory zone in the cell. This refractory zone is maintained until cytosolic and luminal Ca^{2+} are restored to their original levels. The time of this restoration phase will depend on the relative activity of Ca^{2+}-release channels and Ca^{2+}-pumps and on the presence of Ca^{2+}-binding proteins in the cytosol and the lumen (Figure 3). The generation of global Ca^{2+} waves from elementary events will therefore be dependent on the local IP_3R density, the local IP_3 and Ca^{2+} concentrations and the actual diffusion rate for Ca^{2+} ions (Berridge, 1997).

Finally, the regulation of the IP_3R by luminal Ca^{2+} can have a dual purpose: on one hand it helps to modulate the intracellular Ca^{2+} signals, but on the other hand it will protect the endoplasmic reticulum against excessive Ca^{2+} depletion which can lead to abnormal protein synthesis, assembly, folding and sorting (see Michalak et al., this book).

7. SUMMARY

In summary, the available evidence indicates that Ca^{2+} regulates IP_3R1, IP_3R2 and IP_3R3 in a tight and complex way. The fast kinetics of activation and inactivation of the IP_3R is hereby instrumental for the formation of local and global intracellular spatio-temporal Ca^{2+} signals. An important point is the fact that the Ca^{2+} ions released by the IP_3R can act as a potent accelerator and a potent brake on their own release. The type 1 IP_3R, and probably also the type 2 and type 3 IP_3Rs, are therefore critically controlled by Ca^{2+}. The main features are the stimulatory and the inhibitory effects of cytosolic Ca^{2+} and the stimulatory effect of luminal Ca^{2+} on IP_3-induced Ca^{2+} release. The mechanisms of interaction of Ca^{2+} with the IP_3R have not yet been clarified. For IP_3R1, calmodulin has been shown to be an important factor for the inhibitory phase. In addition, multiple Ca^{2+}-binding sites have been identified and the inhibition of IP_3 binding by Ca^{2+} may contribute to the inhibitory phase. For IP_3R2 and IP_3R3 similar mechanisms may be involved, although differences in the nature of the processes have been described. This indicates that the Ca^{2+}-dependent positive and negative feedback systems acting on the IP_3R are important mechanisms for which cells have evolved different, partially redundant, mechanisms. Further work is needed to establish the role of the various Ca^{2+}- and calmodulin-binding sites on the various IP_3R isoforms and to identify other accessory proteins or factors that may be involved in this regulation.

ACKNOWLEDGEMENTS

The authors thank all present and past members of the "IP_3-group" in Leuven for fruitful discussions and cooperation. Stimulating discussions with Drs. S.K. Joseph, C.W. Taylor, M.J. Berridge, and M.D. Bootman are acknowledged. The work performed in the authors laboratory was supported by grants 1.5342.97N and 3.0238.95 of the Fund for Scientific Research-Flanders (FWO), by the Concerted Actions of the K.U.Leuven, and by EC grant BMH4-CT96-0656. IS and PDS are Senior Research Assistants and JBP is Research Associate of the FWO. Finally, the authors want to apologize for all the excellent work on Ca^{2+} regulation of the IP_3R that, due to place constraints could not be cited.

REFERENCES

Adkins, C.E. and Taylor, C.W., 1999, Lateral inhibition of inositol 1,4,5-trisphosphate receptors by cytosolic Ca^{2+}, *Curr. Biol.* 9, 1115–1118.

Adkins, C.E., Morris, S.A., De Smedt, H., Sienaert, I., Török, K. and Taylor, C.W., 2000, Ca^{2+}-calmodulin inhibits Ca^{2+} release mediated by types 1, 2 and 3 inositol trisphosphate receptors, *Biochem. J.* 345, 357–363.

Barritt, G.J., Calcium signalling in liver cells, this book.

Berridge, M.J., 1993, Inositol trisphosphate and calcium signalling, *Nature* 361, 315–325.

Berridge, M.J., 1997, Elementary and global aspects of calcium signalling, *J. Physiol. (Lond.)* 499, 290–306.

Bezprozvanny, I., Watras, J. and Ehrlich, B.E., 1991, Bell-shaped calcium-response curves of Ins(1,4,5)P_3- and calcium-gated channels from endoplasmic reticulum of cerebellum, *Nature* 351, 751–754.

Cardy, T.J.A., Traynor, D. and Taylor, C.W., 1997, Differential regulation of types-1 and -3 inositol trisphophate receptors by cytosolic Ca^{2+}, *Biochem. J.* 328, 785–793.

Coquil, J.F., Picard, L. and Mauger, J.P., 1999, Regulation of cerebellar Ins(1,4,5)P_3 receptor by interaction between Ins(1,4,5)P_3 and Ca^{2+}, *Biochem. J.* 341, 697–704.

Danoff, S.K., Supattapone, S. and Snyder, S.H., 1988, Characterization of a membrane protein from brain mediating the inhibition of inositol 1,4,5-trisphosphate receptor binding by calcium, *Biochem. J.* 254, 701–705.

DeLisle, S. and Welsh, M.J., 1992, Inositol trisphosphate is required for the propagation of calcium waves in *Xenopus* oocytes, *J. Biol. Chem.* 267, 7963–7966.

Finch, E.A., Turner, T.J. and Goldin, S.M., 1991, Calcium as a coagonist of inositol 1,4,5-trisphosphate-induced calcium release, *Science* 252, 443–446.

Iino, M., 1990, Biphasic Ca^{2+} dependence of inositol 1,4,5-trisphosphate-induced Ca release in smooth muscle cells of the guinea pig taenia caeci, *J. Gen. Physiol.* 95, 1103–1122.

Joseph, S.K., Rice, H.L. and Williamson, J.R., 1989, The effect of external calcium and pH on inositol trisphosphate-mediated calcium release from cerebellum microsomal fractions, *Biochem. J.* 258, 261–265.

Lechleiter, J.D. and Clapham, D.E., 1992, Molecular mechanisms of intracellular calcium excitability in *X. laevis* oocytes, *Cell* 69, 283–294.

Lin, C., Widjaja, J. and Joseph, S.K., 2000, The interaction of calmodulin with alternatively spliced isoforms of the type-I inositol trisphosphate receptor, *J. Biol. Chem.* 275, 2305–2311.

Macrez, N. and Mironneau, J., Ca^{2+} release in muscle cells, this book.

Mak, D.O., McBride, S. and Foskett, J.K., 1998, Inositol 1,4,5-tris-phosphate activation of inositol trisphosphate receptor Ca^{2+} channel by ligand tuning of Ca^{2+} inhibition, *Proc. Natl. Acad. Sci. USA* 95, 15821–15825.

Marchant, J.S. and Taylor, C.W., 1997, Cooperative activation of IP_3 receptors by sequential binding of IP_3 and Ca^{2+} safeguards against spontaneous activity, *Curr. Biol.* 7, 510–518.

Michalak, M., Nakamura, K., Papp, S. and Opas, M., Calreticulin and dynamics of the endoplasmic reticulum lumenal environment, this book.

Michikawa, T., Hirota, J., Kawano, S., Hiraoka, M., Yamada, M., Furuichi, T. and Mikoshiba, K., 1999, Calmodulin mediates calcium-dependent inactivation of the cerebellar type 1 inositol 1,4,5-trisphosphate receptor, *Neuron* 23, 799–808.

Missiaen, L., De Smedt, H., Droogmans, G. and Casteels, R., 1992, Ca^{2+} release induced by inositol 1,4,5-trisphosphate is a steady-state phenomenon controlled by luminal Ca^{2+} in permeabilized cells, *Nature* 357, 599–602.

Missiaen, L., De Smedt, H., Parys, J.B. and Casteels, R., 1994, Co-activation of inositol trisphosphate-induced Ca^{2+} release by cytosolic Ca^{2+} is loading-dependent, *J. Biol. Chem.* 269, 7238–7242.

Missiaen, L., Parys, J.B., Weidema, A.F., Sipma, H., Vanlingen, S., De Smet, P., Callewaert, G. and De Smedt, H., 1999, The bell-shaped Ca^{2+} dependence of the inositol 1,4,5-trisphosphate-induced Ca^{2+} release is modulated by Ca^{2+}/calmodulin, *J. Biol. Chem.* 274, 13748–13751.

Parys, J.B., Sernett, S.W., DeLisle, S., Snyder, P.M., Welsh, M.J. and Campbell, K.P., 1992, Isolation, characterization, and localization of the inositol 1,4,5-trisphosphate receptor protein in *Xenopus laevis* oocytes, *J. Biol. Chem.* 267, 18776–18782.

Parys, J.B., Missiaen, L., De Smedt, H., Sienaert, I. and Casteels, R., 1996, Mechanisms responsible for quantal Ca^{2+} release from inositol trisphosphate-sensitive calcium stores, *Pflügers Arch. Eur. J. Physiol.* 432, 359–367.

Patel, S., Joseph, S.K. and Thomas, A.P., 1999, Molecular properties of inositol 1,4,5-trisphosphate receptors, *Cell Calcium* 25, 247–264.

Petersen, O.H., Petersen, C.C. and Kasai, H., 1994, Calcium and hormone action, *Annu. Rev. Physiol.* 56, 297–319.

Picard, L., Coquil, J.F. and Mauger, J.P., 1998, Multiple mechanisms of regulation of the inositol 1,4,5-trisphosphate receptor by calcium, *Cell Calcium* 23, 339–348.

Rossi, D., Barone, V., Simeoni, I. and Sorrentino, V., Ryanodine-sensitive calcium release channels, this book.

Sienaert, I., De Smedt, H., Parys, J.B., Missiaen, L., Vanlingen, S., Sipma, H. and Casteels, R., 1996, Characterization of a cytosolic and a luminal Ca^{2+} binding site in the type I inositol 1,4,5-trisphosphate receptor, *J. Biol. Chem.* 271, 27005–27012.

Sienaert, I., Missiaen, L., De Smedt, H., Parys, J.B., Sipma, H. and Casteels, R., 1997, Molecular and functional evidence for multiple Ca^{2+}-binding domains in the type 1 inositol 1,4,5-trisphosphate receptor, *J. Biol. Chem.* 272, 25899–25906.

Sipma, H., De Smet, P., Sienaert, I., Vanlingen, S., Missiaen, L., Parys, J.B. and De Smedt, H., 1999, Modulation of inositol 1,4,5-trisphosphate binding to the recombinant ligand-binding site of the type-1 inositol 1,4,5-trisphosphate receptor by Ca^{2+} and calmodulin, *J. Biol. Chem.* 274, 12157–12162.

Stricker, S.A., 1999, Comparative biology of calcium signaling during fertilization and egg activation in animals, *Dev. Biol.* 211, 157–176.

Swillens, S., Combettes, L. and Champeil, P., 1994, Transient inositol 1,4,5-trisphosphate-induced Ca^{2+} release: A model based on regulatory Ca^{2+}-binding sites along the permeation pathway, *Proc. Natl. Acad. Sci. USA* 91, 10074–10078.

Tanimura, A. and Turner, R.J., 1996, Calcium release in HSY cells conforms to a steady-state mechanism involving regulation of the inositol 1,4,5-trisphosphate receptor Ca^{2+} channel by luminal [Ca^{2+}], *J. Cell. Biol.* 132, 607–616.

Taylor, C.W., 1998, Inositol trisphosphate receptors: Ca^{2+}-modulated intracellular Ca^{2+} channels, *Biochim. Biophys. Acta* 1436, 19–33.

Vanlingen, S., Sipma, H., De Smet, P., Callewaert, G., Missiaen, L., De Smedt, H. and Parys, J.B., 2000, Ca^{2+} and calmodulin differentially modulate myo-inositol 1,4,5-trisphosphate (IP_3)-binding to the recombinant ligand-binding domains of the various IP_3 receptor isoforms, *Biochem. J.* 346, 275–280.

Yamada, M., Miyawaki, A., Saito, K., Nakajima, T., Yamamoto Hino, M., Ryo, Y., Furuichi, T. and Mikoshiba, K., 1995, The calmodulin-binding domain in the mouse type 1 inositol 1,4,5-trisphosphate receptor, *Biochem. J.* 308, 83–88.

Yoneshima, H., Miyawaki, A., Michikawa, T., Furuichi, T. and Mikoshiba, K., 1997, Ca^{2+} differentially regulates the ligand-affinity states of type 1 and type 3 inositol 1,4,5-trisphosphate receptors, *Biochem. J.* 322, 591–596.

Calreticulin and Dynamics of the Endoplasmic Reticulum Lumenal Environment

Marek Michalak, Kimitoshi Nakamura, Sylvia Papp and Michal Opas

1. INTRODUCTION

It is widely accepted that Ca^{2+} is a universal signaling molecule in the cell (Pozzan et al., 1994). Due to the versatility of Ca^{2+} signaling both, Ca^{2+} storage and release must be tightly controlled in spatiotemporal manner (Petersen and Burdakov, 1999). While several organelles participate in control of Ca^{2+} homeostasis, the endoplasmic reticulum (ER) appears to be the most important organelle controlling many aspects of intracellular Ca^{2+} homeostasis (Pozzan et al., 1994; Meldolesi and Pozzan, 1998). Ca^{2+} is released from the ER by $InsP_3$ receptor and/or RyR Ca^{2+} release channels and it is taken up by the sarcoplasmic/endoplasmic reticulum Ca^{2+}-ATPase (SERCA) (MacLennan et al., 1997 and Parys et al., this book). Alterations in the intracellular and ER lumenal Ca^{2+} concentration regulate a variety of diverse cellular functions including secretion, contraction-relaxation, cell motility, cytoplasmic and mitochondrial metabolism, protein synthesis, modification and folding, gene expression, cell cycle progression and apoptosis (Pozzan et al., 1994). The ER is a site of synthesis of membrane proteins, membrane lipids and secreted proteins. It contains the largest concentration of chaperones involved in protein folding, modification and assembly. The ER and its lumen contain a characteristic set of resident proteins that are involved in every aspect of the ER function. It is also likely that Ca^{2+} plays a role of

---

Marek Michalak and Kimitoshi Nakamura • Canadian Institutes of Health Research Group in Molecular Biology of Membranes and the Department of Biochemistry, University of Alberta, Edmonton, Alberta, Canada. **Sylvia Papp and Michal Opas** • Department of Anatomy and Cell Biology, University of Toronto, Toronto, Ontario, Canada.

R. Pochet, R. Donato, J. Haiech, C. Heizmann and V. Gerke (eds.): Calcium: The Molecular Basis of Calcium Action in Biology and Medicine, 191–204.

signaling molecule in the ER lumen. The ER lumenal Ca^{2+} concentration ($[Ca^{2+}]_{ER}$) affect several processes in the ER lumen including modulation of chaperone-substrate and protein-protein interactions (Corbett et al., 1999). There are several examples of how changes in the $[Ca^{2+}]_{ER}$ may affect the function of this important organelle. It is not surprising, therefore, that the ER proteins have profound effects on many cell functions. In this article we focus on how calreticulin and other ER lumenal Ca^{2+} binding chaperones affect many cellular functions with a special emphasis on their role in pathological conditions (see also Eggleton and Llewellyn, this book).

2. CALRETICULIN

Calreticulin is one of the major ER lumenal proteins (Michalak et al., 1999). It is a 46-kDa protein encoded by a single gene (Michalak et al., 1999). The protein is synthesized with an N-terminal cleavable amino acid signal sequence and a C-terminal KDEL (Lysine – Aspartic acid – Glutamic acid – Leucine) ER retrieval signal (Michalak et al., 1999). These specific amino acid sequences are responsible for targeting and retention of calreticulin in the ER lumen.

Structural predictions of calreticulin suggest that the protein has at least three domains (Michalak et al., 1999) (Figure 1). The N-terminal half of the molecule is predicted to be a highly folded, globular structure containing 8 anti-parallel β-strands connected by protein loops. The N-domain binds Zn^{2+} (Michalak et al., 1999), interacts with the DNA binding domain of the glucocorticoid receptor *in vitro* (Burns et al., 1994), with Rubella virus RNA (Nakhasi et al., 1998), α-integrin (Rojiani et al., 1991), and with PDI and ERp57 (Baksh et al., 1995; Corbett et al., 1999; Molinari and Helenius, 1999; Oliver et al., 1999). These specific protein-protein interactions may be regulated by Ca^{2+} binding to the C-domain of the protein (Corbett et al., 1999). The N-domain of calreticulin also inhibits proliferation of endothelial cells and suppresses angiogenesis (Pike et al., 1998).

Calreticulin has two Ca^{2+} binding sites: a high affinity, low capacity site in the P-domain and a low affinity high capacity site in the C-domain (Michalak et al., 1999). The P-domain of calreticulin comprises a proline-rich sequence which binds Ca^{2+} with high affinity ($K_d = 1\ \mu$M; $B_{\max} = 1$ mole Ca^{2+}/mole protein) (Michalak et al., 1999). This region of the protein is critical for the lectin-like chaperone activity of calreticulin (Michalak et al., 1999) (Figure 1). The C-terminal region of the protein is highly acidic and negatively charged (Michalak et al., 1999). The C-domain of calreticulin binds Ca^{2+} ($K_d = 2$ mM; $B_{\max} = 25$ moles Ca^{2+}/mole protein) (Michalak et al., 1999). This domain binds to blood clotting factors and inhibits injury-induced blood vessel damage (restenosis) (Michalak et al., 1999). The C-domain of calreticulin may play a role of "Ca^{2+}-sensor" as it is responsible for

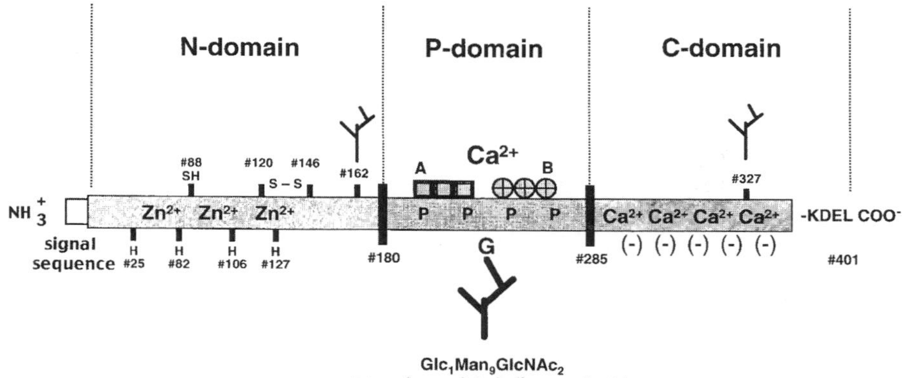

Figure 1. A model of domains of calreticulin. The figure shows schematic representation of calreticulin domains. The protein contains an N-terminal amino acid signal sequence (empty box) and a C-terminal KDEL ER retrieval signal. The location of the disulfide-bridge and four histidines (H) in the N-domain of calreticulin is indicated. The location of potential glycosylation sites (residues 162 and 327) is marked. The N-domain of calreticulin binds Zn^{2+} and involves four of the histidine (H) residues found in this domain. The P-domain of calreticulin comprises a proline-rich sequence (indicated by "P") with three repeats of the amino acid sequence PxxIxDPDAxKPEDWDE (Repeats A) followed by three repeats of the sequence GxWxPPxIxNPxYx (Repeats B) indicated by squares and circles, respectively. The P-domain of calreticulin binds Ca^{2+}. The P-domain also binds, in a Ca^{2+}-dependent manner, $Glc_1Man_9GlcNAc_2$ monoglucosylated oligosaccharides. The C-domain of calreticulin is highly acidic and negatively charged (indicated by "(−)") and binds over 25 moles of Ca^{2+}/mole of protein.

Ca^{2+} sensitivity of interaction between calreticulin and other ER lumenal chaperones (Corbett et al., 1999).

3. FUNCTIONS OF CALRETICULIN

Numerous studies showed that calreticulin is localized to the ER lumen in many diverse species including plants (Michalak et al., 1999, and references therein). Figure 2 shows immunolocalization of calreticulin in mouse embryonic fibroblasts (MEF). The protein is localized to the ER network in wild type MEF and in calreticulin deficient cells transfected with calreticulin expression vector (Figure 2) further conforming ER localization of calreticulin. Yet, calreticulin has been implicated to participate in many cellular functions inside and outside of the ER (Eggleton et al., 1997; Krause and Michalak, 1997; Nakhasi et al., 1998; Ellgaard et al., 1999; Michalak et al., 1999). Therefore, as expected any changes in calreticulin expression and function have profound effects on many cellular functions. Calreticulin deficiency is embryonic lethal (Mesaeli et al., 1999). However, there are numerous physiological and pathological conditions where the expression of calreticulin is significantly increased. These include viral infection, prostate

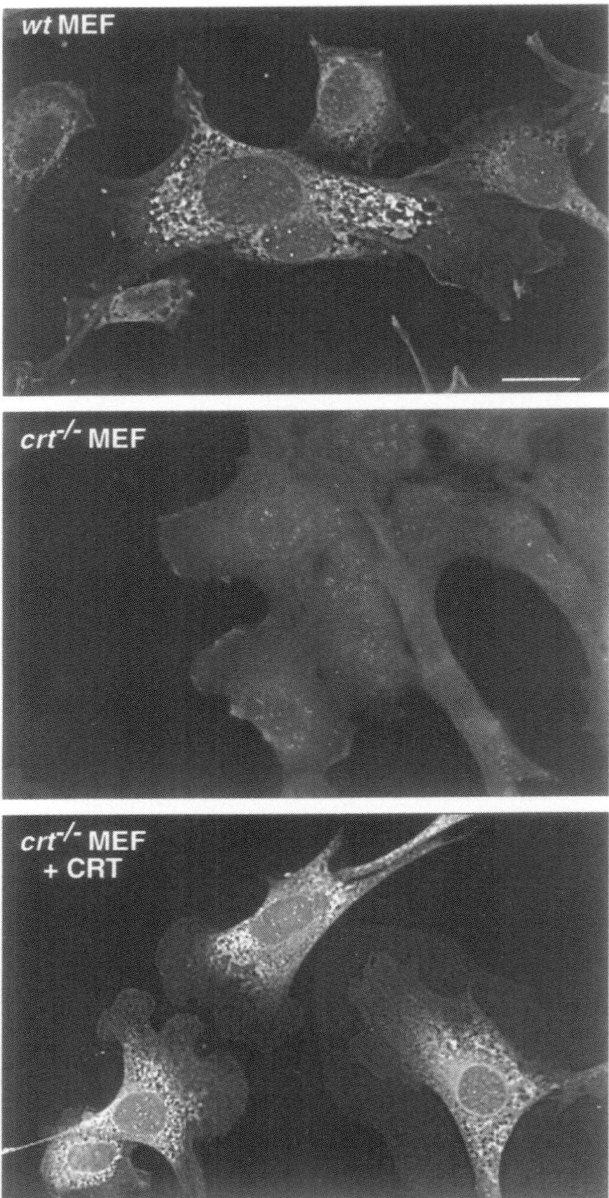

Figure 2. Calreticulin is localized to ER in mouse embryonic fibroblasts. Mouse embryonic fibroblasts (MEF) were isolated from 14.5 days old wild type (*wt MEF*) and calreticulin knockout (*crt*$^{-/-}$ *MEF*) embryos (Mesaeli et al., 1999). MEF derived from the calreticulin knockout embryos were also transfected with calreticulin expression vector (*crt*$^{-/-}$ *MEF* + *CRT*). The figure shows localization of calreticulin to the ER network in wild type cells and calreticulin knockout cells transfected with calreticulin expression vector (*crt*$^{-/-}$ *MEF* + *CRT*). As expected there is no calreticulin positive staining in calreticulin deficient cells (*crt*$^{-/-}$ *MEF*). Scale bar 25 μm.

cancer cells, chemical stress, or starvation, activation of the immune system to name a few (Michalak et al., 1999). Therefore, it is essential to understand the role of calreticulin, and other ER lumenal proteins, in control of many ER functions as those may impinge upon a variety of cellular functions distal to the ER.

3.1. Calreticulin, a lectin-like chaperone

Calreticulin is a molecular chaperone and it plays an important role in preventing the aggregation of partially folded proteins and glycoproteins (Bergeron et al., 1994; Ellgaard et al., 1999; Saito et al., 1999). Its interaction with glycoproteins makes calreticulin a lectin-like chaperone similar to calnexin, an integral ER membrane protein (Bergeron et al., 1994; Ellgaard et al., 1999). Both chaperones are involved in the "quality control" process during the synthesis of basically every glycoprotein (Ellgaard et al., 1999). Calreticulin and calnexin may be the most important chaperones in the ER lumen. They are required for proper folding majority of all cell surface molecules that are glycosylated including as ion channels, surface receptors, integrins and many transporters.

Both proteins may act sequentially and constitute a cycle involved in chaperoning of glycosylated and non-glycosylated proteins (Bergeron et al., 1994; Ellgaard et al., 1999; Ihara et al., 1999; Saito et al., 1999). Both proteins bind to $Glc_1Man_9GlcNAc_2$ oligosaccharides and recognize the terminal glucose and four internal mannose moieties (Michalak et al., 1999). The $Glc_3Man_9GlcNAc_2$ carbohydrate is attached to newly synthesized proteins followed by removal of glucose moieties by glucosidase I and II. If the glycoprotein is not correctly folded, the terminal glucose is once again attached by the UDP-glucose:glycoprotein glucosyltransferase. Unfolded glycoproteins (and non-glycosylated proteins) undergo cycles of interaction with calnexin and calreticulin (Ellgaard et al., 1999; Ihara et al., 1999; Saito et al., 1999). The lectin-binding site of calreticulin and calnexin is localized to the Ca^{2+} binding P-domain of the protein and the bound Ca^{2+} is essential for the lectin-like function of these proteins. Region(s) of calreticulin (or calnexin) involved in chaperone interactions between calreticulin and non-glycosylated proteins have not be determined yet. Interestingly, while glycoproteins are bound to calreticulin or calnexin the disulfide bonds of the substrates are rearranged by the disulfide isomerase activity associated with ERp57, an ER chaperone, a homologue of protein disulphide isomerase (PDI), with thiol-dependent reductase, and cysteine-dependent protease activities (Zapun et al., 1998). This suggests that calreticulin binding to carbohydrates may be a "signal" to recruit other chaperones to assist in protein folding.

3.2. Calreticulin, an ER lumenal "Ca^{2+}-sensor"

Ca^{2+} is released from the ER by the $InsP_3$/RyR receptors (Pozzan et al., 1994) and taken up to the ER lumen by SERCA (MacLennan et al., 1997). The $[Ca^{2+}]_{ER}$ is approximately 400 μM (Meldolesi and Pozzan, 1998). This $[Ca^{2+}]$ is two to three orders of magnitude higher than in the cytosol. Ca^{2+} release from the ER lowers $[Ca^{2+}]_{ER}$ to < 50 μM (Meldolesi and Pozzan, 1998). Calreticulin and other ER lumenal chaperones play important role in Ca^{2+} storage in the ER lumen. Besides calreticulin many other Ca^{2+} binding proteins have been identified; BiP; Grp94 (glucose-regulated protein), BiP (Grp78) (immunoglobulin heavy-chain binding protein), PDI (protein disulphide isomerase, ERp72 (endoplasmic reticulum protein), and ER/calcistorin (Meldolesi and Pozzan, 1998). This indicates that that these ER proteins play important role in intralumenal Ca^{2+} homeostasis. As we may suspect release of Ca^{2+} from the ER lumen has far reaching consequences for cellular functions. Reduction of the $[Ca^{2+}]_{ER}$ (ER Ca^{2+} depletion conditions) leads to accumulation of misfolded proteins, activation of expression of ER chaperones (Kaufman, 1999), activation of ER-nucleus and ER-plasma membrane "signaling", (Kaufman, 1999), inhibition of ER-Golgi trafficking and blocking of the nuclear import and export (Stehno-Bittel et al., 1995). Clearly, changes of $[Ca^{2+}]_{ER}$ have profound effects at multiple cellular sites including the structure and function of the ER lumenal Ca^{2+} binding chaperones.

It has recently become apparent that Ca^{2+}, besides its role as a universal cytosolic signaling molecule (Pozzan et al., 1994), may also play a signaling role within the ER lumen (Corbett and Michalak, 2000). $[Ca^{2+}]_{ER}$ affect several processes in the ER lumen including modulation of chaperone-substrate and protein-protein interactions (Corbett et al., 1999). There are several examples of how changes in the $[Ca^{2+}]_{ER}$ may affect the function of this important organelle. Binding of carbohydrate to calreticulin and calnexin occurs at high $[Ca^{2+}]_{ER}$ (when Ca^{2+} stores are full) and it is inhibited at low $[Ca^{2+}]_{ER}$ under the conditions of Ca^{2+} depletion of the stores. Calreticulin interacts with PDI (protein disulphide isomerase) (Baksh et al., 1995; Corbett et al., 1999) and these interactions are Ca^{2+}-dependent in a way reminiscent of the emptying and refilling of the ER Ca^{2+} stores (Corbett et al., 1999). Calreticulin also interacts with ERp57, a PDI-like ER lumenal chaperone (Corbett et al., 1999). These interactions are also sensitive to Ca^{2+} (Corbett et al., 1999) and affect ERp57 chaperone activity (Zapun et al., 1998). The most important observation is that Ca^{2+}-sensitivity of interactions between calreticulin, PDI and ERp57 is confined to the high capacity Ca^{2+} binding C-domain of calreticulin (Corbett et al., 1999) suggesting that the C-domain of calreticulin may play a role of Ca^{2+} "sensor" in the ER lumen.

Calreticulin forms structural and functional protein complexes with other chaperones including BiP, Grp94, Grp78 (BiP) PDI, ERp72, p50 and a 46-kDa protein (Nigam et al., 1994). These associations and dissociation between ER lumenal chaperones may be regulated by Ca^{2+} binding to calreticulin, as described for PDI and ERp57 (Baksh et al., 1995; Corbett et al., 1999). Moreover, Ca^{2+} in the ER lumen may also play a role in the stability of lumenal chaperones. For example, degradation of PDI and calreticulin by the protease activity of ERp72 is Ca^{2+}-dependent (Urade et al., 1993). PDI degradation is enhanced in the presence of 1 mM Ca^{2+}, but degradation of calreticulin is inhibited by the presence of 1 mM Ca^{2+} (Urade et al., 1993). Therefore, in addition to its role in the cytosol Ca^{2+} may be considered an important signaling molecule in the ER lumen (Corbett and Michalak, 2000).

3.3. Calreticulin, a regulator of Ca^{2+} homeostasis

Overexpression of calreticulin in a variety of cellular systems (Bastianutto et al., 1995; Mery et al., 1996; Opas et al., 1996) does result in an increased amount of Ca^{2+} stored in the ER lumen without any significant effect on the cytoplasmic $[Ca^{2+}]$ (Bastianutto et al., 1995; Mery et al., 1996; Opas et al., 1996). On the other hand, when calreticulin expression is abolished by knocking out the calreticulin gene by homologous recombination technique, Ca^{2+} storage capacity of the ER embryonic stem (ES) cells or mouse embryonic fibroblasts (MEF) is not changed (Coppolino et al., 1997; Mesaeli et al., 1999). This is likely because the other ER lumenal Ca^{2+} binding chaperones (Grp94, BiP, PDI) compensate for the loss of calreticulin. Calreticulin deficient MEF cells have diminished agonist-mediated, $InsP_3$-dependent Ca^{2+} release from the ER (Mesaeli et al., 1999) suggesting that calreticulin, from the ER lumen, affects Ca^{2+} transport across the ER membrane and consequently the $[Ca^{2+}]_{ER}$.

Ca^{2+} is taken up into the ER lumen by SERCA (Pozzan et al., 1994). There are three differentially expressed genes encoding the SERCA protein, SERCA 1, 2 and 3 (MacLennan et al., 1997). SERCA2a is the cardiac/slow-twitch muscle isoform, whereas SERCA2b, with a C-terminal extension, is expressed in smooth muscle and non-muscle tissues (MacLennan et al., 1997). Camacho's group carried out elegant studies on the role of calreticulin in Ca^{2+} homeostasis utilizing the *Xenopus* oocyte model (Camacho and Lechleiter, 1995; John et al., 1998). They demonstrated that co-expression of calreticulin with SERCA2b (but not SERCA1 or 2a) results in a sustained elevation in Ca^{2+} release without concomitant oscillations upon injection of $InsP_3$ (Camacho and Lechleiter, 1995; John et al., 1998). SERCA2b has an additional transmembrane segment and a C-terminal 12 residues tail localized to the ER lumen (MacLennan et al., 1997), containing a putative N-glycosylation site (residue N1036). Site-directed mutagenesis of the N1036 (John et al., 1998) or truncation studies involving removal of N1036 (Verboomen et al., 1994) revealed that this residue is critical for calreticulin-dependent effects on SERCA2b function and for its isoform specific functional differences. Effects of calreticulin on SERCA2b involve the

P-domain of the protein suggesting involvement of the chaperone, lectin-like function of calreticulin (John et al., 1998). Based on these observations John et al. (1998) proposed that the C-terminal tail of SERCA2b may be glycosylated *in vivo* and that calreticulin modulates SERCA2b Ca^{2+} transport activity by a direct interaction with the glycosylated C-terminal tail of the pump (John et al., 1998). This is an exciting hypothesis suggesting that lectin-like region of calreticulin may play a dual role in the ER lumen: chaperoning of newly synthesized integral and secreted proteins and modulation of "functional" conformations of the mature, fully functional integral (and perhaps lumenal) ER glycoproteins (John et al., 1998). An attractive hypothesis is, that, similar to potential SERCA-calreticulin interactions, calreticulin may bind to the glycosylated intralumenal loop(s) of the $InsP_3$ receptor and modulate Ca^{2+}-release. However, as yet there is no evidence for either glycosylation of SERCA2b or for calreticulin-SERCA or calreticulin-$InsP_3$ receptor interactions.

3.4. Calreticulin, an ER signaling molecule

Changes in the level of expression of calreticulin affect cell adhesion (Rojiani et al., 1991; Opas et al., 1996; Coppolino et al., 1997; Fadel et al., 1999), integrin-dependent Ca^{2+} signaling (Coppolino et al., 1997), steroid-sensitive gene expression both *in vitro* and *in vivo* (Michalak, 1999) and store-operated Ca^{2+} influx (Bastianutto et al., 1995; Mery et al., 1996). What is the mechanisms involved in calreticulin-dependent modulation of functions outside of the ER?

Since calreticulin has not been found in the cytoplasm it is conceivable that calreticulin influence cell adhesion indirectly, from the ER lumen, via modulation of gene expression of adhesion-related molecules (Fadel et al., 1999; Opas et al., 1996) and/or by changes in the integrin-dependent Ca^{2+}-signaling (Coppolino et al., 1997). For example, increased expression of calreticulin modulates cell adhesion by coordinating up-regulation of expression of vinculin and N-cadherin (Opas et al., 1996; Fadel et al., 1999). Down-regulation of calreticulin causes inverse effects (Opas et al., 1996). The changes in cell adhesion are also coincident with changes in the levels of protein tyrosine phosphorylation in cells differentially expressing calreticulin (Fadel et al., 1999). It is well documented that protein phosphorylation/dephosphorylation of tyrosine is a major mechanism for regulation of cell adhesion (Burridge and Chrzanowska-Wodnicka, 1996; Hanks and Polte, 1997). One of the proteins undergoing tyrosine dephosphorylation in calreticulin-overexpressing cells is β-catenin (Opas, unpublished observations), a structural component of cadherin-mediated adhesion complexes, a member of the armadillo protein family and a part of the Wnt/Wingless signaling pathway (Barth et al., 1997). Tyrosine dephosphorylated β-catenin is stabilized in junctional complexes (Hazan and Norton, 1998). Phosphorylated β-catenin, when displaced to the cytoplasm, may bind LEF/TCF transcription factors and translocate to the nucleus (Barth et al., 1997; Huttenlocher et al., 1998). Phosphorylation of β-catenin by a serine/threonine kinase, glycogen synthase kinase-3β in a complex with axin and adenomatous polyposis coli protein targets it for ubiquitination and subsequent degradation (Barth et al., 1997; Ikeda et al., 1998). Although the mechanism(s) are still elusive, it is conceivable that the effects of calreticulin overexpression on cell adhesion may be due to calreticulin effects on a signaling pathway, which includes the vinculin/catenin-cadherin protein system and may involve changes in activity of tyrosine kinases and/or phosphatases. A direct implication of this for cell-substratum interactions is that calreticulin effects may target primarily focal contact-mediated adhesion, which indeed was shown to be the case (Fadel et al., 1999).

Calreticulin binds to the DNA binding domain of steroid receptors and transcription factors containing the amino acid sequence KxFF(K/R)R and prevents their interaction with DNA *in vitro* (Burns et al., 1994; Dedhar et al., 1994). Transcriptional activation by glucocorticoid, androgen, retinoic acid and vitamin D_3 receptors *in vivo* is modulated in cells overexpressing calreticulin (Burns et al., 1994; Dedhar et al., 1994; Michalak et al., 1996). Again, these are surprising findings since calreticulin is an ER resident protein and steroid receptors are found in the cytoplasm or in the nucleus. What could be the mechanism(s) by which calreticulin affects functions outside the ER including cell adhesion and gene expression? Despite many years of investigation calreticulin has not been identified in the cytosol. Cytoplasmically-targeted calreticulin does not have any effect on the function of steroid receptors or cell adhesion *in vivo* (Michalak et al., 1996; Opas et al., 1996; Fadel et al., 1999). Calreticulin must modulate cell adhesion and gene expression from the ER lumen. These findings indicate that calreticulin may participate in a signaling network in the lumen of the ER (Kaufman, 1999).

4. CALRETICULIN KNOCKOUT MOUSE

Recently gene targeting by homologous recombination was used to generate calreticulin deficient ES cells and the mouse (Coppolino et al., 1997; Mesaeli et al., 1999). To our knowledge this was the first time homologous recombination was used to knockout a gene encoding an ER lumenal protein. Since the protein is involved in a number of diverse and important functions, it was not surprising that calreticulin knockout mouse was not be viable (Mesaeli et al., 1999). What was surprising was that calreticulin deficient embryos likely die from a lesion in cardiac development. In the adult, calreticulin is expressed mainly in non-muscle and smooth muscle cells, and is only a minor component of the skeletal and cardiac muscle (Mesaeli et al., 1999). However, calreticulin gene is activated during cardiac development concomitant with an elevated expression of the protein, which drops sharply in the newborn heart (Mesaeli et al., 1999). These indicate that calreticulin belongs to the family of cardiac embryonic genes and plays critical role during cardiogenesis. Grp94, another ER lumenal Ca^{2+} binding chaperone, is also up-regulated during cardiomyogenesis (Barnes and Smoak, 1997) suggesting that ER chaperones in general must play an important role in formation of the heart.

What could be the role of calreticulin and other ER lumenal proteins in cardiac development? Cardiac development is an extremely complex process under strict transcriptional control with functions of many of these transcription factors depending on sustained, $InsP_3$-dependent Ca^{2+} release (Olson and Srivastava, 1996). For example, transcriptional activity of GATA-4 is enhanced several fold by the formation of heterodimers with NF-AT and NF-AT/GATA-4/calcineurin synergistically activate marker genes for cardiac hypertrophy (Molkentin et al., 1998). Nuclear import of the NF-AT transcription factors requires dephosphorylation by calcineurin (Rao et al., 1997). Ca^{2+} release by $InsP_3$-dependent pathway, but not Ca^{2+} pulses, is required to activate calcineurin and to maintain the NF-AT transcription factor in the nucleus (Crabtree, 1999). Ca^{2+} release from the ER is impaired in calreticulin deficient MEF suggesting that a role for calreticulin during cardiac development likely relates to its effects on ER Ca^{2+} transport (Mesaeli et al., 1999). An important lesson we have learned from the calreticulin deficient mouse is that SR and ER membrane Ca^{2+} stores may be structurally and functionally distinct compartments in cardiomyocytes. Furthermore, Ca^{2+} pools that signals the developmental (or cardiac pathology) response may be distinct from those involved in excitation-contraction coupling in the SR.

4.1. Endoplasmic reticulum in the heart

ER is a site of synthesis of membrane proteins, membrane lipids and secreted proteins. Cardiac cells, similar to other cell types, require ER membrane for housekeeping functions such as protein/lipid turnover, protein modification and folding. The ER in non-muscle cells and SR in cardiac and skeletal muscle cells are also considered one of the most important and metabolically relevant sources of cellular Ca^{2+} for variety of functions including secretion, contraction-relaxation, cell motility, cytoplasmic and mitochondrial metabolism, protein synthesis, modification and folding, gene expression, cell cycle progression and apoptosis (Pozzan et al., 1994). Several ER-specific proteins have been identified in cardiac muscle, including calnexin, BiP, Grp94, PDI (Fliegel et al., 1990; Cala et al., 1993; Cala and Jones, 1994; Barnes and Smoak, 1997). It is conceivable that these and other ER membrane proteins are found in the muscle ER and the SR membrane. However, this awaits further investigation.

Calreticulin deficient cardiomyocytes develop a functional SR and contract spontaneously, however, their Ca^{2+} homeostasis by ER membrane is impaired (Mesaeli et al., 1999) suggesting that cardiomyocytes can distinguish between SR and ER Ca^{2+}. It is not clear how cardiomyocytes discriminate between elevations in Ca^{2+} associated with development, chronic long-term hypertrophic stimuli and normal fluctuations in Ca^{2+} level during each phase of contraction-relaxation. Work with calreticulin deficient mouse indicates that the Ca^{2+} pool that signals the developmental (or hypertrophy) response are distinct from those involved in excitation-contraction coupling in the SR. It is conceivable that different amplitude and frequency of the Ca^{2+} signals could be generated from cardiac ER or SR to represent different biologically important information. Does calreticulin, and other ER membrane proteins, play a role in cardiac pathology? Solving this question may well hold key answers to many clinical problems in the cardiovascular field.

ACKNOWLEDGEMENTS

Research in the authors' laboratories is supported by the Canadian Institutes of Health Research, the Heart and Stroke Foundation of Alberta, the Heart and Stroke Foundation of Ontario, and the Alberta Heritage Foundation for Medical Research. M.M. is a Senior Scientist of the Canadian Institutes of Health Research and a Medical Scientist of the Alberta Heritage Foundation for Medical Research. K.N. is a Postdoctoral Fellow of the Alberta Heritage Foundation for Medical Research.

REFERENCES

Baksh, S., Burns, K., Andrin, C. and Michalak, M., 1995, Interaction of calreticulin with protein disulfide isomerase, *J. Biol. Chem.* 270, 31338–31344.

Barnes, J.A. and Smoak, I.W., 1997, Immunolocalization and heart levels of GRP94 in the mouse during post-implantation development, *Anat. Embryol.* 196, 335–341.

Barth, A.I., Nathke, I.S. and Nelson, W.J., 1997, Cadherins, catenins and APC protein: Interplay between cytoskeletal complexes and signaling pathways, *Curr. Opin. Cell Biol.* 9, 683–690.

Bastianutto, C., Clementi, E., Codazzi, F., Podini, P., De Giorgi, F., Rizzuto, R., Meldolesi, J. and Pozzan, T., 1995, Overexpression of calreticulin increases the Ca^{2+} capacity of rapidly exchanging Ca^{2+} stores and reveals aspects of their lumenal microenvironment and function, *J. Cell Biol.* 130, 847–855.

Bergeron, J.J., Brenner, M.B., Thomas, D.Y. and Williams, D.B., 1994, Calnexin: A membrane-bound chaperone of the endoplasmic reticulum, *Trends Biochem. Sci.* 19, 124–128.

Burns, K., Duggan, B., Atkinson, E.A., Famulski, K.S., Nemer, M., Bleackley, R.C. and Michalak, M., 1994, Modulation of gene expression by calreticulin binding to the glucocorticoid receptor, *Nature* 367, 476–480.

Burridge, K. and Chrzanowska-Wodnicka, M., 1996, Focal adhesions, contractility, and signaling, *Annu. Rev. Cell Dev. Biol.* 12, 463–518.

Cala, S.E. and Jones, L.R., 1994, GRP94 resides within cardiac sarcoplasmic reticulum vesicles and is phosphorylated by casein kinase II, *J. Biol. Chem.* 269, 5926–5931.

Cala, S.E., Ulbright, C., Kelley, J.S. and Jones, L.R., 1993, Purification of a 90-kDa protein (Band VII) from cardiac sarcoplasmic reticulum. Identification as calnexin and localization of casein kinase II phosphorylation sites, *J. Biol. Chem.* 268, 2969–2975.

Camacho, P. and Lechleiter, J.D., 1995, Calreticulin inhibits repetitive intracellular Ca^{2+} waves, *Cell* 82, 765–771.

Coppolino, M.G., Woodside, M.J., Demaurex, N., Grinstein, S., St-Arnaud, R. and Dedhar, S., 1997, Calreticulin is essential for integrin-mediated calcium signalling and cell adhesion, *Nature* 386, 843–847.

Corbett, E.F. and Michalak, M., 2000, Calcium, a signaling molecule in the endoplasmic reticulum?, *Trends Biochem. Sci.* 25, 307–311.

Corbett, E.F., Oikawa, K., Francois, P., Tessier, D.C., Kay, C., Bergeron, J.J.M., Thomas, D.Y., Krause, K.H. and Michalak, M., 1999, Ca^{2+} regulation of interactions between endoplasmic reticulum chaperones, *J. Biol. Chem.* 274, 6203–6211.

Crabtree, G.R., 1999, Generic signals and specific outcomes: signaling through Ca^{2+}, calcineurin, and NF-AT, *Cell* 96, 611–614.

Dedhar, S., Rennie, P.S., Shago, M., Hagesteijn, C.Y., Yang, H., Filmus, J., Hawley, R., Bruchovsky, N., Cheng, H., Matusik, R.J. and Giguere, V., 1994, Inhibition of nuclear hormone receptor activity by calreticulin, *Nature* 367, 480–483.

Eggleton, P. and Llewellyn, D.H., Autoimmune disease and calcium-binding proteins, this book.

Eggleton, P., Reid, K.B., Kishore, U. and Sontheimer, R.D., 1997, Clinical relevance of calreticulin in systemic lupus erythematosus, *Lupus* 6, 564–571.

Ellgaard, L., Molinari, M. and Helenius, A., 1999, Setting the standards: Quality control in the secretory pathway, *Science* 286, 1882–1888.

Fadel, M.P., Dziak, E., Lo, C.M., Ferrier, J., Mesaeli, N., Michalak, M. and Opas, M., 1999, Calreticulin affects focal contact-dependent but not close contact-dependent cell-substratum adhesion, *J. Biol. Chem.* 274, 15085–15094.

Fliegel, L., Newton, E., Burns, K. and Michalak, M., 1990, Molecular cloning of cDNA encoding a 55-kDa multifunctional thyroid hormone binding protein of skeletal muscle sarcoplasmic reticulum, *J. Biol. Chem.* 265, 15496–15502.

Ghosh, A. and Greenberg, M.E., 1995, Calcium signaling in neurons: Molecular mechanisms and cellular consequences, *Science* 268, 239–247.

Hanks, S.K. and Polte, T.R., 1997, Signaling through focal adhesion kinase, *Bioessays* 19, 137–145.

Hazan, R.B. and Norton, L., 1998, The epidermal growth factor receptor modulates the interaction of E-cadherin with the actin cytoskeleton, *J. Biol. Chem.* 273, 9078–9084.

Huttenlocher, A., Lakonishok, M., Kinder, M., Wu, S., Truong, T., Knudsen, K.A. and Horwitz, A.F., 1998, Integrin and cadherin synergy regulates contact inhibition of migration and motile activity, *J. Cell Biol.* 141, 515–526.

Ihara, Y., Cohen-Doyle, M.F., Saito, Y. and Williams, D.B., 1999, Calnexin discriminates between protein conformational states and functions as a molecular chaperone *in vitro*, *Molecular Cell* 4, 331–341.

Ikeda, S., Kishida, S., Yamamoto, H., Murai, H., Koyama, S. and Kikuchi, A., 1998, Axin, a negative regulator of the Wnt signaling pathway, forms a complex with GSK-3beta and beta-catenin and promotes GSK-3beta-dependent phosphorylation of beta-catenin, *EMBO J.* 17, 1371–1384.

John, L.M., Lechleiter, J.D. and Camacho, P., 1998, Differential modulation of SERCA2 isoforms by calreticulin, *J. Cell Biol.* 142, 963–973.

Kaufman, R.J., 1999, Stress signaling from the lumen of the endoplasmic reticulum: Coordination of gene transcriptional and translational controls, *Genes & Dev.* 13, 1211–1233.

Krause, K.-H. and Michalak, M., 1997, Calreticulin, *Cell* 88, 439-443.

MacLennan, D.H., Rice, W.J. and Green, N.M., 1997, The mechanism of Ca^{2+} transport by sarco(endo)plasmic reticulum Ca^{2+}-ATPases, *J. Biol. Chem.* 272, 28815–28818.

Meldolesi, J. and Pozzan, T., 1998, The endoplasmic reticulum Ca^{2+} store: A view from the lumen, *Trends Biochem. Sci.* 23, 10–14.

Mery, L., Mesaeli, N., Michalak, M., Opas, M., Lew, D.P. and Krause, K.-H., 1996, Overexpression of calreticulin increases intracellular Ca^{2+} storage and decreases store-operated Ca^{2+} influx, *J. Biol. Chem.* 271, 9332–9339.

Mesaeli, N., Nakamura, K., Zvaritch, E., Dickie, P., Dziak, E., Krause, K.-H., Opas, M., MacLennan, D.H. and Michalak, M., 1999, Calreticulin is essential for cardiac development, *J. Cell Biol.* 144, 857–868.

Michalak, M., Burns, K., Andrin, C., Mesaeli, N., Jass, G.H., Busaan, J.L. and Opas, M., 1996, Endoplasmic reticulum form of calreticulin modulates glucocorticoid-sensitive gene expression, *J. Biol. Chem.* 271, 29436–29445.

Michalak, M., Corbett, E.F., Mesaeli, N., Nakamura, K. and Opas, M., 1999, Calreticulin: one protein, one gene, many functions, *Biochem. J.* 344, 281–292.

Molinari, M. and Helenius, A., 1999, Glycoproteins form mixed disulphides with oxidoreductases during folding in living cells, *Nature* 402, 90–93.

Molkentin, J.D., Lu, J.R., Antos, C.L., Markham, B., Richardson, J., Robbins, J., Grant, S.R. and Olson, E.N., 1998, A calcineurin-dependent transcriptional pathway for cardiac hypertrophy, *Cell* 93, 215–228.

Nakhasi, H.L., Pogue, G.P., Duncan, R.C., Joshi, M., Atreya, C.D., Lee, N.S. and Dwyer, D.M., 1998, Implications of calreticulin function in parasite biology, *Parasitol. Today* 14, 157–160.

Nigam, S.K., Goldberg, A.L., Ho, S., Rohde, M.F., Bush, K.T. and Sherman, M., 1994, A set of endoplasmic reticulum proteins possessing properties of molecular chaperones includes Ca^{2+}-binding proteins and members of the thioredoxin superfamily, *J. Biol. Chem.* 269, 1744–1749.

Oliver, J.D., Roderick, H.L., Llewellyn, D.H. and High, S., 1999, ERp57 Functions as a subunit of specific complexes formed with the ER lectins calreticulin and calnexin, *Mol. Biol. Cell* 10, 2573–2582.

Olson, E.N. and Srivastava, D., 1996, Molecular pathways controlling heart development, *Science* 272, 671–676.

Opas, M., Szewczenko-Pawlikowski, M., Jass, G.K., Mesaeli, N. and Michalak, M., 1996, Calreticulin modulates cell adhesiveness via regulation of vinculin expression, *J. Cell Biol.* 135, 1913–1923.

Parys, J.B., Sienaert, I., Vanlingen, S., Callewaert, G., De Smet, P., Missiaen, L. and De Smedt, H., Regulation of inositol 1,4,5-trisphosphate-induced Ca^{2+} release by Ca^{2+}, this book.

Petersen, O.H. and Burdakov, D., 1999, Polarity in intracellular calcium signaling, *Bioessays* 21, 851–860.

Pike, S.E., Yao, L., Jones, K.D., Cherney, B., Appella, E., Sakaguchi, K., Nakhasi, H., Teruya-Feldstein, J., Wirth, P., Gupta, G. and Tosato, G., 1998, Vasostatin, a calreticulin fragment, inhibits angiogenesis and suppresses tumor growth, *J. Exp. Med.* 188, 2349–2356.

Pozzan, T., Rizzuto, R., Volpe, P. and Meldolesi, J., 1994, Molecular and cellular physiology of intracellular calcium stores, *Physiol. Rev.* 74, 595–636.

Rao, A., Luo, C. and Hogan, P.G., 1997, Transcription factors of the NFAT family: Regulation and function, *Annu. Rev. Immunol.* 15, 707–747.

Rojiani, M.V., Finlay, B.B., Gray, V. and Dedhar, S., 1991, *In vitro* interaction of a polypeptide homologous to human Ro/SS-A antigen (calreticulin) with a highly conserved amino acid sequence in the cytoplasmic domain of integrin alpha subunits, *Biochemistry* 30, 9859–9866.

Saito, Y., Ihara, Y., Leach, M.R., Cohen-Doyle, M.F. and Williams, D.B., 1999, Calreticulin functions *in vitro* as a molecular chaperone for both glycosylated and non-glycosylated proteins, *EMBO J.* 18, 6718–6729.

Stehno-Bittel, L., Luckhoff, A. and Clapham, D.E., 1995, Calcium release from the nucleus by $InsP_3$ receptor channels, *Neuron* 14, 163–167.

Urade, R., Takenaka, Y. and Kito, M., 1993, Protein degradation by ERp72 from rat and mouse liver endoplasmic reticulum, *J. Biol. Chem.* 268, 22004–22009.

Verboomen, H., Wuytack, F., Van Den Bosch, L., Mertens, L. and Casteels, R., 1994, The functional importance of the extreme C-terminal tail of the gene 2 organellar Ca^{2+}-transport ATPase (SERCA2a/b), *Biochem. J.* 303, 979–984.

Vitadello, M., Colpo, P. and Gorza, L., 1998, Rabbit cardiac and skeletal interaction myocytes differ in constitutive and inducible expression of the glucose-regulated protein GRP94, *Biochem. J.* 332, 351–359.

Zapun, A., Darby, N.J., Tessier, D.C., Michalak, M., Bergeron, J.J.M. and Thomas, D.Y., 1998, Enhanced catalysis of ribonuclease B folding by the of calnexin or calreticulin with ERp57, *J. Biol. Chem.* 273, 6009–6012.

Ryanodine-Sensitive Calcium Release Channels

Daniela Rossi, Virginia Barone, Ilenia Simeoni and Vincenzo Sorrentino

1. INTRODUCTION

Variations in the intracellular calcium concentration $[Ca^{2+}]_i$ play an important role in a variety of intracellular processes, including secretion, contraction and cell proliferation (Berridge, 1993). Increases in $[Ca^{2+}]_i$ can be exerted by an influx of ions from the extracellular medium and/or by a release of Ca^{2+} from intracellular stores. Two classes of intracellular channels that play a key role in regulating Ca^{2+} release from intracellular stores have been so far identified: the inositol 1,4,5-trisphosphate ($InsP_3$) receptors and the Ryanodine Receptors (RyRs) (Berridge, 1993; Sorrentino and Volpe, 1993). $InsP_3$ receptors are described in Parys et al. (this book). This chapter will focus on the second class of intracellular Ca^{2+} release channels: the Ryanodine Receptors.

RyRs, so named for their ability to bind with high affinity the plant alkaloid ryanodine, have originally been described in skeletal and cardiac muscles, where they are mainly known for their role in regulating Ca^{2+} release from the sarcoplasmic reticulum in order to trigger contraction. Electron microscopy studies on muscle fibers have shown the presence of periodically arranged electron densities, called "feet" or "foot structures", later identified as the cytoplasmic domain of ryanodine receptors (Franzini-Armstrong and Protasi, 1997). Feet are lined in a special domain of the sarcoplasmic reticulum membrane, the junctional membrane, facing the sarcolemmal T-tubules. The highly ordered organization of RyRs in muscle fibers is strictly

Daniela Rossi, Ilenia Simeoni and Vincenzo Sorrentino • Section of Molecular Medicine, Department of Neurosciences, University of Siena, Italy. **Virginia Barone and Vincenzo Sorrentino** • DIBIT, Istituto Scientifico San Raffaele, Milan, Italy.

R. Pochet, R. Donato, J. Haiech, C. Heizmann and V. Gerke (eds.): Calcium: The Molecular Basis of Calcium Action in Biology and Medicine, 205–219.

associated with their function in regulating Ca^{2+} release for muscle contraction. Indeed, RyRs are strategically organized on the junctional sarcoplasmic to allow interaction with dihydropyridine receptors (DHPRs) voltage dependent calcium channels, localized on the plasma membrane. Activation of muscle contraction occurs through a mechanism referred to as "excitation-contraction coupling" that involves the coordinate activation of DHPRs and RyRs (Powell et al., 1996; Takeshima et al., 1994, 1995; Nakai et al., 1996). There are two mechanisms that mediate excitation-contraction coupling. In skeletal muscle, a direct coupling model has been described. According to this model RyRs are physically coupled with DHPRs and open in relation to conformational changes of the DHPRs induced by membrane depolarization. In cardiac fibers, by contrast, RyRs are not in physical association with DHPRs and are activated by a Ca^{2+}-Induced Ca^{2+} Release (CICR) mechanism. More precisely, cardiac DHPRs open following membrane depolarization and allow a Ca^{2+} influx from the extracellular enviroment to occur. This sarcolemmal Ca^{2+} current is not sufficient to activate contraction, but can induce the opening of cardiac RyRs which promote more sustained release of Ca^{2+} from the sarcoplasmic reticulum (Fabiato, 1983).

In addition to striated muscles, recent evidence has shown that RyRs are also present in the central nervous system and many other tissues (Giannini et al., 1995; Ledbetter et al., 1994) providing grounds for a view of intracellular Ca^{2+} release mechanisms which now include RyRs as potential participants of the Ca^{2+} release machinery in many eukaryotic cells, in association with $InsP_3$ receptors.

In the following sections, the main features of the ryanodine receptor gene family will be considered, with particular attention to the overall protein structure, the expression pattern and modulation of RyR activity. The functional role of RyR channels in intracellular Ca^{2+} signalling will be discussed in the final sections, taking into consideration the more recent insights obtained with RyR knockout animal models.

2. THE RyR GENE FAMILY

RyRs have been purified, cloned and sequenced from a variety of species. In mammals, three isoforms of RyRs (RyR1, RyR2 and RyR3) have been identified and they are encoded by different genes. RyR1, RyR2 and RyR3 have been mapped in the human genome on chromosomes 19q13.1, 1 and 15q14-q15, respectively. The three isoforms include about 5000 amino acid residues and share an overall amino acid sequence identity of about 70%, although in three regions, named divergency (D) regions, the three isoforms differ significantly among them. With reference to the RyR1 sequence, region D1 spans amino acids 4254–4631, region D2 amino acids 1342–1403 and region D3, a glutamate rich sequence, lies between residues 1872 and 1923.

Two RyR isoforms, known as α-RyR and β-RyR have been identified in fish, amphibian and avian skeletal muscles and they have been found to be the homologues of mammalian RyR1 and RyR3, respectively (Oyamada et al., 1994; Ottini et al., 1996; Airey et al., 1990; Lai et al., 1992; Olivares et al., 1991).

Alternative splice variants have been described for the three isoforms. Two insertions of 5 and 6 amino acids, respectively, have been localised in human and mouse RyR1 mRNA (Futatsugi et al., 1995; Zhang et al., 1993). Two putative alternative splicing sites have been postulated in RyR2 mRNA (Nakai et al., 1990) and three alternative splicing sites have been described in RyR3 (Marziali et al., 1996). Therefore, different mRNAs can be transcribed from the same gene, suggesting that Ca^{2+} release events may involve combinations of channels derived not only from different RyR genes but also from the same gene.

3. RYANODINE RECEPTOR STRUCTURE

RyR channels are formed by the assembly of four identical subunits, each of about 550 kDa. Based on sequence analysis, each subunit is predicted to have a large (about 4000 amino acids) N-terminal domain protruding into the cytosol followed by a region containing 4 to 10 transmembrane segments and by a short cytoplasmic domain of about 100 amino acids. The transmembrane regions are highly conserved at the protein sequence level, with the exception of domains 3 and 4, that show the lower degree of homology.

A considerable improvement in understanding ryanodine receptor structure has come from cryo-electron microscopy and three-dimensional reconstruction studies. Reconstruction of the ryanodine receptor from skeletal and cardiac muscle have confirmed the fourfold symmetry of the channels with a large cytoplasmic assembly and a small transmembrane region that protrudes 7 nm from one of its faces. The same studies have revealed the presence of a cylindrical domain in the transmembrane region, with a diameter of about 2 to 3 nm that has been proposed to correspond to the Ca^{2+}-conducting pore. A globular mass density, referred to as the “channel plug” is also located in the centre of the channel, nearest to the cytoplasmic assembly (Radermacher et al., 1994; Serysheva et al., 1995; Sharma et al., 1998; Wagenknecht et al., 1997). The membrane spanning regions of the four subunits make a group forming the pore of the homotetrameric channel. Recently, it has been proposed that a conserved sequence (GVRAGGGIGD) in one of the transmembrane domain is part of the pore-forming segment of RyRs (Zhao et al., 1999).

The cytoplasmic assembly is constructed from 10 or more domains that are loosely packed together and some of them are themselves multidomain structures (Radermacher et al., 1994; Serysheva et al., 1995; Sharma et al., 1998; Wagenknecht et al., 1997). The large cytoplasmic domain represents the modulatory region of the receptor and contains several binding sites for nucleotide, calmodulin, FKBP12, high and low affinity binding sites for Ca^{2+}, as well as phosphorylation and glycosylation sites. Indeed, electron microscopy reconstructions have identified the three-dimensional location of calmodulin and FKBP12 on the cytoplasmic assembly of RyR, close to the transmembrane ion channel (Wagenknecht et al., 1997).

4. PATTERN OF EXPRESSION OF RYANODINE RECEPTORS

Ryanodine receptors were first identified by the characteristic action of the plant alkaloid ryanodine on vertebrate striated muscles. Actually, RyR1 and RyR2 isoforms are predominantly expressed in skeletal and in cardiac muscle, respectively. RyR2 represents also the major brain isoform, while RyR3 has been found to be ubiquitously expressed at low levels. However, RNase protection, RT-PCR analysis and *in situ* hybridisation have shown that RyR1 and RyR2 are also widely detected in different tissues although at lower levels than in striated muscle (Giannini et al., 1995; Ledbetter et al., 1994).

In particular, RyR3 is expressed in different organs such as the esophagus, stomach, testis, ovary, urinary bladder, uterus, spleen, lung and kidney (Giannini et al., 1992, 1995; Hakamata et al., 1992). RyR1 and RyR2 mRNA have been detected, in addition to skeletal and cardiac muscles respectively, in the esophagus, gut, stomach, thymus, adrenal gland and ovary (Giannini et al., 1995; Ledbetter et al., 1994).

It is noteworthy that several tissues have been shown to express more than one isoform of RyR, but the role of each isoform has still to be clarified. For example all the three RyR isoforms are expressed in the brain. *In situ* hybridisation on adult mouse and rabbit brain (Furuichi et al., 1994; Giannini et al., 1995) confirmed RyR2 as the predominant brain isoform (McPherson and Campbell, 1993), and showed a heterogeneous distribution of the three receptors in different areas. RyR1 is expressed in the Purkinje cells, in the dentate gyrus of the hippocampus, in the CA3 and CA1 cells of the Ammon's horn and in the olfactory bulb. RyR2 expression is localised in the granular cell layer of the cerebellum, in the dentate gyrus, in the amygdala, in the cortex and in the granular cell layer of the olfactory bulb. RyR3 is present in the granular cell layer of the cerebellum, in the CA1 region of the Ammon's horn, in the caudate/putamen nuclei, and in the mitral and granular cell layer of the olfactory bulb. Finally RyR3 is also present in the lateral septum (Giannini et al., 1995).

Another interesting feature is represented by the expression pattern of different RyR isoforms in skeletal muscle. In most avian, amphibian and fish skeletal muscles, two isoforms of RyRs, named α and β, that correspond to mammalian RyR1 and RyR3, are expressed (Oyamada et al., 1994; Ottini et al., 1996; Airey et al., 1990; Lai et al., 1992; Olivares et al., 1991). Recent evidence has indicated that also in mammalian skeletal muscles, in addition to RyR1, the RyR3 isoform is present on the junctional sarcoplasmic reticulum (Sorrentino and Reggiani, 1999). In particular, it has been shown that the RyR3 isoform is mainly associated with muscle development. In fact, Western blot assays on adult skeletal muscles from different mammalian vertebrates show relatively high levels of RyR3 protein only in the diaphragm muscle (Conti et al., 1996; Tarroni et al., 1997), while in other muscles low to undetectable expression levels have been described. By contrast, during mouse muscle development, the RyR3 isoform is expressed in all muscles, from the late embryonic stage and during the first two weeks after birth. At variance with RyR1, which reaches the highest level of expression in the adult, RyR3 expression is downregulated in most muscles to start from 2–3 weeks of post-natal life (Bertocchini et al., 1997; Flucher et al., 1999).

5. MODULATION OF RYANODINE RECEPTOR ACTIVITY

Different putative modulatory sites have been identified in the N-terminal domain of the sequence, including calmodulin binding sites, phosphorylation and glycosylation sites, high affinity nucleotide-binding sequences and Ca^{2+}-binding domains of the EF-hand type (Takeshima et al., 1989; Zorzato et al., 1990; Nakai et al., 1990; Otsu et al., 1990; Hakamata et al., 1992; Marziali et al., 1996).

Indeed different approaches including Ca^{2+} release experiments, single channel analysis or [^{3}H]ryanodine binding, have shown that many substances can modulate RyR activity. In particular both endogenous and exogenous or pharmacological regulators have been studied. The first include ions, (Ca^{2+}, Mg^{2+}, H^+, Fe^{2+}, inorganic phosphate), adenine nucletide, cyclic adenosine diphosphate ribose (cADPR), calmodulin, Protein Kinase A (PKA) and Protein Kinase C (PKC) phosphorylation. Exogenous regulators are represented by different pharmacological agents including ryanodine, methylxanthines, caffeine, suramin, halothane, ruthenium red, procaine.

In this section the effect of some endogenous regulators will be considered. Several sites for calmodulin-dependent protein kinase (CaM kinase) have been predicted to be on the primary sequence of RyRs. Experimental evidence has shown that CaM kinase preferentially phosphorylates the serine 2809 (Ser2809) in the RyR2 (Witcher et al., 1991). The corresponding site in RyR1 (Ser2843) seems to be phosphorylated by CaM protein kinases and by Protein Kinase A (PKA) (Suko et al., 1993).

Calmodulin binding sites have been identified on RyRs (Hakamata et al., 1992; Nakai et al., 1990; Zorzato et al., 1990) and a direct interaction has been reported (Yang et al., 1994). The effect of calmodulin on RyR1 activity has been shown to be biphasic, increasing Ca^{2+} release at low Ca^{2+} concentrations and inhibiting it at higher Ca^{2+} concentrations (Ikemoto et al., 1995, 1998; Tripathy et al., 1995).

Another important potential endogenous regulator of RyR activity is Ca^{2+}. Ca^{2+} at nanomolar to micromolar concentration activates the channel, whereas at micromolar to millimolar concentration it inhibits the receptor (Smith et al., 1985; Ma and Zhao, 1994). Functional Ca^{2+} binding sites have been identified on RyR1 sequence, between amino acids 4014 and 4765. Antibodies against this region increase the Ca^{2+} sensitivity of channels reconstituted into lipid bilayers, indicating that this domain may be involved in a CICR process (Chen et al., 1992). RyR activation by Ca^{2+} is considered to be mediated by high affinity Ca^{2+} binding sites in the protein. Site directed mutagenesis of glutamate 3885 to alanine in the transmembrane sequence M2 of rabbit RyR3 reduced Ca^{2+} sensitivity, indicating that this amino acid may act as the Ca^{2+} sensor of the channel (Chen et al., 1998). By contrast, Ca^{2+} inactivation of RyR seems to be mediated by low affinity Ca^{2+} binding sites, which have recently been located between amino acids 3726 and 5037 at the COOH terminus of RyR1 (Du et al., 1999).

Cyclic adenosine diphosphate ribose (cADPR) has been shown to be a Ca^{2+} mobilising metabolite of β-NAD$^+$. It mediates Ca^{2+} release in many cell types, like sea urchin eggs, rat pituitary cells, dorsal root ganglion cells and pancreatic β-cells (Galione et al., 1991; Currie et al., 1992, Koshiyama et al., 1991; Takasawa et al., 1993). In addition, the ADP-ribosyl cyclase, the enzyme that catalyses cADPR synthesis, has been found to be widespread, suggesting that cADPR could play an important role in cell physiology (Rusinko and Lee, 1989).

cADPR has originally been found to mobilise Ca^{2+} by a mechanism independent of the $InsP_3Rs$ in sea urchin eggs. Actually, heparin, an inhibitor of $InsP_3Rs$, does not affect cADPR induced Ca^{2+} release (Dargie et al., 1990). On the other hand, cADPR induced Ca^{2+} release can be inhibited by procaine and ruthenium red, two blockers of RyRs. Sea urchin egg homogenates treated with ryanodine and caffeine, two agonists of ryanodine receptors, were desensitised to further addition of cADPR, although $InsP_3Rs$ could still release Ca^{2+} from intracellular stores. Conversely, cADPR desensitises the homogenates to further addition of caffeine or ryanodine, indicating that cADPR can mobilise Ca^{2+} through RyRs (Galione et al., 1991). In addition, cADPR-mediated Ca^{2+} release has been found to be enhanced by Ca^{2+} and sensitive to pharmacological regulators of ryanodine receptors (Galione et al., 1991). However, studies on the ability of cADPR to directly modulate skeletal and cardiac isoforms of the RyRs in planar lipid bilayers have given discordant results (Lee, 1997). Studies on Ca^{2+} fluxes and [^{3}H]ryanodine binding have shown that cADPR can activate RyR2 but seems to inhibit RyR1 channels (Meszaros et al., 1993). By contrast, Fruen et al. (1994) demonstrated that neither cADPR nor the related metabolite β-NAD$^+$ affect RyR2 activity as determined by [^{3}H]ryanodine binding. Similarly, cADPR failed to activate single channels in planar lipid bilayers; and furthermore RyR1 [^{3}H]ryanodine binding was unaffected by cADPR (Fruen et al., 1994). On the contrary, activation of sheep RyR1 channels incorporated into planar bilayers has been observed both by cADPR and β-NAD$^+$ (Sitsapesan and Williams, 1995). In addition, single channel recordings indicated that cADPR stimulates Ca^{2+} release through RyR3, while RyR1 was insensitive (Sonnleitner et al., 1998). Moreover, in sea urchin eggs, cADPR-mediated Ca^{2+} release is dependent on the presence of calmodulin (Lee et al., 1994), suggesting that cADPR activation may involve one or more mediator proteins.

6. ANIMAL MODELS

Muscle contraction is triggered by the release of Ca^{2+} from the sarcoplasmic reticulum after depolarisation of transverse tubules. The ryanodine receptor Ca^{2+} release channels type 1 and type 2 are responsible for the excitation-contraction coupling in skeletal and cardiac muscles respectively. In skeletal muscle the RyR3 isoform is also expressed and is localised in the terminal cisternae of the sarcoplasmic reticulum together with the preponderant RyR1 isoform. As mentioned in the previous sections, co-expression of two distinct RyR isoforms named α and β is commonly described in non-mammalian vertebrate skeletal muscles from chickens, frogs, and fish (Oyamada et al., 1994; Ottini et al., 1996; Airey et al., 1990; Lai et al., 1992; Olivares et al., 1991).

The advantage of the expression of two RyR isoforms is still unclear. Several observation indicate that avian α and β RyRs display different gating properties (Percival et al., 1994). And also that mammalian RyR1 and RyR3 are characterized by different functional properties (Sutko and Airey, 1996), suggesting that they may play different roles in muscle physiology.

In the last few years, this question has been considered using different approaches. In particular, the use of genetic techniques, like homologous recombination to generate knockout mice has considerably contributed to elucidating the role in *in vivo* muscle physiology as well as to investigating RyR functional activity in tissues other than muscle.

Until now knockout mice for each of the three ryanodine receptor isoforms (RyR1-RyR2-RyR3) and mice doubly knockedout for RyR1 and RyR3 have been generated and characterised. The main features of RyR knockout mice will be considered in the following sections.

7. STUDIES ON KNOCKOUT MICE FOR THE RyR1 GENE

In mammalian skeletal muscle, RyR1 is the predominant isoform. RyRs are lined on the junctional sarcoplasmic membrane and DHPRs form tetrads, groups of four receptors, themselves organized in arrays that face arrays of RyRs (Franzini-Armstrong and Protasi, 1997). The development of RyR1 knockout mice confirmed the essential role of this isoform in excitation-contraction coupling and further indicates a direct involment in tetrads formation.

Actually, mice carrying a targeted disruption of the RyR1 gene show complete loss of the skeletal muscle excitation-contraction coupling and die perinatally due to respiratory failure (Takeshima et al., 1994). The mutant neonates display also an abnormal curvature of the spine, thin limbs and a thick neck area, and the skeleton preparations show an abnormal rib cage and an arched vertebral column (Takeshima et al., 1994).

Skeletal muscle fibres from neonate knockout mice are small and fragmented and their nuclei remain centrally located. However, no abnormalities in the structure of the sarcomere is observed. Triad assembly occurs normally even in the absence of RyR1, and other triadic proteins like FKBP12, triadin, calsequestrin and SERCA are apparently normally expressed (Buck et al., 1997). Interestingly, DHPR protein expression has been found to be diminished in RyR1 knockout mice (Buck et al., 1997). Further, the formation of tetrads, the specific arrangement of four DHPRs on the plasma membrane in proximity to RyRs is prevented, indicating that anchoring to RyR1 is necessary for tetrads organisation (Protasi et al., 1998).

Skeletal muscle from RyR1 knockout mice fail to respond to electrical stimulation, although they retain the ability to release Ca^{2+} in response to caffeine, even if at lower levels than wild type muscles (Takeshima et al., 1994). As the RyR3 isoform is expressed in skeletal muscles, it has been proposed that this residual Ca^{2+} release could be mediated by this isoform (Takeshima et al., 1995), first indicating a possible role of RyR3 in the regulation of skeletal muscle contraction. This was later confirmed using RyR3 knockout mice, as described in the following section.

8. STUDIES ON KNOCKOUT MICE FOR THE RyR2 GENE

The ryanodine receptor type 2 functions as a CICR channel. RyR2 is predominantly expressed in cardiac muscle, where it has an essential role in the cardiac type of excitation-contraction coupling. In this case, excitation-contraction coupling is started by a small increase of cytosolic Ca^{2+} due to influx through the DHPR that, secondarily, can induce a massive Ca^{2+} release via RyR2 by a CICR mechanism, thus causing cardiac muscle contraction (Fabiato, 1985; Nabauer et al., 1989). No physical coupling is established between RyR2 and cardiac DHPRs.

Generations of mice carrying a targeted disruption of the RyR2 gene indicate a role of this isoform during myocardial development. Knockout mice die at embryonic day (E) 10 and show morphological abnormalities in the heart tube, like irregular organisation of the myocardium, the trabeculae and the epicardium. Mutant cardiac myocytes lose functional channel activity and no residual caffeine response can be detected indicating that no other ryanodine receptor isoforms are expressed. In addition, cardiac myocytes present ultrastructural defects as large vacuoles of the sarcoplasmic reticulum and abnormal mitochondria. These abnormalities have not been found in other tissues and it has been suggested that they may be due to complete loss of the RyR2 Ca^{2+} release channels which may act as a valve for intracellular Ca^{2+} stores (Takeshima et al., 1998). Actually, dense Ca^{2+} precipitates have been observed by electron microscopy in RyR2 knockout myocytes. In this respect, in the absence of RyR2, Ca^{2+} stores may become overloaded and this may cause the formation of large vacuoles. Ca^{2+} not sequestered from the sarcoplasmic reticulum may flow into mitochondria and into other organelles. These data indicate that during myocardial development, RyR2 is required for intracellular Ca^{2+} homeostasis in myocytes (Takeshima et al., 1998).

9. STUDIES ON KNOCKOUT MICE FOR THE RyR3 GENE

The RyR3 isoform shows a wide pattern of expression but lacks a preferential association with one tissue. The knockout mice lacking RyR3 show apparently normal growth and reproduction with no gross abnormalities. Histological examination of skeletal muscles of RyR3 knockout mice fails to reveal abnormalities in sarcomere and triad organisation.

Skeletal muscles from RyR3 deficient mice retain normal excitation-contraction coupling (Takeshima et al., 1996; Bertocchini et al., 1997). However, the CICR rate has been found to be much lower than in wild type muscles, indicating that RyR3 may function as a CICR channel (Takeshima et al., 1996).

In skeletal muscle RyR3 is present in all muscles in the late stages of fetal development and between 2–3 weeks after birth. Later RyR3 levels progressively decrease and this isoform is no longer detected in adult muscles with the exception of the diaphragm muscle. Studies on neonatal muscles from RyR3 knockout mice have shown that they are 35% less efficient in translating electrical stimulation into force generation than muscles from control mice; furthermore caffeine-induced contracture is reduced to 20% (Bertocchini et al., 1997). These results provide the first evidence for a specific contribution of RyR3 to contractile performance regulation and for a functional relevance of this isoform in the neonatal phase of skeletal muscle development.

To further investigate the role of RyR3 in excitation-contraction coupling of skeletal muscle, studies have been performed using RyR3 knockout mice, to verify how the RyR3 channels may contribute to elementary events of Ca^{2+} release called “sparks”. Analysis of multiple Ca^{2+} sparks parameters indicates that RyR3 has a pronounced contribution to sparks during the embryonic stage of muscle cell development (Conklin et al., 1999; Shirokova et al., 1999).

In the central nervous system, RyR2 is the most abundant isoform, while type 1 isoform expression is restricted in Purkinje cells. In the central nervous system, RyR3 is preferentially expressed in the hyppocampus and in the striatum. The role of the RyR3 channel in the central nervous system has been investigated using the RyR3 knockout mouse.

The RyR3-null mice were analysed at morphological, functional and behavioural levels. Results show that mutant mice homozygous for a non-functional allele of the RyR3 gene display specific changes in hippocampal synaptic plasticity without noticeable alterations in hippocampal morphology, basal synaptic transmission and presynaptic function. Robust Long Term Potentiation (LTP) induced by repeated, strong tetanisation in the CA1 region and in the dentate gyrus is unaltered, whereas weak forms of potentiation generated by either a single weak tetanus or by depotentiation of a robust LTP are impaired. These physiological deficits are paralleled by a reduced flexibility in re-learning a new target in the water maze test. In contrast, learning performance in the acquisition phase and during probe trial does not differ between mutants and their wild type littermates (Balschun et al., 1999).

An increased speed of locomotion and a mild tendency to circular running is also found in RyR3 deficient mice, suggesting that Ca^{2+} release via RyR3 is essential for the function of specific neurons in the central nervous system (Takeshima et al., 1996; Balschun et al., 1999). Locomotor activity is controlled by complex neuronal circuits in the brain that include frontoparietal cortex, basal ganglia, and thalamus where the RyR3 protein is expressed. Abnormal Ca^{2+} signalling of certain neurons in these regions caused by lack of RyR3 may result in the locomotor hyperactivity. Mutant mice possess no significant defects in their digestive or circulation systems indicating that the loss of RyR3 does not affect the physiological functions of smooth muscle.

10. STUDIES ON DOUBLE KNOCKOUT MICE FOR RyR1 AND RyR3 GENES

The doubly mutant mice do not actively move and die after birth as was the case with RyR1 deficient mice. The doubly mutant neonates have an external appearance indistinguishable from that of RyR1 deficient mice (Ikemoto et al., 1997).

Double knockout mice confirm the functional data obtained from single knockout mice, showing a complete loss of excitation-contraction coupling and contraction in response to caffeine and ryanodine stimulation indicating the absence of all ryanodine/caffeine sensitive pathways of Ca^{2+} release. Indeed, in permeabilized myocytes lacking both RyRs, Ikemoto et al. (1997) observed that the CICR mechanism is completely lost, and caffeine fails to induce Ca^{2+} release (Ikemoto et al., 1997).

Morphological analysis of double knockout muscles shows a severe muscular degeneration with the myofibrils often branched and poorly developed and the cross striation misaligned. In addition, they are hardly able to develop tension when directly activated with micromolar Ca^{2+} after membrane permeabilisation. This loss of contractile response seems to be proportional to the reduction in myofibrillar protein content. In fact, a significant reduction in Myosin heavy chain and Troponin T and Troponin I content has been observed in $RyR1^{-/-}$ muscle fibers and even more so in double knockout mice (Barone et al., 1998).

In double mutants, triads and diads are present, although less frequent than in normal muscle, and they all lack feet. This indicates that neither RyR1 nor RyR3 are essential for the formation of triadic junctions suggesting the presence of other molecular components responsible for the triad formation in skeletal muscle cells.

11. CONCLUSION

In the last five years, studies with genetically modified mice have confirmed that RyR1 and RyR2 are essential for excitation-contraction coupling in skeletal and cardiac muscle respectively. Contemporary analysis of RyR3 knockout mice has provided a series of interesting observations which indicate that additional control of cellular functions can be obtained by co-expression of more than one isoform of Ca^{2+} release channel in cells such as skeletal muscles and CNS neurons.

REFERENCES

Airey, J.A., Beck, C.F., Murakami, K., Tanksley, S.J., Deerinck, T.J., Ellisman, M.H. and Sutko, J.L., 1990, Identification and localization of two triad junctional foot protein isoforms in mature avian fast twitch skeletal muscle, *J. Biol. Chem.* 265, 14187–14194.

Balschun, D., Wolfer, D.P., Bertocchini, F., Barone, V., Conti, A., Zuschratter, W., Missiaen, L., Lipp, H.P., Frey, U. and Sorrentino, V., 1999, Deletion of the ryanodine receptor type 3 (RyR3) impairs forms of synaptic plasticity and spatial learning, *EMBO J.* 18, 5264–5273.

Barone, V., Bertocchini, F., Bottinelli, R., Protasi, F., Allen, P.D., Franzini Armstrong, C., Reggiani, C. and Sorrentino, V., 1998, Contractile impairment and structural alterations of skeletal muscles from knockout mice lacking type 1 and type 3 ryanodine receptors, *FEBS Lett.* 422, 160–164.

Berridge, J.B., 1993, Inositol trisphosphte and calcium signalling. Nature, 361, 315-325.

Bertocchini, F., Ovitt, C.E., Conti, A., Barone, V., Schöler, H.R., Bottinelli, R., Reggiani, C. and Sorrentino, V., 1997, Requirement for the ryanodine receptor type 3 for efficient contraction in neonatal skeletal muscle, *EMBO J.* 16, 6956–6963.

Buck, E.D., Nguyen, H., Pessah, I. and Allen, P.D., 1997, Dyspedic mouse skeletal muscle expresses major elements of the triadic junction but lacks detectable ryanodine receptor protein and function, *J. Cell Biol.* 272, 7360–7367.

Chen, S.R.W., Zhang, L. and MacLennan, D.H., 1992, Characterization of a Ca^{2+} binding and regulatory site in the Ca^{2+} release channel (ryanodine receptor) of rabbit skeletal muscle sarcoplasmic reticulum, *J. Biol. Chem.* 267, 23318–23326.

Chen, S.R.W., Ebisawa, K., Li, X. and Zhang, L., 1998, Molecular identification of the ryanodine receptor Ca^{2+} sensor, *J. Biol. Chem.* 273, 14675–14678.

Conklin, M.W., Barone, V., Sorrentino, V. and Coronado, R, 1999, Contribution of ryanodine receptor type 3 to Ca^{2+} sparks in embryonic mouse skeletal muscle, *Biophys. J.* 77(3), 1394–1403.

Conti, A., Gorza, L. and Sorrentino, V., 1996, Differential distribution of ryanodine receptor type 3 (RyR3) gene product in mammalian skeletal muscles, *Biochem. J.* 316, 19–23.

Currie, K.P., Swann, K., Galione, A. and Scott, R.H., 1992, Activation of Ca(2+)-dependent currents in cultured rat dorsal root ganglion neurones by a sperm factor and cyclic ADP-ribose, *Mol. Biol. Cell.* 3, 1415–1425.

Dargie, P.J., Agre, M.C. and Lee, H.C., 1990, Comparison of Ca^{2+} mobilizing activities of cyclic ADP-ribose and inositol trisphosphate, *Cell Regul.* 1, 279–290.

Du, G.G. and MacLennan, D.H., 1999, Ca^{2+} inactivation sites are located in the COOH-terminal quarter of recombinant rabbit skeletal muscle Ca^{2+} release channels (ryanodine receptors), *J. Biol. Chem.* 274, 26120–26126.

Fabiato, A., 1983, Calcium-induced release of calcium from the cardiac sarcoplasmic reticulum, *Am. J. Physiol.* 245, C1–C14.

Fabiato, A., 1985, Simulated calcium current can both cause calcium loading in and trigger calcium release from the sarcoplasmic reticulum of a skinned canine cardiac Purkinje cell, *J. Gen. Physiol.* 85, 291–320.

Flucher, B.E., Conti, A., Takeshima, H. and Sorrentino, V., 1999, Type 3 and Type 1 ryanodine receptors are localized in triads of the same mammalian skeletal muscle fibers, *J. Cell Biol.* 146, 621–629.

Franzini-Armstrong, C. and Protasi, F., 1997, Ryanodine receptors of striated muscles: A complex channel capable of multiple interactions, *Physiol. Rev.* 77, 699–729.

Fruen, B.R., Mickelson, J.R., Shomer, N.H., Velez, P. and Louis, C.F., 1994, Cyclic ADP-ribose does not affect cardiac or skeletal muscle ryanodine receptors, *FEBS Lett.* 352, 123–126.

Furuichi, T., Furutama, D., Hakamata, Y., Nakai, J., Takeshima, H. and Mikoshiba, K., 1994, Multiple types of ryanodine receptor/Ca^{2+} release channels are differentially expressed in rabbit brain, *J. Neurosci.* 14, 4794–4805.

Futatsugi, A., Kuwajima, G. and Mikoshiba, K., 1995, Tissue-specific and developmentally regulated alternative splicing in mouse skeletal muscle ryanodine receptor mRNA, *Biochem J.* 305, 373–378.

Galione, A., Lee, H.C. and Busa, W., 1991, Ca^{2+}-induced Ca^{2+} release in sea urchin egg homogenates: Modulation by cyclic ADP-ribose, *Science* 253, 1143–1146.

Giannini, G., Clementi, E., Ceci, R., Marziali, G. and Sorrentino, V., 1992, Expression of a ryanodine receptor-Ca^{2+} channel that is regulated by TGF-β, *Science* 257, 91–94.

Giannini, G., Conti, A., Mammarella, S., Scrobogna, M. and Sorrentino, V., 1995, The ryanodine receptor/Calcium release channel genes are widely and differentially expressed in murine brain and peripheral tissues, *J. Cell Biol.* 128, 893–904.

Hakamata, Y., Nakai, J., Takeshima, H. and Imoto, K., 1992, Primary structure and distribution of a novel ryanodine receptor/calcium release channel from rabbit brain, *FEBS Lett.* 312, 229–235.

Ikemoto, T., Iino, M. and Endo, M., 1995, Enhancing effect of calmodulin on Ca^{2+}-induced-Ca^{2+} release in the sarcoplasimc reticulum of skeletal muscle fibers, *J. Physiol.* 487, 573–582.

Ikemoto, T., Komazaki, S., Takeshima, H., Nishi, M., Noda, T., Iino, M. and Endo, M., 1997, Functional and morphological features of skeletal muscle from mutant mice lacking both type 1 and type 3 ryanodine receptors, *J. Physiol.* 501, 305–312.

Ikemoto, T., Takeshima, H., Iino, M. and Endo, M., 1998, Effect of calmodulin on Ca^{2+} induced Ca^{2+} release of skeletal muscle from mutant mice expressing either ryanodine receptor type 1 or type 3, *Pflügers Arch. – Eur. J. Physiol.* 437, 43–48.

Koshiyama, H., Lee, H.C. and Tashjian, A.H., 1991, Novel mechanism of intracellular calcium release in pituitary cells, *J. Biol. Chem.* 266, 16985–16988.

Lai, F.A., Liu, Q.-L., Xu, L., El-Hashem, A., Kramarcy, N.R., Sealock, R. and Meissner, G., 1992, Amphibian ryanodine receptor isoforms are related to those of mammalian skeletal or cardiac muscle, *Am. J. Physiol.* 263, C365–C372.

Ledbetter, M.W., Preiner, J.K., Louis, C.F. and Mickelson, J.R., 1994, Tissue distribution of ryanodine receptor isoforms and alleles determined by reverse transcription polymerase chain reaction, *J. Biol. Chem.* 269, 31544–31551.

Lee, H.C., 1997, Mechanisms of calcium signaling by cyclic ADP-ribose and NAADP, *Physiol. Rev.* 77, 1133–1164.

Lee, H.C., Aarhus, R., Graeff, R., Gurnack, M.E. and Walseth, T.F., 1994, Cyclic ADP ribose activation of the ryanodine receptor is mediated by calmodulin, *Nature* 370, 307–309.

Ma, J. and Zhao, J.Y., 1994, Highly cooperative and hysteretic response of the skeletal muscle ryanodine receptor to changes in proton concentrations, *Biophys. J.* 67, 626–633.

Marziali, G., Rossi, D., Giannini, G., Charlesworth, A. and Sorrentino, V., 1996, cDNA cloning reveals a tissue specific expression of alternatively spliced transcripts of the ryanodine receptor type 3 (RyR3) calcium release channel, *FEBS Lett.* 394, 76–82.

McPherson, P.S. and Campbell, K.P., 1993, The ryanodine receptor/Ca^{2+} release channel, *J. Biol. Chem.* 268, 13765–13768.

Meszaros, L.G., Bak, J. and Chu, A., 1993, Cyclic ADP-ribose as an endogenous regulator of the non-skeletal type ryanodine receptor Ca^{2+} channel, *Nature* 364, 76–79.

Nabauer, M., Callewaert, G., Cleemann, L. and Morad, M., 1989, Regulation of calcium release is gated by calcium current, not gating charge, in cardiac myocytes, *Science* 244, 800–803.

Nakai, J., Imagawa, T., Hakamata, Y., Shigekawa, M., Takeshima, H. and Numa, S., 1990, Primary structure and functional expression from cDNA of the cardiac ryanodine receptor/calcium release channel, *FEBS Lett.* 271, 169–177.

Nakai, J., Dirksen, R.T., Nguyen, H.T., Pessah, I.N., Beam, K.G. and Allen, P.D., 1996, Enhanced dihydropyridine receptor channel activity in the presence of ryanodine receptor, *Nature* 380, 72–75.

Olivares, E.B., Tanksley, S.J., Airey, J.A., Beck, C.F., Ouyang, Y., Deerinck, T.J., Ellisman, M.H. and Sutko, J.L., 1991, Multiple foot protein isoforms in amphibian, avian and piscine skeletal muscles, *Biophys. J.* 59, 1153–1163

Otsu, K., Willard, H.F., Khanna, V.K., Zorzato, F., Green, N.M. and MacLennan, D.H., 1990, Molecular cloning of cDNA encoding the Ca^{2+} release channel (ryanodine receptor) of rabbit cardiac muscle sarcoplasmic reticulum, *J. Biol. Chem.* 265, 13472–13483.

Ottini, L., Marziali, G., Conti, A., Charlesworth, A. and Sorrentino, V., 1996, α and β isoforms of ryanodine receptors from chicken skeletal muscle are the homologues of mammalian RyR1 and RyR3, *Biochem. J.* 315, 207–215.

Oyamada, H., Murayama, T., Takagi, T., Iino, M., Iwabe, N., Miyata, T., Ogawa, Y. and Endo, M., 1994, Primary structures and distribution of ryanodine-binding protein isoforms of the bullfrog skeletal muscle, *J. Biol. Chem.* 269, 17206–17214.

Percival, A., Airey, J.A., Grinsell, M.M., Kenyon, J.L., Williams, A.J. and Sutko, J.L., 1994, Chicken skeletal muscle ryanodine receptor isoforms: Ion channel properties, *Biophys. J.* 67, 1834–1850.

Powell, J.A., Petherbridge, L. and Flucher, B.E., 1996, Formation of triads without the dihydropyridine receptor alpha subunits in cell lines from dysgenic skeletal muscle, *J. Cell Biol.* 134, 375–387.

Protasi, F., Franzini-Armstrong, C. and Allen, P.D., 1998, Role of ryanodine receptors in assembly of calcium release units in skeletal muscle, *J. Cell Biol.* 140, 831–842.

Radermacher, M., Rao, V., Grassucci, R., Frank, J., Timerman, A.P., Fleischer, S. and Wagenknecht, T., 1994, Cryo-electron microscopy and three-dimensional reconstruction of the calcium release channel/ryanodine receptor from skeletal muscle, *J. Cell Biol.* 127, 411–423.

Rusinko, N. and Lee, H.C., 1989, Widespread occurrence in animal tissues of an enzyme catalyzing the conversion of NAD^+ into a cyclic metabolite with intracellular Ca^{2+} mobilizing activity, *J. Biol. Chem.* 64, 11725–11731.

Serysheva, I., Orlova, E.V., Chiu, W., Sherman, M.B., Hamilton, S.L.H. and van Heel, M., 1995, Electron cryomicroscopy and angular reconstitution used to visualize the skeletal muscle calcium release channel, *Nature Struct. Biol.* 2, 18–24.

Sharma, M.R., Penczek, P., Grassucci, R., Xin, H., Fleischer, S. and Wagenknecht, T., 1998, Cryoelectron microscopy and image analysis of the cardiac ryanodine receptor, *J. Biol. Chem.* 273, 18429–18434.

Shirokova, M., Shirokov, R., Rossi, D., Gonzales, A., Kirsch, W.G., Garcia, J., Sorrentino, V. and Rios, E., 1999, Spatially segregated control of Ca^{2+} release in developing skeletal muscle of mice, *J. Physiol. (Lond.)* 521(2), 483–495.

Sitsapesan, R. and Williams, A.J., 1995, Cyclic ADP-ribose and related compounds activate sheep skeletal sarcoplasmic reticulum Ca^{2+} release channel, *Am. J. Physiol.* 268, C1235–C1240.

Smith, J.S., Coronado, R. and Meissner, G., 1985, Sarcoplasmic reticulum contains adenine nucleotide-activated calcium channels, *Nature* 316, 446–449.

Sonnleitner, A., Conti, A., Bertocchini, F., Schindler, H. and Sorrentino, V., 1998, Functional properties of the Ryanodine receptor type 3 (RyR3) Ca^{2+} release channel, *EMBO J.* 17, 2790–2798.

Sorrentino, V. and Volpe, P., 1993, Ryanodine receptors: How many, where and why?, *Trends Pharmacol. Sci.* 14, 98–103.

Sorrentino, V. and Reggiani, C., 1999, Expression of the ryanodine receptor type 3 in skeletal muscle. A new partner in excitation-contraction coupling?, *Trends Card. Med.* 9, 54–61.

Suko, J., Maurer-Fogy, I., Plank, B., Bertel, O., Wyskovsky, W., Hohenegger, M. and Hellmann, G., 1993, Phosphorilation of serine 2843 in ryanodine receptor-calcium release channel of skeletal muscle by cAMP, cGMP and CAM-dependent protein kinase, *Biochim. Biophys. Acta* 1175, 193–206.

Sutko, J.L. and Airey, J.A., 1996, Ryanodine receptors Ca^{2+} release channels: Does diversity in form equal diversity in function, *Physiol. Rev.* 76, 1027–1071.

Takasawa, S., Natsa, K., Yonekura, H. and Okamoto, H., 1993, Cyclic ADP-ribose in insulin secretion from pancreatic beta cells, *Science* 259, 370–373.

Takeshima, H., Nishimura, S., Matsumoto, T., Ishida, H., Kangawa, K., Minamino, N., Matsuo, H., Ueda, M., Hanaoka, M., Hirose, T. and Numa, S., 1989, Primary structure and expression from complementary DNA of skeletal muscle ryanodine receptor, *Nature* 339, 439–445.

Takeshima, H., Iino, M., Takekura, H., Nishi, M., Kuno, J., Minowa, O., Takano, H. and Noda, T., 1994, Excitation-contraction uncoupling and muscular degeneration in mice lacking functional skeletal muscle ryanodine-receptor gene, *Nature* 369, 556-559.

Takeshima, H., Yamazawa, T., Ikemoto, T., Takekura, H., Nishi, M., Noda, T. and Iino, M., 1995, Ca^{2+}-induced Ca^{2+} release in myocites from dyspedic mice lacking the type-1 ryanodine receptor, *EMBO J.* 14, 2999–3006.

Takeshima, H., Ikemoto, T., Nishi, M., Nishiyama, N., Shimuta, M., Sugitani, Y., Kuno, J., Saito, I., Saito, H., Endo, M., Iino, M. and Noda, T., 1996, Generation and characterization of mutant mice lacking ryanodine receptor type 3, *J. Cell Biol.* 271, 19649–19652.

Takeshima, H., Komazaki, S., Hirose, K., Nishi, M., Noda, T. and Iino, M., 1998, Embryonic lethality and abnormal cardiac myocytes in mice lacking ryanodine receptor type 2, *EMBO J.* 17, 3309–3316.

Tarroni, P., Rossi, D., Conti, A. and Sorrentino, V., 1997, Expression of the ryanodine receptor type 3 calcium release channel during development and differentiation of mammalian skeletal muscle cells, *J. Biol. Chem.* 272, 19808–19813.

Tripathy, A., Xu, L., Mann, G. and Meissner, G., 1995, Calmodulin activation and inhibition of skeletal muscle Ca^{2+} release channel (ryanodine receptor), *Biophys. J.* 69, 106–119.

Wagenknecht, T., Radermacher, M., Grassucci, R., Berkowitz, J., Xin, H.B. and Fleisher, S., 1997, Locations of calmodulin and FK506-binding protein on the three-dimensional architecture of the skeletal muscle ryanodine receptor, *J. Biol. Chem.* 272, 32463– 32471.

Witcher, D.R., Kovacs, R.J., Schulman, H., Cefali, D.C. and Jones, L.R., 1991, Unique phosphorylation site on the cardiac ryanodine receptor regulates calcium channel activity, *J. Biol. Chem.* 266, 11144–11152.

Yang, H.C., Reedy, M.M., Bruke, C.L. and Strasburg, G.M., 1994, Calmodulin interaction with the skeletal muscle sarcoplasmic reticulum calcium channel protein, *Biochemistry* 33, 518–525.

Zhang, Y., Chen, H.S., Khanna, V.K., De Leon, S., Phillips, M.S., Schappert, K., Britt, B.A., Brownell, A.K.W. and MacLennan, D.H., 1993, A mutation in the human ryanodine receptor gene associated with central core desease, *Nat. Genet.* 5, 46–50.

Zhao, M., Li, P., Li, X., Zhang, L., Winkfein, R., Wayne Chen, S.R., 1999, Molecular identification of the ryanodine receptor pore-forming segment, *J. Biol. Chem.* 274, 25971–25974.

Zorzato, F., Fujii, J., Otsu, K., Phillips, M., Green, N.M., Lai, F.A., Meissner, G. and MacLennan, D.H., 1990, Molecular cloning of cDNA encoding human and rabbit forms of the Ca^{2+} release channel (ryanodine receptor) of skeletal muscle sarcoplasmic reticulum, *J. Biol. Chem.* 265, 2244–2256.

Calcium Inhibition of Cytoplasmic Streaming

Hozumi Kawamichi, Akio Nakamura and Kazuhiro Kohama

1. INTRODUCTION

Cytoplasmic streaming is commonly observed in plant cells, where organelles associated with myosin move along the actin cables that run beneath the cell membrane. The driving force for the streaming is produced by the interaction of myosin with actin in a similar way to muscle contraction. In muscle cells, cytoplasmic Ca^{2+} concentration ($[Ca^{2+}]_i$) is mostly kept at low levels, i.e., submicromolar levels. When the cell is excited, $[Ca^{2+}]_i$ increases up to micromolar levels. The increase in $[Ca^{2+}]_i$ causes the actin-myosin interaction so that the muscle contracts. When $[Ca^{2+}]_i$ returns to submicromolar levels, the interaction is abolished and the muscle relaxes (Ebashi and Endo, 1968).

Thus, the distinguishing feature between cytoplasmic streaming and muscle contraction is that the former occurs continuously at low $[Ca^{2+}]_i$, whereas the latter is resting. Upon the elevation of $[Ca^{2+}]_i$ by cellular excitation, the streaming is abolished (Table 1). These observations suggest that Ca^{2+} works as an inhibitor for the streaming, a regulatory mode that is quite distinct from the mode of muscle contraction.

An alternative organism that shows cytoplasmic streaming is the plasmodium of lower eukaryote of *Physarum polycephalum*. The streaming has been demonstrated to be induced by the actin-myosin interaction, which can be monitored by the physiological methods to measure muscle contraction (Kamiya, 1981). Further, the plasmodium can be cultured in a quantity large enough for biochemical study (Kohama et al., 1998). This feature enabled actin, myosin and actomyosin-related proteins to be purified by modifying the procedures for muscle protein (Hatano, 1973). In this article, we will review

Hozumi Kawamichi, Akio Nakamura and Kazuhiro Kohama • Department of Pharmacology, Gunma University School of Medicine, Maebashi, Gunma, Japan.

R. Pochet, R. Donato, J. Haiech, C. Heizmann and V. Gerke (eds.): Calcium: The Molecular Basis of Calcium Action in Biology and Medicine, 221–244.

Table 1. Regulation of actin-myosin interaction by Ca^{2+}

Cells	Resting ($Ca^{2+} < \mu M$)	Excited ($Ca^{2+} > \mu M$)
Plant	+	−
Animal	−	+

+, active interaction between actin and myosin; –, no active interaction. Cytoplasmic concentration of Ca^{2+} ($[Ca^{2+}]_i$) is kept as low as possible by the extrusion of Ca^{2+} through cell membrane and by the sequestration of Ca^{2+} into endoplasmic reticulum. When cell is excited, $[Ca^{2+}]_i$ increases due to the entry of extracellular Ca^{2+} and to the release of sequestred Ca^{2+}. The increase is unstable, because the above devices for keeping $[Ca^{2+}]_i$ low operate again. The cell utilizes such a transient increase in $[Ca^{2+}]_i$ as a second messenger to regulate the actin-myosin interaction. The utilization is quite distinct, $[Ca^{2+}]_i$ works as an inhibitor for plant cells and an activator for animal cells. The interaction of *Physarum* can be classified as plant type.

how the actin-myosin interaction of *Physarum polycephalum* is regulated by Ca^{2+} based upon biochemical analysis.

2. CALCIUM INHIBITION IN PLANT CELLS

The intermodal cells of higher plants of *Chara* and *Nitella* are popular materials for the observation of cytoplasmic streaming. When they are excited by the electrical stimulation, $[Ca^{2+}]_i$ is elevated transiently (Kikuyama and Tazawa, 1982; Williamson and Ashley, 1982). Concomitantly, abolition of the streaming is detectable (Kishimoto and Akahori, 1959; Tazawa and Kishimoto, 1968). Such an inhibitory role of Ca^{2+} was further analyzed by changing $[Ca^{2+}]_i$, i.e., the cells were perfused by artificial solutions containing various concentrations of Ca^{2+} into the intermodal cells, or by soaking the cells in these solutions after their cytoplasmic membrane was destroyed (Williamson, 1975; Shimmen and Yano, 1984; Tominaga et al., 1983).

In *Vallisneria* leaf cells, red light induces efflux of Ca^{2+} to, and infrared light influx of Ca^{2+} from, the medium outside the cell (Takagi, 1993). Accordingly, the cytoplasmic streaming is detectable when observed under red light and is abolished by the application of infrared light (Takagi and Nagai, 1985, 1986), providing another example of the calcium inhibition of cytoplasmic streaming.

Biochemical approaches for myosin in higher plants have been hampered by the small quantity of cytoplasm. In spite of this difficulty, myosin was purified recently from the pollen tubes of *Lilium*, where we can observe similar calcium inhibition of cytoplasmic streaming (Kohno and Shimmen 1988a, b). The myosin is composed of 170 kDa protein as its heavy chain and calmodulin (CaM) as its light chain. The effect of μM levels of Ca^{2+} on the myosin ATPase activity in the presence of actin is inhibitory, and the *in vitro* motility of actin and myosin was also inhibited by Ca^{2+} (Yokota et al., 1999).

3. CALCIUM INHIBITION OF THE CYTOPLASMIC STREAMING OF PLASMODIA OF PHYSARUM POLYCEPHALUM

In *Physarum* plasmodia, we observe the vigorous shuttle movement of organelles. In contrast to plant cells, plasmodial actin does not form actin cables but rather forms a complex with myosin to build up "vessels", in which cytoplasm streams passively due to the contraction of the actomyosin vessel wall (Kamiya, 1981). $[Ca^{2+}]_i$ changes within μM levels, as can be observed by injecting aequorin into plasmodia as a Ca^{2+} probe (Ridgway and Durham, 1976). Yoshimoto et al. (1981a, b) permeabilized the membrane of a plasmodial strand to alter the cytoplasmic concentration of Ca^{2+} directly and related its contraction to the concentration of Ca^{2+}. They concluded that the actin-myosin interaction, as measured by the contraction, is maximal in the absence of Ca^{2+} and reduced with an increase in the Ca^{2+} concentration. This inhibitory effect of Ca^{2+} is confirmed with similar cell-free models under various concentrations of cytoplasmic Ca^{2+} (Achenbach and Wohlfarth-Bottermann, 1986a, b; Pies and Wohlfarth-Bottermann, 1986) (Table 2). Using a small fragment of plasmodium that shows contraction-relaxation movement rather than the cytoplasmic streaming, Ishigami et al. (1996) were able to relate this movement to the changes in cytoplasmic concentration of Ca^{2+}. They concluded that the increase in local $[Ca^{2+}]$ acts as a trigger to make the plasmodium relax (Table 2).

4. CALCIUM INHIBITION OF ACTIN-MYOSIN INTERACTION OF PHYSARUM AS EXAMINED IN VITRO

4.1. Calcium inhibition of native actomyosin

As reviewed by Ebashi and Endo (1968), the first step to analyze how Ca^{2+} regulates the actin-myosin interaction *in vitro* is to examine the effect of Ca^{2+} on the natural actomyosin, which is composed of actin, myosin and their relaxed protein(s). Native actomyosin was prepared from the homogenate of *Physarum* plasmodia by utilizing the fact that it is extractable at high salt and precipitable at low salt (Kohama et al.,1980, Ogihara et al.,1983). Its ATPase activity was reduced with the increase in Ca^{2+} concentrations in the assay medium. This effect is in good conformity with calcium inhibition of the

Table 2. Calcium inhibition of actin-myosin interaction of *Physarum* plasmodium

	Reference
Actin-activated ATPase activity of myosin	Ogihara et al. (1983), Kohama and Kendrick-Jones (1986), Kohama et al. (1991a) and Ishikawa et al. (1991)
Superprecipitation of actomyosin	Kohama et al. (1980) and Ogihara et al. (1983)
In vitro motility assay	
Nitella-basal motility assay	Kohama and Shimmen (1985)
Myosin-coated surface assay	Okagaki et al. (1989)
Tension development of actomyosin threads	Sugino and Matsumura (1983)
Contraction of cell-free model	Yoshimoto et al. (1981a), Yoshimoto and Kamiya (1984), Achenbach and Wohlfarth-Bottermann (1986a, b)
Intact cell	Ishigami et al. (1996)

superprecipitation of the native actomyosin, a classical method for detecting the actin-myosin interaction.

4.2. Calcium inhibition of hybrid actomyosins

Ca^{2+} regulates the actin-myosin interaction of skeletal muscle through troponin (Ebashi and Endo, 1968). Therefore, the site of action of Ca^{2+} is contained neither in actin nor in myosin of skeletal muscle. To utilize such a property, we purified actin and myosin from the native actomyosin preparation and then reconstituted actomyosins from actin and myosin of *Physarum* and those of skeletal muscle. The respective actomyosins were then subjected to the measure of ATPase activity (Kohama and Kendrick-Jones, 1986), finding that the activity of the actomyosin was inhibited by Ca^{2+} only when myosin was from *Physarum*.

However, the ATPase activity of the actomyosin produced from skeletal muscle myosin was not affected by Ca^{2+} no matter whether actin was from *Physarum* or skeletal muscle. Thus, Ca^{2+} is demonstrated to exert an inhibitory effect through myosin.

In the absence of actin, the ATPase activity of *Physarum* myosin is also inhibited by Ca^{2+}, although the effect is slight (Kohama and Kendrick-Jones, 1986). This observation also indicates that the site of action of Ca^{2+} is myosin and suggests that the effect of Ca^{2+} should be amplified by actin. (*N.B.*

Table 3. Myosin movement along actin cables (μm/sec)

Solution	*Physarum*	Scallop
EGTA	1.40 ± 0.08 ($n = 18$)	0 ($n = 3$)
Ca^{2+}	0.40 ± 0.08 ($n = 11$)	1.22 ± 0.39 ($n = 3$)

Beads coated with *Physarum* or scallop myosin were introduced into *Nitella* cells together with EGTA-containing solution ($pCa^{2+} < 8$) or with Ca^{2+} solution ($pCa^{2+} \approx 6$) and observed with a Nomarski microscope (Shimmen and Kohama, 1984; Kohama and Shimmen, 1985).

Myosin is not the sole site of Ca^{2+}; the role of the actin-binding proteins in calcium inhibition will be explained in Section 6.)

4.3. Calcium inhibition as detected by in vitro Motility Assays

The ATPase measurement in the presence of actin is not the only way to monitor actin-myosin interaction. To demonstrate the myosin-linked nature of calcium inhibition by an alternative method, small latex beads were coated with *Physarum* myosin. They were then allowed to move along actin cables of *Nitella* internodal cells as described by Shimmen and Yano (1984). As expected from the inhibitory effect of Ca^{2+} on the ATPase activity of myosin, the beads moved much faster in the absence of Ca^{2+} than in its presence (Table 3, Shimmen and Kohama, 1984; Kohama and Shimmen, 1985).

Calcium regulation of the ATPase activities for myosin from scallop muscle is well characterized as will be described in Section 4.6. The effect of Ca^{2+} on scallop myosin is diametrically different, stimulating the activity (Vale et al., 1984). Accordingly, the beads with scallop myosin moved in the presence of Ca^{2+} (Table 3). The different effect of Ca^{2+} using the same actin cables in this *Nitella*-based motility assay indicates that myosin in the site for Ca^{2+} inhibition of the actin-myosin interaction in *Physarum*.

To try another *in vitro* motility assay (Kron and Spudich, 1986; Harada et al., 1987), we prepared coverslips coated with either *Physarum* myosin or scallop myosin. Fluorescent actin from skeletal muscle was mounted on the coverslips in the presence of ATP and was observed with a fluorescence microscope (Okagaki et al., 1989). As shown in Figure 2, actin moved in the absence of Ca^{2+} on the *Physarum* myosin coverslip, and movement was inhibited by Ca^{2+}. Conversely, the same actin moved on the surface coated with scallop muscle myosin in the presence of actin but not in the absence of Ca^{2+}. These differences confirm the myosin-linked nature of calcium inhibition.

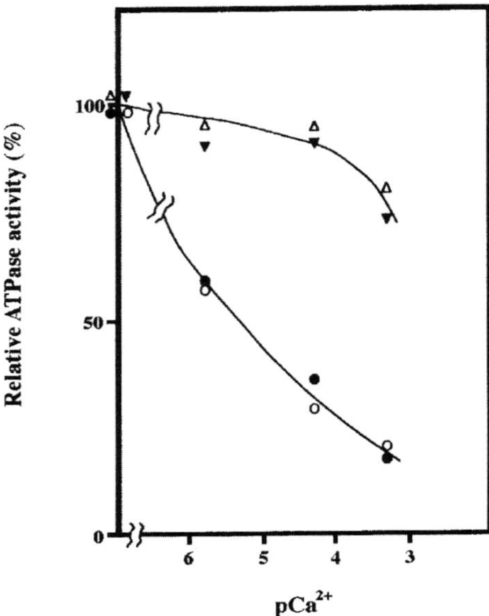

Figure 1. Effect of Ca^{2+} on the ATPase activity of hybrid actomyosin formed from skeletal muscle and *Physarum* plasmodia (Kohama and Kendrick-Jones, 1986). Actomyosin reconstituted from *Physarum* actin and *Physarum* myosin (●), actomyosin reconstituted from skeletal muscle actin and *Physarum* myosin (○), actomyosin reconstituted from *Physarum* actin and skeletal muscle myosin (▲), and actomyosin reconstituted from skeletal muscle actin and skeletal muscle myosin (△). $pCa^{2+} = -\log[Ca^{2+}]M$.

4.4. Ca-binding properties of Physarum myosin

The role of Ca^{2+} in the assays for actin-myosin interaction as described above is in accordance with the idea that Ca^{2+} binds to the myosin molecule. In order to demonstrate this, we measured Ca-binding activity of the myosin under the conditions comparable to the above assays, and found that the myosin bound Ca^{2+} with a high affinity at μM levels (Kohama and Kendrick-Jones, 1986) and with a binding capacity of 2 mol Ca^{2+} per mol myosin (Table 4). These data are comparable with those of myosin from scallop muscle, a typical myosin that is in an active form when it binds in Ca-containing solution. Release of Ca^{2+} from scallop myosin upon the withdrawal of Ca^{2+} inactivates its activity (Kendrick-Jones et al., 1970). The Ca-binding properties of *Physarum* myosin suggest that its calcium switch resembles that of scallop myosin (see Section 4.6). However, the effect of Ca^{2+} for the interaction is quite distinct; *Physarum* myosin is in an active form when it releases Ca^{2+} and in an inactive form when it binds Ca^{2+} (Figure 2).

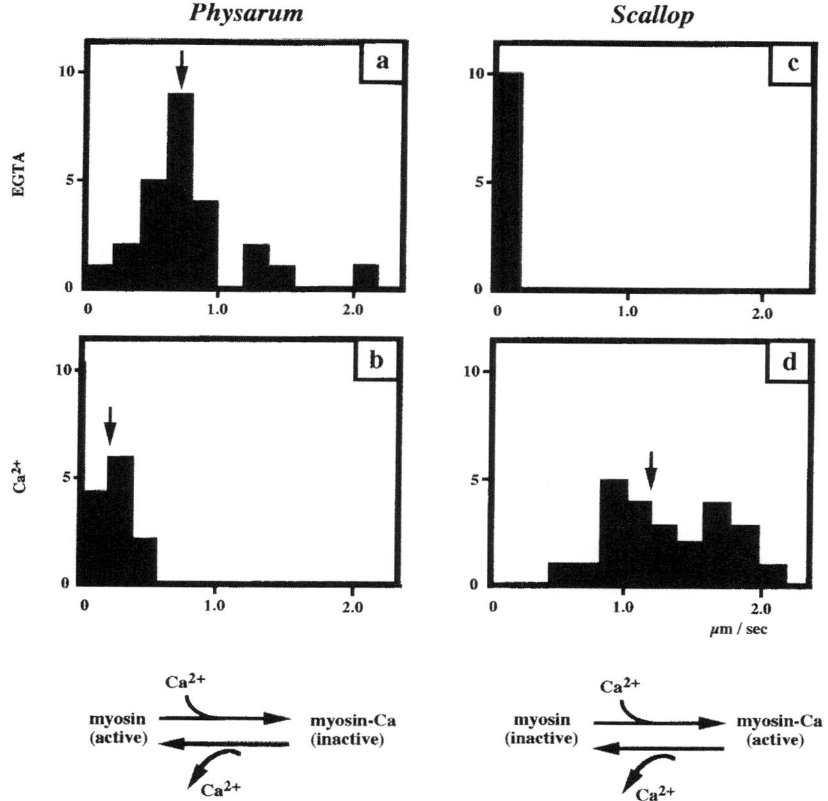

Figure 2. Inhibitory (a, b) and stimulatory (c, d) effects of Ca^{2+} as measured by the myosin-coated surface assay (Okagaki et al., 1989). Actin filament was labeled with rhodamine-phalloidine and mounted on a coverslip coated with *Physarum* myosin (a, b) or scallop myosin (c, d). ATP-dependent movement of actin filament was observed in the presence of 0.1 mM EGTA (a, c) or 0.1 mM Ca^{2+} (b, d) with a fluorescent microscope equipped with a video camera. Ordinate, numbers of moving actin filament; abscissa, velocities (μm/sec). Arrows indicate average velocity of the movement.

Table 4. Calcium binding to *Physarum* myosin and its subunits

Myosin (phosphorylated)	1.27 ± 0.16 mol/mol ($n = 3$)
Myosin (dephosphorylated)	1.21 ± 0.07 mol/mol ($n = 3$)
Calcium-binding light chain (CaLc)	0.37 ± 0.092 mol/mol ($n = 6$)
Phosphorylatable light chain (PLc)	ND

The binding of Ca^{2+} to myosin and its subunits was measured in 0.5 M KCl, 1 mM $MgCl_2$, 20 mM Tris-HCl (pH 7.5), and 30 μM Ca^{2+} containing ^{45}Ca by equilibrium dialysis (Kohama et al., 1991a, b). ND, not detectable. *Note*. CaLc and PLc are classified into essential and regulatory light chains of vertebrate myosin, respectively (Section 4.5).

Physarum CaM

```
          10        20        30        40        50        60
Ac-VDSLTEEQIAEFKEAFSLFDKDGDGNITTKELGTVMRSLGQNPTEAELQDMINEVDADGN

          70        80        90       100       110       120
GTIDFPEFLTMMARKMADTDTEEEIREAFKVFDKDGNGFISAAELRHVMTNLGEKLSDEE

         130       140
VDEMIREADVDGDGQVNYDEFVKMMLSK
```

Physarum CaLc

```
          10        20        30        40        50        60
Ac-TASADQIQECFQIFDKDNDGKVSIEELGSALRSLGKNPTNAELNTIKGQLNAKEFDLATF

          70        80        90       100       110       120
KTVYRKPIKTPTEQSKEMLDAFRALDKEGNGTIQEAELRQLLLNLGDALTSSEVEELMKE

         130       140
VSVSGDGAINYESFVDMLVTGYPLASA
```

Figure 3. Amino acid sequence of calcium-binding light chain (CaLc) (EMBL/Gene Bank accession number J03499) of *Physarum* myosin (Kobayashi et al., 1988) and *Physarum* calmodulin (CaM) (DDBJ/EMBL/Gene Bank accession number AB022702) (Toda et al., 1990). Molecular weights and statiscal p*I* for CaLc are 16,080 and 4.33 respectively, and those for CaM are 16,606 and 3.92.

4.5. Calcium-binding light chain (CaLc) as a Ca-receptive subunit of Physarum myosin

Physarum myosin is composed of a pair of heavy chains (230 kDa in SDS-PAGE) and two pairs of light chains (18 and 14 kDa in SDS-PAGE). The domain structure of heavy chain has been examined; the binding sites for ATP, actin and light chains are all in the heavy chain (Kohama et al., 1988b). The primary structure of Calcium-binding light chain (CaLc) was determined by both peptide analysis and cDNA cloning as shown in Figure 3 (Kobayashi et al., 1988), and its molecular weight is calculated to be 16,084 Da. The EF-hand structure, which is a consensus sequence for Ca-binding proteins (Kretsinger, 1980), is identified at one position at the N-terminal. The highest homology of CaLc is found in calmodulin of bovine brain. On a biochemical basis, CaLc shows calmodulin-like activity that it activates phosphodiesterase activity through CaLc in the reverse manner (Figure 4) as Ca^{2+} inhibition of the ATPase activity of *Physarum* myosin via CaLc (Figure 1). Thus, CaLc is not the site of activation or inhibition by Ca^{2+}. CaLc appears to work merely as a Ca-receptive subunit in *Physarum* myosin (Kohama et al., 1991b). As will be described in Section 5, the 18 kDa myosin light chain is phos-

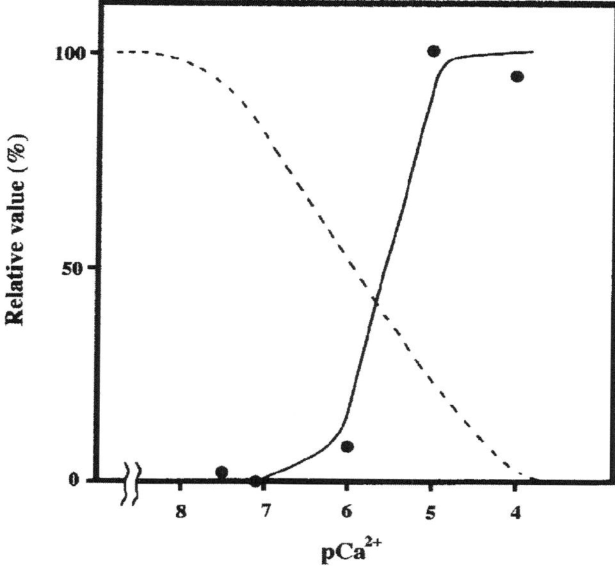

Figure 4. CaLc stimulates phosphodiesterase activity in the presence of Ca^{2+}. Phosphodiesterase activities (continuous line) (Kohama et al., 1991b) were measured in the presence of CaLc under conditions that allowed direct comparison with the actin-activated myosin ATPase activity (the broken line).

phorylable and is called the phosphorylatable light chain (PLc). The isolated PLc binds to skeletal muscle heavy chain as a substitute for 5,5′-dithiobis (2-nitrobenzonic acid) (DTNB) light chain (Kohama et al., 1991b), but does not bind Ca^{2+} (Table 4).

4.6. Speculative mode of Ca-binding of CaLc based on the knowledge of scallop myosin

The crystal structure of the regulatory domain of scallop myosin, which is composed of the essential and regulatory light chains and the light chain-binding fragment of the heavy chain, shows that essential light chain sequesters Ca^{2+} at its N-terminal EF-hand structure, where the regulatory light chain is closely associated (Xie et al., 1994). The isolated essential light chain is unable to bind Ca^{2+}. However, the regulatory domain can bind Ca^{2+} as strongly as parent scallop myosin (Kwon et al., 1990). The association of regulatory light chain allows the regulatory domain to bind Ca^{2+} by stabilizing the bond.

Chimeras of the essential light chain of Ca^{2+} binding scallop myosin and that of non-Ca-binding cardiac myosin were produced in *E. coli* as recombinant proteins, and it was demonstrated that the third EF-hand structure of the scallop essential light-chain was required for the myosin to bind Ca^{2+} (Jancso and Szent-Gyorgyi, 1994). Similar analysis was carried out with chimeras between regulatory light chain of scallop and skeletal muscle myosins. The crucial portion for Ca^{2+} binding by scallop myosin is at the C terminal of scallop regulatory light chain (Fromherz and Szent-Gyorgyi, 1995). Analysis with amino acid replacement demonstrated that Gly117 at the C terminal of the regulatory light chain is of primary importance to stabilise Ca^{2+} bound to the scallop essential light chain (Jancso and Szent-Gyorgyi, 1994).

Physarum CaLc and PLc are classified into essential and regulatory light chains, respectively, as shown in Table 4 (Kohama et al., 1991b). An EF-hand structure is identified at the N-terminal of CaLc of *Physarum* myosin as described previously (Figure 3). Resembling scallop myosin, Ca-binding to CaLc is too weak to explain the Ca-binding activity of *Physarum* myosin. We speculate that *Physarum* myosin binds Ca^{2+} at its N-terminal in a mode similar to that for scallop myosin and that the binding is stabilized by PLc. It would be intriguing to know whether PLc has a Gly residue analogous to Gly117 in its C-terminal portion.

5. PHOSPHORYLATION AND DEPHOSPHORYLATION OF PHYSARUM MYOSIN

5.1. The sites of phosphorylation in physarum myosin

Myosins from vertebrate muscle and nonmuscle sources are purified in the dephosphorylated state. However, those from lower eukaryotes, such as *Physarum*, *Dictyostelium* (Kuczmarski and Spudich, 1980; Maruta et al., 1983a), and *Acanthamoeba* (Collins and Korn, 1981), are prepared in the phosphorylated form. In the case of *Physarum* myosin, the total phosphate content as determined after assignment was 4.0–6.8 mol *Pi* per mole myosin (Kohama and Kendrich-Jones, 1986). We cultured *Physarum* plasmodia in the presence of $H_3[^{32}P]O_4$ and then prepared crude myosin. Autoradiography after SDS-PAGE showed that the major site of phosphorylation is in the heavy chain, but PLc is also partly phosphorylated (Kohama and Kendrick-Jones, 1986).

5.2. Dephosphorylation and calcium inhibition

We dephosphorylated *Physarum* myosin with phosphatase (Ogihara et al., 1983) and compared its actin-activated ATPase activity with that of untreated, phosphorylated myosin (Kohama et al., 1991a). As shown in Figure 5a, the activity of phosphorylated myosin was high in the absence of Ca^{2+} and decreased with an increase in the Ca^{2+} concentrations. In contrast, the activity of dephosphorylated myosin was low irrespective of Ca^{2+} concentration. We examined the Ca-binding activity of dephosphorylated myosin. As shown in

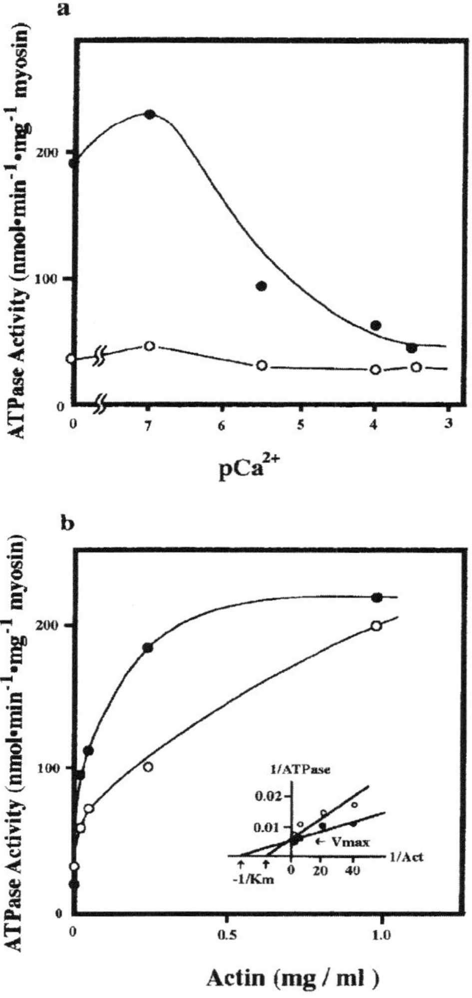

Figure 5. (a) *Physarum myosin* is purified in the phosphorylated form, and then is dephosphorylated by the phosphatase treatment. Dephosphorylation reduced phosphate to 1.6 mol P_i per mole of 500 kDa myosin. Actin-activated ATPase activities of phosphorylated myosin (•), and dephosphorylated myosin (○), in the presence of various consentrations of Ca^{2+} (Kohama et al., 1991a). (b) Effect of actin on ATPase activities of phosphorylated (•) and dephosphorylated (○) myosins (Kohama et al., 1991a). Double-reciprocal plots (inset) showed that the dephosphorylation procedure reduced K_m; $V_{\max}$ was not significantly different between phosphorylated and dephosphorylated myosins.

Table 4, dephosphorylated myosin bound Ca^{2+} as well as the phosphorylated form. Thus, dephosphorylation of *Physarum* myosin minimizes myosin ATPase activity so that there is no activity to be inhibited by Ca^{2+} (Kohama et al., 1991a).

We measured the ATPase activities in the various concentrations of actin (Figure 5b) and found that dephosphorylation reduces the affinity of myosin for actin, whereas ATPase maximum activity (V_{max}) remains unaffected (Kohama et al., 1991a). Therefore, the crucial factor in determining which of the two modes (i.e., Ca-binding mode and phosphorylating mode) is dominant *in vivo* is the concentration of actin in *Physarum* cells. In muscular tissues, the concentration of actin is comparable to that of myosin. In non-muscle tissues, however, actin greatly exceeds myosin in concentration (Pollard and Weihing, 1974). This is true in *Physarum* cells (Ishikawa et al., 1991), and hence the mode of Ca^{2+} binding should be physiological. Change in the phosphorylated state of myosin may not play a major role *in vivo*.

5.3. Dependence on Ca^{2+} of protein kinases and phosphatases that act on myosin

Kinases specific for the myosin heavy chain and PLc have molecular weights of 76 and 55 kDa, respectively, as indicated by SDS-PAGE (Okagaki et al., 1991a, b). The effect of Ca^{2+} on their activities of vertebrate is similar to its effect on myosin, i.e., these activities are most pronounced in the absence of Ca^{2+} and are reduced with an increase in Ca^{2+} concentration. A sole Ca-binding protein is involved in the inhibitory effect of Ca^{2+} (see Section 7).

The phosphatase activities for myosin are much lower in *Physarum* cells than the kinase activities. However, native actomyosin preparation (see Section 4.1) contained not only the kinase activities but also the phosphatase activities. We incubated the preparation with $[^{32}P]\gamma$-ATP to phosphorylate myosin; the phosphorylation was terminated by the addition of the kinase inhibitor staurosporine (Okagaki et al., 1991a). We observed that the radioactivity incorporated into these proteins was gradually lost due to the phosphatase activity. The decrease in radioactivity was abolished by the phosphatase inhibitor okadaic acid. The phosphatase activities were low in the presence of Ca^{2+} and increased with an increase in $[Ca^{2+}]$ to μM levels. Furthermore, the calmodulin inhibitor trifluoperazine also inhibited the phosphatase activities, suggesting the involvement of calmodulin in these activities (Kohama et al., 1993).

Figure 6 summarizes the role of Ca^{2+} and phosphorylation in the relationship between Ca-binding and phosphorylation of *Physarum* myosin (Kohama, 1990; Kohama et al., 1993). Step 2 is mediated by phosphatase activity, which works at high Ca^{2+} levels. Kinase reactions are expressed by step 4, which occurs at low Ca^{2+} concentrations. Steps 1 and 3 are mediated by direct Ca-binding to myosin. The Ca-binding activity remains similar regardless of the phosphorylated state of myosin (Table 4). As discussed in Section 5.2, dephosphorylation of myosin reduces its affinity to actin. In *Physarum* cells, the concentration of actin exceeds that of myosin. Therefore, myosins in both phosphorylated and dephosphorylated forms are in the active form in low $[Ca^{2+}]$. Ca-binding to both myosins inhibits their activities.

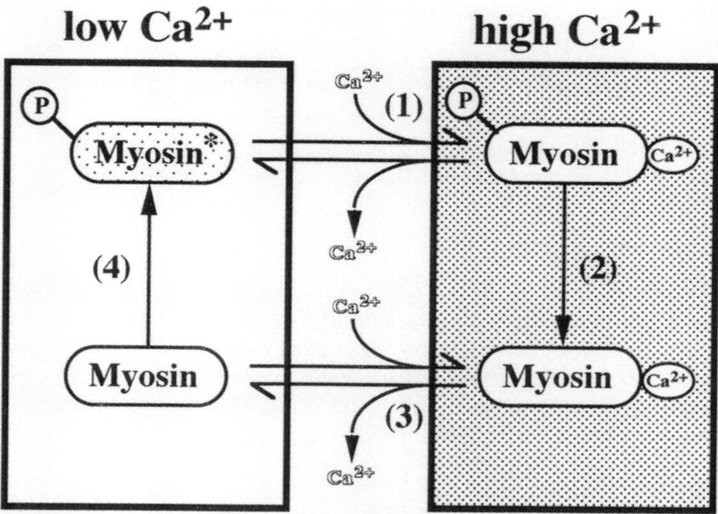

Figure 6. A model of the role of Ca^{2+} and phosphorylation in *Physarum* myosin. Myosin is active only in the phosphorylated form at low Ca^{2+} concentrations (asterisk). Binding of Ca^{2+} occurs irrespective of the extent of phosphorylation at high [Ca^{2+}] (see steps 1 and 3). Myosin can be dephosphorylated at high [Ca^{2+}] (step 2) and phosphorylated at low Ca^{2+} (step 4). The Ca-binding proteins are calcium-binding light chain for steps 1 and 3, calmodulin for step 2, and calcium-dependent inhibitory factor for step 4. The phosphorylated form of myosin is denoted by Ⓟ. *N.B.* Myosin with asterisk is the sole active form. However, Figure 5b indicates that dephosphorylated myosin is as active as phosphorylated myosin under V_{max} conditions. As descaibed in the text, the concentration of actin in *Physarum* plasmodia is much higher than that of myosin, providing V_{max} conditions *in vivo*. Therefore, binding and release of Ca^{2+}, rather than phosphorylation and dephosphorylation of myosin, should be the physiological mode of regulation.

In other words, the physiological switch that determines whether *Physarum* myosin is active or inactive is Ca-binding to myosin.

6. ACTIN-BINDING PROTEINS OF PHYSARUM THAT ARE INVOLVED IN CALCIUM INHIBITION

The actin-linked mode of calcium inhibition has been indicated in an earlier stage of our studies (Kohama, 1981; Kohama and Shimmen, 1985). When ATPase activity of native actomyosin (see Section 4.1) was compared with actomyosin reconstituted from purified actin and myosin, the inhibitory effect of Ca^{2+} of the former was more pronounced than that of the latter (Kohama and Kohama, 1984). If the inhibition was exerted exclusively through myosin, the effect should be the same. Thus, the inhibitory factor(s) is speculated to be lost during the purification.

Table 5. Velocities of movement of actin filaments on coverslips coated with *Physarum* myosin

Actin filament	Velocity (μm/sec, $n = 30$)
Control (0.1 mM EGTA)	1.39 ± 0.49
+Caldesmon-like protein (0.1 mM EGTA)	1.95 ± 0.66
Control (0.1 mM Ca^{2+})	0.99 ± 0.42
+Caldesmon-like protein (0.1 mM Ca^{2+})	1.26 ± 0.60
+Caldesmon-like protein + caldesmon (0.1mM Ca^{2+})	1.08 ± 0.50

Caldesmon-like protein was purified from *Physarum* plasmodia as a heat-stable, actin-binding protein. Unlike smooth muscle caldesmon, caldesmon-like protein stimulated the velocity of movement. Because calmodulin in the presence of Ca^{2+} abolishes binding to actin, caldesmon-like protein works to argument increase calcium inhibition (Ishikawa et al., 1991).

6.1. Caldesmon-like protein of Physarum

We purified a 210 kDa, heat-stable protein from *Physarum* that reacts with an antibody against caldesmon, an actin-binding, regulatory protein of smooth muscle (Ishikawa et al., 1991, 1992). This caldesmon-like protein binds to actin and stimulates the actin-activated ATPase activity of *Physarum* myosin (Table 5). The stimulation is abolished when calmodulin is mixed with the calmodulin-like protein in the presence of Ca^{2+}. Furthermore, calmodulin (Figure 3) is found in *Physarum* (Ishikawa et al., 1991). Therefore, caldesmon-like protein is able to produce calcium inhibition together with calmodulin.

It must be noted that the stimulatory effect of caldesmon-like protein is distinct from the inhibitory effect reported for smooth muscle caldesmon (Sobue and Sellers, 1991). However, smooth muscle caldesmon is also shown to stimulate its actin-myosin interaction under the specified conditions. Because this stimulation is related to the myosin-binding property of caldesmon (Lin et al., 1994), we need to reexamine the regulatory mode of *Physarum* caldesmon-like protein.

6.2. Other actin-binding proteins Physarum

When CaLc was isolated from *Physarum* myosin, it was found to interact with actin (Kohama and Shimmen, 1985) and was further confirmed by the measure of viscosity and flow birefringence of actin (Kohama et al., 1988a). We also examined the effect of CaLc on the actin-activated ATPase activity of *Physarum* myosin and found that CaLc worked so as to enhance calcium inhibition (Kohama et al., 1985). This finding is in accordance with our observation that a significant amount of CaLc is present in *Physarum* plasmodium without associating myosin (Kohama et al., 1985). These previous experiments were carried out using CaLc purified from *Physarum*. Now that recombinant CaLc is available, we need to confirm the results and to identify the sequence responsible by mutating CaLc.

Fragmin binds to actin filaments at their barbed end and severs them in the presence of Ca^{2+} (Hasegawa et al., 1980). Furthermore, it forms a complex with monomeric actin, and the complex becomes a nucleus for polymerization of actin into filaments (Hasegawa et al., 1980; Hinssen, 1981a,b; Sugino and Hatano, 1982; Maruta et al., 1983b). Fragmin is considered to be an analog of vertebrate gelsolin (Yin and Stossel, 1980). Isoforms of fragmin were purified by Uyeda et al.(1988) and Furuhashi and Hatano (1989). Profilin was also identified in *Physarum* as another protein that forms a complex with monomeric actin (Ozaki et al., 1983).

The actin-binding proteins that may be involved in the organization of *Physarum* plasmodium are high-molecular-weight actin-binding protein (Sutoh et al., 1984), a homolog of smooth muscle filamin (Wang, 1977); connectin / titin (Ozaki and Maruyama, 1980; Gassner et al., 1985), an elastic protein originally found in muscle tissue (Maruyama et al., 1976); a 52 kDa protein that bundles actin filaments (Itano and Hatano, 1991); and caldesmon-like protein (Ishikawa et al., 1991, 1992). Ishikawa et al. (1995) also purified an ATP-dependent actin-binding protein from *Physarum* plasmodium.

As described in Section 5.1 caldesmon-like protein together with CaM takes part in the calcium inhibition of the actin-myosin interaction (Ishikawa et al., 1991). Sugino and Matsumura (1983) showed that the tension of the actomyosin thread of *Physarum* was reduced by fragmin in the presence of Ca^{2+} (Table 2). Connectin/titin, although prepared from skeletal muscle, binds to actin in a calcium-dependent manner and inhibits the actin-myosin interaction (Kellermayer and Granzier, 1996). This effect remains to be confirmed with *Physarum* actomyosin.

7. Ca-BINDING PROTEINS IN PHYSARUM

7.1. Three Ca-binding proteins that support calcium inhibition

In this review, we have described three Ca-binding proteins in *Physarum* as summarized in Figure 7. The most characterized one is CaLc (Kohama et al., 1992); not only does CaLc work as a Ca-binding subunit of *Physarum* myosin but it also binds to actin and both properties contribute to produce calcium inhibition.

The amino acid sequence of *Physarum* calmodulin (Figure 3) is very similar to that of brain calmodulin (88% identity). Calmodulin binds to caldesmon-like protein in the presence of Ca^{2+} to abolish the stimulatory effect of the protein on the actin-myosin interaction (see Table 5). The abolition enhances calcium inhibition of the actin-myosin interaction.

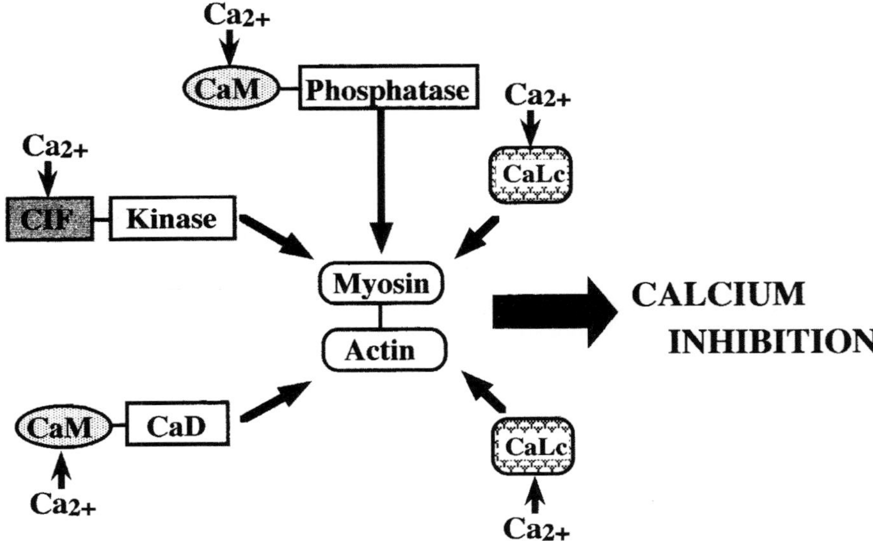

Figure 7. Three Ca-binding proteins involved in calcium inhibition of the *Physarum* actomyosin system. CIF, calcium-dependent inhibitory factor for kinases (Okagaki et al., 1991a, b); CaLc, calcium-binding light chain of *Physarum* myosin (Kobayashi et al., 1988; Kohama et al., 1991b); CaM, *Physarum* calmodulin (Tada et al., 1990); CaD, caldesmon-like protein of *Physarum* (Ishikawa et al., 1991). CaLc works as a sole Ca^{2+}-receptive protein in *Physarum* myosin, producing calcium inhibition. When CIF binds Ca^{2+}, CIF inhibits kinase for *Physarum* myosin. The phosphorylated form interacts with actin more effectively than when dephosphorylated. Ca^{2+} stimulates myosin phosphatase activity by binding to CaM (Kohama et al., 1993) to dephosphorylate myosin. As explained in Figure 6, these phosphorylation/dephosphorylation of myosin should not be a major regulatory mode under physiological conditions, however. CaD enhances calcium inhibition by binding to actin, which is negated by CaM in the presence of Ca^{2+}. CaLc binds actin and allows Ca^{2+} to cause inhibition (Kohama et al., 1985).

In the presence of Ca^{2+}, calmodulin stimulates myosin phosphatase activity (Kohama et al., 1993). In the absence of Ca^{2+}, myosin phosphatase remains low to keep myosin phosphorylated (Figure 6). Therefore, calmodulin indirectly contributes to calcium inhibition of myosin.

As described in Section 5.3, the activities of the kinases for the heavy (76 kDa in SDS-PAGE) and light (55 kDa) chains of *Physarum* myosin are also calcium inhibitory, i.e., at low Ca^{2+} concentration *Physarum* myosin tends to be phosphorylated. Calcium inhibition of kinase activities also contributes to the inhibition of the actin-myosin interaction. A Ca-binding protein of 38 kDa in SDS-PAGE has been shown to be involved in the calcium inhibition of kinase activities and is called calcium-dependent inhibitory factor (CIF) (Okagaki et al., 1991a, b). Actin kinase of *Physarum* has been shown to be subject to calcium inhibition (Okagaki et al., 1991a). When actin forms a complex with fragmin, the kinase is able to phosphorylate actin (Furuhashi and Hatano, 1990, 1992; Furuhashi et al., 1992; Gettemans et al., 1992, 1993).

7.2. Preliminary characterization of recombinant 40 kDa protein

A 40 kDa Ca-binding protein was purified from *Physarum* (Nakamura et al., 1994). The cloning of its cDNA demonstrated that it is identical to LAV1-2 cloned by Laroche et al., (1989) as an abundant mRNA specific to *Physarum* plasmodia. We expected 40 kDa protein protein to work as the CIF described in Sections 5.3 and 7.1, because CIF has a similar molecular mass (38 kDa) (Okagaki et al., 1991a, b). However, the 40 kDa protein obtained as a recombinant protein in *E. coli* failed to exert an inhibitory effect on myosin heavy- and light-chain kinases.We interpreted this to mean that CIF differs from 40 kDa protein. The other interpretation is that posttranscriptional modifications such as glycosylation and phosphorylation may be required for 40 kDa protein to inhibit the kinases.

7.2.1. Structure of 40 kDa protein

As shown in Figure 8, the 40 kDa protein consists of 355 amino acid residues (Nakamura et al., 1994; Nakamura and Kohama, 1999) The calculated molecular mass and p*I* are 40,508 Da and 4.97, respectively. Four EF-hand Ca-binding consensus sequences (Kretsinger, 1980) were found in the C-terminal half, which also contained a consensus sequence for nuclear proteins as reported by Robbins et al. (1991). The sequence is 226 RKIDTNSNGTLS-RKEFR 242. The N-terminal half contained an α-helical structure predicted by MacDNASIS Pro 1993. Residues 1–32 formed aggregates of the 40 kDa protein.

7.2.2. Ca-binding properties of the 40 kDa protein

Recombinant 40 kDa protein was obtained as a soluble form in low ionic strength. However, elevation to μM levels allowed the 40 kDa protein to form large aggregates (Nakamura et al., 1994, 2000; Nakamura and Kohama, 1995). The ability to form aggregates was abolished by producing a mutant that was deleted for residues 1–32. When this truncated protein was subjected to the Ca-binding assay with a flow dialysis apparatus (Womack and Colowic, 1973), 4 mol Ca/mol of 40 kDa protein was bound with half-maximal binding at a concentration of 29 μM. The concentration is much lower than that of *Physarum* calmodulin (5.7 μM), which was also obtained in the recombinant form (Figure 8). The C-terminal, but not N-terminal half of the 40 kDa protein is highly homologous to calmodulin. We speculate that the secret of the higher Ca-binding activity of the 40 kDa protein is in its N-terminal half (Nakamura and Kohama, 1999; Nakamura et al., 2000). It will be of interest to see whether a fusion protein of calmodulin with N-terminal half 40 kDa protein has increased Ca-binding properties.

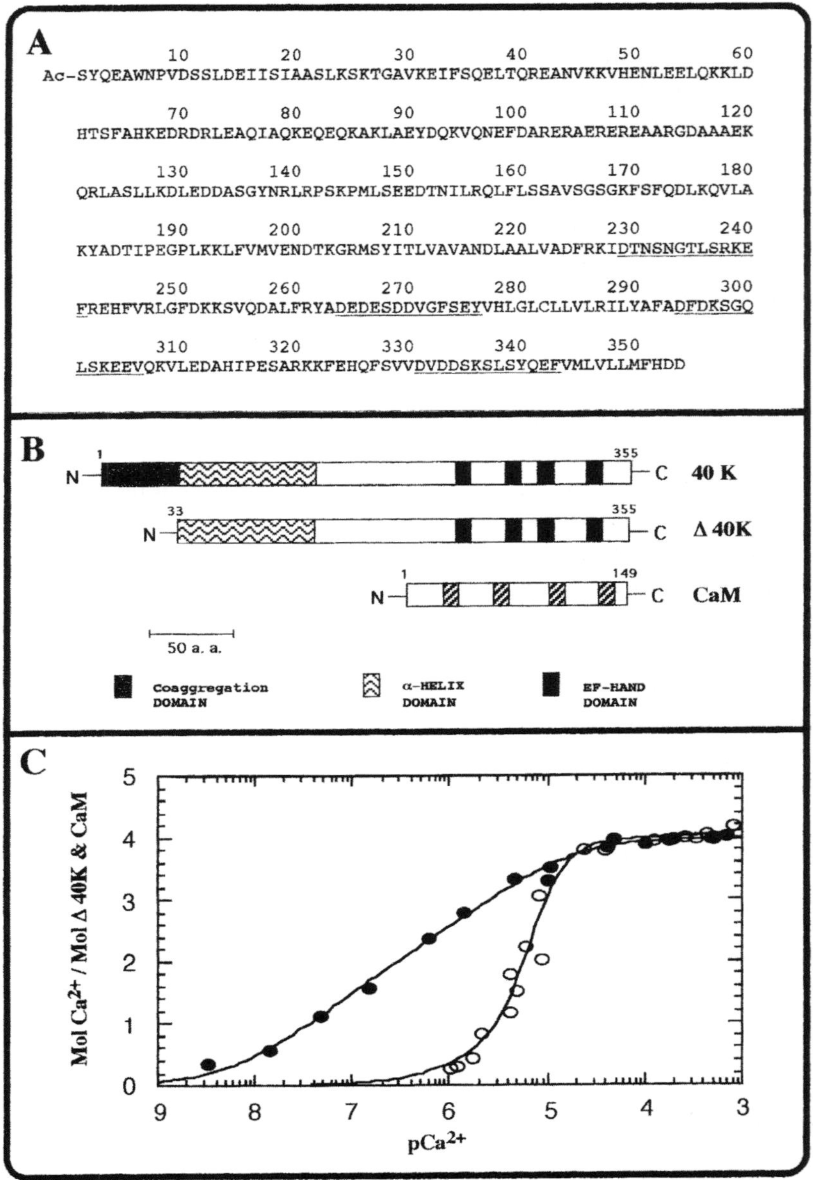

Figure 8. Structure and function of the 40 kDa protein (Nakamura and Kohama, 1999). (A) Amino acid sequence of the 40 kDa protein deduced from its nucleotide sequence (EMBL/Gene Bank accession number X14502). Molecular weight, 40,508 Da; statistical p*I* 4.97. (B) Domain structures of the parent 40 kDa protein and the truncated form of the 40 kDa protein 40K compared with that of calmodulin (CaM). (C) We expressed the 40 kDa protein, its truncated form, and *Physarum* calmodulin (CaM) in *E. coli* and purified them. The measurement of Ca-binding to the 40 kDa protein was hampered by its Ca-dependent aggregation. Therefore, we used the truncated form as shown (•). We also measured the Ca-binding activities of CaM (○).

8. PERSPECTIVES

In this review, the inhibitory effect of Ca^{2+} on the actin-myosin interaction that causes cytoplasmic streaming was analyzed. In the case of *Physarum*, myosin is the major site of Ca^{2+} action; myosin is in an active form when Ca^{2+} in absent and is inactivated when the calcium-binding light-chain of myosin binds Ca^{2+} (Figure 1). Such calcium inhibition is augmented through actin-binding proteins (Table 5). Role of phosphorylation/dephosphorylation of myosin in the intention were also described (Figure 5).

The myosins explained in this review, i.e., myosins from *Physarum* plasmodia, scallop muscle, and skeletal muscle, are conventional with the specific light chains. However, plant myosin (Section 2) purified by Yokota et al. (1999) is unconventional with CaM as it light chain. Unconventional myosins have been identified from vertebrate cells. Myosin I from brush border consists of a 119 kDa heavy chain and calmodulin. Ca^{2+} dissociates calmodulin from the heavy chain and prevents the interaction of myosin I with actin (Collins et al., 1990; Swanljung-Collins and Collins, 1991). Myosin V, a myosin-like protein from brain, is another calmodulin-containing protein. The effect of Ca^{2+} as monitored by the *in vitro* motility assay is inhibitory (Cheney et al., 1993). However, actin-activated ATPase activity of the protein is stimulated by Ca^{2+}, a discrepancy that must be solved in the near future.

The kinase activities responsible for phosphorylation of myosin are detectable for both phosphorytable light chain and heavy chain and are inhibitied in a Ca-containing solution. Calcium-dependent inhibitory factor is expected to work as a Ca-binding protein (Figures 6 and 7). It is important to note that the kinases are active without requiring Ca^{2+}. Such enzymes are expected to be involved in general cell maintenance and thus be exempted from the regulation by Ca^{2+}. However, the discovery of calcium inhibition indicates that the idea of exemption is an erroneous idea and suggests that some of the housekeeping enzymes may be under the control of Ca^{2+} if they are associated with novel Ca-binding proteins that should exert an inhibitory effect. We hope that our review may inspire scientists to further research such proteins.

REFERENCES

Achenbach, F. and Wohlfarth-Bottermann, K.-E., 1986a, Reactivation of cell-free models of endoplasmic drops from Physarum polycephalum after glycerol extraction at low ionic strength, *Eur. J. Cell Biol.* 40, 135–138.

Achenbach, F. and Wohlfarth-Bottermann, K.-E., 1986b, Successive contraction-relaxation cycles experimentally induced in cell-free models of *Physarum polycephalum*, *Eur. J. Cell Biol.* 42, 111–117.

Cheney, R.E., O'Shea, M.K., Heuser, J.E., Coelho, M.V., Wolenski, J.S. and Mooseker, M.S., 1993, Brain myosin-V is a two-headed unconventional myosin with motor activity, *Cell* 75, 13–23.

Collins, J.H. and Korn, E.D., 1981, Purification and characterization of actin-activatable, Ca^{2+}-sensitive myosin II from *Acanthamoeba*, *J. Biol. Chem.* 256, 2586–2595.

Collins, K., Sellers, J.R. and Matsudaira, P., 1990, Calmodulin dissociation regulates brush border Myosin I (110-KD-Calmodulin) mechanochemical activity in vitro, *J. Cell. Biol.* 110, 1137–1147.

Ebashi, S. and Endo, M., 1968, Calcium ion and muscle contraction, *Progr. Biophys. Mol. Biol.* 18, 123–183.

Fromherz, S. and Szent-Gyorgyi, A.G., 1995, Role of essential light chain EF hand domains in clacium binding and regulation of scallop myosin, *Proc. Natil. Acad. Sci. USA* 92, 7652–7656.

Furuhashi, K. and Hatano, S., 1989, A fragmin-like protein from *Physarum polycephalum* that severs F-actin and caps the barbed end of F-actin in a Ca^{2+}-sensitive way, *J. Biochem.* 106, 311–318.

Furuhashi, K. and Hatano, S., 1990, Control of actin filament length by phosphorylation of fragmin-actin complex, *J. Cell. Biol.* 111, 1081–1087.

Furuhashi, K. and Hatano, S., 1992, Actin kinase: A protein kinase that phosphorylates actin of fragmin-actin complex, *J. Biochem.* 111, 366–370.

Furuhashi, K., Hatano, S., Ando, S., Nishizawa, K. and Inagaki, M., 1992, Phosphorylation by actin kinase of pointed end-domain on the actin molecule, *J. Biol. Chem.* 267, 9326–9330.

Gassner, D., Shraideh, Z. and Wohlfarth-Bottermann, K.E., 1985, A giant titin-like protein in *Physarum polycephalum*: Evidence for its candidacy as a major component of an elastic cytoskeletal superthin filament lattice, *Eur. J. Cell Biol.* 37, 44–62.

Gettemans, J., De Ville, Y., Vandekerckhove, J. and Waelkens, E., 1992, *Physarum* actin is phosphorylated as the actin-fragmin complex at residues Thr203 and Thr202 by a specific 80 kDa kinase, *EMBO J.* 11, 3185–3191.

Gettemans, J., De Ville, Y., Vandekerckhove, J. and Waelkens, E., 1993, Purification and partial amino acid sequence of the actin-fragmin kinase from *Physarum polycephalum*, *Eur. J. Biochem.* 214, 111–119.

Harada, Y., Noguchi, A., Kishino, A. and Yanagida, T., 1987, Sliding movement of single actin filaments on one-headed myosin filaments, *Nature* 326, 805–808.

Hasegawa, T., Takahashi, S., Hayashi, H. and Hatano, S., 1980, Fragmin: A calcium ion sensitive regulatory factor on the formation of actin filaments, *Biochemistry* 19, 2677–2683.

Hatano, S., 1973, Contractile proteins from the myxomycete plasmodium, *Adv. Biophys.* 5, 143–176.

Hinssen, H., 1981a, An actin-modulating protein from *Physarum polycephalum*. I. Isolation and purification, *Eur. J. Cell Biol.* 23, 225–233.

Hinssen, H., 1981b, An actin-modulating protein from *Physarum polycephalum*. II. Ca^{++}-dependence and other properties, *Eur. J. Cell Biol.* 23, 234–240.

Ishigami, M., Yoshiyama, S. and Furuhashi, K., 1996, Calcium waves corresponding to the contraction-relaxation cycle of Physarum polycephalum, *Cell. Struct. Funct.* 21, 628 (Abstract).

Ishikawa, R., Okagaki, T., Higashi-Fujime, S. and Kohama, K., 1991, Stimulation of the interaction between actin and myosin by *Physarum* caldesmon-line protein and smooth muscle caldesmon, *J. Biol. Chem.* 266, 21784–21790.

Ishikawa, R., Okagaki, T. and Kohama, K., 1992, Regulation by Ca^{2+}-calmodulin of the actin-bundling activity of *Physarum* 210-kDa protein, *J. Muscle Res. Cell Motil.* 13, 321–328.

Ishikawa, R., Sasaki, Y., Nakamura, A., Takagi, T. and Kohama, K., 1995, Purification of an ATP-dependent actin-binding protein from a lower eukaryote, *Physarum polycephalum*, *Biochem. Biophys. Res. Commun.* 212, 347–352.

Itano, N. and Hatano, S., 1991, F-actin bundling protein from *Physarum polycephalum*: Purification and its capacity for co-bundling of actin filaments and microtubules, *Cell Motil. Cytoskel.* 19, 244–254.

Jancso, A. and Szent-Gyorgyi, A.G., 1994, Regulation of scallop myosin by the regulatory light chain depends on a single glycine residere, *Proc. Natl. Acad. Sci. USA* 91, 8762–8766.

Kamiya, N., 1981, Physical and chemical basis of cytoplasmic streaming, *Annu. Rev. Plant Physiol.* 32, 205–236.

Kellermayer, M.S. and Granzier, H.L., 1996, Calcium-dependent inhibition of in vitro thin-filament motility by native titin, *FEBS Lett.* 380, 281–286.

Kendrick-Jones, J., Lehman, W. and Szent-Gyorgyi, A.G., 1970, Regulation in molluscan muscles, *J. Mol. Biol.* 54, 313–326.

Kikuyama, M. and Tazawa, M., 1982, Ca^{2+} ion reversibly inhibits the cytoplasmic streaming after K^{+} induced cessation, *Protoplasma* 113, 241–243.

Kishimoto, U. and Akahori, H., 1959. Protoplasmic streaming of an internodal cell of *Nitella flexilis*, *J. Gen. Physiol.* 42, 1167–1183.

Kobayashi, T., Takagi K., Konishi, K., Hamada, Y., Kawaguchi, M. and Kohama, K., 1988, Amino acid sequence of the calcium-binding light chain of myosin from the lower eukaryote, *Physarum polycephalum*, *J. Biol. Chem.* 263, 305–313.

Kohama, K., 1981, Ca-dependent inhibitory factor for the actin-myosin-ATP interaction of *Physarum polycephalum*, *J. Biochem.* 90, 1829–1832.

Kohama, K., 1990, Inhibitory mode for Ca^{2+} regulation, *Trends Pharmacol. Sci.* 11, 433–435.

Kohama, K. and Kendrick-Jones, J., 1986, The inhibitory Ca^{2+}-regulation of the actin-activated Mg-ATPase activity of myosin from *Physarum polycephalum* plasmodia, *J. Biochem.* 99, 1433–1446.

Kohama, K. and Kohama, T., 1984, Myosin confers inhibitory Ca^{2+}-sensitivity on actin-myosin-ATP interaction of *Physarum polycephalum* under physiological conditions, *Proc. Jpn. Acad.* 60B, 435–439.

Kohama, K. and Shimmen, T., 1985, Inhibitory Ca^{2+}-control of movement of beads coated with *Physarum* myosin along actin-cables in *Chara* internodal cells, *Protoplasma* 129, 88–91.

Kohama, K., Kobayashi, K. and Mitani, S., 1980, Effects of Ca ion and ADP on superprecipitation of myosin B from slime mold, *Physarum polycephalum*, *Proc. Jpn. Acad.* 56B, 591–596.

Kohama, K., Uyeda, T.Q.P., Takano-Ohmuro, H., Tanaka, T., Yamaguchi, T., Maruyama, K. and Kohama, T., 1985, Ca^{2+}-binding light chain of *Physarum* myosin confers inhibitory Ca^{2+}-sensitivity on actin-myosin-ATP interaction via actin, *Proc. Jpn. Acad.* 61B, 501–505.

Kohama, K., Oosawa, M., Ito, T. and Maruyama, K., 1988a, Physarum myosin light chain interacts with actin in a Ca^{2+}-dependent manner, *J. Biochem.* 104, 995–998.

Kohama, K., Sohda, M., Murayama, K. and Okamoto, Y., 1988b, Domain structure of Physarum myosin heavy chain, *Protoplasma* (Suppl. 2), 37–47.

Kohama, K., Kohno, T., Okagaki, T. and Shimmen, T., 1991a, Role of actin in the myosin-linked Ca^{2+}-regulation of ATP-dependent interaction between actin and myosin of a lower eukaryote, *Physarum polycephalum*, *J. Biochem.* 110, 508–513.

Kohama, K., Okagaki, T., Takano-Ohmuro, H. and Ishikawa, R., 1991b, Characterization of calcium-binding light chain as a Ca^{2+}-receptive subunit of *Physarum* myosin, *J. Biochem.* 110, 566–570.

Kohama, K., Ishikawa, R. and Okagaki, T., 1992, Calcium inhibition of *Physarum* actomyosin system: Myosin-linked and actin-linked natures, in *Calcium Inhibiton*, K. Kohama (ed.), Japan Sci. Soc. Press, Tokyo/CRC Press, Boca Raton, pp. 91–107.

Kohama, K., Ye, L.-H. and Nakamura, A., 1993, Calcium-binding proteins that are involved in the calcium inhibition of the actomyosin system of a lower eukaryote, *Physarum polycephalum*, *Biomed. Res.* 14 (Suppl. 2), 57–62.

Kohama, K., Ishikawa, R. and Ishigami, M., 1998, Large scale culture of *Physarum*: A simple way of growing plasmodia to purify actomyosin and myosin, in *Cell Biology Hand Book*, J.E. Celis (ed.), 2nd edn., Vol. 1, Academic Press, San Diego, pp. 466–471.

Kohno, T. and Shimmen T., 1988a, Accelerated sliding of pollen tube organelles along *Characeae* actin bundles regulated by Ca^{2+}, *J. Cell. Biol.* 106, 1539–1543.

Kohno, T. and Shimmen T., 1988b, Mechanism of Ca^{2+} inhibition of cytoplasmic streaming in lily pollen tubes, *J. Cell. Sci.* 91, 501–509.

Kretsinger, R.H., 1980, Structure and evolution of calcium-modulated proteins, *CRC Crit. Rev. Biochem.* 8, 119–174.

Kron, S.J. and Spudich, J.A., 1986, Fluorescent actin filaments move on myosin fixed to a glass surface, *Proc. Natl. Acad. Sci. USA* 83, 6272–6276.

Kuczmarski, E.R. and Spudich, J.A., 1980, Regulation of myosin self-assembly: Phosphorylation of Dictyostelium heavy chain inhibits formation of thick filaments, *Proc. Natl. Acad. Sci. USA* 77, 7292–7296.

Kwon, H., Goodwin, E.B., Nyitray, L., Berliner, E., O'Neall-Hennessey, E., Medandri, F.D. and Szent-Gyorgyi, A.G., 1990, Isolation of the regulatory domain of scallop myosin: Role of the essential light chain in calcium binding, *Proc. Natl. Acad. Sci. USA* 87, 4771–4775.

Laroche, A., Lemieux, G. and Pollotta, D., 1989, The nucleotide sequence of a developmentally regulated cDNA for Physarum polycephalum, *Neucleic Acids Res.* 17, 10502.

Lin, Y., Ishikawa, R., Okagaki, T., Ye, L.-H. and Kohama, K., 1994, Stimulation of the ATP-dependent interaction between actin and myosin by a myosin-binding fragment of smooth muscle caldesmon, *Cell Motil. Cytoskel.* 29, 250–258.

Maruta, H., Baltes, W., Dieter, P., Marme, D. and Gerisch, G., 1983a, Myosin heavy chain kinase inactivated by Ca^{2+}/calmodulin from aggregating cells of *Dictyostelium discoideum*, *EMBO J* 2, 535–542.

Maruta, H., Isenberg, G., Schreckenbach, T., Hallmann, H., Risse, G., Shibayama, T. and Hesse, J., 1983b, Ca^{2+}-dependent actin-binding phosphoprotein in *Physarum polycephalum*. I. Ca^{2+}/actin-dependent inhibition of its phosphorylation, *J. Biol. Chem.* 258, 10144–10150.

Maruyama, K., Natori, R. and Nonomura, Y., 1976, New elastic protein from muscle, *Nature (London)* 262, 58–59.

Nakamura, A. and Kohama, K., 1995, Calcium inhibition of actin-myosin interaction, in *Calcium as Cell Signal*, K. Maruyama, Y. Nonomura and K. Kohama (eds.), Igaku-shoin, Tokyo, pp. 270–276.

Nakamura, A. and Kohama, K., 1999, Calcium regulation of the actin-myosin interaction of Physarum policephalum, *Int. Rev. Cytol.* 19, 53–98.

Nakamura, A., Okagaki, T., Takagi, T., Tanaka, T. and Kohama, K., 1994, Molecular cloning and expression of Ca^{2+}-binding protein that inhibits myosin light chain kinase activity in lower eukaryote *Physarum polycephalum*, *Jpn. J. Pharmacol.* 64, 112 (Abstract).

Nakamura, A., Okagaki, T., Takagi, T., Nakashima, K.-I., Yazawa, M. and Kohama, K., 2000, Calcium binding properties of recombinant calcium binding protein 40, a major calcium binding protein of lower eukaryote *Physarum Polycephalum*, *Biochemistry* 39, 3827–3834.

Ogihara, S., Ikebe, M., Takahashi, K. and Tonomura, Y., 1983, Requirement of phosphorylation of *Physarum* myosin heavy chain for thick filament formation, actin activation of Mg^{2+}-ATPase activity, and Ca^{2+}-inhibitory superprecipitation, *J. Biochem.* 93, 205–223.

Okagaki, T., Higashi-Fujime, S. and Kohama, K., 1989, Ca^{2+} activates actin-filament sliding on scallop myosin but inhibits that on *Physarum* myosin, *J. Biochem.* 106, 955–957.

Okagaki, T., Ishikawa, R. and Kohama, K., 1991a, Inhibitory Ca^{2+}-regulation of myosin light chain kinase in the lower eukaryote, *Physarum polycephalum*: Role of a Ca^{2+}-dependent inhibitory factor, *Eur. J. Cell Biol.* 56,113-122.

Okagaki, T., Ishikawa, R. and Kohama, K., 1991b, Purification of a novel Ca-binding protein that inhibits myosin light chain kinase activity in lower eukaryote *Physarum polycephalum*, *Biochem. Biophys. Res. Commun.* 176, 564–570.

Ozaki, K. and Maruyama, K., 1980, Connection, an elastic protein of muscle. A connectin-like protein from the plasmodium *Physarum polycephalum*, *J. Biochem.* 88, 883–888.

Ozaki, K., Sugino, H., Hasegawa, T., Takahashi, S. and Hatano, S., 1983, Isolation and characterization of *Physarum* profilin, *J. Biochem.* 93, 295–298.

Pies, N.J. and Wohlfarth-Bottermann, K.E., 1986, Reactivation of a cell-free model from *Physarum polycephalum*: Studies on cryosections indicate an inhibitory effect of Ca^{++} on cytoplasmic actomyosin contraction, *Eur. J. Cell Biol.* 40, 139–149.

Pollard, T.D. and Weihing, R.R., 1974, Actin and myosin and cell motility, *CRC Crit. Rev. Biochem.* 2, 1–65.

Ridgway, E.B. and Durham, A.C.H., 1976, Oscillations of calcium-ion concentrations in *Physarum polycephalum*, *J. Cell Biol.* 69, 223–226.

Robbins, J., Dilorth, S.M., Laskey, R.A. and Dingwall, C., 1991, Two interdependent basic domains in nucleoplasmin nuclear targeting sequence: Identification of a class of bipartite nuclear targeting sequence, *Cell* 64, 615–623.

Shimmen, T. and Kohama, K., 1984, Ca^{2+}-sensitive sliding of latex beads coated with *Physarum* or scallop myosin along actin bundles in *Characeae* cells. Abstracts of the papers presented at the Third International Congress on Cell Biology, Tokyo, p. 504 (Abstract).

Shimmen, T. and Yano, M., 1984, Active sliding movement of latex beads coated with skeletal muscle myosin on *Chara* actin bundles, *Protoplasma* 121, 132–137.

Sobue, K. and Sellers, J.R., 1991, Caldesmon, a novel regulatory protein in smooth muscle and nonmuscle actomyosin system, *J. Biol. Chem.* 266, 12115–12118.

Sugino, H. and Hatano, S., 1982, Effect of fragmin on actin polymerization: Evidence for enhancement of nucleation and capping of the barbed end, *Cell Motil. Cytoskel.* 2, 457–470.

Sugino, H. and Matsumura, F., 1983, Fragmin induces tension reduction of actomyosin threads in the presence of micromolar levels of Ca^{2+}, *J. Cell Biol.* 96, 199–203.

Sutoh, K., Iwane, M., Matsuzaki, F., Kikuchi, M. and Ikai, A., 1984, Isolation and characterization of a high molecular weight actin-binding protein from *Physarum polycephalum* plasodia, *J. Cell Biol.* 98, 1611–1618.

Swanljung-Collins, H. and Collins, J.H., 1991, Ca^{2+} stimulates the Mg^{2+}-ATPase activity of brush boder Myosin I with three or four calmodulin light chains but inhibits with less than two bound, *J. Biol. Chem.* 266, 1312–1319.

Takagi, S., 1993, Photoregulation of cytoplasmic streaming, *Cell. Struct. Funct.* 18, 498. (abstract)

Takagi, S. and Nagai, R., 1985, Light-controlled cytoplasmic streaming in *Vallisneria* mesophyll cells, *Plant & Cell Physiol.* 26, 941–951.

Takagi, S. and Nagai, R., 1986, Intracellular Ca^{2+} concentration and cytoplasmic streaming in *Valisneria* mesophyll cells, *Plant & Cell Physiol.* 27, 953–959.

Tazawa, M. and Kishimoto, U., 1968, Cessation of cytoplasmic streaming of *Chara* internodes during action potential, *Plant & Cell Physiol.* 9, 361–368.

Toda, H., Okagaki, T. and Kohama, K., 1990, Amino acid sequence of calmodulin from lower eukaryote; *Physarum polycephalum*, in *The Biology and Medicine of Signal Trunsduction*, Y. Nishizuka, M. Endo and T. Tanaka (eds.), Advances in Second Messenger and Phosphoprotein Research, Vol. 24, Raven Press, New York, p. 614 (Abstract).

Tominaga, Y., Shimmen, T. and Tazawa, M., 1983, Control of cytoplasmic streaming by extracellular Ca^{2+} in permeabilized *Nitella* cells, *Protoplasma* 116, 75–77.

Uyeda, T.Q.P., Hatano, S., Kohama, K. and Furuya, M., 1988, Purification of myxamoebal fragmin, and switching of myxamoebal fragmin to plasmodia fragmin during differentiation of Physarum polycephalum, *J. Muscle Res. Cell Motil.* 9, 233–240.

Vale, R.P., Szent-Gyorgyi, A.G. and Sheetz, M.P., 1984, Movement of scallop on *Nitella* actin filaments: Regulation by calcium, *Proc. Natl. Acad. Sci. USA* 81, 6775–6778.

Wang, K., 1977, Filamin, a new high-molecular weight protein found in smooth muscle and non muscle cells. Purifications and properties of chicken gizzard filamin, *Biochemistry* 16, 1857–1865.

Williamson, R.E., 1975, Cytoplasmic streaming in *Chara*: A cell model activated by ATP and inhibited by cytocharasin B, *J. Cell Sci.* 17, 655–688.

Williamson, R.E. and Ashley, C.C., 1982, Free Ca^{2+} and cytoplasmic streaming in the alga *Chara*, *Nature* 296, 647–651.

Womack, F.C. and Colowic, S.P., 1973. Rapid mesurement of binding of ligands by rate of dialysis, *Methods in Enzymology* 27, 464–471.

Xie, X., Harrison, I., Schlichting, D.H., Sweet, R.M., Kalabokis, V.N., Szent-Gyorgyi, A.G. and Cohen, C., 1994, Structure of the regulatory domain of scallop myosin at 2.8 Å resolution, *Nature* 368, 304–312.

Yin, H.L. and Stossel, T.P., 1980, Purification and structural properties of gelsolin, a Ca^{2+}-activated regulatory protein of macrophages, *J. Biol. Chem.* 255, 9490–9493.

Yokota, E. Muto, S. and Shimmen, T., 1999, Inhibitory regulation of higher-plant myosin by Ca^{2+} ions, *Plant Physiol.* 119, 231–239.

Yoshimoto, Y. and Kamiya, N., 1984, ATP and calcium-controlled contraction in a saponin model of *Physarum polycephalum*, *Cell Struct. Funct.* 9, 135–141.

Yoshimoto, Y., Matsumura, F. and Kamiya, N., 1981a, Simultaneous oscillations of Ca^{2+} efflux and tension generation in the premeabilized plasmodial strand of *Physarum*, *Cell Motil.* 1, 432–443.

Yoshimoto, Y., Sakai, T. and Kamiya, N., 1981b, ATP oscillation in *Physarum* plasmodia, *Protoplasma* 109, 159–168.

Penta-EF-Hand (PEF) Proteins and Calsenilin/DREAM: Involvement of the New EF-Hand Calcium-Binding Proteins in Apoptosis and Signal Transduction

Masatoshi Maki

1. INTRODUCTION

Calcium ions control a variety of cellular phenomena, including muscle contraction, adhesion, secretion, motility, growth, differentiation, gene expression, etc. Alterations in intracellular Ca^{2+} homeostasis are also commonly observed during apoptosis or programmed cell death by various stimuli such as glucocorticoid treatment, Interleukin-3 withdrawal, T cell receptor cross-linking, Fas/CD95 stimulation and oxidative stress (see reviews by McConkey and Orrenius, 1997, by Krebs, 1998, and references therein). The duration and extent of Ca^{2+} influx may determine whether cells survive, die by apoptosis or undergo necrosis. Ca^{2+}-binding proteins are the mediators of the signals, and play pivotal roles in the above-mentioned cellular phenomena through a variety of different mechanisms.

Some enzymes, such as protein kinase C and phospholipase C contain Ca^{2+}-binding sites within the enzyme molecules, whereas calmodulin kinases require Ca^{2+}-bound calmodulin for their activations. Based on primary and 3-D structures of the Ca^{2+}-binding domains, the intracellular Ca^{2+}-binding proteins are classified into four groups: EF-hand, C2-domain, annexins and acidic Ca^{2+}-storage proteins. The EF-hand motif, the Ca^{2+}-binding helix-loop-helix structure, has been identified in numerous Ca^{2+}-binding proteins (Kawasaki et al., 1998). The number of repetitive EF-hand motifs in protein molecules, regardless of whether they are capable of Ca^{2+}-binding, ranges

Masatoshi Maki • Department of Applied Molecular Biosciences, Graduate School of Bioagricultural Sciences, Nagoya University, Japan.

R. Pochet, R. Donato, J. Haiech, C. Heizmann and V. Gerke (eds.): Calcium: The Molecular Basis of Calcium Action in Biology and Medicine, 245–258.

from two to eight. In this review, I will focus on the novel Ca^{2+}-binding proteins with five or four EF-hand motifs which draw attention for possible involvement in apoptosis and signal transduction.

2. ALG-2

Stimulation of T cells via CD3/T cell receptor complex induces apoptosis in a Ca^{2+}-dependent manner (McConkey et al., 1989). In 1996, Vito and his co-workers reported the selection of genes involved in apoptosis of mouse T cell hybridoma 3DO cells induced by T cell receptor cross-linking. They employed the "death trap" method where a cDNA library constructed in a mammalian expression vector was used to transfect cells to protect them from death by expecting inactivation of apoptotic factors through different mechanisms: (i) antisense RNAs which inhibit the synthesis of pro-apoptotic proteins; (ii) sense RNAs containing either (a) a complete open reading frame (ORF) for an anti-apoptotic protein or (b) a partial ORF producing either a dominant negative protein of a pro-apoptotic gene or a gain of function of an anti-apoptotic protein (D'Adamio et al., 1997). One of the isolated apoptosis-linked genes named ALG-2 was expressed from the vector as an antisense transcript, and another clone named ALG-3 behaved as a dominant negative protein as described later in this chapter. The ALG-2 cDNA turned out to be identical with a previously reported partial cDNA of the putative EF-hand Ca^{2+}-binding protein pMP42 (Kageyama et al., 1989). The open reading frame of the full-length cDNA encodes a protein of 191 amino acid residues (Figure 1). ALG-2 mRNA is expressed in various mouse tissues. Overexpression of ALG-2 sensitizes NIH3T3 fibroblast cells to apoptosis induced by the combination of the phorbol 12-myristate 13-acetate (PMA) and calcium ionophore ionomycin which mimicked the T cell receptor signaling (D'Adamio et al.,1997). Depletion of ALG-2 by antisense transcription also protected the T cell hybridoma from apoptosis by other stimuli, such as glucocorticoid and Fas/CD95. Caspases were activated in the ALG-2 depleted cells upon stimulation with these reagents, indicating that ALG-2 function is required downstream or is independent of the protease activity for cell death to occur (Lacanà et al.,1997). *In situ* hybridization histochemistry of the rat brain reveals high levels of ALG-2 mRNAs in the granule and pyramidal cell layers of the hippocampus, choroid plexus, area postrema, and a number of hindbrain nuclei (Venn and Conway, 1998). ALG-2 mRNA levels in aged rats are not significantly different from those in young animals. The chromosomal localization of the human ALG-2 gene has been assigned to 5p15.3 (GenBank accession No. AF035606), but no hereditary diseases linked to this locus has been reported yet.

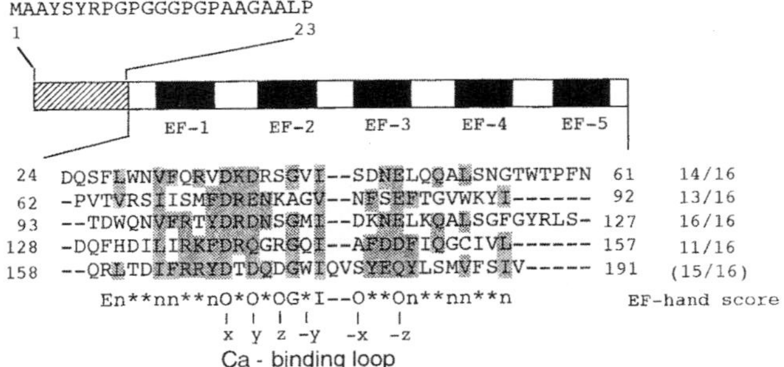

Figure 1. Five repetitive EF-hand motifs in ALG-2. Identical or similar residues in at least three repeats are stippled. The canonical EF-hand sequence contains 16 preferred residues: E, acidic; n, hydrophobic; O, oxygen-containing; G, glycine, I, aliphatic side chains; asterisks, variable; hyphens, gaps. EF-hand score: number of matched residue. (Taken from Maki et al., *J. Biochem.* 124, 1170–1177, 1998. Reproduced with permission from the Japanese Biochemical Society.)

3. PEF PROTEIN FAMILY

Comparing with calmodulin, the well-known four-EF-hand type Ca^{2+}-binding protein, ALG-2 has two unique features. First, it has an additional EF-hand like sequence at the *C*-terminus. Because of the presence of a two-residue insertion in the corresponding Ca^{2+}-binding loop, the fifth EF-hand (EF-5) may not bind Ca^{2+}. Second, ALG-2 has a stretch of glycine and hydrophobic residues in the *N*-terminus. These features as well as similarity in the amino acid sequence, ALG-2 can be classified into a five-EF-hand (penta-EF-hand, PEF) protein family or the calpain small subunit family (Maki et al., 1997). The lengths of the *N*-terminal hydrophobic domains are variable among the members (Figure 2).

Calpains, the Ca^{2+}-dependent cysteine proteases widely distributed in animal tissues and cells, are involved in various physiological and pathological reactions coupled with Ca^{2+}-mobilization by cleaving certain enzymes, receptors, cytoskeletal proteins, transcription factors, etc. (Ono et al., 1998). Both μ-calpain (low Ca^{2+}-requiring form) and m-calpain (high Ca^{2+}-requiring form), expressed ubiquitously in mammals and avians, consist of a mutually distinct catalytic large subunit (80 kDa) and a common regulatory small subunit (30 kDa). Both large and small subunits possess PEF domains (or named calmodulin-like domains) at the *C*-terminal regions. Some invertebrate calpains lack the PEF domains. The typical PEF-domain-containing calpains as well as the small subunit appeared rather new in the evolution (Jekely and Friedrich, 1999). The PEF domains function not only as Ca^{2+}-dependent regulatory domains for calpain activation but also as substrate recognition sites such as for IκBα (Shumway et al., 1999). Endogenous calpain inhibitor protein calpastatin also interacts with these domains (Takano

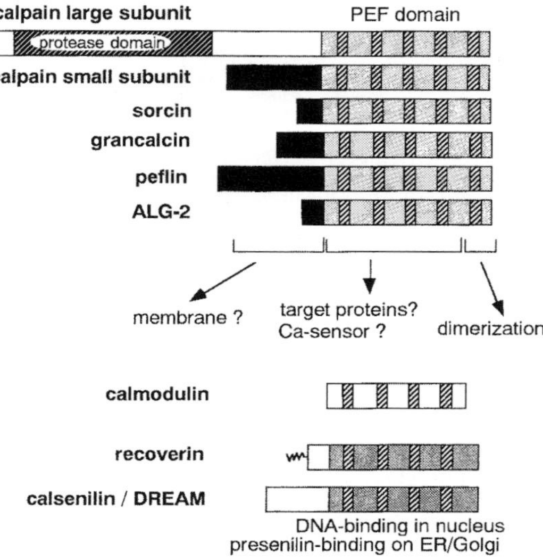

Figure 2. Schematic structures of penta-EF-hand (PEF) proteins and four EF-hand proteins. Hatched boxes indicate potential Ca^{2+}-binding loops of EF-hands. Closed boxes of PEF proteins indicate the N-terminal glycine-rich hydrophobic regions. The N-terminus of recoverin is linked to a myristoyl or related acyl group.

et al., 1995). Calpains are suggested to be involved in certain types of cell death (Squier et al., 1999; Chi et al., 1999).

Sorcin was first found as a gene amplified together with the P-glycoprotein gene in multidrug-resistant cancer cells. However, the role of sorcin in drug resistance is not clear because its level of expression does not correlate with the degree of resistance. Sorcin has been identified in a wide variety of cells and seems involved in Ca^{2+} homeostasis by modulating ryanodine receptors (reviewed by Valdivia, 1998). Ca^{2+}-bound form of sorcin inhibits cardiac ryanodine receptors, allowing feed back regulation. Sorcin phosphorylated by protein kinase A (PKA) is inactive in inhibing the receptors. Although physiological function is not clear, interaction of sorcin with C-terminal region of α_1 subunit of L-type Ca^{2+} channel is also reported (Meyers et al., 1998). Grancalcin is the close homolog of sorcin, and is supposed to be associated with granule-membrane fusion and degranulation of neutrophils (Boyhan et al., 1992). Sorcin and grancalcin form dimers, respectively, as revealed by gel filtration. Dimers seem to be formed through EF-5's as demonstrated by X-ray crystallography of the calpain small subunit dimer (Lin et al., 1997). Dimerization of ALG-2 has been also shown by the yeast two-hybrid system and the co-immunoprecipitation method (Missotten et al., 1999). Gel filtration of the recombinant protein expressed in *E. coli*, however, shows a monomeric form, and suggests that the interaction between EF-5's of ALG-2 molecules is weak (Maki et al., 1998; Lo et al., 1999). Mutational analyses of the recombinant protein reveal that EF-1 and EF-3 are essential for Ca^{2+} binding (Vito et al., 1996b; Lo et al., 1999). Ca^{2+} induces conformational change of ALG-2 to expose a hydrophobic surface, which may permit interaction with target proteins (Maki et al., 1998). Recently we have identified a new human PEF protein named peflin, which has the longest *N*-terminal hydrophobic domain in the family but most similar to ALG-2 (Kitaura et al., 1999). Homology search reveals occurrence of PEF proteins not only in the animal kingdom but also in fungi and plants (Maki, 2000).

4. ALG-2 INTERACTING PROTEIN

Using the yeast two-hybrid interaction screening system, a cDNA for the ALG-2 binding protein named AIP1 (ALG-2 interacting protein 1) was isolated from a mouse liver cDNA library (Vito et al., 1999). Both ALG-2 and AIP1 are co-localized in the cytosol and the presence of Ca^{2+} is an indispensable requisite for their association. Overexpression of the *N*-terminal truncated form of AIP1 protects HeLa and COS cells from death induced by withdrawal of trophic factors. AIP1 is expressed in various tissues including brain, heart, liver, spleen, testis, etc. By the similar method, another group independently isolated a cDNA clone named Alix (ALG-2 interacting protein X) from a mouse adult brain cDNA library (Missotten et al., 1999). Although there are some differences in the deduced amino acid sequences due to a frame-shift in nucleotide sequences (GenBank accession numbers: AIP1, AF119955; Alix, MMAJ5073) or single nucleotide substitution, AIP1 and Alix are essentially derived from the same gene. Note that AIP1/Alix is absolutely different from actin interacting protein 1 or atrophin interacting protein 1 (also abbreviated as AIP1). To avoid confusion, the term Alix is used in the following sections.

Alix contains 869 amino acid residues, and the entire sequence except the *C*-terminal region of Alix (1-723) is similar to several proteins predicted as signal transducing factors such as *Xenopus* Xp95 (81% identity, Che et al., 1999), *C. elegans* YNK1 (42% identity, Che et al., 1997), *Aspergillus nidulans* PalA (25% identity, Negrete-Urtasun et al., 1997), *Arabidopsis thaliana* hypothetical protein (23% identiy, GenBank accession No. AC007591, protein-id AAD39642.1) and budding yeast BRO1 (17% identity, Nickas and Yaffe, 1996) (Figure 3). All these proteins contain consensus motifs (R/K)X(2,3)(D/E)X(2,3)Y for potential tyrosine phosphorylation sites (313-KDNDFIY in Alix). Moreover, they contain common features in their *C*-terminal regions: rich in proline residues and presence of PXXP motifs which represent potential binding sites for SH3 domains (*src* homology domain 3). Indeed, an SH3-containing protein named SETA has been shown to interact with Alix (Chen et al., 2000). The phylogenetic tree analysis suggests that Xp95 and YNK1 are *Xenopus* and *C. elegans* orthologs of mouse Alix, respectively (Maki, 2000). Xp95 is a 95 kDa protein in *Xenopus* oocytes and phosphorylated from the first through second meiotic divisions during progesterone-induced oocyte maturation. Xp95 is phosphorylated by

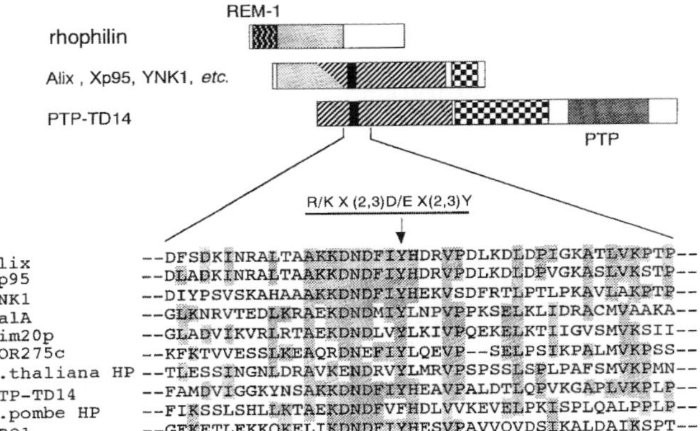

Figure 3. Schematic structures of ALG-2 interacting protein Alix and related proteins. Amino acid sequences containing potential tyrosine-phosphorylation sites are shown. REM-1, Rho-effector motif class 1; PTP, protein tyrosine phosphatase domain. Checkered boxes indicate proline-rich domain. *N*-terminal regions of Alix and its homologs are similar to rhophilin, but there is no similarity between rhophilin and PTP-TD14.

Src kinase and to a lesser extent by Fyn kinase, but not by mitotic Cdc2 kinase *in vitro*. Function of *C. elegans* YNK1 is not known, but this protein is expressed *in vivo* throughout development of the worm.

Implications of Alix function in signal transductions are derived from genetic studies of budding yeast. *BRO1* mutant yeast cells exhibit a temperature-sensitive growth defect that is suppressed by the addition of osmotic stabilizers (sorbitol) or Ca^{2+} to the growth medium or by additional copies of the *BCK1* gene, which encodes a MEK kinase that functions in a mitogen-activated protein kinase pathway mediating maintenance of cell integrity. Involvement of the yeast hypothetical PEF protein YG-25 in this system remains unknown. Alkaline adaptation of fungi *Aspergillus* (*Emericella*) *nidulans* and pH regulation of gene expression is mediated by the zinc finger transcription factor PacC, which activates transcription of genes expressed at alkaline pH and prevent transcription of genes expressed at acid pH. The products of six genes, *palA*, *-B*, *-C*, *-F*, *-H*, and *-I* form a signal transduction pathway through which alkaline ambient pH is able to elicit the conversion of the full-length form of PacC to the functional proteolyzed form. The *palB* gene product PalB is a cysteine protease of the calpain superfamily lacking the PEF domain (Denison et al., 1995), but it is not the protease responsible for the final conversion of PacC to its functional form. It is not yet known whether ALG-2 like protein is involved in this pathway.

Interestingly, a recently identified novel rat putative protein tyrosine phosphatase PTP-TD14 has a BRO1-like domain and shows a significant similarity with Alix (191-716, 17% identity) (Cao et al., 1998) in addition to the Pro-rich region. Expression of PTP-TD14 in NIH-3T3 cells inhibits Ha-*ras*-mediated transformation more than three folds. This inhibitory activity is not observed in the case of the phosphatase domain mutant. PTP-TD14 is localized in the cytoplasm in association with vesicle-like structures. This association appears not to require the catalytic or C-terminal region of its PTPase domain. Thus, the BRO1 domain is suggested to determine its intracellular localization. Rhophilin, an interacting protein with Ras-related small GTPase Rho (Watanabe et al., 1996), shows a significant similarity with the N-terminal regions of the Alix family members except PTP-TD14, but lacks the conserved potential Tyr-phosphorylation site. Rho functions as a molecular switch in regulation of focal adhesions, stress fibers, contractile ring, cytokinesis, etc., through interactions with different classes of proteins (Narumiya et al., 1997). Because the Alix family members lack the Rho-effector motif class 1 (REM-1), these proteins may not, at least directly, interact with the small GTPase.

At present, actions of Alix on apoptosis are not clear and further studies are required to understand the functions of this protein in signal transductions associated with Tyr-phosphorylation, if it occurs. Ubiquitous presence of homologs of both ALG-2 and Alix in plants and fungi as well as in animals suggests that these proteins are common components in signal transduction pathways conserved in all eukaryotic cells. Abnormalities in their functions may lead various pathological phenomena in human, and ALG-2 and/or Alix might become good targets for diagnostic markers and development of therapeutic drugs in the future.

5. PRESENILINS

Mutations in the highly homologous genes presenilin 1 (PS1) and presenilin 2 (PS2) alter the processing of the β-amyloid precursor protein (APP) by γ-secretase and lead to increase the level of neurotoxic peptide $A\beta_{1-42}$, causing most early-onset of familial Alzheimer's disease (FAD) (Selkoe, 1998; Monteiro and Stabler, this book). Presenilins are polytopic integral membrane proteins with six to eight membrane spanning regions, and processed at the cytoplasmic hydrophilic loop to the 30 kDa N-terminal fragment and the 20 kDa C-terminal fragment to form a complex probably associated with other proteins. Presenilins are also known to be involved in the processing of Notch, which is a single transmembrane receptor and determines cell fate in development. Recent evidence that PS1 itself has proteolytic properties could explain many of the biological and biochemical alterations caused by PS1 deficiency or clinical mutations in PS1. Since presenilins are primarily localized to the endoplasmic reticulum and the Golgi apparatus, the apparent difference in subcellular localization between presenilins and APP or Notch faces a "spatial paradox" (Annaert and De Strooper, 1999).

In addition to their roles in processing of APP and producing the neurotoxic peptide $A\beta$, presenilins are suggested to be involved in sensitization of apoptosis by as yet unclarified mechanisms (Figure 4). Overexpression

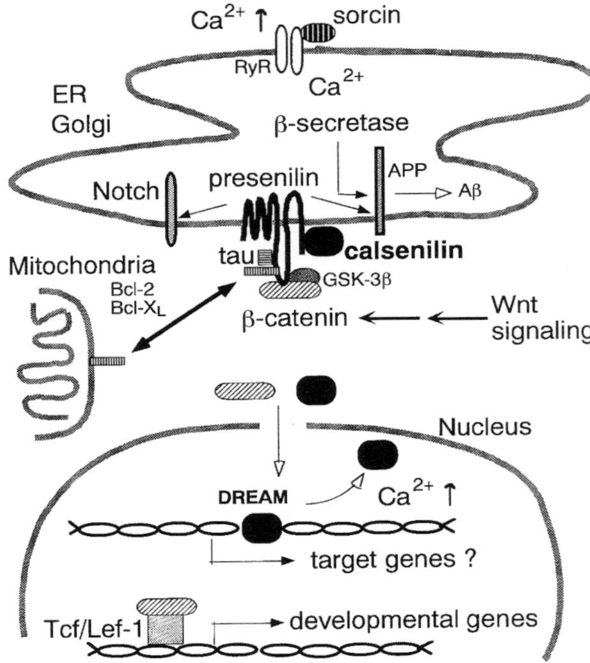

Figure 4. Presenilin-interacting proteins. Calsenilin, identical with the Ca^{2+}-dependent transcriptional repressor DREAM, binds to the *C*-terminal tails of presenilins 1 and 2. APP, β-amyloid precursor protein; ER, endoplasmic reticulum; GSK-3β, glycogen synthase kinase 3β; RyR, ryanodine receptor; Tcf/Lef-1, T cell factor/lymphocyte enhancer-binding factor 1.

of PS2 increases cell death induced by various apoptotic stimuli, whereas FAD-associated presenilin mutations produce molecules with enhanced basal apoptotic activity (Vito et al., 1996a; Wolozin et al., 1996). ALG-3 is an experimentally obtained truncated form of mouse PS2, corresponding to the *C*-terminal 103 amino acids, and rescues cells from T cell receptor-induced apoptosis by inhibiting Fas-mediated death signal (Vito et al., 1996b). Under physiological condition, *C*-terminal fragments of PS2 are generated by translation of a short form of the PS2 mRNA and also by cleavage of PS2 protein with unknown protease. Caspase-3 also generates a shorter *C*-terminal fragment. These ALG-3 like fragments protect cells from apoptotic stimuli (Vito et al., 1997). Mechanisms of the modulation of apoptosis by presenilins are not clarified yet, but reports of various proteins interacting with presenilins are now accumulating.

β-catenin, originally identified as a protein binding to the cytoplasmic domain of cell adhesion molecule cadherin, is now recognized as a transcriptional co-factor functioning in nucleus and binds to a T cell factor/lymphocyte enhancer-binding factor 1 (Tcf/Lef-1). Stabilization of β-catenin is essential for Wnt/Wingless signaling (Eastman and Grosschedl, 1999). Although the consequence is not clear, presenilins are shown to interact with β-catenin.

Destabilization of β-catenin by mutations in the PS1 gene potentiates neuronal apoptosis by inactivation of Tcf/Lef-1 (Zhang et al., 1998). The protein level of β-catenin is lowered in the brain of FAD patients. A recent report by Kang et al. in 1999, however, is in apparent disagreement where they show that β-catenin is rather destabilized by wild type PS1 but not by PS1 mutations. Since PS1 also associates with GSK-3β, a Ser/Thr-kinase which phosphorylates β-catenin as well as glycogen synthase and the microtubule-associated protein tau, it may act as a molecular tether connecting the kinase and substrates (Takashima et al., 1998).

Recently, anti-apoptotic factor Bcl-X_L has been shown to associate with PS1 and PS2 (Passer et al., 1999). Presenilins influence mitochondrial-dependent apoptotic activities, such as cytochrome c release and Bax-mediated apoptosis by modulating the balance between the anti- and pro-apoptotic Bcl-2 family members. Although physiological relevance to apoptosis is not known, Ca^{2+}-binding proteins are also reported to interact with PS2. Among them are the carboxy-terminal region of the μ-calpain large subunit (Shinozaki et al., 1998), sorcin (Pack-Chung et al., 2000) and calsenilin as described below.

6. CALSENILIN/DREAM

Calsenilin was identified as a PS-2-binding protein in 1998 by Buxbaum and his co-workers. While other presenilin binding proteins interact with the large cytoplasmic loop, calsenilin interacts with the last C-terminal tail hanging in the cytoplasm. It also binds to PS1. Human calsenilin, 256 amino acid residues, has four EF-hands and belongs to the recoverin family (Figure 2). It is expressed at its highest level in brain, but scarcely in other tissues such as liver, lung, heart, etc. In transfection experiments, immunostaining of the epitope-tagged calsenilin exhibits a diffuse cytoplasmic pattern in COS-7 cells, fibroblast-like cells containing SV40 T antigen. After co-expression with PS2, however, the staining pattern shifts to an ER-membrane distribution, similar to that for PS2. Overexpression of calsenilin in human neuroglioma cells leads to alternative processing of PS2, similar to that observed after caspase activation during apoptosis. The aforementioned anti-apoptotic effects of ALG-3 or the physiological C-terminal fragments of presenilins may be explained by sequestering excess cytoplasmic pool of calsenilin or leaving the C-terminal tails of presenilins free to access to other factors. It is not yet known, however, whether the interactions between calsenilin and presenilins are adversely modified by FAD mutations and contribute to disrupted Ca^{2+} homeostasis. Effects of Ca^{2+} on the calsenilin–presenilin interaction remain to be seen.

DREAM, a 31 kDa protein, is a transcriptional repressor which was identified through its ability to bind to the downstream regulatory element (DRE) of the prodynorphin gene (Carrión et al., 1999). DRE functions as a transcriptional silencer and it can repress transcription in an orientation-independent manner when placed downstream from the transcription start site of a heterologous promoter (Carrión et al., 1998). However, DRE shows a position-dependent repressor activity, being totally ineffective when placed upstream of the TATA box. DREAM contains four EF-hands and also belongs to the recoverin family. In the absence of Ca^{2+}, DREAM remains tightly bound to DRE and represses the transcription. Binding of Ca^{2+} to DREAM induces conformational change of the protein and release it from DRE. The DRE sequence (putative assignment, GNARYYRAG) is found in at least one other gene regulated by Ca^{2+}, the immediate-early gene *c-fos*. The direct effect of Ca^{2+} on the transcription factor is a new mechanism of calcium signaling.

More surprisingly, DREAM is identical with calsenilin. Although the identity of the two proteins (GenBank accession numbers are: calsenilin, AF120102; DREAM, AJ1730) was not mentioned in the respective papers by the authors, database search revealed the surprising fact. The coincidental discovery, as is often the case in science, of the identical protein by independent groups working in different research fields will shed a new light on the biological functions of calsenilin/DREAM. What is the subcellular distribution of the nuclear factor DREAM? The same question should be given to calsenilin again because Buxbaum and his co-workers analysed epitope-tagged calsenilin overexpressed in the transfected COS-7 cells. Native calsenilin in neuronal cells may have different subcellular distribution. DREAM in a tetrameric form (110 kDa) has a much higher affinity to DRE than in a monomeric form. Would ALG-3 or the *C*-terminal fragments of presenilins modulate the transcriptional repression activity of DREAM and change the gene expression levels of survival or pro-apoptotic factors? The discoveries of calsenilin and DREAM have raised more questions than the authors of the respective papers have imagined, and the link of presenilins with the Ca^{2+}-regulated transcriptional repressor would allow us to gain more insights into the molecular mechanism of apoptosis associated with FAD.

7. SUMMARY

The penta-EF-hand (PEF) protein family includes the calpain small subunit, sorcin, grancalcin, peflin and ALG-2 in mammals. The family members contain five EF-hand motifs in their *C*-termini as well as stretches of glycine and hydrophobic residues in their *N*-termini. ALG-2 is a new 22 kDa intracellular Ca^{2+}-binding protein involved in apoptosis. Forced reduction of ALG-2 gene expression by the antisense-RNA strategy makes the transfected T cell-lines resistant to apoptosis induced by various stimuli. ALG-2 binds to a newly identified protein named AIP1 or Alix. The overall structure of AIP1/Alix is similar to *Xenopus* oocyte phosphoprotein Xp95, *Aspergillus nidulans* alkaline adaptation factor PalA and the budding yeast protein named

BRO1, which is involved in a MAP kinase cascade. All these proteins contain potential tyrosine-phosphorylation sites and a few PXXP motifs (potential SH3-binding sites), suggesting involvement of ALG-2 or its homologs in the basic signal transductions common to eukaryotic cells. On the other hand, calsenilin is a newly identified member of the recoverin family and contains four-EF-hand motifs. It interacts with the C-terminal regions of presenilins whose mutations are linked to the familial Alzheimer's disease. Overexpression of the C-terminal fragments of presenilins suppresses apoptosis of the transfected cells induced by the neurotoxic peptide $A\beta$ in a transdominant negative fashion. Surprisingly, calsenilin is identical with a recently identified Ca^{2+}-dependent transcriptional repressor named DREAM, suggesting dual roles of the novel Ca^{2+}-binding protein in cytoplasm and nucleus.

REFERENCES

Annaert, W. and De Strooper, B., 1999, Presenilins: Molecular switches between proteolysis and signal transduction, *Trends Neurosci.* 22, 439–443.

Boyhan, A., Casimir, C.M., French, J.K., Teahan, C.G. and Segal, A.W., 1992, Molecular cloning and characterization of grancalcin, a novel EF-hand calcium-binding protein abundant in neutrophils and monocytes, *J. Biol. Chem.* 267, 2928–2933.

Buxbaum, J.D., Choi, E.K., Luo, Y., Lilliehook, C., Crowley, A.C., Merriam, D.E. and Wasco, W., 1998, Calsenilin: A calcium-binding protein that interacts with the presenilins and regulates the levels of a presenilin fragment, *Nature Med.* 4, 1177–1181.

Cao, L., Zhang, L., Ruiz-Lozano, P., Yang, Q., Chien, K.R., Graham, R.M. and Zhou, M., 1998, A novel putative protein-tyrosine phosphatase contains a BRO1-like domain and suppresses Ha-*ras*-mediated transformation, *J. Biol. Chem.* 273, 21077–21083.

Carrión, A.M., Mellström, B. and Naranjo, J.R., 1998, Protein kinase A-dependent derepression of the human prodynorphin gene via differential binding to an intragenic silencer element, *Mol. Cell. Biol.* 18, 6921–6929.

Carrión, A.M., Wolfgang, A.L., Ledo, F., Mellström, B. and Naranjo, J.R., 1999, DREAM is a Ca^{2+}-regulated transcriptional repressor, *Nature* 398, 80–84.

Che, S., Weil, M.M., Etkin, L.D., Epstein, H. and Kuang, J., 1997, Molecular cloning of a splice variant of *Caenorhabditis elegans* YNK1, a putative element in signal transduction *Biochim. Biophys. Acta* 1354, 231–240.

Che, S., El-Hodiri, H.M., Wu, C.F., Nelman-Gonzalez, M., Wei, M.M., Etkin, L.D., Clark, R.B. and Kuang, J., 1999, Identification and cloning of Xp95, a putative signal transduction protein in *Xenopus* oocytes, *J. Biol. Chem.* 274, 5522–5531.

Chen, B., Borinstein, S.C., Gillis, J., Sykes, V.W. and Bogler, O., 2000, The glioma-associated protein SETA interacts with AIP1/Alix and ALG-2 and modulates apoptosis in astrocytes, *J. Biol. Chem.* 275, 19275–19281.

Chi, S., Hiwasa, T., Maki, M., Sugaya, S., Nomura, J., Kita, K. and Suzuki, N., 1999, Suppression of okadaic acid-induced apoptosis by overexpression of calpastatin in human UVr-1 cells, *FEBS Lett.* 459, 391–394.

D'Adamio, L., Lacanà, E. and Vito, P., 1997, Functional cloning of genes involved in T-cell receptor-induced programmed cell death, *Seminars Immunol.* 9, 17–23.

Denison, S.H., Orejas, M. and Arst, H.N. Jr., 1995, Signaling of ambient pH in Aspergillus involves a cysteine protease, *J. Biol. Chem.* 270, 28519–28522.

Eastman, Q. and Grosschedl, R., 1999, Regulation of LEF-1/TCF transcription factors by Wnt and other signals, *Curr. Opinion Cell Biol.* 11, 233–240.

Jekely, G. and Friedrich, P., 1999, The evolution of the calpain family as reflected in paralogous chromosome regions, *J. Mol. Evol.* 49, 272–281.

Kageyama, H., Shimizu, M., Tokunaga, K., Hiwasa, T. and Sakiyama, S., 1989, A partial cDNA for a novel protein which has a typical EF-hand structure, *Biochim. Biophys. Acta* 1008, 255–257.

Kang, D.E., Soriano, S., Frosch, M.P., Collins, T., Naruse, S., Sisodia, S.S., Leibowitz, G., Levine, F. and Koo, E.H., 1999, Presenilin 1 facilitates the constitutive turnover of β-catenin: Differential activity of Alzheimer's disease-linked PS1 mutants in the β-catenin-signaling pathway, *J. Neurosci.* 19, 4229–4237.

Kawasaki, H., Nakayama, S. and Kretsinger, R.H., 1998, Classification and evolution of EF-hand proteins, *Biometals* 11, 277–295.

Kitaura, Y., Watanabe, M., Satoh, H., Kawai, T., Hitomi, K. and Maki, M., 1999, Peflin, a novel member of the five-EF-hand-protein family, is similar to the Apoptosis-Linked Gene 2 (ALG-2) protein but possesses nonapeptide repeats in the N-terminal hydrophobic region, *Biochem. Biophys. Res. Commun.* 263, 68–75.

Krebs, J., 1998, The role of calcium in apoptosis, *Biometals* 11, 375–382.

Lacanà, E., Ganjei, K.J., Vito, P. and D'Adamio, L., 1997, Dissociation of apoptosis and activation of IL-1β-converting enzyme/Ced-3 proteases by ALG-2 and the truncated Alzheimer's gene ALG-3. *J. Immunol.* 158, 5129–5135.

Lin, G., Chattopadhyay, D., Maki, M., Wang, K.K.W., Carson, M., Jin, L., Yuen, P., Takano, E., Hatanaka, M., DeLucas, L.J. and Narayana, S.V.L., 1997, Crystal structure of calcium bound domain VI of calpain at 1.9Å resolution and its role in enzyme assembly, regulation, and inhibitor binding, *Nature Struct. Biol.* 4, 539–547.

Lo, K.W.-H., Zhang, Q., Li, M. and Zhang, M., 1999, Apoptosis-linked gene product ALG-2 is a new member of the calpain small subunit subfamily of Ca^{2+}-binding proteins, *Biochem.* 38, 7498–7508.

Maki, M., 2000, Roles of ALG-2 and its interacting protein in apoptosis and signal transduction, *Electr. J. Pathol. Histol.* 6, 001-08.

Maki, M., Narayana, S.V.L. and Hitomi, K., 1997, A growing family of the Ca^{2+}-binding proteins with five EF-hand motifs, *Biochem. J.* 328, 718–720.

Maki, M., Yamaguchi, K., Kitaura, Y., Satoh, H. and Hitomi, K., 1998, Calcium-induced exposure of a hydrophobic surface of mouse ALG-2, which is a member of the penta-EF-hand protein family, *J. Biochem.* 124, 1170–1177.

McConkey, D.J. and Orrenius, S., 1997, The role of calcium in the regulation of apoptosis, *Biochem. Biophys. Res. Commun.* 239, 357–366.

McConkey, D.J., Hartzell, P., Amador-Perez, J.F., Orrenius, S. and Jondal, M., 1989, Calcium-dependent killing of immature thymocytes by stimulation via the CD3/T cell receptor complex, *J. Immunol.* 143, 1801–1806.

Meyers, M.B., Puri, T.S., Chien, A.J., Gao, T., Hsu, P.H., Hosey, M.M. and Fishman, G.I., 1998, Sorcin associates with the pore-forming subunit of voltage-dependent L-type Ca^{2+} channels, *J. Biol. Chem.* 273, 18930–18935.

Missotten, M., Nichols, A., Rieger, K. and Sadoul, R., 1999, Alix, a novel mouse protein undergoing calcium-dependent interaction with the apoptosis-linked-gene 2 (ALG-2) protein, *Cell Death Diff.* 6, 124–129.

Monteiro, M.J. and Stabler, S., 2000, Genetic factors and the role of calcium in Alzheimer's disease pathogenesis, this book.

Narumiya, S., Ishizaki, T. and Watanabe, N., 1997, Rho effectors and reorganization of actin cytoskeleton, *FEBS Lett.* 410, 68–72.

Negrete-Urtasun, S., Denison, S.H. and Arst, H.N. Jr., 1997, Characterization of the pH signal transduction pathway gene palA of *Aspergillus nidulans* and identification of possible homologs, *J. Bact.* 179, 1832–1835.

Nickas, M.E. and Yaffe, M.P., 1996, *BRO1*, a novel gene that interacts with components of the Pkc1p-mitogen-activated protein kinase pathway in *Saccharomyces cerevisiae*, *Mol. Cell. Biol.* 16, 2585–2593.

Ono, Y., Sorimachi, H. and Suzuki, K., 1998, Structure and physiology of calpain, an enigmatic protease, *Biochem. Biophys. Res. Commun.* 245, 289–294.

Pack-Chung, E., Meyers, M.B., Pettngell, W.P., Moir, R.D., Brownawell, A.M., Cheng, I., Tanzi, R.E. and Kim, T.W., 2000, Presenilin 2 interacts with sorcin, a modulator of the ryanodine receptor, *J. Biol. Chem.* 275, 14440–14445.

Passer, B.J., Pellegrini, L., Vito, P., Ganjei, J.K. and D'Adamio, L., 1999, Interaction of Alzheimer's presenilin-1 and presenilin-2 with Bcl-X_L. A potential role in modulating the threshold of cell death, *J. Biol. Chem.* 274, 24007–24013.

Selkoe, D.J., 1998, The cell biology of β-amyloid precursor protein and presenilin in Alzheimer's disease, *Trends Cell Biol.* 8, 447–453.

Shinozaki, K., Maruyama, K., Kume, H., Tomita, T., Saido, T.C., Iwatsubo, T. and Obata, K., 1998, The presenilin 2 loop domain interacts with the μ-calpain C-terminal region, *Int. J. Mol. Med.* 1, 797–799.

Shumway, S.D., Maki, M. and Miyamoto, S., 1999, The PEST domain of IκBα is necessay and sufficient for *in vitro* degradation by μ-calpain, *J. Biol. Chem.* 274, 30874–30881.

Squier, M.K., Sehnert, A.J., Sellins, K.S., Malkinson, A.M., Takano, E. and Cohen, J.J., 1999, Calpain and calpastatin regulate neutrophil apoptosis, *J. Cell. Physiol.* 178, 311–319.

Takano, E., Ma, H., Yang, H.Q., Maki, M. and Hatanaka, M., 1995, Preference of calcium-dependent interactions between calmodulin-like domains of calpain and calpastatin subdomains, *FEBS Lett.* 362, 93–97.

Takashima, A., Murayama, M., Murayama, O., Kohno, T., Honda, T., Yasutake, K., Nihonmatsu, N., Mercken, M., Yamaguchi, H., Sugihara, S. and Wolozin, B., 1998, Presenilin 1 associates with glycogen synthase kinase-3β and its substrate tau, *Proc. Natl. Acad. Sci. USA* 95, 9637–9641.

Valdivia, H.H., 1998, Modulation of intracellular Ca^{2+} levels in the heart by sorcin and FKBP12, two accessory proteins of ryanodine receptors, *Trends Pharmacol. Sci.* 19, 479–482.

Venn, M.K. and Conway, E.L., 1998. Localization of mRNA for the apoptosis-linked gene ALG-2 in young and aged rat brain, *Neuroreport* 9, 1981–1985.

Vito, P., Wolozin, B., Ganjei, J.K., Iwasaki, K., Lacanà, E. and D'Adamio, L., 1996a, Requirement of the familial Alzheimer's disease gene PS2 for apoptosis. Opposing effect of ALG-3, *J. Biol. Chem.* 271, 31025–31028.

Vito, P., Lacanà, E. and D'Adamio, L., 1996b, Interfering with Apoptosis: Ca^{2+}-binding protein ALG-2 and Alzheimer's disease gene ALG-3, *Science* 271, 521–525.

Vito, P., Ghayur, T. and D'Adamio, L., 1997, Generation of anti-apoptotic presenilin-2 polypeptides by alternative transcription, proteolysis, and caspase-3 cleavage, *J. Biol. Chem.* 272, 28315–28320.

Vito, P., Pellegrini, L., Guiet, C. and D'Adamio, L., 1999, Cloning of AIPI, a novel protein that associates with the apoptosis-linked gene ALG-2 in a Ca^{2+}-dependent reaction, *J. Biol. Chem.* 274, 1533–1540.

Watanabe, G., Saito, Y., Madaule, P., Ishizaki, T., Fujisawa, K., Morii, N., Mukai, H., Ono, Y., Kakizuka, A. and Narumiya, S., 1996, Protein kinase N (PKN) and Rho-related protein rhophilin as targests of small GTPase Rho, *Science* 271, 645–649.

Wolozin, B., Iwasaki, K., Vito, P., Ganjei, J.K., Lacanà, E., Sunderland, T., Zhao, B., Kusiak, J.W., Wasco, W. and D'Adamio, L., 1996, Participation of presenilin 2 in apoptosis: Enhanced basal activity conferred by an Alzheimer mutation, *Science* 274, 1710–1713.

Zhang, Z., Hartmann, H., Do, V.M., Abramowski, D., Sturchler-Pierrat, C., Staufenbiel, M., Sommer, B., van de Wetering, M., Clevers, H., Saftig, P., De Strooper, B., He, X. and Yankner, B.A., 1998, Destabilization of beta-catenin by mutations in presenilin-1 potentiates neuronal apoptosis, *Nature* 395, 698–702.

Activation of Programmed Cell Death by Calcium: Protection against Cell Death by the Calcium Binding Protein, Calbindin-D_{28k}

Sylvia Christakos, Frank Barletta, Michael Huening, Jody Kohut and Mihali Raval-Pandya

1. INTRODUCTION

Calcium has been known to be a potent second messenger for a wide range of cellular processes from fertilization to cell death. It has been implicated in the regulation of protein kinases, phosphatases, protease activity, chromatin structure and transcription as well as in the regulation of muscle contraction, nerve transmission, cytoskeletal organization, cell cycle progression and differentiation (Berridge, 1997; Berridge et al., 1998). Calcium homeostasis is tightly regulated such that any exogenous or internally generated calcium load is rapidly controlled to maintain calcium balance. Calcium ions signal from outside to inside by raising the intracellular cytosolic calcium concentration. An increase in cytosolic calcium can also occur inside the cell by release from stores. The mitochondria and the endoplasmic reticulum cross talk with each other to modulate the cytosolic calcium concentration. Thus an interplay among membrane components, intracellular organelles, calcium pumps and ion channels exists. Although calcium signaling is complex and incorporates multiple factors, it has been suggested that the ability of the calcium ion to interact with a family of calcium binding proteins (Kd = 10^{-8}–10^{-10} M), known as EF-hand proteins, can play an important role in the transduction of the calcium signal into a biological response (Christakos et al., 1989, 1997; Heizmann and Braun, 1992; Heizmann and Hunziker, 1991; Schafer and Heizmann, 1996; Zimmer et al., 1995). This family of calcium

Sylvia Christakos, Frank Barletta, Michael Huening, Jody Kohut and Mihali Raval-Pandya • Department of Biochemistry and Molecular Biology, UMDNJ-New Jersey Medical School, Newark, NJ, U.S.A.

R. Pochet, R. Donato, J. Haiech, C. Heizmann and V. Gerke (eds.): Calcium: The Molecular Basis of Calcium Action in Biology and Medicine, 259–275.

binding proteins consists of over 200 members and is characterized by the EF-hand structural motif. The EF-hand domain is an octahedral structure consisting of two α helices separated by a 12 amino acid loop that contains side chain oxygens necessary for orienting the divalent calcium ion (Strynadka and James, 1989; Tufty and Kretsinger, 1975; see Figure 1). Calcium binding proteins belonging to this family include calmodulin, parvalbumin, troponin C, calretinin, calcineurin, calpain, Spec I, myosin light chains, the vitamin D dependent calcium binding proteins, calbindin-D_{9k} (localized primarily in mammalian intestine) and calbindin-D_{28k} (localized primarily in avian and mammalian kidney, in avian intestine, in mammalian and avian pancreas, and in molluskan to mammalian brain; Christakos et al., 1989, 1997) and members of the S100 protein family (Schafer and Heizmann, 1996; Zimmer et al., 1995; Donato, 1999). Calmodulin, which has 4 calcium binding domains, is a ubiquitous, highly conserved calcium binding protein (from yeast to mammals) which acts as a calcium mediated activator of a number of different intracellular enzymes including phosphodiesterase, myosin light chain kinase, CaATPase, adenylate cyclase, the serine/threonine phosphatase, calcineurin and phosphorylase kinase (James et al., 1995). Troponin C, found in skeletal muscle, also has 4 calcium binding domains (2 of higher affinity, Kd $\sim$ 10^{-7} M and 2 of lower affinity, Kd $\sim$ 10^{-5} M) and plays a key role in Ca^{2+} mediated regulation of muscle contraction (Malhotra, 1994). Parvalbumin, the S100 proteins and calbindin-D_{9k} have two calcium binding domains. Parvalbumin, present in highest concentrations in fast twitch muscles, has been suggested to be involved in relaxation following muscle contraction and is also present in the central nervous system in a subpopulation of neurons containing the inhibitory neurotransmitter γ-aminobutyric acid (Andressen et al., 1993). S100B is present in glia, acts extracellularly and is known to be involved in neurite outgrowth and astrocyte proliferation (Reeves, 1994; Kligman and Marshak, 1985; Winningham-Major et al., 1989; Selinfreund et al., 1991). Calbindin-D_{28k} and calretinin have 6 helix-loop-helix domains (4 of the 6 domains are functional and bind calcium; see Figure 2 for the structure of calbindin-D_{28k} and Minghetti et al., 1988). Calretinin, unlike calbindin-D_{28k}, is expressed mainly in nerve cells (Rogers, 1987). Although they share 58% amino acid identity, calretinin and calbindin-D_{28k} are expressed largely in separate subpopulations of neurons. Both calbindin-D_{28k} and calretinin have been suggested to act by buffering calcium and, at least in the brain, these proteins have been suggested to have a role in protection from excitotoxic insults (Andressen et al., 1993; Keller, this book). Besides calbindin-D_{28k} and calretinin, additional roles in cell death or protection against cell death have also been proposed for other members of the EF-hand family of proteins including calmodulin, parvalbumin, S100B and calcineurin. Calmodulin antagonists have been shown to protect against glucocorticoid induced cell death in lymphocytes (Dowd et al., 1991). cAMP has been reported to promote cell death in various cells. It has been suggested that calmodulin, an activator of adenylyl cyclase, may have a role in

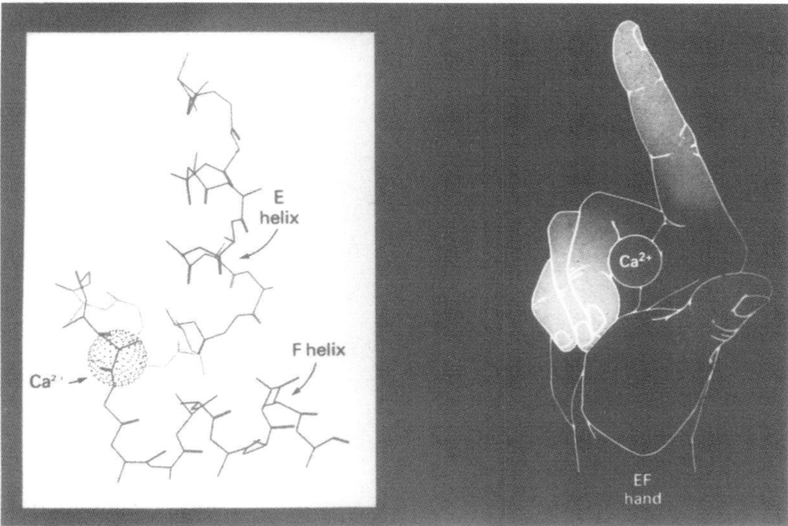

Figure 1. EF-hand structural motif (helix-loop-helix). In the EF-hand configuration two helices are represented by the extended forefinger and the thumb. The clenched middle finger represents the loop that contains the oxygen ligands of the calcium ion. The EF hand is a recurring motif in calbindin and other calcium binding proteins. Reprinted with permission from L. Stryer, 1995, *Biochemistry*, Freeman, San Francisco, p. 1064.

causing an increase in intracellular cAMP, activation of protein kinase A and subsequent cell death (Dowd and Miesfeld, 1992). S100B has been reported to induce apoptosis in PC12 cells (adrenal pheochromocytoma cells which can differentiate to cells with neuronal properties) and to induce cell death in neurons through nitric oxide release from astrocytes (Mariggio et al., 1994; Hu et al., 1997). In addition, high levels of calcineurin have been reported to predispose neuronal cells to degenerate (Asai et al., 1999). Although calmodulin, S100B and calcineurin have been reported to promote cell death, it has been suggested that parvalbumin, similar to calbindin-D_{28k} and calretinin, may be important for cell survival (Heizmann and Braun, 1992).

The role of calmodulin as a calcium mediated enzyme activator and the role of troponin C in muscle contraction have been known for some time. However, the association of a number of the calcium binding proteins with cell death or protection from cell death suggests that a clearer understanding of the interrelationship between calcium binding proteins and programmed cell death should provide new insight into the role of calcium in cell death and the role of programmed cell death in normal development, degeneration and malignant cell growth. Thus, this chapter will focus on the role of calcium in programmed cell death or apoptosis and specifically on the role of the calcium binding protein calbindin-D_{28k} in protection against cell death and the possible mechanisms involved in cytoprotection by calbindin.

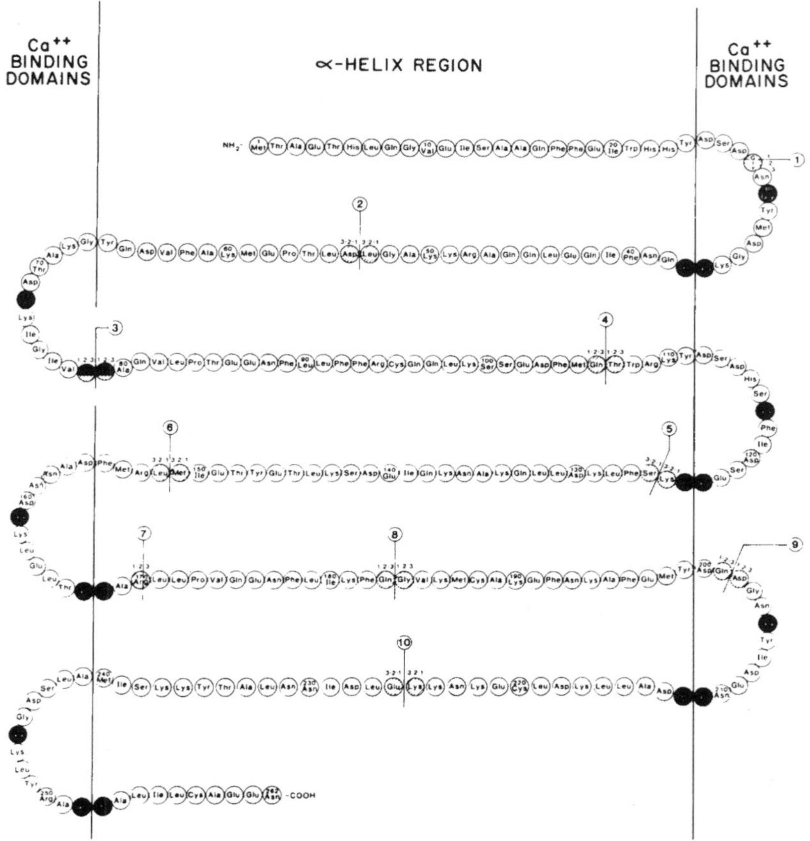

Figure 2. Position of intervening sequences within the structure of chicken calbindin-D_{28K}. Locations of introns are indicted by circled numbers. Numbers above amino acids indicate codon positions. Invariant Glu/leu and Gly amino acids are indicated by black circles. Calcium binding domains are separated from the α-helix region by vertical lines. Reprinted with permission from Minghetti et al., 1988.

2. APOPTOSIS: GENERAL CONSIDERATIONS

Apoptosis is a term meant to define a biological process of programmed cell death. Apoptosis occurs during normal physiology and development of a multicellular eukaryote as a means to maintain tissue or cellular homeostasis. Aberration of apoptosis control leads to pathological conditions. Increased apoptosis has been implicated in a number of human diseases including neurodegenerative disorders such as Alzheimer's disease, Parkinson's disease as well as in myocardial and brain ischemia following stroke. Decreased apoptosis has been implicated in cancer (Li and Yuan, 1999). Apoptosis is characterized by morphological alterations including condensation and fragmentation of nuclear chromatin, plasma and nuclear membrane budding and cytoplasmic

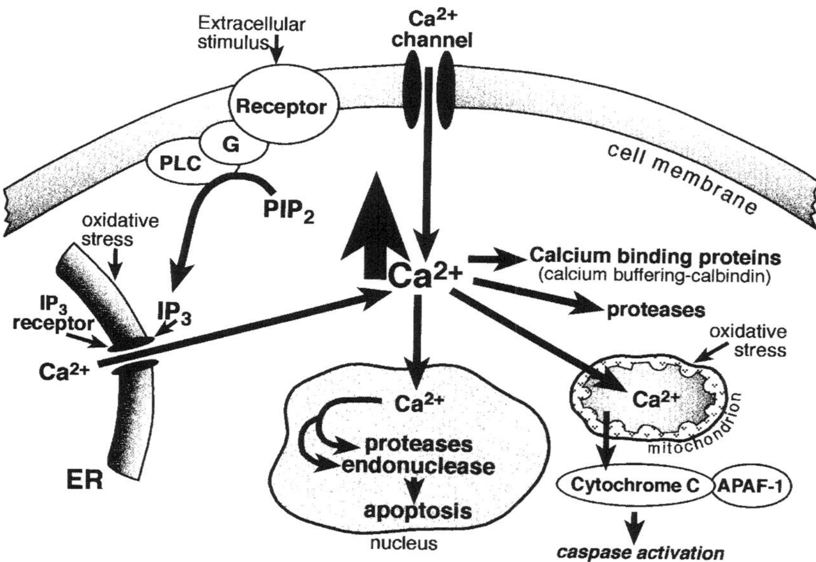

Figure 3. Activation of programmed cell death by calcium. Increased levels of intracellular calcium are derived from sources either outside the cell of from stores within the endoplasmic reticulum (ER). In response to the action of an extracellular stimulus inositol 1,4,5 triphosphate (IP_3) is generated by the action of the enzyme phospholipase C (PLC) on phophatidylinositol 4,5 bisphosphate (PIP_2). IP_3 acts on receptors in the ER which cause release of calcium from ER stores. Calcium can be taken up by the mitochondria. Overloading the mitochondria with calcium results in mitochondrial dysfuntion and release of cytochrome c. Cytochrome c binding to apoptosis protease activating factor 1 (APAF-1) results in activation of caspases and apoptosis. Sustained calcium elevations can also result in apoptosis by activation of calcium dependent proteases and endonuclease. Oxidative stress can disrupt intracellular calcium homeostasis, resulting in adverse effects on ER and mitochondria. Calcium binding proteins can promote cell death (calmodulin and S100β) or they can be important for cell survival by acting as calcium buffers (calbindin-D_{28k}, calretinin, parvalbumin). Recent studies in our lab have suggested mechanisms other than calcium buffering for protection against cell death by calbindin-D_{28k}.

contraction. Inducers of apoptosis include cytokines, ultraviolet irradiation, growth factor deprivation and tumor necrosis factor. Downstream events induce apoptosis and the emerging theme is that a cascade of cysteine proteases known as caspases (15 have currently been identified) is activated which enables the disintegration of cellular morphology and results, in addition to the effect of endonuclease, in the breakdown of nuclear chromatin and DNA (Ashkanezi and Dixit, 1998; Cohen, 1997; Green, 1998).

3. CALCIUM AND APOPTOSIS

Calcium is thought to play a critical regulatory role in apoptosis. Increased levels of intracellular calcium can be derived in part from outside the cell through calcium channels. The drug nifedipine, which blocks L type voltage dependent calcium channels in the plasma membrane, can, in certain cases, protect against apoptosis, suggesting a role for calcium influx in apoptosis (Guo et al., 1997). Increases in intracellular calcium can also be derived from stores within the endoplasmic reticulum (ER). In response to the action of external stimuli (growth factors, hormones, neurotransmitters or other receptor agonists) inositol 1,4,5-triphosphate (IP_3) is produced by the action of phospholipase on phosphatidylinositol 4,5 bisphosphate (PIP_2) at the plasma membrane. IP_3 acts on receptors in the endoplasmic reticulum resulting in the release of calcium from the ER (Figure 3). Also associated with the ER are pumps for calcium uptake (SERCAs) and luminal calcium binding proteins (Meldolesi and Pozzan, 1998). Abnormal ER calcium regulation has also been linked to apoptosis. Increased levels of the IP_3 receptor were reported in apoptotic lymphocytes and IP_3 antisense oligonucleotides were reported to block apoptosis (Khan et al., 1996). An association of ER calcium regulation and apoptosis is also suggested by studies which have shown that thapsigargin, which releases calcium from ER stores, can induce apoptosis and dantrolene, an agent that blocks calcium release from ER stores, prevents apoptosis (Guo et al., 1997). Under conditions of normal calcium signaling, some of the calcium released from the ER is rapidly taken up by the mitochondria and then returned to the endoplasmic reticulum (Berridge et al., 1998). Sustained elevations in intracellular calcium, however, have adverse effects on mitochondria. A reduction in inner mitochondrial membrane potential, thought to be mediated by the opening of the mitochondrial permeability transition pore (PT), has been reported to accompany early apoptosis. It has been suggested that the PT, which involves a multiprotein complex formed in the contact site between the inner and outer mitochondrial membranes, can function as a sensor for damage and stress, resulting in the uncoupling of the respiratory chain (Kroemer et al., 1998). Inhibition of PT by pharmacological intervention prevents apoptosis, further suggesting the involvement of PT in the apoptotic process (Zamzami et al., 1996). As a consequence of mitochondrial dysfunction cytochrome c is released from the mitochondria, and this release has been suggested to be an apoptotic trigger (Li et al.,1997). The release of cyochrome c is not necessarily associated with a reduction in the inner mitochondrial membrane potential (Bossy-Wetzel et al., 1998). Cytochrome c binds to apoptosis protease activating factor 1 (APAF-1) and this has been reported to result in the activation of caspases, the enzymes that play a central role in apoptosis (Li et al., 1997). Thus, cross talk between calcium signaling in the cytosol and the shuttling of calcium between endoplasmic reticulum and mitochondria takes place (Figure 3). An effect of calcium on the mitochrondria results in several changes that initiate apoptotic mechanisms. Although calcium influx and abnormal ER calcium regulation have each been linked to apoptosis, calcium signaling initiated by external stimuli or those that arise due to secondary mechanisms inside the cell, can both converge on the mitochondria. Thus, it has been suggested that the mitochondria can assess the converged calcium signals and either decrease or amplify its potential via the mitochondrial membrane associated components to dictate life or death decisions of the calcium signal (Kroemer et al., 1998).

Calcium sensitive proteases including calpain and a lamin protease, known as nuclear scaffold protease, as well as calcium activated endonuclease (NUC18) have been reported to be targets for sustained calcium elevations in apoptosis (Molinari and Carafoli, 1997; Neamati et al., 1995; Gaido and Cidlowski, 1991). Calpain has been reported to irreversibly degrade critical cytoskeletal and membrane proteins. Calpain substrates include lamin A and B and α spectrin. Calpain inhibitors have been reported to protect against apoptosis (Wang et al., 1996; Ono et al., this book).

In addition to calcium dependent proteases and calcium activated endonuclease, the calcium calmodulin dependent threonine phosphatase, calcineurin has been reported to be involved in apoptosis. High levels of calcineurin have been shown to render cells more susceptible to apoptosis induced by serum reduction or brief exposure to calcium ionophore (Asai et al., 1999), suggesting a possible mechanism for inhibition of apoptosis by calmodulin antagonists. One suggested mechanism for the action of calcineurin, which is still controversial, is activation of nitric oxide synthase by dephosphorylation (nitric oxide has been reported to result in cellular toxicity by reacting with superoxide anion resulting in the formation of the destructive peroxynitrite radical) (Dawson et al., 1993).

Thus although transient increases in intracellular calcium are important for the regulation of normal cellular processes, prolonged elevations in intracellular calcium elicited by a variety of metabolic insults including oxidative stress can result in programmed cell death. Calcium appears to damage the cell in part by activating proteases, endonuclease, the calcium dependent phosphatase, calcineurin, calsenilin and the new PEF protein family (Maki, this book).

4. PROTECTION AGAINST CALCIUM MEDIATED CELL DEATH: ROLE OF THE CALCIUM BINDING PROTEIN CALBINDIN-D_{28k}

Calbindin-D_{28k}, previously known as the vitamin D dependent calcium binding protein (CaBP), became officially known as calbindin-D_{28k} in 1985 (Wasserman, 1985). Initially identified in avian intestine in 1967, calbindin-D_{28k} has since been reported in many other tissues including mammalian and avian kidney, mammalian and avian pancreas and bone and mammalian to molluskan brain (Christakos et al., 1989, 1997). Calbindin-D_{28k} contains 261 amino acid residues (Figure 2), has a molecular weight of approximately 28,000 (based on migration on sodium dodecyl sulfate polyacrylamide gels and 30,000 based on amino acid sequence) and is blocked at the amino terminus (Minghetti et al., 1988; Parmentier et al., 1987). In avian intestine and in avian and mammalian kidney calbindin-D_{28k} is induced by vitamin D and has been suggested to be involved in transcellular calcium transport. In brain calbindin has a widespread but restricted distribution. It is present in most neuronal cell groups and fiber tracts. Neurons containing calbindin-D_{28k} are found in the hippocampus, cerebral cortex, hypothalamus, amygdala, pyriform region and the thalamus. Purkinje cells of the cerebellum stain most intensely for calbindin-D_{28k} (Feldman and Christakos, 1983; Celio, 1990). In the brain calbindin is not regulated by vitamin D (Christakos et al., 1989, 1997; Pasteels et al., 1986).

Studies using neuronal cells in culture (embryonic rat hippocampal cells) showed a direct relationship between calbindin positive neurons and protection against damage induced by glutamate or calcium ionophore (Mattson et al., 1991). These studies were the first to provide evidence for an excitoprotective role for calbindin and suggested that calbindin, by buffering calcium, could prevent calcium mediated neuronal cell death that could result from excitotoxic insults. Since that time studies by us and others have suggested either indirectly by correlative evidence or directly by stable transfection and expression in cells, a role for calbindin in protection against cytotoxicity. Studies of calbindin in relation to several human neurodegenerative conditions have provided indirect correlative evidence for a protective role of calbindin. For example, in Parkinson's disease, a neurodegenerative disease characterized by loss of dopaminergic (DA) neurons, it has been suggested that an increase in oxygen free radicals in the substantia nigra is responsible for initiating the process that leads to the loss of DA neurons and that the final effector of neuronal degeneration is a rise in intracellular calcium concentration possibly resulting from the oxidative stress (Hirsch, 1992). In immunocytochemical studies the selectivity of the DA neuron loss has been found to be inversely related to the presence of calbindin-D_{28k}. For example, the loss of neurons in the substantia nigra is about 80%. Only 3.8% of those cells contain calbindin. In the central gray matter where 92.6% of the cells express calbindin-D_{28k} the loss of cells is a mere 3% (Hirsch, 1992). In Alzheimer's disease, a syndrome characterized neuropathologically by the presence of neurofibrillary tangles, amyloid plaques and the loss of neurons in specific brain regions involved in learning and memory, altered calcium homeostasis has been implicated at least in part for the pathogenesis of the neuronal degeneration (for review see Mattson et al., 1996; Mattson, 1997). Increased levels of calcium in neurofibrillary tangles (Garruto et al., 1984) and increased activation of calcium dependent proteases in vulnerable neurons (Nixon et al., 1994) have been reported. In addition, amyloid β peptide ($A\beta$), the cleavage product of the amyloid precursor protein and the major component of plaques in Alzheimer's disease, has been suggested to damage neurons by a mechanism involving induction of oxidative stress and disruption of calcium homeostasis (Mattson, 1997). Immunocytochemical and in situ hybridization or Northern analysis of postmortem Alzheimer's brain tissue have indicated a loss of calbindin immunoreactive neurons and a decrease in calbindin mRNA in vulnerable neurons (Chan-Palay et al., 1993; Iacopino and Christakos, 1990a; Maguire-Zeiss et al., 1995; Sutherland et al., 1993). In addition, cases with high levels of neurofibrillary tangles have been reported to have less calbindin mRNA than low neurofibrillary tangles cases (Maguire-Zeiss et al., 1995). These findings suggest that calbindin containing neurons are affected in Alzheimer's disease, resulting in perturbed calcium homeostasis and a cascade of events resulting in neuronal cell death.

Besides Parkinson's disease and Alzheimer's disease, immunohistochemical studies have also demonstrated the loss of calbindin from granule cells in the hippocampus from temporal lobe epilepsy patients (Magloczky et al., 1997). Changes in intracellular calcium homeostasis have been reported to be involved in the development of hyperactivity in the epileptic hippocampus (Baimbridge et al., 1985). A decrease in calbindin in the hippocampus has also been reported in animal models of epilepsy (Baimbridge et al., 1985) and a decrease in calbindin gene expression was observed in affected brain regions or associated areas in the genetically epilepsy prone rat (Montpied et al., 1995). Nagerl and Mody (1998), in electrophysiological studies using dentate granule cells from surgical specimens from temporal lobe epilepsy patients, suggested that calbindin may act by binding calcium and restricting free calcium entry.

In addition to neurodegenerative diseases and epilepsy, alterations in calcium homeostasis have also been reported in ischemia. Ischemic death of vulnerable neurons has been reported to be preceded by a pathological accumulation of intracellular calcium, and differences in calbindin-D_{28k} levels in CA1 and CA2 pyramidal cells of the hippocampus have been reported to be inversely related to degeneration after ischemia (Rami et al., 1992). The dentate granule cells of the hippocampus, which possess calbindin, are the least degenerated cells in the hippocampus following induced ischemia (Freund et al., 1990; Goodman et al., 1993). However, it should be noted with regard to ischemia that the presence and absence of calbindin does not always correlate with protection and degeneration respectively (Freund et al., 1990).

Direct evidence of a protective role of calbindin, providing more conclusive evidence than correlative studies, has been demonstrated in studies using cells lines as well primary neuronal cultures in which the calbindin gene has been transfected resulting in the overexpression of calbindin. Overexpression of calbindin has been reported to increase cell survival in response to a variety of insults which involve calcium dependent events including exposure to calcium ionophore, cAMP, glucocorticoid, IgG from amyotrophic lateral sclerosis patients, hypoglycemia and Aβ (Dowd et al., 1992; Guo et al., 1998; Ho et al., 1996; Meier et al., 1997). Recent studies by Phillips et al. (1999) indicated that infection of hippocampal cells with calbindin-D_{28k} herpes simplex viral vector increased survival of hippocampal cells treated with antimetabolite 3-acetylpyridine (3AP) (which damages these neurons by uncoupling mitochondrial electron transport) and of hippocampal cells following treatment with kainic acid which results in neurotoxicity. This study is important since it suggests that calbindin can protect against toxins that directly alter calcium homeostasis (kainic acid results in a massive influx of extracellular calcium by activation of kainic acid preferring glutamate receptors) as well as against trauma that indirectly alters intracellular calcium following a cascade of events (uncoupling of electron transport results in a decrease in available energy and an increase in cytosolic calcium due to an imbalance of metabolic costs of calcium sequestration and extrusion).

Studies in PC12 cells (Guo et al., 1998), glial cells (Wernyj et al., 1999), lymphocytes (Dowd et al., 1992) and motor neuron hybrid cells (Ho et al., 1996) indicate that the type of cell death protected by calbindin is apoptotic. Recent findings using PC12 cells suggest a mechanism involved in the antiapoptotic action of calbindin (Guo et al., 1998). Presenilin-1 (PS-1) is an integral membrane protein localized in the ER and expressed in neurons throughout the brain. Mutations in PS-1 are causally linked to approximately 50% of the cases of early-onset familial Alzheimer's disease (Nishimura et al., 1999). Expression of mutant PS-1 in PC-12 cells (by transfection of the gene and subsequent expression of the protein) was found to sensitize these cells to apoptosis induced by Aβ and to potentiate elevations in the intracellular calcium concentration as well as the generation of reactive oxygen species induced by Aβ. Impairment of mitochondrial function by Aβ was also exacerbated by the expression of mutant PS-1. Overexpression of calbindin suppressed the proapoptotic action of mutant PS-1, attenuated the increase in intracellular calcium (Figure 4) and largely prevented the impairment of mitochondrial function. These findings suggest (similar to the findings of Phillips et al. using 3AP) an important role for calbindin in preventing apoptotic cell death by stabilizing intracellular calcium levels and preventing apoptotic mitochondrial alterations. Thus, although it remains to be determined, one mechanism whereby calbindin may protect against apoptotic cell death would be by preventing calcium mediated mitochondrial damage and the subequent release of cytochrome c, which has been suggested to be an apoptotic trigger (Li et al., 1997).

Further evidence for a neuroprotective role for calbindin comes from results showing that calbindin can be induced by brain injury (Mattson et al., 1995) and by increased levels of glucocorticoids (Iacopino and Christakos, 1990b). High levels of corticosterone have been reported to result in an increase in intracellular calcium in hippocampal neurons and to cause cell death mainly in areas where calbindin is not localized (Elliott and Sapolsky, 1992, 1993). The induction of calbindin may be a compensatory mechanism to promote neuronal survival. In addition, studies have indicated that neurotrophin 3, brain derived neurotropic factor and fibroblast growth factor can induce calbindin in rat hippocampal cultures (Collazo et al., 1992; Ip et al., 1993). These neurotrophic factors have been reported to protect neurons against neuronal toxicity. Thus the induction of calbindin by these factors further suggests a role for calbindin in the process of protection against cytotoxicity.

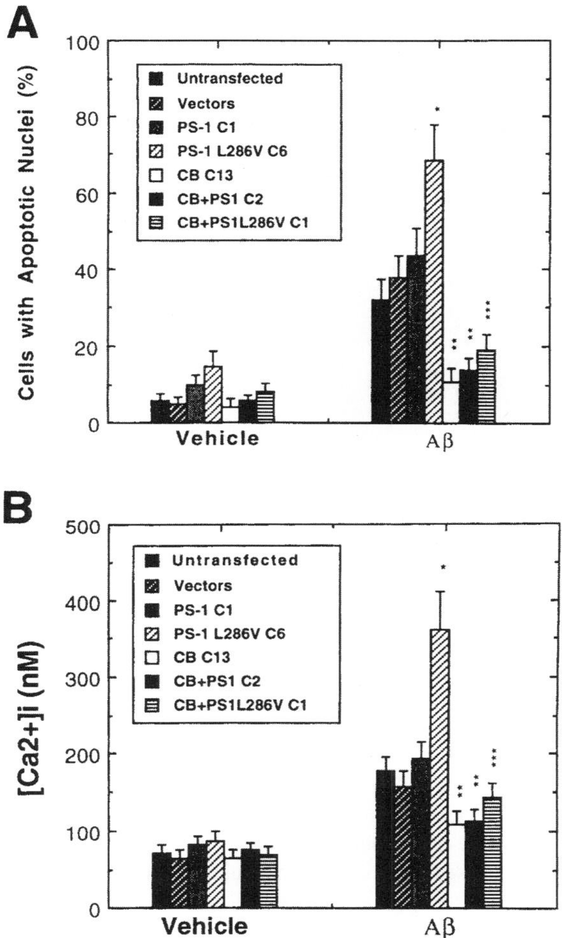

Figure 4. Calbindin-D_{28k} attenuates apoptosis (A) and increases in intracellular calcium (B) induced by Aβ and enhanced in PC12 cells expressing mutant PS-1. Vectors, PC12 cells transfected with empty vectors; PS-1, PC12 cells transfected with presenilin 1; PS-1 L286V, PC12 cells transfected with mutant presenilin-1; CB, PC12 cells transfected with calbindin-D_{28k}. (C1, C2, C6 and C13 refer to clone numbers.) Adapted from Guo et al. (1998) with permission.

It should be noted that calbindin protects against apoptotic cell death in cells in addition to lymphocytes and cells of the central nervous system. Recent studies have indicated that cytokine mediated destruction of pancreatic β cells, a cause of insulin dependent diabetes, can be inhibited by calbindin-D_{28k}. Studies were done transfecting the pancreatic β cell line βTC-3 with the calbindin-D_{28k} gene. In the calbindin transfected cells stimulation of free radical formation by cytokines was inhibited (Suarez-Pinzon et al., 1999). These findings suggest that calbindin is an important regulator of apoptosis which can protect β cells from autoimmune destruction in type 1 diabetes.

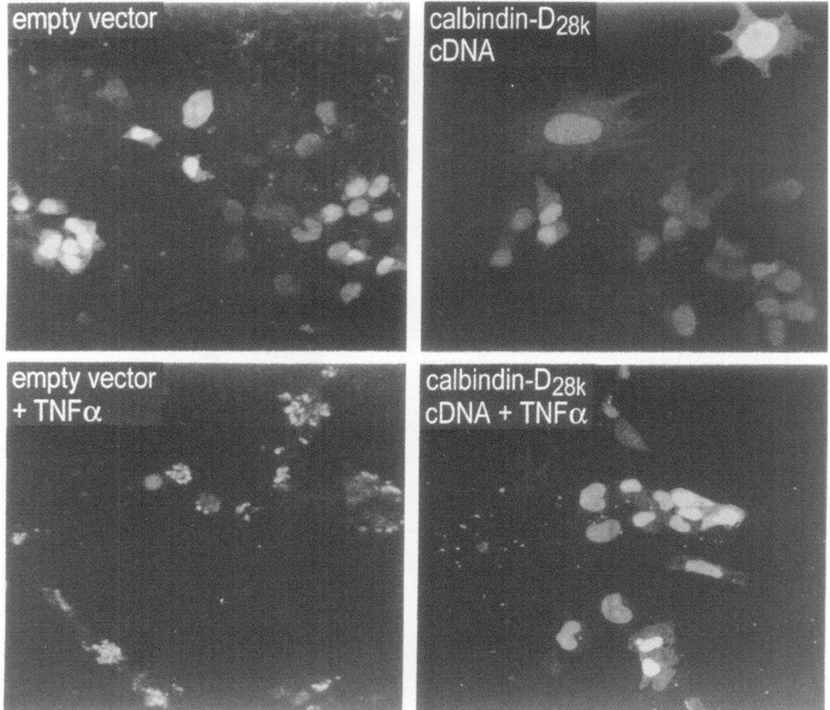

Figure 5. Overexpression of calbindin-D_{28k} suppresses nuclear fragmentation of osteoblastic cells induced by TNFα. Cells were transfected with the expression vector rep4 alone (empty vector) or containing the cDNA for calbindin-D_{28k} (calbindin-D_{28k}) together with an expression vector containing the coding sequence of green fluorescent protein with a nuclear localization sequence. Forty-eight hours after transfection, cells were exposed to 1 nM TNFα for 16 hours. Cells were fixed, mounted and examined with a Zeiss confocal laser scanning microscope. Note the presence of apoptotic nuclei in TNFα treated vector transfected cells but not in the calbindin transfected cells similarly treated. Quantitative analysis indicated that TNFα increased the percentage of apoptotic cells from 10.8 to 78.1% in vector transfected cells and from 0 to 20.1% in calbindin-D_{28k} transfected cells. Similar results were observed in an additional experiment.

In addition, in recent studies we found that calbindin can also protect against apoptosis of osteoblast cells which are involved in bone formation. The rate of osteoblast apoptosis influences the rate of bone formation. Tumor necrosis factor (TNF) has been reported to have proapoptotic effects in osteoblasts (Jilka et al., 1998). Stable transfection of calbindin in osteoblastic MC3T3-E1 cells was found to block TNF induced apoptosis as determined by cell viability and nuclear morphology of cells cotransfected with the green fluorescent protein targeted to the nucleus (Figure 5). Recently, work in our laboratory has suggested a novel mechanism by which calbindin prevents apoptosis in bone cells. As indicated previously, a number of different types of proteases are involved in effecting the morphological changes of apoptosis. One major protein family important in apoptosis is the interleukin-1β-converting enzyme(ICE)-like protease family, also known as the caspase family. Some members of this family (caspases 2, 8, 9 and 10) function as initiator caspases that can self-activate by autocatalysis, while others (caspases 3, 4, 5, 6, 7, 11, 12 and 13) act as effector caspases that cleave an ever increasing list of cellular target proteins (Chan and Mattson, 1999). Caspase-3 is a key mediator of apoptosis and is a common downstream effector of multiple apoptotic signaling pathways, including cytochrome c-mediated activation of caspase 9 (Chan and Mattson, 1999). Our findings suggest that calbindin is capable of directly inhibiting the activity of caspase-3 *in vitro*, and that this inhibition results in an inhibition of TNFα-induced apoptosis in MC3T3-E1 osteoblastic cells in culture (Bellido et al., 1998). These data suggest that while at least part of calbindin's protective effect may result from the buffering of rises in intracellular calcium, other mechanisms of action, such as inhibition of caspase activity, may also have a significant role to play in the prevention of apoptosis by calbindin-D_{28k}.

In summary, results support the direct involvement of calbindin in protecting against cell death not only in apoptosis susceptible cells of the central nervous system and in lymphocytes but also in pancreatic β cells and in osteoblasts involved in bone formation. A further understanding of the mechanisms whereby calbindin-D_{28k} protects against apoptotic cell death will have important therapeutic implications for the prevention of cellular degeneration.

REFERENCES

Andressen, C., Blumcke, I. and Celio, M.R., 1993, Calcium binding proteins: Selective markers of nerve cells, *Cell Tissue Res.* 271, 181–208.

Asai, A., Qiu, J.-H., Narita, Y., Chi, S., Saito, N., Shinoura, N., Hamada, H., Kuchino, Y. and Kirino, T., 1999, High level calcineurin activity predisposes neuronal cells to apoptosis, *J. Biol. Chem.* 274, 34450–34458.

Ashkanezi, A. and Dixit, V.M., 1998, Death receptors: Signaling and modulation, *Science* 281, 1305–1308.

Baimbridge, K.G., Mody, I. and Miller, J.J., 1985, Reduction of rat hippocampal calcium binding protein following commissural, amygdala, septal, perforant path and olfactory bulb kindling, *Epilepsia* 26, 460–465.

Bellido, T., Han, L., Huening, M., Barger, S.W., Manolagas, S.C. and Christakos, S., 1998, Calbindin-D_{28k} is expressed in osteoblastic cells and suppresses their apoptosis by inhibiting caspase 3 activity, *J. Bone Min. Res.* 23 (Suppl. 1), S177.

Berridge, M.J., 1997, The AM and FM of calcium signaling, *Nature* 386, 759–760.

Berridge, M.J., Bootman, M.D. and Lipp, P., 1998, Calcium – A life and death signal, *Nature* 395, 645–648.

Bossy-Wetzel, E., Newmeyer, D.D. and Green, D.R., 1998, Mitochondrial cytochrome c release in apoptois occurs upstream of DEVD-specific caspase activation and independently of mitochondrial transmembrane depolarization, *Embo J.* 17, 37–49.

Celio, M.R., 1990, Calbindin and parvalbumin in the rat nervous system, *Neuroscience* 35, 375–475.

Chan, S.L. and Mattson, M.P., 1999, Caspase and calpain substrates: Roles in synaptic plasticity and cell death, *J. Neurosci. Res.* 58, 167–190.

Chan-Palay, V., Hochli, M., Savaskan, E. and Hungerecker, G., 1993, Calbindin-D_{28k} and monoamine oxidase A immunoreactive neurons in the nucleus basalis of Meynert in senile dementia of the Alzheimer type and Parkinson's disease, *Dementia* 4, 1–15.

Christakos, S., Gabrielides, C. and Rhoten W.B., 1989, Vitamin D-dependent calcium binding proteins: Chemistry, distribution, functional considerations and molecular biology, *Endo. Rev.* 10, 84–107.

Christakos, S., Beck, J.D. and Hyllner, S.J., 1997, Calbindin-D_{28k}, in *Vitamin D*, D. Feldman, F. Glorieux and J.W. Pike (eds.), Academic Press, San Diego, CA, pp. 209–221.

Cohen, G.M., 1997, Caspases: The executioners of apoptosis, *Biochem. J.* 326, 1–16.

Collazo, D., Takahashi, H. and McKay, R.D.G., 1992, Cellular targets and trophic functions of neurotrophin-3 in the developing rat hippocampus, *Neuron* 9, 643–656.

Dawson, T.M., Steiner, J.P., Dawson, V.L., Dinerman, J.L., Uhl, G.R. and Snyder, S.H., 1993, Immunosuppressent FK 506 enhances phosphorylation of nitric oxide synthase and protects against glutamate neurotoxicity, *Proc. Natl. Acad. Sci. USA* 90, 9808–9812.

Donato, R., 1999, Functional roles of S100 proteins, calcium binding proteins of the EF-hand type, *Biochim. Biophys. Acta* 1450, 191–231.

Dowd, D.R. and Miesfeld, R.L., 1992, Cyclic AMP-induced apoptotic pathways in lymphocytes share distal events, *Mol. Cell. Biol.* 12, 3600–3608.

Dowd, D.R., MacDonald, P.N., Komm, B.S., Haussler, M.R. and Miesfeld, R.L., 1991, Evidence for early induction of calmodulin gene expression in lymphocytes undergoing glucocorticoid-mediated apoptosis, *J. Biol. Chem.* 266, 18423–18426.

Dowd, D.R., MacDonald, P.N., Komm, B.S., Haussler, M.R. and Miesfeld, R.L., 1992, Stable expression of calbindin-D_{28k} complementary DNA interferes with the apoptotic pathway in lymphocytes, *Mol. Endo.* 6, 1843–1848.

Elliott, E.M. and Sapolsky, R.M., 1992, Corticosterone enhances kainic acid-induced calcium elevation in cultured hippocampal neurons, *J. Neurochem.* 59, 1033–1040.

Elliott, E.M. and Sapolsky, R.M., 1993, Corticosterone impairs hippocampal neuronal calcium regulation – Possible mediating mechanisms, *Brain Res.* 602, 84–90.

Feldman, S.C. and Christakos, S., 1983, Vitamin D-dependent calcium binding protein in rat brain: Biochemical and immunocytochemical characterization, *Endocrinology* 112, 290–302.

Freund, T.F., Buzsaki, G., Leon, A., Baimbridge, K.G. and Somogyi, P., 1990, Relationship of neuronal vulnerability and calcium binding protein immunoreactivity in ischemia, *Exp. Brain Res.* 83, 55–66.

Gaido, M.L. and Cidlowski, J.A., 1991, Identification, purification and characterization of a calcium-dependent endonuclease (NUC 18) from apoptotic rat thymocytes, *J. Biol. Chem.* 266, 18580–18585.

Garruto, R.M., Fukatsu, R., Yanagihara, R., Gajdusek, D.C., Hook, G. and Fiori, C.E., 1984, Imaging of calcium and aluminum in neurofibrillary tangle-bearing neurons in parkinsonism-dementia of Guam, *Proc. Natl. Acad. Sci. USA* 81, 1875–1879.

Goodman, J.H., Wasterlain, C.G., Massarweh, W.F., Dean, E., Sollas, A.L. and Sloviter, R.S., 1993, Calbindin-D_{28k} immunoreactivity and selective vulnerability to ischemia in dentate gyrus of developing rat, *Brain Res.* 606, 309–314.

Green, D.R., 1998, Apoptotic pathways: The roads to ruin, *Cell* 94, 695–698.

Guo, Q., Sopher, B.L., Pham, D.G., Furukawa, K., Robinson, N., Martin, G.M. and Mattson, M.P., 1997, Alzheimer's presenilin mutation sensitizes neural cells to apoptosis induced by trophic factor withdrawal and amyloid β-peptide: Involvement of calcium and oxyradicals, *J. Neurosci.* 17, 4212–4222.

Guo, Q., Christakos, S., Robinson, N. and Mattson, M.P., 1998, Calbindin-D_{28k} blocks the proapoptotic actions of mutant presenilin: Reduced oxidative stress and preserved mitochondrial function, *Proc. Natl. Acad. Sci. USA* 95, 3227–3232.

Heizmann, C.W. and Braun, K., 1992, Changes in Ca^{2+}-binding proteins in human neurodegenerative disorders, *Trends Neurosci. Res.* 15, 259–264.

Heizmann, C.W. and Hunziker, W., 1991, Intracellular calcium binding proteins: More sites than insights, *Trends Biochem. Sci.* 16, 98–103.

Hirsch, E.C., 1992, Why are nigral catecholaminergic neurons more vulnerable than other cells in Parkinson's disease?, *Ann. Neurol.* 32, S88–S93.

Ho, B.-K., Alexianu, M.E., Colom, L.V., Mohamed, A.H., Serrano, F. and Appel, S.H., 1996, Expression of calbindin-D_{28k} cDNA prevents amyotropohic lateral sclerosis IgG-mediated cytotoxicity, *Proc. Natl. Acad. Sci. USA* 93, 6796–6801.

Hu, J., Ferreira, A. and Van Eldik, L.J., 1997, S100 induces neuronal cell death through nitric oxide release from astrocytes, *J. Neurochem.* 69, 2294–2300.

Iacopino, A.M. and Christakos, S., 1990a, Specific reduction of neuronal calcium binding protein (calbindin-D_{28k}) gene expression in aging and neurodegenerative diseases, *Proc. Natl. Acad. Sci. USA* 87, 4078–4082.

Iacopino, A.M. and Christakos, S., 1990b, Corticosterone regulates calbindin-D_{28k} mRNA and protein levels in rat hippocampus, *J. Biol. Chem.* 265, 10177–10180.

Ip, N.Y., Li, Y., Yancopoulos, G.D. and Lindsay, R.M., 1993, Cultured hippocampal neurons show responses to BDNF, NT-3 and NT-4 but not NGF, *J. Neurosci.* 13, 3394–3405.

James, P., Vorherr, T. and Carafoli, E., 1995, Calmodulin-binding domains: Just two faced or multi-faceted?, *Trends Biochem. Sci.* 20, 38–42.

Jilka, R.L., Weinstein, R.S., Bellido, T., Parfitt, A.M. and Manolagas, S.C., 1998, Osteoblast programmed cell death (apoptosis): Modulation by growth factors and cytokines, *J. Bone Miner. Res.* 13, 793–802.

Keller, B.U., The role of intracellular calcium signaling in motoneuron function and disease, this book.

Khan, A.A., Soloski, M.J., Sharp, A.H., Schilling, G., Sabatini, D.M., Li, S.H., Ross, C.A. and Snyder, S.H., 1996, Lymphocyte apoptosis: Medition by increased type 3 inositol 1,4,5-triphosphate receptor, *Science* 273, 503–507.

Kligman, D. and Marshak, D.R., 1985, Purification and characterization of a neurite extension factor from bovine brain, *Proc. Natl. Acad. Sci. USA* 82, 7136–7139.

Kroemer, G., Dallaporta, B. and Resche-Rigon, M., 1998, The mitochondrial death/life regulator in apoptosis and necrosis, *Annu. Rev. Physiol.* 60, 619–642.

Li, H. and Yuan, J., 1999, Decifering the pathways of life and death. *Curr. Opin. Cell Biol.* 11, 261–266.

Li, P., Nijhawan, D., Budihardjo, I., Srinivasula, S.M., Ahmad, M., Alnemri, E.S. and Wang, X., 1997, Cytochrome c and dATP-dependent formation of Apaf-1/caspase-9 complex initiates an apoptosis protease cascade, *Cell* 91, 479–489.

Magloczky, Z.S., Halasz, P., Vajda, J., Czirjak, S. and Freund, T.F., 1997, Loss of calbindin-D_{28k} immunoreactivity from dentate granule cells in human temporal lobe epilepsy, *Neuroscience* 76, 377–385.

Maguire-Zeiss, K.A., Li, Z.W., Shimoda, L.M.N. and Hamill, R.W., 1995, Calbindin D_{28k} mRNA in hippocampus, superior temoral gyrus and cerebellum: Comparison between control and Alzheimer's disease subjects, *Mol. Brain Res.* 30, 362–366.

Maki, M., Penta-EF-hand (PEF) proteins and calsenilin/DREAM: Involvement of the new EF-hand calcium-binding proteins in apoptosis and signal transduction, this book.

Malhotra, A., 1994, Role of regulatory proteins (troponin-tropomyosin) in pathological states, *Mol. Cell. Biochem.* 135, 43–50.

Mariggio, M.A., Fulle, S., Calissano, P., Nicoletti, I. and Fano, G., 1994, The brain protein S100ab inuces apoptosis in PC12 cells, *Neurosci.* 60, 29–35.

Mattson, M.P., 1997, Cellular actions of β-amyloid precursor protein and its soluble and fibrillogenic peptide derivatives, *Physiol. Rev.* 77, 1081–1132.

Mattson, M.P., Rychlik, B., Chu, C. and Christakos S., 1991, Evidence for calcium reducing and excitoprotective roles for the calcium binding protein calbindin-D_{28k} in hippocampal neurons, *Neuron* 6, 41–51.

Mattson, M.P., Chang, B., Baldwin, S., Smith-Swintosky, V.L., Keller, J., Geddes, J.V., Scheff, S.W. and Christakos, S., 1995, Brain injury and tumor necrosis factors induce calbindin-D_{28k} in astrocytes: Evidence for a cytoprotective response, *J. Neurosci. Res.* 42, 357–370.

Mattson, M.P., Furukawa, K., Bruce, A.J., Mark, R.J. and Blanc, E.M., 1996, Calcium homeostasis and free radical metabolism as convergence points in the pathophysiology of dementia, in *Molecular Mechanisms of Dementia*, W. Wasco and R. E. Tanzi (eds.), Humana Press, Totowa, NJ, pp. 103–143.

Meier, T.J., Ho, D.Y. and Sapolsky, R.M., 1997, Increased expression of calbindin-D_{28k} via hepes simplex virus amplicon vector decreases calcium ion mobilization and enhances neuronal survival after hypoglycemic challenge, . *J. Neurochem.* 69, 1039–1047.

Meldolesi, J. and Pozzan, T., 1998, The endoplasmic reticulum Ca^{2+} store: A view from the lumen, *Trends Biol. Sci.* 23, 10–14.

Minghetti, P.P., Cancela, L., Fujisawa, Y., Theofan, G. and Norman, A.W., 1988, Molecular structure of the chicken vitamin D-induced calbindin-D_{28k} gene reveals eleven exons, six Ca^{2+}-binding domains and numerous promoter regulatory elements, *Mol. Endocrinol.* 2, 355–367.

Molinari, M. and Carafoli, E., 1997, Calpain: A cytosolic proteinase active at the membrane, *J. Membr. Biol.* 156, 1–8.

Montpied, P., Winsky, L., Dailey, J.W., Jobe, P.C. and Jacobowitz, D.M., 1995, Alteration in levels of expression of brain calbindin-D_{28k} and calretinin mRNA in genetically epilepsy-prone rates, *Epilepsia* 36, 911–921.

Nagerl, U.V. and Mody, I., 1998, Calcium-dependent inactivation of high-threshold calcium currents in human dentate gyrus granule cells, *J. Physiol.* 509(1), 39–45.

Neamati, N., Fernandez, A., Wright, S., Kiefer, J. and McConkey, D.J., 1995, Degradation of lamin B1 precedes oligonucleosomal DNA in apoptotic thymocytes and isolated thymocyte nuclei, *J. Immunol.* 154, 3788–3795.

Nishimura, M., Yu, G. and St George-Hyslop, P.H., 1999, Biology of presenilins as causitive molecules for Alzheimer disease, *Clin. Genet.* 55, 219–225.

Nixon, R.A., Saito, K.I., Grynspan, F., Griffin, W.R., Katayama, S., Honda, T., Mohan, P.S., Shea, T.B. and Beermann, M., 1994, Calcium-activated neutral proteinase (calpain) system in aging and Alzheimer's disease, *Ann. NY Acad. Sci.* 747, 77–91.

Ono, Y., Hata, S., Sorimachi, H. and Suzuki, K., Calcium and muscle disease: Pathophysiology of calpains and limb-girdle muscular distrophy Type 2a (LGMD2A), this book.

Parmentier, M., Lawson, D.E. and Vassart, G., 1987, Human 27-kDa calbindin complementary DNA sequence. Evolutionary and functional implications, *Eur. J. Biochem.* 170, 207–215.

Pasteels, J.L., Pochet, R., Surardt, L., Hubeau, C., Chirnoaga, M., Parmentier, M. and Lawson, D.E., 1986, Ultrastructural localization of brain 'vitamin D-dependent' calcium binding proteins, *Brain Res.* 384, 294–303.

Phillips, R.C., Meier, T.J., Giuli, L.C., McLaughlin, J.R., Ho, D.Y. and Sapolsky, R.M., 1999, Calbindin D_{28k} gene transfer via herpes simplex virus amplicon vector decreases hippocampal damage *in vivo* following neurotoxic insults, *J. Neurochem.* 73, 1200–1205.

Rami, A., Rabie, A., Thomasset, M. and Krieglstein, J., 1992, Calbindin-D_{28k} and ischemic damage of pyramidal cells in rat hippocampus, *J. Neurosci. Res.* 31, 89–95.

Reeves, R.H., 1994, Astrocytosis and axonal proliferation in the hippocampus of S100b transgenic mice, *Proc. Natl. Acad. Sci. USA* 91, 5359–5363.

Rogers, J.H., 1987, Calretinin: A gene for a novel calcium-binding protein expressed principally in neurons, *J. Cell Biol.* 105, 1343–1353.

Schafer, B.W. and Heizmann, C.W., 1996, The S100 family of EF-hand calcium binding proteins: Functions and pathology, *Trends Biochem. Sci.* 21, 134–140.

Selinfreund, R.H., Barger, S.W., Pledger, W.J. and Van Eldik, L.J., 1991, Neurotrophic protein S100 stimulates glial cell proliferation, *Proc. Natl. Acad. Sci. USA* 88, 3554–3558.

Strynadka, N.C.J. and James, M.N.G., 1989, Crystal structures of the helix-loop-helix calcium binding proteins, *Annu. Rev. Biochem.* 58, 951–998.

Suarez-Pinzon, W., Rabinovitch, A., Strynadka, K., Sooy, K. and Christakos, S., 1999, Cytokine mediated apoptotic destruction of pancreatic β cells, a cause of insulin dependent diabetes, is inhibited by calbindin-D_{28k}, *J. Bone Miner. Res.* 14 (Suppl. 1), S327.

Sutherland, M.K., Wong, L., Somerville, M.J., Yoong, L.K.K., Bergeron, C., Parmentier, M. and McLachlan, D.R., 1993, Reduction of calbindin-28k mRNA levels in Alzheimer as compared to Huntington hippocampus, *Mol. Brain Res.* 18, 32–42.

Tufty, R.M. and Kretsinger, R.H., 1975, Troponin and parvalbumin calcium binding regions predicted in myosin light chain and T4 lysozyme, *Science* 187, 167–169.

Wang, K.K., Nath, R. and Posner, A., 1996, An alpha-mercapto acrylic acid derivative is a selective nonpeptide cell-permeable calpain inhibitor and is neuroprotective, *Proc. Natl. Acad. Sci. USA* 93, 6687–6692.

Wasserman, R.H., 1985, Nomenclature of the vitamin D induced calcium binding proteins, in *Vitamin D: Chemical, Biochemical and Clinical Update*, A.W. Norman, K. Schaefer, H.G. Grigoleit and D.V. Herrath (eds.), de Gruyter, Berlin, pp. 321–322.

Wernyj, R.P., Mattson, M.P. and Christakos, S., 1999, Expression of calbindin-D_{28k} in C6 glial cells stabilizes intracellular calcium levels and protects against apoptosis induced by calcium ionophore and amyloid β-peptide, *Mol. Brain Res.* 64, 69–79.

Winningham-Major, F., Staecker, J.L., Barger, S.W., Coats, S. and Van Eldik, L.J., 1989, Neurite extension and neuronal survival activities of recombinant S100 proteins that differ in the content and position of cysteine residues, *J. Cell. Biol.* 109, 3036–3071.

Zamzami, N., Marchetti, P., Castedo, M., Hirsch, T., Susin, S.A., Mose, B. and Kroemer, G., 1996, Inhibitors of permeability transition interfere with the disruption of the mitochondrial transmembrane potential during apoptosis, *FEBS Lett.* 384, 53–57.

Zimmer, D.B., Cornwall, E.H., Landar, A. and Song, W., 1995, S100 protein family: Function and expression, *Brain Res. Bull.* 37, 417–429.

Calcium and Cellular Ageing

Alexej Verkhratsky and Emil Toescu

1. INTRODUCTION

Calcium, an ubiquitous cytoplasmic second messenger, controls numerous physiological events and is also involved in cellular pathology. Either excess or deficit of cytoplasmic calcium could initiate cellular malfunction and result in cellular death. As will be highlighted in this volume, intracellular calcium is regulated by co-ordinated activity of several molecular cascades, represented by calcium transporters (i.e. ion channels, pumps and exchangers) and calcium binding proteins. Any unbalance of this delicate ensemble of calcium handling proteins may result in fatal consequences for living tissues.

Dysregulation of calcium homeostasis is important for acute pathological processes such as calcium mediated excitotoxicity (Choi, 1994; Zipfel et al., 1999) and for chronic cellular alterations. In particular calcium mishandling may (at least partially) be responsible for a progressive decline in cellular function associated with the ageing process (Verkhratsky and Toescu, 1998a; Thibault et al., 1998). Many theories have been proposed which try to determine the general mechanism of ageing. However, none of them has yet been able to explain the multitude of changes occurring in senescence. Particularly mysterious, remains the process of brain ageing. It is no secret that years take their inevitable toll from all systems and tissues, making them generally weaker. Brain is no exception, and ageing is usually associated with decline in mental capabilities, though the extent of this decline varies enormously. Decline of human mental capabilities in senescence presents one of the most exciting challenges to modern gerontology. Indeed, the increase of longevity faced by the developed world drastically increases the number of old people and the social pressure of an ageing population. In the present chapter we shall view current knowledge about the relations between cellular calcium homeostasis and the process of ageing.

Alexej Verkhratsky • Manchester University, School of Biological Sciences, Manchester, U.K. **Emil Toescu** • Department of Physiology, Birmingham University, Birmingham, U.K.

R. Pochet, R. Donato, J. Haiech, C. Heizmann and V. Gerke (eds.): Calcium: The Molecular Basis of Calcium Action in Biology and Medicine, 277–286.

2. CALCIUM AND CELLULAR AGEING

The first notion of the possible link between the process of ageing and calcium metabolism was made at the beginning of this century (Novi, 1912). Since then this issue has been examined rather intensively (Gibson and Peterson, 1987; Thibault et al., 1998; Verkhratsky and Toescu, 1998a) resulting in the formation of a general view of calcium dysregulation as one of the possible mechanisms of age-dependent decline of cellular function. In fact these views are most coherently presented by the so-called "calcium hypothesis of neuronal ageing" (Khachaturian, 1991; Khachaturian et al., 1989; Verkhratsky and Toescu, 1998a), which link the age-associated mental decline with disturbances in calcium homeostasis. The initial hypothesis postulated that the age-dependent decline of Ca^{2+} homeostatic systems causes chronic $[Ca^{2+}]_i$ elevation which eventually results in cellular death. The main support for this hypothesis was provided by extensive investigation into Alzheimer's disease (AD) which increases nearly exponentially with age. The leading mechanism of AD-associated dementia is considered to be profound loss of neurones. As has been demonstrated by several groups (Mattson et al., 1993; Zipfel and Choi, 1999), AD coincides with severe disturbances in cellular calcium transport and overloading of cells with Ca^{2+}, which may account for cellular death. Recent morphological studies, however, have demonstrated that decrease in neuronal number does not accompany normal ageing of the brain: only when the brain is affected by AD does the neuronal loss become detectable (Gomez-Isla et al., 1996; West et al., 1994). Thus, ageing of the central nervous system may take two distinct routes – either pathological (AD and other neurodegenerative processes) or physiological, which is not associated with an actual decrease in the number of cells, though still producing decline in mental capabilities. Is calcium dysregulation involved in both types of ageing? This important question challenges the cellular physiology of ageing.

3. CALCIUM HOMEOSTATIC SYSTEMS IN AGEING

Investigations of age-dependent changes in calcium homeostatic systems present a substantial experimental problem. Progress in experimental techniques has made possible direct measurement of $[Ca^{2+}]_i$ in cells and subcellular compartments (see methodological chapters in this book); electrophysiological and biochemical techniques also provide a sensitive means for monitoring transmembrane ion fluxes and ion binding to various molecules. Nevertheless, the old tissues present a special challenge for cell isolation/culturing, and experimental investigation of tissue fragments (e.g. brain slices) is technically very demanding and slows down accumulation of new data. This has to be kept in mind when discussing the experimental data on age-dependent changes in cellular calcium regulation.

3.1. Calcium uptake

The studies of calcium uptake were probably the first to directly address the question of calcium handling in aged preparations. Using the $^{45}Ca^{2+}$ technique it was demonstrated that the uptake of Ca^{2+} into rat brain synaptosomes, different brain regions and cultured fibroblasts declined with age (Gibson et al., 1986; Peterson et al., 1985; Peterson and Gibson, 1983). These initial data were subsequently confirmed by several groups which essentially found the same age-dependent decline in Ca^{2+} uptake in a wide variety of preparations (Leslie et al., 1985; Satrustegui et al., 1996).

3.2. Calcium buffering

All cells contain in their cytosol a variety of Ca^{2+} binding proteins (CaBPs), with affinities ranging from hundreds of micromolar (e.g., calpain) to submicromolar (e.g., parvalbumin) (for more details about the calcium buffering capacity, see Keller in this book). The major part of Ca^{2+} ions entering the cell are almost instantly buffered by these cytoplasmic calcium-binding sites. The expression of various intracellular Ca^{2+} binding proteins is affected by ageing. The level of expression of major cytosolic Ca^{2+} buffers, calbindin and calretinin, was substantially decreased in the hippocampus, but not in the cerebellum and cortex of old rats and rabbits (De Jong et al., 1996; Papazafiri et al., 1995; Villa et al., 1994). The levels of calbindin-28 were also decreased in aged retinal preparations (Papazafiri et al., 1995). Reflecting this down-regulation of expression of calcium-binding proteins, the number of calbindin-immunoreactive neurones was markedly decreased in old rat hippocampus (Krzywkowski et al., 1995, 1996). Intracellular Ca^{2+} buffers were also reduced in peripheral adrenergic nerves of the old rats (Duckles et al., 1996). The age-dependent changes in expression of calcium-binding proteins varies in different brain regions: in medial septum and striatum the calbindin expression remained either unchanged or slightly reduced, whereas the number of neurones positive for another Ca^{2+} binding protein, parvalbumin, decreased markedly (Krzywkowski et al., 1996). Functional experiments on neurones isolated from the medial septum and from the nucleus of the diagonal band revealed an opposite trend: in these neurones the overall cytoplasmic calcium buffer capacity was actually increased with ageing (Murchison and Griffith, 1998). These data once more suggest the regional variations of age-dependent changes in calcium homeostatic cascades.

3.3. Plasmalemmal Ca^{2+} influx

The major pathways for plasmalemmal calcium influx are formed by two families of ion channels, the voltage-operated calcium channels (VOCCs) and the ligand-operated calcium permeable channels (ionotropic receptors).

The age-dependent changes of VOCCs were the most extensively studied. It appears that ageing does not affect the properties of channels protein as the elementary calcium currents did not change in aged neurones (Kostyuk et al., 1993; Thibault and Landfield, 1996; Thibault et al., 1995). Nevertheless, the density of VOCCs undergoes substantial age-dependent changes; and moreover they appear to be region-specific. In other words, in sensory neurones ageing led to a substantial decrease in density of high-threshold (L-type) Ca^{2+} channels and in complete disappearance of low-threshold VOCC (Kostyuk et al., 1993). In contrast, in aged hippocampal neurones a very substantial (2–3 fold) increase in L-type calcium channel density was observed (Campbell et al., 1996; Thibault and Landfield, 1996; Thibault et al., 1998). This increase in channel density coincided with an age-dependent up-regulation of the expression of mRNA for α_{1D}-VOCC subunit in hippocampal tissue (Thibault et al., 1998), suggesting therefore that synthesis of Ca^{2+} channels may be increased in old cells. In basal forebrain neurones, ageing does not affect significantly the density of L-type VOCCs, but increases the number of T-type channels (Murchison and Griffith, 1995, 1996). These contradictory results further stress a substantial heterogeneity in ageing in various cell types.

The functional properties of ionotropic receptors in old cells is virtually unknown, though there are some indirect data indicating that their molecular composition could be affected by ageing. For instance, it has been suggested that hippocampal neurones in old animals may express glutamate receptors lacking GluR-B subunit (Pagliusi et al., 1994). Such a peculiar down-regulation of GluR-B expression may result in an increase of Ca^{2+} permeability of glutamate ionotropic receptors.

3.4. Intracellular Ca^{2+} stores

An important part of calcium homeostasis is associated with specialised intracellular organelles known as intracellular calcium stores. Two types of these organelles, the endoplasmic reticulum (ER) and the mitochondria, play the major role in cytoplasmic calcium dynamics. Both these types of calcium storage organelles fulfil two main functions: they may act as a Ca^{2+} buffering system (Friel and Tsien, 1992; Neher, 1998; Toescu, 1998b) as well as a source of Ca^{2+} (Pozzan et al., 1994; Verkhratsky and Petersen, 1998; Verkhratsky and Shmigol, 1996; Lips and Keller, 1998). The accumulation of Ca^{2+} into an ER calcium store is mediated by SERCA pumps, whereas Ca^{2+} is released from the ER lumen via intracellular Ca^{2+} channels represented by $InsP_3$-gated ($InsP_3$ receptor) and Ca^{2+}-gated (ryanodine receptor) Ca^{2+} channels. The mitochondrial Ca^{2+} accumulation occurs via Ca^{2+} uniporter due to the existence of a strong electro-driving force between cytoplasm and the mitochondrial matrix. Similarly, the mitochondrial Ca^{2+} could also be rapidly released via several pathways (Ichas and Mazat, 1998).

Studies of the age-dependent status of intracellular calcium stores are very limited. Biochemical studies revealed a significant decrease in the density of $InsP_3$ receptors in membrane fractions isolated from the cerebellum (Igwe and Ning, 1993) and cerebrum (Martini et al., 1994) of aged rats. Conversely, the density of ryanodine receptors seems to remain constant during ageing (Martini et al., 1994). Age-dependent decrease in the density of $InsP_3$-receptors coincides with greater $InsP_3$ production in old cells (Igwe and Filla, 1995, 1997), possibly therefore representing a compensatory response. Interestingly, Ca^{2+} release from the $InsP_3$-sensitive ER stores was markedly enhanced in AD fibroblasts, due to both increase in $InsP_3$ production and increase in sensitivity of $InsP_3$ receptors to this second messenger (Etcheberrigaray et al., 1994; Ito et al., 1994), although the density of $InsP_3$ receptors was diminished (Kurumatani et al., 1998).

Functional studies demonstrated that the sensitivity of ryanodine receptors to caffeine does not change with ageing (Murchison and Griffith, 1999; Verkhratsky et al., 1994). Nevertheless the amount of releasable calcium within the ER lumen appears to be decreased in aged neurones (Kirischuk and Verkhratsky, 1996; Kirischuk et al., 1996; Murchison and Griffith, 1999; Verkhratsky et al., 1994). This reduction could result from decreased activity of SERCA pumps or decreased buffering capacity of ER lumen. Alternatively, as was suggested by Murchison and Griffith (1999) the amount of Ca^{2+} sequestered by the ER store could be reduced due to the increased cytoplasmic Ca^{2+}-buffer capacity. However, this might be a specific property of the medial septal neurones investigated by these authors.

Mitochondrial calcium transport also could be affected by ageing. Indeed, as was demonstrated by Satrustegui et al. (1996) mitochondrial Ca^{2+} accumulation is reduced in old neurones, which might reflect the chronic mitochondrial depolarization which occurs during ageing (Toescu and Verkhratsky, unpublished observations).

3.5. Calcium extrusion

The Ca^{2+} extrusion to the extracellular space is mediated by either an energy-dependent plasmalemmal Ca^{2+} pump (plasmalemmal Ca^{2+} ATPases – PCMA) or by an electrochemically driven Na^+/Ca^{2+} exchanger. The calcium extrusion seems to suffer most during cellular ageing. Detailed investigation of PMCA activity has shown that even under optimal conditions of saturating Ca^{2+} and ATP, the neuronal Ca^{2+} ATPase activity decreases with age (Michaelis et al., 1996). Sodium/calcium exchanger in contrast seems to be unaffected by age (Colvin et al., 1996).

Physiological experiments also demonstrated that ageing is associated with a remarkable increase in recovery time needed to reduce the transient Ca^{2+} increase due to membrane depolarization. Using acutely isolated cells and cerebellar granule neurones from brain slices, both obtained from aged animals, it has been shown that, regardless of the fact that the amplitude of the Ca^{2+} signal evoked by KCl depolarization (50 mM) was substantially diminished, the time taken to recover the resting $[Ca^{2+}]_i$ levels increased significantly (Kirischuk et al., 1992, 1996; Kirischuk and Verkhratsky, 1996; Verkhratsky et al., 1994). The half-time of $[Ca^{2+}]_i$ recovery measured in adult neocortical neurones was 10 ± 1 s (mean $\pm$ SEM), whereas in old neurones this value increased to 27 ± 3 s (Verkhratsky et al., 1994). Similarly, in DRG neurones, the value for half-time recovery increased from 25 ± 3 s in adult neurones to 43 ± 5 s in old cells. Furthermore, in experiments performed in brain slices, it was found that larger Ca^{2+} loads, as evoked by longer depolarization of the neurones and from which adult neurones invariably recover, frequently resulted, in aged slices, in irreversible increases in $[Ca^{2+}]_i$. This deterioration of PMCA-driven Ca^{2+} extrusion may result either from decrease in number/activity of plasmalemmal calcium pumps or from diminished ATP supply.

4. PHYSIOLOGICAL IMPLICATIONS OF Ca^{2+} DYSREGULATION IN AGEING

Based on the experimental data viewed above we may conclude that ageing is indeed associated with certain changes in performance of calcium homeostatic systems. However, these changes are rather subtle. It is still unclear whether age-dependent alterations of calcium handling may have substantial functional consequences.

The amount of experimental data describing actual changes in $[Ca^{2+}]_i$ dynamics in old cells is quite limited. It is still unclear whether ageing causes significant changes in resting cytoplasmic Ca^{2+} concentration. The $[Ca^{2+}]_i$ measurements on synaptosomes showed that resting $[Ca^{2+}]_i$ was somewhat increased in old preparations (Giovannelli and Pepeu, 1989; Martinez et al., 1988). When fura-2 $[Ca^{2+}]_i$ recordings were performed on aged peripheral and central neurones in both tissue culture and in acutely prepared brain slices, it was found that ageing coincides with a consistent increase in resting $[Ca^{2+}]_i$ (Kirischuk et al., 1992; Kirischuk and Verkhratsky, 1996; Verkhratsky et al., 1994). However, as the amount of direct $[Ca^{2+}]_i$ measurements is very limited it is still impossible to unequivocally conclude that ageing is associated with increase in neuronal resting $[Ca^{2+}]_i$ in all brain regions. The final conclusion on this matter awaits substantially more experimental evidence.

Nevertheless it seems to be generally accepted that some forms of Ca^{2+} dysregulation exists in aged cells, and it might be considered as a potentially crucial factor for the development of age-dependent cell malfunction (Figure 1). These subtle changes in calcium homeostasis might be particularly important for the performance of the central nervous system, the latter being critically dependent on Ca^{2+}-dependent synaptic transmission: the long-term

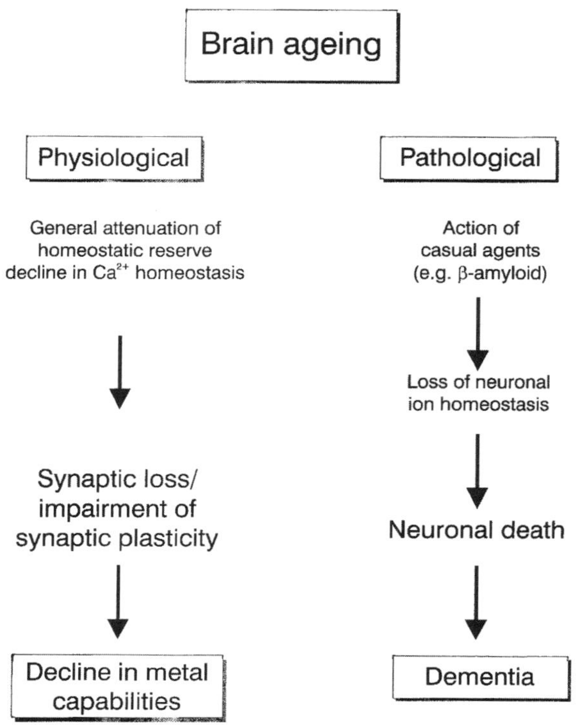

Figure 1. A schematic representation of the possible involvement of calcium dysregulation in the ageing process of the brain.

potentiation, LTP, which is currently regarded as a cellular model of memory and which is highly Ca^{2+}-dependent is altered with ageing (Barnes, 1988; Foster and Norris, 1997; Landfield, 1988). It seems obvious that disruption of $[Ca^{2+}]_i$ handling may significantly weaken the information processing in the brain as synaptic transmission is heavily Ca^{2+} dependent. The age-dependent decrease of $[Ca^{2+}]_i$ recovery following neuronal excitation (which today seems to be the most consistent experimental finding) would promote Ca^{2+} accumulation in the presynaptic terminal, which certainly may affect the efficacy of synaptic transmission. The chronic increase in $[Ca^{2+}]_i$ in the limited volume of a presynaptic terminal may also trigger a local excitotoxic effect, leading to the elimination of the most active synapses. Does such a mechanism indeed operate in the ageing process? This remains the intriguing and challenging question for the cellular physiology of ageing.

ACKNOWLEDGEMENT

This project work was supported by the Biotechnology and Biological Sciences Research Council, U.K.

REFERENCES

Barnes, C.A., 1988, Aging and the physiology of spatial memory, *Neurobiol. Aging* 9, 563–568.

Campbell, L.W., Hao, S.-Y., Thibault, O., Blalock, E.M. and Landfield, P.W., 1996, Aging changes in voltage-gated calcium currents in hippocampal CA1 neurones, *J. Neurosci.* 16, 6286–6295.

Choi, D.W., 1994, Calcium and excitotoxic neuronal injury, *Ann NY Acad Sci.* 747, 162–171.

Colvin, R.A., Walker, J.P., Schummers, J. and Davis, N., 1996, Aging does not affect steady-state expression of the Na+/Ca^{2+} exchanger in rat brain, *Cell Mol. Neurobiol.* 16, 11–19.

De Jong, G.I., Naber, P.A., Van den Zee, E.A., Thompson, L.T., Disterhoft, J.F. and Luiten, P.G.M., 1996. Age-related loss of calcium binding proteins in rabbit hippocampus, *Neurobiol. Aging* 17, 459–465.

Duckles, S.P., Tsai, H. and Buchholz, J.N., 1996, Evidence for decline in intracellular calcium buffering in adrenergic nerves of aged rats, *Life Sci.* 58, 2029–2035.

Etcheberrigaray, E., Gibson, G.E. and Alkon, D.L., 1994, Molecular mechanisms of memory and the pathophysiology of Alzheimer's disease, *Ann. NY Acad. Sci.* 747, 245–255.

Foster, T.C. and Norris, C.M., 1997, Age-associated changes in Ca^{2+}-dependent processes: Relation to hippocampal synaptic plasticity, *Hippocampus* 7, 602–612.

Friel, D.D. and Tsien, R.W., 1992, A caffeine- and ryanodine-sensitive Ca^{2+} store in bullfrog sympathetic neurones modulates effects of Ca^{2+} entry on $[Ca^{2+}]_i$, *J. Physiol. Lond.* 450, 217–246.

Gibson, G.E. and Peterson, C., 1987, Calcium and the aging nervous system, *Neurobiol. Aging* 8, 329–343.

Gibson, G.E., Perrino, P. and Dienel, G.A., 1986, In vivo brain calcium homeostasis during aging, *Mech. Ageing. Dev.* 37, 1–12.

Giovannelli, L. and Pepeu, G., 1989, Effect of age on K^+-induced cytosolic Ca^{2+} changes in rat cortical synaptosomes, *J. Neurochem.* 53, 392–398.

Gomez-Isla, T., Price, J.L., McKell, D.W., Jr., Morris, J.C., Growdon, J.H. and Hyman, B.T., 1996, Profound loss of layer II entorhinal cortex neurones occurs in very mild Alzheimer's disease, *J. Neurosci.* 16, 4491–4500.

Ichas, F. and Mazat, J.-P., 1998, From calcium signaling to cell death: Two conformations for the mitochondrial permeability transition pore. Switching from low- to high-conductance state, *Biochim. Biophys. Acta* 1366, 33–50.

Igwe, O.J. and Filla, M.B., 1995, Regulaton of phosphainositide transduction system in the rat spinal cord during aging, *Neuroscience* 69, 1239–1251.

Igwe, O.J. and Filla, M.B., 1997, Aging-related regulation of two myo-inositol 1,4,5-trisphosphate signal transduction pathway in the rat striatum, *Brain Res. Mol. Brain. Res* 46, 39–53.

Igwe, O.J. and Ning, L., 1993, Inositol 1,4,5-trisphosphate arm of the phosphatidylinositide signal transduction pathway in the rat cerebellum during aging, *Neurosci. Lett.* 164, 167–170.

Ito, E., Oka, K., Etcheberrigaray, R., Nelson, T.J., McPhie, D.L., Tofel-Grehl, B., Gibson, G.E. and Alkon, D.L., 1994, Internal Ca^{2+} mobilization is altered in fibroblasts from patients with Alzheimer diseasem *Proc. Natl. Acad. Sci, USA* 91, 534–538.

Keller, B.U., 2000, The role of intracellular calcium signaling in motoneuron function and disease, this book.

Khachaturian, Z.S., 1991, Calcium and the aging brain: Upsetting a delicate balance?, *Geriatrics* 46, 78–79.

Khachaturian, Z.S., Cotman, C.W. and Pettegrew, W., 1989, Calcium, membranes, aging, and Alzheimer's disease, *Ann. NY Acad. Sci.* 568, 1–292.

Kirischuk, S. and Verkhratsky, A., 1996, Calcium homeostasis in aged neurones, *Life Sci.* 59, 451–459.

Kirischuk, S., Pronchuk, N. and Verkhratsky, A., 1992, Measurements of intracellular calcium in sensory neurons of adult and old rats, *Neuroscience* 50, 947–951.

Kirischuk, S., Voitenko, N., Kostyuk, P. and Verkhratsky, A., 1996, Age associated changes of cytoplasmic calcium homeostasis in cerebellar granule neurons in situ: Investigation on thin cerebellar slices, *Exp. Gerontology* 31, 475–487.

Kostyuk, P., Pronchuk, N., Savchenko, A. and Verkhratsky, A., 1993, Calcium currents in aged rat dorsal root ganglion neurones, *J. Physiol. Lond.* 461, 467–483.

Krzywkowski, P., De Bilbao, F., Senut, M.C. and Lamour, Y., 1995, Age-related changes in parvalbumin- and GABA-immunoreactive cells in the rat septum, *Neurobiol. Aging* 16, 29–40.

Krzywkowski, P., Potier, B., Billard, J.M., Dutar, P. and Lamour, Y., 1996, Synaptic mechanisms and calcium binding proteins in the aged rat brain, *Life Sci.* 59, 421–428.

Kurumatani, T., Fastbom, J., Bonkale, W.L., Bogdanovic, N., Winblad, B., Ohm, T.G. and Cowburn, R.F., 1998, Loss of inositol 1,4,5-trisphosphate receptor sites and decreased PKC levels correlate with staging of Alzheimer's disease neurofibrillary pathology, *Brain Res.* 796, 209–221.

Landfield, P.W., 1988, Hippocampal neurobiological mechanisms of age-related memory dysfunction, *Neurobiol. Aging* 9, 571–579.

Leslie, S.W., Chandler, L.J., Barr, E. and Farrar, R.P., 1985, Reduced calcium uptake by rat brain mitochondrial and synaptosomes in responses to aging, *Brain Res.* 329, 177–183.

Lips, M.B. and Keller B.U., 1998, Endogenous calcium buffering in motoneurons of the nucleus hypoglossus from mouse, *J. Physiol. Lond.* 511, 105–117.

Martinez, A., Vitorica, J. and Satrustegui, J., 1988, Cytosolic free calcium levels increase with age in rat brain synaptosomes, *Neurosci. Lett.* 88, 336–342.

Martini, A., Battaini, F., Govoni, S. and Volpe, P., 1994, Inositol 1,4,5-trisphosphate receptor and ryanodine receptor in the aging brain of Wistar rats, *Neurobiol. Aging* 15, 203–206.

Mattson, M.P., Barger, S.W., Cheng, B., Lieberburg, I., Smith Swintosky, V.L. and Rydel, R.E., 1993, β-Amyloid precursor protein metabolites and loss of neuronal Ca^{2+} homeostasis in Alzheimer's disease, *Trends Neurosci.* 16, 409–414.

Michaelis, M.L., Bigelow, D.J., Schoneich, C., Williams, T.D., Ramonda, L., Yin, D., Huhmer, A.F., Yao, Y., Gao, J. and Squier, T.C., 1996, Decreased plasma membrane calcium transport activity in aging brain, *Life Sci.* 59, 405–412.

Murchison, D. and Griffith, W.H., 1995, Low-voltage activated calcium currents increase in basal forebrain neurones from aged rats, *J. Neurophysiol.* 74, 876–887.

Murchison, D. and Griffith, W.H., 1996, High-voltage-activated calcium currents in basal forebrain neurons during aging, *J. Neurophysiol.* 76, 158–174.

Murchison, D. and Griffith, W.H., 1998, Increased calcium buffering in basal forebrain neurons during aging, *J. Neurophysiol.* 76, 350–364.

Murchison, D. and Griffith, W.H., 1999, Age-related alterations in caffeine-sensitive calcium stores and mitochondrial buffering in rat basal forebrain, *Cell Calcium* 25, 439–452.

Neher, E., 1998, Usefulness and limitations of linear approximations to the understanding of Ca^{2+} signals, *Cell Calcium* 24, 345–357.

Novi, I., 1912, Le calcium et le magnésium du cerveau dans les différents âges, *Arch. Ital. Biol.* 58, 333–336.

Pagliusi, S.R., Gerrard, P., Abdallah, M., Talabot, D. and Catsicas, S., 1994, Age-related changes in expression of AMPA-selective glutamate receptor subunits: Is calcium-permeability altered in hippocampal neurons?, *Neuroscience* 61, 429–433.

Papazafiri, P., Podini, P., Meldolesi, J. and Yamaguchi, T., 1995, Ageing affects cytosolic Ca^{2+} binding proteins and synaptic markers in the retina but not in cerebral cortex neurons of the rat, *Neurosci. Lett.* 186, 65–68.

Peterson, C. and Gibson, G.E., 1983, 3,4-diaminopyridine alter synaptosomal calcium uptake, *J. Biol. Chem.* 258, 11482–11486.

Peterson, C., Gibson, G.E. and Blass, J.P., 1985, *Altered calcium uptake in cultured skin fibroblasts from patients with Alzheimer's disease*, *New Engl. J. Med.* 312, 1063–1064.

Pozzan, T., Rizzuto, R., Volpe, P. and Meldolesi, J., 1994, Molecular and cellular physiology of intracellular calcium stores, *Physiol. Rev.* 74, 595–636.

Satrustegui, J., Villalba, M., Pereira, R., Bogonez, E. and Martinez Serrano, A., 1996, Cytosolic and mitochondrial calcium in synaptosomes during aging, *Life Sci.* 59, 429–434.

Thibault, O. and Landfield, P.W., 1996, Increase in single L-type calcium channels in hippocampal neurons during aging, *Science* 272, 1017–1020.

Thibault, O., Mazzanti, M.L., Blalock, E.M., Porter, N.M. and Landfield, P.W., 1995, Single-channel and whole-cell studies of calcium currents in young and aged rat hippocampal slice neurons, *J. Neurosci. Methods* 59, 77–83.

Thibault, O., Porter, N.M., Chen, K.C., Blalock, E.M., Kaminker, P.G., Clodfelter, G.V., Brewer, L.D. and Landfield, P.W., 1998, Calcium dysregulation in neuronal aging and Alzheimer's disease: history and new directions, *Cell Calcium* 24, 417–433.

Toescu, E.C., 1998a, Apoptosis and cell death in neuronal cells: Where does calcium fit in?, *Cell Calcium* 24, 387–403.

Toescu, E.C., 1998b, Intraneuronal Ca^{2+} stores act mainly as a 'Ca^{2+} sink' in cerebellar granule neurones, *Neuroreport* 9, 1227–1231.

Verkhratsky, A. and Petersen, O.H., 1998, Neuronal calcium stores, *Cell Calcium* 24, 333–343.

Verkhratsky, A. and Shmigol, A., 1996, Calcium-induced calcium release in neurones, *Cell Calcium* 19, 1–14.

Verkhratsky, A., Shmigol, A., Kirischuk, S., Pronchuk, N. and Kostyuk, P., 1994, Age-dependent changes in calcium currents and calcium homeostasis in mammalian neurons, *Ann. NY Acad. Sci.* 747, 365–381.

Verkhratsky, A. and Toescu, E.C., 1998a, Calcium and neuronal ageing, *Trends Neurosci.* 21, 2–7.

Verkhratsky, A.N. and Toescu, E.C., 1998b, *Integrative Aspects of Calcium Signalling*, Plenum Press, London.

Villa, A., Podini, P., Panzeri, M.C., Racchetti, G. and Meldolesi, J., 1994, Cytosolic Ca^{2+} binding proteins during rat brain ageing: Loss of calbindin and calretinin in the hippocampus, with no change in the cerebellum, *Eur. J. Neurosci.* 6, 1491–1499.

West, M.J., Coleman, P.D., Flood, D.G. and Troncoso, J.C., 1994, Differences in the pattern of hippocampal neuronal loss in normal ageing and Alzheimer's disease, *Lancet* 344, 769–772.

Zipfel, G.J. and Choi, D.W., 1999, The changing landscape of ischaemic brain injury mechanisms, *Nature* 399 (Suppl. 6738), A7–A14.

Zipfel, G.J., Lee, J.M. and Choi, D.W., 1999, Reducing calcium overload in the ischemic brain, *New Engl. J. Med.* 341, 1543–1544.

The Epidermal Growth Factor Receptor and the Calcium Signal

Antonio Villalobo, María José Ruano, Paloma I. Palomo-Jiménez, Hongbing Li and José Martín-Nieto

1. INTRODUCTION

Among the multiple systems that become operative during the mitogenic activation of a cell there is an early signal mediated by calcium. Thus, the cytosolic concentration of this ubiquitous second messenger transiently increases when a variety of mitogenic receptors that belong to the superfamily of receptors with tyrosine kinase activity are activated by their ligands (Rozengurt, 1986). Among these receptors is the epidermal growth factor receptor (EGFR/ErbB1), a 170 kDa plasma membrane glycoprotein that, alike other related members of the ErbB family, is involved in proliferation, differentiation and even the control of apoptotic processes (Carpenter, 1987; Ullrich and Schlessinger, 1990; Alroy and Yarden, 1997). In this review we will analyze how the EGFR generates the calcium signal, the subsequent activation of diverse Ca^{2+}-dependent systems, and the role that these systems play on the regulation and fate of the receptor.

2. THE CALCIUM TRANSIENT INDUCED BY THE EGFR

The activation of the EGFR induces a transient increase in the cytosolic concentration of free calcium ($[Ca^{2+}]_c$) that is generally preceded by an increase

---

Antonio Villalobo, María José Ruano, Paloma I. Palomo-Jiménez and Hongbing Li • Instituto de Investigaciones Biomédicas, Consejo Superior de Investigaciones Científicas and Universidad Autónoma de Madrid, Madrid, Spain. **José Martín-Nieto** • División de Genética, Departamento de Fisiología, Genética y Microbiología, Universidad de Alicante, Alicante, Spain.

R. Pochet, R. Donato, J. Haiech, C. Heizmann and V. Gerke (eds.): Calcium: The Molecular Basis of Calcium Action in Biology and Medicine, 287–303.

in the turnover of phosphoinositides in a variety of cell types (Moolenaar et al., 1986; Pandiella et al., 1988; Gilligan et al., 1988; Cheyette and Gross, 1991; Himpens et al., 1993). This cytosolic calcium rise is also followed by an increase in the concentration of free calcium in the nucleus (Himpens et al., 1993; Elliget et al., 1996).

This calcium signal appears to be a phylogenetically ancient mechanism, as it is evoked by the interaction of Lin-3 and Let-23, the EGF and EGFR homologs, respectively, in *Caenorhabditis elegans* (Clandinin et al., 1998). Nevertheless, in some cell types, such as in Swiss 3T3 fibroblasts, EGF by itself fails to induce a significant calcium transient, although this growth factor induces a cytosolic calcium rise in these cells after bradykinin stimulation (Olsen et al., 1988; Bierman et al., 1990).

The EGF-induced increase in $[Ca^{2+}]_c$ exhibits two components: a release of calcium from intracellular stores, and a net calcium influx from the extracellular medium (Gilligan et al., 1988; Tinhofer et al., 1996). These two events can also be kinetically dissected using the human EGFR expressed in *Xenopus* oocytes (Browaeys-Poly et al., 1998). Interestingly, the entry of calcium in the cell by the opening of Ca^{2+}-channels is highly regulated since it is inhibited by several systems such as sphingosine (Hudson et al., 1994), and annexin VI, a Ca^{2+}/phospholipid-binding protein of yet unknown function (Fleet et al., 1999). On the other hand, both Ras and Rac activate the EGF-induced calcium entry (Tinhofer et al., 1996; Peppelenbosch et al., 1996), whereas an increase in the level of extracellular calcium inhibits EGF-induced Ras activation (Medema et al., 1994). Moreover, trimeric G proteins appear to control as well the EGF-induced calcium influx (Hughes et al., 1987; Kuryshev et al., 1993).

The mechanism by which the EGFR elicits an increase in the $[Ca^{2+}]_c$ is related to the activation of phospholipase Cγ (PLCγ) and phospholipase A_2 (PLA2) and the subsequent production of a series of lipidic second messengers that act as effectors of Ca^{2+}-channels (see Figure 1).

Thus, PLCγ binds to phospho-Tyr_{992} in the transactivated EGFR via its SH2 domains and results phosphorylated and thereby activated (Wahl et al., 1988; Margolis et al., 1989; Nishibe et al., 1989, 1990; Chang et al., 1991; Vega et al., 1992). This phospholipase hydrolyzes phosphatidylinositol-4,5-bisphosphate (PtdIns-4,5-P2) to inositol-1,4,5-trisphosphate (Ins-1,4,5-P3) plus 1,2-diacylglycerol (1,2-DAG). Ins-1,4,5-P3 in turn opens Ca^{2+}-channels located in the endo/sarcoplasmic reticulum releasing Ca^{2+} from its lumen (Hughes et al., 1991; Mikoshiba, 1993), which together with 1,2-DAG activate a number of protein kinase C (PKC) isoforms (Azzi et al., 1992) (see Figures 1 and 2).

Moreover, PLA2 also becomes activated in cells stimulated with EGF by an indirect mechanism that involves its phosphorylation at serine residues (Peppelenbosch et al., 1992; Schalkwijk et al., 1995). This phospholipase hydrolyzes membrane phospholipids (PL) liberating arachidonic acid (AA), which is transformed in leukotriene C_4 (LC4) through the 5-lipoxygenase

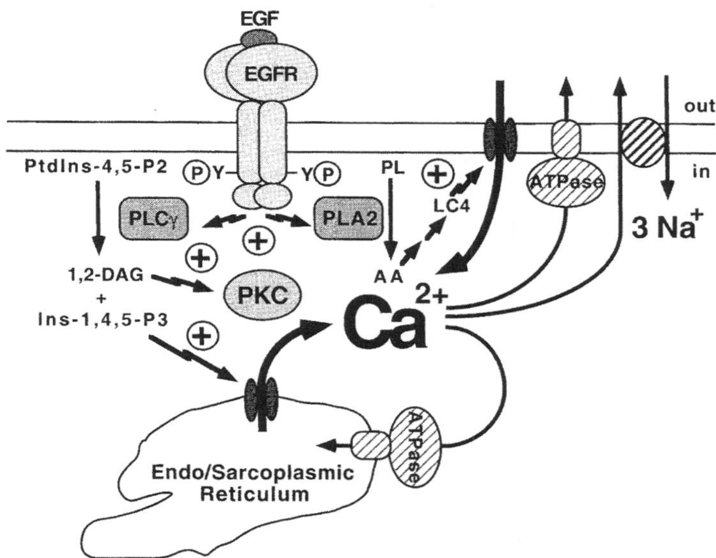

Figure 1. The transient calcium signal generated by the EGFR. The transactivation of the EGFR elicits the activation of PLCγ and PLA2. The $[Ca^{2+}]_c$ rises because Ins-1,4,5-P3 activates Ca^{2+}- channels located in the endo/sarcoplasmic reticulum and LC4 activates additional Ca^{2+}-channels located in the plasma membrane. The generated calcium signal is promptly dissipated by the concurrent operation of Ca^{2+}-ATPases and the Na^+/Ca^{2+} exchanger. Phosphate groups are represented by encircled P. See text for details.

pathway. This compound is a potent effector that opens voltage-insensitive Ca^{2+}-channels located in the plasma membrane (Peppelenbosch et al., 1992) (see Figure 1). The activation of cells by EGF also produces a hyperpolarizing response, since the subsequent increase in the $[Ca^{2+}]_c$ activates Ca^{2+}-dependent K^+ channels (Enomoto et al., 1986; Peppelenbosch et al., 1991). This change in membrane potential further activates hyperpolarization-sensitive Ca^{2+}-channels contributing to potentiate the ensuing Ca^{2+} influx. This cascade of events appears to be negatively regulated by PKC (Peppelenbosch et al., 1991).

Figure 1 also shows the extinction of the calcium signal by the cooperative work of the endo/sarcoplasmic reticulum Ca^{2+}-ATPase, and the calmodulin-dependent Ca^{2+}-ATPase and Na^+/Ca^{2+} exchanger from the plasma membrane. Interestingly, EGF augments the level of Na^+/Ca^{2+} exchanger mRNA, suggesting a prominent role of this transport system in the dissipation of the calcium signal (Smith and Smith, 1994).

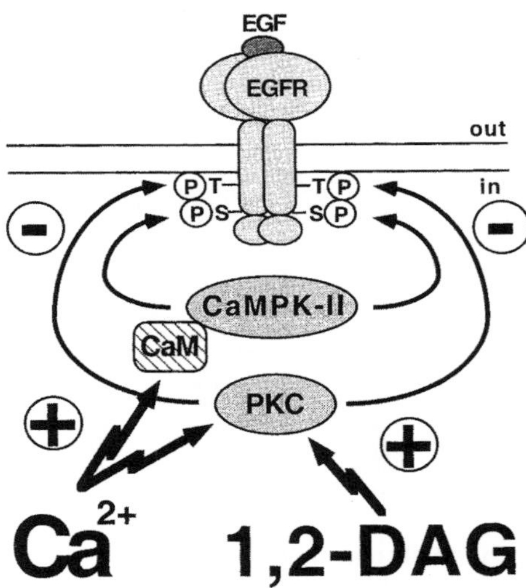

Figure 2. Retroinhibition of the EGFR by calcium-dependent protein kinases. The activation of PKC and CaMPK-II by the generated calcium signal results in the phosphorylation of the EGFR retroinhibiting its activity. Phosphate groups are represented by encircled P. See text for details.

3. REGULATION OF THE EGFR BY CALCIUM-DEPENDENT PROTEIN KINASES

The calcium signal generated by the EGFR activates two Ca^{2+}-dependent protein kinases that regulate the activity of the receptor. These are PKC, which is activated by calcium and 1,2-DAG, and calmodulin-dependent protein kinase-II (CaMPK-II), which is activated by the Ca^{2+}/calmodulin (CaM) complex formed by the ensuing EGFR-mediated $[Ca^{2+}]_c$ rise (Figure 2).

Phosphorylation of the EGFR by PKC at Thr_{654}, located in its cytosolic juxtamembrane region, inhibits its tyrosine kinase activity (Hunter et al., 1984; Davis, 1988; Livneh et al., 1988; Countaway et al., 1990; Lund et al., 1990) (see Figure 2). Moreover, this phosphorylation has been shown to prevent receptor internalization (Logsdon and Williams, 1984; Lund et al., 1990). However, the regulatory role of PKC on the receptor appears to be more complex, since Thr_{669}, the site of phosphorylation in the receptor by a mitogen-activated protein kinase (MAPK), is also under the control of PKC (Morrison et al., 1993, 1996).

The phosphorylation of Ser_{1046} and Ser_{1047} in the EGFR by CaMPK-II also inhibits the tyrosine kinase activity of the receptor (Countaway et al., 1992; Theroux et al., 1992; Feinmesser et al., 1999) (see Figure 2). However, this kinase also phosphorylates the receptor at Ser_{744}, located in the tyrosine kinase domain, perhaps inducing a further regulation of the receptor (Feinmesser et al., 1999). Moreover, ErbB2/Neu, a cousin of the EGFR, is also phosphorylated by CaMPK-II at Thr_{1172} what could have some regulatory implications (Feinmesser et al., 1996).

4. DEPHOSPHORYLATION OF THE EGFR BY CALCIUM-DEPENDENT PROTEIN PHOSPHATASES

Calcium appears to be implicated as well in the dephosphorylation of the EGFR, although some uncertainty remains in this area. In this context, it has been shown that the artificial increase in the $[Ca^{2+}]_c$ induced by calcium ionophores augments the activity of phosphotyrosine phosphatases in whole cells. However, the activation of the EGFR induces a marked inhibition of this activity, which has led to the conclusion that the Ca^{2+} signal generated by the receptor cannot be responsible for such inhibition (Mishra and Hamburger, 1993). Nevertheless, although calcineurin, a Ca^{2+}/calmodulin-dependent phosphatase, mainly dephosphorylates serine/threonine residues, both the autophosphorylated EGFR and cellular proteins phosphorylated by the receptor at tyrosine residues have been shown to be dephosphorylated *in vitro* by this phosphatase (Chernoff et al., 1984; Pallen et al., 1985). However, the physiological meaning of these observations is not yet clear, as the presence of a contaminating *bona fide* tyrosine phosphatase has not been explicitly ruled out. Nevertheless, more recently it has been shown that calcineurin rather acts *in vivo* as a regulator of downstream signaling pathways activated by the receptor, such as catalyzing the dephosphorylation of the transcription factor Elk-1 at Ser_{383} inducing its inactivation (Sugimoto et al., 1997).

5. CALCIUM-DEPENDENT TRANSACTIVATION OF THE EGFR

In excitable cells a series of mechanisms are able to generate an increase in the $[Ca^{2+}]_c$, and it has been recently shown that this calcium signal induces the transactivation of the EGFR. Thus, membrane depolarization activating voltage-sensitive Ca^{2+}-channels, and bradykinin stimulation lead to the phosphorylation of the EGFR in a ligand-independent manner in pheochromocytoma PC12 cells, perhaps because its phosphorylation by the non-receptor tyrosine kinase Src. This process results in the subsequent activation of the Ras/MAPK pathway via the phosphorylation of Shc and the formation of the Grb2-Sos complex (Rosen and Greenberg, 1996; Zwick et al., 1997). Similar observations have been done by the stimulation of these cells with extracellular UTP, an effect that is mediated by G protein-coupled P_{2Y2} receptors (Soltoff, 1998). Moreover, the transactivation of the EGFR following cell depolarization induced by KCl treatment has been shown to be mediated by a CaMPK, whereas the transactivation induced by bradykinin treatment is not mediated by this kinase (Zwick et al., 1999).

In addition, the activation of cardiac fibroblasts and vascular smooth muscle cells by angiotensin II also produces a Ca^{2+}-dependent transactivation of the EGFR that is mediated by Src and appears to be sensitive to calmodulin inhibitors. Similarly, this pathway results in the activation of the MAPK cascade (Murasawa et al., 1998; Eguchi et al., 1998).

6. PHOSPHORYLATION OF CALMODULIN BY THE EGFR

An interesting new development in our understanding of the possible interplay between the Ca^{2+} signal generated by the receptor and its regulation, came when it was shown that the receptor phosphorylates calmodulin *in vitro* and that very low concentrations of this cation prevent this phosphorylation ($K'_{i[Ca^{2+}]} \approx 0.3\ \mu M$) (San José et al., 1992; Benguría and Villalobo, 1993; Benguría et al., 1994, 1995; De Frutos et al., 1997; Benaim et al., 1998; Palomo-Jiménez et al., 1999). The phosphorylation of calmodulin by the receptor mainly occurs at Tyr_{99}, a residue located in its third Ca^{2+}-binding site (Benguría et al., 1994; Benaim et al., 1998; Palomo-Jiménez et al., 1999). Since the carbonyl oxygen of Tyr_{99} intervenes in the coordination of Ca^{2+} at this site (Strynadka and James, 1989), this process, and/or the major conformational changes occurring in the protein upon calcium binding, could hinder the accessibility of the –OH group of Tyr_{99} to the kinase domain of the EGFR. Moreover, the phosphorylation of calmodulin was strictly dependent on the presence of a positively-charged polypeptide in the assay system (San José et al., 1992; Benguría and Villalobo, 1993; Benguría et al., 1994, 1995; De Frutos et al., 1997; Benaim et al., 1998; Palomo-Jiménez et al., 1999). This suggests that a physiological cofactor with these characteristics may be necessary for calmodulin phosphorylation in whole cells. Apparently, the formation of a complex between such cationic cofactor and the negatively-charged calmodulin could induce the exposure of its phosphorylation site (Benguría et al., 1994).

The phosphorylation of calmodulin by the EGFR could has profound consequences on its biological activity. Thus, we have shown that phospho(Tyr)calmodulin (P-CaM) fails to significantly activate the phosphodiesterase for cyclic nucleotides from bovine heart, in contrast to the activation of this enzyme exerted by non-phosphorylated calmodulin (Palomo-Jiménez et al., 1999). However, it has been reported that the phosphodiesterase for cyclic nucleotides from bovine brain was fully activated by a similar preparation of phospho(Tyr)calmodulin (Corti et al., 1999). It remains to be determined wether this apparent discrepancy is due to the distinct origin of the phosphodiesterases used, or to some unrecognized differences on the phosphorylation states of calmodulin.

An additional intriguing observation is that when the EGFR phosphorylates calmodulin, the tyrosine kinase of the receptor becomes further activated (Villalobo et al., 1997; Villalobo and Gabius, 1998). This suggests that the accumulated phospho(Tyr)calmodulin exerts this potentiation. This extra activation of the EGFR is observed in the presence but not in the absence of EGF, is hindered by an anti-calmodulin antibody that prevents its

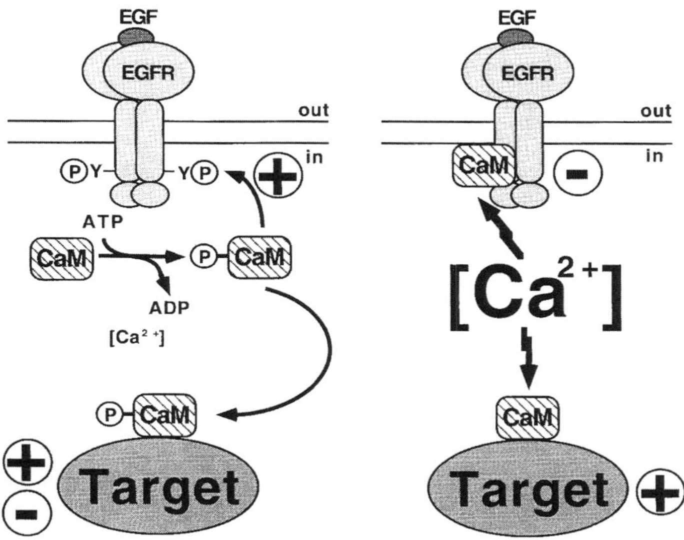

Figure 3. Regulatory interactions between calmodulin and the EGFR. (*Left panel*) When the $[Ca^{2+}]_c$ is low, at the start of the mitogenic signal, the EGFR phosphorylates calmodulin. This phosphorylation modifies its capacity to activate calmodulin-dependent target systems, and results in further activation of the receptor itself. Phosphate groups are represented by encircled P. (*Right panel*) When the $[Ca^{2+}]_c$ is high, because the generated receptor-dependent calcium signal, the Ca^{2+}/calmodulin complex is formed activating target systems and retroinhibiting the EGFR tyrosine kinase. See text for details.

phosphorylation, and is not observed when mammalian calmodulin is substituted by trypanosomatid calmodulin, which has a phenylalanine instead of a tyrosine at position 99 (Villalobo et al., 1997). Therefore, we have postulated that phospho(Tyr)calmodulin could be an intracellular activator of the EGFR in whole cells, exerting a stimulatory effect by potentiating the activation induced by the extracellular ligand (Villalobo et al., 1997; Villalobo and Gabius, 1998).

In the left panel of Figure 3 it is shown that when the $[Ca^{2+}]_c$ is low, calmodulin becomes phosphorylated upon EGFR activation, and that the resulting phospho(Tyr)calmodulin exerts both a potentiation of the activity of the receptor and a differential action on different calmodulin-dependent target systems as compared to non-phosphorylated calmodulin. Calmodulin phosphorylation could thus constitute a fine tuning system to modify the activity of this modulator, as opposed to the gross tuning exerted by the rise in $[Ca^{2+}]_c$, as the latter should activate all calmodulin-dependent enzymes without any discrimination, while the former could spare some calmodulin-dependent systems from this activation (Palomo-Jiménez et al., 1999).

7. REGULATORY ROLE OF CALMODULIN ON THE EGFR

The first indication that calmodulin could directly interact with the EGFR, thereby exerting a modulatory role on its activity, was attained by the isolation of the receptor from solubilized rat liver plasma membrane using Ca^{2+}-dependent calmodulin-affinity chromatography (San José et al., 1992). The isolated receptor is able to undergo autophosphorylation and exhibits a very active tyrosine kinase activity toward exogenous substrates that is stimulated either by EGF or transforming growth factor-α, another EGFR ligand (San José et al., 1992; Benguría and Villalobo, 1993; Benguría et al., 1994, 1995; Benaim et al., 1998; Palomo-Jiménez et al., 1999). However, the human receptor overexpressed in EGFR-T17 fibroblasts and isolated by this method, undergoes autophosphorylation but this process is not stimulated by EGF (Martín-Nieto and Villalobo, 1998). This suggests that only a subpopulation of the receptor, which may have some post-translational modification(s), is able to bind calmodulin in whole cells (Martín-Nieto and Villalobo, 1998).

Sequence analysis of the human EGFR shows that its cytosolic juxtamembrane region contains a potential calmodulin-binding site consisting of a cluster of basic and hydrophobic amino acids that could form a positively-charged amphiphilic α-helix (San José et al., 1992; Benguría and Villalobo, 1993; Martín-Nieto and Villalobo, 1998). To ascertain that calmodulin could bind to this site in the receptor, a recombinant fusion protein harboring a segment of the human EGFR comprised between amino acids 645 and 660 was prepared (Martín-Nieto and Villalobo, 1998). Using this system, it was demonstrated that calmodulin indeed binds to this segment of the receptor in a Ca^{2+}-dependent manner, exhibiting a $K'_{d[\mathrm{CaM}]}$ of 0.4 μM (Martín-Nieto and Villalobo, 1998).

The identified calmodulin-binding site in the EGFR is identical in the receptor from human, mouse and chicken, and has two amino acid substitutions in rat. The conservation of this calmodulin-binding site along evolution underscores its possible physiological relevance. In contrast, this region presents significant amino acid substitutions in other members of the EGFR family, such as ErbB2/Neu, ErbB3, and ErbB4, whereas in the EGFR counterparts from invertebrates very low sequence homology is found (Martín-Nieto and Villalobo, 1998).

Using the purified EGFR from rat liver, we have demonstrated that calmodulin inhibits the tyrosine kinase activity of the receptor in a Ca^{2+}-dependent manner (San José et al., 1992; Benguría and Villalobo, 1993; Benguría et al., 1995). The $K'_{i[\mathrm{CaM}]}$ for this inhibition process was determined to be approximately 1 μM (San José et al., 1992). This value is close to the $K'_{d[\mathrm{CaM}]}$ calculated for the fusion protein (0.4 μM), suggesting that calmodulin exerts its inhibitory action by binding to this juxtamembrane site in the whole receptor (Martín-Nieto and Villalobo, 1998).

In the right panel of Figure 3 the interaction of the Ca^{2+}/calmodulin complex with the EGFR is depicted together with the subsequent inhibition of the

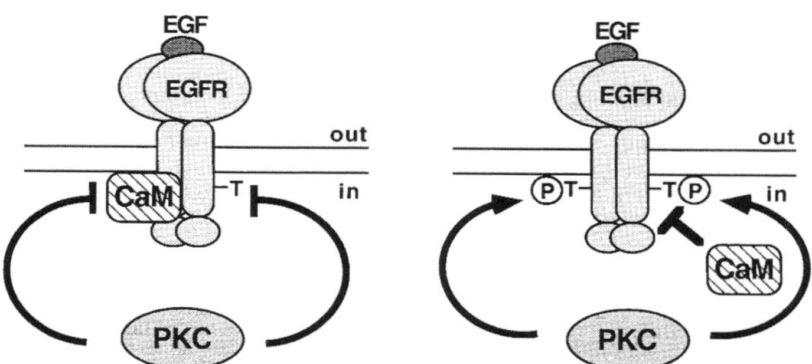

Figure 4. Mutually incompatible regulation of the EGFR by calmodulin and PKC. (*Left panel*) Binding of calmodulin to the EGFR prevents its phosphorylation by PKC. (*Right panel*) Phosphorylation of the EGFR by PKC prevents calmodulin binding to the receptor. Phosphate groups are represented by encircled P. See text for details.

receptor tyrosine kinase activity, which should lead to the termination of the mitogenic signal. In contrast, the Ca^{2+}/calmodulin complex activates a series of calmodulin-dependent target systems, some of which could be relevant for the restoration of the basal $[Ca^{2+}]_c$, such as the calmodulin-dependent plasma membrane Ca^{2+}-ATPase (see Figure 1).

The calmodulin binding site of the human EGFR has the sequence Arg_{645}-Arg-Arg-His-Ile-Val-Arg-Lys-Arg-Thr_{654}-Leu-Arg-Arg-Leu-Leu-Gln_{660} (San José et al., 1992; Benguría and Villalobo, 1993; Martín-Nieto and Villalobo, 1998), and Thr_{654}, located in the center of this sequence, is known to be the major phosphorylation site by PKC in the receptor as discussed earlier (Hunter et al., 1984; Davis, 1988; Livneh et al., 1988; Countaway et al., 1990; Lund et al., 1990). Indeed, we have shown that PKC phosphorylates as well this threonine residue in a fusion protein and that binding of calmodulin prevents this phosphorylation with a $K'_{i[CaM]}$ of 0.5–1 μM (Martín-Nieto and Villalobo, 1998). Conversely, the phosphorylation of this residue by this kinase inhibits calmodulin binding to the fusion protein (Martín-Nieto and Villalobo, 1998). The left panel of Figure 4 represents the incapacity of PKC to phosphorylate the EGFR when calmodulin is bound to the receptor probably because the occlusion of the phosphorylation site (Thr_{654}), and the right panel of this figure represents the inability of calmodulin to bind to the receptor when this has been phosphorylated at Thr_{654} by PKC.

As discussed in a previous section, phosphorylation of Thr_{654} by PKC in the EGFR inhibits its tyrosine kinase activity (Hunter et al., 1984; Davis, 1988; Livneh et al., 1988; Countaway et al., 1990; Lund et al., 1990), similarly to the inhibitory action exerted by the binding of the Ca^{2+}-calmodulin complex to the isolated receptor (San José et al., 1992; Benguría and Villalobo, 1993; Benguría et al., 1995). Therefore, the confluence of two mutually incompatible regulatory systems acting on the same site of the EGFR is intriguing. As phosphorylation by PKC also prevents receptor internalization

(Logsdon and Williams, 1984; Lund et al., 1990), this apparently redundant regulation could be explained assuming that the binding of calmodulin exerts an effect on the fate of the receptor distinct from that elicited by PKC phosphorylation. In this context, we have postulated that binding of calmodulin to the EGFR could be an internalization signal (Martín-Nieto and Villalobo, 1998), in contrast to the anti-internalization effect imposed by PKC phosphorylation (Logsdon and Williams, 1984; Lund et al., 1990).

8. PHOSPHORYLATION OF CALMODULIN-BINDING PROTEINS BY THE EGFR

Another interesting interconnection between the EGFR and the generated calcium signal is established when the receptor phosphorylates Ca^{2+}/calmodulin-dependent proteins, subsequently altering their biological activities. Not too many examples exist of phosphorylation of this sort of proteins by the receptor, but certainly these phosphorylations may have important physiological implications, particularly those relevant for cytoskeletal organization. Thus, the EGFR has been shown to phosphorylate the β subunit of fodrin, a non-erythroid spectrin present in brain (Akiyama et al., 1986). Another example of this sort of regulation is given by a mutant EGFR encoded by the S3-v-*erb*B oncogene that phosphorylates caldesmon, a Ca^{2+}/calmodulin-dependent regulatory protein involved in both inhibition of smooth muscle contraction and the organization of the actin cytoskeletal network (McManus et al., 1997; Wang et al., 1999). Moreover, the phosphorylation of caldesmon enhances its interaction with the Shc-Grb2-Sos complex, which could have important signaling consequences for sarcoma cells expressing this oncogene by inducing actin stress fiber disassembly (McManus et al., 1997; Wang et al., 1999).

An additional target for EGFR-mediated phosphorylation is connexin32, a calmodulin-binding protein that forms gap junction channels. This protein has been shown to become phosphorylated *in vitro* by the receptor at a tyrosine residue that appears to be located in its calmodulin-binding site, since calmodulin inhibits its phosphorylation (Díez et al., 1995, 1998). Nevertheless, in whole cells the phosphorylation of connexin43, another member of the connexin family, becomes phosphorylated at serine residues by a MAPK activated by the EGFR (Kanemitsu and Lau, 1993). In any event, the direct or indirect phosphorylation of connexins by the receptor could be of importance for the transient closing and/or disassembly of intercellular channels during mitogenesis, which could be of special relevance in tumor cells overexpressing the EGFR or expressing receptor variants with hyperactive tyrosine kinase.

9. THE CALCIUM SIGNAL AND THE PROTEOLYTIC PROCESSING OF THE EGFR

After signaling is achieved, the EGFR undergoes internalization in endosomes that form from clathrin-coated pits, followed by its proteolytic degradation in lysosomes (Todderud and Carpenter, 1989; Ullrich and Schlessinger, 1990). As exposed below, calcium appears to be implicated as well in the proteolytic processing of the receptor.

The EGFR has in its cytosolic C-terminal region, adjacent to and distal from the tyrosine kinase domain, a 48 amino acid segment that is involved in both endocytosis and the generation of the calcium signal. This receptor segment is denoted as the CAIN domain (for $\underline{Ca}^{2+}$/$\underline{In}$ternalization domain) (Chen et al., 1989; Chang et al., 1991). However, distinct sub-domains within this segment have been shown to be responsible for each of the two events, that do not appear to be obligatorily coupled (Chang et al., 1991). In fact, it has been demonstrated that Tyr_{992}, present in the CAIN domain, is the critical element for the generation of the calcium signal by the receptor, since such residue when phosphorylated appears to be the docking site for PLCγ (Chang et al., 1991; Vega et al., 1992). Interestingly, the tandem duplication of a region comprising both the tyrosine kinase and the CAIN domains generates a 190 kDa EGFR mutant form that is coexpressed with the wild-type 170 kDa receptor in astrocytoma cells, which could contribute to its oncogenic potential (Fenstermaker et al., 1998).

Although most of the EGFR molecules become proteolytically degraded in lysosomes after internalization, the cytosolic Ca^{2+}-dependent protease calpain appears to play as well a role in its degradation, transforming the native 170 kDa receptor into a major membrane-bound 150 kDa fragment both *in vitro* and *in vivo* (Cassel and Glaser, 1982; Gates and King, 1983, 1985; Yeaton et al., 1983; Stoscheck et al., 1988). Using a purified recombinant EGFR cytoplasmic domain, it has been possible to identify seven proteolytic cleavage sites in the EGFR, two of which are located between the tyrosine kinase domain and the C-terminal segment containing the tyrosine residues that become autophosphorylated (Gregoriou et al., 1994). Most interestingly, EGF also induces the activation of calpain, a process that could be mediated by the Ca^{2+} signal generated by the own receptor. The activation of this protease brings about the subsequent detachment of the rear edge of the cell during EGF-induced motility, a process that is inhibited by IP-10, an interferon-inducible protein which thus plays an important counterbalancing role by diminishing fibroblast migration during wound healing (Shiraha et al., 1999).

10. CONCLUDING REMARKS

As it has been briefly outlined in the preceding sections, the interplay between the calcium signal and the regulation of the activity and fate of the EGFR is quite complex, and in this elaborated network some questions remain unanswered. We believe that the most urgent problems to be solved in the near future in this context are: (i) to establish the mechanism for the Ca^{2+}/calmodulin-dependent transactivation of the EGFR mediated by membrane depolarization and G protein-coupled receptors; (ii) to establish whether calmodulin is phosphorylated in whole cells by the EGFR and other ErbB family members, and to identify the role that phospho(Tyr)calmodulin may play on these receptors and the myriad of calmodulin-dependent systems present in the cell; (iii) to identify new Ca^{2+}-dependent systems which become directly or indirectly phosphorylated by the EGFR and to establish any change in their functions, and (iv) to identify the consequences of the alteration of these regulatory systems in tumor cells.

ACKNOWLEDGMENTS

The research in the authors' laboratory was financed by grants (to A.V.) from the Comisión Interministerial de Ciencia y Tecnología (SAF99-0052), the Consejería de Educación de la Comunidad de Madrid (08.5/0019/1997), and the Agencia Española de Cooperación Internacional (99CN0011). We apologize to all colleagues whose original works have not been cited because of space limitations.

REFERENCES

Akiyama, T., Kadowaki, T., Nishida, E., Kadooka, T., Ogawara, H., Fukami, Y., Sakai, H., Takaku, F. and Kasuga, M., 1986, Substrate specificities of tyrosine-specific protein kinases toward cytoskeletal proteins *in vitro*, *J. Biol. Chem.* 261, 14797–14803.

Alroy, I. and Yarden, Y., 1997, The ErbB signaling network in embryogenesis and oncogenesis: signal diversification through combinatorial ligand-receptor interactions. *FEBS Lett.* 410, 83–86.

Azzi, A., Boscoboinik, D. and Hensey, C., 1992, The protein kinase C family, *Eur. J. Biochem.* 208, 547–557.

Benaim, G., Cervino, V. and Villalobo, A., 1998, Comparative phosphorylation of calmodulin from trypanosomatids and bovine brain by calmodulin-binding protein kinases, *Comp. Biochem. Physiol. Part C* 120, 57–65.

Benguría, A. and Villalobo, A., 1993, Calmodulin and the epidermal growth factor receptor: A reciprocal regulation?, *Bio-Reguladores* 2, 74–85.

Benguría, A., Hernández-Perera, O., Martínez-Pastor, M.T., Sacks, D.B. and Villalobo, A., 1994, Phosphorylation of calmodulin by the epidermal-growth-factor-receptor tyrosine kinase, *Eur. J. Biochem.* 224, 909–916.

Benguría, A., Martín-Nieto, J., Benaim, G. and Villalobo, A., 1995, Regulatory interaction between calmodulin and the epidermal growth factor receptor, *Ann. N.Y. Acad. Sci.* 766, 472–476.

Bierman, A.J., Koenderman, L., Tool, A.J. and de Laat, S.W., 1990. Epidermal growth factor and bombesin differ strikingly in the induction of early responses in Swiss 3T3 cells, *J. Cell. Physiol.* 142, 441–448.

Browaeys-Poly, E., Cailliau, K. and Vilain, J.P., 1998, Fibroblast and epidermal growth factor receptor expression in *Xenopus* oocytes displays distinct calcium oscillatory patterns, *Biochim. Biophys. Acta* 1404, 484–489.

Carpenter, G., 1987, Receptors for epidermal growth factor and other polypeptide mitogens, *Annu. Rev. Biochem.* 56, 881–914.

Cassel, D. and Glaser, L., 1982, Proteolytic cleavage of epidermal growth factor receptor: A Ca^{2+}-dependent, sulfhydryl-sensitive proteolytic system in A431 cells, *J. Biol. Chem.* 257, 9845–9848.

Chang, C.-P., Kao, J.P.Y., Lazar, C.S., Walsh, B.J., Wells, A., Wiley, H.S., Gill, G.N. and Rosenfeld, M.G., 1991, Ligand-induced internalization and increased cell calcium are mediated via distinct structural elements in the carboxyl terminus of the epidermal growth factor receptor, *J. Biol. Chem.* 266, 23467–23470.

Chen, W.S., Lazar, C.S., Lund, K.A., Welsh, J.B., Chang, C.-P., Walton, G.M., Der, C.J., Wiley, H.S., Gill, G.N. and Rosenfeld, M.G., 1989, Functional independence of the epidermal growth factor receptor from a domain required for ligand-induced internalization and calcium regulation, *Cell* 59, 33–43.

Chernoff, J., Sells, M.A. and Li, H.C., 1984, Characterization of phosphotyrosyl-protein phosphatase activity associated with calcineurin, *Biochem. Biophys. Res. Commun.* 121, 141–148.

Cheyette, T.E. and Gross, D.J., 1991, Epidermal growth factor-stimulated calcium ion transients in individual A431 cells: Initiation kinetics and ligand concentration dependence, *Cell Regul.* 2, 827–840.

Clandinin, T.R., DeModena, J.A. and Sternberg, P.W., 1998, Inositol trisphosphate mediates a RAS-independent response to LET-23 receptor tyrosine kinase activation in *C. elegans*, *Cell* 92, 523–533.

Corti, C., LeClerc L'Hostis, E., Quadroni, M., Schmid, H., Durussel, Y., Cox, J., Hatt, P.D., James, P. and Carafoli, E., 1999, Tyrosine phosphorylation modulates the interaction of calmodulin with its target proteins, *Eur. J. Biochem.* 262, 790–802.

Countaway, J.L., McQuilkin, P., Girones, N. and Davis, R.J., 1990, Multisite phosphorylation of the epidermal growth factor receptor: Use of site-directed mutagenesis to examine the role of serine/threonine phosphorylationm, *J. Biol. Chem.* 265, 3407–3416.

Countaway, J.L., Nairn, A.C. and Davis, R.J., 1992, Mechanism of desensitization of the epidermal growth factor receptor protein-tyrosine kinase, *J. Biol. Chem.* 267, 1129–1140.

Davis, R.J., 1988, Independent mechanisms account for the regulation by protein kinase C of the epidermal growth factor receptor affinity and tyrosine-protein kinase activity, *J. Biol. Chem.* 263, 9462–9469.

De Frutos, T., Martín-Nieto, J. and Villalobo, A., 1997, Phosphorylation of calmodulin by permeabilized fibroblasts overexpressing the human epidermal growth factor receptor, *Biol. Chem.* 378, 31–38.

Díez, J.A., Elvira, M. and Villalobo, A., 1995, Phosphorylation of connexin-32 by the epidermal growth factor receptor tyrosine kinase, *Ann. N.Y. Acad. Sci.* 766, 477–480.

Díez, J.A., Elvira, M. and Villalobo, A., 1998, The epidermal growth factor receptor tyrosine kinase phosphorylates connexin32, *Mol. Cell. Biochem.* 187, 201–210.

Eguchi, S., Numaguchi, K., Iwasaki, H., Matsumoto, T., Yamakawa, T., Utsunomiya, H., Motley, E.D., Kawakatsu, H., Owada, K.M., Hirata, Y., Marumo, F. and Inagami, T., 1998, Calcium-dependent epidermal growth factor receptor transactivation mediates the angiotensin II-induced mitogen-activated protein kinase activation in vascular smooth muscle cells, *J. Biol. Chem.* 273, 8890–8896.

Elliget, K.A., Phelps, P.C. and Smith, M.W., 1996, Transforming growth factor beta modulation of the epidermal growth factor Ca^{2+} signal and c-Fos oncoprotein levels in A431 human epidermoid carcinoma cells, *Cell Growth Differ.* 7, 461–468.

Enomoto, K., Cossu, M.F., Maeno, T., Edwards, C. and Oka, T., 1986, Involvement of the Ca^{2+}-dependent K^+ channel activity in the hyperpolarizing response induced by epidermal growth factor in mammary epithelial cells, *FEBS Lett.* 203, 181–184.

Feinmesser, R.L., Gray, K., Means, A.R. and Chantry, A., 1996, HER-2/c-erbB2 is phosphorylated by calmodulin-dependent protein kinase II on a single site in the cytoplasmic tail at threonine-1172, *Oncogene* 12, 2725–2730.

Feinmesser, R.L., Wicks, S.J., Taverner, C.J. and Chantry, A., 1999, Ca^{2+}/calmodulin-dependent kinase II phosphorylates the epidermal growth factor receptor on multiple sites in the cytoplasmic tail and serine 744 within the kinase domain to regulate signal generation, *J. Biol. Chem.* 274, 16168–16173.

Fenstermaker, R.A., Ciesielski, M.J. and Castiglia, G.J., 1998, Tandem duplication of the epidermal growth factor receptor tyrosine kinase and calcium internalization domains in A-172 glioma cells, *Oncogene* 16, 3435–3443.

Fleet, A., Ashworth, R., Kubista, H., Edwards, H., Bolsover, S., Mobbs, P. and Moss, S.E., 1999, Inhibition of EGF-dependent calcium influx by annexin VI is splice form-specific, *Biochem. Biophys. Res. Commun.* 260, 540–546.

Gates, R.E. and King, L.E., Jr., 1983, Proteolysis of the epidermal growth factor receptor by endogenous calcium-activated neutral protease from rat liver, *Biochem. Biophys. Res. Commun.* 113, 255–261.

Gates, R.E. and King, L.E., Jr., 1985, Different forms of the epidermal growth factor receptor kinase have different autophosphorylation sites, *Biochemistry* 24, 5209–5215.

Gilligan, A., Prentki, M., Glennon, C. and Knowles, B.B., 1988, Epidermal growth factor-induced increases in inositol trisphosphates, inositol tetrakisphosphates, and cytosolic Ca^{2+} in a human hepatocellular carcinoma-derived cell line, *FEBS Lett.* 233, 41–46.

Gregoriou, M., Willis, A.C., Pearson, M.A. and Crawford, C., 1994, The calpain cleavage sites in the epidermal growth factor receptor kinase domain, *Eur. J. Biochem.* 223, 455–464.

Himpens, B., de Smedt, H. and Casteels, R., 1993, Intracellular Ca^{2+} signaling induced by vasopressin, ATP, and epidermal growth factor in epithelial LLC-PK1 cells, *Am. J. Physiol.* 265, C966–C975.

Hudson, P.L., Pedersen, W.A., Saltsman, W.S., Liscovitch, M., MacLaughlin, D.T., Donahoe, P.K. and Blusztajn, J.K., 1994, Modulation by sphingolipids of calcium signals evoked by epidermal growth factor, *J. Biol. Chem.* 269, 21885–21890.

Hughes, B.P., Crofts, J.N., Auld, A.M., Read, L.C. and Barritt, G.J., 1987, Evidence that a pertussis-toxin-sensitive substrate is involved in the stimulation by epidermal growth factor and vasopressin of plasma-membrane Ca^{2+} inflow in hepatocytes, *Biochem. J.* 248, 911–918.

Hughes, A.R., Bird, G.S., Obie, J.F., Thastrup, O. and Putney, J.W., Jr., 1991, Role of inositol (1,4,5)trisphosphate in epidermal growth factor-induced Ca^{2+} signaling in A431 cells, *Mol. Pharmacol.* 40, 254–262.

Hunter, T., Ling, N. and Cooper, J.A., 1984, Protein kinase C phosphorylation of the EGF receptor at a threonine residue close to the cytoplasmic face of the plasma membrane, *Nature* 311, 480–483.

Kanemitsu, M.Y. and Lau, A.F., 1993, Epidermal growth factor stimulates the disruption of gap junctional communication and connexin43 phosphorylation independent of 12-*O*-tetradecanoylphorbol-13-acetate-sensitive protein kinase C: The possible involvement of mitogen-activated protein kinase, *Mol. Biol. Cell* 4, 837–848.

Kuryshev, Y.A., Naumov, A.P., Avdonin, P.V. and Mozhayeva, G.N., 1993, Evidence for involvement of a GTP-binding protein in activation of Ca^{2+} influx by epidermal growth factor in A431 cells: Effects of fluoride and bacterial toxins, *Cell Signal* 5, 555–564.

Livneh, E., Dull, T.J., Berent, E., Prywes, R., Ullrich, A. and Schlessinger, J., 1988, Release of a phorbol ester-induced mitogenic block by mutation at Thr-654 of the epidermal growth factor receptor, *Mol. Cell. Biol.* 8, 2302–2308.

Logsdon, C.D. and Williams, J.A., 1984, Intracellular Ca^{2+} and phorbol esters synergistically inhibit internalization of epidermal growth factor in pancreatic acini, *Biochem. J.* 223, 893–900.

Lund, K.A., Lazar, C.S., Chen, W.S., Walsh, B.J., Welsh, J.B., Herbst, J.J., Walton, G.M., Rosenfeld, M.G., Gill, G.N. and Wiley, H.S., 1990, Phosphorylation of the epidermal growth factor receptor at threonine 654 inhibits ligand-induced internalization and down-regulation, *J. Biol. Chem.* 265, 20517–20523.

Margolis, B., Rhee, S.G., Felder, S., Mervic, M., Lyall, R., Levitzki, A., Ullrich, A., Zilberstein, A. and Schlessinger, J., 1989, EGF induces tyrosine phosphorylation of phospholipase C-II: A potential mechanism for EGF receptor signaling, *Cell* 57, 1101–1107.

Martín-Nieto, J. and Villalobo, A., 1998, The human epidermal growth factor receptor contains a juxtamembrane calmodulin-binding site, *Biochemistry* 37, 227–236.

McManus, M.J., Lingle, W.L., Salisbury, J.L. and Maihle, N.J., 1997, A transformation-associated complex involving tyrosine kinase signal adapter proteins and caldesmon links v-erbB signaling to actin stress fiber disassembly, *Proc. Natl. Acad. Sci. USA* 94, 11351–11356.

Medema, J.P., Sark, M.W., Backendorf, C. and Bos, J.L., 1994, Calcium inhibits epidermal growth factor-induced activation of $p21^{ras}$ in human primary keratinocytes, *Mol. Cell. Biol.* 14, 7078–7085.

Mikoshiba, K., 1993, Inositol 1,4,5-trisphosphate receptor, *Trends Pharmacol. Sci* 14, 86–89.

Mishra, S. and Hamburger, A.W., 1993, Role of intracellular Ca^{2+} in the epidermal growth factor induced inhibition of protein tyrosine phosphatase activity in a breast cancer cell line, *Biochem. Biophys. Res. Commun* 191, 1066–1072.

Moolenaar, W.H., Aerts, R.J., Tertoolen, L.G.J. and de Laat, S.W., 1986, The epidermal growth factor-induced calcium signal in A431 cells, *J. Biol. Chem.* 261, 279–284.

Morrison, P., Takishima, K. and Rosner, M.R., 1993, Role of threonine residues in regulation of the epidermal growth factor receptor by protein kinase C and mitogen-activated protein kinase, *J. Biol. Chem.* 268, 15536–15543.

Morrison, P., Saltiel, A.R. and Rosner, M.R., 1996, Role of mitogen-activated protein kinase kinase in regulation of the epidermal growth factor receptor by protein kinase C, *J. Biol. Chem.* 271, 12891–12896.

Murasawa, S., Mori, Y., Nozawa, Y., Gotoh, N., Shibuya, M., Masaki, H., Maruyama, K., Tsutsumi, Y., Moriguchi, Y., Shibazaki, Y., Tanaka, Y., Iwasaki, T., Inada, M. and Matsubara, H., 1998, Angiotensin II type 1 receptor-induced extracellular signal-regulated protein kinase activation is mediated by Ca^{2+}/calmodulin-dependent transactivation of epidermal growth factor receptor, *Circ. Res.* 82, 1338–1348.

Nishibe, S., Wahl, M.I., Rhee, S.G. and Carpenter, G., 1989, Tyrosine phosphorylation of phospholipase C-II *in vitro* by the epidermal growth factor receptor, *J. Biol. Chem.* 264, 10335–10338.

Nishibe, S., Wahl, M.I., Hernández-Sotomayor, S.M.T., Tonks, N.K., Rhee, S.G. and Carpenter, G., 1990, Increase of the catalytic activity of phospholipase C-γ1 by tyrosine phosphorylation, *Science* 250, 1253–1256.

Olsen, R., Santone, K., Melder, D., Oakes, S.G., Abraham, R. and Powis, G., 1988, An increase in intracellular free Ca^{2+} associated with serum-free growth stimulation of Swiss 3T3 fibroblasts by epidermal growth factor in the presence of bradykinin, *J. Biol. Chem.* 263, 18030–18035.

Pallen, C.J., Valentine, K.A., Wang, J.H. and Hollenberg, M.D., 1985, Calcineurin-mediated dephosphorylation of the human placental membrane receptor for epidermal growth factor urogastrone, *Biochemistry* 24, 4727–4730.

Palomo-Jiménez, P.I., Hernández-Hernando, S., García-Nieto, R.M. and Villalobo, A., 1999, A method for the purification of phospho(Tyr)calmodulin free of non-phosphorylated calmodulin, *Prot. Express. Purif.* 16, 388–395.

Pandiella, A., Beguinot, L., Velu, T.J. and Meldolesi, J., 1988, Transmembrane signalling at epidermal growth factor receptors overexpressed in NIH 3T3 cells: Phosphoinositide hydrolysis, cytosolic Ca^{2+} increase and alkalinization correlate with epidermal-growth-factor-induced cell proliferation, *Biochem. J.* 254, 223–228.

Peppelenbosch, M.P., Tertoolen, L.G.J. and de Laat, S.W., 1991, Epidermal growth factor-activated calcium and potassium channels, *J. Biol. Chem.* 266, 19938–19944.

Peppelenbosch, M.P., Tertoolen, L.G.J., den Hertog, J. and de Laat, S.W., 1992, Epidermal growth factor activates calcium channels by phospholipase A2/5-lipoxygenase-mediated leukotriene C4 production, *Cell* 69, 295–303.

Peppelenbosch, M.P., Tertoolen, L.G.J., de Vries-Smits, A.M.M., Qiu, R.-G., M'Rabet, L., Symons, M.H., de Laat, S.W. and Bos, J.L., 1996, Rac-dependent and -independent pathways mediate growth factor-induced Ca^{2+} influx, *J. Biol. Chem.* 271, 7883–7886.

Rosen, L.B. and Greenberg, M.E., 1996, Stimulation of growth factor receptor signal transduction by activation of voltage-sensitive calcium channels, *Proc. Natl. Acad. Sci. USA* 93, 1113–1118.

Rozengurt, E., 1986, Early signals in the mitogenic response, *Science* 234, 161–166.

San José, E., Benguría, A., Geller, P. and Villalobo, A., 1992, Calmodulin inhibits the epidermal growth factor receptor tyrosine kinase, *J. Biol. Chem.* 267, 15237–15245.

Schalkwijk, C.G., Spaargaren, M., Defize, L.H., Verkleij, A.J., van den Bosch, H. and Boonstra, J., 1995, Epidermal growth factor (EGF) induces serine phosphorylation-dependent activation and calcium-dependent translocation of the cytosolic phospholipase A2, *Eur. J. Biochem.* 231, 593–601.

Shiraha, H., Glading, A., Gupta, K. and Wells, A., 1999, IP-10 inhibits epidermal growth factor-induced motility decreasing epidermal growth factor receptor-mediated calpain activity, *J. Cell. Biol.* 146, 243–254.

Smith, L. and Smith, J.B., 1994, Regulation of sodium-calcium exchanger by glucocorticoids and growth factors in vascular smooth muscle, *J. Biol. Chem.* 269, 27527–27531.

Soltoff, S.P., 1998, Related adhesion focal tyrosine kinase and the epidermal growth factor receptor mediate the stimulation of mitogen-activated protein kinase by the G-protein-coupled P2Y2 receptor: Phorbol ester or $[Ca^{2+}]_i$ elevation substitute for receptor activation, *J. Biol. Chem.* 273, 23110–23117.

Stoscheck, C.M., Gates, R.E. and King, L.E., Jr., 1988, A search for EGF-elicited degradation products of the EGF receptor, *J. Cell Biochem.* 38, 51–63.

Strynadka, N.C.J. and James, M.N.G., 1989. Crystal structures of the helix-loop-helix calcium-binding proteins, *Annu. Rev. Biochem.* 58, 951–998.

Sugimoto, T., Stewart, S. and Guan, K.L., 1997, The calcium/calmodulin-dependent protein phosphatase calcineurin is the major Elk-1 phosphatase, *J. Biol. Chem.* 272, 29415–29418.

Theroux, S.J., Latour, D.A., Stanley, K., Raden, D.L. and Davis, R.J., 1992, Signal transduction by the epidermal growth factor receptor is attenuated by a COOH-terminal domain serine phosphorylation site, *J. Biol. Chem.* 267, 16620–16626.

Tinhofer, I., Maly, K., Dietl, P., Hochholdinger, F., Mayr, S., Obermeier, A. and Grunicke, H.H., 1996, Differential Ca^{2+} signaling induced by activation of the epidermal growth factor and nerve growth factor receptors, *J. Biol. Chem.* 271, 30505–30509.

Todderud, G. and Carpenter, G., 1989, Epidermal growth factor: The receptor and its function, *BioFactors* 2, 11–15.

Ullrich, A. and Schlessinger, J., 1990, Signal transduction by receptors with tyrosine kinase activity, *Cell* 61, 203–212.

Vega, Q.C., Cochet, C., Filhol, O., Chang, C.-P., Rhee, S.G. and Gill, G.N., 1992, A site of tyrosine phosphorylation in the C terminus of the epidermal growth factor receptor is required to activate phospholipase C, *Mol. Cell. Biol.* 12, 128–135.

Villalobo, A. and Gabius, H.-J., 1998, Signaling pathways for transduction of the initial message of the glycocode into cellular responses, *Acta Anat* 161, 110–129.

Villalobo, A., Suju, M., Benaim, G., Palomo-Jiménez, P.I., Salerno, M., Martín-Nieto, J. and Benguría, A., 1997, Phosphocalmodulin activates the epidermal growth factor receptor, *Tenth International Symposium on Calcium-Binding Proteins and Calcium Function in Health and Disease*, Lund, Sweden, Abstracts, p. 78.

Wahl, M.I., Daniel, T.O. and Carpenter, G., 1988, Antiphosphotyrosine recovery of phospholipase C activity after EGF treatment of A-431 cells, *Science* 241, 968–970.

Wang, Z., Danielsen, A.J., Maihle, N.J. and McManus, M.J., 1999, Tyrosine phosphorylation of caldesmon is required for binding to the Shc.Grb2 complex, *J. Biol. Chem.* 274, 33807–33813.

Yeaton, R.W., Lipari, M.T. and Fox, C.F., 1983, Calcium-mediated degradation of epidermal growth factor receptor in dislodged A431 cells and membrane preparations, *J. Biol. Chem.* 258, 9254–9261.

Zwick, E., Daub, H., Aoki, N., Yamaguchi-Aoki, Y., Tinhofer, I., Maly, K. and Ullrich, A., 1997, Critical role of calcium-dependent epidermal growth factor receptor transactivation in PC12 cell membrane depolarization and bradykinin signaling, *J. Biol. Chem.* 272, 24767–24770.

Zwick, E., Wallasch, C., Daub, H. and Ullrich, A., 1999, Distinct calcium-dependent pathways of epidermal growth factor receptor transactivation and PYK2 tyrosine phosphorylation in PC12 cells, *J. Biol. Chem.* 274, 20989–20996.

PART TWO

Calcium Related Diseases

Annexinopathies

Harvey B. Pollard and Meera Srivastava

1. INTRODUCTION

Annexins constitute an extensive gene family of Ca^{2+}-dependent phospholipid binding proteins, whose existence has been appreciated for nearly a quarter of a century. With one exception, members of the annexin gene family (abbreviated ANX) are characterized by a characteristic C-terminal tetrad repeat of approximately 70 residues each. The exception is annexin 6, which has eight such repeats. The gene family is highly conserved and the gene is also very ancient (Morgan and Fernandez, 1997). A primitive form of one mammalian family member (ANX7) is found in the slime mold *Dictyostelium* (Doring et al., 1991; Gerke, 1991), in the ascomycete *Neurospora crassa* (Braun et al., 1998), and in the dysenteric protozoan, *Giardia lamblia* (Morgan and Fernandez, 1995). *In vitro*, different mammalian members of the annexin gene family specialize in mediating inflammation (ANX1); blocking blood coagulation (ANX2, 3, 4, 5, 6, and 8); driving membrane fusion (ANX1 and 7); functioning as a Ca^{2+}-activated GTPase (ANX7); blocking phospholipase A2 activity (many ANX's); behaving as a Ca^{2+} channel in membranes (ANX1, 5, 6, 7). A number of annexins are targets for protein kinases, and so have been implicated in cell proliferentiation and differentiation. These problems have been amply documented in several extensive, though aging, reviews (Moss, 1992; Raynal and Pollard, 1994), and will not be further pursued here.

Most of the members of this gene family were discovered using biochemical rather than molecular methods, and by investigators with quite diverse interests. As a consequence the field became riddled with a nomenclatorial chaos of names representing functions, real or anticipated, rather than anything these gene products had in common. Indeed, it was only in 1990 that

Harvey B. Pollard and Meera Srivastava • Department of Anatomy, Physiology and Genetics, and Institute for Molecular Medicine, U.S.U. School of Medicine (USUHS), Bethesda, MD, U.S.A.

R. Pochet, R. Donato, J. Haiech, C. Heizmann and V. Gerke (eds.): Calcium: The Molecular Basis of Calcium Action in Biology and Medicine, 307–316.

the name "annexin" was agreed upon to represent members of this remarkable gene family (Crumpton and Dedman, 1990). Some of these early names, such as lipocortin, placental anticoagulant, calcimedin, synexin, calelectrin, and others, summon up the enthusiasm and hopes of the discoverers, and the extensive range of biological processes that the investigators hoped to explain. However, this enthusiasm has been accompanied by a most insidious problem. While the *in vitro* properties of many of these annexins have clearly announced relevance to important biological problems, demonstrating this relevance, *in vivo*, continues to be a real challenge.

It has been an article of faith among annexinologists that their favorite annexins would eventually prove to be important for fundamental biological processes. But 25 years is a long time to wait. However, as anticipated by the true believers, this faith has been finally found to be fully justified. As a mark of this progress, Jacob Rand (1999) published an editorial in the *New England Journal of Medicine* entitled "Annexinopathies: A New Class of Diseases". In recognition of this paradigm shift, this chapter is entitled "Annexinopathies", and is focussed on those few but robust examples in which dysfunction of specific mammalian annexins unambiguously cause specific types of disease. Thus, what comes next is not a review but an update, which the author hopes will shortly be out of date. The annexinopathies to be described fall into the disease categories of disorders of coagulation (annexins 2 and 5), cardiac development (annexin 6), and insulin secretion (annexin 7).

2. COAGULATION DISORDERS: ANNEXINS 2 AND 5

2.1. Annexin 2

A series of studies by Hajjar and colleagues over the past several years has clearly identified the central role of annexin 2 in binding plasminogen and tissue plasminogen activator (t-PA) to cells. The result of the binding process is the conversion of plasminogen to the clot-dissolving protease plasmin (Hajjar and Krishnan, 1999). Thus, under physiological conditions on endothelial cells, the interaction of plasminogen and t-PA results in promoting fibrinolysis (Hajjar et al., 1994). As a co-factor for t-PA, annexin 2 increases the efficiency of plasmin formation by a factor of 60 (Cesarman et al., 1994). The process is termed "fibrinolytic surveillance", and is central to moment-to-moment survival of mankind.

Tissue plasminogen activator (t-PA) binds to endothelial cells via the N-terminal tail of annexin 2, at a site quite distinct from that where plasminogen binds (Hajjar et al., 1998). Strong evidence supporting this concept lies in the fact that the hexapeptide LCKLSL, corresponding to [7-12]-ANX2, substantially blocks t-PA binding to full length ANX2. The details of this interaction between ANX2 and t-PA have further important epidemiological and clinical consequences for atherosclerosis and myocardial infarction. For example, it is known that high levels of homocysteine count as a positive risk factor for myocardial infarction. Fortunately. folic acid supplements reduce both homocysteine levels and the risk. Interestingly, homocysteine, at physiologically relevant concentrations, prevents the binding of t-PA to ANX2, thereby suppressing the capacity of the plasminogen/t-PA system to promote fibrinolysis (Hajjar and Jacovina, 1998). Theoretically, this is the mechanism by which homocysteine becomes a risk factor for atherothrombosis. A second risk factor for atherothrombosis is the level of the LDL-like particle, Lp(a). Interestingly, Lp(a) inhibits the binding of plasminogen to ANX2 (Hajjar and Jacovina, 1998). Thus the two epidemiologically important risk factors for atherothrombosis, homocysteine and Lp(a), have sites of action on distinct plasminogen and t-PA binding sites on a common molecule, ANX2.

In addition to these physiological functions, Hajjar and colleagues have recently shown that bleeding in acute promyelocytic leukemia (APL) is due to hyperexpression of ANX2 on the surfaces of APL tumor cells (Menell et al., 1999). APL is described as due to a clonal expansion of immature promyelocytes bearing the characteristic balanced translocation, t(15;17)(q22-24; q12-21). APL is associated with a hemorrhagic diathesis, resulting from disseminated intravascular coagulopathy (DIC), abnormal fibrinolysis, or both, and can be treated with all-*trans* retinoic acid.

Menell et al. (1999) show with fluorescein-tagged antibody that APL cells have higher levels of ANX2 that do other types of leukemic cells. In addition, the APL cells stimulate t-PA-dependent plasmin generation twice as efficiently as non-APL cells. Mennel et al. (1999) then show that anti-ANX2 antibody blocks plasmin increases driven by APL cells. They also show that the plasmin increase can be induced in a non-APL cell by tranfecting in the ANX2 gene. Finally, treatment of APL cells with all-*trans* retinoic acid results in a decrease in ANX2 mRNA, and a reduction in plasmin increases. Najjar and colleagues suggest that the mechanism by which humans with APL have such profound bleeding tendencies is the overexpression of cellular ANX2, with resulting hyperactivation of fibrinolytic plasmin activity. These data are summarized in Figure 1.

2.2. Annexin 5

Annexin 5 was discovered initially on the basis of its ability to block blood coagulation (Tait et al., 1988; Reutelsingsperger et al., 1988). Tait and associates isolated their ANX5 from placenta, and presciently called it placental anticoagulant protein I. Reutelsingsperger and colleagues had isolated their ANX5 from aorta, and concluded that endothelial cells were the origin. They termed ANX5 as vascular anticoagulant α. The mechanism of coagulation inhibition involves inhibition of Factor Xa-dependent conversion of prothrombin to thrombin, and the specific action of ANX5 is believed to be interference with the ability of the phospholipid component to support the process. The

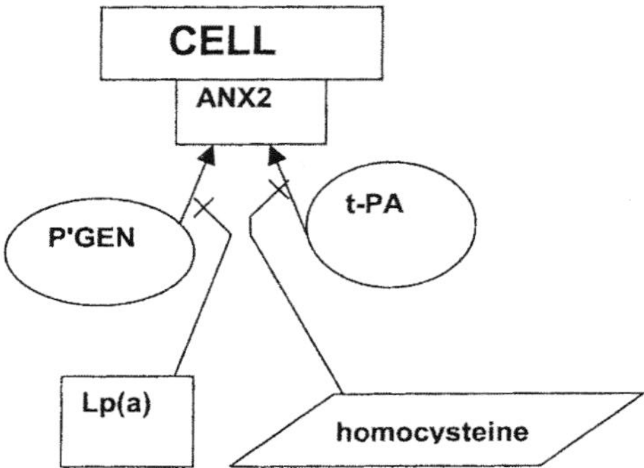

Figure 1. Coagulation processes mediated by annexin 2. Annexin 2 provides independent sites for binding of plasminogen (P'GEN) and tissue plasminogen activator (t-PA). t-PA activates the process of generation of plasmin from plasminogen 60-fold. Each site is antagonized by lipoprotein(a) or by homocysteine, respectively, which are risk factors for atherothrombotic processes.

target for ANX5 is in fact phosphatidylserine (Thiagarajan and Tait, 1990), which becomes exposed on the external leaflet of the plasma membrane as a consequence of damage to the cell. Andree et al. (1992) have proposed that the inhibitory action of ANX5 on coagulation was through displacement of Factor Va, a cofactor for Factor Xa, which is required for Factor Xa to activate prothrombin. Students of the annexin field will be aware of a controversy, sometimes still not resolved, over whether the immediate target of annexin action is the phospholipid or the enzyme(s) whose activities are affected by the annexin (Raynal and Pollard, 1994). However, the commonly accepted mechanism by which ANX5 blocks coagulation is by the formation of a protective shield over exposed acidic phospholipids, thereby displacing coagulation factors from the required phospholipid cofactors.

The application of this information to human disease has come to us in terms of the antiphospholipid (aPL) syndrome, in which patients present with instances of thromboembolism or recurrent pregnancy loss, in association with antibodies against anionic phospholipids (Rand and Wu, 1999). Understanding of this process in terms of ANX5 was initiated by Krikun et al. (1994), who reported that ANX5 is localized to the apical surface of villous trophoblasts in the placenta. Rand et al. (1994) also correlated displacement of ANX7 by these antiphospholipid antibodies with recurrent spontaneous abortion. Subsequently, Vogt et al. (1997) showed that an antibodies to phosphatidylserine could displace ANX5 from this location, thereby permitting coagulation factors such as Factor X and substrate prothrombin access to the lipidic surface. Rand et al. (1998) demonstrated that the specific action of the antiphospholipid antibodies is to inhibit ANX5 binding to phospholipids.

Yet, the mechanism by which this process occurs is still not completely understood (Rand and Wu, 1999). In the specific case of the aPL syndrome, Rand et al. (1994) had shown by immunocytochemistry that placentas from

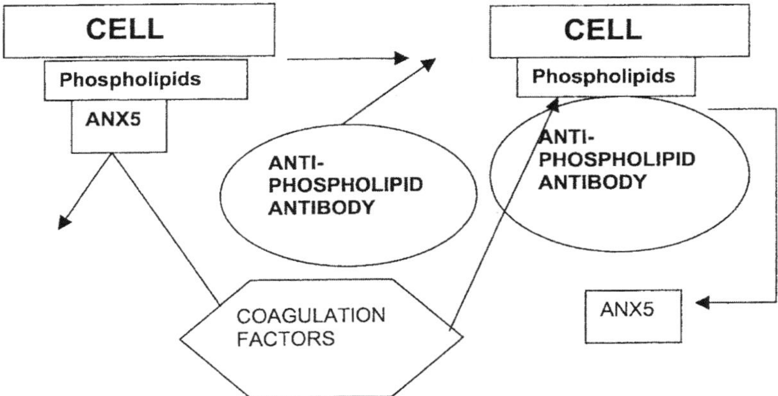

Figure 2. Coagulation processes mediated by annexin 5. Annexin 5 provides a barrier against the binding of coagulation factors to the cell surface. This barrier is antagonized by antiphospholipid antibodies, and is the pathophysiological basis of the antiphospholipid syndrome.

women with aPL syndrome have marked reductions of ANX5 on the apical membranes of syncytiotrophoblasts. However, ANX5 remains on the basal membrane of the syncytiotrophoblasts, and within the underlying cytotrophoblasts. Thus, removal of ANX5 seems limited to those sites where circulating antiphospholipid antibody can reach. The procoagulant effect of the antiphospholipid antibody is therefore proposed to be the removal the protective shield of ANX5 from the phospholipid surface. It is unclear how the antibody to the phospholipid is powerful enough to remove the ANX5, yet is itself sterically insufficient to deny the coagulation factors access to the very surface the antibodies themselves now occupy. These data are summarized in Figure 2.

3. CARDIOVASCULAR CONSEQUENCES OF ANNEXIN 6 EXPRESSION

Annexin 6 is the only known annexin to have eight, not four repeats, and was originally discovered as an inhibitor of ANX7 (Pollard and Scott, 1982). ANX6 has also been shown to have regulatory effects on the ryanodine-sensitive calcium release channel in skeletal muscle, and on the cardiac Na^+/Ca^{2+} exchanger. In an effort to probe the actions of ANX6 on the heart, Gunteski-Hamblin et al. (1996) created a transgenic mouse in which the ANX6 was driven alpha myosin heavy chain promotor. The results were transgenic lines in which the ANX6 levels in the heart were up to ten-fold elevated in both atria and ventricles compared to controls. *In vivo*, the mice exhibited enlarged, dilated hearts, acute diffuse myocarditis, lymphocytic infiltration, moderate to severe fibrosis throughout the heart, and mild fibrosis around the pulmonary veins in the lung. Studies with cardiomyocytes isolated from transgenic tissue revealed lower than normal basal levels of Ca^{2+}, and a reduced rise in free Ca^{2+} following depolarization.

Dedman et al. (1999) have also described the properties of a transgenic mouse in which a dominant negative form of ANX6 is expressed in the heart. This construct consists of a truncated [1-129]-ANX6, with a MW of 16 kDa. These animals reportedly develop ventricular hypertrophy and display signs of dilated cardiomyopathy at 5–6 months of age. At 8 months of age, these animals cannot sustain vigorous exercise, and by 12–14 months of age they develop congestive heart failure and die. Electrophysiologically, cardiomyocytes from these mutant mice have attenuated free Ca^{2+} transients, and reduced rates of inactivation of L-type calcium channels.

Given the interesting cardiovascular consequences of over-expressed full length and dominant negative ANX6, it has been quite a surprise to learn from Moss and colleagues that a nullizygous ANX6 (-/-) knockout mouse has a normal immunological and cardiovascular function (Hawkins et al., 1999). One can only anticipate that other choices for the knockout location in the ANX6 gene may yield a different phenotype.

4. SECRETORY DYSFUNCTION IN THE ANX7 KNOCKOUT MOUSE

Annexin 7 was the first annexin to be isolated, but the seventh to be sequenced (Creutz et al., 1978). The authors had been searching for a calcium-dependent mediator of exocytotic membrane fusion, and named it synexin, after Greek words meaning a "meeting" or a "coming together". This protein has had a history of interesting *in vitro* properties, which have, until recently, been difficult to show relevance to function *in vivo*. For example, *in vitro*, ANX7 forms highly active Ca^{2+} channels in planar lipid bilayers, and drives membrane fusion at nearly diffusion limited rates. However, only recently was it discovered that ANX7 depends for maximal activity on not only Ca^{2+} but also GTP (Caohuy et al., 1996) and Protein Kinase C (Caohuy and Pollard, 2000). Thus, for many years studies by ourselves and others on ANX7 had been inadvertently performed on the less active non-phosphorylated, GDP-form of the protein.

In an effort to determine in some fundamental manner the *in vivo* role of ANX7 for living processes, we prepared a knockout mouse for this gene (Srivastava et al., 1999). Given our anticipation regarding the importance of the ANX7 gene, we were not too surprised to find that the nullizygous Anx7 (-/-) mouse is lethal after embryonic day 10. The cause of death appears to be due to cerebral hemorrhage. No placental pathology was apparent. By contrast, the Anx7(+/-) heterozygote is viable and fertile. However, levels of Anx7 protein are lower than in normal mice, and these mutants also exhibit a variety of dysfunctional phenotypes. Due to an interest in diabetes, we initially directed our attention to insulin secretion by beta cells in the islets of

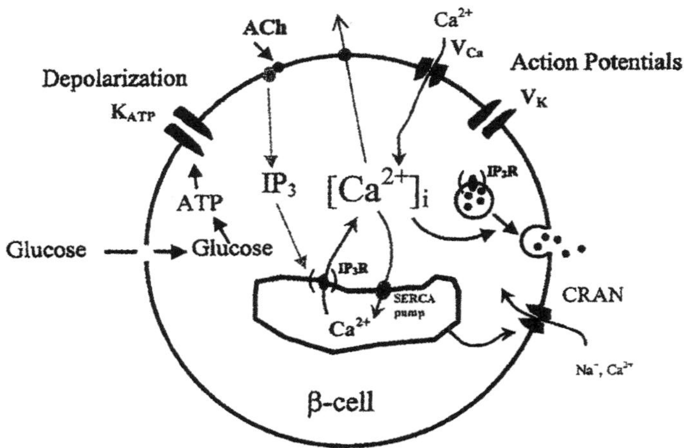

Figure 3. Inhibition of insulin secretion by reduced annexin 7. Insulin secretion is mediated by a complex process involving entry of extracellular Ca^{2+} through voltage-sensitive calcium channels, controlled by glucose-dependent depolarization of K_{ATP}, and recruitment of intracellular Ca^{2+} from intracellular calcium stores. Reduced Anx7 gene dosage in the Anx7(+/-) knockout mouse results in lower levels of Anx7 protein, a ten-fold reduction in expression of IP3Receptor protein, and inhibition of insulin secretion. Figure is from Srivastava et al. (1999).

Langerhans. Histologically, individual islets of Langerhans in these mutant mice are enormous, sometimes encompassing the entire width of the murine pancreas. The hyperplastic islets are made up of hyperplastic beta cells. And within these large beta cells are insulin granules containing up to twice the amounts of insulin found in insulin granules of normal littermate controls. The rate of insulin secretion is attenuated at physiological calcium concentrations, but can be elevated by raising calcium levels to supraphysiological levels. Recruitment of cytosolic Ca^{2+} from intracellular calcium stores is substantially reduced, and store-operated calcium channel activity is virtually undetectable. The mechanism of these secretory difficulties seems to be a substantial (ten-fold) reduction in the levels of IP3 Receptor expressed both in the ER and in secretory granules. These data are summarized in Figure 3. The ability of the heterozygous animals to survive appears to depend on substantial compensatory responses. The beta cells do not function well; so the mouse has responded by increasing the number and insulin content of beta cells. Such changes have been noted in human type 2 diabetes, in which defective secretion is initially accompanied by islet hyperplasia.

5. CONCLUSION AND PROLOGUE

This chapter is ended with the term "prologue" to indicate that these examples of annexinopathies are probably just the beginning. Hopefully, insightful clinicians will continue to look carefully at mysterious annexin-like maladies with the new molecular toolkits now available. It is clear that knockout and transgenic mouse models of the different annexins may also provide functional suggestions which will hopefully cross species lines. It is very likely that nature has had important reasons for vesting massive amounts of effort in maintaining the annexin gene family intact across many species and for millions of years. The review by Raynal and Pollard (1994) summarized the possible realms of actions of the annexins in terms of exocytosis and membrane trafficking, inflammation and anticoagulant properties, and cell proliferation and differentiation. With the advent of the new millenium, it is encouraging that some of these possibilities actually appear to be concretely realized in terms of good science and applications to human disease. We cannot ask for more.

REFERENCES

Andree, H.A.M., Hermens, W.T., Hemker, H.C. and Willems, G.M., 1992, Displacement of Factor Va by annexin V, in *Phospholipid Binding and Anticoagulant Action of Annexin V*, H.A.M. Andree (ed.), Universitaire Pers Maastricht, Maastricht.

Braun, E.l., Kang, S., Nelson, M.A. and Natvig, D.O., 1998, Identification of the first fungal annexin: Analysis of annexin gene duplications and implications for eukaryotic evolution, *J. Mol. Evol.* 47, 531–543.

Cesarman, G.M., Guevara, C.A. and Hajjar, K.A., 1994, An endothelial cell receptor for plasminogen/tissue plasminogen activator(t-PA) II. Annexin II-mediated enhancement of t-PA-dependent plasminogen activation, *J. Biol. Chem.* 269, 21198–21203.

Caohuy, H. and Pollard, H.B., 2000, Activation of annexin 7 by protein kinase C, *in vitro* and *in vivo*, *J. Biol. Chem.*, in press.

Caohuy, H., Srivastava, M. and Pollard, H.B., 1996, Membrane fusion protein synexin (Annexin VII) as a Ca^{2+}/GTP sensor in exocytotic secretion, *Proc. Natl. Acad. Sci. USA* 93, 10797–10802.

Creutz, C.E., Pazoles, C.J. and Pollard, H.B., 1978, Identification and purification of an adrenal medullary protein (synexin) that causes calcium dependent aggregation of isolated chomaffin granules, *J. Biol. Chem.* 253, 2858–2866.

Crumpton, M.J. and Dedman, J.R., 1990, Protein terminology tangle, *Nature* 345, 212.

Dedman, J.R., Kaetzel, M.A., Gebhardt, B., Choi, D., Grupp, I., Yatani, SA. and Matlib, M.A., 1999, Annexins: In search of function, in *Fiftieth Harden Conference: The Annexins*, The Biochemical Society, London.

Doring, V., Schleichter, M. and Noegel, A..A., 1991, *Dictyostelium* annexin VII (synexin). cDNA sequence and isolation of gene disruption mutant, *J. Biol. Chem.* 266, 17509–17515.

Gerke, V., 1991, Identification of a homologue for annexin VII (synexin) in Dictyostelium discoideum, *J. Biol. Chem.* 266, 1697–1700.

Gunteski-Hamblin, A.M., Song, G., Walsh, R.A., Frenzke, M., Boivin, G.P. Dorn, G.W., 2nd, Kaetzel, M.A., Horseman, N.D. and Dedman, J.R., 1996, Annexin VII overexpression targeted to heart alters cardiomyocyre function in transgenic mice, *Am. J. Physiol.* 270, H1091–H1100.

Hajjar, K.A. and Jacovina, A.T., 1998, Modulation of annexin II by homocysteine: Implications for atherothrombosis, *J. Invest. Med.* 46, 364–369.

Hajjar, K.A. and Krishmnan, S., 1999, Annexin II: A mediator of the plasmin/plasminogen activator system, *Trends Cardiovasc. Med.* 9, 128–138.

Hajjar, K.A., Jacovina, A.T. and Chacko, J., 1994, An endothelial cell receptor for plasminogen/tissue plasminogen activator. I. Identity with annexin II, *J. Biol. Chem.* 269, 21191–21197.

Hajjar, K.A., Mauri, L., Jacovina, A.T., Zhong, F., Mirza, U.A., Padovan, J.C. and Chait, B.T., 1998, Tissue plasminogen activator binding to the annexin II tail domain. Direct modulation by homocysteine, *J. Biol. Chem.* 273, 9987–9993.

Hawkins, T.E., Roes, J., Rees, D., Monkhouse, J. and Moss, S.E., 1999, Immunological development and cardiovascular function are normal in annexin VI null mutant mice, *Mol. Cell Biol.* 19, 8028–8032.

Krikun, G., Lockwood, C.J., Wu, X.X., Zhou, X.D., Guller, S., Calandri, C., Guha, A., Nemerson, Y. and Rand, J.H., 1994, The expression of the placental anticoagulant protein annexin V, by villous trophoblasts: Immunolocalizayioon and in vitro regulation, *Placenta* 15, 601–612.

Menell, J.S., Cesarman, G.M., Jacovina, A.T., McLaughlin, M.A., Lev, E.A. and Hajjar, K.A., 1999, Annexin II and bleeding in acute promyelocytic leukemia, *New Eng. J. Med.* 340, 994–1004.

Morgan, R.O. and Fernandez, M.P., 1995, Molecular phylogeny of annexins and identification of a primitive homologue in Giardia lamblia, *Mol. Biol. Evol.* 12, 967–979.

Morgan, R.O. and Fernandez, M.P., 1997, Distinct annexin subfamilies in plants and protists diverged prior to animal annexins and from a common ancestor, *J. Mol. Evol.* 44, P178–188.

Moss, S.E. (ed.), 1992, *The Annexins*, Portland Press, London.

Plow, E.F., Herren, T., Redlitz, A., Miles, L.A. and Hoover-Plow, J.L., 1995, The cell biology of the plasminogen system, *FASEB J.* 9, 939–945.

Pollard, H.B. and Scott, J.H., 1982, Synhibin: A new calcium dependent membrane binding protein that inhibits synexin-induced chromaffin granule aggregation and fusion, *FEBS Lett.* 150, 201–206.

Rand, J.H., 1999, "Annexinopathies" – A new class of diseases, *New Eng. J. Med.* 340, 1035–1036.

Rand, J.H. and Wu, X.X., 1999, Antibody-mediated disruption of the annexin V antithrobotic shield: A new mechanism for thrombosis in the antiphospholipid syndrome, *Thrombosis and Haemostasis* 82, 649–655.

Rand, J.H., Wu, X.X., Guller, S., et al., 1994, Reduction of annexin V (placental anticoagulant protein-I) on placental villi of women with antiphospholipid antibodies and recurrent spontaneous abortion, *Am. J. Obstet. Gynecol.* 171, 1566–1572.

Rand, J.H., Wu, X.X., Andree, H.A.M., et al., 1997, Pregnancy loss in the antiphospholipid-antibody syndrome-a possible thrombogenic mechanism, *New Eng. J. Med.* 337, 154–160 (see Erratum, ibid. 337, 1327).

Rand, J.H., Wu, X.X., Andree, H.A.M., Ross, J.B., Rosinova, E., Gascon-Lema, M.G., Calandri, C. and Harpel, P.C., 1998, Antiphospholipid antibodies accelerate plasma coagulation by inhibiting annexin V binding to phospholipids: A "lupus procoagulant" phenomenon, *Blood* 92, 1652–1660.

Raynal, P. and Pollard, H.B., 1994, Annexins: the problem of assessing the biological role of a gene family of multifunctional calcium-and phospholipid-binding poroteins, *Biochim. Biophys. Acta* 1197, 63–93.

Reutelsingsperger, C.P., Kop. J.M., Hornstra, G. and Hemker, H.C., 1988, Purification and characterization of a novel protein from bovine aorta that inhibits coagulation: Inhibition of the phospholipid dependent Factor Xa-catalyzed prothrombin activation, through a high affinity binding of the anticoagulant to the phospholipids, *Eur. J. Biochem.* 173, 171–178.

Srivastava, M., Atwater, I., Glasman, M., Leighton, X., Goping,G., Caohuy, H., Mears, D., Rojas, E., Westfal, H., Pichil, J. and Pollard, H.B., 1999, Defects in IP3 receptor expression, Ca^{2+}-signaling and insulin secretion in the anx7 (+/-) knockout mouse, *Proc. Natl. Acad. Sci. USA* 96, 13783–13788.

Tait, J.F., Sakata, M., McMullen, B.A., Miao, C.H., Funakoshi, T., Hendrickson, L.E. and Fujikawa, K., 1988, Placental anticoagulant proteins: Isolation and comparative characterization of four members of the lipocortin family, *Biochemistry* 27, 6268–6276.

Thiagarajan, P. and Tait, J.F., 1990, Binding of Annexin V/placental anticoagulant protein I to platelets: evidence for phosphatidylserine exposure in the procoagulant response of activated platelets, *J. Biol. Chem.* 265, 17420–17423.

Vogt, E., Ng, A.K. and Rote, N.S., 1997, Antiphosphatidylserine antibody removes annexin V and facilitates binding of prothrombin at the surface of a choriocarcinoma model of trophoblast differentiation, *Am. J. Obstet. Gynecol.* 177, 964–972.

Autoimmune Disease and Calcium Binding Proteins

P. Eggleton and D.H. Llewellyn

1. INTRODUCTION

Calcium is a crucially important regulatory ion in biological systems. Hence, regulation of its concentration within cells is tightly controlled by a combination of storage proteins, pumps, and ion channels, whilst its functions are mediated through effector proteins that bind it, such as calmodulin. It is becoming increasingly clear that all these types of proteins involved in the control of intracellular Ca^{2+} signalling can be targeted by the body's own immune system, leading to a range of responses from autoantibody production, inactivation of complement, tissue destruction and interference with cell physiology. The production of autoantibodies against calcium binding proteins (CBPs) in patients with autoimmune diseases such as systemic lupus erythematosus (SLE) and rheumatoid arthritis may perhaps simply reflect the non-specific increase in autoantibody levels observed in these patients. Alternatively, several of these CBPs, such as calreticulin and members of the S100/calgranulin C family appear to have homologues on or near the surface of human parasites that may lead to a hyperactive autoimmune response due to some pathological mechanism, involving cross-reactivity of pathogens and host protein. Irrespective of whether these responses are generated specifically, or, result from more general autoimmune mechanisms, it is clear that they can have effects upon Ca^{2+}-dependent physiological responses. In some cases autoantibodies are generated in neurodegenerative disease and lung carcinomas against a number of calcium channels. In amyotrophic lateral sclerosis (ALS), antibodies to neuronal channels trigger the entry of calcium into the cells and activates apoptosis. Calcium channel antibodies also act as

P. Eggleton • MRC Immunochemistry Unit, Department of Biochemistry, University of Oxford, Oxford, U.K., **D.H. Llewellyn** • Department of Medical Biochemistry, University of Wales College of Medicine, Cardiff, Wales, U.K.

R. Pochet, R. Donato, J. Haiech, C. Heizmann and V. Gerke (eds.): Calcium: The Molecular Basis of Calcium Action in Biology and Medicine, 317–331.

clinical markers of some neoplastic syndromes. Rather than attempt to catalogue every autoimmune response that has been documented against proteins involved in intracellular Ca^{2+} regulation, the aim of this chapter is to consider using specific examples, the mechanisms by which autoimmune responses are generated against these proteins, together with the pathophysiological implications of such responses.

2. ER/SR Ca^{2+} BINDING PROTEINS

The ability of cells to store Ca^{2+} in the endoplasmic reticulum (ER) (or in muscle, the sarcoplasmic reticulum, SR) is important since it serves as a pool that can be released upon appropriate stimulation e.g. by the action of inositol tris-phosphate (IP_3), and acts as a sump into which Ca^{2+} can be pumped from the cytosol in order to control the cytosolic Ca^{2+} concentration (see Michalak et al., this book and Parys et al., this book). Within the lumen of the ER, there are proteins with acidic C termini constituting high-capacity, low-affinity Ca^{2+} binding sites, which are considered to act as Ca^{2+}-storage proteins, particularly calreticulin (Corbett et al., 1999), grp78 (or BiP) (Lievremont et al., 1997), and Protein Disulphide Isomerase (PDI) (Lucero et al., 1998). They are also involved in promoting the correct folding and assembly of proteins in the ER, which is the primary site for the processing of membrane-bound and secretory proteins. These proteins, especially calreticulin, are targets for autoimmunity in patients suffering from SLE and rheumatoid disease, whilst antibodies against the major Ca^{2+}-storage protein in muscle cells, calsequestrin, may possibly be involved in eye muscle damage in autoimmune thyroid-associated opthalomopathy (Gunji et al., 1999).

Autoantibodies to calreticulin and grp78, as well as many other autoantigens of diverse structure and function, including double stranded (ds) DNA, ribosomal P protein, histone-DNA complexes and Ro and La complexes, have been found in the sera of patients with SLE and rheumatoid disease. Unfortunately, as yet, it is not completely understood why certain self-components are targeted by the immune system in sufferers of autoimmune disease. Moreover, the question of whether antibody responses to certain intracellular components is indicative of a greater pathological role played by certain antigens, or, is simply a marker of the autoimmune disease in progress, remains debatable. After all, low levels of autoantibody production to host components occurs in all individuals to some degree. Nevertheless, autoantibodies against calreticulin are found in 40–80% of unselected SLE sera and the protein has been implicated in pathological processes such as epitope spreading and complement inactivation.

2.1. The involvement of calreticulin and grp78 in epitope spreading

Recent detailed studies on the autoimmune and physiological properties of calreticulin have begun to provide evidence that suggests it has a specific pathological role in SLE and other autoimmune diseases. Unlike many other host components targeted by the immune system, calreticulin and other chaperones are directly involved in immune-mediated recognition processes i.e. the regulation of the folding and hetero-dimer formation of the major histocompatibility complex (MHC) class I and II molecules involved in antigen presentation. These molecules bind to cytoplasmic peptides derived from either "self", viruses, or tumours and present them on the cell surface for subsequent recognition by CD^{8+} CD^{4+} T cells. Several peptides of calreticulin have been identified in the peptide binding groves of DR molecules of the MHC. Whether these peptides actually trigger a cell-mediated autoimmune response remains unclear.

Epitope spreading is a phenomenon in which a potential autoantigen forms a complex with other host molecules such that the whole becomes antigenic leading to the development of antibody responses directed against other components of the complex. Both grp78 and calreticulin are subject to this phenomenon. For example, in patients with Sjögrens syndrome (SS) and SLE, high titre circulating antibodies are frequently found against linked sets of cell components. Autoantibodies are often directed against the Ro/SS-A complex, a human rheumatic disease autoantigen of unknown cellular function. The complex is composed of 46, 52 and 60 kDa proteins and four small cytoplasmic RNA components designated hyRNA 1, 3, 4 and 5. In the 1980s, calreticulin was proposed to also be a member of the Ro/SS-A complex, which remained a controversial idea until yeast two-hybrid analysis *in vivo* confirmed that calreticulin interacts with Ro52 polypeptide (Cheng et al., 1996). These studies also confirmed that calreticulin can play a supportive role in the formation of the Ro/SS-A complex and bind directly to human cytosolic (hy) RNA. However, calreticulin normally resides in the ER while the Ro polypeptides are associated with cytosolic-based hyRNA. Under normal conditions, these self-antigens must remain non-antigenic to the immune system. However, they do come into contact with one another under specific stress conditions, such as apoptosis, where they co-localise in rough ER apoptotic blebs, which, in keratinocytes, are known to contain Ro52, Ro60, grp78 and calreticulin. During the increased inflammation and apoptosis that occurs during autoimmune disease, presumably the antigens are exposed to the cell surface, taken up by antigen presenting cells and presented to naïve T cells that co-operate with B cells to eventually generate specific autoantibody against these clusters of host proteins.

If this is the case, one should be able to demonstrate specific immunity against clusters of proteins experimentally. McCluskey and colleagues evaluated this possibility by experimental immunisation of mice with one component of the Ro/SS-A complex at a time to determine if there is an immune response against any of the other components (Keech et al., 1996). Immunisation of normal inbred mice with either the 46, 52 and 60 kDa proteins led to a consistent pattern of response, in which high titre production of antibody against the immunised protein occurred first. This was followed 7 to 14 days later by lower and delayed autoantibody production of the other components of the apoptotic ER bleb cluster of proteins. Immunisation of mice with Ro52 and Ro62 leads to the generation of autoantibodies against calreticulin (Kinoshita et al., 1998) and grp78 (Kinoshita et al., 1999). Since grp78 does not associate with the Ro/SS-A complex in non-stressed cells, these immunisation studies provide strong evidence that epitope spreading to the four proteins, Ro52, Ro60, grp78 and calreticulin probably occurs as a result of their association within the same apoptotic compartment. Grp78 contains a binding motif for Ro52 and several potential binding motifs for Ro60 (Kinoshita et al., 1999), while calreticulin binds to Ro52 directly as demonstrated by *in vivo* by yeast two-hybrid analysis (Cheng et al., 1996). Other ER-resident chaperones, such as grp94/96 are also targets of autoimmunity, and suggest these proteins play a general role enhancing the immunogenicity of clusters of antigens with which they associate. This is of concern given that a single immunogenic peptide determinant can readily initiate spreading of an autoimmune response to whole clusters of host molecules in a genetically susceptible host (Figure 1).

The current understanding of epitope spreading requires further investigation to explain why particular host proteins evoke an autoimmune response in the first place. Epitopes buried deep within the folds of the three-dimensional structure a protein are not naturally presented to antigen presenting cells and are termed cryptic. However, some inflammatory trigger or specific cellular activation may render such sites visible to autoimmune T cells. Alternatively, host proteins may share homology with proteins or peptides from pathogens. There are numerous examples of immunodominant homologues of calreticulin in parasites that trigger an antibody response in patients infected with various forms of nematodes, blood flukes and protozoa (Nakhashi et al., 1998). The amino acid sequence homology between the parasitic forms of calreticulin and human calreticulin is approximately 60–70%. Recognition of small peptide epitopes of calreticulin presented by antigen presenting cells to T cells would result in cross-reactivity with host calreticulin homologues by autoreactive T cells. The autoimmune dominant epitopes recognised by antiserum against calreticulin from SLE patients have recently been mapped (Eggleton et al., 2000). Similar mapping studies of anti-calreticulin antibodies from patients with parasitic diseases will reveal if common peptide sequences induce similar B-cell mediated antibody response in different disease states. However, another possibility is that self-peptides that share very little structural homology between various antigens, as is the case with calreticulin, Ro60 and Ro52, can result in epitope spreading due to recognition of the autoantibodies recognising conformational epitopes shared amongst these autoantigens.

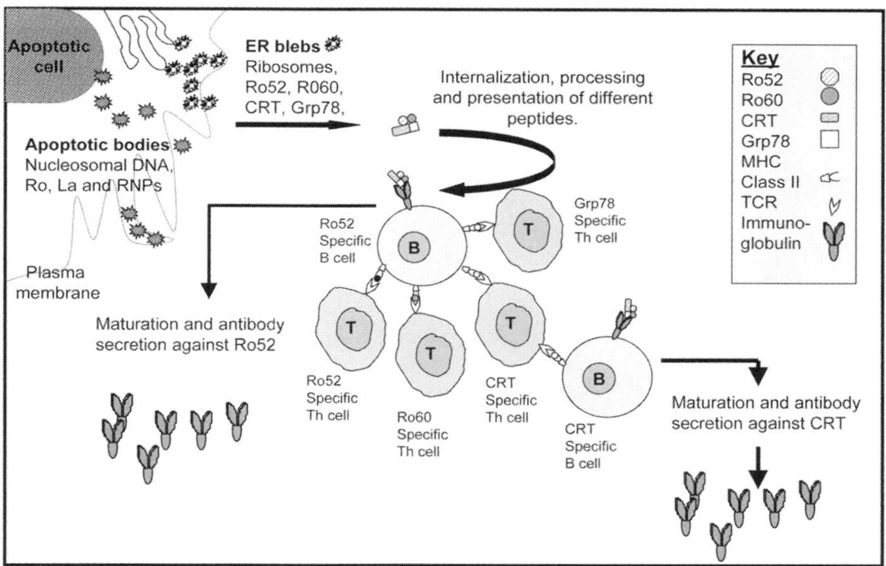

Figure 1. Model showing possible mechanism of B and T-cell co-operation during epitope spreading to calcium binding proteins, calreticulin and grp78. Clusters of autoantigens including the calcium binding proteins – calreticulin and grp78 (BiP) once release into the extracellular environment may be internalised by B cells specific for Ro52 or Ro60. After antigen processing, MHC class II molecules with peptides from the ER cluster of proteins are presented on the cell surface. These are presented to naive Ro52, Ro60, GRP78 and calreticulin Th-cells, leading to their priming and activation. If the initial peptide presented to a specific Th-cell is Ro52, then the B cell will mature and produce Ro52 antibody. If the initial B cell has primed several Th-cells for specificity against the other proteins, then a calreticulin specific B cell can get cognate help from the calreticulin specific TH-cell, resulting in maturation and production of a second antibody.

2.2. The interaction of calreticulin with the first component of complement – Protective or pathological?

As well as being an autoantigen, calreticulin is known to bind C1q, a subunit of the first component of complement (C1), which provides the initial trigger for the activation of the classical complement cascade. In classical complement activation, the C-terminal globular head region of C1q binds to the C_{H2} domains of immune complex (IC)-fixed immunoglobulin. Complement activation is often regarded as an inflammatory process and is most definitely activated during microbial infections. However, it is also a protective process essential in certain autoimmune diseases, where it is involved in the prevention of formation of large insoluble ICs, which might cause tissue injury. In SLE and other autoimmune diseases, both the prevention of antigen-complex formation and their consequent clearance are impaired. Disease symptoms often correlate with decreased levels of complement components, including C1q.

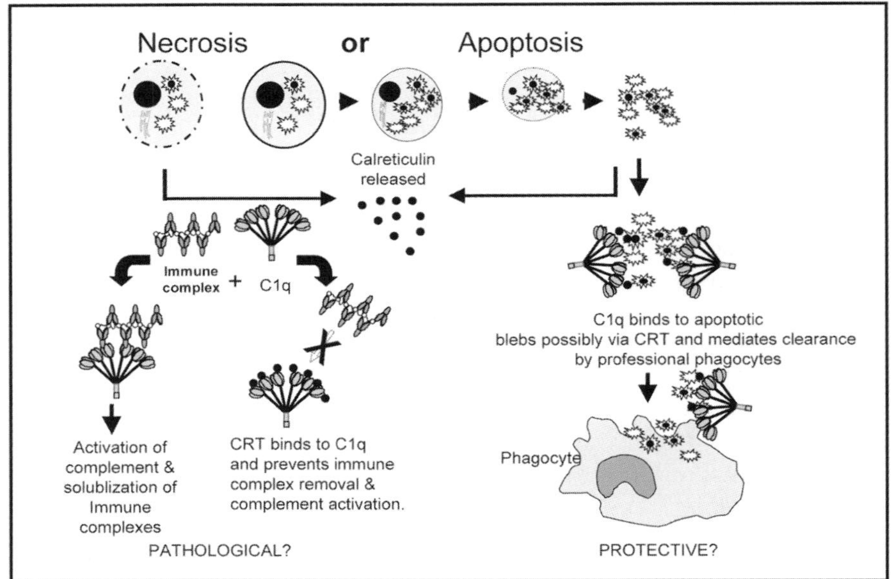

Figure 2. Proposed pathological and physiological interactions between calreticulin and C1q. Once released into the extracellular environment, calreticulin may compete with immune complexes for C1q inhibiting complement activation and immune complex phagocytosis and solubilisation. Alternatively, C1q binds to ER blebs containing calreticulin and other potential autoantigens and may enhance their controlled removal by macrophages.

During pathological flares of activity in SLE, a number of organs are targeted by the immune system leading to localised inflammation and cell death. Calreticulin released from cells either by apoptosis or necrosis is available to bind to C1q (Figure 2). In a comparative study, calreticulin was detected more frequently in SLE sera and in higher amounts than found in control sera (Kishore et al., 1997b). *In vitro*, the binding sites of calreticulin for C1q were identified as being in the N-terminal portion of this 400 amino acid protein (Kishore et al., 1997a). More detailed studies using a series of overlapping calreticulin synthetic peptides revealed that calreticulin contains up to six immunoglobulin C_{H2}-like motifs similar to the ExKxKx C1q binding motif found in the C_{H2} domain of IgG (Kovacs et al., 1998). A number of these peptides were shown to inhibit binding of C1q to IgG. Moreover, several of these peptides and recombinant fragments of calreticulin containing four C_{H2}-like motifs were capable of inhibiting the classical pathway of complement activation. Furthermore, the N-terminal portion of the protein containing the anti-complement activation binding sites is highly resistant to proteolysis by proteolytic enzymes released from leukocytes (Eggleton et al., 2000). This suggests that release of calreticulin from cells in sites of local inflammation, may generate a proteolytic stable fragment capable of interfering with the ability of C1q to associate with immune complexes in autoimmune-related disorders. As for a protective role of C1q-calreticulin interaction, calreticulin-containing blebs of apoptotic keratinocytes directly bind to C1q in the absence of antibody (Korb and Ahearn, 1997). One can speculate that C1q binding to ER blebs possibly via calreticulin, targets the blebs for phagocytosis and removal in order to prevent the immune responses described in detail above, such as epitope spreading (Figure 2).

2.3. Stress-induced expression of ER lumenal proteins and autoimmune disease

It is now clear that expression of grp78, calreticulin, and other ER lumenal Ca^{2+} binding proteins can be upregulated under a variety of cellular stresses, including perturbation of normal ER function (ER stress) (Llewellyn et al., 1996; Waser et al., 1997), amino acid deprivation (Plakidoudymock and McGivan, 1994), heat shock (Conway et al., 1995; Nguyen et al., 1996) and exposure to heavy metals (Nguyen et al., 1996). Such enhanced expression may have a role in autoimmune disease. Viruses have been implicated in the aetiopathogenesis of autoimmune disorders with some such as SS, RA and SLE having been shown to follow infection by cytomegalovirus (CMV) (Einsele et al., 1992; Thorn et al., 1988; Vasquez et al., 1992). It remains an intriguing possibility that viral infection may alter ER function e.g. by CMV (Ahn et al., 1996) such that a stress response is triggered thereby inducing the expression of these proteins. In rheumatoid neutrophils, the expression of calreticulin is enhanced although the mechanism is unclear (Sheikh and Llewellyn, unpublished results). Interestingly, the Ca^{2+} storage capacity of rheumatoid cells has been demonstrated previously to be greatly elevated (Hallgren et al., 1985), whilst abnormalities in Ca^{2+} store release have been reported in neutrophils from patients with rheumatoid and other inflammatory diseases (Davies et al., 1991, 1994).

Stress-induced expression of calreticulin may also have a role in SLE, which is a photosensitive disorder in which symptoms can be exacerbated by exposure to sunlight. UV-B irradiation of transformed keratinocytes can increase the expression of calreticulin at the cell surface, but as yet it remains unknown whether this results from an actual increase in the transcriptional activity of the calreticulin gene.

3. S-100 PROTEINS

The S-100 proteins are a family of acidic low molecular weight (10–12 kDa) EF-hand Ca^{2+} binding proteins, encoded mostly by genes located in a clustered region on human chromosome 1q21. They are believed to be activator proteins, involved in regulating signal transduction mechanisms across the cell membrane. A number of them have been implicated in autoimmune diseases, including S100B (S-100β) in autoimmune disease of the inner ear

(Gloddek et al., 1999), and S100A12 (calgranulin C) in Mooren's disease (Gottsch et al., 1997).

S100A12 is a protein of unknown function comprised of 91-amino acids with a predicted molecular weight of 10.6 kDa (Wicki et al., 1996). It is abundant in neutrophils where, upon secretion from the cell, it is thought to play a role in innate immunity by binding to and targeting foreign antigens for removal by the phagocytic system. Mooren's disease is an autoimmune eye disorder that often presents as a progressive ulceration and destruction of the keratocytes (corneal fibroblasts) in the peripheral cornea. The disease has been observed in patients from India and Nigeria suffering from helminthic infections. In some cases, there is evidence to support an autoimmune process, in which the human neutrophil calcium binding proteins associated with parasite proteins mounts an immune response to similar proteins present in the cornea of the eye. These observations have led for a search for proteins that may be responsible for host-parasite interactions. In 1996, a human protein secreted from activated neutrophils designated calgranulin-related protein (CGRP) was located on the surface of *Onchocerca volvulus* extracts in human subcutaneous nodules. *O. volvulus* is a filarial nematode and the causative agent for river blindness. It was proposed that S100A12, as it was subsequently named, together with defensins, may be involved in attacking and killing mirofilariae once they enter the skin via the bites of *Simulium* flies.

Interestingly, autoantibodies to both defensins and S100A12 have been found in patients with *O. volvulus* infections. It has been speculated that these host molecules are not normally antigenic, but, once they associate with the surface of nematodes, their interaction with parasitic proteins leads to epitope spreading, resulting in the induction of an autoimmune response against both the parasitic and host proteins. The adult stages of *O. volvulus* can live for many years in subcutaneous nodules, multiplying and living in the dermal layers of the skin. Over a long period of time, autoantibodies against S100A12 may be generated eventually coming into contact with the protein in other sites, including the cornea of the eye. This is a possibility based on the recent identification and determination of the primary structure of S100A12 from human and bovine corneal cDNA libraries (Gottsch et al., 1997). Recombinant S100A12, which was subsequently expressed, was recognised by autoantibodies in the sera of 6 of 15 patients with Mooren's ulcer but none of 14 normal control sera by Western blots (Gottsch and Liu, 1998). If an autoantibody response to S100A12 is an integral part of the pathogenesis of the Mooren's ulceration, one might expect upregulation of this calcium binding protein in keratocytes in the presence of pro-inflammatory cytokines. This has been recently demonstrated in cultured bovine keratocytes stimulated with interleukin-1α (IL-1α) and tumour necrosis factor-α (TNF-α). S100A12 was found to be present in the keratocytes in the stromal matrix of the cornea and mRNA levels were upregulated after between 3 and 6 hours of cytokine exposure (Gottsch et al., 1999). Increased levels of this protein locally within

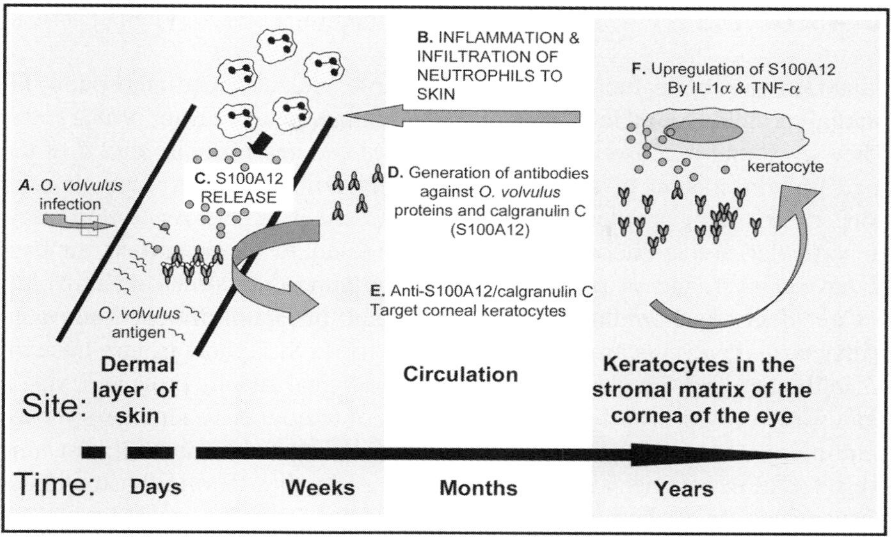

Figure 3. Proposed mechanism of autoimmunity induced by S100A12 (calgranulin C) in the eye disease – Mooren's corneal ulceration. (A) Infection of the skin with the filarial nematode – *Onchocercus volvulus* results in (B) a low-grade inflammatory reaction over months and years. During this period, neutrophils migrate to subcutaneous nodules that contain the adult nematodes. Neutrophils release defensins including S100A12 which (C) associates with the parasite proteins. (D) Eventually, antibodies against parasite antigens and S100A12 are generated. (E) These antibodies circulate for many years and cross-react with S100A12 in other tissues including keranocytes in the eye resulting in Mooren's corneal ulceration. (F) Cytokines upregulated at the site of autoimmunity, upregulate the production of S100A12, which in turn exasperates the overall pathology.

the cornea in association with a hyper-reactive autoimmune response may in part play a pathological role in the destruction of the cornea characteristic of Mooren's ulcer (Figure 3).

A number of neurological human diseases have also been linked to S100B, including Alzheimer's disease, epilepsy and Down's syndrome. In the context of autoimmune disease, there have been reports that S100B can induce an aggressive autoimmune T cell mediated response. Recent experimental evidence involving injection of T cells specific for S100B in Lewis rats, resulted in inflammation of the brain and manifest as autoimmune encephalomyelitis. In the same animals there was also evidence of inflammation in the inner ear cavities leading to labyrinthitis and impaired hearing seven days post transfer of T cells (Gloddek et al., 1999). Interestingly, inflammatory changes induced by S100B have only been reported in the inner ear, eye and brain, yet S100B is expressed in many other organs and tissues. What is more puzzling is how autoreactive T helper cells target S100B antigen across blood-brain and blood labyrinth barriers. This work provides an ongoing model in which a S100 protein is implicated in central nervous system pathology and autoimmune disease.

4. ANNEXINS

Annexins are multi-functional water soluble, calcium and phospholipid-binding proteins found in all most eukaryotes, but not in yeast or prokaryotes. They are found in a diverse variety of tissues and can make up to 2% of the total protein content of the cell (Raynal and Pollard, 1994). Thus, it is of concern that some annexins have been implicated as targets of autoimmunity. In particular, autoantibodies against annexin I and, to a lesser extent, annexin V have been found in patients with autoimmune disease. Annexin V has a Ca^{2+}-dependent binding affinity for anionic phospholipids and activated platelets, and prevents pro-thrombinase activity. In SLE patients, anti-annexin V antibodies may block some of the physiological functions of annexin V. For example, significantly higher incidences of arterial or venous thrombosis, intra-uterine foetal loss, and prolonged activated partial thromboplastin time were found in patients with anti-annexin V antibodies than in those without anti-annexin V antibodies. Consequently, these antibodies may be associated with the pathogenesis of thrombotic events observed in these patients (Kaburaki et al., 1997).

Crystallographic studies of several mammalian annexins in the early 1990s revealed that they all possess a similar domain topology in which five α-helices (A to E) compactly folded to give an overall structure of one domain. The axes of the helices A, B, D, E are orientated in a anti-parallel manner, while helix C runs perpendicular through the middle of the molecule. The result is a partially curved molecule with a convex and concave face. (The details of the structure and functions of annexins in health and disease are dealt with in the chapter by Pollard and Srivastava, this book.) Four of these domains I, II, III and IV are orientated in a cyclic array, forming tight hydrophobic contacts with one another to give the overall structure that contains a total of 20 helices. Both the Ca^{2+}-binding and membrane binding sites of the annexins are located on the convex side of the molecule. However, the crystal structures provide very few clues as to why these proteins provoke an autoimmune response. The more frequent prevalence of autoantibodies against annexin I rather than annexin V in the past has been thought to reflect the fact that the former is expressed both intra- and extracellularly whereas the annexin V is predominantly an intracellular protein (Bastian, 1997). However, recently annexin V as well as annexin I has been shown to be released from cells, but the regulatory mechanism of release for both annexins are not known. However, annexin I has been shown to be secreted by rat neutrophils during inflammation. The comparative concentrations of extracellular annexin I and V released under autoimmune conditions have not been determined, but this may play a part in the more prevalent autoimmune response directed against annexin I. Given the highly similar three-dimensional structure of annexin I and V, different autoantigenic characteristics may reflect the conformational stability of the proteins in an extracellular environment. Thermodynamic calorimetry and protein unfolding studies have shown significant energetic differences exists between annexin I and annexin V despite their overall structural homology which could provide a basis for their functional (Rosengarth et al., 1999) and autoimmune characteristics.

Annexins have been reported to have an number of anti-inflammatory effects. Annexin I has been observed to have the ability to inhibit MAP Kinase activation of phospholipase A_2 activity and neutrophil migration, fever and ischemic brain damage (Donato and Russo-Marie, 1999). In 1981, Hirata and colleagues observed that sera from patients with rheumatoid disease were able to cancel the inhibitory action of lipocortin on phospholipase A2 activity. This effect decreased when the IgM fractions were removed from the sera, suggesting IgM antibodies against annexin I were present in the sera (Hirata et al., 1981), but since annexin I appears to have a number of anti-inflammatory roles, the original conclusions of this early study may be open to further interpretation. In addition, IgM and IgG antibodies to annexin I have been found in the sera of patients with SLE and RA, as well as in the sera of non-autoimmune disease patients. It is still not known for certain whether antibodies to annexins are a direct cause of disease in certain autoimmune disorders, or are simply prevalent in patients who raise a number of autoantibodies to other host antigens. However, as more information on annexin structure and function becomes available, a pathological role for autoantibodies against these proteins is becoming more evident.

5. AUTOIMMUNITY INVOLVING Ca^{2+} CHANNELS

The redistribution or release of cellular components that control cell effector mechanisms often occurs in response to an action potential, which in turn is regulated by the influx of ions through voltage-gated channels. These Ca^{2+} channels are classified at the molecular level according to DNA sequence differences in their α_1 subunits. Four subtypes of calcium channels are known, T, L, N and P/Q-types. The Ca^{2+} channels regulate the movement of Ca^{2+} in a number of neuromuscular operations but are also a target for autoimmunity in a number of paraneoplastic conditions. This in itself is not surprising, since most tumour cells are targeted by the immune system. Host antibodies target specific calcium channels in neurons of the peripheral and central nervous system in various neurological disorders. All subtypes of Ca^{2+} channel are affected by autoantibodies. These antibodies act as clinical markers for a number of neuromuscular disorders such as Lambert–Eaton syndrome, sensory neuronopathy brain-stem encephalopathy and encephalomyelopathy. Whether the antibodies are a cause of the neurological disorders has been a focus of research over the past decade. The best characterised of these disorders is the Lambert–Eaton myasthenic syndrome (LEMS). Normally, acetylcholine is stored in membrane-bound vesicles in pre-synaptic axon terminals near the plasma membrane. Once Ca^{2+} enters an axon terminal through voltage-gated channels the level of Ca^{2+} in the cytosol rise from $< 0.1\ \mu M$ to several micromolar. The Ca^{2+} influx leads to the association of the synaptic vesicle with the plasma membrane, inducing membrane fusion and release of acetylcholine into the synaptic cleft. This neurotransmitter combines with nicotine acetylcholine receptor molecules in the membrane of the post-synaptic cell, resulting in transient increases in both Na^+ and K^+ ions. In LEMS, antibodies against these Ca^{2+} channels are thought to interfere with the initial calcium influx. *In vitro* radiolabelling studies have also revealed that acetylcholine release is inhibited from rat cortical synaptosomes. A large number of other studies by researchers suggest LEMS antibodies downregulate Ca^{2+} channels in both rodent and human carcinoma cell lines (Sher et al., 1998), as well as inhibit hormone release and proliferation by various cells. The precise inhibitory mechanism of LEMS autoantibodies on cellular processes is not understood, but it is clearly more than just an epi-phenomenon.

LEMS autoantibodies do not appear to target Ca^{2+} channels in muscle or cardiac cells, but a number of autoantibodies against a wide spectrum of myocardial antigens do occur in susceptible individuals. Adenine nucleotide translocator (ANT) is one antigen that is a target for autoantibody generation in myocarditis patients. This protein located on the inner mitochondrial membrane in these patients produces antibodies that cross-react with a myolemmal Ca^{2+} channel in isolated myocytes and alter Ca^{2+} fluxes. Recently, antibodies against ANT raised in guinea pigs by immunisation with the isolated protein were shown to markedly affect the isolated hearts of these animals. In particular, the beating hearts presented with impaired myocardial energy metabolism and altered calcium homeostasis (Schulze et al., 1999).

Amyotrophic lateral sclerosis (ALS) is a fatal neurodegenerative disorder in which autoantibodies against L-type and P-type Ca^{2+} channels have been implicated. However, the existence of anti-Ca^{2+} channel antibodies is controversial. Independent researchers have suggested as many as 70–80% of all patients with ALS have anti-L type Ca^{2+} channel antibodies, while others have failed to detect anti-Ca^{2+} channel antibodies at all, suggesting instead that the antibodies are non-specific. Recent research on a small number of ALS patients confirmed the presence of specific anti-L type Ca^{2+} channel antibodies which once purified from patient sera can inhibit the release of dopamine from PC12 cells, providing evidence that autoantibodies to calcium channels specifically target physiological alterations to Ca^{2+} channel-dependent mechanisms (Offen et al., 1998). However it must be borne in mind that such *in vitro* evidence does not provide conclusive evidence as yet that such antibodies are a major causative factor of neuro-degenerative disorders.

6. CONCLUSIONS

In this chapter we have demonstrated that autoimmune responses can occur against almost every type of Ca^{2+} binding protein, ranging from those involved in the regulation of intracellular Ca^{2+} signalling such as calsequestrin, to those whose function is as yet still unclear, e.g. some of the S-100 proteins. The mechanisms by which such proteins become autoantigenic include epitope spreading, conformational similarity and possibly abnormal or aberrant expression of cyptic epitopes. In some cases, it remains unclear why such proteins become targeted by the immune system, but small peptides as well as whole proteins can act as potent autoantigens *in vivo*. However, it may be that as further research reveals the pathophysiological conseqences of the autoimmune responses involving these proteins, then we may be able to discern why that is the case. Moreover, further understanding of the pathogenesis of some autoimmune diseases could help in determining the functions of those Ca^{2+} binding proteins that remain obscure.

ACKNOWLEDGEMENTS

We thank the Arthritis Research Campaign for their generous support (P.E., grant EO521) and the Welsh scheme for Health and Social Research and Nuffield Foundation for financial support (D.H.L.).

REFERENCES

Ahn, K.S., Angulo, A., Ghazal, P., Peterson, P.A., Yang, Y. and Fruh, K., 1996, Human cytomegalovirus inhibits antigen presentation by a sequential multistep process, *Proc. Natl. Acad. Sci. USA* 93, 10990–10995.

Bastian, B.C., 1997, Annexins in cancer and autoimmune disease, *Cell. Mol. Life Sci.* 53, 554–556.

Cheng, S.T., Nguyen, T.Q., Yang, Y.S., Capra, J.D. and Sontheimer, R.D., 1996, Calreticulin binds hYRNA and the 52-kDa polypeptide component of the Ro/SS-A ribonucleoprotein autoantigen, *J. Immunol.* 156, 4484–4491.

Conway, E.M., Liu, L.L., Nowakowski, B., Steinermosonyi, M., Ribeiro, S.P. and Michalak, M., 1995, Heat shock-sensitive expression of calreticulin – *in-vitro* and *in-vivo* up-regulation, *J. Biol. Chem.* 270, 17011–17016.

Corbett, E.F., Oikawa, K., Francois, P., Tessier, D.C., Kay, C., Bergeron, J.J., Thomas, D.Y., Krause, K.H. and Michalak, M., 1999, Ca^{2+} regulation of interactions between endoplasmic reticulum chaperones, *J. Biol. Chem.* 274, 6203–6211.

Davies, E.V., Campbell, A.K., Williams, B.D. and Hallett, M.B., 1991, Single cell imaging reveals abnormal intracellular calcium signals within rheumatoid synovial neutrophils, *Br. J. Rheumatol.* 30, 443–448.

Davies, E.V., Williams, B.D., Whiston, R.J., Cooper, A.M., Campbell, A.K. and Hallett, M.B., 1994, Altered Ca^{2+} signalling in human neutrophils from inflammatory sites, *Ann. Rheum. Dis.* 53, 446–449.

Donato, R. and Russo-Marie, F., 1999, The annexins: Structure and functions, *Cell Calcium* 26, 85–89.

Eggleton, P., Ward, F.J., Johnson, S., Khamashta, M.A., Hughes, G.R.V., Hajela, V., Michalak, M., Corbett, E.F., Staines, N.A. and Reid, K.B.M., 2000, Fine specificity of autoantibodies to calreticulin: Epitope mapping and characterisation, *Clin. Exp. Immunol.* 120, 384–391.

Einsele, H., Steidle, M., Muller, C.A., Fritz, P., Zacher, J. and Schmidt, H., 1992, Demonstration of cytomegalovirus (CMV) DNA and anti-CMV response in the synovial membrane and serum of patients with rheumatoid arthritis, *J. Rheumatol.* 19, 677–681.

Gloddek, B., Lassmann, S., Gloddek, J. and Arnold, W., 1999, Role of S-100b as potential autoantigen in an autoimmune disease of the inner ear, *J. Neuroimmunol.* 101, 39–46.

Gottsch, J.D. and Liu, S.H., 1998, Cloning and expression of human corneal calgranulin C (CO-Ag), *Curr. Eye Res.* 17, 870–874.

Gottsch, J.D., Stark, W.J. and Liu, S.M., 1997, Cloning and sequence analysis of human and bovine corneal antigen (CO-Ag) cDNA: Identification of host-parasite protein Calgranulin C., *Tr. Am. Ophth. Soc.* XCV, 111–125.

Gottsch, J.D., Li, Q., Ashraf, F., O'Brien, T.P., Stark, W.J. and Liu, S.H., 1999, Cytokine-induced calgranulin C expression in keratocytes, *Clin. Immunol.* 91, 34–40.

Gunji, K., Kubota, S., Stolarski, C., Wengrowicz, S., Kennerdell, J.S. and Wall, J.R., 1999, A 63 kDa skeletal muscle protein associated with eye muscle inflammation in Graves' disease is identified as the calcium binding protein calsequestrin, *Autoimmunity* 29, 1–9.

Hallgren, R., Svenson, K., Johansson, E. and Lindh, U., 1985, Abnormal calcium and magnesium stores in erythrocytes and granulocytes from patients with inflammatory connective tissue diseases. Relationship to inflammatory activity and effect of corticosteroid therapy, *Arthritis Rheum.* 28, 169–173.

Hirata, F., del Carmine, R., Nelson, C.A., Axelrod, J., Schiffmann, E. and Warabi, A., 1981, Presence of autoantibody for phospholipase inhibitory protein, lipomodulin, in patients with rheumatic disease, *Proc. Natl. Acad. Sci. USA* 78, 3190–3194.

Kaburaki, J., Kuwana, M., Yamamoto, M., Kawai, S. and Ikeda, Y., 1997, Clinical significance of anti-annexin V antibodies in patients with systemic lupus erythematosus, *Am. J. Hematol.* 54, 209–213.

Keech, C.L., Gordon, T.P. and McCluskey, J., 1996, The immune response to 52-kDa and 60-kDa Ro is linked in experimental autoimmunity, *J. Immunol.* 157, 3694–3699.

Kinoshita, G., Keech, C.L., Sontheimer, R.D., Purcell, A., McCluskey, J. and Gordon, T.P., 1998, Spreading of the immune response from 52 kDaRo and 60 kDaRo to calreticulin in experimental autoimmunity, *Lupus* 7, 7–11.

Kinoshita, G., Purcell, A.W., Keech, C.L., Farris, A.D., McCluskey, J. and Gordon, T.P., 1999, Molecular chaperones are targets of autoimmunity in Ro(SS-A) immune mice, *Clin. Exp. Immunol.* 115, 268–274.

Kishore, U., Sontheimer, R.D., Sastry, K.N., Zaner, K.S., Zappi, E.G., Hughes, G.R., Khamashta, M.A., Strong, P., Reid, K.B. and Eggleton, P., 1997a, Release of calreticulin from neutrophils may alter C1q-mediated immune functions, *Biochem. J.* 322, 543–550.

Kishore, U., Sontheimer, R.D., Sastry, K.N., Zappi, E.G., Hughes, G.R., Khamashta, M.A., Reid, K.B. and Eggleton, P., 1997b, The systemic lupus erythematosus (SLE) disease autoantigen-calreticulin can inhibit C1q association with immune complexes, *Clin. Exp. Immunol.* 108, 181–190.

Korb, L.C. and Ahearn, J.M., 1997, C1q binds directly and specifically to surface blebs of apoptotic human keratinocytes: Complement deficiency and systemic lupus erythematosus revisited, *J. Immunol.* 158, 4525–4528.

Kovacs, H., Campbell, I.D., Strong, P., Johnson, S., Ward, F.J., Reid, K.B. and Eggleton, P., 1998, Evidence that C1q binds specifically to CH2-like immunoglobulin gamma motifs present in the autoantigen calreticulin and interferes with complement activation, *Biochemistry* 37, 17865–17874.

Lievremont, J.P., Rizzuto, R., Hendershot, L. and Meldolesi, J., 1997, BiP, a major chaperone protein of the endoplasmic reticulum lumen, plays a direct and important role in the storage of the rapidly exchanging pool of Ca^{2+}, *J. Biol. Chem.* 272, 30873–30879.

Llewellyn, D.H., Kendall, J.M., Sheikh, F.N. and Campbell, A.K., 1996, Induction of calreticulin expression in hela-cells by depletion of the endoplasmic-reticulum Ca^{2+} store and inhibition of N-linked glycosylation, *Biochem. J.* 318, 555–560.

Lucero, H.A., Lebeche, D. and Kaminer, B., 1998, ERcalcistorin/protein-disulfide isomerase acts as a calcium storage protein in the endoplasmic reticulum of a living cell. Comparison with calreticulin and calsequestrin, *J. Biol. Chem.* 273, 9857–9863.

Michalak, M., Nakamura, K., Papp, S. and Opas, S., Calreticulin and dynamics of the endoplasmic reticulum lumenal environment, this book.

Nakhashi, H.L., Pogue, G.P., Duncan, R.C., Joshi, M., Atreya, C.D., Lee, N.S. and Dwyer, D.M., 1998, Implications of calreticulin function in parasite biology, *Parasitology Today* 14, 157–160.

Nguyen, T.O., Capra, J.D. and Sontheimer, R.D., 1996, Calreticulin is transcriptionally upregulated by heat shock, calcium and heavy metals, *Mol. Immunol.* 33, 379–386.

Offen, D., Halevi, S., Orion, D., Mosberg, R., Stern-Goldberg, H., Melamed, E. and Atlas, D., 1998, Antibodies from ALS patients inhibit dopamine release mediated by L-type calcium channels, *Neurology* 51, 1100–1103.

Parys, J.B., Sienaert, I., Vanlingen, S., Callewaert, G., De Smet, P., Missiaen, L. and De Smedt, H., Regulation of inositol 1,4,5-triphosphate-induced Ca^{2+} release by Ca^{2+}, this book.

Plakidoudymock, S. and McGivan, J.D., 1994, Calreticulin – A stress protein-induced in the renal epithelial-cell line Nbl-1 by amino-acid deprivation, *Cell Calcium* 16, 1–8.

Pollard, H.B. and Srivastava, M., Annexinopathies, this book.

Raynal, C. and Pollard, H.B., 1994, Annexins: The problems of assessing the biological role for a gene family of multifunctional calcium- and phospholipid-binding proteins, *Biochim. Biophys. Acta* 1197, 63–93.

Rosengarth, A., Rosgen, J., Hinz, H.-J. and Gerke, V., 1999, A comparison of the energetics of Annexin I and Annexin V, *J. Mol. Biol.* 288, 1013–1025.

Schulze, K., Heineman, F.W., Schultheiss, H.P. and Balaban, R.S., 1999, Impairment of myocardial calcium homeostasis by antibodies against the adenine nucleotide translocator, *Cell Calcium* 25, 361–370.

Sher, E., Codignola, A., Passafaro, M., Tarroni, P., Magnelli, V., Carbone, E. and Clementi, F., 1998, Nicotinic receptors and calcium channels in small cell lung carcinoma. Functional role, modulation and autoimmunity, *Ann. N.Y. Acad. Sci.* 841, 606–624.

Thorn, J.J., Oxholm, P. and Andersen, H.K., 1988, High levels of complement fixing antibodies against cytomegalovirus in patients with primary Sjögren's syndrome, *Clin. Exp. Rheum.* 6, 71–74.

Vasquez, V., Barzaga, R.A. and Cunha, B.A., 1992, Cytomegalovirus-induced flare of systemic lupus erythematosus, *Heart Lung.* 21, 407–408.

Waser, M., Mesaeli, N., Spencer, C. and Michalak, M., 1997, Regulation of calreticulin gene expression by calcium, *J. Cell. Biol.* 138, 547–557.

Wicki, R., Marenholz, I., Mischke, D., Schafer, B.W. and Heizmann, C.W., 1996, Characterization of the human S-100 A12 (calgranulin C, p6, CAAF1, CGRP) gene, a new member of the S-100 gene classification on chromosome 1q21, *Cell Calcium* 20, 459–464.

Calcium Antagonists and Calcium Sensitizers

P.A. van Zwieten

1. HISTORICAL BACKGROUNDS

The concept of calcium antagonism was postulated independently and almost simultaneously by Godfraind and Kaba (1969) and Fleckenstein and Fleckenstein-Grün (1980). The term calcium antagonists for the drugs which cause vasodilatation by blockade of calcium entry in depolarized arteries was firmly established by Godfraind and Kaba in 1969 and it has been used ever since. The term calcium antagonist has been defined as official nomenclature by the International Union of Pharmacology (IUPHAR). Other terms sometimes encountered in the literature are calcium entry blockers, calcium slow channel blockers or calcium blockers. The term calcium antagonists should be adhered to as the official nomenclature.

The now well-known calcium antagonist (CA) verapamil was originally but incorrectly believed to be a β-blocker, also because of its chemical structure (Figure 1) which indeed resembles that of a β-blocker. In the 1960s verapamil was recognized as an inhibitor of the transmembranous influx of calcium ions in myocardial and vascular smooth muscle cells (Fleckenstein and Fleckenstein-Grün, 1980). Similar findings were described for drugs such as cinnarizine and lidoflazine (Godfraind and Kaba, 1969). Calcium antagonists (CA) as selective inhibitors of calcium influx in depolarized smooth muscle have been recognized as useful therapeutics in the treatment of hypertension and angina pectoris. Verapamil may also be used as an antiarrhythmic agent or in the treatment of obstructive cardiomyopathy.

This chapter will be dealing with the pharmacology and therapeutic use of the CA.

P.A. van Zwieten • Departments of Pharmacotherapy, Cardiology, and Cardiopulmonary Surgery, Academic Medical Centre, University of Amsterdam, Amsterdam, The Netherlands.

R. Pochet, R. Donato, J. Haiech, C. Heizmann and V. Gerke (eds.): Calcium: The Molecular Basis of Calcium Action in Biology and Medicine, 333–363.

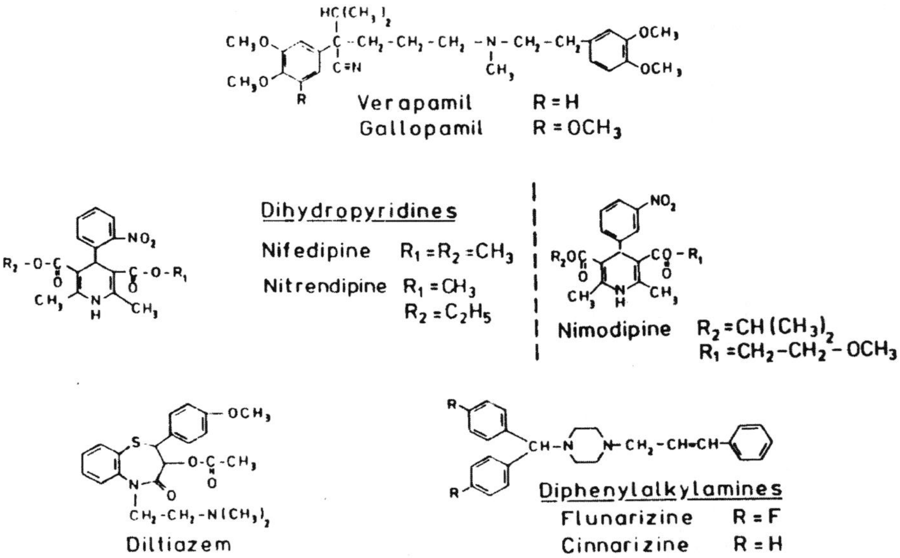

Figure 1. Chemical structures of the major subgroups of calcium antagonists: phenylalkylamines (verapamil and related drugs), benzothiazepines (diltiazem and related drugs), dihydropyridines (nifedipine and related drugs).

2. MODE OF ACTION AT THE CELLULAR LEVEL

Ion channels in the cell membrane which are selective for calcium, potassium and sodium ions, respectively, have been identified and defined (Hess et al., 1984, 1986; Tsien and Tsien, 1990). Calcium channels may be identified by combined electrophysiological techniques. Over the years different types of calcium channels have been identified and called L, T, and N-channels, respectively (Sher et al., 1991). The differentiation of these channels is based on their sensitivity to membrane potential variation and to the time required to reach inactivation.

For the various actions of the CA the L-type channels are by far the most important ones. L-type channels are known to occur in myocardial cells, nodal tissues, and in vascular smooth muscle. Their chemical structure, that is their amino acid sequence has been clarified by means of cloning techniques (Catterall, 1993). Their binding sites for calcium ions have been identified as well (Godfraind, 1997).

L-channels are activated by depolarization, thus facilitating the influx of extracellular calcium ions into the cell. This influx causes, partly via the stimulated release of calcium ions from intracellular sources (sarcoplasmic reticulum, mitochondria, etc.), the activation of the contractile proteins, such as actine, myosin and troponin (Catterall, 1993; Godfraind, 1997). Conversely, the blockade of the L-type channels by CA will cause vascular relaxation and a decrease in myocardial contractile activity (see Figure 2).

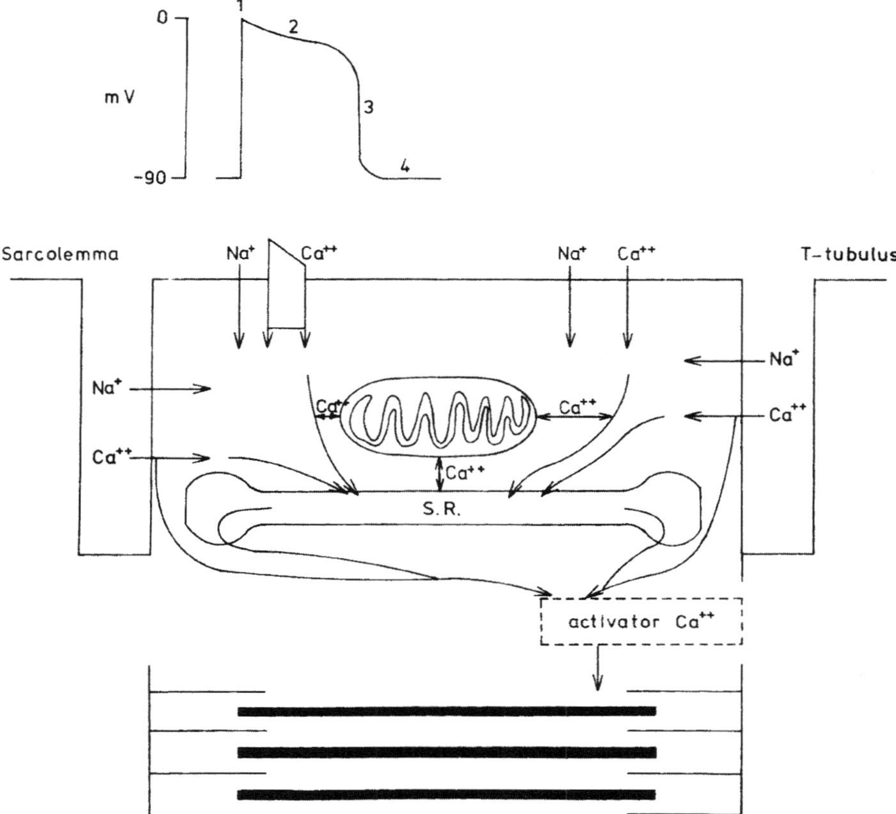

Figure 2. Effect of calcium antagonists on a cardiac cell, with a typical cardiac action potential (top). The calcium (slow) inward current flows during the characteristic plateau phase (phase 2) of the action potential. This calcium influx is selectively inhibited by CA. Activation of the sarcoplasmic reticulum (SR) and other cellular calcium pools occur via Ca^{2+} and Na^{+} which flow into the cell. The sarcoplasmic reticulum and other pools donate activator Ca^{2+} which stimulate the contractile proteins. The presence of tubular systems (invaginations), characteristic of cardiac tissues, causes considerable enlargement of the cellular surface, thus allowing an effective influx of Na^{+} and Ca^{2+}. Inhibition of the calcium inward flux by a CA causes diminished activation of the contractile proteins.

Certain CA (but not all) such as verapamil, mibefradil and diltiazem depress the activity of the nodal tissues, as reflected by impaired A-V conduction and reduced heart rate.

T-channels have recently been identified in nodal cardiac tissues (Mishra and Hermsmeyer, 1994). Their electrophysiological characteristics are different from those of the L-channels: they require a different voltage to be opened, their time of opening is shorter than that of the L-channels, and their capacitance is lower, which means that they can transport fewer calcium ions per unit of time than the L-channels.

T-channels have been proposed as the target of the newer CA mibefradil, which also has affinity for the L-channel (Mehrke et al., 1994; Lüscher et al., 1997). It seems likely, however, that most of the beneficial vasodilator effects of mibefradil can be explained by L-channel blockade (van der Lee et al., 1997). It can be imagined, however, that the negative chronotropic activity is mediated in part by T-channel blockade (Lüscher et al., 1997). More recently some interest has been developed for calcium channels of the N-type, which are predominantly located at the level of various types of neurons (Catterall, 1993; Sher et al., 1991). The release of the neurotransmitter noradrenaline from postganglionic sympathetic neurons is calcium-dependent, and N-type calcium channels are known to be involved in this process. Conversely, the blockade of N-type Ca-channels by particular CA such as mibefradil (van der Lee et al., 1997) and cilnidipine is known to diminish the release of noradrenaline from presynaptic sites at the sympathetic nerve endings.

Other mechanisms than L-channel blockade may contribute to the vasodilator effect of the CA. It has been demonstrated that in vascular smooth muscle the stimulation of both α-adrenoceptors and angiotensin II-receptors with their endogenous agonists (catecholamines and angiotensin II, respectively) will cause the influx of calcium ions into the cell. Conversely, the blockade of L-type calcium channels will reduce this effect. This subtle interaction between CA, α-adrenoceptors and angiotensin II-receptors is assumed to play a part in the vasodilator action of the CA (van Heiningen and van Zwieten, 1988; van Meel et al., 1981; Timmermans and van Zwieten, 1982; van Zwieten et al., 1987), in particular when the sympathetic nervous system and/or the renin-angiotensin-aldosterone system is activated.

3. CLASSIFICATION OF CALCIUM ANTAGONISTS

Blockade of the voltage operated, L-type calcium channel is the common mechanism shared by all subtypes of CA. So far the classification of the various subtypes of CA is based upon their chemical structures (Figure 1). Accordingly, the three major subtypes of CA to be distinguished are:

1. verapamil, gallopamil and mibefradil (phenylalkylamines and derivatives);
2. nifedipine and related drugs (dihydropyridine-CA);
3. diltiazem and related drugs (benzothiazepines).

The group of the dihydropyridine-CA has been expanded recently by several new agents, to be discussed in a separate section. Apart from the three major groups of CA mentioned a few drugs with aberrant structures have been introduced as therapeutics which are formally spoken CA. Flunarizine and bepridil are examples of such agents. Since they do not play an important role in cardiovascular medicine they will not be discussed in this chapter.

4. HAEMODYNAMIC ACTIONS

4.1. Antihypertensive action

Vasodilation is a major property of the CA, underlying most of their therapeutic actions. Vasodilation occurs predominantly at the level of the resistance vessels (precapillary arterioles), thus causing a reduction in elevated peripheral resistance, as in essential and other forms of hypertension (Nayler, 1988a). In therapeutic doses, the CA have no primary dilator effect on the venous system, thus explaining the absence of orthostatic hypotension during treatment with these agents (Gray Ellrodt and Singh, 1983; Nayler, 1988a; Opie, 1990; Struyker Boudier et al., 1989), in spite of their potent antihypertensive activity.

The fall in blood pressure and vascular resistance trigger a sinoaortic baroreflex-mediated rise in sympathetic activity, as reflected by the transient tachycardia associated with nifedipine and other dihydropyridine CA (Gray Ellrodt and Singh, 1983; Nayler, 1988a; Opie, 1990; Struyker Boudier et al., 1989). The direct cardiac actions of verapamil and diltiazem, in addition to their interference with the baroreflex mechanism, prevent a rise in heart rate and cardiac output. As a result of a long-term adaptation process, heart rate and cardiac output return to values close to (dihydropyridines) or slightly below (verapamil, diltiazem) control levels (Figure 3) (Gray Ellrodt and Singh, 1983; Nayler, 1988a; Opie, 1990; Struyker Boudier et al., 1989).

Vasodilation occurs especially in the skeletal muscle and coronary vascular beds. The gastrointestinal, cerebral and renal vascular beds are also dilated, whereas the skin is only slightly affected, although facial flushes may occur as an adverse reaction to dihydropyridine CA.

Vasodilation in the coronary system, in particular the relaxation of coronary spasm, is a major component of the anti-ischaemic activity of CA in coronary heart disease. A mild natriuretic effect, probably at the renal distal tubular level may explain why the CA, although potent vasodilators, do not cause fluid retention. Peripheral (ankle) oedema, a well-known side effect of the dihydro-pyridine CA does not reflect systemic fluid retention, but rather a direct effect on the microcirculation, and possibly also on the lymphatic circulation (Gray Ellrodt and Singh, 1983).

From a haemodynamic point of view, established hypertension is characterized by a rise in peripheral vascular resistance and slowly decreasing cardiac output over the course of several years (Struyker Boudier et al., 1989). Since elevated peripheral resistance is the most consistent haemodynamic change in hypertensive disease, vasodilation at the level of the arterioles (resistance vessels) appears to be a logical approach in treatment. All three different categories of classic CA are, indeed, vasodilators which cause a reduction in peripheral resistance and a concomitant fall in blood pressure in hypertensives (Nayler, 1988a; Opie, 1990; Muiesan et al., 1982). For instance, diltiazem treatment in hypertensives largely maintains the diurnal variations

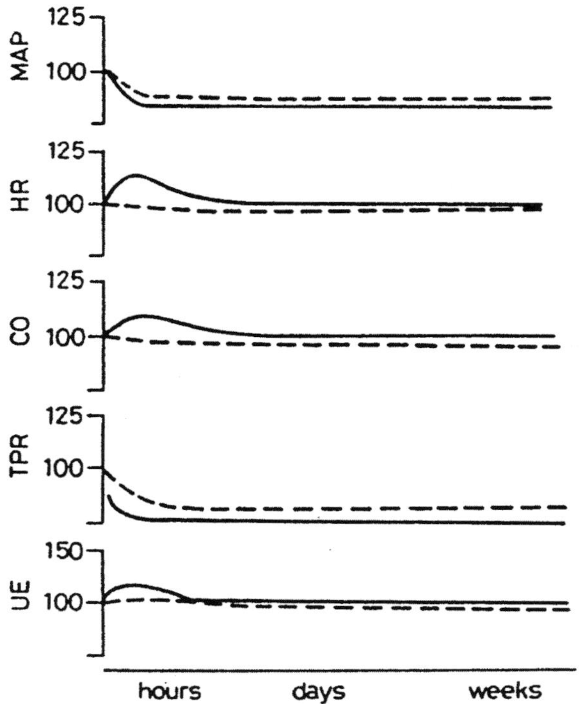

Figure 3. Differential haemodynamic patterns of dihydropyridine CA (nifedipine) and verapamil (or diltiazem). Solid lines, dihydropyridines: dotted lines, verapamil or diltiazem. Reflex tachycardia is evoked by the dihydropyridine CA and also a rise in sodium excretion.

in pressure, though at a lower level than in normotensive subjects (Kiowski et al., 1990). The degree of antihypertensive activity caused by various types of CA is very similar at appropriate doses, but different types have different effects on heart rate, as mentioned above. With long-term use, CA may induce a regression of myocardial and vascular hypertrophy (Agabiti Rosei et al., 1994; Sever, 1985).

Several attempts have been made to establish a relationship between calcium metabolism/fluxes and the pathogenesis of hypertension, but no relationship has so far been demonstrated convincingly.

4.2. Anti-anginal activity

The imbalance between myocardial oxygen supply and demand, which underlies myocardial ischaemia and its sequelae in angina pectoris, may be improved by CA by a reduction in peripheral vascular resistance and, in consequence, a reduction in cardiac afterload and left ventricular wall tension; these mechanisms will reduce myocardial oxygen consumption (van Zwieten, 1985, 1989). A further action of CA is coronary vasodilation, in

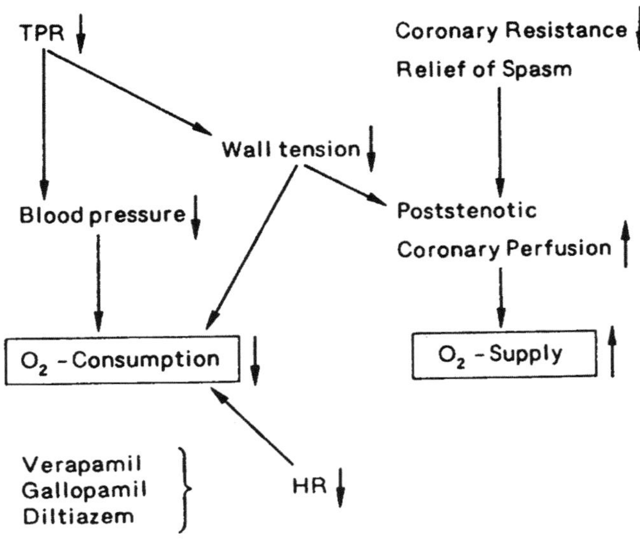

Figure 4. Diagram of the potential mechanism of CA, showing their beneficial effects in angina pectoris. The final result is an improvement in the imbalance between myocardial oxygen demand and supply. TPR: total peripheral resistance.

particular the relaxation of coronary spasm, which improves the myocardial oxygen supply. These two mechanisms are brought about by all groups of CA (dihydropyridines, verapamil, and diltiazem; Figure 4).

A reduction in heart rate, as induced by verapamil or diltiazem, further contributes to the antianginal activity of these two compounds, by reducing the oxygen consumption of the myocardium (Figures 3 and 4). The dihydropyridines, as mentioned, either cause reflex tachycardia (nifedipine in the non-retarded formulation) or leave the heart rate unchanged.

At the cellular level, most CA are thought to preserve ATP levels in the ischaemic heart, as demonstrated in animal and biochemical experiments. However, there is no evidence that this cellular anti-ischaemic mechanism is of clinical importance in the treatment of angina (Nayler, 1988b).

4.3. Secondary prevention after acute coronary syndromes (myocardial infarction, unstable angina)

Verapamil is the only CA for which a clear protective effect has been shown in patients who have suffered an acute coronary syndrome. Increased survival and a lower re-infarction rate were found in the 2nd Danish Verapamil Infarction Trial (DAVIT II) study among patients treated for 12–18 months after a myocardial infarction (verapamil *versus* placebo) (The Danish Study Group, 1990; Fischer Hansen, 1992). In a smaller study with diltiazem, the Diltiazem Multicenter Postinfarction Research Trial, a favourable trend was seen following a myocardial infarction in patients without pulmonary congestion after long-term treatment with diltiazem (compared with placebo) (Multicenter DPTR Group, 1988).

Nifedipine has no protective effect in patients who have survived an acute coronary syndrome. In patients who had experienced unstable angina the incidence of subsequent myocardial infarction was higher in those treated with nifedipine, compared with placebo. This negative effect of nifedipine was suppressed by simultaneous treatment with a β-blocker (HINT Research Group, 1986).

The reason for the cardioprotective effect of verapamil in the DAVIT II study has not been determined. It seems likely, however, that the reduction in the heart rate may have been an important beneficial factor. Conversely, the reflex tachycardia provoked by nifedipine may explain the unfavourable effect of this CA in the HINT study (HINT Research Group, 1986). The favourable effect of adding a β-blocker in this study was associated with the suppression of the nifedipine-induced reflex tachycardia. Taken together the suppression of tachycardia by verapamil is likely to have been an important component of the protective action of this CA in the DAVIT-II trial.

4.4. Antiarrhytmic activity

Verapamil and sometimes diltiazem may be used as antiarrhythmics in the treatment of supraventricular tachy-arrhythmias. The antiarrhythmic activity is based upon an impairment of the electrical activity in the cardiac nodal tissues, causing impaired A-V conduction and a reduction in heart rate. In contrast, the dihydropyridine-CA (nifedipine and related drugs) display no useful antiarrhythmic activity.

In clinical practice verapamil may be used both intravenously and via the oral route. Atrial fibrillation is a major indication, including the atrial fibrillation frequently observed post-operatively after surgical valve reconstruction or coronary bypass surgery (CABG).

5. ADVERSE REACTIONS

The short-term adverse reactions clearly differ with respect to the subtype (chemical class) of CA involved. The most important side-effects may be summarized as follows (van Zwieten, 1992; van Zwieten et al., 1995):

- verapamil and related drugs: constipation; impaired A-V conductance; vasodilatation, flush, headache; reduced cardiac contractile force (cave in patients with heart failure);
- mibefradil does not cause constipation and displays no negative inotropic activity; the adverse reactions to diltiazem are very similar to

those of verapamil, whereas constipation probably occurs less frequently than with verapamil;
- nifedipine and related dihydropyridines frequently cause ankle oedema, probably reflecting a direct action on the local microcirculation, and possibly also the lymphe circulation;
- headache and flush are caused by vasodilation;
- reflex tachycardia, induced by vasodilatation, may be felt by the patient as palpitations.

The side-effect profile of the newer CA, to be discussed in a separate section, appears to be more favourable than that of the non-retarded nifedipine, which is rapid and short acting.

6. LONG-TERM SAFETY

The safety of long-term treatment with CA has been challenged by three studies, two case-control studies in hypertensives (Pahor et al., 1995; Psaty et al., 1995), and a meta-analysis of 16 studies on nifedipine aiming at secondary prevention subsequent to acute coronary syndromes (Furberg et al., 1995). From these studies it was concluded that long-term treatment with CA may be associated with a higher risk of cardiovascular morbidity and mortality. It has also been submitted that long-term use of CA may elevate the risk of cancer in elderly populations (Pahor et al., 1996a). From smaller studies it has been concluded by certain authors that CA would enhance the risk of bleeding in the digestive tract (Pahor et al., 1996b), or during surgical interventions (Wagenknecht et al., 1995). Most of these studies have been subject to emotional and biased, premature publication in the media before they were published in appropriate scientific journals.

In the case-control studies in hypertension, CA appeared to cause a modest increase in the risk of myocardial infarction while the meta-analysis appeared to show that nifedipine was associated with increased mortality in patients with symptomatic coronary heart disease. However, all three studies have severe methodological flaws, and their interpretation has been strongly criticized (Tijssen and Hugenholz, 1996; Mancia and van Zwieten, 1996; WHO-ISH Study Group, 1997) with respect to the following issues: erroneous calculations; conclusions drawn too "heavy" on the basis of too meagre statistics; confounding by indication. Furthermore, it should be realized that the relevance of case control studies is limited, especially if the differences between cases and controls is as modest (increase of risk by not more than 60%) as in the aforementioned cardiovascular studies. As a rule the relevance of case control studies is considered as valid if the difference between cases and controls is at least 300%, as encountered for instance in the large retrospective studies on the association between smoking cigarettes and lung cancer in the 1960s.

There is considerable uncertainty about the negative conclusions drawn from the cardiovascular studies, as reflected by the recent United States Food and Drug Administration (FDA) decision not to change the general policy towards the use of CA as antihypertensive agents. Conversely, more and more data are now beginning to emerge which severely challenge the validity and relevance of the aforementioned studies on CA.

In a case control study by Aursnes et al. (1995), designed similarly as those already mentioned CA did not at all increase the risk of cardiovascular morbidity or mortality and proved even protective. In a cohort study by Braun et al. (1996) in 15502 patients with coronary heart disease the risk ratio of death was not influenced by CA, neither in the positive nor in the negative sense.

In a case-control study by Alderman et al. (1997) hypertensive patients with a history of at least one major cardiovascular event were compared with correctly matched controls. The rapidly acting, non-retarded nifedipine increased the risk of death (RR = 3.88) when compared with a β-blocker. However, the more recently introduced slow- and long-acting CA significantly reduced the risk of death (RR = 0.76) when compared with a β-blocker.

Both in the STONE- and SYST-EUR-studies (Gong et al., 1996; Staessen et al., 1997) total mortality was unchanged by CA-treatment, whereas protection against cerebro- and cardiovascular events was obvious. In neither of the two studies an increased risk of cancer or gastrointestinal bleeding was observed. In a recent communication by Lever (1997) it was reported that in the well-known WOSCOP trial (Sheperd et al., 1995) the survival, cardiovascular morbidity/mortality and incidence of cancer proved uninfluenced by the long-term use of CA. The recently published HOT-study did not show any evidence for noxious effects of felodipine on mortality, MI or cancer.

In conclusion, the challenge of the long-term safety of CA is controversial and not convincing. The major, randomized ongoing studies (STOP-2, ELSA, VHAS, NORDIL, and others) are the only ones which can offer a decisive answer concerning the long-term safety of CA. The data so far available do not require us to condemn all CA. In contrast, once the aforementioned ongoing studies will have been completed and published the CA will be the antihypertensive agents of which the long-term safety and protective action will be known better than those of all other antihypertensives.

Considerable doubt has been cast upon the safety of rapidly and short-acting dihydropyridines such as non-retarded nifedipine. It seems wise to replace this agent by the slow- and long-acting CA discussed in the forthcoming section on the new CA. The validity and safety of these newer CA are subject to critical tests in the aforementioned ongoing trials.

7. DRUG INTERACTIONS (Hansten, 1992; Stockley, 1994)

We only mention a few clinically relevant interactions.

- verapamil or diltiazem + β-blocker: additive depression of A-V conduction, bradycardia. This problem may arise in particular if a patient treated with a β-blocker (long term, orally), receives an i.v. injection of verapamil;
- verapamil + digoxin: verapamil impairs the renal excretion of digoxin and hence raises the plasma level of the cardiac glycoside, thus increasing its toxicity;
- plasma levels and the risk of neurotoxicity of carbamazepine are enhanced by verapamil and diltiazem;
- plasma levels of cyclosporin are increased by verapamil, diltiazem and nicardipine.

8. OUTCOME STUDIES ON THE USE OF CALCIUM ANTAGONISTS

Nifedipine (Adalat®) was introduced as an antihypertensive agent in the 1970s. Somewhat later the antihypertensive efficacy of verapamil and diltiazem were recognized. Since then numerous newer CA have been introduced as antihypertensives, to be discussed in a separate section (Section 12) of this chapter.

The effective reduction of elevated blood pressure by the CA has been described in numerous communications and review papers (see, for instance, Bühler, 1995; Müller et al., 1985; Pedrinelli et al., 1986; Ribstein et al., 1985).

The haemodynamic effects caused by the CA in hypertensives largely follow the principles discussed in a preceding section (Section 4.1). The same holds for the adverse reactions.

In spite of several theoretic discussions there is no specific category of hypertensive patients which should preferably be treated with CA, or rather not. An earlier hypothesis that CA should preferably be used in elderly hypertensive patients (Bühler, 1995) has been dismissed. Accordingly, CA are effective antihypertensives in the following categories of hypertensive patients:

- patients with essential hypertension (Müller et al., 1985; Pedrinelli et al., 1986; Ribstein et al., 1985),
- patients with renovascular hypertension (Rodicio et al., 1993),
- black hypertensive patients (Dustan, 1987),
- patients with both hypertension and diabetes mellitus (since the CA do not influence glucose tolerance or plasma lipids) (Henry, 1990).

CA should be avoided in pregnant women with hypertension. Teratogenic effects of CA in humans are uncertain and have not been clearly described, but animal experiments have demonstrated noxious effects of the CA on the fetus (Rubin, 1995).

There exists no doubt concerning the efficacy of CA in lowering elevated blood pressure, also on long-term use. However, large-scale outcome studies concerning the protection by CA against the complications of hypertension (stroke, myocardial infarction, heart failure, renal disease) have been scarce so far, in spite of the widespread use of these agents in antihypertensive drug treatment for many years. Outcome studies on the effects of CA in angina and heart failure have been scarce but these are now on the way.

8.1. Outcome studies

8.1.1. Treatment of hypertension

The Treatment of Mild Hypertension Study (TOMHS) evaluated 5 classes of antihypertensive monotherapy and placebo, in comparison with life style intervention (Liebson et al., 1995). Accordingly, 902 mild hypertensives were followed for 4 years. Left ventricular mass (LVM) was monitored in 844 of the 902 patients. In this study the following drugs were used: chlorthalidone (diuretic); acebutolol (β-blocker); doxazosin (α_1-adrenoceptor antagonist); amlodipine (Ca-antagonist). These drugs were compared with non-pharmacological treatment, consisting of nutritional-hygienic intervention plus placebo. The drugs used (including the Ca-antagonist amlodipine) were virtually equi-effective in lowering blood pressure. The drugs were somewhat more effective in lowering blood pressure than the non-pharmacological treatment.

The Shanghai Trial of Nifedipine in the Elderly (STONE) (Gong et al., 1996) was a single-blind study on the use of slow-release nifedipine *versus* placebo in elderly Chinese hypertensives. In the active treatment arm of this study nifedipine significantly reduced the overall risk of cardiovascular evenst (including stroke) by almost 60%. The rarity of myocardial infarction in the Chinese patient population may be considered as a weak point of the STONE-study from a Western point of view. However, it is one of the very few placebo-controlled studies where the beneficial effect of a CA against placebo has been clearly demonstrated.

The SYST-EUR-Study was performed in 4695 patients of 60 years and older with isolated systolic hypertension (Staessen et al., 1997). The dihydropyridine-CA nitrendipine was compared with placebo. If necessary enalapril or hydrochlorothiazide were added. The study was stopped prematurely after 2 years of follow up, because the active treatment with nitrendipine proved clearly beneficial when compared with placebo. In particular the incidence of stroke proved significantly and substantially (−42%) reduced by the treatment with nitrendipine.

Hypertension Optimal Treatment Study (HOT) In this very large study it was shown that felodipine caused an intensive lowering of blood pressure which was associated with a lowered rate of cardiovascular events (Hansson et al., 1998). Total mortality was lower than in the treated patients of the Glasgow clinic and there was certainly no indication of increased mortality due to treatment with CA.

Important ongoing studies on CA in hypertensives are for instance:

- STOP-2 (Dahlöf et al., 1993): diuretics, β-blockers compared with ACE-inhibitors and CA, in elderly hypertensives;
- ELSA (Bond et al., 1993): lacidipine, (a lipophilic CA), compared with atenolol, on carotid wall thickness and atherosclerosis in elderly hypertensives;
- VHAS (Zanchetti, 1996): comparison of verapamil (slow release) with chlorthalidone, on carotid wall thickness and atherosclerosis, in hypertensives;
- NORDIL (The Nordil Study Group, 1993): diltiazem compared with a β-blocker and/or diuretic in hypertensives aged 50–60 years;
- ALLHAT (Davis et al., 1996): amlodipine in hypertensives; blood pressure and lipid profile. Most of these studies will be completed within the forthcoming 1–3 years and their results are anxiously awaited.

8.1.2. Ischaemic heart disease and chronic heart failure

International Nifedipine Trial on Antiatherosclerotic Therapy (INTACT) In patients with well-established coronary heart disease, nifedipine treatment significantly reduced the formation of new atherosclerotic lesions in the coronary arterial system (Lichtlen et al., 1990). The number of MI's was not reduced by nifedipine treatment.

Montréal Heart Study The design and outcome were practially the same as in the INTACT study. The patients were treated with nicardipine (Waters et al., 1990).

Canadian Amlodipine/Atenolol in Silent Ischaemia Study (CASIS) In this study patients with stable coronary artery disease and ischaemia during treadmill testing and ambulatory monitoring were treated with amlodipine and atenolol, respectively, as well as with the combination of both drugs (Davies et al., 1995). Amlodipine prolonged exercise time by 29 *versus* 3% prolongation by atenolol. The combination amlodipine/atenolol prolonged exercise time by 34%. Accordingly, ischaemia during treadmill testing was more effectively suppressed by amlodipine than by the β_1-blocker atenolol. During ambulatory monitoring, the frequency of ischaemic episodes decreased by 28% with amlodipine, by 57% with atenolol and by 72% with combined treatment with amlodipine and atenolol.

Consequently, ischaemia during ambulatory monitoring was more effectively suppressed by atenolol than by amlodipine. The combination of both drugs was more effective than either of the two individual drugs in both types of investigations concerning the antianginal activity. There was no evidence whatsoever concerning safety problems with either of the two drugs.

Angina Prognosis Study in Stockholm (APSIS) A randomized, double-blind trial with either metoprolol or verapamil was undertaken in 809 patients under 70 years of age with stable angina pectoris (Rehnqvist et al., 1996). The patients were followed between 6 and 75 months (median 3–4 years). Combined cardiovascular events did not differ and occurred in 30.8 and 29.3% of metoprolol- and verapamil-treated patients, respectively. Total mortality in metoprolol- and verapamil-treated patients was 5.4 and 6.2%, respectively. Cardiovascular mortality was 4.7% in both groups. Non-fatal cardiovascular events occurred in 26.1 and 24.3% of metoprolol- and verapamil-treated patients, respectively.

Total Ischaemic Burden European Trial (TIBET) A randomized, double-blind study of atenolol, nifedipine (SR) and their combination (Dargie et al., 1996) with ambulatory monitoring was undertaken in 682 men and women with chronic stable angina who were followed up for two years. No differences were shown on hard endpoints such as cardiac death, non-fatal myocardial infarction and unstable angina. Furthermore, the study also showed no evidence of an association between the presence, frequency or total duration of ischaemic evenst on Holter monitoring treatment, and the main outcome measures.

Prospective Randomized Evaluation of the Vasular Effects of Norvasc Trial (PREVENT) PREVENT is a multicentre, randomized, placebo-controlled, double-blind clinical trial of 825 patients with predefined angiographic evidence of coronary artery disease (Byington et al., 1997). Patients were treated with amlodipine (5–10 mg daily) or placebo for three years; both quantitative coronary angiography and B-mode ultrasonography were performed in order to establish the progression of coronary artery disease as well as the rate of progression of atherosclerosis in the carotid arteries. Furthermore, several important clinical outcome criteria were monitored, such as mortality, major cardiovascular events (MI, sudden death), congestive heart failure and stroke, as well as the incidence of cancer.

Preliminary data were presented at a recent American Heart Association Meeting (Dallas, November 1998), and a detailed publication will follow in the near future (PREVENT Trial Group, 1998). The major finding was the significant reduction (31%) in cardiovascular and cerebrovascular morbidity and mortality events caused by amlodipine treatment. These events included MI and angina attacks, stroke, cardiovascular death, hospitalizations for severe angina and heart failure. The reduction in the requirement for PTCA and CABG even amounted to 46%. In spite of these highly relevant clinical findings, the coronary angiographic measurements did not show a discernible difference in the progression of coronary atherosclerosis between amlodipine- and placebo-treated patients. In contrast, amlodipine caused a significant retardation of carotid plaque formation when compared with placebo. The apparent discrepancy between the findings in the coronary and carotid arteries may be explained by the different techniques used, taking into account that ultrasonography is clearly more sensitive than coronary angiography. In the course of the study no problems whatsoever were observed concerning the safety of amlodipine.

8.2. Prospective Randomized Amlodipine Survival Evaluation (PRAISE)

Patients with chronic heart failure were treated with amlodipine. A moderate improvement was seen in patients with heart failure caused by idiopathic cardiomyopathy. Amlodipine had no influence, either beneficial or detrimental, on the course of heart failure caused by ischaemic heart disease (Packer et al., 1996).

9. CALCIUM ANTAGONISTS AND LEFT VENTRICULAR HYPERTROPHY

Left ventricular hypertrophy (LVH) is considered as an important, virtually independent risk factor in hypertensives, already recognized in the Framingham Study (Levy et al., 1989). It is therefore an important issue whether treatment with antihypertensive drugs will cause regression of LVH.

Numerous animal experiments have indicated that CA may cause regression of the LVH associated with hypertension. In clinical studies this issue has been followed up and most of these investigations have shown that various types of CA indeed cause regression of LVH on long-term treatment (Agabiti Rosei et al., 1994; Liebson et al., 1995; Thürmann et al., 1996). Attempts have been made to compare the efficacy of different types of antihypertensives in this respect (Devereux, 1997; Malbantgil et al., 1996; Matsuzaki et al., 1997; Schmieder et al., 1987). The impression is obtained that ACE-inhibitors and CA are the most efficacious antihypertensive drugs with respect to the regression of LVH in hypertensives, and probably more effective than β-blockers or diuretics (Devereux, 1997). The studies so far performed do not allow a quantitative and definite conclusion with respect to the different efficacies of various types of drugs, also because of methodological difficulties in the quantitative determination of LVH in patients (Devereux, 1997). Trials designed in such a manner that quantitative comparisons between various drugs can be made are on the way.

Taken together there are sound reasons to assume that long-term antihypertensive treatment (at least for several months) with CA will cause significant and relevant reduction of LVH. For review on this issue see (Agabiti Rosei et al., 1994; Devereux, 1997). In studies with isolated vascular smooth muscle cells in isolated vessel preparations CA have been shown to impair vascular smooth muscle proliferation (Catapano, 1997). This potentially beneficial effect may also occur in the treatment of hypertensive patients. Because of methodological difficulties such an effect has so far not been convincingly demonstrated in human hypertensives treated with CA.

10. CALCIUM ANTAGONISTS AND ATHEROSCLEROSIS

In vitro and animal studies have strongly suggested that CA may impair lipid accumulation in the aorta. CA have been demonstrated to impair a variety of processes which underly atherosclerotic plaque formations, such as cholesterol deposition, cellular proliferation and migration, increased cellular matrix, calcium overload and platelet aggregation (Catapano, 1997). The antiatherogenic activity of CA is not mediated by a reduction in plasma lipid levels, which are not influenced by CA in therapeutically active concentrations. The impression is obtained that lipophilic CA such as lacidipine and lercanidipine are the more active antiatherogenic agents, at least in animal and biochemical experiments. So far it has been very difficult to demonstrate an antiatherogenic effect of CA in human patients.

Nifedipine was shown to impair the formation of new lesions in the coronary arterial system in patients with ischaemic heart disease. Large, established lesions, however, were not affected by nifedipine treatment, and nifedipine did not protect against acute coronary syndromes in these patients (Lichtlen et al., 1990). Comparable findings were obtained for nicardipine in the Montréal Heart Study (Waters et al., 1990). In the MIDAS-Trial isradipine and hydrochlorothiazide were compared in mild to moderate hypertensives. Atherosclerotic plaque formation was monitored by means of an ECHO-Doppler procedure. Unfortunately, the data concerning an antiatherogenic action of isradipine were inconclusive because of methodological difficulties (Borhani et al., 1996).

We already mentioned the ELSA-Study (European Lacidipine Study on Atherosclerosis) where about 2300 hypertensive patients were randomized to lacidipine or atenolol (Bond et al., 1993). Both the antihypertensive effect and a possible influence on plaque formation in the carotid region are monitored. The study is ongoing. In the similarly designed VHAS-Study verapamil and captopril are compared (Zanchetti, 1996).

In conclusion, the potential antiatherogenic activity of CA is a most important issue, but it remains to be demonstrated whether it indeed occurs in hypertensive patients when treated with these drugs.

11. CALCIUM ANTAGONISTS AND PERIOPERATIVE HYPERTENSION

Perioperative hypertension, predominantly caused by sympathetic stimulation, occurs frequently during thoracic surgery. Usually, this acute form of hypertension is suppressed by using short-acting, intravenously administered antihypertensive drugs. Nitroglycerin, sodium nitroprusside, clonidine, urapidil, ketanserin, and also a few short-acting β-blockers or CA may be used for this purpose (van Zwieten and van Wezel, 1993). A theoretical argument in favour of using CA in this condition is the anti-ischaemic activity of this class of drugs, which may be helpful in patients with coronary artery disease undergoing surgery (Underwood et al., 1991). This anti-ischaemic activity is not offered by other vasodilators like sodium nitroprusside, which may even cause additional problems as a result of coronary "steal". However, the anti-ischaemic activity of CA, when given intravenously, has been challenged by reports of a substantial risk of pro-ischaemic effects with the dihydropyridine group (Waters, 1991).

Among the various types of CA, only dihydropyridines, which are predominantly vasodilator agents with little or no cardiodepressant activity, have been studied in detail in surgical patients in an effort to counteract or prevent perioperative hypertension. Among the numerous dihydropyridines available at present, most of the studies have been limited to nifedipine and nicardipine. A few smaller studies have been performed with isradipine (a newer dihydropyridine) and with diltiazem. As a whole, nifedipine and nicardipine are considered as useful, short-acting dihydropyridine-CA in the treatment of perioperative hypertension (van Zwieten and van Wezel, 1993; Kaplan, 1989; Visser et al., 1986).

12. NEW CALCIUM ANTAGONISTS

12.1. General trends

Several new CA have been introduced as potential therapeutic agents in cardiovascular medicine. The majority of these compounds are dihydropyridines (Figure 5), although an interesting verapamil derivative (mibefradil; Figure 6) has been developed as well.

Nifedipine, the prototype of the dihydropyridine-CA has been used on a very large scale for many years but is known to have a number of deficiencies, including negative inotropic activity and a short duration of action, thus requiring at least three daily doses in long-term treatment for hypertension or angina pectoris. In particular, the rapid development of its haemodynamic effects, reflecting its kinetic characteristics, implies that reflex tachycardia triggered by peripheral vasodilation will occur, which is potentially harmful as a pro-ischaemic process.

These disadvantages of nifedipine stimulated the development of dihydropyridine-CA with less negative inotropic activity and a more favourable pharmacokinetic profile. A certain degree of vasoselectivity claimed for these newer compounds implies that they may have less or possibly no depressant action on cardiac contractile force, while the therapeutic efficacy is

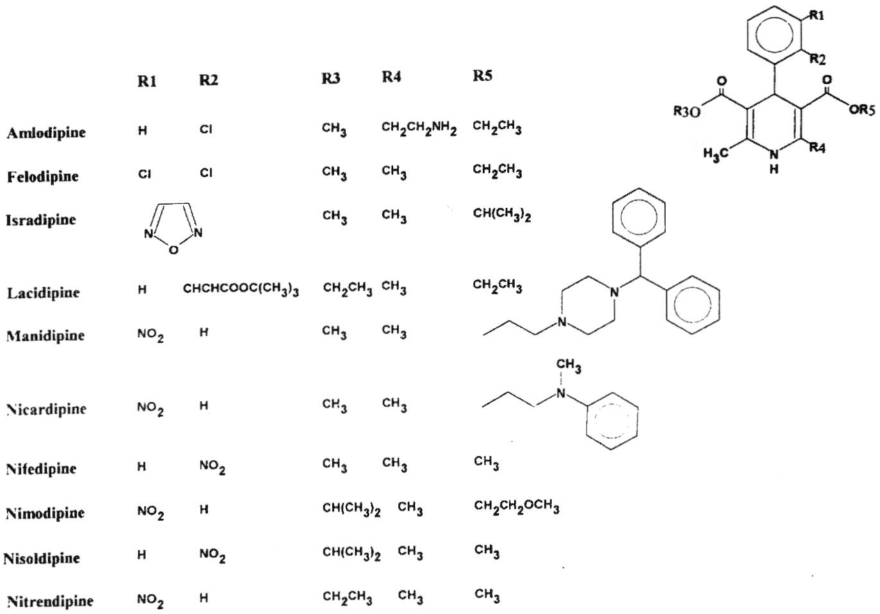

	R1	R2	R3	R4	R5
Amlodipine	H	Cl	CH_3	$CH_2CH_2NH_2$	CH_2CH_3
Felodipine	Cl	Cl	CH_3	CH_3	CH_2CH_3
Isradipine			CH_3	CH_3	$CH(CH_3)_2$
Lacidipine	H	$CHCHCOOC(CH_3)_3$	CH_2CH_3	CH_3	CH_2CH_3
Manidipine	NO_2	H	CH_3	CH_3	
Nicardipine	NO_2	H	CH_3	CH_3	
Nifedipine	H	NO_2	CH_3	CH_3	CH_3
Nimodipine	NO_2	H	$CH(CH_3)_2$	CH_3	$CH_2CH_2OCH_3$
Nisoldipine	H	NO_2	$CH(CH_3)_2$	CH_3	CH_3
Nitrendipine	NO_2	H	CH_2CH_3	CH_3	CH_3

Figure 5. Chemical structures of dihydropyridine CA, including several newer compounds.

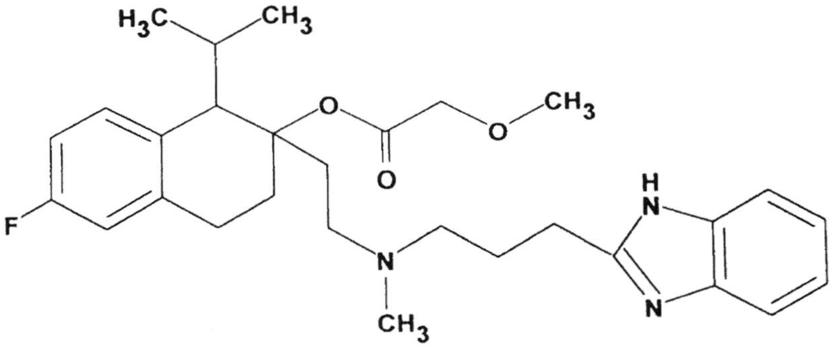

Figure 6. Chemical structure of mibefradil (Ro 40-5967), a benzimidazole derivative. This CA, a vasodilator, which also reduces heart rate, is virtually devoid of negative inotropic activity.

at least the same as or even better than that of nifedipine. Indeed, several of the newer dihydropyridine-CA as well as the verapamil-like CA mibefradil (compound Ro 40.5967) appear to have a certain degree of vascular selectivity, which implies that in therapeutic doses they induce little or no cardiodepressant activity, unlike nifedipine, verapamil or diltiazem.

An even more sophisticated approach has been followed in current attempts to develop CA with selectivity for a particular vascular bed. Furthermore, pharmacokinetic improvements of newer CA have indeed been achieved for several dihydropyridines, but also for verapamil, diltiazem and mibefradil. These pharmacokinetic improvements are based upon two different principles:

a) the development of slow release (retarded) preparations of CA which are as such rapidly and short acting;
b) the development of CA-molecules with a slow and long duration of action of their own.

For a survey of the various new CA, see Table 1.

12.2. Pharmacokinetic improvements

12.2.1. Slow release (retarded) preparations

Nifedipine was originally introduced as capsules, non-retarded, and characterized by a rapid onset and short action. Accordingly, this preparation had to be administered 3–4 times daily in order to obtain and maintain sufficient therapeutic activity over a period of 24 h. Furthermore, substantial reflex tachycardia occurred, triggered by the rapid vasodilator effect of this preparation. It is the general feeling that this preparation, although still available, should no more be used, not even in case of a hypertensive emergency where several alternative medications can be thought of. The non-retarded nifedipine capsules were replaced by a slow release preparation which has to be administered twice daily in order to obtain an acceptable control of blood pressure in hypertensives, which is associated with less intensive reflex tachycardia when compared with the non-retarded preparation. More recently a more sophisticated retarded preparation of nifedipine has been introduced as the nifedipine-gastrointestinal therapeutic system (GITS) (Zanchetti, 1994; Brogden and McTavish, 1995). This preparation is based upon the principle of an ALZET-minipump, which after oral ingestion is compressed via an osmotic mechanism and hence slowly squeezes out the solution of nifedipine with zero-order kinetics into the intestinal lumen. Consequently, a constant, slow release of the active drug (nifedipine) is achieved, thus leading to a slowly developing, long lasting effect, high bioavailability, stable blood levels, as well as a favourable trough:peak ratio (Zanchetti, 1994; Brogden and McTavish, 1995).

Felodipine, isradipine and nicardipine are also available as slow release preparations, as well as the so-called coat core formulation of nisoldipine (Fodor, 1997). Similarly as for these dihydropyridine-CA both verapamil and diltiazem are now available as slow-release preparations. All of these preparations allow a once daily administration with acceptable or good trough to peak ratio, slow onset of action, and the virtual absence of reflex tachycardia or neuro-endocrine activation.

12.2.2. Slow- and long-acting CA

A second possibility to obtain CA with an improved kinetic profile is based upon the use of molecules which as such slowly develop their effects and

Table 1. Survey of the newer calcium antagonists (CA). Commercial names as used in the Netherlands

1. *Slow release preparations of classic and newer CA*:
   verapamil-SR (Isoptin®)
   diltiazem-SR (Tildiem®)
   nifedipine-SR (Adalat®)
   nifedipine-GITS (Adalat-OROS®)
   felodipine-SR (Plendil®)
   isradipine-SR (Lomir®)
   nicardipine-SR (Cardene®)
   nisoldipine-Coat Core (CC) (Sular®)
2. *Newer dihydropyridines with a slow onset and long duration of action*:
   amlodipine (Norvasc®)
   lacidipine (Motens®)
   lercanidipine (Lerdip®)
   barnidipine (Cyress®)
   manidipine
3. *Lipophilic CA*:
   lacidipine (Motens®)
   lercanidipine (Lerdip®)
   mibefradil (Posicor®)
   manidipine
   barnidipine (Cyress®)
4. *Vasoselective Ca*:
   lacidipine (Motens®)
   lercanidipine (Lerdip®)
   manidipine
   barnidipine (Cyress®)
   (amlodipine, Norvasc®: only partially)
5. *Mibefradil (Posicor®)*:
   vasodilator with negative chronotropic and dromotropic activity; no negative influence on cardiac contractile force. Mibefradil blocks L-, T-, and N-type calcium channels.
6. *Cilnidipine*:
   Dihydropyridine-CA which blocks both L- and N-type channels. Cilnidipine inhibits the release of noradrenaline from the sympathetic nerve endings, via the blockade of N-type channels.

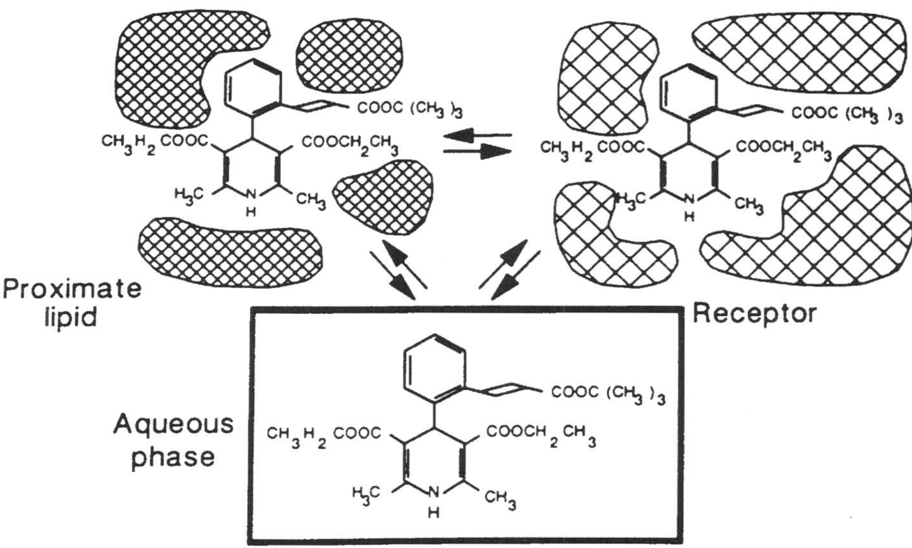

Figure 7. A diagrammatic representation of the 3-compartment model of lacidipine which may explain its slower onset and longer duration of action than that of most other calcium antagonists (from van Zwieten, 1998, with permission).

maintain their action sufficiently long to allow a once daily administration for the treatment of hypertension or angina. Amlodipine, a dihydropyridine, was one of the first examples of such slow- and long-acting CA. The kinetic characteristics of this molecule are explained by its strong affinity for the membranes of vascular smooth muscle. Furthermore the molecule is rather stable and hepatic degradation occurs but slowly (Murdoch and Heel, 1991).

12.2.3. Lipophilic CA

Several newer dihydropyridine-CA may be characterized as lipophilic molecules. The lipid (fat) solubility of such compounds implies that they readily dissolve in lipid-rich depots, including those within the cell membrane (see Figure 7). From these depots they are slowly released to subsequently reach their receptors, that is the L-type calcium channel. This phenomenon explains their slow onset of action and, therefore, the virtual absence of reflex tachycardia and sympathetic activation. The persistence of such lipophilic CA within cellular and other lipid depots also explains their long duration of action (Leonetti, 1991; Pfaffendorf et al., 1993). The interaction between the lipophilic CA and the cell membrane, as illustrated schematically in Figure 7 has been explored in detail for lacidipine and lercanidipine (Cafiero et al., 1997; Guarneri et al., 1997; Leonetti, 1991; Pfaffendorf et al., 1993). The slow onset of the effects of such drugs as well as their lack of reflex tachycardia in antihypertensive doses is illustrated in Figure 8. This figure demonstrates the antihypertensive action of lacidipine in patients over

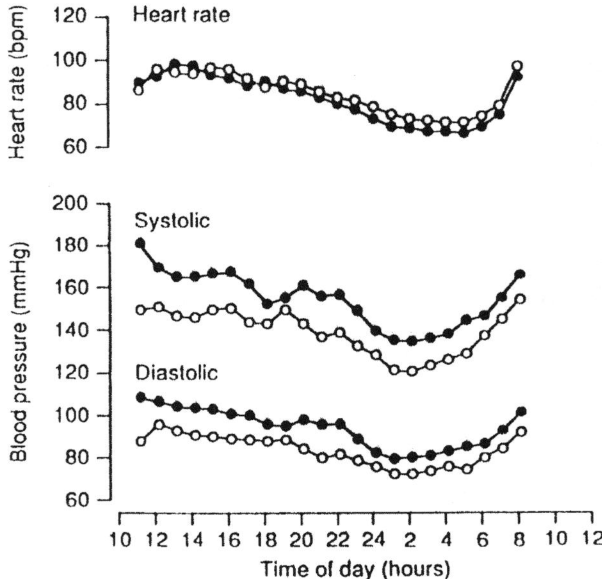

Figure 8. Circadian curves of blood pressure and heart rate in hypertensive subjects, before (•) and after (○) 4–8 weeks of treatment with lacidipine 4–8 mg. Note the sustained 24-hour effect of lacidipine and the absence of reflex tachycardia (from van Zwieten and Pfaffendorf, 1994, with permission).

a period of 24 h. The antihypertensive action after a single dose persists for 24 h and the circadian rhythm of blood pressure is maintained, although at a lower level. Interestingly, heart rate remains unchanged, in spite of the vasodilator/antihypertensive action of lacidipine, indicating that no reflex tachycardia is triggered. Similar findings have been obtained with lercanidipine, also a lipophilic CA (Guarneri et al., 1997; Cafiero and Giasi, 1997).

12.3. Vascular selectivity

Amlodipine, barnidipine, felodipine, isradipine, lacidipine, manidipine, nicardipine, and nisoldipine are examples of dihydropyridine-CA with a certain degree of vascular selectivity. The stronger effect of these agents on the vascular system than on the heart has been attributed to subtle differences between L-type calcium channels in the various tissues and organs. The vascular selectivity of nicardipine has been well demonstrated in a study in patients with coronary heart disease; when the drug was injected into a left main coronary artery it did not depress cardiac contractile force, in spite of a clear dilator effect in the coronary system (van Zwieten, 1998; Plosker and Faulds, 1996). Nisoldipine has been claimed, among the dihydropyridine-CA to display the highest degree of vascular selectivity, but this claim is

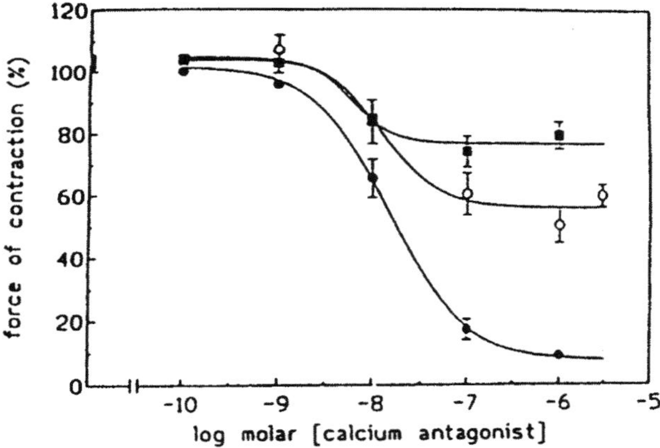

Figure 9. Concentration-response curves for the inhibitory effects of nifedipine (open circle), nisoldipine (closed square), and manidipine (closed circle) on serotonin-induced contractions in isolated rat renal artery preparations. Note the strong antagonistic effect of manidipine (from van Zwieten, 1998, with permission).

only based upon *in vitro* experiments with isolated human and animal blood vessels (Plosker and Faulds, 1996).

A few CA have been claimed to display moderate selectivity for a particular, specialized vascular bed. The evidence for such claims is in most cases rather meagre. A certain degree of selectivity for a particular vascular bed is probably best substantiated for the renovascular selectivity of manidipine, which indeed shows most potent dilator effects on renal vessels in intact animals and *in vitro* (Figure 9), whereas improved renal function in conditions of renal insufficiency have been demonstrated in animal models (van Zwieten and Pfaffendorf, 1994). Clinical data which do not contradict the claim of renal selectivity of manidipine are beginning to emerge (Rodicio, 1996). It goes without saying that a renal selective CA would be of great clinical and scientific interest.

12.4. Mibefradil

Mibefradil (Ro 40-5967) has been derived chemically from verapamil, although its molecule contains a few rather unusual substituents (Figure 6). Mibefradil appears to bind to the same [^{3}H]-desmethoxy-verapamil binding sites as verapamil in cardiac membrane homogenates. The haemodynamic profile of mibefradil may be briefly characterized as follows (Lüscher et al., 1997; Mehrke et al., 1994): Mibefradil is a vasodilator agent, predominantly at arteriolar sites (resistance vessels) with significant negative chronotropic and negative dromotropic activity, whereas contractile force is not reduced by this agent, as shown in isolated organs, animal models, but also in patients (Bernink et al., 1996; Portegies et al., 1991). The vasodilator activity

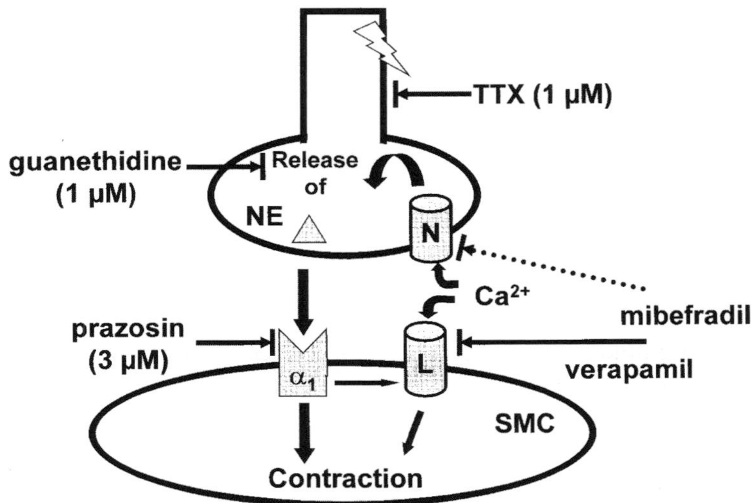

Figure 10. Schematic presentation of L- and N-type calcium channels in the synaps associated with vascular smooth muscle. At postsynaptic sites the α_1-adrenoceptor is coupled to the L-channel, and stimulation of the α-receptor by noradrenaline induces the influx of calcium ions into the cell. The α_1-adrenoceptor can be blocked by the antagonist prazosin (and related drugs), the influx of calcium ions can be inhibited by all classic calcium antagonists. At presynaptic sites the noradrenaline release is calcium-sensitive, and modulated by calcium N-type channels, which can be blocked by mibefradil.

of mibefradil is fully explained by its blocking effect on L-type calcium channels (van der Lee et al., 1997). Mibefradil has a stronger effect on T-type than on L-type calcium channels, and it may well be that its negative chronotropic and dromotropic activities are brought about, at least in part, by T-channel blockade in the sinus node and in the A-V conduction system (Lüscher et al., 1997; Mehrke et al., 1994). Interestingly, it was recently demonstrated in isolated human atrial tissue preparations that mibefradil in therapeutically relevant preparations also blocks N-type calcium channels. Consequently, noradrenaline release from cardiac sympathetic nerves is suppressed by mibefradil via N-channel blockade (Göthert et al., 1997). Similar findings have been obtained in isolated rat vessels (Göthert et al., 1997; van der Lee et al., 2000) (Figure 10). Mibefradil has been withdrawn because of its numerous interactions with other drugs, and also because it tends to impair A-V conduction in therapeutic doses. However, it must be considered as an interesting example of a CA which simultaneously blocks L, T, and N-type calcium channels. Its haemodynamic profile is potentially attractive.

12.5. Cilnidipine

Cilnidipine, a newer dihydropyridine was recently put forward as a CA with strong N-channel blocking activity in addition to its blocking effect on L-channels (Fujii et al., 1997; Minami et al., 1998). The N-channel blocking activity implies that cilnidipine inhibits the release of noradrenaline from sympathetic nerve endings. This may be considered as a potentially beneficial action.

13. CALCIUM SENSITIZERS

13.1. General aspects

It has been attempted repeatedly to develop inotropic drugs with modes of action which differ from those of classical agents, such as calcium ions, cardiac glycosides, β-adrenoceptor agonists (catecholamines and analogues), and phosphodiesterase inhibitors. A new mode of action is offered by the calcium sensitizers (CS), which sensitize the contractile proteins in the myofilaments to calcium ions and hence increase cardiac contractile force (Ruegg and Morano, 1989). CS will increase contractility without changing the intracellular calcium concentration. Increased sensitivity of the contractile proteins can for instance be achieved by an augmented affinity of troponin C for Ca^{2+}. Accordingly, more Ca^{2+} is bound and more force is thus generated at a given cytosolic Ca^{2+}-concentration. However, on theoretical grounds (Nielsen Kudsk and Addershvile, 1995) CS may be expected to inhibit cardiac relaxation, since Ca^{2+} might possibly be detached more slowly from troponin C. Impaired cardiac relaxation may lead to diastolic dysfunction.

Potential advantages of CS are, at least on theoretical grounds:

- an increased contractility which is not associated with an increased myocardial oxygen consumption;
- the inotropic effect is independent of β-adrenoceptors and therefore not impaired by the down-regulation of these receptors as in congestive heart failure;
- the pro-arrhythmogenic potential of the CS is lower than that of classic inotropic agents, such as β-adrenoceptor agonists, PDE-inhibitors or cardiac glycosides.

13.2. Calcium sensitizers so far developed

Over the past decade several experimental CS have been developed by pharmaceutical industry. Examples are adibendan, levosimendan, pimobendan, sulmazole, MCI-154, DPI-201.206, EMD 53998, and EMD 57033 (Nielsen Kudsk and Addershvile, 1995). Unfortunately, all of the compounds are not only CS, but also, to varying degrees, PDE-inhibitors. Accordingly, the therapeutic value of the process of calcium sensitization by these agents is difficult to judge. As a result of PDE III-inhibition cyclic AMP accumulation and secondary increased activity of phospholamban may contribute to the improved cardiac contractility. On the other hand the pro-arrhythmogenic activity is increased as a result of PDE III-inhibition.

Levosimendan appears to be the most promising compound of this category from a clinical point of view. Levosimendan binds to troponin C and stabilizes the conformational change of this contractile protein by calcium ions. The positive inotropic effect of levosimendan in acute heart failure, as during cardiac surgery (Nijhawan et al., 1999), is well established. Heart rate remains virtually unchanged and a pro-arrhythmogenic effect has so far not been observed. Interestingly, levosimendan appears to slightly accelerate diastolic relaxation instead of impairing it, as might be feared on theoretical grounds. Its use in cardiac surgery an anaesthesiology is subject to various clinical studies.

REFERENCES

Agabiti-Rosei, E., Muiesan, M.L., Rizzoni, D., Zulli, R., Calebich, S., Castellano, M., et al., 1994, Cardiovascular structural changes and calcium antagonist therapy in patients with hypertension, *J. Cardiovasc. Pharmacol.* 24 (Suppl. A), S37–S43.

Alderman, M.H., Cohen, H., Roqune, R. and Medhaven, S., 1997, Effect of long-acting and short-acting calcium antagonists on cardiovascular outcomes in hypertensive patients, *Lancet* 349, 594–598.

Aursnes, I., Litleskare, I., Froyland, H. and Abdelnoor, M., 1995, Association between various drugs used for hypertension and risk of acute myocardial infarction, *Blood Press.* 4, 157–163.

Bernink, P.J.L.M., Prager, G., Schelling, A. and Kobrin, I., 1996, On behalf of the Mibefradil International Study Group. 1996, Antihypertensive properties of the novel calcium antagonist mibefradil (Ro 40-5967), *Hypertens.* 27 (Part I), 426–432.

Bond, G., Dal Palu, C., Hansson, L., Magnani, B., Mancia, G., et al.. 1993, European Lacidipine Study on Atherosclerosis (ELSA), *J. Hypertens.* 11 (Suppl. 5), S405.

Borhani, N.O., Mercuri, M., Borhani, P.A., et al., 1996, Final outcome results of the Multicenter Isradipine Diuretic Atherosclerosis Study (MIDAS), *JAMA* 276, 785–791.

Braun, S., Boyko, V., Behar, S., Reicher-Reiss, H., Shotan, A., et al., 1996, Calcium antagonists and mortality in patients with coronary artery disease: A cohort study of 11,575 patients, *J. Am. Coll. Cardiol.* 28, 7–11.

Brogden, R.N. and McTavish, D., 1995, Nifedipine gastrointestinal therapeutic system (GITS), *Drugs* 50, 495–512.

Bühler, F R., 1995, Antihypertensive care with calcium antagonists, in *Hypertension. Pathophysiology, Diagnosis and Management*, 2nd edn., J.H. Laragh and B.M. Brenner (eds.), Raven Press, New York, pp. 2801–2814.

Byington, R.P., Miller, M.E., Herrington, D., et al., 1997, Rationale, design, and baseline characteristics of the prospective randomized evaluation of the vascular effects of Norvasc Trial (PREVENT), *Am. J. Cardiol.* 80, 1087–1090.

Cafiero, M. and Giasi, M., 1997, Long-term (12-month) treatment with lercanidipine in patients with mild to moderate hypertension, *J. Cardiovasc. Pharmacol.* 29 (Suppl. 2), S45–S49.

Catapano, A.L., 1997, Calcium antagonists and atherosclerosis. Experimental evidence, *Eur. Heart J.* 18 (Suppl. A), A80–A86.

Catterall, W.A., 1993, Structure and function of voltage-gated ion channels, *Trends Neurosci.* 16, 500–506.

Dahlöf, B., Hansson, L., Lindholm, L.H., Schersten, B., Wester, P.O., Ekbom, Y., et al., 1993, STOP-Hypertension-2: A prospective intervention trial of 'newer' versus 'older' treatment alternatives in old patients with hypertension, *Blood Press.* 2, 136–141.

The Danish Study Group on Verapamil in Myocardial Infarction, 1990, Effect of verapamil on mortality and major events after acute myocardial infarction (The Danish Verapamil Infarction Trial II – DAVIT II), *Am. J. Cardiol.* 66, 779–785.

Dargie, H.J., Ford, I., Fox, K.M., on behalf of the TIBET (Total Ischaemic Burden European Trial) study group, 1996, Effects of ischaemia and treatment with atenolol, nifedipine SR and their combination on outcome in patients with chronic stable angina, *Eur. Heart J.* 17, 104–112.

Davies, R.F., Hbibi, H., Davies, R., et al., 1995, Effects of amlodipine, atenolol and their combination on myocardial ischaemia during treadmill exercise and ambulatory monitoring, *J. Am. Coll. Card.* 25, 619–625.

Davis, B.R., Cutler, J.A., Gordon, D.J., et al., for the ALLHAT- Research Group, 1996, Rationale and design for the antihypertensive and lipid lowering treatment to prevent heart attack (ALLHAT), *Am. J. Hypertens.* 9, 342–360.

Devereux, R.B., 1997, Do antihypertensive drugs differ in their ability to regress left ventricular hypertrophy?, *Circulation* 95, 1983–1985.

Dustan, H.P., 1987, Nitrendipine in black US patients, *J. Cardiovasc. Pharmacol.* 9 (Suppl. 4), 267–271.

Fischer Hansen, J., 1992, Secondary prevention with calcium antagonists after acute myocardial infarction, *Drugs* 44, 33–43.

Fleckenstein, A. and Fleckenstein-Grün, G., 1980, Cardiovascular protection by CA-antagonists, *Eur. Heart J.* 1, 15–21.

Fodor, J.G., 1997, Nisoldipine CC: Efficacy and tolerability in hypertension and ischemic heart disease, *Cardiovasc. Drugs Ther.* 10, 873–879.

Fujii, S., Kameyama, K., Hosono, M., Hayashi, Y. and Kitamura, K., 1997, Effect of cilnidipine, a novel dihydropyridine Ca^{2+}-channel antagonist, on N-type Ca^{2+}-channel in rat dorsal root ganglion neurons, *J. Pharmacol. Exp. Ther.* 280, 1184–1191.

Furberg, C.D., Psaty, B.M. and Meyer, J.V., 1995, Nifedipine. Dose-related increase in mortality in patients with coronary heart disease, *Circulation* 92, 1326–1330.

Godfraind, T. 1997, Vasodilators and calcium antagonists, in *Antihypertensive Drugs*, P.A. van Zwieten and W.J. Greenlee (eds.), Harwood Academic Publishers, Amsterdam, pp. 313–375.

Godfraind, T. and Kaba, A., 1969, Blockade or reversal of contraction induced by calcium and adrenaline in depolarized arterial smooth muscle, *Br. J. Pharmacol.* 36, 549–560.

Godfraind, T., Miller, R.C. and Wibo, M., 1986, Calcium antagonism and calcium entry blockade, *Pharmac. Rev.* 38, 321–416.

Gong, L., Zhang, W., Zhy, Y. and Zhu, J., 1996, Shanghai Trial of Nifedipine in the Elderly (STONE), *J. Hypertens.* 14, 1237–1245.

Göthert, M. and Molderings, G.J., 1997, Mibefradil and ω-conotoxin GVIA-induced inhibition of noradrenaline release from the sympathetic nerves of the human heart, *Naunyn-Schmiedeb. Arch. Pharmacol.* 356, 860–863.

Gray Ellrodt, A. and Singh, B.N., 1983, Clinical applications of slow channel blocking compounds, *Clin. Pharmacol. Ther.* 23, 1–43.

Guarneri, L., Sironi, G., Angelico, P., Ibba, M., Greto, L., et al., 1997, In vitro and in vivo vascular selectivity of lercanidipine and its enantiomers, *J. Cardiovasc. Pharmacol.* 29 (Suppl. 1), S25–S32.

Hansson, L., Zanchetti, A. and Carruthers, S.G., 1998, Effects of intensive blood pressure lowering and low-dose aspirin in patients with hypertension: Principal results of the Hypertension Optimal Treatment (HOT) randomised trial, *Lancet* 351, 1755–1762.

Hansten, Ph.D., 1992, Important drug interactions, in *Basic and Clinical Pharmacology*, 5th edn., B.G. Katzung (ed.), Prentice-Hall, Englewoods Cliffs, NJ, pp. 931–942.

Henry, P.D., 1990, Calcium antagonists as anti-atherosclerotic agents, *Arteriosclerosis* 10, 963–965.

Hess, P., Lansman, J.B. and Tsien, R.W., 1984, Different modes of Ca-channel gating behaviour favoured by dihydropyridine Ca-agonists and -antagonists, *Nature* 311, 538–544.

Hess, P., Lansman, J.B. and Tsien, R.W., 1986, Calcium channel selectivity for divalent and monovalent cations, *J. Gen. Physiol.* 88, 293–319.

Holland Interuniversity Nifedipine-metoprolol Trial (HINT) Research Group, 1986, Early treatment of unstable angina in the coronary care unit: A randomised, double blind, placebo-controlled comparison of recurrent ischemia in patients treated with nifedipine or metoprolol or both, *Br. Heart J.* 56, 400–413.

Kaplan, J.A., 1989, The role of nicardipine during anesthesia and surgery, *Clin. Ther.* 11, 84–93.

Kiowski, W., Linder, L. and Bühler, F.R., 1990, Arterial vasodilator and antihypertensive effects of diltiazem, *J. Cardiovasc. Pharmacol.* 16 (Suppl. 6), S7–S10.

Leonetti, G., 1991, Clinical position of lacidipine, a new dihydro-pyridine calcium antagonist, in the treatment of hypertension, *J. Cardiovasc. Pharmacol.* 18 (Suppl. 11), S18–S21.

Lever, A.F., 1997, Calcium antagonists and cancer, *8th European Meeting on Hypertension*, Milan, June 1997.

Levy, D., Garrison, R.J., Savage, D.D., Kannel, W.B. and Castelli, W.P., 1989, Left ventricular mass and incidence of coronary heart disease in an elderly cohort: The Framingham Heart Study, *Ann. Intern. Med.* 110, 101–107.

Lichtlen, P.R., Hugenholtz, P.G., Rafflenbeul, W., Hecker, H., Jost, S. and Deckers, J.W., 1990, Retardation of angiographic progression of coronary artery disease by nifedipine. Results of an international nifedipine trial on anti-atherosclerotic therapy (INTACT), *Lancet* 335, 1109–1113.

Liebson, P.R., Grandits, G.A., Dianzumba, S., Prineas, R.J., Grimm, R.H., et al., 1995, for the Treatment of Hypertension Study Research Group. Comparison of five antihypertensive monotherapies and placebo for change in left ventricular mass in patients receiving nutritional-hygienic therapy in the Treatment of Mild Hypertension Study (TOHMS), *Circulation* 91, 698–706.

Lüscher, T.F., Clozel, J.P. and Noll, G., 1997, Pharmacology of the calcium antagonist mibefradil, *J. Hypertens.* 15 (Suppl. 3), S11–S18.

Malbantgil, I., Önder, R., Killiçcioglu, B., Boydak, B., Terzioglu, U. and Yilmaz, H., 1996, The efficacy of felodipine ER on regression of left ventricular hypertrophy in patients with primary hypertension, *Blood Press.* 5, 285–291.

Mancia, G. and van Zwieten, P.A., 1996, How safe are calcium antagonists in hypertension and coronary heart disease?, *J. Hypertens.* 14, 13–17.

Matsuzaki, K., Mukai, M., Sumimoto, T. and Murakami, E., 1997, Effects of ACE-inhibitors versus calcium antagonists on left ventricular morphology and function in patients with essential hypertension, *Hypertens. Res.* 20, 7–10.

Mehrke, G., Zong, X.G., Flockerzi, V. and Hofmann, F., 1994, The Ca^{++} channel blocker Ro 40-5967 blocks differently T-type and L-type Ca^{++} channels, *J. Pharmacol. Exp. Ther.* 271, 1483–1488.

Minami, J., Ishimitsu, T., Kawano, Y., Numabe, A. and Matsuoka, H., 1998, Comparison of 24-hour blood pressure, heart rate, and autonomic nerve activity in hypertensive patients treated with cilnidipine or nifedipine retard, *J. Cardiovasc. Pharmacol.* 32, 331–336.

Mishra, S.K. and Hermsmeyer, K., 1994, Selective inhibition of T-type Ca^{2+}-channels by Ro 40-5967, *Circ. Res.* 75, 144–148.

Muiesan, G., Agabiti-Rosei, E., Castellano, M., Alicandri, C.L., Corea, L., Fariello, R., et al., 1982, Antihypertensive and humoral effects of verapamil and nifedipine in essential hypertension, *J. Cardiovasc. Pharmacol.* 4 (Suppl. 3), S325–S329.

Müller, F.B., Ha, H.R., Hotz, M., Schmidlin, O., Follath, F. and Bühler, F., 1985, Once a day verapamil in essential hypertension, *Br. J. Clin. Pharmacol.* 21 (Suppl. 2), 143S–147S.

The Multicenter Diltiazem Postinfarction Trial Research Group, 1988, The effect of diltiazem on mortality and reinfarction after myocardial infarction, *New Engl. J. Med.* 319, 385–392.

Murdoch, D. and Heel, R.C., 1991, Amlodipine, *Drugs* 41, 478–505.

Nayler, W.G., 1988a, *Calcium Antagonists* Academic Press, Harcourt Brace Jovanovich Publishers, London, pp. 1–347.

Nayler, W.G., 1988b, Calcium antagonists and myocardial ischaemia, in *Calcium Antagonists*, W.G. Nayler (ed.), Academic Press, New York, pp. 157–176.

Nielsen Kudsk, J.E. and Addershvile, J., 1995, Will calcium sensitizers play a role in the treatment of heart failure?, *J. Cardiovasc. Pharmacol.* 26 (Suppl. 1), S77–S84.

Nijhawan, N., Nicolosi, A.C., Montgomery, M.N., Aggarwal, A., Pagel, P.S. and Warltier, D.C., 1999, Levosimendan enhances cardiac performance after cardiopulmonary bypass: A prospective, randomized, placebo-controlled trial, *J. Cardiovasc. Pharmacol.* 34, 219–228.

NORDIL Study Group, 1993, The NORDIC Diltiazem Study: An intervention study in hypertension comparing calcium antagonist based treatment with conventional therapy, *Blood Press.* 2, 312–321.

Opie, L.H., 1990, *Clinical Use of Calcium Channel Antagonistic Drugs*, 2nd edn., Kluwer Academic Publishers, Boston, pp. 70–130.

Packer, M., O'Connor, C.M., Ghali, J.K., et al., 1996, Effects of amlodipine on morbidity and mortality in severe chronic heart failure, *New Engl. J. Med.* 335, 1107–1114.

Pahor, M., Guralnik, J.M., Corti, M., Foley, D.J., Carbonin, P. and Havlik, J.R., 1995, Long-term survival and use of antihypertensive medications in older persons, *J. Am. Ger. Soc.* 43, 1–7.

Pahor, M., Guralnik, J., Salive, M., Corti, M., Carbonin, P. and Havlik, R., 1996a, Do calcium channel blockers increase the risk of cancer?, *J. Hypertens.* 9, 695–699.

Pahor, M., Guralnik, J., Furberg, C., Carbonin, P. and Havlik, R., 1996b, Risk of gastrointestinal haemorrhage with calcium antagonists in hypertensive persons over 67 years old, *Lancet* 347, 1061–1065.

Pedrinelli, R., Fouad, F.M., Tarazi, R.C., Bravo, E.L. and Textor, S.C., 1986, Nitrendipine, a calcium entry blocker. Renal and humoral effects in human arterial hypertension, *Arch. Intern. Med.* 146, 62–65.

Pfaffendorf, M., Mathy, M.J. and van Zwieten, P.A., 1993, In vitro effects of nifedipine, nisoldipine, and lacidipine on rat isolated coronary small arteries, *J. Cardiovasc. Pharmacol.* 21, 496–502.

Plosker, G.L. and Faulds, D., 1996, Nisoldipine Coat-Core. A review of its pharmacology and therapeutic efficacy in hypertension, *Drugs* 52, 232–253.

Portegies, M.C.M., Schmitt, R., Kraaij, C.J., et al., 1991, Lack of negative inotropic effect of a new calcium antagonist Ro 40-5967 in patients with stable angina pectoris, *J. Cardiovasc. Pharmacol.* 18, 746–751.

PREVENT Trial, 1998, Communication at the 71st Scientific Sessions of the American Heart Association, Dallas USA, November 1998.

Psaty, B.M., Heckbert, S.R., Koepsell, T.D., Sisovick, D.S., Raghunathan, T.E., Weiss, N.S., et al., 1995, The risk of myocardial infarction associated with antihypertensive drug therapies, *JAMA* 274, 620–625.

Rehnqvist, N., Hjemdahl, P., Billing, E., et al., 1996, Effects of metoprolol vs verapamil in patients with stable angina pectoris. The Angina Prognosis Study in Stockholm (APSIS), *Eur. Heart. J.* 17, 76–81.

Ribstein, J., de Treglode, D., Mimran, A., 1985, Acute effects of nifedipine on arterial pressure in healthy subjects and hypertensives, *Arch. Mal. Coeur* 78, 29–32.

Rodicio, J.L., 1996, Renal effects of calcium antagonists with special reference to manidipine hydrochloride, *Blood Press.* 5 (Suppl. 5), 10–15.

Rodicio, J.L., Morales, J.M., Alcazar, M. and Ruilope, L.M., 1993, Calcium antagonists and renal protection, *J. Hypertens.* 11, S49–S53.

Rubin, P., 1995, *Prescribing in Pregnancy*, 2nd edn., BMJ Publishing Group, London, pp. 99–102.

Ruegg, J.C. and Morano, I., 1989, Calcium sensitivity modulation of cardiac myofibrillar proteins, *J. Cardiovasc. Pharmacol.* 14 (Suppl. 3), S20–S23.

Schmieder, R.E., Messuli, F.H., Garavglio, G.E. and Nunes, B.D., 1987, Cardiovascular effects of verapamil in essential hypertension, *Circulation* 76, 1143–1155.

Sever, P., 1986, 1985, The year of the hypertension trials, *Trends Pharmacol. Sci.* 6, 134–139.

Sheperd, J., et al., 1995, Prevention of coronary heart disease with pravastatin in men with hypercholesterolaemia, *New Engl. J. Med.* 333, 1301–1307.

Sher, E., Biancardi, E., Passafaro, M. and Clementi, F., 1991, Physiopathology of neuronal voltage-operated calcium channels, *FASEB J.* 5, 2677–2683.

Staessen, J.A., et al., for the Systolic Hypertension in Europe Trial (SYST-EUR), 1997, Randomised double-blind comparison of placebo and active treatment for older patients with isolated systolic hypertension, *Lancet* 350, 757–764.

Stockley, I.H., 1994, *Drug Interactions*, 3rd edn., Blackwell, Oxford.

Struyker Boudier, H.A.J., de Mey, J.G., Smits, J.F.M. and Nievelstein, H.M.N.W., 1989, Hemodynamic actions of calcium entry blockers, in *Clinical Aspects of Calcium Entry Blockers*, P.A. van Zwieten (ed.), Karger Verlag, Basel, pp. 21–66.

Thürmann, P.A., Stephens, N., Heagerty, A.M., Kenedi, P., Weidinger, G. and Rietbrock, N., 1996, Influence of isradipine and spirapril on left ventricular hypertrophy and resistance arteries, *Hypertension* 28, 450–456.

Tijssen, J.G.P. and Hugenholtz, P.G., 1996, Critical appraisal of recent studies on nifedipine and other calcium channel blockers in coronary heart disease and hypertension, *Eur. Heart J.* 17, 1152–1157.

Timmermans, P.B.M.W.M. and van Zwieten, P.A., 1982, α_2-Adrenoceptors. Classification, localisation, mechanisms and targets for drugs, *J. Med. Chem.* 25, 1389–1401.

Tsien, R.W. and Tsien, R.Y., 1990, Calcium channels, stores and oscillations, *Ann. Rev. Cell. Biol.* 6, 715–760.

Underwood, S.M., Davies, S.W., Feneck, R.O., Lunnon, M.W. and Walesby, R.K., 1991, Comparison of isradipine with nitroprusside for control of blood pressure following myocardial revascularization: Effects on hemodynamics, cardiac metabolism and coronary blood flow, *J. Cardiothorac. Vasc. Anesth.* 5, 348–356.

Van Heiningen, P.N.M. and van Zwieten, P.A., 1988, Differential sensitivity to calcium entry blockade of angiotensin II-induced contractions of rat and guinea-pig aorta, *Arch. Int. Pharmacodyn.* 296, 118–130.

Van der Lee, R., Pfaffendorf, M. and van Zwieten, P.A., 1997, Effects of mibefradil and other calcium antagonists on microvessels of different end organs, *8th European Meeting on Hypertension*, Milan, June 1997.

Van der Lee, R., Pfaffendorf, M., De Mey, J.G.R. and van Zwieten, P.A., 2000, Inhibitory effect of mibefradil on contractions induced by sympathetic neurotransmitter release in rat tail artery, *Naunyn–Schmiedeb. Arch. Pharmacol.* 361, 74–79.

Van Meel, J.C.A., de Jonge, A.J., Kalkman, H.O., Wilffert, B., Timmermans, P.B.M.W.M. and van Zwieten, P.A., 1981, Vascular smooth muscle contraction initiated by postsynaptic α_2-adrenoceptor activation is induced by an influx of extracellular calcium, *Eur. J. Pharmacol.* 69, 205–208.

Van Zwieten, P.A., 1985, Drug targets in unstable angina, in *Unstable Angina, Current Concepts and Management*, P.G. Hugenholtz (ed.), Schattauer Verlag, Stuttgart, pp. 151–157.

Van Zwieten, P.A., 1989, Clinical aspects of calcium entry blockers, *Progr. Basic Clin. Pharmacol.* 2, 1–20.

Van Zwieten, P.A., 1992, Therapy update – What is new? Calcium antagonists, *Neth. J. Cardiol.* 5, 17–22.

Van Zwieten, P.A., 1998, The newer calcium antagonists, *Cardiologie* 5, 4–13.

Van Zwieten, P.A. and Lie, K.I., 1995, Long term efficacy and safety of calcium antagonists, *Cardiologie* 2, 457–460.

Van Zwieten, P.A. and Pfaffendorf, M., 1994, New aspects of the pharmacology of dihydropyridine calcium antagonists, *JAMA SE Asia* 15 (Suppl.), 9–19.

Van Zwieten, P.A. and van Wezel, H.B., 1993, Antihypertensive drug treatment in the perioperative period, *J. Cardiothorac. Vasc. Anesth.* 7, 213–226.

Van Zwieten, P.A., Timmermans, P.B.M.W.M. and van Heiningen, P.N.M., 1987, Receptor subtypes involved in the action of calcium antagonists, *J. Hypertens.* 5 (Suppl. 4), S21–S28.

Visser, C.A., Koolen, J.J., van Wezel, H.B., Jonges, R., Hoedemaker, G. and Dunning, A.J., 1986, Effects of intracoronary nicardipine and nifedipine on left ventricular function and coronary sinus blood flow, *Br. J. Clin. Pharmacol.* 22 (Suppl. 2), 313S–318S.

Wagenknecht, L., Furberg, C., Hammon, J., Legault, C. and Troost, B., 1995, Surgical bleeding: Unexpected effect of a calcium antagonist, *Br. Med. J.* 310, 776–777.

Waters, D., 1991, Proischemic complications of dihydropyridine calcium channel blockers, *Circulation* 84, 2598–2600.

Waters, D., Lespérence, J., Francetich, M., et al., 1990, A controlled clinical trial to assess the effects of a calcium channel blocker on the progression of coronary atherosclerosis, *Circulation* 82, 1940–1953.

WHO-ISH Study, *Ad hoc* subcommittee of the Liaison Committee of the World Health Organisation and the International Society of Hypertension, 1997, Effects of calcium antagonists on the risks of coronary heart disease, cancer and bleeding, *J. Hypertens.* 15, 105–115.

Zanchetti, A., on behalf of the Italian nifedipine-GITS study group, 1994, The 24-hour efficacy of a new once-daily formulation of nifedipine, *Drugs* 49 (Suppl. 1), 23–31.

Zanchetti, A., 1996, Antiatherosclerotic effects of antihypertensive drugs: Recent evidence and ongoing trial, *Clin. Exp. Hypertens.* 18, 489–499.

Calcium-Binding Proteins in Type I Allergy: Elicitors and Vaccines

Rudolf Valenta, Anna Twardosz, Ines Swoboda, Brigitte Hayek, Susanne Spitzauer and Dietrich Kraft

1. INTRODUCTION

Type I allergy is an immunologically-mediated hypersensitivity disease with complex genetic background affecting almost 25% of the population (Kay, 1997; Lockey and Bukantz, 1998). As a major feature of their disease, allergic patients produce IgE antibodies against *per se* mostly harmless antigens (i.e., allergens) which, after allergen-binding, can activate effector and inducer cells of the atopic immune response (Ravetch and Kinet, 1991; Beaven and Metzger, 1993; Bieber, 1996; Stingl and Maurer, 1997). Depending on the site and duration of allergen contact and the type of immune cells involved, the manifestations of Type I allergy may greatly vary (e.g., allergic rhinitis, conjunctivitis, asthma, dermatitis, gastrointestinal disease). Progress made in the field of molecular allergen characterization has revealed that calcium-binding proteins from many sources are frequent targets for IgE antibodies of allergic patients and thus can act as widely spread elicitors of Type I allergy (Valenta et al., 1998). Calcium-binding allergens from different sources share sequence and structural similarities. Therefore, patients who are cross-sensitized to calcium-binding allergens can exhibit allergic symptoms after exposure to many allergen sources. Calcium-depletion experiments indicate that IgE antibodies of sensitized individuals recognize preferentially the calcium-bound forms of the allergens whereas the apoforms are poorly recognized. This fact opens possibilities to employ genetic engineering and synthetic peptide chemistry for the production of hypoallergenic apoforms

Rudolf Valenta, Anna Twardosz, Brigitte Hayek and Dietrich Kraft • Department of Pathophysiology, Vienna General Hospital, University of Vienna, Austria. **Ines Swoboda and Susanne Spitzauer** • Department of Medical and Clinical Chemistry, Vienna General Hospital, University of Vienna, Austria.

R. Pochet, R. Donato, J. Haiech, C. Heizmann and V. Gerke (eds.): Calcium: The Molecular Basis of Calcium Action in Biology and Medicine, 365–377.

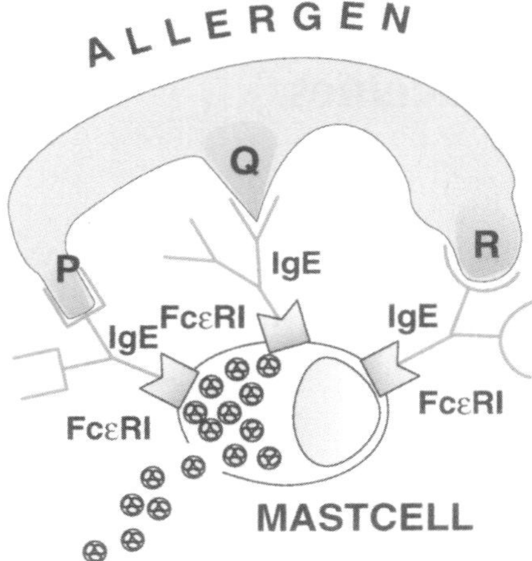

Figure 1. Allergen-induced crosslinking of mast cell-bound IgE antibodies leads to degranulation and mediator release.

which may be used for therapeutic vaccination against Type I allergy with a reduced rate of anaphylactic side effects. This chapter summarizes current knowledge about calcium-binding proteins as elicitors of Type I allergy and discusses strategies for their conversion into hypoallergenic allergy vaccines.

2. PATHOPHYSIOLOGY OF TYPE I ALLERGY

More than 25% of the population suffer from the symptoms of Type I allergy (allergic rhinoconjunctivitis, asthma, dermatitis, food intolerance, anaphylactic shock) (Kay, 1997; Lockey and Bukantz, 1998). As a major hallmark of their disease, allergic patients exhibit an increased tendency to produce IgE antibodies against *per se* harmless antigens (i.e., allergens). Allergic individuals become sensitized against allergens early in life and, in contrast to non-atopic individuals, respond to allergen contact with the production of specific IgE antibodies. B cells from allergic patients preferentially undergo class switch to IgE production. This process is controlled by a variety of cytokines (e.g., IL-4, IL-13, IL-5) which are secreted by T cells but also by effector cells (e.g., basophils, mast cells) of the immune system (Vercelli et al., 1998; Haas et al., 1999). The cytokine milieu which controls the class switch to IgE production against allergens during the early sensitization phase seems to be regulated by a complex, as yet not fully understood, genetic background (Casolaro et al., 1996). However, the allergen-specific IgE production by B-epsilon memory cells in the course of the disease is triggered primarily by allergen exposure itself (Henderson et al., 1975).

The presence of allergen-specific IgE antibodies in serum and other body fluids by itself would not be able to cause allergic disease because IgE represents the least abundant class of antibodies and cannot activate the complement system (Ishizaka et al., 1966; Johansson and Bennich, 1967). However, IgE-allergen immune complexes can bind to low and high affinity receptors for IgE on effector as well as inducer cells of the immune system and thereby cause symptoms of disease (Ravetch and Kinet, 1991; Beaven and Metzger, 1993; Bieber, 1996; Stingl and Maurer, 1997). When allergens crosslink effector cell (e.g., mast cell, basophil)-bound IgE antibodies, degranulation of preformed mediators (e.g., histamine, leukotriens) is induced which causes the immediate symptoms of Type I allergy (allergic rhinoconjunctivitis, acute allergic asthma) (Ravetch and Kinet, 1991; Beaven and Metzger, 1993). Figure 1 illustrates the allergen-induced crosslinking of mast cell-bound IgE antibodies. Allergens contain different epitopes for IgE antibodies (Figure 1: P, Q, R) and thus can crosslink the corresponding IgE antibodies which are bound to mast cells via the high affinity IgE receptor (Figure 1: FcεRI). It has been demonstrated that dimerization of effector cell-bound IgE is sufficient for the induction of degranulation (Segal et al., 1977). Chronic inflammation (e.g., atopic dermatitis, chronic asthma) can be induced by allergen-induced activation of T cells and eosinophils (Bieber, 1996; Stingl and Maurer, 1997; Desreumaux and Capron, 1996). In this context it was demonstrated that IgE-mediated presentation of allergens can strongly induce T cell activation (Mudde et al., 1990; Maurer et al., 1995, 1996) and it is now also well established that eosinophils express the high affinity receptor for IgE (Desreumaux and Capron, 1996) and thus may become activated after allergen contact. The symptoms of allergic disease may strongly vary depending on the exposed organ sites (eyes, respiratory tract, gastrointestinal tract) and the cell types involved (mast cells, basophils, T cells, eosinophils). It is however generally accepted that the IgE recognition of allergens (exogenous and probably also endogenous proteins/glycoproteins, see Valenta et al., 2000) represents the initiating pathophysiological event in the disease which is then orchestered and augmented by the consecutive activation of effector and inducer cells of atopy.

3. DIAGNOSIS AND TREATMENT OF TYPE I ALLERGY

From what has been said about the pathophysiology of Type I allergy we may continue to mention a few words regarding diagnosis and therapy of the disease. In the year 1921 Prausnitz and Küstner showed in an elegant experiment that the symptoms of Type I allergy depend on two specific elements which may vary from individual to individual and a "tissue component" which is present also in non-atopic persons (Prausnitz and Küstner, 1921). In their experiment they transferred serum from a fish allergic patient into the skin of a non-atopic individual and showed that subsequent exposure of this skin area to fish lead to an allergic reaction. The allergen-specific components present in serum were termed "reagins" and in 1967 were identified as IgE antibodies (Ishizaka et al., 1966; Johansson and Bennich, 1967). The elements which may vary among allergic subjects are therefore the spectrum of the disease eliciting allergens and the corresponding allergen-specific IgE antibodies whereas allergic as well as non-atopic individuals contain the cellular machinery required for the induction of allergic symptoms. At present, the diagnosis of Type I allergy involves the identification of the disease eliciting allergen, the detection of allergen-specific IgE antibodies and the demonstration that provocation with the given allergen will induce an allergic reaction (reviewed in Valenta et al., 1999a). Allergic patients may exhibit greatly varying sensitization profiles and thus may be divided into groups of patients who are sensitized against few allergen components (oligosensitized individuals) and into groups of patients who react to multiple allergen components (polysensitized individuals).

While the symptoms of Type I allergy can be controlled by pharmacotherapy (i.e., using immunosuppressive and anti-inflammatory agents), there are only a few causative therapy approaches towards Type I allergy. These approaches comprise (1) allergen-specific immunotherapy (Bousquet et al., 1998a, b), (2) attempts to reduce the levels of IgE antibodies, (3) interference with the interaction of IgE with its receptors, and (4) specific blockade of IgE-mediated signal transduction. Although suited for general therapy of atopy, the latter three approaches are at present either experimental in nature or not yet clinically evaluated. By contrast, allergen-specific immunotherapy which is based on the administration of allergens in order to induce a state of unresponsiveness represents the traditional and well established form of causative treatment (Bousquet et al., 1998a, b). Numerous clinical studies have documented the clinical efficacy of immunotherapy but there is some controversy regarding the immunological mechanisms underlying the treatment (Durham and Till, 1998). First classical studies demonstrated that allergen-specific immunotherapy induces "blocking IgG antibodies" which protect patients by inhibiting IgE-mediated effector cell activation and mediator release (Cooke et al., 1935) but several other immunological changes were observed in the context with successful immunotherapy. They include

1. Reduction of mediator concentrations and numbers of effector cells (mast cells, eosinophils) in the target organs of atopy (Creticos et al., 1985; Otsuka et al., 1991; Durham et al., 1996; Furin et al., 1991; Rak et al., 1998).
2. Inhibition of the recruitment of T cells and eosinophils to sites of allergen contact (Durham et al., 1996).
3. Induction of suppressor cells (Rocklin et al., 1980).
4. Decreases of IL-4 production by T cells and increases of Th1 type cells in peripheral blood of treated patients (Secrist et al., 1993, 1995; Jutel et al., 1995).
5. Induction of anergy in allergen-specific T cells (Akdis and Blaser, 1999).
6. Induction of allergen-specific IgG responses against new epitopes documenting vaccination character of immunotherapy (Ball et al., 1999).

One major disadvantage of allergen-specific immunotherapy is that the administration of allergens can cause severe anaphylactic side effects. Another problem is that treatment is currently performed with allergen-extracts consisting of difficult to standardize mixtures of allergens and other undefined materials and therefore cannot be adapted to the patients individual sensitization profile (Bousquet et al., 1998a, 1998b).

Recent advances in the field of molecular allergen characterization and regarding the production of defined recombinant allergens give now rise to the hope that in the near future safer forms of immunotherapy can be developed (Valenta and Kraft, 1995). The new forms of treatment may be based on the use of defined allergen-components which are selected according to the patients sensitization profile and which are produced as hypoallergenic derivatives by recombinant DNA technology or synthetic peptide chemistry (reviewed in Valenta et al., 1999b).

4. CALCIUM-BINDING ALLERGENS

In the last decade, the number of available allergen sequences has rapidly increased due to the introduction of molecular biology techniques for allergen characterization (reviewed in Valenta and Kraft, 1995). Sequence analysis of allergen-encoding cDNAs revealed the presence of typical calcium-binding motifs, termed EF-hands, within allergens from various sources (reviewed in Valenta et al., 1998). Parvalbumin represented the first calcium-binding allergen described (Aas and Jebsen, 1967; Elsayed and Aas, 1971). It represents the major allergen in fish and belongs to a subfamily of closely related calcium-binding proteins that contain two functional and one silent EF-hand motif (Coffee and Bradshaw, 1973; Kretsinger and Nockolds, 1973). In countries with high fish consumption almost 1 per 1000 of the population is allergic to fish with most of the IgE antibodies directed against parvalbumin. Because of sequence and structural similarities among parvalbumins, fish allergic individuals exhibit allergic reactions after contact and/or consumption of various fish species. Next after parvalbumin, EF-hand domains were identified in the deduced amino acid sequences of two birch pollen allergens (Seiberler et al., 1992) and the allergens were subsequently identified as a three and a two EF-hand allergen designated, Bet v 3 (Seiberler et al., 1994) and Bet v 4 (Engel et al., 1997; Twardosz et al., 1997), respectively. It was found that IgE recognition of Bet v 3 was strongly influenced by the presence or absence of protein-bound calcium (Seiberler et al., 1994), a property which was shared by most of the other calcium-binding allergens described later on. Table 1 provides an overview of the calcium-binding allergens described so far. The allergens were grouped according to the number of calcium-binding domains. Allergen sources, molecular weights, accession numbers for the sequences and references are also displayed. As is evident from Table 1, pollens from various plant species turned out to be major sources of calcium-binding allergens. The latter may be related to the fact that calcium ions are important for pollen tube growth and perhaps for the self-incompatibility response in plants (Brewbaker and Kwack, 1963). Many calcium-binding allergens are therefore abundantly expressed in pollen tissue and, due to the fact that pollen from wind pollinated plants becomes airborn in large amounts, can sensitize patients via the respiratory mucosa. In this context it is noteworthy that calcium-binding allergens, although representing intracellular proteins, are rapidly eluted when pollen becomes hydrated as it happens after contact with the mucosa of patients (Grote et al., 1999). Bet v 3 from birch pollen was the first described calcium-binding pollen allergen (Seiberler et al., 1994). It was isolated by immunoscreening of a birch pollen cDNA library using serum IgE from a pollen allergic patient. Within its sequence, Bet v 3 contains three typical calcium-binding motifs and was found to be highly expressed in mature pollen. Another family of calcium binding allergens representing two EF-hand proteins of 8–9 kDa was identified in tree-, weed-, and grass pollens (Table 1). Jun o 2 and Ole e 8 represent four EF-hand calcium-binding allergens which were isolated from cypress and olive pollen (Tinghino et al., 1998; Ledesma et al., 2000).

Calcium-binding allergens were however also described in parasites (Santiago et al., 1998; Pritchard et al., 1999; McGibbon and Lee, 1995), which are potent inducers of IgE responses, and in animal hair/dander (Rautiainen et al., 1995). Recently even a calcium-binding autoallergen was isolated using serum IgE from atopic dermatitis patients for screening of a human epithelial expression cDNA library (Natter et al., 1998).

5. CALCIUM-BINDING PROTEINS AS CROSSREACTIVE ALLERGENS

The classical EF-hand domain found also in calcium-binding allergens consists of an α-helix, a loop coordinating the Ca^{2+} ion and a second α-helix (Kawasaki and Kretsinger, 1994; Kawasaki et al., 1998). In the EF-hand, calcium ions are usually bound through four carboxylate or carboxamide groups and a single backbone carbonyl oxygen placed in the loop with a specific spacing. Although EF-hand domains represent highly conserved sequence motifs also in calcium-binding allergens, it seems that crossreactivity of IgE antibodies from allergic patients who are sensitized against a particular calcium-binding allergen is rather limited to closely related proteins. There is extensive IgE crossreactivity of fish allergic individuals with parvalbumins from various fish species which explains why parvalbumin-sensitized individuals exhibit allergic symptoms after ingestion of various fish species (Bugajska-Schretter et al., 1998, 2000) (Table 1). Likewise, there is strong crossreactivity among the members of the two-EF-hand pollen allergens (Table 1). Patients who are sensitized to the two-EF hand allergen from birch always exhibit IgE crossreactivity with the homologous proteins in weed and

Table 1. Calcium-binding allergens

No. of Ca^{2+}-binding sites	Designation	MW kDa	Source	Tissue	References	Genebank Accession No.
1	Sj22.6	22.6	schistosome (*Schistosoma japonicum*)	tegument	Santiago et al (1998)	AF030404
2	BDA11 (S100A7, psoriasin)	11.6	cow (*Bos taurus*)	skin	Rautiainen et al (1995)	L39834
	calreticulin	46	hookworm (*Necator americanus*)		Pritchard et al (1999)	AJ006790
	ABA-1 (member of S100 family)	10	ascaris (*Ascaris lumbricoides*)		McGibbon et al (1995)	A37188
	Ara CALC (autoantigen)	54.5	man (*Homo sapiens*)	A431	Natter et al (1998)	Y17711
				keratinocytes		D29420
				lung		T90587
				brain		Z43480, AL117423
				fetal liver		R87799
				heart		T32765
	Aln g 4	9.3	alder (*Alnus glutinosa*)	pollen	Hayek et al (1998)	Y17713
	Bet v 4	9.3	birch (*Betula verrucosa*)	pollen	Seiberler et al (1992); Twardosz et al (1997); Engel et al (1997)	Y12560, X87153
	Ole e 3	9.4	olive (*Olea europea*)	pollen	Batanero et al (1996); Ledesma et al (1998)	AF015810
	Cyn d 7	8.7	bermuda grass (*Cynodon dactylon*)	pollen	Smith et al (1997); Suphioglu et al (1997)	X91256
	Phl p 7	8.6	timothy grass (*Phleum pratense*)	pollen	Niederberger et al (1999)	Y17835
	Bra r 1	8.6	rape (*Brassica rapa*)	pollen	Toriyama et al (1995)	D63153
	Bra n 1	8.6	rape (*Brassica napus*)	pollen	Toriyama et al (1995)	D63351
	Bra r 2	9.1	rape (*Brassica rapa*)	pollen	Toriyama et al (1995)	D63154
	Bra n 2	9.0	rape (*Brassica napus*)	pollen	Toriyama et al (1995)	D63152
3	Bet v 3	23.7	birch (*Betula verrucosa*)	pollen	Seiberler et al (1992); Seiberler et al (1994)	X79267
	allergen M (Gad c 1)	12.3	cod (*Gadus callarias*)	muscle	Elsayed et al (1971)	
	parvalbumin	12	carp (*Cyprinus carpio*)	muscle	Coffee et al (1973)	
	parvalbumin	12	salmon (*Salmo salar*)	muscle	Lindstroem et al (1996)	X97824, X97825
4	Jun o 2	18	cypress (*Juniperus oxycedrus*)	pollen	Tinghino et al (1998)	AF031471
	Ole e 8	18.9	olive (*Olea europea*)	pollen	Ledesma et al (2000)	AF078680

grasspollens and therefore constitute a group of allergic patients who can exhibit allergic reactions after contact with pollens from many, even unrelated plant species (Engel et al., 1997; Twardosz et al., 1997). While there is extensive crossreactivity to calcium-binding allergens within certain allergen families (parvalbumins, 2 EF-hand pollen allergens) there seems to be little or no crossreactivity among the various families of calcium-binding allergens in

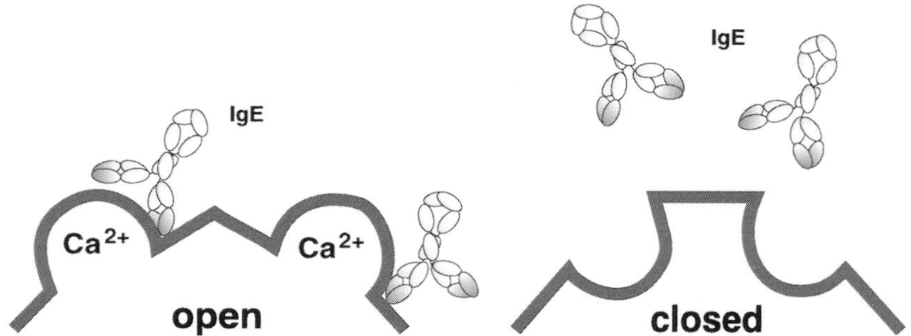

Figure 2. IgE antibodies recognize preferentially epitopes on the open (i.e., calcium-bound) form but not on the closed (i.e., apo-) form of calcium-binding allergens. For a colour version of this figure, see page xxiv.

general. The latter fact indicates that the less conserved sequence motifs and, perhaps conformational epitopes which are unique to a given protein family, play an important role for IgE recognition.

Crossreactivities of IgE antibodies within certain families of calcium-binding proteins may have important implications for diagnosis as well as for therapy. If an allergic patient contains for instance IgE antibodies against a particular 2 EF-hand pollen allergen (e.g., birch: Bet v 4) these IgE antibodies will crossreact with 2 EF-hand allergens in grass-, weed- and other tree pollens leading to allergic reactions against most plant pollens. Calcium-binding allergens can thus be used as marker allergens which when used for diagnostic testing allow to predict broad sensitization of a given patient (reviewed in Valenta et al., 1999a). On the other hand it seems possible to treat sensitized patients by specific immunotherapy only with a few representative members containing the crossreactive epitopes in order to cover the complete family of crossreactive calcium-binding allergens.

6. CONFORMATION-DEPENDENT IgE RECOGNITION OF CALCIUM-BINDING ALLERGENS

It is well established that calcium-binding proteins can undergo dramatic open-closed conformational changes depending on the presence or absence of protein-bound calcium (Herzberg et al., 1986; Heinzmann and Hunziker, 1991; Strynadka and James, 1991; Trave et al., 1995; Ikura, 1996; Laberge et al., 1997). It is also well established that these conformational changes can strongly affect the physochemical properties of calcium-binding proteins. E.g., it is known that the calcium-bound (open conformations) are more stable than the apoforms (closed conformations) and that the calcium-bound (open conformations) forms can expose hydrophobic surfaces which are not available in the closed conformation (Figure 2). Calcium can thus strongly influence the ability of calcium-binding proteins to interact with other ligands and thereby represents a physiological calcium-dependent regulation mechanism for ligand interactions (Ikura, 1996). Likewise it was discovered that the IgE-binding capacity of calcium-binding allergens is affected by calcium (Seiberler et al., 1992; Seiberler et al., 1994) (Figure 2). A calcium-dependent modulation of the IgE-binding capacity was reported for calcium-binding birch pollen allergens, Bet v 3 and Bet v 4, as well as for other calcium-binding pollen allergens and the major fish allergen, parvalbumin (Seiberler et al., 1992, 1994; Engel et al., 1997; Twardosz et al., 1997; Bugajska-Schretter et al., 1998). Most sensitized patients contain IgE antibodies which recognize preferentially the calcium-bound form and exhibit strongly reduced IgE binding to the calcium-free apoform. This observation indicates that patients who are allergic against calcium-binding allergens were preferentially sensitized against the calcium-bound forms which are also more stable than the apoforms (Hayek et al., 1998; Bugajska-Schretter et al., 2000).

7. CONVERSION OF CALCIUM-BINDING ALLERGENS INTO HYPOALLERGENIC MOLECULES FOR VACCINATION AGAINST ALLERGY

A major disadvantage of allergen-specific immunotherapy is that administration of allergens to patients can cause anaphylactic side effects (Bousquet et al., 1998a, b). Therefore attempts are made to convert active allergens into molecules with reduced allergenic activity for immunotherapy with reduced rate of anaphylactic side effects. The finding that IgE recognition of calcium-binding allergens depends on protein-bound calcium has stimulated ideas to convert these allergens into allergen variants with reduced IgE binding capacity by mutation of the calcium-binding sites (Engel et al., 1997; Okada et al., 1998). It was demonstrated that mutated versions of the birch pollen allergen Bet v 4 and the 2 EF-hand allergens from rape containing amino acid exchanges within the calcium-binding domains lost their calcium-binding properties and exhibited a strongly reduced IgE-binding capacity. It was also demonstrated that deletion variants and fragments of calcium-binding allergens exhibited reduced IgE binding capacity (Twardosz et al., 1997; Hayek et al., 1998).This opens now possibilities to produce by genetic engineering hypoallergenic versions of calcium-binding allergens which may be used for allergen-specific immunotherapy with reduced risk of anaphylactic side effects (reviewed in Singh et al., 1999; Valenta et al., 1999b).

ACKNOWLEDGEMENTS

This study was supported by grant F0506 of the Austrian Science Fund and by the ICP project of the Austrian Ministry of Science. We thank Anton Jäger for skillful preparation of the artwork.

REFERENCES

Aas, K. and Jebsen, J.W., 1967, Studies of hypersensitivity to fish. Partial purification and crystallization of a major allergenic component of cod, *Int. Arch. Allergy* 32, 1–20.

Akdis, C.A. and Blaser, K., 1999, IL-10-induced anergy in peripheral T cell and reactivation by microenvironmental cytokines: Two key steps in specific immunotherapy, *FASEB J.* 13, 603–609.

Ball, T., Sperr, W.R., Valent, P., Lidholm, J., Spitzauer, S., Ebner, C., Kraft, D. and Valenta, R., 1999, Induction of antibody responses to new B cell epitopes indicates vaccination character of allergen immunotherapy, *Eur. J. Immunol.* 29, 2026–2036.

Batanero, E., Villalba, M., Ledesma, A., Puente, X.S. and Rodriguez, R., 1996, Ole e 3, an olive tree allergen, belongs to a widespread family of pollen proteins, *Eur. J. Biochem.* 241, 772–778.

Beaven, M.A. and Metzger, H., 1993, Signal transduction by Fc receptors: The Fc epsilon RI case, *Immunol Today* 14, 222–226.

Bieber, T., 1996, Fc epsilon RI on antigen-presenting cells, *Curr. Opin. Immunol.* 8, 773–777.

Bousquet, J., Lockey, R., Malling, H.J. and the WHO Panel Members, 1998a, Allergen immunotherapy: Therapeutic vaccines for allergic diseases. A WHO position paper, *J. Allergy Clin. Immunol.* 102, 558–562.

Bousquet, J., Lockey, R. and Malling, H.J., 1998b, WHO position paper. Allergen immunotherapy: Therapeutic vaccines for allergic diseases, *Allergy* 53, 1–42.

Brewbaker, J.L. and Kwack, B.H., 1963, The essential role of calcium ion in pollen germination and pollen tube growth, *Am. J. Both.* 50, 859–865.

Bugajska-Schretter, A., Elfman, L., Fuchs, T., Kapiotis, S., Rumpold, H., Valenta, R. and Spitzauer, S., 1998, Parvalbumin, a cross-reactive fish allergen, contains IgE-binding epitopes sensitive to periodate treatment and Ca^{2+} depletion, *J. Allergy Clin. Immunol.* 101, 67–74.

Bugajska-Schretter, A., Grote, M., Vangelista, L., Valent, P., Sperr, W.R., Rumpold, H., Pastore, A., Reichelt, R., Valenta, R. and Spitzauer, S., 2000, Purification, biochemical, and immunological characterization of a major food allergen: Different immunoglobulin E recognition of the apo- and calcium-bound form of carp parvalbumin, *Gut* 46, 661–669.

Casolaro, S., Georas, S.N., Song, Z. and Ono, S.J., 1996, Biology and genetics of atopic disease, *Curr. Opin. Immunol.* 8, 796–803.

Coffee, C.J. and Bradshaw, R.A., 1973, Carp muscle calcium-binding protein. I. Characterization of the tryptic peptides and the complete amino acid sequence of component B, *J. Biol. Chem.* 248, 3302–3312.

Cooke, R., Barnard, J., Hebald, S. and Stull, A., 1935, Serological evidence on immunity with co-existing sensitization in a type of human allergy, hay fever, *J. Exp. Med.* 62, 733–750.

Creticos, P.S., Adkinson, N.F., Jr., Kagey-Sobotka, A., Proud, D., Meier, H.L., Naclerio, R.M., Lichtenstein, L.M. and Norman, P.S., 1985, Nasal challenge with ragweed in hay fever patients: Effect of immunotherapy, *J. Clin. Invest.* 76, 2247–2253.

Desreumaux, P. and Capron, M., 1996, Eosinophils in allergic reactions, *Curr. Opin. Immunol.* 8, 790–795.

Durham, S.R. and Till, S.J., 1998, Immunological changes associated with allergen immunotherapy, *J. Allergy Clin. Immunol.* 102, 157–164.

Durham, S.R., Ying, S., Varney, V.A., Jacobson, M.R., Sudderick, R.M., Mackay, I.S., Kay, A.B. and Hamid, Q.A., 1996, Grass pollen immunotherapy inhibits allergen-induced infiltration of CD4+ T lymphocytes and eosinophils in the nasal mucosa and increases the number of cells expressing messenger RNA for interferon-γ, *J. Allergy Clin. Immunol.* 97, 1356–1365.

Engel, E., Richter, K., Obermeyer, G., Briza, P., Kungl, A.J., Simon, B., Auer, M., Ebner, C., Rheinberger, H.J., Breitenbach, M. and Ferreira, F., 1997, Immunological and biological properties of Bet v 4, a novel birch pollen allergen with two EF-hand calcium-binding domains, *J. Biol. Chem.* 272, 28630–28637.

Elsayed, S. and Aas, K., 1971, Characterization of a major allergen (cod). Observation on effect of denaturation on the allergenic activity, *J. Allergy* 47, 283–291.

Furin, M.M., Norman, P.S., Creticos, P.S., Proud, D., Kagey-Sobotka, A., Lichtenstein, L.M. and Naclerio, R.M., 1991, Immunotherapy decreases antigen-induced eosinophil migration into the nasal cavity, *J. Allergy Clin. Immunol.* 88, 27–32.

Grote, M., Hayek, B., Reichelt, R., Kraft, D. and Valenta, R., 1999, Immunogold electron microscopic localization of the crossreactive two EF-hand calcium-binding birch pollen allergen Bet v 4 in dry and rehydrated birch pollen, *Int. Arch. Allergy Immunol.* 120, 287–294.

Haas, H., Falcone, F.H., Holland, M.J., Schramm, G., Haisch, K., Gibbs, B.F., Bufe, A. and Schlaak, M., 1999, Early interleukin-4: ist role in the switch towards a Th2 response and IgE-mediated allergy, *Int. Arch. Allergy Immunol.* 119, 86–94.

Hayek, B., Vangelista, L., Pastore, A., Sperr, W.R., Valent, P., Vrtala, S., Niederberger, V., Twardosz, A., Kraft, D. and Valenta, R., 1998, Molecular and immunologic characterization of a highly cross-reactive two EF-hand calcium-binding alder pollen allergen, Aln g 4: Structural basis for calcium-modulated IgE recognition, *J. Immunol.* 161, 7031–7039.

Heinzmann, C.W. and Hunziker, W., 1991, Intracellular calcium-binding proteins: more sites than insights, *Trends Biochem. Sci.* 16, 98–103.

Henderson, L.L., Larson, J.B. and Gleich, G.L., 1975, Maximal rise in IgE antibody following ragweed pollination season, *J. Allergy Clin. Immunol.* 55, 10–15.

Herzberg, O., Moult, J. and James, M.N., 1986, A model for the Ca^{2+} induced conformational transition of troponin C. A trigger for muscle contraction, *J. Biol. Chem.* 261, 2638–2644.

Ikura, M., 1996, Calcium binding and conformational response in EF-hand proteins, *Trends Biochem. Sci.* 21, 14–17.

Ishizaka, K., Ishizaka, T. and Hornbrook, M.M., 1966, Physicochemical properties of human reaginic antibody, *J. Immunol.* 97, 840–845.

Johansson, S.G.O. and Bennich, H., 1967, Immunological studies of an atypical (myeloma) immunoglobulin, *Immunology* 13, 381–394.

Jutel, M., Pichler, W.J., Skrbic, D., Urwyler, A., Dahinden, C. and Müller, U.R., 1995, Bee venom immunotherapy results in decrease of IL-4 and IL-5 and increase of IFN-γ secretion in specific allergen-stimulated T cell cultures, *J. Immunol.* 154, 4187–4194.

Kawasaki, H. and Kretsinger, R.H., 1994, Calcium-binding proteins, in *Protein Profile 1*, H. Kawasaki and R.H. Kretsinger (eds.), Academic Press, San Diego, pp. 343–390.

Kawasaki, H., Nakayama, S. and Kretsinger, R.H., 1998, Classification and evolution of EF-hand proteins, *Biometals* 11, 277–295.

Kay, A.B., 1997, *Allergy and Allergic Diseases*, Blackwell Science, Oxford.

Kretsinger, R.H. and Nockolds, C.E., 1973, Carp muscle calcium-binding protein. II. Structure determination and general description, *J. Biol. Chem.* 248, 3313–3326.

Laberge, M., Wright, W.W., Sudhakar, K., Liebman, P.A. and Vanderkooi, J.M., 1997, Conformational effects of calcium release from parvalbumin: Comparison of computational simulations with spectroscopic investigations, *Biochemistry* 36, 5363–5371.

Ledesma, A., Villalba, M., Batanero, E. and Rodriguez, R., 1998, Molecular cloning and expression of active Ole e 3, a major allergen from olive-tree pollen and member of a novel family of Ca^{2+}-binding proteins (polcalcins) involved in allergy, *Eur. J. Biochem.* 258, 454–459.

Ledesma, A., Villalba, M. and Rodriguez, R., 2000, Cloning, expression and characterization of a novel four EF-hand Ca^{2+}-binding protein from olive pollen with allergenic activity, *FEBS Lett.* 466, 192–196.

Lindstroem, C.D.V., van Do, T., Hordvik, I., Endresen, C. and Elsayed, S., 1996, Cloning of two distinct cDNAs encoding parvalbumin, the major allergen of atlantic salmon (Salmo salar), *Scand. J. Immunol.* 44, 335–344.

Lockey, R.F. and Bukantz, S.C. (eds.), 1998, *Allergen Immunotherapy*, 2nd edition, Marcel Dekker, New York.

Maurer, D., Ebner, C., Reininger, B., Fiebiger, E., Kraft, D., Kinet, J.P. and Stingl, G., 1995, The high affinity IgE receptor (Fc epsilon RI) mediates IgE-dependent allergen presentation, *J. Immunol.* 154, 6285–6290.

Maurer, D., Fiebiger, E., Ebner, C., Reininger, B., Fischer, G.F., Wichlas, S., Jouvin, M.H., Schmitt-Egenolf, M., Kraft, D., Kinet, J.P. and Stingl, G., 1996, Peripheral blood dendritic cells express Fc epsilon RI as a complex composed of Fc epsilon RI alpha- and Fc epsilon RI gamma-chains and can use this receptor for IgE-mediated allergen presentation, *J. Immunol.* 157, 607–616.

McGibbon, A.M. and Lee, T.D., 1995, Structural characteristics of the Ascaris allergen, ABA-1, *Parasite* 2, 41–48.

Mudde, G.C., van Reijsen, F.C., Boland, G.J., de Gast, G.C., Bruijnzeel, P.L. and Bruijnzeel-Koomen, C.A., 1990, Allergen presentation by epidermal Langerhans cells from patients with atopic dermatitis is mediated by IgE, *Immunology* 69, 335–341.

Natter, S., Seiberler, S., Hufnagl, P., Binder, B.R., Hirschl, A.M., Ring, J., Abeck, D., Schmidt, T., Valent, P. and Valenta, R., 1998, Isolation of cDNA clones coding for IgE autoantigens with serum IgE from atopic dermatitis patients, *FASEB J.* 12, 1559–1569.

Niederberger, V., Hayek, B., Vrtala, S., Laffer, S., Twardosz, A., Vangelista, L., Sperr, W.R., Valent, P., Rumpold, H., Kraft, D., Ehrenberger, K., Valenta, R. and Spitzauer, S., 1999, Calcium-dependent immunoglobulin E recognition of the apo- and calcium-bound form of a cross-reactive two EF-hand timothy grass pollen allergen, Phl p 7, *FASEB J.* 13, 843–856.

Okada, T., Swoboda, I., Bhalla, P.L., Toriyama, D. and Singh, M.B., 1998, Engineering of hypoallergenic mutants of the Brassica pollen allergen, Bra r 1, for immunotherapy, *FEBS Lett.* 434, 255–260.

Otsuka, H., Mezawa, A., Ohnishi, M., Okubo, K., Sehi, H. and Okuda, M., 1991, Changes in nasal metachromatic cells during allergen immunotherapy, *Clin. Exp. Allergy* 21, 115–120.

Prausnitz, C. and Küstner, H., 1921. Studien über die Überempfindlichkeit, *Zentralbl. Bakteriol.* 86, 160–169.

Pritchard, D.I., Brown, A., Kasper, G., McElroy, P., Loukas, A., Hewitt, C., Berry, C., Fullkrug, R. and Beck, E., 1999, A hookworm allergen which strongly resembles calreticulin, *Parasite Immunol.* 21, 439–450.

Rak, S., Rowhagen, O. and Venge, P., 1998, The effect of immunotherapy on bronchial hyperresponsiveness and eosinophil cationic protein in pollen allergic patients, *J. Allergy Clin. Immunol.* 82, 470–480.

Rautiainen, J., Rytkonen, M., Parkkinen, S., Pentikainen, J., Linnala-Kamkkunen, A., Virtanen, T., Pelkonen, J. and Mantyjarvi, R., 1995, cDNA cloning and protein analysis of a bovine dermal allergen with homology to psoriasin, *J. Invest. Dermatol.* 105, 660–663.

Ravetch, J.V. and Kinet, J.P., 1991, Fc receptors, *Annu. Rev. Immunol.* 9, 457–492.

Rocklin, R.E., Sheffer, A., Greineder, D.K. and Melmon, K.L., 1980, Generation of antigen-specific supressor cells during allergy desensitization, *New Engl. J. Med.* 302, 1213–1219.

Santiago, M.L., Hafalla, J.C.R., Kurtis, J.D., Aligui, G.L., Wiest, P.M., Olveda, R.M., Olds, G.R., Dunne, D.W. and Ramirez, B.L., 1998, Identification of the *Schistosoma japonicum* 22.6 kD antigen as a major target of the human IgE response: Similarity of IgE-binding epitopes to allergen peptides, *Int. Arch. Allergy Immunol.* 117, 94–106.

Secrist, H., Chelen, C.J., Wen, Y., Marshall, J.D. and Umetsu, D.T., 1993, Allergen immunotherapy decreases interleukin 4 production in CD4+ T cells from allergic individuals, *J. Exp. Med.* 178, 2123–2130.

Secrist, H., DeKruyff, R.H. and Umetsu, D.T., 1995, Interleukin 4 production by CD4+ T cells from allergic individuals is modulated by antigen concentration and antigen-presenting cell type, *J. Exp. Med.* 181, 1081–1089.

Segal, D.M., Taurog, J.D. and Metzger, H., 1977, Dimeric immunoglobulin E serves as a unit signal for mast cell degranulation, *Proc. Natl. Acad. Sci. USA* 74, 2993–2997.

Seiberler, S., Scheiner, O., Kraft, D. and Valenta, R., 1992, Homology of two cDNAs coding for birch pollen allergens with calmodulin: Protein-bound Ca^{2+} affects the IgE-binding capacity, *Int. Arch. Allergy Immunol.* 99, 380–381.

Seiberler, S., Scheiner, O., Kraft, D., Lonsdale, D. and Valenta, R., 1994, Characterization of a birch pollen allergen, Bet v III, representing a novel class of Ca^{2+} binding proteins: Specific expression in mature pollen and dependence of patients' IgE binding on protein-bound Ca^{2+}, *EMBO J.* 13, 3481–3486.

Singh, M.B., de-Weerd, N. and Bhalla, P.L., 1999, Genetically engineered plant allergens with reduced anaphylactic activity, *Int. Arch. Allergy Immunol.* 119, 75–85.

Smith, P.M., Xu, H., Swoboda, I. and Singh, M.B., 1997, Identification of a Ca^{2+} binding protein as a new Bermuda grass pollen allergen Cyn d 7: IgE cross-reactivity with oilseed rape pollen allergen Bra r 1, *Int. Arch. Allergy Immunol.* 114, 265–271.

Stingl, G. and Maurer, D., 1997, IgE-mediated allergen presentation via Fc epsilon RI on antigen-presenting cells, *Int. Arch. Allergy Immunol.* 113, 24–29.

Strynadka, N.C.J. and James, M.N.G., 1991, Towards an understanding of the effects of calcium on protein structure and function, *Curr. Opin. Struct. Biol.* 1, 905–914.

Suphioglu, C., Ferreira, F. and Knox, R.B., 1997, Molecular cloning and immunological characterization of Cyn d 7, a novel calcium-binding allergen from Bermuda grass pollen, *FEBS Lett.* 402, 167–172.

Tinghino, R., Barletta, B., Palumbo, S., Afferni, C., Iacovacci, P., Mari, A., Di Felici, G. and Pini, C., 1998, Molecular characterization of a cross-reactive Juniperus oxycedrus pollen allergen, Jun o 2: A novel calcium-binding allergen, *J. Allergy Clin. Immunol.* 101, 772–777.

Toriyama, K., Okada, T., Watanabe, M., Ide, T., Ashida, T., Xu, H. and Singh, M.B., 1995, A cDNA clone encoding an IgE-binding protein from Brassica anther has significant sequence similarity to Ca^{2+}-binding proteins, *Plant. Mol. Biol.* 29, 1157–1165.

Trave, G., Lacombe, P.J., Pfuhl, M., Saraste, M. and Pastore, A., 1995, Molecular mechanism of the calcium-induced conformational change in the spectrin EF-hands, *EMBO J.* 14, 4922–4931.

Twardosz, A., Hayek, B., Seiberler, S., Vangelista, L., Elfman, L., Grönlund, H., Kraft, D. and Valenta, R., 1997, Molecular characterization, expression in *Escherichia* coli and epitope analysis of a two EF-hand calcium-binding birch pollen allergen, Bet v 4, *Biochem. Biophys. Res. Comm.* 239, 197–204.

Valenta, R. and Kraft, D., 1995, Recombinant allergens for diagnosis and therapy of allergic diseases, *Curr. Opin. Immunol.* 7, 751–756.

Valenta, R., Hayek, B., Seiberler, S., Bugajska-Schretter, A., Niederberger, V., Twardosz, A., Natter, S., Vangelista, L., Pastore, A., Spitzauer, S. and Kraft, D., 1998, Calcium-binding allergens: From plants to man, *Int. Arch. Allergy Immunol.* 117, 160–166.

Valenta, R., Lidholm, J., Niederberger, V., Hayek, B., Kraft, D. and Grönlund, H., 1999a, The recombinant allergen-based concept of component-resolved diagnostics and immunotherapy (CRD and CRIT), *Clin. Exp. Allergy* 29, 896–904.

Valenta, R., Vrtala, S., Focke-Tejkl, M., Bugajska-Schretter, A., Ball, T., Twardosz, A., Spitzauer, S., Grönlund, H. and Kraft, D, 1999b, Genetically engineered and synthetic allergen derivatives: Candidates for vaccination against Type I allergy, *Biol. Chem.* 380, 815–824.

Valenta, R., Seiberler, S., Natter, S., Mahler, V., Mossabeb, R., Ring, J. and Stingl, G., 2000, Autoallergy: A pathogenetic factor in atopic dermatitis?, *J. Allergy Clin. Immunol.* 105, 432–437.

Vercelli, D., De Monte, L., Monticelli, S., Di Bartolo, C. and Agresti, A., 1998, To E or not to E? Can IL-4-induced B cell choose between IgE and IgG4?, *Int. Arch. Allergy Immunol.* 116, 1–4.

Calcium Channelopathies in Nervous System

Daniela Pietrobon

1. CALCIUM CHANNELOPATHIES

What do epilepsy, migraine headache, episodic ataxia, peryodic paralysis, malignant hyperthermia, night blindness have in common? These human neurological disorders can be caused by mutations in genes encoding ion channels, and therefore can be considered as "channelopathies", a term used here to name inherited ion channel pathologies. More specifically, they can be considered as calcium channelopathies since they can be caused by mutations in genes encoding calcium channels. This chapter deals with known human and mouse calcium channelopathies of the central nervous system (CNS). The human diseases comprise (i) a recessive retinal disorder, X-linked congenital stationary night blindness (xlCSNB), associated with mutations in the CACNA1F gene, encoding the pore-forming subunit of an L-type voltage-dependent calcium channel expressed only in the retina, and (ii) a group of rare allelic autosomal dominant human neurological disorders including familial hemiplegic migraine (FHM), episodic ataxia type 2 (EA-2) and spinocerebellar ataxia type 6 (SCA6), all associated with mutations in the CACNA1A gene, encoding the pore-forming subunit of neuronal P/Q-type voltage-dependent calcium channels. Mutations at the mouse orthologue of the CACNA1A gene cause a group of recessive neurological disorders, including the tottering and leaner phenotypes with ataxia and absence epilepsy, and the rolling Nagoya phenotype with ataxia without seizures. Two other spontaneous mouse mutants with ataxia and absence epilepsy, lethargic and stargazer, have mutations in genes encoding a calcium channel auxiliary β subunit and a putative calcium channel auxiliary γ subunit. The skeletal muscle human calcium channelopathies hypokalemic peryodic paralysis and malignant hyperthermia will not be dealt with here.

Daniela Pietrobon • Department of Biomedical Sciences, University of Padova, Padova, Italy.

R. Pochet, R. Donato, J. Haiech, C. Heizmann and V. Gerke (eds.): Calcium: The Molecular Basis of Calcium Action in Biology and Medicine, 379–400.

Understanding the phenotypic expression of these monogenic disorders may shed light on similar mechanisms in other prevalent and polygenic disorders such as migraine and epilepsy. Currently, limited treatment can be offered to patients with calcium channelopathies. Analysis of the functional consequences of calcium channel mutations may help to develop rational treatment for these diseases. Moreover, it provides an important approach for identifying the poorly understood roles that calcium channels play in different neuronal populations in the CNS.

2. ION CHANNELS AND CELLULAR EXCITABILITY

Ion channels are ubiquitous membrane proteins forming pores that allow ions to move rapidly through cell membranes down their electrochemical gradients. Most ion channels are gated, i.e capable of making transitions between conducting and non-conducting conformations. Gating of ion channels can be dependent on membrane potential, extracellular ligands, intracellular second messengers and metabolites, protein-protein interactions, phosphorylation and other factors. The flow of ions through ion channels create electrical currents that produce rapid changes in the membrane potential. Opening and closing of ion channels generate a complex system of electrical signaling, and form the basis of cellular excitability in neurons and muscle cells. Molecular cloning has revealed a large number of channel genes, that most likely reflects a diversity of specific signaling needs. The specific properties of individual channel types, their localization, and level of expression are primary determinants of cellular excitability. Secondary determinants are cellular proteins and external environmental factors modulating ion channels.

In neurons and muscle cells membrane excitability is tightly regulated by a balance of excitatory (membrane depolarizing) and inhibitory (membrane hyperpolarizing) signals. Ion channel defects are predicted to tip this delicate balance leading to abnormal cellular excitability. Indeed, most channelopathies share the clinical feature of episodic occurrence along with manifestations suggesting membrane hyperexcitability. The biological abnormality is sufficiently mild that patients may be totally normal between attacks; however, under certain circumstances some factor pushes patients past some boundary and precipitate the attack. Precipitating factors in many of these disorders include stress, fatigue, exercise and certain foods (Ptacek, 1998).

3. NEURONAL VOLTAGE-DEPENDENT CALCIUM CHANNELS

Since Ca^{2+} ions (unlike K^+, Na^+, Cl^- ions) are intracellular second messengers, calcium channels are unique among ion channels because of the diversity of additional cellular functions, besides cellular excitability, they regulate. Voltage-dependent calcium channels play the unique role of transducers

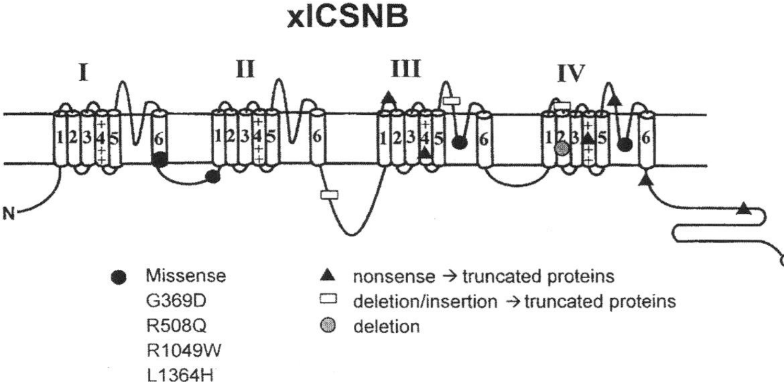

Figure 1. Mutations of the CACNA1F gene associated with X-linked congenital stationary night blindness (xlCSNB): location in the secondary structure of the calcium channel α_{1F} subunit.

of electrical signals into chemical signals. Neuronal voltage-dependent calcium channels control a broad array of functions including neurotransmitter release, neurite outgrowth, synaptogenesis, neuronal excitability, activity-dependent gene expression, as well as neuronal survival, differentiation and plasticity.

Voltage-dependent Ca^{2+} channels are multisubunit complexes composed of a pore-forming and voltage-sensing α_1 subunit and several auxiliary subunits, including $\alpha_2\delta$ and β subunits. They constitute a complex family of channels comprising a large number of different subtypes, which have in common a steep voltage dependence of the open probability and a very high selectivity for Ca^{2+} over Na^+ and K^+ ions in physiological solutions. The structure of the α_1 subunit shares a basic design with other voltage-gated ion channels, consisting of six membrane spanning segments (S1 to S6) flanked by cytoplasmic and extracellular loops, with the loop between S5 and S6 folded into the membrane to form part of the pore. This basic design is repeated in four homologous domains (Figure 1). The S4 segment has been shown to play a major role in voltage sensing. High selectivity for Ca^{2+} ions is achieved through a high affinity binding site formed by a ring of four glutamates, located within the pore close to the external mouth (Sather et al., 1994; Armstrong and Hille, 1998).

Ca^{2+} channels α_1, α_2-δ and β subunits are encoded by at least ten ($\alpha_{1A,B,C,D,E,F,G,I,H,S}$), three ($\alpha_2\delta$-1, -2, -3) and four ($\beta_{1,2,3,4}$) different genes, respectively (Dunlap et al., 1995; Randall and Benham, 1999). Further molecular diversity is created by the existence of multiple splice variants for each gene. The auxiliary subunits, in particular the β subunits, have major functional effects on both membrane targeting and modulation of calcium channels containing $\alpha_{1A,B,C,D,E,S}$ subunits (Walker and De Waard, 1998). In heterologous expression systems, different β subunits in combination with a given α_1 subunit give rise to calcium channels with different biophysical properties. Moreover, functionally different calcium channels can be formed by different splice variants of a given α_1 subunit (Lin et al., 1997; Bourinet et al., 1999; Tottene et al., 2000). All the different Ca^{2+} channel subunits, except α_{1S}, are expressed in the brain, with a differential distribution in different neuronal populations. Therefore, the potential for combinatorial structural and functional heterogeneity of brain calcium channels is enormous.

According to pharmacological criteria, native neuronal high-voltage-activated Ca^{2+} channels have been classified as dihydropyridine-sensitive channels (L-type), ω-conotoxin-GVIA-sensitive channels (N-type) and ω-agatoxin-IVA-sensitive channels (P/Q-type). An additional component of current (R-type) has been identified as the calcium current resistant to the specific inhibitors of L-, N- and P/Q-type channels (Dunlap et al., 1995; Randall and Benham, 1999). Ca^{2+} channels subtypes with distinct biophysical properties have been identified within each of the four pharmacological classes of high-voltage-activated calcium channels. The following correlation between native neuronal calcium channels and cloned α_1 subunits has been established: α_{1B}, α_{1A} and α_{1E} subunits are the pore-forming subunits of N-, P/Q- and R-type calcium channels, respectively; α_{1C}, α_{1D} and α_{1F} subunits are pore-forming subunits of different L-type calcium channels; α_{1G}, α_{1I} and α_{1H} subunits are pore-forming subunits of different low-voltage activated T-type Ca^{2+} channels. Different combinations with auxiliary subunits and/or alternative splicing of α_1 subunits most likely account for the large functional diversity of native L-, N-, P/Q- and R-type calcium channels. Indeed, coimmunoprecipitation studies have shown that native L-, N- and P/Q-type channels can contain each of the β subunits (Walker and De Waard, 1998).

4. X-LINKED CONGENITAL STATIONARY NIGHT BLINDNESS: A HUMAN α_{1F} CALCIUM CHANNELOPATHY

xlCSNB is a recessive non-progressive human eye disease characterized by night blindness, variable reduced day vision, decreased visual acuity, myopia, nystagmus and strabismus. xlCSNB is thought to result from decreased effectivness of synaptic transmission between photoreceptors and second-order neurons in the retina. Two distinct clinical entities have been proposed, complete and incomplete xlCSNB, whose loci have been mapped to chromosome Xp11.4 and Xp11.23, respectively. A gene, called CACNA1F, which encodes a calcium channel α_{1F} subunit, has been mapped to Xp11.23, and mutations in this gene have been found in the majority of screened families with incomplete xlCSNB (Strom et al., 1998; Torben Bech-Hansen et al., 1998). CACNA1F is a new member of the family of genes encoding L-type calcium channel α_1 subunits, since α_{1F} subunits have 60–70% aminoacids homology with α_{1C}, α_{1D} and α_{1S} subunits, and contain the conserved dihydropyridine binding site of L-type channels. α_{1F} subunits are expressed only in the retina, in both the outer nuclear cell layer, containing the photoreceptor cell bodies, and the inner nuclear cell layer, consisting of horizontal, bipolar and amacrine cells. Of the 15 mutations associated to xlCSNB identified so far, eleven are nonsense mutations or deletion/insertions causing truncated and deleted proteins, and four are missense mutations leading to substitution of conserved aminoacids and a change in net charge (Figure 1). Three of the four aminoacid substitutions are in regions, the P loops of domains III and IV and the S6 segment of domain I, that are thought to form part of the pore. Although the consequences of the mutations on channel function have not been investigated, the truncated mutant proteins are unlikely to form functional channels, and therefore the association of xlCSNB with loss-of-function mutations appears most likely. Since L-type channels control neurotransmitter release from photoreceptor presynaptic terminals, loss-of-function mutations in presynaptic L-type channels would decrease presynaptic calcium influx and tonic glutamate release in darkness, with consequent relative depolarization of bipolar cells. The expression of α_{1F} subunits, not only in photoreceptors, but also in other neurons of the retina suggests that additional mechanisms are probably involved in the disease.

5. HUMAN α_{1A} CALCIUM CHANNELOPATHIES: GENOTYPE AND PHENOTYPE

Migraine is a frequent and clinically heterogeneous neurological disorder, affecting up to 15% of females and 6% of males in Caucasian populations. Migraine attacks, typically lasting one to three days, are characterized by severe, unilateral pounding head pain associated with nausea, vomiting, and sensitivity to light and sound (migraine without aura). In about 15% of patients, the attacks are preceded by transient neurologic abnormalities, such as visual, sensory, motor, or cognitive impairment (migraine with aura). Family-, twin- and population-based studies suggest that genetic factors are involved, most likely as part of a multifactorial mechanism. FHM (Familial Hemiplegic Migraine) is a rare autosomal dominant subtype of migraine with aura of childhood onset, characterized by intermittent unilateral weakness or paralysis lasting for hours to days. Some patients show nystagmus (i.e. abnormal eye movements, an earliest sign of ataxia) and develop a slowly progressive ataxia (i.e. imbalance and clumsiness) with evidence of cerebellar atrophy later in life. A gene for FHM has been assigned to chromosome 19p13 in about 50% of families tested. Some evidence suggests that the same locus may also be involved in more frequent forms of migraine. A fraction of the FHM families linked to chromosome 19 and none of the unlinked families had progressive cerebellar atrophy and ataxia. Cerebellar ataxia may be diagnosed prior to the first migraine attack and progresses independently of the frequency and/or severity of attacks. Unilateral weakness is suggestive of brainstem, subcortical, or possibly cortical involvement, whereas ataxia points to cerebellar involvement (Ophoff et al., 1996, 1998).

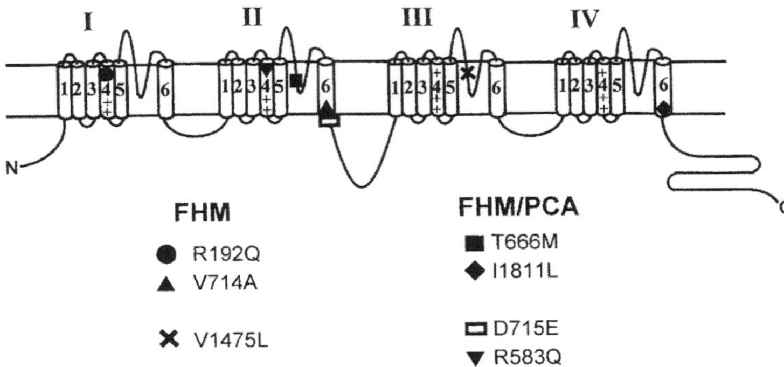

Figure 2. Mutations of the CACNA1A gene associated with pure familial hemiplegic migraine (FHM) and FHM with progressive ataxia (FHM/PCA): location in the secondary structure of the calcium channel α_{1A} subunit.

The gene on chromosome 19p13 responsible for FHM was identified in 1996 as being CACNA1A, which encodes α_{1A} subunits, the pore-forming subunits of P/Q-type calcium channels (Ophoff et al., 1996). α_{1A} subunits are expressed through the human and murine brain in most presynaptic terminals and also in the cell body and dendrites of many neurons (Volsen et al., 1995; Westenbroek et al., 1995). P/Q type calcium channels play a prominent role in controlling neurotransmitter release in many synapses (Dunlap et al., 1995). Their localisation also in dendrites and cell bodies suggests additional post-synaptic roles (Llinas et al., 1992). In both humans and mice, the expression of α_{1A} subunits is particularly high in the cerebellum, in both Purkinje and granule cells and also in the molecular layer (Volsen et al., 1995; Fletcher et al., 1996; Burgess et al., 1999). In the molecular layer α_{1A}-immunostaining appears at the level of dendrites of Purkinje cells, at terminals of parallel fibers of granule cells and basket cells (Volsen et al., 1995; Burgess et al., 1999). Double immunolabelling experiments have provided direct evidence for the presence of α_{1A} subunits in the majority of synaptic terminals in the rat cerebellum (Westenbroek et al., 1995). The basic neuronal circuit in the cerebellum is composed of the Purkinje cells, the only output element of the cerebellar cortex, and two inputs: a monosynaptic input to the Purkinje cells, the climbing fibers, and a disynaptic input, the mossy fiber-granule cell-Purkinje cell system. Most of the Ca^{2+} current of Purkinje cells and a large fraction of the Ca^{2+} current of cerebellar granule cells is inhibited by ω-AgaIVA, the spider toxin that specifically inhibits P/Q-type Ca^{2+} channels (Mintz et al., 1992; Randall and Tsien, 1995; Tottene et al., 1996). In the rat, the same toxin inhibits most of the excitatory synaptic transmission onto Purkinje cells, at both parallel fibers and climbing fibers synapses (Mintz et al., 1995). Moreover, it inhibits also most of the inhibitory synaptic transmission between Purkinje cells and deep cerebellar nuclei (Iwasaki et al., 2000).

Four different missense mutations were identified in five unrelated FHM families, all resulting in substitutions of conserved aminoacids in important functional regions of the α_{1A} subunit (Figure 2): R192Q, a substitution of a positively charged arginine with a neutral glutamine in the S4 segment of domain I, which forms part of the voltage sensor; T666M, a substitution of threonine for methionine in the pore-lining segment (P loop) between segments S5 and S6 of domain II, in close proximity to one of the key glutamates that form the high-affinity binding site for divalent ions in the selectivity filter; V714A and I1811L, substitutions of two conserved aminoacids located at the intracellular end of segments S6 of domains II and IV: these segments are thought to contribute to the lining of the part of the pore internal to the selectivity filter. Interestingly, at least some of the affected members of FHM families with mutations T666M and I1811L showed progressive cerebellar ataxia (FHM/PCA), whereas patients with mutations R192Q and V714A did not show cerebellar symptoms (pure FHM). A strong correlation between the T666M genotype and the FHM/PCA phenotype has been recently established (Ducros et al., 1999). Moreover, three new FHM missense mutations have been reported (Battistini et al., 1999; Carrera et al., 1999; Ducros et al., 1999), two of which lead to substitution of conserved aminoacids in pore regions (V1457L in the P loop of domain III and D715E next to V714A in IIS6) and the third (R583Q) to substitution of arginine for glutamine in the voltage sensor IIS4. Two of these mutations (D715E and R583Q) were associated with a FHM/PCA phenotype.

The functional consequences of the four FHM mutations described by Ophoff et al. (1996) have been investigated in heterologous expression systems expressing calcium channels containing either human (Hans et al., 1999) or rabbit α_{1A} subunits (Kraus et al., 1998). The FHM mutations altered the biophysical properties of recombinant P/Q-type channels. Single channel patch-clamp recordings revealed that two mutations located in pore regions (T666M and V714A) reduced the single channel conductance of human P/Q type calcium channels (Figure 3A). Surprisingly, a minor fraction of mutant channels had the wild-type conductance, suggesting that the abnormal channel may switch on and off, perhaps depending on some unknown factor (Hans et al., 1999). Two mutations located in S6 segments (V714A and I1811L) increased the single-channel open probability at all voltages and shifted the voltage-range of channel activation towards more negative voltages. The single channel open probability was increased also by mutation R192Q located in the voltage sensor region (Figure 3B). The rate of recovery from inactivation of both human and rabbit P/Q-type channels was increased by both mutations V714A and I811L, and decreased by mutation T666M (Kraus et al., 1998; Hans et al., 1999). The FHM mutations also produced changes in the density of functional channels in the membrane (Hans et al., 1999). The three mutations in the pore region decreased the density of functional channels. Strikingly, mutation R192Q had the opposite effect (Figure 3C).

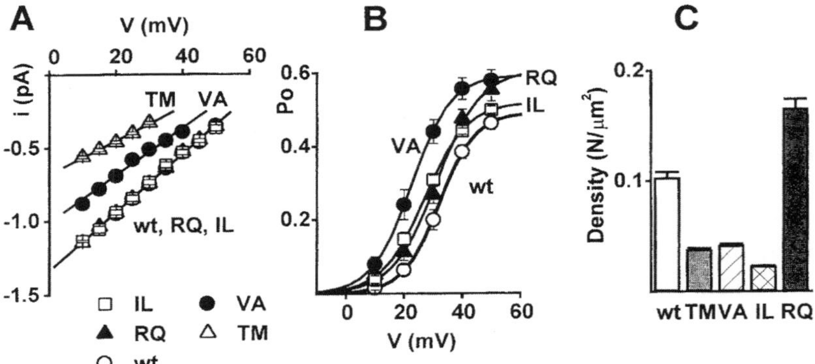

Figure 3. Effect of FHM mutations on single channel current and conductance, open probability and density of functional human recombinant P/Q-type calcium channels. (A) Unitary current-voltage, i-V, relationships of calcium channels containing wild-type (wt) and mutant human α_{1A-2} subunits: T666M (TM), I1811L (IL), R192Q (RQ), V714A (VA). (B) Voltage dependence of the open probability, po, of single calcium channels containing wt or mutant α_{1A-2} subunits. (C) Density of functional calcium channels containing wt and mutant α_{1A-2} subunits. The density was calculated from the average number of channels per patch and the average patch area in cell attached patch-clamp recordings (Hans et al., 1999).

Table 1. Effect of FHM mutations on Ca^{2+} influx. The table shows how the changes in functional properties of human recombinant Ca^{2+} channels produced by FHM mutations are predicted to affect Ca^{2+} influx into neurons. An increased, a decreased or an unchanged Ca^{2+} influx with respect to wt wild type are indicated with ↑, ↓, and –, respectively. The smaller symbols in parenthesis refer to the function of a minority of mutants

	Channel Density	Single Channel Current	Open Probability	Recovery from Inactivation
T666M	↓↓	↓↓ (-)	–	↓
V714A	↓↓	↓ (-)	↑↑	↑↑
I1811L	↓↓↓	– (↓)	↑	↑↑
R192Q	↑↑	–	↑	–

Table 1 shows how the changes in functional properties of human recombinant channels produced by the FHM mutations are predicted to affect calcium influx into neurons. The question of whether the four FHM mutations lead to gain- or loss-of-function in terms of Ca^{2+} influx does not have a simple and univocal answer. Mutation T666M should lead to a reduction of Ca^{2+} influx (loss-of-function) and mutation R192Q to an increase of Ca^{2+} influx (gain-of-function), whether the function of the single calcium channel or the density of functional channels are considered. On the other hand, mutations V714A and I1811L would lead to an overall gain-of-function at the single channel level, given the higher open probability and the faster rate of recovery from inactivation, while they may lead to an overall loss-of-function at the level of the whole-cell calcium current, given the decreased density of functional channels (particularly large for I811L). Considering the possibility that the FHM mutations may differentially affect the expression of P/Q-type calcium channels in different neurons, then one can predict either an increased or a decreased Ca^{2+} influx through mutant channels (especially V714A and I1811L) depending on the type of neuron. Moreover, the FHM mutations might affect differently P/Q-type calcium influx in different neurons also as a consequence of expression of different α_{1A} splice variants and/or different auxiliary subunits. For example, the predicted effect of the increased rate of recovery from inactivation of the V714A and I1811L mutants would be enhancement of calcium influx during repetitive activity in neurons expressing inactivating α_{1A} variants (as shown in Table 1) or no change in calcium influx during repetitive activity in neurons expressing non-inactivating variants. For the same reasons, the FHM mutations might affect differently calcium influx in different compartments of the same neuron (e.g. dendrites vs synaptic terminals).

EA-2 (Episodic Ataxia type 2) is a rare dominantly inherited neurological disorder, characterized by interictal nystagmus and episodes (lasting hours to days) of ataxia, i.e. truncal instability, unsteady gait, loss of limb coordination, and sometimes vertigo or dizziness, that may be precipitated by stress or fatigue. Clinical onset generally occurs in childhood or early adulthood. The symptoms suggest intermittent derangement of cerebellar function. But some patients develop progressive cerebellar ataxia and cerebellar atrophy predominating on the anterior vermis, and about 50% of the patients report migraine symptoms. Weakness and/or confusion are other symptoms often associated with episodic ataxia (Denier et al., 1999; Jen, 1999).

EA-2 has been mapped in the same interval as the FHM locus and mutations in the same CACNA1A gene have been found in a large fraction of familial and sporadic cases (Ophoff et al., 1996; Denier et al., 1999; Jen, 1999). The majority of mutations (but not all) disrupt the open reading frame leading to truncation, exon skipping or intron inclusion of the gene product (Figure 4). The mutant protein is unlikely to form functional channels, but since it retains association sites for auxiliary subunits, might interfere with channel targeting and function by sequestering associated proteins. Alternatively, the aberrant transcripts may be unstable and degraded, resulting in loss-of-function. Great intra- and interfamilial variability exists in the symptoms (both episodic and permanent) experienced by EA2 patients with this type of mutations. A severe progressive ataxia in the absence of paroxysmal episodes may be part of the clinical spectrum of EA-2. A missense mutation leading to substitution of a highly conserved neutral glycine with a positively

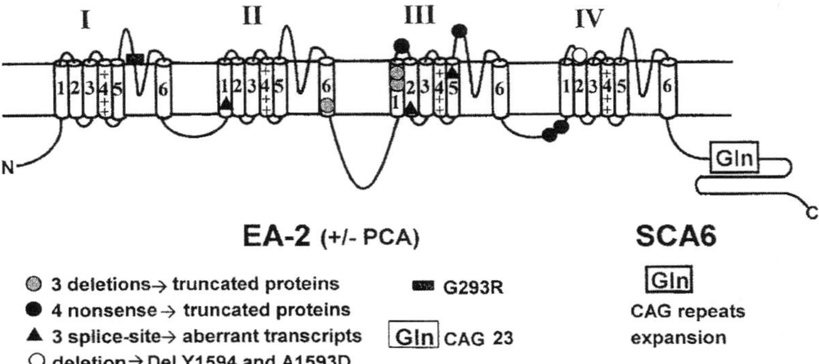

Figure 4. Mutations of the CACNA1A gene associated with episodic ataxia type 2 (EA-2) with and without progressive ataxia (PCA) and with spinocerebellar ataxia type 6 (SCA6): location in the secondary structure of the calcium channel α_{1A} subunit.

charged arginine in the S5–S6 linker of domain I was identified in a family with affected members showing either a combination of progressive and episodic ataxia or very severe early-onset progressive cerebellar ataxia without episodic features (Figure 4). Although not yet investigated, this mutation is expected to result in major changes in the channel permeation properties, most likely leading to reduced calcium influx.

The autosomal dominant spinocerebellar ataxias are a group of inherited neurodegenerative disorders characterized by progressive ataxia and cerebellar degeneration, caused by expansions of CAG trinucleotide repeats coding for an extended polyglutamine sequence (35 to 135 repeats). Zhuchenko et al. (1997) identified small CAG expansions, ranging from 21 to 27 repeat units, in the 3′ end of CACNA1A in patients with a late-onset autosomal dominant slowly progressive ataxic syndrome that they named SCA6 (Spinocerebellar Ataxia type 6). Patients develop permanent balance and coordination difficulties progressively leading to impairment of gait that may cause them to become wheel-chair bound. There is evidence of marked cerebellar atrophy especially in the superior vermis, with more severe loss of Purkinje cells than granule cells, and variable mild atrophy of the brain stem. Episodic features have been reported in SCA6 patients, leading to the suggestion that SCA6 and EA-2 represent a clinical continuum (Jen, 1999).

Alternative splicing of the CACNA1A gene results in at least six mRNA isoforms with different 3′ end, in which the CAG repeat is either part of the noncoding or part of the coding region (in 3 isoforms). In the latter isoforms, the CAG repeat encodes a polyglutamine stretch at the C-terminus. Immunohistochemistry in human brain sections showed the expression of isoforms containing a polyglutamine stretch in both control and SCA6 patients (Ishikawa et al., 1999a). The most notable difference with respect to control was the presence of cytoplasmic aggregations of the α_{1A} protein exclusively in the cytoplasm of SCA6 Purkinje cells, together with a mild reduction of immunoreactivity in the Purkinje and molecular layers, suggestive of reduced α_{1A} expression. There are some indirect evidences that selective neuronal degeneration in SCA6 might be associated with aggregation of α_{1A} protein. Expanded polyglutamines were found to directly alter the biophysical properties of heterologously expressed P/Q-type channels. Recombinant channels containing rabbit α_{1A} subunits with 30 or 40 (but not 24) polyglutamines had abnormal voltage-dependence of steady state-inactivation, that would reduce the number of channels available to open from resting membrane potential and predict a reduced calcium influx (Matsuyama et al., 1999). Current densities of wild-type and mutant channels were similar in this study. However, these findings are rather controversial since increased current density with respect to wild-type has been recently reported in cells expressing human α_{1A} subunits with 23, 27 and 72 polyglutamines (Piedras-Renteria et al., 1999).

6. MOUSE CALCIUM CHANNELOPATHIES: GENOTYPE AND PHENOTYPE

Spontaneously arising mutations in the mouse orthologue of CACNA1A were identified in three recessive neurological mouse mutants: tottering (tg), leaner (tg^{la}) and rolling Nagoya (tg^{rol}) (Fletcher et al., 1996; Burgess and Noebels, 1999; Mori et al., 1999). Both tottering (tg/tg) and leaner (tg^{la}/tg^{la}) mice exhibit intermittent seizures very similar to human absence epilepsy, a generalized, non-convulsive epileptic disorder that is most common among children and is characterized by cortical spike-wave discharges recorded by EEG concomitant with behavioural immobility. The neurological phenotype of tottering mice is in addition characterized by mild ataxia (wobbly gait) and infrequent spontaneous motor seizures (involuntary movements of limb and trunk, more appropriately classified as episodic dyskinesia, a form of dystonia involving only specific muscles). The leaner phenotype is usually described as being characterized by much more severe ataxia than tottering, without episodic dyskinesia. However, the typical stiff posture with extended limbs and severe impairment in the ability to walk of leaner mice can be more appropriately described as dystonia (Fletcher, personal communication). Neurological symptoms appear in the second week after birth in leaner and later, at 3–4 weeks after birth, in tottering. In the leaner cerebellum there is extensive degeneration of granule, Golgi and Purkinje cells. Granule cell loss begins in the second week after birth, and progresses slowly over a period of months, showing an anterior-posterior gradient, with more severe loss in the anterior lobe. Purkinje cell loss begins later at about 4 weeks after birth, and occurs in parasagittal stripes separated by areas of normal cells. Interestingly, the pattern of surviving Purkinje cells is essentially coextensive with the striped pattern of zebrin (a cerebellum marker) staining and tyrosine hydroxylase expression. In both leaner and tottering mice, the normally transient expression of tyrosine hydroxylase is not suppressed, resulting in expression in the adult. In contrast with leaner, cerebellar atrophy, not accompanied by loss of Purkinje cells, has been reported only in some old tottering mice. A characteristic feature of tottering brain is the aberrant synaptogenesis of locus ceruleus neurons, with consequent hyperinnervation by noradrenergic terminals of all major locus ceruleus targets, including the cerebellum and the hippocampus (Fletcher et al., 1996; Burgess and Noebels, 1999). Rolling Nagoya mice (tg^{rol}/tg^{rol}) exhibit severe ataxia, beginning at 10–14 days after birth, but not episodic dyskinesia nor seizures. Locus ceruleus hyperinnervation, cerebellar granule cell loss, especially in the anterior lobe, and increased tyrosine hydroxylase activity in the cerebellum have been reported (Rhyu et al., 1999a). At the ultrastructural level, the three allelic mouse mutants show an altered cerebellar phenotype consisting in enlarged parallel fibers varicosities synapsing on several Purkinje cell dendritic spines, in addition to the common monosynaptic contacts (Rhyu et al., 1999b).

The mutation in the CACNA1A gene causing the tottering phenotype is a missense mutation resulting in substitution of proline with leucine (P601L) in the S5-S6 linker of domain II (located about 20 aminoacids from the threonine to leucine substitution in FHM), whereas the tg^{la} phenotype is caused by a splice site mutation producing a frameshift in the reading frame and two aberrant splice products with altered C-terminal sequences. The mutation causing the Rolling Nagoya phenotype is an arginine to proline substitution in S4 of domain III (Mori et al., 1999). Immnunocytochemistry and *in situ* hybridization in leaner cerebellar slices showed no difference in α_{1A} protein and mRNA level with respect to wild-type (Lau et al., 1998). However, Northern blot analysis revealed selective reduced expression of one of the two major α_{1A} transcript already at 9 days after birth (P9) (Doyle et al., 1997). No differences in α_{1A} mRNA levels were detected in adult tottering cerebellum, but an interesting upregulation of α_{1C} mRNA level in Purkinje cells was observed (Doyle et al., 1997; Campbell and Hess, 1999).

The functional consequences of the tg and tg^{la} mutations have been investigated in both heterologous expression systems (Wakamori et al., 1998) and native Purkinje cells (Dove et al., 1998; Lorenzon et al., 1998; Wakamori et al., 1998). The mutations caused a reduction of P/Q-type calcium current density of 70–80 and 15–50% in freshly dissociated Purkinje cells of leaner and tottering mice (10 to 35 days in age), respectively, with no change in the non-P/Q-type current. A 50–60% decrease in current density was also measured after transient expression of rabbit α_{1A} subunits containing the tottering mutation or one of the two possible abnormal leaner C-termini (whereas the same current density was found for the other abnormal splicing product). Neither in the Purkinje cells nor in the heterologous expression system there was evidence of changes in the kinetics or voltage-dependent properties of mutant tottering P/Q-type channels. A small shift of the voltage range of both activation and inactivation towards more positive voltages has been reported for leaner by some authors (Wakamori et al., 1998) but not by others (Dove et al., 1998; Lorenzon et al., 1998). Neither mutation apparently changed the single channel conductance. A reduction of P/Q-type current density has been reported also in tg^{rol}/tg^{rol} Purkinje cells (Mori et al., 1999). Overall, the electrophysiological data suggest that the main consequence of the allelic tottering mutations is a decrease in the number of functional P/Q-type channels in the membrane. Very recently, a strong reduction of the contribution of P/Q-type channels to calcium influx into presynaptic terminals of CA3 pyramidal cells was measured in hippocampal slices of adult tottering (Qian and Noebels, 2000). The reduced presynaptic P/Q-type calcium influx was at least partially compensated by an increased influx through presynaptic N-type calcium channels, with a consequent much lower reduction of synaptic transmission than expected. Interestingly, a reduced excitatory but not inhibitory synaptic transmission has been measured at synapses on neurons of the ventrobasal nucleus in the tottering thalamus (Caddick et al., 1999).

Given the correlation between extent of reduction of P/Q-type calcium current in Purkinje cells and severity of the ataxia phenotype in leaner and tottering, it is quite interesting to consider the phenotype of the recently generated α_{1A}-deficient mice ($\alpha_{1A}^{-/-}$) (Jun et al., 1999; Fletcher et al., 2000). Knockout mice ($\alpha_{1A}^{-/-}$) exhibit a neurological phenotype very similar to that of leaner. Interestingly, histological examination of $\alpha_{1A}^{-/-}$ brains at 15 weeks of age revealed selective degeneration of the cerebellum in a specific pattern, with loss of Purkinje cells in parasagittal stripes and graded loss of granule cells more severe in the anterior lobe, like in leaner (Fletcher et al., 2000). A selective histopathological involvement of the cerebellum might be predicted from the known high level of α_{1A} expression in the cerebellum. However, the highly selective pattern of neurodegeneration is surprising; it implies the existence of two types of cerebellar neurons: one with absolute dependence of P/Q calcium channel function for survival, and another that can tolerate lack of P/Q channels. A similar (70%) reduction in total calcium current was measured in Purkinje cells dissociated from leaner and $\alpha_{1A}^{-/-}$ mice. In $\alpha_{1A}^{-/-}$ Purkinje cells, the ablation of the P/Q-type current was partially compensated by an increase of the L- and N-type calcium currents (unlike leaner) (Jun et al., 1999). A partially compensating increase of L- and N-type calcium current, with no change of the R-type component, can also be observed in $\alpha_{1A}^{-/-}$ cerebellar granule cells (Fletcher et al., 2000; but see different findings in Jun et al., 1999). The null allele behaves as a strictly recessive mutation, since heterozygous $\alpha_{1A}^{-/+}$ mice have no phenotype in terms of motor performance, cerebellar neuroanatomy or EEG, but show a 50% reduction of P/Q-type calcium current in cerebellar granule cells (Fletcher et al., 2000).

Lethargic mice (lh/lh) have a clinical phenotype almost identical to that of tottering, except that lh/lh exhibit earlier (onset at P15) and more pronounced ataxia, and also suffer from reduced body weight and transient defects in the immune system (Burgess and Noebels, 1999). The mutation causing the lethargic phenotype is in the Cacnb4 gene encoding the auxiliary calcium channel subunit β_4, and is equivalent to a null mutation (Burgess et al., 1997).

The level of expression of different β subunits vary considerably in different regions of mouse brain and in different cells within the same region. In both humans and mice the expression of β_4 subunits, like that of α_{1A} subunits, is particularly high in the cerebellum in both Purkinje and granule cells (Volsen et al., 1997; Burgess et al., 1999). In lethargic mutants there is evidence for an increased level of expression of β_{1b} subunits in both forebrain and cerebellum, which partially compensates for the lack of β_4 subunit (McEnery et al., 1998). Moreover, coimmunoprecipitation experiments show an increase in the fraction of α_{1A} and α_{1B} subunits associated with β_{1b} and β_3 in the brain, and suggest that there are no channels lacking an auxiliary β subunit in the mutant (Burgess et al., 1999). Thus, the β_4 subunit can be replaced by the remaining coexpressed β subunits in the lethargic brain, a process termed "subunit reshuffling". There are conflicting data on the effect of the lethargic mutation on the level of expression of α_1 subunits (no change in either α_{1A} or α_{1B} subunits in Burgess et al. 1999, but decreased expression of α_{1B} in McEnery et al., 1998). The kinetics and voltage-dependent properties of the P/Q-type calcium current in wild-type and lh/lh Purkinje cells were very similar, and, although quantitative values were not provided, there was no evidence for significantly different current densities (Burgess et al., 1999). Synaptic transmission and the pharmacological profile of presynaptic calcium influx at CA3-CA1 synapses in wild-type and lh/lh hippocampal slices were also very similar (Qian and Noebels, 2000). In contrast, as in tottering, excitatory but not inhibitory synaptic transmission was reduced at synapses on the ventrobasal nucleus in the lethargic thalamus (Caddick et al., 1999).

A mutation in a gene encoding a brain specific protein called stargazin, that shares a modest partial sequence similarity and predicted secondary structure with the auxiliary γ subunit of the skeletal muscle calcium channel, is the genetic defect in stargazer (stg/stg) and allelic waggler (stg^{wag}/stg^{wag}) mice. These mouse mutants exhibit recurrent seizures similar to human absence epilepsy, distinctive head-tossing and severe ataxia beginning at P15 (Letts et al., 1998; Burgess and Noebels, 1999). Complex alterations are present in the stargazer brain, including remarkable outgrowth of dentate granule axons in the hippocampus. The adult cerebellum is morphologically normal, but there are evidences for delayed granule cell migration and maturation, most likely as a consequence of striking downregulation of brain-derived neurotrophic factor expression in cerebellar granule cells. Recently, it has been shown that cerebellar granule cells of adult stargazer and waggler lack functional α-amino-3-hydroxy-5-methyl-4-isoxazolepropionic acid (AMPA) receptors, which implies a virtual absence of massive afferent information from mossy fibers, since at the mossy fiber-granule cell synapse usual excitatory synaptic transmission at low frequency is mediated mainly by AMPA receptors (Chen et al., 1999; Hashimoto et al., 1999). Moreover, synaptic transmission was reduced also at the parallel-fiber-Purkinje cell synapse, most likely as a consequence of lower transmitter release.

To date, the only reported evidence linking stargazin to calcium channel function is a small hyperpolarizing shift in the steady state inactivation curve of recombinant calcium channels containing rabbit α_{1A} subunits upon coexpression with stargazin (Letts et al., 1998). Thus, there is still considerable debate as to whether stargazin and its human orthologue γ_2 subunit (and other γ isoforms) really form part of neuronal voltage-dependent calcium channels (Randall and Benham, 1999). The mutation in stargazer leads to premature termination of the transcript and therefore most likely to a non functional protein. Assuming stargazin is a neuronal γ subunit affecting steady state inactivation of calcium channels, then one would predict enhanced calcium influx in stg neuronal cells as a consequence of increased channel availability in the absence of the γ subunit. However, the predicted increased calcium entry cannot account for the functional deficits measured in stg/stg cerebellar synapses, showing a decreased (not increased) excitatory transmission, and it is unclear how it would lead to lack of functional AMPA receptors.

7. HOW DO CALCIUM CHANNEL MUTATIONS PRODUCE DISEASE?

Researchers are beginning to answer the question of how specific calcium channel mutations alter channel function. The more challenging question of how the alterations in channel function lead to selective cellular dysfunction and to the episodic as well as persistent neurological symptoms of calcium channelopathies remains unanswered. To answer this question, parallel studies at the molecular, cellular, neuronal network and behavioral levels in mouse genetic models will be extremely important.

Since Ca^{2+} ions regulate numerous intracellular signalling pathways, including those regulating gene expression, the activity of calcium channels is inextricably linked to a broad array of functions in the developing and mature nervous system, including cell proliferation, differentiation, survival, neurite outgrowth, synaptogenesis, neuronal excitability, transmitter release, and plasticity. Although the primary functional defect may be abnormal Ca^{2+} entry into neurones, any of these secondary processes could contribute substantially to the disease phenotype. It is tempting to speculate that abnormal neuronal excitability due to dysfunctional calcium channels may trigger the episodic neurologic symptoms (e.g. attacks of migraine, ataxia in humans, absence epilepsy in mouse mutants), whereas chronic abnormal calcium homeostasis may lead to progressive neuronal degeneration and to progressive fixed dysfunction of the tissue involved.

Different mutations in the gene CACNA1A, encoding the pore-forming subunit of P/Q-type calcium channels, cause in humans a spectrum of disorders (FHM, EA-2 and SCA6) with overlapping symptoms, including progressive ataxia and cerebellar degeneration (Table 2). Progressive ataxia and cerebellar atrophy are present only in a fraction of FHM families linked to CACNA1A and in none of the unlinked families, and are strongly correlated with the T666M genotype. Mutation T666M is the only FHM mutation, of the four analyzed, that leads to a predicted decrease of P/Q-type Ca^{2+} influx into cerebellar neurons at both the single channel (cf. lower single channel conductance) and whole-cell level (cf. lower density of functional channels). A reduction of Ca^{2+} influx mainly as a consequence of reduced expression of functional channels can be predicted also for mutation I1811L, the other mutation causing FHM associated to progressive cerebellar ataxia. These findings suggest that mutations in P/Q-type calcium channels cause progressive ataxia and cerebellar atrophy through a loss of function mechanism.

This conclusion is supported by the severe ataxia (and/or dystonia) exhibited by both α_{1A}-deficient and leaner mice (with either absent or strongly reduced P/Q-type calcium current) and by the highly selective pattern of cerebellar degeneration of both leaner and $\alpha_{1A}^{-/-}$ mice, which resembles the cerebellar atrophy seen in patients with expanded CAG repeats (Ishikawa et al., 1999b) and in certain EA-2 patients (Denier et al., 1999). Since neither tottering nor $\alpha_{1A}^{-/+}$ mice with 50% reduced P/Q current density show loss of cerebellar Purkinje cells, it appears that, in mice, calcium influx must be reduced below a critical threshold to cause cell death and consequent cerebellar degeneration. It remains unknown why, when reduction of calcium influx exceeds this threshold, Purkinje cell death occurs in a striped pattern in contrast with the uniform expression of α_{1A} subunits.

Remarkably, the human mutations are inherited as autosomal dominants, causing cerebellar degeneration in the heterozygous state. Assuming that they are loss-of-function mutations, haploinsufficiency could be the underlying pathological mechanism, but this would be dissimilar from the recessive mouse mutations. Another possibility is that, as a consequence of the human mutations, other functions are disrupted. For example, death of Purkinje cells in SCA6 might be caused by mechanisms similar to those of other CAG repeat disorders and be associated with aggregation of the α_{1A} protein.

The recognized role of the cerebellum is to provide for coordination and fine control of movements. One attractive idea to explain the fundamental operation of the cerebellum is that it acts as a "comparator" that compensates for errors in movement by comparing intention with performance. Among the typical deficits produced by lesions of the cerebellum are a variety of abnormalities in the execution of voluntary movements which can be globally referred to as ataxia. In mice with mutations in calcium channel subunits, ataxia is not correlated with Purkinje cell loss. Indeed both tottering and lethargic mice are ataxic but do not show Purkinje cell loss. Moreover, in both leaner and knockout mice the neurological symptoms appear before the death of Purkinje cells. Moreover, ataxia does not appear to be correlated in a simple manner with the reduction of P/Q-type calcium current in Purkinje cells, since a 50% reduction of P/Q current in heterozygous $\alpha_{1A}^{-/+}$ mice does not cause ataxia, whereas a similar or lower reduction in tottering

Table 2. Calcium channelopathies in the CNS. Symbols ↑, ↓, and – indicate increased, decreased and unchanged calcium entry, respectively, as predicted on the basis of the functional consequences of the mutations measured on recombinant calcium channels in the case of the human diseases, and as measured in dissociated Purkinje cells in the case of the mouse mutants

	Clinical Phenotype	Gene (chromosome)	Gene product	Mutations	Consequence on calcium entry
Human	**CSNB**: night blindness	**CACNA1F** (Xp11.23)	α_{1F}	Premature stops Deletions Missense	unknown (?↓)
	FHM: migraine	**CACNA1A** (19p13)	α_{1A}	R192Q V714A V1475L	↑ ↑↓ unknown
	FHM/PCA: migraine + progressive ataxia	**CACNA1A** (19p13)	α_{1A}	T666M I1811L D715E R583Q	↓↓ ↓↑ unknown unknown
	EA2: episodic ataxia ± progressive ataxia ± migraine	**CACNA1A** (19p13)	α_{1A}	Premature stops Splicing errors Deletions Missense CAG_{23}	unknown (?↓)
	SCA6: progressive ataxia ± episodic ataxia	**CACNA1A** (19p13)	α_{1A}	CAG repeats expansion	↓ (or ↑ ?)
Mouse	**Tottering**: ataxia+seizures	**Cacna1a** (8)	α_{1A}	P601L	↓
	Leaner: ataxia+seizures	**Cacna1a** (8)	α_{1A}	Splicing errors → novel C-terminus	↓↓
	Rolling Nagoya: ataxia	**Cacna1a** (8)	α_{1A}	Missense	↓
	Lethargic: ataxia+seizures	**Cacnb4** (2)	β_4	Truncation	-
	Stargazer: ataxia+seizures	**Cacnγ2** (2)	γ_2	Truncation	unknown

and lethargic mice (with no change in kinetic properties of the current) is associated with ataxia. The ataxia phenotype is probably linked in a complex manner to dysfunction of cerebellar neuronal circuits that involve P/Q-type calcium channels at multiple synapses. As shown in tottering and lethargic, the same mutation in P/Q channels may affect differently neurotransmitter release at different synapses. Many mechanisms could lead to differential effect at different synapses, including different intrinsic contributions of P/Q-type channels in controlling release, different mutation-induced changes in expression of P/Q-type channels and different compensatory changes of other calcium channel α_1 and/or auxiliary subunits.

The clinical distinction between the different human and mouse α_{1A} calcium channelopathies is based mainly on the different episodic neurological symptoms, ranging from migraine in FHM to episodic ataxia in EA-2 and to absence epilepsy in the mouse mutants (Table 2). It remains unclear how different mutations in the same gene can lead to such varied clinical phenotypes and, on the other hand, how mutations that are predicted to affect calcium influx in the opposite direction all lead to the same migraine phenotype. Also unclear is how in general a permanent mutation leads to episodic disorders.

Given the delicate balance between excitatory and inhibitory signals that regulates neuronal excitability, even relatively small changes in channel activity may tip this delicate balance. Thus, channelopathies may be considered as defects of cellular excitability. It has been suggested that the differential effect of the tottering and lethargic mutations on excitatory and inhibitory transmission in the thalamus may tip the balance between excitation and inhibition towards inhibition of thalamic relay neurons, that can lead to synchronization of these neurons into a burst-firing mode and generation of spike wave discharges typical of absence epilepsy (Caddick et al., 1999). Although an established model that explains migraine attacks is still lacking, a favoured hypothesis considers a persistent state of hyperexcitability of neurons in the cerebral cortex as the basis for susceptibility to migraine (Welch, 1998). This state would favor the onset of cortical spreading depression, which is believed to initiate the attacks of migraine with aura. Today, one can only speculate about different mechanisms with which neuronal excitability can be increased and/or the development of cortical spreading depression can be facilitated by either loss- or gain-of-function variants of presynaptic and/or postsynaptic P/Q-type calcium channels. However, to explain the similar FHM phenotype produced by mutations that are either gain- or loss-of-function in heterologous expression systems, the assumption of differential changes in calcium influx produced by a given mutation in different neurons, and/or differential up- and down-regulation of inhibitory and excitatory neuronal elements, appears necessary.

The episodic nature of the neurological symptoms might perhaps be explained by the mutant channels providing a continous background of neuronal instability and by the intervention of an internal modulatory factor and/or an external environmental factor that pushes the system past some boundary and precipitates an attack. The observation that anxiety or stress triggers almost instantaneously attacks of ataxia in many EA-2 patients suggests potentially important modulatory action of noradrenergic or serotonergic inputs to the cerebellum. Interestingly, lesions of central noradrenergic axons early in development obliterate seizure expression in tottering mutants (Burgess and Noebels, 1999). The finding that the functional effect of the FHM mutations on single channel conductance was not present in some mutant channels or periods of activity, suggests the interesting possibility that some unknown factor can precipitate an attack by directly switching the abnormal channel on or off (Hans et al., 1999).

In human calcium channelopathies there is remarkable phenotypic variability in terms of frequency of attacks, features and severity of symptoms among affected individuals in all reported pedigrees, suggesting that environmental, metabolic and other genetic factors contribute to the phenotypic expression of the mutations. Specific alleles or external factors that influence the expression of proteins involved in signalling, including phosphatases, kinases, G proteins, neurotransmitter receptors and ligand gated or other ion channel types, may be expected to play significant modifying roles in the phenotypic expression of calcium channel defects.

REFERENCES

Armstrong, C.M. and Hille, B., 1998, Voltage-gated ion channels and electrical excitability, *Neuron* 20, 371–380.

Battistini, S., Stenirri, S., Piatti, M., Gelfi, C., Righetti, P.G., Rocchi, R., Giannini, F., Battistini, N., Guazzi, G.C., Ferrari, M. and Carrera, P., 1999, A new CACNA1A gene mutation in acetazolamide-responsive familial hemiplegic migraine and ataxia, *Neurology* 53, 38–43.

Bourinet, E., Soong, T.W., Sutton, K., Slaymaker, S., Mathews, E., Monteil, A., Zamponi, G.W., Nargeot, J. and Snutch, T.P., 1999, Splicing of α_{1A} subunit gene generates phenotypic variants of P- and Q-type calcium channels, *Nature Neuroscience* 2, 407–415.

Burgess, D.L., Biddlecome, G.H., McDonough, S.I., Diaz, M.E., Zilinski, C.A., Bean, B.P., Campbell, K.P. and Noebels, J.L., 1999, β reshuffling modifies N- and P/Q-type Ca^{2+} channel subunit compositions in lethargic mouse brain, *Mol. Cell. Neurosci.* 13, 293–311.

Burgess, D.L., Jones, J.M., Meisler, M.H. and Noebels, J.L., 1997, Mutation of the Ca^{2+} channel β subunit gene Cchb4 is associated with ataxia and seizures in the lethargic (lh) mouse, *Cell* 88, 385–392.

Burgess, D.L. and Noebels, J.L., 1999, Single gene defects in mice: The role of voltage-dependent calcium channels in absence models, *Epilepsy Res.* 36, 111–122.

Caddick, S.J., Wang, C., Fletcher, C.F., Jenkins, N.A., Copeland, N.G. and Hosford, D.A., 1999, Excitatory but not inhibitory synaptic transmission is reduced in lethargic ($Cacnb4^{lh}$) and tottering ($Cacna1a^{tg}$) mouse thalami, *J. Neurophysiol.* 81, 2066–2074.

Campbell, D.B. and Hess, E.J., 1999, L-type calcium channels contribute to the tottering mouse dystonic episodes, *Mol. Pharmacol.* 55, 23–31.

Carrera, P., Piatti, M., Stenirri, S., Grimaldi, L.M.E., G., Marchioni, E., Curcio, M., Righetti, P.G., Ferrari, M. and Gelfi, C., 1999, Genetic heterogeneity in Italian families with familial hemiplegic migraine, *Neurology* 53, 26–32.

Chen, L., Bao, S., Qiao, X. and Thompson, R.F., 1999, Impaired cerebellar synapse maturation in waggler, a mutant mouse with a disrupted neuronal calcium channel g subunit, *Proc. Natl. Acad. Sci.* 96, 12132–12137.

Denier, C., Ducros, A., Vahedi, K., Joutel, A., Thierry, P., Ritz, A., Castelnovo, G., Deonna, T., Gerard, P., Devoize, J.L., Gayou, A., Perrouty, B., Soisson, T., Autret, A., Warter, J.M., Vighetto, A., Van Bogaert, P., Alamowitch, S., Roullet, E. and Tournier-Lasserve, E., 1999, High prevalence of CACNA1A truncations and broader clinical spectrum in episodic ataxia type 2, *Neurology* 52, 1816–1821.

Dove, L.S., Abbott, L.C. and Griffith, W.H., 1998, Whole-cell and single-channel analysis of P-type calcium currents in cerebellar Purkinje cells of leaner mutant mice, *J. Neurosci.* 18, 7687–7699.

Doyle, J., Ren, X., Lennon, G. and Stubbs, L., 1997, Mutations in the CACNL1A4 calcium channel gene are associated with seizures, cerebellar degeneration, and ataxia in tottering and leaner mutant mice, *Mamm. Genome* 8, 113–120.

Ducros, A., Denier, C., Joutel, A., Vahedi, K., Michel, A., Darcel, F., Madigan, M., Guerouaou, D., Tison, F., Julien, J., Hirsch, E., Chedru, F., Bisgard, C., Lucotte, G., Despres, P., Billard, C., Barthez, M.A., Ponsot, G., Bousser, M.G. and Tournier-Lasserve, E., 1999, Recurrence of the T666M calcium channel CACNA1A gene mutation in familial hemiplegic migraine with progressive cerebellar ataxia, *Am. J. Hum. Genet.* 64, 89–98.

Dunlap, K., Luebke, J.I. and Turner, T.J., 1995, Exocytotic Ca^{2+} channels in mammalian central neurons, *Trends Neurosci.* 18, 89–98.

Fletcher, C.F., Lutz, C.M., O'Sullivan, T.N., Shaughnessy, J.D.J., Hawkes, R.H., Frankel, W.N., Copeland, N.G. and Jenkins, N.A., 1996, Absence epilepsy in Tottering mutant mice is associated with calcium channel defects, *Cell* 87, 607–617.

Fletcher, C.F., Tottene, A., Wilson, S., Dubel, S.J., Paylor, R., Hosford, D., Tessarollo, L., Lennon, V., McEnery, M.W., Pietrobon, D., Copeland, N.G. and Jenkins, N.A., 2000, Cerebellar atrophy in mice lacking the P/Q voltage-dependent calcium channel, submitted.

Hans, M., Luvisetto, S., Williams, M.E., Spagnolo, M., Urrutia, A., Tottene, A., Brust, P.F., Johnson, E.C., Harpold, M.M., Stauderman, K.A. and Pietrobon, D., 1999, Functional consequences of mutations in the human α_{1A} calcium channel subunit linked to familial hemiplegic migraine, *J. Neurosci.* 19, 1610–1619.

Hashimoto, K., Fukaya, M., Qiao, X., Sakimura, K., Watanabe, M. and Kano, M., 1999, Impairment of AMPA receptor function in cerebellar granule cells of ataxic mutant mouse Stargazer, *J. Neurosci.* 19, 6027–6036.

Ishikawa, K., Fujigasaki, H., Saegusa, H., Ohwada, K., Fujita, T., Iwamoto, H., Komatsuzaki, Y., Toru, S., Toriyama, H., Watanabe, M., Ohkoshi, N., Shoji, S., Kanazawa, I., Tanabe, T. and Mizusawa, H., 1999a, Abundant expression and cytoplasmic aggregations of α_{1A} voltage-dependent calcium channel protein associated with neurodegeneration in spinocerebellar ataxia type 6, *Hum. Mol. Genet.* 8, 1185–1193.

Ishikawa, K., Watanabe, M., Yoshizawa, K., Fujita, T., Iwamoto, H., Yoshizawa, T., Harada, K., Nakamagoe, K., Yomatsuzaki, Y., Satoh, A., Doi, M., Ogata, T., Kanazawa, I., Shoji, S. and Mizusawa, H., 1999b, Clinical, neuropathological, and molecular study in two families with spinocerebellar ataxia type 6 (SCA6), *J. Neurol. Neurosurg. Psychiatry* 67, 86–89.

Iwasaki, S., Momiyama, A., Uchitel, O.D. and Takahashi, T., 2000, Developmental changes in calcium channel types mediating central synaptic transmission, *J. Neurosci.* 20, 59–65.

Jen, J., 1999, Calcium channelopathies in the central nervous system, *Curr. Opin. Neurobiol.* 9, 274–280.

Jun, K., Piedras-Renteria, E.S., Smith, S.M., Wheeler, D.B., Lee, S.B., Lee, T.G., Chin, H., Adams, M.E., Scheller, R.H., Tsien, R.W. and Shin, H.-P., 1999, Ablation of P/Q-type Ca^{2+} channel currents, altered synaptic transmission, and progressive ataxia in mice lacking the α_{1A}-subunit, *Proc. Natl. Acad. Sci.* 96, 15245–15250.

Kraus, R.L., Sinnegger, M.J., Glossmann, H., Hering, S. and Striessnig, J., 1998, Familial hemiplegic migraine mutations change α_{1A} Ca^{2+} channel kinetics, *J. Biol. Chem.* 273, 5586–5590.

Lau, F.C., Abbott, L.C., Rhyu, I.J., Kim, D.S. and Chin, H., 1998, Expression of calcium channel a1A mRNA and protein in the leaner mouse (tg^{1a}/tg^{1a}) cerebellum, *Mol. Brain Res.* 59, 93–99.

Letts, V.A., Felix, R., Biddlecome, G.H., Arikkath, J., Mahaffey, C.L., Valenzuela, A., Bartlett II, F.S., Mori, Y., Campbell, K.P. and Frankel, W.N., 1998, The mouse stargazer gene encodes a neuronal Ca^{2+}-channel γ subunit, *Nature Genetics* 19, 340–347.

Lin, Z., Haus, S., Edgerton, J. and Lipscombe, D., 1997, Identification of functionally distinct isoforms of the N-type Ca^{2+} channel in rat sympathetic ganglia and brain, *Neuron* 18, 153-166.

Llinas, R., Sugimori, M., Hillman, D.E. and Cherksey, B., 1992, Distribution and functional significance of the P-type, voltage-dependent Ca^{2+} channels in the mammalian central nervous system, *Trends Neurosci.* 15, 351–355.

Lorenzon, N.M., Lutz, C.M., Frankel, W.N. and Beam, K.G., 1998, Altered calcium channel currents in Purkinje cells of the neurological mutant mouse leaner, *J. Neurosci.* 18, 4482–4489.

Matsuyama, Z., Wakamori, M., Mori, Y., Kawakami, H., Nakamura, S. and Imoto, K., 1999, Direct alteration of the P/Q-type Ca^{2+} channel property by polyglutamine expansion in spinocerebellar ataxia 6, *J. Neurosci.* 19, RC14 (11—15).

McEnery, M.W., Copeland, T.D. and Vance, C.L., 1998, Altered expression and assembly of N-type calcium channel α_{1B} and β subunits in epileptic lethargic (lh/lh) mouse, *J. Biol. Chem.* 273, 21435–21438.

Mintz, I.M., Venema, V.J., Swiderek, K.M., Lee, T.D., Bean, B.P. and Adams, M.E., 1992, P-type calcium channels blocked by the spider toxin w-Aga-IVA, *Nature* 355, 827–829.

Mintz, I.M., Sabatini, B.L. and Regehr, W.G., 1995, Calcium control of transmitter release at a cerebellar synapse, *Neuron* 15, 675–688.

Mori, Y., Wakamori, M., Matsuyama, Z., Fletcher, C., Copeland, N.G., Jenkins, N.A., Oda, S. and Imoto, K., 1999, A defect in voltage sensor of P/Q-type Ca^{2+} channel is associated with the ataxic mouse mutation Rolling Nagoya (tg^{rol}), in *Society for Neuroscience Abstracts*, Miami Beach, FL, p. 721.

Ophoff, R.A., Terwindt, G.M., Vergouwe, M.N., van Eijk, R., Oefner, P.J., Hoffman, S.M.G., Lamerdin, J.E., Mohrenweiser, H.W., Bulman, D.E., Ferrari, M., Haan, J., Lindhout, D., van Hommen, G.-J.B., Hofker, M.H., Ferrari, M.D. and Frants, R.R., 1996, Familial hemiplegic migraine and episodic ataxia type-2 are caused by mutations in the Ca^{2+} channel gene CACNL1A4, *Cell* 87, 543–552.

Ophoff, R.A., Terwindt, G.M., Frants, R.R. and Ferrari, M.D., 1998, P/Q-type Ca^{2+} channel defects in migraine, ataxia and epilepsy, *Trends Pharmacol. Sci.* 19, 121–127.

Piedras-Renteria, E.S., Watase, K., Zoghbi, H.Y., Lee, C.C. and Tsien, R.W., 1999, Alteration of expressed α_{1A} Ca^{2+} channel currents arising from expanded trinucleotide repeats in spinocerebellar ataxia type 6 (SCA6), in *Society for Neuroscience Abstracts*, Miami Beach, FL, p. 1056.

Ptacek, L.J., 1998, The place of migraine as a channelopathy, *Curr. Opin. Neurobiol.* 11, 217–226.

Qian, J. and Noebels, J.L., 2000, Presynaptic Ca^{2+} influx at a mouse central synapse with Ca^{2+} channel subunit mutations, *J. Neurosci.* 20, 163–170.

Randall, A. and Benham, C.D., 1999, Recent advances in the molecular understanding of voltage-gated Ca^{2+} channels, *Mol. Cell. Neurosci.* 14, 255–272.

Randall, A. and Tsien, R.W., 1995, Pharmacological dissection of multiple types of Ca^{2+} channel currents in rat cerebellar granule neurons, *J. Neurosci.* 15, 2995–3012.

Rhyu, I.J., Oda, S., Uhm, C.-S., Kim, H., Suh, Y.-S. and Abbott, L.C., 1999a, Morphological investigation of rolling mouse Nagoya (tg^{rol}/tg^{rol}) cerebellar Purkinje cells: An ataxic mutant revisited, *Neurosci. Lett.* 266, 49–52.

Rhyu, I.J., Abbott, L.C., Walker, D.B. and Sotelo, C., 1999b, An ultrastructural study of granule cell/Purkinje cell synapses in tottering (tg/tg), leaner (tg^{1a}/tg^{1a}) and compound heterozygous tottering/leaner (tg/tg^{1a}) mice, *Neuroscience* 90, 717–728.

Sather, W.A., Yang, J. and Tsien, R.W., 1994, Structural basis of ion channel permeation and selectivity, *Curr. Opin. Neurobiol.* 4, 313–323.

Strom, T.M., Nyakatura, G., Apfelstedt-Sylla, E., Hellebrand, H., Lorenz, B., Weber, B.H.F., Wutz, K., Gutwillinger, N., Ruther, K., Dresher, B., Sauer, C., Zrenner, E., Meitinger, T., Rosenthal, A. and Meindl, A., 1998, An L-type calcium-channel gene mutated in incomplete X-linked congenital stationary night blindness, *Nature Genetics* 19, 260–263.

Torben Bech-Hansen, N., Naylor, M., Maybaum, T.A., Pearce, W.G., Koop, B., Fishman, G.A., Mets, M., Musarella, M.A. and Boycott, K.M., 1998, Loss-of-function mutations in a calcium-channel α_1-subunit gene in Xp11.23 cause X-linked congenital stationary night blindness, *Nature Genetics* 19, 264–267.

Tottene, A., Moretti, A. and Pietrobon, D., 1996, Functional diversity of P-type and R-type calcium channels in rat cerebellar neurons, *J. Neurosci.* 16, 6353–6363.

Tottene, A., Volsen, S. and Pietrobon, D., 2000, α_{1E} subunits form the pore of three cerebellar R-type calcium channels with different pharmacological and permeation properties, *J. Neurosci.* 1, 171–178.

Volsen, S.G., Day, N.C., McCormack, A.L., Smith, W., Craig, P.J., Beattie, R., Ince, P.G., Shaw, P.J., Ellis, S.B., Gillespie, A., Harpold, M.M. and Lodge, D., 1995, The expression of neuronal voltage-dependent calcium channels in human cerebellum, *Mol. Brain Res.* 34, 271–282.

Volsen, S.G., Day, N.C., McCormack, A.L., Smith, W., Craig, P.J., Beattie, R.E., Smith, D., Ince, P.G., Shaw, P.J., Ellis, S.B., Mayne, N., Burnett, J.P., Gillespie, A. and Harpold, M.M., 1997, The expression of voltage-dependent calcium channel beta subunits in human cerebellum, *Neuroscience* 80, 161–174.

Wakamori, M., Yamazaki, K., Matsunodaira, H., Teramoto, T., Tanaka, I., Niidome, T., Sawada, K., Nishizawa, Y., Sekiguchi, N., Mori, E., Mori, Y. and Imoto, K., 1998, Single tottering mutations responsible for the neuropathic phenotype of the P-type calcium channel, *J. Biol. Chem.* 273, 34857–34867.

Walker, D. and De Waard, M., 1998, Subunit interaction sites in voltage-dependent Ca^{2+} channels: Role in channel function, *Trends Neurosci.* 21, 148–154.

Welch, K.M.A., 1998, Current opininons in headache pathogenesis: Introduction and synthesis, *Curr. Opin. Neurol.* 11, 193–197.

Westenbroek, R.E., Sakurai, T., Elliott, E.M., Hell, J.W., Starr, T.V.B., Snutch, T.P. and Catterall, W.A., 1995, Immunochemical identification and subcellular distribution of the α_{1A} subunits of brain calcium channels, *J. Neurosci.* 15, 6403–6418.

Zhuchenko, O., Bailey, J., Bonnen, P., Ashizawa, T., Stockton, D.W., Amos, C., Dobyns, W.B., Subramony, S.H., Zoghbi, H.Y. and Lee, C., 1997, Autosomal dominant cerebellar ataxia (SCA6) associated with small polyglutamine expansions in the α_{1A}-voltage-dependent calcium channel, *Nature Genet.* 15, 62–69.

Calcium and Diabetes

Shahidul Islam

1. INTRODUCTION

Diabetes mellitus is a heterogeneous group of syndromes characterised by chronic hyperglycemia, perturbed lipid metabolism as well as many hormonal and metabolic disturbances. Poorly controlled diabetes leads to debilitating complications e.g. accelerated atherosclerosis, microangiopathy, nephropathy, neuropathy, foot diseases and blindness. Throughout the world, diabetes afflicts $\sim$ 140 million people and the number may double by the year 2025. Diabetic syndromes result from complex interactions between many predisposing genetic factors and environmental ones. Type 1 diabetes is due to specific destruction of pancreatic β-cells mainly by autoimmune mechanisms. Type 2 diabetes is phenotypic expression of a multitude of defects that perturb lipid metabolism, reduce the ability of insulin to lower plasma glucose effectively, and cause varying degrees of insulin deficiency. Type 2 diabetes is the commonest form of diabetic syndromes: in Sweden, annual incidences of type 1 and type 2 diabetes per 100,000 inhabitants in 1991–1995 being 14.7 and 265.6 respectively (Berger et al., 1999).

Pancreatic β-cells and skeletal muscle are two of the main tissues involved in glucose homeostasis. These two tissues are also the main targets for common antidiabetic medicines. Long-term complications of diabetes involve many other cells and tissues e.g. basement membranes, endothelium, vascular smooth muscle cells, platelets and monocytes. Ca^{2+} homeostasis and normal Ca^{2+} signalling are essential for optimal function of all of these cells. Because of crucial roles of $[Ca^{2+}]_i$ in secretion in many cells, as well as in cell-proliferation and cell-death, numerous investigators are exploring diverse aspects of Ca^{2+}-signalling in β-cells and roles of intracellular Ca^{2+} homeostasis in diabetes (reviewed by Levy, 1999). It is generally assumed that in type 2 diabetes, β-cell defects may reside in the pathways that link

Shahidul Islam • Department of Molecular Medicine, Karolinska Institutet, Karolinska Hospital, Stockholm, Sweden.

R. Pochet, R. Donato, J. Haiech, C. Heizmann and V. Gerke (eds.): Calcium: The Molecular Basis of Calcium Action in Biology and Medicine, 401–413.

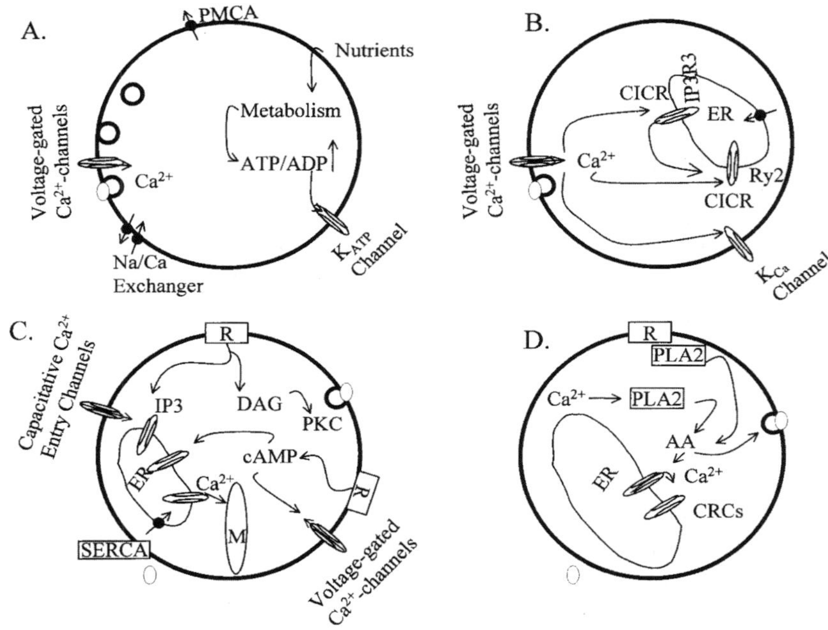

Figure 1. This figure illustrates some of the molecules and processes involved in Ca^{2+} homeostasis and Ca^{2+}-dependent stimulus secretion coupling in β-cells. (A) Nutrient metabolism increases cytoplasmic ATP/ADP ratio leading to closure of K_{ATP} Channel, membrane depolarisation and Ca^{2+} entry through voltage-gated Ca^{2+} channels. PMCA, plasma membrane Ca^{2+} ATPase. (B) Ca^{2+} signalling is amplified by Ca^{2+} induced Ca^{2+} release through intracellular Ca^{2+} channels. Calcium activated potassium channels (K_{ca} channel) mediate membrane repolarisation. IP3R3, type 3 inositol 1,4,5-trisphosphate receptor; Ry2, type 2 ryanodine receptor. (C) Depletion of intracellular Ca^{2+} stores activates capacitative Ca^{2+} entry channels in the plasma membrane. Ca^{2+} released from the ER increases mitochondrial (M) Ca^{2+} and activates mitochondrial metabolism. cAMP-dependent pathways modulate Ca^{2+} signalling by phosphorylating intracellular as well as plasma membrane Ca^{2+} channels. (D) Ca^{2+} activates PLA2 leading to formation of arachidonic acid (AA). Multiple mechanisms are involved in the formation of AA which directly or through its metabolites release further Ca^{2+} by activating the ER Ca^{2+} release channels (CRCs).

metabolism of nutrients and ligand-receptor interactions to biosynthesis and exocytosis of insulin, as well as survival of these cells. In this chapter, I shall give an overview of the roles of intracellular Ca^{2+}, placing emphasis on studies that have used diabetic patients or diabetic animal models to understand pathogenesis of the condition, its complications, and mechanisms of action of antidiabetic drugs.

2.

β-cells like most cells, share some basic mechanisms of Ca^{2+} handling which will not be elaborated in this chapter (see chapters by Guse and Nadal and

Soria in this book). Some features of Ca^{2+} signalling in electrically excitable cells, e.g. muscle and β-cells have been outlined by Macrez and Mironneau and Marsh et al. in this book. β-cells are nested in the pancreas together with other cell types, in numerous tiny islets. These cells are difficult to study since pure and functioning β-cells cannot be easily obtained in adequate numbers. These cells are specialised as fuel sensors: signals generated from nutrient metabolism increase cytoplasmic Ca^{2+} concentration ($[Ca^{2+}]_i$) by multiple mechanisms (Nadal and Soria, 1997, and this book). It is widely accepted that an increase in the cytoplasmic ATP/ADP ratio leads to closure of ATP-sensitive potassium channels resulting in membrane-depolarisation, opening of voltage gated Ca^{2+} channels in the plasma membrane and Ca^{2+} entry (Figure 1A). Other signals that are generated from nutrient metabolism and that may be involved in metabolism-secretion coupling include malonyl CoA (Antinozzi et al., 1998), arachidonic acid (Figure 1D) (Ma et al., 1998) and cyclic ADP ribose (cADPR) (Okamoto, 1999; Guse, this book). Ca^{2+} entering through the plasma membrane Ca^{2+} channels may trigger further Ca^{2+} release through the intracellular Ca^{2+} channels i.e. type 3 and type 2 inositol 1,4,5-trisphosphate receptors (Lee et al., 1999; Parys et al., this book) and type 2 ryanodine receptor (Islam et al., 1992, 1998) (Figure 1B). Gut hormones and neurotransmitters potentiate glucose-induced insulin secretion. These agents increase $[Ca^{2+}]_i$ in β-cells by releasing the ion from endoplasmic reticulum (ER) and stimulating capacitative Ca^{2+} entry (Miura et al., 1997b). Cytoplasmic Ca^{2+} increases mitochondrial Ca^{2+} which stimulates mitochondrial metabolism resulting in increased production of not only ATP but also a putative signalling substance that may enhance exocytosis (Maechler and Wollheim, 1999). β-cells also have multiple molecular mechanisms for extrusion or sequestration of Ca^{2+}. These include several isoforms (e.g. 1b, 2b and 4b) of plasma membrane Ca^{2+} ATPase (Varadi et al., 1996a), different isoforms of thapsigargin-sensitive ER Ca^{2+} ATPases e.g. SERCA 2b and SERCA 3 (Islam and Berggren, 1993; Varadi et al., 1996b; and for a review on SERCA see Paterlini-Bréchot et al., this book) and Na^+/Ca^{2+} exchanger (Van Eylen et al., 1998). Cross-talks between Ca^{2+}- and cAMP-signalling pathways take place at multiple levels with dramatic effects on insulin secretion (Grapengeisser et al., 1991; Abdel-Halim et al., 1996; Holz et al., 1999) (Figure 1C).

Insulin secretion is associated with an increase in $[Ca^{2+}]_i$ in β-cells (Nadal and Soria, 1997). Glucose does not stimulate insulin secretion if $[Ca^{2+}]_i$ is not increased (Pertusa et al., 1999). *In vivo*, insulin is released in regular pulses ($\sim$ every 8–15 min in large mammals), superimposed on basal secretion (Goodner et al., 1977). *In vitro*, insulin pulses may be more frequent. In perfused human- and monkey-pancreata, the pulse interval is $\sim$ 5–7 min (Goodner et al., 1991). Mechanisms that generate these oscillations reside in the islets (Marchetti et al., 1994). Glucose causes oscillation of metabolism (e.g. oscillation in the intracellular ATP/ADP ratio) in β-cell by inducing oscillation in the activity of many enzymes and of concentration of their effectors (MacDonald et al., 1997). In *in vitro* experiments, application of glucose to islets or β-cells results in conspicuous $[Ca^{2+}]_i$ oscillations. When $[Ca^{2+}]_i$ and insulin secretion is measured in single islets, the two oscillations appear to be superimposable (Bergsten, 1995). Metabolic oscillations lead to rhythmic membrane depolarisation-repolarisation and consequent opening of the voltage-gated Ca^{2+} channels (Martin et al., 1997). Metabolic oscillations can also cause periodic release of Ca^{2+} from the ER (Corkey et al., 1988). Apparently, oscillation in metabolism drives oscillation of $[Ca^{2+}]_i$, the latter being more effective in triggering insulin exocytosis (Ravier et al., 1999). The two oscillations, however, cooperate to produce pulsatile insulin release. Cellular energy and phosphorylation status determine effectiveness of Ca^{2+}-triggered exocytosis. In type 2 diabetes and obesity, regular oscillations of insulin secretion is lost even before onset of hyperglycemia, suggesting that metabolic and Ca^{2+}-oscillatory mechanisms that underlie pulsatile insulin release become deranged early in course of the pathogenesis of the disease (O'Rahilly et al., 1988; Polonsky et al., 1998).

Because of difficulty in obtaining human islets, specially human diabetic islets, most studies have used islets from many rodent models of diabetes for studying roles of Ca^{2+} in the pathogenesis of the disease. These models probably do not represent human diabetic syndromes well, but important insights have been obtained by studying them. Neonatal rats exposed to streptozotocin develop a syndrome mimicking type 2 diabetes when they become adult. Triphenyltin, an organic tin compound, produce diabetes in hamsters without causing obvious morphological changes in the islet cells. *db/db* mice and Zucker diabetic fatty (ZDF) rats lack functional leptin receptors, as a consequence of which they develop hyperphagia, obesity and diabetes resembling type 2 human diabetes. Transgenic overexpression of calmodulin, a Ca^{2+} binding protein, in β-cells results in an insulin dependent diabetes in the mice (Epstein et al., 1989). Two other models of type 2 diabetes are genetically diabetic hamsters and Goto Kakizaki (GK) rats, the latter becoming increasingly popular as a model of non-obese type 2 diabetes. β-cells from streptozotocin-induced diabetic rats (Tsuji et al., 1993), ZDF rats (Roe et al., 1996), *db/db* mice (Roe et al., 1994), genetically diabetic or triphenyltin-induced diabetic hamsters (Lindström et al., 1996; Miura et al., 1997a), all show diminished $[Ca^{2+}]_i$ response to glucose. Typically, fewer β-cells from diabetic islets respond by an elevation of $[Ca^{2+}]_i$ when challenged with stimulating concentration of glucose. $[Ca^{2+}]_i$-increase is delayed, the rate of rise of $[Ca^{2+}]_i$ is slower and magnitude of maximal increase of $[Ca^{2+}]_i$ is diminished in diabetic β-cells. More importantly, in diabetic islets $[Ca^{2+}]_i$ increase tends to be persistent rather than oscillatory as in normal islets (Figure 2) (Roe et al., 1994, 1996). When stimulated by glucose, some preparations of GK rat islets also show a delayed and slower rise of $[Ca^{2+}]_i$ (Zaitsev et al., 1997). Furthermore, maximal increase of $[Ca^{2+}]_i$ in GK rat β-cells may be lower than that in the controls (Kato et al., 1996). It should be noted that, impairment of $[Ca^{2+}]_i$ response in diabetic β-cells is generally selective for

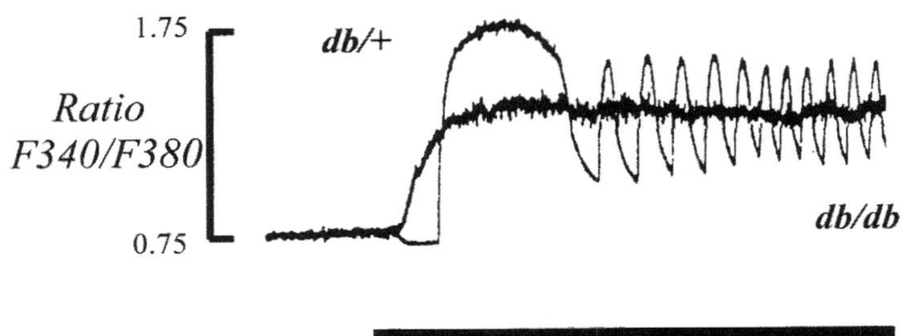

Figure 2. Cytoplasmic free Ca^{2+} concentration oscillates in normal islets of Langerhans and such oscillation is lost in diabetic islets. The figure shows changes in cytoplasmic free Ca^{2+} concentration measured by fura-2 technique, in single islets obtained from a diabetic model (*db/db* mice) or its normal counterparts (*db/+*). Initial glucose concentration was 2 mM which was increased to 12 mM during the period indicated by the thick line. (Reproduced with permission from M.W. Roe et al., 1994.)

glucose. $[Ca^{2+}]_i$ response of β-cells from neonatally streptozotocin-induced diabetic rats and GK rats, to KCl or arginine is not reduced but is rather augmented (Kato et al., 1994; Zaitsev et al., 1997). Such augmented $[Ca^{2+}]_i$ response may be due to increased L-type and T-type voltage-gated Ca^{2+} channel currents (Kato et al., 1994, 1996).

β-cell specific over-expression of calmodulin in transgenic mice reduces basal $[Ca^{2+}]_i$ and increases Ca^{2+} buffering in these cells. As a result of such alterations in Ca^{2+} handling in β-cells, these mice develop severely impaired insulin secretion and insulin-dependent diabetes (Epstein et al., 1989). Islets from these mice show markedly reduced $[Ca^{2+}]_i$ response to glucose (Ribar et al., 1995). In these islets glycolysis is impaired and β-cell mass is also grossly reduced (Ribar et al., 1995). The calmodulin-transgenic mice illustrate essential roles of Ca^{2+} in β-cells for nutrient metabolism, stimulus secretion coupling as well as proliferation of the cells. However, Ca^{2+}-independent mechanisms may also be involved in producing some of the defects observed in these mice. The reduced β-cell mass in these animals could be due to increased apoptosis. Calmodulin activates adenylyl cyclase thereby increasing intracellular cAMP and PKA activity, which are known to induce apoptosis in some cell types (Dowd and Miesfeld, 1992). $[Ca^{2+}]_i$ response of human diabetic islets has been examined only rarely and the results are difficult to interpret because of small sample size and variability of responses (Kindmark et al., 1994).

A variety of molecular mechanisms that may cause perturbations of Ca^{2+} signalling and Ca^{2+} homeostasis in diabetic β-cells have been identified. In *db/db* islets there is about five fold reduction of ER Ca^{2+} ATPase proteins (Roe et al., 1994). In GK rat islets also SERCA 3 isoform of ER Ca^{2+} ATPase is significantly reduced (Varadi et al., 1996b). In ZDF rats there is decreased expression of genes encoding the α_1-subunits of the β-cell L-type voltage-gated Ca^{2+} channel (Roe et al., 1996). In many diabetic islets, as in many other cells in diabetes, the basal $[Ca^{2+}]_i$ is often slightly elevated (Wang et al., 1996). In β-cells of GK rats and non-obese diabetic mice (a model of type 1 diabetes), increased basal $[Ca^{2+}]_i$ may be due to increased expression or activity of voltage gated Ca^{2+}channels in the plasma membrane (Kato et al., 1996; Wang et al., 1996).

CD38 is a multifunctional protein bound to the plasma membrane of many cells including β-cells. It produces small amount ($< 2\%$) of a potent Ca^{2+} releasing second messenger, cADPR and large amount of ADP ribose from NAD^+ (Lee et al., 1999; Guse, this book). cADPR activates ryanodine receptor Ca^{2+} release channel in many cells and it has been postulated to play a role in insulin secretion from β-cells (Okamoto, 1999; Islam and Berggren, 1997). CD38 knock out mice show impaired glucose-induced increase of $[Ca^{2+}]_i$ and insulin secretion (Okamoto, 1999). In diabetic GK rats the level of CD38 (and many other proteins) is reduced (Matsuoka et al., 1995). In some diabetic patients, mutation in the CD38 gene (Yagui et al., 1998) and autoantibodies against CD38 (Pupilli et al., 1999; Ikehata et al., 1998) have been described.

When islets or β-cells are exposed to high concentrations of glucose for long period of time, they become unresponsive to subsequent challenges with stimulatory concentrations of glucose. This phenomenon called "glucose toxicity" is an important component of diabetic syndrome. In human islets exposed to high glucose for 48 hours, the basal $[Ca^{2+}]_i$ is markedly elevated and stimulation by glucose does not increase $[Ca^{2+}]_i$ further (Björklund et al., 2000).

Alterations in cytoplasmic Ca^{2+} homeostasis may play a role in causing certain forms of β-cell-death e.g. apoptosis (see chapters by Christakos et al. and Maki in this book for mechanisms of Ca^{2+} mediated apoptosis). It has been postulated that hyperactivity of voltage-gated Ca^{2+} channels in the plasma membrane and resulting Ca^{2+} overload over a long period of time, may cause progressive loss of normal β-cells in some models of diabetes (Kato et al., 1996). In type 1 diabetes, proinflammatory cytokines can mediate apoptosis by Ca^{2+}-dependent mechanisms (Suarez-Pinzon et al., 1999). According to one report, serum from patients with newly diagnosed type 1 diabetes contains factors that increase activity of voltage-activated Ca^{2+} channels and thereby induce β-cell apoptosis (Juntti-Berggren et al., 1993). Exposure to cytokines induce a low voltage-activated Ca^{2+} current in mouse β-cells causing elevation of basal $[Ca^{2+}]_i$ and apoptosis (Wang et al., 1996). In the β-cells of non-obese diabetic mice, and tumour cells derived from them (NIT-1 cells), there is increased activity of a low-voltage-activated Ca^{2+} channel in the plasma membrane, which increases basal $[Ca^{2+}]_i$ and causes cell death (Wang et al., 1996). Agents that cause prolonged increase in $[Ca^{2+}]_i$ in β-cells may cause apoptosis of these cells, a mechanism that is thought to be relevant also in the pathogenesis or progression of type 2 diabetes (Efanova et al., 1998b).

Ca^{2+} mediates actions of commonly used antidiabetic drugs on β-cells. Hypoglycemic drugs or drug-candidates belonging to both sulfonylurea (Abrahamsson et al., 1985) and non-sulfonylurea (Fujitani et al., 1997) types induce insulin secretion mainly by increasing $[Ca^{2+}]_i$. An increase in $[Ca^{2+}]_i$ in β-cells is essential also for insulin-releasing effects of imidazolines (Efanova et al., 1998a; Shepherd et al., 1996). Glucagon like peptide-1, a hormone that sensitizes β-cells to stimulation by glucose, shows great potential for treatment of type 2 diabetes. It promotes insulin secretion by orchestrating an interplay between Ca^{2+}- and cAMP-dependent pathways leading to characteristic changes in $[Ca^{2+}]_i$ (Fridolf and Ahren, 1993; Holz et al., 1995). Some thiazolidinedions also stimulate insulin secretion by stimulating Ca^{2+} entry in to the β-cells (Ohtani et al., 1996).

While physiological roles of Ca^{2+} in insulin secretion is widely known, its role in insulin action is less clear. There is some evidence that $[Ca^{2+}]_i$ may be involved in mediating some of the actions of insulin and may contribute to the pathogenesis of insulin resistance in type 2 diabetes. Insulin and insulin-like growth factor-1 increase $[Ca^{2+}]_i$ in muscle cells, the main site of glucose disposal (Bruton et al., 1999; Semsarian et al., 1999). High $[Ca^{2+}]_i$ in muscle, reduces phosphoserine phosphatase activity and thereby reduces normal dephosphorylation of glycogen synthetase and GLUT-4 which may contribute to insulin resistance (Begum et al., 1993; Sowers and Draznin, 1998). Activation of some isoforms of PKC by diacylglycerol may mediate insulin resistance in muscle (Bossenmaier et al., 1997). Elevated extracellular Ca^{2+}, as occurs in patients with hyperparathyroidism, often results in insulin resistance, hyperinsulinemia, and glucose intolerance (Richards and Thompson, 1999).

Long-term complications of diabetes are due to a vicious interplay between many secondary biochemical disturbances e.g. increased sorbitol production, non-enzymic glycosylation of proteins, oxidative stress, increased protein kinase C (PKC) and MAP kinase activity. Impaired cellular Ca^{2+} homeostasis may be another factor contributing to the pathogenesis of such complications (reviewed by Massry and Smogorzeski, 1997). Advanced glycation end products (AGE) inhibit agonist stimulated $[Ca^{2+}]_i$ increase in human glomerular mesangial cells (Mene et al., 1999). In diabetes and hypertension, basal $[Ca^{2+}]_i$ is elevated and Ca^{2+} signalling is impaired in many cells including platelets (Takaya et al., 1997; Vicari et al., 1996), endothelial cells and vascular smooth muscle cells. One action of insulin is to decrease vascular resistance by stimulating endothelial nitric oxide production and by reducing $[Ca^{2+}]_i$ in vascular smooth muscle cells (Cleland et al., 1998). Diabetic conditions alter subcellular distribution of $[Ca^{2+}]_i$ in vascular smooth muscle cells resulting in increased contractility of the cells and hypertension (Fleischhacker et al., 1999). High glucose also impairs Ca^{2+} signalling in vascular endothelial cells and may thus contribute to defective endothelium-dependent relaxation associated with type 2 diabetes (Salameh and Dhein, 1998). High $[Ca^{2+}]_i$ in the endothelial cells may contribute to thickening of basement membrane, a hall mark of diabetic small vessel disease, by increasing secretion of basement membrane proteins. Increased $[Ca^{2+}]_i$ in endothelial cells may also promote leakiness of the capillaries possibly by stimulating contraction of the endothelial cell cytoplasm (Michel and Curry, 1999). Heart cells cultured in high glucose show a slower clearance of intracellular Ca^{2+} (Ren et al., 1999). Persistent activation of the diacylglycerol-PKC pathway under conditions of hyperglycemia is implicated in the pathogenesis of vascular complications of diabetes (Park et al., 1999). PKC activation in monocytes may contribute to atherosclerosis in diabetes (Ceolotto et al., 1999). Mechanisms by which high glucose increases basal $[Ca^{2+}]_i$ in many cells, may include non-enzymic glycosylation of Ca^{2+}-ATPase and increased PKC activity (Gonzalez Flecha et al., 1999).

3. CONCLUDING REMARKS

$[Ca^{2+}]_i$ signalling is crucial for insulin secretion and is possibly also involved in insulin action. It must be emphasised that diabetes and its complications are unlikely to be due to defect in any one single signalling molecule or pathway. It is, however, remarkable that numerous studies have consistently demonstrated impaired Ca^{2+} signalling in β-cells in nearly all animal models of type 2 diabetes. Such impairments may be symptomatic of primary metabolic perturbations or secondary complications of hyperglycemia, but may also be due to defects in molecules involved in Ca^{2+} signalling and Ca^{2+} homeostasis. Diabetic conditions alter Ca^{2+} handling in many cells thus elevating basal $[Ca^{2+}]_i$ or altering subcellular Ca^{2+} distribution, mechanisms that may be among the factors that lead to long-term complications of the disease. The role of perturbed Ca^{2+} signalling and Ca^{2+} homeostasis in the pathogenesis of diabetes needs to be appreciated in the wider context of many other metabolic and signalling defects that have been reported in these complex group of syndromes.

ACKNOWLEDGEMENTS

The author is recepient of a Juvenile Diabetes Foundations International Award. Research at the author's laboratory is supported by the Swedish Medical Research Council, the Swedish Council for Natural Science Research, the Swedish Medical Society, the Novo-Nordisk Foundation, the Swedish Diabetic Society and the European Commission.

REFERENCES

Abdel-Halim, S.M., Guenifi, A., Khan, A., Larsson, O., Berggren, P.O., Östenson, C.G. and Efendić S., 1996, Impaired coupling of glucose signal to the exocytotic machinery in diabetic GK rats: A defect ameliorated by cAMP, *Diabetes* 45, 934–940.

Abrahamsson, H., Berggren, P.O. and Rorsman, P., 1985, Direct measurements of increased free cytoplasmic Ca^{2+} in mouse pancreatic β-cells following stimulation by hypoglycemic sulfonylureas, *FEBS Lett.* 190, 21–24.

Antinozzi, P.A., Segall, L., Prentki, M., McGarry, J.D. and Newgard, C.B., 1998, Molecular or pharmacologic perturbation of the link between glucose and lipid metabolism is without effect on glucose-stimulated insulin secretion. A re-evaluation of the long-chain acyl-CoA hypothesis, *J. Biol. Chem.* 273, 16146–16154.

Begum, N., Leitner, W., Reusch, J.E., Sussman, K.E. and Draznin, B., 1993, GLUT-4 phosphorylation and its intrinsic activity. Mechanism of Ca^{2+}-induced inhibition of insulin-stimulated glucose transport, *J. Biol. Chem.* 268, 3352–3356.

Berger, B., Stenström, G. and Sundkvist, G., 1999, Incidence, prevalence, and mortality of diabetes in a large population. A report from the Skaraborg Diabetes Registry, *Diabetes Care* 22, 773–738.

Bergsten, P., 1995, Slow and fast oscillations of cytoplasmic Ca^{2+} in pancreatic islets correspond to pulsatile insulin release, *Am. J. Physiol.* 268, E282–E287.

Björklund, A., Lansner, A. and Grill, V., 2000, Glucose induced $[Ca^{2+}]_i$ abnormalities in human pancreatic islets: Important role of overstimulation. (Manuscript).

Bossenmaier, B., Mosthaf, L., Mischak, H., Ullrich, A. and Haring, H.U., 1997, Protein kinase C isoforms beta 1 and beta 2 inhibit the tyrosine kinase activity of the insulin receptor, *Diabetologia* 40, 863–466.

Bruton, J.D., Katz, A., Westerblad, H., 1999, Insulin increases near-membrane but not global Ca^{2+} in isolated skeletal muscle, *Proc. Natl. Acad. Sci, USA* 96, 3281–3286.

Ceolotto, G., Gallo, A., Miola, M., Sartori, M., Trevisan, R., Del Prato, S., Semplicini, A. and Avogaro, A., 1999, Protein kinase C activity is acutely regulated by plasma glucose concentration in human monocytes in vivo, *Diabetes* 48, 1316–1322.

Cleland, S.J., Petrie, J.R., Ueda, S., Elliott, H.L. and Connell, J.M., 1998, Insulin as a vascular hormone: Implications for the pathophysiology of cardiovascular disease, *Clin. Exp. Pharmacol. Physiol.* 25, 175–184.

Corkey, B.E., Tornheim, K., Deeney, J.T., Glennon, M.C., Parker, J.C., Matschinsky, F.M., Ruderman, N.B. and Prentki, M., 1988, Linked oscillations of free Ca^{2+} and the ATP/ADP ratio in permeabilized RINm5F insulinoma cells supplemented with a glycolyzing cell-free muscle extract, *J. Biol. Chem.* 263, 4254–4258.

Dowd, D.R. and Miesfeld, R.L., 1992, Evidence that glucocorticoid- and cyclic AMP-induced apoptotic pathways in lymphocytes share distal events, *Mol. Cell. Biol.* 12, 3600–3608.

Efanova, I.B., Zaitsev, S.V., Brown, G., Berggren, P.O. and Efendić, S., 1998a, RX871024 induces Ca^{2+} mobilization from thapsigargin-sensitive stores in mouse pancreatic β-cells, *Diabetes* 47, 211–218.

Efanova, I.B., Zaitsev, S.V., Zhivotovsky, B., Kohler, M., Efendić, S., Orrenius, S. and Berggren, P.O., 1998b, Glucose and tolbutamide induce apoptosis in pancreatic β-cells. A process dependent on intracellular Ca^{2+} concentration, *J. Biol. Chem.* 273, 33501–33507.

Epstein, P.N., Overbeek, P.A. and Means, A.R., 1989, Calmodulin-induced early-onset diabetes in transgenic mice, *Cell* 58, 1067–1073.

Fleischhacker, E., Esenabhalu, V.E., Spitaler, M., Holzmann, S., Skrabal, F., Koidl, B., Kostner, G.M. and Graier, W.F., 1999, Human diabetes is associated with hyperreactivity of vascular smooth muscle cells due to altered subcellular Ca^{2+} distribution, *Diabetes* 48, 1323–1330.

Fridolf, T. and Ahren, B., 1993, Effects of glucagon like peptide-1(7-36) amide on the cytoplasmic Ca^{2+}-concentration in rat islet cells, *Mol. Cell. Endocrinol.* 96, 85–90.

Fujitani, S., Okazaki, K. and Yada, T., 1997, The ability of a new hypoglycaemic agent, A-4166, compared to sulphonylureas, to increase cytosolic Ca^{2+} in pancreatic β-cells under metabolic inhibition, *Brit. J. Pharmacol.* 120, 1191–1198.

Gonzalez Flecha, F.L., Castello, P.R., Gagliardino, J.J. and Rossi, J.P., 1999, Molecular characterization of the glycated plasma membrane calcium pump, *J. Membr. Biol.* 171, 25–34.

Goodner, C.J., Walike, B.C., Koerker, D.J., Ensinck, J.W., Brown, A.C., Chideckel, E.W., Palmer, J. and Kalnasy, L., 1977, Insulin, glucagon, and glucose exhibit synchronous, sustained oscillations in fasting monkeys, *Science* 195, 177–179.

Goodner, C.J., Koerker, D.J., Stagner, J.I., Samols, E., 1991, *In vitro* pancreatic hormonal pulses are less regular and more frequent than *in vivo*, *Am. J. Physiol.* 260, E422–E429.

Grapengiesser, E., Gylfe, E. and Hellman, B., 1991, Cyclic AMP as a determinant for glucose induction of fast Ca^{2+} oscillations in isolated pancreatic β-cells, *J. Biol. Chem.* 266, 12207–12210.

Holz, G.G., 4th, Leech, C.A. and Habener, J.F., 1995, Activation of a cAMP-regulated Ca^{2+}-signaling pathway in pancreatic beta-cells by the insulinotropic hormone glucagon-like peptide-1, *J. Biol. Chem.* 270, 17749–17757.

Holz, G.G., Leech, C.A., Heller, R.S., Castonguay, M. and Habener, J.F., 1999, cAMP-dependent mobilization of intracellular Ca^{2+} stores by activation of ryanodine receptors in pancreatic β-cell. A Ca^{2+} signaling system stimulated by the insulinotropic hormone glucagon-like peptide-1-(7-37), *J. Biol. Chem.* 274, 14147–14156.

Ikehata, F., Satoh, J., Nata, K., Tohgo, A., Nakazawa, T., Kato, I., Kobayashi, S., Akiyama, T., Takasawa, S., Toyota, T. and Okamoto, H., 1998, Autoantibodies against CD38 (ADP-ribosyl cyclase/cyclic ADP-ribose hydrolase) that impair glucose-induced insulin secretion in noninsulin-dependent diabetes patients, *J. Clin. Invest.* 102, 395–401.

Islam, M.S. and Berggren, P.O., 1993, Mobilization of Ca^{2+} by thapsigargin and 2,5-di-(t-butyl)-1,4-benzohydroquinone in permeabilized insulin-secreting RINm5F cells: Evidence for separate uptake and release compartments in inositol 1,4,5-trisphosphate-sensitive Ca^{2+} pool, *Biochem. J.* 293, 423–429.

Islam, M.S. and Berggren, P.O., 1997, Cyclic ADP-ribose and the pancreatic β-cell: Where do we stand?, *Diabetologia* 40, 1480–1484.

Islam, M.S., Rorsman, P. and Berggren, P.O., 1992, Ca^{2+}-induced Ca^{2+} release in insulin-secreting cells, *FEBS Lett.* 296, 287–291.

Islam, M.S., Leibiger, I., Leibiger, B., Rossi, D., Sorrentino, V., Ekström, T.J., Westerblad, H., Andrade, F.H. and Berggren, P.O., 1998, *In situ* activation of the type 2 ryanodine receptor in pancreatic β-cells requires cAMP-dependent phosphorylation, *Proc. Natl. Acad. Sci. USA* 95, 6145–6150.

Juntti-Berggren, L., Larsson, O., Rorsman, P., Ämmala, C., Bokvist, K., Wahlander, K., Nicotera, P., Dypbukt, J., Orrenius, S., Hallberg, A., et al., 1993, Increased activity of L-type Ca^{2+} channels exposed to serum from patients with type I diabetes, *Science* 261, 86–90.

Kato, S., Ishida, H., Tsuura, Y., Okamoto, Y., Tsuji, K., Horie, M., Okada, Y. and Seino, Y., 1994, Increased calcium-channel currents of pancreatic β-cells in neonatally streptozocin-induced diabetic rats, *Metabolism* 43, 1395–1400.

Kato, S., Ishida, H., Tsuura, Y., Tsuji, K., Nishimura, M., Horie, M., Taminato, T., Ikehara, S., Odaka, H., Ikeda, I., Okada, Y. and Seino, Y., 1996, Alterations in basal and glucose-stimulated voltage-dependent Ca^{2+} channel activities in pancreatic β-cells of non-insulin-dependent diabetes mellitus GK rats, *J. Clin. Invest.* 97, 2417–2425.

Kindmark, H., Köhler, M., Arkhammar, P., Efendić, S., Larsson, O., Linder, S., Nilsson, T. and Berggren, P.O., 1994, Oscillations in cytoplasmic free calcium concentration in human pancreatic islets from subjects with normal and impaired glucose tolerance, *Diabetologia*, 37, 1121–1131.

Lee, B., Jonas, J.C., Weir, G.C. and Laychock, S.G., 1999, Glucose regulates expression of inositol 1,4,5-trisphosphate receptor isoforms in isolated rat pancreatic islets, *Endocrinology* 140, 2173–2182.

Lee, H.C., Munshi, C. and Graeff, R., 1999, Structures and activities of cyclic ADP-ribose, NAADP and their metabolic enzymes, *Mol. Cell. Biochem.* 193, 89–98.

Levy, J., 1999, Abnormal cell calcium homeostasis in type 2 diabetes mellitus – A new look on an old disease, *Endocrine* 10, 1–6.

Lindström, P., Sehlin, J. and Frankel, B.J., 1996, Glucose-stimulated elevation of cytoplasmic calcium is defective in the diabetic Chinese hamster islet B cell, *Eur. J. Endocrinol.* 134, 617–625.

Ma, Z., Ramanadham, S., Hu, Z. and Turk, J., 1998, Cloning and expression of a group IV cytosolic Ca^{2+}-dependent phospholipase A2 from rat pancreatic islets. Comparison of the expressed activity with that of an islet group VI cytosolic Ca^{2+}-independent phospholipase A2, *Biochim. Biophys. Acta* 1391, 384–400.

MacDonald, M.J., Al-Masri, H., Jumelle-Laclau, M. and Cruz, M.O., 1997, Oscillations in activities of enzymes in pancreatic islet subcellular fractions induced by physiological concentrations of effectors, *Diabetes* 46, 1996–2001.

Maechler, P. and Wollheim, C.B., 1999, Mitochondrial glutamate acts as a messenger in glucose-induced insulin exocytosis, *Nature* 402, 685–689.

Marchetti, P., Scharp, D.W., Mclear, M., Gingerich, R., Finke, E., Olack, B., Swanson, C., Giannarelli, R., Navalesi, R. and Lacy, P.E., 1994, Pulsatile insulin secretion from isolated human pancreatic islets, *Diabetes* 43, 827–830.

Martin, F., Pertusa, J.A. and Soria, B., 1997, Oscillations of cytosolic Ca^{2+} in pancreatic islets of Langerhans, *Adv. Exp. Med. Biol.* 426, 195–202.

Massry, S.G. and Smogorzewski, M., 1997, Role of elevated cytosolic calcium in the pathogenesis of complications in diabetes mellitus, *Miner. Electrolyte Metab.* 23, 253–260.

Matsuoka, T., Kajimoto, Y., Watada, H., Umayahara, Y., Kubota, M., Kawamori, R., Yamasaki, Y. and Kamada, T., 1995, Expression of CD38 gene, but not of mitochondrial glycerol-3-phosphate dehydrogenase gene, is impaired in pancreatic islets of GK rats, *Biochem. Biophys. Res. Commun.* 214, 239–246.

Mene, P., Pascale, C., Teti, A., Bernardini, S., Cinotti, G.A. and Pugliese, F., 1999, Effects of advanced glycation end products on cytosolic Ca^{2+} signaling of cultured human mesangial cells, *J. Am. Soc. Nephrol.* 10, 1478–1486.

Michel, C.C. and Curry, F.E., 1999, Microvascular permeability, *Physiol. Rev.* 79, 703–761.

Miura, Y., Kato, M., Ogino, K. and Matsui, H., 1997, Impaired cytosolic Ca^{2+} response to glucose and gastric inhibitory polypeptide in pancreatic β-cells from triphenyltin-induced diabetic hamster, *Endocrinology* 138, 2769–2775.

Miura, Y., Henquin, J.C. and Gilon, P., 1997a, Emptying of intracellular Ca^{2+} stores stimulates Ca^{2+} entry in mouse pancreatic β-cells by both direct and indirect mechanisms, *J. Physiol.* 503, 387–398.

Nadal, A. and Soria, B., 1997, Glucose metabolism regulates cytosolic Ca^{2+} in the pancreatic β-cell by three different mechanisms, *Adv. Exp. Med. Biol.* 426, 235–243.

Ohtani, K., Shimizu, H., Tanaka, Y., Sato, N. and Mori, M., 1996, Pioglitazone hydrochloride stimulates insulin secretion in HIT-T 15 cells by inducing Ca^{2+} influx, *J. Endocrinol.* 150, 107–111.

Okamoto, H., 1999, The CD38-cyclic ADP-ribose signaling system in insulin secretion, *Mol. Cell. Biochem.* 193, 115–118.

O'Rahilly, S., Turner, R.C. and Matthews, D.R., 1988, Impaired pulsatile secretion of insulin in relatives of patients with non-insulin-dependent diabetes, *New. Engl. J. Med.* 318, 1225–1230.

Park, J.Y., Ha, S.W. and King, G.L., 1999, The role of protein kinase C activation in the pathogenesis of diabetic vascular complications, *Perit. Dial. Int.* 19, S222–S227.

Pertusa, J.A., Sanchez-Andres, J.V., Martin, F. and Soria, B., 1999, Effects of calcium buffering on glucose-induced insulin release in mouse pancreatic islets: An approximation to the calcium sensor, *J. Physiol.* 520, 473–483.

Polonsky, K.S., Sturis, J. and Van Cauter, E., 1998, Temporal profiles and clinical significance of pulsatile insulin secretion, *Horm. Res.* 49, 178–184.

Pupilli, C., Giannini, S., Marchetti, P., Lupi, R., Antonelli, A., Malavasi, F., Takasawa, S., Okamoto, H. and Ferrannini, E., 1999, Autoantibodies to CD38 (ADP-ribosyl cyclase/cyclic ADP-ribose hydrolase) in Caucasian patients with diabetes: effects on insulin release from human islets, *Diabetes* 48, 2309–2315.

Ravier, M.A., Gilon, P. and Henquin, J.C., 1999, Oscillations of insulin secretion can be triggered by imposed oscillations of cytoplasmic Ca^{2+} or metabolism in normal mouse islets, *Diabetes* 48, 2374–2382.

Ren, J., Dominguez, L.J., Sowers, J.R. and Davidoff, A.J., 1999, Metformin but not glyburide prevents high glucose-induced abnormalities in relaxation and intracellular Ca^{2+} transients in adult rat ventricular myocytes, *Diabetes* 48, 2059–2065.

Ribar, T.J., Jan, C.R., Augustine, G.J. and Means, A.R., 1995, Defective glycolysis and calcium signaling underlie impaired insulin secretion in a transgenic mouse, *J. Biol. Chem.* 270, 28688–28695.

Richards, M.L. and Thompson, N.W., 1999, Diabetes mellitus with hyperparathyroidism: Another indication for parathyroidectomy?, *Surgery*, 126, 1160–1166.

Roe, M.W., Philipson, L.H., Frangakis, C.J., Kuznetsov, A., Mertz, R.J., Lancaster, M.E., Spencer, B., Worley, J.F., 3rd and Dukes, I.D., 1994, Defective glucose-dependent endoplasmic reticulum Ca^{2+} sequestration in diabetic mouse islets of Langerhans, *J. Biol. Chem.* 269, 18279–18282.

Roe, M.W., Worley, J.F., 3rd, Tokuyama, Y., Philipson, L.H., Sturis, J., Tang, J., Dukes, I.D., Bell, G.I. and Polonsky, K.S., 1996, NIDDM is associated with loss of pancreatic β-cell L-type Ca^{2+} channel activity, *Am. J. Physiol.* 270, E133–E140.

Salameh, A. and Dhein, S., 1998, Influence of chronic exposure to high concentrations of D-glucose and long-term beta-blocker treatment on intracellular calcium concentrations of porcine aortic endothelial cells, *Diabetes* 47, 407–413.

Semsarian, C., Wu, M.J., Ju, Y.K., Marciniec, T., Yeoh, T., Allen, D.G., Harvey, R.P. and Graham, R.M., 1999, Skeletal muscle hypertrophy is mediated by a Ca^{2+}-dependent calcineurin signalling pathway, *Nature* 400, 576–581.

Shepherd, R.M., Hashmi, M.N., Kane, C., Squires, P.E. and Dunne, M.J., 1996, Elevation of cytosolic calcium by imidazolines in mouse islets of Langerhans: Implications for stimulus-response coupling of insulin release, *Brit. J. Pharmacol.* 119, 911–916.

Sowers, J.R. and Draznin, B., 1998, Insulin, cation metabolism and insulin resistance, *J. Basic Clin. Physiol. Pharmacol.* 9, 223–233.

Suarez-Pinzon, W., Sorensen, O., Bleackley, R.C., Elliott, J.F., Rajotte, R.V. and Rabinovitch, A., 1999, β-cell destruction in NOD mice correlates with Fas (CD95) expression on β-cells and proinflammatory cytokine expression in islets, *Diabetes* 48, 21–28.

Takaya, J., Iwamoto, Y., Higashino, H., Ishihara, R. and Kobayashi, Y., 1997, Increased intracellular calcium and altered phorbol dibutyrate binding to intact platelets in young subjects with insulin-dependent and non-insulin-dependent diabetes mellitus, *Metabolism* 46, 949–953.

Tsuji, K., Taminato, T., Ishida, H., Okamoto, Y., Tsuura, Y., Kato, S., Kurose, T., Okada, Y., Imura, H. and Seino, Y., 1993, Selective impairment of the cytoplasmic Ca^{2+} response to glucose in pancreatic β-cells of streptozotocin-induced non-insulin-dependent diabetic rats, *Metabolism* 42, 1424–1428.

Van Eylen, F., Lebeau, C., Albuquerque-Silva, J. and Herchuelz, A., 1998, Contribution of Na^{+}/Ca^{2+} exchange to Ca^{2+} outflow and entry in the rat pancreatic β-cell: Studies with antisense oligonucleotides, *Diabetes* 47, 1873–1880.

Varadi, A., Molnar, E. and Ashcroft, S.J., 1996a, A unique combination of plasma membrane Ca^{2+}-ATPase isoforms is expressed in islets of Langerhans and pancreatic β-cell lines, *Biochem. J.* 314, 663–669.

Varadi, A., Molnar, E., Östenson, C.G. and Ashcroft, S.J., 1996b, Isoforms of endoplasmic reticulum Ca^{2+}-ATPase are differentially expressed in normal and diabetic islets of Langerhans, *Biochem. J.* 319, 521–527.

Vicari, A.M., Taglietti, M.V., Pellegatta, F., Spotti, D., Melandri, M., Galli, L., Ronchi, P. and Folli, F., 1996, Deranged platelet calcium homeostasis in diabetic patients with end-stage renal failure. A possible link to increased cardiovascular mortality?, *Diabetes Care* 19, 1062–1066.

Wang, L., Bhattacharjee, A., Fu, J. and Li, M., 1996, Abnormally expressed low-voltage-activated calcium channels in β-cells from NOD mice and a related clonal cell line, *Diabetes* 45, 1678–1683.

Yagui, K., Shimada, F., Mimura, M., Hashimoto, N., Suzuki, Y., Tokuyama, Y., Nata, K., Tohgo, A., Ikehata, F., Takasawa, S., Okamoto, H., Makino, H., Saito, Y. and Kanatsuka, A., 1998, A missense mutation in the CD38 gene, a novel factor for insulin secretion: Association with Type II diabetes mellitus in Japanese subjects and evidence of abnormal function when expressed in vitro, *Diabetologia* 41, 1024–1028.

Zaitsev, S., Efanova, I., Östenson, C.G., Efendić, S. and Berggren, P.O., 1997, Delayed Ca^{2+} response to glucose in diabetic GK rat, *Biochem. Biophys. Res. Commun.* 239, 129–133.

Inherited and Acquired Disorders of Extracellular Calcium (Ca_0^{2+})-Sensing

Edward M. Brown, Naibedya Chattopadhyay and Mei Bai

1. INTRODUCTION

The cloning of a G protein-coupled, extracellular calcium (Ca_0^{2+})-sensing receptor (CaR) from parathyroid, kidney and other cell types has elucidated how Ca_0^{2+} exerts its direct actions on various cells and tissues, especially those involved in maintaining systemic Ca_0^{2+} homeostasis (for a review, see Brown, 1999). The parathyroid cell is the prototypical Ca_0^{2+}-sensing cell. Its exquisite sensitivity to changes in Ca_0^{2+} enables it to serve as a primary regulator of systemic calcium metabolism (Brown, 1991). That is, high Ca_0^{2+}-evoked changes in parathyroid hormone (PTH) secretion modulate the translocation of Ca_0^{2+} ions into or out of the extracellular fluid (ECF) through the actions of PTH on the effector elements of the homeostatic mechanism (e.g., kidney, bone and intestine) in ways that restore normocalcemia. For instance, in response to a reduction in Ca_0^{2+}, PTH secretion increases. PTH then acts on the kidney to increase tubular reabsorption of Ca^{2+} and, if the hypocalcemia persists for more than a few hours, to stimulate proximal tubular synthesis of $1,25(OH)_2D_3$ (Bringhurst et al., 1998). The latter enhances intestinal absorption of Ca^{2+} and acts synergistically with PTH to promote net release of Ca^{2+} from bone. The combination of enhanced renal Ca^{2+} conservation, mobilization of Ca^{2+} from bone and increased intestinal Ca^{2+} absorption restores Ca_0^{2+} to its normal level (Brown, 1991).

The availability of the cloned CaR has also made it possible to elucidate this receptor's role in mediating known actions of Ca_0^{2+} on additional cells

Edward M. Brown, Naibedya Chattopadhyay and Mei Bai • Endocrine-Hypertension Division and Membrane Biology Program, Department of Medicine, Brigham and Women's Hospital and Harvard Medical School, Boston, MA, U.S.A.

R. Pochet, R. Donato, J. Haiech, C. Heizmann and V. Gerke (eds.): Calcium: The Molecular Basis of Calcium Action in Biology and Medicine, 415–442.

involved in systemic Ca_0^{2+} homeostasis that also respond to changes in Ca_0^{2+} in physiologically relevant ways. The thyroidal C-cells, for instance, secrete calcitonin (CT) – a hormone that promotes hypocalcemia, principally by inhibiting bone resorption – when Ca_0^{2+} rises (Bringhurst et al., 1998). Indeed, these cells express a CaR identical to that present in the parathyroid, which is thought to mediate Ca_0^{2+}-induced changes in CT release (Freichel et al., 1996; Garrett et al., 1995).

The synthesis of $1,25(OH)_2D_3$ by the renal proximal tubule, in addition to being indirectly regulated by Ca_0^{2+} through the associated changes in PTH secretion, is also directly inhibited by high Ca_0^{2+} (Brown, 1991; Weisinger et al., 1989). Although the CaR is expressed on the apical membrane of the proximal tubular epithelial cells (Riccardi et al., 1998), it is not currently known whether it mediates this direct action of Ca_0^{2+} on the 1-hydroxylation of 25-hydroxyvitamin D_3. The CaR is also expressed further along the nephron in the cortical thick ascending limb (CTAL) (Riccardi et al., 1998), where elevating peritubular but not luminal Ca_0^{2+} inhibits Ca^{2+} and Mg^{2+} reabsorption (De Rouffignac and Quamme, 1994). This direct action of Ca_0^{2+} on tubular function provides a mechanism for autoregulating renal divalent cation handling. As discussed in more detail later, inherited disorders of Ca_0^{2+} homeostasis caused by mutations in the CaR gene have established that this action of Ca_0^{2+} on the CTAL is almost certainly CaR-mediated (for a review, see Brown, 1999). Finally, elevated Ca_0^{2+} has actions on bone-forming osteoblasts and bone-resorbing osteoclasts that could contribute to maintaining Ca_0^{2+} homeostasis – stimulating and inhibiting, respectively, their differentiation from their respective progenitors as well as their bone-forming and/or bone-resorbing activities (Quarles, 1997; Zaidi et al., 1999). Precursors of both osteoblasts and osteoclasts, most fully differentiated osteoblasts (Chang et al., 1999) and some mature, multinucleated osteoclasts (Chang et al., 1999) express the CaR (for a review, see Yamaguchi et al., 1999), although further work is needed to determine whether this receptor mediates the known actions of Ca_0^{2+} on these two cell types and/or their precursors. Thus, in addition to directly regulating the secretion of the Ca_0^{2+}-elevating hormone, PTH, and the Ca_0^{2+} lowering hormone, CT, Ca_0^{2+} itself serves as a key Ca_0^{2+}-lowering "hormone" in humans. It does so, at least in part, through its own cell surface, G protein-coupled, Ca_0^{2+}-sensing receptor that is expressed on most, if not all, of the cells involved in maintaining systemic mineral ion metabolism (Brown et al., 1999). For additional details on the CaR's role in normal mineral ion metabolism, see Ward and Riccardi (this book).

The identification of naturally-occurring loss- or gain-of-function mutations in the CaR gene in human hyper- and hypocalcemic disorders (Brown, 1999), respectively, has considerably elucidated the physiology of systemic Ca_0^{2+} homeostasis by proving the CaR's central, non-redundant role in maintaining near constancy of Ca_0^{2+} by the mechanisms just outlined. From a pathophysiological perspective, however, the cloning of the CaR has also provided a new way of categorizing disorders of Ca_0^{2+} homeostasis – viz., by enabling the identification of disorders caused by abnormal Ca_0^{2+}-sensing. Moreover, while some of these disorders are generalized abnormalities in Ca_0^{2+}-sensing, in others the defect is "tissue-selective" (e.g., limited to a single tissue, such as the parathyroid). Moreover, in addition to inherited disorders arising from CaR mutations, other conditions have been identified in which the defect in Ca_0^{2+}-sensing is acquired and does not involve such mutations (Brown, 1999). The purpose of this chapter is to review these recent advances in our understanding of inherited and acquired, generalized and tissue-selective abnormalities in Ca_0^{2+}-sensing.

2. INHERITED ABNORMALITIES IN EXTRACELLULAR-CALCIUM SENSING

2.1. Disorders with generalized resistance to Ca_0^{2+}

2.1.1. Familial hypocalciuric hypercalcemia (FHH)

FHH is a genetic syndrome with autosomal dominant inheritance of generally asymptomatic, mild to moderate hypercalcemia accompanied by mild hypermagnesemia or serum magnesium levels in the upper part of the normal range, as well as generally normal serum levels of PTH, phosphorus, 25-hydroxyvitamin D_3 and $1,25(OH)_2D_3$ (for reviews, see Heath, 1994; Law and Heath, 1985; Marx et al., 1981a). A very characteristic finding in most persons with FHH is an unexpectedly low rate of urinary Ca^{2+} excretion given that there is coexistent hypercalcemia (e.g., there is "relative" hypocalciuria, which is inappropriate in the setting of hypercalcemia) (Heath, 1994; Law and Heath, 1985; Marx et al., 1981a). Studies utilizing induced hyper- and hypocalcemia have documented an increase in the set-point of the parathyroid glands for Ca_0^{2+}-regulated PTH secretion (i.e., the level of Ca_0^{2+} that half-maximally suppresses PTH secretion) in patients with FHH (Auwerx et al., 1984; Khosla et al., 1993). Thus the parathyroid glands in FHH are "resistant" to Ca_0^{2+}. Furthermore, the relative hypocalciuria in FHH persists even after parathyroidectomy – indicating that it is not simply the result of blunted suppression of PTH secretion by elevated Ca_0^{2+} but rather represents an additional, intrinsic renal tubular defect (Attie et al., 1983; Davies et al., 1984). Of the maneuvers that normally increase urinary Ca^{2+} excretion by direct renal actions, only administration of the loop diuretic, ethacrynic acid, which inhibits the Na/K/2Cl cotransporter in the CTAL, enhances urinary Ca^{2+} excretion in hypoparathyroid patients with FHH (Attie et al., 1983). This finding indicates the presence of excessively avid renal tubular Ca^{2+} reabsorption, likely occurring within the CTAL in FHH, which is the major nephron segment in which this cotransporter is expressed. Finally, persons with FHH concentrate their urine normally despite their hypercalcemia, unlike patients with primary hyperparathyroidism (PHPT) who have an equivalent degree of hypercal-

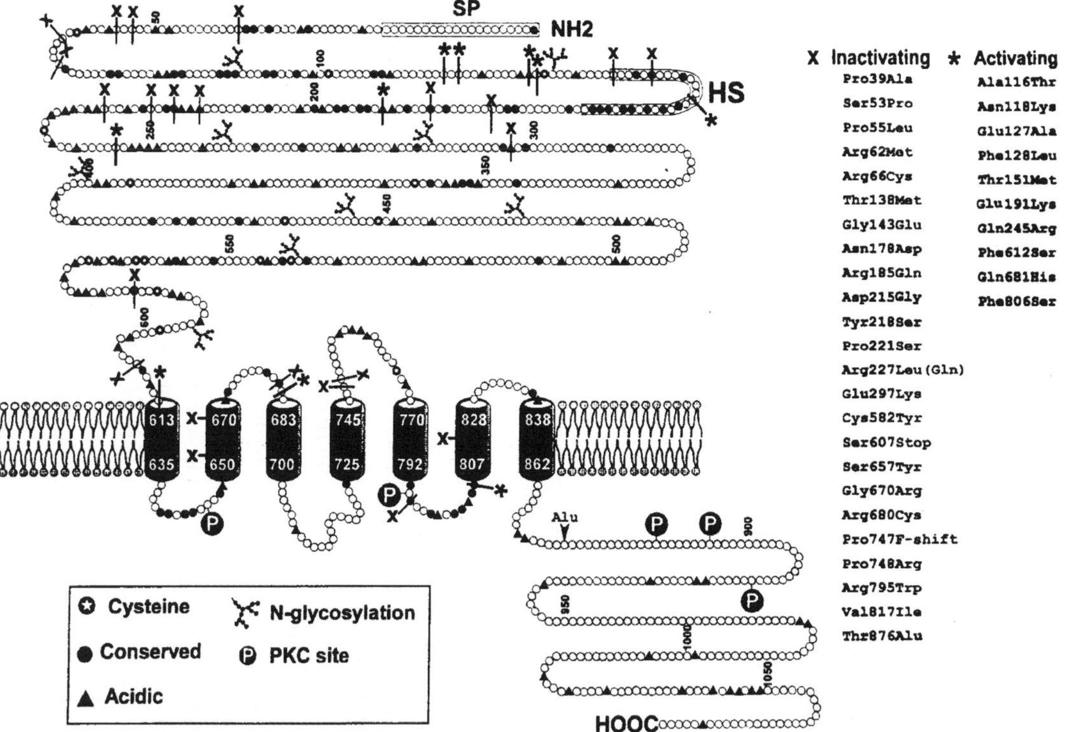

Figure 1. Schematic representation of the topology of the Ca_0^{2+}-sensing receptor cloned from human parathyroid, indicating locations of activating and inactivating mutations. Abbreviations are as follows: SP = signal peptide; HS = hydrophobic segment. Also delineated are the positions of various missense and nonsense mutations that cause familial hypocalciuric hypercalcemia (FHH) or autosomal dominant hypocalcemia; the latter are shown using the three letter amino acid code. The normal amino acid is indicated before and the mutant amino acid after the relevant codon. From Brown et al. (1997), with permission.

cemia, whose maximal urinary concentrating power is modestly diminished (Marx et al., 1981b). Thus individuals with FHH exhibit an apparent "resistance" to the known, direct actions of hypercalcemia on the parathyroid gland and on several aspects of renal function, namely urinary calcium excretion and concentrating ability.

What is the basis for the extracellular calcium resistance in FHH? The FHH gene was first mapped to the long arm of chromosome 3 (band q21–24) in four large families (Chou et al., 1992). Linkage analysis also formally proved that persons with FHH are heterozygous for this disease gene. FHH in at least 90% of cases is associated with a disease gene on chromosome 3, although a phenotypically similar disorder has recently been linked to two different loci on chromosome 19 [one on the long arm (Heath et al., 1992) and one on the short arm (Lloyd et al., 1999)] – indicating that FHH is genetically heterogeneous. Among families with the disease gene linked to chromosome 3, about two thirds harbor heterozygous inactivating mutations within the CaR's coding regions (e.g., see Figure 1). It is presumed that the remaining families that exhibit linkage to chromosome 3 have defects elsewhere in the CaR gene that reduce its level of expression, although this has not been formally proven.

Altogether, about 40 different CaR mutations have been identified to date in persons with FHH (Aida et al., 1995; Chou et al., 1995; Heath et al., 1996; Janicic et al., 1995; Kobayashi et al., 1997; Pearce et al., 1995; Ward et al., 1997). Each family generally has its own unique mutation, although a few unrelated families share identical mutations (e.g., Arg185Gln; see Bai et al., 1997b; Pollak et al., 1993). As illustrated in Figure 1, most FHH mutations are missense mutations clustered within (a) the first half of the receptor's large amino-terminal extracellular domain (ECD), (b) the region of the ECD immediately before the first transmembrane domain (TMD1) or (c) the CaR's TMDs, intracellular loops (ICLs), extracellular loops (ECLs) or carboxyl (C)-terminal tail. When expressed in heterologous mammalian expression systems, such as human embryonic kidney (HEK293) cells, these missense mutations inactivate the CaR to varying extents (Bai et al., 1996, 1997a, b; Pearce et al., 1996a) (Figure 2). Some mutations cause only modest (e.g., less than 2-fold) rightward shifts in the CaR's EC_{50} (the level of Ca_0^{2+} half-maximally activating it), without producing any change in its maximal activity at high Ca_0^{2+}, while others essentially abolish its biological activity. In nearly all cases, the mutant receptors would be activated little, if at all, by the modestly elevated levels of Ca_0^{2+} present *in vivo* in persons with FHH.

Although the levels of CaR protein expression in the parathyroid glands of persons with FHH have not been quantified, in mice heterozygous for targeted disruption of the CaR gene – an animal model of FHH (see below) – there is about a 50% reduction in the CaR's level of expression as assessed by immunohistochemistry with an anti-CaR antiserum (Ho et al., 1995). This result suggests that there is little upregulation of the production of the CaR protein by the remaining normal gene. Thus, if the same situation were true

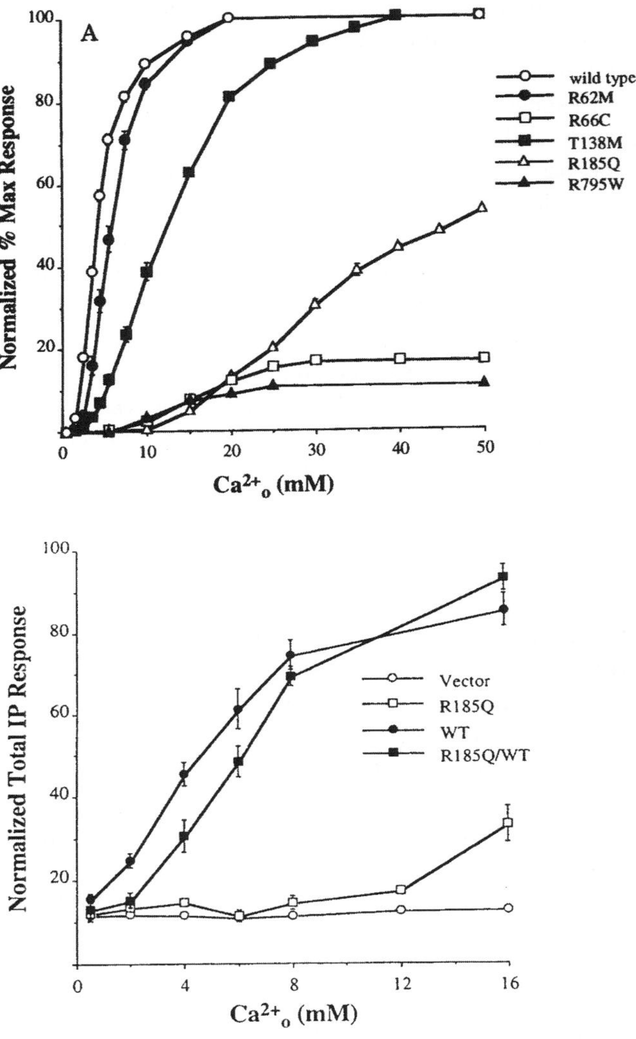

Figure 2. Expression of CaRs with FHH mutations in HEK293 cells. (A) The results illustrate the effects of varying levels of Ca_0^{2+} on Ca_i^{2+} in HEK293 cells that have been transiently transfected with wild type CaR or the indicated mutant CaRs. Results are normalized to percent of the maximal response of the wild type receptor. Reproduced from Bai et al. (1996), with permission. (B) Coexpression of mutant (arg185gln) and wild type CaRs in HEK293 cells. The results demonstrate the high Ca_0^{2+}-elicited increases in total cellular inositol phosphates (an indicator of activation of phoshpolipase C) in HEK 293 cells transiently transfected with empty vector (e.g., not containing any CaR cDNA), wild type CaR, a mutant CaR bearing arg185gln, or both the mutant and wild type CaRs. Note the "dominant negative" action of the CaR containing arg185gln when it is cotransfected with the wild type CaR, which right-shifts the high Ca_0^{2+}-induced activation of the wild type receptor and presumably contributes to the greater degree of hypercalcemia present in the neonate with *de novo* heterozygous FHH as a result of this mutation. From Bai et al. (1997b), with permission.

in FHH, it would imply that about a 50% reduction in the level of the CaR protein in the parathyroid produces the very modest, 10–20% rightward shift in the set-point for Ca_0^{2+}-regulated PTH release that is observed in most persons with FHH in whom this parameter has been quantified (Auwerx et al., 1984; Khosla et al., 1993). Thus the "resistance" to Ca_0^{2+} in FHH most likely results from a reduced level of expression of normally functioning CaRs. The tendency of individuals with FHH to exhibit mild hypermagnesemia (Marx et al., 1981a) may indicate that the CaR also contributes to "setting" the level of Mg_0^{2+} in blood (Brown, 1999; Strewler, 1994). That is, because it is activated by both Ca_0^{2+} and Mg_0^{2+}, the CaR in parathyroid and/or kidney may contribute importantly to Mg_0^{2+} homeostasis (Hebert et al., 1997). Indeed, families with FHH that exhibit more severe hypercalcemia tend to have greater degrees of elevation of Mg_0^{2+}, probably because the CaR contributes to maintaining the levels of both ions within their respective narrow ranges of normality (Aida et al., 1995; Marx et al., 1981a). Because most families with FHH have a very benign clinical course, with little in the way of symptoms or complications characteristic of hypercalcemia, there is a general consensus that parathyroid surgery should be avoided. In effect, these patients have "reset" their homeostatic systems to a new, mildly elevated level at which they are quite comfortable, in most cases, to remain.

Some families with missense mutations of the CaR, as just pointed out, exhibit more severe hypercalcemia than is the norm in this condition. For instance, while most families have total serum calcium concentrations averaging around 11 mg/dl, two families have been described with serum total calcium concentrations that average 12.5 and 13.4 mg/dl, respectively (Bai et al., 1996, 1997b). What accounts for the greater degree of hypercalcemia in these latter two families? Recent studies have shed some light on this matter. When the cell surface expression of the two mutations (R795W and R185Q respectively) in these two families was examined in transiently transfected HEK293 cells, it was relatively robust in both cases – compared to that of several other missense mutations – and was not dissimilar from that of the wild type CaR (Bai et al., 1996). The biological activity of these two receptors, in contrast, was severely compromised. When expressed by transient transfection in HEK293 cells, they showed, respectively, 3- and 7-fold increases in their EC_{50}'s, and their maximal activities were only about 15 and 40–50% of that of the wild type CaR (Bai et al., 1996). The recent recognition that the CaR resides on the cell surface principally in the form of a disulfide-linked dimer (Bai et al., 1998; Ward et al., 1998) provides a potential explanation for how these two mutated CaRs exert their dominant negative actions on the wild type CaR. That is, the mutant CaRs form heterodimers with the wild type receptor, and the presence of the mutant CaR in the heterodimer interferes in some fashion with the capacity of high Ca_0^{2+} to activate the wild type partner (Bai et al., 1997b). The apparent EC_{50} of the heterodimer is raised to a sufficiently high level that the serum calcium concentrations present in these FHH patients *in vivo* are inadequate to stimulate its activity (Figure 2). The only remaining normally functioning receptors, therefore, are the wild type homodimers, which on a purely statistical basis would comprise only 25% of the cell surface receptor (e.g., one would expect a 1:2:1 ratio of wild type homodimer, wild type-mutant heterodimer and mutant homodimer, respectively). It will be of interest as additional families with FHH are described with unusually high serum calcium concentrations to determine whether their mutant CaRs also exhibit severely compromised, high Ca_0^{2+}-evoked activation of intracellular signalling combined with robust cell surface expression, thereby encouraging the formation of wild type-mutant CaR heterodimers.

Additional types of mutations have recently been described within the CaR gene in FHH. Several families harbor mutations – such as nonsense mutations, deletions or insertions within the receptor's coding sequence – that generate premature stop codons, thereby producing truncated and, in most cases, biologically inert receptors (for a review, see Brown, 1999). An unusual mutation of this general type, which has so far been described in only one kindred – a family in Nova Scotia – is the insertion of a 383 bp Alu repetitive element at codon 876 within the receptor's C-terminal tail (Janicic et al., 1995). This Alu sequence harbors stop codons within all three of its reading frames, thereby producing truncated CaR proteins that contain long stretches of phenylalanines within their C-terminal tails (e.g., coded for by the element's long poly(A/T) tract). As might have been anticipated from the earlier discussion of the dominant negative actions of some mutant CaRs, families with truncated, inactive mutant CaRs tend to have very mild hypercalcemia. This likely occurs because the truncated receptor is unable to interact with the wild type CaR and, therefore, acts as a true "null" mutant (e.g., as if the mutant allele were totally absent). As a consequence, the number of wild type receptors is probably only reduced by about 50%, as in the heterozygous CaR knockout mice. Indeed, these mice mild only exhibit about a 10% increase in mean serum calcium concentration (Ho et al., 1995), as described in more detail below.

Based on the functional properties of the various FHH mutations that have been studied to date, the following conclusions can be tentatively drawn from these "experiments-in-nature" concerning the CaR's structure-function relationships and its roles in regulating mineral ion and water metabolism (for a review, see Hebert et al., 1997): (1) Some missense mutations in the CaR's ECD alter its apparent affinity for Ca_0^{2+} without changing its level of expression (at least as determined by Western analysis, which does not distinguish intracellular forms of the receptor from those residing on the cell surface). These mutated amino acids may modify the receptor's affinity for Ca_0^{2+} through direct or indirect mechanisms. Elucidation of the three dimensional structure of the receptor's ECD will eventually be required to determine precisely how the CaR binds Ca_0^{2+} and its other polycationic agonists. (2) Some additional mutations probably interfere with the receptor's capacity to activate its respective G-protein(s) (i.e., arg795trp) or perhaps other processes that are necessary for initiating signal transduction (e.g., conformational changes in the receptor's ECD, TMs and/or ICLs). (3) The elevated set-point of the parathyroid gland for Ca_0^{2+} is in most cases the result of reduced cell surface expression of normally functioning CaRs arising from the wild type CaR allele (i.e., a simple gene-dosage effect). In some cases, however, the "resistance" of the parathyroid glands to Ca_0^{2+} in FHH is further exacerbated by a "dominant negative" interaction between wild type and mutant receptors. The latter most likely results from the formation of heterodimers of the wild type and mutant CaRs, thereby reducing the normal partner's biological activity as well as the density of normally functioning CaRs (e.g., wild type homodimers) on the cell surface. (4) The overly avid renal tubular reabsorption of Ca^{2+} and Mg^{2+} that take place in the CTAL in FHH likely reflects "resistance" of the renal tubule to the normally calciuric action of elevated Ca_0^{2+} and document the CaR's key role in regulating renal tubular Ca^{2+} reabsorption. (5) The persistence of this tubular defect even following parathyroidectomy emphasizes another important point related to renal Ca^{2+} handling in the CTAL. Since multiple hormones stimulate the Na/K/2Cl cotransporter in a cAMP-dependent manner in this nephron segment (De Rouffignac and Quamme, 1994), the high Ca_0^{2+}-induced decrease in PTH secretion may by itself be insufficient to substantially reduce renal tubular Ca^{2+} reabsorption in the CTAL. Instead, direct CaR-mediated inhibition of divalent cation reabsorption in the CTAL is probably crucial for the increase in urinary Ca^{2+} excretion that normally occurs in response to hypercalcemia, which can still occur independent of circulating PTH (e.g., in hypoparathyroidism). (6) Finally, the capacity of persons with FHH to concentrate their urine normally provides further indirect support for the CaR's mediatory role in inhibiting urinary concentrating ability.

2.1.2. Neonatal severe hypercalcemia (NSHPT)

Neonatal severe hyperparathyroidism (NSHPT) is a condition encountered in newborn infants within the first 6 months of life in which there is much more severe resistance to Ca_0^{2+} than in FHH. Levels of serum calcium range from moderately elevated in the mildest cases (e.g., 12–13 mg/dl) to values as high as 30.8 mg/dl in the most severely affected ones (Brown, 1999; Chattopadhyay et al., 1996; Heath, 1994). These infants often exhibit hypotonia and poor feeding as well as the clinical sequellae of multiple fractures caused by their severe hyperparathyroidism. Despite the severe hypercalcemia, relative hypocalciuria can be present in NSHPT, as can hypermagnesemia (Heath, 1994). PTH levels are generally markedly elevated – in the range of 10-fold higher than the upper limit of normal – thereby causing overt hyperparathyroid bone disease, a hallmark of this condition. As a consequence, there are often multiple fractures of long bones. Sometimes the presence of numerous rib fractures produce a "flail chest" syndrome that causes respiratory embarrassment because of the infant's inability to expand its chest wall effectively (Heath, 1994).

The parathyroid glands in NSHPT are usually enlarged many-fold in mass and exhibit chief cell hyperplasia. Ca_0^{2+}-regulated PTH secretion has been studied *in vitro* using parathyroid tissue resected at the time of total parathyroidectomy from several infants with NSHPT (Bai et al., 1997b; Cooper et al., 1986; Marx et al., 1986). Dispersed parathyroid cells prepared from these hyperplastic glands have exhibited substantial, e.g., ~2-fold or greater, increases in set-point. Traditionally it has been recommended that infants with NSHPT undergo total parathyroidectomy, owing to the substantial mortality associated with this condition in early clinical descriptions of NSHPT. More recent series, however, have emphasized the greater spectrum of severity encountered in patients with NSHPT in the modem era (Heath, 1994). In many cases, particularly when the serum calcium concentrations is less than 15 mg/dl at presentation, aggressive medical treatment with modem therapeutic modalities, e.g., bisphosphonates, permits a more conservative approach to clinical management. After several weeks or months of such vigorous medical management and careful monitoring of the infant's clinical and biochemical parameters, there is not infrequently healing of the bone disease and reversion to a clinical state more closely resembling FHH (Heath, 1994).

With the discovery that NSHPT is caused in several cases by mutations in the CaR gene, it has been possible to identify and characterize infants with a wider spectrum of clinical severity than previously recognized and to begin to describe relationships between genotype and phenotype that will eventually simplify the diagnosis and management of this condition in the future, as described in more detail below.

NSHPT due to homozygous or compound heterozygous inactivating CaR mutations NSHPT in some but not all cases represents the homozygous form of FHH, usually owing to the consanguineous union of two individuals with FHH. Pollak et al. (1994b) showed in 11 families in which the FHH gene mapped to chromosome 3q that consanguineous marriages of affected individuals in four such families produced children with NSHPT. Subsequent studies have confirmed that the inheritance of two abnormal copies of the CaR gene can produce NSHPT (Aida et al., 1995; Chou et al., 1995; Janicic et al., 1995). Since they lack any normal copies of the CaR gene, these infants generally exhibit substantially more severe biochemical and clinical manifestations than in FHH, owing to severe resistance of CaR-expressing tissues, especially the parathyroid glands, to Ca_0^{2+}. In some cases, however, mutations that cause more modest functional defects in the CaR produce a less severe clinical and biochemical phenotype. Indeed, in some such cases, the condition may escape clinical detection until later in life. For instance, a woman inheriting two copies of the missense mutation, pro39ala, from her two related parents – each of whom harbored the same mutation – was only identified as having hypercalcemia at age 35 (Aida et al., 1995). Remarkably, she was seemingly asymptomatic despite having serum calcium concentrations of 15–17 mg/dl, PTH levels at the upper limit of normal and quite substantial (~50%) elevations in serum magnesium level. This case reiterates the probable role of the CaR in "setting" the level of Mg_0^{2+}. Furthermore, it points out that several of the symptoms and complications of hypercalcemia may be CaR-mediated, since this individual was asymptomatic despite a level of hypercalcemia that would be expected to produce severe compromise of renal and mental functions in otherwise normal people.

Another cause of FHH resulting from loss of both normal CaR alleles – albeit even more uncommon that the occurrence of homozygous FHH – was recently described in an infant with a serum calcium concentration in the range of 25 mg/dl (Kobayashi et al., 1997). In this case, the two unrelated parents each had different CaR mutations – a truncation mutation (arg185stop) in the father and a missense mutation (gly670glu) in the mother – producing a severely affected neonate who was a compound heterozygote. As is not infrequently the case in families in which there are infants with NSHPT, the biochemical findings in the parents were very subtle. Indeed, both parents were biochemically "unaffected" in the sense that their serum calcium concentrations were within the upper part of the normal range, thereby escaping earlier clinical detection. The mild phenotype in the two parents points out that in some instances the abnormality in Ca_0^{2+}-sensing in FHH can be so mild that the serum calcium concentration remains within the normal range. In our experience, however, persons with serum calcium concentrations persistently near the upper limit of normal (e.g., above 10 mg/dl but below the nominal upper normal limit) usually have some type of abnormality in Ca_0^{2+} homeostasis, which is not infrequently primary hyperparathyroidism or, less commonly, FHH, particularly if this finding has been present for several years in the former case or indefinitely in the latter.

NSHPT due to heterozygous inactivating CaR mutations Not all cases of NSHPT, however, are caused by homozygous or compound heterozygous FHH. In fact, most cases occur in a sporadic form or in FHH families in which there is only one affected parent (Chattopadhyay et al., 1996; Heath, 1994). One explanation for the occurrence of NSHPT in these cases is the presence of CaRs that exert a dominant negative action on the wild type CaR (Bai et al., 1996, 1997a), thereby producing more severe biochemical and clinical features in the affected infants than is the case in most FHH families. In other cases, gestation of the affected fetus in a normal mother may aggravate the situation further by exposing the parathyroid glands of the fetus to what is perceived to be "hypocalcemia" owing to their resistance to Ca_0^{2+}, which promotes sufficiently severe additional "secondary" parathyroid hyperplasia to cause NSHPT. It should be recognized, however, that there is no obvious increased incidence of NSHPT in affected infants born of normal mothers; thus, there must be additional, currently poorly understood factors that contribute to the development of NSHPT in some infants heterozygous for inactivating CaR mutations.

NSHPT can also occur as a result of de novo heterozygous CaR mutations (Bai et al., 1997b; Pearce et al., 1995) (i.e., there is a germline CaR mutation that occurs in the offspring of normal parents). Two such infants exhibited hyperparathyroid bone disease and hypercalcemia that was less severe than that seen in NSHPT due to homozygous or compound heterozygous CaR mutations (Pearce et al., 1995). We recently identified another case of NSHPT in an infant harboring a *de novo* CaR mutation identical to the one described earlier in a kindred with unusually severe hypercalcemia as well as several infants with NSHPT (Bai et al., 1997b). In both the infant with the *de novo* mutation and the family harboring this mutation, the relatively large discrepancy between the set-points of the mothers' and infants' parathyroid glands in utero and this receptor's dominant negative action may have produced more severe hyperparathyroidism in both pre- and postnatal life, increasing, therefore, the risk of NSHPT.

The severe hypercalcemia in NSHPT demonstrates further the CaR's central, non-redundant role in Ca_0^{2+}-regulated PTH secretion and in Ca_0^{2+} homeostasis more generally. Furthermore, the Ca_0^{2+} resistance that is present in both FHH and NSHPT are analogous to the resistance to more classical hormones that is caused by inactivating mutations in other guanine nucleotide regulatory protein (G protein)-coupled receptors – including the thyrotropin and adrenocorticotropin receptors (Spiegel, 1996) – as well as in other types of receptors – for instance, the thyroid hormone and androgen receptors (Jameson, 1999; Refetoff, 1982).

2.1.3. The broadening clinical spectrum of FHH and NSHPT

The recent availability of the cloned CaR has made it possible to document a broader clinical spectrum of PTH-dependent, hypercalcemic conditions than was previously appreciated arising from inactivating mutations of the CaR gene. For instance, the mildly affected homozygote described earlier with the pro39ala mutation (Aida et al., 1995) escaped clinical detection in the neonatal period despite having "homozygous FHH", which in most cases presents as NSHPT. Another, similar case first presenting in adulthood has been described recently in preliminary form, in which the parathyroid glands were unusually high in their content of fat cells (Fukumoto et al., 1998). Thus this case was described as exhibiting "lipohyperlasia" of the parathyroid glands. Another cases with a heterozygous inactivating mutation in the CaR gene also presented with sufficiently severe biochemical abnormalities to warrant parathyroid surgery and was found to have lipohyperplasia of all four parathyroid glands (Fukumoto et al., 1998). In both of the latter two cases, the two CaR mutations that were described (gln27arg and pro55leu) showed only mildly increased EC_{50}s when expressed transiently (5.2 and 7.4 mM Ca_0^{2+}, respectively, vs. 3.9 mM Ca_0^{2+} for the wild type receptor). Thus while the mild phenotype of the homozygote is explainable based on the underlying mild abnormality in Ca_0^{2+}-sensing by the mutant CaR, it is not clear why the second patient developed overt hyperparathyroidism despite being heterozygous for a mutant receptor that also exhibited a relatively mild functional defect – at least in the assay employed. These cases, particularly when taken in the context of instances of NSHPT arising from compound heterozygous, *de novo* heterozygous or familial heterozygous CaR mutations, point out the heterogeneity of the genetic defects in the CaR gene that can result in the same clinical condition, albeit one with an increasingly broad range of clinical severity.

Two additional kindreds in which affected individuals were heterozygous for missense mutations in the CaR, exhibit clinical features that broaden the phenotype of FHH toward that encountered in primary hyperparathyroidism. In both, the families described have – in some affected members – overt hypercalciuria that caused kidney stones. Moreover, one of the families have frank elevations in serum level of PTH (Carling et al., 2000; Soei et al., 1999). These cases are important to identify, because unlike the usual family with FHH, the hypercalciuria puts affected family members at risk for stones and progressive bone loss – complications of even mild primary hyperparathyroidism that are considered clear indications for surgical intervention. Therefore, unlike the majority of families with FHH, in whom surgical intervention is contraindicated, in cases in which PTH levels are frankly elevated, the degree of elevation in the serum calcium concentration is unusually high or in which there is overt hypercalciuria, surgical intervention may be warranted. Given the tendency of FHH to recur biochemically following anything less than total parathyroidectomy, the optimal surgical approach to these cases (e.g., subtotal parathyroidectomy vs. total parathyroidectomy with or without transplantation of parathyroid tissue to the forearm in the latter instance) is a challenging problem in surgical management.

3. CaR KNOCKOUT MICE AS A MODEL FOR HUMAN DISEASE AND EXPERIMENTAL TOOL

The development of mice heterozygous or homozygous for targeted disruption (e.g., knockout) of the CaR gene has generated useful animal models of FHH and NSHPT, respectively. Heterozygous mice are phenotypically unremarkable and they exhibit normal fertility and longevity. Their serum calcium concentrations average 10.4 mg/dl – about 10% higher than that of their normal littermates – while their circulating PTH levels are approximately 50% higher than in normal mice (Ho et al., 1995). Urinary calcium excretion in the heterozygous mice is modestly lower than in normal mice despite their hypercalcemia – similar to the situation in FHH. Immunocytochemistry and western analysis using specific anti-CaR antisera have documented that the level of CaR protein expression in the parathyroid glands and kidneys of the heterozygous CaR knockout mice are both diminished by about 50% (Ho et al., 1995). This reduced number of otherwise normal CaRs probably is the major cause of the "resistance" of parathyroid and kidney to Ca_0^{2+}, similar to the pathophysiology discussed earlier in FHH.

Mice that are homozygous for CaR gene inactivation, in contrast, although close to normal size at birth, grow at a markedly reduced rate thereafter compared to their normal or heterozygous littermates (Ho et al., 1995). Most of the homozygotes die within the first few weeks after birth, and only occasional ones survive for as long as 3–4 weeks. Homozygous CaR knockout mice exhibit severe hypercalcemia, with serum calcium concentrations that average 14.8 mg/dl. Their serum Mg^{2+} concentrations are slightly higher than in the heterozygous CaR knockout mice – though not elevated to the same extent as their serum levels of Ca_0^{2+}. The increased levels of Mg_0^{2+} in mice heterozygous and homozygous for targeted disruption of the CaR (Ho et al., 1995) – similar to those present in FHH (Marx et al., 1981a) and NSHPT (Aida et al., 1995) – supports the CaR's involvement in "setting" the serum level of Mg_0^{2+} (Brown et al., 1999; Strewler, 1994). Serum PTH levels are about 10-fold higher in homozygous than in wild type mice – an increase that is comparable to that present in NSHPT, particularly that caused by homozygous inactivating CaR mutations. In spite of their severely elevated serum calcium concentrations, the urinary Ca^{2+} concentration in homozygous mice is lower than in normal mice. Mice homozygous for CaR knockout, similar to infants with NSHPT, exhibit marked parathyroid hyperplasia, supporting the CaR's role in tonically suppressing parathyroid cellular proliferation (Ho et al., 1995). Skeletal x-rays reveal substantial reductions in bone mineral density, bowing of long bones and kyphoscoliosis. Mice that are homozygous for targeted disruption of the CaR gene, therefore, are similar in several of their phenotypic and biochemical features to infants with the forms of NSHPT caused by homozygous or compound heterozygous inactivating mutations of the CaR gene (e.g., with no normally functioning CaRs in parathyroid and kidney).

3.1. Disorders with generalized oversensitivity to extracellular calcium

3.1.1. Autosomal dominant hypocalcemia (hypercalciuric hypocalcemia)

A form of autosomal dominant hypocalcemia and, occasionally, sporadic cases of hypocalcemia can occur as a consequence of heterozygous activating CaR mutations (Brown, 1999). The biochemical and clinical features of these individuals resemble those of patients with hypoparathyroidism in the sense that circulating PTH levels are insufficient to maintain normocalcemia. The reduced levels of PTH, in the former, however, result not from a diminished parathyroid cell mass, as is the case in the latter, but rather from "resetting" downward of the set-points of parathyroid and kidney, respectively, for regulation of PTH secretion and urinary Ca^{2+} excretion by Ca_0^{2+} (Brown, 1999). In other words, CaR-expressing cells are overly sensitive to Ca_0^{2+} owing to their expression of CaRs harboring activating mutations. Therefore, this condition is the prototypical example of generalized "overresponsiveness" to Ca_0^{2+} of CaR-expressing tissues and is just the converse of the generalized resistance to Ca_0^{2+} in FHH and NSHPT.

Individuals with activating mutations of the CaR exhibit mild to moderate, and occasionally severe, hypocalcemia combined with relative hypercalciuria (e.g., they have hypercalciuric hypocalcemia) (Baron et al., 1996; Pearce et al., 1996a; Pollak et al., 1994a). They are not infrequently asymptomatic but can experience seizures and other symptoms of hypocalcemia and are susceptible to renal complications of treatment of their hypocalcemia with vitamin D and calcium. These complications include nephrolithiasis, nephrocalcinosis, and reversible (and in some cases irreversible) renal impairment, which most likely are a consequence of the marked hypercalciuria occurring in affected individuals as their levels of Ca_0^{2+} are elevated toward normal, owing to the presence of activated renal CaRs (Pearce et al., 1996a). Therapy with vitamin D and calcium supplementation, therefore, must be administered with great care in these cases and only when symptoms are clearly related to hypocalcemia (e.g. seizures).

This syndrome was documented to be linked to the CaR locus on chromosome 3 by Finegold et al. (1994). Pollak et al., (1994a) then documented the presence of a missense mutation of the CaR at residue 127 in a family with autosomal dominant hypocalcemia that increased the CaR's activity at low Ca_0^{2+} when it was studied by expression in *Xenopus laevis* oocytes. Subsequently, about a dozen missense mutations have been characterized in families with autosomal dominant hypocalcemia (Baron et al., 1996; Lienhardt et al., 2000; Lovlie et al., 1996; Mancilla et al., 1997, 1998; Pearce et al., 1996a; Perry et al., 1994; Pollak et al., 1994a); for reviews, see Chattopadhyay et al. (1996) and De Luca and Baron (1998); and in several cases of sporadic hypocalcemia (Baron et al., 1996; De Luca et al., 1997; Mancilla et al., 1998). Most of them are missense mutations that are present within the CaR's ECD, although some reside in the regions encompassed by the receptor's TMDs. The predominance of activating mutations in the CaR's ECD provides further support for the extracellular domain's role in activation of the receptor by Ca_0^{2+}. A novel type of activating mutation has recently been identified in a French family (Lienhardt et al., 2000). Affected members of this family exhibit a large deletion of 181 amino acids within the CaR's C-tail. Interestingly, one family member who is homozygous for this mutation has clinical and biochemical findings that are no more severe than those in heterozygous family members (Lienhardt et al., 2000). Therefore, it appears that, in contrast to the situation in FHH/NSHPT, the presence of one abnormal copy of an activated CaR in families with autosomal dominant hypocalcemia may be sufficient to "reset" the homeostatic system to the level determined by that receptor's biochemical properties and that the presence of a second abnormal allele has little in the way of additional biochemical consequences.

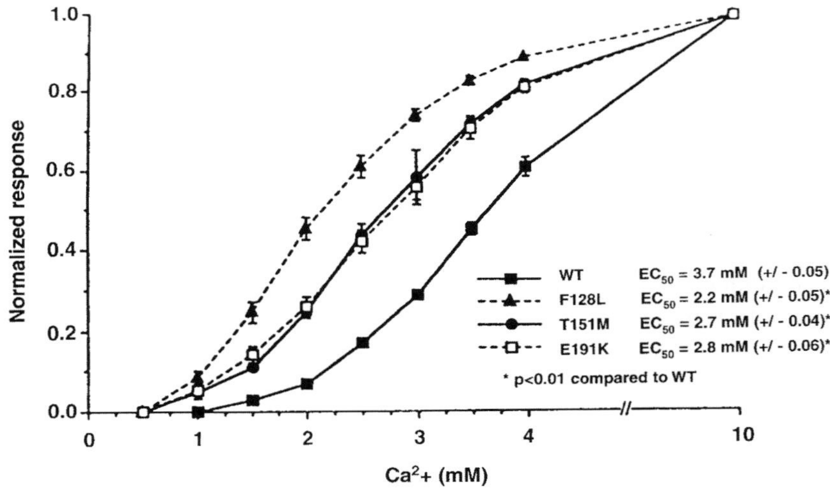

Figure 3. Expression of mutant CaRs containing activating mutations in HEK293 cells. Results indicate Ca_0^{2+}-evoked elevations in Ca_i^{2+} in HEK293 cells transiently transfected with the wild type human CaR or CaRs containing the activating mutations indicated. Reproduced in modified form with permission from Pearce et al. (1996).

Expression of several of the known activating mutations of the CaR in HEK293 cells has shown a clear left-shift in the receptors' activation by Ca_0^{2+} (Bai et al., 1996; De Luca et al., 1997; Lienhardt et al., 2000; Mancilla et al., 1997; Pearce et al., 1996a, b) (Figure 3) and, in some cases, an increase in their maximal activities, particularly when changes in inositol phosphates were assessed (De Luca et al., 1997; Mancilla et al., 1997). This observation appears to provide a clear explanation for how these receptors "reset" the level of Ca_0^{2+} *in vivo* in families with autosomal dominant hypocalcemia. The mechanism underlying the apparent increase in affinity of CaRs harboring activating mutations, however, is not yet clear. In some cases, as in the CaR with the deletion within its C-tail, there can be an increase in cell surface expression of the mutant CaR, which might reduce the level of Ca_0^{2+} needed to achieve sufficient occupancy of the receptor by its agonist to initiate receptor signaling (Lienhardt et al., 2000). Other mutations might alter the conformation of the CaR in such a way as to perturb the equilibrium between its active and inactive forms or increase the intrinsic affinity of the receptor's binding site(s) for Ca_0^{2+}. Much further work is needed to understand more fully the processes underlying the activation of both normal and abnormal CaRs by Ca_0^{2+}. Activating mutations of the CaR do not in most cases increase the basal or "constitutive" activity of the receptor (e.g., that present in the total absence of extracellular calcium). In contrast, activating mutations of other GPCRs, such as those in the PTH, TSH and LH receptors, which are more commonly present in the respective receptors' TMDs than within their ECDs, generally produce some degree of constitutive activation of these receptors (Parma et al., 1995; Schipani et al., 1995; Spiegel, 1996).

The increased level of urinary Ca^{2+} excretion in autosomal dominant or sporadic hypocalcemia owing to activating mutations of the CaR, even in the presence of overt hypocalcemia (Baron et al., 1996; Pearce et al., 1996a), probably reflects direct inhibition of renal tubular reabsorption of Ca^{2+} (and Mg^{2+}) by mutant CaRs activated at inappropriately low levels of Ca_0^{2+}. This latter effect is the opposite of that observed with inactivating CaR mutations in FHH (Attie et al., 1983), lending additional indirect support to the notion that the CaR directly regulates renal tubular Ca^{2+} handling in addition to its indirect regulation mediated via changes in PTH secretion. There may also be a "resetting" downward of the inhibitory action of Ca_0^{2+} on vasopressin-stimulated reabsorption of water in persons with autosomal dominant hypocalcemia (Hebert et al., 1997). That is, during treatment of hypocalcemia with vitamin D and calcium supplementation, these individuals not infrequently develop polyuria and polydipsia, probably as a result of defective urinary concentration and/or increased thirst at normal or even low levels of Ca_0^{2+} (Pearce et al., 1996a). These are symptoms typical of hypercalcemia and generally, of course, do not develop in otherwise normal persons until they become frankly hypercalcemic. Therefore, because of the various renal complications that can develop in persons with activating mutations of the CaR during well-intentioned treatment of their hypocalcemia with vitamin D and calcium, it is important (1) to limit such therapy to persons with symptoms of hypocalcemia that are clearly caused by the hypocalcemia, (2) to raise the level of serum calcium to the smallest extent that ameliorates their symptoms, and (3) to carefully monitor serum and urinary calcium concentrations and indices of renal function during follow up. In the future, it is possible that the use of a suitable CaR antagonist, which would "reset" the overresponsiveness of parathyroid and kidney – and, therefore, serum and urine Ca_0^{2+} – toward normal, would represent an effective treatment for symptomatic patients with this condition (see Section 5).

4. ACQUIRED DISORDERS OF EXTRACELLULAR CALCIUM-SENSING

4.1. Disorders with generalized abnormalities in extracellular calcium-sensing

4.1.1. Antibodies to the CaR

A substantial fraction of endocrine disorders are the result of aberrant immune function leading to autoimmune activation, inhibition and/or frank destruction of the function of endocrine tissues. For instance, these disorders include autoimmune forms of Graves' disease, primary hypothyroidism, type I diabetes, Addison's disease and some forms of ovarian failure (Eisenbarth and Verge, 1998; Neufeld et al., 1980). In some cases, multiple autoimmune endocrine disorders can coexist in the same patient. In several cases, the associated antibodies perturb endocrine function by binding to hormone receptors. For instance, Grave's disease is caused by antibodies that activate the G-protein-coupled receptor for thyrotropin hormone stimulating hormone (TSH), thereby mimicking the normal stimulatory action of TSH on thyrocytes (for a review, see Paschke et al., 1996). Another, less common example of autoimmune-mediated disruption of endocrine cell function mediated by anti-receptor antibodies results from blocking or activating antibodies to the insulin receptor (Flier et al., 1978).

It would not be surprising, therefore, if naturally-occurring antibodies existed that were directed at the CaR. Autoantibodies to the parathyroid glands were first reported by Blizzard et al. (1966) thereby suggesting an autoimmune etiology for acquired hypoparathyroidism. The nature of the antigen(s) to which these anti-parathyroid antibodies were directed, however, has remained obscure until recent studies revealed that in some cases there can be anti-CaR autoantibodies in patients with acquired hypoparathyroidism. Li et al. (1996) showed that the sera of 5 out of 20 patients with autoimmune hypoparathyroidism contained antibodies that bound to the CaR. This autoantigen appeared to be disease-specific since it was only recognized by the sera from the patients with acquired hypoparathyroidism but not by those from patients with other autoimmune diseases. The epitope recognized by the autoantibodies was exclusively localized to the receptor's extracellular domain. These authors, however, were unable to demonstrate any functional consequences of the binding of the antibodies to the CaR when the latter was expressed in HEK293 cells, suggesting that the hypoparathyroidism in these cases did not result from activation of the CaR but more likely from autoimmune destruction of the parathyroid glands (Li et al., 1996). In contrast, in preliminary studies, we have recently studied two patients with hypoparathyroidism in whom anti-CaR antibodies inhibited PTH secretion from dispersed parathyroid cells (Kifor et al., manuscript in preparation), suggesting that anti-CaR antibodies can in some cases produce hypoparathyroidism by mimicking the stimulatory action of the CaR's normal endogenous ligand, Ca_0^{2+}. It should be pointed out that hypoparathyroidism caused by anti-CaR antibodies that activate the receptor would be expected to produce a clinical disorder more similar to that caused by activating mutations in the CaR than that resulting simply from deficiency of PTH, since antibody-mediated activation of CaRs in the CTAL of the kidney would be expected to produce excessive hypercalciuria at any given level of Ca_0^{2+}.

4.2. Tissue-selective resistance to extracellular calcium

4.2.1. Primary and uremic/tertiary hyperparathyroidism

A decrease in the CaR's level of expression in parathyroid cells has been observed in pathological parathyroid glands resected from patients with primary hyperparathyroidism (PHPT) as well as those having severe secondary hyperparathyroidism (SHPT) that is caused by chronic renal insufficiency in most (Farnebo et al., 1997; Gogusev et al., 1997; Kaneko et al., 1999; Kifor et al., 1996), but not all studies (Garner et al., 1997). The latter study utilized quantitative RT-PCR to compare the expression of CaR transcripts in normal and pathological parathyroid glands. One likely difficulty in the interpretation of the results using this latter approach (Garner et al., 1997) would be that normal but not hyperparathyroid parathyroid glands contain substantial numbers of fat cells, which account for about 50% of the volume of the normal parathyroid gland. Studies employing immunocytochemistry or *in situ* hybridization have not demonstrated expression of the CaR in the fat cells present in normal parathyroid glands (Farnebo et al., 1997; Gogusev et al., 1997; Kaneko et al., 1999; Kifor et al., 1996). Extraction of RNA from normal parathyroid glands (Garner et al., 1997), therefore, could "dilute" CaR transcripts derived from parathyroid chief cells with RNA devoid of these transcripts originating from fat cells, thereby reducing the apparent level of CaR mRNA expression in normal parathyroid glands to one comparable to those in pathological parathyroid cells.

What is the basis for the reduced level of CaR expression in PHPT and SHPT? Several studies have documented the absence of mutations in the CaR's coding region in these two conditions. Therefore, the CaRs expressed in these hyperparathyroid glands are normal, at least in their primary structures. The mechanism producing the reduced CaR expression in PHPT and SHPT and its relationship, if any, to the concomitant parathyroid cellular hyperplasia in these pathological parathyroid cells remains to be determined. It should be noted, however, that the parathyroid proliferation observed in mice with knockout of the CaR gene (Ho et al., 1995) as well as in patients homozygous for inactivating mutations of the human CaR gene (Brown, 1999; Chattopadhyay et al., 1996) strongly support the CaR's involvement in tonically suppressing parathyroid cellular proliferation. Finally, the observed reduction in CaR expression in pathological parathyroid tissue likely contributes to the increase in the set-point for Ca_0^{2+}-regulated PTH secretion that is commonly observed not only in PHPT but also in severe, uremic secondary hyperparathyroidism (Brown, 1983). Of interest, the average reduction to the level of CaR expression, which is about 60% (Kifor et al., 1996), is similar to that observed in the parathyroid glands of mice heterozygous for targeted disruption of the CaR (~50%) (Ho et al., 1995), in whom the elevation in serum calcium concentration is not dissimilar from that observed in patients with mild primary hyperparathyroidism (e.g., 10–20%).

4.2.2. Reduced renal CaR expression in renal insufficiency

The level of expression of the CaR is also reduced in the remaining kidney tissue in rats with chronic renal insufficiency induced by subtotal nephrectomy (Mathias et al., 1997). This decrease expression of the CaR may contribute to the associated reduction in urinary Ca^{2+} excretion that occurs in this setting, because of the inverse relationship between the activity of the CaR and the concomitant rate of urinary calcium excretion that is present in persons with inactivating mutations in the CaR (Hebert, 1997). That is, the rats develop tissue-selective resistance of the remaining kidney tissue to the normally calciuric action of extracellular calcium. Since, as described later, 1,25-dihydroxyvitamin D_3 [$1,25(OH)_2D_3$] is known to increase the CaR's level of expression in the kidney (Brown et al., 1996), the decrease in CaR expression with impaired renal function could result, in part, from the concomitant diminution in circulating $1,25(OH)_2D_3$ concentration that occurs universally during the progression of renal insufficiency (Bringhurst et al., 1998). Alternatively or in addition, the elevation in circulating PTH levels that also takes place with chronic renal failure could also participate in promoting the associated reduction in CaR gene expression (Mathias et al., 1997).

5. CaR-BASED THERAPEUTICS

Calcimimetics are a class of small organic molecules that have been shown to bind within the CaR's transmembrane domains (Nemeth et al., 1996, 1998b). They act as allosteric modifiers – activating the receptor in the presence, but not in the absence of extracellular calcium, thereby reducing the CaR's EC_{50} for activation by Ca_0^{2+}. They are useful, therefore, when the activity and/or expression of the receptor is reduced, e.g., by decreasing an elevated set-point of the parathyroid gland toward normal. Indeed, calcimimetics are currently undergoing clinical trials for the treatment of PHPT as well as for the severe SHPT encountered in patients being treated for end stage renal insufficiency by hemodialysis or peritoneal dialysis (Nemeth, 1996; Ott, 1998; Silverberg et al., 1996, 1997). Initial results indicate that they produce rapid, dose-dependent reductions in serum PTH followed temporally by decreases in serum calcium concentration with repeated administration, especially at higher doses. Calcimimetics have also been shown to inhibit parathyroid cellular proliferation in an experimental model of renal insufficiency induced in rats by partial nephrectomy – an action that should mitigate the progression of uremic secondary HPT. Activation of the CaR in the CTAL would be expected to increase urinary calcium excretion, and this action may explain, in part, an initial increase in urinary calcium levels following initial dosing of patients with PHPT with calcimimetics. An important contributory factor to this increase in urinary calcium excretion, however, may also be the calcimimetic-evoked reduction in PTH secretion. In fact, available data do not suggest that increased urinary calcium excretion will be a significant clinical problem during treatment with this class of agents. If effective for treating primary and secondary hyperparathyroidism, calcimimetics would represent the first effective medical form of therapy for the hyperparathyroid state. Indeed, in a single patient with parathyroid carcinoma refractory to other forms of treatment for severe and progressive hypercalcemia and hyperparathyroidism, the use of a calcimimetic promoted substantial biochemical improvement over a period of more than a year (Collins et al., 1998).

Studies are also underway attempting to develop rapidly acting CaR antagonists that would evoke a "pulse" in plasma PTH levels as a treatment for osteoporosis (Nemeth et al., 1998a). Available data indicate that when administered exogenously in a pulsatile manner (i.e., a once daily injection), PTH can exert an anabolic action on bone, particularly on the cancellous bone of the spine. Therefore, the use of a CaR antagonist could potentially provide an alternative means of producing a transient rise in circulating levels of PTH by blocking the tonic inhibitory action of Ca_0^{2+} on PTH secretion (basal PTH levels are only about 20–25% of the maximal PTH levels that can be achieved by acutely induced hypocalcemia). A kidney-specific CaR antagonist might also provide an effective means of treating patients with recurrent Ca^{2+}-containing renal stones owing to excessively high levels of urinary calcium excretion. In this setting, reducing the activity of the CaR in the CTAL would be expected to reduce urinary Ca^{2+} excretion by two-fold or more (e.g., analogous to the impact that the reduced number of normally functioning CaRs in the CTAL in FHH has on urinary Ca^{2+} excretion; that is, a kidney selective CaR antagonist would induce "FHH of the kidney" without promoting concomitant excessive PTH secretion. The use of a CaR antagonist might also prove very useful in symptomatic patients with activating mutations of the CaR, in whom attempts at correction of hypocalcemia by supplementation with calcium and vitamin D promote excessive hypercalciuria and attendant renal damage and/or stones. In this circumstance, "resetting" the calciostat upward might restore ostensibly "normal" Ca_0^{2+} homeostasis.

Finally, in addition to the potential implications of CaR-based therapeutics in parathyroid and renal disorders, it appears likely that the agents could also be useful a wider range of diseases of CaR-expressing tissues in which modulating the receptor's activity would have desirable consequences from a therapeutic perspective. For example, if the CaR were the mediator of the stimulatory effect of Ca_0^{2+} on osteoblast function (Yamaguchi et al., 1999), a "bone-specific" CaR activator could represent an effective anabolic agent for treating conditions with decreased skeletal mass, especially post-menopausal and other forms of osteoporosis. As the CaR's roles in the numerous tissues in which it is expressed are better defined, CaR-based therapeutics may prove applicable to human diseases other than those perturbing systemic Ca_0^{2+} homeostasis.

6. SUMMARY AND CONCLUSIONS

The cloning of the CaR, in addition to shedding considerable light on the normal control of mineral ion homeostasis, has also shed light on the pathophysiology of Ca_0^{2+} metabolism. It has permitted the identification of inherited conditions in which the responsiveness of CaR-expressing cells involved in maintaining the normal, near constancy of Ca_0^{2+} is either reduced or increased. In the case of inactivating mutations in the CaR, the loss of approximately half of the normal complement of CaRs through the inactivation of one allele of the receptor in FHH produces a state of mild-to-moderate resistance of CaR-expressing tissues to Ca_0^{2+} with resultant mid hypercalcemia. Because the homeostatic system is reset to maintain a higher than normal level of Ca_0^{2+}, it is likely that the level of the CaR's activity in individual tissues regulated by this receptor is relatively normal. In contrast, with loss of both alleles of the receptor in NSHPT caused by homozygous or compound heterozygous FHH, there is severe Ca_0^{2+}-resistance, which is some cases is incompatible with life unless the hypercalcemia is corrected by total parathyroidectomy. Conversely, oversensitivity of the CaR to Ca_0^{2+} in autosomal dominant hypocalcemia or persons with sporadic hypoparathyroidism resets the "calciostat" downward, with resultant hypocalcemia and a tendency toward excessive rates of urinary calcium excretion (hypercalciuric hypocalcemia).

In addition to inherited abnormalities in Ca_0^{2+}-sensing, there is acquired resistance of the parathyroid gland(s) to Ca_0^{2+} in primary and some cases of severe secondary hyperparathyroidism (SHPT) encountered in patients with chronic renal insufficiency. In this setting a reduced level of expression of an apparently structurally normal CaR – of unknown etiology – likely contributes, at least in part, to the "tissue-selective" resistance of the abnormal parathyroid gland to Ca_0^{2+}. Calcimimetic CaR activators, which increase the sensitivity of the CaR to Ca_0^{2+} through an allosteric mechanism will likely provide an effective form of medical treatment for the Ca_0^{2+}-resistance of the parathyroid in PHPT and SHPT. Further studies will reveal the extent to which activating antibodies directed at the CaR's extracellular domain are a clinical significant cause of hypoparathyroidism in the population of patients with autoimmune hypoparathyroidism. A functional defect arising from this mechanism – in contrast to a permanent ablation of parathyroid function owing to parathyroid cellular destruction mediated by immune mechanisms – could be amenable to treatment with a CaR antagonist, which would reset the calciostat upward, thereby mitigating or even completely normalizing the hypocalcemia. It remains to be determined whether there are cases of "autoimmune FHH" caused by anti-CaR antibodies that reduce the CaR's activity. Finally, tissue-specific defects in the level of expression and/or function of the CaR in tissues other than the parathyroid gland, e.g., in the kidney, could lead to states in which the deranged function of the receptor in that tissue has pathophysiological consequences (i.e., hypercalciuria). The application of CaR-based therapeutics could, in turn, provide novel therapeutic approaches for clinical conditions for which there currently are no truly effective medical therapies.

ACKNOWLEDGEMENTS

The authors gratefully acknowledge generous grant support from the National Institutes of Health (DK41415, DK48330 and DK52005 to E.M.B.), NPS Pharmaceuticals, Inc., the National Space Bioscience Research Institute (NSBRI), and the St. Giles Foundation.

REFERENCES

Aida, K., Koishi, S., Inoue, M., Nakazato, M., Tawata, M. and Onaya, T., 1995, Familial hypocalciuric hypercalcemia associated with mutation in the human $Ca^{(2+)}$-sensing receptor gene, *J. Clin. Endocrinol. Metab.* 80, 2594–2598.

Attie, M.F., Gill, Jr., J., Stock, J.L., Spiegel, A.M., Downs, Jr., R.W., Levine, M.A. and Marx, S.J., 1983, Urinary calcium excretion in familial hypocalciuric hypercalcemia. Persistence of relative hypocalciuria after induction of hypoparathyroidism, *J. Clin. Invest.* 72, 667–676.

Auwerx, J., Demedts, M. and Bouillon, R., 1984, Altered parathyroid set point to calcium in familial hypocalciuric hypercalcaemia, *Acta Endocrinologica (Copenh.)* 106, 215–218.

Bai, M., Janicic, N., Trivedi, S., Quinn, S.J., Cole, D.E., Brown, E.M. and Hendy, G.N., 1997a, Markedly reduced activity of mutant calcium-sensing receptor with an inserted Alu element from a kindred with familial hypocalciuric hypercalcemia and neonatal severe hyperparathyroidism, *J. Clin. Invest.* 99, 1917–1925.

Bai, M., Pearce, S.H., Kifor, O., Trivedi, S., Stauffer, U.G., Thakker, R.V., Brown, E.M. and Steinmann, B., 1997b, In vivo and in vitro characterization of neonatal hyperparathyroidism resulting from a de novo, heterozygous mutation in the Ca^{2+}-sensing receptor gene: Normal maternal calcium homeostasis as a cause of secondary hyperparathyroidism in familial benign hypocalciuric hypercalcemia, *J. Clin. Invest.* 99, 88–96.

Bai, M., Quinn, S., Trivedi, S., Kifor, O., Pearce, S.H.S., Pollak, M.R., Krapcho, K., Hebert, S.C. and Brown, E.M., 1996, Expression and characterization of inactivating and activating mutations in the human Ca_0^{2+}-sensing receptor, *J. Biol. Chem.* 271, 19537–19545.

Bai, M., Trivedi, S. and Brown, E.M., 1998, Dimerization of the extracellular calcium-sensing receptor (CaR) on the cell surface of CaR-transfected HEK293 cells, *J. Biol. Chem.* 273, 23605–23610.

Baron, J., Winer, K.K., Yanovski, J.A., Cunningham, A.W., Laue, L., Zimmerman, D. and Cutler, G.B., Jr., 1996, Mutations in the $Ca^{(2+)}$-sensing receptor gene cause autosomal dominant and sporadic hypoparathyroidism, *Hum. Mol. Genet.* 5, 601–606.

Blizzard, R.M., Chee, D. and Davis, W., 1966, The incidence of parathyroid and other antibodies in the sera of patients with idiopathic hypoparathyroidism, *Clin. Exp. Immunol.* 1, 119–128.

Bringhurst, F.R., Demay, M.B. and Kronenberg, H.M., 1998, Hormones and disorders of mineral metabolism, in *Williams Textbook of Endocrinology*, J.D. Wilson, D.W. Foster, H.M. Kronenberg and P.R. Larsen (eds.), 9th edn., W.B. Saunders, Philadelphia, pp. 1155–1209.

Brown, A.J., Zhong, M., Finch, J., Ritter, C., McCracken, R., Morrissey, J. and Slatopolsky, E., 1996, Rat calcium-sensing receptor is regulated by vitamin D but not by calcium, *Am. J. Physiol.* 270, F454–460.

Brown, E.M., 1991, Extracellular Ca^{2+} sensing, regulation of parathyroid cell function, and role of Ca^{2+} and other ions as extracellular (first) messengers, *Physiol. Rev.* 71, 371–411.

Brown E.M., 1983, Four parameter model of the sigmoidal relationship between parathyroid hormone release and extracellular calcium concentration in normal and abnormal parathyroid tissue, *J. Clin. Endocrinol. Metab.* 56, 572–581.

Brown, E.M., 1999, Physiology and pathophysiology of the extracellular calcium-sensing receptor, *Am. J. Med.* 106, 238–253.

Brown, E.M., Vassilev, P.M., Quinn, S. and Hebert, S.C., 1999, G-protein-coupled, extracellular $Ca^{(2+)}$-sensing receptor: A versatile regulator of diverse cellular functions, *Vitam. Horm.* 55, 1–71.

Carling, T., Szabo, E., Bai, M., Westin, G., Gustavsson, P., Trivedi, S., Hellman, P., Brown, E.M., Dahl, N. and Rastad, J., 2000, Autosomal dominant mild hyperparathyroidism. A novel hypercalcemic disorder caused by a mutation in the cytoplasmic tail of the calcium receptor, *J. Clin. Endocrinol. Metab.* 85, 2042–2047.

Chang, W., Tu, C., Chen, T.-H., Komuves, L., Oda, Y., Pratt, S., Miller, S. and Shoback, D., 1999, Expression and signal transduction of calcium-sensing receptors in cartilage and bone, *Endocrinology* 140, 5883–5893.

Chattopadhyay, N., Mithal, A. and Brown, E.M., 1996, The calcium-sensing receptor: A window into the physiology and pathophysiology of mineral ion metabolism, *Endocr. Rev.* 17, 289–307.

Chou, Y.H., Brown, E.M., Levi, T., Crowe, G., Atkinson, A.B., Arnqvist, H.J., Toss, G., Fuleihan, G.E., Seidman, J.G. and Seidman, C.E., 1992, The gene responsible for familial hypocalciuric hypercalcemia maps to chromosome 3q in four unrelated families, *Nat. Genet.* 1, 295–300.

Chou, Y.H., Pollak, M.R., Brandi, M.L., Toss, G., Arnqvist, H., Atkinson, A.B., Papapoulos, S.E., Marx, S., Brown, E.M., Seidman, J.G., et al., 1995, Mutations in the human $Ca^{(2+)}$-sensing-receptor gene that cause familial hypocalciuric hypercalcemia, *Am. J. Hum. Genet.* 56, 1075–1079.

Collins, M.T., Skarulis, M.C., Bilezikian, J.P., Silverberg, S.J., Spiegel, A.M. and Marx, S.J., 1998, Treatment of hypercalcemia secondary to parathyroid carcinoma with a novel calcimimetic agent, *J. Clin. Endocrinol. Metab.* 83, 1083–1088.

Cooper, L., Wertheimer, J., Levey, R., Brown, E., Leboff, M., Wilkinson, R. and Anast, C.S., 1986, Severe primary hyperparathyroidism in a neonate with two hypercalcemic parents: Management with parathyroidectomy and heterotopic auto transplantation, *Pediatrics* 78, 263–268.

Davies, M., Adams, P.H., Lumb, G.A., Berry, J.L. and Loveridge, N., 1984, Familial hypocalciuric hypercalcaemia: Evidence for continued enhanced renal tubular reabsorption of calcium following total parathyroidectomy, *Acta Endocrinol (Copenh.)* 106, 499–504.

De Luca, F. and Baron, J. 1998, Molecular biology and clinical importance of the $Ca^{(2+)}$-sensing receptor, *Curr. Opin. Pediatr.* 10, 435–440.

De Luca, F., Ray, K., Mancilla, E.E., Fan, G.F., Winer, K.K., Gore, P., Spiegel, A.M. and Baron, J., 1997, Sporadic hypoparathyroidism caused by de novo gain-of-function mutations of the $Ca^{(2+)}$-sensing receptor, *J. Clin. Endocrinol. Metab.* 82, 2710–2715.

De Rouffignac, C. and Quamme, G.A., 1994, Renal magnesium handling and its hormonal control, *Physiol. Rev.* 74, 305–322.

Eisenbarth, G.S. and Verge, C.F., 1998, Immunoendocrinopathy syndromes, in *Williams Textbook of Endocrinology*, J.D. Wilson, D.W. Foster, H.M. Kronenberg and P.R. Larsen (eds.), 9th edn., W.B. Saunders, Philadelphia, pp. 1651–1662.

Farnebo, F., Enberg, U., Grimelius, L., Backdahl, M., Schalling, M., Larsson, C. and Farnebo, L.O., 1997, Tumor-specific decreased expression of calcium sensing receptor messenger ribonucleic acid in sporadic primary hyperparathyroidism, *J. Clin. Endocrinol. Metab.* 82, 3481–3486.

Finegold, D.N., Armitage, M.M., Galiani, M., Matise, T.C., Pandian, M.R., Perry, Y.M., Deka, R. and Ferrell, R.E., 1994, Preliminary localization of a gene for autosomal dominant hypoparathyroidism to chromosome 3q13, *Pediatr. Res.* 36, 414–417.

Flier, J.S., Bar, R.S., Muggeo, M., Kahn, C.R., Roth, J. and Gorden, P., 1978, The evolving clinical course of patients with insulin receptor autoantibodies: Spontaneous remission or receptor proliferation with hypoglycemia, *J. Clin. Endocrinol. Metab.* 47, 985–995.

Freichel, M., Zink-Lorenz, A., Holloschi, A., Hafner, M., Flockerzi, V. and Raue, F., 1996, Expression of a calcium-sensing receptor in a human medullary thyroid carcinoma cell line and its contribution to calcitonin secretion, *Endocrinology* 137, 3842–3848.

Fukumoto, S., Chikatsu, N., Okazaki, R., Suzawa, M., Tamura, Y., Takeda, S., Takeuchi, Y., Obara, T. and Fujita, T., 1998, Parathyroid lipohyperplasia is caused by mutations in calcium-sensing receptor (CaSR), *Bone* 23, S283 (Abstract T346).

Garner, S.C., Hinson, T.K., McCarty, K.S., Leight, M., Leight, G.S., Jr. and Quarles, L.D., 1997, Quantitative analysis of the calcium-sensing receptor messenger RNA in parathyroid adenomas, *Surgery* 122, 1166–1175,

Garrett, J.E., Tamir, H., Kifor, O., Simin, R.T., Rogers, K.V., Mithal, A., Gagel, R.F. and Brown, E.M., 1995, Calcitonin-secreting cells of the thyroid express an extracellular calcium receptor gene, *Endocrinology* 136, 5202–5211.

Gogusev, J., Duchambon, P., Hory, B., Giovannini, M., Goureau, Y., Sarfati, E. and Drueke, T.B., 1997, Depressed expression of calcium receptor in parathyroid gland tissue of patients with hyperparathyroidism, *Kidney Int.* 51, 328–336.

Heath, D.A., 1994, Familial hypocalciuric hypercalcemia, in *The Parathyroids*, J.P. Bilezikian, R. Marcus and M.A. Levine (eds.), Raven Press, New York, pp. 699–710.

Heath, H., III, Odelberg, S., Jackson, C.E., Teh, B.T., Hayward, N., Larsson, C., Buist, N.R., Krapcho, K.J., Hung, B.C., Capuano, I.V., Garrett, J.E. and Leppert, M.F., 1996, Clustered inactivating mutations and benign polymorphisms of the calcium receptor gene in familial benign hypocalciuric hypercalcemia suggest receptor functional domains, *J. Clin. Endocrinol. Metab.* 81, 1312–1317.

Heath, H., III, Leppert, M.F., Lifton, R.P. and Penniston, J.T., 1992, Genetic linkage analysis in familial benign hypercalcemia using a candidate gene strategy. I. Studies in four families, *J. Clin. Endocrinol. Metab.* 75, 846–851.

Hebert, S.C., Brown, E.M. and Harris, H.W., 1997, Role of the Ca^{2+}-sensing receptor in divalent mineral ion homeostasis, *J. Exp. Biol.* 200, 295–302.

Ho, C., Conner, D.A., Pollak, M.R., Ladd, D.J., Kifor, O., Warren, H.B., Brown, E.M., Seidman, J.G. and Seidman, C.E., 1995, A mouse model of human familial hypocalciuric hypercalcemia and neonatal severe hyperparathyroidism, *Nat. Genet.* 11, 389–394.

Jameson, L., 1999, *Hormone Resistance Syndromes*, Humana Press, Towata, NJ.

Janicic, N., Pausova, Z., Cole, D.E. and Hendy, G.N., 1995, Insertion of an Alu sequence in the $Ca^{(2+)}$-sensing receptor gene in familial hypocalciuric hypercalcemia and neonatal severe hyperparathyroidism, *Am. J. Hum. Genet.* 56, 880–886.

Kaneko, C., Mizunashi, K., Tanaka, M., Uzuki, M., Kikuchi, M., Sawai, T., and Goto, M.M., 1999, Relationship between Ca^{2+}-dependent change of serum PTH and extracellular Ca^{2+}-sensing receptor expression in parathyroid adenoma, *Calcif. Tissue Int.* 64, 271–272.

Khosla, S., Ebeling, P.R., Firek, A.F., Burritt, M.M., Kao, P.C. and Heath III, H., 1993, Calcium infusion suggests a "set-point" abnormality of parathyroid gland function in familial benign hypercalcemia and more complex disturbances in primary hyperparathyroidism, *J. Clin. Endocrinol. Metab.* 76, 715–720.

Kifor, O., Moore, F.D., Jr., Wang, P., Goldstein, M., Vassilev, P., Kifor, I., Hebert, S.C. and Brown, E.M., 1996, Reduced immunostaining for the extracellular Ca^{2+}-sensing receptor in primary and uremic secondary hyperparathyroidism, *J. Clin. Endocrinol. Metab.* 81, 1598–1606.

Kobayashi, M., Tanaka, H., Tsuzuki, K., Tsuyuki, M., Igaki, H., Ichinose, Y., Aya, K., Nishioka, N. and Seino, Y., 1997, Two novel missense mutations in calcium-sensing receptor gene associated with neonatal severe hyperparathyroidism, *J. Clin. Endocrinol. Metab.* 82, 2716–2719.

Law, Jr., W.M. and Heath, III, H., 1985, Familial benign hypercalcemia (hypocalciuric hypercalcemia). Clinical and pathogenetic studies in 21 families, *Ann. Int. Med.* 105, 511–519.

Li, Y., Song, Y.H., Rais, N., Connor, E., Schatz, D., Muir, A. and Maclaren, N., 1996, Autoantibodies to the extracellular domain of the calcium sensing receptor in patients with acquired hypoparathyroidism, *J. Clin. Invest.* 97, 910–914.

Lienhardt, A., Bai, M., Garabedian, M., Sinding, C., Boulesteix, J., Rigaud, I.M., Brown, E.M. and Kottler, M.L., 2000, A large homozygous or heterozygous in-frame deletion within the calcium-sensing receptor's carboxyl-terminal cytoplasmic tail that causes autosomal dominant hypocalcemia, *J. Clin. Endocrinol. Metab.*, in press.

Lloyd, S.E., Parmett, A.A., Dixon, P.H., Whyte, M.P. and Thakker, R.V., 1999, Localization of familial benign hypercalcemia, Oklahoma vanant (FBHOk), to chromosome 19q13, *Am. J. Hum. Genet.* 64, 189–195.

Lovlie, R., Eiken, H.G., Sorheim, J.I. and Boman, H., 1996, The $Ca^{(2+)}$-sensing receptor gene (PCAR1) mutation T151M in isolated autosomal dominant hypoparathyroidism, *Hum. Genet.* 98, 129–133,

Mancilla, E.E., De Luca, F. and Baron, J. 1998, Activating mutations of the Ca^{2+}-sensing receptor, *Mol. Genet. Metab.* 64, 198–204.

Mancilla, E.E., De Luca, F., Ray, K., Winer, K.K., Fan, G.-F. and Baron, J., 1997, A Ca^{2+}-sensing receptor mutation causes hypoparathyroidism by increasing receptor sensitivity to Ca^{2+} and maximal signal transduction, *Pediatr. Res.* 42, 443–447.

Marx, S.J., Attie, M.F., Levine, M.A., Spiegel, A.M., Downs, Jr., R.W. and Lasker, R.D., 1981a, The hypocalciuric or benign variant of familial hypercalcemia: Clinical and biochemical features in fifteen kindreds, *Medicine (Baltimore)* 60, 397–412.

Marx, S.J., Attie, M.F., Stock, J.L., Spiegel, A.M. and Levine, M.A., 1981b, Maximal urine-concentrating ability: Familial hypocalciuric hypercalcemia versus typical primary hyperparathyroidism, *J. Clin. Endocrinol. Metab.* 52, 736–740.

Marx, S., Lasker, R., Brown, E., Fitzpatrick, L., Sweezey, N., Goldbloom, R., Gillis, D. and Cole, D., 1986, Secretory dysfunction in parathyroid cells from a neonate with severe primary hyperparathyroidism, *J. Clin. Endocrinol. Metab.* 62, 445–449.

Mathias, R., Nguyen, H., Zhang, M. and Portale, A., 1997, Expression of the renal calcium-sensing receptor is reduced in rats with experimental chronic renal insufficiency, *J Bone Miner. Res.* 12, S326 (Abstract F400).

Nemeth, E.F., 1996, Calcium receptors as novel drug targets, in *Principles of Bone Biology*, J.P. Bilezikian, L.G. Raisz and G.A. Rodan (eds.), Academic Press, San Diego, pp. 1019–1035.

Nemeth, E.F., Fox, J., Delmar, E.G., Steffey, M.E., Lambert, L.D., Conklin, R.L., Bhatnagar, P.K. and Gowen, M., 1998a, Stimulation of parathyroid hormone secretion by a small molecule antagonist of the calcium receptor, *J. Bone Miner. Res.* 23, S156 (Abstract #1030).

Nemeth, E.F., Steffey, M.E., Hammerland, L.G., Hung, B.C., Van Wagenen, B.C., DelMar, E.G. and Balandrin, M.F., 1998b, Calcimimetics with potent and selective activity on the parathyroid calcium receptor, *Proc. Natl. Acad. Sci. USA* 95, 4040–4045.

Neufeld, M., Maclaren, N. and Blizzard, R., 1980, Autoimmune polyglandular syndromes, *Pediatr. Ann.* 9, 154–162.

Ott, S., 1998, Editorial: Calcimimetics – New drugs with the potential to control hyperparathyroidism, *J. Clin. Endocrinol. Metab.* 83, 1080–1082.

Parma, J., Van Sande, J., Swillens, S., Tonacchera, M., Dumont, J. and Vassart, G., 1995, Somatic mutations causing constitutive activity of the thyrotropin receptor are the major cause of hyperfunctioning thyroid adenomas: Identification of additional mutations activating both the cyclic adenosine 3′,5′-monophosphate and inositol phosphate-Ca^{2+} cascades, *Mol. Endocrinol.* 9, 725–733.

Paschke, R., Van Sande, J., Parma, J. and Vassart, G. 1996, The TSH receptor and thyroid diseases, *Baillieres Clin. Endocrinol. Metab.* 10, 9–27.

Pearce, S.H., Trump, D., Wooding, C., Besser, G.M., Chew, S.L., Grant, D.B., Heath, D.A., Hughes, I.A., Paterson, C.R., Whyte, M.P. and Thakker, R.V., 1995, Calcium-sensing receptor mutations in familial benign hypercalcemia and neonatal hyperparathyroidism, *J. Clin. Invest.* 96, 2683–2692.

Pearce, S.H., Bai, M., Quinn, S.J., Kifor, O., Brown, E.M. and Thakker, R.V., 1996a, Functional characterization of calcium-sensing receptor mutations expressed in human embryonic kidney cells, *J. Clin. Invest.* 98, 1860–1866.

Pearce, S.H., Williamson, C., Kifor, O., Bai, M., Coulthard, M.G., Davies, M., Lewis-Bamed, N., McCredie, D., Powell, H., Kendall-Taylor, P., Brown, E.M. and Thakker, R.V., 1996b, A familial syndrome of hypocalcemia with hypercalciuria due to mutations in the calcium-sensing receptor, *New Engl. J. Med.* 335, 1115–1122.

Perry, Y.M., Finegold, D.M., Annitage, M.M. and Ferrell, R.E., 1994, A missense mutation in the Ca-sensing receptor causes familial autosomal dominant hypoparathyroidism, *Am. J. Human Genet.* 55 (Suppl.), A17 (Abstract).

Pollak, M.R., Brown, E.M., Chou, Y.H., Hebert, S.C., Marx, S.J., Steinmann, B., Levi, T., Seidman, C.E. and Seidman, J.G., 1993, Mutations in the human $Ca^{(2+)}$-sensing receptor gene cause familial hypocalciuric hypercalcemia and neonatal severe hyperparathyroidism, *Cell* 75, 1297–1303.

Pollak, M.R., Brown, E.M., Estep, H.L., McLaine, P.N., Kifor, O., Park, J., Hebert, S.C., Seidman, C.E. and Seidman, J.G., 1994a, Autosomal dominant hypocalcaemia caused by a $Ca^{(2+)}$-sensing receptor gene mutation, *Nat. Genet.* 8, 303–307.

Pollak, M.R., Chou, Y.H., Marx, S.J., Steinmann, B., Cole, D.E., Brandi, M.L., Papapoulos, S.E., Menko, F.H., Hendy, G.N., Brown, E.M., et al., 1994b, Familial hypocalciuric hypercalcemia and neonatal severe hyperparathyroidism. Effects of mutant gene dosage on phenotype, *J. Clin. Invest.* 93, 1108–1112.

Quarles, L.D., 1997, Cation-sensing receptors in bone: A novel paradigm for regulating bone remodeling?, *J. Bone Miner. Res.* 12, 1971–1974.

Refetoff, S., 1982, Syndromes of thyroid hormone resistance, *Am. J. Physiol.* 243, E88–E98.

Riccardi, D., Hall, A.E., Chattopadhyay, N., Xu, J.Z., Brown, E.M. and Hebert, S.C., 1998, Localization of the extracellular Ca^{2+}/polyvalent cation-sensing protein in rat kidney, *Am. J. Physiol.* 274, F611–622.

Schipani, E., Kruse, K. and Juppner, H., 1995, A constitutively active mutant PTH-PTHrP receptor in Jansen-type metaphyseal chondrodysplasia, *Science* 268, 98–100.

Silverberg, S.J., Bone, III, H.G., Marriott, T.B., Locker, F.G., Thys-Jacobs, S., Dziern, G., Kaatz, S., Sanguinetti, E.L. and Bilezikian, J.P., 1997, Short-term inhibition of parathyroid hormone secretion by a calcium-receptor agonist in patients with primary hyperparathyroidism, *New Engl. J. Med.* 337, 1506–1510.

Silverberg, S.J., Thys-Jacobs, S., Locker, F.G., Sanguinetti, E.L., Marriott, T.B. and Bilezikian, J.P., 1996, The effects of the calcimimetic drug NPS R-568 on parathyroid hormone secretion in primary hyperparathyroidism, *J. Bone Miner. Res.* 11 (Suppl. 1), S116 (Abstract 187).

Soei, Y.L., Karperien, M., Bakker, B., Breuning, M.B., Hendy, G.N. and Papapoulos, S.E., 1999, Familial benign hypercalcemia (FBH) with age-associated hypercalciuria and a missense mutation in the calcium-sensing receptor (CaSR) expands the spectrum of the syndrome toward primary hyperparathyroidism, *J. Bone Mineral Res.* 14, S447 (abstract SU062).

Spiegel, A.M., 1996, Defects in G protein-coupled signal transduction in human disease, *Annu. Rev. Physiol.* 58, 143–170.

Spiegel, A.M., 1996, Mutations in G protein and G protein-coupled receptors in endocrine disease, *J. Clin. Endocrinol. Metab.* 81, 2434–2442.

Strewler, G.J., 1994, Familial benign hypocalciuric hypercalcemia – From the clinic to the calcium sensor, *West. J. Med.* 160, 579–580.

Ward, B.K., Stuckey, B.G., Gutteridge, D.H., Laing, N.G., Pullan, P.T. and Ratajczak, T., 1997, A novel mutation (L174R) in the Ca^{2+}-sensing receptor gene associated with familial hypocalciuric hypercalcemia, *Hum. Mutat.* 10, 233–235.

Ward, D.T., Brown, E.M. and Harris, H.W., 1998, Disulfide bonds in the extracellular calcium-polyvalent cation-sensing receptor correlate with dimer formation and its response to divalent cations in vitro, *J. Biol. Chem.* 273, 14476–14483.

Weisinger, J.R., Favus, M.J., Langman, C.B. and Bushinsky, D., 1989, Regulation of 1,25-dihydroxyvitamin D_3 by calcium in the parathyroidectomized, parathyroid hormone-replete rat, *J. Bone Miner. Res.* 4, 929–935.

Yamaguchi, T., Chattopadhyay, N. and Brown, E.M., 1999, G protein-coupled extracellular Ca^{2+} (Ca_0^{2+})-sensing receptor (CaR): Roles in cell signaling and control of diverse cellular functions, *Adv. Pharmacol.* 47, 209–253.

Zaidi, M., Adebanjo, O.A., Moonga, B.S., Sun, L. and Huang, C.L., 1999, Emerging insights into the role of calcium ions in osteoclast regulation, *J. Bone Miner. Res.* 14, 669–674.

Calcium and Muscle Disease: Pathophysiology of Calpains and Limb-Girdle Muscular Dystrophy Type 2A (LGMD2A)

Yasuko Ono, Shoji Hata, Hiroyuki Sorimachi and Koichi Suzuki

1. INTRODUCTION

Precise regulation of numerous cellular events is essential for cell function and survival. Proteolysis in the cytosol is one of the important systems that modulate cellular functions internally. In muscle tissues, the involvement of several classes of proteases in various cellular events has been studied extensively, including their roles in myofibrillar protein turnover, myoblast fusion, cell cycle progression, and so on. Calpain, an intracellular Ca^{2+}-dependent cysteine protease, is considered one of the most significant proteases in muscle tissues in both normal and pathological conditions. Muscular dystrophy is a group of neuromuscular disorders that present with necrosis and irregularly sized muscle fibers. Excessive degradation of myofibrillar proteins is responsible for this pathological change. Since degenerative muscle is often accompanied by an elevated cytosolic Ca^{2+} concentration, it is almost certain that proteolysis by calpain initiates the fiber necrosis. Therefore, the study of calpain in muscular dystrophy has aimed principally at developing means of suppressing calpain activity. Recent studies have revealed another aspect of the relationship between calpain and muscular dystrophy. A mutation in p94, also called calpain 3, is responsible for limb-girdle muscular dystrophy type

Yasuko Ono, Shoji Hata, Hiroyuki Sorimachi and Koichi Suzuki • Laboratory of Molecular Structure and Function, Department of Molecular Biology, Institute of Molecular and Cellular Biosciences, University of Tokyo, Tokyo, Japan. **Hiroyuki Sorimachi** • Department of Applied Biological Chemistry, Graduate School of Agricultural and Life Sciences, University of Tokyo, Tokyo, Japan.

R. Pochet, R. Donato, J. Haiech, C. Heizmann and V. Gerke (eds.): Calcium: The Molecular Basis of Calcium Action in Biology and Medicine, 443–464.

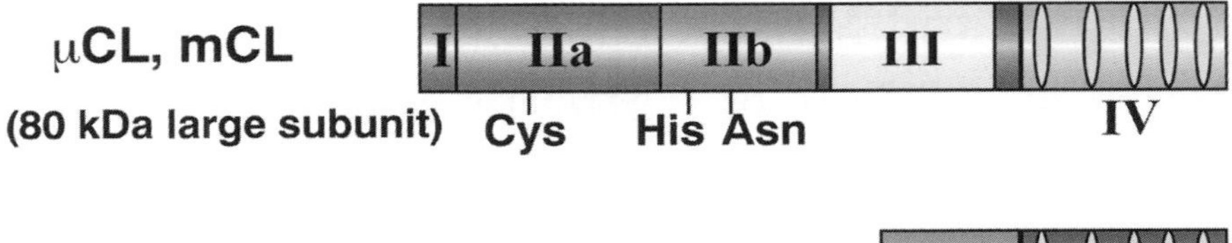

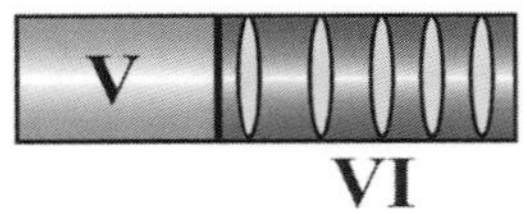

Figure 1. Schematic representation of the molecular structure of calpain. Conventional calpains, μ-calpain and m-calpain, are composed of a large catalytic subunit (μCL or mCL) and a small subunit (30K). Domain II is a cysteine protease domain. IIa and IIb correspond to the crystal structure shown in Figure 2. Cys, His, and Asn are catalytic residues. Domains IV and VI are Ca^{2+}-binding domains containing five EF-hand motifs. For a colour version of this figure, see page xxvii.

2A (LGMD2A). This provides definitive evidence that the calpain system is vital for normal muscle function. Moreover, it suggests for the first time that the activity of one calpain species might oppose the effect of another. In this chapter, we briefly review the molecular structure of calpain and focus on the uncertain, but likely, pathophysiological roles of calpain in muscle diseases based on current knowledge.

2. OVERVIEW OF THE MOLECULAR STRUCTURE AND DIVERSITY OF CALPAIN

Calpain (EC 3.4.22.17; Family C02: clan CA) is an intracellular Ca^{2+}-dependent protease, representing one of the most important groups of peptidases, the cysteine peptidase (Murachi et al., 1981; Carafoli and Molinari, 1998; Suzuki and Sorimachi, 1998). Calpain plays indispensable roles in several physiological processes by modulating the activity, structure, and localization of its target proteins, including protein kinase, transcriptional factors, cytoskeletal proteins, and muscle proteins (Sorimachi et al., 1997). In mammals, two groups of isozymes, the μ- and m-calpains, have been thoroughly studied. In the following paragraphs, the term "calpain" is used to represent both isozymes, unless otherwise specified. Their characteristics can be summarized as follows:

(i) Subunit composition and structure: Both μ- and m-calpains are heterodimers composed of distinct 80 kDa large subunits and a common 30 kDa small subunit (Ishiura et al., 1978) (Figure 1). The large subunit (abbreviated μCL and mCL, respectively) and small subunit (abbreviated 30K) are divided into four (I–IV) and two (V–VI) domains, respectively (Ohno et al., 1984; Sakihama et al., 1985). Domain II is a protease domain that includes the catalytic triad residues. Domains IV and VI are homologous Ca^{2+}-binding domains that contain five EF-hand motifs (EF-1-5). The crystal structure of domain VI revealed Ca^{2+}-binding to EF-1-4, but not to EF-5, and homodimerization as a result of the interaction between the EF-5 of each molecule in the presence and absence of Ca^{2+} (Blanchard et al., 1997; Lin et al., 1997). More recently, the overall structure of m-calpain in the absence of Ca^{2+} was resolved, proving that heterodimerization of mCL and 30K was EF-5 dependent (Hosfield et al., 1999; Strobl et al., 2000) (Figure 2). In addition, the N-terminal alpha-helical region of domain I contacts domain VI in several places, suggesting involvement of domain I in heterodimer formation. Interestingly, the folding of domain III is very similar to that of the C2-domain, a Ca^{2+} and phospholipid-binding module, in spite of the lack of sequence similarity. Therefore, at least three domains of calpain seem capable of binding Ca^{2+}.

(ii) Tissue distribution: Both isozymes are expressed in most mammalian tissues. The relative proportions vary from tissue to tissue. For example, μ-calpain predominates in erythrocytes, whereas m-calpain is more abundant in cardiac muscle.

(iii) Activation and Regulation: The Ca^{2+}-concentrations required for *in vitro* proteolytic activity of μ- and m-calpain are ca. 10^{-5} and 10^{-3} M, respectively. The Ca^{2+} sensitivity of calpain is increased by autolysis, the addition of phospholipids or activator proteins (although this is not reproducible), or subunit dissociation (Suzuki et al., 1981; Saido et al., 1992; Yoshizawa et al., 1995). The crystal structure suggests a novel activation mechanism. In the absence of Ca^{2+}, a catalytic triad of Cys, His, and Asn is not assembled. With an elevated cellular Ca^{2+} concentration, Ca^{2+} binding to domains IV and VI (and probably domain III) causes formation of the triad, and activation (Strobl et al., 2000). In mammalian cells, calpastatin, a calpain-specific inhibitor protein, is also expressed ubiquitously and regulates calpain activity (Takahashi-Nakamura et al., 1981).

With the recent discovery of various calpain large subunit homologues, it has become obvious that calpains comprise an independent superfamily among cysteine proteases called the "calpain superfamily" (Sorimachi et al., 1997) (Table 1). Another important finding is that the five-EF hand (FEF) motif structure found in domains IV and VI (thus, they are called FEF domains) form a distinct subfamily of EF-hand proteins together with other proteins with FEF domains. The reviews of Sorimachi et al. (1997) and Kawasaki et al. (1998) contain details on the FEF family of proteins.

The vertebrate calpain homologues can be divided into two groups depending on the mode of expression. They are ubiquitous species, such as the μ- and m-calpains, and tissue-specific species, of which p94 is a prominent example. It is generally assumed that tissue-specific species are indispensable for the function of the corresponding tissue. This notion was recently proved with the discovery that a mutation in the p94 gene causes LGMD2A (Richard et al., 1995).

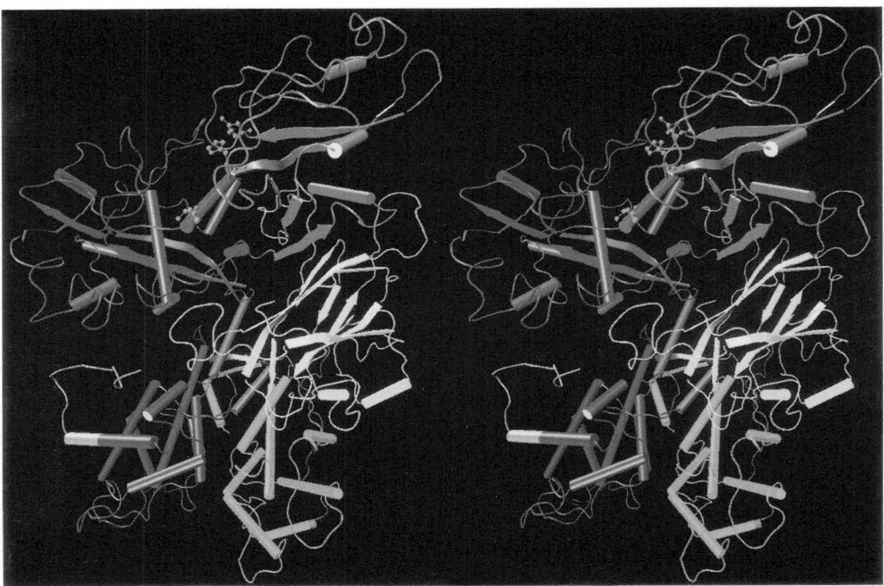

Figure 2. Ribbon structure of human m-calpain in the absence of Ca^{2+}. The colour of each domain corresponds to that in Figure 1. The cylinders represent helical structures. The N-terminus of mCL resides in the center of the molecule. For a colour version of this figure, see page xxvii.

Table 1. Calpain and related molecules

Calpain superfamily			**FEF protein family**	
Calpain homologues			**Vertebrate**	
Vertebrate			30K	
μCL, mCL, μ/mCL		(ubiquitous expression)	Calpain homologues	
p94, Lp82, nCL-2, nCL-4		(tissue-specific expression)	ALG-2	
Other			Grancalcin	
Dm	Dm-calpain (CALPA), CALPB		Sorcin	
Sm	Sm-calpain		**Other**	
Sj	Sj-calpain		*Sc*	FEF1p(YGR058w)
Calpain-like proteases				
Vertebrate				
nCL-2'[1]				
hTRA-3(CAPN5), CANPX(CAPN6)[2]				
SOLH, PalBH				
Other				
Dm	CALPA'[1], SOL			
Ce	TRA-3, Ce-pXXXs[3]			
En	PalB			
Ao	$PalB^{ory}$			
Sc	Cpl1p			

Dm: *Drosophila melanogaster*, *Sm*: *Schistosoma mansoni*, *Sj*: *Schistosoma japonica*, *Ce*: *Caenorhabditis elegans*, *En*: *Emericella nidulans*, *Ao*: *Aspergillus oryzae*, *Sc*: *Saccharomyces cerevisiae*.

[1]nCL-2′ and CALPA′ are alternative splice variant of nCL-2 and CALPA, respectively. [2]Active site residues are substituted. [3]XXX represents molecular weight estimated from the genome sequences. Thirteen species have been identified to date.

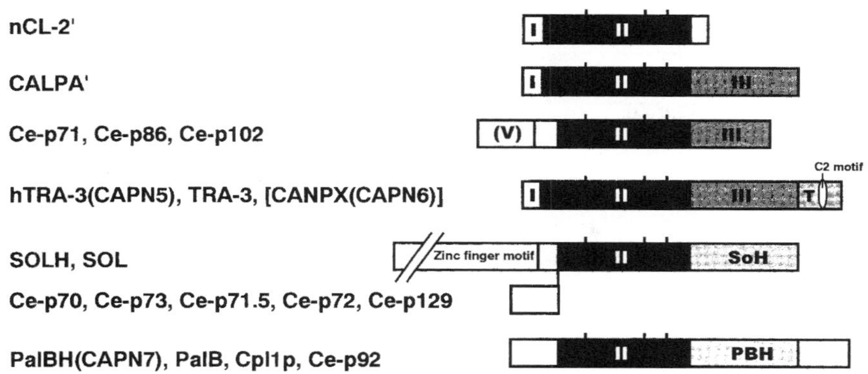

Figure 3. Domain structures of calpain-like proteases. The calpain-like proteases presented in Table 1 are categorized according to structural characteristics. In the CANPX(CAPN6), active site, Cys and/or His is substituted.

It is intriguing to categorize all the homologues by domain structure. We designate homologues that share similarity only to domain II of μCL and mCL as "calpain-like proteases" (Sorimachi et al., 1997). One of the interesting features of these molecules is that they can be further classified by the specific C-terminal structures conserved between mammalian and non-vertebrate orthologues, such as domains T, PBH, and SoH (Delaney et al, 1991; Denison et al., 1995; Barnes and Hodgkin, 1996; Mugita et al., 1997; Dear et al., 1997; Kamei et al., 1998; Futai et al., 1999; Franz et al., 1999) (Figure 3). It is now speculated that they are regulated differently from the species containing an FEF structure.

Since the major topic of this chapter is the pathophysiological roles of calpain in muscle diseases, the characteristics of p94, or calpain 3, will be described in more detail. For details on other members of the calpain superfamily, please refer to the reviews of Sorimachi et al. (1997) and Ono et al. (1998a) and the papers they cite.

In 1989, p94, a skeletal muscle-specific calpain homologue, became the first tissue-specific calpain identified (Sorimachi et al., 1989). It is expressed predominantly in skeletal muscle, where its mRNA level is 10 times higher than those of μCL and mCL. Little is expressed in other tissues. One of the features of p94 is that it contains three unique insertions (NS, IS1, and IS2), while it is still very similar to μCL and mCL (Figure 4). Recent studies have revealed the characteristics of p94 (Kinbara et al., 1997; Sorimachi et al., 1993, 1995; Richard et al., 1995; Ma et al., 1997).

(i) p94 autolyzes very rapidly (its half-life is less than 10 min *in vitro*) and this activity requires both IS1 and IS2.

(ii) Inhibitors effective in stopping the protease activity of calpain, such as E-64, leupeptin, and EDTA, do not inhibit the autolysis of p94. Calpastatin is proteolyzed by p94 and therefore is not effective.

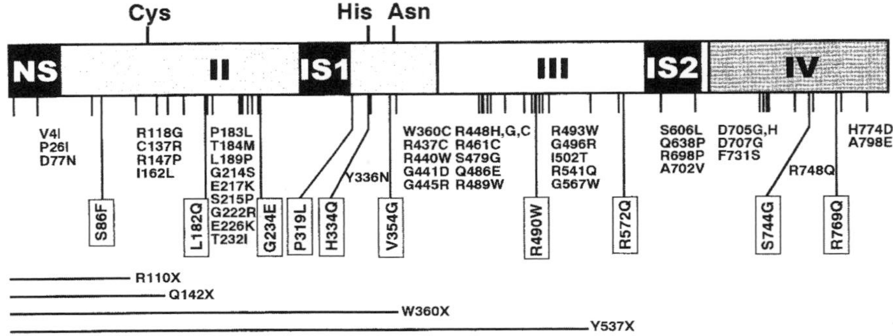

Figure 4. Structure of p94 and the positions of LGMD2A mutations. The locations of missense and nonsense mutations are indicated. The boxed mutants are characterized in Table 3.

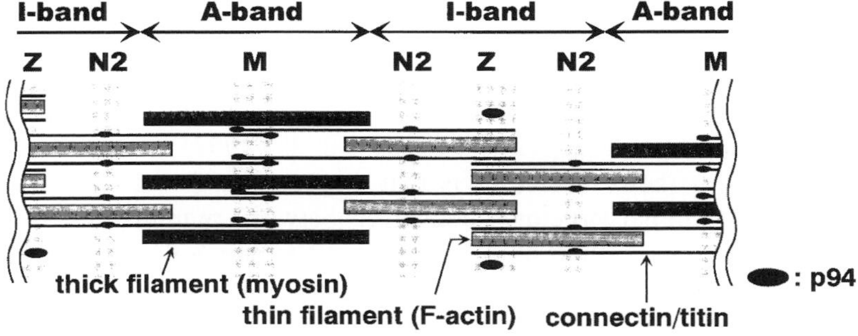

Figure 5. Localization of p94 in myofibril. A skeletal muscle sarcomere is represented schematically. Circles correspond to the location demonstrated by both the yeast two-hybrid system and immunofluorescence. Z, N2, and M mean Z-, N_2-, and M-lines, respectively.

(iii) p94 localizes both in the nucleus and cytosol when expressed in cultured cells.

(iv) The protease activity of p94 is apparently Ca^{2+}-independent, or EDTA and EGTA-resistant, even though the FEF structure in domain IV is very similar to those of μCL and mCL.

(v) p94 specifically binds to connectin/titin, a gigantic filamentous protein spanning the M- to Z-lines of the muscle sarcomere, but not to 30K. Connectin/titin binds to p94 at the N2A sequences in the N_2-line region and Mis7 in the C-terminal region.

(vi) In immunostained myofibril, p94 is located along the N_2- and Z-lines of the sarcomere (Figure 5).

(vii) p94 was identified as the gene product responsible for LGMD2A by positional cloning. (A detailed description of the roles of p94 in LGMD2A is given in the following sections.)

(viii) Splice variants of p94 are expressed specifically in the rat, mouse, pig, and cow, and rabbit lens.

Table 2. Genes and/or loci responsible for muscular dystrophies

Type	Locus (D/R[1])	Gene product	Note
DMD/BMD	Xp21.1 (R)	dystrophin	model animal: mdx mouse
CMD	6q22-23 (R)	laminin a2	model animal: dy mouse
FCMD	9q31 (R)	fukutin	secretory protein
EDMD	Xq28 (R)	emerin	nuclear membrane
AD-EDMD	1q11-23 (D)	lamin A/C	nuclear lamina
TMD	2q31 (D)	n.d. (titin?)	spanning a half-sarcomere
DM	19q13.2-13.3 (D)	DMPK	binding to MKBP[2]/HSPB2
DM-2	3q (D)	n.d.	
LGMD1A	5q31 (D)	n.d.	
LGMD1B	1q11-21 (D)	n.d.	
LGMD1C	3p25 (D)	caveolin-3	also responsible for MM[3]
LGMD2A	15q15.1(R)	p94	DG and SG complexes remain
LGMD2B	2p1 3(R)	dysferlin	similar to Ce FERp[4]
LGMD2C	13q12 (R)	γ-SG[5]	a component of SG-complex
LGMD2D	17q12-21 (R)	α-SG	a component of SG-complex
LGMD2E	4q12 (R)	β-SG	a component of 1f SG-complex
LGMD2F	5q33 (R)	d-SG	a component of SG-complex
LGMD2G	17q11-12 (R)	T-cap (telethonin)	binding to the N-terminus of titin
LGMD2H	9q31-33 (R)	n.d.	

[1]D: dominant, R: recessive; [2]Myotonic dystrophy protein kinase binding protein; [3]Miyoshi myopathy; [4]*Caenorhabditis elegans* spermatogenesis factor protein; [5]Sarcoglycan

Recent progress in the pathological studies of calpain involvement in muscle tissues, including the role of p94 in LGMD2A, is reviewed in the following sections.

3. CALPAIN AND MUSCLE DISEASE

Initially, studies on the relationship between calpain and muscle disease focused on the contribution of the protease activity of calpain to the increased protein degradation observed in muscular dystrophy. Muscular dystrophy is a group of neuromuscular disorders characterized by necrosis and degeneration of muscle fibers (Bonnemann et al., 1996; Emery, 1998; Ozawa et al., 1998; Bushby, 1999; Toniolo and Minetti, 1999). Patients suffer from progressive muscle wasting and weakness, which is occasionally lethal. Various types of muscular dystrophy have been defined genetically (Table 2).

The role of the proteolytic system in the pathology of muscular dystrophy was first investigated in Duchenne-type muscular dystrophy (DMD) (Rabbani et al., 1984; Sugita et al., 1980). DMD is one of the most disabilitating muscular dystrophies, with early onset and rapid progression of muscle atrophy. The gene product responsible for DMD is dystrophin. Dystrophin is a major component of the dystrophin-glycoprotein complex in skeletal muscle fibers, and is found on the cytoplasmic side of the sarcolemma. Dystrophin disappears in DMD muscle and decreases in Becker-type muscular dystrophy (BMD), which has milder symptoms (Hoffman et al., 1987).

In the course of searching for a molecular explanation for the pathology of DMD, an elevated intracellular Ca^{2+} concentration and the consequent hyper-degradation of muscle proteins by calpain appeared crucial for muscle fiber necrosis (Turner et al., 1988). Actually, cytoskeletal/structural proteins represent a group of good calpain substrates (Goll et al., 1999). Identification of an essential enzyme in the disease condition suggested that calpain activity in necrotic fibers might be a diagnostic marker, or a target for therapy. Subsequently, great effort has been made to control calpain activity, using cell-penetrating calpain-specific inhibitors, although clinically useful means have not yet been developed.

The mechanism of excessive calpain activation is now explained in the following way. A defect in dystrophin causes complete or partial disappearance of dystrophin-associated complex and destabilizes the membrane structure, rendering dystrophic muscle more susceptible to stress (Menke and Joskusch, 1991). The subsequent elevation of the cytosolic Ca^{2+} concentration triggers muscle contraction and the activation of calpain. Then, muscle cells are damaged from the inside as calpain proteolyzes various muscle proteins, such as alpha-actinin, myosin, connectin/titin, etc. The increased probability of Ca^{2+}-leak channels and/or stretch-sensitive Ca^{2+}-channels is considered the major cause of elevated Ca^{2+} influx (Fong et al., 1990; Hopf et al., 1996; Imbert et al., 1996). It has also been shown that proteolysis by calpain increases the activity of Ca^{2+}-leak channels (Turner et al., 1993). In addition, nitric oxide synthase (NOS), which opposes contraction, is a possible calpain substrate and disappears from DMD muscle (Brenman et al., 1995). Therefore, it is suspected that positive feedback accelerates muscle fiber necrosis.

This model describes the process occurring in DMD; however, the irregular activation of calpain could destroy any kind of cell from the inside. In fact, calpain proteolysis has been identified as a key event in many pathological states, including cell damage in ischaemic areas of brain and myocardium, tumorinogenesis in neurofibromatosis type 2A, and cataract formation (Saido et al., 1994; David and Shearer, 1993; Yoshida and Harada, 1997; Kimura et al., 1998). Mutations in Ca^{2+} channels associated with a number of neuromuscular diseases might also cause hyperactivation of calpain (Greenberg, 1999). It is possible that therapeutic reagents developed for one disease might be useful in others.

4. p94 AND LGMD2A

In 1995, the gene product responsible for limb-girdle muscular dystrophy type 2A (LGMD2A) was identified as p94 (Richard et al., 1995). This finding was remarkable for at least two reasons: it was the first report that a defect in a protease caused a muscular dystrophy, and p94 was the first calpain homologue whose mutation was identified as causing an inherited disorder. In an attempt to investigate the molecular mechanism of LGMD2A, it was shown that the loss of p94 proteolytic activity caused LGMD2A. This is consistent with the recessive inheritance of LGMD2A and completely opposed the previous notion that muscular dystrophies were linked to the hyper-activation of calpain(s) (Ono et al., 1998b). In this section, the current knowledge of limb-girdle muscular dystrophy is reviewed, focusing on the relationship between p94 and LGMD2A.

4.1. Limb-girdle muscular dystrophy

Limb-girdle muscular dystrophy (LGMD) is a genetically heterogeneous group of inherited progressive neuromuscular disorders with both dominant and recessive inheritance (Beckmann and Bushby, 1996; Ozawa et al., 1998). Early onset and symmetrical atrophy of the proximal limb and trunk muscle are general characteristics of LGMD patients. Until very recently, however, clinical diagnosis of LGMD was primarily a process of exclusion, a patient without any symptoms characteristic of other well-known diseases such as DMD is diagnosed as an LGMD patient. As shown in Table 2, the responsible gene/locus has been determined for four autosomal dominant and eight autosomal recessive forms of LGMD. The forms for which the genes responsible have been revealed can be diagnosed definitively. Another advantage of identifying the gene responsible is that the pathology of the disease can be studied on a molecular basis, which will in turn, hopefully, accelerate establishing a therapy.

LGMD2C, 2D, 2E, and 2F are caused by mutations in the gamma-, alpha-, beta-, and delta-sarcoglycan component of the sarcoglycan complex, respectively. Generically, they are called sarcoglycanopathies (Ozawa et al., 1998) (Figure 6). The sarcoglycan complex is localized at the sarcolemma, and stabilizes its structure in combination with other dystrophin-associated proteins. A defect in one sarcoglycan destabilizes the complex and is suspected of triggering the degeneration of muscle cells by a mechanism analogous to that of DMD/BMD. One of the remarkable features of sarcoglycanopathy is that the dystrophin-dystroglycan complex remains at the sarcolemma, even in the absence of the sarcoglycan complex (Beckmann and Bushby, 1996). Therefore, either the sarcoglycan complex or the coexistence of both the sarcoglycan and dystrophin-dystroglycan complexes may be critical for maintaining the function of the dystrophin-glycoprotein complex. The physiological function

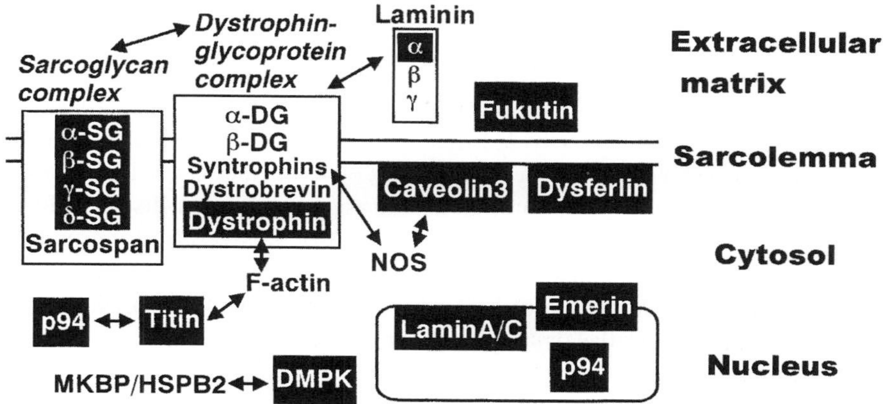

Figure 6. Locations of the molecules related to muscular dystrophies and/or sarcolemma structure. Proteins probably involved in muscular dystrophies are shown with open letters. Arrows mean interactions between these proteins (complexes).

of the sarcoglycan complex at the molecular level is not completely understood. Recently, knockout mice have been generated for each sarcoglycan species (Hack et al., 1998; Duclos et al., 1998; Coral-Vazquez et al., 1999; Araishi et al., 1999). Microscopic observation of these mice suggested that differences in the tissue distribution outside skeletal muscle contributed to the variation in symptoms observed for each type of knockout. The identification of novel sarcoglycan homologues and sarcoglycan interacting proteins has also revealed new aspects of the sarcoglycan complex (Liu and Enfvall, 1999; Thompson et al., 2000). The cytosolic Ca^{2+} concentration and level of calpain activity in these LGMD's should be examined in further studies.

LGMD2A is unique in that the responsible gene product is p94, a skeletal muscle-specific isoform of calpain, a cytosolic protease (Sorimachi et al., 2000). Therefore, the molecular mechanism of LGMD2A appears totally different from that of other muscular dystrophies, which result from defects in proteins without enzymatic activities, e.g., structural proteins.

In contrast to the sarcoglycanopathies, all the sarcoglycans, as well as dystrophin, are normally detected in LGMD2A biopsies. In addition to frequent muscle necrosis-regeneration and wide variation in muscle fiber size, both of which are general histological features of LGMD, many lobulated fibers are recognized in the end stage of LGMD2A (Fardeau et al., 1996). However, it is difficult to clinically distinguish LGMD2A from other LGMD's. It is an open question why the final symptoms are very similar to those of sarcoglycanopathy in spite of the fact that the sarcoglycan complex remains in LGMD2A patients.

Statistically, LGMD2A comprises 30–60% of the cases of autosomal recessive LGMD, the highest proportion of muscular dystrophies regardless of race (Sorimachi et al., 2000). Originally, the p94 gene was identified as responsible for familial cases of LGMD2A; however, mutations in the p94 gene were found in sporadic cases of LGMD with similar incidence (Minami et al., 1999). Nearly 100 distinct pathogenic mutations of the p94 gene have been identified (Kawai et al., 1998; Chou et al., 1999; Minami et al., 1999; Richard et al., 1999) (Fig. 4). The mutations are distributed along the entire p94 gene, without any particular hot spots. The ambiguity in the genetics of LGMD2A, the so-called "Réunion paradox", which many of the researchers involved admit, is that different mutations that probably descended from the same genetic isolate are found in consanguineous families (Richard et al., 1995). In other words, why is the frequency of mutations in the p94 gene so high in particular genetic isolates? This remains to be answered.

4.2. Defects of p94 in LGMD2A

It was predicted that hyper-activation of p94 is unlikely to cause LGMD2A since LGMD2A shows recessive inheritance and some of the mutations identified clearly abolish protease activity. Although the physiological function of p94 is unknown, what occurs in LGMD2A is opposite to the ordinary concept that the activation of proteases, including calpains, causes the final degradation of muscle proteins.

Previous studies have revealed several properties specific to p94, and the implications for the physiological function of p94 are of great interest (see Section 2). To verify that loss of p94 function causes LGMD2A, we attempted to identify the properties of p94 that are altered by any of the LGMD2A-associated mutations. We focused on missense point mutations of the p94 gene. In theory, point mutations should affect a few, but not all, of these properties. As summarized in Table 3, 10 point mutants were constructed and examined for autolytic activity, inter-molecular autolytic activity to p94:C129S, connectin/titin binding ability, Ca^{2+}-dependency of proteolytic activity, and *in vivo* ability to proteolyze fodrin and calpastatin.

The results were convincing and surprising. The autolytic activities and connectin/titin binding abilities of the mutants were not necessarily disrupted. This suggests that not all of the LGMD2A mutants disrupt protease activity and that dissociation from connectin/titin is not essential to LGMD2A. When the wild-type p94 is expressed in COS7 cells, it causes proteolysis of endogenous fodrin and calpastatin. In contrast, all the mutants lost this effect almost completely (Ono et al., 1998b). This strongly supports the hypothesis that LGMD2A mutations are pathogenic because they suppress proteolysis by p94. Thus, identification of the *in vivo* substrates of p94 and investigation of their significance has become an urgent topic. We have identified calpastatin, DMPK (myotonin protein kinase), and other candidate p94 substrates (Brook et al., 1992; Sorimachi et al., 1995). Their *in vivo* susceptibility to p94 proteolysis and its consequence are now under investigation.

Table 3. Properties of LGMD2A mutants of p94

Type	Domain	Autolysis	*in vivo* proteolysis of		Binding to	
			C129S[1]	Fodrin[2]	N2A[3]	C-ter[3]
Wild type	–	+	+	+	+	+
C129S	II	–	–	–	+	+
S86F	I	–	–	–	–	–
L182Q	II	–	–	–	–	–
G234E	II	–	–	–	+	–
H334Q	II	–	–	–	+	+
V354G	II	–	–	–	–	–
P319L	IS1	–	+	–	–	+
R490W	III	– (Ca^{2+}-dependent)	+	–	+	+
R572Q	III	– (Ca^{2+}-dependent)	N.D.	–	+	–
S744G	IV	++	+	–	+	–
R769Q	IV	++	+	–	+	–

[1]p94:C129S coexpressed with each construct in COS7 cells; [2] Proteolysis of COS7 cell endogeneous fodrin; [3]N2A or C-terminal region of connectin/titin.

4.3. Molecular characterization of p94

Considering the skeletal muscle-specific expression of p94, it is possible that p94 is involved in the organization of the muscle system. Previously, antisense treatment targeting the mCL subunit and p94 revealed their inhibitory effects on myoblast fusion and myofibrillar integration, respectively (Poussard et al., 1996). Inhibition of p94 translation results in the absence of mature Z-lines and the diffuse distribution of alpha-actinin. In this system, it is not clear which property of p94, e.g., protease enzymatic activity or structure as a connectin/titin binding protein, is necessary for reverting to normal appearance. If the loss of p94 protease activity impairs myofibrillar integrity, this contradicts the fact that embryonic myogenesis is not impaired in LGMD2A. This might be reasonable, since even an embryonically lethal genotype, such as alpha5 integrin $-/-$, allows normal *in vitro* differentiation into myotubes at the level of ES cells or myoblasts, possibly by dystrophic muscle apparatus (Taverna et al., 1998). It may be that p94 is not always required solely as a protease. For example, p94 might be an element of a structural device binding to connectin/titin at a certain developmental stage.

The state of p94 *in vivo* has been examined at various levels. As mentioned earlier, in immunostained myofibrils, p94 is located along the N_2- and Z-lines. The former corresponds to the region where the N2A sequence, one of the binding sites of connectin/titin to p94, is localized. The latter substance has not yet been identified, although coexistence of novel p94 interacting molecules has been predicted. Another group studied the fiber type-specific distribution of p94 protein (Jones et al., 1999). They showed that p94 was present in both fast and slow fibers, and that the relative content of p94 to actin is much higher in fast fibers. The expression level or stability of p94 is not simply correlated to the distribution of connectin/titin splice variants with the binding sites for p94. This appears to challenge the current hypothesis that p94 is stabilized by binding to connectin/titin. It would be interesting to clarify the determinants of the level of p94 in independent fibers.

The spatiotemporal pattern of p94 expression during human and mouse development has been studied with *in situ* hybridization experiments (Fougerousse et al., 2000). The results can be summarized as follows. (1) Expression of p94 mRNA in skeletal muscle was observed in both human and mouse embryos; (2) besides skeletal muscle, human embryos express p94 in the heart and smooth muscle of the digestive tract; and (3) the lens-specific splice variant of p94 is only detected in mouse embryos. Developmental changes in p94 expression were also investigated, focusing on alternative splicing (Herasse et al., 1999). Various p94 splice variants without the exon(s) corresponding to p94 specific insertion sequences, NS, IS1, and IS2, have been identified. Comparison of these variants at the protein level confirmed that these insertion sequences are responsible for p94-specific features, such as autolytic activity and connectin/titin binding. The biological significance of these observations should be substantiated by further investigations. The mechanism regulating p94 expression and its relation to muscle development are of great interest. Moreover, the possibility that each splice variant plays a specific role, or is regulated differently, is of interest.

4.4. Evaluation of p94 in pathogenic states

The central question is why a deficiency of p94 causes the muscle abnormality observed in LGMD2A. We have not successfully ascertained the consequence of p94-mediated proteolysis *in vivo*. In terms of the physiological relevance of proteolysis in muscle, calpain activity, mainly that of m-calpain, has been demonstrated in the proliferation of myogenic cells, myotube formation, and myofibrillar protein turnover (Goll et al., 1999). In these cellular events, calpain activity triggers reorganization of the cell system (the cytoskeleton, composition of cytosolic enzymes, or gene expression) in response to elevation of the cytosolic Ca^{2+} concentration.

The availability of protease inhibitors for m-calpain, such as E-64 and calpastatin, was advantageous for defining the involvement of its protease activity and identifying the substrates targeted. However, there is no effective means to selectively suppress p94 protease activity. In other words, it is necessary to prove that degradation of muscle proteins by other proteolytic systems, calpains, proteasomes, and so on, is enhanced when the protease activity of p94 is inhibited. Effective p94 inhibitors should be sought and the state of p94 under pathogenic conditions investigated.

The importance of developing a methodology that enables the systematic diagnosis of muscular dystrophy has been considered (Anderson and Davison, 1999). One strategy is to detect abnormal p94 at the protein level using p94-specific antibodies. Several lines of evidence have shown that in LGMD2A, the expression of p94 protein is abnormal, and usually lowered (Spencer et al., 1997; Anderson et al., 1998). The variability of identified LGMD2A mutations means that the identification of pathogenic features using genetic analyses will require a lot of effort. Although it is not clear why the amount of mutant p94 protein decreases, quantification of p94 protein might assist in clinical diagnosis. If the alteration of the amount of p94 is caused by other proteolytic systems, such as proteasome and other calpains, identification of the responsible protease activity will provide insight into the linkage of p94 protease activity and other proteases.

In line with emerging concern about the contribution of apoptosis to muscle degeneration, it was recently reported that LGMD2A muscle biopsies show apoptotic myonuclei more frequently than other types of muscular dystrophies, such as DMD and sarcoglycanopathies ($0.40\% \pm 0.16\%$, $n = 13$ vs. $0.0071\% \pm 0.0049\%$, $n = 8$) (Mukasa, et al., 1999; Baghdiguian et al., 1999). It was also demonstrated that IkBalpha is a potent p94 substrate. As a possible molecular basis for LGMD2A, it was suggested that nuclear accumulation of IkBalpha caused by the loss of p94 activity results in apoptotic myonuclei. In LGMD2A, apoptotic myonuclei are found mainly in multinucleated myofibers, coexisting with other non-apoptotic nuclei. The experimental model of myotubes derived from C2C12 cells has shown that postmitotic multinucleated skeletal muscle cells also undergo the process of apoptosis, i.e., caspase activation and DNA degradation, as mononuclear cells do (McArdle et al., 1999). Nevertheless, it is unknown whether a muscle cell dies and is removed, as occurs in the process of apoptosis in other cell systems. It should be determined whether an apoptotic change in part of the nuclei determines the fate of a whole cell.

A study of the effect of interleukin 6 (IL-6) suggested a negative correlation between muscle atrophy and p94 (Tsujinaka et al., 1996). Administering IL-6 to the rat promotes proteolysis, which augments atrophy. Consistent with this observation, transgenic mice overexpressing IL-6 develop atrophy. In these mice, p94 decreases to 50% of the level in control mice at the mRNA level, whereas cathepsins, proteasome subunits, and μ- and m-calpains increase. The changes in the mRNA levels can be eliminated when the atrophy is blocked by the administration of anti-IL-6 antibody. This observation implies that transduction of a certain signal through the IL-6 receptor regulates p94 at the transcriptional level, and that down-regulation of p94 could cause atrophy.

5. CONTROL OF PROTEASE ACTIVITY

We have discussed p94 based on the premise that p94 protease activity is vital for maintaining muscle cell structure or function. Although it has not been observed *in vivo*, hyper-activation of p94 must be harmful for cells, as is the case with other calpains. In fact, when wild-type p94 is overexpressed in COS7 cells, many more cells are detached from the culture dish than with untransfected cells or cells overexpressing μ- or m-calpains (Sorimachi et al., 1993). The explanation is that many cytoskeletal proteins identified as calpain substrates, such as fodrin, are proteolyzed by p94.

It is not yet determined whether p94 is active completely independent of Ca^{2+}, but proteolysis by p94 proceeds at a Ca^{2+} concentration that does not permit μ- and m-calpain activation. One remarkable finding is that two LGMD2A mutations in domain III of p94, R490W and R572Q, cause Ca^{2+}-dependency. These two mutant p94 are hardly active at cellular Ca^{2+} concentrations, while the addition of excessive Ca^{2+} restores the autolytic activity of these mutants (Ono et al., 1998b). Contradictory to the authors' observation, it was also reported that recombinant p94 expressed using a baculovirus expression system requires Ca^{2+} for autolytic activity (Branca et al., 1999). Further study is required to attain a unified hypothesis. However, it is worth noting that p94 is active at cellular Ca^{2+} concentrations or, in extreme cases, is Ca^{2+}-independent. Considering that the mutations in domain III alter the Ca^{2+}-sensitivity of p94, and that the recently resolved crystal structure of m-calpain reveals the involvement of domain III in the Ca^{2+}-dependent activation mechanism, studies of the structure of p94 will benefit discussion on the structural bases of calpain activity as well as that of p94 itself.

It seems as though p94 is regulated by association with connectin/titin and autolysis (Kinbara et al., 1998). It is also possible that other unidentified factors interact with p94 and suppress or stabilize its activity. The authors' hypothesis is as follows (Sorimachi et al., 2000). The role of connectin/titin is to localize each p94 molecule at a certain position to suppress inter-molecular autolysis or disrupt the conformation for autolysis (Kinbara et al., 1998). At the same time, p94 is prevented from hydrolyzing cellular proteins. Thus, the state in which p94 binds to connectin/titin is not cytotoxic. Furthermore, p94 protects connectin/titin and possibly neighboring cytoskeletal proteins by its protease activity or steric hindrance of other proteases.

By yet unknown signals, p94 dissociates from connectin/titin and translocates to the nucleus or other cytoplasmic region to hydrolyze its substrates. In this case, nuclear proteins, such as transcription factors, and other enzymes, such as kinase, phosphatases, and proteases, could be substrates. Autolytic activity of p94 is advantageous for limiting the duration of proteolysis by p94. In our experimental system, COS7 cells, and *in vitro* translation, p94 itself serves as its best substrate. Excessive p94 activity would be utilized to clear itself away.

In addition, studies of DMPK, another enzyme whose gene is responsible for muscular dystrophy, have shown the involvement of the stress response cascade in maintaining muscle cell function (Suzuki and Sorimachi, 1998; Sugiyama et al., 2000). Therefore, proteins involved in this cascade, such as heat shock proteins (HSP) as well as DMPK itself, may be subjected to proteolytic regulation by p94.

The missing factor in this speculative scheme is the signal dissociating p94 from connectin/titin and subsequent activation. Mechanical stresses generated by tension or cell-cell interaction, which is usually followed by activation of a certain group of ion channels, or signal transduction through transmembrane receptors, such as the IL-6 receptor, have not been excluded as the signal. If other structural proteins, e.g., dystrophin-glycoprotein complex, many of which are gene products responsible for muscular dystrophy, regulate the input from such factor(s), p94 might be damaged as a secondary effect in other muscular dystrophies. In such a case, a defect in p94 protease activity would contribute, at least partially, to the muscle degeneration caused by various stimuli.

Moreover, if connectin/titin cannot serve as a platform for p94, the level of potentially functional p94 would decrease due to the lack of a stabilizer. Such a condition may exist in two distinct myopathies recently reported; tibial muscular dystrophy (TMD) linked to 2q31, and a subtype of hypertrophic cardiomyopathy (HCM), where connectin/titin may be disrupted (Haravuori et al., 1998; Satoh et al., 1999). In addition, very recently, T-cap, also called telethonin, was identified as a responsible gene product for LGMD2G (Moreira et al., 2000). T-cap binds to the N-terminus of connectin/titin and is speculated to be a "bolt" facilitating the assembly of connectin/titin into Z-line (Gregorio et al., 1998). It would be informative to examine the state of p94 in these diseases.

6. PERSPECTIVES

In the past few years, it has been revealed that many types of muscular dystrophy and myopathies are caused by mutations of structural proteins. These defects have been explained as causing an elevated cytosolic Ca^{2+} concentration by perturbing the function of Ca^{2+} channels and vice versa. It is almost certain that the mutations found in LGMD2A cause p94 to lose its protease activity, and the novelty of its molecular mechanism is of particular interest. The characterization of p94 has been attempted from many angles, as described in this chapter, but the essential event that is deeply involved in the mechanism of LGMD2A remains to be found.

The contribution of Ca^{2+} to the activation of p94 *in vivo* has not been of as much concern, due to the apparent Ca^{2+}-independency of p94 observed in multiple experimental systems. However, it is still possible that p94 has a very high Ca^{2+}-sensitivity. In addition, it might be interesting to examine whether p94 is involved in Ca^{2+}-homeostasis of muscle cells (Shevchenko et al., 1998).

Although no study has examined whether the loss of p94 protease activity impairs the Ca^{2+}-handling ability and thus permits hyper-activation of calpains, this might be an alternative mechanism for LGMD2A. We prefer the term "different mechanism" for LGMD2A pathology. The variety of responsible gene products means that there are multiple entrances for pathogenic sequences. Hopefully, studies of these muscle diseases will teach us how the muscle system is organized and maintained to fulfill its function.

REFERENCES

Anderson, L.V.B. and Davison, K., 1999, Multiplex Western blotting system for the analysis of muscular dystrophy proteins, *Am. J. Pathol.* 154, 1017–1022.

Anderson, L.V., Davison, K., Mossa, J.A., Richard, I., Fardeau, M., Tome, F.M., Hubner, C., Lasa, A., Colomer, J. and Beckmann, J.S., 1998, Characterization of monoclonal antibodies to calpain 3 and protein expression in muscle from patients with limb-girdle muscular dystrophy type 2A, *Am. J. Pathol.* 153, 1169–1179.

Araishi, K., Sasaoka, T., Imamura, M., Noguchi, S., Hama, H., Wakabayashi, E., Yoshida, M., Hori, T. and Ozawa, E., 1999, Loss of the sarcoglycan complex and sarcospan leads to muscular dystrophy in β-sarcoglycan-deficient mice, *Hum. Mol. Genet.* 8, 1589–1598.

Baghdiguian, S., Martin, M., Richard, I., Pons, F., Astier, C., Bourg, N., Hay, R.T., Chemaly, R., Halaby, G., Loiselet, J., Anderson, L.V.B., de Munain, A.L., Fardeau, M., Mangeat, P., Beckmann, J.S. and Lefranc, G., 1999, Calpain 3 deficiency is associated with myonuclear apoptosis and profound perturbation of the IkBalpha/NF-kB pathway in limb-girdle muscular dystrophy type 2A, *Nat. Med.* 5, 503–511.

Barnes, T.M. and Hodgkin, J., 1996, The tra-3 sex determination gene of *Caenorhabditis elegans* encodes a member of the calpain regulatory protease family, *EMBO J.* 15, 4477–4484.

Beckmann, J.S. and Bushby, K.M., 1996, Advances in the molecular genetics of the limb-girdle type of autosomal recessive progressive muscular dystrophy, *Curr. Opin. Neurol.* 9, 389–393.

Blanchard, H., Grochulski, P., Li, Y., Arthur, J.S., Davies, P.L., Elce, J.S. and Cygler, M., 1997, Structure of a calpain Ca^{2+}-binding domain reveals a novel EF-hand and Ca^{2+}-induced conformational changes, *Nat. Struct. Bio.* 4, 532–538.

Bonnemann, C.G., McNally, E.M. and Kunkel, L.M., 1996, Beyond dystrophin: Current progress in the muscular dystrophies, *Curr. Opin. Pediatr.* 8, 569–582.

Branca, D., Gugliucci, A., Bano, D., Brini, M. and Carafoli, E., 1999, Expression, partial purification and functional properties of the muscle-specific calpain isoform p94, *Eur. J. Biochem.* 265, 839–846.

Brenman, J.E., Chao, D.S., Xia, H., Aldape, K. and Bredt, D.S., 1995, Nitric oxide synthase complexed with dystrophin and absent from skeletal muscle sarcolemma in Duchenne muscular dystrophy, *Cell* 82, 743–752.

Brook, J.D., McCurrach, M.E., Harley, H.G., Buckler, A.J., Church, D., Aburatani, H., Hunter, K., Stanton, V.P., Thirion, J.P., Hudson, T., Sohn, R., Zemelman, B., Snell, R.G., Rundle, S.A., Crow, S., Davies, J., Shelbourne, P., Buxton, J., Jones, C., Juvonen, V., Johnson, K., Harper, P.S., Shaw, D.J. and Housman, D.E., 1992, Molecular basis of myotonic dystrophy: Expansion of a trinucleotide (CTG) repeat at the 3$'$ end of a transcript encoding a protein kinase family member, *Cell* 68, 799–808.

Bushby, K.M.D., 1999, Making sense of the limb-girdle muscular dystrophies, *Brain* 122, 1403–1420.

Carafoli, E. and Molinari, M., 1998, Calpain: A protease in search of a function?, *Biochem. Biophys. Res. Commun.* 247, 193–203.

Chou, F.L., Angelini, C., Daentl, D., Garcia, C., Greco, C., Hausmanowa-Petrusewicz, I., Fidzianska, A., Wessel, H. and Hoffman, E.P., 1999, Calpain III mutation analysis of a heterogeneous limb-girdle muscular dystrophy population, *Neurol.* 52, 1015–1020.

Coral-Vazquez, R., Cohn, R.D., Moore, S.A., Hill, J.A., Weiss, R.M., Davisson, R.L., Straub, V., Barresi, R., Bansal, D., Hrstka, R.F., Williamson, R. and Campbell, K.P., 1999, Disruption of the sarcoglycan-sarcospan complex in vascular smooth muscle: A novel mechanism for cardiomyopathy and muscular dystrophy, *Cell* 98, 465–474.

David, L.L. and Shearer, T.R., 1993, Beta-crystallins insolubilized by calpain II in vitro contain cleavage sites similar to beta-crystallins insolubilized during cataract, *FEBS Lett.* 324, 265–270.

Dear, N., Matena, K., Vingron, M. and Boehm, T., 1997, A new subfamily of vertebrate calpains lacking a calmodulin-like domain: Implications for calpain regulation and evolution, *Genomics* 45, 175–184.

Delaney, S.J., Hayward, D.C., Barleben, F., Fischbach, K.F. and Gabor-Miklos, G.L., 1991, Molecular cloning and analysis of small optic lobes, a structural brain gene of *Drosophila melanogaster*, *Proc. Natl. Acad. Sci. USA* 88, 7213–7218.

Denison, S.H., Orejas, M. and Arst. Jr., H.N., 1995, Signaling of ambient pH in *Aspergillus* involves a cysteine protease, *J. Biol. Chem.* 270, 28519–28522.

Duclos, F., Straub, V., Moore, S.A., Venzke, D.P., Hrstka, R.F., Crosbie, R.H., Durbeej, M., Lebakken, C.S., Ettinger, A.J., van der Meulen, J., Holt, K.H., Lim, L.E., Sanes, J.R., Davidson, B.L., Faulkner, J.A., Williamson, R. and Campbell, K.P., 1998, Progressive muscular dystrophy in alpha-sarcoglycan-deficient mice, *J. Cell. Biol.* 142, 1461–1471.

Emery, A.E.H., 1998, The muscular dystrophies, *BMJ* 317, 991–995.

Fardeau, M., Hillaire, D., Mignard, C., Feingold, N., Feingold, J., Mignard, D., deUbeda, B., Collin, H., Tome, F.M.S., Richard, I. and Beckmann, J.S., 1996, Juvenile limb-girdle muscular dystrophy: Clinical, histopathological, and genetic data from a small community living in the Reunion Island, *Brain* 119, 295–308.

Fong, P.Y., Turner, P.R., Denetclaw, W.F. and Steinhardt, R.A., 1990, Increased activity of calcium leak channels in myotubes of Duchenne human and mdx mouse origin, *Science* 250, 673–676.

Fougerousse, F., Bullen, P., Herasse, M., Lindsay, S., Richard, I., Lee, S., Suel, L., McMahon, A., Durand, D., Robson, S., Wilson, D., Abitbol, M., Beckmann, J.S. and Strachan, T., 2000, Human-mouse differences in the embryonic expression patterns of developmental control genes and disease genes, *Hum. Mol. Genet.* 9, 165–173.

Franz, T., Vingron, M., Boehm, T. and Dear, T.N., 1999, Capn7: A highly divergent vertebrate calpain with a novel C-terminal domain, *Mamm. Genome* 10, 318–321.

Futai, E., Maeda, T., Sorimachi, H., Kitamoto, K., Ishiura, S. and Suzuki, K., 1999, The protease activity of a calpain-like cysteine protease in *Saccharomyces cerevisiae* is required for alkaline adaptation and sporulation, *Mol. Gen. Genet.* 260, 559–568.

Goll, D.E., Thompson, V.F., Taylo, R.G., Ouali, A. and Chou, R.G.R., 1999, *Calpain: Pharmacology and Toxicology of Calcium-Dependent Protease*, Taylor & Francis, Philadelphia, pp. 127–160.

Greenberg, D.A., 1999, Neuromuscular disease and calcium channels, *Muscle Nerve* 22, 1341–1349.

Gregorio, C.C., Trombitas, K., Centner, T., Kolmerer, B., Stier, G., Kunke, K., Suzuki, K., Obermayr, F., Herrmann, B., Granzier, H., Sorimachi, H. and Labeit, S., 1998, The NH_2 terminus of titin spans the Z-disc: Its interaction with a novel 19-kD ligand (T-cap) is required for sarcomeric integrity, *J. Cell. Biol.* 143, 1013–1027.

Hack, A.A., Ly, C.T., Jiang, F., Clendenin, C.J., Sigrist, K.S., Wollmann, R.L. and McNally, E.M., 1998, Gamma-sarcoglycan deficiency leads to muscle membrane defects and apoptosis independent of dystrophin, *J. Cell. Biol.* 142, 1279–1287.

Haravuori, H., Makela-Bengs, P., Udd, B., Partanen, J., Pulkkinen, L., Somer, H. and Peltonen, L., 1998, Assignment of the tibial muscular dystrophy locus to chromosome 2q31, *Am. J. Hum. Genet.* 62, 620–626.

Herasse, M., Ono, Y., Fougerousse, F., Kimura, E., Stockholm, D., Beley, C., Montarras, D., Pinset, C., Sorimachi, H., Suzuki, K., Beckmann, J.S. and Richard, I., 1999, Expression and functional characteristics of calpain 3 isoforms generated through tissue-specific transcriptional and posttranscriptional events, *Mol. Cell. Biol.* 19, 4047–4055.

Hoffman, E.P., Brown, R.H., Jr. and Kunkel, L.M., 1987, Dystrophin: The protein product of the Duchenne muscular dystrophy locus, *Cell* 51, 919–928.

Hopf, F.W., Turner, P.R., Denetclaw, W.F., Jr., Reddy, P. and Steinhardt, R.A., 1996, A critical evaluation of resting intracellular free calcium regulation in dystrophic mdx muscle, *Am. J. Physiol.* 271, 1325–1339.

Hosfield, C.H., Elce, J.S., Davies, P.L. and Jia, Z., 1999, Crystal structure of calpain reveals the structural basis for Ca^{2+}-dependent protease activity and a novel mode of enzyme activation, *EMBO J.* 18, 6880–6889.

Imbert, N., Vandebrouck, C., Constantin, B., Duport, G., Guillou, C., Cognard, C. and Raymond, G., 1996, Hypo-osmotic shocks induce elevation of resting calcium level in Duchenne muscular dystrophy myotubes contracting in vitro, *Neuromuscul. Disord.* 6, 351–360.

Ishiura, S., Murofushi, H., Suzuki, K. and Imahori, K., 1978, Studies of a calcium-activated neutral protease from chicken skeletal muscle. I. Purification and characterization, *J. Biochem.* 84, 225–230.

Jones, S.W., Parr, T., Sensky, P.L., Scothern, G.P., Bardsley, R.G. and Buttery, P.J., 1999, Fiber type-specific expression of p94, a skeletal muscle-specific calpain, *J. Muscle Res. Cell Motil.* 20, 417–424.

Kamei, M., Webb, G.C., Young, I.G. and Campbell, H.D., 1998, SOLH, a human homologue of the Drosophila melanogaster small optic lobes gene is a member of the calpain and zinc-finger gene families and maps to human chromosome 16p13.3 near CATM (cataract with microphtalmia), *Genomics* 51, 197–206.

Kawai, H., Akaike, M., Kunishige, M., Inui, T., Adachi, K., Kimura, C., Kawajiri, M., Nishida, Y., Endo, I., Kashiwagi, S., Nishino, H., Fujiwara, T., Okuno, S., Roudaut, C., Richard, I., Beckmann, J.S., Miyoshi, K. and Matsumoto, T., 1998, Clinical, pathological, and genetic features of limb-girdle muscular dystrophy type 2A with new calpain 3 gene mutations in seven patients from three Japanese families, *Muscle Nerve* 21, 1493–1501.

Kawasaki, H., Nakayama, S. and Kretsinger, R.H., 1998, Classification and evolution of EF-hand proteins, *Biometals* 11, 277–295.

Kimura, Y., Koga, H., Araki, N., Mugita, N., Fujita, N., Takeshima, H., Nishi, T., Yamashima, T., Saido, T.C., Yamasaki, T., Moritake, K., Saya, H. and Nakao, M., 1998, The involvement of calpain-dependent proteolysis of the tumor suppressor NF2 (merlin) in schwannomas and meningiomas, *Nat. Med.* 4, 915–922.

Kinbara, K., Ishiura, S., Tomioka, S., Sorimachi, H., Jeong, S.Y., Amano, S., Kawasaki, H., Kolmerer, B., Kimura, S., Labeit, S. and Suzuki, K., 1998, Purification of native p94, a muscle-specific calpain, and characterization of its autolysis, *Biochem. J* 335, 589–596.

Kinbara, K., Sorimachi, H., Ishiura, S. and Suzuki, K., 1997, Muscle-specific calpain, p94, interacts with the extreme C-terminal region of connectin, a unique region flanked by two immunoglobulin C2 motifs, *Arch. Biochem. Biophys.* 342, 99–107.

Lin, G.D., Chattopadhyay, D., Maki, M., Wang, K.K.W., Carson, M., Jin, L., Yuen, P.W., Takano, E., Hatanaka, M., DeLucas, L.J. and Narayana, S.V., 1997, Crystal structure of calcium bound domain VI of calpain at 1.9Å resolution and its role in enzyme assembly, regulation, and inhibitor binding, *Nat. Struct. Biol.* 4, 539–547.

Liu, L.A. and Enfvall, E., 1999, Sarcoglycan isoforms in skeletal muscle, *J. Biol. Chem.* 274, 38171–38176.

Ma, H., Fukiage, C., Azuma, M. and Shearer, T.R., 1997, Cloning and expression of mRNA for calpain Lp82 from rat lens: Spice variant of p94, *Invest. Ophthalmol. Vis. Sci.* 39, 454–461.

Menke, A. and Joskusch, H., 1991, Decreased osmotic stability of dystrophin-less muscle cells from the mdx mouse, *Nature* 349, 69–71.

McArdle, A., Maglara, A., Appleton, P., Watson, A.J., Grierson, I. and Jackson, M.J., 1999, Apoptosis in multinucleated skeletal muscle myotubes, *Lab. Invest.* 79, 1069–1076.

Minami, N., Nishino, I., Kobayashi, O., Ikezoe, K., Goto, Y. and Nonaka, I., 1999, Mutations of calpain 3 gene in patients with sporadic limb-girdle muscular dystrophy in Japan, *J. Neurol. Sci.* 171, 31–37.

Moreira, E.S., Wiltshire, T.J., Faulkner, G., Nilforoushan, A., Vainzof, M., Suzuki, O.T., Valle, G., Reeves, R., Zatz, M., Passos-Bueno, M.R. and Jenne, D.E., 2000, Limb-girdle muscular dystrophy type 2G is caused by mutations in the gene encoding the sarcomeric protein telethonin, *Nat. Genet.* 24, 163–166.

Mugita, N., Kimura, Y., Ogawa, M., Saa, H. and Nakao, M., 1997, Identification of a novel, tissue-specific calpain htra-3: A human homologue of the Caenorhabditis elegans sex determination gene, *Biochem. Biophys. Res. Commun.* 239, 845–850.

Mukasa, T., Momoi, T. and Momoi, M.Y., 1999, Activation of caspase-3 apoptotic pathways in skeletal muscle fibers in laminin alpha2-deficient mice, *Biochem. Biophys. Res. Commun.* 260, 139–142.

Murachi, T., Tanaka, K., Hatanaka, M. and Murakami, T., 1981, Intracellular Ca^{2+}-dependent protease (calpain) and its high-molecular-weight endogenous inhibitor (calpastatin), *Adv. Enz. Reg.* 19, 407–424.

Ohno, S., Emori, Y., Imajoh, S., Kawasaki, H., Kisaragi, M. and Suzuki, K., 1984, Evolutionary origin of a calcium-dependent protease by fusion of genes for a thiol protease and a calcium-binding protein?, *Nature* 312, 566–570.

Ono, Y., Sorimachi, H. and Suzuki, K., 1998a, Structure and physiology of calpain, an enigmatic protease, *Biochem. Biophys. Res. Commun.* 245, 289–294.

Ono, Y., Shimada, H., Sorimachi, H., Richard, I., Saido, T.C., Beckmann, J.S., Ishiura, S. and Suzuki, K., 1998b, Functional defects of a muscle-specific calpain, p94, caused by mutations associated with limb-girdle muscular dystrophy type 2A, *J. Biol. Chem.* 273, 17073–17078.

Ozawa, E., Noguchi, S., Mizuno, Y., Hagiwara, Y. and Yoshida, M., 1998, From dystrophinopathy to sarcoglycanopathy: Evolution of a concept of muscular dystrophy, *Muscle Nerve* 21, 431–438.

Poussard, S., Duvert, M., Balcerzak, D., Ramassamy, S., Brustis, J.J., Cottin, P. and Ducastaing, A., 1996, Evidence for implication of muscle-specific calpain (p94) in myofibrillar integrity, *Cell Growth Differ.* 7, 1461–1469.

Rabbani, N., Moses, L., Anandavalli, T.E. and Anandaraj, M.P., 1984, Calcium-activated neutral protease from muscle and platelets of Duchenne muscular dystrophy cases, *Clin. Chim. Acta* 143, 163–168.

Richard, I., Broux, O., Allamand, V., Fougerousse, F., Chiannilkulchai, N., Bourg, N., Brenguier, L., Davaud, C., Pasturaud, P., Roudaut, C., Hillaire, D., Passos-Bueno, M.R., Zatz, M., Tischrield, J.A., Fardeu, M., Jackson, C.E., Cohen, D. and Beckmann, J.S., 1995, Mutations in the proteolytic enzyme calpain 3 cause limb-girdle muscular dystrophy type 2A, *Cell* 81, 27–40.

Richard, I., Roudaut, C., Saenz, A., Pogue, R., Grimbergen, J.E.M.A., Anderson, L.V.B., Beley, C., Cobo, A., de Diego, C., Eymard, B., Gallano, P., Ginjaar, H.B., Lasa, A., Pollitt, C., Topaloglu, H., Urtizverea, J.A., de Visser, M., van der Kooi, A., Bushby, K., Bakker, E., Lopez de Munain, A., Fardeau, M. and Beckmann, J.S., 1999, Calpainopathy – A survey of mutations and polymorphisms, *Am. J. Hum. Genet.* 64, 1524–1540.

Saido, T.C., Shibata, M., Takenawa, T., Murofushi, H. and Suzuki, K., 1992, Positive regulation of μ-calpain action by polyphosphoinositides, *J. Biol. Chem.* 267, 24585–24590.

Saido, T.C., Sorimachi, H. and Suzuki, K., 1994, Calpain: New perspective in molecular diversity and physiological-pathological involvement, *FASEB J.* 8, 814–822.

Sakihama, T., Kakidani, H., Zenita, K., Yumoto, N., Kikuchi, T., Sasaki, T., Kannagi, R., Nakanishi, S., Ohmori, M., Takio, K. and Murachi, T., 1985, A putative Ca^{2+}-binding protein: Structure of the light subunit of porcine calpain elucidated by molecular cloning and protein sequence analysis, *Proc. Natl. Acad. Sci. USA* 82, 6075–6079.

Satoh, M., Takahashi, M., Sakamoto, T., Hiroe, M., Marumo, F. and Kimura, A., 1999, Structural analysis of the titin gene in hypertrophic cardiomyopathy: Identification of a novel disease gene, *Biochem. Biophys. Res. Commun* 262, 411–417.

Shevchenko, S., Feng, W., Varsanyi, M. and Shoshan-Barmatz, B., 1998, Identification, characterization and partial purification of a thiol-protease which cleaves specifically the skeletal muscle ryanodine receptor/Ca^{2+} release channel, *J. Membrane Biol.* 161, 33–43.

Sorimachi, H., Imajoh-Ohmi, S., Emori, Y., Kawasaki, H., Ohno, S., Minami, Y. and Suzuki, K., 1989, Molecular cloning of a novel mammalian calcium-dependent protease distinct from both m- and M-types, *J. Biol. Chem.* 264, 20106–20111.

Sorimachi, H., Toyama-Sorimachi, N., Saido, T.C., Kawasaki, H., Sugita, H., Miyasaka, M., Arahata, K., Ishiura, S. and Suzuki, K., 1993, Muscle-specific calpain, p94, is degraded by autolysis immediately after translation, resulting in disappearance from muscle, *J. Biol. Chem.* 268, 10593–10605.

Sorimachi, H., Kinbara, K., Kimura, S., Takahashi, M., Ishiura, S., Sasagawa, S., Sorimachi, N., Shimada, H., Tagawa, K., Maruyama, K. and Suzuki, K., 1995, Muscle-specific calpain, p94, responsible for limb-girdle muscular dystrophy type 2A, associates with connectin through IS2, a p94-specific sequence, *J. Biol. Chem.* 270, 31158–31162.

Sorimachi, H., Ishiura, S. and Suzuki, K., 1997, Structure and physiological function of calpains, *Biochem. J.* 328, 721–732.

Sorimachi, H., Minami, N., Ono, Y., Suzuki, K. and Nonaka, I., 2000, Limb-girdle muscular dystrophy with calpain 3 (p94) gene mutations (calpainopathy), *Neurosci. News* 3, 20–27.

Spencer, M.J., Tidball, J.G., Anderson, L.V., Bushby, K.M., Harris, J.B., Passos-Bueno, M.R., Somer, H., Vainzof, M. and Zatz, M., 1997, Absence of calpain 3 in a form of limb-girdle muscular dystrophy (LGMD2A), *J. Neurol. Sci.* 146, 173–178.

Strobl, S., Fernandez-Catalan, C., Braun, M., Huber, R., Masumoto, H., Nakagawa, K., Irie, A., Sorimachi, H., Bourenkow, G., Bartunik, H., Suzuki, K. and Bode, W., 2000, The crystal structure of calcium-free human m-calpain suggests an electrostatic switch mechanism for activation by calcium, *Proc. Natl. Acad. Sci. USA* 97, 588–592.

Sugita, H., Ishiura, S., Suzuki, K. and Imahori, K., 1980, Ca^{2+}-activated neutral protease and its inhibitors: In vitro effect on intact myofibrils, *Muscle Nerve* 3, 335–339.

Sugiyama, Y., Suzuki, A., Kishikawa, M., Akutsu, R., Hirose, T., Waye, M.M.Y., Tsui, S.K.W., Yoshida, S. and Ohno, S., 2000, Muscle develops a specific form of small heat shock protein complex composed of MKBP/HSPB2 and HSPB3 during myogenic differentiation, *J. Biol. Chem.* 275, 1095–1104.

Suzuki, K. and Sorimachi, H., 1998, A novel aspect of calpain activation, *FEBS Lett.* 433, 1–4.

Suzuki, K., Tsuji, S., Kubota, S., Kimura, Y. and Imahori, K., 1981, Limited autolysis of Ca^{2+}-activated neutral protease (CANP) changes its sensitivity to Ca^{2+} ions, *J. Biochem.* 90, 275–278.

Takahashi-Nakamura, M., Tsuji, S., Suzuki, K. and Imahori, K., 1981, Purification and characterization of an inhibitor of calcium-activated neutral protease from rabbit skeletal muscle, *J. Biochem.* 90, 1538–1589.

Taverna, D., Disatnik, M.H., Rayburn, H., Bronson, R.T., Yang, J., Rando, T.A. and Hynes, R.O., 1998, Dystrophic muscle in mice chimeric for expression of alpha5 integrin, *J. Cell. Biol.* 143, 849–859.

Thompson, T.G., Chan, Y.M., Hack, A.A., Brosius, M., Rajala, M., Lidov, H.G., McNally, E.M., Watkins, S. and Kunkel, L.M., 2000, Filamin 2 (FLN2). A muscle-specific sarcoglycan interacting protein, *J. Cell. Biol.* 148, 115–126.

Toniolo, D. and Minetti, C., 1999, Muscular dystrophies: Alterations in a limited number of cellular pathways?, *Curr. Opin. Genet. Dev.* 9, 275–282.

Tsujinaka, T., Fujita, J., Ebisui, C., Yano, M., Kominaimi, E., Suzuki, K., Tanaka, K., Katsume, A., Ohsugi, Y., Shiozaki, H. and Monden, M., 1996, Interleukin 6 receptor antibody inhibits muscle atrophy and modulates proteolytic systems in interleukin 6 transgenic mice, *J. Clin. Invest.* 97, 244–249.

Turner, P.R., Westwood, T., Regen, C.M. and Steinhardt, R.A., 1988, Increased protein degradation results from elevated free calcium levels found in muscle from mdx mice, *Nature* 335, 735–738.

Turner, P.R., Schults, R., Ganguly, B. and Steinhardt, R.A., 1993, Proteolysis results in altered leak channel kinetics and elevated free calcium in mdx muscle, *J. Membr. Biol.* 133, 243–251.

Yoshida, K. and Harada, K., 1997, Proteolysis of erythrocyte-type and brain-type ankyrins in rat heart after postischemic reperfusion, *J. Biochem.* 133, 279–285.

Yoshizawa, T., Sorimachi, H., Tomioka, S., Ishiura, S. and Suzuki, K., 1995, A catalytic subunit of calpain possesses full proteolytic activity, *FEBS Lett.* 358, 101–103.

Role of Calcium Channels in Drug Dependence

Fernando Rodríguez de Fonseca, Miguel Angel Gorriti and Miguel Navarro

1. INTRODUCTION: ROLE OF CALCIUM IN DRUG ADDICTION

Drug addiction is a medical illness that has devastating health, social, political and economic consequences. Indeed, it is one of the most important health problems facing the world today. Addiction has this devastating impact despite extensive political effort and social/behavioral research aimed at reducing addiction. Social and psychological approaches to the treatment of addiction – while beneficial to many – have been only partially successful in alleviating the worldwide burden of substance abuse (Koob and Le Moal, 1997; Nestler and Aghajanian, 1997; O'Brien, 1997). Based on recent scientific advances, attitudes toward addiction and approaches to the treatment and prevention of addiction are rapidly changing. It is now clear that addiction should be conceptualized as a brain disease that occurs in susceptible individuals as a consequence of cellular and molecular changes in nervous system function (Koob et al., 1998). It is well established that certain brain circuits, like those involving mesencephalic dopaminergic neurons, are particularly important for mediating the behavioral and psychological actions of drugs abuse (Koob and Le Moal, 1997; Rodríguez de Fonseca and Navarro, 1998). This has led to the development of more sophisticated molecular hypotheses to explain critical features of addiction such as sensitization, tolerance, withdrawal and dependence (Nestler and Aghajanian, 1997; Koob et al., 1998). Molecular pharmacology research as well as genetic studies in inbred strains of rodents and the achievement of genetically modified animals have allowed to find those molecular targets of major abused drugs (Guppy et al., 1995; Koob

---

Fernando Rodríguez de Fonseca, Miguel Angel Gorriti and Miguel Navarro • Instituto Universitario de Drogodependencias (Departamento de Psicobiología, Facultad de Psicología), Universidad Complutense, Madrid, Spain.

R. Pochet, R. Donato, J. Haiech, C. Heizmann and V. Gerke (eds.): Calcium: The Molecular Basis of Calcium Action in Biology and Medicine, 465–476.

et al., 1998; Crabbe et al., 1999; Ledent et al., 1999). Among these targets, Ca^{2+}-dependent signaling events play a central role (Dolin et al.,1987; Bongianni et al., 1988; Herman et al., 1995; Miyakawa et al., 1997; Gerstin et al., 1998). The present review is intended to outline the main contribution of Ca^{2+} signaling in the mediation of acute and chronic drug effects. Understanding the acute and chronic actions of drugs such as ethanol, opiates, cannabinoids or psychostimulants on the functionality of voltage-gated, ligand-activated or G-protein-coupled receptor-regulated calcium channels as well as on the activity of Ca^{2+}-dependent transducers such as protein kinases will undoubtfully help to identify the cellular responses underlying addiction-related processes such as drug tolerance, drug-induced sensitization or withdrawal syndromes. Additionally, this information will help to establish new strategies for the treatment of drug abuse. In fact, new therapies based on these Ca^{2+} signaling mechanism are already available (Tokuyama et al., 1995; Johnson et al., 1999). This is the case of the antagonist of the glutamate-activated Ca^{2+} channel termed NMDA receptor which has provided new drugs for the treatment of ethanol dependence in humans (Spanagel and Zieglgänsberger, 1997; Zerning et al., 1997).

2. ADDICTION AS A BRAIN DISEASE

Addiction or drug dependence is a chronically relapsing disorder characterized by a non-controlled compulsion to obtain and take the drug, as well as by the induction of neuroadaptions as results of repeated drug exposure (American Psychiatric Association, 1994; Koob and Le Moal, 1997). These adaptive responses are revealed as a negative emotional state (withdrawal syndrome composed of characteristic physical and behavioral manifestations) when access to the drug is prevented. The phenotype derived of such drug-induced neuroadaptions is equivalent to the status of drug dependence and occur in neurons belonging to the structurally and functionally-related elements of the brain that provides the anatomical substrate for motivated behaviors, the limbic system (Rodríguez de Fonseca and Navarro, 1998). Thus, drugs of abuse converge in a rather well defined limbic neural system, the *reward system*, involved in the prediction and evaluation of the natural positive and negative reinforcing signals associated to relevant experiences. Drug and drug-associated experiences converge to modify the reward system and configure the phenotype of the drug-vulnerable subject (Koob and Le Moal, 1997). The physical and dynamical alterations in the circuits that processes motivation and emotion are likely the substrate for the long-term changes in behavior that are a characteristic of drug users: conditioning, reward expectancy, loss of control of drug intake, drug craving and relapse (Koob and Le Moal, 1997; Rodríguez de Fonseca and Navarro, 1998). The main limbic subcircuits that process these different elements of addictive behavior have been elucidated. The contribution of each element of the limbic system to drug addiction is far from the aim of the present work and can be found elsewhere (Pich et al., 1995; Rodríguez de Fonseca et al., 1997; Rodríguez de Fonseca and Navarro, 1998). The existence of convergent actions of abused drugs on the functionality of limbic neurons has been demonstrated in animal models (Pich et al., 1995; Navarro et al., 1998; Pabello et al., 1998; Ledent et al., 1999; Rodríguez de Fonseca et al., 1999). This convergence is based on the existence of shared mechanisms for abused drugs in individual cells of the reward system. Although there are specific cellular targets for each class of abused drugs, as depicted in Figure 1, most drugs abused activate common cellular and molecular mechanisms that ultimately are the responsible of the transition between occasional drug use and the loss of control over drug self-administration (Nestler and Aghajanian, 1997; Koob et al., 1998). These common mechanisms include signaling mediated by second messengers such as cAMP or Ca^{2+}, and their correspondent transducers, protein kinases (PK) such as PKA, PKC, Ca^{2+} -calmodulin-PK or Fyn-kinase (Tokuyama et al., 1995; Miyakawa et al., 1997; Nestler and Aghajanian, 1997; Gerstin et al., 1998; Koob et al., 1998). Acute exposure to a certain drug is able of setting in motion short-term adaptive responses mediated by phosphorylation of membrane/cytoplasmic proteins. However, chronic exposure to the drug will result in protein kinase-dependent activation of transcription factors whose activity will lead to long-term neuroadaptions that characterize the dependent phenotype. Calcium-dependent signaling also undergoes this two-step process of adaptive responses characteristic of acute and chronic drug exposure (Dolin and Little, 1989; Goto et al., 1993; Bernstein and Welch, 1995).

3. MOLECULAR MECHANISMS MEDIATING ACUTE ACTIONS OF ABUSED DRUGS: THE PLACE OF Ca^{2+} SIGNALING

Figure 1 illustrates how drugs abused activate endogenous signaling mechanisms and converge on intracellular messengers, including Ca^{2+}, to produce their effects. It is important to note that acute actions of drugs abuse on Ca^{2+} signaling are just the starting point of a complex set of neuroadaptions that will be relevant for establishing the dependent phenotype after chronic drug exposure (Nestler and Aghajanian, 1997; Koob et al., 1998). The drug targets depicted in the scheme may not be located in the same cells. In fact, the different acute behavioral/physiological actions of abused drugs will be dependent on the distribution of that primary neuropharmacological target, whose distribution often goes beyond the neural circuits of the reward system. However, at the end all drugs abuse will activate the ascending mesocorticolimbic dopaminergic cells and their immediate connections (Koob et al., 1998). Thus, we can assume that all the cellular mechanisms activated by drugs abused will converge in a restricted set of neural cells by both, the direct action on drug targets present in reward-relevant neurons, or by the

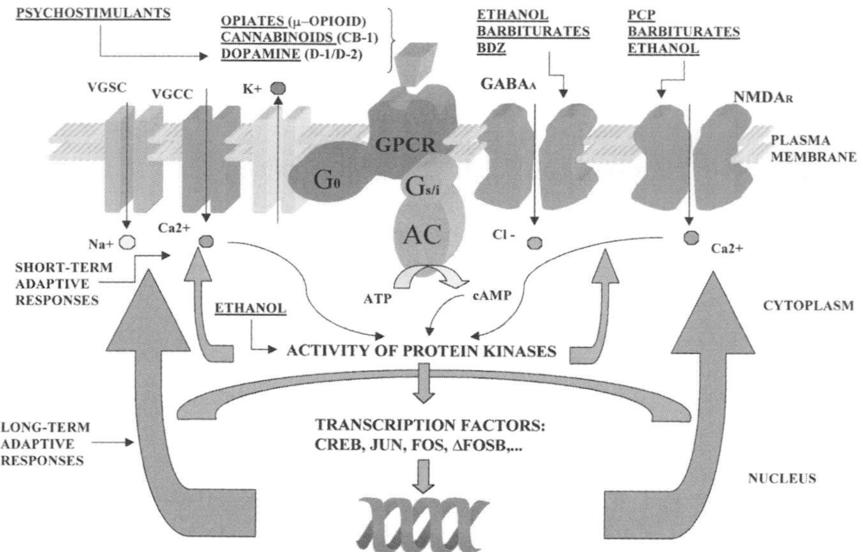

Figure 1. Place of calcium in molecular events associated with acute and chronic drug exposure. Four different types of cellular targets have been described for abused drugs: uptake systems, G-protein-coupled receptors (GPCR), ligand-gated ion channels and cytoplasmic protein kinases. Psychostimulants (cocaine, amphetamine) affect the functionality of monoamine uptake systems, increasing the synaptic concentrations of these transmitters, specially of dopamine, in reward-relevant synapses. Dopamine, as well as opiates or cannabinoids activate specific GPCR such as dopamine D-1 and D-2 families of receptors, the μ-opioid or the cannabinoid CB-1 receptor respectively. GPCR activation results in dynamic changes in second messengers such as cAMP and Ca^{2+}, through the G-protein mediated regulation of effectors such as adenylate cyclase (AC) or Ca^{2+} and K^+ channels. Ethanol, barbiturates, phencyclidine and benzodiazepines (BDZ) affect the functionality of $GABA_A$ or glutamate NMDA receptors ($NMDA_R$), leading to cellular hyperpolarization and to an elevation in cytoplasmic Ca^{2+} levels, respectively. Changes in cAMP and Ca^{2+} levels activate protein kinases, the later effect also directly elicited by ethanol. Protein kinases modulate the function of plasma membrane (i.e. voltage-gated calcium channels (VGCC) and membrane receptors such as GCPR, $GABA_A$ or $NMDA_R$) and transcription factors. Protein-kinase-mediated phosphorylation of plasma membrane proteins constitute the first stage in cellular homeostatic responses counteracting acute effects of drugs abuse. For instance, short-term adaptive cellular responses underlying acute ethanol tolerance are mediated by protein kinases (PK) such as PKC, PKA or Fyn-kinase, which directly modulate $GABA_A$ receptor activity and the desensitization of $NMDA_R$. Activation of transcription factors lead to long-term cellular changes responsible for the neuroadaptions associated with chronic tolerance, sensitization or dependence. These cellular neuroadaptions associated to long-term exposure to abused drugs include, among others, the up-regulation of the cAMP-PKA signaling pathway in opiate and cannabinoid-dependent animals or the up-regulation of L-type VGCC and modifications in subunit composition of glutamate $NMDA_R$ induced by chronic ethanol and opiates. Additional cellular proteins such as voltage-gated sodium channels (VGSC), G-proteins, neurotransmitter receptors, enzymes involved in transmitter synthesis, etc., can also be affected by chronic drug exposure. Other abused drugs such as nicotine or methylxantines (caffeine, etc.), not depicted here to improve clarity of the scheme, similarly modulate the activaty of specific endogenous signaling systems (cholinergic and purinergic respectively). For a colour version of this figure, see page xxvi.

transynaptic activation of mesolimbic dopaminergic neurons (and associated systems such as corticotropin-releasing factor neurons) elicited by the drug.

With the exception of ethanol that can also act intracellularly (Miyakawa et al., 1997), most common abused drugs act primarily by either activation G-protein-coupled receptors for neurotransmitters (GPCR), or modulation of ligand-activated ion channels. In the first case, this activation is produced by two mechanisms: (a) direct interaction with GPCR, as it is the case of opiates and cannabinoids, or (b) by increasing the synaptic availability of endogenous transmitters, such as dopamine, by blocking its uptake or promoting reverse transport from synaptic vesicle stores and cytoplasmic pool, as it is the case of psychostimulants (Nestler and Aghajanian, 1997; Koob et al., 1998). In the first situation drugs will act then as exogenous GPCR agonists, while in the second one they will work as indirect GPCR agonists (i.e. through the dopamine-mediated activation of dopamine D_1 and D_2 receptors). The GPCRs involved in the action of drugs abuse are preferentially coupled to the adenylate cyclase-protein kinase A transduction system (Nestler and Aghajanian, 1997). Other drugs, including ethanol, benzodiazepines, barbiturates or toluene act through ligand-activated ion channels such as the $GABA_A$ receptor complex (a Cl^- channel) and the glutamate NMDA receptor (a Ca^{2+} channel) (Littleton et al., 1992; Suzuki et al., 1993; Tabakoff and Hoffman, 1996; Miyakawa et al., 1997). Their acute effects are either a potentiation of the ionic conductance for Cl^- of the $GABA_A$ channel, resulting in sedation and anxiolysis, or a blockade of NMDA-Ca^{2+} currents. Lastly, ethanol has been proposed also to directly inhibit L-type of Ca^{2+} channels (Wang et al., 1994) and to modulate the activity of Ca^{2+}-dependent protein kinase C and Fyn-kinase which participate in the rapid desensitization of both $GABA_A$ and NMDA receptors (Miyakawa et al., 1997; Gerstin et al., 1998).

What is the repercussion of the activation of primary drug targets on Ca^{2+} signaling? Drugs acting directly or indirectly at GCPRs inhibit high-threshold Ca^{2+} currents through a G_0-mediated regulation of N-, P- and Q-type Ca^{2+} channels. These actions are rapidly counteracted by phosphorylation processes mediated by PKA, PKC or β-arrestin proteins (Nestler and Aghajanian, 1997; Koob et al., 1998). Additionally, these receptors can modulate the activity of glutamate NMDA receptors through its phosphorylation mediated by PKA. The activity of this kinase is dependent on the GCPR-modulation of intracellular levels of cAMP. Changes in cytoplasmic Ca^{2+} affect the activity of Ca^{2+}-dependent kinases, such as Ca^{2+}/calmodulin-dependent kinase II or PKC. The decrease in the activity of these proteins derived of the inhibition of Ca^{2+} influx is followed by a rapid up-regulation that likely participates in the neuroadaptions associated with chronic drug exposure (Nestler and Aghajanian, 1997; Gerstin et al., 1998; Koob et al., 1998).

Besides their effects on GABA-activated Cl^- channels, ethanol and barbiturates inhibit NMDA Ca^{2+} currents (Miyakawa et al., 1997; Wirkner et al., 1999). Ethanol effects on NMDA receptors is dependent on the expression of

Table 1. Effects of acute and chronic ethanol exposure on selected calcium signaling events

Effect	Reference
Acute Exposure	
Inhibition of $^{45}Ca^{2+}$ uptake	(Stokes and Harris, 1982)
Inhibition of voltage-dependent Ca^{2+} influx (L-channels)	(Wang et al., 1994)
Inhibition of glutamate NMDA receptors	(Miyakawa et al., 1997)
Chronic Exposure	
Increase of $^{45}Ca^{2+}$ uptake	(Lynch and Littleton, 1983)
Increase of voltage-dependent Ca^{2+} influx (L-channels)	(Goto et al., 1993)
Up-regulation of L-type Ca^{2+} channels	(Gerstin et al., 1998)
Reversal of ethanol-mediated inhibition of NMDA receptors	(Miyakawa et al., 1997)
Up-regulation of glutamate NMDA receptors	(Tabakoff and Hoffman, 1996)

a particular subunit combination of the Ca^{2+} channel. Of them, NR1/NR2A and NR1/NR2B combinations are preferentially sensitive to ethanol inhibition. The result of this NMDA inhibition is the depression of excitatory synaptic potentials mediated by this type of excitatory aminoacid receptor (Tabakoff and Hoffman, 1996; Wirkner et al. 1999). This effect can be added to the hyperpolarization elicited by the ethanol- and barbiturate-induced potentiation of Cl^- influx through the $GABA_A$ complex. The inhibitory effects of barbiturates and ethanol are rapidly blocked by phosphorylation of the NMDA receptor by several kinases, including the Fyn-kinase (Miyakawa et al., 1997; Gerstin et al., 1998). This phosphorylation is responsible of the development of acute tolerance described for ethanol. Lastly, acute ethanol exposure induces a characteristic inhibition of K^+-evoked Ca^{2+} uptake and L-type Ca^{2+} currents (see Table 1).

The overall analysis of the effects of drugs abuse in Ca^{2+} signaling points to a widespread inhibition of Ca^{2+} influx through voltage-gated Ca^{2+} channels and NMDA receptors. They contribute to the acute reinforcing processes elicited by the drugs and natural reinforcers (Calcagnetti and Schechter, 1992; Koob and Le Moal, 1997). These effects are mediated mainly through a direct actions or through GCPRs, and are rapidly counteracted by phosphorylation mechanisms mediated by a wide set of intracellular kinases (Fan et al., 1999).

4. NEUROADAPTIONS ASSOCIATED WITH CHRONIC DRUG EXPOSURE: L-TYPE Ca^{2+} CHANNELS AND GLUTAMATE NMDA RECEPTORS

As we have discussed above, several molecular mechanisms, including Ca^{2+} signaling, are recruited by drug stimulation as a second stage in cellular signaling. They have been found to contribute not only to the acute effects of drugs but to the long-term changes induced by repeated drug exposure. In this line, second messengers such as cAMP or specific transcriptions factors such as Jun, Fos, ΔFosB or CREB participate in the transition to the drug-dependence status (Nestler and Aghajanian, 1997). The actions of these transcription factors on specific genes such as those coding for voltage-gated Ca^{2+} channels or local mechanisms regulating synaptic transmission such as those involved in neurotransmitter storage and release contribute to the adaptive plastic alterations in synaptic transmission in reward-relevant circuits characterized in drug-dependent animals.

Chronic neuroadaptions induced by drugs abuse characteristically oppose the acute effects induced by the drug (Koob and Le Moal, 1997). A classical example is the up-regulation of the cAMP/PKA signaling pathway found after chronic exposure to opiates or cannabinoids. As we can see in Table 1 and Figure 1, this counteregulatory response accounts also for Ca^{2+} signaling processes. In fact, the interference with Ca^{2+} signaling neuroadaptions involving L-type Ca^{2+} channels and NMDA receptors has been reported to prevent the development of opiate, ethanol and barbiturate tolerance and dependence, a finding of important therapeutic implications. Most neuroadaptions described in Ca^{2+} signaling mechanisms reverse the drug-induced acute inhibition of Ca^{2+} influx through voltage-dependent Ca^{2+} channels and NMDA receptors as well as the decrease in the activity of Ca^{2+}-dependent kinases. These adaptions include the up-regulation of the expression of the genes coding for these proteins. For instance, chronic ethanol induce the up-regulation of L-type Ca^{2+} channels (an effect mediated by PKCdδ) the up-regulation of NMDA receptor, changes in subunit composition of $GABA_A$ and NMDA receptors and up-regulation of PKC isozymes (see Tables 1 and 2). All these adaptive responses are closely associated with an increase in the amount of ethanol drunk by dependent subjects. Similar mechanisms have been described in opiate and barbiturate dependence.

One of the important consequences of the up-regulation of Ca^{2+} channels associated with chronic drug exposure is the well characterized increase in electrical excitability associated with drug withdrawal (Caro et al., 1988; Pérez-Vázquez et al., 1994; Ripley and Little, 1995; Krystal et al., 1996; Bailey et al., 1998). Cessation of drug use removes the inhibitory action of the drug and releases the opposed nature of the neuroadaptive processes. This phenomenon is markedly present during withdrawal associated to central nervous system depressants such as ethanol and barbiturates. In fact, seizures are a common feature of ethanol or barbiturate withdrawal syn-

Table 2. Effects of L-type Ca^{2+} channel blockers and NMDA receptor antagonists in human addicts and in animal models of drug addiction

Effect	Reference
L-type Ca^{2+} channel blockers	
Prevention of ethanol tolerance	(Dolin and Little, 1989)
Blockade of ethanol withdrawal	(Whittington and Little, 1991; Watson et al., 1994)
Blockade of ethanol self-administration	(Kuzmin et al., 1999)
Prevention of opiate dependence	(Tokuyama et al., 1995)
Reduction of opiate withdrawal	(Barrios and Baeyens, 1991)
Blockade of amphetamine reward in humans	(Johnson et al., 1999)
Reduction of nicotine self-administration	(Martellota et al., 1995)
Prevention of barbiturate tolerance	(Suzuki et al., 1993)
NMDA receptor antagonists	
Decrease of ethanol intake and craving	(Spanagel and Zieglgänsberger, 1997)
Decrease of cocaine self-administration	(Pulvirenti et al., 1992)
Decrease of psychostimulant-induced hyperactivity	(Pulvirenti et al., 1994)
Prevention of opiate tolerance and reduction of opiate withdrawal symptoms	(Herman et al., 1995)

dromes. Supporting this, antagonists of the L-type of Ca^{2+} channels can attenuate ethanol and barbiturate withdrawal symptomatology (Table 2). This electrical excitability is also contributing to the overt signs associated with opiate withdrawal, since several symptoms associated with opiate cessation can also be antagonized by dihydropyridines (Schnur et al., 1992; Zharkovsky et al., 1993; Kishioka et al., 1996) . Neuropharmacological markers of drug withdrawal such as the depression in mesolimbic dopamine outflow (Rosetti et al., 1999), or the increased release of glutamate in the locus coeruleus, can also be reversed by L-type Ca^{2+} channel antagonists (Tokuyama and Ho, 1996).

Lastly, Ca^{2+} signaling is also associated with the development of sensitization to psychostimulants (Kalivas et al., 1992). Sensitization, the increase in the behavioral response to psychostimulants derived of previous exposures to the drug, is thought to contribute to cocaine and amphetamine addiction. Although apparently opposed to the development of tolerance described before, sensitization processes converge through the same signaling mechanisms altered by other abused drugs. Thus, L-type and N-type of Ca^{2+} channel antagonists injected in the mesolimbic circuit attenuates the expression of behavioral sensitization to cocaine. The same effect could be observed by inhibiting Ca^{2+}/calmodulin-dependent protein kinase II and calcium-dependent protein kinase C (Pierce et al., 1998).

5. THERAPEUTIC IMPLICATIONS OF CALCIUM CHANNELS IN DRUG ADDICTION

What is the clinical relevance of Ca^{2+} signaling neuroadaptions associated with drug abuse? From the above described findings, Ca^{2+} signaling contribute to several processes that constitute the natural history of drug addiction: acute and chronic tolerance, sensitization, dependence and withdrawal syndromes. There are many examples derived from basic research that point to the utility of drugs that modulate Ca^{2+} channels in the therapy of addiction. Table 2 shows the potential utility of L-type Ca^{2+} channels antagonists as well as the already in use applications of NMDA receptor antagonists for the prevention of tolerance and dependence, the attenuation of withdrawal syndromes and the suppression of drug craving. Thus, dihydropyridines attenuates psychostimulant, opiate, ethanol and nicotine self-administration, a finding reproduced with NMDA antagonists such as MK 801 or acamprosate (see Table 2). This latest drug is already under study for the treatment of alcoholism. Acamprosate reduced ethanol intake in rodents.This finding, together with its ability to influence the activity of voltage-gated Ca^{2+} channels, as well as its antagonistic effect on NMDA receptor function and on the transcription of NMDA receptor subunit prompted the first clinical trials in humans (Spanagel and Zieglgänsberger, 1997; Zerning et al., 1997). Following this rationale, recent studies have demonstrated that pretreatment with the dihydropyridine isradipine, a L-type channel antagonist attenuated some of the rewarding effects of d-methamphetamine in humans (Bankole et al., 1999). Although there are only few trials already in curse, we should expect that future studies will explore these basic findings to identify the real contribution of calcium channels-based therapeutic approaches to the treatment of drug addiction.

ACKNOWLEDGEMENTS

This work has been supported by Comunidad de Madrid (grant 08.5/0013/98), DGICYT (grant PM-96/0047) and Delegación del Gobierno para el Plan Nacional Sobre Drogas.

REFERENCES

American Psychiatric Association, 1994, *Diagnostic and Statistical Manual of Mental Disorders*, 4th edn., American Psychiatric Press, Washington.

Bailey, C.P., Molleman, A. and Little, H.J., 1998, Comparison of the effects of drugs on hyperexcitability induced in hippocampal slices by withdrawal from chronic ethanol comsumption, *Br. J. Pharmacol.* 123, 215–222.

Bankole, A.J., Roache, J.D., Bordnick, P.S. and Ait-Daoud, N., 1999, Isradipine, a dihydropyridine-class calcium channel antagonist, attenuates some of d-methamphetamine's positive subjective effects: A preliminary study, *Psychopharmacology* 144, 295–300.

Barrios, M. and Baeyens, J.M., 1991, Different effects of L-type calcium channel blockers and stimulants on naloxone-precipitated withdrawal in mice acutely dependent on morphine, *Psychopharmacology* 104, 397–403.

Bernstein, M.A. and Welch, S.P., 1995, Alterations in L-type calcium channels in the brain and spinal chord of acutely treated and morphine-tolerant mice, *Brain Res.* 696, 83–88.

Bongianni, F., Carla, V., Moroni, F. and Pellegrini-Gianpietro, D.E., 1988, Calcium channels inhibitors suppress the morphine-withdrawal syndrome in rats, *Br. J. Pharmacol.* 88, 561–567.

Calcagnetti, D.J. and Schechter, M.D., 1992, Attenuation of drinking sweetened water following calcium channel blockade, *Brain Res. Bull.* 28, 967–973.

Caro, G., Barrios, M. and Baeyens, J.M., 1988, Dose-dependent and stereoselective antagonism by diltiazem of naloxone-precipitated morphine abstinence after acute morphine-dependence in vivo and in vitro, *Life Sci.* 43, 1523–1527.

Crabbe, J.C., Phillips, T.J., Buck, K.J., Cunningham, C.L. and Belknap, J.K., 1999, Identifying genes for alcohol and drug sensitivity: Recent progress and future directions, *Trends Neurosci.* 22, 173–179.

Dolin, S.J. and Little, H.J., 1989, Are changes in neuronal calcium channels involved in ethanol tolerance?, *J. Pharmacol. Exp. Ther.* 250, 985–991.

Dolin, S., Little, H., Hudspith, M., Pagonis, C. and Littleton, J., 1987, Increased dihydropyridine-sensitive calcium channels in rat brain may underlie ethanol physical dependence, *Neuropharmacology* 26, 275–279.

Fan, G.H., Wang, L.Z., Qiu, H.C., Ma, L. and Pei, G., 1999, Inhibition of calcium/calmodulin-dependent protein kinase II in rat hippocampus attenuates morphine tolerance and dependence, *Mol. Pharmacol.* 56, 39–45.

Gerstin, E.H., McMahon, T., Dadgar, J. and Messing, R.O., 1998, Protein kinase C mediates ethanol-induced up-regulation of L-type calcium channels, *J. Biol. Chem.* 273, 16409–16414.

Goto, M., Lemasters, J.J. and Thurman, R.G., 1993, Activation of voltage-dependent calcium channels in kupfer cells by chronic treatment with alcohol in the rat, *J. Pharmacol. Exp. Ther.* 267, 1264–1268.

Guppy, L.J., Crabbe, J.C., Littleton, J.M., 1995, Time course and genetic variation in the regulation of calcium channel antagonist binding sites in rodent tissues during the induction of ethanol physical dependence and withdrawal, *Alcohol Alcohol* 30, 607–615.

Herman B.H., Vocci, F. and Bridge, P., 1995, The effects of NMDA receptor antagonists and nitric oxide synthase inhibitors on opioid tolerance and withdrawal. Medication development issues for opiate addiction, *Neuropsychopharmacology* 13, 269–293.

Johnson, B.A., Roache, J.D., Bordnick P.S. and Ait-Daoud, N., 1999, Isradipine, a dihydropyridine-class calcium channel antagonist, attenuates some of d-amphetamine's positive subjective effects: A preliminary study, *Psychopharmacology* 144, 295–300.

Kalivas, P.W., Striplin, C.D., Steketee, J.D., Klitenick, M.A. and Duffy, P., 1992, Cellular mechanisms of behavioral sensitization, *Ann. N.Y. Acad. Sci.* 654, 128–134.

Kishioka, S., Inoue, N., Nishida, S., Fukunaga, Y. and Yamamoto, H., 1996, Diltiazem inhibits naloxone-precipitated and spontaneous morphine withdrawal in rats, *Eur. J. Pharmacol.* 361, 7–14.

Koob, G.F. and Le Moal, M., 1997, Drug abuse: Hedonic homeostatic dysregulation, *Science* 278, 52–57.

Koob, G.F., Sanna, P.P. and Bloom, F.E., 1998, Neuroscience of addiction, *Neuron* 21, 467–476.

Krystal, J.H., Compere, S., Nestler, E.J. and Rasmussen, K., 1996, Nimodipine reduction of naltrexone-precipitated locus coeruleus activation and abstinence behavior in morphine-dependent rats, *Physiol. Behav.* 59, 863–866.

Kuzmin, A., Semenova, S., Zvartau, E. and De Very, J., 1999, Effects of calcium channel blockade on intravenous self-administration of ethanol in rats, *Eur. Neuropsychopharmacol.* 9, 197–203.

Ledent, C., Valverde, O., Cossu, G., Petitet, F., Aubert, J.F., Beslot, F., Bohme, G.A., Imperato, A., Pedrazzini, T., Roques, B.P., Vassart, G., Fratta, W. and Parmentier, M., 1999, Unresponsiveness to cannabinoids and reduced addictive effects of opiates in CB1 receptor knockout mice, *Science* 283, 401–404.

Littleton, J., Little, H. and Laverty, R., 1992, Role of neuronal calcium channels in ethanol dependence: From cells cultures to the intact animal, *Ann. N.Y. Acad. Sci.* 654, 324–334.

Lynch, M.A. and Littleton, J.M., 1983, Possible association of alcohol tolerance with increased synaptic Ca^{2+} sensitivity, *Nature* 303, 175–177.

Martellota, M.C., Kuzmin, A., Zvartan, Z., Cossu, G., Gessa, G.L. and Fratta, W., 1995, Isradipine inhibits nicotine intravenous self-administration in drug naive mice, *Pharmacol. Biochem. Behav.* 52, 271–274.

Miyakawa, T., Yagi, T., Kitazawa, H., Yasuda, M., Kawai, N., Tsuboi, K. and Niki, H., 1997, Fyn-kinase as a determinant of ethanol sensitivity: Relation to NMDA-receptor function, *Science* 278, 698–701.

Navarro, M., Chowen, J.A., Carrera, M.R.A., Del Arco, I., Villanúa, M.A., Martín, Y., Roberts, A., Koob, G.F. and Rodríguez de Fonseca, F.A., 1998, CB-1 cannabinoid receptor antagonist-induced opiate withdrawal in morphine-dependent rats. *NeuroReport* 9, 3397–3402.

Nestler, E.J. and Aghajanian, G.K., 1997, Molecular and cellular basis of addiction, *Science* 278, 58–62.

O'Brien, C.P., 1997, A range of research-based pharmacotherapies for addiction, *Science* 278, 66–70.

Pabello, N.G., Hubbell, C.L., Cavallaro, C.A., Barringer, T.M., Mendez, J.J. and Reid, L.D., 1998, Responding for rewarding brain stimulation: Cocaine and isradipine plus naltrexone, *Pharmacol. Biochem. Behav.* 61, 181–192.

Pérez-Vázquez, J.L., Valiante, T.A. and Carlen, P.L., 1994, Changes in calcium currents during ethanol withdrawal in a genetic mouse model, *Brain Res.* 649, 305–309.

Pich, E.M., Lorang, M., Yeganeh, M., Rodríguez de Fonseca, F., Raber, J., Koob, G.F. and Weiss, F., 1995, Increase of extracellular corticotropin-releasing factor-like inmunoreactivity levels in the amygdala of awake rats during restraint stress and ethanol withdrawal as measured by mycrodialysis, *J. Neurosci.* 15, 5439–5447.

Pierce, R.C., Quick, E.A., Reeder, D.C., Morgan, Z.R. and Kalivas, P.W., 1998, Calcium-mediated second messengers modulate the expression of behavioural sensitization to cocaine, *J. Pharmacol. Exp. Ther.* 286, 1171–1176.

Pulvirenti, L., Maldonado, R. and Koob, G.F., 1992, NMDA receptors in the nucleus accumbens modulate intravenous cocaine but not heroin self-administration in the rat, *Brain Res.* 594, 327–330.

Pulvirenti, L., Berrier, R., Kriefeldt, M. and Koob, G.F., 1994, Modulation of locomotor activity by NMDA receptors in the nucleus accumbens core and shell regions of the rat, *Brain Res.* 664, 231–236.

Ripley, T.L. and Little, H.J., 1995, Effects on ethanol withdrawal hyperexcitability of chronic treatment with a competitive N-methyl-D-Aspartate receptor antagonist, *J. Pharmacol. Exp. Ther.* 272, 112–118.

Rodríguez de Fonseca, F. and Navarro, M., 1998, Role of the limbic system in dependence on drugs, *Ann. Med.* 30, 397–405.

Rodríguez de Fonseca, F., Carrera, M.R., Navarro, M., Koob, G.F. and Weiss, F., 1997, Activation of corticotropin-releasing factor in the limbic system during cannabinoid withdrawal, *Science* 276, 2050–2054.

Rodríguez de Fonseca, F., Roberts, A.J., Bilbao, A.J., Koob, G.F. and Navarro, M., 1999, Cannabinoid receptor antagonist SR141716A decreases operant ethanol self administration in rats exposed to ethanol-vapor chambers, *Acta Pharmacol. Sinica* 20, 1109–1114.

Rosseti, Z.L., Isola, D., De Very, J. and Fadda, F., 1999, Effects of nimodipine on extracellular dopamine levels in the rat nucleus accumbens in ethanol withdrawal, *Neuropsychopharmacology* 38, 1361–1369.

Schnur, P., Espinoza, M., Flores, R., Ortiz, S., Vallejos, S. and Wainwright, M., 1992, Blocking naloxone-precipitated withdrawal in rats and hamsters, *Pharmacol. Biochem. Behav.* 43, 1093–1098.

Spanagel, R. and Zieglgänsberger, W., 1997, Anti-craving compounds for ethanol: New pharmacological tools to study addictive process, *Trends Pharmacol. Sci.* 18, 54–59.

Spanagel, R., Sillaber, I., Zieglgänsberger, W., Corrigall, W.A., Stewart, J. and Shaham, Y., 1998, Acamprosate suppresses the expression of morphine-induced sensitization in rats but does not affect heroin self-administration or relapse induced by heroin or stress, *Psychopharmacology* 139, 391–401.

Stokes, J.A. and Harris, R.A., 1982, Alcohols and synaptosomal calcium transport, *Mol. Pharmacol.* 22, 99–104.

Suzuki, T., Mizoguchi, H., Noguchi, H., Yoshii, T. and Misawa, M., 1993, Effects of flunarizine and diltiazem on physical dependence on barbital in rats, *Pharmacol. Biochem. Behav.* 45, 703–712.

Tabakoff, B. and Hoffman, P.L., 1996, Alcohol addiction: An enigma among us, *Neuron* 16, 909–912.

Tokuyama, S. and Ho, I., 1996, Inhibitory effects of diltiazem, an L-type Ca^{2+} channel blocker, on naloxone-increased glutamate levels in the locus coeruleus of opioid-dependent rats, *Brain Res.* 722, 212–216.

Tokuyama, S., Feng, Y., Wakabayashi, H. and Ho, I.K., 1995, Ca^{2+} channel bloker diltiazem, prevents physical dependence and the enhancement of protein kinase C activity by opioid infusion in rats, *Eur. J. Pharmacol.* 279, 92–98.

Twitchell, W., Brown, S. and Mackie, K., 1997, Cannabinoids inhibit N- and P/Q-type calcium channels in cultured rat hippocampal neurons, *J. Neurophysiol.* 78, 43–50.

Wang, X., Wang, G., Lemos, J.R. and Treitsman, S.N., 1994, Ethanol directly modulates gating of a dyhidropyridine-sensitive Ca^{2+} channel in neurohypophysial terminals, *J. Neurosci.* 14, 5443–5460.

Watson, W.P., Misra, A., Cross, A.J., Green, A.R. and Little, H.J., 1994, The differential effects of felodipine and nitrendipine on cerebral dihydropyridine binding ex vivo and the ethanol withdrawal syndrome in mice, *Br. J. Pharmacol.* 112, 1017–1024.

Whittington, M.A. and Little, H.J., 1991, A calcium channel antagonist stereoselectively decreases ethanol withdrawal hyperexcitability but not that due to biccuculine, in hippocampal slices, *Br. J. Pharmacol.* 103, 1313–1320.

Wirkner, K., Poelchen, W., Koles, L., Muhlberg, K., Scheibler, P., Allgaier, C. and Illes, P., 1999, Ethanol-induced inhibition of NMDA receptor channels, *Neurochem. Int.* 35, 153–162.

Zerning, G., Fabisch, H. and Fabish, H., 1997, Pharmacotherapy of alcohol dependence, *Trends Pharmacol. Sci.* 18, 229–231.

Zharkovsky, A., Totterman, A.M., Moisio, J. and Ahtee, L., 1993, Concurrent nimodipine attenuates the withdrawal signs and the increase in cerebral dihydropiridine binding sites after chronic morphine treatment in rats, *Naunyn Schmiedeberg Arch. Pharmacol.* 347, 483–486.

S100 Proteins and Fatty Acid Transport Are Altered in Skin Diseases

Gerry Hagens and Georges Siegenthaler

1. INTRODUCTION

Fatty acids (FAs) are essential components for every living cell. FAs are used e.g. as metabolic fuels, as precursors of membrane lipids, or as bioactive molecules that participate in cell signaling. Disorders of FA-levels result in many diseases, indicating the existence of mechanisms that regulate FA-levels and the many diverse FA-tasks. Such regulatory mechanisms include the trafficking of FAs. In fact, FAs are highly hydrophobic and unstable molecules. They form poorly soluble salts (soap) with Ca^{2+} and possess surfactant properties that can become harmful to the cell. Therefore, FAs must be transported, solubilized and protected by FA-carriers.

FA-transport in the blood circulation and FA-translocation across the cell membrane involve albumin, albumin receptors (Antohe et al., 1993), and proteins anchored in the plasma membrane (Glatz et al., 1997). Translocation of FAs might also be carried out via a passive process of FA-diffusion and -partition between the membrane leaflets (Kamp and Hamilton, 1993). Once internalized, FAs are transported by small molecular weight cytoplasmic fatty acid-binding proteins (FABPs) (Glatz and van der Vusse, 1996) or the very recently described FA-p34 complex (Siegenthaler et al., 1997).

FABPs comprise a family of at least 8 proteins with a molecular mass of about 15 kDa and the members of that family are named according to the tissue of their first isolation (for a review, see Veerkamp and Maatman, 1995). FABPs are widely distributed among different cell types and several observations suggest that the major function of FABPs is the intracellular transport of FAs and their targeting to specific metabolic pathways. Each FABP type displays a characteristic pattern of tissue expression and distinct

---

Gerry Hagens and Georges Siegenthaler • Clinique de Dermatologie and DHURDV, University Hospital Geneva, Geneva, Switzerland.

R. Pochet, R. Donato, J. Haiech, C. Heizmann and V. Gerke (eds.): Calcium: The Molecular Basis of Calcium Action in Biology and Medicine, 477–492.

ligand-binding properties, suggesting specific and complementary functions for each protein. Analyses on the tertiary structure of members of this family revealed that FABPs resemble a clamp, and that the FA-binding site lays within the clamp in a hydrophobic pocket (for a review, see Glatz and van der Vusse, 1996).

Findings that might indicate the existence of regulatory mechanisms for FABP functions have emerged from the characterization of the epidermal-type FABP (E-FABP), which associates with the Ca^{2+}-binding protein (CaBPs) S100A7. The discovery of this E-FABP/S100A7 complex led to the hypothesis that FA-transport might be generally regulated by CaBPs and that a Ca^{2+}/FA cross-talk might exist.

This hypothesis has been reinforced by the identification of the heterocomplex FA-p34 complex of 34 kDa that is composed of the CaBPs S100A8 and S100A9. These S100 proteins belong to a multigenic family of Ca^{2+}-binding proteins of the EF-hand type that are differentially expressed in a large number of cell types. Members of these proteins have been implicated in many diverse Ca^{2+}-dependent processes like regulation of protein phosphorylation, enzyme activities, cell proliferation, differentiation, cytoskeletal interactions, or inflammation (for a review, see Donato, 1999).

The next sections will describe the two complexes in more detail and present mechanisms that control complex formation. We will also discuss the expression of the complexes in certain biological systems, that are characterized by a cross-talk between FAs and Ca^{2+}, and what role the complexes might play in these systems.

2. THE E-FABP/S100A7 COMPLEX

Most S100 proteins are known to interact with other proteins (for a review, see Heizmann and Cox, 1998). This is also true for S100A7 (also called psoriasin, Madsen et al., 1991), which binds non-covalently to the epidermal-type fatty acid-binding protein E-FABP (Siegenthaler et al., 1994). First evidences for a complex formation between the two proteins were obtained by purifying E-FABP under native conditions, whereby S100A7 co-purified with E-FABP (Hagens et al., 1999b). Further investigations revealed that S100A7 co-immunoprecipitated with E-FABP using an antiserum directed against E-FABP. The high specificity of the E-FABP/S100A7 complex formation was proven by showing that other S100 proteins such as S100A8 and S100A9 did not significantly bind to E-FABP (Hagens et al., 1999a).

E-FABP was purified from human keratinocytes and transports saturated as well as unsaturated FAs in these cells. Stearic acid, followed by linoleic and oleic acid, has the highest affinity for E-FABP (Siegenthaler et al., 1994). S100A7 is devoid of any FA-binding capacity (Hagens et al., 1999a). Keratinocytes express a cytoplasmic and a membrane-attached form (Bürgisser et al., 1995). Both forms are expressed as homodimers with a molecular mass of about 22 kDa (Bürgisser et al., 1995; Brodersen et al., 1999) and can be distinguished by SDS-PAGE. Surprisingly, the two forms do not differ in their molecular masses as revealed by mass spectrometry, suggesting that conformational differences might be responsible for the different electrophoretic mobilities. The presence of two different S100A7 forms also indicates that the protein might be able to translocate from the cytosol to the membrane, which goes along with important conformational changes.

In vitro, the E-FABP/S100A7 complex is reconstituted only in presence of the chelator EDTA, suggesting that bivalent cations might regulate complex formation. As for most S100 proteins, binding of the bivalent cations Ca^{2+} or Zn^{2+} leads to conformational changes on the proteins. Ca^{2+} induces only minor conformational changes on S100A7, whereas Zn^{2+}-binding has profound effects on S100A7 conformation (Brodersen et al., 1999). This is in line with our observation, showing that Zn^{2+} strongly inhibits formation of or disrupts the E-FABP/S100A7 complex, whereas Ca^{2+} has only minor effects. Taken together, E-FABP/S100A7 complex formation at least *in vitro* is regulated by a balance of Ca^{2+} and Zn^{2+} concentrations. Binding of FAs to E-FABP/S100A7 complex does not affect complex stability and it is thought that in keratinocytes FA-transport is performed by free E-FABP as well as by the E-FABP/S100A7 complex.

3. THE FA-p34 COMPLEX

FA-p34 is a heterocomplex composed by the non-covalent association of S100A8 and S100A9 (also called MRP8 and MRP14, respectively). The complex has been characterized in human psoriatic keratinocytes (kFA-p34) (Siegenthaler et al., 1997) and in human neutrophils (nFA-p34) (Roulin et al., 1999).

The separate subunits S100A8 and S100A9 do not exhibit FA-binding properties, whereas the reconstitution of the FA-p34 heterocomplex leads to the acquisition of the FA-binding capacity. Further *in vitro* analyses revealed that FA-p34 has a high affinity for unsaturated FAs like arachidonic acid and linoleic acid, both precursors and mediators of inflammation. FA-p34 shows poor affinity for saturated FAs, which have mainly structural and energetic functions, and no binding to retinoids. These *in vitro* data were confirmed by the identification of an *in vivo* ligand for FA-p34. Indeed, linoleic acid was the predominant ligand bound to purified kFA-p34 as shown by GC analysis (Roulin et al., 1999). However, FA-p34 might most probably also bind other unsaturated FAs *in vivo*.

Densitometrical analysis of k and nFA-p34 subunits separated by 2D gel electrophoresis suggests that FA-p34 associates with a $(S100A8)_2S100A9$ stoichiometry. This proposed stoichiometry is in line with the molecular mass of 34 kDa measured by gel filtration (Siegenthaler et al., 1997) and reports describing the existence of a trimer composed of two S100A8 and one S100A9 protein (Edgeworth et al., 1991; Teigelkamp et al., 1991). However, S100A8 and S100A9 preferably form heterodimers (Hunter and Chazin, 1998), and a murine heterodimer composed of recombinant S100A8 and S100A9 has been found to bind arachidonic acid (Klempt et al., 1997). Thus, the exact stoichiometry of the FA-p34 complex remains to be determined.

2D gel analysis of purified kFA-p34 and nFA-p34 further revealed that each of the two complexes should be considered as a heterogenous family of complexes (Roulin et al., 1999). In fact, due to an alternative translation initiation site, the S100A9 subunit exists as a truncated as well as a native S100A9 form and both forms can be phosphorylated in a protein kinase C (PKC)-dependent or -independent way (Edgeworth et al., 1989; Roth et al., 1993; Guignard et al., 1996). Also, S100A8 has been found to be expressed as a phosphorylated protein (Guignard et al., 1996). Concerning FA-p34, the biggest part of the complexes seems to be composed of two particular S100A8 and two particular S100A9 isoforms, that still have to be identified unambiguously. However, it is not excluded that other subunit isoforms, present in lower quantities, become involved in complex formation (Roulin et al., 1999).

The domains that are responsible for the proper assembly of the FA-p34 complex must still be identified. Since S100 proteins contain two hydrophobic and two ionic domains (Kligman and Hilt, 1988), the complementary association of hydrophobic and ionic domains on S100 proteins might result in a complex, and the specific juxtaposition of hydrophobic domains creates a hydrophobic pocket that binds FAs.

FA-p34 complex formation might be controlled by several mechanisms. Since it is unlikely that post-translational modifications significantly contribute to the control of complex formation, purified FA-p34 contains all subunit isoforms in more or less high amounts. Post-translational modifications might rather control the function of the complexes as will be discussed later. S100 proteins are characterized by two Ca^{2+}-binding structural motifs, called EF-hands (Kligman and Hilt, 1988). Binding of Ca^{2+} to S100A8 and S100A9 results in conformational changes that regulate interactions with other proteins (Edgeworth et al., 1991, Teigelkamp et al., 1991). Since the presence of Ca^{2+} is a prerequisite for FA-p34 complex formation, Ca^{2+}-mediated conformational changes of the subunits allow their right association, thereby creating a FA-binding site. Whether other bivalent cations influence complex formation has still to be determined. Fact is that most S100 proteins bind also Zn^{2+} and Mg^{2+} reversibly at specific sites and some S100 proteins bind these cations in a cooperative way (Heizmann and Cox, 1998). Thus, Zn^{2+} concentration might also be another parameter involved in the control of complex formation and/or function. Other investigations will show whether this is the case and whether FA-p34 formation/disruption is dependent also on other cations.

4. FUNCTIONS OF THE FA-p34 AND E-FABP/S100A7 COMPLEXES

Investigating the expression patterns of the complexes and their subunits under normal and pathological conditions might provide important clues about the physiological significance of such complex formations. This section will introduce biological systems that are characterized by a Ca^{2+}/FA cross-talk and discuss the roles the complexes might have in these systems.

4.1. The Ca^{2+}/FA cross-talk: Examples from human skin and neutrophils

In normal skin and in culture, keratinocyte differentiation is recognized by the formation of several epidermal cell layers (stratification). Starting from the first cell layer (proliferative compartment), keratinocytes differentiate and participate in the formation of the stratum spinosum and the stratum granulosum cell layers and finally become part of the cornified cell envelope (stratum corneum) at the skin surface. One of the key factors that control keratinocyte differentiation is the extracellular Ca^{2+}-gradient. In normal skin Ca^{2+}-levels are low in the lower cell layers and increase progressively reaching the highest level in the outermost cell layers. It is now established that Ca^{2+} is intimately involved in complex mechanisms regulating keratinocyte growth and differentiation, and Ca^{2+} may be an important trigger for several terminal events associated with cornification. Ca^{2+} has been shown to act in the late stage of the S phase, just prior to DNA synthesis, inhibiting cell proliferation and initiating differentiation (Eglijo et al., 1986). In addition, Ca^{2+} activates several enzymatic systems like the Ca^{2+}-dependent transglutaminase, which is responsible for the cross-linking of cornified envelope precursors (Dale et al., 1994). How the epidermal Ca^{2+}-gradient is established and stabilized remains to be determined but it is generally believed that an array of CaBPs, among which figure S100 proteins (Robinson et al., 1997), play an important role in these tasks.

One of the most important functions of skin is to protect the body from water evaporation. Transepidermal water loss is controlled by lipids that are secreted by differentiating keratinocytes via lamellar bodies and arranged in the intercellular spaces of epidermis. The formation of this protective lipid barrier is tightly controlled by mechanisms involved in keratinocyte differentiation. In fact, marked changes in lipid composition of successive epidermal cell layers indicate that FA-metabolism and transport are modulated during Ca^{2+}-dependent differentiation of keratinocytes (Lampe et al., 1983). Removing the epidermal lipid barrier affects the epidermal Ca^{2+}-gradient indicating the presence of a Ca^{2+}/FA cross-talk. The immediate consequence of a lipid barrier disruption is an alteration of the Ca^{2+}-gradient with higher cell layers containing decreased Ca^{2+}-levels. This in turn, leads to an accelerated lipid synthesis and secretion (Menon et al., 1994). However, not only lipid, but also DNA synthesis is stimulated upon disruption of the lipid barrier (Proksch et al., 1993), and it is thought that stimulation of DNA synthesis leading to epidermal hyperplasia may be a second mechanism by which the epidermis tries to repair defects in barrier function.

Psoriasis is a skin disease characterized by hyperproliferative and abnormally differentiated keratinocytes. The epidermal Ca^{2+}-gradient is strongly altered in psoriatic skin (Forslind et al., 1999), and abnormally high Zn^{2+}-levels were recorded in the stratum granulosum and corneum regions of psoriatic lesions (Proksch et al., 1993). The hyperproliferative and abnormally differentiated phenotype of psoriatic keratinocytes reflects most probably consequences of altered Ca^{2+}-concentrations in the different cell layers. These changes go along with an impaired epidermal lipid barrier function (Motta et al., 1994), which is accompanied by an altered and accelerated FA-metabolism, and an increased FA-release (Kragballe and Voorhees, 1987; Fogh et al., 1989; Menon et al., 1994). In this context, it has been suggested that diseased cells accelerate lipid metabolism in order to re-establish a functional lipid barrier (Menon et al., 1994). Additionally, the altered lipid composition indicates that different metabolic pathways are active in diseased skin compared to normal skin (Duell et al., 1988). The molecular mechanisms underlying these diseases remain to be elucidated. However, important changes in Ca^{2+}-, Zn^{2+}-, and FA-homeostasis are thought at least to contribute to the physiopathology of this skin disease.

Many reports describe the roles of unsaturated FAs in keratinocytes and neutrophils under normal and pathologic conditions (for references, see Smith, 1992, Goetzl et al., 1995). Arachidonic acid is a precursor for a family of bioactive lipid mediators known as eicosanoids that includes prostaglandins, thromboxanes, and leukotrienes. Arachidonic acid and its metabolites have been shown to act as second messengers. Thereby, arachidonic acid and other unsaturated FAs seem to play an essential role in activation of NADPH oxidase involved in superoxide anion (O_2^-) generation, which is essential for neutrophil-mediated killing of microbial pathogens (Badwey et al., 1981, Hendersen et al., 1993). Arachidonic acid and its metabolites released into the extracellular space by keratinocytes or neutrophils, have been shown to possess a chemotactic potential and thus contribute to the establishment of an inflammatory state (for a review, see Conrad, 1999). Such a FA-release might contribute to the epidermal inflammatory response observed in psoriasis.

Beside the fact that arachidonic acid regulates activation of several protein kinases (Blobe et al., 1995; Rao et al., 1994), or interferes with the synthesis of certain proteins (Jurivich et al., 1994), this FA has been shown to play an important role in the mobilization of intracellular calcium and activation of calcium channels (Ordway et al., 1989; Soliven et al., 1993). Such changes have been shown to activate several Ca^{2+}-dependent enzymes like phospholipase A2 (PLA2), which is responsible for the release of arachidonic acid or other unsaturated FAs from phospholipid pools (for a review, see Lambeau and Lazdunski, 1999). A good example of the Ca^{2+}/FA cross-talk in keratinocytes and neutrophils is certainly the transport of FA by the E-FABP/S100A7 and FA-p34 complexes, respectively. The roles and functions of these two FA transporters will be discussed below. However, the various other roles of S100A8 and S100A9 associated with inflammatory processes have been intensively reviewed by others (Passey et al., 1999; Donato, 1999) and will not be discussed in detail here.

4.2. E-FABP/S100A7 in human skin under normal and pathological conditions

E-FABP is the predominant FA-carrier in normal epidermis. Its expression is observed in the last layers of the stratum spinosum and throughout the stratum granulosum (Hagens et al., 1999a). The expression of E-FABP in differentiating cell layers indicates that these layers contain an active FA-metabolism. Differentiating keratinocytes produce and secret high amounts of FAs via lamellar bodies in order to maintain a functional lipid barrier. Thus, E-FABP might be involved in this differentiation-dependent FA-metabolism by transporting FAs to their site of metabolism or secretion. In addition, removing the lipid barrier by treatment of skin with detergent increases expression of E-FABP (Le et al., 1996).

In contrast, S100A7 expression is confined to a few cells of the last living cell layer of the stratum granulosum (Hagens et al., 1999a) explaining the scarce detection of the E-FABP/S100A7 complex in normal human skin. The presence of S100A7 in normal skin might contribute to the maintenance of normal Ca^{2+}-concentrations in the uppermost cell layer. This particular expression pattern also indicates that S100A7 and E-FABP expression are differently regulated.

Ca^{2+}- and FA-homeostasis are completely altered in lesional psoriatic skin, which is also characterized by an impaired epidermal lipid barrier and hyperproliferative keratinocytes. The severe changes in lesional psoriatic skin go along with the overexpression of E-FABP and S100A7 (Hagens et al., 1999a; Siegenthaler et al., 1994; Bürgisser et al., 1995) and the E-FABP/S100A7 complex (Hagens et al., 1999a). Table 1 compares the levels of calcium, FA, E-FABP, S100A7 and S100A8/S100A9 in normal epidermis and shows how these levels are altered in psoriatic epidermis. The modified expression of the proteins might reflect the skin's efforts in order to re-establish the lipid barrier or to satisfy the FA-need to assure enhanced cellular turnover. The fact that E-FABP and S100A7 are co-expressed in all suprabasal cell layers explains the high amounts of E-FABP/S100A7 complex in psoriatic skin. This co-expression might also reflect the altered FA-metabolism in these cells compared to keratinocytes from normal skin (Duell et al., 1988) that express only E-FABP and low amounts of S100A7. Whether association

Table 1. Levels of various important actors involved in the homeostasis of human epidermis. Epidermis is constituted by stratum corneum (sc), granulosum (sg), spinosum (ss) and basale (sb) The data for S100A8/S100A9 levels was obtained by immunohistochemistry with an antibody against FA-p34 (unpublished data). +++ very abundant, ++ abundant, + detectable, – not detected.

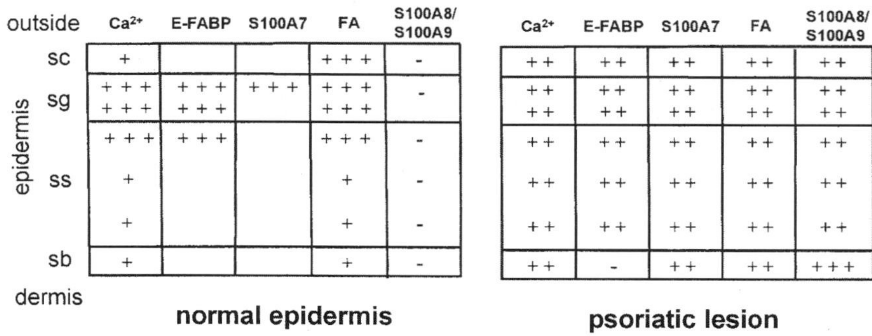

outside		normal epidermis: Ca^{2+}	E-FABP	S100A7	FA	S100A8/ S100A9	psoriatic lesion: Ca^{2+}	E-FABP	S100A7	FA	S100A8/ S100A9
epidermis	sc	+			+++	-	++	++	++	++	++
	sg	+++ +++	+++ +++	+++	+++ +++	-	++ ++	++ ++	++ ++	++ ++	++ ++
		+++	+++		+++	-	++	++	++	++	++
	ss	+			+	-	++	++	++	++	++
		+			+	-	++	++	++	++	++
	sb	+			+	-	++	-	++	++	+++
dermis											

of E-FABP with S100A7 might change E-FABP characteristics (e.g. ligand specificity, affinity, targeting of ligands, etc.) remains to be determined.

In lesional psoriatic epidermis S100A7 is also detected in the cytosol and in the nucleus of cells from the basal layer, where E-FABP is scarcely expressed. Again, this shows that S100A7 expression is differently regulated when compared to E-FABP and also suggests specific and separate functions for the two proteins. The nuclear localization of S100A7 in some psoriatic keratinocytes (Figure 1) is an interesting observation, however, its nuclear function remains to be determined.

In the non-keratinized epidermis like oral mucosa a high expression of E-FABP and S100A7 has been observed (Masouyé et al., 1997, and unpublished observations, respectively). Expression of the proteins in these cells might depend on the same mechanisms that induce the expression of E-FABP and S100A7 in psoriatic scales. Indeed, oral mucosa, similar to psoriatic lesions, is characterized by a high keratinocyte turnover compared to normal skin (Squir et al., 1975) and an important lipid metabolism (Lekholm and Svennerholm, 1977). Thus, oral mucosa seems to demand high expression of free E-FABP, and eventually of the E-FABP/S100A7 complex, in order to assure proper FA-transport.

Confocal microscopy allows a closer look and a more precise analysis of the intracellular localization of proteins. Beside their cytosolic co-localization in the upper cell layers of lesional psoriatic epidermis (see Figure 1), confocal microscopy revealed that E-FABP and S100A7 are also detected at a perinuclear structure, most probably the endoplasmic reticulum (Hagens et al., 1999a). Since S100A7 exists as a cytosolic as well as a membrane-attached form, this finding indicates that S100A7 might act as a docking protein for E-FABP. Thus, E-FABP might be guided and anchored via

(a) (b)

(c)

Figure 1. Confocal microscopical analysis of S100A7 (a), E-FABP (b) and double staining (c) of lesional psoriatic skin. Picture (c) shows that S100A7 and E-FABP co-localize in all suprabasal cell layers (yellow fluorescence). The predominant green coloured basal layer (c) confirms the very low expression of E-FABP in basal keratinocytes (b). For a colour version of this figure, see page xxxii.

S100A7 to sites were FAs are needed. The mechanisms for such a guidance or anchorage are unknown, but might depend on bivalent cations. For example, a locally high Ca^{2+}-concentration might trigger the translocation of S100A7 to the membrane fraction where E-FABP can dock to S100A7. Once arrived at the site, cation-induced conformational changes on membrane-attached S100A7 might disrupt the complex so that E-FABP is able to return to the cytosolic fraction. Such a mechanism could be active at the endoplasmic reticulum. This organelle and sub-compartments thereof are well-characterized Ca^{2+}-pools, in which saturated FAs have been shown to promote Ca^{2+}-sequestration by acting as Ca^{2+}-complexing agents (Rys-Sikora and Gill, 1996). Increased Ca^{2+}-concentrations might induce the translocation of S100A7 to the endoplasmic reticulum. E-FABP might dock to S100A7 anchored in the membrane of the endoplasmic reticulum and release its FA-ligand for Ca^{2+} sequestration. Similarly, S100A7 might dock E-FABP in the vicinity of the nucleus (perinuclear) in order to assure the release of the FA-ligand near the nucleus. Since FAs have been shown to interact with nuclear receptors (for a review, see Wahli et al., 1999) the transport of FAs into the nuclear neighbourhood might facilitate the meeting of FAs with their corresponding receptors.

4.3. The FA-p34 complex in human skin under normal and pathological conditions

kFA-p34 was first identified and characterized in the particulate fraction of lesional psoriatic skin. In contrast, neither kFA-p34 nor its subunits S100A8 and S100A9 were detected in normal epidermis, indicating that expression of kFA-p34 and the separate subunits might be triggered by abnormal differentiation, hyperproliferation of keratinocytes and eventually also by inflammation. Similar to the E-FABP/S100A7 complex, the high expression of kFA-p34 might again reflect the diseased cells' efforts to transport enough FAs for the renovation of the defective skin lipid barrier or to satisfy the increased cellular turnover. Psoriatic lesions possess also an altered FA-metabolism and FA-composition compared to normal skin. In fact, increased amounts of arachidonic acid and its precursors and metabolites are produced in skin regions affected by psoriasis compared to normal skin (Duell et al., 1988). These unsaturated FAs are mediators of inflammation and possess chemotactic potential, which partially explains the strong infiltration of neutrophils into psoriatic lesions. The facts that kFA-p34 is highly overexpressed in psoriatic lesions and that the complex has a high affinity for unsaturated FAs, suggests that at least a part of unsaturated FA-transport in diseased cells is performed by kFA-p34. The complex might also contribute to the establishment of the epidermal inflammatory response evidenced by the infiltration of neutrophils into the lesions. Such a contribution might be achieved by shuttling unsaturated FAs with chemotactic potential from the cytosol to the membrane fraction, from where the FAs are released into the extracellular space. Indeed, psoriatic keratinocytes express the majority of kFA-p34 in the membrane fraction (Siegenthaler et al., 1997).

How kFA-p34 translocates from the cytosolic to the membrane fraction remains to be answered. However, observations made with the FA-p34 complex from neutrophils clearly demonstrated a cation-dependent translocation of the neutrophil complex, as will be described in the next section.

4.4. The FA-p34 complex in resting and activated human neutrophils

Subunit composition of FA-p34 from human polymorphonuclear leukocytes (PMNL) does not differ significantly from the complex isolated from human keratinocytes, as revealed by 2D gel analysis, and affinity for unsaturated FAs are very similar in both complexes.

PMNL express considerable amounts of nFA-p34, reaching up to 2.6% of total cytosolic proteins. It seems that this complex is the predominant FA-carrier in these cells, since no other FA-carrier activity has been measured using several different techniques (Roulin et al., 1999). The importance of an abundant FA-carrier specific for unsaturated FAs is underlined by the fact that neutrophils generate and release large amounts of arachidonic acid and its metabolites upon stimulation. These FAs have to be transported to their sites of action in order to be processed by enzymes, to activate transcription factors, or to fulfill their chemotactic function.

Upon activation of neutrophils using extracellular Ca^{2+}-dependent stimuli, one observes a decrease of cytosolic nFA-p34 in parallel to a strong increase of nFA-p34 in the membrane fraction (Roulin et al., 1999). The increase in the membrane fraction is explained by translocation of the complex and by translocation of the separate subunits and their subsequent assembly to nFA-p34 at the membrane. As such a translocation has never been observed for any FA-carrier, it is suggested that nFA-p34 might play an important role in the shuttling, up-take, or release of unsaturated FAs.

The mechanisms that control translocation have still to be investigated, but the Ca^{2+}-binding properties of the complex subunits might give important clues on this point. Both proteins, S100A8 as well as S100A9, contain two EF-hands that differ in their affinities for Ca^{2+}. The high affinity EF-hands might bind the cation in order to permit complex assembly. Local increases in Ca^{2+}-level, which are observed upon neutrophil activation, might engage the low affinity EF-hand that drives the protein to sites containing high Ca^{2+}-levels. Since both proteins contain separate Zn^{2+}-binding sites it is not excluded that this cation might act similar to Ca^{2+}.

The roles of the post-translational modifications identified on the subunits remain to be elucidated, but phosphorylation of subunits might control translocation. Indeed, a recent report describes that predominantly phosphorylated S100A9 is found in the membrane fraction of activated monocytes (van den Bos et al., 1996) and it seems that phosphorylation increases the protein's affinity for Ca^{2+}. Increased Ca^{2+}-affinity might, in turn, trigger translocation as explained above. Post-translational modifications might also control functions of the complex, i.e. ligand affinities, interactions with targets, etc. In this context, nFA-p34 complexes that differ in subunits composition might have different affinities for a particular unsaturated FA. In addition, one complex might be involved in the secretion of its FA ligand, whereas the other complex might be responsible for delivering its ligand to a membrane anchored enzyme.

S100A8 and S100A9 are abundantly expressed in neutrophils and monocytes/macrophages. The proteins that are involved in nFA-p34 complex formation represent only a minor part of the subunit pool, indicating that S100A8 as well as S100A9, and complexes thereof have several different functions. For example, it has been described that S100A8 is a potent chemoattractant for neutrophils and monocytes/macrophages and causes a sustained inflammatory reaction *in vivo* (Lackmann et al., 1993). The C-terminal 26 amino acids of S100A9 are identical to a peptide known as neutrophil-immobilizing factor (NIF). In order to exert these functions, the proteins have to be secreted. In fact, S100A8 and S100A9 have been shown to be secreted from activated monocytes via a novel pathway that involves tubulin and PKC (Rammes et al., 1997). It has been suggested that secretion of the proteins from activated monocytes might involve the formation of a FA-binding complex (Klempt et al., 1997). Such a complex should be able to anchor itself at the plasma membrane via its FA-binding capacity in order to facilitate its secretion by an unknown mechanism. Once externalized, the proteins might exert their chemotactic functions that contribute to the pathophysiology of inflammatory diseases. In this context, unsaturated chemotactic FAs bound to FA-p34 might be secreted together with the complex and further enhance the chemotactic function of the FA-p34 subunits.

Whether other cells of myeloid origin express the complex is unknown. Monocytes express large amounts of S100A8 and S100A9 and the complex might also be formed in these cells, however, differentiation of monocytes into macrophages is associated with loss of both proteins (Zwaldo et al., 1988). On the other hand, at sites of chronic inflammation in patients suffering of, e.g. rheumatoid arthritis or tuberculosis, macrophages express both proteins, and it would be interesting to determine whether FA-p34 is also present in these cells. In contrast, macrophages located at acutely inflamed tissues express only S100A9. It is established that differentiation-dependent processes regulate the expression of S100A8, S100A9, but how nFA-p34 expression is regulated in cells of myeloid origin is uncertain. However, the close link between expression and disease suggests that inflammatory stimuli might contribute to the regulation of the expression of the complex and its subunits.

5. CONCLUDING REMARKS

The control of a balanced FA-level is crucial for the maintenance of a healthy/normal state of many tissues or cells. FA-transport definitely plays an important role in this control. Our data suggest that FA-transport in turn might be controlled by a Ca^{2+}/FA cross-talk. The FA-carrier complexes E-FABP/S100A7 and FA-p34 and the mechanisms that control their formation/action exemplify this cross-talk. The involvement of S100A7, S100A8, and S100A9 in FA-transport represents a new function for these proteins and it remains to be determined whether other S100 proteins participate in the control of cellular FA-levels.

Increased formation of the E-FABP/S100A7 and FA-p34 complexes is accompanied by a diseased or activated state of cells or tissues expressing the complexes. The increased expression of the complexes might reflect a consequence or a cause of the diseased/activated state. Whatever the answer, furthering the knowledge about the functions of the complexes will contribute to a better understanding of the physiopathology of certain diseases and allow developments aiming at the treatment of diseases characterized by an altered Ca^{2+}/FA cross-talk, as it is the case in psoriasis or activated neutrophils.

ACKNOWLEDGEMENT

This work was supported by the Swiss National Science Foundation, grant No. 32-56812.99 to Dr. G. Siegenthaler.

REFERENCES

Antohe, F., Dobrila, L., Heltianu, C., Simionescu, N. and Simionescu, M., 1993, Albumin binding proteins function in the receptor-mediated binding and transcytosis of albumin across cultured endothelial cells, *Eur. J. Cell. Biol.* 60, 268–275.

Badwey, J., Curnutte, J. and Karnovsky, M., 1981, cis-Polyunsaturated fatty acids induce high levels of superoxide production by human neutrophils, *J. Biol. Chem.* 256, 12640–12643.

Blobe, G.C., Khan, W.A. and Hannun, Y.A., 1995, Protein kinase C: Cellular target of the second messenger arachidonic acid, *Prostaglandins Leukot. Essent. Fatty Acids* 52, 129–135.

Brodersen, D.E., Nyborg, J. and Kjieldgaard, M., 1999, Zinc-binding sites of an S100 protein revealed. Two crystal structures of Ca^{2+}-bound human psoriasin (S100A7) in the Zn^{2+}-loaded and Zn^{2+}-free states, *Biochemistry* 38, 1695–1704.

Bürgisser, D., Siegenthaler, G., Kuster, T., Hellman, U., Hunziker, P., Birchler, N. and Heizmann, C.W., 1995, Amino acid sequence analysis of human S100A7 (Psoriasin) by tandem mass spectrometry, *Biochem. Biophys. Res. Comm.* 217, 257–263.

Conrad, D.J., 1999, The arachidonate 12/15 lipoxygenases. A review of tissue expression and biologic function, *Clin. Rev. Allergy Immunol.* 17, 71–89.

Dale, B.A., Resing, K.A. and Presland, R.B., 1994, Keratohyalin granule proteins, in *The Keratinocyte Handbook*, E.B. Lane and F.M. Watt (eds.), Cambridge University Press, Cambridge, pp. 323–350.

Donato, R., 1999, Functional roles of S100A proteins, calcium-binding proteins of the EF-hand type, *Biochim. Biophys. Acta* 1450, 191–231.

Duell, E.A., Ellis, C.N. and Voorhees, J.J., 1988, Determination of 5,12, and 15-lipoxygenase products in keratomed biopsies of normal and psoriatic skin, *J. Invest. Dermatol.* 91, 446–450.

Edgeworth, J., Freemont, P. and Hogg, N., 1989, Ionomycin-regulated phosphorylation of the myeloid calcium-binding protein p14, *Nature* 342, 189–192.

Edgeworth, J., Gorman, M., Bennet, R., Freemont, P. and Hogg, N., 1991, Identification of p8,14 as a highly abundant heterodimeric calcium-binding protein complex of myeloid cells, *J. Biol. Chem.* 266, 7706–7713.

Egljio, K., Hennings, H. and Clausen, O.P.F., 1986, Altered growth kinetics proceed Ca^{2+}-induced differentiation in mouse epidermal cells, *In Vitro* 22, 332–336.

Fogh, K., Herlin, T. and Kragballe, K., 1989, Eicosanoids in acute and chronic psoriatic lesions: Leukotriene B4, but not 12-hydroxy-eicosatetraenoic acid, is present in biologically active amounts in acute guttate lesions, *J. Invest. Dermatol.* 92, 837–841.

Forslind, B., Werner-Linde, Y., Lindberg, M. and Pallon, J., 1999, Elemental analysis mirrors epidermal differentiation, *Acta Derm. Venereol.* 79, 12–17.

Glatz, J.F.C. and van der Vusse, G.J., 1996, Cellular fatty acid-binding proteins: Their function and physiological significance, *Prog. Lipid Res.* 35, 243–282.

Glatz, J.F.C., Luiken, J.J.F.P., van Nieuwenhoven, F.A. and van der Vusse, G.J., 1997, Molecular mechanisms of cellular uptake and intracellular translocation of fatty acids, *Prostaglandins Leukot. Essent. Fatty Acids* 57, 3–9.

Goetzl, E.J., An, S. and Smith, W.L., 1995, Specificity of expression and effects of eicosanoid mediators in normal physiology and human diseases, *FASEB J.* 9,1051–1058.

Guignard, F., Mauel, J. and Markert, M., 1996, Phosphorylation of Myeloid-related proteins MRP14 and MRP8 during human neutrophil activation, *Eur. J. Biochem.* 241, 265–271.

Hagens, G., Masouyé, I., Augsburger, E., Hotz, R., Saurat, J.H. and Siegenthaler, G., 1999a, Calcium-binding protein S100A7 and epidermal-type fatty acid-binding protein are associated in the cytosol of human keratinocytes, *Biochem. J.* 339, 419–427.

Hagens, G., Roulin, K., Hotz, R., Saurat, J.H., Hellman, U. and Siegenthaler, G., 1999b, Probable interaction between S100A7 and E-FABP in the cytosol of human keratinocytes from psoriatic scales, *Mol. Cell. Biochem.* 192, 123–128.

Heizmann, C.W. and Cox, J.A., 1998, New perspectives on S100 proteins: A multi-functional Ca(2+)-, Zn(2+)- and Cu(2+)-binding protein family, *Biometals* 11, 383–397.

Henderson, L.M., Moule, S.K. and Chappell, J.B., 1993, The immediate activator of the NADPH oxidase is arachidonate not phosphorylation, *Eur. J. Biochem.* 211, 157–162.

Hunter, M.J. and Chazin, W.J., 1998, High level expression and dimer characterization of the S100 EF-hand proteins, migration inhibitory factor-related proteins 8 and 14, *J. Biol. Chem.* 273, 12427–12435.

Jurivich, D., Sistonen, L., Sarge, K. and Morimoto, R., 1994, Arachidonate is a potent modulator of human heat shock gene transcription, *Proc. Natl. Acad. Sci. USA* 91, 2280–2284.

Kamp, F. and Hamilton, J.A., 1993, Movement of fatty acids, fatty acid analogues, and bile acids across phospholipid bilayers, *Biochemistry* 32, 11074–11085.

Klempt, M., Melkonyan, H., Nacken, W., Wiesmann, D., Holtkemper, U. and Sorg, C., 1997, The heterodimer of the Ca^{2+}-binding proteins MRP8 and MRP14 binds arachidonic acid, *FEBS Lett.* 408, 81–84.

Kligman, D. and Hilt, D.C., 1988. The S100 protein family, *Trends Biochem. Sci.* 13, 437–443.

Kragballe, K. and Voorhees, J.J., 1987, Eicosanoids in psoriasis-15-HETE on the stage, *Dermatologica* 174, 209–213.

Lackmann, M., Rajasekariah, P., Iismaa, S.E., Jones, G., Cornish, C.J., Hu, S.P., Simpson, R.J., Moritz, R.L. and Geczy, C.L., 1993, Identification of a chemotactic domain of the pro-inflammatory S100 protein CP-10, *J. Immunol.* 150, 2981–2991.

Lambeau, G. and Lazdunski, M., 1999, Receptors for a growing family of secreted phospholipases A2, *Trends Pharmacol. Sci.* 20, 162–170.

Lampe, M.A., Williams, M.L. and Elias, P.M., 1983, Human epidermal lipids: Characterization and modulation during differentiation, *J. Lipid Res.* 24, 131–140.

Le, M., Schalwijk, J., Siegenthaler, G., van de Kerkhof, P.C., Veerkamp, J.H. and van de Valk, P.G., 1996, Changes in keratinocyte differentiation following mild irritation by sodium dodecyl sulphate, *Arch. Dermatol. Res.* 288, 684–690.

Lekholm, U. and Svennerholm, L., 1977, Lipid pattern and fatty acid composition of human palatal oral epithelium, *Scand. J. Dent. Res.* 85, 279–290.

Madsen, P., Rasmussen, H.H., Leffers, H., Honoré, B., Dejgaard, K., Olsen, E., Kiil, J., Walbum, E., Andersen, A.H., Basse, B., Lauridsen, J.B., Ratz, G.P., Celis, A., Vanderkerckhove, J. and Celis, J.E., 1991, Molecular cloning, occurrence, and expression of a novel partially secreted protein "psoriasin" that is highly up-regulated in psoriatic skin, *J. Invest. Dermatol.* 97, 701–712.

Masouyé, I., Hagens, G., van Kuppevelt, T.H., Madsen, P., Saurat, J.H., Veerkamp, J.H., Pepper, M.S. and Siegenthaler, G., 1997, Endothelial cells of the human microvasculature express epidermal fatty acid-binding protein (E-FABP), *Circul. Res.* 81, 297–303.

Menon, G.K., Price, L.F., Bommannan, B., Elias, P.M. and Feingold, K.R., 1994, Selective obliteration of the epidermal Ca^{2+}-gradient leads to enhanced lamellar body secretion, *J. Invest. Dermatol.* 102, 789–795.

Motta, S., Monti, M., Sesana, S., Mellesi, L., Ghidoni, R. and Caputo, R., 1994, Abnormality of the water barrier function in psoriasis. Role of ceramide fractions, *Arch. Dermatol.* 130, 452–456.

Ordway, R.W., Walsh, J.V. and Singer, J.J., 1989, Arachidonic acid and other fatty acids directly activate potassium channels in smooth muscle cells, *Science* 244, 1176–1179.

Passey, R.J., Xu, K., Hume, D.A. and Geczy, C.L., 1999, S100A8: Emerging functions and regulation, *J. Leukocyte Biol.* 66, 549–556.

Proksch, E., Holleran, W.M., Menon, G.K., Elias, P.M. and Feingold, K.R., 1993, Barrier function regulates epidermal lipid and DNA synthesis, *Br. J. Dermatol.* 128, 473–482

Rammes, A., Roth, J., Goebeler, M., Klempt, M., Hartmann, M. and Sorg, C., 1997, Myeloid-related protein (MRP) 8 and MRP 14, calcium-binding proteins of the S100 family are secreted by activated monocytes via a novel, tubulin-dependent pathway, *J. Biol. Chem.* 272, 9496–9502.

Rao, G.M., Baas, A., Glasgow, W., Eling, T., Runge, M. and Alexander, R., 1994, Activation of mitogen-activated protein kinases by arachidonic acid and its metabolites in vascular smooth muscle cells, *J. Biol. Chem.* 269, 32586–32591.

Robinson, N.A., Lapic, S., Welter, J.F. and Eckert, R.L., 1997, S100A11, S100A10, annexin I, desmosomal proteins, small proline-rich proteins, plasminogen activator inhibitor-2, and involucrin are components of the cornified envelope of cultured human epidermal keratinocytes, *J. Biol. Chem.* 272, 12035–12046

Roth, J., Burwinkel, F., van den Bos, C., Goebeler, M., Vollmer, E. and Sorg, C., 1993, MRP8 and MRP14, S-100 like proteins associated with myeloid differentiation, are translocated to plasma membrane and intermediate filaments in a calcium-dependent manner, *Blood* 82, 1875–1883.

Roulin, K., Hagens, G., Hotz, R., Saurat, J.H., Veerkamp, J.H. and Siegenthaler, G., 1999, The fatty acid-binding heterocomplex FA-p34 formed by S100A8 and S100A9 is the major fatty acid carrier in neutrophils and translocates from the cytosol to the membrane upon stimulation, *Exp. Cell Res.* 247, 410–421.

Rys-Sikora, K.E. and Gill, D.L., 1996, The role of fatty acids within endoplasmic reticulum calcium pools, in *Frontiers in Bioactive Lipids*, Vanderhoek (ed.), Plenum Press, New York, pp. 31–38.

Schafer, B.W. and Heizmann, C.W., 1996, The S100 family of EF-hand Ca^{2+}-binding proteins: Functions and pathology, *Trends Biochem. Sci.* 21, 134–140.

Siegenthaler, G., Hotz, R., Chatellard-Gruaz, D., Didierjean, L., Hellman, U. and Saurat, J.H., 1994, Purification and characterization of the human epidermal fatty acid-binding protein: Localization during epidermal differentiation in vivo and in vitro, *Biochem. J.* 302, 363–371.

Siegenthaler, G., Roulin, K., Chatellard-Gruaz, D., Hotz, R., Saurat, J.H., Hellman, U. and Hagens, G., 1997, A heterocomplex formed by the calcium-binding proteins MRP8 (S100A8) and MRP14 (S100A9) binds unsaturated fatty acids with high affinity, *J. Biol. Chem.* 272, 9371–9377.

Smith, W.L., 1992, Prostanoid biosynthesis and mechanisms of action, *Am. J. Physiol.* 263, 181–191.

Soliven, G., Takeda, M., Shandy, T. and Nelson, D., 1993, Arachidonic acid and its metabolites increase Ca(i) in cultured rat oligodendrocytes, *Am. J. Physiol.* 264, 632–640.

Squir, C.A., Johnson, N.W. and Hackemann, M., 1975, *Structure and Function of Normal Human Oral Mucosa in Health and Disease*, Blackwell, Oxford.

Teigelkamp, S., Bhardwaj, R.S., Roth, J., Meinhardus-Hager, G., Karas, M. and Sorg, C., 1991, Calcium-binding complex assembly of the myeloic differentiation proteins MRP8 and MRP14, *J. Biol. Chem.* 266, 13462–13467.

van den Bos, C., Roth, J., Koch, H.G., Hartmann, M. and Sorg, C., 1996, Phosphorylation of MRP14, an S100 protein expressed during monocytic differentiation, modulates Ca(2+)-dependent translocation from cytoplasm to membranes and cytoskeleton, *J. Immunol.* 156, 1247–1254.

Veerkamp, J.H. and Maatman, R.G.H.J., 1995, Cytoplasmic fatty acid-binding proteins: their structure and genes, *Progr. Lipid Res.* 34, 17–52.

Wahli, W., Devchand, P.R., Ijpenberg, A. and Desvergne, B., 1999, Fatty acids, eicosanoids, and hypolipidemic agents regulate gene expression through direct binding to peroxisome proliferator-activated receptors, *Adv. Exp. Med. Biol.* 447, 199–209.

Zwaldo, G., Brüggen, J., Gerhards, G., Schlegel, R. and Sorg, C., 1988, Two calcium-binding proteins associated with specific stages of myeloid differentiation are expressed by subsets of macrophages in inflammatory tissues, *Clin. Exp. Immunol.* 72, 510–515.

Calcium-Binding Proteins in Degenerative and Cancer-Related Diseases of the Eye

Ricardo L. Gee, Lalita Subramanian, Teresa M. Walker, Paul R. van Ginkel and Arthur S. Polans

Calcium-binding proteins purportedly are involved in a variety of human pathologies, but the supporting evidence often is circumstantial. In contrast, studies of visual function have provided clear examples of how defects in either the expression or function of a calcium-binding protein can manifest as a human disease (Polans et al., 1996).

Phototransduction initiates in the outer segments of vertebrate photoreceptors through a series of chemical reactions that couples light to the generation of an electrical signal (Polans et al., 1996; Koutalos and Yau, 1996). Figure 1 illustrates how light triggers a biochemical cascade leading to the activation of a cGMP phosphodiesterase, which hydrolyzes cGMP and thereby reduces a circulating current that flows into the cell through cGMP-gated cation channels in the outer segment plasma membrane. While the current entering the cell is carried primarily by sodium ions, approximately 15% of the current is comprised of calcium ions. Closure of the cGMP-gated channels during the light response reduces calcium influx but not its extrusion from the cell via a sodium/calcium, potassium exchanger. The resulting decrease in internal calcium from approximately 500–700 nM in the dark to 30 nM in the light stimulates the enzyme guanylate cyclase, promoting the resynthesis of cGMP, the re-opening of the cation channels, and recovery of the dark current. The effect of calcium on guanylate cyclase is mediated

Ricardo L. Gee, Lalita Subramanian, Teresa M. Walker, Paul R. van Ginkel and Arthur S. Polans • Department of Ophthalmology and Visual Sciences and the Department of Biomolecular Chemistry, University of Wisconsin, Madison, WI, U.S.A.

R. Pochet, R. Donato, J. Haiech, C. Heizmann and V. Gerke (eds.): Calcium: The Molecular Basis of Calcium Action in Biology and Medicine, 493–504.

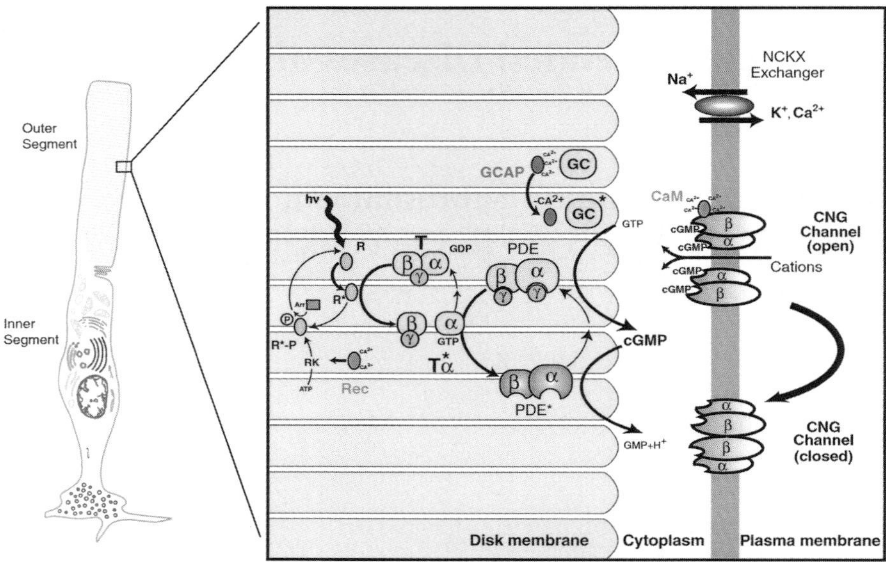

Figure 1. Phototransduction and the role of calcium-binding proteins. The outer segment of the photoreceptor is the primary site of phototransduction. The outer segment consists of a series of flattened, double-membranous disks surrounded by a plasma membrane. Light causes the isomerization of the visual chromophore, 11-cis retinal, to its all-trans configuration. 11-cis retinal is normally coupled to the protein opsin, the complex referred to as rhodopsin (R), and isomerization leads to the activation of rhodopsin (R*) in the disk membranes and the subsequent biochemical reactions depicted in the figure. A trimeric G protein called transducin (T) is activated (T*), causing the stimulation of a phosphodiesterase (PDE*) by removing two inhibitory gamma subunits; this leads to the hydrolysis of cGMP. Cyclic nucleotide gated (CNG) channels in the outer segment plasma membrane are open in the dark when cGMP levels are elevated and close upon illumination as the concentration of cGMP decreases, thus altering the flux of cations into the photoreceptor. The change in membrane potential diminishes the release of chemical transmitter at the synapse so that the reception of light in the outer segment is conveyed to second order neurons in the retina. While sodium ions carry most of the current entering the outer segment, calcium ions also flow through the CNG channels. The flux of calcium ions through the channels is reduced by illumination, while their extrusion via the sodium-calcium-potassium exchanger (NCKX) continues, thus reducing the intracellular concentration of calcium in the light. This reduction stimulates a guanylate cyclase (GC) through the action of GCAP (guanylate cyclase activating protein) in order to resynthesize cGMP, reopen the CNG channels, and reestablish the dark current. Calcium also modifies the phototransduction cycle through the action of recoverin (Rec), which is thought to interact with rhodopsin kinase, inhibiting the phosphorylation of rhodopsin, which otherwise is a major step required for the inactivation of R*. Calmodulin (CaM) also modulates the phototransduction cycle by modifying the affinity of the CNG channels towards cGMP. As light lowers the concentration of intracellular calcium, the binding of calmodulin to the channels changes, increasing their affinity to cGMP, thereby enhancing the return to the dark-adapted state. The light-dependent changes in the concentration of calcium, mediated by calcium-binding proteins, play a critical role in determining the adaptation mechanisms of the photoreceptor. In the absence of such changes in the concentration of calcium, photoreceptor cells could not respond to the wide range of light intensities that we normally experience (Palczewski et al., 2000). For a colour version of this figure, see page xxviii.

by a small family of calcium-binding proteins of the EF-hand superfamily referred to as GCAPs (guanylate cyclase activator proteins) (Palczewski et al., 1994; Dizhoor et al., 1995; Haeseleer et al., 1999). GCAPs are expressed specifically in the photosensitive tissues of the eye, and are unique modulators of calcium in that they activate their effector molecule while in their calcium-free form (for further review, see Palczewski et al., 2000). An additional activator of guanylate cyclase, identified as S100B, has been shown to activate guanylate cyclase in its calcium-occupied form *in vitro* (Duda et al., 1996). While some immunocytochemical studies have demonstrated the presence of S100B in photoreceptor cells (Rambotti et al., 1999), this laboratory was unable to localize S100B beyond Müller cells and the astrocytes and glia of the nerve fiber layer (unpublished observations). In addition, S100B has not been detected biochemically in preparations of purified rod outer segments, the site of phototransduction.

Degenerative events occur in photoreceptor cells as the result of mutations in the gene encoding GCAP1. A mutation (Y99C) results in a tyrosine to cysteine change found in humans afflicted with an autosomal dominant cone dystrophy (Payne et al., 1998; Sokal et al., 1998). The primary effect in cone cells is consistent with the elevated expression of GCAP1 in cones compared to rod cells. The Y99C mutation has been shown to alter the calcium sensitivity of GCAP1 in a manner that promotes the constitutive activation of guanylate cyclase at high calcium concentrations, limiting its ability to fully inactivate guanylate cyclase under physiological dark conditions. An increase in the concentration of cGMP ensues, and such alterations have been linked in other studies to the degeneration of photoreceptor cells (Ulshafer et al., 1980; Bowes et al., 1989; Pittler and Baehr, 1991). Elevated levels of cGMP cause further cation channels to open in the photosensitive outer segments, leading to a persistent rise in the intracellular concentration of calcium as the flux of calcium ions through the channels is enhanced. Apoptosis likely occurs as a consequence of these pathological concentrations of calcium by pathways that are both dependent and independent of other calcium-binding proteins (Fain and Lisman, 1999; Palczewski et al., 2000).

Recoverin is another member of the EF-hand superfamily of calcium-binding proteins that plays a critical role in the adaptation of retinal rods and cones, supposedly by regulating the phosphorylation of the photopigment molecule rhodopsin (Figure 1; for a review, see Polans et al., 1996). Recoverin contains two canonical EF hands and is modified at its amino terminus by a heterogeneous acylation in which myristate (C14:0) and related acyl groups (C14:1, C14:2, and C12:0) are linked to the amino terminal glycine residue (Dizhoor et al., 1992). This modification to the amino terminus of recoverin is thought to enhance membrane binding in the presence of calcium, a so-called calcium-myristoyl switch (Zozulya and Stryer, 1992), although other studies indicate that calcium-induced conformational changes of recoverin do not alter its partitioning between the cytoplasmic and membrane compartments of the photoreceptor cell (Johnson et al., 1997).

Recoverin normally is expressed in the rod and cone photoreceptor cells of the retina. In some circumstances, however, recoverin is expressed in primary neoplasms originating outside of the eye (Polans et al., 1995). This leads to an immune response, which then inadvertently destroys retinal photoreceptor cells, eventually causing vision impairment or complete blindness. The disease, cancer-associated retinopathy (CAR), is an example of a series of so-called paraneoplastic diseases, whereby the development of a tumor has an effect on a distant site in the nervous system. In the case of CAR, the loss of vision often precedes the diagnosis of cancer so that the detection of recoverin antibodies in the patient's serum acts as an early warning for an arising cancer. CAR can be reproduced in an animal model either by direct inoculation with recoverin or by the transfer of lymphocytes from an immunized animal to a naive recipient (Adamus et al., 1994). The immunodominant and immunopathological regions of recoverin have been mapped and overlap at the first α-helix of EF-hand 2 (Adamus et al., 1994). The basis for recoverin expression in tumor tissues remains a mystery, but it may be due to the location of the recoverin gene on chromosome 17 near the site of p53, a major tumor suppressor gene.

Aside from these examples, the over expression of several calcium-binding proteins has been implicated in establishing the malignant and metastatic phenotypes of various tumors (Schafer and Heizmann, 1996; Takenaga et al., 1997; Chen et al., 1997; Donato, 1999). S100B, for example, is of interest because of its expression in tumor tissues, its potential role in metastatic processes, and especially its use as a marker to identify malignant melanomas and other types of tumors (Ilg et al., 1996). S100A4, also referred to as CAPL, mts1 and various clonal names, has been associated with the metastatic propensity of different tumor cell lines (Ebralidze et al., 1989; Lloyd et al., 1998).

Early studies also postulated the involvement of a novel calcium-binding protein in the development of an ocular tumor, uveal melanoma (Kan-Mitchell et al., 1990, 1993). Uveal melanoma is the principal ocular tumor arising in adults. The uvea consists of the iris, ciliary body and choroid, and the transformation of melanocytes within these structures leads to a proliferating mass that can alter the structure of the uvea and retina, as well as metastasize to other sites. The etiology of the disease is unknown and the subject of limited investigation. There are few prognostic indicators that can be observed clinically, contributing to enucleation as the principal method of treatment. Unfortunately, systemic metastases may have occurred by the time the ocular symptoms are recognized, and death, often due to hepatic disease, occurs in about half of the afflicted individuals. There are few similarities between uveal and cutaneous melanoma; they differ in their metastatic profiles, susceptibilities to treatment, and thus far their genetic basis (for a review, see Albert, 1997).

In the remainder of this chapter we will describe a series of calcium-binding proteins associated with uveal melanoma and examine their differ-

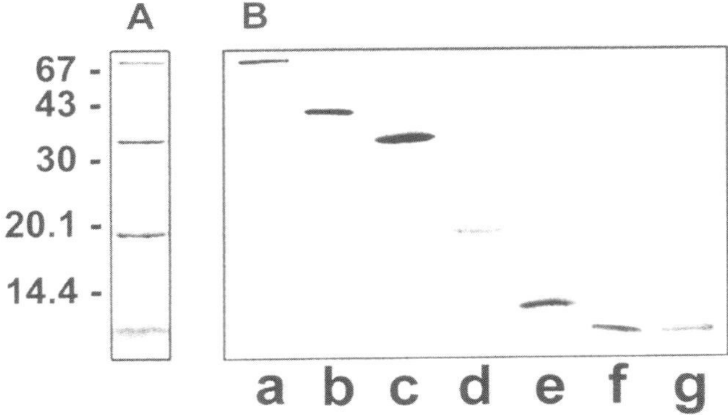

Figure 2. Identification of calcium-binding proteins associated with ocular melanoma cell lines. Homogenates of cell lines were extracted with EDTA, and a soluble fraction was adjusted with calcium, applied to a phenyl-Sepharose column, and bound material eluted with EDTA. Aliquots were subjected to SDS-PAGE and stained for protein (A). An aliquot of the sample depicted in panel A also was transferred to Immobilon and immunostained with antibodies specific for: annexin VI (lane a), capg (lane b), annexin V (lane c), calmodulin (lane d), S100A11 (lane e), S100B (lane f), and S100A6 (lane g). Reprinted from *BBA*, 1448, 290–297, 1998, with permission from Elsevier Science.

ential expression in both cultured melanoma cells and in primary tumors. We will also discuss methodologies to elucidate the function of these calcium-binding proteins and ultimately to provide possible drug treatments or an avenue for gene therapy in order to intervene in the normal progression of the disease.

There are a variety of biochemical and molecular methods for identifying calcium-binding proteins. One convenient approach takes advantage of the feature that some calcium-binding proteins expose hydrophobic regions upon binding calcium, resulting in their affinity for hydrophobic matrices, such as phenyl-Sepharose (Polans et al., 1993). Figure 2A demonstrates the phenyl-Sepharose eluate of a cell lysate derived from an established cell line of a human uveal melanoma. These enriched proteins were further purified by RP-HPLC using a C_4 column and unambiguously identified by a combination of Edman analysis and mass spectrometry (van Ginkel et al., 1998). They are annexin VI (67 kD), capg (39 kD), annexin V (35 kD), calmodulin (18 kD – in the presence of EGTA), S100A11 (formerly S100C, 12 kD), S100B (9.5 kD), and S100A6 (formerly calcyclin, 9.5 kD). Figure 2B illustrates a Western blot analysis of this phenyl-Sepharose fraction, demonstrating that antibodies specific for each of these calcium-binding proteins can be generated. These antibodies are useful for localization studies and for determining the relative level of expression of each protein in cultures of normal uveal melanocytes, and established lines of malignant uveal melanoma.

As shown in Figure 3 the melanoma cell line expressed substantially more S100A11 and capg, along with modest elevation of annexin VI and S100A6 (calcyclin) compared to normal uveal melanocytes. No differences

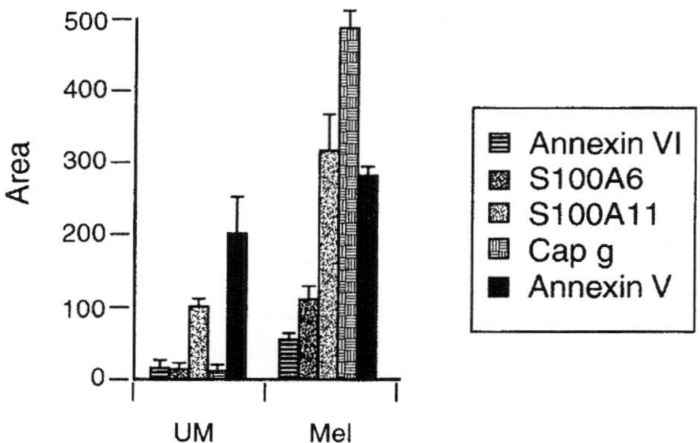

Figure 3. Differential expression of calcium-binding proteins. Immunoblots from at least three separate experiments were quantified by densitometry and area integration. Values are the mean ± S.D. UM, normal uveal melanocytes; Mel, malignant uveal melanoma cell line.

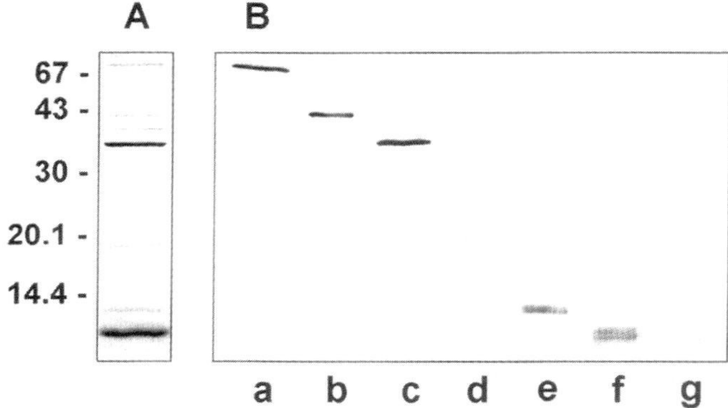

Figure 4. Calcium-binding proteins in a primary ocular tumor. Within 30 min of enucleation, melanoma tumor tissue was dissected and an extract processed for phenyl-Sepharose chromatography (A) and immunoblotting (B) as described in the legend of Figure 1. Cytogenetic information pertaining to this tumor tissue is given in the text. Reprinted from *BBA*, 1448, 290–297, 1998, with permission from Elsevier Science.

in the expression of annexin V and S100B were detected (van Ginkel et al., 1998). Multiple cell lines of uveal melanoma demonstrate the same finding. These observations also support previous reports, for example, that S100A6 is related to malignancy and metastasis in other types of tumors (Maelandsmo et al., 1997; Boni et al., 1997), including cutaneous melanoma (Weterman et al., 1992).

We have identified a unique set of calcium-binding proteins and demonstrated their differential expression in uveal melanoma cell lines. In order to avoid *in vitro* artifacts, however, findings derived from studies of cell lines must be validated using primary tumor tissues. As shown in Figure 4, the same calcium-binding proteins found in cell lines of uveal melanoma also are

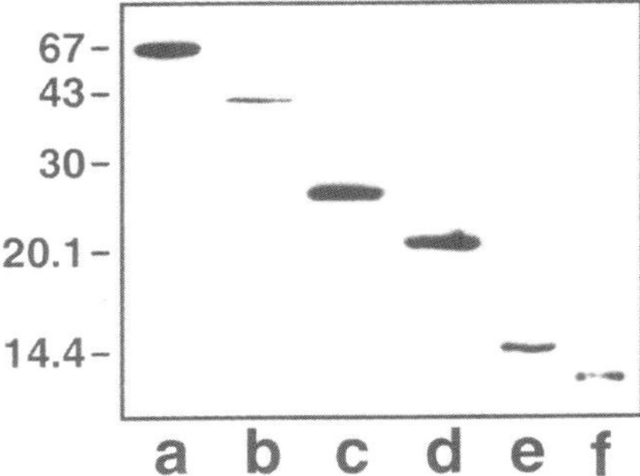

Figure 5. Calcium-binding proteins in an established cell line of uveal melanoma. A cell line with different morphological characteristics (spindle subtype A) than the cells used to generate the data of Figure 1 (spindle subtype B) and Figure 3 (epithelioid) was homogenized and processed for phenyl-Sepharose chromatography as described in the legend for Figure 1. An immunoblot of the material eluted from the column with EDTA was stained with antibodies (except anti-S100A6) as shown in the legend for Figure 1. Cytogenetic information pertaining to this established cell line is given in the text.

present in fresh biopsies of uveal melanoma obtained within 30 minutes of enucleation. It is worth noting that S100A6 was expressed at lower levels in primary tumor tissues, which probably reflects the fact that malignant cells comprising the primary tumor have a much slower doubling time than cells selected in culture, which are provided with growth factors and special media.

Recent cytogenetic (Prescher et al., 1994; Singh et al., 1994; Speicher et al., 1994; Prescher et al., 1995; Luyten et al., 1996; McNamara and Kennedy, 1997; McNamara et al., 1997; Mitelman et al., 1997; White et al., 1998) and biochemical (Coupland et al., 1998; Mouriaux et al., 1998) evidence indicate that uveal melanomas may not originate as the result of a single type of genetic event, rather that disparate mutations may lead to a common phenotype, namely uncontrolled proliferation. Therefore, it may be more prudent to understand the shared cellular pathways that contribute to the malignant and metastatic phenotypes of different tumors and to develop treatments based on these pathways rather than to devise therapies based on intervening in a host of genetic aberrations. With this in mind, it is interesting that the set of calcium-binding proteins associated with uveal melanoma are expressed in both cell lines and primary tumor tissues regardless of their cytogenetic profiles. Figure 5 shows the Western blot analysis of a different cell line from that depicted in either Figures 2 or 4; again the same series of calcium-binding proteins are detected. The cytogenetics, however, differ considerably:

Figure 4: 46,X,der(X)t(X;10)(q13;q11.2),der(1)t(1;13)(p13;q14),-3,i(8q),+I(8q),der(9)t(9;10)(q34;q11.2),-10,-13

Figure 5: 44,XY,der(3)t(3;8)(q23;q22),I(6p),der(7)t(7;17)(q22;q11.2), -9,der(10)t(10;17)(q22;q11.2),der(13)t(12;13)(p11.2;p12), 16,der(17)t(7;17)(q22;q11.2),der(19)t(11;19)(q21;p13.3),del(19)(q13.1)

As shown here, the chromosomal aberrations associated with these two tumors, incorporating the initial transformation events as well as later mutations, share little in common. Therefore, regardless of the genetic events that initiate the transformation of normal uveal melanocytes into malignant melanomas, certain gene products, in this case a specific set of calcium-binding proteins, may be necessary to support the complete cancer phenotype. With this in mind, uveal melanoma cell lines are being stably transfected with both sense and antisense constructs for each of the calcium-binding proteins thus far identified, and these lines will be used for intraocular inoculation of athymic nude mice. Normally, uveal melanoma cell lines grow in the eye and disseminate to the liver in this animal model, paralleling the progression of the tumor in humans. We can now determine whether by interfering with either the expression or function of these calcium-binding proteins we also can alter the progression of the tumor.

Recent advances in both biochemical and molecular methods make it feasible to study the cellular pathways that endow tumor cells with special properties related to their malignant and metastatic phenotypes. Such studies provide significant information about the course of the disease and about effective treatments, independent of knowing the genetic mutations that cause the disease. In this laboratory we have compared normal uveal melanocytes with malignant melanomas using such techniques as suppression subtractive hybridization (measuring differences in the levels of mRNAs; Diatchenko et al., 1996) and 2-D gel electrophoresis coupled with mass spectrometry (measuring differences in protein products; Clauser et al., 1995; Matsui et al., 1997). Such methodologies have revealed interesting expression patterns for calcium-binding proteins in addition to the ones described thus far. As an example, ALG-2 (the **A**poptosis-**L**inked **G**ene **2**) was detected by subtractive hybridization. (ALG-2 is described in more detail in the chapter by Maki, this book). In contrast to the calcium-binding proteins already discussed, ALG-2 expression is diminished in more malignant melanoma cell lines. ALG-2 was first detected in studies of a T cell hybridoma in which its expression was found to be necessary for cell death in response to Fas or glucocorticoid treatment, as well as in response to other apoptotic signals (Vito et al., 1996). By down-regulating a necessary component of the cell death pathway, in addition to altering regulators of the cell cycle, melanoma cells may enhance their capacity for cell division and survival, hence their malignancy and ability to disseminate. It will be critical in future experiments to further characterize the ALG-2 pathway in uveal melanoma tumor cells and to determine whether the activation of this pathway will limit the progression of this type of tumor.

Ocular tissues have provided a unique opportunity to study the structure, function, and pathologies associated with a novel group of calcium-binding proteins. Effector molecules have been identified for some of these proteins so that detailed studies of their molecular interactions can be initiated. It is likely that lessons learned by the study of these ocular proteins will be relevant to other tissues and to other human diseases.

ACKNOWLEDGEMENTS

This research was supported by grants from Research to Prevent Blindness, Inc. (RPB) to the Department of Ophthalmology and Visual Sciences of the University of Wisconsin, and from the Retina Research Foundation. ASP is the recipient of a Jules and Doris Stein Professorship from RPB.

REFERENCES

Adamus, G.H., Ortega, D., Witkowska, D. and Polans, A., 1994, Recoverin: A potent uveitogen for the induction of photoreceptor degeneration in Lewis rats, *Exp. Eye Res.* 59, 447–456.

Albert, D.M., 1997, The ocular melanoma story. LIII Edward Jackson Memorial Lecture: Part II, *Am. J. Ophthalmol.* 123(6), 729–741.

Boni, R., Burg, G., Doguoglu, A., Ilg, E.C., Schafer, B.W., Muller, B. and Heizmann, C.W., 1997, Immunohistochemical localization of the Ca^{2+} binding s100 proteins in normal human skin and malanocytic lesions, *Br. J. Dermatol.* 137, 39–43.

Bowes, C., Danciger, M., Kozak, C.A. and Farber, D.B., 1989, Isolation of a candidate cDNA for the gene causing retinal degeneration in the rd mouse, *Proc. Natl. Acad. Sci. USA* 86, 9722–9726.

Chen, D., Davies, M.P., Rudland, P.S. and Barraclough, R., 1997, Transcriptional down-regulation of the metastasis-inducing S100A4 (p9Ka) in benign but not in malignant rat mammary epithelial cells by GC-factor, *J. Biol. Chem.* 272(32), 20283–20290.

Clauser, K.R., Hall, S.R., Smith, D.M., Webb, J.W., Andrews, L.E., Tran, H.M., Epstein, L.B. and Burlingame, A.L., 1995, Rapid mass spectrometric peptide sequencing and mass matching for characterization of human melanoma proteins isolated by two-dimensional PAGE, *Proc. Natl. Acad. Sci. USA* 92, 5072–5076.

Coupland, S.E., Bechrakis, N., Schüler, A., Anagnostopoulos, I., Hummel, M., Bornfeld, N. and Stein, H., 1998, Expression patterns of cyclin D1 and related proteins regulating G1-S phase transition in uveal melanoma and retinoblastoma, *Br. J. Ophthalmol.* 82, 961–970.

Dasher, A.M., Boikov, S.G. and Olshevskaya, E.V., 1998, Constitutive activation of photoreceptor guanylate cyclase by Y99C mutant of GCAP-1. Possible role in causing human autosomal dominant cone degeneration, *J. Biol. Chem.* 273, 17311–17314.

Diatchenko, L., Lau Y.C., Campbell, A.P., Chenckik, A., Mooadam, F., Huang, B., Lukyanov, S., Lukyanov, K., Gurskaya, N., Sverdlov, E.D. and Siebert, P.D., 1996, Suppression subtractive hybridization: A method for generating differentially regulated or tissue-specific cDNA probes and libraries, *Proc. Natl. Acad. Sci. USA* 93, 6025–6030.

Dizhoor, A.M., Ericsson, L.H., Johnson, R.S., Kumar, S., Olshevskaya, E., Zozulya, S., Neubert, T.A., Stryer, L., Hurley, J.B. and Walsh, K.A., 1992, The NH2 terminus of retinal recoverin is acylated by a small family of fatty acids, *J. Biol. Chem.* 267, 16033–16036.

Dizhoor, A.M., Olshevskaya, E.V., Wong, S.C., Stults, J.T., Ankoudinova, I., Hurley, J.B. and Henzel, W.J., 1995, Cloning, sequencing, and expression of a 24-kDa calcium-binding protein activating photoreceptor guanylyl cyclase, *J. Biol. Chem.* 270, 25200–25206.

Donato, R., 1999, Functional roles of S100 proteins, calcium-binding proteins of the EF-hand type, *Biochim. Biophys. Acta* 1450, 191–231.

Duda, T., Goraczniak, R.M. and Sharma, R.K., 1996, Molecular characterization of S100A1-S100B protein in retina and its activation mechanism of bovine photoreceptor guanylate cyclase, *Biochemistry* 35, 6263–6266.

Ebralidze, A., Tulchinsky, E., Grigorian, M., Afanasyeve, A., Senin, V., Revazzova, E. and Lukanidin, E., 1989, Isolation and characterization of the gene specifically expressed in different metastatic cells and whose deduced gene product has a high homology to a Ca^{2+}-binding protein family, *Genes Dev.* 3(7), 1086–1093.

Fain, G.L. and Lisman, J.E., 1999, Light, Ca^{2+}, and photoreceptor death: New evidence for the equivalent-light hypothesis from arrestin knockout mice, *Invest. Ophthalmol. Vis. Sci.* 40(12), 2770–2772.

Haeseleer, F., Sokal, I. and Li, N., 1999, Molecular characterization of a third member of the guanylyl cyclase-activating regulator of phototransduction, *J. Biol. Chem.* 274, 6526–6535.

Ilg, E.C., Schafer, B.W. and Heizmann, C.W., 1996, Expression pattern of S100 calcium-binding proteins in human tumors, *Int. J. Cancer* 68(3), 325–32.

Johnson, W.C., Palczewski, K., Gorczyca, W.A., Riazance-Lawrence, J.H., Witkowska, D. and Polans, A.S., 1997, Calcium binding to recoverin: Implications for secondary structure and membrane association, *Biochim. Biophys. Acta* 1342, 164–174.

Kan-Mitchell, J., Ras, N., Albert, D., Van Eldik, L. and Taylor, C.R., 1990, S100 Immunophenotypes of uveal melonomas, *Invest. Ophthalmol. Vis. Sci.* 31, 1492–1496.

Kan-Mitchell, J., Liggett, P.E., Taylor, C.R., Rao, N., Granada, E.S.V., Danenberg, K.D., White, W.L., Van Eldik, J., Horikoshi, T. and Danberg, P.V., 1993, Differential S100B expression in choroidal and skin melanomas: Quantitation by the polymerase chain reaction, *Invest. Ophthalmol. Vis. Sci.* 345, 3366–3375.

Koutalos, Y. and Yau, K.-W., 1996, Regulation of sensitivity in vertebrate rod photoreceptors by calcium, *Trends Neurosci.* 19, 73–81.

Lloyd, B.H., Platt-Higgins, A., Rudland, P.S. and Barraclough, R., 1998, Human S100A4 (p9Ka) induces the metastiatic phenotype upon benign tumour cells, *Oncogene* 17, 465–473.

Luyten, G.P., Mooy, C.M., Post, J., Jensen, O.A., Luider, T.M. and De Jong, P.T., 1996, Metastatic uveal melanoma – A morphologic and immunohistochemical analysis, *Cancer* 78, 1967–1971.

Maelandsmo, G.M., Florenes, V.A., Mellingsaeter, T., Hovig, E., Kerbel, R.S. and Fodstad, O., 1997, Differential expression patterns of S100A2, S100A4 and S100A6 during progression of human malignant melanoma, *Int. J. Cancer* 74, 464–469.

Matsui, N.M., Smith, D.M., Clauser, K.R., Fichmann, J., Andrews, L.E., Sullivan, C.M., Burlingame, A.L. and Epstein, L.B., 1997, Immobilized pH gradient two-dimensional gel electrophoresis and mass spectrometric identification of cytokine-regulated proteins in ME-180 cervical carcinoma cells, *Electrophoresis* 18, 409–417.

McNamara, M. and Kennedy, S.M., 1997, Successful establishment of uveal and conjunctival melanoma *in vitro*, *In Vitro Cell. Dev. Biol. Anim.* 33, 236–239.

McNamara, M., Felix, C., Davison, E.V., Fenton, M. and Kennedy, S.M., 1997, Assessment of chromosome 3 copy number in ocular melanoma using fluorescence in situ hybridization, *Cancer Genet. Cytogenet.* 98, 4–8.

Mitelman, F., Mertens, F. and Johansson, B., 1997, A breakpoint map of recurrent chromosomal rearrangements in human neoplasia, *Nat. Genet.* 15, 417–474.

Mouriaux, F., Casagrande, F., Pillaire, M., Manenti, S., Malecaze, F. and Darbon, J., 1998, Differential expression of G1 cyclins and cyclin-dependent kinase inhibitors in normal and transformed melanocytes, *Invest. Ophthalmol. Vis. Sci.* 39, 876–884.

Palczewski, K., Subbaraya, I. and Gorczyca, W.A., 1994, Molecular cloning and characterization of retinal photoreceptor guanylyl cyclase activating protein (GCAP), *Neuron* 13, 395–404.

Palczewski, K., Polans, A.S., Baehr, W. and Ames, J.B., 2000, Calcium-binding proteins in the retina: Structure, function, and the etiology of human retinal diseases, *Bioessays*, in press.

Payne, A.M., Downes, S.M., Bessant, D.A., Taylor, R., Holder, G.E., Warren, M.J., Bird, A.C. and Bhattacharya, S.S., 1998, A mutation in guanylate cyclase activator 1A (GUCA1A) in an autosomal dominant cone dystrophy pedigree mapping to a new locus on chromosome 6p21.1, *Hum. Mol. Genet.* 7(2), 273–277.

Pittler, S.J. and Baehr, W., 1991, Identification of the nonsense mutation in the rod photoreceptor cGMP PDE beta subunit gene of the rd mouse, *Proc. Natl. Acad. Sci. USA* 88, 8322–8326.

Polans, A.S., Crab, J. and Palczewski, K., 1993, Calcium-binding proteins in the retina, *Methods Neurosci.* 15, 248–260.

Polans, A.S., Witkowska, D., Haley, T.L., Amundson, D., Baizer, L. and Adamus, G., 1995, Recoverin, a photoreceptor-specific calcium-binding protein, is expressed by the tumor of a patient with cancer-associated retinopathy, *Proc. Natl. Acad. Sci. USA* 92(20), 9176–9180.

Polans, A., Baehr, W. and Palczewski, K., 1996, Turned on by Ca^{2+}: The physiology and pathology of Ca(2+)-binding proteins in the retina, *Trends Neurosci.* 19(12), 547–54.

Prescher, G., Bornfeld, N. and Becher, R., 1994, Two subclones in a case of uveal melanoma: Relevance of monosomy 3 and multiplication of chromosome 8q, *Cancer Genet. Cytogenet.* 77, 144–146.

Prescher, G., Bornfeld, N., Friedrichs, W., Seeber, S. and Becher, R., 1995, Cytogenetics of twelve cases of uveal melanoma and patterns of nonrandom anomalies and isochromosome formation, *Cancer Genet. Cytogenet.* 80, 40–46.

Rambotti, M.G., Giambanco, I., Spreca, A. and Donato, R., 1999, S100B and S100A1 proteins in bovine retina: Their calcium-dependent stimulation of a membrane bound guanylate cyclase activity as investigated by ultracytochemistry, *Neurosci.* 92, 1089–1101.

Schafer, B.W. and Heizmann, C.W., 1996, The S100 family of EF-hand calcium-binding proteins: Functions and pathology, *Trends Biochem. Sci.* 21(4), 134–140.

Singh, A.D., Boghosian-Sell, L., Wary, K.K., Shields, C.L., De Potter, P., Donoso, L.A., Shields, J.A. and Cannizzaro, L.A., 1994, Cytogenetic findings in primary uveal melanoma, *Cancer Genet. Cytogenet.* 72, 109–115.

Sokal, I., Li, N., Surgucheva, I., Warren, M.J., Payne, A.M., Bhattacharya, S.S., Baehr, W. and Palczewski, K., 1998, GCAP1 (Y99C) mutant is constitutively active in autosomal dominant cone dystrophy, *Mol. Cell.* 2(1), 129–133.

Speicher, M.R., Prescher, G., Du Manoir, S., Jauch, A., Horsthemke, B., Bornfeld, N., Becher, R. and Cremer, T., 1994, Chromosomal gains and losses in uveal melanomas detected by comparative genomic hybridization, *Cancer Res.* 54, 3817–3823.

Takenaga, K., Nakamura, Y. and Sakiyama, S., 1997, Expression of antisense RNA to S100A4 gene encoding an S100-related calcium-binding protein suppresses metastatic potential of high-metastatic Lewis lung carcinoma cells, *Oncogene* 14(3), 331–337.

Ulshafer, R.J., Garcial, C.A. and Hollyfield, J.G., 1980, Sensitivity of photoreceptors to elevated levels of cGMP in the human retina, *Invest. Ophthalmol. Vis. Sci.* 19, 1236–1241.

van Ginkel, P.R., Gee, R.L., Walker, T.M., Hu, D.N., Heizmann, C.W., and Polans, A.S., 1998, The identification and differential expression of calcium-binding proteins associated with ocular melanoma, *Biochim. Biophys. Acta* 1448, 290–297.

Vito, P., Lacana, E., and D'Adamio, L., 1996, Interfering with apoptosis: Ca^{2+}-binding protein ALG-2 and Alzheimer's disease gene ALG-3, *Science* 271, 521–525.

Weterman, M.A.J., Stoopen, G.M., Van Muijen, G.N.P., Kuznicki, J. and Ruiter, D.J., 1992, Expression of calcyclin in human melanoma cell lines correlates with metastatic behavior in nude mice, *Cancer Res.* 52(5), 1291–1296.

White, V.A., Chambers, J.D., Courtright, P.D., Chang, W.Y. and Horsman, D.E., 1998, Correlation of cytogenetic abnormalities with the outcome of patients with uveal melanoma, *Cancer* 83, 354–359.

Zozulya, S. and Stryer, L., 1992, Calcium-myristoyl protein switch, *Biochemistry* 89, 11569–11573.

Calcium ATPases Genes and Cell Transformation

Patrizia Paterlini-Bréchot, Mounia Chami and Devrim Gozuacik

1. INTRODUCTION

Although several important studies have pointed out the general and versatile role of calcium in cell growth and differentiation (Berridge et al., 1998), evidence that Ca^{2+}-governing genes may be selectively mutated and involved in cell transformation is, as yet, lacking.

Following on from a recent *in vivo* observation which focused our attention on a SERCA gene (Chami et al., 2000), we aim here to present direct and indirect evidence to support the possible role of SERCA genes in cell transformation.

Sarco/Endoplasmic Reticulum Calcium ATPases (SERCA) account for most Ca^{2+} uptake into rapidly exchanging pools of all types of cells. They couple ATP hydrolysis with cation transport, and maintain a high level of free Ca^{2+} within the ER/SR lumen, in equilibrium with the Ca^{2+} bound to lumenal proteins, such as calsequestrin, calreticulin and BiP (Pozzan et al., 1994; MacLennan et al., 1997). They are mainly localized in the SR/ER membranes, although some recent results have suggested the existence of SERCA on Golgi membranes (Lin et al., 1999).

Three distinct SERCA genes, located on chromosomes 16p12 (SERCA1), 12q23-24 (SERCA2) and 17p13.3 (SERCA3) encode a number of differentially expressed isoforms (Wu et al., 1995). SERCA1a (adult form) and 1b (fetal form) are mainly found in fast-twitch skeletal muscle and SERCA 2a in slow-twitch skeletal and cardiac muscle, while SERCA2b is widespread and often coexists with SERCA3, which also displays isoforms (SERCA3a and

Patrizia Paterlini-Bréchot, Mounia Chami and Devrim Gozuacik • Unité INSERM 370, Institut Pasteur/Necker, Faculté de Médecine Necker, Paris, France.

R. Pochet, R. Donato, J. Haiech, C. Heizmann and V. Gerke (eds.): Calcium: The Molecular Basis of Calcium Action in Biology and Medicine, 505–519.

SERCA3b, (Bobe et al., 1998)) and is expressed in several tissues, including the intestine, thymus, cerebellum, spleen, lymph nodes, lung, platelets, endothelial and secretory epithelial cells (Wu et al., 1995). All SERCA proteins have a similar structure-function relationship. They are transmembrane proteins with 10 putative transmembrane domains, a large hydrophilic cytosolic domain containing the ATP-binding sequence and the Asp-351 residue which undergo phosporylation upon Ca^{2+} binding (Figure 1). Binding of two Ca^{2+} by SERCA, which implicates six transmembrane Ca^{2+} binding residues and the cytoplasmic loop between transmembrane segments 6 and 7 (Falson et al., 1997; MacLennan et al., 1997), is followed by ATP-dependent phosphorylation of Asp 351, and by a change in the conformation of the enzyme which releases Ca^{2+} into the ER lumen (Figure 2). Electron microscopy analyses of crystallized rabbit SERCA1 have confirmed this structure. Density cross-sections of the SERCA1 crystal form through the transmembrane region suggest the existence of a large cavity, between transmembrane domains M4, M5 and M6, consistent with a path for calcium release to the lumen of the SR/ER (Zhang et al., 1998) (Figure 1). This evidence corroborates the findings of site-directed mutagenesis analyses that have localized calcium binding residues in transmembrane segments M4, M5 and M6 (MacLennan et al., 1997).

All SERCA proteins are specifically inhibited by thapsigargin (Lytton et al., 1991), a plant-derived sesquiterpene lactone tumor promoter. SERCA3 exhibits a lower Ca^{2+} affinity and a higher optimum pH (Lytton et al., 1992) than other SERCA, and SERCA1 has a higher turnover rate than SERCA2b and SERCA2a (Sumbilla et al., 1999). In spite of these biochemical differences, SERCA1a can functionally substitute for SERCA2a in transgenic mouse hearts (Ji et al., 1999). SERCA activity is regulated by phospholamban, a homopentamer of 6 kDa subunits which is expressed in cardiac, slow-twitch skeletal smooth muscle and endothelial cells (Paul, 1998; Toyofuku et al., 1993). Dephosphorylated phospholamban binds to the SERCA pump, shifting its Kd for Ca^{2+} to a lower affinity value and thereby inhibiting the pump. Phosphorylation of phospholamban by the c-AMP and Ca^{2+}/calmodulin-dependent protein kinases dissociates the protein from the SERCA and relieves this inhibition. Sarcolipin inhibits SERCA1-dependent Ca^{2+} uptake at low Ca^{2+} concentrations, and stimulates Ca^{2+} uptake at saturating Ca^{2+} concentrations (Odermatt et al., 1998). Although the significance of the existence of different isoforms is not completely understood, molecular, biochemical, tissue and cell distribution data suggest a modulation of the expression of individual calcium pumps as a function of fine modulation of calcium events in different cellular regions and at different times during the cell differentiation program. For example, in salivary gland acinar cells, SERCA3 is expressed at the basal pole, whereas SERCA2b, which has higher Ca^{2+} affinity and a lower optimum pH, is localized at the luminal pole (Lee et al., 1997). A switch from one SERCA isoform to another has been described under different physiological conditions. *In vitro*

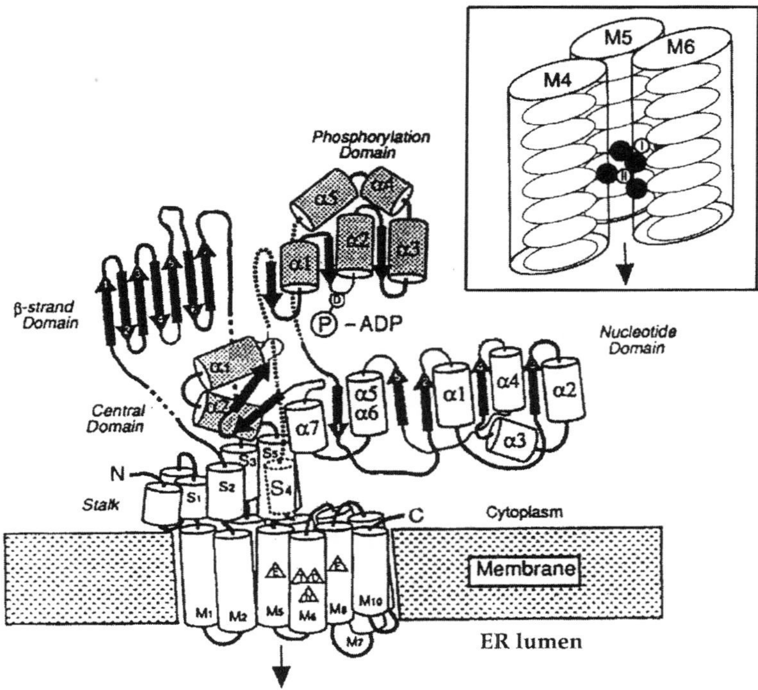

Figure 1. Ca^{2+} pump ATPase tertiary structure prediction (Green and Stokes, 1992). This model includes 10 transmembrane helices (M1 to M10), the α-helical stalk sector (S1–S5) and globular subdomains in the cytoplasmic region which form β-strand, phosphorylation, nucleotide binding domains (MacLennan et al., 1985). The arrow pointing to the ER lumen indicates the release of calcium ions from Ca^{2+} binding sites in M4, M5, and M6. *Insert*: Helices M4, M5 and M6 are shown in oblique section. Positions of Ca^{2+} binding residues (Glu309 in M4, Glu771 in M5, Asn796 and Asp800 in M6) are indicated as filled black circles. Ca^{2+} ions binding sites I and II are shown as empty circles (MacLennan et al., 1998).

differentiation of a human myeloid/promyelocytic cell line along the neutrophil/granulocytic lineage was associated with a decrease in SERCA2b and a parallel increase in SERCA PL/IM (SERCA3b) (Launay et al., 1999). On the other hand, the pharmacological activation of several T lymphocyte cell lines resulted in the down-regulation of SERCA PL/IM and the up-regulation of SERCA2b (Launay et al., 1997). Isoform switching and an increase in the SERCA2a pump was observed in PDGF-treated, aortic smooth muscle cells undergoing cell proliferation (Magnier et al., 1992). It was subsequently shown that this SERCA2a up-regulation was related to G1/S transition in the cell cycle (Magnier-Gaubil et al., 1996). The myogenic differentiation of undifferentiated mouse tumor cells BC3H1 was associated with a switch from predominant SERCA2b to predominant SERCA2a (De Smedt et al., 1991). In vascular endothelial cells, the pattern of SERCA expression was shown to change during cell proliferation: SERCA3 messenger levels fell, with IP3R isoform switching occurring in parallel (Mountian et al., 1999).

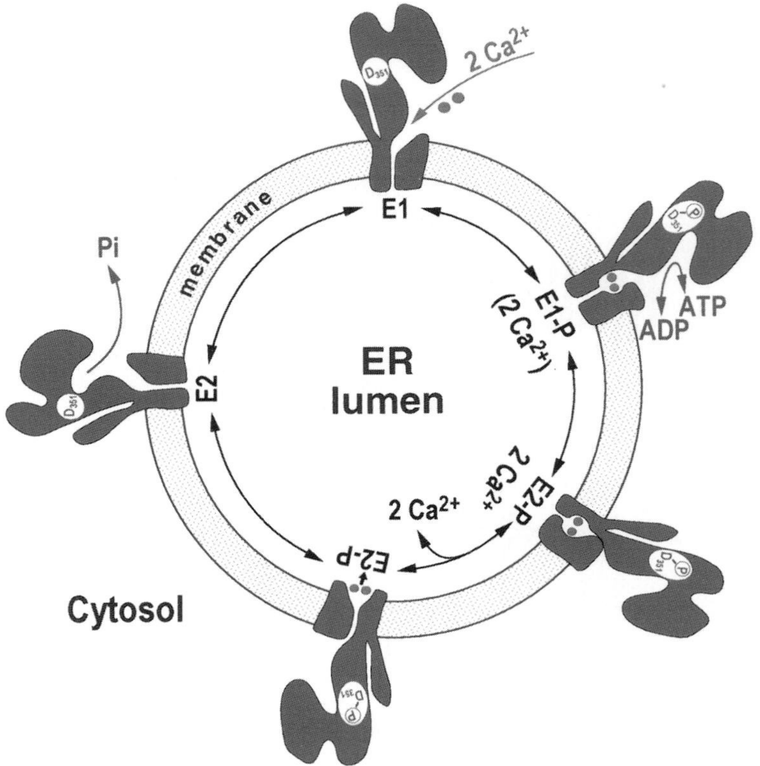

Figure 2. Reaction cycle for Ca^{2+} transport by SERCA pumps. In the E1 conformation, high affinity Ca^{2+} binding sites located near the center of helices M4, M5, and M6 are accessible to cytoplasmic Ca^{2+} but not to lumenal Ca^{2+}. Phosphorylation from ATP, following the occupation of both sites by Ca^{2+} (E1-P ($2Ca^{2+}$)), leads to linked movements of both cytoplasmic and transmembrane domains, resulting in occlusion through closure of the entry gate (E2-P $2Ca^{2+}$). Ca^{2+} is now contained in a polar cavity formed near the center of the transmembrane domains. Further conformational changes open the gate allowing the exit of two Ca^{2+} ions to the lumen (E2-P). This conformation, which Ca^{2+} affinity is very low, also activates dephosphorylation (E2) and returns the pump to the high Ca^{2+} affinity form (E1), completing the cycle (MacLennan et al., 1997; Yonekura et al., 1997). For a colour version of this figure, see page xxix.

In thapsigargin-resistant cells, a novel intracellular Ca^{2+} pump is triggered, which is barely sensitive to thapsigargin and present at very low levels in parental thapsigargin-sensitive cells (Waldron et al., 1997). SERCA switching may also occur in a pathological setting. Although both SERCA2 and SERCA3 are expressed in β cells in the pancreas, rats in a spontaneous model of non-insulin-dependent diabetes mellitus exhibited a marked reduction in SERCA 3 expression in the islets of Langerhans (Varadi et al., 1996).

2. SERCA1 GENE IS A TARGET FOR HBV-RELATED INSERTIONAL MUTAGENESIS

We have recently described the clonal mutation of SERCA1 gene selectively occurring in a human hepatocellular carcinoma, but not in the adjacent non-tumorous liver (Chami et al., 2000). This mutation was due to integration of the Hepatitis B Virus (HBV) DNA into SERCA1 gene in the liver cell genome. Viral-related insertional mutagenesis, i.e. the integration of viral DNA into or in the proximity of a cellular gene, is one of the molecular mechanisms involved in Hepadnavirus-related liver cell transformation (Buendia, 1992). The following sub-section deals with this issue and with its impact in human liver carcinogenesis.

2.1. HBV-related insertional mutagenesis in human hepatocellular carcinomas

Chronic HBV infection is etiologically related to human hepatocellular carcinoma (HCC) (Paterlini and Bréchot, 1994). Most HCCs contain integrated HBV DNA in the liver cellular DNA, suggesting that the integration may be directly involved in carcinogenesis. Studies addressing this issue are rare in the litterature and include analysis of HBV-DNA integration sites in tumorous liver tissues, cloning of mutated cDNAs from the tumorous tissue, analysis of their *in vitro* effects in transiently transfected and stably transfected cell lines, *in vitro* transformation assays and *in vivo* (nude mice) tests of tumorigenic activity.

In one HCC, the viral genome was found to be integrated into the Retinoic acid receptor β gene (RAR beta) (Dejean et al., 1986), a gene involved in cell differentiation and proliferation, which has been discovered through this original approach. The viral integration was potentially encoding a chimeric HBV/RAR beta transcript. It has been subsequently proved that the expression of this chimeric contruct is transforming. In fact, infection of erythrocytic progenitor cells by a retrovirus expressing the HBV/RAR beta construct induced their transformation (Garcia et al., 1993). In addition, the identification of the RAR beta gene led to the discovery of a family of genes and in particular the RAR alpha receptor, which has subsequently been shown to be rearranged in acute promyelocytic leukemia (De Thé et al., 1990).

In a second liver tumor, the HBV genome was found integrated into the human Cyclin A gene (Wang, et al., 1990). This gene plays a major role in both the S and G2/M check-points of the cell cycle and was cloned for the first time thanks to these viral integration studies. Chimeric HBV/Cyclin A transcripts were found to be expressed in the tumor. It has been then proved that cells stably transfected with the HBV/Cyclin A chimeric construct undergo cell transformation and are tumorigenic when injected to nude mice (Berasain et al., 1998). Moreover, following the first report on cyclin A mutation in a human liver cancer, mutations of genes of the same family were looked for in different tumors and amplification of human cyclin D gene was discovered in different types of carcinoma (Buschges et al., 1999; Jiang et al., 1992).

In addition to these two fully documented studies, other cases have been reported. Integration of HBV-DNA has been found in the gene encoding Mevalonate Kinase in the PLC/PRF/5 cell line (Graef et al., 1994), and in the Carboxypeptidase N locus in a human HCC (Pineau et al., 1996). In both instances, chimeric transcripts were identified in the tumor cells, but the effect of their *in vitro* expression on cell transformation has not yet been investigated.

Up to now, studies have failed to identify preferential cellular target sequences for HBV DNA integration. Moreover, insertional mutagenesis has been considered a rare event in human liver carcinogenesis (Koike, 1998). However, it is worthwhile to note that, due to the long and cumbersome cloning methods used in the past to isolate HBV/cellular DNA junctions, few tumors have been analyzed so far (around 30 HCCs in a 20-years period). In addition to that, data bank analyses to search for cellular genes can take now advantage from the recent and huge advances in human genome sequencing.

Some years ago, we have developed a new approach, based on an HBV-Alu PCR assay, to isolate viral/cellular DNA junctions (Minami et al., 1995). This method has allowed us to isolate 18 HBV/cellular DNA junction fragments; among these, new target genes, including SERCA1, have been identified (work in progress). Thus, recent results suggest that the actual prevalence of HBV-related insertional mutagenesis should be reassessed.

Taken overall, these data strongly support the idea that studies of HBV-related insertional mutagenesis are likely to reveal new genes and mechanisms involved in cell transformation. On the same line, using a large-scale approach, retrovirus-related insertional mutagenesis has recently been used to discover genes involved in leukemic disease (Li et al., 1999).

2.2. HBV-DNA integration into the SERCA1 gene

We isolated an HBV-Alu PCR fragment from a liver tumorous DNA (86T). In this fragment, the HBV X (Weil et al., 1999) gene was fused in frame with SERCA1 gene exon 3 (Zhang et al., 1995). Northern blot analysis showed a 3.2 kb transcript, driven by the viral X promoter, hybridizing with both HBV and SERCA1 probes, and expressed in the tumor but not in adjacent, non-tumorous liver tissue (Figure 3A). A cDNA library, constructed using tumor polyA$^+$ mRNAs, yielded eight recombinant clones (Figure 3B). Their 5′ ends exhibited the same HBV X sequence, fused in frame to SERCA1 exon 3. Interestingly, SERCA1 cDNA was characterized by the splicing of exon 11, leading to a 22 codon frameshift and a premature stop codon in exon 12. Three of the eight clones exhibited an additional splicing of SERCA1 exon 4. These cDNA clones potentially encode two chimeric proteins, the structure

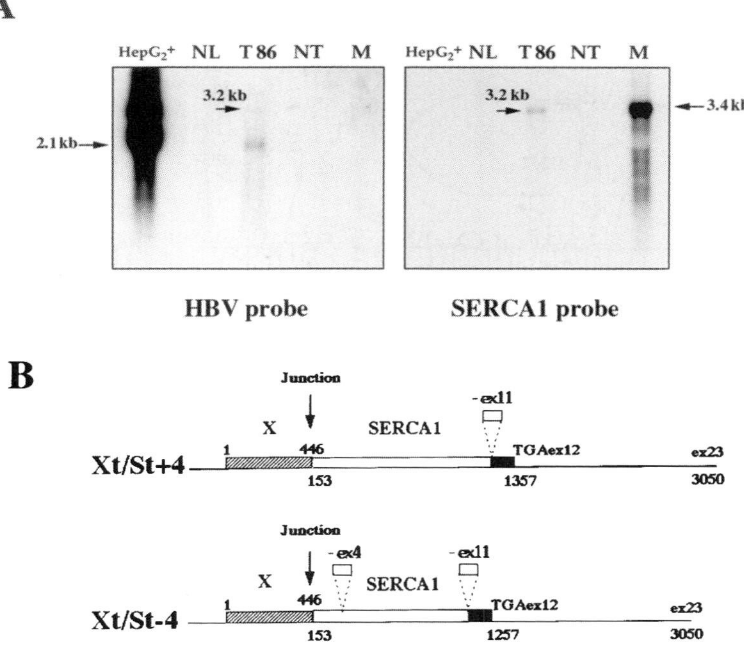

Figure 3. (A) Expression of chimeric HBV X/SERCA1 mRNA in a tumorous liver tissue (T86). The same Northern blot of poly A^+ RNA from various tissues (M, muscle (positive control for SERCA1 probe); T86, tumorous tissue T86; NT, adjacent non-tumorous tissue NT86; NL, normal liver and $HepG2^+$, HepG2 cells transfected with HBV (positive control for HBV probe)) was hybridized with SERCA1 and HBV probe. A 3.2 Kb transcript hybridizing with both HBV and SERCA1 probe is present in T86 (arrow), but not in NT. (B) Hybrid cDNAs cloned from T86. Xt/St+4, clones with splicing of SERCA1 exon 11 (-ex11) leading to a frameshift (black box) and a stop codon in exon 12. Xt/St-4, clones with exon 11 and exon 4 (-ex4) splicing. X, HBV X sequence.

of which is shown in Figure 4A. SERCA1 N-terminal end (51 aminoacids) is replaced by a nearly complete HBV X protein (148 out of 154 aminoacids), and a mutated peptide is encoded at the C-terminal end. The two proteins lack six putative SERCA1 transmembrane segments (M5-M10), including five of the six Ca^{2+} binding residues (MacLennan et al., 1997) and the cytoplasmic loop between transmembrane segments 6 and 7 which also controls Ca^{2+} binding (Falson et al., 1997) (Figure 4A). In addition, one chimeric protein also lacks a peptide encoding the second putative transmembrane segment (M2) and the last 5 C-terminal residues of M1 (Figure 4A). Thus, consistent with previously reported mutants (MacLennan et al., 1998), these chimeric proteins cannot function as calcium pumps. The expression of these two chimeric proteins was selectively detected, by Western blot, in the tumor but not in the corresponding non-tumorous tissue (Figure 4B). When transiently expressed in liver-derived cell lines, the chimeric proteins localize to ER, induce ER Ca^{2+} depletion and apoptosis (Chami et al., 2000). Moreover,

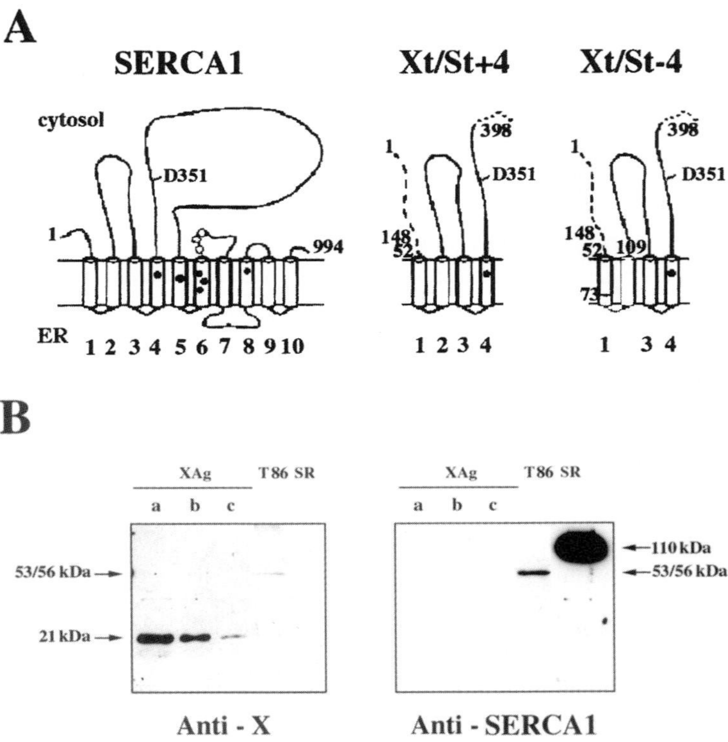

Figure 4. (A) A schematic representation of SERCA1 (left) and chimeric proteins: numbers below drawings indicate transmembrane domains. D351: phosphorylatable aspartic acid 351. (•) transmembrane and (○) cytoplasmic calcium binding residues. The HBV X is depicted as broken line, the peptide created by frameshift is shown as a dotted line and the peptide encoded by exon 4 and deleted in Xt/St-4 is in grey. (B) Chimeric protein (X/SERCA) is expressed in T86. The same Western blot was successively hybridized with anti-HBV X and anti-SERCA1 antibodies. X Ag (21 kDa), purified recombinant HBV X protein (a: 100 ng, b: 50 ng, c: 50 ng mixed with 10 μg of normal liver microsomes); T86, T86 microsomes (50 μg); SR, SERCA1 from rabbit muscle microsomes (50 ng).

cell death was shown to be due to the SERCA rather then the X part of the protein. Namely, X protein, which has been described as having a proapototic effect, was shown to loose not only its proapoptotic but also its transactivating properties when fused to the SERCA protein and therefore targeted to the ER.

Further results have shown that non-chimeric SERCA1 transcripts encoding truncated SERCA1 proteins, which are C-terminally truncated and cannot function as calcium pumps, are expressed in normal liver. This was assessed by cloning SERCA1 transcripts from normal liver. Using SERCA1 specific primers on exons 10 and 13, we found that the SERCA1 transcripts lacking exon 11 were in fact expressed in adult heart, spleen, and thymus, as well as in liver tissues, but not in muscle, thus ruling out a general phenomenon of illegitimate transcription (Chelly et al., 1989). When expressed *in vitro* under the control of a strong heterologous promoter, these truncated SERCA1 proteins induce apoptosis (Chami et al., manuscript in preparation).

Thus, in the liver tumorous tissue, the HBV X promoter cis-activates mRNAs, normally expressed at low levels in non-tumorous livers, as chimeric transcripts.

3. SERCA GENES AND APOPTOSIS

The involvement of SERCA proteins in apoptosis has been inferred from studies using the SERCA-specific inhibitor, thapsigargin (Ghosh et al., 1991; Graber et al., 1996; Short et al., 1993). Empting of ER Ca^{2+} with either thapsigargin, or 2,5-di-*tert*-butylhydroquinone (DBHQ) (Short et al., 1993), causes cells to undergo either growth arrest in G0-like state (Ghosh et al., 1991) or apoptotic death (Furuya et al., 1994). Apoptosis follows the ER calcium depletion induced by SERCA inhibition, and it is noteworthy that the anti-apoptotic protein Bcl-2 has been reported to counteract ER calcium depletion (reviewed by McConkey and Orrenius, 1997). It has also been suggested that the expression of SERCA may be modulated by the expression of Bcl-2 (Kuo et al., 1998). However, it is difficult to assess the direct impact of SERCA genes on cell growth and death, since we can hardly distinguish between the effect of thapsigargin, by itself, and the effect of thapsigargin due to its Ca^{2+} pump inhibitory effect, on cell viability and growth. Moreover, ER calcium depletion has also been reported to follow ER overload and unfolded protein response, without Ca^{2+} pump inhibition, and eventually lead to dysfunctions of protein synthesis, growth arrest and apoptosis (Kaufman, 1999).

In this setting, chimeric SERCA proteins provide an *in vivo*-derived model to study the direct role of SERCA genes on cell viability and growth. Chimeric SERCA proteins have been shown to be targeted to the ER, and their structure makes it possible to predict that they are inserted into the ER membrane (Chami et al., 2000). Overexpression of both chimeric SERCAs and SERCA moieties induces apoptosis. In addition to that, overexpression of the viral X moiety, targeted to the ER membrane by a prolactin cassette, is unable to induce apoptosis to a significant degree. In line with these findings, the overexpression of truncated, non-chimeric SERCAs also induces apoptosis. In this model, apoptosis is therefore related to truncated SERCA proteins rather than to ER overload or response to the accumulation of mutated, unfolded proteins (Chami et al., 2000). Consistently with these data, SERCA overexpression, obtained by transduction of COS cells with SERCA-expressing adenovirus vectors, has been shown to induce apoptotic cell death (Ma et al., 1999). Taken together, these results are consistent with the direct implication of SERCA genes in the control of apoptosis.

4. SERCA GENES, CELL GROWTH AND CELL TRANSFORMATION

Many experimental findings converge to show the prominent role of calcium signalling in cell proliferation (Berridge, 1995). Previous reports also demonstrated the role of SERCA genes in the exit from G0 into the cell cycle (Cheng et al., 1996) and in the G1/S transition (Magnier-Gaubil et al., 1996). Modulation of SERCA isoforms with an increased ratio of SERCA2b/SERCA3 has been reported in vascular endothelial cells (Mountian et al., 1999) and in activated lymphocytes (Launay, et al., 1997). By handling Ca^{2+} content in the cytosol and ER, SERCA genes have a direct impact on calcium signalling, including the induction and propagation of calcium waves (Camacho and Lechleiter, 1993, Lee et al., 1997, Morgan and Jacob, 1998). In turn, calcium wave amplitude and duration have been shown to differentially activate certain transcription factors (Dolmetsch et al., 1997). The potential role of calcium waves in cell transformation has been raised by showing that long lasting calcium oscillations are displayed by transformed but not normal Madin–Darby kidney cells (Wojnowski et al., 1994). Therefore, several experimental data converge to show that SERCA genes expression participates directly in the control of cell growth. In this context, it seems plausible that a disfuction of this mechanism might play a role in cell trasformation. Furthermore, cell transformation and apoptosis may share common pathways. For example, the occurrence of cell proliferation or death depends, in some documented cases, from the expression level of cellular oncogenes, such as c-myc, the supply of survival factors and the activity of different cellular partners (Harrington et al., 1994; Philips et al., 1997).

SERCA2 gene mutations have recently been reported in Darier–White disease (Sakuntabhai et al., 1999), a rare dermatosis with dominant inheritance, characterized by focal areas of separation between suprabasal epidermal cells and abnormal keratinocyte differentiation, indicating that modifications to intracellular calcium levels, directly related to SERCA2 gene mutations, may lead to a distinct program of altered cell growth and differentiation (Peacocke and Christiano, 1999). Mutations of SERCA3 gene have been identified in patients with type II diabetes indicating that the SERCA3 locus possibly contributes to the genetic susceptibility to this disease (Varadi et al., 1999). Consistently with these observations, the impaired function and expression of SERCA3 has been associated to beta cell apoptosis in diabetic animal models (Varadi et al., 1996).

In the model of chimeric SERCA, it is not clear why the cis-activation of chimeric SERCA1 proteins, unable to pump calcium, has been selectively found, *in vivo*, in clonally proliferating transformed cells, while *in vitro* overexpression of the same chimeric proteins massively induces apoptosis. One hypothesis is that pre-apoptotic cells are characterized by DNA mutations and breaks. If, for an unpredictable reason, one of these cells escapes cell death, it may generate a tumor cell clone. A further, non-exclusive, possibility is that the fine modulation of chimeric SERCA1 protein expression has the potential to change the balance of cell proliferation and apoptosis. In the tumorous tissue, chimeric SERCAs are expressed under the control of the viral X promoter, which is of moderate strength, while their *in vitro* overexpression is governed by the CMV promoter. Transgenic experiments in mice, with floxed chimeric constructs under control of the HBV X promoter, are currently under way in an attempt to answer these questions.

The autosomal, recessive inheritance of SERCA1 gene mutations, leading to mutated SERCA1 proteins unable to pump calcium, has been reported in Brody disease, being characterized by impaired muscle relaxation (MacLennan et al., 1998). At variance with this syndrome, the tumorous tissue we described displays overexpression of mutated SERCA1 proteins while endogenous SERCA liver-specific expression (SERCA2b) is maintained. The interplay of the levels of expression of different SERCA isoforms in different cell types, and the impact of their fine regulation on the basic mechanisms of cell death, growth and differentiation is as yet unclear.

Although the direct implication of SERCA genes in cell transformation is, at present, only a working hypothesis, evidence is growing that their impact may go well beyond the limits of Ca^{2+} transport across the ER membrane. In this respect, combined studies on cell physiology and pathological conditions may provide new pieces in the puzzle. This view should stimulate further analysis of SERCA gene mutations in different types of cancer, and efforts to treat cancer disease and modify tumor cell invasion through the pharmacological modulation of intracellular calcium (Kohn and Liotta, 1995).

ACKNOWLEDGEMENTS

We wish to thank Christian Bréchot and Pierre Falson for helpful discussions.

REFERENCES

Berasain, C., Patil, D., Perera, E., Huang, S., Mouly, H. and Bréchot, C., 1998, Oncogenic activation of a human cyclin A2 targeted to the endoplasmic reticulum upon Hepatitis B virus genome insertion, *Oncogene* 16, 1277–1288.

Berridge, M.J., 1995, Calcium signalling and cell proliferation, *Bioessays* 17, 491–500.

Berridge, M.J., Bootman, M.D. and Lipp, P., 1998, Calcium – A life and death signal [news], *Nature* 395, 645–648.

Bobe, R., Lacabaratz-Porret, C., Bredoux, R., Martin, V., Ozog, A., Launay, S., Corvazier, E., Kovacs, T., Papp, B. and Enouf, J., 1998, Expression of two isoforms of the third sarco/endoplasmic reticulum Ca^{2+} ATPase (SERCA3) in platelets. Possible recognition of the SERCA3b isoform by the PL/IM430 monoclonal antibody, *FEBS Lett.* 423, 259–264.

Buendia, M.A., 1992, Hepatitis B viruses and hepatocellular carcinoma, *Adv. Cancer Res.* 59, 167–226.

Buschges, R., Weber, R., Actor, B., Lichter, P., Collins, V. and Reifenberger, G., 1999, Amplification and expression of cyclin D (CCND1, CCND2 and CCND3) in human malignant gliomas, *Brain Pathol.* 9, 435–442.

Camacho, P. and Lechleiter, J.D., 1993, Increased frequency of calcium waves in Xenopus laevis oocytes that express a calcium-ATPase, *Science* 260, 226–229.

Chami, M., Gozuacik, D., Saigo, K., Capiod, T., Falson, P., Lecoeur, H., Urashima, T., Beckmann, J., Gougeon, M.-L., Claret, M., le Maire, M., Bréchot, C. and Paterlini-Bréchot, P., 2000, Hepatitis B Virus-related insertional mutagenesis implicates SERCA1 gene in the control of apoptosis, *Oncogene* 19, 2877–2886.

Chelly, J., Concordet, J.-P., Kaplan, J.-C. and Kahn, A., 1989, Illegitimate transcription: Transcription of any gene in any cell type, *Proc. Natl. Acad. Sci. USA* 86, 2617–2621.

Cheng, G., Liu, B., Yu, Y., Diglio, C. and Kuo, T., 1996, The exit from G0 into the cell cycle requires and is controlled by Sarco(endo)plasmic Reticulum Ca^{2+} pump, *Arch. Biochem. Biophys.* 329, 65–72.

Dejean, A., Bougueleret, L., Grzeschik, K.H. and Tiollais, P., 1986, Hepatitis B virus DNA integration in a sequence homologous to v-erbA and steroid receptor genes in a hepatocellular carcinoma, *Nature* 322, 70–72.

De Smedt, H., Eggermont, J., Wuytack, F., Parys, J., Van Den Bosch, L., Missiaen, L., Verbist, J. and Casteels, R., 1991, Isoform switching of the Sarco(endo)plasmic Reticulum Ca^{2+} pump during differentiation of BC3H1 myoblasts, *J. Biol. Chem.* 266, 7092–7095.

De Thé, H., Chomienne, C., Lanotte, M., Degos, L. and Dejean, A., 1990, The t(15;17) translocation of acute promyelocytic leukemia fuses the retinoic acid receptor alpha gene to a novel transcribed locus, *Nature* 347, 558–561.

Dolmetsch, R.E., Lewis, R.S., Goodnow, C.C. and Healy, J.I., 1997, Differential activation of transcription factors induced by Ca^{2+} response amplitude and duration [published erratum appears in *Nature* 1997, Jul 17; 388(6639):308], *Nature* 386, 855–858.

Falson, P., Menguy, T., Corre, F., Bouneau, L., de Gracia, A.G., Soulie, S., Centeno, F., Moller, J.V., Champeil, P. and le Maire, M., 1997, The cytoplasmic loop between putative transmembrane segments 6 and 7 in sarcoplasmic reticulum Ca^{2+}-ATPase binds Ca^{2+} and is functionally important, *J. Biol. Chem.* 272, 17258–17262.

Furuya, Y., Lundmo, P., Short, A., Gill, D. and Isaacs, J., 1994, The role of calcium, pH, and cell proliferation in the programmed (apoptotic) death of androgen-independent prostatic cancer cells induced by thapsigargin, *Cancer Res.* 54, 6167–6175.

Garcia, M., De Thé, H., Tiollais, P., Samarut, J. and Dejean, A., 1993, A hepatitis B virus pre-S-retinoic acid receptor béta chimera transforms erythrocytic progenitor cells in vitro, *Cell Biology* 90, 89–93.

Ghosh, T., Bian, J., Short, A., Rybak, S. and Gill, D., 1991, Persistent intracellular calcium pool depletion by thapsigargin and its influence on cell growth, *J. Biol. Chem.* 266, 24690–24697.

Graber, M.N., Alfonso, A. and Gill, D.L., 1996, Ca^{2+} pools and cell growth: Arachidonic acid induces recovery of cells growth-arrested by Ca^{2+} pool depletion, *J. Biol. Chem.* 271, 883–888.

Graef, E., Caselmann, W., Wells, J. and Koshy, R., 1994, Insertional activation of mevalonate kinase by hepatitis B virus DNA in a human hepatoma cell line, *Oncogene* 9, 81–87.

Green, N. and Stokes, D., 1992, Structural modelling of P-type ion pumps, *Acta Physiol. Scand.* 146, 59–68.

Harrington, E., Bennett, M., Fanidi, A. and Evan, G., 1994, C-myc induced apoptosis in fibroblasts is inhibited by specific cytokines, *EMBO J.* 13, 3286–3295.

Ji, Y., Loukianov, E., Loukianova, T., Jones, L. and Periasamy, M., 1999, SERCA1a can functionally substitute for SERCA2a in the heart, *Am. J. Physiol.* 276, H89–97.

Jiang, W., Kahn, S., Tomita, N., Zhang, Y., Lu, S. and Weinstein, I., 1992, Amplification and expression of the human cyclin D gne in esophageal cancer, *Cancer Res.* 52, 2980–2983.

Kaufman, R.J., 1999, Stress signaling from the lumen of the endoplasmic reticulum: Coordination of gene transcriptional and translational controls, *Genes Dev.* 13, 1211–1233.

Kohn, E. and Liotta, L., 1995, Molecular insights into cancer invasion: Strategies for prevention and intervention, *Cancer Res.* 55, 1856–1862.

Koike, K., 1998, HBV DNA integration and insertional mutagenesis, in *Hepatitis B Virus: Molecular Mechanisms in Disease and Novel Strategies for Therapy*, R. Koshy and W. Caselmann (eds.), London Imperial College Press, London, pp. 133–160.

Kuo, T.H., Kim, H.R., Zhu, L., Yu, Y., Lin, H.M. and Tsang, W., 1998, Modulation of endoplasmic reticulum calcium pump by Bcl-2, *Oncogene* 17, 1903–1910.

Launay, S., Bobe, R., Lacabaratz-Porret, C., Bredoux, R., Kovacs, T., Enouf, J. and Papp, B., 1997, Modulation of endoplasmic reticulum calcium pump expression during T lymphocyte activation, *J. Biol. Chem.* 272, 10746–10750.

Launay, S., Giann, M., Kovacs, T., Bredoux, R., Bruel, A., Gelebart, P., Zassadowski, F., Chomienne, C., Enouf, J. and Papp, B., 1999, Lineage-specific modulation of calcium pump expression during myeloid differentiation, *Blood* 93, 4395–4405.

Lee, M., Xu, X., Zeng, W., Diaz, J., Kuo, T., Wuytack, F., Racymaekers, L. and Muallem, S., 1997, Polarized expression of Ca^{2+} pumps in pancreatic and salivary gland cells. Role in initiation and propagation of $(Ca^{2+})_i$ waves, *J. Biol. Chem.* 272, 15771–15776.

Li, J., Shen, H., Himmel, K., Dupuy, A., Largaespada, D., Nakamura, T., Shaughnessy Jr., J., Jenkins, N. and Copeland, N., 1999, Laukaemia disease genes: Large-scale cloning and pathway predictions, *Nat. Genet.* 23, 348–353.

Lin, P., Yao, Y., Hofmeister, R., Tsien, R. and Gist Farquhar, M., 1999, Overexpression of CALNUC (Nucleobindin) increases agonist and thapsigargin releasable Ca^{2+} storage in the Golgi, *J. Cell Biol.* 145, 279–289.

Lytton, J., Westlin, M. and Hanley, M., 1991, Thapsigargin inhibits the sarcoplasmic or endoplasmic reticulum Ca-ATPase family of calcium pumps, *J. Biol. Chem.* 266, 17067–17071.

Lytton, J., Westlin, M., Burk, S., Shull, G. and MacLennan, D., 1992, Functional comparison between isoforms of the sarcoplasmic or endoplasmic reticulum family of calcium pumps, *J. Biol. Chem.* 267, 14483–14489.

Ma, T., Mann, D., Lee, J. and Gallinghouse, G., 1999, SR compartment calcium and cell apoptosis in SERCA overexpression, *Cell Calcium* 26, 25–36.

MacLennan, D., Brandl, C., Korczak, B. and Green, N., 1985, Amino-acid sequence of a Ca^{2+} + Mg^{2+}-dependent ATPase from rabbit muscle sarcoplasmic reticulum, deduced from its complementary DNA sequence, *Nature* 316, 696–700.

MacLennan, D.H., Rice, W.J. and Green, N.M., 1997, The mechanism of Ca^{2+} transport by sarco(endo)plasmic reticulum Ca^{2+}-ATPases, *J. Biol. Chem.* 272, 28815–28818.

MacLennan, D.H., Rice, W.J., Odermatt, A. and Green, N.M., 1998, Structure-function relationships in the Ca(2+)-binding and translocation domain of SERCA1: Physiological correlates in Brody disease, *Acta Physiol. Scand.* Suppl., 643, 55–67.

Magnier, C., Papp, B., Corvazier, E., Bredoux, R., Wuytack, F., Eggermont, J., Maclouf, J. and Enouf, J., 1992, Regulation of sarco-endoplasmic reticulum Ca(2+)-ATPases during platelet-derived growth factor-induced smooth muscle cell proliferation, *J. Biol. Chem.* 267, 15808–15815.

Magnier-Gaubil, C., Herbert, J.M., Quarck, R., Papp, B., Corvazier, E., Wuytack, F., Levy-Toledano, S. and Enouf, J., 1996, Smooth muscle cell cycle and proliferation. Relationship between calcium influx and sarco-endoplasmic reticulum Ca^{2+}ATPase regulation, *J. Biol. Chem.* 271, 27788–27794.

McConkey, D. and Orrenius, S., 1997, The role of calcium in the regulation of apoptosis, *Biochem. Biophys. Res. Comm.* 239, 357–366.

Minami, M., Poussin, K., Bréchot, C. and Paterlini, P., 1995, A novel PCR technique using Alu-specific primers to identify unknown flanking sequences from the human genome, *Genomics* 29, 403–408.

Morgan, A.J. and Jacob, R., 1998, Differential modulation of the phases of a Ca^{2+} spike by the store Ca^{2+}-ATPase in human umbilical vein endothelial cells, *J. Physiol. (Lond.)* 513, 83–101.

Mountian, I., Manolopoulos, V., De Smedt, H., Parys, J., Missiaen, L. and Wuytack, F., 1999, Expression patterns of sarco/endoplasmic reticulum Ca^{2+}-ATPase and inositol 1,4,5-triphosphate receptor isoforms in vascular endothelial cells, *Cell Calcium* 25, 371–380.

Odermatt, A., Becker, S., Khannan, V., Kurzydlowski, K., Leisner, E., Pette, D. and MacLennan, D., 1998, Sarcolipin regulates the activity of SERCA1, the fast-twitch skeletal muscle sarcoplasmic Ca^{2+}-ATPase, *J. Biol. Chem.* 273, 12360–12369.

Paterlini, P. and Bréchot, C., 1994, Hepatitis B virus and primar liver cancer in hepatitis B surface antigen-positive and negative patients, in *Primary Liver Cancer, Etiological and Progression Factors*, C. Bréchot (ed.), CRC Press, London, pp. 167–190.

Paul, R., 1998, The role of phospholamban and SERCA3 in regulation of smooth muscle-endothelial cell signalling mechanisms: Evidence from gene-ablated mice, *Acta Physiol. Scand.* 164, 589–597.

Peacocke, M. and Christiano, A.M., 1999, Bumps and pumps, SERCA 1999 [news; comment], *Nat. Genet.* 21, 252–253.

Philips, A., Bates, S., Ryan, K., Helin, K. and Vousden, K., 1997, Induction of DNA synthesis and apoptosis are separable functions, *Genes Dev.* 11, 1853–1863.

Pineau, P., Marchio, A., Terris, B., Mattei, M.G., Tu, Z.X., Tiollais, P. and Dejean, A., 1996, A t(3;8) chromosomal translocation associated with hepatitis B virus intergration involves the carboxypeptidase N locus, *J. Virol.* 70, 7280–7284.

Pozzan, T., Rizzuto, R., Volpe, P. and Meldolesi, J., 1994, Molecular and cellular physiology of intracellular calcium stores, *Physiol. Rev.* 74, 595–636.

Sakuntabhai, A., Ruiz-Perez, V., Carter, S., Jacobsen, N., Burge, S., Monk, S., Smith, M., Munro, C.S., O'Donovan, M., Craddock, N., Kucherlapati, R., Rees, J.L., Owen, M., Lathrop, G.M., Monaco, A.P., Strachan, T. and Hovnanian, A., 1999, Mutations in ATP2A2, encoding a Ca^{2+} pump, cause Darier disease [see comments], *Nat. Genet.* 21, 271–277.

Short, A.D., Bian, J., Ghosh, T.K., Waldron, R.T., Rybak, S.L. and Gill, D.L., 1993, Intracellular Ca^{2+} pool content is linked to control of cell growth, *Proc. Natl Acad. Sci. USA* 90, 4986–4990.

Sumbilla, C., Cavagna, M., Zhong, L., Ma, H., Lewis, D., Farrance, I. and Inesi, G., 1999, Comparison of SERCA1 and SERCA2a expressed in COS-1 cells and cardiac myocytes, *Am. J. Physiol.* 277, H2381–2391.

Toyofuku, T., Kurzydlowski, K., Tada, M. and MacLennan, D., 1993, Identification of regions in the Ca^{2+}-ATPase of Sarcoplasmic Reticulum that affect functional association with Phospholamban, *J. Biol. Chem.* 268, 2809–2815.

Varadi, A., Molnar, E., Ostenson, C. and Ashcroft, S., 1996, Isoforms of endoplasmic reticulum Ca(2+)-ATPase are differentially expressed in normal and diabetic islets of Langerhans, *Biochem. J.* 319, 521–527.

Varadi, A., Lebel, L., Hashim, Y., Mehta, Z., Ashcroft, S. and Turner, R., 1999, Sequence variants of the sarco(endo)plasmic reticulum Ca(2+)-transport ATPase 3 gene (SERCA 3) in caucasian type II diabetic patients (UK prospective diabetes study 48), *Diabetologica* 42, 1240–1243.

Waldron, R., Short, A. and Gill, D., 1997, Store-operated Ca^{2+} entry and coupling to Ca^{2+} pool depletion in Thapsigargin-resistant cells, *J. Biol. Chem.* 272, 6440–6447.

Wang, J., Chenivesse, X., Henglein, B. and Bréchot, C., 1990, Hepatitis B virus integration in a cyclin A gene in a hepatocellular carcinoma, *Nature* 343, 555–557.

Weil, R., Sirma, H., Giannini, C., Kremsdorf, D., Bessia, C., Dargemont, C., Brechot, C. and Israel, A., 1999, Direct association and nuclear import of the hepatitis B virus X protein with the NF-kappaB inhibitor IkappaBalpha, *Mol. Cell. Biol.* 19, 6345–6354.

Wojnowski, L., Hoyland, J., Mason, W., Schwab, A., Westphale, H. and Oberleithner, H., 1994, Cell transformation induces cytoplasmic Ca^{2+} oscillator in Madin–Darby canine kidney cells, *Pflügers Arch.* 426, 89–94.

Wu, K.D., Lee, W.S., Wey, J., Bungard, D. and Lytton, J., 1995, Localization and quantification of endoplasmic reticulum Ca(2+)-ATPase isoform transcripts, *Am. J. Physiol.* 269, C775–784.

Yonekura, K., Stokes, D., Sasabe, H. and Toyoshima, C., 1997, The ATP-binding site of Ca^{2+}-ATPase revealed by electron image analysis, *Biophysical J.* 72, 997–1005.

Zhang, Y., Fujii, J., Phillips, M., Chen, H.-S., Karpati, G., Yee, W.-C., Schrank, B., Cornblath, D., Bobylan, K. and MacLennan, D., 1995, Characterization of cDNA and genomic DNA encoding SERCA1, the Ca^{2+}-ATPase of human fast-twitch skeletal muscle sarcoplasmic reticulum, and its elimination as a candidate gene for Brody disease, *Genomics* 30, 415–424.

Zhang, P., Toyoshima, C., Yonekura, K., Green, N.M. and Stokes, D.L., 1998, Structure of the calcium pump from sarcoplasmic reticulum at 8-A resolution, *Nature* 392, 835–839.

Interaction of Dimeric S100B($\beta\beta$) with the Tumor Suppressor Protein p53: A Model for Ca^{2+}-Dependent S100-Target Protein Interactions

David J. Weber, Richard R. Rustandi, France Carrier and Danna B. Zimmer

1. INTRODUCTION

The tumor suppressor protein p53 interacts with a number of proteins to mediate its pleiotropic effects (Giaccia and Kastan, 1998; Levine, 1997). The Ca^{2+}-dependent interaction of p53 with members of the S100 calcium binding protein family is of particular interest since like p53, these dimeric proteins affect cell cycle progression and are overexpressed in numerous tumor cells (Donato, 1991, 1999; Heizman, 1999; Ilg, 1996; Kligman and Hilt, 1988; Schafer and Heizmann, 1996; Zimmer et al., 1995). For example, reactive gliomas have as much as 20 times more S100B($\beta\beta$) than in non-transforming cell lines (Donato, 1991). Increased levels of S100B($\beta\beta$) are also found in renal cell tumors and malignant mature T-cells (such as doubly negative $CD4^-$/$CD8^-$ adult T-cells in leukemia patients). As is the case for S100B($\beta\beta$), a number of other S100 proteins are often upregulated in cancer (Kligman and Hilt, 1988). For example, S100A1, S100A6, and S100B($\beta\beta$) are elevated significantly in metastatic human mammary epithelial cells (Schafer and Heizmann, 1996), and increased levels of CAPL (S100A4) in transgenic mice induce metastatic mammary tumors (Barraclough, 1998; Sherbet and Lakshmi, 1998). In the case of S100A4, expression of antisense

David J. Weber, Richard R. Rustandi and France Carrier • Deparment of Biochemistry and Molecular Biology, University of Maryland School of Medicine, Baltimore, MD, U.S.A.
Danna B. Zimmer • Deparment of Pharmacology, University of South Alabama College of Medicine, Mobile, AL, U.S.A.

R. Pochet, R. Donato, J. Haiech, C. Heizmann and V. Gerke (eds.): Calcium: The Molecular Basis of Calcium Action in Biology and Medicine, 521–539.

Table 1. Comparison of the amino acid sequences of the two putative Ca^{2+}-binding sites in each S100β subunit with a consensus EF-hand Ca^{2+}-binding site

Position[1]	1	2	3	4	5	6	7	8	9	10	11	12
Consensus[2]	**D**	*B*	**D** **N**	G	**D** **N**	G	*X*	*H*	**O**	*H*	*A*	**E**
Ligand coordinate[3]	**X**		**Y**		**Z**		**–Y**		**–X**			**–Z**
S100β												
			R				H					
Site I[4]	**S**	G	**E**	G	**D**	K	**K**	L	K	K	S	**E**
(ψ-EF-hand)												
Site II[5]	**D**	E	**D**	G	**D**	G	**E**	C	D	F	Q	**E**
(typical EF-hand)												

[1]The position number is based on the consensus sequence for a normal EF-hand (Strynadka and James, 1989).

[2]Abbreviations include *A*, acidic residue; *B*, basic residue; *H*, hydrophobic; *X*, any residue; *O*, Asp, Glu, Thr, or Ser (Strynadka and James, 1989).

[3]The coordinate of the Ca^{2+} ligands are shown in bold, and the underlined coordinate has a carbonyl oxygen as the Ca^{2+} ligand based on previously determined structures of EF-hand-containing proteins (Kretsinger, 1980). In most EF-hand proteins a water oxygen coordinates at position 9; parvalbumin is the exception in which case a side chain carboxylate oxygen from a glutamate coordinates at this position.

[4]Underlined residues have a carbonyl oxygen as the Ca^{2+} ligand and the sidechain carboxylate oxygens of E32 form a bidentate ligand based on the X-ray structure of Ca^{2+}-bound S100B($\beta\beta$) (Matsumara et al., 1998). Lanthanide luminescence data, however, indicates that only four ligands contribute to binding in the pseudo-EF-hand (Chaudhuri et al., 1997), which is consistent with the NMR structure (Drohat et al., 1998), which has E32 as monodentate and does not include the carbonyl oxygen of D21 as a ligand.

[5]The residues that ligand Ca^{2+} are shown in bold, and the underlined residue has the carbonyl oxygen as the Ca^{2+} ligand based on the X-ray and NMR structures and on lanthanide luminescence spectroscopy (Chaudhuri et al., 1997; Drohat et al., 1998; Matsumura et al., 1998).

RNA to S100A4 suppresses metastatic potential for a high-metastatic Lewis lung carcinoma (Barraclough, 1998; Sherbet and Lakshmi, 1998). In addition, protein levels of S100B($\beta\beta$), S100L (S100A2), S100A4, and S100A6 correlate with malignant melanoma as detected by antibodies specific for each protein (McNutt, 1998). In fact, S100 antibodies are used clinically to identify and classify tumors in brain, lung, bladder, intestine, kidney, cervix, breast, skin, head and neck, lymph, testes, larynx, and mouth among others. These results, together with the knowledge that a p53-S100B($\beta\beta$) interaction occurs *in vitro* (Baudier et al., 1992; Delphin et al., 1999; Garbuglia et al., 1999; Rustandi et al., 1999; Rustandi et al., 1998; Wilder et al., 1998) and *in vivo* (Carrier, 1999; Scotto et al., 1998, 1999), provides a basis for the detailed examination of the binding between S100B($\beta\beta$) and p53. This Ca^{2+}-dependent complex formation can also be used as a model for a number of S100-target protein interactions.

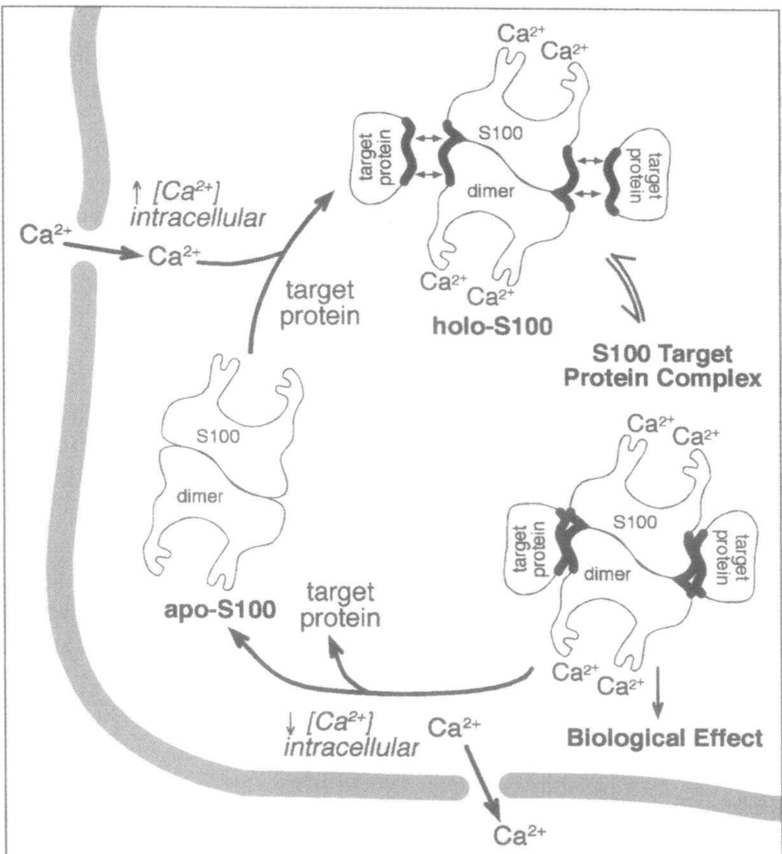

Figure 1. Schematic model of the Ca^{2+}-dependent interaction of S100 and target proteins. As intracellular Ca^{2+} increases, members of the EF-hand containing S100 protein family bind Ca^{2+} and change in conformation. This conformational change is necessary to expose a site on the S100 protein that interacts with specific target proteins to elicit a biological effect. When intracellular Ca^{2+} concentrations decrease the metal and target protein dissociate from the S100 protein and the biological effect is stopped.

2. PROPERTIES OF S100 PROTEINS

Presumably, the biological functions of S100B($\beta\beta$) and other S100 proteins are related to tightly regulated intra- and extracellular Ca^{2+} ion concentrations (Figure 1) (Donato, 1991, 1999). Based on sequence homologies to other EF-hand containing proteins, S100 proteins contain two helix-loop-helix Ca^{2+} ion binding domains per subunit now referred to as the "typical" and "pseudo" EF-hand motifs, respectively (Kligman and Hilt, 1988; Strynadka and James, 1989) (Table 1). The "pseudo" EF-hand has 14 rather than 12 residues, and generally binds Ca^{2+} with lower affinity ($K_d > 50\ \mu$M) than the typical EF-hand ($K_d < 50\ \mu$M; Table 1) (Kligman and Hilt, 1988; Rustandi et al., 1998). These *in vitro* binding constants may approximate *in vivo* affinities; however, the effective binding in cells depends upon the microenvironment of the S100 protein, and thus may be different (Kligman and Hilt, 1988). For example, S100B($\beta\beta$) binds Ca^{2+} with a lower affinity at elevated sodium or potassium concentrations; thus, fluctuations in monovalent cation concentrations profoundly affect the Ca^{2+} binding state of S100 proteins (Donato, 1991, 1999). Furthermore, if *in vitro* binding is a good first approximation for Ca^{2+} binding *in vivo*, then both Ca^{2+} binding sites on each S100β subunit would be saturated only in locations where Ca^{2+} concentrations reach a relatively high level, such as in subcellular compartments or in extracellular space. On the other hand, membrane-bound S100B($\beta\beta$) has a 15- to 18-fold higher affinity for Ca^{2+} (Donato, 1991, 1999), and the presence of a peptide derived from the C-terminus of p53 increases the Ca^{2+}-binding affinity by at least 3-fold (Rustandi et al., 1998). Furthermore, Ca^{2+} ion binding affinity is increased for S100B($\beta\beta$) by 5- to 10-fold in the presence of Zn^{2+} which binds ($K_d = 0.1$–$1.0\ \mu$M) at sites different from those of Ca^{2+} (Donato, 1991, 1999; Rustandi et al., 1998). Under these conditions, Ca^{2+} binding to most S100 proteins is physiologically relevant at a large number of locations throughout the cell, including the cytoplasm where Ca^{2+} ion concentrations are generally low (0.1 to 10 μM) (Berridge and Irvine, 1989).

A number of *in vitro* studies based on sulfhydryl reactivity, fluorescence spectroscopy, UV differential absorbance, circular dichroism (CD), and 1D NMR demonstrate that Ca^{2+} ion binding induces a conformational change in most S100 proteins (Donato, 1991, 1999). This change in conformation is required to interact with their protein targets (Figure 1) (Donato, 1991, 1999). Kyte–Doolittle hydropathy plots of the S100 family predict that a hydrophobic patch exists in loop 2 (the "hinge" region) and the C-terminal region of S100 proteins that may be exposed upon Ca^{2+}-binding and interact with membranes and/or specific effector proteins (Kligman and Hilt, 1988). The lack of sequence homology in the "hinge" and C-terminal loop with other members of the S100 family can explain why specificity in target protein binding is observed among many of the S100 family members (Kligman and Hilt, 1988). The 3D structures of several S100 proteins described in this chapter confirm this to be the case.

Recently, NMR and/or X-ray crystallography were used to solve the high-resolution 3D structures of several S100 proteins in the apo-(Ca^{2+}-free), holo-(Ca^{2+}-bound), and target protein-bound states (Table 2). A detailed comparison of accurately defined structures among family members is critical for understanding how specific S100 proteins bind unique targets to elicit different biological effects. Since S100B($\beta\beta$) is the only S100 protein for which all three complexes (apo-, holo-, and target protein-bound) are solved at high resolution, a comparison of these three structures will be discussed. This comparison illustrates how S100B($\beta\beta$) binds p53 in a Ca^{2+}-dependent manner, which is relevant to the Ca^{2+}-dependent target protein interactions of several S100 family members (Groves et al., 1998).

3. 3D STRUCTURE OF APO-S100B($\beta\beta$)

The structure of apo-S100B($\beta\beta$) was first reported by two groups with minor differences (Drohat et al., 1996; Kilby et al., 1996), and later refined using dipolar coupling restraints to give a very accurate high resolution structure (Figure 2) (Drohat et al., 1999). As found for S100A6 (Maler et al., 1999; Potts et al., 1995), S100B was found to exist in solution as a symmetric dimer, termed S100B($\beta\beta$), held together by noncovalent interactions under reducing conditions. This is quite a different structure from the disulfide-linked form of the dimer, S100B($\beta_{S-S}\beta$), that is responsible for many of the extracellular activities reported for this protein (Donato, 1991, 1999; Kligman and Hilt, 1988; Zimmer et al., 1995).

3.1. Is the non-covalent dimer physiologically relevant?

Gel filtration, dynamic light scattering, and ultracentrifugation studies of S100B($\beta\beta$) were done in the presence and absence of Ca^{2+} (Drohat et al., 1997; Landar et al., 1997). Together these data confirm that wild-type, C68S, C68V, C84S, C84A, C68S + C84S, C68V + C84S, and C68V + C84A S100B($\beta\beta$) proteins are each dimeric (Landar et al., 1997). It is clear that the dimer interface must be held together by non-covalent forces since double mutant proteins, which lacked cysteine residues, remained dimeric (Landar et al., 1997). Also, large-zone analytical gel filtration chromatography with ^{35}S-labeled protein showed that S100B exists exclusively as a dimer ($> 99\%$) both in the presence or absence of Ca^{2+} at concentrations as low as 1 nM. This concentration is significantly below ($> 1 \times 10^4$) the concentration of S100B($\beta\beta$) in glial cells (μM levels) (Donato, 1991). Therefore, dimeric S100B($\beta\beta$), rather than a single S100β subunit, is likely the form of S100B presented to target proteins at physiologically relevant protein concentrations inside the cell. However, it is imperative to determine the oligomerization state of each S100B-target protein complex since target protein binding may affect S100B($\beta\beta$) dissociation (Drohat, et al., 1997).

The overall fold of apo-S100B($\beta\beta$) has a distinct hydrophobic core and a very hydrophilic solvent exposed surface area which is consistent with its high solubility (Drohat et al., 1996). The dimer interface of S100B($\beta\beta$) is defined by the nearly antiparallel alignment of helices 1 and 1′ (170°) and helices 4 and 4′ (170°), respectively (Table 3). In addition, helices 1 and 1′ of S100B($\beta\beta$) are nearly perpendicular to helices 4 and 4′, respectively, to give an X-type four helical bundle as observed for all of the dimeric S100 proteins for which a 3D structure is reported (Figure 2; Table 2). Furthermore, the sta-

Table 2. RMSD (NMR) or Resolution/R-factor (X-ray) of Dimeric S100 protein structures[1]

S100 protein	apo-	apo-refined	Ca^{2+}-bound	Ca^{2+}, Zn^{2+}-bound		target protein-bound
S100B						
Rat	**1.04** Med. (1SYM) NMR	**0.39** High (1B4C) NMR	**1.27** Med. (1QLK) NMR		–	**0.49**[2] High (1DT7) NMR
Human	–	–	**1.61** Med. (1UWO) NMR		–	–
Bovine	**1.26** Med. (1CFP) NMR	–	**2.0/19.5%** High (1MHO) X-ray		–	–
S100A6						
Human	**2.72** Low (1CNP) NMR	**0.73** High (2CNP) NMR	**2.70** Low (1AO3) NMR		–	–
S100A7						
Human	–	–	**1.05/10.9%**[3] High (1PSR) X-ray	**2.05/21.6%**[4] High (2PSR) X-ray	**2.50/22.3%**[5] Med. (3PSR) X-ray	–
S100A10						
Human	**2.25/24.6%** High (1A4P) X-ray	–	–	–	**2.44/23.3%** Med. (1BT6) X-ray[6]	–

[1]Shown in bold is the RMSD (in Å) from the mean structure for all backbone residues in a family of NMR structures or the resolution (in Å)/R-factor for the X-ray crystal structures. Also listed are the overall resolution of the structure (i.e. high, medium, or low), the protein data bank (PDB) accession code in parentheses, and the method of the structure determination. Details are described for each structure in the primary literature for (i) *rat* apo-S100B (Drohat et al., 1996), refined apo-S100B (Drohat et al., 1999), Ca^{2+}-bound S100B (Drohat et al., 1998), p53-bound S100B (Rustandi et al., 2000), (ii) *human* Ca^{2+}-bound S100B (Smith and Shaw, 1998), (iii) *bovine* apo-S100B (Kilby et al., 1996), Ca^{2+}-bound S100B (Matsumara et al., 1998), (iv) *human* apo-S100A6 (Potts et al., 1995), refined apo-S100A6 (Maler et al., 1999), Ca^{2+}-bound S100A6 (Sastry et al., 1998), (v) *human* Ho^{3+}-bound S100A7 (Brodersen et al., 1999a), Ca^{2+}, Zn^{2+}-bound S100A7 (Brodersen et al., 1999b), and (vi) *human* apo-S100A10 (to be published), annexin II-bound S100A10 (Rety et al., 1999).

[2]Bound to a peptide derived from the C-terminus of p53 (residues 367–388).

[3]Ho^{3+} is substituted for Ca^{2+}.

[4]Ca^{2+} and Zn^{2+} are bound to both subunits.

[5]One subunit is bound with Ca^{2+} and Zn^{2+}, and the other subunit is bound with only Ca^{2+}.

[6]Bound to a peptide derived from annexin II.

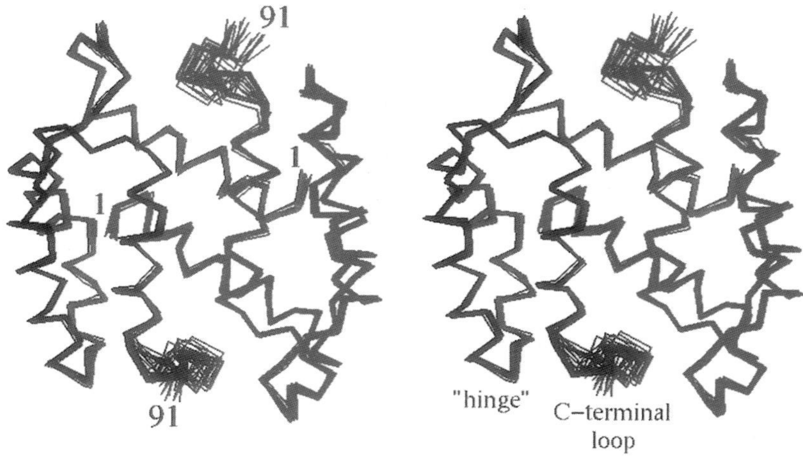

Figure 2. Stereoview of 20 apo-S100B($\beta\beta$) structures refined to high resolution using dipolar coupling restraints as described previously (Drohat et al., 1999). Residues in the C-terminus of apo-S100B($\beta\beta$) and Ca^{2+}-loaded S100B($\beta\beta$) (Drohat et al., 1998) (Figure 3) are in conformational exchange, and are not well defined. The conformation in the C-terminal region of S100B($\beta\beta$) is stabilized when bound to a peptide derived from p53 (Rustandi et al., 1999). The proximity of non-conserved residues in loop 2 (the "hinge") and the C-terminal loop ultimately confer specificity to S100-target protein interactions. For a colour version of this figure, see page xxx.

bility of the S100B dimer ($K_d < 1$ nM) found both in presence and absence of Ca^{2+} is due to the similar orientation of a large number of hydrophobic residues at the subunit interface (Drohat et al., 1996, 1998, 1999). These residues are highly conserved within the S100 protein family (Drohat, et al., 1997); calbindinD_{9K} is the only S100 protein that lacks these hydrophobic residues, and it is monomeric in solution.

3.2. Ca^{2+}-binding loops in apo-S100B($\beta\beta$)

In the absence of Ca^{2+}, each S100β subunit is highly structured with two helix-loop-helix Ca^{2+}-binding domains. In a manner similar to other calcium binding proteins, the two Ca^{2+} ion binding domains of S100B($\beta\beta$) are brought into close proximity with a short two-stranded antiparallel β-sheet. In addition, the overall fold and interhelical angle between helices 1 and 2 of its S100-hand are similar to those of the S100-hand present in apo-calbindinD_{9K} and apo-calcyclin (Yap et al., 1999). However, the interhelical angle between helices 3 and 4 differs significantly from classical EF-hand motifs in other Ca^{2+} ion binding proteins such as apo-calmodulin and apo-calbindinD_{9K} (Table 3) (Yap et al., 1999). This difference results from a novel

Table 3. Interhelical angles in S100B complexes[1]

Helices	rat S100B-Ca^{2+}-p53[2,3]	rat Ca^{2+}-S100B[3,4]	bovine Ca^{2+}-S100B[5]	rat apo-S100B[3,4]
I–II	$120 \pm 3^{\perp}$	$137 \pm 5^{\perp}$	$135^{\perp}$	$133 \pm 1^{\perp}$
I–III	$-123 \pm 2^{\perp}$	$-118 \pm 5^{\perp}$	–	$-46 \pm 1^{\parallel}$
I–IV	$125 \pm 1^{\perp}$	$128 \pm 4^{\perp}$	$127^{\perp}$	$120 \pm 1^{\perp}$
II–III	$115 \pm 2^{\perp}$	$104 \pm 3^{\perp}$	$97^{\perp}$	$149 \pm 1^{\parallel}$
II–IV	$-26 \pm 2^{\parallel}$	$-35 \pm 4^{\parallel}$	–	$-40 \pm 1^{\parallel}$
III–IV	$110 \pm 1^{\perp}$	$106 \pm 4^{\perp}$	$101^{\perp}$	$-166 \pm 1^{\parallel}$
I–I′	$-152 \pm 2^{\perp}$	$-155 \pm 1^{\perp}$	–	$-153 \pm 1^{\perp}$
I–IV′	$-66 \pm 1^{\perp}$	$-58 \pm 3^{\perp}$	–	$-66 \pm 1^{\perp}$
IV–IV′	$138 \pm 2^{\perp}$	$159 \pm 5^{\perp}$	–	$155 \pm 1^{\perp}$

[1]Interhelical angles (Ω) range from -180 to $+180°$, and are classified as either: parallel ($\parallel$) for $0° \leq |\Omega| \leq 40°$ and $140° \leq |\Omega| \leq 180°$, or as perpendicular ($\perp$) for $40° < |\Omega| < 140°$, as described (Harris et al., 1994). The sign (+ or −) of Ω can be determined by: (1) orienting the molecule such that the two helices of interest (i, j) are in planes parallel to the screen with the first helix (i) in front of the second (j) and the first helix (i) aligned vertically (0°, with its N→C vector pointing upwards), (2) aligning an imaginary vector vertically (0°) with its tail placed at the N-terminus of the second helix (j) and rotating it (by $\leq 180°$ clockwise or counterclockwise) to align with the N→C vector of the second helix (j) where a clockwise rotation gives a positive (+) Ω value and a counter-clockwise rotation gives a negative (−) Ω value.
[2]The interhelical angles from the p53 peptide to helices I–IV are 139 ± 3 (I), 100 ± 3 (II), -16 ± 3 (III), and 95 ± 3 (IV), respectively.
[3]Interhelical angles were calculated using the program Iha 1.4.
[4]Helices in rat apo-S100B are: helix I (2–18), helix II (29–40), helix III (50–63), and helix IV (70–83) (Drohat et al., 1999), and helices in Ca^{2+}-loaded S100B are: helix I (2–20), helix II (29–38), helix III (50–61), and helix IV (70–83).
[5]Taken from the X-ray crystal structure of Matsumura et al. (1998).

orientation of helix 3 in the second helix-loop-helix Ca^{2+} binding domain of apo-S100B($\beta\beta$) (Figures 2 and 3).

4. 3D STRUCTURE OF Ca^{2+}-BOUND S100B($\beta\beta$)

A comparison of the Ca^{2+}-bound structures to those of the apo-form becomes important for understanding how S100 proteins function in solution (Groves et al., 1998). However, a discrepancy regarding the position of helix 3 in the apo-structures of S100B($\beta\beta$) from different sources (rat, bovine) and calcyclin (S100A6) led to confusion early on regarding the extent of the conformational change that occurs for dimeric S100 proteins upon the addition of Ca^{2+} (Drohat et al., 1996; Kilby et al., 1996; Potts et al., 1995). These

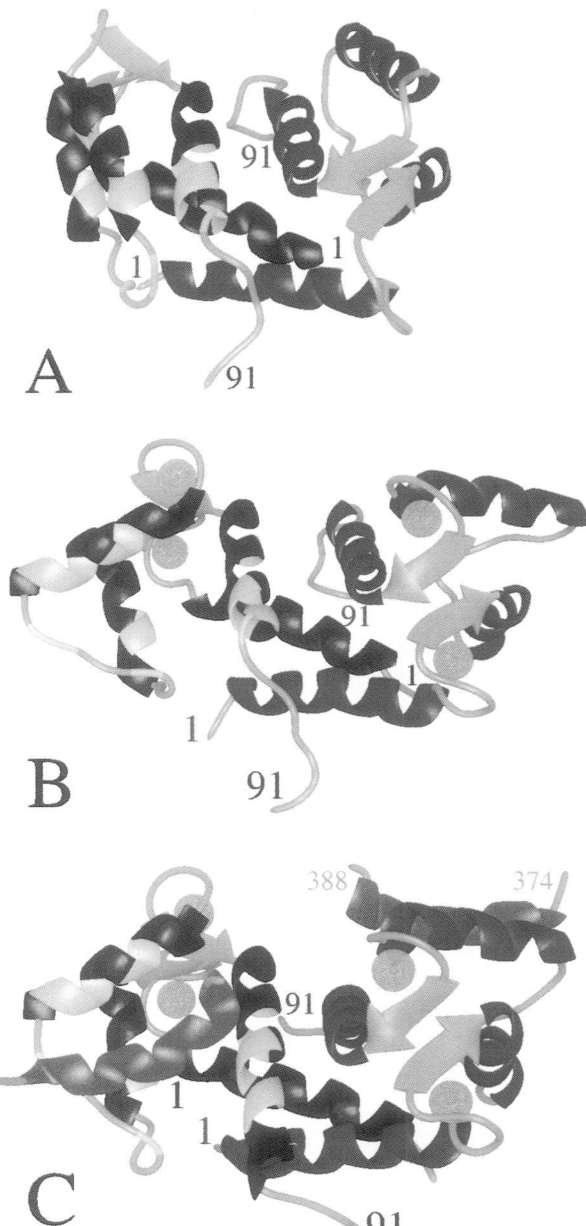

Figure 3. Ribbon diagrams illustrating the 3D structures of (A) apo-S100B($\beta\beta$), (B) Ca^{2+}-bound S100B($\beta\beta$), and (C) p53 peptide-bound S100B($\beta\beta$). Residues that interact with the p53 peptide (in C) are shown in purple on one subunit of the apo- and Ca^{2+}-bound S100B($\beta\beta$) structures to illustrate the Ca^{2+}-dependence of the p53-S100B interaction. Residues that interact with p53 are buried in (A) apo-S100B, but are exposed to solvent after a Ca^{2+}-dependent conformational change (in B). This change in protein structure is required for p53 binding (Rustandi et al., 1998, 1999). For a colour version of this figure, see page xxxi.

issues were resolved, however, when the apo-structures from rat S100B($\beta\beta$) and S100A6 were refined to higher resolution (Drohat et al., 1999; Maler et al., 1999), and when the Ca^{2+}-bound structures for both of these S100 proteins were determined (Table 2) (Drohat et al., 1998; Matsumura et al., 1998; Sastry et al., 1998; Smith and Shaw, 1998).

The secondary structure of Ca^{2+}-bound S100B($\beta\beta$) is consistent with the presence of two helix-loop-helix EF-hand Ca^{2+}-binding domains as determined by NMR (Drohat, et al., 1998; Matsumura et al., 1998; Sastry et al., 1998; Smith and Shaw, 1998) and X-ray crystallography (Drohat et al., 1998; Matsumura et al., 1998; Sastry et al., 1998; Smith and Shaw, 1998). The overall geometry of the Ca^{2+}-binding loops (ϕ, φ angles) and the interhelical angles of each EF-hand in Ca^{2+}-bound S100B($\beta\beta$) are similar to those of several EF-hand containing proteins including Ca^{2+}-bound calbindinD$_{9K}$, troponin C, calmodulin, and parvalbumin (Strynadka and James, 1989; Yap et al., 1999). The residues involved in Ca^{2+}-ion coordination observed in the X-ray structure of S100B($\beta\beta$) (Drohat et al., 1998; Matsumura et al., 1998; Sastry et al., 1998; Smith and Shaw, 1998) are identical to those in both sites of Ca^{2+}-bound calbindinD$_{9K}$ (Table 1), which is supported for the most part by the NMR solution structure (Drohat et al., 1998) and by data from lanthanide luminescence spectroscopy (Chaudhuri et al., 1997). The only difference being that lanthanide luminescence data is consistent with only four ligands provided by the protein in the pseudo-EF-hand, and yet the X-ray crystal structure has six (Table 1).

Comparison of the apo- and Ca^{2+}-bound structures of S100B($\beta\beta$) showed no differences in the orientation of helices (1, 1′, 4, or 4′) in the X-type four helical bundle dimer interface. These findings confirm the gel filtration data which indicates that the S100B($\beta\beta$) dimer is stable below 1 nM concentrations both in the presence and absence of Ca^{2+} (Drohat et al., 1997; Landar et al., 1997). Another similarity is that the interhelical angles between helices 1 and 2 in the pseudo-EF-hand are identical in the two structures (apo-: $133 \pm 1°$; Ca^{2+}-bound: $137 \pm 5°$; Table 3). The absence of a conformational change in the pseudo-EF-hand is not surprising, since the position of helix 1 is severely restricted by core interactions at the dimer interface in both the presence and absence of Ca^{2+}.

4.1. The conformational change

In the second EF-hand (the typical EF-hand), however, there is a *large change in the position of helix 3* upon the addition of Ca^{2+} (Figure 3). In this EF-hand, a $\sim$ 90° change is observed in the interhelical angles between helices 3 and 4 upon the addition of Ca^{2+} (Ca^{2+}-bound: $\Omega_{3-4} = 106 \pm 4°$; apo-: $\Omega_{3-4} = 194 \pm 1°$; Table 3). It is the entering helix (helix 3) in this EF-hand rather than the exiting helix (helix 4) that reorients significantly upon Ca^{2+} binding because of the involvement of helix 4 in the dimer interface of

S100B($\beta\beta$). This is unlike the conformational changes observed in most EF-hand containing proteins such as calmodulin and troponin C where the exiting helix rotates when Ca^{2+} binds (Yap et al., 1999). For S100B($\beta\beta$), the Ca^{2+}-dependent conformational change exposes a cleft defined by residues in the hinge region, the C-terminal loop, and helix 3 (Figures 2 and 3). This surface on Ca^{2+}-bound S100B($\beta\beta$) is important for target protein binding (Figure 3).

5. 3D STRUCTURE OF S100B($\beta\beta$) BOUND TO THE NEGATIVE REGULATORY DOMAIN OF p53

A tightly associated dimer is also observed for S100B($\beta\beta$) in the p53 peptide complex (Figure 3). The only difference in this dimer interface from that observed in the apo- and Ca^{2+}-loaded S100B($\beta\beta$) is the IV-IV′ interhelical angle ($\Delta\Omega_{IV-IV'} = 19 \pm 2°$), but this is mostly due to an extension of these two helices when the p53 peptide binds. While the addition of Ca^{2+} to apo-S100B($\beta\beta$) causes a large change in the orientation of helix 3 ($\Delta\Omega_{III-IV} = 90°$; Figures 3B, 3C), no additional change in this interhelical angle is observed when the p53 peptide binds (Table 3). In fact, each subunit of S100B($\beta\beta$) has interhelical angles similar to those of several other EF-hand containing proteins when bound to the p53 peptide (Rustandi et al., 1998, 2000). Yet the geometry of Ca^{2+} coordination is improved with the addition of the p53 peptide as judged by small differences in the position of helix 4, which affects the Ca^{2+} ion coordination, and by a 3-fold increase in Ca^{2+} ion binding affinity (Rustandi et al., 1998, 1999).

The most notable structural changes occurs in the C-terminus and in loop 2 of S100B($\beta\beta$) when the p53 peptide binds. Specifically, helix 4 is extended by 5 residues with a concomitant loss of conformational exchange averaging (Rustandi et al., 1998, 1999), and loop 2, termed the "hinge" region, is considerably rotated when the structure of S100B($\beta\beta$) in the presence and absence of peptide are compared. In fact, the "hinge" and helix 4 clamp down on the p53 peptide to give a number of important hydrophobic and salt-bridge type interactions (Figure 4). Also, slower amide exchange rates are observed for many of these residues in the binding pocket when compared to those in the absence of peptide (Rustandi et al., 1998, 1999). It is indeed these regions of S100B($\beta\beta$) that have the least amount of sequence homology with other members of the S100 protein family and gives rise to specificity in the S100-target protein interactions as previously suggested (Drohat et al., 1996; Kligman and Hilt, 1988).

In the absence of S100B($\beta\beta$), the p53 peptide (S367–E388) exists as a random coil as judged by the lack of NOE correlations and a very narrow range of NMR spectral dispersion. When bound to Ca^{2+}-loaded S100B($\beta\beta$), the C-terminal p53 peptide (residues S376–T387) adopts a helical conformation. Additionally, a number of interactions between the C-terminal region of the peptide and S100B($\beta\beta$) contribute to the stable helical conformation

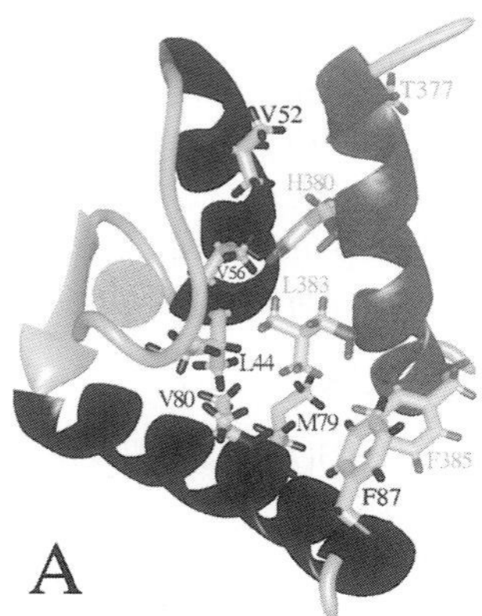

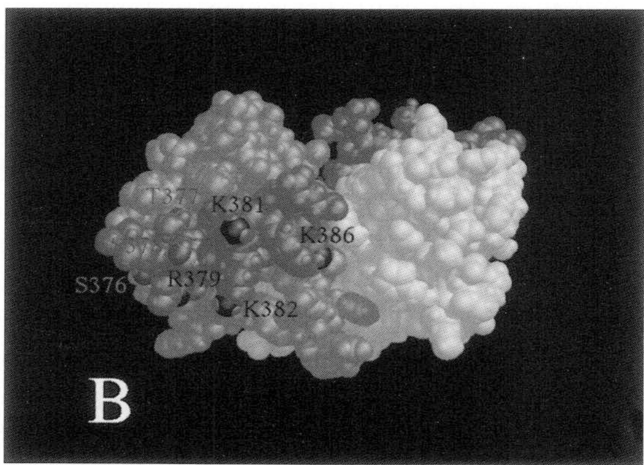

Figure 4. Ribbon diagram and space-filling model showing the positioning of sidechain residues of S100B($\beta\beta$) and p53. (A) Ribbon diagram displaying the sidechains involved in a hydrophobic binding pocket. These residues include L44, V52, V56, M79, V80, F87 (colored in blue) from S100($\beta\beta$) and H380, L383, and F385 (colored in green) from the p53 peptide. (B) Space-filling model of Ca^{2+}-loaded S100B($\beta\beta$) bound to the p53 peptide illustrating the positions of the PKC-phosphorylation sites, S376, T377, and S378 (red) and several p300-acetylation sites (in blue) on p53. For a colour version of this figure, see page xxx.

of the p53 peptide in this complex, while the N-terminal region of the p53 peptide (S367 to Q375) remains relatively unstructured in solution with only a few NMR-derived constraints. The binding surface for the p53 peptide on S100B($\beta\beta$) is defined by several residues from loop 2, helix 3, and helix 4 on S100B($\beta\beta$) and by residues Q375, R379, H380, L383, M384, K386, F385, T387, and E388 of the p53 peptide. Two hydrophobic residues (L383, F385) of the peptide interact extensively with a hydrophobic patch on S100B($\beta\beta$) (Figure 4). There are likely salt bridges between residues R379 and K386 from the peptide to residues E45 on the "hinge" and E86 on helix 4 of S100B($\beta\beta$) based on this NMR structure.

The Ca^{2+}-dependence of the S100B($\beta\beta$)-p53 interaction can easily be observed by comparing the structural data of apo-, Ca^{2+}-bound, and the p53 peptide complexes of S100B($\beta\beta$) (Figure 3). Most of the residues that interact

Table 4. Amino acid sequence of protein targets recognized by S100B($\beta\beta$)[1]

Position		1	2	3	4	5	6	7	8
S100 binding[2]		**K**	L	X	W	X	X	I	L
Human p53 (367–388)	S H L **K** S **K** **K** G Q *S* *T* *S* **R** H **K**	**K**	L	M	F	**K**	T	E	
CapZ (265–276)	T **R** T	**K**	I	D	W	N	**K**	I	L

[1]Lysine and arginine residues are shown in bold; sites for phosphorylation are shown in italics, and residues that match the S100B binding sequence are underlined. In parenthesis are the residue numbers of the protein from which peptides were prepared, and shown to bind S100B($\beta\beta$).

[2]The S100B-binding domain was determined by Dimlich and coworkers (Ivanenkov et al., 1995) by screening a peptide library. The CapZ peptide was found to be the tightest binding peptide of those screened.

with p53 (18 of 21) are buried in the apo-S100B($\beta\beta$) structure (Figure 3A). When Ca^{2+} binds to S100B($\beta\beta$), however, these same residues become exposed to solvent due to a large change in the position of helix 3 (Figure 3B); this conformational change is required for the interaction with the p53 peptide (Figure 3C) (Rustandi et al., 1998, 2000). The direct involvement of the "hinge" and the C-terminus of S100B($\beta\beta$) in the p53 interaction supports numerous mutagenesis and biochemical studies done previously that implicate these two regions in S100 proteins as being critical for binding their targets (Donato, 1999). Likewise, much of the consensus sequence found for S100B($\beta\beta$) protein targets, as determined by screening large peptide libraries, is found in the C-terminus of p53 (Table 4) (Ivanenkov et al., 1995; Wilder et al., 1998). In fact, when the p53 peptide is mutated (F385W) to better match this "S100B-binding consensus sequence", the affinity of the p53 peptide for S100B($\beta\beta$) goes up by approximately 5-fold (Rustandi et al., 1998).

6. BIOLOGICAL IMPLICATIONS OF S100B($\beta\beta$) BINDING p53

It is now clear that p53 acts as a transcription factor, and that its activity can be upregulated by modifications to the C-terminus of the protein often termed the "negative regulatory domain" (Chiarugi et al., 1998; Giaccia and Kastan, 1998; Hupp et al., 1992; Levine, 1997). It is generally thought that the negative regulatory domain of p53 interacts with its own DNA binding domain to allosterically inhibit its activity, and that phosphorylation and/or acetylation of this domain activates p53 by disrupting this intramolecular interaction (Chiarugi et al., 1998; Hupp et al., 1992). In fact, deletion of the C-terminus or binding of artificial oligonucleotides, peptides, and antibodies such as Pab421 can stimulate p53 DNA binding, presumably by a similar mechanism, but the physiological relevance of these modifications has not been shown. Peptides derived from the C-terminal negative regulatory domain also activate DNA-binding of both wild-type and mutant p53, which may have therapeutic ramifications (Selivanova et al., 1999).

The opposite effect was observed in reporter gene assays done with transient co-transfections of S100B($\beta\beta$) and p53 (Carrier et al., 1999). In this case, the binding of S100B($\beta\beta$) to the C-terminal negative regulatory domain is not sufficient to activate p53-dependent transcription, but rather inhibits p53 activity by about 50% (Carrier et al., 1999). This inhibition is most likely related to an observation made by Baudier and colleagues in 1992 (Baudier et al., 1992). In these studies, it was demonstrated that phosphorylation of p53 by protein kinase C (PKC) occurs *in vivo* and *in vitro*, and that this post-translational modification is inhibited by S100B($\beta\beta$) in a Ca^{2+}-dependent manner (Baudier et al., 1992; Wilder et al., 1998). They initially proposed that a direct interaction between the negative regulatory domain of p53 and S100B($\beta\beta$) is responsible for this inhibition (Baudier et al., 1992), which was later confirmed in studies with a peptide derived from the C-terminus of p53 (residues 367–388) (Rustandi et al., 1998, 1999; Wilder et al., 1998). Thus, the S100B-dependent inhibition of p53 phosphorylation is the result of a Ca^{2+}-dependent interaction between S100B($\beta\beta$) and the PKC substrate (p53), rather than with the kinase itself (Wilder et al., 1998).

The NMR structure of the p53 peptide bound to S100B($\beta\beta$) also gives some direct insights into how S100B($\beta\beta$) may inhibit p53-dependent transcription activation. Two important PKC phosphorylation sites (S376 and T377) and two acetylation sites (R379, K386) on p53 are sterically blocked by S100B($\beta\beta$) binding (Figure 4B). The identity of these particular phosphorylation sites of p53 is particularly important because mutations at either one of these residues (S376A, T377A) decreases p53 activity to nearly the same level (38 and 50% inhibition, respectively) (Youmell et al., 1998) as observed in the assays done with transient co-transfections of S100B($\beta\beta$) and p53 (50% inhibition) (Carrier et al., 1999). Thus, the inhibition p53 activity by S100B($\beta\beta$) is very likely related to the ability of S100B($\beta\beta$) to block post-translational modifications in the C-terminal negative regulatory domain. Perhaps a general mechanism for regulating signal transduction is emerging based on results with S100B($\beta\beta$) (Wilder et al., 1998) and similar results with calmodulin (Sheu et al., 1995). In this model, Ca^{2+}-binding proteins provide a preemptive control mechanism by binding at or near the site(s) of post-translational modification(s) to sterically inhibit catalysis. This is, in some sense, opposite to the control mechanism provided by protein phosphatases, which react after a signal is transduced (Rustandi et al., 1998; Wilder et al., 1998).

The interaction between S100B($\beta\beta$) and p53 also involves the oligomerization domain of p53 (Delphin et al., 1999) and promotes p53 tetramer dissociation (Baudier, et al., 1992). Since tetrameric p53 is more active than dimeric or monomeric forms of the tumor suppressor, it is also likely that the disruption of p53 tetramers contributes to the S100B($\beta\beta$)-dependent loss of p53 activity *in vivo* (Baudier et al., 1992; Delphin et al., 1999; Rustandi et al., 1998, 1999; Wilder et al., 1998). Actually, some combination of both mechanisms is likely because phosphorylation in the C-terminal negative regulatory domain stabilizes the p53 tetramer (Sakaguchi et al., 1997). In summary, the binding of Ca^{2+} to S100B($\beta\beta$) induces a conformational change that is required for binding to p53 (Figure 3). As a result of this binding, post-translational modifications on p53 are blocked (Baudier et al., 1992; Delphin et al., 1999; Rustandi et al., 1998, 1999), the p53 tetramer is disrupted (Baudier et al., 1992), and p53-dependent transcription activation is inhibited (Carrier et al., 1999). The structure of the S100B($\beta\beta$)-p53 complex reported here can be used to develop inhibitors that block the Ca^{2+}-dependent interaction between these two proteins. Such a drug could possibly restore p53 activity in astrocytomas and gliomas, which have significantly elevated levels of S100B($\beta\beta$) and p53 (Mueller et al., 1999; Rubio et al., 1993).

7. DO ALL S100 PROTEINS REGULATE THEIR TARGETS IN A Ca^{2+}-DEPENDENT MANNER?

In some cases, the Ca^{2+}-dependent interaction of S100B($\beta\beta$) can be used as model for other S100-target protein interactions. For example, S100A1 binds a peptide derived from the actin capping protein (CapZ) via a Ca^{2+}-dependent interaction involving many similar residues from the "hinge" and C-terminus as found in the S100B($\beta\beta$)-p53 peptide complex (Garbuglia et al., 1999; Landar et al., 1998; Osterloh et al., 1998). Also, most S100 proteins are dimeric, bind Ca^{2+}, and show spectroscopic evidence for a conformational change prior to target protein binding (Donato, 1999). On the other hand, S100A6 has much smaller structural changes than S100B($\beta\beta$) upon binding Ca^{2+}, and it is not clear whether Ca^{2+}-binding is necessary for binding its protein targets. Thus, it is important to examine each S100 protein in order to determine details involved in their mechanism of action.

In some cases, Ca^{2+}-binding is clearly not necessary for target protein binding and activation. For example, a peptide derived from annexin II binds S100A10 in a very similar orientation as the p53 peptide bound to S100B($\beta\beta$) (Rety et al., 1999); however, for S100A10, the interaction *is not* Ca^{2+}-dependent. In fact, S100A10 does not bind Ca^{2+} because the EF-hand motifs in this protein depart significantly from the consensus EF-hand sequence (Table 1). Yet, the 3D structure for S100A10 in the absence of Ca^{2+} is very similar to most EF-hand containing proteins in the Ca^{2+}-bound state. In another case, S100B($\beta\beta$) and S100A1 can activate a few protein targets in a Ca^{2+}-independent manner (Zimmer et al., 1995). For these interactions, however, the target protein binding site does not involve the C-terminus of the protein as found in the S100B($\beta\beta$)-p53 peptide complex (Landar et al., 1998), so this site must be distinct from the surface of S100B($\beta\beta$) that interacts with p53.

8. Zn^{2+}-BINDING TO S100 PROTEINS

Several S100 proteins, including S100B($\beta\beta$), also bind Zn^{2+} ions ($K_d = 0.1$–$1.0\ \mu$M) at sites different from those of Ca^{2+} (Donato, 1991, 1999; Rustandi et al., 1998), and for S100B($\beta\beta$), Ca^{2+} ion binding affinity is increased by 5- to 10-fold in the presence of Zn^{2+} (Donato, 1991, 1999). Data from fluorescence, UV differential spectroscopy, and CD studies indicate that Zn^{2+} ion binding, like Ca^{2+}, induces a conformational change in S100B($\beta\beta$) (Chaudhuri et al., 1997; Donato, 1991, 1999). However, a different protein conformation occurs with Zn^{2+} (*versus* Ca^{2+}) because different reactivities with sulfhydryl reagents are found in the two different metal ion complexes (Baudier and Cole, 1988). In addition, Zn^{2+} ions prevent disulfide-linked dimer formation, S100B($\beta_{S-S}\beta$); whereas Ca^{2+} ions induce the accumulation of the covalent dimer. Titrations of Ca^{2+}-bound S100B($\beta\beta$) with Zn^{2+} indicate that three histidine residues (H42, H86, and H90) are candidate ligands for the Zn^{2+} site, since they exhibited large changes in backbone ^{1}H and/or ^{15}N chemical shift (as measured by NMR). Interestingly, these histidine residues are located in the hinge and the C-terminal loops of S100B($\beta\beta$), which are regions that are involved in target protein binding. Interestingly, Zn^{2+}-binding to S100B($\beta\beta$) increases the binding affinity of another peptide target (CapZ peptide) by 5-fold (Barber et al., 1999). These results, together with the finding that S100A7 has a Zn^{2+} binding site located in a similar position (Brodersen, et al., 1999a, b), suggest that Zn^{2+} binding may be important in the biological function of several S100 proteins.

9. SUMMARY

A large change in the position of helix 3 in S100B($\beta\beta$) upon binding Ca^{2+} is required for S100B($\beta\beta$) to bind the tumor suppressor protein, p53. The 3D structures of S100B($\beta\beta$) in the apo-, Ca^{2+}-bound, and p53 complexes clearly illustrate that the p53 binding site is buried in the apo-state and that several hydrophobic residues that are critical for binding p53 are exposed upon binding Ca^{2+}. This interaction between p53 and S100B($\beta\beta$) sterically blocks sites of post-translational modifications that occur *in vivo* (i.e. phosphorylation, acetylation) to activate p53. Thus, inhibiting the S100B-p53 interaction could have an important therapeutic value. In the future, it is important to determine whether other S100 proteins regulate p53, and if metal ions other than Ca^{2+}, such as Cu^{2+} and Zn^{2+}, affect the S100-p53 interaction.

REFERENCES

Barber, K.R., McClintock, K.A., Jamieson, G.A., R.V.W, D., and Shaw, G.S., 1999, Specificity and Zn enhancement of S100B binding epitope TRTK-12, *J. Biol. Chem.* 274, 1502–1508.

Barraclough, R., 1998, Calcium-binding protein S100A4 in health and disease, *Biochem. Biophys. Acta* 1448, 190–199.

Baudier, J. and Cole, R.D., 1988, Interaction between the microtubule-associated τ proteins and S100b regulate τ protein phosphorylation by the Ca^{2+}/calmodulin-dependent protein kinase II, *J. Biol. Chem.* 263, 5876–5883.

Baudier, J., Delphin, C., Grundwald, D., Khochbin, S. and Lawrence, J.J., 1992, Characterization of the tumor suppresor protein p53 as a protein kinase C substrate and a S100B-binding protein, *Proc. Natl. Acad. Sci. USA* 89, 11627–11631.

Berridge, M.J. and Irvine, R.F., 1989, Inositol phosphates and cell signalling, *Nature* 341, 197–204.

Brodersen, D.E., Etzerodt, M., Madsen, P., J.E., C., Thogersen, H.C., Nyborg, J. and Kjeldgaard, M., 1999a, EF-hands at atomic resolution: The structure fo human psoriasin (S100A7) solved by MAD phasing, *Structure* 6, 477–489.

Brodersen, D.E., Nyborg, J. and Kjeldgaard, M., 1999b, Zinc-binding site of an S100 protein revealed. Two crystal structures of Ca-bound human psoriasin (S100A7) in the Zn-loaded and Zn-free states, *Biochemistry* 38, 1695–1704.

Carrier, F., Blake, M., Zimmer, D., Rustandi, R.R. and Weber, D.J., 1999, Abrogation of p53 transcriptional activity by the S100 calcium binding proteins: Possible implication in angiogenesis, *Proc. of the AACR* 40, 102.

Chaudhuri, D., Horrocks, W.W., Amburgey, J.C. and Weber, D.J., 1997, Characterization of lanthanide ion binding to the EF hand protein S100b by luminescence spectroscopy, *Biochemistry* 36, 9674–9680.

Chiarugi, V., Cinelli, M. and Magnelli, L., 1998, Acetylation and phosphorylation of the carboxy-terminal domain of p53: Regulative significance, *Oncology Res.* 10, 55–57.

Delphin, C., Ronjat, M., Deloulme, J.C., Garin, G., Debussche, L., Higashimoto, Y., Sakaguchi, K. and Baudier, J., 1999, Calcium-dependent interaction of S100B with the C-terminal domain of the tumor suppressor p53, *J. Biol. Chem.* 274, 10539–10544.

Donato, R., 1991, Perspectives in S-100 protein biology, *Cell Calcium* 12, 713–726.

Donato, R., 1999, Functional roles of S100 proteins, calcium-binding proteins of the EF-hand type, *Biochem. Biophys. Acta* 1450, 191–231.

Drohat, A.C., Amburgey, J.C., Abildgaard, F., Starich, M.R., Baldisseri, D. and Weber, D.J., 1996, Solution structure of rat apo-S100B($\beta\beta$) as determined by NMR spectroscopy, *Biochemistry* 35, 11577–11588.

Drohat, A.C., Nenortas, E., Beckett, D. and Weber, D.J., 1997, Oligomerization state of S100B at nanomolar concentration determined by large-zone analytical gel filtration chromatography. *Protein Sci.* 6, 1577–1582.

Drohat, A.C., Baldisseri, D.M., Rustandi, R.R. and Weber, D.J., 1998, Solution structure of calcium-bound rat S100B($\beta\beta$) as determined by NMR spectroscopy, *Biochemistry* 37, 2729–2740.

Drohat, A.C., Tjandra, N., Baldisseri, D.M. and Weber, D.J., 1999, The use of dipolar couplings for determining the solution structure of rat apo-S100B($\beta\beta$), *Protein Sci.* 8, 800–809.

Garbuglia, M., Verzini, M., Rustandi, R.R., Osterloh, D., Weber, D.J., Gerke, V. and Donato, R., 1999, Role of the C-terminal extension in the interaction of S100A1 and GFAP, tubilin, S100A1- and S100B-inhibitory peptide, TRTK-12, and a peptide derived from p53, and the S100A1 inhibitory effect on GFAP polymerization, *Biochem. Biophys. Res. Commun.* 254, 36–41.

Giaccia, A.J. and Kastan, M.B., 1998, The complexity of p53 modulation: Emerging patterns from divergent signals, *Genes and Develop.* 12, 2973–2983.

Groves, P., Finn, B.E., Kuznicki, J. and Forsen, S., 1998, A model for target protein binding to calcium-activated S100 dimers, *FEBS Lett.* 421, 175–179.

Harris, N.L., Resnell, S.R. and Cohen, F.E., 1994, Four helix bundle diversity in globular proteins, *J. Mol. Biol.* 236, 1356–1368.

Heizman, C.W., 1999, Ca^{2+}-binding S100 proteins in the central nervous system, *Neurochem. Res.* 24, 1097–1100.

Hupp, T.R., Meek, D.W., Midgley, C.A. and Lane, D.P., 1992, Regulation of the specific DNA binding function of p53, *Cell* 71, 875–886.

Ilg, E.G., Schafer, B.W. and Heizmann, C.W., 1996, Expression pattern of S100 Ca^{2+}-binding proteins in human tumors, *Int. J. Cancer* 68, 325–332.

Ivanenkov, V.V., Jamieson, G.A.J., Gruenstein, E. and Dimlich, R.V.W., 1995, Characterization of S100B binding epitiopes. Identification of a novel target, the actin capping protein, CapZ, *J. Biol. Chem.* 270, 14651–14658.

Kilby, P.M., Van Eldik, L.J. and Roberts, G.C.K., 1996, The solution structure of the bovine S100B protein dimer in the calcium-free state, *Structure* 4, 1041–1052.

Kligman, D. and Hilt, D., 1988, The S100 protein family, *Trends Biochem. Sci.* 13, 437–443.

Kretsinger, R.H., 1980, Structure and evolution of calcium-modulated proteins, *CRC Crit. Rev. Biochem.* 8, 119–174.

Landar, A., Hall, T.L., Cornwall, E.H., Correia, J.J., Drohat, A.C., Weber, D.J. and Zimmer, D.B., 1997, The role of cysteine residues in S100B dimerization and regulation of target protein activity, *Biochim. Biophys. Acta* 1343, 117–129.

Landar, A., Rustandi, R.R., Weber, D.J. and Zimmer, D.B., 1998, S100A1 utilizes different mechanisms for interacting with calcium-dependent and calcium independent target proteins, *Biochemistry* 37, 17429–17438.

Levine, A.J., 1997, p53, the cellular gatekeeper for growth and division, *Cell* 88, 323–331.

Maler, L., Potts, B.C.M. and Chazin, W.J., 1999, High resolution solution structure of apo-calcyclin and structural variations in the S100 family of calcium-binding proteins, *J. Biomol. NMR* 13, 233–247.

Matsumura, H., Shiba, T., Inoue, T., Harada, S. and Kai, Y., 1998, A novel mode of target recognition suggested by the 2.0 Å structure of holo S100B from bovine brain, *Structure* 6, 233–241.

McNutt, N.S., 1998, The S100 family of multipurpose calcium-binding proteins, *J. Cut. Path.* 25, 521–529.

Mueller, A., Bachi, T., Hochli, M., Schafer, B.W. and Heizmann, C.W., 1999, Subcellular distribution of S100 proteins in tumor cells and their relocation in response to calcium activation, *Hist. Chem. Biol.* 111, 453–459.

Osterloh, D., Ivanenkov, V.V. and Gerke, V., 1998, Hydrophobic residues in the C-terminal region of S100A1 are essential for target protein binding, but not for dimerization, *Cell Calcium* 24, 137–151.

Potts, B.C.M., Smith, J., Akke, M., Macke, T.J., Okazaki, K., Hidaka, H., Case, D.A. and Chazin, W.J., 1995, The structure of calcyclin reveals a novel homodimeric fold for S100 Ca-binding proteins, *Nat. Struct. Biol.* 2, 790–796.

Rety, S., Sopkaova, J., Renouard, M., Osterloh, D., Gerke, V., Tabaries, S., Russo-Marie, F. and Lewit-Bentley, A., 1999, The crystal structure of a complex of p11 with the annexin II N-terminal peptide, *Nat. Struct. Biol.* 6, 89–95.

Rubio, M.-P., von Deimling, A., Yandell, D.W., Wiestler, O.D., Gusella, J.F. and Louis, D.N., 1993, Accumulation of wild-type p53 protein in human astrocytomas, *Cancer Res.* 53, 3465–3467.

Rustandi, R.R., Drohat, A.D., Baldisseri, D.M., Wilder, P.T. and Weber, D.J., 1998, The Ca^{2+}-dependent interaction of S100B($\beta\beta$) with a peptide derived from p53, *Biochemistry* 37, 1951–1960.

Rustandi, R.R., Baldisseri, D.M., Drohat, A.C. and Weber, D.J., 1999, Structural changes in the C-terminus of Ca-bound rat S100B($\beta\beta$) upon binding the a peptide derived from the C-terminal regulatory domain of p53, *Protein Sci.* 8, 1743–1751.

Rustandi, R.R., Baldisseri, D.M. and Weber, D.J., 2000, Structure of the negative regulatory domain of p53 bound to S100B($\beta\beta$), *Nat. Struct. Biol.* 7, 570–574.

Sakaguchi, K., Sakamoto, H., Lewis, M.S., Anderson, C.W., Erickson, J.W., Appela, E. and Xie, D., 1997, Phosphorylation of ser-392 stabilizes the tetramer formation of tumor suppressor protein p53, *Biochemistry* 36, 10117–10124.

Sastry, M., Ketchem, R.R., Crescenzi, O., Weber, C., Lubienski, M.J., Hidaka, H. and Chazin, W.J., 1998, The three-dimensional structure of Ca^{2+}-bound calcyclin: Implications for Ca^{2+}-signal transduction by S100 proteins, *Structure* 6, 223–231.

Schafer, B.W. and Heizmann, C.W., 1996, The S100 family of EF-hand calcium-binding proteins: Functions and pathology, *Trends Biochem. Sci.* 21, 134–140.

Scotto, C., Deloulme, J.C., Rousseau, D., Chambaz, E. and Baudier, J., 1998, Calcium and S100B regulation of p53-dependent cell growth arrest and apoptosis, *Mol. Cell. Biol.* 18, 4272–4281.

Scotto, C., Delphin, C., Deloulme, J.C. and Baudier, J., 1999, Concerted regulation of wild-type p53 nuclear accumulation and activation by S100B and calcium-dependent protein kinase C, *Mol. Cell. Biol.* 19, 7168–7180.

Selivanova, G., Iotsova, V., Okan, I., Fritsche, M., Strom, M., Groner, B., Grafstrom, R.C. and Wiman, K.G., 1999, Restoration of the growth suppression function of mutant p53 by a synthetic peptide derived from the p53 C-terminal domain, *Nature Med.* 3, 632–638.

Sherbet, G.V. and Lakshmi, M.S., 1998, S100A4 (MTS1) calcium binding protein in cancer growth, invasion and metastasis, *Anticancer Res.* 18, 2415–2422.

Sheu, F.-S., Huang, F.L. and Huang, K.-P., 1995, Differential responses of protein kinase C substrates (MARCKS, neuromodulin, and neurogranin) phosphorylation to calmodulin and S100, *Arch. Biochem. Biophys.* 316, 335–342.

Smith, S.P. and Shaw, G.S., 1998, A novel calcium-sensitive switch revealed by the structure of human S100B in the calcium-bound form, *Structure* 6, 211–222.

Strynadka, N.C.J. and James, M.N.G., 1989, Crystal structures of the helix-loop-helix calcium- binding proteins, *Annu. Rev. Biochem.* 58, 951–998.

Wilder, P.T., Rustandi, R.R., Drohat, A.C. and Weber, D.J., 1998, S100B($\beta\beta$) inhibits the protein kinase C-dependent phosphorylation of a peptide derived from p53 in a Ca^{2+}-dependent manner, *Protein Sci.* 7, 794–798.

Yap, K.L., Ames, J.B., Swindells, M.B. and Ikura, M., 1999, Diversity of conformational states and changes within the EF-hand protein superfamily, *Proteins: Struct. Funct. Genet.* 15, 499–507.

Youmell, M., Park, S.J., Basu, S. and Price, B.D., 1998, Regulation of the p53 protein by protein kinase C and protein kinase ζ, *Biochem. Biophys. Res. Commun.* 245, 514–518.

Zimmer, D.B., Cornwall, E.H., Landar, A. and Song, W., 1995, The S100 protein family: History, function, and expression, *Brain Res. Bull.* 37, 417–429.

The Interaction of Calmodulin with Novel Target Proteins

Kelly Y. Chun and David B. Sacks

1. INTRODUCTION

Ionized calcium (Ca^{2+}) is the most common signal transduction element in cells. Intracellular free Ca^{2+} concentrations ($[Ca^{2+}]_i$), which average $\sim$ 100 nM in resting cells, are 20,000-fold lower than the extracellular concentrations (Clapham, 1995; Berridge, 1993). Unlike other signaling molecules, Ca^{2+} cannot be metabolized. Therefore, cells contain numerous specialized extrusion proteins (e.g., pumps and channels) and binding proteins that tightly regulate $[Ca^{2+}]_i$. Ca^{2+} is required for proliferation and survival of mammalian cells (Takuwa et al., 1995), yet paradoxically, prolonged high $[Ca^{2+}]_i$ leads to cell death (Nicotera et al., 1994).

Many proteins that bind Ca^{2+} contain the EF hand motif (Kretsinger et al., 1986). This is a helix-loop-helix motif that coordinates Ca^{2+} through $\sim$ 6 oxygen atoms, which are provided by side chains of conserved glutamate and aspartate amino acid residues. Ca^{2+}-binding proteins can be divided into two groups; those that buffer Ca^{2+} by sequestration as $[Ca^{2+}]_i$ rise (e.g., parvalbumin and calsequestrin) and those that are activated by Ca^{2+} (e.g., calmodulin) (for reviews, see Clapham, 1995; Berridge, 1993).

The primary mediator of Ca^{2+}-dependent signaling in eukaryotic cells is calmodulin (Cohen and Klee, 1988). Calmodulin is a highly acidic, ubiquitous protein with a molecular mass of 16.7 kDa. Its tertiary structure resembles a dumbbell, with two globular domains connected by a long α-helical linker with no contacts between the two globular domains. The two Ca^{2+}-binding domains in each lobe (i.e., total of four Ca^{2+}-binding sites per molecule) are similar. The effects of calmodulin are produced by direct interaction with target enzymes, or indirectly via multiple specific kinases.

Kelly Y. Chun and David B. Sacks • Department of Pathology, Brigham and Women's Hospital and Harvard Medical School, Boston, MA, U.S.A.

R. Pochet, R. Donato, J. Haiech, C. Heizmann and V. Gerke (eds.): Calcium: The Molecular Basis of Calcium Action in Biology and Medicine, 541–563.

Upon ligation of Ca^{2+}, calmodulin undergoes a conformational change enabling hydrophobic and ion pairing interactions to cooperate synergistically with binding regions on specific target proteins. The calmodulin-binding domain is a short region of 14–26 amino acid residues that is not conserved among targets. However, there are two general structural features. The calmodulin binding region in "classic" targets has a propensity to form a basic amphiphilic α-helix (O'Neil and DeGrado, 1990). The second calmodulin-binding domain is the more recently identified IQ motif (Cheney and Mooseker, 1992). In addition, calmodulin can bind to proteins that do not contain the general structural features described. For most target enzymes, binding of Ca^{2+}/calmodulin displaces an auto-inhibitory domain, thereby initiating intracellular signaling events that regulate diverse, fundamental cell functions, ranging from cell motility and division to highly specialized functions in learning and memory.

Calmodulin binds to and regulates multiple proteins, including over 30 protein kinases, molecular motors, nitric oxide synthase and adenylate cyclase. Better-characterized "classic" calmodulin targets include (Ca^{2+} + Mg^{2+})-ATPase, response-specific kinases that target specific substrates (e.g., phosphodiesterase, myosin light chain kinase, phosphorylase kinase and eEF-2 kinase), multifunctional Ca^{2+}/calmodulin-dependent protein kinases (CaM-kinase I, II and IV), and a phosphatase (calcineurin). Detailed information on "classic" calmodulin target proteins can be found elsewhere (Cohen and Klee, 1988; Carafoli and Klee, 1999; Van Eldik and Watterson, 1998). The last decade produced a significant expansion in the number of "novel" or "non-classic" calmodulin target proteins that have been identified. These categories of calmodulin-dependent signaling proteins significantly expand the repertoire of cellular processes in which calmodulin participates. The purpose of this review is to highlight selected areas pertaining to those novel calmodulin target proteins in which recent progress has been notable. The emphasis will be on small GTPases and the involvement of calmodulin in the Rho signaling pathways.

2. IQ-CONTAINING PROTEINS

Gene sequence data obtained over the past few years has revealed the presence of protein modules that regulate signal transduction by mediating protein-protein interactions (Pawson and Scott, 1997). These conserved modules are used to construct a complex network of interacting proteins that organize and control intracellular signaling pathways. For example, Src homology 2 (SH2) and phosphotyrosine-binding (PTB) domains bind to phosphotyrosine-containing sequences, pleckstrin homology (PH) domains bind phosphatidylinositol phosphates, and Src homology 3 (SH3) and WW domains bind proline-rich sequences (Pawson and Scott, 1997). Analogous to these protein modules, IQ motifs are targeted by calmodulin. The first recog-

Table 1. Families of IQ-containing proteins

Proteins	Function(s)
Unconventional myosins	• Actin-based motility, cell growth and development, organelle movement, membrane trafficking, signal transduction
Neuronal proteins	
Neurogranin	• Neurite growth
Neuromodulin	• Cytoskeletal rearrangement
N-WASP	• Actin depolarization
Ion channels	
Voltage-gated Ca^{2+} channels	• Heart muscle contraction, hormone secretion
Intracellular signaling	
IRS-1, -2	• Insulin, growth factor and cytokine signaling
Ras-associated proteins	
IQGAP1, -2	• Reorganization of the cytoskeleton
Ras-GRF	• Intracellular signaling
EEA1	• Trafficking, vesicular transport to late endosomes

nition of IQ motifs was in tandem repeats in the neck region of members of the family of unconventional (non-muscle) myosins (Cheney and Mooseker, 1992). The myosin family consists of at least 13 distinct classes (Hasson and Mooseker, 1996). These proteins are mechanoenzymes that hydrolyze ATP to generate mechanical force. Class II is conventional myosin. The remaining 6 vertebrate unconventional myosin classes participate in a wide range of functions (Wolenski, 1995). All vertebrate unconventional myosins contain: (i) a conserved N-terminal "head" motor domain that binds to actin and hydrolyzes ATP, (ii) a "neck" domain that contains 1–6 repeats of IQ motifs, and (iii) a "tail" domain that serves a specific function at the C-terminus (e.g., dimerization, membrane binding, SH3 binding).

Subsequently, IQ motifs have been identified in numerous proteins, ranging from proteins involved in signaling to ion channels (see Table 1 for representative examples). IQ motifs are usually present in 2 to 6 tandem repeats, each comprising approximately 25 amino acid residues with the core fitting the consensus IQXXXRGXXXR (X represents any amino acid; the first residue need not be isoleucine, but is usually hydrophobic). The sequence was designated the IQ motif on the basis of its conserved core (Cheney and Mooseker, 1992). IQ motifs are not obligatory calmodulin-binding sequences and have been shown to also bind myosin-associated light chains. Compared to classic calmodulin-binding peptides (e.g., those found in myosin light chain kinase or Ca^{2+}/calmodulin-dependent kinase II), relatively little is known about the interaction of calmodulin with IQ-motifs. The crystal structure of a calmodulin-IQ complex has not been solved. Insight into how calmodulin binds to IQ-containing targets was derived by molecular modeling from the crystal structure of the IQ-containing heavy chain of scallop myosin bound to myosin-associated light chain (Houdusse and Cohen, 1995). The structure revealed that the conserved portion of the IQ motif (IQXXXR) is the most important region that dictates both the conformation and positioning of the C-terminal lobe (Houdusse and Cohen, 1995). In addition to the initial conserved residues, a "complete" IQ-motif also contains a second part (GXXXR), not found in "incomplete" IQ motifs. The C-terminal arginine is the crucial residue in a complete IQ motif that determines the position of the N-terminal lobe of the calmodulin and eliminates the Ca^{2+} requirement for calmodulin binding (Houdusse and Cohen, 1995; Houdusse et al., 1996). An "incomplete" IQ motif requires Ca^{2+} for calmodulin binding.

3. "NON-CLASSIC" CALMODULIN TARGET PROTEINS

The diversity and complexity of Ca^{2+}/calmodulin-regulated cellular processes are daunting and the large list of proteins regulated by calmodulin grows almost weekly. Obviously it is impossible to write a comprehensive review discussing all of the targets of calmodulin. Our intent is to address selected calmodulin-binding proteins that are not thought of as "classic" calmodulin targets. In fact, none of these proteins was included in a comprehensive book on calmodulin published in 1988 (Cohen and Klee, 1988). Some of these "novel" targets are listed in Table 2.

3.1. The role of Ca^{2+}/calmodulin in receptor signaling pathways

Most extracellular agonists exert their effects on cells via transmembrane signaling systems. Many growth factors, such as the epidermal growth factor (EGF), fibroblast growth factor (FGF), platelet-derived growth factor (PDGF) and nerve growth factor (NGF), bind to plasma membrane receptors, thereby activating intrinsic intracellular tyrosine kinases (Schlessinger and Ullrich, 1992). Cross-talk between the tyrosine kinase and Ca^{2+} second messenger system has been demonstrated in several signaling cascades (Huckle et al., 1992; Brady and Palfrey, 1993). For example, PDGF, EGF and NGF activate phospholipase Cγ, which initiates Ca^{2+} release from intracellular stores. Furthermore, Ca^{2+} (Rozengurt, 1986) and calmodulin (Katori et al., 1994) participate in PDGF and FGF signaling, and both NGF and EGF enhance phosphorylation of a calmodulin-binding protein in PC12 cells (Brady and

Table 2. "Non-classic" calmodulin target proteins

Categories	Calmodulin targets
Cell surface receptors and channels	• Insulin receptor • EGF receptor • Angiotensin II receptor • G protein-coupled receptors • Metabotropic glutamate receptor • N-methyl-D-aspartate receptor • Polymeric immunoglobulin receptor • Small- and intermediate conductance Ca^{2+}-activated potassium (SK, IK) channels • P/Q type Ca^{2+} channels • Cyclic nucleotide-gated channels
Peripheral membrane-associated proteins	• G protein-coupled receptor kinases (GRKs) • Heterotrimeric G protein (β subunit)
Intracellular signaling proteins	• PI3K • IRS-1, -2 • IQGAP1, -2 • N-WASP • Ras-GRF • A kinase-anchoring protein79 (AKAP79) • 14-3-3ε
Small GTPases	• Rab3A • Rad, Kir, Gem • Ral-A
Nuclear proteins	• Estrogen receptor • $p21^{Cip1}$ (cyclin-dependent kinase inhibitor) • Rin1 (small GTPase) • basic helix-loop-helix (bHLH) proteins • p68 RNA helicase

Palfrey, 1993). The participation of calmodulin in EGF signaling has been characterized in some detail. The EGF receptor interacts with calmodulin in a Ca^{2+}-dependent manner (Benguria et al., 1995; San Jose et al., 1992; Martin-Nieto and Villalobo, 1998) and phosphocalmodulin increases EGF-stimulated activity of the kinase (San Jose et al., 1992). In addition, EGF stimulates the phosphorylation of calmodulin by the EGF receptor tyrosine kinase (Benguria et al., 1994).

3.2. Insulin action

Analogous to growth factors, insulin binds to the extracellular domain of specific transmembrane receptors, inducing tyrosine phosphorylation of selected intracellular proteins. The insulin receptor substrates IRS-1 and IRS-2 (and the more recently identified IRS-3 and IRS-4) are the major targets of the insulin receptor tyrosine kinase and function by transducing the insulin signal to downstream effectors (White and Kahn, 1994). Briefly, the mechanism is as follows: following tyrosine phosphorylation, IRS proteins associate with SH2 domains on target proteins, including phosphatidylinositol 3-kinase (PI3K), GRB2 and SHP-2 (White and Kahn, 1994). This binding alters the activity of the SH2-containing proteins, propagating the insulin signal via a cascade involving multiple enzymes.

Although considerable evidence supports a role for Ca^{2+} in insulin action (Levy et al., 1994), this concept is controversial. Nevertheless, calmodulin has been implicated in insulin action. For example, the insulin receptor contains a calmodulin-binding domain (Graves et al., 1985; Sacks et al., 1989) and calmodulin increases insulin receptor tyrosine kinase activity in a Ca^{2+}-dependent manner (Graves et al., 1985). Moreover, insulin stimulates the phosphorylation of calmodulin *in vitro* (Sacks and McDonald, 1988) and in intact cells (Sacks et al., 1992; Joyal and Sacks, 1994). In addition, it has been reported by several groups that antagonism of calmodulin inhibits insulin-stimulated glucose transport (Shechter, 1984; Shashkin et al., 1995; Yang et al., 2000). Despite these accumulated data, the molecular mechanism and role of calmodulin in insulin signaling remain to be established.

More recently, calmodulin has been observed to interact with proteins distal to the insulin receptor. Both IRS-1 and IRS-2 co-immunoprecipitate with calmodulin from Chinese hamster ovary (CHO) cells (Munshi et al., 1996). In contrast to other IRS-1 targets, binding of calmodulin is independent of tyrosine phosphorylation of IRS-1. The association is enhanced by Ca^{2+} and inhibited by trifluoperazine, a cell-permeable calmodulin antagonist. IRS-1 is a docking protein that contains several interaction modules, including PH, PTB and multiple phosphorylation motifs (Sun et al., 1995). In addition, several IQ-like motifs were recognized in both IRS-1 and IRS-2 (Munshi et al., 1996). Synthetic peptides corresponding to the IQ motifs significantly inhibited the association between IRS-1 and calmodulin (Munshi et al., 1996). Subsequent investigation has revealed that the interaction between calmodulin and IRS proteins is enhanced in several models of insulin resistance in rats (Li et al., 2000). Interestingly, overexpression of calmodulin in CHO cells reduces insulin-stimulated tyrosine phosphorylation of IRS-1, with a concomitant decrease in insulin-stimulated binding of PI3K to IRS-1. Thus, calmodulin appears to impair the ability of IRS-1 to interact with downstream targets and may contribute to insulin resistance.

A second provocative interaction between calmodulin and downstream targets of the insulin receptor was described recently. Calmodulin bound directly to the SH2 domains in the p85 regulatory subunit of PI3K (Joyal et al., 1997b). Binding was phophotyrosine-independent and regulated by Ca^{2+}. This observation was extremely surprising because SH2 domains were believed to bind exclusively to specific sequences that are phosphorylated on tyrosine residues (Pawson and Scott, 1997). The specificity of calmodulin binding was validated by competition with a phosphotyrosine-containing peptide (Joyal et al., 1997b). Moreover, calmodulin significantly enhanced PI3K activity both *in vitro* and in intact cells. Interestingly, the specific phosphatidylinositol phosphates generated in intact cells by Ca^{2+}/calmodulin differed from those induced by incubating cells with insulin or growth factors. It was subsequently shown that calmodulin also binds to the p110γ isoform of PI3K, but no functional sequelae were demonstrated (Fischer et al., 1998). Additional data supporting a role for calmodulin in PI3K function have been obtained recently. Calmodulin antagonists inhibited insulin-stimulated translocation of the insulin-sensitive glucose transporter 4 (GLUT4), with concomitant reduction in Akt (an enzyme downstream of PI3K) (Yang et al., 2000). Analysis of the mechanism revealed that antagonism of calmodulin prevented insulin from stimulating the formation of phosphatidylinositol 3,4,5-trisphosphate (a product of PI3K) *in vivo*. Although the physiological significance of calmodulin in normal PI3K function is not known, these observations expand the multiple interconnections that occur between signaling pathways, providing an additional locus of communication between the Ca^{2+} and growth factor receptor pathways.

3.3. Nuclear receptors

Estrogen receptors, like all steroid hormone receptors, are transcription factors that alter gene transcription. Several reports couple Ca^{2+}/calmodulin to estrogen receptor function. These include: (i) calmodulin binds to estrogen receptors in a Ca^{2+}-dependent manner (Castoria et al., 1988) and thus increases the K_d of estradiol binding (Bouhoute and Leclercq, 1995), (ii) Ca^{2+}/calmodulin stimulates activation of estradiol binding to the estrogen receptor (Migliaccio et al., 1984), (iii) Ca^{2+}/calmodulin stimulates tyrosine phosphorylation of the estrogen receptor (Migliaccio et al., 1984), (iv) association of calmodulin with activated estrogen-estrogen receptor complex enhances the ability of the complex to interact with the estrogen response element (Bouhoute and Leclercq, 1995), and (v) estrogen receptor co-immunoprecipitates with endogenous calmodulin in an estradiol-independent manner (Joyal et al., unpublished observation). Together these data suggest that calmodulin may participate in estrogen receptor-induced transcriptional activity.

3.4. G protein-coupled receptors and heterotrimeric G proteins

While most growth factors elicit their effects in cells via tyrosine-containing receptors, numerous other factors transmit signals from the extracellular environment to the cytoplasm via receptors that are coupled to heterotrimeric guanine nucleotide-binding proteins (G proteins) (for a review, see Gutkind, 1998). Heterotrimeric G proteins consist of α, β, and γ subunits (G$\alpha\beta\gamma$), each of which is encoded by a separate gene. Additional complexity is provided by the multiplicity of isoforms: 16 Gα, 5 Gβ and 11 Gγ subunits have been identified. G proteins are regulated by the binding of guanine nucleotides to Gα. In the inactive, GDP-bound state, Gα associates with $\beta\gamma$ subunits (G$\beta\gamma$) forming heterotrimers. Receptor-stimulated guanine nucleotide exchange results in the release of GDP and binding of GTP to Gα. GTP binding to Gα dissociates Gα from G$\beta\gamma$. GTP-bound Gα and free G$\beta\gamma$ initiate intracellular signaling by activating downstream effectors (for a review, see Hamm, 1998).

Calmodulin regulates both G protein-coupled receptors (GPCRs) and heterotrimeric G proteins using at least three distinct mechanisms. Firstly, calmodulin mediates a Ca^{2+}-dependent negative regulation of GPCRs. For example, calmodulin binds directly to the G protein-coupling domain of opioid receptors, reducing activation of the G protein by GPCR. Calmodulin competes with G proteins for binding to the opioid receptor, implying that calmodulin acts as a second messenger that inhibits GPCR activation (Wang et al., 1999). Similarly, calmodulin deactivates light responses in rhodopsin, a GPCR, *in vivo* (Scott et al., 1997). Light stimulation activates phospholipase C via a G protein. The resultant increase in $[Ca^{2+}]_i$ induces the binding of Ca^{2+}/calmodulin to rhodopsin, deactivating the signal (Scott et al., 1997).

A second mechanism of GPCR inactivation by Ca^{2+}/calmodulin proceeds via a two-step process involving G protein-coupled receptor kinases (GRKs). GRK is a family of serine/threonine kinases that phosphorylates agonist-activated GPCRs, thereby inducing binding of arrestin protein, which uncouples GPCR signaling from the G protein (for a review, see Lefkowitz, 1998). The family of GRKs includes six members, GRK1 to 6. Several EF-hand containing Ca^{2+}-sensor proteins selectively regulate the activity of individual GRK subtypes. GRK1 is inhibited by the binding of recoverin, a photoreceptor-specific Ca^{2+}-binding protein (Levay et al., 1998; Iacovelli et al., 1999). All the other GRKs are inhibited by calmodulin, with GRK5 ($IC_{50} \sim 50$ nM) being the most sensitive (Pronin et al., 1997; Chuang et al., 1996; Levay et al., 1998). This high affinity interaction between GRK5 and Ca^{2+}/calmodulin did not influence the catalytic activity of the kinase, but reduced GRK5 binding to the GPCR and phospholipid membrane (Pronin et al., 1997; Iacovelli et al., 1999).

In the third mechanism, Ca^{2+}/calmodulin regulates G protein action directly by binding to G$\beta\gamma$, thereby preventing heterotrimer formation and maintaining the Gα and G$\beta\gamma$ in the active state (Liu et al., 1997; Wu et

Table 3. Major subfamilies and representative members of the small GTPases

Subfamilies	Members	Functions
Ras	• H-, K-, N-Ras • Rap1A, -B, Rap2A, -B	• Cell growth and proliferation
Rho	• RhoA, -B, -C • Rac1, -2 • Cdc42Hs, -G25K	• Remodeling of cytoskeleton, mitogenesis, cell transformation, cell-cell adhesion
Arf	• Arf1, -2, -3 • Arf4, -5 • Arf6	• Vesicle trafficking • Plasma membrane transport
Rab	• Rab (> 30 isoforms)	• Vesicle docking
Ran	• Ran	• Nuclear transport
Ral	• RalA, -B	• Cell proliferation
Rad	• Rad • Gem • Kir	• Cytoskeletal organization • Signal transduction • Invasion and metastasis
Rin	• Rin • Rit	• Neuronal signaling?

al., 1992). However, the physiologic function of calmodulin binding to the heterotrimeric G protein is unclear, as direct binding of calmodulin to $G\beta\gamma$ did not alter its ability to activate phospholipase $C\gamma$, a major $G\beta\gamma$ target (Liu et al., 1997). Nevertheless, these data demonstrate that Ca^{2+}/calmodulin plays a central role in regulating G protein-mediated signal transduction at multiple levels by interacting directly with the GPCRs, heterotrimeric G proteins and GRKs.

3.5. Small GTPases

Distinct from the heterotrimeric G proteins is a group of small GTPases that constitute the Ras superfamily. The mammalian Ras superfamily contains > 60 members, which can be grouped on the basis of sequence into subfamilies (Exton, 1998; Hall, 1998) (Table 3). These small GTPases act as molecular switches for the selection, amplification, timing and delivery of signals from diverse sources and participate in multiple fundamental cellular processes. Analogous to the $G\alpha$ subunit of heterotrimeric G proteins, the small GTPases cycle between two conformational states, active GTP-bound and inactive GDP-bound (Figure 1). Two distinct classes of regulatory proteins modulate this cycle. Guanine nucleotide exchange factors (GEFs) induce the release

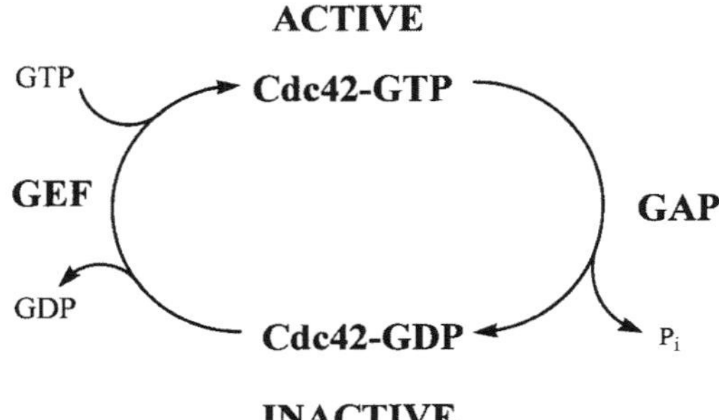

Figure 1. Regulation of Cdc42. Small GTPases are activated by a guanine nucleotide exchange factor (GEF) that induces the release of GDP, allowing GTP to bind. A GTPase activating protein (GAP) enhances the rate of intrinsic GTP hydrolysis, returning the GTPase to the GDP-bound conformation.

of GDP, allowing GTP – in excess in the cell – to bind. Families of GTPase activating proteins (GAPs) enhance the slow intrinsic Ras GTPase activity, returning the small GTPase to the inactive GDP-bound form (Figure 1).

3.6. Ca^{2+} calmodulin signaling and the Ras pathway

The Ca^{2+} and Ras signaling pathways intersect at multiple levels. It has been shown that Ca^{2+} influx by voltage-sensitive Ca^{2+} channels leads to activation of mitogen activated protein (MAP) kinase (Lev et al., 1995). One of the proteins that links this Ca^{2+}-mediated signal to the MAP kinase pathway is the protein tyrosine kinase Pyk2 (Lev et al., 1995). Investigation has revealed that Pyk2 activates the MAP kinase pathway in a Ca^{2+}-dependent manner in response to extracellular signals via GPCRs (Dikic et al., 1996), metabotropic glutamate receptors (Siciliano et al., 1996), and the immune receptor, FcεRI (Okazaki et al., 1997).

Calmodulin also participates in Ras signaling. Ca^{2+}/calmodulin binds to an IQ motif in Ras-GRF (a Ras-specific guanine nucleotide exchange factor, also known as $Cdc25^{Mm}$), activating Ras *in vivo* (Farnsworth et al., 1995). Mutation of the IQ motif, which abrogated binding of calmodulin to Ras-GRF, abolished Ras activation (Farnsworth et al., 1995). The N-terminal domain of Ras-GRF acts cooperatively with the IQ domain to facilitate Ca^{2+}/calmodulin-mediated activation of the exchange factor *in vivo* (Buchsbaum et al., 1996). In addition, Ca^{2+}-induced oligomerization of Ras-GRF is necessary for full activity (Anborgh et al., 1999). In contrast, Ca^{2+}/calmodulin inhibited the activation of Ras by Ras-GRF *in vitro* (Baouz et al., 1997), suggesting that other components may affect the *in vivo* interaction of calmodulin with Ras-GRF. Therefore, the regulation of Ras-GRF by calmodulin appears to be complex, requiring cooperative interactions among multiple regions of Ras-GRF, perhaps in conjunction with other as yet unidentified molecules. In support of this speculation, GRFβ, which has an N-terminus identical to Ras-GRF, negatively regulates Ca^{2+}-dependent Ras signaling in beta cells of the pancreas (Arava et al., 1999).

Ca^{2+} signaling is coupled to virtually all identified Ras-related pathways at some level. Many of the small GTPases (Table 3) have been observed to bind calmodulin. RalA, a downstream target of Ras, binds to calmodulin in a Ca^{2+}-dependent manner and this enhances GTP binding to RalA (Wang et al., 1997; Wang and Roufogalis, 1999). Rab3A is associated with secretory vesicles of neuronal and endocrine cells and controls the Ca^{2+}-triggered release of neurotransmitters and hormones. The interaction with Ca^{2+}/calmodulin induces the dissociation of Rab3A from synaptic membranes (Park et al., 1997) and modulates Rab3 function (Coppola et al., 1999). Calmodulin also binds to all members of the Rad subfamily, eliciting functional alteration. Ca^{2+}/calmodulin binding to Kir and Gem inhibits GTP binding (Moyers et al., 1997; Fischer et al., 1996), while calmodulin appears to be important for the correct subcellular localization of Rad (Moyers et al., 1997). Rin (Lee et al., 1996) also associates directly with calmodulin, but the physiological relevance has not been established.

3.7. Rho subfamily

The mammalian Rho GTPase subfamily includes at least 10 members, which interact with a variety of effectors to regulate cellular processes such as cell movement, axonal guidance, and changes in cell shape. The best-characterized members of the Rho family of proteins are RhoA, Rac1 and Cdc42. Activation of Rho produces stress fibers and adhesion plaques, Rac stimulates the formation of lamellipodia and membrane ruffling at the cell periphery, and Cdc42 triggers formation of filopodia or microspikes (Hall, 1998). These cellular effects are mediated by several groups of proteins that interact with the Rho family. Downstream targets for Rac and/or Cdc42 include p21-activated kinases (PAKs), PI3K, WASP (Wiskott–Aldrich syndrome protein), phospholipase D, and IQGAP1.

4. IQGAP1

4.1. Functional domains

IQGAP1 was cloned in 1994 by RNA-PCR from metastatic human osteosarcoma tissue (Weissbach et al., 1994). The protein was originally designated as a member of the Ras-GAP protein family based on the presence in the C-terminal half of a region with significant sequence similarity to the catalytic domain of Ras-GAPs (Figure 2). Examination of the sequence of IQGAP1 reveals the presence of a calponin homology domain (CHD), a WW domain,

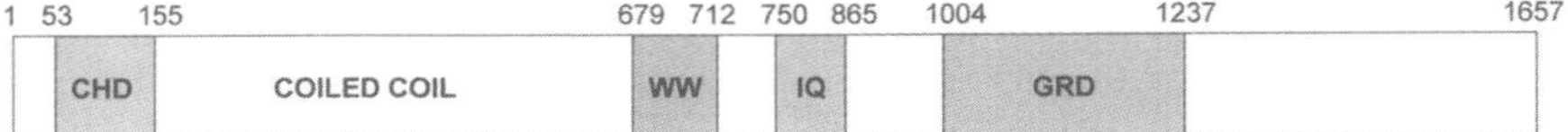

Figure 2. Schematic representation of domain organization of IQGAP1. The major indentified domains are shown, with the amino acid residue numbers indicated. CHD (calponin homology domain); coiled coil (presumptive α-helical domain with significant sequence identity to myosins); WW (a poly-proline binding domain; IQ (four tandem calmodulin-binding motifs); GRD (Ras-GAP related domain).

an N-terminal repeat region, which is predicted to form a coiled-coil, and four IQ motifs. The multiple conserved modules present in IQGAP1 imply that IQGAP1 participates in protein-protein interactions and may be an effector for several signaling pathways. Subsequent analysis by several groups documented that IQGAP1 binds activated Cdc42 and Rac (Hart et al., 1996; Fukata et al., 1999), actin (Erickson et al., 1997; Ho et al., 1999; Bashour et al., 1997), calmodulin (Joyal et al., 1997a; Ho et al., 1999), E-cadherin (Li et al., 1999), (Kuroda et al., 1998), and β-catenin (Fukata et al., 1999; Kuroda et al., 1998). The interaction and functional effects of IQGAP1 with each of these targets will be briefly described below.

4.2. Ca^{2+}/calmodulin regulates the binding of IQGAP1 to Cdc42 and F-actin

Analysis by nanoelectrospray tandem mass spectrometry revealed that IQGAP1 binds calmodulin (Joyal et al., 1997a). It was observed that IQGAP1 binds to both apocalmodulin and Ca^{2+}/calmodulin, with the latter exhibiting a 2.4-fold greater affinity. *In vitro* analysis localized the binding site to the IQ region, with some low-affinity binding of Ca^{2+}/calmodulin to the CHD (Ho et al., 1999). This interaction was hypothesized to have physiological relevance since IQGAP1 is the major calmodulin-binding protein in Ca^{2+}-free breast epithelial cell lysates (Joyal et al., 1997a). Furthermore, immunodepletion of calmodulin from cell lysates in the presence of Ca^{2+} reduced by 50% the amount of IQGAP1, suggesting that a high fraction of endogenous IQGAP1 is associated with calmodulin in intact cells (Ho et al., 1999). In support of this, immunocytochemical analysis of unstimulated MCF-7 cells revealed considerable overlap in the intracellular localization of endogenous calmodulin and IQGAP1 (Figure 5). Both IQGAP1 and calmodulin exhibit a diffuse distribution throughout the cytoplasm, with some localization at the plasma membrane and in the perinuclear area. Note that calmodulin, but not IQGAP1, is also present in the nucleus. Because calmodulin elicits effects by regulating target proteins, it was thought that the binding of calmodulin to IQGAP1 may modulate IQGAP1 function.

Activated (GTP-bound) Cdc42 and Rac bind to the GRD region of IQGAP1 (Hart et al., 1996). Based on its sequence similarity to Ras-GAP,

IQGAP1 was hypothesized to be a GAP for Cdc42 (Weissbach et al., 1994). Surprisingly, IQGAP1 was observed by two groups to inhibit the intrinsic GTPase activity of Cdc42 (Hart et al., 1996; Ho et al., 1999) and maintain Cdc42 in the active state. Calmodulin modulates this interaction by inhibiting the association of IQGAP1 with Cdc42 (Joyal et al., 1997a). Interestingly, the inhibitory effect of calmodulin on the Cdc42-IQGAP1 interaction requires physiological Ca^{2+} concentrations; apocalmodulin has no effect (Joyal et al., 1997a). Because the Cdc42 and calmodulin binding regions in IQGAP1 are some distance apart, binding of Ca^{2+}/calmodulin to the IQ region presumably alters the tertiary conformation of IQGAP1, preventing the association of Cdc42.

Ca^{2+}, calmodulin, and IQGAP1 are believed to be components that regulate and link Cdc42 signaling to downstream effects, including cytoskeletal organization. IQGAP1 colocalizes with microfilaments in lamellipodia and membrane ruffles (Bashour et al., 1997). Two independent studies suggest that IQGAP1 is a direct downstream target of Cdc42 (Erickson et al., 1997, Ho et al., 1999) and serves as a scaffolding protein that brings Cdc42 in close proximity to microfilaments. IQGAP1 can form a ternary complex with activated Cdc42 and F-actin (Erickson et al., 1997). Both F-actin and calmodulin bind to the CHD of IQGAP1. *In vitro* analysis demonstrates that F-actin and Ca^{2+}/calmodulin compete for binding to the CHD of IQGAP1 (Ho et al., 1999). Ca^{2+}/calmodulin probably acts as a negative regulator of Cdc42 signaling to the cytoskeleton by inhibiting IQGAP1 function as a scaffolding protein. By competing with F-actin for binding to IQGAP1, calmodulin prevents F-actin association with IQGAP1. Furthermore, Ca^{2+}/calmodulin binding to IQGAP1 also prevents Cdc42 from associating with IQGAP1 (Joyal et al., 1997a), further ensuring a separation of Cdc42 from microfilaments.

4.3. Ca^{2+}/calmodulin and IQGAP1 regulate E-cadherin function

In addition to the effects on the actin cytoskeleton, the Rho family members have more recently been shown to modulate Ca^{2+}-dependent cell-cell adhesion via E-cadherin (Braga et al., 1997). E-cadherin, a member of the cadherin family of cell surface adhesion molecules, is a transmembrane protein that is required for epithelial cell-cell adhesion and epithelial cell polarity (Koch et al., this book; Jiang, 1996; Takeichi, 1991). Ca^{2+}-dependent cell-cell contact induces the accumulation of E-cadherins to the regions of cell-cell contact, where homophilic interactions with E-cadherin on adjacent cells are thought to act as a cell adhesion zipper (Shapiro et al., 1995). The cytoplasmic domain of E-cadherin binds directly to either β-catenin or γ-catenin (also known as plakoglobin). E-cadherin complexed to β- or γ-catenin binds to α-catenin, which links the cadherin-catenin complex to actin.

Recent studies have implicated IQGAP1 and calmodulin as mediators of cadherin-catenin cell adhesion (Li et al., 1999; Fukata et al., 1999; Kuroda et al., 1998). Kuroda et al. (1998) demonstrated that IQGAP1 binds directly to E-cadherin and β-catenin *in vitro* and *in vivo*. IQGAP1 inhibits E-cadherin mediated cell-cell adhesion and this negative regulation by IQGAP1 on the cadherin-catenin complex is relieved when activated Cdc42 binds IQGAP1 (Fukata et al., 1999). IQGAP1 presumably alters the interaction between the cytoplasmic filament system and the transmembrane cadherin protein, thereby dissociating the cell adhesion zipper. An additional level of complexity is added by calmodulin, which modulates the interaction between E-cadherin and IQGAP1 (Li et al., 1999). Inhibition of calmodulin function induced the translocation of IQGAP1 to cell-cell junctions, with concomitant movement of E-cadherin away from the cell-cell junctions. The consequence was significantly impaired E-cadherin homophilic adhesion (Li et al., 1999). The molecular mechanism proposed is that disruption of the binding of calmodulin to IQGAP1 increases the association of IQGAP1 with the E-cadherin-catenin complex, thereby attenuating E-cadherin function. Together, these data suggest that IQGAP1 functions as a scaffolding protein, providing a molecular link that couples Ca^{2+}/calmodulin to Cdc42 and E-cadherin function.

5. RHO, CALMODULIN AND IQGAP1 IN MALIGNANCY

5.1. Rho pathways and malignancy

The Rho proteins participate in mitogenesis, transformation, protein kinase cascades, adhesion and transcriptional regulation (Macara et al., 1996). Rac and Rho are essential downstream components of oncogenic transformation by Ras (Qiu et al., 1995; Symons, 1996). Activation of Rac by Tiam-1, a GEF, promotes lymphocyte invasion (Michiels et al., 1995). Expression of constitutively active Rac or Cdc42 disrupts the normal polarization of mammary epithelial cells and induces, via PI3K, motility and invasion (Keely et al., 1997). Moreover, Rac and Cdc42 activate both c-Jun kinase (JNK) – leading to transcriptional activation of c-Jun – and transcription by serum response factor (Symons, 1996). The importance of Rho GTPases in transformation is emphasized by the large number of GEFs for this family (e.g., Dbl, Vav, Tiam, and OST) that are oncogenes which mediate malignant transformation through activation of Rho GTPases (Hall, 1998). The essential role of the Rho GTPases in transformation underscores the critical connection between the cytoskeleton and transformation (Hunter, 1997).

5.2. Calmodulin in malignant cells

A substantial body of evidence implicates Ca^{2+} and calmodulin in carcinogenesis. Non-transformed cells require extracellular Ca^{2+} to proliferate,

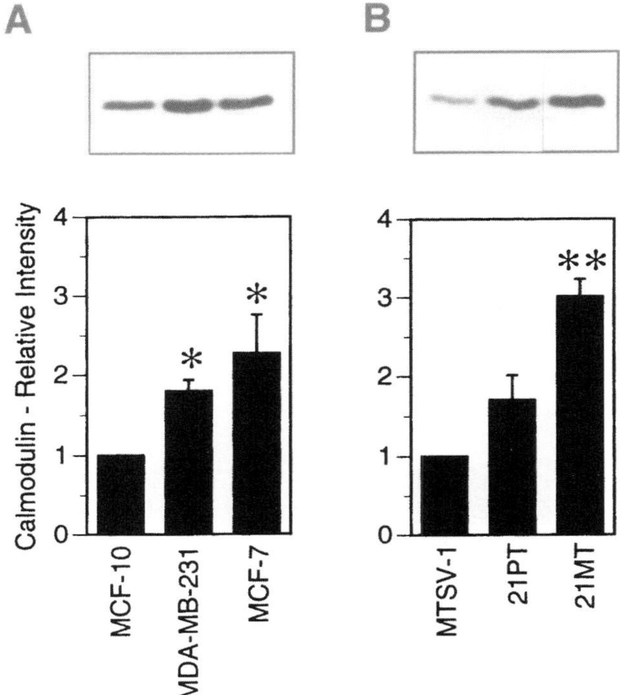

Figure 3. Calmodulin levels in breast cell lines. Cells were grown to equal confluence. For each experiment, equal amounts of protein lysate from the indicated cell lines were resolved by SDS-PAGE and transferred to PVDF. Blots were probed with anti-calmodulin antibody and developed with ECL. Lower panels, densitometry of the calmodulin bands (expressed as mean $\pm$ SE, $n = 3$) relative to MCF-10 (A) or MTSV-1 (B). *, significantly different from MCF-10 ($p < 0.005$); **, significantly different from MTSV-1 ($p < 0.001$).

while transformed cells proliferate independently of Ca^{2+} (Takuwa et al., 1995). Moreover, there is an inverse correlation between the ability of transformed cells to form tumors *in vivo* and their Ca^{2+}-dependence in culture (Boynton et al., 1977). Calmodulin has a fundamental role in cell proliferation and cell cycle progression (for a review, see Liu et al., 1991). The calmodulin concentrations are increased 2-fold at the G_1/S boundary and anti-calmodulin antibody inhibits DNA replication (Reddy et al., 1992). The essential role of calmodulin in normal cellular proliferation implies that abnormal cellular proliferation should alter the amount of calmodulin or its interactions with target proteins (Hait and Lazo, 1986).

The calmodulin content in several transformed cells in culture and tumor tissue is increased significantly (Takuwa et al., 1995). For example, chicken embryo fibroblasts transformed by Rous sarcoma virus (Van Eldik and Burgess, 1983; Watterson et al., 1976) and a human acute lymphoblastic leukemia cell line (Rainteau et al., 1987) has 2- to 3-fold more calmodulin than normal cells. Using a highly specific anti-calmodulin monoclonal an-

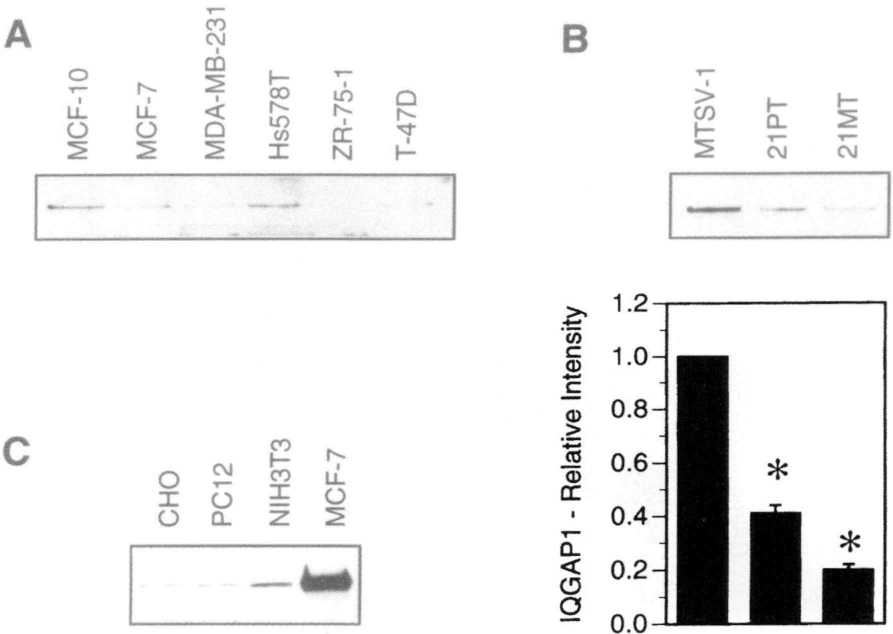

Figure 4. Comparison of the amount of IQGAP1 among cells. Equal amounts of protein lysate from the indicated cells lines were processed as described in the legend to Figure 3. Blots were probed with anti-IQGAP1 antibody and developed by ECL. (B) Lower panel, densitometry of the IQGAP1 bands (expressed as mean $\pm$ SE, $n = 3$) relative to MTSV-1. *, significantly different from from MTSV-1 ($p < 0.0001$).

tibody (Sacks et al., 1991), we compared the concentration of calmodulin among several cultured human breast epithelial cells (Figure 3). Malignant MCF-7 and MDA-MB-231 cells had approximately 2-fold more calmodulin than MCF-10 (Figure 3A), the closest cell line to normal breast epithelium (Russo et al., 1993). Further analysis was performed by comparing 21T cells with MTSV-1 breast epithelial cells. MTSV-1 cells are non-tumorigenic and retain many features of normal luminal epithelial cells (Taylor-Papadimitriou et al., 1993). The 21T malignant breast epithelial cell lines were derived from a single patient and consist of primary tumor (21PT) and metastatic tumor (21MT). Non-tumorigenic MTSV-1 cells had lower calmodulin concentrations than the 21T series (Figure 3B). Moreover, higher calmodulin content correlated with increasing malignancy. It is not possible to determine from these data whether the increased calmodulin is a cause or a result of the malignancy. Nevertheless, the increased concentrations of calmodulin may contribute to neoplastic transformation.

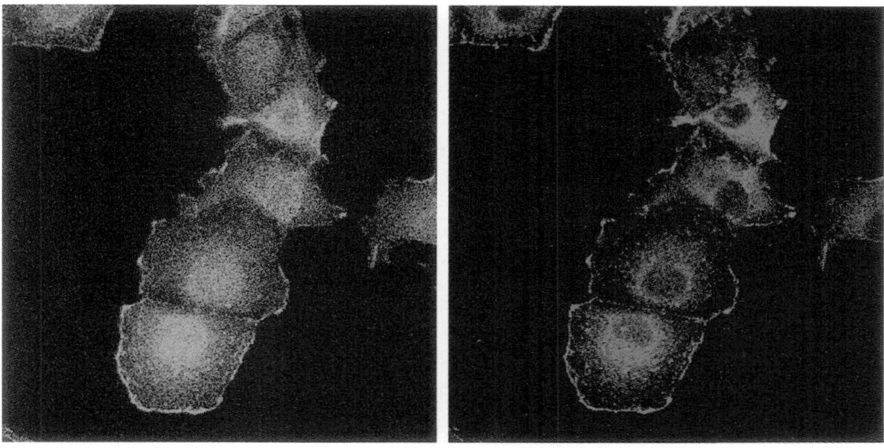

Figure 5. Subcellular location of calmodulin and IQGAP1 by confocal microscopy. MCF-7 cells were grown on plastic slides, fixed and probed with anti-calmodulin and anti-IQGAP1 antibodies. Primary antibodies were visualized with the appropriate fluorescent labeled secondary antibodies. Calmodulin is depicted in green and IQGAP1 in red. For a colour version of this figure, see page xxxvii.

5.3. IQGAP1 in malignant cells

IQGAP1 exhibits sequence similarity to Sar1 and neurofibromin. Neurofibromin is a tumor suppressor (Bernards, 1995) and Sar1, the closest relative of IQGAP1, is a suppressor of a mutationally activated form of Ras (Weissbach et al., 1994). These observations led to the proposal that IQGAP1 may also be a tumor suppressor (Weissbach et al., 1994).

The relative levels of expression of IQGAP1 protein in cell lines derived from several different tissues are shown in Figure 4. Five human breast carcinoma cell lines – MCF-7, ZR-75-1, T-47D, MDA-MB-231 and Hs578T – and non-tumorigenic MCF-10 were compared (Figure 4A). The IQGAP1 content in all the malignant cell lines was significantly lower than in the non-transformed cells (Figures 4A and 4B). Analysis of the 21T series revealed a significant decrease in IQGAP1 concentrations with increasing malignancy (Figure 4B). Analysis of selected cell lines demonstrated that IQGAP1 is expressed in a variety of tissues (Figure 4C). Note that there is a high level of expression of IQGAP1 in breast epithelium. These data, coupled with the report of a reduction of IQGAP1 expression in some human primary lung tumors (Mitchell et al., 1996), demonstrate that certain malignant tumors have decreased amounts of IQGAP1. Whether IQGAP1 participates in carcinogenesis and, if so, whether its interaction with calmodulin is a component, remains to be determined.

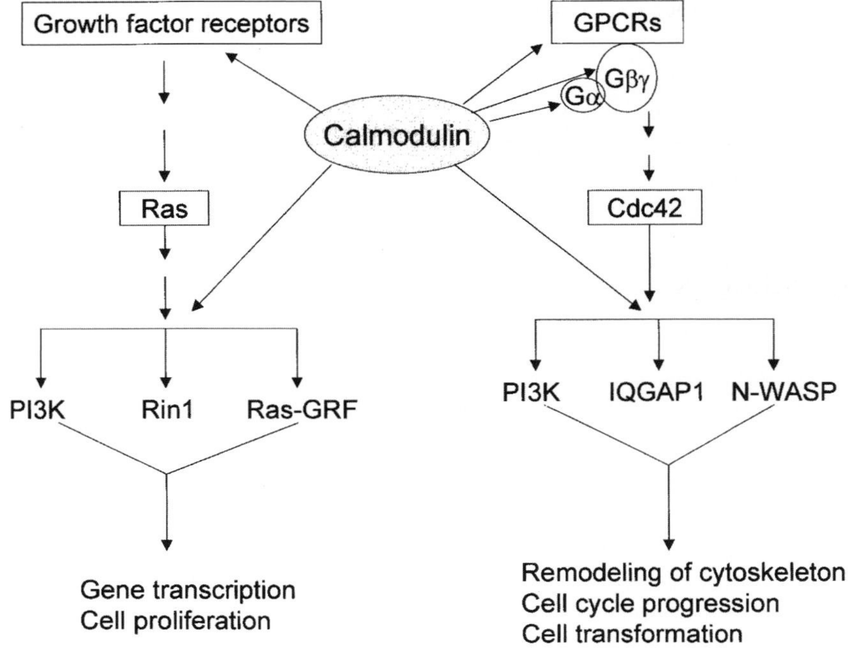

Figure 6. Model of the participation of calmodulin in Ras and Cdc42 signaling pathways. Selected proteins involved in the Ras and Cdc42 signaling pathways are depicted. With the exception of Ras and Cdc42, calmodulin has been demonstrated to interact directly with all of the proteins shown. The communication between Ras and Cdc42 signaling has been omitted for clarity. Details are in the text.

6. CONCLUSION

A model of the identified targets of calmodulin in Ras signaling is depicted in Figure 6. As described above, calmodulin intersects at multiple loci in the Ras and the Cdc42 signaling cascades. Although it does not bind directly to Ras or Cdc42, calmodulin can modulate events both upstream and downstream of these small GTPases and is thus a potentially important regulatory component of the functions elicited by these proteins. It is now evident that the earlier notion that calmodulin is dependent on Ca^{2+} to interact with target proteins was too restrictive and must be expanded; apocalmodulin can modulate diverse pathways. The fundamental physiological processes in which calmodulin participates has expanded dramatically over the last decade. The development of novel and advanced techniques to study normal cellular homeostatic mechanisms and the disruptions that occur in disease states is likely to continue to enhance our comprehension of the diverse biological functions in which calmodulin is a crucial element.

ACKNOWLEDGEMENTS

This work was supported in part by National Institutes of Health Grant CA75205. We thank Stella H. Kim for performing the immunocytochemistry and providing the graphics. We are grateful to A. Pardee and J. Taylor-Papadimitriou for generously providing cell lines.

REFERENCES

Anborgh, P.H., Qian, X., Papageorge, A.G., Vass, W.C., DeClue, J.E. and Lowy, D.R., 1999, Ras-specific exchange factor GRF: Oligomerization through its dbl homology domain and calcium-dependent activation of Raf, *Mol. Cell. Biol.* 19, 6411–6422.

Arava, Y., Seger, R. and Walker, M.D., 1999, GRFβ, a novel regulator of calcium signaling, is expressed in pancreatic beta cells and brain, *J. Biol. Chem.* 274, 24449–24452.

Baouz, S., Jacquet, E., Bernardi, A. and Parmeggiani, A., 1997, The N-terminal moiety of CDC25Mm, a GDP/GTP exchange factor of Ras proteins, controls the activity of the catalytic domain, *J. Biol. Chem.* 272, 6671–6676.

Bashour, A.-M., Fullerton, A.T., Hart, M.J. and Bloom, G.S., 1997, IQGAP1, a Rac- and Cdc42-binding protein, directly binds and cross-links microfilaments, *J. Cell Biol.* 137, 1555–1566.

Benguria, A., Perera, O.H., Pastor, M.T., Sacks, D.B. and Villalobo, A., 1994, Phosphorylation of calmodulin by the epidermal-growth-factor-receptor tyrosine kinase, *Eur. J. Biochem.* 224, 909–916.

Benguria, A., Soriano, M., Joyal, J.L., Sacks, D.B. and Villalobo, A., 1995, Phosphorylation of calmodulin by plasma-membrane-associated protein kinase(s), *Eur. J. Biochem.* 234, 50–58.

Bernards, A., 1995, Neurofibromatosis type 1 and Ras-mediated signaling: Filling in the GAPs, *Biochim. Biophys. Acta* 1242, 43–59.

Berridge, M.J., 1993, Inositol trisphosphate and calcium signalling, *Nature* 361, 315–325.

Bouhoute, A. and Leclercq, G., 1995, Modulation of estradiol and DNA binding to estrogen receptor upon association with calmodulin, *Biochem. Biophys. Res. Commun.* 208, 748–755.

Boynton, A.L., Whitfield, J.F., Isaacs, R.J. and Tremblay, R.G., 1977, Different extracellular calcium requirements for proliferation of nonneoplastic, preneoplastic, and neoplastic mouse cells, *Cancer Res.* 37, 2657–2661.

Brady, M.J. and Palfrey, H.C., 1993, Rapid and sustained phosphorylation of a calmodulin-binding protein (CaM-BP100) in NGF-treated PC12 cells, *J. Biol. Chem.* 268, 17951–17958.

Braga, V.M.M., Machesky, L.M., Hall, A.A. and Hotchin, N.A., 1997, The small GTPases Rho and Rac are required for the establishment of cadherin-dependent cell-cell contacts, *J. Cell Biol.* 137, 1421–1431.

Buchsbaum, R., Telliez, J.-P., Goonesekera, S. and Feig, L.A., 1996, The N-terminal pleckstrin, coiled-coil, and IQ domains of the exchange factor Ras-GRF act cooperatively to facilitate activation by calcium, *Mol. Cell. Biol.* 16, 4888–4896.

Carafoli, E. and Klee, C., 1999, *Calcium as a Cellular Regulator* Oxford University Press, New York.

Castoria, G., Migliaccio, A., Nola, E. and Auricchio, F., 1988, *In vitro* interaction of estradiol receptor with Ca^{2+}-calmodulin, *Mol. Endo.* 2, 167–174.

Cheney, R.E. and Mooseker, M.S., 1992, Unconventional myosins, *Curr. Opin. Cell Biol.* 4, 27–35.

Chuang, T.T., Paolucci, L. and DeBlasi, A., 1996, Inhibition of G protein-coupled receptor kinase subtypes by Ca^{2+}/calmodulin, *J. Biol. Chem.* 271, 28691–28696.

Clapham, D.E., 1995, Calcium signaling, *Cell* 80, 259–268.

Cohen, P. and Klee, C., 1988, *Calmodulin*, Elsevier, New York.

Coppola, T., Perret-Menoud, V., Luthi, S., Farnsworh, C.C., Glomset, J.A. and Regazzi, R., 1999, Disruption of Rab3-calmodulin interaction, but not other effector interactions, prevents Rab3 inhibition of exocytosis, *EMBO J.* 18, 5885–5891.

Dikic, I., Tokiwa, G., Lev, S., Courtneidge, S.A. and Schlessinger, J., 1996, A role for Pyk2 and Src in linking G-protein-coupled receptor with MAP kinase activation, *Nature* 383, 547–550.

Erickson, J.W., Cerione, R.A. and Hart, M.J., 1997, Identification of an actin cytoskeletal complex that includes IQGAP and the Cdc42 GTPase, *J. Biol. Chem.* 272, 24443–24447.

Exton, J., 1998, Small GTPase minireview series, *J. Biol. Chem.* 273, 7.

Farnsworth, C.L., Freshney, N.W., Rosen, L.B., Ghosh, A., Greenberg, M.E. and Feig, L.A., 1995, Calcium activation of Ras mediated by neuronal exchange factor Ras-GRF, *Nature* 376, 524–527.

Fischer, R., Wei, Y., Anagli, J. and Berchtold, M.W., 1996, Calmodulin binds to and inhibits GTP binding of the Ras-like GTPase Kir/Gem, *J. Biol. Chem.* 271, 25067–25070.

Fischer, R., Julsgart, J. and Berchtold, M.W., 1998, High affinity calmodulin target sequence in the signalling molecule PI3-kinase, *FEBS Lett.* 425, 175–177.

Fukata, M., Kuroda, S., Nakagawa, M., Kawajiri, A., Itoh, N., Shoji, I., Matsuura, Y., Yonehara, S., Fujisawa, H., Kikuchi, A. and Kaibuchi, K., 1999, Cdc42 and Rac1 regulate the interaction of IQGAP1 with beta-catenin, *J. Biol. Chem.* 274, 26044–26050.

Graves, C.B., Goewert, R.R. and McDonald, J.M., 1985, The insulin receptor contains a calmodulin-binding domain, *Science* 230, 827–829.

Gutkind, J.S., 1998, The pathways conecting G protein-coupled receptors to the nucleus through divergent mitogen-activated protein kinase cascades, *J. Biol. Chem.* 273, 1839–1842.

Hait, W.N. and Lazo, J.S., 1986, Calmodulin: A potential target for cancer chemotherapeutic agents, *J. Clin. Oncol.* 4, 994–1012.

Hall, A., 1998, Rho GTPases and the actin cytoskeleton, *Science* 279, 509–514.

Hamm, H.E., 1998, The many faces of G protein signaling, *J. Biol. Chem.* 273, 669– 672.

Hart, M.J., Callow, M.G., Souza, B. and Polakis, P., 1996, IQGAP1, a calmodulin-binding protein with a rasGAP-related domain, is a potential effector for cdc42Hs, *EMBO J.* 15, 2997–3005.

Hasson, T. and Mooseker, M.S., 1996, Vertebrate unconventional myosins, *J. Biol. Chem.* 271, 16431–16434.

Ho, Y.-D., Joyal, J.L., Li, Z. and Sacks, D.B., 1999, IQGAP1 integrates Ca^{2+}/calmodulin and Cdc42 signaling, *J. Biol. Chem.* 274, 464–470.

Houdusse, A. and Cohen, C., 1995, Target sequence recognition by the calmodulin superfamily: Implication from light chain binding to the regulatory domain of scallop myosin, *Proc. Natl. Acad. Sci. USA* 92, 10644–10647.

Houdusse, A., Silver, M. and Cohen, C., 1996, A model of Ca^{2+}-free calmodulin binding to unconventional myosins reveals how calmodulin acts as a regulatory switch, *Structure* 4, 1475–1490.

Huckle, W.R., Dy, R.C. and Earp, S.H., 1992, Calcium-dependent increase in tyrosine kinase activity stimulated by angiotensin II, *Proc. Natl. Acad. Sci. USA* 89, 8837–8841.

Hunter, T., 1997, Oncoprotein networks, *Cell* 88, 333–346.

Iacovelli, L., Sallese, M., Mariggio, S. and deBlasi, A., 1999, Regulation of G protein-coupled receptor kinase subtype by calcium sensor proteins, *FASEB J.* 1, 1–8.

Jiang, W.G., 1996, E-cadherin and its associated protein catenins, cancer invasion and metastasis, *Brit. J. Surg.* 83, 437–446.

Joyal, J.L. and Sacks, D.B., 1994, Insulin-dependent phosphorylation of calmodulin in rat hepatocytes, *J. Biol. Chem.* 269, 30039–30048.

Joyal, J.L., Annan, R.S., Ho, Y.D., Huddleston, M.E., Carr, S.A., Hart, M.J., and Sacks, D.B., 1997a, Calmodulin modulates the interaction between IQGAP1 and Cdc42. Identification of IQGAP1 by nanoelectrospray tandem mass spectrometry, *J. Biol. Chem.* 272, 15419–15425.

Joyal, J.L., Burks, D.J., Pons, S., Matter, W.F., Vlahos, C.J., White, M.F. and Sacks, D.B., 1997b, Calmodulin activates phosphatidylinositol 3-kinase, *J. Biol. Chem.* 272, 28183–28186.

Katori, T., Yasuda, H., Fukuda, H. and Kimura, S., 1994, Involvement of Ca^{2+}-calmodulin in platelet-derived growth factor-, fibroblast growth factor-, and insulin-induced ornithine decarboxylase in NIH-3T3 cells, *Metabolism* 43, 4–10.

Keely, P.J., Westwick, J.K., Whitehead, I.P., Der, C.J. and Parise, L.V., 1997, Cdc42 and Rac1 induce integrin-mediated cell motility and invasiveness through PI(3)K, *Nature* 390, 632–636.

Koch, A., Engel, J. and Maurer, P., Calcium binding to extracellular matrix proteins, functional and pathological effects, this book.

Kretsinger, R.H., Rudnick, S.E. and Weissman, L.J., 1986, Crystal structure of calmodulin, *J. Inorg. Biochem.* 28, 289–302.

Kuroda, S., Fukata, M., Nakagawa, M., Fujii, K., Nakamura, T., Ookubo, T., Izawa, I., Nagase, T., Nomura, N., Tani, H., Shoji, I., Matsuura, Y., Yonehard, S. and Kaibuchi, K., 1998, Role of IQGAP1, a target of the small GTPases Cdc42 and Rac1, in regulation of E-cadherin-mediated cell-cell adhesion, *Science* 281, 832–835.

Lee, C.J., Della, N.G., Chew, C.E. and Zack, D.J., 1996, Rin, a neuron-specific and calmodulin-binding small G-protein, and Rit define a novel subfamily of Ras proteins, *J. Neuroscience* 16, 6784–6794.

Lefkowitz, R.J., 1998, G protein-coupled receptors, *J. Biol. Chem.* 273, 18677–18680.

Lev, S., Moreno, H., Martinez, R., Canoll, P., Peles, E., Musacchio, J.M., Plowman, G.D., Rudy, B. and Schlessinger, J., 1995, Protein tyrosine kinase PYK2 involved in Ca^{2+}-induced regulation of ion channel and MAP kinase functions, *Nature* 376, 737-745.

Levay, K., Satpaev, D.K., Pronin, A.N., Benovic, J.L. and Slepak, V.Z., 1998, Localization of the sites for Ca^{2+}-binding proteins on G protein-coupled receptor kinases, *Biochemistry* 37, 13650–13659.

Levy, J., Gavin, J.R. and Sowers, J.R., 1994, Diabetes mellitus: A disease of abnormal cellular calcium metabolism, *Am. J. Med.* 96, 260–273.

Li, Z., Lee, S., Higgins, J., Brenner, M. and Sacks, D.B., 1999, Calmodulin and IQGAP1 modulate E-cadherin function, *J. Biol. Chem.* 274, 37885–37892.

Li, Z., Joyal, J.L. and Sacks, D.B., 2000, Binding of IRS proteins to calmodulin is enhanced in insulin resistance, *Biochemistry* 39, 5089–5096.

Liu, J., Farmer, J.D., Jr., Lane, W.S., Friedman, J., Weissman, I. and Schreiber, S.L., 1991, Calcineurin is a common target of cyclophilin-cyclosporin A and FKBP-FK506 complexes, *Cell* 66, 807–815.

Liu, M., Yu, B., Nakanishi, O., Wieland, T. and Simon, M., 1997, The Ca^{2+}-dependent binding of calmodulin to an N-terminal motif of the heterotrimeric G protein beta subunit, *J. Biol. Chem.* 272, 18801–18807.

Macara, I.G., Lounsbury, K., Richards, S.A., McKiernan, C. and Bar-Sagi, D., 1996, The Ras superfamily of GTPases, *FASEB J.* 10, 625–630.

Martin-Nieto, J. and Villalobo, A., 1998, The human epidermal growth factor receptor contains a juxtamembrane calmodulin-binding site, *Biochemistry* 37, 227–236.

Michiels, F., Habets, G.G.M., Stam, J.C. van der Kammen, R.A. and Collard, J.G., 1995, A role for Rac in Tiam1-induced membrane ruffling and invasion, *Nature* 375, 338–340.

Migliaccio, A., Rotondi, A. and Auricchio, F., 1984, Calmodulin-stimulated phosphorylation of 17β-estradiol receptor on tyrosine, *Proc. Natl. Acad. Sci. USA* 81, 5921–5925.

Mitchell, C.E., Palmisano, W.A., Lechner, J.F., Belinsky, S.A., Bernards, A. and Weissbach, L., 1996, Altered expression of the IQGAP1 gene in human lung cancer cell lines, *Proc. Annu. Meet. Am. Assoc. Cancer. Res.* 37, A3563.

Moyers, J.S., Bilan, P.J., Zhu, J. and Kahn, C.R., 1997, Rad and Rad-related GTPases interact with calmodulin and calmodulin-dependent protein kinase II, *J. Biol. Chem.* 272, 11832–11839.

Munshi, H.G., Burks, D.J., Joyal, J.L., White, M.F. and Sacks, D.B., 1996, Ca^{2+} regulates calmodulin binding to IQ motifs in IRS-1, *Biochemistry* 35, 15883–15889.

Nicotera, P., Zhivotovsky, B. and Orrenius, S., 1994, Nuclear calcium transport and the role of calcium in apoptosis, *Cell Calcium* 16, 279–288.

Okazaki, H., Zhang, J., Hamawy, M.M. and Siraganian, R.P., 1997, Activation of protein-tyrosine kinase Pyk2 is downstream of Syk in FcεRI signaling, *J. Biol. Chem.* 272, 32443–32447.

O'Neil, K.T. and DeGrado, W.F., 1990, How calmodulin binds its targets: Sequence independent recognition of amphiphilic α-helices, *Trends Biochem. Sci.* 15, 59–64.

Park, J.B., Farnsworth, C.C. and Glomset, J.A., 1997, Ca^{2+}/calmodulin causes Rab3A to dissociate from synaptic membranes, *J. Biol. Chem.* 272, 20857–20865.

Pawson, T. and Scott J.D., 1997, Signaling through scaffold, anchoring and adaptor proteins, *Science* 278, 2075–2080.

Pronin, A.N., Satpaev, D.K., Slepak, V.Z. and Benovic, J.L., 1997, Regulation of G protein-coupled receptor kinases by calmodulin and localization of the calmodulin binding domain, *J. Biol. Chem.* 272, 18273–18280.

Qiu, R.-G., Chen, J., Kirn, D., McCormick, F. and Symons, M., 1995, An essential role for Rac in Ras transformation, *Nature* 374, 457–459.

Rainteau, D., Sharif, A., Bourrillon, R. and Weinman, S., 1987, Calmodulin in lymphocyte mitogenic stimulation and in lymphoid cell line growth, *Exp. Cell Res.* 168, 546–554.

Reddy, G.P.V., Reed, W.C., Sheehan, E.L. and Sacks, D.B., 1992, Calmodulin-specific monoclonal antibodies inhibit DNA replication in mammalian cells, *Biochemistry* 31, 10426–10430.

Rozengurt, E., 1986, Early signals in the mitogenic response, *Science* 234, 161–166.

Russo, J., Calaf, G. and Russo, I.H., 1993, A critical approach to the malignant transformation of human breast epithelial cells with chemical carcinogens, *Crit. Rev. Oncog.* 4, 403–417.

Sacks, D.B. and McDonald, J.M., 1988, Insulin-stimulated phosphorylation of calmodulin by rat liver insulin receptor preparations, *J. Biol. Chem.* 263, 2377–2383.

Sacks, D.B., Fujita-Yamaguchi, Y., Gale, R.D. and McDonald, J.M., 1989, Tyrosine-specific phosphorylation of calmodulin by the insulin receptor kinase purified from human placenta, *Biochem. J.* 263, 803–812.

Sacks, D.B., Porter, S.E., Ladenson, J.H. and McDonald, J.M., 1991, Monoclonal antibody to calmodulin: Development, characterization and comparison with polyclonal anti-calmodulin antibodies, *Anal. Biochem.* 194, 369–377.

Sacks, D.B., Davis, H.W., Crimmins, D.L. and McDonald, J.M., 1992, Insulin-stimulated phosphorylation of calmodulin, *Biochem. J.* 286, 211–216.

San Jose, E., Benguria, A., Geller, P. and Villalobo, A., 1992, Calmodulin inhibits the epidermal growth factor receptor tyrosine kinase, *J. Biol. Chem.* 267, 15237–15245.

Schlessinger, J. and Ullrich, A., 1992, Growth factor signaling by receptor tyrosine kinases, *Neuron* 9, 383–391.

Scott, K., Sun, Y., Beckingham, K.A. and Zuker, C.S., 1997, Calmodulin regulation of Drosophila light-activated channels and receptor function mediates termination of the light response in vivo, *Cell* 91, 375–383.

Shapiro, L., Fannon, A.M., Kwong, P.D., Thompson, A., Lehmann, M.S., Grubel, G., Legerand, J., Als-Nielsen, J. and Hendrickson, W.A., 1995, Structural basis of cell-cell adhesion by cadherins, *Nature* 374, 327–337.

Shashkin, P., Koshkin, A., Langley, D., Ren, J.M., Westerblad, H. and Katz, A., 1995, Effects of CGS 9343B (a putative calmodulin antagonist) on isolated skeletal muscle, *J. Biol. Chem.* 270, 25613–25618.

Shechter, Y., 1984, Trifluoperazine inhibits insulin action on glucose metabolism in fat cells without affecting inhibition of lipolysis, *Proc. Natl. Acad. Sci. USA* 81, 327–331.

Siciliano, J.C., Toutant, M., Derkinderen, P., Sasaki, T. and Girault, J., 1996, Differential regulation of proline-rich tyrosine kinase 2/cell adhesion kinase beta (PYK2/CAKβ) and pp125FAK by glutamate and depolarization in rat hippocampus, *J. Biol. Chem.* 271, 28942–28946.

Sun, X.J., Wang, L.M., Zhang, Y., Yenush, L., Myers, M.G., Jr., Glasheen, E., Lane, W.S., Pierce, J.H. and White, M.F., 1995, Role of IRS-2 in insulin and cytokine signalling, *Nature* 377, 173–177.

Symons, M., 1996, Rho family GTPases: The cytoskeleton and beyond, *Trends Biochem. Sci.* 21, 178–181.

Takeichi, M., 1991, Cadherin cell adhesion receptors as a morphogenetic regulator, *Science* 251, 1451–1455.

Takuwa, N., Zhou, W. and Takuwa, Y., 1995, Calcium, calmodulin and cell cycle progression, *Cell Signalling* 7, 93–104.

Taylor-Papadimitriou, J., Berdichevsky, F., D'Souza, B. and Burchell, J., 1993, Human models of breast cancer, *Cancer Surveys* 16, 59–78.

Van Eldik, L.J. and Burgess, W.H., 1983, Analytical subcellular distribution of calmodulin and calmodulin-binding proteins in normal and virus-transformed fibroblasts, *J. Biol. Chem.* 258, 4539–4547.

Van Eldik, L.J. and Watterson, D.M., 1998, *Calmodulin and Signal Transduction*, Academic Press, Orlando.

Wang, D., Sadee, W. and Quillan, J.M., 1999, Calmodulin binding to G protein-copling domain of opiod receptor, *J. Biol. Chem.* 274, 22081–22088.

Wang, K.L. and Roufogalis, B.D., 1999, Ca^{2+}/calmodulin stimulates GTP binding to the Ras-related protein Ral-A, *J. Biol. Chem.* 274, 14525–14528.

Wang, K.L., Khan, M.T. and Roufogalis, B.D., 1997, Identification and characterization of calmodulin-binding domain in Ral-A, a Ras-related GTP-binding protein purified from human erythrocyte membrane, *J. Biol. Chem.* 272, 16002–16009.

Watterson, D.M., Van Eldik, L.J., Smith, R.E. and Vanaman, T.C., 1976, Calcium-dependent regulatory protein of cyclic nucleotide metabolism in normal and transformed chicken embryo fibroblasts, *Proc. Natl. Acad. Sci. USA* 73, 2711–2715.

Weissbach, L., Settleman, J., Kalady, M.F., Snijders, A.J., Murthy, A.E., Yan, Y.X. and Bernards, A., 1994, Identification of a human RasGAP-related protein containing calmodulin-binding motifs, *J. Biol. Chem.* 269, 20517–20521.

White, M.F. and Kahn, C.R., 1994, The insulin signaling system, *J. Biol. Chem.* 269, 1–4.

Wolenski, J.S., 1995, Regulation of calmodulin-binding myosins, *Trends Cell Biol.* 5, 310–316.

Wu, K., Nigam, S.K., Le Doux, M., Huang, Y.Y., Aoki, C. and Siekevitz, P., 1992, Occurrence of the α subunits of G proteins in cerebral cortex synaptic membrane and postsynaptic density fractions: Modulation of ADP-ribosylation by Ca^{2+}/calmodulin, *Proc. Natl. Acad. Sci. USA* 89, 8686–8690.

Yang, C., Watson, R.T., Elmendorf, J.S., Sacks, D.B. and Pessin, J.E., 2000, Calmodulin antagonists inhibit insulin-stimulated GLUT4 (glucose transporter 4) translocation by preventing the formation of phosphatidylinositol 3,4,5-trisphosphate in 3T3L1 adipocytes, *Mol. Endocrinol.* 14, 317–326.

Type II Citrullinemia (Citrin Deficiency): A Mysterious Disease caused by a Defect of Calcium-Binding Mitochondrial Carrier Protein

Keiko Kobayashi, Mikio Iijima, Tomotsugu Yasuda, David S. Sinasac, Naoki Yamaguchi, Lap-Chee Tsui, Stephen W. Scherer and Takeyori Saheki

1. INTRODUCTION

Citrullinemia (OMIM 215700) (McKusick, 1998) is an autosomal recessive disease that is caused by a deficiency of argininosuccinate synthetase (ASS; EC 6.3.4.5). The clinical, biochemical and molecular aspects of citrullinemia have been reviewed elsewhere (Walser, 1983; Saheki et al., 1987a; McKusick, 1998). So far, we have analyzed almost 200 patients with citrullinemia in our laboratory and have classified them into three types according to enzyme abnormality and into two forms according to pathogenesis (Figure 1) (Saheki et al., 1981, 1985a, 1987a, b; Kobayashi et al., 1993, 1999). The first form is the classical form (CTLN1) found in most patients with neonatal/infantile-onset citrullinemia (type I and type III), first described by McMurray et al. (1962); the second form is the adult-onset type II citrullinemia (CTLN2) caused by a liver-specific ASS deficiency. In CTLN1, the enzyme defect is found in all tissues and cells in which ASS is expressed (Saheki et al., 1980, 1981, 1982, 1983a, 1985a, b, 1987a, b). To date, we have identified 36 mutations in the ASS gene located on chromosome 9q34 and have clarified the pathogenesis

Keiko Kobayashi, Mikio Iijima, Tomotsugu Yasuda, Naoki Yamaguchi and Takeyori Saheki • Department of Biochemistry, Faculty of Medicine, Kagoshima University, Kagoshima, Japan. **David S. Sinasac, Lap-Chee Tsui and Stephen W. Scherer** • Department of Molecular and Medical Genetics, University of Toronto, and Department of Genetics, The Hospital for Sick Children, Toronto, Ontario, Canada.

R. Pochet, R. Donato, J. Haiech, C. Heizmann and V. Gerke (eds.): Calcium: The Molecular Basis of Calcium Action in Biology and Medicine, 565–587.

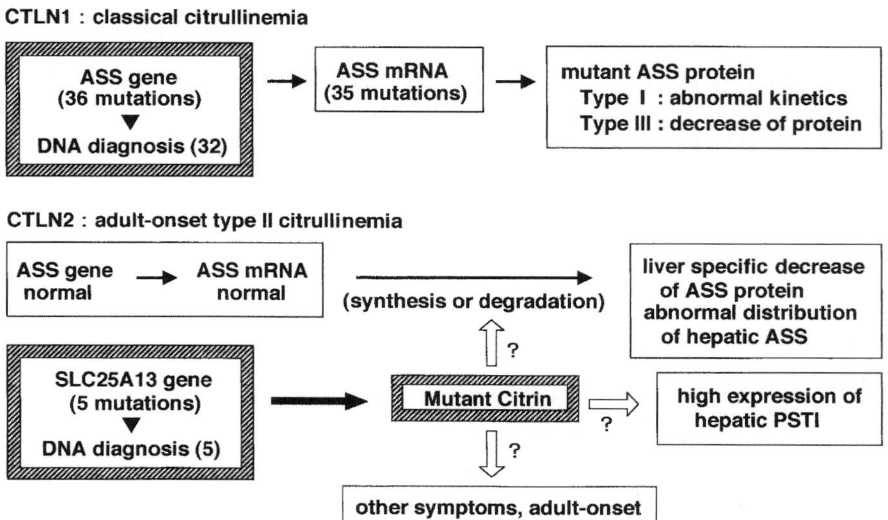

Figure 1. Classification and characterization of citrullinemia.

of most CTLN1 patients at the molecular level (Kobayashi et al., 1987, 1990, 1991, 1994, 1995a; Kakinoki et al., 1997).

In 1968, Miyakoshi et al. (1968) reported that blood citrulline levels were increased in hyperammonemia patients with a special type of chronic recurrent hepatocerebral degeneration described by Inose (1952). This special type of hepatocerebral degeneration differed from a portal-systemic encephalopathy, and came to be designated as pseudoulegyric hepatocerebral disease on the basis of pathological brain changes, or as nutritional hepatocerebral disease on the basis of the metabolic disorders caused by a highly unbalanced diet or developmental disturbances caused by endocrinal abnormalities. However, from the enzymological and immunological studies of late-onset patients with citrullinemia in Japan, Saheki and co-workers (Saheki et al., 1980, 1981, 1982, 1983a, 1985a, b, 1987a, b; Imamura et al., 1987) found that this type of hepatocerebral disease is a type of citrullinemia with a quantitative decrease of ASS protein specifically in the liver. Saheki classified Japanese citrullinemia patients into three types and named the quantitative type "adult-onset type II citrullinemia" to distinguish it from classical citrullinemia type I (qualitative type with abnormal kinetics caused by missense mutation) and type III (mainly caused by abnormal splicing of the ASS gene) (Saheki et al., 1981, 1985a, 1987a, b; Kobayashi et al., 1994, 1995a).

CTLN2 is characterized by a liver-specific deficiency of ASS protein with normal kinetic properties (Saheki et al., 1980, 1981, 1982, 1983a, 1985a, b, 1987a; Imamura et al., 1987). However, there are no apparent abnormalities in hepatic ASS mRNA or within the ASS gene locus of CTLN2 patients (Sase et al., 1985; Kobayashi et al., 1986, 1993) and the primary defect of CTLN2 has been unknown for a long time. We have found that 26 of 132 CTLN2

[A]	SLC25A13	SLC25A12
Chromosome	7q21.3	2q24
Gene	160 kb (18 exons)	not identified
mRNA	3.4 kb (liver, kidney)	2.8, 3.3, 6.6 kb (muscle, brain)
Protein	Citrin (675 aa, 74 kDa)	Aralar (678 aa, 74 kDa)
Deficiency	CTLN2	unknown

[B]

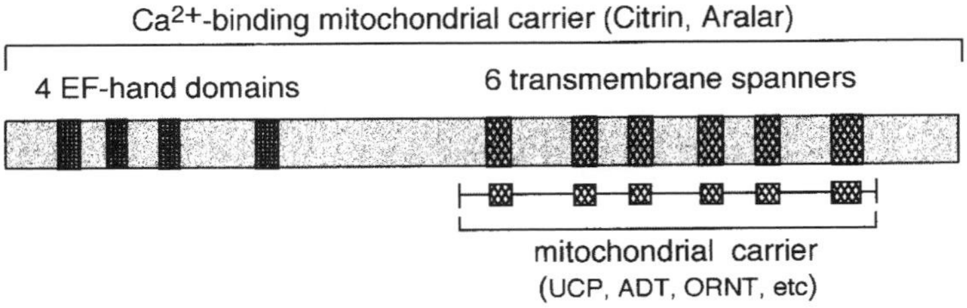

Ca2+-binding mitochondrial carrier (Citrin, Aralar)

4 EF-hand domains

6 transmembrane spanners

mitochondrial carrier

(UCP, ADT, ORNT, etc)

Figure 2. Comparison of SLC25A13 and SLC25A12 (A) and schematic protein structure of Ca^{2+}-binding mitochondrial carrier and other mitochondrial carrier (B). The figure summarizes the data from Camacho et al. (1999), Crackower et al. (1999), Kobayashi et al. (1999), Sinasac et al. (1999), Walker (1992), del Arco and Satrústegui (1998) and del Arco et al. (2000). UCP, mitochondrial uncoupling protein; ADT, ADP-ATP translocase; ORNT, ornithine transporter.

patients (24 of 126 families) are apparently from consanguineous parents (Kobayashi et al., 1993) and sibling cases have been affected (Tsujii et al., 1976), indicating that CTLN2 is probably an autosomal recessive disorder. Most recently, we identified the CTLN2 locus (Kobayashi et al., 1999) on chromosome 7q21.3 by homozygosity mapping analysis of individuals from 18 consanguineous unions, a novel SLC25A13 gene (Kobayashi et al., 1999; Sinasac et al., 1999) that encodes a putative mitochondrial carrier protein by positional cloning (Figure 2), and 5 different DNA sequence alterations (Kobayashi et al., 1999) that account for the mutations in all CTLN2 patients examined (Figure 3). The SLC25A13 gene encodes a 3.4 kb transcript expressed most abundantly in the liver. The predicted SLC25A13 protein, designated citrin, encodes a polypeptide of 675 amino acids, and is bipartite in structures, the amino-terminal half harboring four EF-hand domains and the carboxyl-terminal half having the characteristic features of a mitochondrial carrier; it is a sort of calcium-binding mitochondrial solute carrier protein.

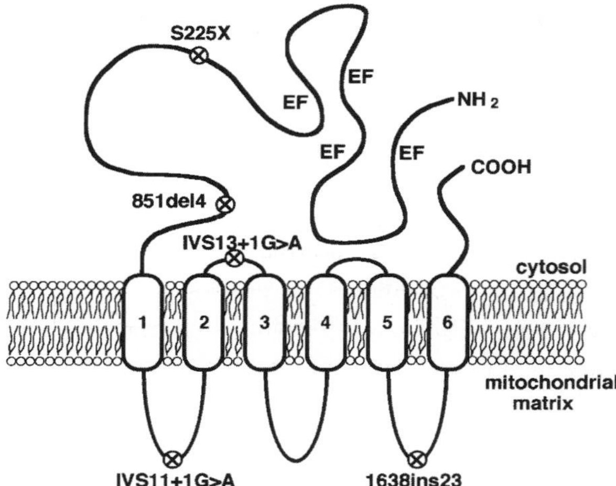

Figure 3. Topological model of human citrin and five mutations in CTLN2. A schematic representation of human citrin is depicted based on the prediction of the amino acid sequence (Kobayashi et al., 1999). The approximate positions of alterations in citrin caused by the mutations in CTLN2 patients are indicated.

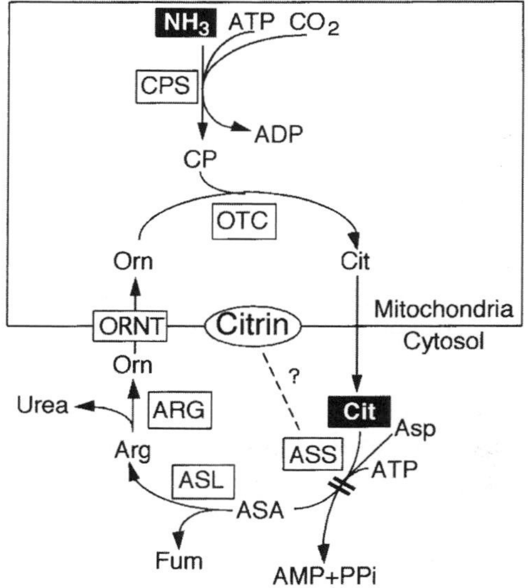

Figure 4. Urea cycle and citrin. CPS, carbamoylphosphate synthetase; OTC, ornithine transcarbamoylase; ASS, argininosuccinate synthetase; ASL, argininosuccinate lyase; ARG, arginase; ORNT, ornithine transporter; CP, carbamoylphosphate; Cit, citrulline; Asp, aspartate; ASA, argininosuccinate; Fum, fumarate; Arg, arginine; Orn, ornithine.

2. THE ROLE OF MITOCHONDRIA IN UREA CYCLE METABOLISM

2.1. Components of the urea cycle

In mammals, the urea cycle is the main mechanism for detoxifying waste nitrogen (Figure 4). It involves five enzymes; carbamoylphosphate synthetase (CPS), ornithine transcarbamoylase (OTC), ASS, argininosuccinate lyase (ASL) and arginase (ARG), and a protein, ornithine transporter (ORNT) which transports ornithine from the cytosol to the mitochondria. These proteins function in different subcellular compartments: the mitochondrial matrix (CPS and OTC), cytosol (ASS, ASL and ARG) and the mitochondrial inner membrane (ORNT). The liver is the only organ which contains all the enzymes and plays a role in detoxifying ammonia. Some of the enzymes are expressed in the intestine (CPS and OTC) and kidney (ASS, ASL and ARG but which differs from hepatic ARG), serving to provide a biosynthetic pathway for arginine (Funahashi et al., 1981; Windmueller and Spaeth, 1981). All the urea cycle enzymes except ARG are expressed in the small intestine of the suckling rat (de Jonge et al., 1998), suggesting that before weaning, arginine biosynthesis from citrulline takes place in the small intestine rather than in the kidney. N-acetylglutamate synthase is another enzyme which is essential for the proper operation of the cycle. Generally, malfunction of any of these enzymes or proteins results in hyperammonemia.

The urea cycle is also known as an ornithine cycle. Ornithine is one of the most important regulatory factors for the operation of the urea cycle, and the concentration of ornithine in the liver changes in response to changes in urea synthesis (Saheki et al., 1980; Li et al., 1999). Other factors necessary for urea synthesis are the electron transport system that generates ATP, the TCA cycle and aspartate aminotransferase which supply aspartate, and the dibasic amino acid transport system at the plasma membrane of intestine and kidney cells that supplies arginine and ornithine, the latter being a carrier of nitrogen.

2.2. Mitochondrial transport of metabolites in the urea cycle

Ornithine, which is formed in the cytosol, must enter the mitochondrial matrix where it is carbamoylated to citrulline; and citrulline must leave the mitochondria in order to regenerate ornithine. Early investigations into the mode of transport of ornithine and citrulline across the mitochondrial inner membrane suggested the existence of: (a) two independent transport systems (Gamble and Lehninger, 1973), one for ornithine cation and the other electroneutral for citrulline, and (b) an ornithine/citrulline exchange system (Bradford and McGivan, 1980). In yeast, the ARG11 gene has been identified as a mitochondrial ornithine carrier involved in arginine biosynthesis (Palmieri et al., 1997). While the 32 kDa protein transports ornithine from mitochondrial matrix to cytosol in exchange for protons, it apparently does not transport citrulline, thereby supporting hypothesis (a). The Arg11p sequence is most similar to that of *Neurospora crassa* Arg13p, which transports ornithine from cytosol to mitochondrial matrix. Support for hypothesis (b) is found in the purification of a putative ornithine/citrulline carrier protein from rat liver mitochondria (Indiveri et al., 1997).

The existence of an ornithine transport molecule in man is also suggested by the presence of a mutant protein with abnormal kinetics in patients with hyperornithinemia, hyperammonemia and homocitrullinuria (HHH) (Inoue et al., 1988). Recently, the human ornithine transporter gene (ORNT1) mapped to chromosome 13q14 was isolated by using the sequences of two fungal proteins, Arg11p and Arg13p; and the mutations were identified in the ORNT1 gene of patients with HHH syndrome (Camacho et al., 1999). The human and mouse ornithine transporters composed of 301 amino acids (Figure 2) exhibit 27% sequence identity with the fungal proteins and are highly expressed in the liver.

2.3. Mitochondrial carrier family

The mitochondrial inner membrane harbors a set of carrier proteins for metabolite transport that constitute a superfamily of related proteins (Walker, 1992). Many mitochondrial metabolite transporters or carriers, such as the mitochondrial uncoupling protein (UCP), ADP-ATP translocase (ADT) and ORNT, have the features of the mitochondrial carrier family (MCF). As shown in Figure 2, the MCF proteins are composed of approximately 300 amino acids and consist of 3 repeated regions, each about 100 amino acids long, that contain a conserved sequence motif, each with two hydrophobic α-helical segments connected by an extensive hydrophilic sequence, and two putative transmembrane (TM) domains. A topological model predicts six TM spanners, with amino and carboxyl termini on one side of the membrane and hydrophilic loops on the other.

3. CLASSIFICATION AND CHARACTERIZATION OF CITRULLINEMIA

3.1. Argininosuccinate synthetase (ASS)

ASS is a cytosolic enzyme which catalyzes the formation of argininosuccinate from citrulline and aspartate with the released energy of ATP cleavage to AMP and pyrophosphate (Figure 4). In ureotelic animals, such as man, the enzyme is expressed at high levels in the liver where it functions as part of the urea cycle. ASS is thought to be one of the regulatory enzymes of the urea cycle, because its total activity in the liver is the lowest among the urea cycle members. The liver is not thought to contribute to circulating citrulline or arginine levels because of its high levels of ASS, ASL and ARG. Citrulline synthesized from glutamine in the small intestine is the main source of circulating arginine, and ASS in the kidney plays a role in the synthesis of arginine by utilization of the circulating citrulline (Funahashi et al., 1981; Windmueller and Spaeth, 1981). ASS is highly expressed in the suckling small intestine to enhance arginine synthesis and declines to a hardly detectable level in the second postnatal week (de Jonge et al., 1998). It may be of significance in the regulation of citrulline-arginine-nitric oxide cycle that ASS is expressed in most tissues where the activity is lower but detectable (Nakagawa et al., 1991; Isayama et al., 1997; Nakata et al., 1997).

The ASS purified from mammalian tissues is a homotetramer with a molecular mass of 180 kDa. The ASS cDNA is cloned for human, rat, bovine, mouse and lower organisms. The human processed ASS mRNA of 1.67 kb encodes a 412 amino acid sequence with a subunit molecular weight of 46 kDa. Analyses of genomic DNA indicate the presence of multiple dispersed processed pseudogenes (at least 14) and a single large expressed ASS gene. The expressed ASS gene is located in human chromosome 9q34, spans 63 kb and comprises 16 exons, but the nucleotide sequences of the introns flanking each exon have not yet been completely identified. It is difficult to identify ASS abnormality by analysis of genomic DNA blots using ASS cDNA as a probe, because hybridization of the probe to pseudogenes makes the band patterns complex.

3.2. Classical citrullinemia (CTLN1)

3.2.1. Clinical features

CTLN1 is found in most patients with neonatal/infantile-onset citrullinemia (Walser, 1983; Saheki et al., 1985a, 1987a, b; McKusick, 1998). In the severe neonatal form, irritability, lethargy, poor feeding and tachypnea appear within the first few days of life and usually proceed to rigidity, sometimes with opisthotonus, apnea, convulsion, coma and death (Walser, 1983). Many patients die in the neonatal period, but some cases survive for more than 10 years with treatment. Some children with mild deficiencies have been reported showing episodic hyperammonemia and physical and mental retardation. A few cases are also found in adult-onset CTLN1 patients (oldest onset age 62 years) who suffered from disturbance of consciousness, including mental confusion and restlessness. Some of the patients suffered from hyperammonemia during pregnancy or at the begining of breast-feeding following delivery when they were 24–26 years old (Saheki et al., 1980, 1981, 1982, 1983a, 1985b; Kobayashi et al., 1994, 1995a).

3.2.2. Pathogenesis, biochemicals, and molecular genetics

As summarized in Figure 1, the pathogenesis of most CTLN1 cases has been clarified at the molecular level mainly by Kobayashi and co-workers (Kobayashi et al., 1987, 1990, 1991, 1994, 1995a; Kakinoki, 1997). By sequence analysis of ASS cDNA and/or genomic DNA from approximately 60 CTLN1 patients, we have to date identified 36 mutations in the ASS gene (35 mutations in ASS mRNA): 25 missense mutations, 2 nonsense mutations, 5 deletion mutations, 1 insertion mutation and 3 point mutations leading to abnormal splicing.

CTLN1 is caused by a mutation of the ASS gene which results in a mutant protein (type I) and a very low amount of the enzyme (type III). Type I citrullinemia with missense mutations in at least one allele was found in neonates/infants and also in adults; in contrast, type III patients were homozygotes or compound heterozygotes with deletion mutations of some exons within ASS mRNA derived from abnormal splicing, but no missense mutations: they were all neonates/infants (Kobayashi et al., 1987, 1994, 1995a).

Type I is qualitative: ASS displays abnormal kinetic properties, such as higher Km values for the substrates, and abnormal cooperative properties (Saheki et al., 1980, 1981, 1982, 1983a, 1985a, b, 1987b). Kinetically abnormal enzymes have been found in the liver, kidney and cultured skin fibroblasts of type I patients. The specific activity of the hepatic enzyme in type I patients was calculated to be lower than that of controls from the data on the amount of ASS protein determined by enzyme immunoassay (Imamura et al., 1987; Saheki et al., 1987b; Kobayashi et al., 1995a). In type III citrullinemia almost no ASS activity was detected in the liver, kidney or cultured skin fibroblasts even when much higher concentrations of the substrates were used for the assay (Saheki et al., 1985a, 1987b; Kobayashi et al., 1995a). A very small amount of the CRIM (cross-reactive immune materials) having the same size as normal ASS protein, however, could be detected in the liver of type III patients by means of a sensitive enzyme immunoassay (Kobayashi et al., 1987, 1995a; Imamura et al., 1987), suggesting that ASS mRNA of normal size is produced by the normal splicing process in the tissues of patients with abnormal splicing mutation (Kobayashi et al., 1994, 1995a).

3.2.3. Diagnosis

In general, CTLN1 is characterized by laboratory findings such as a marked elevation of serum/plasma citrulline (2,500 $\pm$ 1,040 nmol/ml, control range: 20–40), a decrease of plasma arginine (58 $\pm$ 31 nmol/ml, control: 80–130) and hyperammonemia. CTLN1 is diagnosed by the findings of 40 to 200 times normal plasma citrulline, a decrease of plasma arginine, and a decreased ASS activity in all tissues and cells. At present, we are able to clinically detect 32 mutations in the ASS gene of CTLN1 patients using genomic DNA by PCR/Southern (a combination of PCR and restriction enzyme digestion) and are performing prenatal diagnosis and carrier detection.

3.3. Adult-onset type II citrullinemia (CTLN2)

3.3.1. Clinical features

The most characteristic feature of CTLN2 is the late onset of serious and recurring symptoms, varying from 11 to 72 years old (34.4 ± 12.8, $n = 102$) (Kobayashi et al., 1997). We represent the age of onset as the age when a significant clinical abnormality first appeared (Inose, 1952; Miyakoshi et al., 1968; Tsujii et al., 1976; Saheki et al., 1980, 1981, 1982, 1983a, b, 1985b, 1986, 1987c; Yajima et al., 1982; Walser, 1983; Sase et al., 1985; Kobayashi et al., 1986, 1993, 1995b, 1997, 1999; Imamura et al., 1987; Yagi et al., 1988; Todo et al., 1992; Yazaki et al., 1996; Kawamoto et al., 1997), although it is somewhat arbitrary since several CTLN2 patients have some sort of difficulty during infancy or early childhood. Most CTLN2 individuals suffered from a sudden disturbance of consciousness associated with flapping tremor, disorientation, restlessness, drowsiness and coma, and the majority died mainly of cerebral edema within a few years of onset.

Many patients have been found at clinics for psychiatry and neurology because they show mental and neurological symptoms such as nocturnal delirium, behavioral aberrations, convulsive seizures, delusions, and/or loss of memory. In infancy, most of them show no symptoms. Delayed mental or physical development has occurred in some. The progress of the psychoneurotic symptoms can generally be divided into three periods (Inose, 1952). In the first period, there are recurring disturbances of consciousness lasting from one night to several days. In the second period, the episodic disturbances become more frequent and bizarre behavior begins to appear, including manic episodes, echolalia and frank psychosis. At this stage motor weakness or paralysis of the limbs is often seen, as well as dysarthria. Patients become difficult to arouse. Muscle spasms, myoclonus and hyperreflexia as well as pathologic reflexes occur. In acute cases, systemic convulsive seizures already appear in this period. In the third period, a highly wasted condition occurs and the brain shows neurological changes of the pseudoulegyric or ischemic type.

3.3.2. Other findings

Many CTLN2 patients show an unbalanced diet. In half of the patients, a peculiar fondness for beans, peanuts, egg, fish and/or meat has been noted from early childhood, usually associated with a dislike for rice, vegetables, fruits and sweets. None have been alcoholics; in fact, several have shown intolerance for alcohol. Many patients are thin (low body weight), several have emaciation without other physical abnormalities, and some show endocrine abnormality in the form of a slight pituitary dwarfism. The ratio of male to female was 77 to 26 in 103 patients examined in our laboratory (Kobayashi et al., 1997). It has also been found that a few CTLN2 patients are associated with pancreatitis (10% of the patients examined in our laboratory), hepatoma (8%) or hyperlipidemia (5%).

CTLN2 patients show almost no or, if any, mild liver disfunctions. Portal-systemic shunts have been excluded by angiography in all of the CTLN2 patients. Pathological findings in the liver include fatty infiltration and mild fibrosis. Patients present abnormal behaviour with neurological symptoms that closely resemble those of hepatic encephalopathy, but brain CT is normal. Electroencephalograms (EEG) show diffuse slow waves with the occasional appearance of triphasic waves. Neuropathologically, characteristic findings at autopsy are brain edema with cerebellar tonsilar herniation, laminoid necrosis with spongy formation in the cerebral cortex, and Alzheimer type II glia.

3.3.3. Biochemical investigations

The most prominent characteristics of CTLN2 are that a quantitative decrease of ASS protein ($15.4 \pm 12.6\%$ of control, $n = 96$) (Kobayashi et al., 1997) is observed specifically in the liver but not in the kidney or cultured skin fibroblasts (Saheki et al., 1980, 1981, 1982, 1983a, 1985b), resulting in hyperammonemia, an increase of serum citrulline (521 ± 290 nmol/ml, $n = 96$), a slight increase of serum arginine (232 ± 167 nmol/ml, $n = 91$), and a significant elevation of urinary argininosuccinate excretion (106 ± 40 nmol/mg of creatinine, $n = 9$; control: 21.3 ± 6.4, $n = 11$) (Saheki et al., 1986, 1987c; Kobayashi et al., 1997).

It has been reported that hyperammonemia is not present consistently in citrullinemia and, if present at all, occurs only during the postprandial period. In CTLN2 patients, blood ammonia frequently shows diurnal rhythms, probably due to dietary intake: it is low in the morning and increases in the evening, and the diurnal rhythms disappear during fasting (Yajima et al., 1982). Serum citrulline levels of CTLN2 patients (5–30 times those of the controls) are not as high as those of CTLN1 patients (usually more than 40 times those of controls) and often fluctuate during their clinical courses (Saheki et al., 1986).

One big difference between CTLN2 and CTLN1 is that the serum arginine levels of CTLN2 patients are higher than in controls, whereas CTLN1 patients are arginine-deficient (Saheki et al., 1986, 1987a; Kobayashi et al., 1997). This phenomenon can be explained by the difference in ASS activity in the kidney between CTLN2 and CTLN1, and by a widely accepted theory about the origin of the circulating arginine (Funahashi et al., 1981; Windmueller and Spaeth, 1981). According to the theory, circulating arginine is derived from arginine synthesized by the action of ASS and ASL in the kidney from citrulline, which is synthesized from glutamate via ornithine in the intestine under normal conditions. Under citrullinemic conditions, citrulline is released in large quantities from the liver owing to the urea cycle defect. Citrulline accumulated in the circulation cannot be converted to arginine in the case of CTLN1 patients because they have a defect of ASS in the kidney, but citrulline is converted efficiently to circulating arginine in the kidney of CTLN2 patients because they have normal renal ASS activity. This is supported by the finding that the concentration of argininosuccinate, the product of the defective enzyme (ASS), was found to be elevated in the urine of CTLN2 patients (Saheki et al., 1987c).

It is also a prominent characteristic of CTLN2 that the serum levels of alanine, serine, glycine and branched-chain amino acids are significantly lower than in the controls (Saheki et al., 1986), in spite of the fact that serum alanine levels are usually higher in hyperammonemia. As a result, the ratio of threonine to serine is characteristically higher (2.43 ± 1.12, $n = 83$) (Saheki et al., 1986; Kobayashi et al., 1997), and the ratio of branched-chain amino acids to aromatic amino acids lower than the controls (Saheki et al., 1986). The mechanisms underlying such changes in serum amino acids except citrulline and arginine remain unclear.

3.3.4. Abnormal localization of ASS in the lobulus

In the controls, ASS was located only in hepatocytes, not in bile duct cells or endothelial cells, and was distributed almost homogeneously in the lobulus with slightly denser staining in the periportal region than in the pericentral region where glutamine synthetase is located (Saheki et al., 1983b, 1987a; Yagi et al., 1988). This distinct distribution of the urea cycle enzymes and glutamine synthetase may play a role in the nitrogen metabolism of the liver (Häusinger, 1990).

CTLN2 is characterized by a liver-specific deficiency of ASS protein. One question was whether the decreased ASS content in the liver of CTLN2 patients represented an even decrease in all hepatocytes or an uneven decrease. Immunohistochemical analysis with the antisera to ASS revealed that the reduction in ASS protein does not invariably occur homogeneously in the liver of CTLN2 patients. There were two types of ASS protein distribution (Saheki et al., 1983b, 1987a; Yagi et al., 1988); one homogeneous and the other clustered. The clustered type of ASS distribution was seen in the liver of CTLN2, but not in CTLN1, hepatitis, liver cirrhosis or hepatoma. Furthermore, the clustered enzyme distribution was observed only with ASS, not with arginase or aldolase B in CTLN2 livers. The ASS-specific heterogeneous or clustered distribution was observed in 14 out of 25 CTLN2 patients examined (Yagi et al., 1988). There were no differences between the homogeneous and clustered types in age, sex, hepatic ASS activity or serum citrulline level. Nevertheless, the clinical prognosis was quite different: the fatality of the CTLN2 patients who belong to the clustered heterogeneous type was much higher than that of the homogeneous type (Yagi et al., 1988). The cause of the uneven distribution of ASS in the liver of CTLN2 patients and its relation to the pathogenesis of CTLN2 remain unresolved.

3.3.5. Increased expression of hepatic PSTI

Differential mRNA display analysis, which was introduced in order to understand the detailed pathophysiology and pathogenesis, revealed that the pancreatic secretory trypsin inhibitor (PSTI) gene was highly expressed in the liver of CTLN2 patients (Kobayashi et al., 1995b). We also found that the concentration of PSTI protein was higher in the liver of CTLN2 patients (0.368 ± 0.954 microg/g liver, $n = 52$) than controls (0.002 ± 0.001, $n = 28$); and there was a significant increase in serum PSTI level (73.2 ± 66.2 ng/ml, $n = 29$, control: 11.5 ± 2.5, $n = 28$) with no change in the other serum markers of pancreatitis, cancer or acute-phase response, such as elastase 1, trypsin, phospholipase A2, α-fetoprotein, CA19-9 or C-reactive protein (Kobayashi et al., 1997). In addition to the serum amino acid pattern, increase of serum PSTI level is useful as a diagnostic marker for CTLN2. However, the mechanisms underlying the increase in hepatic PSTI mRNA remain unresolved.

3.3.6. Liver transplantation

Most CTLN2 patients present mental and neurological symptoms such as the sudden appearance of behavioral aberrations, tremor, disorientation, restlessness and coma. The biochemical findings, which include hyperammonemia, hypercitrullinemia, a mild elevation in serum arginine and an increase in urinary argininosuccinate excretion, result from a specific decrease in hepatic ASS level. However, there have been no effective pharmacological treatments; and the disease management is difficult because the clinical course of the patients is unpredictable and many patients undergo sudden deterioration with progressive cerebral edema. Liver transplantation remains the only therapeutic option for the majority of CTLN2 patients, who would otherwise face possible death from brain edema (Todo et al., 1992; Yazaki et al., 1996; Kawamoto et al., 1997). Since the first liver transplantation was performed in 1988 at the University of Pittsburgh (Todo et al., 1992), 15 CTLN2 patients have been treated with liver transplantation. All the metabolic abnormalities (ammonia, citrulline, arginine and PSTI levels in the blood) were corrected and the neurological symptoms disappeared (Todo et al., 1992; Yazaki et al., 1996; Kobayashi et al., 1993, 1997; Kawamoto et al., 1997), supporting the hypothesis that the CTLN2 is a liver-specific disorder. The first patient treated with the liver transplantation is still alive, in employment, and has no symptoms (Kumashiro, personal communication).

3.4. Molecular genetics of CTLN2

3.4.1. Identification of the disease gene

As summarized in Figure 1, CTLN2 occurs in association with decreased ASS activity and protein in the liver, however, there are no apparent abnormalities in the hepatic ASS mRNA of CTLN2 patients (Kobayashi et al., 1986, 1993; Sase et al., 1985). RFLP analysis of 16 affected CTLN2 patients from consanguineous marriages suggested that the abnormality is not within the ASS gene locus (Kobayashi et al., 1993). We have found that 20% of patients are apparently from consanguineous parents, and siblings have been affected, indicating that CTLN2 is probably an autosomal recessive disorder.

Since 1977, we have analyzed over 130 CTLN2 patients; the availability of DNA samples in CTLN2 patients from 18 consanguineous families greatly facilitated our genetic analyses, allowing us to use homozygosity mapping to delimit the critical region for the disease (Kobayashi et al., 1999). The SLC25A13 gene causing CTLN2 was identified in chromosome 7q21.3 by positional cloning and found to encode a putative calcium-binding mitochondrial carrier protein (Kobayashi et al., 1999). The human SLC25A13 gene spanned 160 kb of genomic DNA organized into 18 exons (Figure 2) (Kobayashi et al., 1999; Sinasac et al., 1999). SLC25A13 is predicted to encode a 74 kDa protein comprising 675 amino acid residues, which we have named citrin (Kobayashi et al., 1999). In an autosomal recessive disease, homozygosity mapping is a powerful method for identifying the disease gene, since only a small number of patients from consanguineous marriages are required for the analysis.

3.4.2. Mutations in SLC25A13 gene of CTLN2

As shown in Figure 3, we identified five distinct mutations in SLC25A13 gene of CTLN2 patients and confirmed their causative role in the disease (Kobayashi et al., 1999). The mutations include: (i) 851del4 – a 4 bp deletion in exon 9, predicting a frameshift and introduction of a stop codon at position 286, leading to premature truncation of the protein; (ii) IVS11 + 1G > A – a substitution at the 5′-end of intron 11 resulting in abnormal splicing and deletion of exon 11 in mRNA. This causes a loss of 53 amino acids (codons 340–392) within the first hydrophilic loop between the TM1 and TM2 domains of citrin; (iii) 1638ins23 – a 23 bp insertion in exon 16 that results in a frameshift at codon 554 and the addition of 16 new amino acids. A stop codon is introduced at position 570, leading to premature truncation of the C-terminus of citrin; (iv) S225X – a substitution at position 674 in exon 7 that changes serine to a stop codon at position 225 and predicts premature truncation of the protein; and (v) IVS13 + 1G > A – a substitution at the 5′-end of intron 13 resulting in abnormal splicing and the deletion of 27 amino acids (codons 411–437) between TM2 and TM3.

These mutant citrins could not locate into the mitochondria membrane after losing their C-terminal half and extension or after having their mitochondrial transmembrane spanning structures destroyed (Figure 3). In a preliminary experiment by Western blot analysis with anti-human citrin antibody, we detected no CRIM band in the liver of CTLN2 patients. This, together with the biochemical data showing decreased ASS protein in CTLN2 patients, suggests that citrin not only functions as a carrier protein but also plays a role in stabilizing ASS protein (Figure 4).

3.4.3. DNA diagnosis of CTLN2

Before identifying the SLC25A13 gene, we had diagnosed patients suffering from type II citrullinemia under criteria described previously (Saheki et al., 1980, 1981, 1982, 1983a, 1985a, b, 1986, 1987a, b, c; Kobayashi et al., 1997) (including their symptoms, hyperammonemia, increased serum citrulline, arginine, ratio of threonine to serine and PSTI levels, and decreased hepatic ASS protein levels). Now, we have found 5 mutations in the SLC25A13 gene of CTLN2 patients and have established the diagnosis for each mutation using genomic DNA (Kobayashi et al., 1999).

Using diagnostic methods for the 5 mutations of SLC25A13 gene, we tested about 100 CTLN2 patients who had been diagnosed with biochemical and enzymatic experiments. We found that 90% of the patients had one or more of the 5 mutations and 85% were diagnosed as compound heterozygotes or homozygotes (Yasuda et al., manuscript in preparation). The results show without doubt that the SLC25A13 gene is the cause of adult-onset type II citrullinemia, and this is clearly in agreement with the classification of CTLN2 as diagnosed under our previous criteria. Through these findings, we are now able to say conclusively that CTLN2 is a genetic disease.

Surprisingly, 5 out of 21 patients from consanguineous union were found to be compound heterozygotes (Kobayashi et al., 1999). This suggests a very high incidence of the mutant genes among the Japanese population. On the basis of the proportion of consanguinity (about 20%), the incidence of CTLN2 has been calculated to be approximately 1 in 100,000 (Kobayashi et al., 1993). In recent preliminary searching of the general Japanese population, we found that the frequency of heterozygotes is 1–2 in 100 (Yamaguchi et al., manuscript in preparation). From the rates of carrier detected, the frequency of homozygotes with abnormal SLC25A13 genes is calculated to be 1–2 in 20,000.

4. CITRIN, A NEW CALCIUM BINDING PROTEIN

4.1. Comparison of citrin with aralar

4.1.1. Protein structure

The predicted SLC25A13 protein, designated citrin, consists of a polypeptide of 675 amino acids (74 kDa) and a bipartite structure: a C-terminal half with the characteristic features of the MCF members and an N-terminal extension harboring four EF-hand domains (Figures 2 and 3) (Kobayashi et al., 1999); it is closely related to a calcium-binding mitochondrial solute carrier protein, aralar (del Arco and Satrústegui, 1998). The amino acid sequence of citrin is most similar to the human aralar (678 aa, 74 kDa) with 77.8% identity and to the *C. elegans* protein (Q21153) with 53.7% identity (Kobayashi et al., 1999). The alignment of citrin with aralar revealed a high degree of sequence conservation between both proteins with the exception of the N-terminal half and C-terminal end: in the N-terminal half, the residue 1-325 of citrin is 69.9% identical to 1–323 of aralar; in the C-terminal half, 326–625 of citrin is 86.7% identical to 324–623 of aralar; in the C-terminal end, 626–675 of citrin is 61.8% identical to 624–678 of aralar.

4.1.2. Tissue distribution

The primary structure of citrin (SLC25A13, chromosome 7q21.3) (Kobayashi et al., 1999; Sinasac et al., 1999), is most similar to that of aralar (SLC25A12, chromosome 2q24) (Crackower et al., 1999), however, the expression pattern of SLC25A13 mRNA encoding citrin largely differed from that of SLC25A12 mRNA encoding aralar in tissue distribution and also in mRNA size (Figure 2) (Kobayashi et al., 1999). The 981 bp fragment derived from the 5′-region of the citrin cDNA was used as a specific probe for Northern blot analysis, of which the nucleotide sequence was 67% identical to the 975 bp sequence of aralar cDNA. Northern blot analysis using a blot of poly(A)$^{+}$RNA from human tissues revealed that SLC25A13 mRNA as a transcript of approximately 3.4 kb was expressed predominantly in the liver, moderately in the kidney, the heart, the pancreas and the placenta, slightly in the brain and skeletal muscle, but not in the lung. The same result was obtained with the 934 bp fragment from the 3′-noncoding region of SLC25A13 cDNA as probe, and the nucleotide sequence was 49.8% identical to the 900 bp of the aralar noncoding region. In contrast, SLC25A12 mRNA expression was not observed in the liver but was more prevalent in the heart, brain and skeletal muscle having two major bands of 2.8 and 3.3 kb in size. A less intense but apparently specific hybridization band of 6.6 kb in size was also observed in Northern blots for aralar.

4.2. Mitochondrial localization

The C-terminal half of citrin as well as of aralar has also a substantial similarity with MCF proteins (20–30% identity). All MCF members appear to consist of a tripartite structure with each of the three repeated segments being about 100 residues in length. Each repeat contains two TM spanners. The typical mitochondrial carrier signature, $GX_3GX_8PX(D/E)X(I/L/V)(K/R)X(R/K)XQX_{20-30}GX_4(Y/W)(R/K)GX_9P$, was included in the C-terminal half of citrin as well as in aralar (Kobayashi et al., 1999; del Arco and Satrústegui, 1998; del Arco et al., 2000). Since the hydrophobic profile of the mitochondrial carriers shows six potential membrane-spanning helices, with both N- and C-termini of the protein and the loops between TM regions 2–3 and 4–5 possibly facing the cytosol, we were able to predict the structure of citrin localized in the mitochondria (Figures 3 and 4) on the basis of the sequence homology and the hydrophobic profile.

It is well known that most MCF members are located in the inner membrane of the mitochondria (Walker, 1992). We detected citrin in the mitochondria by using the expression of GFP-fusion protein in cultured cells and in the mitochondrial inner membrane obtained from subfractionation of mouse and rat liver by Western blot analysis using anti-citrin antibody (Iijima et al., manuscript in preparation). Our results agree with the demonstration by del Arco et al. (2000) using expression of Flag-tagged aralar2/citrin in HEK-293T cells. The mitochondrial localization of aralar was also clarified by a transfection experiment using COS cells and detected in the brain by anti-aralar anibody (del Arco and Satrústegui, 1998).

4.3. Calcium binding property

There is a striking difference in the N- and C-terminal extension (Figures 2 and 3), especially in the extremely long extension in the N-terminal half, between the citrin and aralar proteins and the other MCF proteins (28–38 kDa). Aralar was originally cloned as a member of a subfamily of mitochondrial carrier that binds calcium (del Arco and Satrústegui, 1998). Members of the calcium-binding mitochondrial solute carrier subfamily have a bipartite structure: the N-terminal half harbours four EF-hand domains whereas the C-terminal half has the characterictic features of MCF members. The presence of calcium binding domains in citrin, aralar and other calcium-binding mitochondrial carrier members (Weber et al., 1997) indicates that these proteins may be involved in calcium-regulated metabolite transporters.

On the other hand, searching the protein databases with the amino acid sequence of citrin revealed relationships with different Ca^{2+}-binding proteins (Kawasaki and Kretsinger, 1994), such as troponin C (P21798 and P02590), calcineurin B subunit (P42322), neuron specific calcium-binding protein (P41211) and vitamin D-dependent calcium-binding protein (P29377). The sequence of residues 68–78 in citrin was found in EF-hand Ca^{2+}-binding domain protein (BL00018) and the sequences 58–90 and 92–124 were in the S-100/ICaBP type Ca^{2+}-binding protein (BL00303B). The N-terminal half of citrin contains four possible EF-hand domains (residues 28–39, 66–77, 100–111 and 171–182), which are comparable to those in aralar and conserved in the other calcium-binding proteins.

To examine the possibility that the N-terminal half of citrin could bind calcium, we expressed recombinant His-tagged proteins using the pET system and tested their ability to bind $^{45}Ca^{2+}$ in an overlay assay. As shown in Figure 5, one major protein band of around 35–38 kDa corresponding to the predicted size of N-half citrin (35 kDa, 1–284) and aralar (38 kDa, 1–312), appears in lysate of *E. coli* and purified samples, but not in C-half citrin (44 kDa, 285–675). Similar results have been demonstrated by del Arco et al. (2000) using His-tagged N-terminal regions of aralar2/citrin (9–278). Furthermore, del Arco et al. (2000) reported that by eliminating the first EF-hand domain of aralar2/citrin, the protein (37–278) loses its calcium binding capacity. These results indicate that the N-terminal half of citrin is able to bind calcium and that the first EF-hand domain is required. This supports the notion that neither the third nor fourth EF-hand binds calcium and that calcium binding by EF-hands 1 and 2 is prevented by eliminating one of these EF-hands, as observed in other calcium binding proteins (Vito et al.,

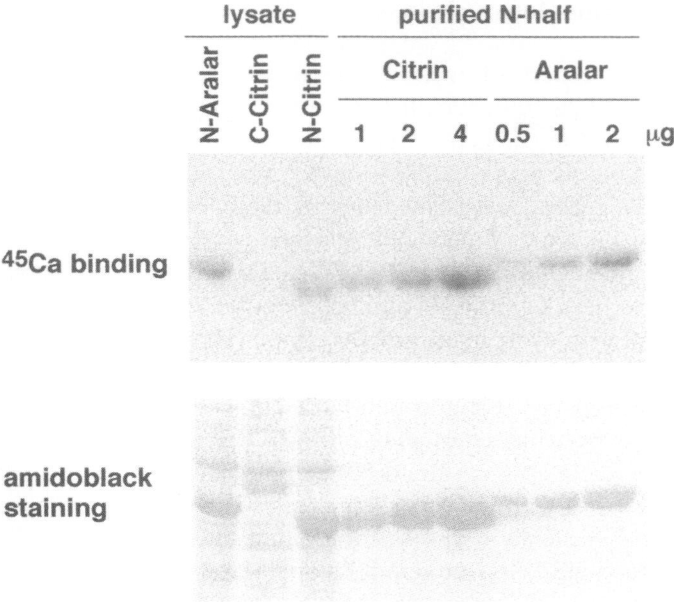

Figure 5. Calcium binding activity by ^{45}Ca overlay assay. The His-tagged proteins; N-half citrin (1–284, 35 kDa), N-half aralar (1–312, 38 kDa) and C-half of citrin (285–675, 44 kDa) were expressed in *E. coli*, and purified. Aliquots of lysates or purified samples were separated by SDS-polyacrylamide gel electrophoresis, transferred to nitrocellulose membranes, and assayed for ^{45}Ca-binding. After exposure of the membrane, the transferred proteins were detected by amidoblack staining.

1996). These results suggest that citrin shares with aralar the function of a calcium-binding mitochondrial solute carrier.

5. PERSPECTIVES

5.1. Are CTLN2 patients found only among Japanese?

CTLN2 appears to be almost exclusively found in Japan – with a frequency of more than 1 in 100,000 (Kobayashi et al., 1993); but possible cases from Europe (Vidailhet et al., 1971; Roerdink et al., 1973) and China (Chow et al., 1996) have also been reported. After publication of our study on the SLC25A13 gene (Kobayashi et al., 1999), we were able to analyze two Chinese patients with citrullinemia and found that the patients were diagnosed as CTLN2 with the same mutation as identified in Japanese CTLN2 patients (Hwu et al., submitted). So it seems that CTLN2 is not restricted to Japan but exists at least in Asia. This suggests the possibility of a common ancestor among these CTLN2 patients. It is now possible to detect CTLN2 patients in other countries without difficulty, because we have sequenced the SLC25A13 gene (Kobayashi et al., 1999; Sinasac et al., 1999).

5.2. Possible function of citrin

An interesting aspect of urea cycle function is the fact that enzymatic components in both mitochondrial matrix and cytosol appear to cluster in association with the mitochondrial membrane (Cheung et al., 1989). This has been proposed as an explanation for the apparent preferential use of endogenously generated intermediates by the enzymes of the urea cycle, a phenomenon known as "metabolic channeling". Citrin may have a central role in these interactions (Figure 4). ASS and the two subsequent activities of the urea cycle may be arranged around the mitochondria, contributing to the channeling of urea cycle intermediates. On the other hand, Demarquoy et al. (1994) reported that ASS is associated with the mitochondrial membrane, and the activity was mainly in mitochondrial fraction of fetal and newborn rat liver and in cytoplasmic fraction of adult rat liver. It is possible that citrin forms a complex in some way with the three "soluble" cytoplasmic enzymes (ASS, ASL and ARG) of the urea cycle, especially ASS within the hepatocytes. If so, the loss of organization in the channeling by the abnormal citrin would lead to a reduction of ASS protein possibly through its destabilization and/or degradation.

5.3. Role of Ca^{2+} in regulating citrin transporter

The supposition that SLC25A13 is indeed a calcium-binding solute carrier is based primarily on its sequence characteristics (Kobayashi et al., 1999). Citrin contains a high degree of amino acid identity with membrane-bound proteins of the mitochondrial solute carrier family and the same four EF-hand domains closest conserved in other calcium-binding proteins. Neither of these proteins have an obvious N-terminal mitochondrial import sequence. As observed with aralar (del Arco and Satrústegui, 1998) and efinal (Weber et al., 1997), citrin may also act as a calcium-binding/transducing protein whereby its N-terminal half confers calcium sensitivity to the activity of its C-terminal half or to other proteins with which it may interact. Citrin and aralar, although structurally similar, have markedly different expression patterns. SLC25A13 is mainly expressed in the liver, kidney and pancreas, while SLC25A12 mRNA appears to be primarily expressed in the heart, brain and skeletal muscle (Kobayashi et al., 1999). Even if citrin and aralar serve to transport the same mitochondrial solutes, the specificity of CTLN2 pathology to the liver would still be explainable by the different tissue expression profiles of the genes.

The enzymes involved in citrulline-arginine-nitric oxide metabolism, ASS, ASL and nitric oxide synthase are expressed in rat pancreas islets and neurons (Nakagawa et al., 1991; Isayama et al., 1997; Nakata et al., 1997). It is interesting that in the presence of stimulatory glucose, citrulline and argininosuccinate at physiological concentrations increase cytosolic

Ca^{2+} concentration in rat β-cells (Nakata et al., 1997). On the other hand, mitochondrial Ca^{2+} transport is mediated by a complex system comprising at least three separate mechanisms (Gunter et al., 1994). Ca^{2+} influx is mediated primarily via a very fast Ca^{2+} uniporter which is energetically downhill, and efflux is mediated via both Na^+-independent and Na^+-dependent efflux mechanisms which are energetically uphill. Both Na^+-independent and Na^+-dependent transport takes place in mitochondria from a wide variety of tissues. However, Na^+-independent transport dominates in the liver and kidneys, whereas Na^+-dependent transport dominates in the heart, skeletal muscle and brain (Gunter et al., 1994). It is noteworthy that two independent Ca^{2+} efflux mechanisms exist in the mitochondrial inner membrane and the different tissue distribution between these two transport systems is similar to the difference of expression between citrin and alarar.

5.4. Unresolved phenomena

Although SLC25A13 gene has now been identified as the disease gene of CTLN2 (Kobayashi et al., 1999), many questions remain unresolved: (1) What is the function of citrin? (2) How do mutant citrin result in CTLN2? (3) Is there any effect of environmental factors, dietary and/or hormonal, on the formation of pathophysiology in CTLN2? What are those factors and mechanisms? (4) Why does non-liver-specific citrin cause liver-specific ASS deficiency? (5) What is the mechanism which increases the expression of PSTI gene? Furthermore, the cause of the clustered distribution of ASS protein in the liver of CTLN2 patients and the reason why CTLN2 is late-onset disease remain unresolved.

The identification of the disease gene for adult-onset type II citrullinemia provides us with a powerful molecular tool to clarify the mechanisms of various phenomena found in CTLN2 patients. The identification of the mutations in the SLC25A13 gene of CTLN2 patients now allows us to establish simple and cost-effective carrier detection and diagnostic screening tests.

ACKNOWLEDGEMENTS

We thank M. Gore for editorial assistance and S. Nagata and A. Sumiyoshi for secretarial assistance. This work was supported in part by Grants-in-Aid for Scientific Research from the Ministry of Education, Science and Culture, the Kodama Foundation for Research in Medical Science, and Health Sciences Research Grants in Research on Human Genome and Gene Therapy from the Ministry of Health and Welfare in Japan.

REFERENCES

Bradford, N.M. and McGivan, J.D., 1980, Evidence for the existence of an ornithine/citrulline antiporter in rat liver mitochondria, *FEBS Lett.* 113, 294–298.

Camacho, J.A., Obie, C., Biery, B., Goodman, B.K., Hu, C.-A., Almashanu, S., Steel, G., Casey, R., Lambert, M., Mitchell, G.A. and Valle, D., 1999, Hyperornithinaemia-hyperammonaemia-homocitrullinuria syndrome is caused by mutations in a gene encoding a mitochondrial ornithine transporter, *Nature Genet.* 22, 151–158.

Cheung, C.-W., Cohen, N.S. and Raijman, L., 1989, Channeling of urea cycle intermediates in situ in permeabilized hepatocytes, *J. Biol. Chem.* 264, 4038–4044.

Chow, W.C., Ng, H.S., Tan, I.K. and Thum, T.Y., 1996, Case report: Recurrent hyperammonaemic encephalopathy due to citrullinaemia in a 52 year old man, *J. Gastroenterol. Hepatol.* 11, 621–625.

Crackower, M.A., Sinasac, D.S., Lee, J.R., Herbrick, J.-A., Tsui, L.-C. and Scherer, S.W., 1999, Assignment of the SLC25A12 gene coding for the human calcium-binding mitochondrial solute carrier protein aralar to human chromosome 2q24, *Cytogenet. Cell Genet.* 87, 197–198.

de Jonge, W.J., Dingemanse, M.A., de Boer, P.A.J., Lamers, W.H. and Moorman, A.F.M., 1998, Arginine-metabolizing enzymes in the developing rat small intestine, *Pediatr Res.* 43, 442–451.

del Arco, A. and Satrústegui, J., 1998, Molecular cloning of aralar, a new member of the mitochondrial carrier superfamily that binds calcium and is present in human muscle and brain, *J. Biol. Chem.* 273, 23327–23334.

del Arco, A., Agudo, M. and Satrústegui, J., 2000, Characterization of a second member of the subfamily of calcium-binding mitochondrial carriers expressed in human non-excitable tissues, *Biochem J.* 345, 725–732.

Demarquoy, J., Fairand, A., Gautier, C. and Vaillant, R., 1994, Demonstration of argininosuccinate synthetase activity associated with mitochondrial membrane: Characterization and hormonal regulation, *Mol. Cell Biochem.* 136, 145–155.

Funahashi, M., Kato, H., Shiosaka, S. and Nakagawa, H., 1981, Formation of arginine and guanidinoacetic acid in the kidney in vivo: Their relations with the liver and their regulation, *J. Biochem.* 89, 1347–1356.

Gamble, J.G. and Lehninger, A.L., 1973, Transport of ornithine and citrulline across the mitochondrial membrane, *J. Biol. Chem.* 248, 610–618.

Gunter, T.E., Gunter, K.K., Sheu, S.-S. and Gavin, C.E., 1994, Mitochondrial calcium transport: Physiological and pathological relevance, *Am. J. Physiol.* 267, C313–C339.

Häusinger, D., 1990, Nitrogen metabolism in liver: Structural and functional organization and physiological relevance, *Biochem. J.* 267, 281–290.

Imamura, Y., Kobayashi, K., Yamashita, T., Saheki, T., Ichiki, H., Hashida, S. and Ishikawa, E., 1987, Clinical application of enzyme immunoassay in the analysis of citrullinemia, *Clin. Chim. Acta* 164, 201–208.

Indiveri, C., Tonazzi, A., Stipani, I. and Palmieri, F., 1997, The purified and reconstituted ornithine/citrulline carrier from rat liver mitochondria: Electrical nature and coupling of the exchange reaction with H^+ translocation, *Biochem. J.* 327, 349–356.

Inose, T., 1952, Hepatocerebral degeneration, a special type, *J. Neuropath. Exp. Neurol.* 11, 401–408.

Inoue, I., Saheki, T., Kayanuma, K., Uono, M., Nakajima, M., Takeshita, K., Koike, R., Yuasa, T., Miyatake, T. and Sakoda, K., 1988, Biochemical analysis of decreased ornithine transport activity in the liver mitochondria from patients with hyperornithinemia, hyperammonemia and homocitrullinuria, *Biochim. Biophys. Acta* 964, 90–95.

Isayama, H., Nakamura, H., Kanemaru, H., Kobayashi, K., Emson, P.C., Kawabuchi, M. and Tashiro, N., 1997, Distribution and co-localization of nitric oxide synthase and argininosuccinate synthetase in the cat hypothalamus, *Arch. Histol. Cytol.* 60, 477–492.

Kakinoki, H., Kobayashi, K., Terazono, H., Nagata, Y. and Saheki, T., 1997, Mutations and DNA diagnoses of classical citrullinemia, *Hum. Mutat.* 9, 250–259.

Kawamoto, S., Strong, R.W., Kerlin, P., Lynch, S.V., Steadman, C., Kobayashi, K., Nakagawa, S., Matsunami, H., Akatsu, T. and Saheki, T., 1997, Orthotopic liver transplantation for adult-onset type II citrullinemia, *Clin. Transplantation* 11, 453–458.

Kawasaki, H. and Kretsinger, R.H., 1994, Calcium-binding proteins, *Protein Profile* 1, 343–517.

Kobayashi, K., Saheki, T., Imamura, Y., Noda, T., Inoue, I., Matuo, S., Hagihara, S., Nomiyama, H., Jinno, Y. and Shimada, K., 1986, Messenger RNA coding for argininosuccinate synthetase in citrullinemia, *Am. J. Hum. Genet.* 38, 667–680.

Kobayashi, K., Ichiki, H., Saheki, T., Tatsuno, M., Uchiyama, C., Nukada, O. and Yoda, T., 1987, Structure of an abnormal messenger RNA for argininosuccinate synthetase in citrullinemia, *Hum. Genet.* 76, 27–32.

Kobayashi, K., Jackson, M.J., Tick, D.B., O'Brien, W.E. and Beaudet, A.L., 1990, Heterogeneity of mutations in argininosuccinate synthetase causing human citrullinemia, *J. Biol. Chem.* 265, 11361–11367.

Kobayashi, K., Rosenbloom, C., Beaudet, A.L. and O'Brien, W.E., 1991, Additional mutations in argininosuccinate synthetase causing citrullinemia, *Mol. Biol. Med.* 8, 95–100.

Kobayashi, K., Shaheen, N., Kumashiro, R., Tanikawa, K., O'Brien, W.E., Beaudet, A.L. and Saheki, T., 1993, A search for the primary abnormality in adult-onset type II citrullinemia, *Am. J. Hum. Genet.* 53, 1024–1030.

Kobayashi, K., Shaheen, N., Terazono, H. and Saheki, T., 1994, Mutations in argininosuccinate synthetase mRNA of Japanese patients, causing classical citrullinemia, *Am. J. Hum. Genet.* 55, 1103–1112.

Kobayashi, K., Kakinoki, H., Fukushige, T., Shaheen, N., Terazono, H. and Saheki, T., 1995a, Nature and frequency of mutations in the argininosuccinate synthetase gene that cause classical citrullinemia, *Hum. Genet.* 96, 454–463.

Kobayashi, K., Nakata, M., Terazono, H., Shinsato, T. and Saheki, T., 1995b, Pancreatic secretory trypsin inhibitor gene is highly expressed in the liver of adult-onset type II citrullinemia, *FEBS Lett.* 372, 69–73.

Kobayashi, K., Horiuchi, M. and Saheki, T., 1997, Pancreatic secretory trypsin inhibitor as a diagnostic marker for adult-onset type II citrullinemia, *Hepatology* 25, 1160–1165.

Kobayashi, K., Sinasac, D.S., Iijima, M., Boright, A.P., Begum, L., Lee, J.R., Yasuda, T., Ikeda, S., Hirano, R., Terazono, H., Crackower, M.A., Kondo, I., Tsui, L.-C., Scherer, S.W. and Saheki, T., 1999, The gene mutated in adult-onset type II citrullinaemia encodes a putative mitochondrial carrier protein, *Nature Genet.* 22, 159–163.

Li, M.X., Nakajima, T., Fukushige, T., Kobayashi, K., Seiler, N. and Saheki, T., 1999, Aberrations of ammonia metabolism in ornithine carbamoyltransferase-deficienct spf-ash mice and their prevention by treatment with urea cycle intermediate amino acids and an ornithine aminotransferase inactivator, *Biochim. Biophys. Acta* 1455, 1–11.

McKusick, V.A., 1998, Citrullinemia, in *Mendelian Inheritance in Man*, Vol. 3, V.A. McKusick (ed.), Johns Hopkins University Press, Baltimore, pp. 2093–2095.

McMurray, W.C., Mohyuddin, F., Rossiter, R.J., Rathbun, J.C., Valentine, G.H., Koegler, S.J. and Zarfas, D.E., 1962, Citrullinuria: A new aminoaciduria associated with mental retardation, *Lancet* i, 138.

Miyakoshi, T., Takahashi, T., Kato, M., Watanabe, M. and Ito, C., 1968, Abnormal citrulline metabolism of Inose-type hepatocerebral disease, *Shinkeikagaku (Japanese)* 7, 88–91.

Nakagawa, S., Mizuma, M., Ichiki, H. and Saheki, T., 1991, Immunocytochemical demonstration of argininosuccinate synthetase in the neuronal structures of the intestinal tract and pancreas of the Japanese monkey, *Acta Histochem. Cytochem.* 24, 209–213.

Nakata, M., Yada, T., Nakagawa, S., Kobayashi, K. and Maruyama, I., 1997, Citrulline-argininosuccinate-arginine cycle coupled to Ca^{2+}-signaling in rat pancreatic β-cells, *Biochem. Biophys. Res. Commun.* 235, 619–624.

Palmieri, L., de Marco, V., Iacobazzi, V., Palmieri, F., Runswick, M.J. and Walker, J.E., 1997, Identification of the yeast ARG-11 gene as a mitochondrial ornithine carrier involved in arginine biosynthesis, *FEBS Lett.* 410, 447–451.

Roerdink, F.H., Gouw, W.L.M., Okken, A., Van der Blij, J.F., Haan, G.L., Hommes, F.A. and Huisjes, H.J., 1973, Citrullinemia, report of a case, with studies on antenatal diagnosis, *Pediatr. Res.* 7, 863–869.

Saheki, T., Tsuda, M., Takada, S., Kusumi, K. and Katsunuma, T., 1980, Role of argininosuccinate synthetase in the regulation of urea synthesis in the rat and argininosuccinate synthetase-associated metabolic disorder in man, *Adv. Enzyme Regulation* 18, 221–238.

Saheki, T., Ueda, A., Hosoya, M., Kusumi, K., Takada, S., Tsuda, M. and Katsunuma, T., 1981, Qualitative and quantitative abnormalities of argininosuccinate synthetase in citrullinemia, *Clin. Chim. Acta* 109, 325–335.

Saheki, T., Ueda, A., Iizima, K., Yamada, N., Kobayashi, K., Takahashi, K. and Katsunuma, T., 1982, Argininosuccinate synthetase activity in cultured skin fibroblasts of citrullinemic patients, *Clin. Chim. Acta* 118, 93–97.

Saheki, T., Ueda, A., Hosoya, M., Sase, M., Nakano, K. and Katsunuma, T., 1983a, Enzymatic analysis of citrullinemia (12 cases) in Japan, *Adv. Exp. Med. Biol.* 153, 63–76.

Saheki, T., Yagi, Y., Sase, M., Nakano, K. and Sato, E., 1983b, Immunohistochemical localization of argininosuccinate synthetase in the liver of control and citrullinemic patients, *Biomed. Res.* 4, 235–238.

Saheki, T., Nakano, K., Kobayashi, K., Imamura, Y., Itakura, Y., Sase, M., Hagihara, S. and Matuo, S., 1985a, Analysis of the enzyme abnormality in eight cases of neonatal and infantile citrullinemia in Japan, *J. Inherit. Metab. Dis.* 8, 155–156.

Saheki, T., Sase, M., Nakano, K. and Yagi, Y., 1985b, Arginine metabolism in citrullinemic patients, in *Guanidines*, A. Mori, B.D. Cohen and A. Lowenthal (eds.), Plenum, New York, pp. 149–158.

Saheki, T., Kobayashi, K., Miura, T., Hashimoto, S., Ueno, Y., Yamasaki, T., Araki, H., Nara, H., Shiozaki, Y., Sameshima, Y., Suzuki, M., Yamauchi, Y., Sakazume, Y., Akiyama, K. and Yamamura, Y., 1986, Serum amino acid pattern of type II citrullinemic patients and effect of oral administration of citrulline, *J. Clin. Biochem. Nutr.* 1, 129–142.

Saheki, T., Kobayashi, K. and Inoue, I., 1987a, Hereditary disorders of the urea cycle in man: Biochemical and molecular approaches, *Rev. Physiol. Biochem. Pharmacol.* 108, 21–68.

Saheki, T., Kobayashi, K., Ichiki, H., Matuo, S., Tatsuno, M., Imamura, Y., Inoue, I., Noda, T. and Hagihara, S., 1987b, Molecular basis of enzyme abnormalities in urea cycle disorders: With special reference to citrullinemia and argininosuccinic aciduria, *Enzyme* 38, 227–232.

Saheki, T., Kobayashi, K., Inoue, I., Matuo, S., Hagihara, S. and Noda, T., 1987c, Increased urinary excretion of argininosuccinate in type II citrullinemia, *Clin. Chim. Acta* 170, 297–304.

Sase, M., Kobayashi, K., Imamura, Y., Saheki, T., Nakano, K., Miura, S. and Mori, M., 1985, Level of translatable messenger RNA coding for argininosuccinate synthetase in the liver of the patients with quantitative-type citrullinemia, *Hum. Genet.* 69, 130–134.

Sinasac, D.S., Crackower, M.A., Lee, J.R., Kobayashi, K., Saheki, T., Scherer, S.W. and Tsui, L.-C., 1999, Genomic structure of the adult-onset type II citrullinemia gene, SLC25A13, and cloning and expression of its mouse homologue, *Genomics* 62, 289–292.

Todo, S., Starzl, T.E., Tzakis, A., Benkov, K.J., Kalousek, F., Saheki, T., Tanikawa, K. and Fenton, W.A., 1992, Orthotopic liver transplantation for urea cycle enzyme deficiency, *Hepatology* 15, 419–422.

Tsujii, T., Morita, T., Matsuyama, Y., Matsui, T., Tamura, M. and Matsuoka, Y., 1976, Sibling cases of chronic recurrent hepatocerebral disease with hypercitrullinemia, *Gastroenterologia Japonica* 11, 328–340.

Vidailhet, M., Levin, B., Dautrevaux, M., Paysant, P., Gelot, S., Badonnel, Y., Pierson, M. and Neimann, N., 1971, Citrullinemie, *Arch. France Ped.* 28, 521–532.

Vito, P., Lacana, E. and D'Adamio, L., 1996, Interfering with apoptosis: Ca^{2+}-binding protein ALG-2 and Alzheimer's disease gene ALG-3, *Science* 271, 521–525.

Walker, J.E., 1992, The mitochondrial transporter family, *Curr. Opin. Struct. Biol.* 2, 519–526.

Walser, M., 1983, Urea cycle disorders and other hereditary hyperammonemic syndrome, in *The Metabolic Basis of Inherited Disease*, J.B. Stanbury, J.B. Wyngaarden, D.S. Frederikson, J.L. Goldstein and M.S. Brown (eds.), McGraw-Hill, New York, pp. 402–438.

Weber, F.E., Minestrini, G., Dyer, J.H., Werder, M., Boffelli, D., Compassi, S., Wehrli, E., Thomas, R.M., Schulthess, G. and Hauser, H., 1997, Molecular cloning of a peroxisomal Ca^{2+}-dependent member of the mitochondrial carrier superfamily, *Proc. Natl. Acad. Sci. USA* 94, 8509–8514.

Windmueller, H.G. and Spaeth, A.E., 1981, Source and fate of circulating citrulline, *Am. J. Physiol.* 241, E473–E480.

Yagi, Y., Saheki, T., Imamura, Y., Kobayashi, K., Sase, M., Nakano, K., Matuo, S., Inoue, I., Hagihara, S., and Noda, T., 1988, The heterogeneous distribution of argininosuccinate synthetase in the liver of type II citrullinemic patients: Its specificity and possible clinical implications, *Am. J. Clin. Pathol.* 89, 735–741.

Yajima, Y., Hirasawa, T. and Saheki, T., 1982, Diurnal fluctuation of blood ammonia levels in adult-type citrullinemia, *Tohoku J. Exp. Med.* 137, 213–220.

Yazaki, M., Ikeda, S., Takei, Y., Yanagisawa, N., Matsunami, H., Hashikura, Y., Kawasaki, S., Makuuchi, M., Kobayashi, K. and Saheki, T., 1996, Complete neurological recovery of an adult patient with type II citrullinaemia after living related partial liver transplantation, *Transplantation* 62, 1679–1681.

High Calcium Concentrations, Calpain Activation and Cytoskeleton Remodeling in Neuronal Regeneration after Axotomy

M.E. Spira, N.E. Ziv, R. Oren, A. Dormann and D. Gitler

1. INTRODUCTION

A vast number of studies has demonstrated causal relations between excessive elevation of the free intra neuronal calcium concentration ($[Ca^{2+}]_i$) and neurodegeneration. Calcium-induced neurodegeneration is believed to occur in acute conditions such as nerve-transection induced Wallerian degeneration (Waller, 1850), mechanical brain trauma, brain ischemia, hypoglycemic coma and status epilepticus. Calcium-induced neurodegeneration is also believed to participate in chronic conditions such as Alzheimer's disease and aging. The degenerative effects of the elevated $[Ca^{2+}]_i$ are thought to be mediated by the unbalanced activation of enzymes that take part in the normal neuronal function. These include proteinases, phospholipases, phosphatases and protein kinases. In turn, the unbalanced activation of these enzymes leads to cytoskeletal damage, membrane dysfunction, enhanced production of free radicals and, finally, neuronal degeneration (reviewed in Choi, 1994; Siesjo, 1994; Rothman and Olney, 1995; Kristian and Siesjo, 1998).

Examination of the literature reveals that the borderlines between "physiological calcium signals" and "pathological calcium concentrations" are not understood. Probably because of technical difficulties in monitoring $[Ca^{2+}]_i$, most studies discuss calcium-induced neurodegeneration in terms of "sustained increases in $[Ca^{2+}]_i$" or "calcium overloading", rather than in quantitative terms of calcium concentration, the duration to which

M.E. Spira, N.E. Ziv, R. Oren, A. Dormann and D. Gitler • Department of Neurobiology, Life Science Institute, The Hebrew University of Jerusalem, Israel; The Interuniversity Institute for Marine Sciences Eilat, Israel.

R. Pochet, R. Donato, J. Haiech, C. Heizmann and V. Gerke (eds.): Calcium: The Molecular Basis of Calcium Action in Biology and Medicine, 589–605.

the calcium levels are elevated, and the specific neuronal compartments which are exposed to the $[Ca^{2+}]_i$ insult. Likewise, the specific modes of the calcium-activated pathological enzymatic actions are not detailed.

Recent studies indicate that $[Ca^{2+}]_i$ levels considered to be pathological (in the range of hundreds of micromollars) participate in normal neuronal functions such as neurotransmitter release (Augustine and Neher, 1992; Llinas et al., 1992). Furthermore, such high calcium concentrations play a role in triggering essential regenerative processes after mechanical trauma in the form of axotomy (Ziv and Spira, 1997; Gitler and Spira, 1998). Recent studies also suggest that the calcium-activated neutral protease, calpain (a Ca^{2+}-activated protease that is thought to play a major role in neurodegenerative processes; see Brorson et al., 1994; Saido et al., 1994; Bednarski et al., 1995), is in fact an essential element in the cascade of events that promote the transformation of a transected axonal tip into a motile growth cone (Gitler and Spira, 1998). Furthermore, calpain activation was found to underlie the recovery of damaged dendrites after excitotoxic injury (Faddis et al., 1997). These findings should be taken into account when evaluating the delicate relations between high $[Ca^{2+}]_i$, the activation of proteases, neuronal regeneration and degeneration.

In the present chapter we describe a cascade of cellular events that leads from nerve injury in the form of axotomy to the formation of a growth cone and regeneration. This description is based on experiments performed in *Aplysia* neurons maintained in primary culture, a preparation which has been used extensively to study different forms of neuronal plasticity (for a review, see Kandel et al., 1991). Using calcium ratio imaging methods we first describe the quantitative relations between $[Ca^{2+}]_i$ gradients formed by axotomy and ultrastructural damage to the axon. We next provide evidence that the ultrastructural modifications induced by axotomy are directly related to the increased levels of the $[Ca^{2+}]_i$ rather than to other injury-related events. We discuss the relations between the $[Ca^{2+}]_i$ at the tip of the transected axon and the crucial formation of a membrane seal over the cut axonal end. In addition, we demonstrate that a localized and transient elevation of the free intra axonal calcium concentration is a sufficient signal to promote growth cone formation and neuritogenesis. Finally, using direct real-time imaging methods, we demonstrate that the activation of calpain at the tip of transected axons is essential for the formation of the growth cone and for regeneration.

2. THE FORMATION OF SPATIOTEMPORAL $[Ca^{2+}]_i$ GRADIENTS FOLLOWING AXOTOMY

The rupturing of the plasma membrane caused by axotomy leads to membrane depolarization which is often associated with the generation of action potentials that propagate antidromically along the proximal axonal segment, and orthodromically along the isolated distal segment. The depolarization and

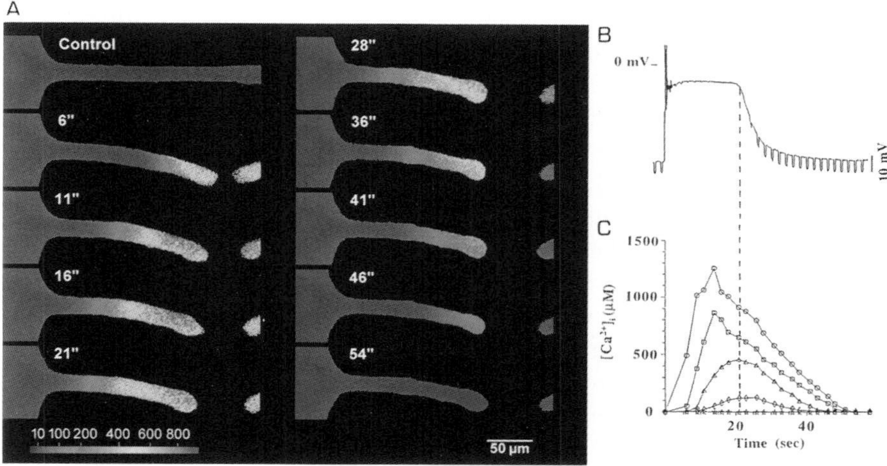

Figure 1. (A) Mag-fura-2 ratio images of the $[Ca^{2+}]_i$ after axotomy of a cultured metacerebral *Aplysia* neuron. Ca^{2+} is observed to diffuse from the cut end towards the cell body (6, 11, 16 seconds), elevating $[Ca^{2+}]_i$ to more than 1000 μM along segments near the cut end of the axon. Following the resealing of the ruptured membrane (between 21 and 28 seconds), the $[Ca^{2+}]_i$ recovers to near control levels. Time is given in seconds from axotomy. Calibration of $[Ca^{2+}]_i$ is given in μM. Correlation between the electrophysiological manifestation of axotomy and the alterations in $[Ca^{2+}]_i$ as obtained from the images shown in (A) are detailed in (B) and (C). (B) Resting potential and the transmembrane voltage drops in response to constant hyperpolaryzing pulses (input resistance measurements) during axonal transection and throughout the recovery period. (C) The $[Ca^{2+}]_i$ as a function of time post axotomy at five points along the axon (from top to bottom: 36 μm from the point of transection, 73, 110, 147 and 211). Note that the recovery of $[Ca^{2+}]_i$ began before the recovery of the membrane potential (dashed line connecting (B) and (C)). From Ziv and Spira (1995). For a colour version of Figure 1A, see page xxxiii.

the generation of action potentials activate voltage gated calcium channels, which in turn elevate the $[Ca^{2+}]_i$ to the micromolar range in the cell body and in neurites which are remote from the site of injury (Ziv and Spira, 1993).

Concomitantly, Ca^{2+} diffuses from the external medium into the axoplasm through the ruptured axonal membrane. The influx of Ca^{2+} along a calcium concentration gradient of 4–5 orders of magnitude (the Ca^{2+} concentration is approximately 100 nM in the axoplasm and 11 mM in the external solution) forms a steep $[Ca^{2+}]_i$ gradient between the axonal cut ends and the remaining parts of the neuron (Figure 1). In cultured *Aplysia* neurons, $[Ca^{2+}]_i$ exceeds 1 mM at the tips of the transected axons and declines to a level of a few hundreds of micromolars 100–200 μm away from the transected tips (Ziv and Spira, 1995). The distance covered by the elevated "$[Ca^{2+}]_i$ front" and the peak $[Ca^{2+}]_i$ level reached before the calcium concentration recovers are determined by a number of factors: (i) the rate of Ca^{2+} diffusion in the axoplasm, a parameter that is mainly controlled by the mobility of the endogenous calcium buffers; (ii) the calcium buffering power of the neuron and the overall Ca^{2+} removal mechanisms of the axon; and (iii) the rate at which a membrane seal is formed over the cut axonal end.

The diffusional spread of Ca^{2+} ions is mainly controlled by intracellular calcium buffers. Fixed buffers tend to retard calcium diffusion, whereas mobile buffers accelerate it (Gabso et al., 1997; Neher, 1995). In cultured *Aplysia* neurons, the majority of the endogenous buffers were found to be stationary. Accordingly, the effective calcium diffusion coefficient in these neurons is low (16 μm^2/sec). This may account for the slow rate at which the calcium front spreads from the point of transection into the axon. However, it should be noted that the ratio between mobile and stationary calcium buffers and their relative concentrations in various types of neurons may differ substantially. For example, frog saccular hair cells contain milimolar amounts of mobile endogenous buffers which possess an estimated diffusion coefficient similar to that of BAPTA (Roberts, 1993). Likewise, Kosaka et al. (1993) detected high concentrations of parvalbumin in the axons of purkinje cells. Thus, it is reasonable to assume that in such neurons axotomy will result in the "flooding" of the neurons by the rapid distribution of Ca^{2+}. As a consequence, the damage inflicted by Ca^{2+} influx in these neurons may be far more pronounced. Furthermore, the calcium buffering power of different types of neurons also varies. Thus, the buffering of calcium influx through the ruptured axonal end may be handled less effectively by some neurons in comparison to others.

In *Aplysia* neurons, a seal is formed over the cut end within 0.5–3 minutes of axotomy. In other preparations that were tested under *semi in-vivo* conditions, complete seal formation was estimated to occur over a substantially longer time course (Borgens et al., 1980; Yawo and Kuno, 1983, 1985; Strautman et al., 1990). Under these conditions, the axoplasm of the transected axon is expected to be exposed to high extracellular Ca^{2+} concentrations for periods of many minutes and even hours.

The recovery of $[Ca^{2+}]_i$ begins prior to the formation of a membrane seal over the cut ends as is indicated by the incomplete restoration of the transmembrane potential and input resistance when the $[Ca^{2+}]_i$ begins to decline (Figure 1). This may be explained by the fact that the formation of a membrane seal over the cut end is formed in a gradual manner (Gallant, 1988; Spira et al., 1993). Accordingly, the rate of Ca^{2+} influx through the cut end of the axon's tip is gradually reduced. The recovery phase of the $[Ca^{2+}]_i$ starts when the rate of Ca^{2+} influx is exceeded by the combined effects of Ca^{2+} extrusion, buffering and diffusion away from the transected axon's tip.

In cultured *Aplysia* neurons, the $[Ca^{2+}]_i$ recovers to control levels in both the proximal and distal axonal segments. This is most likely the case for other neurons transected *in vivo* as evidenced by the fact that isolated axonal segments maintain normal physiological properties for several hours and even days after isolation from the cell body (for references, see George et al., 1995; Bittner, 1991). The relatively rapid recovery of the ruptured mem-

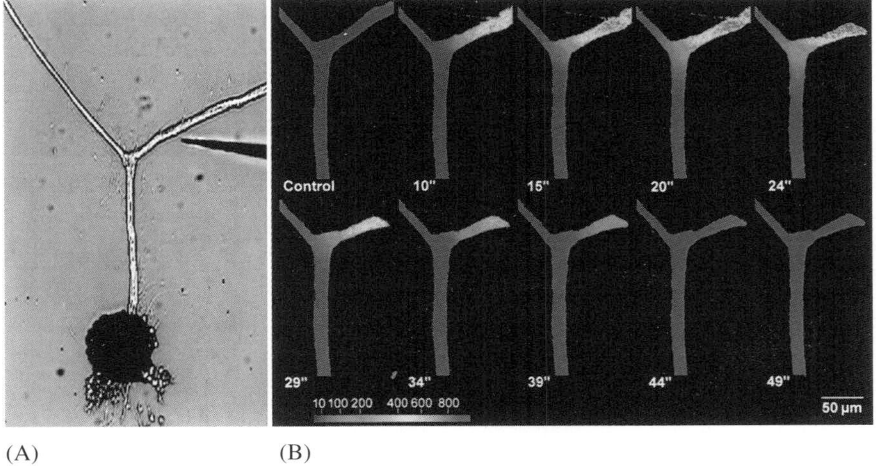

(A) (B)

Figure 2. The spatiotemporal distribution of $[Ca^{2+}]_i$ following axotomy at a point of axonal bifurcation. (A) A brightfield image of the bifurcated axon of a cultured *Aplysia* metacerebral neuron. (B) Mag-fura-2 pseudocolor images of the spatiotemporal distribution of $[Ca^{2+}]_i$ following the transection of the right branch. The front of elevated $[Ca^{2+}]_i$ is observed to reach the branch point, and a significant $[Ca^{2+}]_i$ gradient is formed on both sides of the bifurcation point. The resealing of the membrane occurred 25 seconds after axotomy (not shown). Time is given in seconds from axotomy. Calibration of $[Ca^{2+}]_i$ is given in μM. From Ziv and Spira (1995). For a colour version of this figure, see page xxxiv.

brane protects both the proximal and distal axonal segments from undergoing immediate degeneration after axotomy.

3. RESTRICTION OF THE SPREAD OF CALCIUM FROM AN INJURED AXONAL SEGMENT INTO THE REST OF THE NEURON BY GEOMETRICAL FACTORS

In addition to the aforementioned mechanisms that control the extent of Ca^{2+} spread within a given neuron, we wish to discuss geometrical factors. This is best illustrated by examining the spread of the calcium front at points of axonal bifurcation or at the junction between the axon and the cell body.

The propagation pattern of the $[Ca^{2+}]_i$ front, as it approaches a point of axonal bifurcation, is illustrated in Figure 2. Here a cultured *Aplysia* neuron possessing a bifurcating axon was loaded with the low-affinity calcium indicator mag-fura-2 (Ziv and Spira, 1995). One branch of the bifurcated axon was transected (Figure 2A). The elevated $[Ca^{2+}]_i$ front induced by axotomy propagated along the transected branch. However, as the front reached the point of axonal bifurcation, a sharp $[Ca^{2+}]_i$ gradient of approximately 300 μM was formed between the transected branch and the intact axonal segments beyond the point of bifurcation. Although the cut end of the axon did not reseal, the $[Ca^{2+}]_i$ was not elevated beyond the point of bifurcation. Once the cut end sealed off, as was evidenced by the recovery of the neuron's membrane potential and input resistance (not shown), the $[Ca^{2+}]_i$ returned to pre-axotomy levels, initially at the branch point and then along the transected axon.

This behavior reflects the increased buffering capacity presented to the incoming calcium front by an increased cross sectional area. These observations suggest that both the neuron's enormous capacity for removing excess Ca^{2+} and geometrical factors act to confine extreme increases in $[Ca^{2+}]_i$ caused by axonal transection to the near vicinity of the transection site. This mechanism may effectively isolate injured neuronal segments from other adjacent segments.

4. Ca^{2+} INDUCED RESEALING OF THE CUT END

It is well established that excess intracellular calcium concentrations trigger cascades of pathological alterations. Yet, in the case of axotomy, the influx of Ca^{2+} into the cut axonal end initiates the first "protective" response of the axon by promoting membrane seal formation. It should be noted that in a calcium-free solution or at extracellular calcium levels of a few micromolars, transected axons do not seal and in many cases degenerate. Yawo and Kuno (1983, 1985) demonstrated that extracellular Ca^{2+} concentrations in the range of 0.1–0.5 mM are required for successful membrane sealing of transected giant axons of the cockroach. Extracellular Ca^{2+} concentrations in the range of 100 μM were found to be essential for resealing of transected neurites of cultured rat septal neurons (Xie and Barrett, 1991). Direct measurements of the $[Ca^{2+}]_i$ during the process of membrane seal formation in transected *Aplysia* neurons revealed that the $[Ca^{2+}]_i$ must reach tens of micromolars to promote membrane sealing.

The mechanisms underlying calcium-dependent membrane seal formation after axotomy are not clear. Recent studies suggested that calpain activation is a necessary step in the cascade of events that lead to plasma membrane seal formation. Thus, inhibition of calpain was shown to repress the resealing of the axonal membrane after axotomy. Conversely, supplementing the external medium with calpain accelerated sealing (Xie and Barrett, 1991; Godell et al., 1997; Howard et al., 1999).

A possible role for calpain in this process is the facilitation of the dissociation of the membrane skeleton and of cytoskeletal elements in the axon. This hypothesis is supported by recent observations demonstrating that the rapid depolymerization of microtubules at the transected axonal ends is an essential step in the formation of a membrane seal after axotomy (Khoutorsky and Spira, manuscript in preparation). Here it was found that stabilization of microtubules by taxol (Horwitz, 1994) mechanically impedes the formation of a membrane seal. As a consequence, the transected axon is overloaded with calcium and degenerates.

5. SPATIAL CORRELATION BETWEEN THE $[Ca^{2+}]_i$ GRADIENT AND THE ULTRASTRUCTURAL DAMAGES INDUCED BY AXOTOMY

Early studies reported that ultrastructural alterations associated with axotomy appear in the form of a gradient in which the highest degree of damage occurs near the tip of the cut axon (for example, Ballinger and Bittner, 1980; Gross and Higgins, 1987; Spira et al., 1993). However, the relationship between the ultrastructural changes induced by axotomy and the $[Ca^{2+}]_i$ gradient were only recently determined.

To correlate the spatial distribution pattern of $[Ca^{2+}]_i$ with the ultrastructural damage-gradient induced by axotomy, cultured neurons were loaded with mag-fura-2, their axons were transected, and the alterations in $[Ca^{2+}]_i$ caused by axotomy were recorded. After the recovery of $[Ca^{2+}]_i$ to near-control levels (1–5 minutes after axotomy), the neurons were fixed for EM examination by rapid superfusion with a gluteraldehyde fixation buffer (Benbassat and Spira, 1993; Ziv and Spira, 1997).

Following axotomy, microtubules and neurofilaments that are normally oriented in parallel to the longitudinal axis of the axon are no longer detected at the tip of the transected axon in regions that correspond to peak $[Ca^{2+}]_i$ elevations between 300–1500 μM (Figure 3A). At the very tip of the axon, in regions in which the $[Ca^{2+}]_i$ exceeds 1500 μM, dissociated microtubules and neurofilaments form large electron dense aggregates (Figure 3D). Further away from the cut end, at an axonal segment that corresponds to $[Ca^{2+}]_i$ levels of 300–1500 μM, the dissociated cytoskeletal elements form elongated electron-dense aggregates (Figures 3B and 3C). No intact neurofilaments are observed along this segment.

At larger distances from the transected tip, at regions that correspond to peak $[Ca^{2+}]_i$ of $\sim$ 300 μM (Figure 3A), clusters of relatively short fragments of microtubules can be seen. These fragments are no longer oriented exclusively in parallel to the longitudinal axis of the axon. Another feature characteristic to transected axons is that along segments of $\sim$ 100 μm in which the calcium concentrations transiently exceed 300 μM, the plasma membrane detaches from the axoplasmic core (Figure 3A arrowheads and B asterisk). The space between the core of the axoplasm and the axolema is filled with amorphous material, vesicles and swollen sub-surface cisterns. The detachment of the plasma membrane is most likely the outcome of the calcium induced proteolytic degradation of spectrin, a filamentous key component of the membrane skeleton (see below).

The axotomy-induced alterations to the axon's cytoarchitecture end abruptly 50–150 μm from the cut end, resulting in the formation of a sharp transition zone between the axonal segment in which the cytoarchitecture is altered and the rest of the axon in which the cytoarchitecture appears normal (Figure 3A, asterisk). Examination of the spatiotemporal $[Ca^{2+}]_i$ distribution pattern does not reveal any sharp drop in the $[Ca^{2+}]_i$ gradient that parallels the

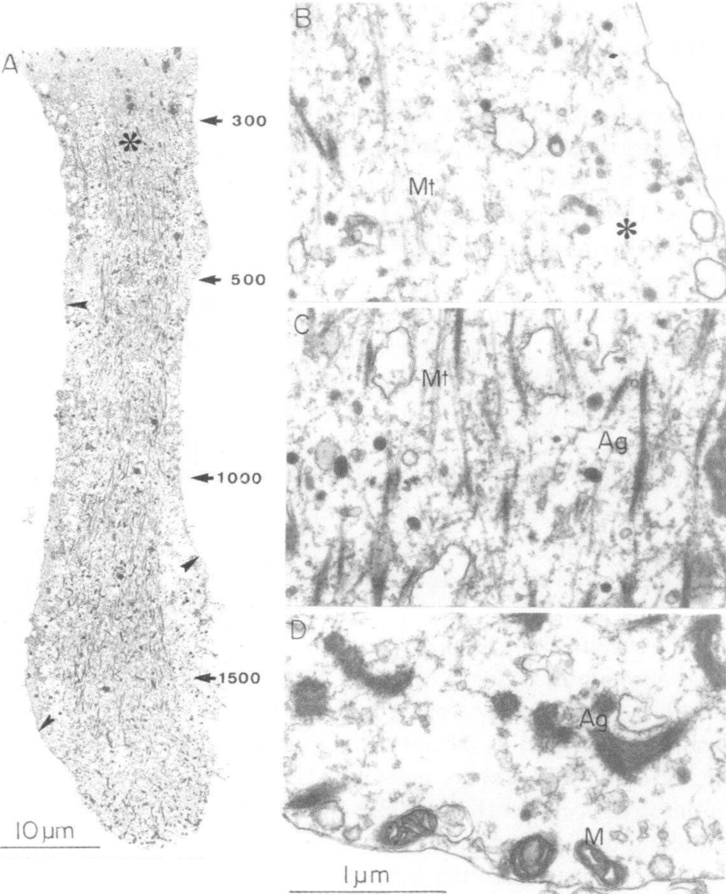

Figure 3. Correlation between the calcium concentration and the ultrastructural damage gradient. (A) An axon was transected while imaging the calcium concentration by mag-fura-2 ratiometric fluorescent microscopy. The axon was fixed for EM after complete recovery of the $[Ca^{2+}]_i$. A low magnification image of the transected axon is shown in (A). Note the disruption of microtubules and neurofilaments in the distal region of the axon, the formation of short fragments of electron dense filamentous material in the core of the axoplasm, and the conspicuous separation of the axolema from the axoplasmic core (arrowheads). In particular, note the sharp transition (asterisk) between the severely altered axoplasm and the unaltered axoplasm of the proximal region in which $[Ca^{2+}]_i$ was elevated to less than 300 μM. The peak Ca^{2+} concentrations (in μM) recorded along the axon after axotomy are indicated on the right-hand side of the figure (arrows). (B) A high magnification of the region adjacent to the axolema reveals a large gap between the axolema and the cytoskeletal core (asterisk), that is filled with amorphous axoplasm and several large vacuoles. (C) A high magnification of the severely altered axoplasmic core reveals that the electron dense filaments are aggregates of amorphous material and short segments of microtubules (Ag). (D) A high magnification of the tip of the cut axon reveals that large electron dense aggregates are formed near the tip as well as vacuoles of unidentified origin. Mt – microtubules. M – mitochondria. From Ziv and Spira (1997).

sharp transition in the axonal cytoarchitecture. In fact, the $[Ca^{2+}]_i$ gradually decreases through the transition region from $\sim$ 300 μM to the micromolar range. The transition zone is characterized by the disappearance of the microtubular aggregates and the reappearance of intact neurofilaments and microtubules. It is of particular interest to note that at this transition zone the detached axolema consistently reattaches to the axoplasmic core (Figure 3A) and that several minutes after axotomy large numbers of vesicles accumulate at the transition zone. This region corresponds to the site from which a growth cone lamellipodium extends after axotomy. To examine whether the elevation in $[Ca^{2+}]_i$ underlies the ultrastructural alterations described above, or whether other injury-related events are involved, we focally applied the calcium ionophore ionomycin to axonal segments of intact *Aplysia* neurons and examined its effects on the axon's cytoarchitecture (Ziv and Spira, 1997). Ionomycin applications that elevated the $[Ca^{2+}]_i$ to 300–500 μM at the application point, and to decreasing values at increasing distances from the site of application, induced ultrastructural alterations identical to those documented in transected axons. Specifically, in regions where $[Ca^{2+}]_i$ was elevated to more than $\sim$ 300 μM, microtubules were disrupted and dissociated microtubules collapsed to form small longitudinal clusters. Neurofilaments were lost and the axolema was detached from the axoplasmic core. In regions in which the $[Ca^{2+}]_i$ was elevated to values below $\sim$ 300 μM, the normal appearance of the axoplasm was retained. In common with the ultrastructure of the transected axons well defined transition zones were observed on both sides of the ionomycin application site. These transition zones were characterized by the disappearance of the electron-dense aggregates, the reappearance of intact microtubules, the reattachment of the plasma membrane to the axoplasmic core and the accumulation of vesicles.

These observations demonstrate that ultrastructural alterations induced by axotomy and by focal applications of ionomycin are identical, and suggest clear correlation between transient increases in $[Ca^{2+}]_i$ and specific ultrastructural consequences.

6. GROWTH CONE FORMATION IS TRIGGERED BY A TRANSIENT ELEVATION OF THE $[Ca^{2+}]_i$ TO 300–500 μM

The calcium-induced ultrastructural alterations described above are usually considered to be pathological in nature. However, our studies revealed that these changes are also part of the process in which the severed axonal segments differentiate into motile growth cones (Ziv and Spira, 1997; Gitler and Spira, 1998).

As mentioned above, the growth cone which is formed after axotomy emerges at some distance from the tip of the cut axon. Correlation between the site of growth cone formation and the $[Ca^{2+}]_i$ gradients recorded after axotomy, revealed that growth cones consistently extend from regions in

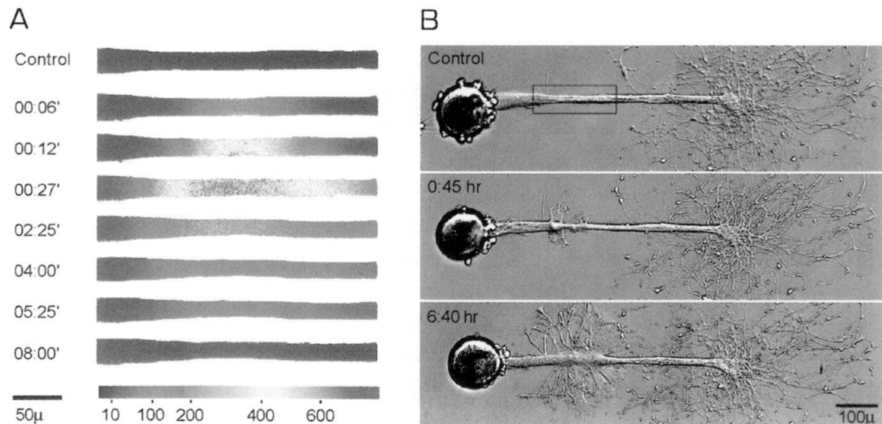

Figure 4. Transient and localized elevation of the $[Ca^{2+}]_i$ to 300–500 μM induces growth cone formation and neuritogenesis in an axon of intact neuron. Mag-fura-2 ratiometric fluorescence microscopy was used to determine the intra-axonal $[Ca^{2+}]_i$ required to induce the transformation of an intact axonal segment into a growth cone. (A) The spatiotemporal alterations in the axonal $[Ca^{2+}]_i$ induced by a focal application of ionomycin. The region shown corresponds to the rectangle in (B), upper panel. $[Ca^{2+}]_i$ is given in μM. (B) The resulting changes in axonal morphology. The transient increase of $[Ca^{2+}]_i$ to $\sim$ 500 μM induced the formation of a growth cone at the application site that subsequently developed into a new neuritic tree. From Ziv and Spira (1997). For a colour version of this figure, see page xxxiv.

which $[Ca^{2+}]_i$ was transiently elevated to 300–500 μM (Ziv and Spira, 1997). The ability of $[Ca^{2+}]_i$ transients to induce growth cone formation was directly established in experiments in which transient elevations of $[Ca^{2+}]_i$ to 300–500 μM induced the formation of ectopic growth cones along axons of intact neurons (Figure 4). Greater Ca^{2+} concentrations were commonly associated with visible damage to the axon, manifested as beading and axonal degeneration. Conversely, lower Ca^{2+} concentrations did not noticeably affect the axonal morphology or ultrastructure.

These findings strongly suggest that the transient elevation of $[Ca^{2+}]_i$ caused by axotomy may directly induce the dedifferentiation of severed axonal segments into growth cones. It is worth noting that the first signs of growth cone formation are only detectable after the axotomy- or ionomycin-induced $[Ca^{2+}]_i$ elevations recover to control levels. We interpret this observation to suggest that the transient elevation in $[Ca^{2+}]_i$ triggers the growth process but is not required for its perpetuation.

7. AXOTOMY INDUCES LOCALIZED AND TRANSIENT ELEVATION OF PROTEOLYTIC ACTIVITY AT THE SEVERED TIP

The findings presented above strongly suggest that a transient elevation in the $[Ca^{2+}]_i$ to 300–500 μM is a signal sufficient to induce growth cone formation. This raises questions as to the nature of the processes and molecules that link elevated Ca^{2+} levels, the massive restructuring of the axon's cytoskeleton and the formation of a growth cone. Recent studies suggest that the transient elevation of the $[Ca^{2+}]_i$ activates the neutral cytosolic protease calpain (a Ca^{2+}-activated protease that is considered to take part in neurodegeneration) and that this protease provides the link between $[Ca^{2+}]_i$ and growth cone formation. Furthermore, the application of calpeptin, a membrane permeable calpain inhibitor, was found to block the dedifferentiation of transected axonal tips into growth cones (Gitler and Spira, 1998).

The spatiotemporal relationship between axotomy, $[Ca^{2+}]_i$ and calpain activation are illustrated in Figure 5. To examine these relations we simultaneously imaged the $[Ca^{2+}]_i$ using mag-fura-2 and proteolytic activity using the fluorogenic membrane permeable proteolysis indicator bis(CBZ-L-Alanyl-L-Alanine amine)-Rhodamine 110 (bCAA-R110). bCAA-R110 is practically non-fluorescent, while the products of the cleavage of its amide bonds, rhodamine 110 (R110) and its monoamides, are highly fluorescent (Leytus et al., 1983a, b). As a result, the proteolytic cleavage of bCAA-R110 is associated with a huge increase in fluorescence.

In these experiments, the basal proteolytic activity in neurons was imaged for approximately 20 minutes and then their axons were transected. As a result, the $[Ca^{2+}]_i$ levels transiently increased, forming a sharp concentration gradient of > 1000 μM at the tip of the axon (Figure 5B) and < 100 μM at a distance of ~ 200 μm from the cut end. The calcium gradient declined to control values within a few minutes. Axotomy was followed by a significant increase in the level of the proteolytic activity (Figure 5C). Such an increase was detected approximately 2 minutes after axotomy. Initially, the proteolytic activity increased uniformly within the axoplasm of the transected axonal tips. Thereafter, discrete peaks of activity developed. The proteolytic activity in the rest of the tip continued to rise in a slower manner. After reaching peak levels the proteolytic activity gradually decreased.

To determine if this proteolytic activity was induced by the transient elevations in $[Ca^{2+}]_i$ associated with axotomy rather than other injury related processes, $[Ca^{2+}]_i$ was focally elevated by applying ionomycin to the axon of an intact neuron while simultaneously imaging $[Ca^{2+}]_i$ and proteolysis. We found that proteolytic activity was induced along restricted axonal segments in which the $[Ca^{2+}]_i$ had been elevated to 300–400 μM. The time course of the $[Ca^{2+}]_i$ transient and that of the proteolytic activity was similar to that observed following axotomy. Eventually, a growth cone emerged from the region where the $[Ca^{2+}]_i$ and the proteolytic activity were increased, within time periods similar to those measured in axotomy experiments. These results illustrate that a transient increase in $[Ca^{2+}]_i$ is sufficient to induce both proteolytic activity and growth cone formation.

Preincubation of the neuron in 100 μM calpeptin completely inhibited both proteolytic activity and growth cone formation induced by axotomy or by ionomycin applications. Calpeptin did not affect the recovery of $[Ca^{2+}]_i$

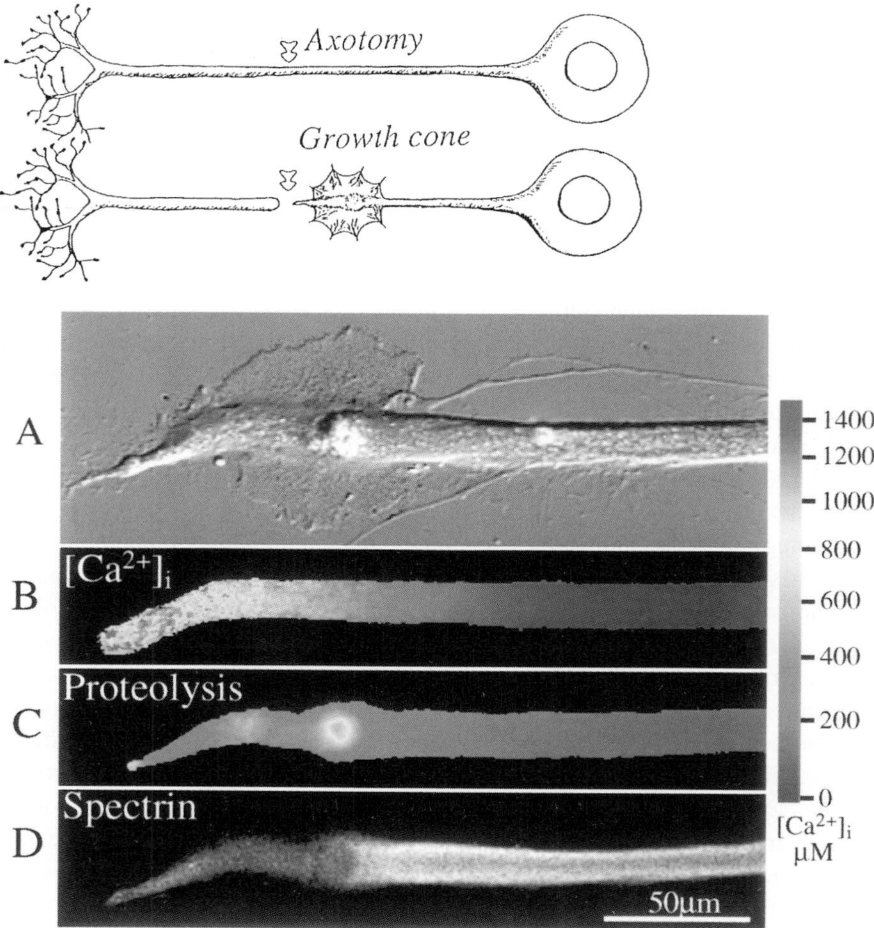

Figure 5. $[Ca^{2+}]_i$, proteolytic activity and spectrin distribution after axotomy. A cultured *Aplysia* buccal neuron was transected while $[Ca^{2+}]_i$ and proteolytic activity were measured. The neuron was fixed and immunolabeled for spectrin 25 minutes after axotomy. (A) Image of the axon prior to fixation. The neuron extended a sizable growth cone lamellipodium following axotomy. (B) The maximal levels of $[Ca^{2+}]_i$ measured during axotomy. The image was acquired 25 seconds after the axotomy was performed. (C) The levels of proteolytic activity, as measured just prior to fixation, 25 minutes after axotomy. (D) Spectrin density is seen to be drastically decreased from the center of the growth cone and up to the transected tip. Notice that the sites from which spectrin was removed correlate very well with the locations in which highest proteolytic activity was measured (C). There exists a very sharp border between the regions in which spectrin density was reduced and those in which spectrin density appears to be unaffected. In contrast, notice that the distribution of spectrin does not conform to the shape of the $[Ca^{2+}]_i$ gradient (B). The location where the growth cone's center was formed correlates to those areas in which $[Ca^{2+}]_i$ was elevated to approximately 350 μM. A $[Ca^{2+}]_i$ scale bar is given to the right. From Gitler and Spira (1998). For a colour version of this figure, see page xxxv.

to control levels once the membrane resealed, suggesting that calpeptin does not alter neuronal Ca^{2+} handling characteristics. The spatio-temporal relations described above and the inhibitory action of calpeptin on growth cone formation, suggested causal relationships between the transient elevation of $[Ca^{2+}]_i$, proteolysis and growth cone formation.

A recent study has linked calpain activation with remodeling of dendrites after sublethal excitotoxic injury *in-vitro*, as assessed by immunolabeling of both MAP2 and calpain generated spectrin fragments (Faddis et al., 1997). This study showed that calpain activation is associated not with the formation of injury induced varicosities but rather with their dissolution, thus supporting the view that calpain activation has a role in recovery from injury.

8. REDUCTION IN SPECTRIN DENSITY SPATIALLY CORRELATES WITH INCREASED LEVELS OF PROTEOLYTIC ACTIVITY AND WITH THE FORMATION OF THE GROWTH CONE

Aunis and Bader (1988) and Perrin et al. (1992) suggested that removal of spectrin makes the inner surface of secretory cells accessible for fusion with intracellular membranes. As spectrin is a substrate of calpain, and since the growth process requires the insertion of intracellular membranes into the neurolema, we studied the relations between calcium-induced proteolysis, the distribution of axonal spectrin and the formation of the growth cone. For this purpose, we measured both the $[Ca^{2+}]_i$ transients and the ensuing proteolytic activity after axonal transection, and then immunolabeled the same neurons with antibodies against spectrin (Figure 5D). Confocal microscope images of control neurons revealed that spectrin is abundant at the submembranal domain, and less so in the inner axonal space. We found that axotomy greatly reduced the density of submembrane spectrin in regions where the $[Ca^{2+}]_i$ was elevated and in which proteolytic activity was induced (Figure 5). Furthermore, calpeptin, which was found to inhibit Ca^{2+} induced proteolytic activity and growth cone formation also inhibited the removal of spectrin from beneath the plasma membrane (Gitler and Spira, 1998).

9. CONCLUDING REMARKS

The results presented here demonstrate that transient and localized elevations of $[Ca^{2+}]_i$ to levels well within ranges currently believed to be toxic to cells (300–500 μM) trigger growth cone formation and neuritogenesis in cultured *Aplysia* neurons. Furthermore, these high calcium concentration activate calpains which are usually considered in the context of neurodegenerative conditions. Yet, under the experimental conditions described above calpains seem to play key roles in the induction of regeneration.

The findings described here on the one hand, and the established relationships between calcium overloading and degeneration on the other, have lead us to hypothesize that growth cone formation and regeneration after axotomy is triggered when the calcium signal is narrowly confined in terms of: (a) the intracellular space which is exposed to the increased calcium concentration; (b) the duration of such elevations; and (c) the values of the calcium concentration. While our experiments revealed that localized elevation of the $[Ca^{2+}]_i$ to levels of 300–500 μM for several minutes induce cytoarchitectural alterations that lead to axonal regeneration, homogeneous elevation of the $[Ca^{2+}]_i$ along the entire neuronal length to these values (for example, by bath application of ionomycin) leads to degeneration. We assume that the different fates of the neuron under these conditions can be explained in the following way: a transient and local elevation of calcium can be efficiently buffered by the "intact" surrounding calcium-buffer systems of the neuron. Thus, under such conditions, the neuron can translate the calcium-induced local molecular alterations into growth and regeneration. In contrast, when the entire buffer system is loaded the calcium buffering system fails to down-regulate the calcium load and thus, the degenerative processes are not only triggered but also amplified.

As to the role of calpain in promoting regeneration rather then degeneration, our findings are consistent with earlier studies that suggest a role for calpain activity in morphological remodeling of neurons in relation to neuronal plasticity (Lynch and Baudry, 1984; Lynch et al., 1990). In these studies it was suggested that calpain activity participates in processes such as long term potentiation (LTP) by perpetuating morphological changes in preexisting structures. This suggestion was based on the detection of calpain-generated spectrin fragments in NMDA treated hippocampal slices (Seubert et al., 1988; del Cerro et al., 1994), as well as in slices which had been stimulated at the theta rhythm, a rhythm which is known to stimulate the production of LTP (Vanderklish et al., 1995).

As mentioned earlier, the recent study of Faddis et al. (1997) has linked calpain activation with remodeling of dendrites after sublethal excitotoxic injury *in-vitro*, thus supporting the view that calpain activation has a role in recovery from injury.

It is premature to generalize our conclusions from the studies of cultured *Aplysia* neurons and from the available information in the literature to other neurons. Nevertheless, it would be beneficial to re-evaluate the current pharmacological concepts taken to prevent acute and chronic neurodegeneration. Recall that the current pharmacological approaches are based on two main concepts: (a) the development of pharmacological agents that control upstream excess increase in the $[Ca^{2+}]_i$, and (b) inhibition of down-stream events by blockage of calcium-activated enzymes such as calpain (Wang and Yuen, 1994). Obviously both calcium and calpain participate under experimental conditions in the recovery processes of neurons after trauma, thus total blockage of them may interfere with recovery and regenerative processes.

ACKNOWLEDGEMENTS

The laboratory of M.E.S. is supported by grants from The Israel Science Foundation (No. 620/98), The German Israel Foundation for Scientific research and Development (No. I-0598-162.01/98) and The US-Israel Bi-National Research Foundation (No. 97-00297-1). M.E. Spira is the Levi Deviali professor in Neurobiology.

REFERENCES

Augustine, G.J. and Neher, E., 1992, Calcium requirements for secretion in bovine chromaffin cells, *J. Physol. Lond.* 450, 247–271.

Aunis, D. and Bader, M.F., 1988, The cytoskeleton as a barrier to exocytosis in secretory cells, *J. Exp. Biol.* 139, 253–266.

Ballinger, M.L. and Bittner, G.D., 1980, Ultrastructural studies of severed medial giant and other CNS axons in crayfish, *Cell Tissue Res.* 208, 123–133.

Bednarski, E., Vanderklish, P., Gall, C., Saido, T.C., Bahr, B.A. and Lynch, G., 1995, Translational suppression of calpain I reduces NMDA-induced spectrin proteolysis and pathophysiology in cultured hippocampal slices, *Brain Res.* 694, 147–157.

Benbassat, D. and Spira, M.E., 1993, Survival of isolated axonal segments in culture: Morphological, ultrastructural, and physiological analysis, *Exp. Neurol.* 122, 295–310.

Bittner, G.D., 1991, Long-term survival of anucleate axons and its implications for nerve regeneration, *Trends Neurosci.* 14, 188–193.

Borgens, R.B., Jaffe, L.F. and Cohen, M.J., 1980, Large and persitent electrical currents enter the transected lamprey spinal cord, *Proc. Natl. Acad. Sci. USA* 77, 1209–1213.

Brorson, J.R., Manzolillo, P.A. and Miller, R.J., 1994, Ca^{2+} entry via AMPA/KA receptors and excitotoxicity in cultured cerebellar Purkinje cells, *J. Neurosci.* 14, 187–197.

Choi, D.W., 1994, Calcium and excitotoxic neuronal injury, *Ann. N.Y. Acad. Sci.* 747, 162–171.

del Cerro, S., Arai, A., Kessler, M., Bahr, B.A., Vanderklish, P., Rivera, S. and Lynch, G., 1994, Stimulation of NMDA receptors activates calpain in cultured hippocampal slices, *Neurosci. Lett.* 167, 149–152.

Faddis, B.T., Hasbani, M.J. and Goldberg, M.P., 1997, Calpain activation contributes to dendritic remodeling after brief excitotoxic injury in vitro, *J. Neurosci.* 17, 951–959.

Gabso, M., Neher, E. and Spira, M.E., 1997, Low mobility of the Ca^{2+} buffers in axons of cultured *Aplysia* neurons, *Neuron* 18, 473–481.

Gallant, P.E., 1988, Effects of the extarnal ions and metabolic poisoning on the constriction of the squid giant axon after axotomy, *J. Neurosci.* 8, 1479–1484.

George, E.B., Glass, J.D. and Griffin, J.W., 1995, Axotomy-induced axonal degeneration is mediated by calcium influx through ion-specific channels, *J. Neurosci.* 15, 6445–6452.

Gitler, D. and Spira, M.E., 1998, Real time imaging of calcium-induced localized proteolytic activity after axotomy and its relation to growth cone formation, *Neuron* 20, 1123–1135.

Godell, C.M., Smyers, M.E., Eddleman, C.S., Ballinger, M.L., Fishman, H.M. and Bittner, G.D., 1997, Calpain activity promotes the sealing of severed giant axons, *Proc. Natl. Acad. Sci. USA* 94, 4751–4756.

Gross, G.W. and Higgins, M.L., 1987, Cytoplasmic damage gradients in dendrites after transection lesions, *Exp. Brain Res.* 67, 52–60.

Horwitz, S.B., 1994, Taxol (paclitaxel): Mechanisms of action, *Ann. Oncol.* 5 (Suppl. 6), S3–S6.

Howard, M.J., David, G. and Barrett, J.N., 1999, Resealing of transected myelinated mammalian axons in vivo: Evidence for involvement of calpain, *Neuroscience* 93, 807–815.

Kandel, E.R., Schwartz, J.H. and Jessell, T.M., 1991, *Principles of Neuronal Science*, Elsevier, New York.

Kosaka, T., Kosaka, K., Nakayama, T., Hunziker, W. and Heizmann, C.W., 1993, Axons and axon terminals of cerebellar Purkinje cells and basket cells have higher levels of parvalbumin immunoreactivity than somata dendrites: Quantitative analysis by immunogold labeling, *Exp. Brain Res.* 93, 483–491.

Kristian, T. and Siesjo, B.K., 1998, Calcium in ischemic cell death, *Stroke* 29, 705–718.

Leytus, S.P., Melhado, L.L. and Mangel, W.F., 1983a, Rhodamine-based compounds as fluorogenic substrates for serine proteinases, *Biochem. J.* 209, 299–307.

Leytus, S.P., Patterson, W.L. and Mangel, W.F., 1983b, New class of sensitive and selective fluorogenic substrates for serine proteinases. Amino acid and dipeptide derivatives of rhodamine, *Biochem. J.* 215, 253–260.

Llinas, R., Sugimori, M. and Silver, R.B., 1992, Microdomains of high calcium concentration in a presynaptic terminal, *Science* 256, 677–679.

Lynch, G. and Baudry, M., 1984, The biochemistry of memory: A new and specific hypothesis, *Science* 224, 1057–1063.

Lynch, G., Kessler, M., Arai, A. and Larson, J., 1990, The nature and causes of hippocampal long-term potentiation, *Prog. Brain Res.* 83, 233–250.

Neher, E., 1995, The use of fura-2 for estimating Ca buffers and Ca fluxes, *Neuropharmacology* 34, 1423–1442.

Perrin, D., Moller, K., Hanke, K. and Soling, H.D., 1992, cAMP and Ca(2+)-mediated secretion in parotid acinar cells is associated with reversible changes in the organization of the cytoskeleton, *J. Cell Biol.* 116, 127–134.

Roberts, W.M., 1993, Spatial calcium buffering in saccular hair cells, *Nature* 363, 74–76.

Rothman, S.M. and Olney, J.W., 1995, Excitotoxicity and the NMDA receptor – Still lethal after eight years, *Trends Neurosci.* 18, 57–58.

Saido, T.C., Sorimachi, H. and Suzuki, K., 1994, Calpain: New perspectives in molecular diversity and physiological-pathological involvement, *FASEB J.* 8, 814–822.

Seubert, P., Larson, J., Oliver, M., Jung, M.W., Baudry, M. and Lynch, G., 1988, Stimulation of NMDA receptors induces proteolysis of spectrin in hippocampus, *Brain Res.* 460, 189–194.

Siesjo, B.K., 1994, Calcium-mediated processes in neuronal degeneration, *Ann. N.Y. Acad. Sci.* 747, 140–161.

Spira, M.E., Benbassat, D. and Dormann, A., 1993, Resealing of the proximal and distal cut ends of transected axons: Electrophysiological and ultrastructural analysis, *J. Neurobiol.* 24, 300–316.

Strautman, A.F., Cork, R.J. and Robinson, K.R., 1990, The distribution of free calcium in transected spinal axons and its modulation by applied electrical fields, *J. Neurosci.* 10, 3564–3575.

Vanderklish, P., Saido, T.C., Gall, C., Arai, A. and Lynch, G., 1995, Proteolysis of spectrin by calpain accompanies theta-burst stimulation in cultured hippocampal slices, *Brain Res. Mol. Brain Res.* 32, 25–35.

Waller, A.V., 1850, Experiments on the section of the glossopharyngeal and hypoglossal nerves of the frog and observations of the alterations produced thereby in the structure of their primitive fibres, *Philos. Trans. Roy. Soc. Lond. (Biol.)* 140, 423–429.

Wang, K.K. and Yuen, P.W., 1994, Calpain inhibition: An overview of its therapeutic potential, *Trends Pharmacol. Sci.* 15, 412–419.

Xie, X.Y. and Barrett, J.N., 1991, Membrane resealing in cultured rat septal neurons after neurite transection: Evidence for enhancement by Ca(2+)-triggered protease activity and cytoskeletal disassembly, *J. Neurosci.* 11, 3257–3267.

Yawo, H. and Kuno, M., 1983, How a nerve fiber repairs its cut end: Involvement of phospholipase A2, *Science* 222, 1351–1353.

Yawo, H. and Kuno, M., 1985, Calcium dependence of membrane sealing at the cut end of the cockroach giant axon, *J. Neurosci.* 5, 1626–1632.

Ziv, N.E. and Spira, M.E., 1993, Spatiotemporal distribution of Ca^{2+} following axotomy and throughout the recovery process of cultured *Aplysia* neurons, *Eur. J. Neurosci.* 5, 657–668.

Ziv, N.E. and Spira, M.E., 1995, Axotomy induces a transient and localized elevation of the free intracellular calcium concentration to the millimolar range, *J. Neurophysol.* 74, 2625–2637.

Ziv, N.E. and Spira, M.E., 1997, Localized and transient elevations of intracellular Ca^{2+} induce the dedifferentiation of axonal segments into growth cones, *J. Neurosci.* 17, 3568–3579.

Genetic Factors and the Role of Calcium in Alzheimer's Disease Pathogenesis

Mervyn J. Monteiro and Stacy M. Stabler

1. INTRODUCTION

Alzheimer's disease (AD) is a progressive neurodegenerative disorder characterized by impaired memory, cognition and altered behavior. Aging increases the incidence of AD from ~10% in people 65 years of age or older to ~47% in persons > 85 years. Although many theories have been proposed regarding the etiology of AD, the exact mechanisms involved are unknown and no effective cure exists. Exploring the role of calcium (Ca^{2+}) in AD has begun to offer some insight into this devastating disease. This review will provide a basic background into AD pathology and genetics, followed by a more in depth focus on the evidence linking altered Ca^{2+} regulation to both the early-onset AD gene products and actual affected tissues.

2. ALZHEIMER'S DISEASE PATHOLOGY AND GENETICS

While AD can be diagnosed with ~90% accuracy, the definitive test is pathological examination of brain at autopsy. Characteristic changes in the brain distinguish AD from other forms of dementia. Brains affected by AD show the presence of neurofibrillary tangles (NFT) within nerve cells and extracellular neuritic plaques in areas important for memory and intellectual functions. NFTs are principally composed of abnormally twisted paired helical filaments (PHFs) assembled from the microtubule-associated protein tau, whereas, plaques contain abnormal deposits of proteolytic fragments of the β-amyloid precursor protein (APP) (Price et al., 1998; Selkoe, 1998). NFTs and

Mervyn J. Monteiro and Stacy M. Stabler • Medical Biotechnology Center, Department of Neurology and Division of Human Genetics University of Maryland, Baltimore, MD, U.S.A.

R. Pochet, R. Donato, J. Haiech, C. Heizmann and V. Gerke (eds.): Calcium: The Molecular Basis of Calcium Action in Biology and Medicine, 607–623.

plaques are thought to be caused by the interplay of several factors including age, genes, and environment.

The majority of AD cases are late-onset, appearing in people over the age of 65. However, a small percentage (~5%) of cases, termed early-onset familial, arise at an unusually young age with some individuals developing the disease as early as their third decade of life. Molecular genetic analysis has linked early-onset familial Alzheimer's disease (FAD) to the autosomal dominant inheritance of mutations in three genes: APP on chromosome 21, and the two homologous genes presenilin 1 and 2 (PS-1 and PS-2) on chromosomes 14 and 1, respectively (reviewed by Hardy, 1997; Price et al., 1998; Selkoe, 1998). So far, nine AD-linked mutations have been identified in APP, four have been found in PS-2, and the vast majority of mutations (> 58) have been mapped to PS-1 (for an update, visit http://molgen-www.uia.ac.be/ADMutations/). It is unclear why PS-1 is subject to such a high rate of mutation. One possibility we speculate is that considering these mutations are compatible with successful early development, they may confer some unknown advantage (e.g. resistance to certain forms of cancer), that later in life inadvertently results in AD. Another feature of almost all FAD mutations mapped to date are they are either missense and/or in-frame deletion mutations, suggesting that gross alteration in these gene products is incompatible with survival in humans.

The genetics of late-onset AD are more complex and a larger number of genes have been implicated as modifiers or risk factors for the disease (Price et al., 1998). The best example of this group is the inheritance of the #4 allele of apolipoprotein epsilon (ApoE4), which increases the lifetime risk and lowers the age of onset for developing AD. The accumulating data, therefore, indicates that AD is genetically heterogeneous. However, numerous other studies suggest involvement of additional non-genetic factors in AD etiology, such as mitochondrial DNA damage, oxidative stress, protein frameshift mutations, inflammatory processes, cell cycle misregulation, and apoptosis (Mattson, 1997; Cotman, 1998; Raina et al., 1999).

3. APP FUNCTION, PROCESSING, AND CALCIUM REGULATION

Multiple APP isoforms generated by alternative splicing are expressed in human brain (e.g. the resulting 695, 751 and 770 amino acids proteins are the most prominent brain isoforms), yet their function has remained enigmatic. Among the functions ascribed to APP are that it acts as a trophic factor, a modulator of cell adhesion, a regulator of intracellular Ca^{2+}, and a neuronal receptor of GTP-binding G_o protein (Mattson, 1997; Selkoe, 1998). However, APP is not essential for survival in mice, as animals disrupted of APP develop and reproduce normally. APP is a cell surface-expressed glycoprotein with a long NH_2-terminal extracellular domain, a single transmembrane domain, and a short cytoplasmic COOH-terminal tail (Figure 1;

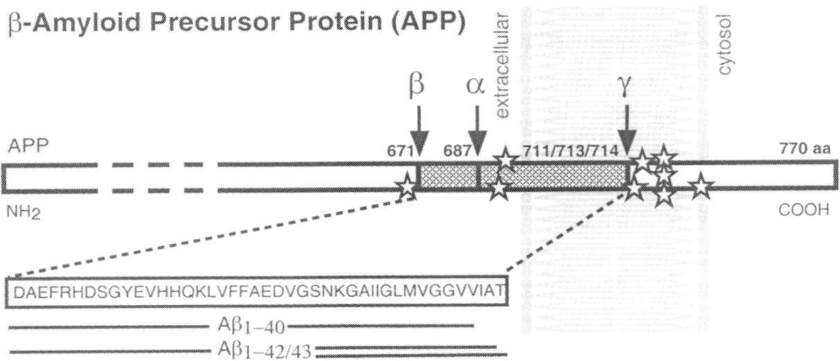

Figure 1. Schematic of APP protein structure. The membrane orientation of the 770 isoform is shown together with the location of the nine FAD mutations (indicated by stars). The proteolytic cleavage sites for α, β, and γ secretases are indicted with arrows. The amino acid sequence of the two major Aβ species, Aβ_{1-40} and A$\beta_{1-42/43}$, that are formed by cleavage of APP at a common NH$_2$-β-secretase site (671) but differ in cleavage at the γ-secretase site (711, or 713/714), are shown.

Mattson, 1997; Selkoe, 1998). APP undergoes a complex pathway of protein trafficking and is proteolytically cleaved at multiple sites generating different fragments (Selkoe, 1998). One fragment containing a portion of the transmembrane segment, named amyloid β-protein (Aβ), is the major component of neuritic and cerebrovascular plaques (Figure 1). Aβ is generated by proteolytic cleavage at the extracellular domain by β-secretase(s) and within the transmembrane domain by γ-secretase(s). Aβ fragments are predominantly 1–40 residues long (Aβ_{40}, although slightly longer 1–42 and 1–43 fragments (which we will collectively refer to as Aβ_{42} for simplicity) that are both more fibrillogenic and amyloidogenic arise due to differences in cleavage at the COOH-terminal γ-site (Figure 1; Selkoe, 1998). Aβ formation is precluded when a third enzyme, α-secretase, cleaves a site within the Aβ region of APP. Cleavage by this more predominant pathway results in the release of the large NH$_2$-terminal secreted form of APP (sAPPα). Production of sAPPα is stimulated by excitatory neurotransmitters, electrical activity and activators of protein kinase C (Mattson, 1997; Selkoe, 1998). sAPPα is believed to be neuroprotective, regulate neuronal excitability, and promote synaptic plasticity (Mattson, 1997; Selkoe, 1998). Enzymes with appropriate α and β cleavage specificities have been identified (Vassar et al., 1999). The identity of the γ-secretase is more controversial. Recent data have indicated that γ-secretase activity is, in large part, influenced by presenilin expression, leading to the suggestion that presenilins are γ-secretase (Haass and De Strooper, 1999). However, direct biochemical evidence that purified presenilin proteins possess proteolytic activity has not been demonstrated. The precise mechanisms controlling APP cleavage to generate Aβ are poorly understood. Agents that enhance production of Aβ include cell stressors (Mattson, 1997), as well, agents that elevate Ca^{2+} levels such as the ionophore A23187 and caffeine which stimulates release of Ca^{2+} from intracellular stores through the ryanodine receptor channels (RyR) (Querfurth and Selkoe, 1994; Querfurth et al., 1997). Furthermore, recent experiments suggest the intriguing possibility that Aβ can be generated by caspase cleavage during apoptosis (Gervais et al., 1999).

A predominant theory for the etiology of AD is the "amyloid cascade hypothesis" which postulates that production and accumulation of Aβ in plaques is the primary event that triggers a neurotoxic cascade leading to neurodegeneration and AD (Hardy, 1997). Alternatively, others have argued that amyloid plaques may, at best, be a by-product of the disease (Neve and Robakis, 1998). Regardless of the controversy surrounding the toxicity of Aβ, both it and sAPP fragments have been reported to modulate Ca^{2+} levels in cells. Extracellular application of sAPP to cultured neuronal cells decreases resting internal Ca^{2+} ($[Ca^{2+}]_i$) due to presumed activation of K^+ channels and suppression of Ca^{2+} influx through voltage-dependent channels and N-methyl-D-aspartate (NMDA) receptors (Mattson, 1997). In contrast, the Aβ-derived fragments cause a slow (hours to days) elevation of resting $[Ca^{2+}]_i$ that is thought to be mediated by Ca^{2+} influx through plasma membrane channels (Mattson, 1997). However, the exact mechanism for how Aβ increases $[Ca^{2+}]_i$ is not known, although evidence exists that Aβ can form Ca^{2+}- and ion-conducting pores upon its insertion into artificial membranes (Pollard et al., 1995). It is generally believed that Aβ disrupts ion homeostasis, causing oxidative damage of the plasma membrane, which directly or indirectly leads to cell death. The relationship between different APP cleavage products and how they affect Ca^{2+} levels is complex and poorly understood. Furthermore, understanding this relationship in the even more complex milieu of the brain will be hugely challenging.

4. PRESENILIN STRUCTURE, LOCALIZATION, AND DEVELOPMENTAL FUNCTIONS

Human PS-1 and PS-2 are multi-transmembrane proteins that share 67% sequence identity. The exact topology of the presenilins is debatable, although the most widely drawn models show proteins that weave through the membrane eight times with the NH_2- and COOH-terminal domains, and a large "loop" between transmembrane domains (TMD) 6 and 7 all orientated towards the cytoplasm (Figure 2; Haass and De Strooper, 1999). Although both proteins share extensive sequence identity along the entire length, their NH_2-terminal domains and the second half of the loop are highly divergent, suggesting that these unique regions could modulate different functions of the two presenilins. An examination of FAD presenilin-linked mutations mapped so far reveals that most are missense mutations in residues that are shared between the two presenilins, suggesting that the conserved residues are important for the function of the proteins. It is interesting to note that the

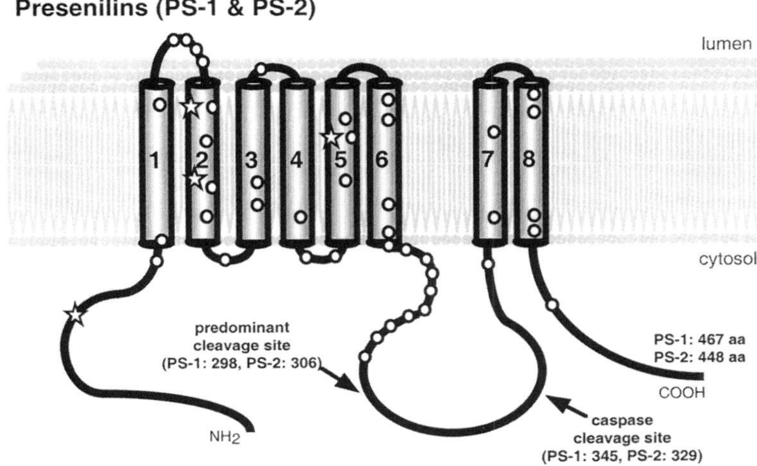

Figure 2. Schematic structure of presenilin proteins. The predicted topology of the presenilins, based on an eight TMD model is depicted. The two proteins (PS-1 467 amino acids long, and PS-2 448 amino acids long) are thought to have a similar structure. Also indicated are the two different proteolytic cleavage sites in the presenilin loop region. FAD-linked mutations in PS-1 are indicated by an "open circle" and those in PS-2 by a "star".

mutations cluster primarily in two locations, the TMD2 segment and the first half of the loop (Figure 2), hinting that these two regions may play important roles in AD pathogenesis.

Both human presenilin genes are ubiquitously expressed in most cell types, but at variable to low levels. In brain the proteins are more highly expressed in neurons than glia (Price et al., 1998). Knowledge of where presenilin proteins localize in cells is important for understanding their function. However, uncertainties exist regarding their precise subcellular localization. In neurons, endogenous PS-1 and PS-2 have been localized to the ER, and to vesicular structures of the somatodentritic compartment and axons (Price et al., 1998). When overexpressed the proteins predominantly localize to the endoplasmic reticulum (ER), the Golgi complex, and nuclear envelope (Haass and De Strooper, 1999). However, in non-neuronal cells endogenous presenilins have been localized to the ER, Golgi complex, centrosomes, centromeres and at the cell surface (Haass and De Strooper, 1999; Raina et al., 1999). It is not known to what extent these diverse locations reflect true sites of presenilin function.

Neither the function of presenilins in human brain nor the mechanisms by which FAD-linked mutations cause AD are known. Nevertheless, clues as to the functions of presenilins have emerged from studies of presenilin homologues in other species. In *Caenorhabditis elegans*, the presenilin homolog, *sel-12*, functions in cell signaling involving Notch-based receptors (Haass and De Strooper, 1999). *Sel-12* mutants are defective in vulva development and egg laying, and can be functionally rescued by either of the two human presenilin genes (Price et al., 1998). Mutation of the *Drosophila melanogaster* presenilin gene produces severe lethal Notch-like phenotypes, again alluding to an important role for presenilins in Notch signaling during development (Haass and De Strooper, 1999). Mammalian presenilins also appear critical for development, as disruption of the mouse PS-1 gene leads to death shortly after birth with embryos displaying central nervous system defects together with abnormal patterning of the axial skeleton and spinal ganglia (Price et al., 1998). Interestingly, PS1−/− mice can be rescued by human transgenes containing FAD-linked mutations, indicating that these FAD mutations do not affect presenilin functions related to embryo development (Price et al., 1998; Haass and De Strooper, 1999). In contrast, disruption of the mouse PS-2 gene produces no obvious defects, but PS-1/PS-2 double knock-out mice die earlier at embryonic day 9.5 and like PS-1−/− mice display severe misexpression of proteins involved in Notch signaling (Donoviel et al., 1999; Herreman et al., 1999). These results indicate that there is functional redundancy between PS-1 and PS-2; PS-1 can compensate for PS-2, but PS-2 cannot compensate for PS-1, at least during early mouse development.

Both PS proteins are frequently observed as cleaved products in protein lysates derived from normal tissues, cultured cells, and transgenic mice (Price et al., 1998; Haass and De Strooper, 1999). The predominant cleavage site is in the loop (Figure 2), but the function, if any, of this cleavage is not known. An alternative cleavage site, mediated by caspase activation, also occurs nearby in the loop (Steiner et al., 1998). In contrast to these cleaved products, full-length presenilin polypeptides accumulate as a major species in certain cells and tissues leaving unresolved which of the two forms (the full length or the cleaved products) are functional (Janicki and Monteiro, 1997; Parkin et al., 1999). It is worth noting that PS mutants deleted of the proteolytic cleavage sites successfully complement the *sel-12* egg laying defect in *C. elegans*, indicating that proteolytic cleavage may not be required for some presenilin functions (Haass and De Strooper, 1999). We proposed that protein cleavage diminishes presenilin functions, since presenilin fragments are less effective in inducing apoptosis than are the full-length proteins (Janicki and Monteiro, 1997).

There is considerable evidence linking presenilins to apoptosis. Initial indication for such a connection was the identification of a COOH-terminal 103 residue fragment of PS-2, termed ALG3 (see also Maki, this book), which rescued T-cells from receptor- and Fas-induced apoptosis in a dominant-negative manner (Vito et al., 1996). It has since been demonstrated that the overexpression of full length and NH_2-terminal fragments of PS-2 induce apoptosis and that cells expressing FAD mutations in either PS-1 or PS-2 have enhanced apoptotic activities relative to cells expressing wild-type presenilins (Vito et al., 1996; Mattson and Guo, 1999; Janicki and Monteiro, 1997). In non-neuronal HeLa cells, overexpression of the FAD PS-2(N141I) mutant potentiates the cell cycle arrest induced by presenilin overexpression, an affect that likely precedes apoptosis (Janicki and Monteiro, 1999). The mechanisms by which presenilins sensitize cells to apoptosis is not clear but speculations have focused on perturbations in Ca^{2+}, oxidative stress, mitochondria misregulation, destabilization of β-catenin, and increased signaling by heterotrimeric GTP-binding proteins (Mattson and Guo, 1999). Intriguingly, FAD-linked mutations in APP have also been linked to increased signaling through heterotrimeric GTP-binding proteins and cause increased apoptosis when overexpressed (Nishimoto et al., 1997). This has lead to speculation that increased apoptosis may be involved in AD pathogenesis (Cotman, 1998). Evidence that apoptosis is increased in AD is beginning to emerge as several apoptotic characteristics (e.g. cell shrinkage, increased DNA fragmentation, increased Bax expression and caspase activation) have been identified in AD brains (reviewed by Stadelmann et al., 1999).

5. THE PRESENILIN CONNECTION TO CALCIUM-BINDING PROTEINS AND CALCIUM MISREGULATION

Recent studies have identified a growing list of presenilin-binding proteins in hopes of obtaining insight into presenilin function. Of particular significance for this review is the interaction of three different Ca^{2+}-binding proteins, calsenilin, calmyrin, and sorcin, with the presenilins.

Calsenilin was recovered in a yeast two-hybrid (Y2H) screen by its interaction with the COOH-terminal 40 residue tail domain of PS-2 (Buxbaum et al., 1998). Binding of the two proteins was confirmed by coimmunoprecipitation of cotransfected calsenilin and presenilins. As expected, since the tail domain sequence is highly conserved between PS-2 and PS-1, both PS-1 and PS-2 coimmunoprecipitate with calsenilin. Calsenilin is a novel protein of 256 amino acids with highest homology to the recoverin family of Ca^{2+}-binding proteins. The protein contains four EF hands and binds radioactive $^{45}Ca^{2+}$ in blot overlay assays. However, unlike recoverin family members, calsenilin lacks a consensus myristoylation sequence at its N-terminus. Interestingly a protein identical to calsenilin called DREAM was subsequently found to be involved in transcriptional repression of the prodynorphin gene involved in memory and pain (Carrion et al., 1999; see also Maki, this book, for further comparison of these two proteins). Calsenilin/DREAM expression is highest in human brain, thyroid, thymus, and testis. Using an epitope-tagged construct the protein was found to localize predominantly to the cytoplasm in transfected cells and colocalized with presenilin when the two genes were cotransfected. Since Carrion et al. (1999) suggest calsenilin/DREAM is involved in transcriptional repression of genes, one might suspect the protein is also present in the nucleus. It will therefore be important to determine the location and intracellular dynamics of endogenous calsenilin. Apart from its role as a transcriptional repressor, how calsenilin influences presenilin function is not known. A putative clue may be its strong binding to the caspase-derived COOH-terminal fragment of PS-2 (Buxbaum et al., 1998). This binding has lead to suggestions that calsenilin either induces production of these caspase-cleaved presenilin fragments, suggesting a role in apoptosis, or that it stabilizes these fragments after their production (Buxbaum et al., 1998). Knock-out studies in mice should help to answer these questions.

A second Ca^{2+}-binding protein, calmyrin, was also isolated in Y2H screens by virtue of its interaction with a 50 amino acid segment comprising the first half of the PS-2 loop (Stabler et al., 1999). Interaction between calmyrin and PS-2 was confirmed by coimmunoprecipitation, affinity-chromatography and by colocalization studies. Deletion, mutagenesis, and Y2H studies narrowed the calmyrin-PS-2 interaction to the 31 NH_2-terminal most residues of the PS-2 loop. This region is highly conserved in sequence between presenilins, yet calmyrin binds PS-1 with approximately 12-fold lower affinity than PS-2. Site-directed mutagenesis of the three residues that differ between PS-1 and PS-2 in the loop segment indicated that all three amino acids contribute to calmyrin binding (Stabler et al., 1999). These results suggest that even small changes in sequence in the presenilin loop lead to structural alterations that affect functional protein-protein interactions. This may in part explain why most FAD mutations map to this region of presenilin (Figure 2).

Calmyrin is a 191 amino acid protein with highest homology to calcineurin B, the regulatory subunit of protein phosphatase 2B (Figure 3), and a recently-identified protein of unknown function termed KIP2 (Seki et al., 1999). Calmyrin was independently isolated in two other Y2H screens and called CIB due to its Ca^{2+}- and integrin α_{IIb}-binding (Naik et al., 1997) and KIP due to its interactions with eukaryotic DNA-dependent protein kinase, DNA-PKcs (Wu and Lieber, 1997). We chose to call the protein calmyrin (for calcium-binding myristoylated protein with homology to calcineurin) as it described the inherent properties of the protein without any bias towards its interacting partners. Calmyrin is highly conserved in mammals with the mouse homolog being 94% identical to the human protein (Figure 3; Saito et al., 1999). Calmyrin is myristoylated, and this fatty acid modification appears to regulate intracellular targeting of the protein (Stabler et al., 1999). Only two of four EF hand motifs in calmyrin are likely to bind Ca^{2+}, as the two NH_2-terminal EF hands have insertions that are predicted to prevent Ca^{2+} binding (Figure 3). Indeed, calmyrin binds Ca^{2+} as revealed in blot assays (Naik et al., 1997). Human calmyrin is located on chromosome 15q25.3-q26.1 (Seki et al., 1998) and is widely expressed with the highest levels in heart, lung and pancreas (Naik et al., 1997; Wu and Lieber, 1997; Stabler et al., 1999; Shock et al., 1999). The protein is expressed at lower, but uniform, levels in all regions of human brain examined (Stabler et al., 1999).

Calmyrin distributes intracellularly to both the nucleus and cytoplasm, although the protein lacks any recognizable nuclear localization signal (Stabler et al., 1999). The protein fractionates in a Triton-sensitive manner, indicating it bind to membranes. The protein localizes to long projections that emanate

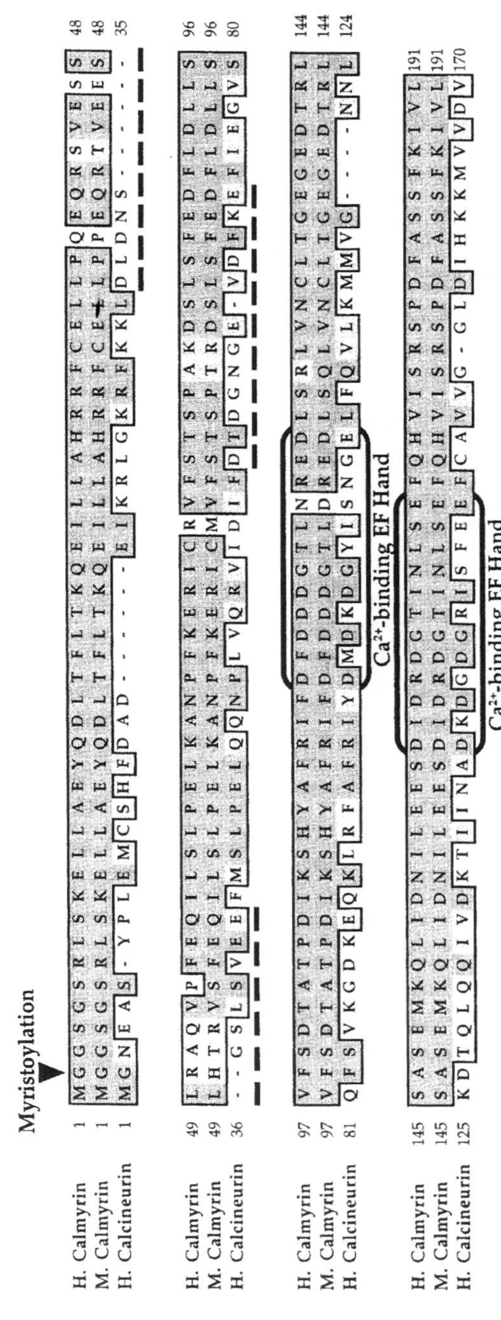

Figure 3. Comparison of human calmyrin, mouse calmyrin, and human calcineurin B proteins. Two 12 amino acid EF hand Ca^{2+}-binding loop sequences that are conserved in calmyrin are circled, whereas two other EF hand sequences in calcineurin that are disrupted in calmyrin due to insertions and indicated by a dashed line.

from the plasma membrane (Stabler et al., 1999), perhaps mediated by its binding to integrins. The binding of calmyrin to integrin α_{IIb} expressed in platelets, appears to be Ca^{2+}-dependent (Shock et al., 1999; see however Vallar et al., 1999). Recently, calmyrin was found to bind to Fnk and Snk, two polo-like serine/thereonine kinases that are induced by stimuli during long-term potentiation (LTP) (Kauselmann et al., 1999). These kinases translocate to dentrites following LTP stimulation suggesting a putative role in regulation of synaptic plasticity. The diverse and interesting interacting partners to which calmyrin binds does not provide precise answers as to its function. Although the protein shares highest homology to calcineurin it is not known if it functions in phosphatase regulation. The protein is likely to function in some signaling pathway since when it is transfected with PS-2 it increases apoptosis (Stabler et al., 1999). In further experiments it will be important to uncover the exact role calmyrin plays in cell signaling, its function in brain, and its relationship, if any, to AD pathogenesis.

Sorcin is the third member of the set of presenilin-interacting Ca^{2+}-binding proteins (Pack-Chung et al., 2000). Binding of sorcin to presenilin was predicted based on homology of sequences in the second half of the PS-2 loop (residues 319–347, just downstream of the calmyrin binding site) with the NH_2-terminal portion of synexin which in turn was known to interact with sorcin. Coimmunoprecipitation and cell fractionation studies confirmed that sorcin binds PS-2, but not PS-1. Surprisingly, sorcin does not bind full-length PS-2, but rather to its proteolytic cleaved COOH-terminal fragment (Pack-Chung et al., 2000). The significance of sorcin interaction with PS-2 fragments is unknown. However, it is interesting that sorcin, a penta EF hand Ca^{2+}-binding protein, is a known modulator of RyRs (Valdivia, 1998) which are major receptors involved in Ca^{2+} release from ER stores and are important for Ca^{2+}-induced Ca^{2+} release, especially in excitable cells such as neurons. Further information of the role of sorcin interaction with presenilin and its relevance to AD pathogenesis remain to be discovered.

Considering that presenilins are now known to interact with several Ca^{2+}-binding proteins, it is especially interesting that overexpressed presenilin FAD mutants alter cellular Ca^{2+} responses and homeostasis under certain conditions (Mattson and Guo, 1999). In fact, there is considerable and growing evidence that suggests both wild-type and mutant presenilins are involved in modulating Ca^{2+} levels, but the precise mechanisms of how this is governed are unclear. PC12 cells overexpressing presenilins bearing FAD mutations exhibit increased Ca^{2+} release when treated with agonists, such as carbachol and bradykinin. These agonists bind cell surface receptors causing stimulation of phospholipase C leading to hydrolysis of phosphadidylinisotol (4,5)-bisphosphate into inositol 1,4,5-triphosphate (IP_3) and diacylglycerol. IP_3 acts as an intracellular messenger by binding to inositol 1,4,5-triphosphate receptors (IP_3Rs) located at the ER, to induce Ca^{2+} release (Clapham, 1995). IP_3R release of Ca^{2+} is regulated by a number of factors, including phosphorylation of the receptor (Snyder et al., 1998), a buildup of Ca^{2+} in the ER lumen, and interestingly, the presence of both IP_3 and Ca^{2+} together (Berridge, 1998). Local increases in Ca^{2+}, therefore, can result in amplification of a Ca^{2+} signal inducing further release of Ca^{2+} from nearby receptors, causing Ca^{2+} sparks and wave propagation (Clapham, 1995; Bootman and Berridge, 1995). Direct evidence that presenilins containing FAD-linked mutations potentiate IP_3 Ca^{2+} release was demonstrated by photo-activation studies of caged IP_3 in *Xenopus laevis* oocytes. Interestingly, both PS-1 and PS-2 FAD mutants expressed in oocytes displayed similar defects in IP_3 responses such that they require lower levels of IP_3 to induce Ca^{2+} responses and the maximal levels of the responses they exhibit are considerably elevated (Leissring et al., 1999a, b). Similar elevations in Ca^{2+} responses were found for presenilin FAD mutants expressed in PC12 cells and treated with the sesquiterpene lactone, thapsigargin. Thapsigargin binds irreversibly to the sarcoplasmic-endoplasmic reticulum Ca^{2+}-ATPases (SERCAs) Ca^{2+} pumps, depleting ER Ca^{2+} stores and raising cytoplasmic Ca^{2+} (Guo et al., 1996, 1997). Collectively these Ca^{2+} studies all indicate that mutant presenilins induce elevations of Ca^{2+} release. Mechanistically this could occur in a number of different ways including the following: (1) the size of the ER Ca^{2+} stores may be increased in cells expressing FAD-linked presenilin mutations, (2) receptors involved in Ca^{2+} release from the ER stores are altered in the FAD mutants and require lower thresholds of either Ca^{2+} or IP_3 to induce Ca^{2+} release, (3) signaling pathways that affect production of Ca^{2+}, IP_3, regulation of Ca^{2+} pumps or their receptors are altered, or that (4) plasma-membrane Ca^{2+} channels may be altered by FAD mutants, especially since presenilin mutants show decreased sensitivity to nifedipine which blocks L type-voltage dependent Ca^{2+} channels (Guo et al., 1996, 1997). In fact, presenilins proteins share weak homology to voltage-dependent Ca^{2+} channels (Sherrington et al., 1995), though there is no evidence to date to suggest that presenilins function as channels. An important consideration with regard to the above defects is the role that Ca^{2+}-binding proteins may play in regulating presenilin functions. In fact coexpression of calbindin28k, an EF hand containing protein involved in calcium buffering, with presenilins, protects cells from apoptosis induced by misregulation of Ca^{2+} homeostasis induced by mutant presenilins (Guo et al., 1998). Collectively these data strongly implicate presenilins in Ca^{2+} regulation, and suggest that FAD mutations in presenilins cause altered Ca^{2+} misregulation.

6. CALCIUM CHANGES IN BRAIN AND CULTURED FIBROBLASTS FROM AD-AFFECTED INDIVIDUALS

Ideally, one would prefer to study the role of Ca^{2+} in AD pathogenesis *in vivo* by examining neurons from affected patients. Unfortunately, as brain biopsies are not common practice, the best alternatives have been to examine post-mortem brain tissue or fibroblast cells cultured from AD affected and presymptomatic individuals. Notably both of these approaches have provided corroborating evidence connecting AD and calcium regulation. It is not surprising that the most compelling evidence of Ca^{2+}-related abnormalities has emerged from studies of individuals carrying autosomal dominant FAD mutations. Fibroblast cell lines cultured from family members carrying FAD mutations in either APP and PS-1 have indeed suggested that Ca^{2+} regulation is abnormal (McCoy et al., 1993; Ito et al., 1994; Hirashima et al., 1996; Gibson et al., 1997; Etcheberrigaray et al., 1998). Cells containing presenilin 1 mutation PS-1(A246E) frequently show elevated receptor-mediated Ca^{2+} responses induced by agonist bradykinin (Gibson et al., 1997; Etcheberrigaray et al., 1998). Similar but less pronounced responses are sometimes found in fibroblasts containing the FAD APP670/671 mutation (Gibson et al., 1997). In fact, fibroblasts with the latter mutation were found to have lower Ca^{2+} responses to bombesin (Gibson et al., 1997), which contrasts to the elevated responses seen in PS-1 FAD fibroblasts (Ito et al., 1994; Hirashima et al., 1996). These data suggest that, although Ca^{2+} misregulation is common to FAD mutations in both APP and presenilin genes, the underlying mechanism leading to disease in the two genes may involve different signal transduction pathways. We should caution however, that not all fibroblasts containing the FAD mutations displayed the Ca^{2+} defects discussed above (Gibson et al., 1997; Etcheberrigaray et al., 1998). The reason for this is at present unclear; maybe other unknown genetic influences or age related changes might be responsible. If fibroblasts are to be useful for AD diagnosis it will be helpful to resolve this issue. The present studies have been restricted to very few families, and therefore it is essential that similar comparisons be made in more families, especially those carrying different mutations in APP, PS-1 and PS-2 genes. Furthermore, if Ca^{2+} responses are truly defective in FAD cases then these alterations should manifest similarly in cells cultured from unrelated individuals carrying identical mutations.

A number of investigators have examined whether changes in expression of Ca^{2+}-binding proteins or Ca^{2+}-release proteins, correlate with disease progression, especially in late-onset AD. Such studies are possible since AD progression occurs with a highly distinctive staging pattern of neuropathological changes in brain (Braak and Braak, 1991). The region of brain most affected during early stages of the disease is the entorhinal cortex, from which neurons project to the hippocampus, a region of the brain known to be important for memory. Such studies have revealed quantitative changes of both Ca^{2+}-binding proteins and RyRs during AD progression. The Ca^{2+}-binding proteins that have been most extensively studied are calbindin-D28k, calretinin, and parvalbumin, all EF hand-containing proteins that are important in Ca^{2+} buffering and homeostasis. During early pathology of AD, parvalbumin immunoreactivity is decreased in certain entorhinal brain subregions, whereas calbindin reactivity is increased (Mikkonen et al., 1999). Other studies have also shown that a subset of parvalbumin-containing neurons become vulnerable to degeneration in AD (Solodkin et al., 1996). Similarly, an initial increase in binding of ryanodine to RyRs, which is followed by a decrease in RyRs, correlates with AD progression (Kelliher et al., 1999). In contrast, calretinin immunoreactive neurons seem to be unaffected during disease progression. Since these changes are correlative it will important to establish if these alterations reflect compensatory changes due to some prior signaling defect or whether changes in these proteins precede neurodegeneration.

7. CONCLUSIONS AND OUTLOOK

As summarized above there is growing evidence that AD is associated with alterations in Ca^{2+} regulation. Ca^{2+} changes have been noted in brain and fibroblasts from AD affected individuals. Additionally, and providing perhaps the strongest evidence, APP and presenilins, genetic factors associated with early-onset AD, are both intimately tied to Ca^{2+} regulation. Mutations in all three genes cause defects in Ca^{2+} responses in cells. The exact mechanisms by which mutations perturb Ca^{2+} levels or its release are less clear. In the case of APP, mutations appear to cause an alteration in a Ca^{2+} signaling pathway (or pathways) or to effect APP cleavage and the subsequent products (sAPP and Aβ), which lead to defects in Ca^{2+} homeostasis or signaling. For presenilins, the evidence linking FAD mutations to Ca^{2+} misregulation in cells is more direct as mutant proteins exhibit defective calcium responses and are proapoptotic when tested in different systems. Moreover, presenilins appear to localized to the ER, an important intracellular site that can function as either a "store" or a "sink" for Ca^{2+}. Finally, presenilins interact with calsenilin, calmyrin, and sorcin, three different Ca^{2+}-binding proteins that are related to members of signal transduction pathways. All of these findings have revealed promising leads into the misregulation of Ca^{2+} in AD, but we must caution that most studies have been performed in cultured cells. How Ca^{2+} responses in individual cells translate in terminally-differentiated neurons and supporting cells in human brain is unknown. For example, it will be interesting to know if similar defects in Ca^{2+} regulation also occur in neurons affected in AD which may express different channels, receptors, and Ca^{2+}-binding proteins. Furthermore, how age-related changes affect these processes needs to be better understood. Clearly the challenge for the future is to determine the precise defects in Ca^{2+} misregulation that participate in AD pathogenesis. The contribution of calcium stores, Ca^{2+}-binding proteins, Ca^{2+} activated channels, and signaling processes will have to be more fully understood if drugs are to be designed to treat and cure this devastating neurodegenerative disease. Based on the rapid progress in AD research during the past few years, it may not be long before the exact mechanisms that lead to the etiology of AD are uncovered and rational therapeutic strategies to prevent and/or ameliorate AD become possible.

ACKNOWLEDGEMENTS

Due to space restrictions many important primary articles could not be referenced, instead reviews are generally quoted. We thank authors who provided information pertinent for this review and Dr. Ann Pluta for critical comments. MJM acknowledges support from the National Institutes on Aging.

REFERENCES

Berridge, M.J., 1998, Neuronal calcium signaling, *Neuron* 21, 13–26.

Bootman, M.D. and Berridge, M.J., 1995, The elemental principles of calcium signaling, *Cell* 83, 675–678.

Braak, H. and Braak, E., 1991, Neuropathological staging of Alzheimer-related changes, *Acta Neuropathol.* 82, 239–259.

Buxbaum, J.D., Choi, E.-K., Luo, Y., Lilliehook, C., Crowley, A.C., Merriam, D.E. and Wasco, W., 1998, Calsenilin: A calcium-binding protein that interacts with the presenilins and regulates the levels of a presenilin fragment, *Nature Med.* 4, 1177–1181.

Carrion, A.M., Link, W.A., Ledo, F., Mellstrom, B. and Naranjo, J.R., 1999, DREAM is a Ca^{2+}-regulated transcriptional repressor, *Nature* 398, 80–84.

Clapham, D.E., 1995, Calcium signaling, *Cell* 80, 259–268.

Cotman, C.W., 1998, Apoptosis decision cascades and neuronal degeneration in Alzheimer's disease, *Neurobiol. Aging* 19, S29–S32.

Donoviel, D.B., Hadjantonakis, A.K., Ikeda, M., Zheng, H., Hyslop, P.S. and Bernstein, A., 1999, Mice lacking both presenilin genes exhibit early embryonic patterning defects, *Genes Dev.* 13, 2801–2810.

Etcheberrigaray, R., Hirashima, N., Nee, L., Prince, J., Govoni, S., Racchi, M., Tanzi, R.E. and Alkon, D.L., 1998, Calcium responses in fibroblasts from asymptomatic members of Alzheimer's disease families, *Neurobiol. Dis.* 5, 37–45.

Gervais, F.G., Xu, D., Robertson, G.S., Vaillancourt, J.P., Zhu, Y., Huang, J., LeBlanc, A., Smith, D., Rigby, M., Shearman, M.S., Clarke, E.E., Zheng, H., Van Der Ploeg, L.H., Ruffolo, S.C., Thornberry, N.A., Xanthoudakis, S., Zamboni, R.J., Roy, S. and Nicholson, D.W., 1999, Involvement of caspases in proteolytic cleavage of Alzheimer's amyloid-beta precursor protein and amyloidogenic A beta peptide formation, *Cell* 97, 395–406.

Gibson, G.E., Vestling, M., Zhang, H., Szolosi, S., Alkon, D., Lannfelt, L., Gandy, S. and Cowburn, R.F., 1997, Abnormalities in Alzheimer's disease fibroblasts bearing the APP670/671 mutation, *Neurobiol. Aging* 18, 573–580.

Guo, Q., Furukawa, K., Sopher, B.L., Pham, D.G., Xie, J., Robinson, N., Martin, G.M. and Mattson, M.P., 1996, Alzheimer's PS-1 mutation perturbs calcium homeostasis and sensitizes PC12 cells to death induced by amyloid beta-peptide, *Neuroreport* 8, 379–383.

Guo, Q., Sopher, B.L., Furukawa, K., Pham, D.G., Robinson, N., Martin, G.M. and Mattson, M.P., 1997, Alzheimer's presenilin mutation sensitizes neural cells to apoptosis induced by trophic factor withdrawal and amyloid beta-peptide: Involvement of calcium and oxyradicals, *J. Neurosci.* 17, 4212–4222.

Guo, Q., Christakos, S., Robinson, N. and Mattson, M.P., 1998, Calbindin D28k blocks the proapoptotic actions of mutant presenilin 1: Reduced oxidative stress and preserved mitochondrial function, *Proc. Natl. Acad. Sci. USA* 95, 3227–3232.

Haass, C. and De Strooper, B., 1999, The presenilins in Alzheimer's disease-proteolysis hold the key, *Science* 286, 916-=919.

Hardy, J., 1997, Amyloid, the presenilins and Alzheimer's disease, *Trends Neurosci.* 20, 154–159.

Herreman, A., Hartmann, D., Annaert, W., Saftig, P., Craessaerts, K., Serneels, L., Umans, L., Schrijvers, V., Checler, F., Vanderstichele, H., Baekelandt, V., Dressel, R., Cupers, P., Huylebroeck, D., Zwijsen, A., Van Leuven, F. and De Strooper, B., 1999, Presenilin 2 deficiency causes a mild pulmonary phenotype and no changes in amyloid precursor protein processing but enhances the embryonic lethal phenotype of presenilin 1 deficiency, *Proc. Natl. Acad. Sci. USA* 96, 11872–11877.

Hirashima, N., Etcheberrigaray, R., Bergamaschi, S., Racchi, M., Battaini, F., Binetti, G., Govoni, S. and Alkon, D.L., 1996, Calcium responses in human fibroblasts: a diagnostic molecular profile for Alzheimer's disease, *Neurobiol. Aging* 17, 549–555.

Ito, E., Oka, K., Etcheberrigaray, R., Nelson, T.J., McPhie, D.L., Tofel-Grehl, B., Gibson, G.E. and Alkon, D.L., 1994, Internal Ca^{2+} mobilization is altered in fibroblasts from patients with Alzheimer disease, *Proc. Natl. Acad. Sci. USA* 91, 534–538.

Janicki, S. and Monteiro, M.J., 1997, Increased apoptosis arising from increased expression of the Alzheimer's disease-associated presenilin-2 mutation (N141I), *J. Cell Biol.* 139, 485–495.

Janicki, S.M. and Monteiro, M.J., 1999, Presenilin overexpression arrests cells in the G1 phase of the cell cycle. Arrest potentiated by the Alzheimer's disease PS2(N141I) mutant, *Am. J. Pathol.* 155, 135–144.

Kauselmann, G., Weiler, M., Wulff, P., Jessberger, S., Konietzko, U., Scafidi, J., Staubli, U., Bereiter-Hahn, J., Strebhardt, K. and Kuhl, D., 1999, The polo-like protein kinases Fnk and Snk associate with a Ca(2+)- and integrin-binding protein and are regulated dynamically with synaptic plasticity, *EMBO J.* 18, 5528–5539.

Kelliher, M., Fastbom, J., Cowburn, R.F., Bonkale, W., Ohm, T.G., Ravid, R., Sorrentino, V. and O'Neill, C., 1999, Alterations in the ryanodine receptor calcium release channel correlate with Alzheimer's disease neurofibrillary and beta-amyloid pathologies, *Neuroscience* 92, 499–513.

Leissring, M.A., Paul, B.A., Parker, I., Cotman, C.W. and LaFerla, F.M., 1999, Alzheimer's presenilin-1 mutation potentiates inositol 1,4,5-trisphosphate-mediated calcium signaling in *Xenopus* oocytes, *J. Neurochem.* 72, 1061–1068.

Leissring, M.A., Parker, I. and LaFerla, F.M., 1999, Presenilin-2 mutations modulate amplitude and kinetics of inositol 1, 4,5-trisphosphate-mediated calcium signals, *J. Biol. Chem.* 274, 32535–32538.

Mattson, M.P., 1997, Cellular actions of β-amyloid precursor protein and its soluble and fibrillogenic derivatives, *Physiol. Rev.* 77, 1081–1132.

Mattson, M.P. and Guo, Q., 1999, The presenilins, *The Neuroscientist* 5, 112–124.

McCoy, K.R., Mullins, R.D., Newcomb, T.G., Ng, G.M., Pavlinkova, G., Polinsky, R.J., Nee, L.E. and Sisken, J.E., 1993, Serum- and bradykinin-induced calcium transients in familial Alzheimer's fibroblasts, *Neurobiol. Aging* 14, 447–455.

Mikkonen, M., Alafuzoff, I., Tapiola, T., Soininen, H. and Miettinen, R., 1999, Subfield- and layer-specific changes in parvalbumin, calretinin and calbindin-D28K immunoreactivity in the entorhinal cortex in Alzheimer's disease, *Neuroscience* 92, 515–532.

Naik, U.P., Patel, P.M. and Parise, L.V., 1997, Identification of a novel calcium-binding protein that interacts with the integrin alphaIIb cytoplasmic domain, *J. Biol. Chem.* 272, 4651–4654.

Neve, R.L. and Robakis, N.K., 1998, Alzheimer's disease: A re-examination of the amyloid hypothesis, *Trends Neurosci.* 21, 15–19.

Nishimoto, I., 1998, A new paradigm for neurotoxicity by FAD mutants of betaAPP: A signaling abnormality, *Neurobiol. Aging* 19, S33–S38.

Pack-Chung, E., Myers, M.B., Pettingell, W.P., Cheng, I., Moir, R.D., Brownawell, A.M., Tanzi, R.E. and Kim, T.-W., 2000, Presenilin 2 interacts with sorcin, a modulator of the ryanodine receptor, *J. Biol. Chem.* 275, 14440–14445.

Parkin, E.T., Hussain, I., Karran, E.H., Turner, A.J. and Hooper, N.M., 1999, Characterization of detergent-insoluble complexes containing the familial Alzheimer's disease-associated presenilins, *J. Neurochem.* 72, 1534–1543.

Pollard, H.B., Arispe, N. and Rojas, E., 1995, Ion channel hypothesis for Alzheimer amyloid peptide neurotoxicity, *Cell Mol. Neurobiol.* 15, 513–526.

Price, D.L., Tanzi, R.E., Borchelt, D.R. and Sisodia, S.S., 1998, Alzheimer's disease: genetic studies and transgenic models, *Ann. Rev. Genet.* 32, 461–493.

Querfurth, H.W. and Selkoe, D.J., 1994, Calcium ionophore increases amyloid beta peptide production by cultured cells, *Biochemistry* 33, 4550–4561.

Querfurth, H.W., Jiang, J., Geiger, J.D. and Selkoe, D.J., 1997, Caffeine stimulates amyloid beta-peptide release from beta-amyloid precursor protein-transfected HEK293 cells, *J. Neurochem.* 69, 1580–1591.

Raina, A.K., Monteiro, M.J., McShea, A. and Smith, M.A., 1999, The role of cell cycle-mediated events in Alzheimer's disease, *Int. J. Exp. Pathol.* 80, 71–76.

Saito, T., Seki, N., Hattori, A., Hayashi, A., Abe, M., Araki, R., Fujimori, A., Fukumura, R., Kozuma, S. and Matsuda, Y., 1999, Structure, expression profile, and chromosomal location of a mouse gene homologous to human DNA-PKcs interacting protein (KIP) gene, *Mamm. Genome* 10, 315–317.

Seki, N., Hayashi, A., Abe, M., Araki, R., Fujimori, A., Fukumura, R., Hattori, A., Kozuma, S., Ohhira, M., Hori, T. and Saito, T., 1998, Chromosomal assignment of the gene for human DNA-PKcs interacting protein (KIP) on chromosome 15q25.3-q26.1 by somatic hybrid analysis and fluorescence in situ hybridization, *J. Hum. Genet.* 43, 275–277.

Seki, N., Hattori, A., Hayashi, A., Kozuma, S., Ohira, M., Hori, T. and Saito, T., 1999, Structure, expression profile and chromosomal location of an isolog of DNA-PKcs interacting protein (KIP) gene, *Biochim. Biophys. Acta* 1444, 143–147.

Selkoe, D.J., 1998, The cell biology of β-amyloid precursor protein and presenilin in Alzheimer's disease, *Trends Cell Biol.* 8, 447–453.

Sherrington, R., Rogaev, E.I., Liang, Y., Rogaeva, E.A., Levesque, G., Ikeda, M., Chi, H., Lin, C., Li, G., Holman, K., et al., 1995, Cloning of a gene bearing missense mutations in early-onset familial Alzheimer's disease, *Nature* 375, 754–760.

Shock, D.D., Naik, U.P., Brittain, J.E., Alahari, S.K., Sondek, J. and Parise, L.V., 1999, Calcium-dependent properties of CIB binding to the integrin alphaIIb cytoplasmic domain and translocation to the platelet cytoskeleton, *Biochem. J.* 342, 729–735.

Snyder, S.H., Lai, M.M. and Burnett, P.E., 1998, Immunophilins in the nervous system, *Neuron* 21, 283–294.

Solodkin, A., Veldhuizen, S.D. and Van Hoesen, G.W., 1996, Contingent vulnerability of entorhinal parvalbumin-containing neurons in Alzheimer's disease, *J. Neurosci.* 16, 3311–3321.

Stabler, S.M., Ostrowski, L.L., Janicki, S.M. and Monteiro, M.J., 1999, A myristoylated calcium-binding protein that preferentially interacts with the Alzheimer's disease presenilin 2 protein, *J. Cell Biol.* 145, 1277–1292.

Stadelmann, C., Deckwerth, T.L., Srinivasan, A., Bancher, C., Bruck, W., Jellinger, K. and Lassmann, H., 1999, Activation of caspase-3 in single neurons and autophagic granules of granulovacuolar degeneration in Alzheimer's disease.Evidence for apoptotic cell death, *Am. J. Pathol.* 155, 1459–1466.

Steiner, H., Capell, A., Pesold, B., Citron, M., Kloetzel, P.M., Selkoe, D.J., Romig, H., Mendla, K. and Haass, C., 1998, Expression of Alzheimer's disease-associated presenilin-1 is controlled by proteolytic degradation and complex formation, *J. Biol. Chem.* 273, 32322–32331.

Valdivia, H.H., 1998, Modulation of intracellular Ca^{2+} levels in the heart by sorcin and FKBP12, two accessory proteins of ryanodine receptors, *Trends Pharmacol. Sci.* 19, 479–482.

Vallar, L., Melchior, C., Plancon, S., Drobecq, H., Lippens, G., Regnault, V. and Kieffer, N., 1999, Divalent cations differentially regulate integrin alphaIIb cytoplasmic tail binding to beta3 and to calcium- and integrin-binding protein, *J. Biol. Chem.* 274, 17257–17266.

Vassar, R., Bennett, B.D., Babu-Khan, S., Kahn, S., Mendiaz, E.A., Denis, P., Teplow, D.B., Ross, S., Amarante, P., Loeloff, R., Luo, Y., Fisher, S., Fuller, J., Edenson, S., Lile, J., Jarosinski, M.A., Biere, A.L., Curran, E., Burgess, T., Louis, J.C., Collins, F., Treanor, J., Rogers, G. and Citron, M., 1999, Beta-secretase cleavage of Alzheimer's amyloid precursor protein by the transmembrane aspartic protease BACE, *Science* 286, 735–741.

Vito, P., Wolozin, B., Ganjei, J.K., Iwasaki, K., Lacana, E. and D'Adamio, L.D., 1996, Requirement of the familial Alzheimer's disease gene PS2 for apoptosis, *J. Biol, Chem.* 271, 31025–31028.

Wu, X. and Lieber, M.R., 1997, Interaction between DNA-dependent protein kinase and a novel protein, KIP, *Mutat. Res.* 385, 13–20.

The Role of Intracellular Calcium Signaling in Motoneuron Function and Disease

Bernhard U. Keller

1. INTRODUCTION

Precise motoneuron performance is essential for many aspects of nervous system function, including swallowing, breathing, locomotion and movement of the tongue. Motoneurones are also particularly vulnerable during human amyotrophic lateral sclerosis (ALS) and corresponding animal models of this neurodegenerative disease. While some motoneuron populations including spinal and brain stem motoneurones are particularly impaired, other populations like oculomotor neurones are largely resistant to ALS-related degeneration. This is a well-known phenomenon in advanced clinical stages of human ALS, but also in related animal models of motoneuron disease (Elliot and Snider, 1995; Ince et al., 1993; Reiner et al., 1995). Selective vulnerability of motoneurones has been closely linked to cell-specific disruptions of calcium signaling, but the underlying cellular and molecular events are only little understood. Recent progress in the experimental analysis of calcium signaling has permitted investigations of calcium regulation in different neuron types, in particular in selectively vulnerable and resistant motoneuron populations in animal models of neurodegenerative disease. This chapter provides an overview of recent advances in this field with an emphasis on calcium signal cascades in ALS-related motoneuron damage. By avoiding technical specialities that can be taken from the reference list, this article is aimed at readers with a background in basic academic or clinical research without specific knowledge about motoneuron physiology or microfluorometric calcium measurements.

Under physiological conditions, calcium signals in motoneurones have been associated with multiple processes, including regulation of action po-

Bernhard U. Keller • Zentrum Physiologie und Pathophysiologie, Universität Göttingen, Göttingen, Germany.

R. Pochet, R. Donato, J. Haiech, C. Heizmann and V. Gerke (eds.): Calcium: The Molecular Basis of Calcium Action in Biology and Medicine, 625–637.

tential (AP) firing rate, synaptic plasticity and responses to neuromodulators such as serotonin (Krieger et al., 1994; Bayliss et al., 1997; Lips and Keller, 1999). On the cellular level, these events are shaped and modulated by calcium influx through membrane ion channel proteins, release from intracellular stores, calcium uptake and extrusion across cellular membranes (McBurny and Neering, 1987; Blaustein, 1988; Baimbridge et al., 1992; Neher, 1995; Lips and Keller, 1998). Under pathophysiological conditions like those found during motoneuron disease, neuronal damage is paralleled by severe disruptions of calcium-related signaling resulting either from excess glutamate-mediated ("excitotoxic") calcium influx or related cellular disturbances (DePaul et al. 1988; Choi, 1988; Rothstein and Kuncl, 1995; Krieger et al., 1996; Shaw and Ince, 1997; Roy et al., 1998). For example, in human amyotrophic lateral sclerosis (ALS) and related animal models, impaired function of excitatory synapses and uncontrolled calcium influx have been directly associated with selective motoneuron degeneration (Medina et al., 1996; Meldrum and Garthwaite, 1990; Rothstein et al., 1992, 1995; Shaw et al., 1999). Also, in genetically determined forms of motoneuron disease both in humans and mice, disruptions of either axonal neurofilaments ("NFL") or the enzyme superoxide dismutase ("SOD1") result in calcium-dependent damage of motoneurones (Tu et al., 1996; Bruijn et al., 1998; Morrison and Morrison, 1998; Siklos et al., 1998; Williamson et al., 1998). In agreement with these elementary analysis, pharmacological reductions of cellular calcium loads provide valuable neuroprotection. This is utilized in recent clinical therapies based on the drug riluzole thought to prevent motoneuron degeneration by reducing calcium influx and excitatory synaptic activity (Smith et al., 1992; Gurney et al., 1996; Roy et al. 1998).

At present, the role of individual cellular parameters for selective neuronal vulnerability is only little understood. For example, studies on hippocampal cells have suggested that low cytosolic concentrations of calcium-binding proteins provide *neuroprotection*, mainly by enhancing calcium-dependent inactivation ("closing") of voltage-dependent calcium channels (Chad, 1989; Abdel-Hamid and Baimbridge, 1997; Nägerl and Mody, 1998). The best studied calcium binding proteins are calbindin and parvalbumin, but also calmodulin and the family of S-100 related proteins (Celio, 1990; Baimbridge et al., 1992; Schafer and Heizmann, 1996; Christakos et al., 2000). The neurodegenerative effect of calbindin has recently received support from studies of transgenic animals, where genetic "knock-out" of this cytosolic calcium-binding protein protected hippocampal cells against ischemia-related damage (Klapstein et al., 1998). On the other hand, several studies have pointed out that *increased* concentrations of cytosolic calcium-binding proteins provide neuroprotection in cellular model systems of ALS and other neurodegenerative diseases (Alexianu et al., 1994, 1998; Reiner et al., 1995; Tymianski et al., 1994; Roy et al., 1998; Christakos et al., this book). In this case, protective effects are mainly attributed to the ability of cytosolic calcium-binding structures ("calcium buffers") to reduce amplitudes of intracellular free calcium levels during glutamate-mediated Ca^{2+} influx and related calcium-dependent cellular disturbances.

This chapter summarizes recent advances in the understanding of endogenous calcium signaling and regulation ("homeostasis") in motoneurones. They are primarily based on the finding that during ALS-related disease brain stem and spinal cord motoneurones are selectively impaired, while others like hippocampal and oculomotor neurones are essentially unaffected. By performing a quantitative analysis of calcium homeostasis, it has recently been possible to compare individual cellular parameters in different neuron types (Lips and Keller, 1998; Palecek et al., 1999; Vanselow and Keller, 2000). In short, such measurements indicate notable differences in calcium signaling in different cell populations, where endogenous calcium buffering is extraordinary low in selectively vulnerable motoneurones.

2. PHYSIOLOGICAL CALCIUM DYNAMICS AND BUFFERING IN MOTONEURONES

Several studies have investigated calcium signaling in motoneurones in their functionally intact state of rhythmic electrical activity (Lev-Tov and O'Donovan, 1995; Ladewig and Keller, 2000). Such processes are thought to reflect the cellular basis of locomotion in the spinal cord and other physiologically relevant mechanisms like breathing and swallowing in the brain stem. A particularly valuable experimental model system is represented by the *in vitro* slice preparation of the mouse brain stem, where the neuronal kernel that generates rhythmic respiratory activity is still intact (Smith et al., 1991; Lips and Keller, 1998; Frermann et al., 1999). In this situation, robust rhythmic respiratory-related activity of brain stem motoneurones in the nucleus hypoglossus is observed by electrophysiological nerve recordings and parallel patch clamp experiments. Under single-cell recording conditions (Keller et al., 1991), rhythmic inspiratory activity of individual hypoglossal motoneurones is represented by repetitive clusters of action potential discharges ("bursts") that are separated by silent ("expiratory") periods lasting several seconds (Figure 1).

In this intact preparation, simultaneous patch clamp and calcium imaging measurements have recently been performed on identified motoneurones (Lips and Keller, 1998, 1999; Ladewig and Keller, 2000). During whole-cell recording, rhythmic-respiratory activity is represented by clusters of excitatory postsynaptic potentials (EPSPs) in association with high frequency burst discharges once the threshold for APs is reached. During inspiratory bursts, rapid rises in somatic and dendritic calcium concentrations are observed using calcium imaging. Somatic and dendritic calcium oscillations display amplitudes between 50 and 200 nM and robust calcium transients are readily observed up to dendritic distances around 100 μm (Figure 1). The physiological role of Ca^{2+} oscillations in motoneurones has been investigated

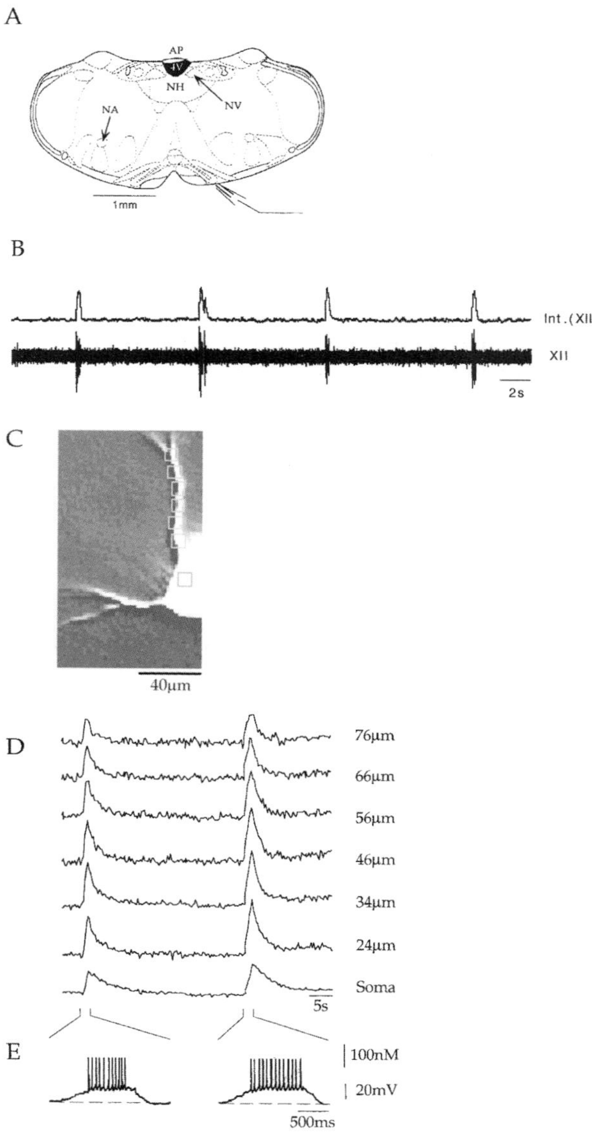

Figure 1. Rhythmic activity and associated calcium oscillations in brain stem motoneurones. (A) Schematic drawing of a transversal section showing the rhythmically active slice preparation of the medulla oblongata including hypoglossal regions. NA: *Nuc ambiguus*, AP: area postrema, NH: *N. hypoglossus*, 4V: 4th ventricle, NV: *N. vagus*. (B) Activity of nerve roots shows compound action potentials during rhythmic-inspiratory discharges. Upper trace shows integrated activity. (C) Micrograph of soma and proximal dendrites of a rhythmically active motoneuron with areas used for calcium measurements. (D) Rhythmic spontaneous discharges and concurrent calcium transients in 6 dendritic compartments at distances from 24–76 μm. (E) Membrane potential recorded in current clamp mode at higher temporal resolution (adapted with permission after Ladewig and Keller, 2000).

in detail by several groups (Bayliss et al., 1997; Viana et al. 1993a, b; Lips and Keller, 1999). In short, burst-related Ca^{2+} elevations are closely linked to action-potential induced openings of voltage dependent Ca^{2+} channels. Subsequently, Ca^{2+}-dependent activation of K^+_{ca}-conductances results in K^+_{ca}-dependent after hyperpolarizations that are a strong determinant of action potential discharge frequency during bursts (Bayliss et al., 1997; Viana et al., 1993a, b). To better define the physiological dynamics of endogenous Ca^{2+} oscillations, several strategies permit a "correction" for experiment-specific parameters like the presence of Ca^{2+} indicator dye and relatively low bath temperatures. By performing such analysis, "physiological" recovery time constants of Ca^{2+} transients are estimated as 70 ms (Lips and Keller, 1999). For a corresponding analysis of Ca^{2+} amplitudes, an average elevation of *several micromolars* is extrapolated for somatic oscillations. Similarly, each action potential is associated with a Ca^{2+}-mediated charge influx of 2 pC into the somatic compartment, which is two orders of magnitude larger compared to 0.01 pC estimated for a single NMDA-receptor mediated excitatory postsynaptic potential (Lips and Keller, 1999).

The "added buffer" approach originally introduced by Neher and Augustine (1992) permits the investigation of motoneuron Ca^{2+} signaling in even more quantitative detail. In essence, cellular Ca^{2+} homeostasis is probed by a competition between "endogenous" (S) and "exogenous" buffers (B) of known quantity that is gradually injected into the cell via the electrophysiological recording pipette (Neher, 1995). Such experiments determine the endogenous Ca^{2+} binding ratio κ_S of a cell, which reflects the ability of the combined set of endogenous cytosolic buffers to bind calcium. For example, a value $\kappa_S = 100$ indicates that only 1 out of 100 calcium ions in the cytosol contributes to the intracellular free calcium concentration, while 99 are bound by endogenous calcium binding molecules. Moreover, the recovery time constant of cytosolic calcium transients depends on κ_S and κ_B the buffering capacity of the calcium indicator dye, where the quantitative model illustrated in Figure 2 predicts a linear relationship

$$\tau = (1 + \kappa'_B + \kappa_S)/\gamma$$

(Neher and Augustine, 1992). In this case, γ denotes the "effective" extrusion rate of the cell, combining the the full set of calcium uptake and extrusion mechanisms (Neher, 1995). Most important, endogenous calcium binding capacities for hypoglossal motoneurones $\kappa_S = 41$ are given by the negative intercept of the x-axis, indicating that 1 out of 41 cytosolic Ca^{2+} ions contribute to the free calcium concentration. Estimates of extrusion rates are obtained from the the linear slope, yielding $\gamma = 60\ s^{-1}$ (Lips and Keller, 1998).

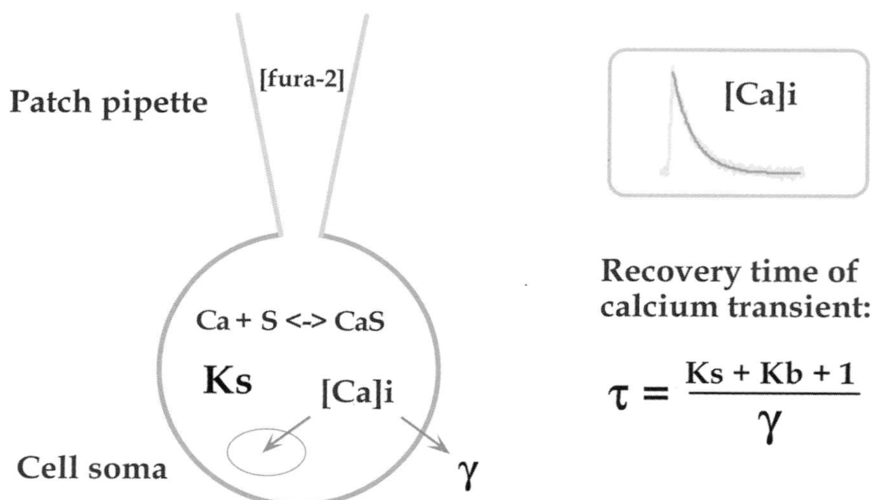

Figure 2. Quantitative model of calcium homeostasis in motoneurones. Scheme of the one-compartment model and the main parameters that determine somatic calcium responses. κ_S denotes the endogenous buffering capacity and γ provides a quantitative description of the "effective" extrusion rate.

3. DIFFERENTIAL CALCIUM BUFFERING CAPACITIES IN MOTONEURONES THAT ARE SELECTIVELY VULNERABLE AND RESISTANT DURING ALS-RELATED MOTONEURON DISEASE

The starting point of this summary was that motoneurones are selectively vulnerable during ALS-related neurodegenerative disease (DePaul et al., 1988; Alexianu et al., 1994; Appel et al., 1995; Reiner et al., 1995; Rothstein and Kuncl, 1995; Krieger et al., 1996; Elliot and Snider, 1995; Ince et al., 1993). By utilizing the quantitative approach, it is now possible to compare in detail individual parameters of cell-specific Ca^{2+} homeostasis and identify potential risk factors that might account for Ca^{2+}-related vulnerability. By implementing this strategy (Lips and Keller, 1998; Palecek et al., 1999; Vanselow and Keller, 2000), comparative studies demonstrated endogenous Ca^{2+} binding ratios of $\kappa_S = 264$ in oculomotor neurones, which is 5–6 times larger compared to $\kappa_S = 41$ and $\kappa_S = 50$ in hypoglossal and spinal motoneurones, respectively (see also Figure 3). Interestingly, values in oculomotor cells are still 3.5 times lower compared to those in Purkinje cells ($\kappa_S = 900$; 6 day old mice; Fierro and Llano, 1996), but comparable to $\kappa_S = 160$–207 in hippocampal CA1 neurones (Helmchen et al., 1996). Extrusion rates in oculomotor neurones are $\gamma = 156\ s^{-1}$, which is similar to $\gamma = 140\ s^{-1}$ found in spinal motoneurones. These rates are slow compared to those in the calyx of Held ($400\ s^{-1}$, 21°C; Helmchen et al., 1997), but several times faster compared to hypoglossal motoneurones ($60\ s^{-1}$) and adrenal chromaffin cells (Neher and

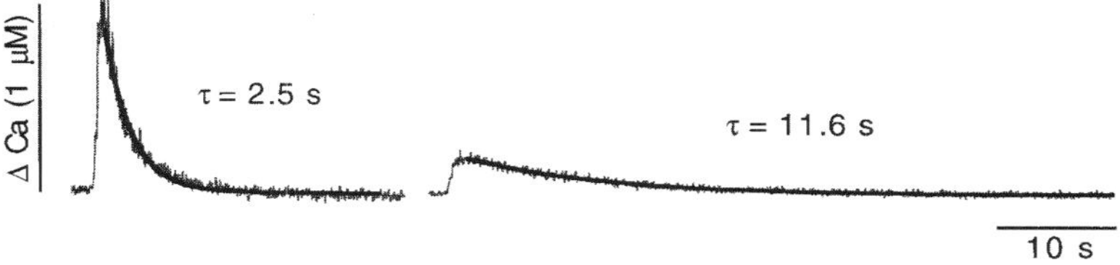

Oculomotor neuron

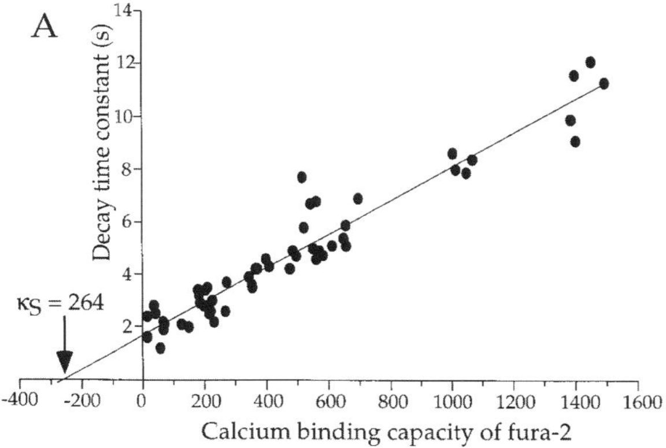

Spinal motoneuron

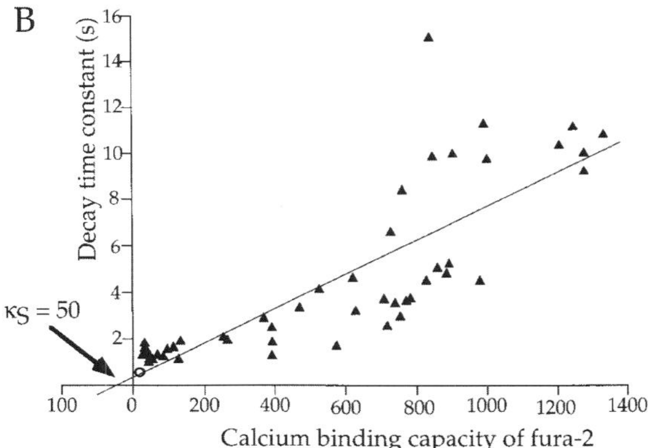

Figure 3. Decay times of calcium transients as a function of calcium binding capacity of fura-2 for oculomotor (left) and spinal (right) motoneurones. Upper trace: calcium transients evoked by short depolarizations to +10 mV. Left trace illustrates a calcium response soon after establishing the whole cell configuration. Right trace illustrates a calcium response after completely filling the neuron with 500 μM fura2. (A), (B) Decay time constants plotted as a function of κ'_B for oculomotor and spinal motoneurones. Negative intercept of the x-axis gives endogenous buffering capacity (adapted after Palecek et al., 1999 and Vanselow and Keller, 2000).

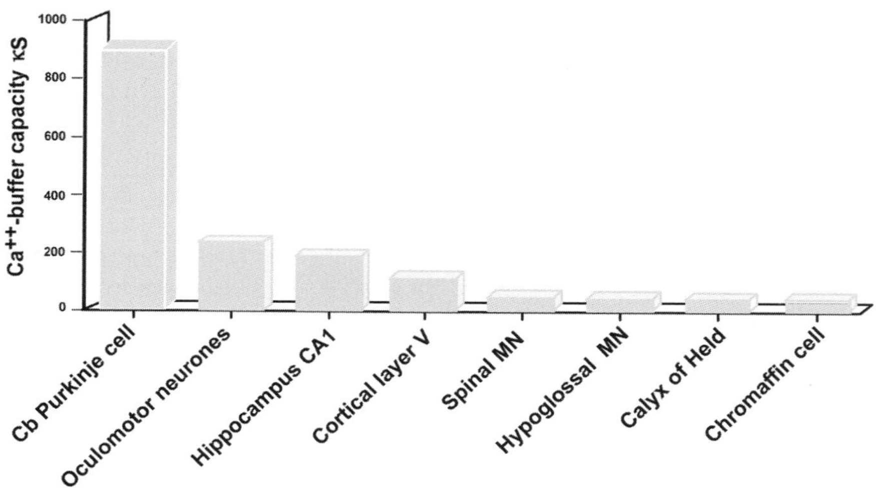

Figure 4. Comparison of endogenous calcium buffering capacities in different cell types. Experimentally determined calcium buffering capacities are illustrated according to the values described by different groups. Values are displayed as given in Lips and Keller (1998). Note the extraordinary low values for hypoglossal and spinal motoneurones.

Augustine, 1992). Most important, these results indicate that differences in endogenous buffering capacity, but not in extrusion rates or voltage dependent calcium influx provide a plausible explanation for selective, Ca^{2+}-related motoneuron vulnerability (Figure 4).

The observation of high buffering capacities in selectively resistant motoneurones is consistent with earlier immunocytochemical studies of endogenous calcium binding proteins (Alexianu et al., 1994; Elliot and Snider, 1995). For example, DePaul et al. (1988) have demonstrated high expression of calcium buffers in oculomotor neurones in a well-studied animal model of ALS-related selective motoneuron degeneration. In this case, high buffering in oculomotor neurones has been linked to elevated expression of parvalbumin, a calcium binding protein that is indeed expressed in low density in hypoglossal and spinal motoneurones. Moreover, *in vitro* cell culture models have shown that elevated cytosolic buffer concentrations reduce ALS-related motoneuron damage, providing further support for the idea that increased buffer concentrations display beneficial protection (Ho et al., 1996; Tymianski et al., 1994; Roy et al., 1998).

Given the broad diversity of molecular mechanisms associated with motoneuron degeneration (Abdel-Hamid and Baimbridge, 1997; Bruijn et al., 1998; Choi, 1988; Reiner et al., 1995; Rothstein et al., 1995; Smith et al., 1992; Morrison and Morrison, 1998), it is unlikely that low endogenous calcium buffering represents the only determinant. For example, our data are perfectly consistent with a model where specialized calcium influx pathways are also critical for selective motoneuron damage. Indeed, several studies suggested a high expression of calcium-permeable glutamate (AMPA/KA) receptors in motoneuron populations that are selectively impaired (Shaw et al., 1999; Shaw and Ince, 1997). More recently, this interpretation has received support from studies on cultured motoneurones, showing that blockers of calcium permeable AMPA/KA receptor channels provide efficient protection (Roy et al., 1998). Another factor could be an ALS-related immune reaction targeted at voltage-dependent calcium channels, where disruption of endogenous calcium homeostasis results from impaired, voltage dependent calcium influx (Appel et al., 1995, Ho et al., 1996). Furthermore, synaptic glutamate transport is also thought to be involved in other forms of ALS-related neurodegenerative disease (Trotti et al., 1999), where resulting overexcitation of glutamatergic synapses is a key stimulus for motoneuron degeneration.

4. SUMMARY AND CONCLUSION

The main conclusion of calcium signal analysis in motoneurones is that *vulnerable* populations also display extraordinary *low* endogenous calcium buffering capacities (Elliot and Snider, 1995; McMahon et al., 1998; Reiner et al., 1995; Ince et al., 1993). Several arguments could account for this phenomenon. One is based on estimates of cytosolic Ca^{2+} transients during "excitotoxic", events. In this case, low buffers evoke large amplitudes of intracellular free Ca^{2+} concentrations, thus increasing the risk for activation of "excitotoxic" *second messenger* cascades and related cellular disturbances (Choi, 1988). A different scenario is obtained for highly buffered oculomotor motoneurones, where the same influx evokes responses with 5–6 fold smaller amplitudes. This presumably decreases the probability for activation of Ca^{2+}-dependent, "apoptotic" lipases and phosphatases known to depend on Ca^{2+} elevations above the micromolar domain (Alexianu et al., 1994; Baimbridge et al., 1992; Choi, 1988; Krieger et al., 1994). Indeed, this risk could be particularly prominent in low buffered, rhythmically active hypoglossal and spinal neurones, where physiological electrical responses are already paralleled by substantial endogenous Ca^{2+} activity (Figure 1). A third argument is provided by the impact of low buffering on spatial Ca^{2+} profiles. As previously described (Neher, 1986; Roberts, 1994; Klingauf and Neher, 1997), high concentrations of mobile buffers accelerate the dispersion of local Ca^{2+} gradients by a process commonly known as "buffered diffusion" (Zhou and Neher, 1993; Roberts, 1994; Klingauf and Neher, 1997). According to this concept, low buffers in spinal and hypoglossal motoneurones provide favourable conditions for localized Ca^{2+} gradients, presumably enhancing the impact of spatially coupled signal cascades. With respect to pathophysiological conditions, this suggests that differential buffering reflects a basic diversity in the spatio/temporal organization of calcium signaling, rather than a singular difference in one cellular parameter.

Finally, it is interesting to note that the low endogenous Ca^{2+} buffering not only provides an increased risk for neuronal damage, but also represents a valuable functional advantage under physiological conditions. As indicated above, low buffers directly accellerate recovery of Ca^{2+} transients if all other parameters are comparable (Neher and Augustine, 1992). This is illustrated by the quantitative model displayed in Figure 2 where calcium decay times increase linearly with endogenous buffering capacity (Neher, 1995; Helmchen et al., 1997). In rhythmically active hypoglossal or spinal motoneurones, where action-potential induced Ca^{2+} oscillations occur at maximum frequencies of 8 Hz (Lips and Keller, 1999; Palecek et al., 1999), rapid recovery of Ca^{2+} transients is an essential requirement for proper cell function. Although fast recovery times of Ca^{2+} transients can be achieved for high buffering by accelerated extrusion rates, this is associated with higher ATP-dependent energy consumption if all other parameters are held constant (Palecek et al., 1999). Taken together, the summarized research therefore indicates that *cellular adaptations that facilitate rapid calcium signaling under physiological conditions enhance a selective vulnerability of brain stem and spinal motoneurones during ALS-related motoneuron disease*.

ACKNOWLEDGEMENTS

I thank D. Crzan for excellent technical assistance and Drs. A.J. Berger, T. Ladewig, P. Lalley, M. Lips, E. Neher, D.W. Richter and B. Vanselow for valuable discussions. This research was supported by DFG grants Ke 403/6-1, Ke 403/12-2. and Sonderforschungsbereich 406.

REFERENCES

Abdel-Hamid, K.M. and Baimbridge, K.G., 1997, The effects of artificial calcium buffers on calcium responses and glutamate-mediated excitotoxicity in cultured hippocampal neurons, *Neuroscience* 81, 673–687.

Alexianu, M.E., Ho, B.K., Mohamed, A.H., La, B.V., Smith, R.G. and Appel, S.H., 1994, The role of calcium-binding proteins in selective motoneuron vulnerability in amyotrophic lateral sclerosis, *Ann. Neurol.* 36, 846–858.

Alexianu, M.E., Robbins, E., Carswell, S. and Appel, S.H., 1998, 1Alpha, 25 dihydroxyvitamin D3-dependent up-regulation of calcium-binding proteins in motoneuron cells, *J. Neurosci. Res.* 51, 58–66.

Appel, S.H., Smith, R.G., Alexianu, M., Siklos, L., Engelhardt, J., Colom, L.V. and Stefani, E., 1995, Increased intracellular calcium triggered by immune mechanisms in amyotrophic lateral sclerosis, *Clin. Neurosci.* 3, 368–374.

Baimbridge, K.G., Celio, M.R. and Rogers, J.H., 1992, Calcium-binding proteins in the nervous system, *Trends Neurosci.* 15, 303–308.

Bayliss, D.A., Viana, F., Talley, E.M. and Berger, A.J., 1997, Neuromodulation of hypoglossal motoneurons: Cellular and developmental mechanisms, *Respir. Physiol.* 110(2-3), 139–150.

Blaustein, M.P., 1988, Calcium transport and buffering in neurons, *Trends Neurosci.* 11, 438–443.

Bruijn, L.I., Housewaert, M.K., Kato, S., Anderson, K.L., Anderson, S.D., Ohama, E., Reaume, A.G., Scott, R.W. and Cleveland, D.W., 1998, Aggregation and motor neuron toxicity of an ALS-linked SOD1 mutant independent from wild-type SOD1, *Science* 281, 1851–1854.

Celio, M.R., 1990, Calbindin and parvalbumin in the rat nervous system, *Neuroscience* 35, 375–475.

Chad, J., 1989, Inactivation of calcium channels, *Comp. Biochem. Physiol. A* 93, 95–105.

Choi, D.W., 1988, Glutamate neurotoxicity and diseases of the nervous system, *Neuron* 1, 623–634.

Christakos, S., Barletta, F., Huening, M., Kohut, J. and Raval-Pandya, M., 2000, Activation of programmed cell death by calcium: Protection against cell death by the calcium binding protein, Calbindin 28K, this book.

DePaul, R., Abbs, J.H., Caligiuri, M., Gracco, V.L. and Brooks, B.R., 1988, Hypoglossal, trigeminal, and facial motoneuron involvement in amyotrophic lateral sclerosis, *Neurology* 38, 281–283.

Elliot, J.L. and Snider, W.D., 1995, Parvalbumin is a marker of ALS-resistant motor neurons, *Neuroreport* 6(3), 449–452.

Fierro, L. and Llano, I., 1996, High endogenous calcium buffering in Purkinje cells from rat cerebellar slices, *J. Physiol.* 496, 617–625.

Frermann, D., Keller, B.U. and Richter, D.W., 1999, Calcium oscillations in rhythmically active respiratory neurones in the brainstem of mouse, *J. Physiol. (Lond.)* 515, 119–131.

Gurney, M.E., Cuttings, F.B., Zhai, P., Doble, A., Taylor, C.P., Andrus, P.K. and Hal, E.D., 1996, Benefit of vitamin E, riluzole, and gabapentin in a transgenic model of familial amyotrophic laterale sclerosis, *Ann. Neurol.* 39, 147–157.

Helmchen, F., Imoto, K. and Sakmann, B., 1996, Ca^{2+} buffering and action potential-evoked Ca^{2+} signaling in dendrites of pyramidal neurons, *Biophys. J.* 70, 1069–1081.

Helmchen, F., Borst, J.G. and Sakmann, B., 1997, Calcium dynamics associated with a single action potential in a CNS presynaptic terminal, *Biophys. J.* 72, 1458–1471.

Ho, B.K., Alexianu, M.E., Colom, L.V., Mohamed, A.H., Serrano, F. and Appel, S.H., 1996, Expression of calbindin-D28K in motoneuron hybrid cells after retroviral infection with calbindin-D28K cDNA prevents amyotrophic lateral sclerosis IgG-mediated cytotoxicity, *Proc. Natl. Acad. Sci. USA* 93, 6796–6801.

Ince, P., Stout, N., Shaw, P., Slade, J., Hunziker, W., Heizmann, C.W. and Baimbridge, K.G., 1993, Parvalbumin and calbindin D-28k in the human motor system and in motoneuron disease, *Neuropathol. Appl. Neurobiol.* 19(4), 291–299.

Keller, B.U., Konnerth, A. and Yaari, Y., 1991, Patch clamp analysis of excitatory synaptic currents in granule cells of rat hippocampus, *J. Physiol.* 435, 275–293.

Klapstein, G.J., Vietla, S., Lieberman, D.N., Gray, P.A., Airaksinen, M.S., Thoenen, H., Meyer, M. and Mody, I., 1998, Calbindin-D28 fails to protect hippocampal neurons against ischemia in spite of its cytoplasmic calcium buffering properties: Evidence from calbindin-D28k knockout mice, *Neuroscience* 85(2), 361–373.

Klingauf, J. and Neher, E., 1997, Modelling buffered Ca^{++} diffusion near the membrane: Implications for secretion in neuroendocrine cells, *Biophys. J.* 72, 674–690.

Krieger, C., Jones, K., Kim, S.U. and Eisen, A.A., 1994, The role of intracellular free calcium in motor neuron disease, *J. Neurol. Sci.* 124, 27–32.

Krieger, C., Lanius, R.A., Pelech, S.L. and Shaw, C.A., 1996, Amyotrophic lateral sclerosis: The involvement of intracellular Ca^{2+} and protein kinase C, *Trends Pharmacol. Sci.* 17, 114–120.

Ladewig, T. and Keller, B.U., 1998, Calcium imaging in rhythmically active motoneurones in the nucleus hypoglossus from mouse, *Pflügers Archi* 435, R62.

Ladewig, T. and Keller, B.U., 2000, Simultaneous patch clamp recording and calcium imaging in a rhythmically active neuronal network in the brainstem slice preparation from mouse, *Pflügers Arch.* 440, 322–332.

Lev-Tov, A. and O'Donovan, M.J., 1995, Calcium imaging of motoneuron activity in the en-bloc spinal cord preparation of the neonatal rat, *J. Neurophysiol.* 74(3), 1324–1334.

Lips, M.B. and Keller, B.U., 1998, Endogenous calcium buffering in motoneurones of the nucleus hypoglossus from mouse, *J. Physiol.* 511, 105–117.

Lips, M.B. and Keller, B.U., 1999, Activity-related calcium dynamics in motoneurones of the nucleus hypoglossus from mouse, *J. Neurophys.* 82(6), 2936–2946.

McBurney, R.N. and Neering, I.R., 1987, Neuronal calcium homeostasis, *Trends Neurosci.* 10, 164–169.

McMahon, A., Wong, B.S., Iacopino, A.M., Ng, M.C., Chi, S. and German, D.C., 1998, Calbindin-D28k buffers intracellular calcium and promotes resistance to degeneration in PC12 cells, *Brain Res. Mol. Brain Res.* 54, 56–63.

Medina, L., Figueredo-Crdenas, G., Rothstein, J.D. and Reiner, A., 1996, Differential abundance of glutamate transporter subtypes in amyotrophic lateral scleroses (ALS)-vulnarable versus ALS-resistant brain stem motor cell groups, *Exp. Neurol.* 142, 287–295.

Meldrum, B. and Garthwaite, J., 1990, Exciatory amino acid neurotoxicity and neurodegenerative disease, *Trends Pharmacol. Sci.* 11, 379–387.

Morrison, B.M. and Morrison, J.H., 1998, Amyotrophic lateal sclerosis associated with mutations in superoxide dismutase: A putative mechanism of degeneration, *Brain Res. Rev.* 29, 121–135.

Nägerl, U.V. and Mody, I., 1998, Calcium-dependent inactivation of high-threshold calcium currents in human dentate gyrus granule cells, *J. Physiol. (Lond.)* 509, 39–45.

Neher, E., 1986, Concentration profiles of intracellular calcium in the presence of a diffusable chelator, *Exp. Brain Res.*, Series 14, 80–96.

Neher, E., 1995, The use of fura-2 for estimating Ca^{2+} buffers and Ca^{2+} fluxes, *Neuropharmacology* 34, 1423–1442.

Neher, E. and Augustine, G.J., 1992, Calcium gradients and buffers in bovine chromaffin cells, *J. Physiology* 450, 273–301.

Palecek, J., Lips, M.B. and Keller, B.U., 1999, Calcium dynamics and buffering in motoneurons of the mouse spinal cord, *J. Physiol.* 520(2), 486–502.

Reiner, A., Medina, L., Figueredo, C.G. and Anfinson, S., 1995, Brainstem motoneuron pools that are selectively resistant in amyotrophic lateral sclerosis are preferentially enriched in parvalbumin: Evidence from monkey brainstem for a calcium-mediated mechanism in sporadic ALS, *Exp. Neurol.* 131, 239–250.

Roberts, W.M., 1994, Localization of calcium signals by a mobile calcium buffer in frog saccular hair cells, *J. Neurosci.* 14, 3246–3262.

Rothstein, J.D. and Kuncl, R.W., 1995, Neuroprotective strategies in a model of chronic glutamate-mediated motor neuron toxicity, *J. Neurochem.* 65, 643–651.

Rothstein, J.D., Martin, L.J. and Kuncl, R.W., 1992, Decreased glutamate transport by the brain and spinal cord in amyotrophic lateral sclerosis, *New Engl. J. Med.* 326, 1464–1468.

Rothstein, J.D., Van Kammen, M., Levey, A.I., Martin, L.J. and Kuncl, R.W., 1995, Selective loss of glial glutamate transporter GLT-1 in amyotrophic lateralsclerosis, *Ann. Neurol.* 38, 73–84.

Roy, J., Minotti, S., Dong, L., Figlewicz, D.D. and Durham, H.D., 1998, Glutamate potentiates the toxicity of mutant Cu/Zn-superoxide dismutase in motor neurones by postsynaptic calcium-dependent mechanisms, *J. Neurosci.* 18(23), 9673–9684.

Schafer, B.W. and Heizmann, C.W., 1996, The S100 family of EF-handed calcium binding proteins: Function and pathology, *Trends Biochem. Sci.* 21, 134–140.

Shaw, P.J. and Ince, P.G., 1997, Glutamate, excitotoxicity and amyotrophic lateral sclerosis, *J. Neurol.* 244, 3–14.

Shaw, P.J., Williams, T.L., Slade, J.Y., Eggett, E.Y. and Ince, P.G., 1999, Low expression of GluR2 Ampa receptor subunit protein by human motor neurons, *Neuroreport* 10(2), 261–265.

Siklos, L., Engelhardt, G.I., Alexianu, M.E., Siddique, T. and Appel, S.H., 1998, Intracellular calcium parallels motoneuron degeneration in SOD-1 mutant mice, *J. Neuropathol. Exp. Neurol.* 57(6), 571–587.

Smith, J.C., Ellenberger, H.H., Ballanyi, K., Richter, D.W. and Feldman, J.L., 1991, Pre-Botzinger complex: A brainstem region that may generate respiratory rhythm in mammals, *Science* 254, 726–729.

Smith, R.G., Hamilton, S., Hofmann, F., Schneider, T., Nastainczyk, W., Birnbaumer, L., Stefani, E. and Appel, S.H., 1992, Serum antibodies to L-type calcium channels in patients with amyotrophic laterale sclerosis, *New Engl. J. Med.* 327, 1721–1728.

Trotti, D., Rolfs, A., Danbolt, N.C., Brown, R.H. and Hediger, M.A., 1999, SOD1 mutants linked to amyotrophic lateral sclerosis selectively inactivate a glial glutamate transporter, *Nat. Neurosci.* 2(5), 427–433.

Tu, P.H., Raju, O., Robinson, K.A., Gurney, M.E., Trojanowski, J.Q. and Lee, M.Y., 1996, Transgenic mice carrying a human mutant superoxid dismutase transgene develop neuronal cytoskeletal pathology resembling human amyotrophic lateral sclerosis lesions, *Proc. Natl. Acad. Sci. USA* 93, 3155–3160.

Tymianski, M., Charlton, M.P., Carlen, P.L. and Tator, C.H., 1994, Properties of neuroprotective cell-permeant Ca^{2+} chelators: Effects on $[Ca^{2+}]_i$ and glutamate neurotoxicity in vitro, *J. Neurophysiol.* 72, 1973–1992,

Vanselow, B. and Keller, B.U., 2000, A quantitative evaluation of calcium dynamics and buffering in oculomotor neurones from mouse, *J. Physiol.* 522, 433–445.

Viana, F., Bayliss, D.A. and Berger, A.J., 1993a, Calcium conductances and their role in the firing behavior of neonatal rat hypoglossal motoneurons, *J. Neurophysiol.* 69, 2137–2149.

Viana, F., Bayliss, D.A. and Berger, A.J., 1993b, Multiple potassium conductances and their role in action potential repolarization and repetitive firing behavior of neonatal rat hypoglossal motoneurons, *J. Neurophysiol.* 69, 2150–2163.

Williamson, T.L., Bruijn, L.I., Zhu, Q., Anderson, K.L., Anderson, S.D., Julien, J. and Cleveland, D.W., 1998, Absence of neurofilaments reduces the selective vulnerability of motor neurons and slows disease caused by a familial amyotrophic lateral sclerosis-linked superoxide dismutase 1 mutant, *Proc. Natl. Acad. Sci. USA* 95, 9631–9636.

Zhou, Z. and Neher, E., 1993, Mobile and immobile calcium buffers in bovine adrenal cells, *J. Physiol.* 469, 245–273.

PART THREE

Methodological Chapters

Pharmaceutical Chemistry and Drug Screening

Jacques Haiech, Jean-Luc Galzi and Marcel Hibert

1. INTRODUCTION

In the early years of pharmacology, the only way to find a new drug was the trial and error method. In the main, natural substances were used to cure human symptoms and were directly tested on patients. Later on, synthetic drugs or purified natural compounds were characterised by their ability to interact with biological targets or cellular pathways.

Now, with the avalanche of sequence data, the pharmaceutical companies have to deal with more than 100 000 putative biological targets (the estimated number of human genes). At the same time, pharmaceutical chemists have been able to increase their productivity in such a way that drug libraries with millions of compounds have been created.

Pharmacologists have now in hand techniques and molecules that allow screening for drugs that interact with non-characterised proteins. However, if we assume that we possess a drug library of 1 000 000 products, that the number of human biological targets is about 100 000 and that the cost for a datapoint in a screening assay is about $1, we would have to spend $100 000 000 000 to get an exhaustive view of what interacts with what. No pharmaceutical firm is ready to invest such an amount of money at once. Moreover, most hits that interact with a target protein do not possess the right properties to become a commercial drug. New and more subtle strategies have to be designed. One of them consists in:

- working on small and smart drug libraries (1000 to 2000 compounds);
- screening virtually drug libraries before doing the real experiments;
- developing general screening assays that do not rely on the function of the orphan and putative biological targets.

Jacques Haiech, Jean-Luc Galzi and Marcel Hibert • IFR Gilbert LAUSTRIAT, Faculté de Pharmacie, Illkirch, France and Ecole Supérieure de Biotechnologie, Illkirch, France.

R. Pochet, R. Donato, J. Haiech, C. Heizmann and V. Gerke (eds.): Calcium: The Molecular Basis of Calcium Action in Biology and Medicine, 641–646.

In the framework of the newly created French Genomic Centre, we have developed in Strasbourg the techniques needed to follow the above strategy.

2. DRUG LIBRARIES

Most molecules with an original structure that have been optimised to become commercial drugs have been derived from natural substances or from synthetic substances built in academic laboratories. We realised that the drawers of the chemical laboratories were full of treasures but useless for the biologists since the compounds were not referenced and not stored in a convenient pattern. In 1998, we decided to gather the different compounds from our laboratory (Pharmacochimie de la Communication Cellulaire, UMR CNRS/ULP 7081) and to create a patrimonial compound library. We used a relational database present in most pharmaceutical firms to reference our compounds and to gather and manage chemical and biological information. We decided to select the compounds archived in the library in order to maximise the structural diversity and the bio-activity potential.

The gathered compounds were stored as powder in normalised tubes. 30 mg of each compound were diluted in DMSO at a concentration of about 10^{-2} M and distributed in 96 multi-tube plates. Those plates (mother plates) were kept at 4°C and used to prepare daughter plates that were in turn used to screen biological targets (the concentration of DMSO is lowered to 1%). Currently, the patrimonial library is composed of about 1500 compounds. It will be extended rapidly. Based on similar ideas, a Smart library composed only of generic commercial drugs has been built by Prestwick Chemical Europe. The use of generic drugs makes it possible to find hits in chemical classes whose biodisponibility and toxicity are well studied.

In 1999, other chemical laboratories decided to join the patrimonial library and the CNRS has decided to support such an initiative in order to create a national academic drug library. A research network working on natural substances is also willing to join this project. More recently, European laboratories started to use the defined standards to gather, reference and store their own compounds. Therefore, through a federation of laboratories using the same protocols to store their in-house compounds, a European academic library is emerging.

3. FLUORESCENCE BASED ASSAY

Fluorescence resonance energy transfer (FRET) takes place when two fluorescent molecules, one energy donor and one energy acceptor, interact with each other. FRET thus allows interactions to be detected (see Figure 1). The requirements for energy transfer are:

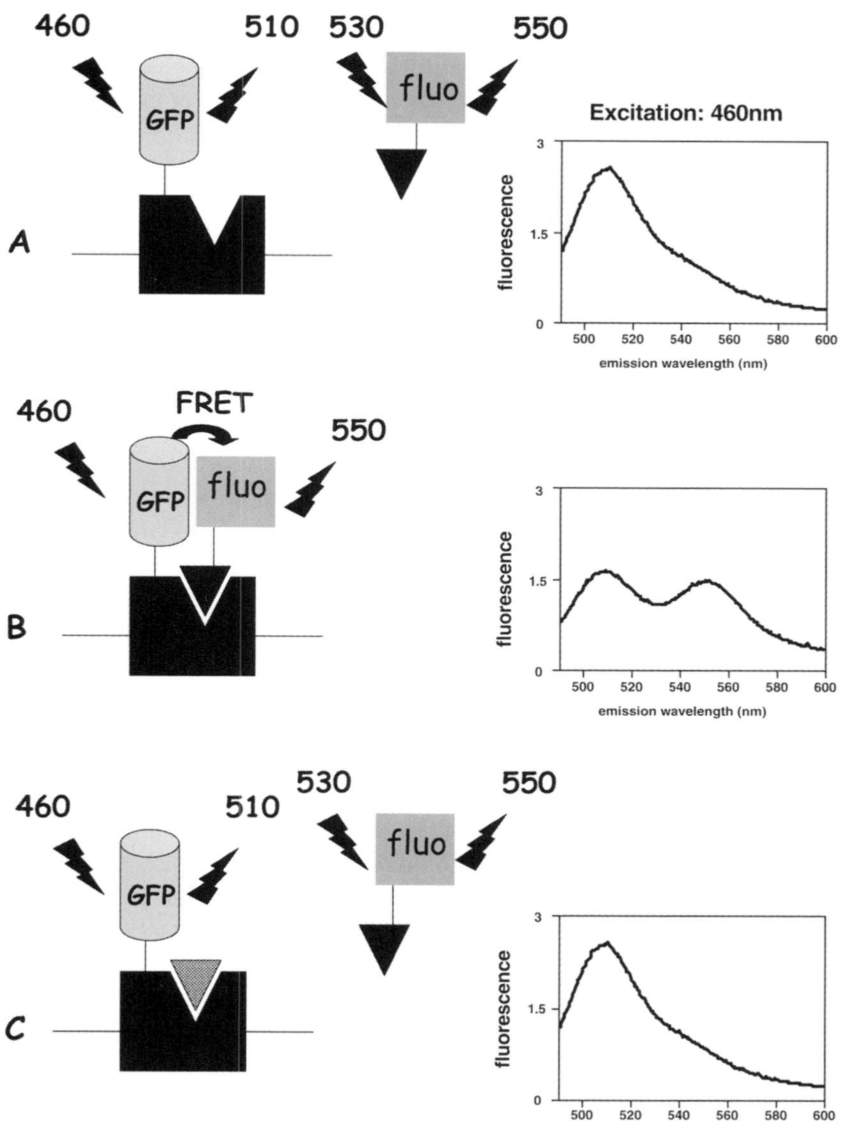

Figure 1. Principle of the assay based on Fluorescence Resonance Energy Transfer (FRET). GFP is Green Fluorescent Protein, "fluo" stands for a relevant fluorescent moiety linked to a ligand. Typical absorption and emission wavelengths are given.

1. the energy acceptor must absorb light at the wavelength of the donor emission, and
2. the two fluorescent molecules must be physically close to each other, generally separated by 20–100A, a distance which, for biological samples, reflects interaction.

The gene encoding GFP, a naturally Green Fluorescent Protein isolated from jellyfish Aequorea victoria (Niwa et al., 1996), can be fused to the gene encoding the target receptor gene to yield a fluorescent receptor (the energy donor) upon expression in cells. One ligand of this target receptor can be rendered fluorescent by chemical means with fluorescent molecules such as Texas Red or rhodamine, to become an energy acceptor. On mixing cells expressing the target fluorescent receptor protein with the fluorescent ligand, the interaction is detected either by a decrease of fluorescence of the receptor or by appearance of fluorescence emission of the ligand, when the preparation is illuminated at the excitation wavelength of the receptor. If an unlabelled molecule subsequently added to this mixture abolishes the fluorescence signal, it will be concluded that it interacts with the target receptor (Vollmer et al., 1999).

We have demonstrated that this approach works for several target receptors belonging to the family of G protein-coupled receptors (Weill et al., 1999; Vollmer et al., 1999), and that it is amenable to automated determinations using pipetting robots and fluorescence plate readers. It is currently used to identify new molecules exhibiting either higher affinity or better selectivity for receptor subtypes of interest for fundamental research as well as for drug discovery.

One important advantage of fluorescence detection over other methods is that measurements can be carried out on single cells (Vollmer et al., 1999; Sayers et al., 1997; see also the chapter from Euroscreen in this book), or even on single molecules (Meseth et al., 1999). Assay miniaturization can thus be envisaged, leading to significant cost reduction, in particular for chemical libraries. The method can also be applied to identify ligands binding to so-called orphan receptors, i.e. gene products of unknown functions.

4. SUPRAMOLECULAR MASS SPECTROMETRY ASSAY

When the protein is soluble and can be produced as a recombinant protein, we use mass spectrometry to follow the interaction with a synthetic molecule. Thus, we have used mass spectrometry to detect the interaction between protein and small ligands or peptides even when the interaction is in the millimolar range. The native protein is dissolved with a given compound at neutral pH and is analysed by "Electro Spray Ionisation" mass spectrometry with a Time Of Flight detector (ESI-TOF apparatus). If the mass that we detect is the sum of the masses of the protein and of the ligand, we know that an interaction occurs. Such a technique allows us to screen thousands of compounds with a relatively small quantity of protein. Then, to further characterise the hits obtained, we use a microcalorimeter to analyse the thermodynamic parameters of the interaction. We have validated such techniques on a calcium binding protein (Calmodulin; Lafitte et al., 1995). Using this approach, we have set up in our research institute a resource facility that allows a research group to screen drug libraries if they have enough available target proteins (a few milligrams). Such a strategy allows us to set up a screening assay without knowledge of the function of the protein and to detect molecules that may be used as pharmacological tools to decipher the role of orphan proteins. This is only one aspect of the use of mass spectrometry in the post genomics era, proteomics representing another important contribution.

5. PERSPECTIVES

The availability of the human genome sequence, probably this year, is going to give access to the whole family of the putative calcium binding proteins. The cellular expression of these proteins under normal and pathological conditions will be checked using "DNA chips" or "DNA arrays". Calcium binding proteins are easy to express in *E. coli* and, using the academic drug library, it will be possible to screen for compounds that specifically bind to one of them. Such molecules are going to be extremely important pharmacological tools and will be used to characterise the calcium signal pathways associated with a specific calcium binding protein. Combined with the ability to analyse the G protein coupled receptor in one cell, it should now be possible to completely characterise the calcium signal pathway in a normal or transformed cell. The recent progress in both Biology and Chemistry opens new avenues for extensively exploring the exquisite complexity of calcium signal pathways.

REFERENCES

Lafitte, D., Capony, J.P., Grassy, G., Haiech, J. and Calas, B., 1995, Analysis of the ion binding sites of calmodulin by electrospray ionization mass spectrometry, *Biochemistry* 34, 13825–13832.

Meseth, U., Wohland, T., Rigler, R. and Vogel, H., 1999, Resolution of fluorescence correlation measurements, *Biophys. J.* 76, 1619–1631.

Niwa, H., Inouye, S., Hirano, T., Matsuno, T., Kojima, S., Kubota, M., Ohashi, M. and Tsuji, F.I., 1996, Chemical nature of the light emitter of the Aequorea green fluorescent protein, *Proc. Natl. Acad. Sci. USA* 93, 13617–13622.

Sayers, L.G., Miyawaki, A., Muto, A., Takeshita, H., Yamamoto, A., Michikawa, T., Furuichi, T. and Mikoshiba, K., 1999, Intracellular targeting and homotetramer formation of a truncated inositol 1,4,5-trisphosphate receptor-green fluorescent protein chimera in Xenopus laevis oocytes: Evidence for the involvement of the transmembrane spanning domain in endoplasmic reticulum targeting and homotetramer complex formation, *Biochem. J.* 323, 273–280.

Vollmer, J.Y., Alix, P., Chollet, A., Takeda, K. and Galzi, J.L., 1999, Subcellular compartmentalization of activation and desensitization of responses mediated by NK2 neurokinin receptors, *J. Biol. Chem.* 274, 37915–37922.

Weill, C., Ilien, B., Goeldner, M. and Galzi, J.L., 1999, Fluorescent muscarinic EGFP-hM1 chimeric receptors: design, ligand binding and functional properties, *J. Recept. Signal Transduct. Res.* 19, 423–436.

Calcium in Drug Screening

Vincent J. Dupriez

1. INTRODUCTION

The recent development of combinatorial chemistry and of automation revolutionized the process of drugs discovery in the pharmaceutical industry. The screening of new therapeutic targets is now performed with tens or hundreds of thousands of compounds and it requires reliable, rapid and affordable screening systems.

For the finding of drugs acting on G-protein-coupled receptors, the classical approach was based on the competition between the tested molecules and a radioligand for their binding to the receptor. This method is expensive, due to the cost of the reagents and of the elimination of radioactive waste, and it does not allow to discriminate between agonists and antagonists of the receptor. Moreover, some compounds bind the receptor at a site different from the binding site of the radioligand and are unable to displace it. These compounds, although they can be good agonists or antagonists of the receptor, can thus be undetectedable by such a binding assay (Thue Schwartz, communication at the 5th meeting of the Society for Biomolecular Screening). Thus the trend is now to use more and more functional assays for the first step of screening of chemical compounds on a receptor. Indeed, functional assays provide direct information on the agonist or antagonist nature of the compound and can be used to quantify the ability of the compound to stimulate or inhibit a second messenger pathway.

The G-protein-coupled receptors (GPCRs) is the widest family of cell membrane receptors. These receptors have an enormous therapeutic potential and are the target of more than half of the drugs currently available on the market. They are specialized in the chemical recognition of various ligands, ranging from bioamines, peptides, glycoproteins, to photons, modified nucleotides, fragrant molecules, etc. Following the recognition of their

Vincent J. Dupriez • Euroscreen S.A., Brussels, Belgium.

R. Pochet, R. Donato, J. Haiech, C. Heizmann and V. Gerke (eds.): Calcium: The Molecular Basis of Calcium Action in Biology and Medicine, 647–659.

ligand, these receptors trigger, by the use of a trimeric protein that binds GTP (G-proteins), the transduction of intracellular signal(s). These signals are usually an increase or a decrease in the level of intracellular cyclic AMP or an increase in intracellular calcium (reviewed by Barritt, 1999). For some receptors, activation can lead to a stimulation of the MAPkinase pathway, an increase in arachidonic acid content, a stimulation or an inhibition of plasma membrane ion channels. As we will explain below, the coupling of virtually any GPCR can be redirected towards an increase in intracellular calcium.

Monitoring of the intracellular calcium concentration can thus be used for the screening of new drugs, acting on any GPCR or on plasma membrane calcium channels. This chapter will focus on the techniques that are used in the pharmaceutical industry to perform this measurement at a high throughput, in order to analyze thousands of compounds for their ability to activate or inhibit a particular receptor.

The two main techniques that are used at a high throughput in the pharmaceutical industry to follow intracellular calcium concentration are calcium-sensitive fluorescent dyes and the photoprotein aequorin.

2. MEASUREMENT OF INTRACELLULAR CALCIUM WITH CALCIUM-SENSITIVE FLUORESCENT DYES

A system that is widely used for drug screening is the use of calcium-sensitive dyes, such as Fluo-3 or 4 or Calcium-Green. These are identical to the one used in confocal microscopy (see Nadal and Soria, this book). The main difference is that, for high-throughput screening, cells (about 65 000 cells/well of a 96-well plate) are placed in the wells of 96- or 384-well plates. A fluorimeter specially adapted for the injection of the compounds to be tested in all the wells at once and for the consequent recording of the fluorescent signal from all the wells is then used (such as the FLIPR from Molecular Devices; see Schroeder and Neagle, 1996). The excitation light comes from an argon laser or from a broadband source such as a Xenon arc lamp and a CCD camera is used to take pictures of the fluorescent light from the bottom of the plate. An optical system allows to focus on the monolayer of cells (in the case of adherent cells) or on the middle of the cell suspension (in the case of non-adherent cells). Before the measurement, cells must be loaded with the dye. The dye present in the extracellular medium needs then to be removed, as the signal yielded by the dye complexed to the calcium present in the extracellular medium would shadow the signal resulting from intracellular calcium increase. To remove this dye, the cells must be washed before the measurement. Measurement must then ideally be performed within 30 minutes; beyond this time, the amount of dye that has gone out of the cells begins to affect the measurement in a significant way. Because there is then less dye in the cell, and because the dye in the extracellular medium highers the basal signal, the increase seen upon stimulation of the cells is less marked and the well-to-well variation increases. For this reason, calcium measurements with fluorescent dyes needs to be performed with batches of cell preparations: one to 3 plates are processed at a time (loading, washing, measurement). This compels the users to prepare the cells all along the experiment.

3. MEASUREMENT OF INTRACELLULAR CALCIUM WITH AEQUORIN-EXPRESSING CELL LINES

Another method allowing to perform measurements of intracellular calcium at a high throughput relies on the use of cell lines expressing aequorin. Aequorin is a photoprotein originating from the jellyfish *Aequorea victoria*, that has been first described by Shimomura et al. (1962) and that has been cloned by Prasher et al. (1985) and Inouye et al. (1985). The apo-enzyme (apoaequorin) is a 21-kDa protein and needs a hydrophobic prosthetic group, coelenterazine, to be converted to aequorin, the active enzyme. This enzyme possesses three "EF-hands", the calcium binding sites which control its activity. Upon Ca^{++} binding, aequorin undergoes an irreversible reaction: the oxidation of coelenterazine into coelenteramide, with production of CO_2 and emission of light. The consumption of aequorin is proportional to the calcium concentration, in the physiological range (from 50 nM to 50 μM) (Brini et al., 1995; Rizutto et al., 1995). Because of this relationship, aequorin can be used in cells as a calcium indicator. A calibration curve can be established, that allows to determine the absolute calcium concentration in the cell by the mean of the amount of light that is emitted from aequorin. The measure of calcium concentration with aequorin gives the same values as the one determined with fluorescent dyes such as, for example, Fura-2 (Brini et al., 1995).

Sheu et al. (1993) and Button and Brownstein (1993) described a system (Figure 1) using cell lines that stably co-express apoaequorin and a GPCR. In this system, cells are incubated with coelenterazine, which is the co-factor of aequorin. During this incubation, coelenterazine enters the cell (it is lipophylic and readily crosses the cell membrane) and conjugates with apoaequorin to form aequorin, which is the active form of the enzyme. When the cells are then exposed to an agonist of the GPCR, intracellular calcium concentration increases. This increase leads to the activation of the catalytic activity of aequorin, which oxidizes coelenterazine and yields apoaequorin, coelenteramide, CO_2 and light. The intensity of light emission is proportional to the increase in intracellular calcium. Thus, in this system, measurement of light emission following agonist addition reflects its ability to activate the GPCR. Because light is emitted only during 20 to 30 seconds after activation of the GPCR, recording of the emitted light must be performed during the few seconds following agonist addition to the cells. This flash-type signal reflects the transient increase in calcium concentration following GPCR stimulation. Stables et al. (1997) showed that, when aequorin is expressed in the mitochondria, the emission of light from cells upon stimulation of a GPCR was higher

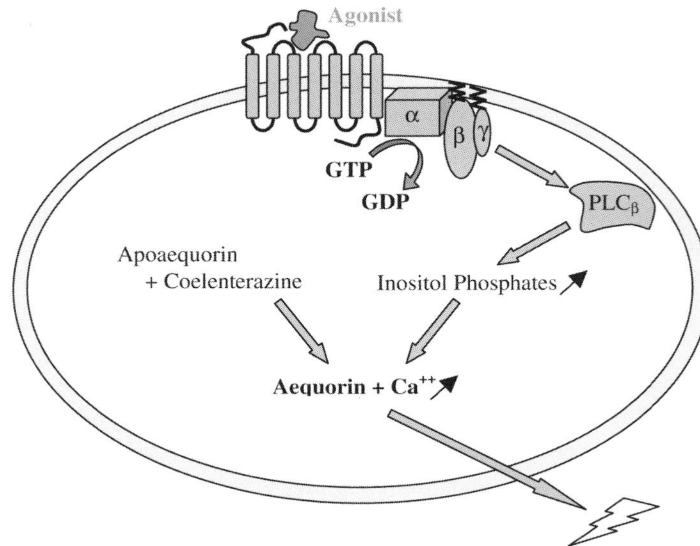

Figure 1. Principle of the measurement of intracellular calcium with cells expressing aequorin. Apoaequorin is expressed at high level by transfecting the cell line with a plasmid coding for aequorin. A GPCR is also expressed by transfecting the cell line with a plasmid coding for it. Before the assay, cells are preincubated with coelenterazine to reconstitute active aequorin. Cells are then incubated for 30 seconds with the tested compound. Compounds acting as agonist will increase intracellular calcium and trigger the emission of a flash light by aequorin. α, β, γ: Gα, Gβ and Gγ proteins; PLCβ: phospholipase Cβ.

than when aequorin is expressed in the cytoplasm. This can be explained by a higher rise of calcium concentration in the mitochondria than in the cytoplasm, because mitochondria are tightly associated with the endoplasmic reticulum, where the local calcium concentration rise is more marked than in the bulk of the cytoplasm. Calcium concentration then rapidly equilibrates between the outside and the inside of the mitochondria (Rizzuto et al., 1998). Aequorin has also been targetted into other subcellular domains (reviewed by Kendall and Badminton, 1998) and could then be used for the screening of drugs able to increase calcium in particular regions of the cell.

Although it has been used for research purposes at small throughput, so far this biological system could not be used at a high-throughput scale. Indeed:

1. the necessity to measure light just after placing the cells in contact with the agonist to be tested compels to use a luminometer equipped with a build-in dispenser;
2. luminometers equipped with a build-in dispenser only allow to inject a single solution into the 96 wells, making it impossible to inject a different drug in each well. Moreover, the washing of the dispenser before each measurement, for the injection of another drug in the next well, is time- and reagent-consuming and thus is not suitable for the high-throughput scale.

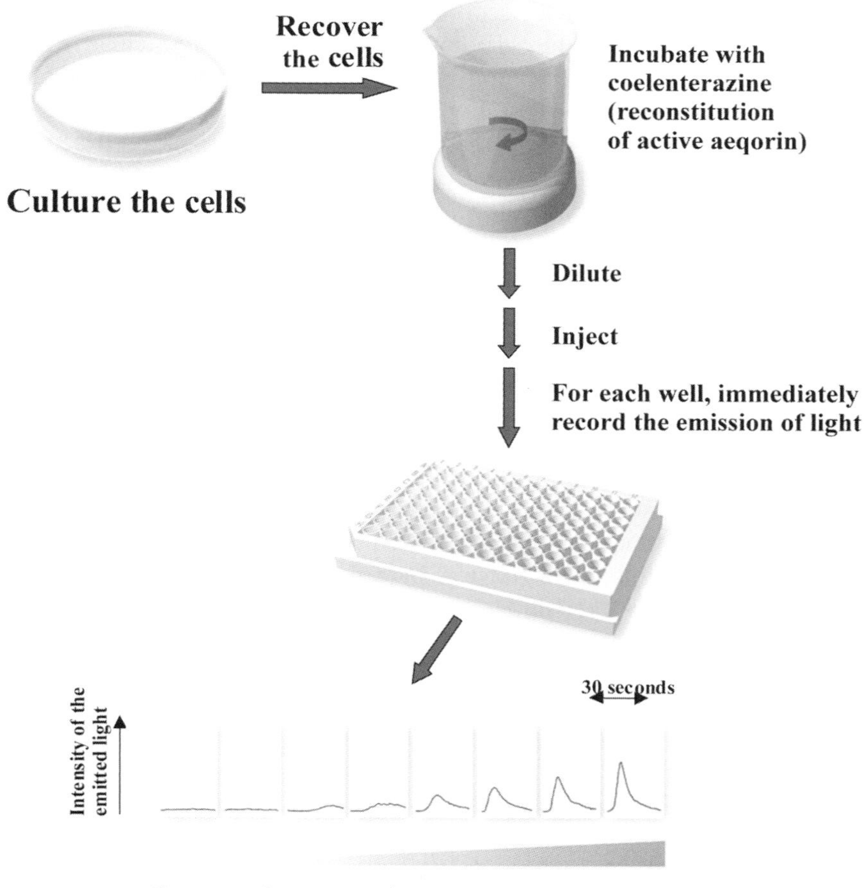

Figure 2. Method developed by Euroscreen to perform high-throughput screening with cell lines co-expressing aequorin and a G-protein coupled receptor.

Euroscreen has developed an aequorin-based system (a patent for this method has been filed by Euroscreen) for high-throughput screening of drugs binding to GPCR offering the advantage of working with conventional luminometers. Following this method (Figure 2), the solutions to be tested for agonistic activities are placed in the wells of a 96-well plate. Cells expressing apoaequorin and a GPCR are detached from the culture plate (if they are adherent), are resuspended in culture medium without serum and are kept in suspension by magnetic stirring and incubated for 4 hours at room temperature, or overnight at 4°C, with coelenterazine to reconstitute active aequorin. Cells are then diluted 10 times with culture medium at room temperature, without serum and are allowed to equilibrate for 30 min. These are then maintained in suspension with a stirrer and the cell suspension is injected, well by well (25 000 cells/well), on the solutions of the compounds to be tested. Light emission is then recorded usually during 10 to 30 seconds, depending on the kinetic of the calcium rise induced by the particular receptor studied. This method, by injecting the same cell suspension in each of the 96 wells, avoids the need of washing the dispenser between each measurement and allows to perform 96 measurements of agonist-induced aequorin light emission in 32 minutes (for 20 seconds/well) with a single dispenser luminometer. Alternatively, it allows to perform 96 measurements of agonist-induced aequorin light emission in 6 minutes with a luminometer equipped with 6 dispensers and detectors (e.g. with the "Microbeta Jet" from EG&G Wallac-Perkin Elmer). This reduces the screening time and the amount of drugs needed for each measurement and renders the aequorin system suitable for high-throughput screening (10 000 samples/day) with mammalian cell lines expressing apoaequorin and a GPCR and a conventional luminometer equipped with 6 injectors and detectors. The fluorimeter usually used for measurements with calcium sensitive fluorescent dyes (FLIPR from Molecular Devices) can also be used for measurements with aequorin. In this case, the LASER excitation light should be turned off and care should be taken to avoid any stray light in the device. Indeed, the level of sensitivity needed for the detection of light emitted from aequorin is higher than in the case of fluorimetry. The use of the FLIPR or of any other CCD camera able to measure the 96- or 384-well plate at one time should increase the throughput of the aequorin system up to hundreds of thousands of measurements per day.

4. EXAMPLES OF RESULTS

This functional screening assay has been tested with success on receptors belonging to different families including serotonin (5-HT_{1B}, 5-HT_{2B}) receptors, the chemokine (CCR1, CCR2, CCR3, CCR4, CCR5, CCR8, CXCR4) receptors, the melanocortin receptor MC4, the TSH receptor, the cannabinoid CB1 receptor, the purinergic $P2Y_2$ receptor, the corticotropin releasing factor CRF_2 receptor, etc. A signal-to-noise ratio above 50 is currently obtained with this system of cell injection. EC_{50} values are similar to affinity values obtain by radioligand binding and/or other functional assays. This method can be used for the detection of agonists and for the detection of antagonists, as detailed below.

4.1. Human chemokine CCR5 receptor

A CHO cell line expressing the chemokine CCR5 receptor and apoaequorin was established. Cells were prepared as described above and were injected on known ligands diluted serially and placed in the wells of a 96-well plate and the emitted light was immediately recorded during 30 seconds. After reading the first well, cells were injected into the next well and emitted light was re-

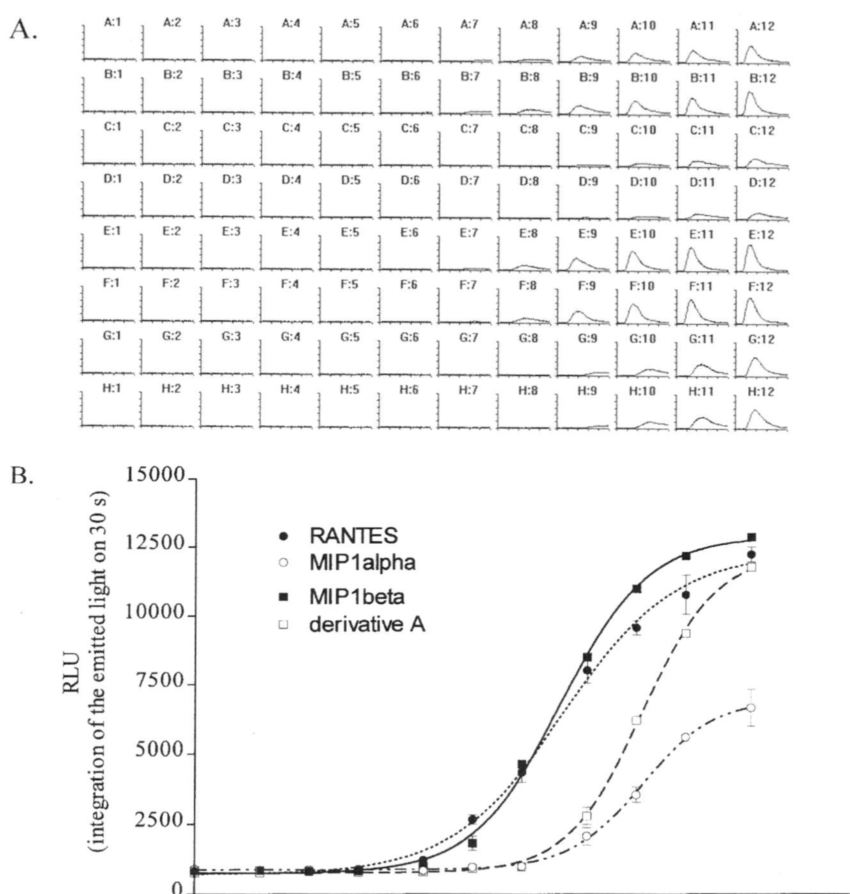

Figure 3. Dose-response curves for agonists of the human chemokine CCR5 receptor with a CHO-CCR5-aequorin cell line. CHO-CCR5-aequorin cells were prepared and used as described in Figure 2 for the detection of dilutions of agonists of CCR5. (A) Curves of the light emission as a function of time are shown for an entire 96-well plate. Increasing (from column 1 to 12) concentrations of RANTES (rows A and B, same dilutions in duplicate), MIP1α (rows C and D, duplicates), MIP1β (rows E and F, duplicates), and an analog of RANTES (rows G and H, duplicates) were disposed in a 96-well plate. Cells were injected in the wells and the emitted light was recorded for 30 seconds. The intensity of the emitted light was plotted as a function of time. The scaling is the same for all the graphs presented here. For this experiment, we used the EG&G Wallac's MicroLumat-Plus microplate luminometer. (B) The emitted light was integrated, yielding a single value for each well and values were plotted against the logarithm of the concentration of agonists used. These results correspond to the experiment shown in Figure 3A. RANTES: CC-chemokine whose full name is "Regulated on activation of normal T cell expressed and secreted"; Mip: CC-chemokine whose full name is "Macrophage inflamatory protein".

corded, etc. For each plate, a series of curves representing the intensity of the emitted light as a function of time for each well was displayed (Figure 3A). The intensity of the emitted light was integrated over 30 seconds using the Winglow software provided with the "MicroLumat-Plus" luminometer (Berthold), yielding for each well one value representative of the emitted light and hence of the stimulation of the CCR5 receptor by the agonist present in the well. These values can be plotted against the logarithm of the ligand concentration to generate dose-response curve as shown in Figure 3B. These allow the determination of half-maximal response doses (EC_{50}) for each ligand.

4.2. Human serotonin $5HT_{2B}$ receptor

A CHO cell line expressing the serotonin $5HT_{2B}$ receptor and apoaequorin was established. Dose-response curves for agonists of the $5HT_{2B}$ receptor were obtained as described above and results are shown in Figure 4A. For the detection of antagonists, CHO-$5HT_{2B}$-aequorin cells were incubated with various concentrations of the compounds to be tested. An agonist (α-methyl-5-HT) at a single concentration was then injected on the mixture of cells and antagonist and the emitted light was recorded and integrated for 30 seconds. Non-surmountable antagonists (Methiothepin, Methysergide) as well as surmountable antagonists (Mesulergine, Mianserin, Ketanserin) prevented in a dose-dependent manner the activation of the 5-HT_{2B} receptor by α-methyl-5-HT (Figure 4B).

One of the main advantages of the aequorin-based assay compared to the use of calcium-sensitive fluorescent dyes is the ease and stability of the loading of the cells with coelenterazine. Indeed, in the aequorin system, a single cell preparation can be used to perform calcium measurements during a whole day or even for several days! In contrast to what happens when using calcium-sensitive fluorescent dyes, once the cells have been incubated with coelenterazine it is not necessary to remove coelenterazine, as it does not emit light by itself (i.e. in the absence of apo-aequorin). After incubating the cells with coelenterazine for 4 hours at room temperature or overnight at 4°C, cells are simply diluted 10 times with fresh culture medium. The coelenterazine that remains in the extracellular medium can then continuously replace the one that has been consumed in the cells; this ensures the stability of the loading of the cells. Figure 5A illustrates the stability of the cell loading for a single cell preparation. To use the same cell preparation for several days, cells can be stored at 4°C before dilution with continuous stirring to keep them in suspension. Each day, an aliquot of this cell suspension is diluted 10 times with culture medium at room temperature and is incubated at room temperature for 1 hour. It can then be used for the aequorin measurements for a whole day. The signal obtained keeps the same intensity for at least 3 days

A.

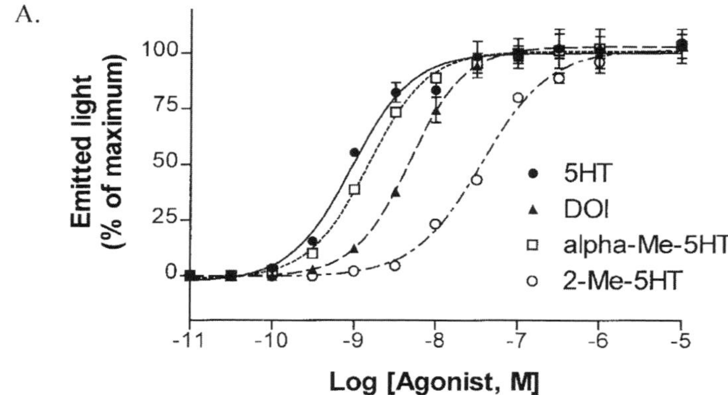

Agonist	EC50 (nM)
5HT	0.8 nM
DOI	4.7 nM
α-methyl-5HT	1.4 nM
2-methyl-5HT	28 nM

B.

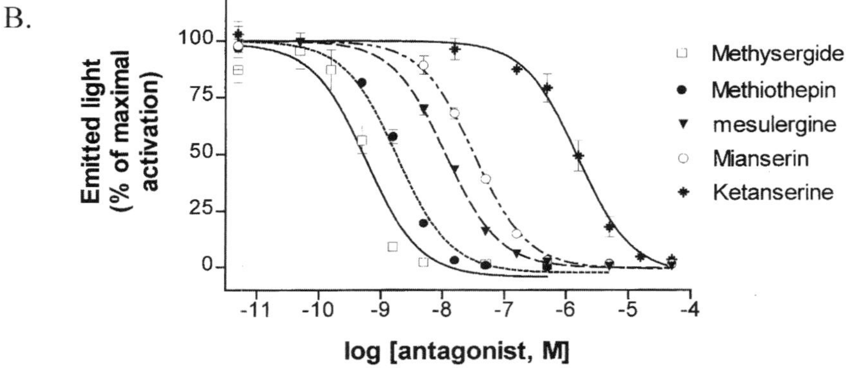

Antagonist	IC50 (nM)
Methysergide	0.6 nM
Methiothepin	2 nM
Mesulergin	10 nM
Mianserin	33 nM
Ketanserin	1.7 μM

Figure 4. Detection of agonists and antagonists of the serotonin $5HT_{2B}$ receptor with CHO-$5HT_{2B}$-aequorin cells. Cells were processed as described in Figure 3. (A) Dose-response curves for agonists of the receptor and EC_{50} values. (B) Dose-response curves for antagonists of the receptor and IC_{50} values. DOI: abbreviation for "(±)-2,5-Dimethoxy-4-iodoamphetamine"; 2-Me-5HT: abbreviation for "2-methyl-5-hydroxytryptamine".

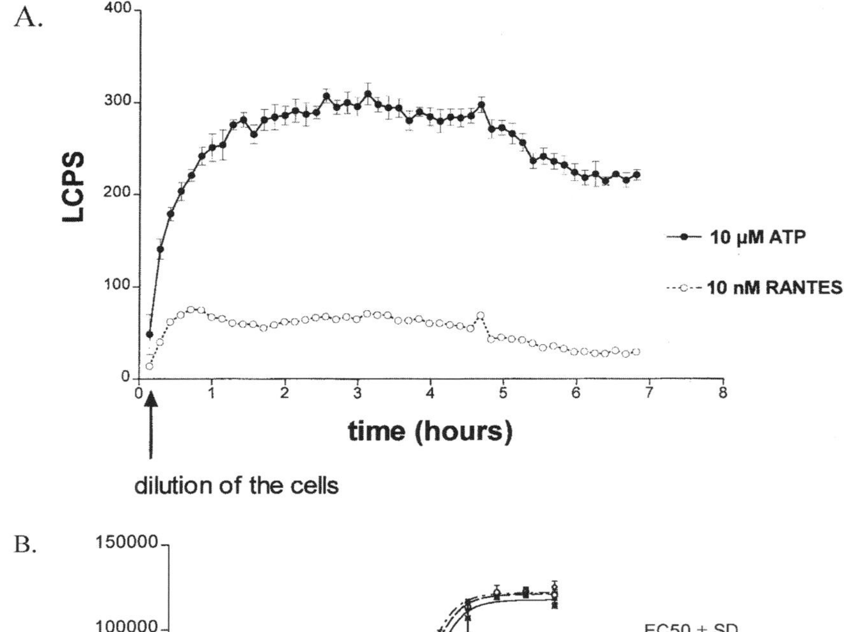

Figure 5. Robustness of calcium measurements with the aequorin system. (A) CHO cells coexpressing aequorin and the human chemokine CCR5 receptor were prepared and active aequorin was reconstituted as described in Figure 2. After loading with coelenterazine, cells were diluted 10 times with culture medium and were maintained in suspension at room temperature by magnetic stirring. At the indicated time, cells were injected on solutions containing 20 μM of ATP, that acts on an endogenous purinergic receptor of the CHO cells, or on solutions containing 10 nM of RANTES, an agonist of the chemokine CCR5 receptor. After each injection, the emitted light was recorded and integrated over the measurement interval. The graph shows that, after 1 hour of equilibration, the value of the signal is stable for at least 4 hours. (B) CHO cells coexpressing aequorin, the melanocortin MC4 receptor and the Gα16 coupling protein were prepared and active aequorin was reconstituted as described in Figure 2. Series of dilution of NDP-α-MSH, an agonist of this receptor, were prepared in culture medium containing 0, 2, 5 and 10% of DMSO and placed in the wells of a 96-well plate. Cells were then injected on these solutions and the emitted light was recorded for 30 seconds. Dose-response curves are shown for each DMSO concentration. The final DMSO concentrations are indicated. These are the starting DMSO concentrations divided by 2 because there is a 2-fold dilution upon addition of the cells on the agonist solutions: 50 μl of cell suspension are added on 50 μl of agonist solutions.

and the cells can still be used, although with a sometimes lower signal, for the rest of the week.

An important feature in the domain of drug screening is the resistance of the test to DMSO. This solvent is indeed the most commonly used to solubilize the compounds to be tested. To perform the assay, the stock solution of the compound to be tested is diluted with the test buffer (culture medium in the case of the aequorin assay) and some DMSO remains at the time of the assay. Figure 5B illustrates the very strong resistance of the aequorin assay to DMSO: it does not induce any non-specific luminescent response, it does not prevent an agonist from stimulating its receptor and does not significantly affect the EC_{50} of this agonist.

5. UNIVERSAL COUPLING OF GPCRs TO THE CALCIUM PATHWAY

As explained above, if the receptor to be screened is not naturally coupled to intracellular calcium increase, it is possible to co-express natural or chimaeric transduction proteins to redirect its coupling towards calcium.

The first possibility is to use the human Gα16 protein, or its murine counterpart Gα15, whose expression is naturally restricted to a subset of haematopoietic cells. The Gα16 and Gα15 proteins have the remarkable property to be able to contact almost any GPCR, with very few exceptions, and to stimulate phospholipase C in response to the binding of an agonist of these receptors. This leads to the coupling of virtually any GPCR to the calcium pathway in cell lines co-expressing the Gα16 or Gα15 protein (Offermanns and Simon, 1995; Milligan et al., 1996).

The second possibility is to use chimaeric G proteins. These are composed of a Gαq protein, where the 5 carboxy-terminal aminoacids have been replaced by those of a Gi or Gs protein, that usually couples GPCRs to an inhibition (Gi) or a stimulation (Gs) of the cAMP pathway. It has been shown that the 5 last aminoacids of the Gα proteins are responsible for the specificity of the interaction between GPCRs and Gα proteins. Thus changing these aminoacids in a Gαq protein changes the specificity of its association with GPCRs. The result is that chimaeric Gαqi and Gαqs proteins redirect the coupling of GPCRs usually coupled to the cAMP pathway towards an intracellular calcium increase (Conklin et al., 1993, 1996; Coward et al., 1999).

The third possibility is to take advantage of the property of the β2 isoform of phospholipase C (PLCβ2) to be stimulated by a great number of Gβ-Gγ dimers (Katz et al., 1992; Camps et al., 1992). The G$\beta\gamma$ dimers that are released from the Gα subunit upon stimulation of a GPCR should then be able to stimulate the activity of PLCβ2. We are currently evaluating the possibility that overexpression of PLCβ2 in cell lines would be able to redirect the coupling of any GPCR towards the calcium pathway. Our preliminary experiments indicate that it is at least able to improve the coupling of the chemokine CCR3 receptor to the calcium pathway.

6. CONCLUSION

Several techniques exist, that allow to perform thousands of measurements of intracellular calcium concentration per day. The possibility to couple almost any GPCR to an increase in intracellular calcium renders this system "universal"; it also allows to screen a target without previous knowledge of the specific signalling pathway. This is very usefull in the case of orphan receptors (receptors whose natural ligand is still unknown). Compared to other functional assays, the measurement of Ca^{++} mobilization is an early event following the binding of an agonist. This limits the risk to have artifacts. The measurement of intracellular calcium is a simple and direct assay, easy to automate, that is widely used in the pharmaceutical industries for the discovery of new drugs.

REFERENCES

Barritt, G.J., 1999, Receptor-activated Ca^{2+} inflow in animal cells: A variety of pathways tailored to meet different intracellular Ca^{2+} signaling requirements, *Biochem. J.* 337, 153–169.

Brini, M., Marsault, R., Bastianutto, C., Alvarez, J., Pozzan, T. and Rizzuto, R., 1995, Transfected aequorin in the measurement of cytosolic Ca^{2+} concentration, *J. Biol. Chem.* 270, 9896–9903.

Button, D. and Brownstein, M., 1993, Aequorin-expressing mammalian cell lines used to report Ca^{2+} mobilization, *Cell Calcium* 14, 663–671.

Camps, M., Carozzi, A., Schnabel, P., Scheer, A., Parker, P.J. and Gierschik, P., 1992, Isozyme-selective stimulation of phospholipase C-beta 2 by G protein beta gamma-subunits, *Nature* 360, 684–686.

Conklin, B.R., Farfel, Z., Lustig, K.D., Julius, D. and Bourne, H.R., 1993, Substitution of three amino acids switches receptor specificity of Gq alpha to that of Gi alpha, *Nature* 363, 274–276.

Conklin, B.R., Herzmark, P. Ishida, S., Voyno-Yasenetskaya, T.A., Sun, Y. and Bourne, H.R., 1996, C-terminal mutations of Gq alpha and Gs alpha that alter the fidelity of receptor activation, *Mol. Pharmacol.* 50, 885–890.

Coward, P., Chan, S., Wada, H., Humphries G. and Conklin, B., 1999, Chimeric G proteins allow a high-throughput signaling assay of Gi-coupled receptors, *Anal. Biochem.* 270, 242–248.

Inouye, S., Noguchi, M., Sakaki, Y., Takagi, Y., Miyata, T., Iwanaga, S. and Tsuji, F.I., 1985, Cloning and sequence analysis of cDNA for the luminescent protein aequorin, *Proc. Natl. Acad. Sci. USA* 82, 3154–3158.

Katz., A., Wu, D. and Simon, M.I., 1992, Subunits beta gamma of heterotrimeric G protein activate beta 2 isoform of phospholipase C, *Nature* 360, 686–689.

Kendall, J.M. and Badminton, M.N., 1998, Aequoria victoria bioluminescence moves into an exciting new era, *Trends Biotechnol.* 16, 216–224.

Milligan, G., Marshall, F. and Rees, S., 1996, $G_{\alpha 16}$ as a universal G protein adapter: Implications for agonist screening strategies, *Trends Pharmacol. Sci.* 17, 235–237.

Offermanns, S. and Simon, M., 1995, G alpha 15 and G alpha 16 couple a wide variety of receptors to phospholipase C, *J. Biol. Chem.* 270, 15175–15180.

Prasher, D.C., McCann, R.O. and Cormier, M.J., 1985, Cloning and expression of the cDNA coding for aequorin, a bioluminescent calcium-binding protein, *Biochem. Biophys. Res. Commun.* 126, 1259–1268.

Rizutto, R., Brini, M., Bastianutto, C., Marsault, R. and Pozzan, T., 1995, Photoprotein-mediated measurement of calcium ion concentration in mitochondria of living cells, *Methods Enzymol.* 260, 417–428.

Rizutto, R., Pinton, P., Carrington, W., Fay, F., Fogarty, K., Lifshitz, L., Tuft, R. and Pozzan, T., 1998. Close contacts with the endoplasmic reticulum as determinants of mitochondrial Ca^{2+} responses, *Science* 280, 1763–1766.

Schroeder, K. and Neagle, B., 1996, FLIPR^{TM}: A new instrument for accurate, high-throughput optical screening, *J. Biomol. Screening* 1, 75–80.

Sheu, Y.-A., Kricka, L.J. and Pritchett, D.B., 1993, Measurement of intracellular calcium using bioluminescent aequorin expressed in human cells, *Anal. Biochem.* 209, 343–347.

Shimomura, O., Johnson, F.H. and Saiga, Y., 1962, Extraction, purification and properties of Aequorin, a bioluminescent protein from the luminous hydromedusan, Aequorea, *J. Cell. Physiol.* 59, 223–240.

Stables, J., Green, A., Marshall, F., Fraser, N., Knight, E., Sautel, M., Milligan, G., Lee, M. and Rees, S., 1997, A bioluminescent assay for agonist activity at potentially any G-protein-coupled receptor, *Anal. Biochem.* 252, 115–126.

Imaging Intracellular Calcium in Living Tissue by Laser-Scanning Confocal Microscopy

Angel Nadal and Bernat Soria

1. INTRODUCTION

For many years physiologists have attempted to watch cells working within organs and to visualize the interactions amongst the different types of cells within them. The development of fluorescence imaging systems and calcium sensitive fluorescent dyes has essentially contributed to revealing many important aspects of cell signaling. Although it has been proven that conventional fluorescence microscopy is an excellent technique for detecting calcium signals in isolated cells, it only provides a very confined insight when used in tissue. When ordinary fluorescence microscopy is used with relatively thick specimens, the main limitations are due to image degradation caused by out of focus flare, further limited by a poor discrimination of depth. In conventional fluorescence, the light that makes up each point of the image spreads out in a solid cone that can reach a significant distance above and below the focus. The spreading cone of light blurs the focused image of the specimen. Also, fluorescent objects that are out of focus introduce unwanted light which highly reduces the contrast of the image. These problems are greatly reduced using laser-scanning confocal microscopy (LSCM), which eliminates out of focus fluorescence, producing well defined thin optical sections out of thick fluorescence specimens (White et al., 1987; Fine et al., 1988; Pawley, 1995). The elimination of out of focus fluorescence increases contrast, clarity and detection, enabling us to do a wide range of investigations that, up until now, have been difficult or impossible to carry out with previous techniques. The combination of confocal microscopy with fluorescent sensit-

Angel Nadal and Bernat Soria • Institute of Bioengineering and Department of Physiology, Miguel Hernández University, Alicante, Spain.

R. Pochet, R. Donato, J. Haiech, C. Heizmann and V. Gerke (eds.): Calcium: The Molecular Basis of Calcium Action in Biology and Medicine, 661–671.

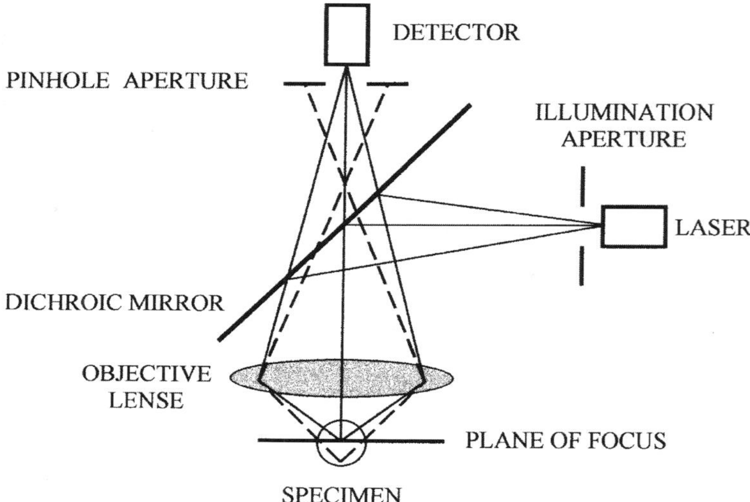

Figure 1. Basic principle of confocal microscopy. A point source of light, usually from a laser illuminates the specimen via a dichroic mirror. Emitted or reflected light from the illuminated point passes through the same pathway as excitation light passing the dichroic and is finally imaged on a pinhole before the photodetector. Such a pinhole excludes out of focus light. The confocal effect depends both on the focussed illumination of a single point in the specimen and the use of a the pinhole before the detector.

ive calcium dyes allows us to image intracellular calcium signals in individual cells within intact living tissues.

2. PRINCIPLES OF CONFOCAL MICROSCOPY

In 1957, Marvin Minsky, a young postdoctoral fellow at Harvard University, invented the confocal scanning microscope and contributed his insight to the potential applications of confocal microscopy (Minsky, 1957, 1988). The physical principle is based on the use of a source of light which is focused on a single point in the specimen. Detection of the emitted or reflected light from the specimen is confined in the same region by a pinhole before a detector. The pinhole aperture allows light that is in focus to pass through to the detector whilst out-of-focus light is thereby eliminated (Figure 1).

In practice, a confocal microscope uses a laser as a powerful point source of light. In order to obtain a two-dimensional image the laser beam scans the specimen by deflecting the light beam with moving mirrors, making an optical section of defined thickness. The optical sectioning depends on several factors: the illumination of a single point in the specimen with a laser beam, the use of a pinhole (of variable aperture) positioned before the detector device, usually a photomultiplier and the use of a high numerical aperture (NA) objective which further decreases the thickness of the optical section. The depth of the optical section is inversely related to the NA and may be as little as 500 nm with a NA 1.4 objective. With modern commercially available laser-scanning confocal microscopes, object features of about 0.2 μm can be seen, and height differences of less that 0.1 μm can be made visible. Confocal microscopy also allows three-dimensional reconstruction. Since optical information is converted into electrical signals by photomultipliers, the image of any object plane can be generated and stored within less than a second. Any object plane of interest can be looked at with a defined focusing (Z axis) movement. A three-dimensional image of a specimen can be reproduced by scanning it in a succession of object planes and producing a stack of slice images. Nevertheless, the time resolution for three-dimensional images is still very poor for measuring $[Ca^{2+}]_i$ movements and this facility is used mainly for morphological studies. Two-photon excitation microscopy is a new alternative to confocal microscopy with many advantages for three-dimensional imaging (Piston, 1999).

3. MONITORING INTRACELLULAR Ca^{2+} MOVEMENTS

In order to measure the spatiotemporal changes of intracellular calcium concentration ($[Ca^{2+}]_i$) in a specimen, it must to be loaded with a calcium sensitive fluorescence dye. Although several Ca^{2+} dyes are available, the most widely used in confocal microscopy are Fluo-3, Rhod-2, Calcium Green-1 and Indo-1. The latter, shows a fine emission shift from 485 to 405 nm with increasing Ca^{2+}, making it suitable for ratiometric laser-scanning (Tsien and Waggoner, 1990). The recent development of high speed confocal microscopes makes Indo-1 a powerful Ca^{2+} indicator, although a UV-laser is necessary for excitation in the 350–365 nm. The two major disadvantages of using UV-lasers are the high cost and the obstacles to correct chromatic aberration which have been described (Lipp and Niggli, 1993). Another problem of using this Ca^{2+} dye is related to the autofluorescence of reduced pyridine nucleotides within the cells, which is well in the range of Indo-1 fluorescence. A final problem is that Indo-1 photobleaches faster than other dyes such as Fura-2 (Tsien and Waggoner, 1990).

Fluo-3 and, to a lesser extent, Rhod-2 and Calcium Green-1 are the most used Ca^{2+} indicators in confocal microscopy, partly because they can be excited using a visible light suitable for low-power visible lasers which is available with the majority of commercial laser scanning confocal microscopes (Minta et al., 1989; Kao et al., 1989). A disadvantage is that none of them present either an excitation or emission shift upon binding of Ca^{2+}. So, it is not possible to perform ratiometric measurements and thus, the calibration of fluorescence signals is difficult. However, this disadvantage has been sorted out by simultaneously loading cells with two Ca^{2+} sensitive fluorescence dyes such as Fluo-3 and Fura Red. Upon binding of Ca^{2+} these two dyes behave reciprocally, fluorescence intensity from fluo-3 increases whilst that of Fura Red decreases, permitting ratiometric measurements of intracellular calcium to be taken (Lipp and Niggli, 1993; Grøndal and Langmoen, 1998).

Cells within tissues can be loaded by two different methods: either injecting the dye in its salt form using a patch pipette or incubating the whole piece of tissue with the acetoxymethyl (AM) derivate of the dye, which is membrane permeable. The former gives a very clear signal and highly detailed two dimensional images can be obtained (Eilers et al., 1995). Nonetheless, only one or, at most few cells are loaded at the same time when they are connected by gap junctions, making it difficult to study interactions between cell populations. Although it is preferable to load with the AM form to study these interactions, this option is not exempt of difficulties. A usual problem experienced is dye penetration into the specimen. As a result, cells close to the surface are normally well loaded but poorer dye loading is observed in cells about 30 μm below the surface (Nadal et al., 1998, Nadal et al., 1999). Although in the particular case of brain slices, histological studies have demonstrated that there is at least a 50 μm area of destroyed tissue at the cut edge, it is possible to find well preserved cells at a depth of around 20–30 μm which responds to NMDA and albumin (Figure 2, Nadal et al., 1998). For this particular preparation, the use of neonatal tissue is essential when loading with the acetoxymethyl ester form of Ca^{2+} dyes (Yuste et al., 1992).

4. EXAMPLES OF Ca^{2+} IMAGING IN LIVING TISSUE

For imaging thin samples such as isolated cells, confocal microscopy might not offer significant advantages over regular fluorescence imaging except in specific cases. Laser scanning confocal microscopy excels when used with thick multicellular preparations such as brain slices, corneal epithelia and the islet of Langerhans, amongst others. LSCM has been successfully combined with electrophysiological techniques, such as patch clamp or with immunocytochemistry after Ca^{2+} measurements were taken to obtain important results concerning different aspects of cell science.

Imaging of intracellular calcium parallelly with patch clamp has been used in brain slices to detect movements of Ca^{2+} in different parts of neurons, such as nerve terminals, cell bodies, dendrites, dendritic spines and axons (Yuste and Tank, 1996; Callewaert et al., 1996). Most of the studies of cellular activity have focussed on synaptic neuronal activity. Recently, however, non-synaptic signaling has received increasing attention. Amongst non-synaptic signaling interaction, glia-glia and bi-directional glia-neuron signaling have been proposed as potential modulators of neuronal functions (Charles, 1994; Bacci et al., 1999). In the last few years, a great number of articles focused on calcium signaling in glial cells have been published (Verkhratsky and Kettenmann, 1996; Verkhratsky et al., 1998). Glial cells respond to most of the current hormones, neurotransmitters and neuromodulators with calcium

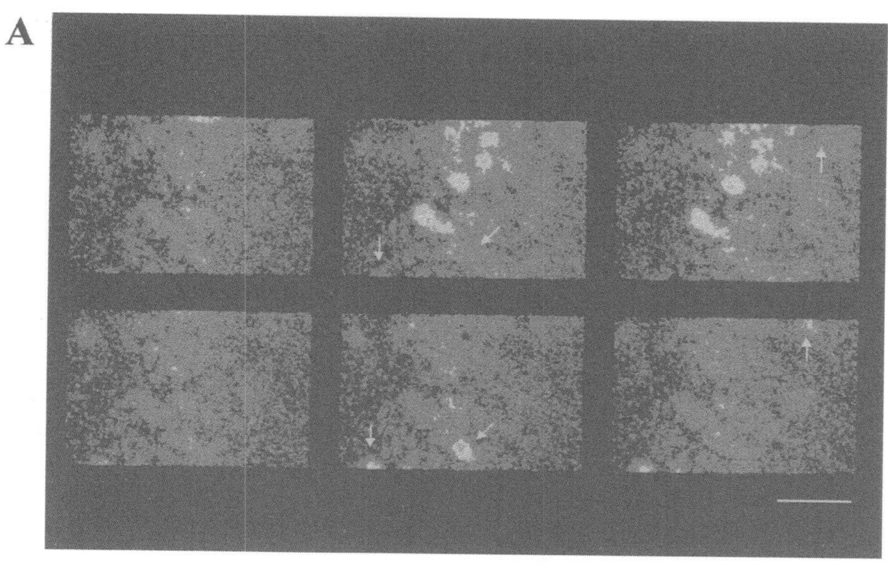

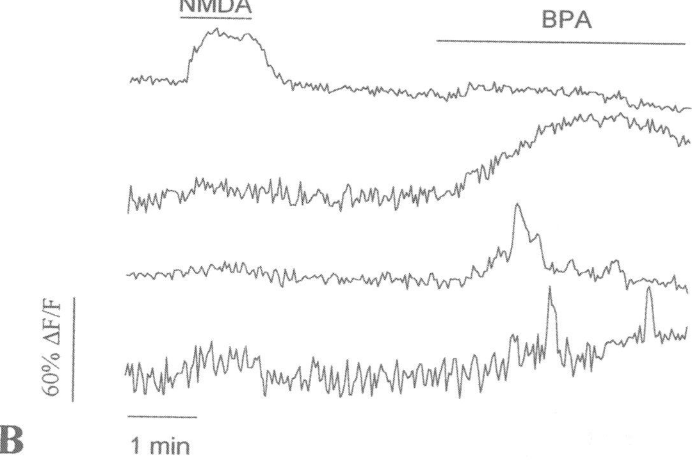

Figure 2. Confocal image of $[Ca^{2+}]_i$ from a fluo-3 loaded cortical brain slice obtained from a neonatal rat. (A) Upper panel: Application of 100 μM NMDA in 0 Mg^{2+} and 10 μM glycine induced Ca^{2+} elevation in a population of cells within the cortical brain slice. From left to right: basal level of fluorescence, 10 s after and 30 s after NMDA application. Lower panel: Effect of 20 mg/ml BPA on the same field of cells. Note that cells indicated by arrows respond to BPA but not to NMDA. Scale bar is 50 μm. (B) Records of fluorescence intensity *versus* time from the experiment in part (A). After Nadal et al. (1998), reproduced with permission. For a colour version of Figure 2A, see page xxxvi.

oscillations associated with intracellular calcium waves and with the intercellular propagation of calcium signals as well (Cornell-Bell et al., 1990; Dani et al., 1992; Nadal et al., 1997; Giaume and Venance, 1998). A consequence of such a calcium elevation is the release of glutamate and the subsequent neuronal stimulation (Purpura et al., 1994), although a crosstalk between glia and neurons due to the presence of gap junction has also been described (Nedergaard, 1994). Studies in cortical slices with confocal microscopy have demonstrated that such an interaction exists in intact brain tissue (Pasti et al., 1997). A difficult task when working with whole tissue is to identify unequivocally the different cell types within it. For instance: how can neurons from astrocytes in brain slices be identified? Although it is possible to perform immunostaining to identify unequivocally each type of cell, it is not only difficult, but time consuming as well and successful experiments are obtained in less than 20% of the cases (Pasti et al., 1997). A second possibility is to take advantage of the absence of an increase of $[Ca^{2+}]_i$ in astrocytes in response to the glutamate receptor agonist N-methyl-D-aspartate (NMDA) which, however, strongly increases $[Ca^{2+}]_i$ in neurons. On the other hand, $[Ca^{2+}]_i$ oscillations are induced by plasma albumin in astrocytes but neurons are unresponsive to this protein (Figure 2 and Nadal et al., 1998).

Identification of cells by their different Ca^{2+} response to a particular stimulus has been proven to be useful in other tissues (Asada et al., 1998; Nadal et al., 1999; Quesada et al., 1999). Another demonstration of the power of LSCM is the imaging of $[Ca^{2+}]_i$ in different cell types within an intact islet of Langerhans.

Pancreatic islets of Langerhans are formed by a heterogeneous population of cells: insulin-releasing β-cells (65–90%), glucagon-releasing α-cells (15–20%), somatostatin-producing δ-cells (3–10%) and pancreatic polypeptide-producing PP-cells (1%). Amongst this population, β-cells have been by far the most studied in terms of stimulus-secretion coupling. In β-cells the stimulus-secretion coupling process involves the closure of K_{ATP} channels as a result of glucose metabolism, membrane depolarization and activation of voltage dependent calcium channels (Prentki and Matchinsky, 1987; Valdeolmillos et al., 1992).

Using conventional Ca^{2+} imaging of the whole islet of Langerhans, it has been demonstrated that stimulatory glucose concentrations induce a synchronous and homogeneous $[Ca^{2+}]_i$ oscillatory pattern as a consequence of the bursting pattern of electrical activity characteristic of pancreatic β-cells (Santos et al., 1991; Valdeolmillos et al., 1993). Less is known about stimulus secretion coupling in non-β-cells. Recent use of confocal microscopy has permitted the investigation of calcium signals in individual cells and the relationship between the different types of cells present within the islet of Langerhans. Islets loaded with the calcium sensitive fluorescent dye Fluo-3 AM look like Figure 3A, where only the periphery of the islet was well loaded, as previously mentioned. Figure 3B shows three different traces, Trace 1 shows a typical record of a pancreatic β-cell, where $[Ca^{2+}]_i$ re-

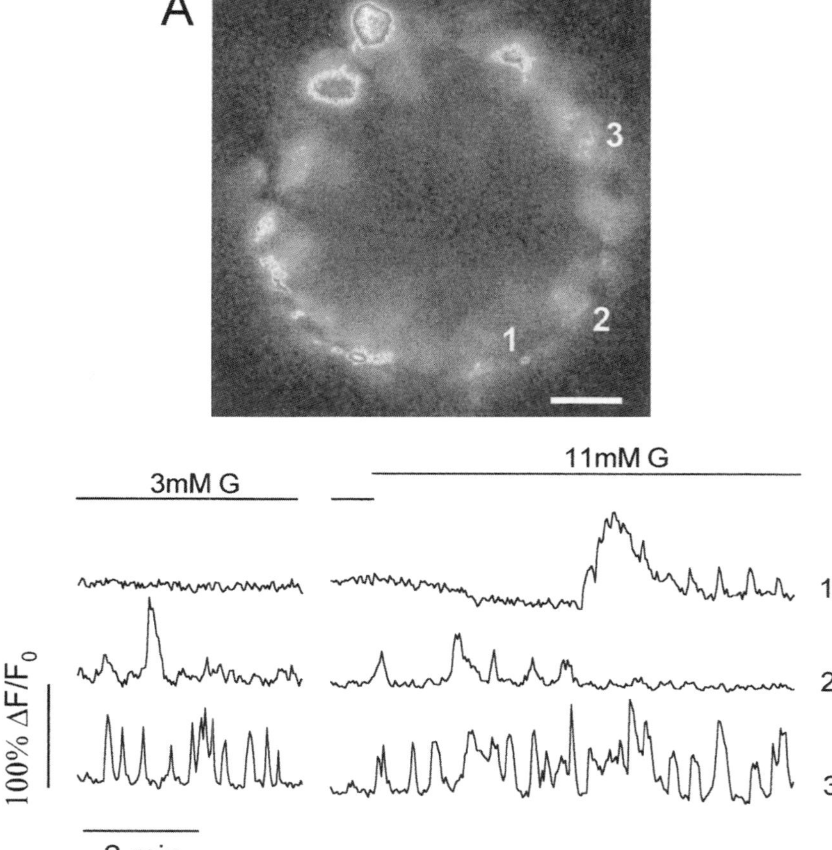

Figure 3. Fluorescence changes measured from individual cells within an intact islet of Langerhans. (A) Colour image of a fluo-3 loaded islet exposed to 3 mM glucose, blue corresponds to low and red to high fluorescence intensity. Scale bar at bottom right represents 15 μm. (B) Records of fluorescence intensity versus time from the islet in (A). Boundaries of each cell were clearly visible from the low fluorescence at the cell edge. Islet exposed to 3 mM glucose were switched to 11 mM glucose as indicated by the bars on the record. Traces 1, 2, 3 corresponded to cells labeled in (A). Time break was 10 minutes. After Nadal et al. (1999), reproduced with permission. For a colour version of Figure 3a, see page xxxvi.

mains silent at low glucose concentrations (3 mM) until glucose stimulation (11 mM) elicits a transient increase of intracellular calcium and a subsequent train of oscillations. In contrast, trace 2 is a record of a cell which displays low frequency oscillations in low glucose, which in turn are abolished by high glucose. These cells were identified as α-cells by immunostaining with an anti-glucagon monoclonal antibody after monitoring $[Ca^{2+}]_i$ signals (Nadal et al., 1999). Trace 3 is characteristic of cells which also oscillated in low glucose, but at a higher frequency; unlike trace 2, 11 mM glucose evoked no change in the oscillatory frequency. Using immunostaining, now with an anti-

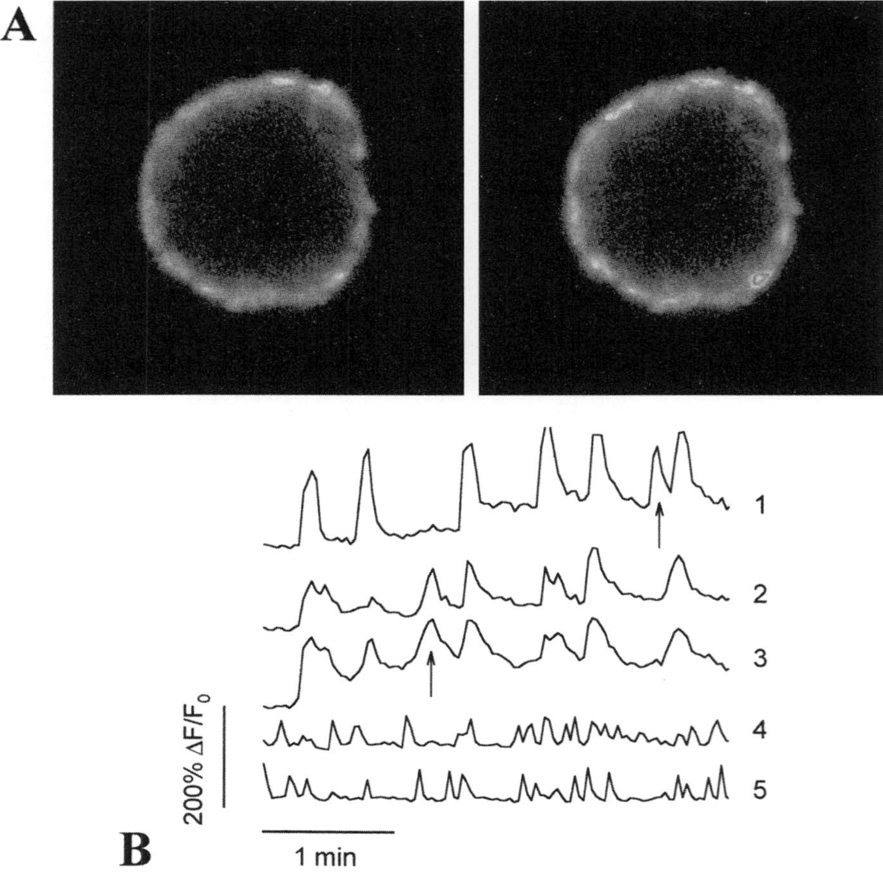

Figure 4. Synchronicity between pancreatic β-cells within an intact islet of Langerhans. (A) Time sequence of a $[Ca^{2+}]_i$ oscillation in an islet of Langerhans exposed to 11 mM glucose imaged with a confocal microscope. From left to right: at the beginning of a $[Ca^{2+}]_i$ oscillation and at the peak of the oscillation. (B) Fluorescence records of 3 β-cells within the islet in (A). Note that when a stimulating glucose concentration (11 mM) are applied $[Ca^{2+}]_i$ oscillations are synchronous in traces 1–3. Traces 4 an 5 represent somatostatin containing δ-cells and are asynchronous. For a colour version of Figure 4a, see page xxxvii.

somatostatin antiserum, these cells were identified as δ-cells (Nadal et al., 1999). This particular preparation is a good model for studying synchronicity amongst calcium signals, since it is well known that dye coupling via gap junctions exists among the same type of cells and between different types of cells (Michaels and Sheridan, 1981; Meda et al., 1982). Also, in terms of calcium signals the whole islet of Langerhans functions as a syncytium in response to stimulatory glucose concentrations (Valdeolmillos et al., 1993).

The experiment in Figure 4 shows an example of individual cells with a typical β-cell $[Ca^{2+}]_i$ pattern, exhibiting a highly synchronous oscillatory $[Ca^{2+}]_i$ behavior (Figure 4B, traces 1 to 3). Nonetheless, some asynchrony is occasionally manifested (note arrows in Figure 4B). On the other hand, δ-cell $[Ca^{2+}]_i$ oscillations, were totally asynchronous among δ-cells and between δ and β cells (Figure 4B, traces 4 and 5). As a consequence, when averaging the signal from cells of the same islet, the $[Ca^{2+}]_i$ derived signal resembles the one obtained using a conventional imaging system. The advantage of this system has been taken to characterize the effect of antidiabetic drugs such as hypoglycemic sulfonylureas in different types of cells within the islet of Langerhans (Quesada et al., 1999).

Heterogeneous responses of different cells have been revealed in other cell systems such as in the corneal epithelial tissue by using LSCM (Kimura et al., 1999). LSCM can be potentially applied to other cell systems as suggested by Satoh et al. (1998), including the smooth muscle of the intestinal wall, the myenteric plexus and the dorsal root ganglia.

5. CONCLUSIONS

Laser scanning confocal microscopy has been proven to be an excellent technique for monitoring Ca^{2+} signals in living tissue. It has not yet been widely used because of the difficulty in preparing living tissue and, more importantly because of the trouble in loading cells in deep regions within the specimen with Ca^{2+}-sensitive fluorescent dyes. Specimen photodamage is another important obstacle which perturbs the sample, specially when fast recording is performed. The development of new Ca^{2+} dyes excitable at visible wavelengths, with a large emission shift to perform ratiometric measurement and resistant to photobleaching would be of great importance. In any case, this problem has been partially solved by the recent invention of two-photon microscopy (Denk et al., 1990; So et al., 1998; Piston, 1999), which allows non-invasive monitoring of biological specimen with confocal-like resolution and should further improve the image of Ca^{2+} signals in living tissue.

REFERENCES

Asada, N., Shibuya, I., Iwanaga, T., Niwa, K. and Kanno, T., 1998, Identification of α- and β-cells in intact isolated islets of Langerhans by their characteristic cytoplasmic Ca^{2+} concentration dynamics and immunocytochemical staining, *Diabetes* 47, 751–757.

Bacci, A., Verderio, C., Pravettoni, E. and Matteoli, M., 1999, The role of glial cells in synaptic function, *Philos. Trans. Roy. Soc. Lond. B* 354, 403–409.

Callewaert, G., Eilers, J. and Konnerth, A., 1996, Axonal calcium entry during fast ‘sodium’ action potentials in rat cerebellar Purkinje neurones, *J. Physiol.* 495, 641–648.

Charles, A.C., 1994, Glia-neurone intercellular calcium signaling, *Dev. Neurosci.* 16, 196–206.

Cornell-Bell, A.H., Finkbeiner, S.M., Cooper, M.S. and Smith, S.J., 1990, Glutamate induces calcium waves in cultured astrocytes: Long-range glial signaling, *Science* 247, 470–473.

Dani, J.W., Chernjavsky, A. and Smith, S.J., 1992, Neuronal activity triggers calcium waves in hippocampal astrocytes networks, *Neuron* 8, 429–440.

Denk, W., Strickler, J.H. and Webb, W.W., 1990, Two-photon laser scanning fluorescence microscopy, *Science* 248, 73–76.

Eilers, J., Callewaert, G., Armstrong, C. and Konnerth, A., 1995, Calcium signaling in a narrow somatic submembrane shell during synaptic activity in cerebellar Purkinje neurones, *Proc. Natl. Acad. Sci. USA* 92, 10272–10276.

Fine, A., Amos, W.B., Durbin, R.M. and McNaughton, P.A., 1988, Confocal microscopy: Applications in neurobiology, *Trends Neurosci.* 11, 346–351.

Giaume, C. and Venance, L., 1998, Intercellular calcium signaling and gap junctional communication in astrocytes, *Glia* 24, 50–64.

Grøndahl, T.Ø. and Langmoen, I.A., 1998, Confocal laser scanning microscopy used to monitor intracellular Ca^{2+} changes in hippocampal CA 1 neurons during energy deprivation, *Brain Res.* 785, 58–65.

Kao, J.P.Y., Harootunian, A.T. and Tsien, R.Y., 1989, Photochemically generated cytosolic calcium pulses and their detection by fluo-3, *J. Biol. Chem.* 264, 8179–8184.

Kimura, K., Yamashita, H., Nishimura, T., Mori, S. and Satoh, Y., 1999, Application of real-time confocal microscopy to observations of ATP-induced Ca^{2+}-oscillatory fluctuations in intact corneal epithelial cells, *Acta Histochem. Cytochem.* 32, 50–63.

Lipp, P. and Niggli, E., 1993, Ratiometric confocal Ca^{2+}-measurements with visible wavelenght indicators in isolated cardiac myocytes, *Cell Calcium* 14, 359–372.

Meda, P., Kohen, E., Kohen, C., Rabinovitch, A. and Orci, L., 1982, Direct communication of homologous and heterologous endocrine islet cells in culture, *J. Cell Biol.* 92, 221–226.

Michaels, R.L. and Sheridan, J.D., 1981, Islets of Langerhans: Dye coupling among immunocytochemically cell types, *Science* 214, 801–803

Minsky, M., 1957, U.S. Patent #3013467, Microscopy Apparatus.

Minsky, M., 1988, Memoir on inventing the confocal scanning microscope, *Scanning* 10, 128–138.

Minta, A., Kao, J.P.Y. and Tsien, R.Y., 1989, Fluorescent indicators for cytosolic calcium based on rhodamine and fluorescein chromophores, *J. Biol. Chem.* 264, 8171–8178.

Nadal, A., Fuentes, E., Pastor, J. and McNaughton, P.A., 1997, Plasma albumin induces calcium waves in rat cortical astrocytes, *Glia* 19, 343–351.

Nadal, A., Sul, J.Y., Valdeolmillos, M. and McNaughton, P.A., 1998, Albumin elicits calcium signals from astrocytes in brain slices from neonatal rat cortex, *J. Physiol.* 509, 711–716.

Nadal, A., Quesada, I. and Soria, B., 1999, Homologous and heterologous asynchronicity between identified α-, β- and δ-cells within intact islets of Langerhans in the mouse, *J. Physiol.* 517, 85–94.

Nedergaard, M., 1994, Direct signaling from astrocytes to neurons in cultures of mammalian astrocytes networks, *Science* 263, 1768–1771.

Pasti, L., Volterra, A., Pozzan, T. and Carmignoto, G., 1997, Intracellular calcium oscillations in astrocytes: A highly plastic, bidirectional form of communication between neurons and astrocytes in situ, *J. Neurosci.* 17, 7817–7830.

Pawley, J.B. (ed.), 1995, *Handbook of Biological Confocal Microscopy*, 2nd edn., Plenum Press, New York.

Piston, D.W., 1999, Imaging living cells and tissues by two-photon excitation microscopy, *Trends Cell Biol.* 9, 66–69.

Prentki, M. and Matchinsky, F.M., 1987, $[Ca^{2+}]_i$, cAMP, and phospholipids-derived messengers in coupling mechanism of insulin secretion, *Physiol. Rev.* 67, 1185–1248.

Purpura, V., Basarsky, T., Liu, F., Jeftinija, K., Jeftinija, S. and Haydon, P., 1994, Glutamate-mediated astrocyte-neuron signalling, *Nature* 369, 744–747.

Quesada, I., Nadal, A. and Soria, B., 1999, Different effects of tolbutamide and diazoxide in α, β and δ cells within intact islets of Langerhans, *Diabetes* 48, 2390–2397.

Santos, R.M., Rosario, L.M., Nadal, A., Garcia-Sancho, J., Soria, B. and Valdeolmillos, M., 1991, Widespread synchronous $[Ca^{2+}]_i$ oscillations due to bursting electrical activity in single pancreatic islets, *Pflügers Arch.* 418, 417–422.

Satoh, Y., Nishimura, T., Kimura, K., Mori, A. and Saino, T., 1998, Application of real-time confocal microscopy for observation of living cells in tissue specimens, *Human Cell* 11, 191–198.

So, P.T.C., König, K., Berland, K., Dong, C.Y., French, T., Bühler, C., Ragan, T. and Gratton, E., 1998, New time-resolved techniques in two-photon microscopy, *Cell. Mol. Biol.* 44, 771–793.

Tsien, R.Y. and Waggoner, A., 1990, Fluorophores for confocal microscopy: Photophysics and photochemistry, in *Handbook of Biological Confocal Microscopy*, 1st edn., J.B. Pawley (ed.), Plenum Press, New York, pp. 169–178.

Valdeolmillos, M., Nadal, A., Contreras, D. and Soria, B., 1992, The relationship between glucose-induced K_{ATP} channel closure and the rise in $[Ca^{2+}]_i$ in single mouse pancreatic β-cells, *J. Physiol.* 455, 173–186.

Valdeolmillos, M., Nadal, A., Soria, B. and Garcia-Sancho, J., 1993, Fluorescence digital image analysis of glucose-induced $[Ca^{2+}]_i$ oscillations in mouse pancreatic islets of Langerhans, *Diabetes* 42, 1210–1214.

Verkhratsky, A. and Kettenmann, H., 1996, Calcium signalling in glial cells, *Trends Neurosci.* 19, 346–352.

Verkhratsky, A., Orkand, R.K. and Kettenmann, H., 1998, Glial calcium: Homeostasis and signaling function, *Physiol. Rev.* 78, 99–141.

White, J.G., Amos, W.B. and Fordham, M., 1987, An evaluation of confocal microscopy versus conventional imaging of biological structures by fluorescence light microscopy, *J. Cell Biol.* 105, 41–48.

Yuste, R. and Tank, D.W., 1996, Dendritic integration in mammalian neurons, a century after Cajal, *Neuron* 16, 701–716.

Yuste, R., Peinado, A. and Katz, L.C., 1992, Neuronal domains in developing neocortex, *Science* 257, 665–669.

Immature Mouse Oocyte as a Model for Imaging Nuclear Calcium Dynamics

Arlette Pesty

1. INTRODUCTION

Free calcium ions (Ca^{2+}) are known in cell physiology to play an ubiquitous role as messengers. Very small and brief fluctuations of their intracellular concentration take part in the transmission of a specific message through the cell. In fact, it is the pulsatility of these fluctuations, designated as "calcium oscillations", which imparts their messenger activity to the calcium ions. Nowadays, more and more teams working on different cellular models are interested in the possible existence of calcium signaling located inside the nucleus (called "germinal vesicle" or GV in the special case of the oocyte), as well as the existence and functionality of a nuclear phosphoinositide cycle.

It is known that the structural components of the nuclear envelope (NE) include a double membrane, i.e. the inner and the outer nuclear membranes, the nuclear pore complexes and the lamina (Stoffler et al.,1999a). Nuclear pore complexes (NPC) are dynamic structures which exhibit a tridimensional architecture whose central framework is an assembly of eight subcomplexes sandwiched between a cytoplasmic ring and a nuclear ring. The latter anchors a kind of basket, assembled from eight thin filaments. In the middle of all this structure is the central pore, possibly plugged with a distinct particle, which acts as a gated channel involved in nucleoplasmic transport. Structural change of the nuclear basket involves its distal ring which may act as an iris-like diaphragm sensible to micromolar calcium fluctuations. In contrast, the cytoplasmic NPC substructure seems rather insensitive to calcium, particularly the position of the central plug (Oberleithner, 1999; Wang and Clapham, 1999; Stoffler et al., 1999b; Bustamante et al., 2000).

Arlette Pesty • INSERM Unité 355, "Maturation Gamétique et Fécondation", Clamart, France.

R. Pochet, R. Donato, J. Haiech, C. Heizmann and V. Gerke (eds.): Calcium: The Molecular Basis of Calcium Action in Biology and Medicine, 673–682.

The NE is contiguous with the endoplasmic reticulum thereby providing a large pool of releasable Ca^{2+} (Lee et al., 1998; Petersen et al., 1998); and the inner membrane of the NE expresses calcium release channels mainly regulated by inositol 1,4,5 trisphosphate receptors (Humbert et al., 1996; Guihard et al., 1997; Santella and Kyozuka, 1997).

The concept of specific nuclear calcium signaling is still a highly controversial topic (Carafoli et al., 1997). First, the difference of fluorescence intensity observed through each cellular compartment could be due to a difference in reactivity of the fluorescent indicator (mainly Fluo-3) in the cytoplasmic and nucleoplasmic environments (Perez-Terzic et al., 1997). But the core of the debate is as follows: some people consider the nuclear envelope as a structure that offers no resistance to calcium ions which would thus freely and immediately equilibrate between the cytosolic and nuclear compartments (Lipp et al., 1997; Shirakawa and Miyazaki, 1996) while others view it as an effective barrier that limits the traffic of calcium between the two compartments (Badminton et al.,1998; Lui et al., 1998; Malviya and Rogue, 1998; Santella et al., 1998).

Within this context, we developed an experimental approach to verify the existence of a specific nuclear calcium signal and to study its potential involvement in the resumption of meiosis. Indeed, micromanipulations and microinjection procedures are easier to practice on this large and isolated cell that is the oocyte, than on any somatic cell; and it constitutes a good model for studying the putative role of specific nuclear calcium signalling during NE disassembly as a physiological process.

2. THE MOUSE OOCYTE AS A CELLULAR MODEL

Briefly, it has been demonstrated that calcium ions play a pivotal role in two steps of meiosis: just before ovulation, at the time of meiosis resumption, characterized by germinal vesicle breakdown (GVB); and later, at the time of fertilization, when gamete fusion occurs inducing meiosis achievement.

Concerning meiosis resumption, the relevance of calcium ions has not been clarified (Whitaker, 1996). However, various authors have agreed with the idea of cytoplasmic calcium oscillations preceding chromatin reorganization and rupture of the nuclear envelope (Figure 1), mainly via the phosphoinositide cycle, in the meiosis process (Pesty et al., 1994; He et al., 1997; Coticchio and Fleming, 1998) as well as in the mitosis process (Whitaker, 1997; Santella, 1998; Hinchcliffe et al., 1999; Silver, 1999). Using confocal microscopy (Figure 2), we observed calcium oscillations not only in the cytoplasmic area of the immature mouse oocyte (i.e. at the GV stage), but in the nucleoplasmic area as well (Lefevre et al., 1995; Pesty et al., 1998). *In vitro*, these calcium oscillations occur spontaneously, that is to say, without exogenous stimulation, in the 10–15 minutes following the mechanical release of the oocyte from the ovarian follicle. Calcium oscillations then occur

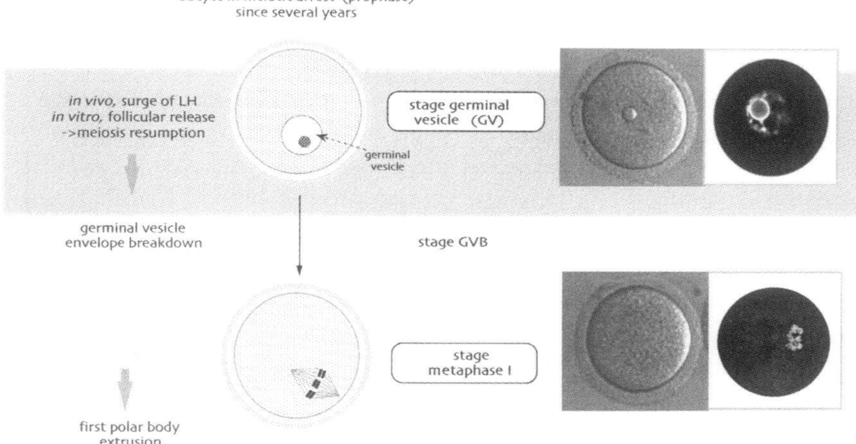

Figure 1. First step of the meiosis resumption: in the immature mouse oocyte, calcium oscillations occur (grey zone) at the time of germinal vesicle breakdown and chromatin re-organisation. *Left inset*: oocyte in phase contrast microscopy. *Right inset*: chromatin is marked with the fluorochrome Syto-15.

over a period of 40–60 minutes with constant frequency and amplitude in 80% of the oocytes. Our purpose was thus to develop an experimental approach for further studying this nuclear calcium signal by selectively loading the nucleus with a Ca^{2+} sensitive fluorescent probe, first maintaining the integrity of the oocyte, then working on the isolated germinal vesicle.

The M2 medium (Fulton and Whittingham, 1978) used for oocyte culture, contains Hepes and, thus, makes it possible to culture oocytes without CO_2 atmosphere with no noticeable fluctuations in pH. To collect oocytes, the antral follicles are disrupted with sterile needles and approximately 15 GV oocytes can be recovered from each mouse ovary.

3. FEATURES OF THE CALCIUM-SENSITIVE FLUOROCHROMES USED

The most often, when experiments are carried out by confocal microscopy on the whole cell, the fluorescent probe Fluo-3 is used (Tsien and Poenie, 1986) This calcium-sensitive probe is characterized by a single excitation wavelength (488 nm) generated by an argon-ion laser source in the visible part of the spectrum. The laser source has a lower phototoxicity than UV; and this is important, considering that we are working on live cells which will be illuminated at least 200 times per experiment. Fluo-3 can be viewed through the standard fluoresceine filters and, between normal resting cytosolic free Ca^{2+} concentration and indicator Ca^{2+} saturation, the enhancement is at least 100-fold, making Fluo-3 particularly useful for measuring the kinetics of Ca^{2+} transients. But the plasma membrane is highly impermeable to this kind of molecule. It is, therefore, necessary to microinject the probe into the oocyte. In fact, it is easier to use the fluorochrome under its acetoxymethyl ester derived form (/AM), rendering the indicator membrane-permeant and insensitive to ions. Once inside the cell, this /AM indicator is hydrolyzed by intracellular esterases, releasing the ion-sensitive indicator which is retained inside the cell. But in view of our purpose, which is to selectively load the nuclei, Fluo-3 offers a disadvantage as it is a small molecule (PM = 1130) which goes through the nuclear pores easily, loading indifferently both cellular compartments of the whole oocyte or leaking from the isolated nucleus.

To improve the calcium signal imaging, Calcium Green indicators are also available, as a family of molecules close to Fluo-3. Their spectral properties offer a higher fluorescence output and make it possible to use a lower fluorochrome concentration. A third kind of calcium probe called Oregon Green 488 BAPTA-1 (Oregon-Green) is still more efficiently excited by the 488 nm spectral line of the argon-ion laser than are Fluo-3 and Calcium Green probes. Additionally, Oregon-Green as Calcium Green-1 is moderately fluorescent in calcium free environment. This property increases the visibility of resting cells and thus facilitates the determination of baseline fluorescence. However, Oregon-Green is also a small molecule. To overcome this limitation, we use the Dextran-conjugated Ca^{2+} indicator form. Dextrans are hydrophilic polysaccharides, characterized by their high molecular weight, good water solubility and low toxicity. They are also biologically inert. OregonGreen Dextran (MW 70000) must be loaded into cells by microinjection. Once loaded, the dextran conjugates are retained well in viable cells.

In our experiments, Fluo-3/acetoxymethyl-ester (Fluo-3/AM) is used when the whole cells are loaded and Oregon Green 488 BAPTA-1 Dextran (Oregon Green Dextran) is used when only the nuclei are loaded. Both products were purchased from Molecular Probes Inc., Eugene, OR, USA.

The cell-permeant Fluo-3/AM is diluted to 1 mM in a 20% Pluronic F-127 (dispersing agent) solution in DMSO (Molecular Probes Inc.) and stored at $-20°C$. It is dissolved extemporarily to a final concentration of 5 μM in a M2 medium. The cells are loaded in the Fluo-3/AM solution at 37°C in the dark for 15 minutes. The cells are then placed on the glass bottom of a culture chamber (Helmut Saur, Reutlingen, Germany), in a drop of M2 medium under mineral oil to be observed by confocalmicroscopy.

The cell-impermeant Oregon Green Dextran is diluted extemporarily in the microinjection medium, then injected into the nuclei (GV) of the cells (Figure 3) under inverted microscope. The microinjection medium is 140 mM KCl, 1 mM $MgCl_2$, 5 mM Hepes, pH 7.2.

To prepare isolated nuclei, the Oregon Green Dextran is first injected into the nucleus of a whole oocyte. Then, the zona pellucida and the plasma membrane of the oocyte are disrupted by aspiration through a thin glass pipette, the diameter of which was a little smaller than that of the oocyte.

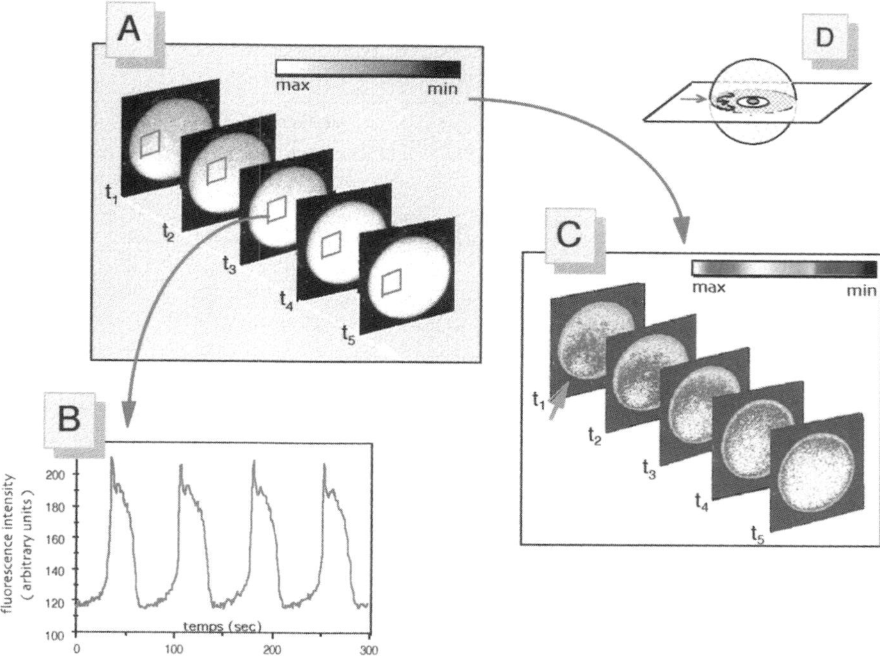

Figure 2. Acquisition and analysis of data recorded in confocal microscopy: (A) sequential grey scale digitalized images are recorded for later analysis; then (B) either a time-course curve of calcium indicator fluorescence emission is constructed on a selected zone of the image to study the kinetics of the Ca^{2+} signal; or (C) a pseudo-colors scale is calibrated on the grey scale of the recorded images, to improve the readability of imaging the spatial progression of the signal. The inset (D) shows the progression of the Ca^{2+} signal in the selected optical plane, through the GV and nucleolus. For a colour version of this figure, see page xxxviii.

The nucleus is immediately put in a drop of medium which mimics the viscosity of the cytosol, to maintain adequate pressure on the nuclear membrane (Figure 4). The nucleus culture medium is 90 mM KCl, 10 mM NaCl, 2 mM $MgCl_2$, 10 mM Hepes, 1.1 mM EGTA, 0.06 $CaCl_2$, pH 7.3, supplemented with 10 mM Polyvinylpyrolidone (MW 10 000) to adjust the viscosity of the medium.

4. CONFOCAL MICROSCOPY AND ACQUISITION OF DATA

Taking into account the shape of the oocyte – a round and single cell of ± 90 μm diameter – measurements and imaging of fluorescence lack precision when carried out with photomultipliers associated with a conventional microscope. The confocal microscopy, instead of measuring the fluorescence on the global oocyte, allows us to select an optical plane through the thickness of the cell, and thus to obtain a sharp and precise image of the distribution of the fluorescence emitted by the calcium indicator. Furthermore, the acquisi-

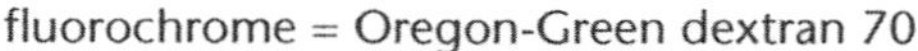

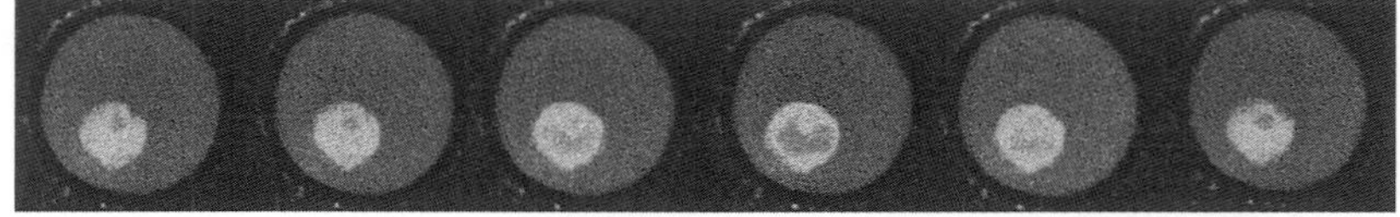

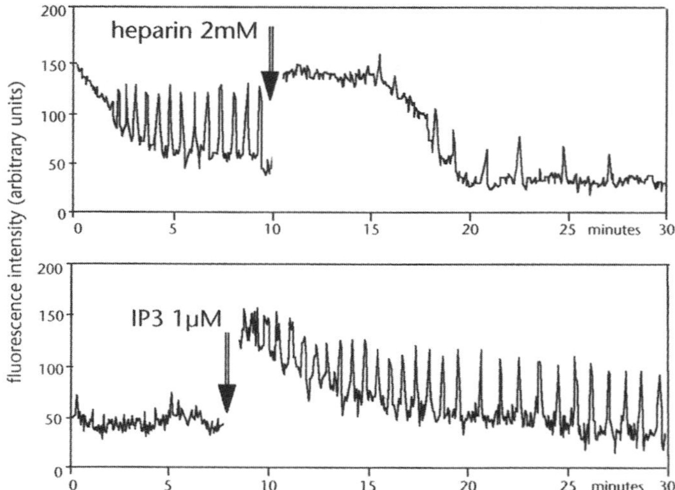

Figure 3. Pseudo-color imaging of nuclear calcium oscillations after microinjection of Oregon Green 488 BAPTA-1 Dextran into the nucleus of a whole oocyte. Below, the time-course curves show the effect of a second microinjection of phosphoinositide cycle antagonists on the spontaneous calcium train of oscillations. For a colour version of this figure, see page xxxviii.

tion of a sequence of images makes it possible to perform a precise study of the spatial distribution of the calcium as a function of time (spatio-temporal study) (Figure 2).

All measurements of fluorescence emission are performed using a confocal laser scanning imaging system interfaced with an inverted microscope. To optimize cell viability, the 25 mW argon-ion laser of the confocal microscope is set at half power and the beam is further filtered through a blue high-sensitive filter (BHS). To perform time-lapse calcium imaging studies, dye-loaded oocytes are viewed in a single optical plane through the GV and nucleolus.

Sequential digitized images are recorded for later analysis using a time-course software specific for the study of Ca^{2+} kinetics. Several experimental parameters such as the screen size of the recorded images, the time lapse between two consecutive images, and the size and position of the interest zones for fluorescence emission analysis are standardized via this software.

Calcium kinetics images are resolved by monitoring the oocytes continuously for up to 60 minutes with a 2-second interval between two consecutive

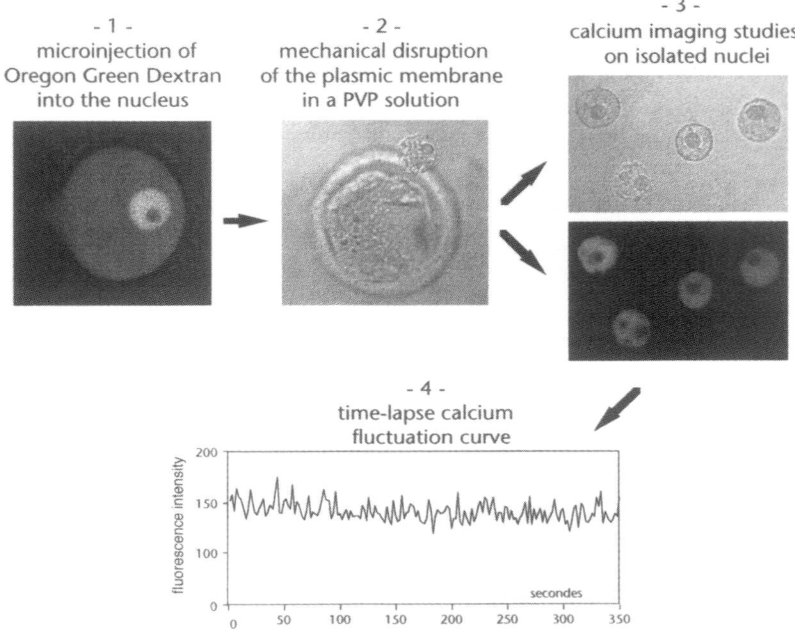

Figure 4. Procedure for obtaining isolated nuclei loaded with Oregon Green 488 BAPTA-1 Dextran. The high molecular weight fluorescent molecule is retained well in viable nuclei and spontaneous Ca^{2+} fluctuations are observable.

acquisitions. Then (A) a time-course curve of calcium indicator fluorescence emission is constructed after subtraction of background values. Fluorescence intensity is measured in rectangles of similar areas; and on the resulting kinetics curves, data are expressed in arbitrary units uncalibrated in terms of calcium concentration (Figure 2). (B) Results are also expressed by imaging fluorescence intensity. In that case, a pseudo-colors scale is calibrated on the grey scale of the recorded images, to improve the readability of imaging (Figure 2).

5. RESULTS AND DISCUSSION

These experimental approaches allowed us to demonstrate the existence of spontaneous nuclear calcium oscillations in the whole mouse oocyte, which can be modulated by nuclear microinjection of inositol tris-phosphate ($InsP_3$), low molecular weight heparin or anti-$InsP_3$-receptor type I monoclonal antibody (Avazeri et al., 1998; Pesty et al., 1998; see Figure 3). The study of Ca^{2+} kinetics analysis on isolated nuclei made it possible to confirm that a specific spontaneous nuclear calcium signal exists, but more as rapid Ca^{2+} fluctuations than as regular "oscillations" (Figure 4). Since the nuclear envelope is contiguous to the endoplasmic reticulum, these results lead us to believe that there is a crosstalk between the cytoplasmic compartment and the nucleus that confers additional properties to the nuclear calcium signal, expressed via its regularity. Such data nourishes the debate about the existence of a particular nuclear Ca^{2+} signal, as observed in starfish oocyte (Santella et al., 1998) and *Xenopus* oocyte (Hennager et al., 1995) but also in other cellular models (Lui et al., 1998a, b).

Working on isolated nuclei, additional technical approaches might be used to verify the integrity and functioning of the nuclear membrane. For example, a calcium-pump ATPase has been described on the outer membrane of liver nuclei. So, measurement of ATP-dependent $^{45}Ca^{2+}$ uptake into nuclei may be useful to confirm the integrity of the NE (Rogue et al., 1998). The same kind of information is provided by assessing the transport of the 10-kDa fluorescent Calcium Green-1 Dextran across the NE (compared to 500-kDa Calcium Green-1 Dextran or 10-kDa Calcium-Insensitive Dextran Lucifer Yellow) (Rogue et al., 1998). Another approach, more sophisticated, consists of the use of specifically targeted aequorin chimeras for selectively monitoring the dynamic changes of calcium concentration inside the nucleus of living cells. The technique is based on the construction of a chimeric cDNA encoding a fusion protein composed of the photoprotein aequorin and a nuclear translocation signal peptide. This modified aequorin is stably expressed in HeLa cells and largely confined to the nucleoplasm (Brini et al., 1993; Badminton et al., 1998). A close technique is utilized with the Green Fluorescent Protein (GFP) for specific labeling of the various subcellular structures, providing powerful tools for visualizing these organelles in living cells (De Giorgi et al., 1996; Subramanian and Meyer, 1997; Strubing and Clapham, 1999; Bustamante et al., 2000).

A large field of investigation will open with the development of these new tools and will perhaps provide some clarification concerning the crosstalk between the different subcellular compartments and the calcium machinery.

REFERENCES

Avazeri, N., Pesty, A. and Lefevre, B., 1998, The germinal vesicle of the mouse oocyte contains elements of the phosphoinositide cycle: What is their role at meiosis resumption?, *Reprod. Nutr. Dev.* 38, 671–682.

Badminton, M.N., Kendall, J.M., Rembold, C.M. and Campbell, A.K., 1998, Current evidence suggests independent regulation of nuclear calcium, *Cell Calcium* 23, 79–86.

Brini, M., Murgia, M., Pasti, L., Picard, D., Pozzan, T. and Rizzuto, R., 1993, Nuclear Ca^{2+} concentration measured with specifically targeted recombinant aequorin, *EMBO J.* 12, 4813–4819.

Bustamante, J.O., Michelette, E.R., Geibel, J.P., Dean, D.A., Hanover, J.A. and McDonnell, T.J., 2000, Calcium, ATP and nuclear pore channel gating, *Pflügers Arch.* 439, 433–444.

Carafoli, E., Nicotera, P. and Santella, L., 1997, Calcium signalling in the cell nucleus, *Cell Calcium* 22, 313–319.

Coticchio, G. and Fleming, S., 1998, Inhibition of phosphoinositide metabolism or chelation of intracellular calcium blocks FSH-induced but not spontaneous meiotic resumption in mouse oocytes, *Dev. Biol.* 203, 201–209.

De Giorgi, F., Brini, M., Bastianutto, C., Marsault, R., Montero, M., Pizzo, P., Rossi, R. and Rizzuto, R., 1996, Targeting aequorin and green fluorescent protein to intracellular organelles, *Gene* 173, 113–117.

Fulton, B.P. and Whittingham, D.G., 1978, Activation of mammalian oocytes by intracellular injection of calcium, *Nature* 273, 149–151.

Guihard, G., Proteau, S. and Rousseau, E., 1997, Does the nuclear envelope contain two types of ligand-gated Ca^{2+} release channels?, *FEBS Lett.* 414, 89–94.

He, C.L., Damiani, P., Parys, J.B. and Fissore, R.A., 1997, Calcium, calcium release receptors, and meiotic resumption in bovine oocytes, *Biol. Reprod.* 57, 1245–1255.

Hennager, D.J., Welsh, M.J. and DeLisle, S., 1995, Changes in either cytosolic or nucleoplasmic inositol 1,4,5 trisphosphate levels can control nuclear Ca^{2+} concentration, *J. Biol. Chem.* 270, 4959–4952.

Hinchcliffe, E.H., Thompson, E.A., Miller, F.J., Yang, J. and Sluder, G., 1999, Nucleo-cytoplasmic interactions that control nuclear envelope breakdown and entry into mitosis in the sea urchin zygote, *J. Cell Sci.* 112, 1139–1148.

Humbert, J.P., Matter, N., Artault, J.C., Koppler, P. and Malviya, A.N., 1996, Inositol 1,4,5-trisphosphate receptor is located to the inner nuclear membrane vindicating regulation of nuclear calcium signaling by inositol 1,4,5-trisphosphate. Discrete distribution of inositol phosphate receptors to inner and outer nuclear membranes, *J. Biol. Chem.* 271, 478–485.

Lee, M.A., Dunn, R.C., Clapham, D.E. and Stehno-Bittel, L., 1998, Calcium regulation of nuclear pore permeability, *Cell Calcium* 23, 91–101.

Lefevre, B., Pesty, A. and Testart, J., 1995, Cytoplasmic and nucleic calcium oscillations in immature mouse oocytes: Evidence of wave polarization by confocal imaging, *Exp. Cell Res.* 218, 166–173.

Lipp, P., Thomas, D., Berridge, M.J. and Bootman, M.D., 1997, Nuclear calcium signalling by individual cytoplasmic calcium puffs, *EMBO J.* 16, 7166–7173.

Lui, P.P., Kong, S.K., Fung, K.P. and Lee, C.Y., 1998a, The rise of nuclear and cytosolic Ca^{2+} can be uncoupled in HeLa cells, *Pflügers Arch.* 436, 371–376.

Lui, P.P., Kong, S.K., Tsang, D. and Lee, C.Y., 1998b, The nuclear envelope of resting C6 glioma cells is able to release and uptake Ca^{2+} in the absence of chemical stimulation, *Pflügers Arch.* 435, 357-361.

Malviya, A.N. and Rogue, P.J., 1998, "Tell me where is calcium bred": Clarifying the roles of nuclear calcium, *Cell* 92, 17–23.

Oberleithner, H., 1999, Aldosterone and nuclear signaling in kidney, *Steroids* 64, 42–50.

Perez-Terzic, C., Jaconi, M. and Stehno-Bittel, L., 1997, Nucleoplasmic and cytoplasmic differences in the fluorescence properties of the calcium indicator Fluo-3, *Cell Calcium* 21, 275–282.

Pesty, A., Lefèvre, B., Kubiak, J., Géraud, G., Tesarik, J. and Maro, B., 1994, Mouse oocyte maturation is affected by lithium via the polyphosphoinositide metabolism and the microtubule network, *Mol. Reprod. Dev.* 38, 187–199.

Pesty, A., Avazeri, N. and Lefevre, B., 1998, Nuclear calcium release by $InsP_3$-receptor channels plays a role in meiosis reinitiation in the mouse oocyte, *Cell Calcium* 24, 239–251.

Petersen, O.H., Gerasimenko, O.V., Gerasimenko, J.V., Mogami, H. and Tepikin, A.V., 1998, The calcium store in the nuclear envelope, *Cell Calcium* 23, 87–90.

Rogue, P.J., Humbert, J.P., Meyer, A., Freyermuth, S., Krady, M.M. and Malviya, A.N., 1998, cAMP-dependent protein kinase phosphorylates and activates nuclear Ca^{2+}-ATPase, *Proc. Natl. Acad. Sci. USA* 95, 9178–9183.

Santella, L., 1998, The role of calcium in the cell cycle: Facts and hypotheses, *Biochem. Biophys. Res. Commun.* 244, 317–324.

Santella, L. and Kyozuka, K., 1997, Effects of 1-methyladenine on nuclear Ca^{2+} transients and meiosis resumption in starfish oocytes are mimicked by the nuclear injection of inositol 1,4,5-trisphosphate and cADP-ribose, *Cell Calcium* 22, 11–20.

Santella, L., De Riso, L., Gragnaniello, G. and Kyozuka, K., 1998, Separate activation of the cytoplasmic and nuclear calcium pools in maturing starfish oocytes, *Biochem. Biophys. Res. Commun.* 252, 1–4.

Shirakawa, H. and Miyazaki, S., 1996, Spatiotemporal analysis of calcium dynamics in the nucleus of hamster oocytes, *J. Physiol. (Lond.)* 494, 29–40.

Silver, R.B., 1999, Imaging structured space-time patterns of Ca^{2+} signals: Essential information for decisions in cell?, *FASEB J.* (Suppl. 2), S209–S215.

Stoffler, D., Fahrenkrog, B. and Aebi, U., 1999a, The nuclear pore complex: From molecular architecture to functional dynamics, *Curr. Opin. Cell Biol.* 11, 391–401.

Stoffler, D., Goldie, K.N., Feja, B. and Aebi, U., 1999b, Calcium-mediated structural changes of native nuclear pore complexes monitored by time-lapse atomic force microscopy, *J. Mol. Biol.* 287, 741–752.

Strubing, C. and Clapham, D.E., 1999, Active nuclear import and export is independent of lumenal Ca^{2+} stores in intact mammalian cells, *J. Gen. Physiol.* 113, 239–248.

Subramanian, K. and Meyer, T., 1997, Calcium-induced restructuring of nuclear envelope and endoplasmic reticulum calcium stores, *Cell* 13, 963–971.

Tsien, R.Y. and Poenie, M., 1986, Fluorescent ratio-imaging: A new window into intracellular ionic signalling, *Trends Pharmacol. Sci.* 11, 450–455.

Wang, H. and Clapham, D.E., 1999, Conformational changes of the in situ nuclear pore complex, *Biophys J.* 77, 241–247.

Whitaker, M., 1996, Control of meiotic arrest, *Rev. Reprod.* 1, 127–138.

Whitaker, M., 1997, Calcium and mitosis, *Progr. Cell Cycle Res.* 3, 261–269.

Single Cell Fluorescence Imaging to Investigate Calcium Signaling in Primary Cultured Neurones

Jennifer M. Pocock and Gareth J.O. Evans

1. INTRODUCTION

This chapter will discuss the use of single cell fluorescence imaging (excluding confocal imaging which is discussed in Nadal and Soria's chapter in this book) to investigate calcium signalling in primary cultured neurones with particular emphasis on cerebellar granule cell neuronal cultures (CGCs).

Calcium has an essential role in neuronal processes, from synaptic transmission to gene regulation. Calcium-dependent mechanisms generally involve an increase in cytoplasmic calcium concentration either via influx across the plasma membrane or efflux from intracellular calcium stores. The synthesis in the mid-1980s of a new generation of fluorescent calcium indicators based upon the structure of EGTA and BAPTA (such as fura-2 and indo-1; see Grynkiewicz et al., 1985) revolutionised the study of neuronal calcium-signaling. When calcium binds to these fluorescent dyes, there is a shift in the wavelength of either their excitation maxima (such as for fura-2) or their emission maxima (such as for indo-1) which can be calibrated with the amount of calcium in the cell. Fluorescence imaging of these dyes offers an alternative or accompaniment to electrophysiological techniques to measure the real time behaviour of calcium ions in cell cultures and tissue slices.

Initial studies employing fluorescent calcium probes to investigate calcium signaling pathways in primary cultured neurons such as cerebellar granule cells used population fluorometry. As the term implies, this allowed the ana-

Jennifer M. Pocock • Cell Signalling Laboratory, Department of Neurochemistry, Institute of Neurology, University College London, London, U.K. **Gareth J.O. Evans** • The Physiological Laboratory, Department of Physiology, University of Liverpool, Liverpool, U.K.

R. Pochet, R. Donato, J. Haiech, C. Heizmann and V. Gerke (eds.): Calcium: The Molecular Basis of Calcium Action in Biology and Medicine, 683–696.

lysis of the calcium responses in neuronal populations. Typically neurons grown at high density on a coverslip are inserted into a cuvette in a fluorimeter. Substances added to the cuvette evoke a response which is recorded as the meaned response of all the cells on the coverslip. Thus addition of 30 mM KCl to populations of cerebellar granule neurons allows the resolution of a transient spike followed by a non-inactivating plateau component (Courtney et al., 1990; Ciardo and Meldolesi, 1991). A component of the sustained calcium plateau could be inhibited by the dihydropyridine, nifedipine, indicating the activation of L-type voltage-dependent calcium channels. The degree to which nifedipine inhibits the plateau component of the KCl-evoked calcium elevation depends on the concentration of KCl used to stimulate the cells; higher KCl concentrations (e.g. 50 mM) result in the additional activation of nifedipine-insensitive calcium channels (Ciardo and Meldolesi, 1991; Pocock et al., 1993).

Fluorometric measurements of calcium responses in neuronal populations reveal information regarding calcium concentrations and a certain level of temporal resolution but no spatial resolution. Cellular calcium signals are complex and consist of a combination of magnitude, spatial and temporal variables. Recording of calcium responses in real time using single cell imaging can provide information concerning all three calcium variables. The use of an imaging system comprising a fluorescence microscope with suitable optics, a low light level CCD camera, filter changer and computer with image processing gives considerably more information (Figure 1). A filter wheel or more conveniently a monochromator which takes less than 10 ms per wavelength change, (a filter wheel can take 100 ms) produces the correct wavelength of light which is reflected onto the cells via a dichroic mirror. Emitted light from the cells goes back to the dichroic mirror and filter set which permit only emitted light of a particular wavelength to reach a digital camera. The exact excitation and emission wavelengths in use will depend on the calcium indicator used. For fura-2, the excitation wavelengths are typically 340 and 380 nm and the emission wavelength is 505–510 nm. The major advantages of an imaging system over a fluorimeter are that calcium responses can be directly observed occurring in the cells rather than inferred from a population response. Recordings can be made from specified cells thus allowing elimination of data from contaminating cell types (such as the few astrocytes or Purkinje cells found in CGC cultures) although an increasingly important area of research is to identify the influence that other cells present in the brain, such as glia, may have on neuronal calcium responses. Single cell imaging allows this contribution to be assessed. Single cell fluorescence imaging allows the attribution of calcium responses to discrete regions of the cell and has advantages over the use of electrophysiological studies in which the somatic response dominates the calcium signal. Neurite calcium responses cannot be measured with either whole-cell voltage clamp or single-channel analysis but can be observed with single cell fluorescence imaging. Furthermore, single cell imaging provides information about whether a re-

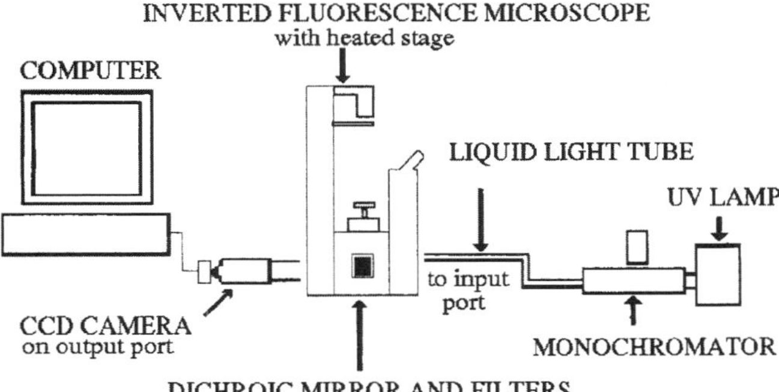

Figure 1. Set up for single cell fluorescence imaging. Cells plated onto a coverslip and loaded with the fluorescent calcium dye are placed in a heated stage on the microscope. A UV lamp provides the excitation whilst the monochromator sets the wavelength of excitation. Between the UV lamp and the monochromator, one of a number of filters with adjustable slits can be inserted to adjust the brightness. The light passes through a light tube to the back of the microscope where it impinges on a dichroic mirror which reflects the light through the objective lens to the cells. Special objectives for fluorescence work must be used to reduce non-specific fluorescence. The emitted fluorescence from the cells is transmitted back to the dichroic which filters out any excitation fluorescence, allowing only emitted light above a certain wavelength to travel to the CCD camera. The CCD chip in the camera consists of a set of light-sensitive diodes, the number of which per chip defines the resolution of the final image. Each diode generates a pixel (picture element) in the final computer image by converting the level of fluorescence it receives to a number (digitisation). The digital output from the camera is stored in the computer as a series of frames which when processed give a false colour image of the fluorescence level detected (see Figure 2) together with a graphical representation of the fluorescence changes (see Figure 4). Imaging is generally carried out under conditions of very low ambient light to prevent dye bleaching and to prevent interference with and possible damage to the camera.

sponse is heterogeneous or homogeneous to the population of cells on the coverslip. It is possible to record the responses of up to 100 CGCs within a given field. When repeated within the same cellular preparation on different coverslips and with different preparations, single cell imaging provides a wealth of information regarding homogeneity and consistency of response. Furthermore, where a compound in use is in limited supply, the reduction in volume of 150–200 μl of cell medium allows maximal economy of use (Pocock et al., 1993). This also applies to the use of neural tissue. Single cell fluorescence imaging of calcium has greater time resolution than can be obtained using current confocal imaging systems.

The method of single cell imaging with fluorescent calcium indicators has a number of inherent pitfalls, although being able to observe the loaded cells is an advantage in detecting some of these. Pitfalls include: *Compartmentalisation*; the acetoxymethyl ester derivatives of calcium indicators are usually employed as they are membrane permeable and are cleaved intracellularly by endogenous esterases to form a membrane impermeable derivative. However, homogeneous distribution throughout the cell is unlikely as the dye may accumulate in membrane-enclosed subcellular structures. Different responses may be obtained depending on whether the fluorescent dye is microinjected into the cell or loaded as the ester due to differences in compartmentalisation (Connor, 1993). *Partial hydrolysis*; in the case of AM-ester forms of dyes, there may be cell to cell variation in esterase activity leading to partial hydrolysis. Only the deesterified dye can bind to calcium. *Leakage*; despite the cleavage of AM-ester dyes producing membrane impermeable derivatives, there will still be some degree of leakage from the cell. *Buffering*; there will be some buffering of cytosolic calcium by the dyes by virtue of their calcium-binding nature. This may not affect large steady state changes in calcium but may affect the kinetics of calcium changes as well as down-stream signaling processes dependent on calcium. Furthermore, the fluorescence dye itself may modulate receptors directly (Taylor and Broad, 1998). *Photobleaching*; during a long experiment, after prolonged exposure to excitation wavelengths, the dyes will photobleach and overall fluorescence will decrease. *Auto-fluorescence*; some cells will have innate fluorescence, however, this should remain constant. *Flare*; this is discussed in Nadel and Soria's chapter in this book. *Fluorescent compounds*; if any compounds used in experiments auto-fluoresce or fluoresce in response to excitation, this can lead to artefactual increases in fluorescence which can be misinterpreted as increases in $[Ca^{2+}]_i$. Many of these artefacts (e.g. leakage, photobleaching, fluorescent compounds) can be detected and overcome by the use of dual wavelength dyes such as fura-2 where the excitation maxima is shifted to a lower wavelength upon calcium binding (from 380 to 340 nm). The results are therefore expressed as a ratio of the emissions obtained from the different excitations and can be calibrated against known concentrations of calcium to give a measurement of $[Ca^{2+}]_i$. Artefactual changes in fluorescence will generally only affect one of the excitation wavelengths and thus be easily detectable. Calibration of fluorescence changes to produce an absolute calcium value may not be reliable (Henke et al., 1996) and often ratio values are used instead.

Single cell imaging of neurons in primary culture allows a more convincing extrapolation of experimental results to the actual *in vivo* situation than could be achieved with a cell line for example. Neurons cultured from different brain regions expectedly possess the properties of the corresponding neurons *in situ*, for example, hippocampal neurons, often the focus of epileptic activity, are highly excitable in culture whereas CGCs, not usually associated with seizures, are not. However, the primary culture of neonatal rat CGCs (see Courtney et al., 1990; Pocock et al., 1995, for details of preparation) is one of the most popular systems for studying neuronal processes related to glutamate neurotransmission and neurotoxicity. The relatively simple preparation yields a highly homogeneous population of neurons that develop in culture, form glutamatergic homosynaptic connections and express a full complement of synaptic machinery including neurotransmitter receptors and voltage-dependent ion channels.

Imaging systems which do not incorporate confocal analysis require the use of monolayers of cells. When plating neurons for single cell fluorescence imaging, a lower density is often more practical than the high density used for population fluorometry. Thus whilst a density of 0.75×10^6 per 13 mm diameter coverslip is typically used for population studies of CGCs, lower cell densities of $0.2–0.3 \times 10^6$ per 13 mm diameter coverslip provide a cell distribution at which resolution of somatic and neurite responses is possible (Pocock et al., 1995) (Figure 2). Such lower densities also allow the presence of any contaminating cells such as glia which tend to underlie the CGCs to be more readily detected.

KCl is the most common reagent used for depolarising excitable membranes in cultured neurons. Application of a millimolar concentration of KCl causes a permanent opening of voltage-dependent K^+ channels generating a clamped depolarisation of the plasma membrane. Since, in the presence of extracellular calcium this stimulation has been shown to trigger neurotransmitter exocytosis, gene transcription and other intracellular signaling events, the study of calcium signaling in response to KCl is likely to be physiologically relevant. Single cell fluorescence imaging of CGCs following KCl-stimulation reveals an enhanced calcium increase at both the soma and neurites (Pocock et al., 1993, 1995). The somatic calcium response of CGCs is much greater than the neuritic response. In order to allow simultaneous capture of the somatic and neuritic responses in the same field, the imaging system must have a sufficient fluorescence detection range. A 12-bit camera provides 2^{12} levels of fluorescence (i.e. 4096 grey levels) which is a sufficient range to resolve both somatic and neuritic responses in the same experiment. Early imaging systems which used 8-bit cameras ($2^8 = 256$ grey levels) meant that to visualise the somata within the fluorescence range of the camera, the neurite fluorescence would be below the level of detection, and to visualise the neurites, the response of the somata would already be saturated (i.e. at the top of the fluorescence scale). Thus a choice had to be made at the beginning of the experiment as to whether the neurite or somatic response was to be followed. Not all cultured neurons or glia require such a range of fluorescence, and it is sometimes a matter of assessing different cameras to ensure that predicted or estimated responses will fall within the range of a camera. Whilst population fluorometric calcium measurements have revealed the activation of multiple calcium channels following high KCl stimulation, single cell fluorescence imaging allows the attribution of these responses to different parts of the neuron. Thus somatic KCl-evoked calcium elevations are dominated by calcium entry through L-type, dihydropyridine-sensitive channels and neurite calcium elevations by a non-L, non-N, non-P-type calcium channel which is sensitive to a spider toxin, Aga-GI (Pocock et al., 1993). These channels may also be inhibited by the snail toxin MVIIC suggesting they fall into the Q class of calcium channels (Figure 3). The neuritic cal-

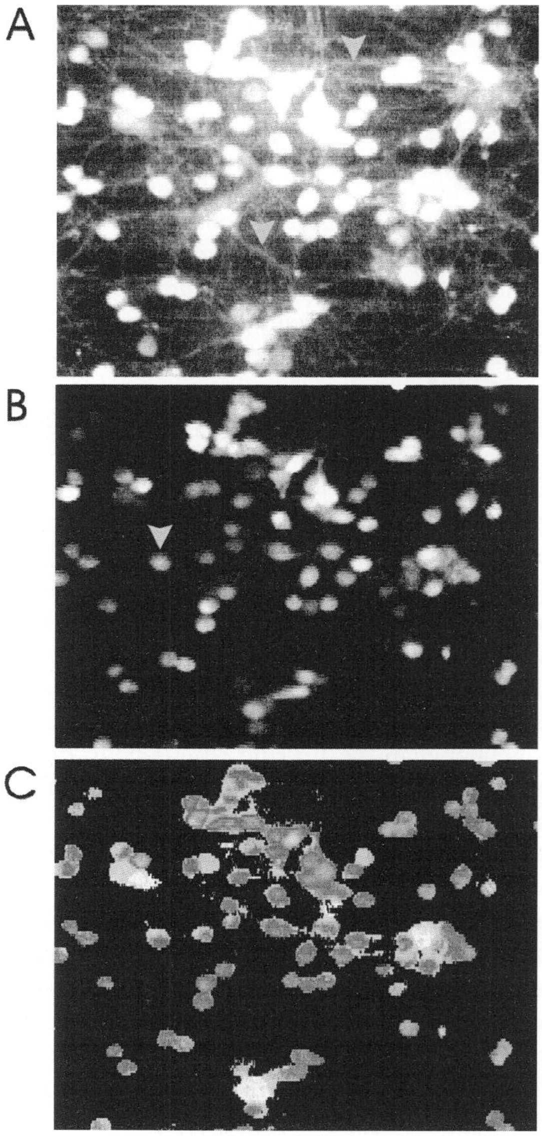

Figure 2. Nine DIV Cerebellar granule neurons plated on small coverslips at low density and loaded for 30 min at 37°C in the presence of 16 μM BSA. The cells were washed and mounted into a heated holder with 150 μM of incubation medium, and placed on the stage of an Olympus IX70 inverted fluorescence microscope. Images were captured using a 12-bit cooled digital CCD camera, Life Science Resources SpectraMASTER High Speed Monochromator, controller and xenon UV lamp, and the output displayed using Life Science Resources Merlin software. Data were collected and analysed off-line to produce (A), (B) and (C). (A) Emission from 380 nm excitation showing field of neurons in which the grey level fluorescence scale has been set so that whilst the somatic response appears saturated (i.e. white), the neurites (arrowheads) are visible. B. The same field of cells in which the fluorescence from the somata is now within scale. (C) 340/380 ratio image of the same cells in (A) and (B) but with a grey level scale set to observe somatic response only. Yellow fluorescence is likely to derive from astrocytes. For a colour version of this figure, see page xxxix.

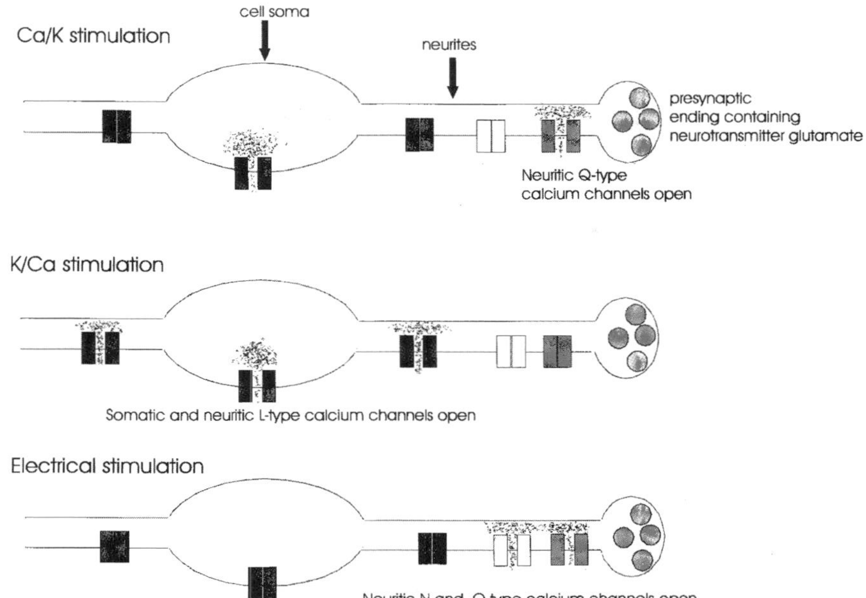

Figure 3. Summary of findings using the spatial resolution of imaging to detect differential modulation of $[Ca^{2+}]_c$ by different voltage-dependent calcium channels at the somata and neurites of cultured cerebellar granule neurons. During Ca/K stimulation, L- and Q-type voltage-dependent calcium channels are activated but only the neuritic Q-type channels (localised at the presynatic endings) modulated glutamate neurotransmitter release. During K/Ca stimulation, somatic and neurite localised L-type calcium channels are activated and L-type channels become coupled to release. During electrical stimulation, L-type calcium channels at the soma or neurites are not activated but N- and Q-type channels are, modulating release.

cium channel is coupled to the release of glutamate and the release of the synaptic vesicle dye FM1-43 (Pocock et al., 1995). Data from various laboratories using specific antagonists of voltage-dependent calcium channels have demonstrated that KCl evoked calcium responses comprise influx through a number of different calcium channel subtypes (Pocock et al., 1993, 1995; Randall and Tsien, 1995). These data (particularly the somatic responses) are in accordance with parallel studies performed using electrophysiological recordings (Forti and Pietrobon, 1993; Pearson et al., 1995).

Altering the type of KCl stimulation protocol can alter the subtypes of calcium channels recruited. Stimulation of CGCs by high KCl in the absence of external calcium followed by the re-addition of calcium triggers neurotransmitter release which is coupled to L-type calcium channels (Pocock et al., 1995; Evans and Pocock, 1999). Analysis of the calcium responses by single cell fluorescence imaging reveals an enhanced calcium influx through L-type calcium channels at the soma whilst neuritic calcium influx shows some sensitivity to nifedipine (Pocock et al., 1995). Thus the recruitment of quiescent calcium channels can occur at both somatic and neuritic loc-

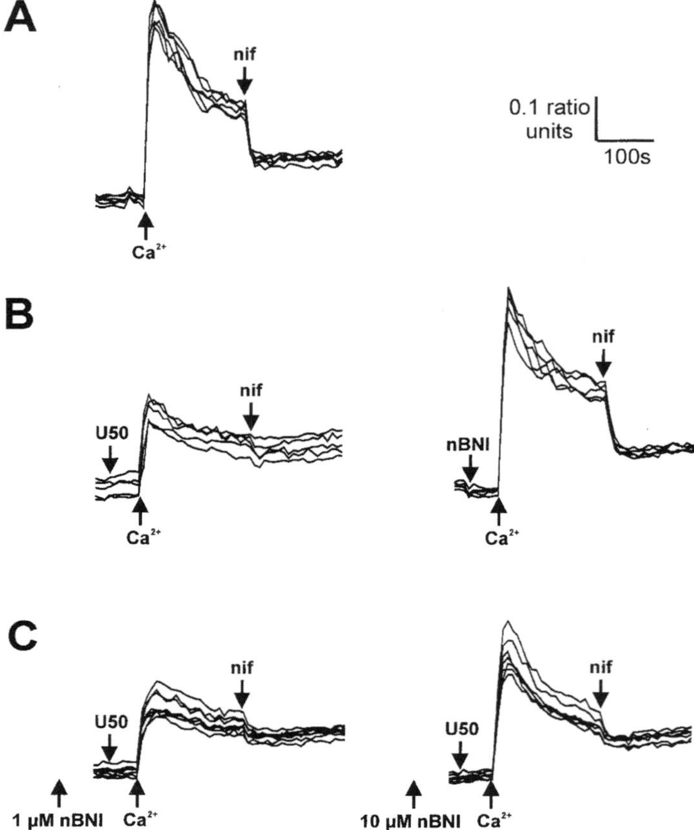

Figure 4. Single cell fluorescence imaging of the effects of U50-488, a dynorphin receptor agonist, or nor-binaltorphimine (nBNI), a dynorphin receptor antagonist, upon K/Ca stimulated cerebellar granule neurons. Somatic $[Ca^{2+}]_c$ increases were measured using fura-2 in the same imaging system as described in the legend of Figure 2. Cells were stimulated by K/Ca under the following conditions (A) control, (B) in the presence of (i) 100 μM U50-488 or (ii) 10 μM nBNI or (C) in the presence of 100 μM U50-488 and nBNI (i), or 10 μM (ii). In (B), the drugs were applied 60 s prior to stimulation by 5 mM Ca^{2+} and in (C), nBNI ws applied 60 s prior to addition of U50-488. Graphs show representative traces from 3 separate experiments. For details of K/Ca stimulation refer to Evans and Pocock (1999).

ations (Figure 3). The physiological relevance of this stimulation is that it may mimic conditions of high synaptic activity such as occurs in plasticity or epilepsy. Under these stimulatory conditions, the recruitment and activation of the L-type calcium channels is controlled by src family kinases since the enhanced nifedipine-sensitive calcium elevation can be inhibited by PP1 (Evans and Pocock, 1999).

Calcium-imaging is frequently used to demonstrate cross-talk between voltage-dependent calcium channels and various plasma membrane receptors such as the NMDA receptor, cholinergic receptors (del Río et al., 1994) or intracellular signaling molecules such as kinases (Evans and Pocock, 1999) using specific agonists and antagonists. An example of using single cell imaging to demonstrate cross-talk between voltage-dependent calcium channels and plasma membrane receptors is shown in Figure 4. The inhibitory κ-opioid neuropeptide, dynorphin, is released from hippocampal neurons and can modulate presynaptic voltage activated calcium channels by binding and activating presynaptic G-protein coupled κ-opioid autoreceptors (Wiley et al., 1997). We wanted to assess whether the dynorphin receptor modulated the K/Ca activated L-type calcium-channel in CGCs. Single cell imaging of fura-2 was used to assess K/Ca evoked calcium responses in the presence of a dynorphin receptor agonist, U50-488H or a covalent antagonist nor-binaltorphimine (nBNI). The contribution of calcium influx through L-type calcium channels was assessed by determining what proportion of the inactivating plateau was inhibited by the L-channel antagonist nifedipine (1 μM). Figure 4 shows that 100 μM U50-488H almost abolished the nifedipine inhibition of the K/Ca evoked calcium response suggesting that activation of the dynorphin receptor leads to an inhibition of calcium influx mediated by L-type calcium channels. In order to confirm that the site of action of the dynorphin agonist was the dynorphin receptor, a specific covalently modifying antagonist of the dynorphin receptor was employed (nBNI). Application of the antagonist prior to the agonist and stimulation reversed a large proportion of the inhibition mediated by the agonist. This experiment demonstrates one of the most common applications of single cell imaging of cultured neurons, the use of a combination of agonists and antagonists to dissect calcium-signalling.

For a more physiological alternative to KCl depolarisation, a technique was developed to stimulate neurons with a uniform electrical field using a platinum electrode placed in the culture dish. Rather than produce a clamped depolarisation of the plasma membrane like KCl, electrical field stimulation generates a series of homogeneous tetrodotoxin-sensitive action potentials that can be adjusted to produce frequencies and amplitudes similar to those found *in vivo* (Beani et al., 1994; Cousin et al., 1997). This form of stimulation would favour the activation of fast inactivating calcium currents, whilst KCl-stimulation of CGCs favours the opening of slow or non-inactivating calcium channels. Thus N-type calcium channels do not contribute to the calcium elevations in CGCs stimulated by high KCl (Pocock et al., 1993, 1995). Analysis of the calcium responses using single cell fluorescence imaging reveals that during electrical stimulation of CGCs, L-type calcium channels are not activated and somatic calcium responses do not occur, confirming the distribution of L-type calcium channels to a somatic location as previously shown for KCl stimulation. Furthermore electrical stimulation reveals that whilst N- and P-type calcium channels are also activated, the same channels which modulate exocytosis at the neurites (i.e. Q-type) are activated whether by KCl stimulation or electrical stimulation (Cousin et al., 1997), and furthermore support current dogma that L-type calcium channels are unlikely to be activated under "normal" stimulatory conditions (Figure 3). The differences between these results and those found using KCl perhaps illustrates how the development of more appropriate *in vitro* models brings us closer to understanding the *in vivo* situation.

Fura-2 fluorescence provides a measurement of the overall intracellular calcium concentration, $[Ca^{2+}]_i$. However, fluctuations in $[Ca^{2+}]_i$ could be due to either influx of calcium through the plasma membrane or efflux from intracellular stores such as calcium-induced calcium release. As an alternative to using inhibitors of intracellular calcium release or antagonists of voltage-dependent calcium channels to dissect the source of a change in $[Ca^{2+}]_i$, a method that allows a distinction to be made between the two types of calcium flux is Mn^{2+} quench (Simpson et al., 1995). The measurement of calcium entry only is made by using low micromolar concentrations of $MnCl_2$ in the extracellular buffer. Mn^{2+} will pass through voltage-dependent calcium channels and ligand gated calcium channels such as the NMDA receptor, but will not stimulate intracellular calcium release or partake in other calcium-signalling processes. Fura-2 can be used to detect the Mn^{2+} entry since Mn^{2+} irreversibly binds fura-2 and at 360 nm excitation (the isosbestic point of fura-2, where there is no change in emission intensity upon binding calcium) the emitted fluorescence is quenched, the degree of quench being proportional to the amount of Mn^{2+} entry. Using this technique, it has been shown in CGCs that NMDA stimulated calcium entry is principally through the receptor channel itself and not through voltage-dependent calcium channels and that AMPA receptors can gate calcium and stimulate intracellular calcium release (Savidge and Bristow, 1997). In another study in which mitochondrial depolarisation in glutamate-exposed CGCs was shown not to affect the rate of Mn^{2+} quench, it was concluded that calcium influx via the NMDA receptor is not inhibited by mitochondrial depolarisation, whilst the activity of calcium efflux pathways is acutely influenced (Castilho et al., 1998).

Due to the fact that cerebellar granule neurons are some of the smallest cells in the brain (5-10 ţm in diameter), studies involving the imaging of their subcellular organelles are somewhat limited. Generally, reports of subcellular organelle calcium imaging have been performed in larger neuronal cells and using more powerful imaging techniques such as confocal microscopy imaging (these techniques are discussed elsewhere in this book). However, in order to selectively image calcium-responses in organelles, cells can be loaded with fura-2-AM and then permeabilised (such as with digitonin) to relieve the cytoplasm of fura-2 and reveal the loading of membrane bound intracellular stores such as the mitochondria and ER (Rizzuto et al., 1993). However, with conventional single cell imaging and a host of specific inhibitors, release of calcium from intracellular stores into the cytoplasm can be easily studied (del Río et al., 1999). Calcium-influx though L-type calcium channels is coupled to changes in neuronal gene transcription. Although very little work has focussed on calcium signaling in the nucleus in these cells, it has been studied in osteoclasts by the differential use of fura-2 and fura-

2-dextran (Parkinson et al., 1998). Fura-2 will load into both the cytoplasm and the nucleus whilst fura-2-dextran will not partition into the nucleus. The involvement of organellar (including Golgi and mitochondrial) or nuclear calcium pools can be detected by the use of different dye loading techniques (Connor, 1993).

The role of intracellular stores such as mitochondria in the buffering of calcium has received increasing interest due to the proposed role of this organelle in neuroexcitotoxicity (Nicholls and Budd, 1998; Keelan et al., 1999). Imaging of CGCs coloaded with fura-2 and the mitochondrial membrane potential sensitive dye rhodamine-123 (Rh 123) reveals that those neurons which respond to glutamate with a large mitochondrial depolarisation subsequently retain an elevated $[Ca^{2+}]_i$ (Khodorov et al., 1996). Similar experiments with cultured hippocampal neurons suggest that mitochondrial dysfunction plays a major role in the deregulation of $[Ca^{2+}]_i$ associated with glutamate toxicity and that the initial acute mitochondrial dysfunction may be induced by the synergistic action of nitric oxide and glutamate (Keelan et al., 1999). The subsequent delayed $[Ca^{2+}]_i$ deregulation which culminates in neuronal death may be a consequence of superoxide production by the damaged mitochondria targetting calcium extrusion mechanisms (Nicholls and Budd, 1998). Direct visualisation of mitochondrial calcium can be achieved with the calcium sensitive fluorescent dye rhod-2, which partitions directly into the mitochondria (Duchen, 1992).

Since CGCs in culture actively develop and form homosynaptic connections, this preparation is a rich source of growth cones, the specialised structures at the growing tips of neurites. Calcium signaling in the growth cone is one of the major mechanisms by which information from the environment is transduced into growth and movement. Indeed, it appears that calcium transients having both positive and negative effects upon growth cones are associated with influx through L- and N-type voltage-dependent calcium channels (Archer et al., 1999; Ohbayashi et al., 1998). The study of calcium signaling in growth cones and other cellular specialisations requires an imaging system since only a small fraction of the total cell volume is of interest. Neurite outgrowth in CGCs can be inhibited by antagonists of voltage-dependent calcium channels, paradoxically however, no calcium-responses that were the target of the antagonists were detectable (Walsh and Doherty, 1997). Archer et al. (1999) have shown with single cell calcium imaging of cultured dorsal root ganglion neurons that FGF and CAMs stimulate small voltage-independent calcium transients through L- and N-type VDCCs. It is proposed that channel "flickerings" are evoked by the generation of fatty acids in response to FGF treatment. These results emphasise that bulk calcium changes need not be necessary for important calcium-dependent processes and may prompt researchers to re-evaluate their results with reference to smaller short-lived calcium responses.

Since CGCs require the presence of KCl/glutamate in order to survive in culture (this stimulation mimics that elicited during development *in vivo* by cerebellar mossy fibres), this system is one of the most widely used to study activity-dependent neuronal death (Milani et al., 1991). The study of neuronal death has implications in the developing brain where there is extensive postnatal apoptosis and also in neurodegeneration. Often, the imaging of calcium-responses is combined with an assay for neuronal death, such as staining with propidium iodide or Hoescht dye to visualise the nuclei of dead or apoptotic cells. The fact that CGCs express the NMDA receptor makes them an ideal model for the investigation of calcium signalling in relation to neurotoxicity (Nicholls and Budd, 1998; Pocock and Nicholls, 1998). Single cell imaging of CGCs exposed to ischaemic conditions reveals that calcium increases firstly in the neurites and then the cell somata, and this is reflected in an initial pulse of neurotransmitter release revealed by co-imaging of calcium and FM1-43 (Pocock and Nicholls, 1998). Inhibition of the plasma membrane Na^+/K^+-ATPase by ouabain reveals an asynchronous calcium increase within a field of CGCs – such asynchronicity is not apparent with population fluorometry (Cousin et al., 1995).

A number of studies have employed single cell calcium imaging to dissect the effects of ethanol upon the survival of CGCs in culture (Iorio et al., 1992; Gruol et al., 1998). This research clearly has physiological implications in the effects of alcohol on the pre- and post-natal brain. Cultures of CGCs acutely treated with ethanol or prepared from rat pups exposed to ethanol show significantly smaller NMDA evoked calcium-responses than controls. The KCl evoked calcium-responses were unaffected suggesting that the NMDA receptor was likely to be the locus for ethanol action (Gruol et al., 1998). Prolonged chronic ethanol exposure affects NMDA receptors by increasing their number, an effect that could relate to withdrawal behaviour (Iorio et al., 1992). This example underlines a major advantage of primary neuronal cultures in that physiological experiments relating to brain function can be performed and the neurons can then be dissected, cultured and analysed *in vitro* for specific effects.

2. SUMMARY

Single cell fluorescence imaging allows a calcium response to be ascribed not only to a particular cell in a field but also to a particular region of that cell, provided that the regions are spacially separated. Single cell fluorescence imaging is temporally fast and this attribute can be used to resolve regional responses within a given cell and within a field of cells. The use of combined dichroics allows dual assessment of both calcium and other cellular parameters such as mitochondrial calcium, membrane potential and neurotransmitter release. The complexity of such measurements can only expand given the number of new fluorescence probes increasingly available and should allow for future correlations between $[Ca^{2+}]_i$ and other cell signaling processes.

REFERENCES

Archer, F.R., Doherty, P., Collins, D. and Bolsover, S.R., 1999, CAMs and FGF cause a local submembrane calcium signal promoting axon outgrowth without a rise in bulk calcium concentration, *Eur. J. Neurosci.* 11, 3565–3573.

Beani, L., Tomasini, C., Govoni, B.M. and Bianchi, C., 1994, Fluorometric determination of electrically evoked increase in intracellular calcium in cultured cerebellar granule cells, *J. Neurosci. Methods* 51, 1–7.

Castilho, R.F., Hansson, O., Ward, M.W., Budd, S.L. and Nicholls, D.G., 1998, Mitochondrial control of acute glutamate excitotoxicity in cultured cerebellar granule cells, *J. Neurosci.* 18, 10277–10286.

Ciardo, A. and Meldolesi, J., 1991, Regulation of intracellular calcium in cerebellar granule neurones: Effect of depolarization and of glutaminergic and cholinergic stimulation, *J. Neurochem.* 56, 184–191.

Connor, J.A., 1993, Intracellular calcium mobilization by inositol 1,4,5-trisphosphate: Intracellular movements and compartmentalization, *Cell Calcium* 14, 185–200.

Courtney, M.J., Lambert, J.J. and Nicholls, D.G., 1990, The interactions between plasma membrane depolarization and glutamate receptor activation in the regulation of cytoplasmic free calcium in cultured cerebellar granule cells, *J. Neurosci.* 10, 3873–3879.

Cousin, M.A., Nicholls, D.G. and Pocock, J.M., 1995, Modulation of ion gradients and glutamate release in cultured cerebellar granule neurons by ouabain, *J. Neurochem.* 64, 2097–2104.

Cousin, M.A., Hurst, H. and Nicholls, D.G., 1997, Presynaptic calcium channels and field-evoked transmitter exocytosis from cultured cerebellar granule cells, *Neuroscience* 81, 151–161.

del Río, E., Nicholls, D.G. and Downes, C.P., 1994, Involvement of calcium influx in muscarinic cholinergic regulation of phospholipase C in cerebellar granule cells, *J. Neurochem.* 63, 535–543.

del Río, E., Mclaughlin, M., Downes, C.P. and Nicholls, D.G., 1999, Differential coupling of G-protein-linked receptors to calcium mobilization through inositol (1,4,5) trisphosphate or ryanodine receptors in cerebellar granule cells in primary culture, *Eur. J. Neurosci.* 11, 3015–3022.

Duchen, M.R., 1992, Contributions of mitochondrial to animal physiology: From homeostatic sensor to calcium signalling and cell death, *J. Physiol.* 516, 1–17.

Evans, G.J.O. and Pocock, J.M., 1999, Modulation of neurotransmitter release by dihydropyridine-sensitive calcium channels involves tyrosine phosphorylation, *Eur. J. Neurosci.* 11, 279–292.

Forti, L. and Pietrobon, D., 1993, Functional diversity of L-type calcium channels in rat cerebellar neurons, *Neuron* 10, 437–450.

Gruol, D.L., Ryabinin, A.E., Parsons, K.L., Cole, M., Wilson, M.C. and Qiu, Z., 1998, Neonatal alcohol exposure reduces NMDA induced calcium signalling in developing cerebellar granule neurons, *Brain Res.* 793, 12–20.

Grynkiewicz, G., Poenie, M. and Tsien, R.Y., 1985, A new generation of calcium indicators with greatly improved fluorescence properties, *J. Biol. Chem.* 260, 3440–3450.

Henke, W., Cetinsoy, C., Jung, K. and Loening, S., 1996, Non-hyperbolic calcium calibration curve of fura-2: Implications for the reliability of quantitative calcium measurements, *Cell Calcium* 20, 287–292.

Iorio, K.R., Reinlib, L., Tabakoff, B. and Hoffman, P.L., 1992, Chronic exposure of cerebellar granule cells to ethanol results in increased N-methyl-D-aspartate receptor function, *Mol. Pharmacol.* 41, 1142–1148.

Keelan, K., Vergun, O. and Duchen, M.R., 1999, Excitotoxic mitochondrial depolarisation requires both calcium and nitric oxide in rat hippocampal neurons, *J. Physiol.* 520, 797–813.

Khodorov, B., Pinelis, V., Vergun, O., Storozhevikh, T. and Vinskaya, N., 1996, Mitochondrial deenergization underlies neuronal calcium overload following a prolonged glutamate challenge, *FEBS Lett.* 397, 230–234.

Milani, D., Guidolin, D., Facci, L., Pozzan, T., Buso, M., Leon, A. and Skaper, S.D., 1991, Excitatory Amino Acid-Induced Alterations of Cytoplasmic Free Calcium in Individual Cerebellar Granule Neurons – Role in Neurotoxicity. *J. Neurosci. Res.* 28, 434–441.

Nicholls, D.G. and Budd, S.L., 1998, Mitochondria and neuronal glutamate excitotoxicity, *Biochim. Biophys. Acta* 1366, 97–112.

Ohbayashi, K., Fukura, H., Inoue, H.K., Komiya, Y. and Igarashi, M., 1998, Stimulation of L-type calcium channel in growth cones activates two independent signaling pathways, *J. Neurosci. Res.* 51, 682–696.

Parkinson, N.A., Bolsover, S. and Mason, W., 1998, Nuclear and cytosolic calcium changes in osteoclasts stimulated with ATP and integrin-binding peptide, *Cell Calcium* 24, 213–221.

Pearson, H.A., Sutton, K.G., Scott, R.H. and Dolphin, A.C., 1995, Characterization of calcium channel currents in cultured rat cerebellar granule neurons, *J. Physiol.* 482, 493–509.

Pocock, J.M., Cousin, M.A. and Nicholls, D.G., 1993, The calcium channel coupled to the exocytosis of L-glutamate from cerebellar granule cells is inhibited by the spider toxin Aga-GI, *Neuropharmacology* 32, 1185–1194.

Pocock, J.M., Cousin, M.A., Parkin, J. and Nicholls, D.G., 1995, Glutamate exocytosis from cerebellar granule cells: the mechanism of a transition to an L-type calcium channel coupling, *Neuroscience* 67, 595–607.

Pocock, J.M. and Nicholls, D.G., 1998, Exocytotic and non-exocytotic modes of glutamate release from cultured cerebellar granule cells during chemical ischaemia, *J. Neurochem.* 70, 806–813.

Randall, A. and Tsien, R.W., 1995, Pharmacological dissection of multiple types of calcium channel currents in rat cerebellar granule neurons, *J. Neurosci.* 15, 2995–3012.

Rizzuto, R., Brini, M., Murgia, M. and Pozzan, T., 1993, Microdomains with high calcium close to IP3-sensitive channels that are sensed by neighboring mitochondria, *Science* 262, 744–747.

Savidge, J.R. and Bristow, D.R., 1997, Distribution of Calcium-permeable AMPA receptors among cultured rat cerebellar granule cells, *Neuroreport* 8, 1877–1882.

Simpson, P.B., Challiss, R.A.J. and Nahorski, S.R., 1995, Divalent-cation entry in cultured rat cerebellar granule cells measured using Mn^{2+} quench of fura-2 fluorescence, *Eur. J. Neurosci.* 7, 831–840.

Taylor, C.W. and Broad, L.M., 1998, Pharmacological analysis of intracellular Ca^{2+} signalling: Problems and pitfalls, *TIPS* 19, 370–374.

Walsh, F.S. and Doherty, P., 1997, Neural cell adhesion molecules of the immunoglobulin superfamily: Role of axon growth and guidance, *Annu. Rev. Cell Dev. Biol.* 13, 425–456.

Wiley, J.W., Moses, H.C., Gross, R.A. and Macdonald, R.L., 1997, Dynorphin A-mediated reduction in multiple calcium currents involves a G(0 alpha)-subtype G protein in rat primary afferent neurons, *J. Neurophys.* 77, 1338–1348.

Calcium Complexities: New Fluorescence Techniques for Probing Mitochondria and Other Subcellular Compartments

Gregory R. Monteith, Vadim N. Dedov and Basil D. Roufogalis

1. INTRODUCTION

As evident in all of the chapters throughout this book, Ca^{2+} is an essential regulator of various cellular functions, including muscle contraction and the release of neurotransmitters and hormones. However, the question can be asked: How can a ubiquitous signal like Ca^{2+} generate so many different physiological responses while maintaining selectivity as a cellular signal? (Berridge, 1997b). Since the advent of Ca^{2+}-sensitive fluorophores and fluorescence microscopy, researchers have been able to visualize Ca^{2+} signaling in living cells. These studies have revealed complexities in temporal and spatial regulation of Ca^{2+} signaling, which appear to hold the key to how Ca^{2+} can act as both a ubiquitous and selective signal (Berridge, 1997b). Furthermore, confocal and electron microscopy studies have shown that the locality of key Ca^{2+} transporters and the concentration of Ca^{2+} in sub-cellular compartments is not homogenous, with consequent physiological implications. In this chapter we will discuss the spatial and temporal complexities of Ca^{2+} signaling, in particular recent studies using confocal, multi-photon and high-speed fluorescence microscopy as well as "caged" regulators of intracellular Ca^{2+}.

Gregory R. Monteith • The School of Pharmacy, The University of Queensland, St. Lucia, Queensland. Australia. **Vadim N. Dedov and Basil D. Roufogalis** • The Faculty of Pharmacy, The University of Sydney, New South Wales. Australia.

R. Pochet, R. Donato, J. Haiech, C. Heizmann and V. Gerke (eds.): Calcium: The Molecular Basis of Calcium Action in Biology and Medicine, 697–713.

2. SPATIAL ASPECTS OF CALCIUM SIGNALING

2.1. Fundamental Ca^{2+} release events: Sparks, puffs, blips and quarks: Small but significant

Perhaps the first and best known example of a non-uniform change in cytosolic free Ca^{2+} ($[Ca^{2+}]_{CYT}$) is a Ca^{2+} wave, where an increase in $[Ca^{2+}]_{CYT}$ is first manifested in a confined region of the cell, followed by an increase in Ca^{2+} which is propagated throughout the cell like a "wave". The observation of fundamental Ca^{2+} release events indicate that Ca^{2+} waves can arise when these confined events act in a coordinated fashion to initiate, then propagate, a Ca^{2+} wave, via calcium induced calcium release activation of ryanodine and/or IP_3-sensitive Ca^{2+} channels (Berridge, 1997a; Marchant et al., 1999). The advent of confocal microscopy with its superior Z axis resolution, and high (ms) temporal resolution in "line scan" mode, enabled researchers to visualize these minute ($\sim 2\ \mu$m) basic events in cellular Ca^{2+} homeostasis (Cannell and Soeller, 1999). The basic Ca^{2+} release event first characterized in cardiac myocytes, and referred to as a Ca^{2+} spark, arises from the opening of ryanodine Ca^{2+} channels on the sarcoplasmic reticulum (SR) Ca^{2+} store (Cheng et al., 1993). There is some debate as to whether a spark is the opening of 1 or more channels (Schneider, 1999). The opening of a single ryanodine channel has been called a "Ca^{2+} quark" (Niggli, 1999), and recent evidence using caged calcium and two-photon microscopy support the existence of this fundamental release event, which is much smaller and of much lower amplitude than a Ca^{2+} spark (Lipp and Niggli, 1998). Ca^{2+} sparks are particularly evident in the T-tubules of cardiac muscle (Niggli, 1999), and as will be discussed latter, the location of these basic elements of Ca^{2+} signaling is of crucial significance.

In cells such as HeLa cells and Xenopus oocytes the fundamental Ca^{2+} release events occur via the opening of IP_3 sensitive Ca^{2+} channels (Berridge, 1997a; Bootman et al., 1997). There have been 3 "levels" proposed for the opening of these channels, namely Ca^{2+} blips (opening of a single IP_3 channel on the endoplasmic reticulum (ER) Ca^{2+} store, perhaps analogous to a quark), Ca^{2+} puffs (opening of a group of IP_3 channels, perhaps analogous to a spark) and global Ca^{2+} signals (which occur after activation of sufficient numbers of Ca^{2+} release channels) (Bootman et al., 1997). The boundaries between these 3 levels (blips, puffs and global signals) are not clearly defined (Sun et al., 1998) and may involve activation of different numbers of IP_3 channels within a group, or activation of all the IP_3 channels in groups consisting of different numbers of IP_3-sensitive channels (Thomas et al., 1998).

Koizumi and colleagues recently described fundamental Ca^{2+} release events from the ER of nerve growth factor differentiated PC12 cells, and primary cultured hippocampal neurons, although the events were less frequent in the hippocampal neurons (Koizumi et al., 1999). The Ca^{2+} release events observed in these cells had some similar features to the Ca^{2+} sparks and Ca^{2+} puffs previously characterized in non-neuronal cells. However, they did have distinct properties, perhaps indicating that the fundamental Ca^{2+} release events in neurons are unique. The fundamental Ca^{2+} release events in the differentiated PC12 cells, and rat cultured hippocampal neurons, were different from Ca^{2+} sparks and puffs, in that they were often produced from clusters consisting of both ryanodine and IP_3 activated Ca^{2+} channels in these neurons (Koizumi et al., 1999). Furthermore, fundamental Ca^{2+} release events in differentiated PC12 cells were more common in areas of neurite branching (Koizumi et al., 1999), indicating spatial regulation of these basic Ca^{2+} signaling events. This is significant, since compartmentalization of the Ca^{2+} signal is particularly evident in neurons, where the Ca^{2+} levels reached in dendritic spines can be dramatically higher than the levels reached in other regions of the neuron (Segal, 1995).

The significance of spatial localization of fundamental Ca^{2+} release events is not confined to differentiated PC12 cells. For example, in cardiac ventricular myocytes Ca^{2+} sparks are most common near T-tubules, where the close proximity between the L-type Ca^{2+} channels of the plasma membrane and the SR plays an important role in excitation-contraction coupling (Niggli, 1999). Likewise, in arterial smooth muscle Ca^{2+} sparks occur most often within a 1 μm distance from the plasma membrane (Nelson et al., 1995). However, in these cells Ca^{2+} sparks are the initiators of muscle relaxation and a single Ca^{2+} spark can activate multiple Ca^{2+} activated K^+ channels, thereby hyperpolarizing the cell, reducing Ca^{2+} inflow through voltage-gated Ca^{2+} channel and dilating an arterial vessel (Fay, 1995; Nelson et al., 1995). Hence the specific location of Ca^{2+} sparks can result in relaxation of a vessel. In *Xenopus* oocytes, spatially distinct regulation of Ca^{2+} puffs is also apparent (Callamaras et al., 1998). In these cells Ca^{2+} puffs are of higher amplitude in the animal hemisphere than the vegetal hemisphere, and the sites for Ca^{2+} release in the animal hemisphere are in closer proximity to each other (Callamaras et al., 1998). Thus, neither Ca^{2+} puffs nor Ca^{2+} sparks occur homogeneously throughout the cytoplasm and this represents a fundamental example of the importance of spatial location in Ca^{2+} signaling.

2.2. Intracellular Ca^{2+} stores and other sub-cellular compartments: More compartmentalization of Ca^{2+}

It is now understood that heterogeneity in the calcium signal is not only generated by fundamental Ca^{2+} release events (Cheng et al., 1993) and Ca^{2+} waves (Simpson and Russell, 1997) but also from cellular organelles such as the S/ER, nucleus and mitochondria. Mitochondrial Ca^{2+} regulation in neuronal and non-neuronal cells will be discussed later in this chapter. Other sub-cellular compartments for Ca^{2+} will be discussed more briefly here.

The importance of the S/ER in Ca^{2+} regulation has been discussed elsewhere in this book, however, some recent studies using the low Ca^{2+} affinity probes mag-fura-2 and fura-2FF (a less Mg^{2+} sensitive probe) report that the IP_3 and ryanodine sensitive stores may be functionally distinct (Golovina and Blaustein, 1997). In some cell types, the S/ER is close to the plasma membrane (Devine et al., 1972), and as described above, the close proximity of the SR to the plasma membrane allows activation of K^+ channels via a Ca^{2+} spark generated from the SR (Fay, 1995; Nelson et al., 1995). The space just below the plasma membrane has long been hypothesized as an area where changes in free Ca^{2+} greatly exceed the levels reached in the bulk cytosol (Rasmussen, 1989). The co-localization of Ca^{2+} influx and Ca^{2+} efflux pathways (Rasmussen, 1989) as well as the close proximity of the S/ER to the plasma membrane (Chen et al., 1992) have been proposed to promote this Ca^{2+} gradient. Until recently, it was impossible to accurately resolve such Ca^{2+} gradients, however, the advent of novel techniques and the design of new Ca^{2+} sensitive fluorophores have provided some evidence for the existence of such gradients in numerous cell types. Aequorin has been targeted to the plasma membrane in the A7r5 rat aortic myocyte cell line using a SNAP-25 aequorin chimera. These studies showed that subplasmalemmal Ca^{2+} ($[Ca^{2+}]_{spm}$) changed to a greater extent than bulk cytosolic Ca^{2+} after stimulation (Marsault et al., 1997). Fura-2 and Indo-1 analogues with lipophillic tails have also been used to assess $[Ca^{2+}]_{spm}$ in neutrophils and skeletal muscle; these studies indicated that changes in $[Ca^{2+}]_{spm}$ exceed those of bulk $[Ca^{2+}]_{CYT}$ (Bruton et al., 1999; Davies and Hallet, 1998). The hypothesis that changes in free Ca^{2+} just beneath the plasma membrane do not match changes in the bulk cytosol is also supported by indirect evidence where the current through Ca^{2+}-activated K^+ channels was used as a qualitative measure of $[Ca^{2+}]_{spm}$ (Ganitkevich and Isenberg, 1996).

A great deal of research is now focused on calcium signaling in the cell nucleus. Readers should consult other reviews (Gerasimenko et al., 1996; Rogue and Malviya, 1999) for a discussion of the sources of nuclear Ca^{2+} and the potential role of $[Ca^{2+}]_N$ in gene transcription and apoptosis. The Golgi apparatus is an organelle which until recently has been largely ignored in terms of Ca^{2+} homeostasis. Recently, experiments with aequorin targeted to the Golgi, has shown that the Golgi sequesters Ca^{2+} via a thapsigargin and 2,5-di-(tert-butyl),1-4-benzohydroquinone sensitive S/ER Ca^{2+}-ATPase (SERCA) and via another active Ca^{2+} transporter, insensitive to these SERCA inhibitors (Pinton et al., 1998). Furthermore, the Golgi was demonstrated to be an IP_3-sensitive store, the free Ca^{2+} in the Golgi declining when HeLa cells were stimulated with histamine (Pinton et al., 1998).

Given the above evidence of spatial localization of Ca^{2+} signaling, it is perhaps not surprising that the location of key Ca^{2+} transporters is not homogeneously distributed in the cell. One such example is the plasma membrane Ca^{2+} ATPase (PMCA) (Monteith and Roufogalis, 1995), which in at least some cell types, does not appear to be evenly dispersed on the plasma membrane, but instead is localized to small plasmalemmal invaginations known as caveolae (Fujimoto 1993). Caveolae are like miniature units of concentrated signal transduction pathways. Caveolae are enriched in caveolin (Schnitzer et al., 1995), protein kinase C enzymes (Mineo et al., 1998), endothelial nitric oxide synthase (Rizzo et al., 1998) and the calcium sensing receptor (Kifor et al., 1998). Furthermore, phosphatidylinositol 4,5 bisphosphate (PIP2) appears to be enriched in caveolae (Pike and Casey, 1996). As discussed previously, the localization of PIP2 together with the PMCA in caveolae makes the hypothesis that PIP2 dynamically regulates PMCA activity more likely (Monteith and Roufogalis, 1995).

Furthermore, highly localized production of IP_3 after PLC-induced hydrolysis of PIP2 (Pike and Casey, 1996) is consistent with the more recent observation that ATP-induced Ca^{2+} waves are initiated in caveolin rich regions of endothelial cells (Isshiki et al., 1998). Furthemore, non-homogenous distribution of Ca^{2+} pumps in cells is also evident by the observation of SERCA2a localization in the luminal pole of pancreatic acini cells, compared with the localization of SERCA2b in the basal pole of the same cells (Lee et al., 1997). This is likely to have implications in the characteristics of the Ca^{2+} waves initiated in this cell type and reinforces the importance of the cellular location of Ca^{2+} transporters (Lee et al., 1997).

3. MITOCHONDRIAL Ca^{2+} REGULATION IN NON-NEURONAL CELLS

Over recent years a plethora of studies have examined mitochondrial matrix free Ca^{2+} ($[Ca^{2+}]_M$) in non-neuronal cells. A comprehensive review of this area is beyond the scope of this chapter (a discussion of recent studies alone and their physiological significance, could easily fill an entire book), and readers are referred to a recent review (Duchen, 1999). However, many recent studies reinforce the concept of spatially defined alterations in free Ca^{2+}, and it will be this aspect which will be the focus of this discussion on mitochondrial Ca^{2+} regulation.

There appear to be two general mechanisms by which Ca^{2+} can be sequestered in mitochondria, involving slow and rapid uptake modes (Gunter et al., 1998; Gunter and Pfeiffer, 1990). In the past mitochondria were thought to influence only $[Ca^{2+}]_{CYT}$ during Ca^{2+} loads in the range likely to be associated with cell death (Carafoli, 1987). However, recent studies using fluorescence digital imaging have demonstrated that mitochondria sequester Ca^{2+} after physiological stimuli in many cell types, although the amplitude of the Ca^{2+} signal and the locality of mitochondria relative to Ca^{2+} release and influx channels greatly influence the degree of Ca^{2+} uptake (Lawrie et al., 1996; Monteith and Blaustein, 1999; Rizzuto et al., 1993, 1998).

In many cell types the uptake of Ca^{2+} by mitochondria is delayed, with peak $[Ca^{2+}]_M$ lagging (in the order of seconds) after the peak level of $[Ca^{2+}]_{CYT}$ has been achieved (Babcock et al., 1997; Jou et al., 1996; Monteith and Blaustein, 1999; Ricken et al., 1998). Although not apparent in all cell types, the recovery of $[Ca^{2+}]_M$ is often slower than the decline in $[Ca^{2+}]_{CYT}$ after stimulation (Monteith and Blaustein, 1999; Ricken et al., 1998), and the uptake of Ca^{2+} into mitochondria is repeatable, indicating that mitochondria can sequester and unload Ca^{2+} after repeat stimuli (Monteith and Blaustein, 1999).

As stated above, the location of mitochondria in the cell is a critical factor in how mitochondria respond during a Ca^{2+} signal. The importance of mitochondrial location in calcium signaling was demonstrated in studies using aequorin (a luminescent Ca^{2+} sensitive protein) targeted to mitochondria in an endothelial cell line (ECV304) and in HeLa cells (Lawrie et al., 1996). In ECV304 cells 14% of mitochondria are within 700 nm of the plasma membrane with very few ($< 4\%$) close to the ER (Lawrie et al., 1996), and these cells undergo greater changes in $[Ca^{2+}]_M$ via Ca^{2+} influx than after activation of Ca^{2+} release from internal stores (Lawrie et al., 1996). In contrast HeLa cells (where most mitochondria are within 700 nm of the ER) undergo greater changes in $[Ca^{2+}]_M$ after Ca^{2+} release from internal stores (Lawrie et al., 1996). Hence, mitochondria appear to be regulated by the high microdomains of Ca^{2+} generated beneath the plasma membrane and close to ER Ca^{2+} release sites (Lawrie et al., 1996). Indeed, heterogeneity in $[Ca^{2+}]_M$ has been demonstrated in activated smooth muscle cells, with some mitochondria not sequestering Ca^{2+} after stimulation (Monteith and Blaustein, 1999), see Figure 1. Mitochondria were also shown to increase their Ca^{2+} (sometimes as a Ca^{2+} wave) when stimulated with ATP. This heterogeneity in changes in $[Ca^{2+}]_M$ may be related to high Ca^{2+} microdomains generated in these cells during ATP stimulation (Monteith and Blaustein, 1999). Recent very elegant studies using green fluorescent protein derivatives targeted to the ER and mitochondria, as well as aequorin targeted to the outer face of the inner mitochondrial membrane, demonstrated directly the close contact between the ER and mitochondria in HeLa cells (Rizzuto et al., 1998). They also demonstrated that the levels of Ca^{2+} which are reached in the microdomain between the ER and mitochondria do indeed exceed the Ca^{2+} levels reached in the rest of the cytoplasm (Rizzuto et al., 1998).

Apart from modulation of Ca^{2+}-sensitive processes in mitochondria, such as ATP production (Hansford, 1985), mitochondria may also regulate Ca^{2+} homeostasis more directly via modulation of Ca^{2+} in microdomains. It has been reported that mitochondria regulate the propagation of $[Ca^{2+}]_{CYT}$ waves (Simpson and Russell, 1996) and that mitochondria suppress Ca^{2+}-mediated feedback activation of IP_3 channels by sequestering Ca^{2+} (Hajnoczky et al., 1999). Indeed "synaptic-like" Ca^{2+} signaling between mitochondria and the ER has been proposed, reinforcing the importance of the spatial localization of mitochondria in regard to Ca^{2+} homeostasis (Csordas et al., 1999). A classic illustration of the importance of mitochondria is the ability of mitochondria surrounding the granule region of pancreatic acinar to stop local IP_3-stimulated Ca^{2+} increases spreading throughout the entire cell (Tinel et al., 1999).

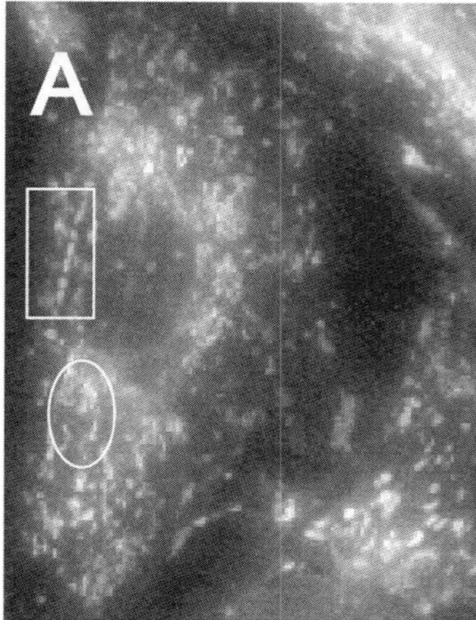

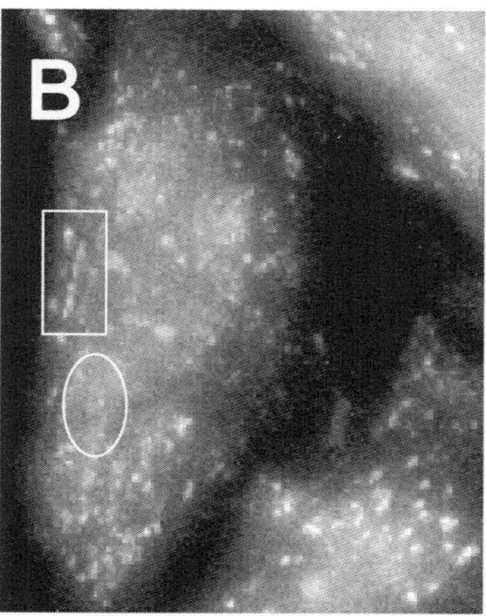

Figure 1. Heterogeneity in the Ca^{2+} response of mitochondria stimulated with 100 μM ATP (Monteith and Blaustein, 1999). Panel A: aortic smooth muscle cells loaded with Mitotracker Green FM (MTG); Panel B: Rhod-2 fluorescence after stimulation with 100 μM ATP. Note that in some regions (e.g. rectangle) almost all mitochondria identified with MTG sequestered Ca^{2+} after ATP, whereas in other regions (e.g., oblong), there was little localized Rhod-2 fluorescence after ATP stimulation. The heterogeneity in mitochondrial Ca^{2+} uptake was even more pronounced after stimulation with 10 μM 5HT (Monteith and Blaustein, 1999).

Once Ca^{2+} is sequestered, it must of course be released, and our understanding of this release process has advanced considerably over recent years. As well as the pathways previously characterized in isolated mitochondria (such as Na^+/Ca^{2+} exchange) (Gunter and Pfeiffer, 1990), it now appears that the mitochondrial permeability transition pore (PTP) can operate in a "low conductance" mode allowing the exit of Ca^{2+} from mitochondria (Ichas and Mazat, 1998). The partial opening of the PTP does not have the dire cellular consequence as full opening of the PTP, where the channel is permeable to proteins < 1500 kDa (Ichas and Mazat, 1998). Instead, the low conductance state of the PTP is associated with a Ca^{2+}-induced Ca^{2+} release phenomena, which may be important in amplifying ER generated Ca^{2+} signals (Ichas et al., 1997) and perhaps initiating or contributing to fundamental Ca^{2+} release events such as Ca^{2+} sparks (Bowser et al., 1998). For an extensive review of the two states of PTP readers should consult a recent review (Ichas and Mazat, 1998).

4. MITOCHONDRIAL Ca^{2+} REGULATION IN NEURONS

As discussed elsewhere in this book, the influx of extracellular Ca^{2+} (eCa^{2+}) through plasma membrane Ca^{2+} channels is a major mechanism by which excitable cells, such as neurons, undergo increases in $[Ca^{2+}]_{CYT}$. The following section of this chapter will primarily focus on mitochondria in Ca^{2+} signaling after activation of neuronal voltage operated Ca^{2+} channels (VOCC) and receptor operated Ca^{2+} channels (ROCC).

The idea that neuronal mitochondria accumulate large amounts of Ca^{2+} originated from studies using isolated mitochondria, including those isolated from brain. Until recently, only pharmacological approaches had been used to characterize mitochondrial Ca^{2+} regulation. These studies involved the application of drugs to manipulate mitochondrial Ca^{2+} transport in isolated mitochondria (for a review, see Budd, 1998). The agents most widely used in these studies include:

(i) ruthenium red, an inhibitor of the mitochondrial uniporter, which facilitates mitochondrial Ca^{2+} influx;

(ii) uncouplers (protonophores), such as cyanide derivatives CCCP and FCCP, which dissipate mitochondrial membrane potential (MMP), and prevent electrogenic mitochondrial Ca^{2+} uptake and stimulate release of accumulated Ca^{2+} from mitochondria;

(iii) CGP-37157, an inhibitor of the mitochondrial Na^{+}/Ca^{2+} exchanger, the major efflux pathway for Ca^{2+} from mitochondria. Removal of Na^{+} from extracellular medium exerts a similar effect by depleting intracellular Na^{+};

(iv) cyclosporin A, an inhibitor of the mitochondrial permeability transition pore. (Mitochondrial pore organization and its biological significance were recently extensively reviewed (Bernardi et al., 1998; Kroemer et al., 1998) and is beyond the scope of this chapter.)

The development of ratiometric fluorescent probes for the measurement of $[Ca^{2+}]_{CYT}$ *in situ*, together with patch-clamp techniques and pharmacological tools, have allowed the study of $[Ca^{2+}]_M$ in individual neurons. Studies have revealed that peak $[Ca^{2+}]_{CYT}$ closely parallels the magnitude of the current evoked by activation of VOCC, with cytosolic Ca^{2+} buffering being more effective for larger Ca^{2+} loads (Thayer and Miller, 1990). The addition of ruthenium red in the patch-clamp pipette prevents such buffering, whereas an uncoupler, if applied after activation of VOCC evokes an additional increase in $[Ca^{2+}]_{CYT}$. The "plateau" phase of 200–600 nM $[Ca^{2+}]_{CYT}$, which followed the spike of $[Ca^{2+}]_{CYT}$ upon activation of VOCC, was considered to be due to the slow release of $[Ca^{2+}]_M$ (Thayer and Miller, 1990). Neuronal uncoupler-sensitive Ca^{2+} stores appeared to be strongly $[Ca^{2+}]_{CYT}$ dependent, because it became influential only when $[Ca^{2+}]_{CYT}$ approached $\sim$ 500 nM, consistent with a mitochondrial Ca^{2+} pool (Friel and Tsien 1994). Increased $[Ca^{2+}]_{CYT}$ upon activation of VOCC, evokes concomitant transient (1–2 minutes) mitochondrial depolarization, and increased respiration (Duchen, 1992). The important role of mitochondria in buffering Ca^{2+} in presynaptic terminals and regulating neurotransmitter release has also now been demonstrated (Peng, 1998; Scotti et al., 1999). Very recently, total mitochondrial Ca^{2+} has been measured in neurons using x-ray microanalysis, and used to demonstrate that increases in the total level of $[Ca^{2+}]_M$ positively correlates with depolarization strength and duration. In these studies,

$[Ca^{2+}]_M$ recovered to prestimulation levels with a time course that paralleled the decline in $[Ca^{2+}]_{CYT}$ (Pivovarova et al., 1999). Mitochondrial Na^+/Ca^{2+} exchanger is considered as the major efflux mechanism for $[Ca^{2+}]_M$ into the cytosol and application of CGP-37157 after activation of VOCC abolishes the "plateau" phase of the $[Ca^{2+}]_{CYT}$ transient, presumably because of the trapping of Ca^{2+} in mitochondria (Pivovarova et al., 1999).

Following a hypoxic-ischemic insult, the collapse of ion gradients results in the inappropriate release of excitatory neurotransmitters, most importantly glutamate, which overstimulates N-methyl-D-aspartate (NMDA) glutamate receptors, which is a major factor in hypoxic cell injury in the CNS (for a review, see Budd, 1998). Using CGP-37157 and FCCP it has been shown that mitochondria not only buffer eCa^{2+} upon stimulation of glutamate receptors but also that mitochondrial buffering of glutamate-induced Ca^{2+} loads becomes progressively more important as the stimulus intensity increases (White and Reynolds, 1996). Mitochondrial calcium uptake is a prerequisite of glutamate-induced neuron death (Stout et al., 1998). Indeed, progressive mitochondrial depolarization is an early signal specific to excitotoxic exposure (White and Reynolds, 1996) and a primary event in glutamate-mediated neurotoxicity (Schinder et al., 1996). The crucial role of mitochondria in neurotoxicity has been recently reviewed and is likely to be related to the degree of $[Ca^{2+}]_{CYT}$ increase (Nicholls and Budd, 1998). Indeed, it is estimated that lethal NMDA exposure occurs at $> 5\ \mu M$ peak $[Ca^{2+}]_{CYT}$, whereas $[Ca^{2+}]_{CYT}$ levels below 1.5 μM are non-lethal (Hyrc et al., 1997). The low-affinity fluorescent Ca^{2+} indicators are required for such studies, as "high-affinity calcium indicators underestimate increases in intracellular calcium concentrations associated with excitotoxic glutamate stimulations" (Stout and Reynolds, 1999).

Differences in total Ca^{2+} permeability, and thus the total amount of Ca^{2+} available for accumulation by mitochondria, may explain the fact that activation of VOCC and some ROCC is not toxic, but activation of glutamate receptors trigger cell death. Using dorsal root ganglion (DRG) neurons as a model for the study of VR-1 (capsaicin) receptors (Caterina et al., 1997) we found that activation of this ROCC evoke distinct Ca^{2+} permeability changes in different subsets of neurons (Dedov and Roufogalis, 1998). We then compared mitochondrial accumulation and $[Ca^{2+}]_{CYT}$ clearance after activation of VOCC and VR-1 receptors in the same neurons. Mitochondrial Ca^{2+} release was found to be responsible for maintaining elevated $[Ca^{2+}]_{CYT}$ levels up to one hour after stimulation of VR-1 receptors, whereas $[Ca^{2+}]_{CYT}$ clearance after activation of VOCC was complete in minutes (Dedov and Roufogalis, 1999). Figure 2 represents a typical $[Ca^{2+}]_{CYT}$ trace for DRG neurons activated by 50 mM KCl (VOCC) or 1 μM capsaicin (a VR-1 receptor agonist). Both KCl and capsaicin elicited rapid activation of Ca^{2+} channels and produced equivalent increases in peak $[Ca^{2+}]_{CYT}$. However, $[Ca^{2+}]_{CYT}$ "plateau" levels and the uncoupler-sensitive $[Ca^{2+}]_{CYT}$ elevation, were much higher after activation of VR-1 receptors. We concluded that mitochondria con-

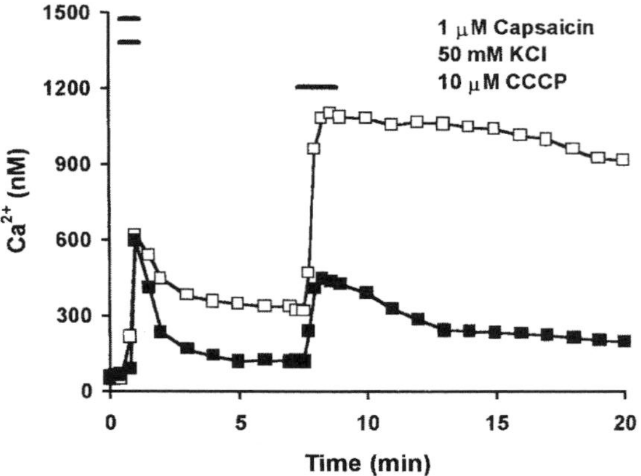

Figure 2. Distinct $[Ca^{2+}]_M$ release patterns upon activation of VOCC or ROCC in DRG neurons. Traces were taken from two different experiments: 50 mM KCl (black squares) or 1 μM capsaicin (white squares) were applied to DRG neurons for 30 s to activate VOCC or ROCC (VR-1), respectively. Neurons were then washed for 7 min and 10 μM CCCP was applied for 1 min. Note that kinetics of activation and amplitudes of peak $[Ca^{2+}]_{CYT}$ transients were similar.

trol $[Ca^{2+}]_{CYT}$ whether VOCC or VR-1 receptors are activated, but that the greater amount of Ca^{2+} accumulated by mitochondria after capsaicin reflects the greater Ca^{2+} influx after VR-1 receptor activation (compared with VOCC activation).

Despite the plethora of studies, the pharmacological approach should be considered carefully because of possible non-specific effects. For example, an increase of $[Ca^{2+}]_{CYT}$ when uncouplers are applied after stimulation of VOCC or glutamate receptors may be due to reduced ATP levels, rather than release of accumulated mitochondrial Ca^{2+} (Budd and Nicholls, 1996a, b). Furthermore, impaired mitochondrial metabolism itself may activate the plasma membrane currents and evoke Ca^{2+} influx (Nowicky and Duchen, 1998). It should also be noted that CGP-37157 can directly inhibit VOCC, and should therefore be used with caution for studies in intact cells (Baron and Thayer, 1997). Some of these difficulties can be overcome by the direct measurement of $[Ca^{2+}]_M$ in neurons.

As discussed above, direct measurement of $[Ca^{2+}]_M$ is now possible using fluorescent probes and mitochondrially targeted aequorin and cameleons. However, transfection of primary cultured neurons with aequorin and cameleons is more difficult than for cell lines. However, such cameleons have already been transfected in primary cultured hippocampal neurons, to assess $[Ca^{2+}]_{CYT}$ (Miyawaki et al., 1999), illustrating the viability of mitochondrially targeting these Ca^{2+} sensors in neurons.

As described previously in this chapter, fluorescent probes such as Rhod-2 have also been used to assess $[Ca^{2+}]_M$ directly in many non-neuronal cells. Simultaneous measurement of $[Ca^{2+}]_{CYT}$ and $[Ca^{2+}]_M$ has revealed that in motor nerve terminals mitochondrial Ca^{2+} uptake contributes to the buffering of presynaptic Ca^{2+} during neuromuscular transmission evoked by brief trains of action potentials (David et al., 1998). However, sustained mitochondrial Ca^{2+} uptake is not always accompanied by progressive elevations in $[Ca^{2+}]_M$ in the same model (David, 1999) and in neurons receiving multiple pulses of NMDA $[Ca^{2+}]_M$ does not always return to baseline, indicating that elevated $[Ca^{2+}]_M$ may persist indefinitely (Peng et al., 1998; Peng, 1998). Moreover, mitochondrial Ca^{2+} uptake in response to increased $[Ca^{2+}]_{CYT}$ is faster and more tightly coupled during NMDA receptor activation than during non-NMDA receptor or VOCC activation (Peng and Greenamyre, 1998). This implies a special relationship between mitochondria and the NMDA receptor. It should be noted that Rhod-2 does have limitations, which may be particularly evident in neurons. Firstly it is not a ratiometric probe, making quantification of $[Ca^{2+}]_M$ (a likely factor in neurotoxicity) more difficult. Furthermore, if selective loading of mitochondria with Rhod-2 is not obtained, then the diffuse pattern of fluorescence in untreated cells and significant contribution of the cytosolic compartment during activation may make assessment of $[Ca^{2+}]_M$ unreliable (Bernardi et al., 1999). Some of these disadvantages may be overcome through the use of two-photon microscopy, as has been used recently in H29 cells (Ricken et al., 1998).

Hence, similarly to other cell types, neuronal mitochondria can buffer Ca^{2+} in situ. Extracellular Ca^{2+}, entering neurons through VOCC or ROCC is a major source for increases in $[Ca^{2+}]_M$ in neurons. The amount of $[Ca^{2+}]_M$ and its metabolic consequences, including neurotoxicity, depend on the nature of the Ca^{2+} channel activated. Activation of VOCC evoke a transient accumulation of $[Ca^{2+}]_M$ and a concomitant drop of MMP, and stimulate mitochondrial metabolism. Activation by glutamate or VR-1 receptors may cause profound Ca^{2+} influx and excessive mitochondrial Ca^{2+} accumulation, the consequence of which is a derangement in mitochondrial function and cell death.

5. THE SIGNIFICANCE OF THE FREQUENCY AND AMPLITUDE OF THE Ca^{2+} SIGNAL: AM AND FM NOT JUST FOR RADIOS

The theory of amplitude and/or frequency modulation has been elegantly described elsewhere (Berridge, 1997b). However, recent studies using novel techniques have demonstrated that amplitude modulation (AM) and frequency modulation (FM) of the Ca^{2+} signal is used by a variety of cells to generate specific cellular responses. Some of these recent studies will be described here. In regard to AM Ca^{2+} signaling, one example discussed above may be Ca^{2+} uptake by mitochondria, which appears to be more predominant during sustained or high Ca^{2+} loads, or where high Ca^{2+} levels are obtained in areas close to Ca^{2+} release channels (Monteith and Blaustein, 1999; Ricken et al., 1998; Rizzuto et al., 1998). Furthermore, the degree of Ca^{2+} uptake by mitochondria may determine whether the Ca^{2+} uptake initiates alterations in ATP production or cytochrome c release and apoptosis (Duchen, 1999; Ichas and Mazat, 1998). The duration and amplitude of the Ca^{2+} response also controls the differential activation of NFAT and NF-κB transcription factors, the latter being selectively activated by higher levels of $[Ca^{2+}]_{CYT}$ (Dolmetsch et al., 1997). The duration of $[Ca^{2+}]_{CYT}$ increases also regulates the activity of cytosolic phospholipase A2 (Hirabayashi et al., 1999), as demonstrated by whether growth or quiescence is induced by hydrogen peroxide in prostate tumor spheroids (Wartenberg et al., 1999). One of the most clear demonstrations of the role of FM in Ca^{2+} signaling for generating specific cellular events came from work using a cell-permeable caged IP_3-analogue (Li et al., 1998). This caged compound entered RBL cells as a cell membrane permeable ester, which once hydrolyzed was trapped in the cytosol. When exposed to a UV flash, the IP_3 analogue was formed (after being released from its inactivating "cage") and mobilized Ca^{2+} from intracellular stores via activation of IP_3 receptors (Li et al., 1998). The unique feature of this study was the alteration of the time between UV flashes, to produce Ca^{2+} oscillations with different frequencies (Li et al., 1998). It was found that the same amount of IP_3 was more efficient at increasing gene expression when given as a series of pulses at a frequency of 1 per minute than at higher or lower frequencies (Li et al., 1998). In these studies, gene expression was monitored in RBL cells, stably transfected with a NF-AT-β-lactamase transcription reporter. Increases in β-lactamase expression were measured using a novel ratio-metric fluorescence technique and the β-lactamase substrate CCF2/AM (Li et al., 1998). It is almost certain that our knowledge of AM and FM Ca^{2+} signaling will expand greatly over the next decade.

6. NEW FRONTIERS

The advances in experimental methodologies, such as two-photon microscopy, confocal microscopy, high speed digital imaging with fluorescent probes for Ca^{2+}, targeted aequorin and caged regulators of Ca^{2+} , have already greatly enhanced our understanding of the complexities of calcium signaling. A major advance has been the recently developed "cameleons" (based on green fluorescent protein and calmodulin) which are fluorescent indicators of Ca^{2+} (with major advantages over aequorin) which can be targeted to specific sub-cellular domains (Miyawaki et al., 1997). These new probes have already, and will no doubt continue, to improve our understanding of Ca^{2+} regulation in intracellular organelles (Miyawaki et al., 1997). The improvement of these "cameleons" by reducing their sensitivity to pH ensures their future use in Ca^{2+} research (Miyawaki et al., 1999). New techniques such as those designed to selectively label specific sites in living cells with fluorescent probes (such as those to measure Golgi pH) may become a focus of research in the forthcoming years (Farinas and Verkman, 1999).

This technique involves sub-cellular targeted expression of single chain antibodies, which can localize hapten-fluorophore conjugates to specific cellular domains (Farinas and Verkman, 1999). In the future this technique may also provide insights into the spatial complexities of Ca^{2+} signaling. More and more innovations such as this are sure to develop in the first decade of the new millennium. From this chapter it is clear that when it comes to understanding Ca^{2+} regulation in cells we do indeed "live in interesting times".

REFERENCES

Babcock, D.F., Herrington, J., Goodwin, P.C., Park, Y.B. and Hille, B., 1997, Mitochondrial participation in the intracellular Ca^{2+} network, *J. Cell. Biol.* 136, 833–844.

Baron, K. and Thayer, S., 1997, CGP37157 modulates mitochondrial Ca^{2+} homeostasis in cultured rat dorsal root ganglion neurons, *Eur. J. Pharmacol.* 340, 295–300.

Bernardi, P., Colonna, R., Costantini, P., Eriksson, O., Fontaine, E., Ichas, F., Massari, S., Nicolli, A., Petronilli, V. and Scorrano, L., 1998. The mitochondrial permeability transition, *Biofactors* 8, 273–281.

Bernardi, P., Scorrano, L., Colonna, R., Petronilli, V. and Di Lisa, F., 1999. Mitochondria and cell death – Mechanistic aspects and methodological issues, *Eur. J. Biochem.* 264, 687–701.

Berridge, M., 1997a, Elementary and global aspects of calcium signalling, *J. Physiol. (Lond.)* 499, 291–306.

Berridge, M.J., 1997b, The AM and FM of calcium signalling, *Nature* 386, 759–760.

Bootman, M., Niggli, E., Berridge, M. and Lipp, P., 1997, Imaging the hierarchical Ca^{2+} signalling system in HeLa cells, *J. Physiol. (Lond).* 499, 307–314.

Bowser, D.N., Minamikawa, T., Nagley, P. and Williams, D.A., 1998, Role of mitochondria in calcium regulation of spontaneously contracting cardiac muscle cells, *Biophys. J.* 75, 2004–2014.

Bruton, J., Katz, A. and Westerblad, H., 1999, Insulin increases near-membrane but not global Ca^{2+} in isolated skeletal muscle, *Proc. Natl. Acad. Sci. USA* 96, 3281–3286.

Budd, S., 1998, Mechanisms of neuronal damage in brain hypoxia/ischemia: Focus on the role of mitochondrial calcium accumulation, *Pharm. Ther.* 80, 203–229.

Budd, S.L. and Nicholls, D.G., 1996a, Mitochondria, calcium regulation, and acute glutamate excitotoxicity in cultured cerebellar granule cells, *J. Neurochem.* 67, 2282–2291.

Budd, S.L. and Nicholls, D.G., 1996b, A reevaluation of the role of mitochondria in neuronal Ca^{2+} homeostasis, *J. Neurochem.* 66, 403–411.

Callamaras, N., Sun, X.P., Ivorra, I. and Parker, I., 1998, Hemispheric asymmetry of macroscopic and elementary calcium signals mediated by InsP3 in Xenopus oocytes, *J. Physiol. (Lond).* 511, 395–405.

Cannell, M. and Soeller, C., 1999, Mechanisms underlying calcium sparks in cardiac muscle, *J. Gen. Physiol.* 113, 373–376.

Carafoli, E., 1987. Intracellular calcium homeostasis, *Annu. Rev. Biochem.* 56, 395–433.

Caterina, M., Schumacher, M., Tominaga, M., Rosen, T., Levine, J. and Julius, D., 1997, The capsaicin receptor: A heat-activated ion channel in the pain pathway, *Nature* 389, 816–824.

Chen, Q., Cannel, M. and Van Breeman, C., 1992, The superficial buffer barier in vascular smooth muscle, *Can. J. Physiol. Pharmacol.* 70, 509–514.

Cheng, H., Lederer, W. and Cannel, M., 1993, Calcium sparks: Elementary events underlying excitation-contraction coupling in heart muscle, *Science* 262, 740–744.

Csordas, G., Thomas, A.P. and Hajnoczky, G., 1999, Quasi-synaptic calcium signal transmission between endoplasmic reticulum and mitochondria, *EMBO J.* 18, 96–108.

David, G., 1999, Mitochondrial clearance of cytosolic Ca^{2+} in stimulated lizard motor nerve terminals proceeds without progressive elevation of mitochondrial matrix [Ca^{2+}], *J. Neurosci.* 19, 7495–7506.

David, G., Barrett, J. and Barret, E., 1998, Evidence that mitochondria buffer physiological Ca^{2+} loads in lizard motor nerve terminals, *J. Physiol (Lond.)* 509, 59–65.

Davies, E. and Hallet, M., 1998, High micromolar Ca^{2+} beneath the plasma membrane in stimulated neutrophils, *Biochem. Biophys. Res. Comm.* 248, 679–683.

Dedov, V.N. and Roufogalis, B.D., 1998, Rat dorsal root ganglion neurones express different capsaicin-evoked Ca^{2+} transients and permeabilities to Mn^{2+}, *Neurosci. Lett.* 248, 151–154.

Dedov, V.N. and Roufogalis, B.D., 1999, Mitochondrial Ca^{2+} accumulation in DRG neurones following activation of capcaicin receptors in DRG neurones, *Neuroscience* 95, 183–188.

Devine, C.E., Somlyo, A.V. and Somlyo, A.P., 1972, Sarcoplasmic reticulum and excitation-contration coupling in mammalian smooth muscle, *J. Cell. Biol.* 52, 690–718.

Dolmetsch, R.E., Lewis, R.S., Goodnow, C.C. and Healy, J.I., 1997, Differential activation of transcription factors induced by Ca^{2+} response amplitude and duration, *Nature* 386, 855–858.

Duchen, M.R., 1992, $Ca^{(2+)}$-dependent changes in the mitochondrial energetics in single dissociated mouse sensory neurons, *Biochem. J.* 283, 41–50.

Duchen, M.R., 1999, Contributions of mitochondria to animal physiology: From homeostatic sensor to calcium signalling and cell death, *J. Physiol. (Lond.)* 516, 1–17.

Farinas, J. and Verkman, A., 1999, Receptor-mediated targeting of fluorescent probes in living cells, *J. Biol. Chem.* 274, 7603–7606.

Fay, F.S., 1995, Calcium sparks in vascular smooth muscle: Relaxation regulators, *Science* 270, 588–589.

Friel, D.D. and Tsien, R.W., 1994, An FCCP-sensitive Ca^{2+} store in bullfrog sympathetic neurons and its participation in stimulus-evoked changes in $[Ca^{2+}]_i$, *J. Neurosci.* 14, 4007–4024.

Fujimoto, T., 1993, Calcium pump of the plasma membrane is localized in caveolae, *J. Cell Biol.* 120, 1147–1157.

Ganitkevich, V.A. and Isenberg, G., 1996, Dissociation of subsarcolemmal from global cytosolic [Ca^{2+}] in myocytes from guinea-pig coronary artery, *J. Physiol. (Lond.)* 490(2), 305–318.

Gerasimenko, O.V., Gerasimenko, J.V., Tepikin, A.V. and Petersen, O.H., 1996, Calcium transport pathways in the nucleus, *Pflügers Arch.* 432, 1–6.

Golovina, V.A. and Blaustein, M.P., 1997, Spatially and functionally distinct Ca^{2+} stores in sarcoplasmic and endoplasmic reticulum, *Science* 275, 1643–1648.

Gunter, T.E. and Pfeiffer, D.R., 1990, Mechanisms by which mitochondria transport calcium, *Am. J. Physiol.* 258, C755–C786.

Gunter, T., Buntinas, L., Sparagna, G. and Gunter, K., 1998, The Ca^{2+} transport mechanisms of mitochondria and Ca^{2+} uptake from physiological-type Ca^{2+} transients, *Biochim. Biophys. Acta* 1366, 5–15.

Hajnoczky, G., Hager, R. and Thomas, A.P., 1999, Mitochondria suppress local feedback activation of inositol 1,4,5-trisphosphate receptors by Ca^{2+}, *J. Biol. Chem.* 274, 14157–14162.

Hansford, G.R., 1985, Regulation between mitochondrial calcium transport and control of energy metabolism, *Rev. Physiol. Biochem. Pharmacol.* 102, 1–17.

Hirabayashi, T., Kume, K., Hirose, K., Yokomizo, T., Iino, M., Itoh, H. and Shimizu, T., 1999, Critical duration of intracellular Ca^{2+} response required for continuous translocation and activation of cytosolic phospholipase A2, *J. Biol. Chem.* 274, 5163–5169.

Hyrc, K., Handran, S.D., Rothman, S.M. and Goldberg, M.P., 1997, Ionized intracellular calcium concentration predicts excitotoxic neuronal death: Observations with low-affinity fluorescent calcium indicators, *J. Neurosci.* 17, 6669–6677.

Ichas, F. and Mazat, J.-P., 1998, From calcium signaling to cell death: Two confomations for the mitochondrial permeability transition pore. Swiching from low- to high-conductance state, *Biochim. Biophys. Acta* 1366, 33–50.

Ichas, F., Jouaville, L.S. and Mazat, J.P., 1997, Mitochondria are excitable organelles capable of generating and conveying electrical and calcium signals, *Cell* 89, 1145–1153.

Isshiki, M., Ando, J., Korenaga, R., Kogo, H., Fujimoto, T., Fujita, T. and Kamiya, A., 1998, Endothelial Ca^{2+} waves prefentially originate at specific loci in caveolin-rich cell edges, *Proc. Natl. Acad. Sci. USA* 95, 5009–5014.

Jou, M.J., Peng, T.I. and Sheu, S.S., 1996, Histamine induces oscillations of mitochondrial free Ca^{2+} concentration in single cultured rat brain astrocytes, *J. Physiol. (Lond.)* 497, 299–308.

Kifor, O., Diaz, R., Butters, R., Kifor, I. and Brown, E., 1998, The calcium-sensing receptor is localized in caveolin-rich plasma membrane domains of bovine parathyroid cells, *J. Biol. Chem.* 273, 21708–21713.

Koizumi, S., Bootman, M.D., Bobanov, L.K., Schell, M.J., Berridge, M.J. and Lipp, P., 1999, Characterization of elementary Ca^{2+} release signals in NGF-differentiated PC12 cells and hippocampal neurons, *Neuron* 22, 125–137.

Kroemer, G., Dallaporta, B. and Resche-Rigon, M., 1998, The mitochondrial death/life regulator in apoptosis and necrosis, *Annu. Rev. Physiol.* 60, 619–642.

Lawrie, A.M., Rizzuto, R., Pozzan, T. and Simpson, A.W., 1996, A role for calcium influx in the regulation of mitochondrial calcium in endothelial cells, *J. Biol. Chem.* 271, 10753–10752.

Lee, M.G., Xu, X., Zeng, W., Diaz, W., Kuo, T.H., Wuytack, F., Racymaekers, L. and Muallem, S., 1997, Polarized expression of Ca^{2+} pumps in pancreatic and salivary gland cells, *J. Biol. Chem.* 272, 15771–15776.

Li, W.-H., Llopis, J., Whitney, M., Zlokarnik, G. and Tsien, R.Y., 1998, Cell-permeant caged InsP3 ester shows that Ca^{2+} spike frequency can optimize gene expression, *Nature* 392, 936–941.

Lipp, P. and Niggli, E., 1998, Fundamental calcium release events revealed by two-photon excitation photolysis of caged calcium in guinea-pig cardiac myocytes, *J. Physiol. (Lond.)* 508(3), 801–809.

Marchant, J., Callamaras, N. and Parker, I., 1999, Initiation of IP_3-mediated Ca^{2+} waves in Xenopus oocytes, *EMBO J.* 18, 5285–5299.

Marsault, R., Murgia, M., Pozzan, T. and Rizzuto, R., 1997, Domains of high Ca^{2+} beneath the plasma membrane of living A7r5 cells, *EMBO J.* 16, 1575–1581.

Mineo, C., Ying, Y.-S., Chapline, C., Jaken, S. and Anderson, R.G.W., 1998, Targeting of protein kinase Calpha to caveolae, *J. Cell. Biol.* 141, 601–610.

Miyawaki, A., Llopis, J., Heim, R., McCaffery, J.M., Adams, J.A., Ikura, M. and Tsien, R.Y., 1997, Fluorescent indicators for Ca^{2+} based on green fluorescent proteins and calmodulin, *Nature* 388, 882–887.

Miyawaki, A., Griesbeck, O., Heim, R. and Tsien, R.Y., 1999, Dynamic and quantitative Ca^{2+} measurements using improved cameleons, *Proc. Natl. Acad. Sci. USA* 96, 2135–2140.

Monteith, G.R. and Blaustein, M.P., 1999, Heterogeneity of mitochondrial matrix free Ca^{2+}: Resolution of Ca^{2+} dynamics in individual mitochondria in situ, *Am. J. Physiol.* 276, C1193–C1204.

Monteith, G.R. and Roufogalis, B.D., 1995. The plasma membrane calcium pump – A physiological perspective on its regulation, *Cell Calcium* 18, 459–470.

Nelson, M.T., Cheng, H., Robart, M., Santana, L., Bonev, A., Knot, H. and Lederer, W., 1995, Relaxation of arterial smooth muscle by calcium sparks, *Science* 270, 633–637.

Nicholls, D.G. and Budd, S.L., 1998, Neuronal excitotoxicity: The role of mitochondria, *Biofactors* 8, 287–299.

Niggli, E., 1999, Localized intracellular calcium signaling in muscle: Calcium sparks and calcium quarks, *Annu. Rev. Physiol.* 61, 311–335.

Nowicky, A.V. and Duchen, M.R., 1998, Changes in $[Ca^{2+}]_i$ and membrane currents during impaired mitochondrial metabolism in dissociated rat hippocampal neurons, *J. Physiol.* 507, 131–145.

Peng, T.I. and Greenamyre, J.T., 1998. Privileged access to mitochondria of calcium influx through N-methyl-D-aspartate receptors, *Mol. Pharmacol.* 53, 974–980.

Peng, T.I., Jou, M.J., Sheu, S.S. and Greenamyre, J.T., 1998, Visualization of NMDA receptor-induced mitochondrial calcium accumulation in striatal neurons, *Exp. Neurol.* 149, 1–12.

Peng, Y.Y., 1998, Effects of mitochondrion on calcium transients at intact presynaptic terminals depend on frequency of nerve firing, *J. Neurophysiol.* 80, 186–195.

Pike, L.J. and Casey, L., 1996, Localization and turnover of phoshatidylinositol 4,5 bisphophate in caveolin-enriched membrane domains, *J. Biol. Chem.* 271, 26453–26456.

Pinton, P., Pozzan, T. and Rizzuto, R., 1998, The Golgi apparatus is an inositiol 1,4,5-trisphosphate-sensitive Ca^{2+} store, with functional properties distinct from those of the endoplasmic reticulum, *EMBO J.* 17, 5298–5308.

Pivovarova, N.B., Hongpaisan, J., Andrews, S.B. and Friel, D.D., 1999, Depolarization-induced mitochondrial Ca accumulation in sympathetic neurons: Spatial and temporal characteristics, *J. Neurosci.* 19, 6372–6384.

Rasmussen, H., 1989, The cycling of calcium as an intracellular messenger, *Scientific American* 10, 44–55.

Ricken, S., Leipziger, J., Greger, R. and Nitschke, R., 1998, Simultaneous measurements of cytosolic and mitochondrial Ca^{2+} transients in HT29 cells, *J. Biol. Chem.* 273, 34961–34969.

Rizzo, V., McIntosh, D.P., Oh, P. and Schnitzer, J.E., 1998, In situ flow activates endothelial nitric oxide synthase in luminal caveolae of endothelium with rapid caveolin dissociation and calmodulin association, *J. Biol. Chem.* 273, 34724–34729.

Rizzuto, R., Brini, M., Murgia, M. and Pozzan, T., 1993, Microdomains with high Ca^{2+} close to IP_3-sensitive channels that are sensed by neighbouring mitochondria, *Science* 262, 744–747.

Rizzuto, R., Pinton, P., Carrington, W., Fay, F., Fogarty, K., Lifshitz, L., Tuft, R. and Pozzan, T., 1998, Close contacts with the endoplsmic reticulum as determinants of mitochondrial Ca^{2+} responses, *Science* 280, 1763–1766.

Rogue, P.J. and Malviya, A.N., 1999, Calcium signals in the cell nucleus, *EMBO J.* 18, 5147–5152.

Schinder, A.F., Olson, E.C., Spitzer, N.C. and Montal, M., 1996, Mitochondrial dysfunction is a primary event in glutamate neurotoxicity, *J. Neurosci.* 16, 6125–6133.

Schneider, M.F., 1999, Ca^{2+} sparks in frog skeletal muscle: Generation by one, some or many SR Ca^{2+} release channels, *J. Gen. Physiol.* 113, 365–371.

Schnitzer, J.E., Oh, P., Jacobson, B.S. and Dvorak, A.M., 1995, Caveolae from luminal plasmalemma of rat lung endothelium: Microdomains enriched in caveolin, Ca^{2+}-ATPase, and inositol trisphosphate receptor, *Proc. Natl. Acad. Sci. USA* 92, 1759–1763.

Scotti, A., Chatton, J. and Reuter, H., 1999, Roles of Na(+)-Ca^{2+} exchange and of mitochondria in the regulation of presynaptic Ca^{2+} and spontaneous glutamate release, *Philos. Trans. Roy. Soc. London, B* 354, 357–364.

Segal, M., 1995, Imaging of calcium variations in living dendritic spines of cultured rat hippocampal neurons, *J. Physiol. (Lond.)* 486, 283–295.

Simpson, P.B. and Russell, J.T., 1996, Mitochondria support inositol 1,4,5-trisphosphate-mediated Ca^{2+} waves in cultured oligodendrocytes, *J. Biol. Chem.* 271, 33493–33501.

Simpson, P.B. and Russell, J.T., 1997, Role of sarcoplasmic/endoplasmic-reticulum Ca^{2+}-ATPases in mediating Ca^{2+} waves and local Ca^{2+}-release microdomains in cultured glia, *Biochem. J.* 325, 239–247.

Stout, A. and Reynolds, I., 1999, High-affinity calcium indicators underestimate increases in intracellular calcium concentrations associated with excitotoxic glutamate stimulations, *Neuroscience* 89, 91–100.

Stout, A.K., Raphael, H.M., Kanterewicz, B.I., Klann, E. and Reynolds, I.J., 1998, Glutamate-induced neuron death requires mitochondrial calcium uptake, *Nature Neurosci.* 1, 366–373.

Sun, X.-P., Callamaras, N., Marchant, J.S. and Parker, I., 1998, A continuum of IP_3 mediated elementary signalling events in Xenopus oocytes, *J. Physiol. (Lond.)* 509(1), 67–80.

Thayer, S.A. and Miller, R.J., 1990, Regulation of the intracellular free calcium concentration in single rat dorsal root ganglion neurones in vitro, *J. Physiol. (Lond.)* 425, 85–115.

Thomas, D., Lipp, P., Berridge, M.J. and Bootman, M.D., 1998, Hormone-evoked elementary Ca^{2+} signals are not stereotypic, but reflect activation of different size clusters and variable recruitment of channels within a cluster, *J. Biol. Chem.* 273, 27130–27136.

Tinel, H., Cancela, J.M., Mogami, H., Gerasimenko, J.V., Gerasimenko, O.V., Tepikin, A.V. and Petersen, O.H., 1999, Active mitochondria surrounding the pancreatic acinar granule region prevent spreading of inositol trisphosphate-evoked local cytosolic Ca^{2+} signals, *EMBO J.* 18, 4999–5008.

Wartenberg, M., Diedershagen, H., Hescheler, J. and Sauer, H., 1999, Growth stimulation versus induction of cell quiescence by hydrogen peroxide in prostate tumor spheroids is encoded by the duration of the Ca^{2+} response, *J. Biol. Chem.* 274, 27759–27767.

White, R.J. and Reynolds, I.J., 1996, Mitochondrial depolarization in glutamate-stimulated neurons: An early signal specific to excitotoxin exposure, *J. Neurosci.* 16, 5688–5697.

Subject Index